NEWCASTLE COLLEGE LIBRARY

01595601

KU-479-857

ref 2/18.

NEWCASTLE
C O L L E G E
L I B R A R Y

REFERENCE BOOK
ONLY

Clay's Handbook of Environmental Health

Nineteenth edition

WITHDRAWN

Clay's Handbook of Environmental Health

Nineteenth edition

EDITED BY

W.H. Bassett
DMA, FCIEH

Series Editor of Clay's Library of
Health and the Environment
Previously Chief Executive and Director of Housing and
Environmental Health, Exeter City Council, UK
Chartered Institute of Environmental Health
Presidential Award 1991

Spon Press
Taylor & Francis Group

LONDON AND NEW YORK

LIBRARY
NEWCASTLE COLLEGE
NEWCASTLE UPON TYNE

Class 614.7

BARCODE 01595601

First published 1933
by H.K. Lewis

Nineteenth edition 2004
by Spon Press
11 New Fetter Lane, London, EC4P 4EE

Simultaneously published in the USA and Canada
by Spon Press
29 West 35th Street, New York, NY 10001

Spon Press is an imprint of the Taylor & Francis Group

© 2004 Spon Press

Typeset in Sabon by
Newgen Imaging Systems (P) Ltd, Chennai, India
Printed and bound in Great Britain by
St. Edmundsbury Press, Bury St. Edmunds, Suffolk

All rights reserved. No part of this book may be reprinted
or reproduced or utilised in any form or by any electronic,
mechanical, or other means, now known or hereafter
invented, including photocopying and recording,
or in any information storage or retrieval system,
without permission in writing from the publishers.

British Library Cataloguing in Publication Data
A catalogue record for this book is available from the British Library

Library of Congress Cataloging in Publication Data
Clay, Henry Hurrell, b. 1882.
   [Handbook of environmental health]
   Clay's handbook of environmental health. – 19th ed.
      p. cm
   Includes bibliographical references and index.
      1. Environmental health – Great Britain – Handbooks, manuals, etc.
   2. Environmental health – Great Britain – Administration – Handbooks, manuals, etc.
   I. Title.
   RA566.5.G7 C53 2004
   614.4–dc22   2003018496

ISBN 0–415–31808–4 (hardback: alk. paper)

This edition of Clay's is dedicated to the memory of Eric W. Foskett OBE who passed away in 2001. A past president of both the Chartered Institute of Environmental Health and the International Federation of Environmental Health, Eric had been a contributor to several editions of Clay's with a chapter based on his detailed study of the history and development of environmental health in the United Kingdom. The chapter appears again in this edition and stands as testament to his enormous drive and enthusiasm on behalf of the environmental health profession.

Contents

Contributors

Ronnie Alexander BA, DMS, MCIEH
Chief Environmental Health Adviser
Welsh Assembly Government
Cardiff

Jon Averns MSc, MCIEH
Director of Port Health Services
London Port Health Authority

Paul C. Belcher BA, MSc, MCIEH,
Principal Lecturer and Head of the Centre for
Environmental Sciences
University of Wales Institute
Cardiff

Alan Blythe FCIEH, M.Inst.WM, DMS, FBIM
Environmental Consultant
Wedmore, Somerset

Robert Butler MSc, BA (Hons), MCIEH
Senior Environmental Health Officer
City of Birmingham

Chris Church
Independent Specialist on Sustainable
Development
London

David Clapham MSc, MCIEH, MRIPH
Principal Enviromental Health Manager
City of Bradford Metropolitan District Council

Jeff Cooper MSc, BSc, M.Inst.WM, FRGS
Producer Responsibility Policy Manager
Environment Agency
Bristol

Philip W. Cox
Research Fellow
Department of Chemical Engineering
University of Birmingham

Richard Elson MSc, MCIEH
Senior Environmental Health Officer
Health Protection Agency
London

Brian Etheridge BSc (Hons), MSc, DMS, MCIEH
Head of Strategy and Intelligence
Health and Safety Executive
London

Liz Falconer MPhil, FIOSH, RSP, MIM
E-learning Development Co-ordinator
Academic Technologies Group
University of the West of England

Rachel J. Flowers BSc (Hons), MCIEH, PGDip, MSc
Head of Health Development
Milton Keynes Primary Care Trust

Norma J. Ford BSc, MCIEH
Senior Lecturer
University of Salford

The late Eric W. Foskett OBE, BSc (Econ.), DPA,
FCIEH, FRSH
Formerly Director of Environmental Health,
City of Manchester
Formerly Past President, Institution of
Environmental Health Officers,
London

Michael J. Gittins C.Eng, FCIEH, M.Inst.E, FRSH, MIOH
Formerly Chief Environmental Health Officer
Leeds City Council

Ian S. Gray MCIEH, AMTSI, AFPH
Policy Officer – Health Development
Chartered Institute of Environmental Health
London

Andrew Griffiths DMS, MCIEH, LTCL
Chartered Institute of Environmental Health
London

Veronica Habgood BSc, MSc, MCIEH
Director of Learning and Quality
University of Greenwich
London

Michael Halls FREHIS, FRSH, MCIWM
International Federation of Environmental
Health
Galashiels
Scotland

Alan Higgins BA, MBA, FCIEH, MIOA
City Environmental Health and Trading
Standards Officer
Portsmouth City Council

Janet Higgitt BSc, MCIEH
Senior Lecturer in Environmental Health
University of Derby

Nigel Horan FCIWEM
Reader in Public Health Engineering
School of Civil Engineering
University of Leeds

Deborah Kendale
Continuing Professional Development
Co-ordinator
Department of Chemical Engineering
University of Birmingham

Sam Knox BSc, MCIEH
Group Chief Environmental Health Officer
Southern Group Environmental Health
Service
Northern Ireland

John F. Leech BSc (Hons), MCIEH
Principal Environmental Health Officer
Exeter City Council

Ian D. MacArthur BSc (Hons), MCIEH
Regional Director Groundwork UK
Formerly Chief Executive UK Public Health
Association
London

Richard T. Mayon-White MB, FFCM
Consultant Epidemiologist
Institute of Health Sciences
University of Oxford

Chris Megainey BSc (Hons), MCIEH
Specialist in Environmental Pollution
London

Chris Melville MCIEH
Principal Environmental Health Officer
North East Lincolnshire Council

Jon Miles BA, MSc, MRPS
Group Leader, Airborne Radio Nuclides
National Radiological Protection Board
Chilton
Didcot
Oxfordshire

Terence Moran JP BSc (Env. Health), LLB, LLM,
MCIEH, MIOA
Head of School of Social Sciences
Leeds Metropolitan University

David Ormandy MCIEH, FRSH
Principal Research Officer
School of Law
University of Warwick

Paul B. Paddock BSc (Hons), MBA, MPhil.,
MCIEH, MIOSH
Principal Lecturer in Environmental Health
The Nottingham Trent University

Richard J. Palfrey MCIEH
Principal Environmental Health Officer
Exeter City Council

Stephen R. Palmer MA, MB, BChir., FFPHM
Director of the Welsh Combined
Centres for Public Health
College of Medicine
University of Wales, Cardiff

Kath Ray BA, MA
Research Assistant
University of Salford

Denise M. Rennie BSc, MSc, MCIEH
Head of Environmental Management
University of Salford

Paul Robinson BA, DMS, MCIEH
Director of Education and Professional Standards
Chartered Institute of Environmental Health
London

Peter Rotheram OBE, BA, FCIEH, FRSH
Executive Secretary
Association of Port Health Authorities
Runcorn
Cheshire

Michael Squires MIOA
Senior Environmental Health Technician
Exeter City Council

Madeleine Smith MSc, BSc, MIFST, MIFFP
Teaching Fellow in Environmental
Health
University of Birmingham

Jill Stewart MSc, BSc (Env. Health)
MCIEH FRSH
Senior Lecturer in Environmental Health and
Housing
University of Greenwich

Amanda Wheway BSc (Hons), MCIEH
Independent Environmental Health
Officer
Formerly Harrogate Borough
Council

John D. Wildsmith BSc (Hons), MSc, MCIEH
Senior Lecturer
Centre for Environmental Sciences
University of Wales Institute
Cardiff

Frank B. Wright LLB (Leeds), LLM (Leicester),
FCIEH, FRSH, FRIPHH
Director of the European Health and Safety
Law Unit
University of Salford

Foreword

For over 70 years, the 18 previous editions of *Clay's Handbook of Environmental Health* have helped both students and practitioners of environmental health throughout the world understand the concept of environmental health and its practical implementation within a variety of management and administrative systems. Clay has justifiably become the informational bedrock of environmental health.

A new edition is always eagerly awaited. This is particularly the case for the nineteenth edition, the first of both a new century and a new millennium. A rapidly changing world demands that environmental health practitioners keep abreast of change and this new edition contains much new material which reflects the problems and pressures that face them today. Old problems sit juxtaposed with new problems and understanding them, and being able to apply solutions to their resolution, remains a constant need for environmental health practitioners.

This new edition has been thoroughly revised at a time when the Chartered Institute of Environmental Health has reviewed and updated its core educational curriculum. It, therefore, incorporates environmental health topics and issues in a way which reflects that review.

In producing the *Handbook*, compromises had to be made in order to take account of the space available and the many new issues that inevitably emerge between one edition and another. In making such compromises there is always a risk that some important issues will be left out or marginalized. Previous experience, careful thought and judicial editing have, however, ensured that no significant issue has been omitted.

There is also, of course, the possibility that such a multi-authored book could lack flow, continuity and internal cohesion. Fortunately, the editor has ensured that the *Handbook* has been structured in order to minimize this potential pitfall. The result, as readers will discover, is a *Handbook* that can be dipped into for information or read in a more comprehensive manner to understand the holistic nature of environmental health.

The *Handbook* also rightly reflects and re-emphasizes the importance of the wider public health agenda. Former generations of environmental health practitioners understood that their contribution was set within a wider public health context. In more recent times, public health has almost ceased to be regarded as a subject in its own right even though substantial public health problems remain to be tackled.

So often debates about 'health' end up as discussions about waiting lists in hospitals rather than the range of preventative measures, which might be applied to produce a healthier nation. It is, therefore, pleasing to see that the institutional and organizational changes, which are currently taking place, both within local government and the NHS, are reflected within the *Handbook* as is the need for sustainable development to become a central feature of our way of life and governance.

Clay's Handbook of Environmental Health will continue to be a faithful servant to both environmental health students and practitioners and will open up the fascinating world of environmental health to them.

It has always 'made a difference' to both those who use it and to the problems that they have had to face. I am confident that this nineteenth edition will maintain that long and proud tradition.

Brian Hanna CBE, FCIEH
President
The Chartered Institute of Environmental Health

Preface

Since the publication of the last edition of *Clay's* in 1999, change in both the breadth and depth of environmental health has continued apace and this has necessitated extensive change to produce this nineteenth edition. The principal changes are outlined here and the extent of them is also reflected in the number of authors, which now totals 42 and of whom 21 are making contributions for the first time.

Part One Environmental Health – Definition and Organization

The introductory chapter has been rewritten to reflect the continuing discussion about the extent of the faculty and its inter-relationship with other disciplines. This material is also presented in a different form to aid study and interpretation.

The extensive changes to the role and monitoring of local government has substantially affected the management of environmental health throughout the United Kingdom and the chapter which deals with this is new. This is supported by appendices which deal with the management of environmental health in Northern Ireland, Scotland and Wales.

The Chartered Institute of Environmental Health has recently redrawn its training and professional standards and this is fully reflected in the chapter contributed by the Director of Education and Professional Standards.

While the coverage of the port health function generally has been updated it is now supported by a case study of the London Port Health Authority in order to give a greater practical appreciation of the subject in view of the increased spotlight being placed on this work, particularly on the controls over imported foodstuffs.

Part Two Environmental Health Law and Administration

The chapter outlining the issues involved with the enforcement of environmental health law now includes consideration of the Human Rights Act and the Regulation of Investigatory Powers Act together with the recent changes to the PACE codes.

Part Three Public Health and Safety

The most significant change here is the introduction of a new chapter which discusses the inter-relationship of public health, environmental health and sustainable development whilst the chapter on risk management has been more strongly focused on its use in environmental health.

Part Four Epidemiology

The establishment of the Health Protection Agency is highlighted in a new chapter on the administration and law of communicable disease.

Part Five Housing

Another new chapter deals with the relationship of housing conditions and health and this also outlines the housing, health and safety rating system (HHSRS) soon to be established in UK law as the replacement for the current fitness standard.

Part Six Occupational Health and Safety
The chapter on toxic and dangerous substances has been rewtitten as a result of new COSSH regulations in this area.

Part Seven Food Safety and Hygiene
This part has been restructured and substantially rewritten in order to focus on those topics of current importance. The introductory chapter is new as are the, now separate, chapters that deal with milk and milk products and fish and shellfish. The chapter on meat hygiene has been refocused to deal more strongly with meat hygiene issues outside of the abattoir and the responsibilities of the environmental health service for them.

Part Eight Environmental Protection
The emergence of contaminated land as a major fuction to be dealt with within the environmental health service is recognized by a new chapter covering this area while those dealing with air pollution, radon and waste management have been extensively revised. Another new chapter encapsulates the role of environmental health in respect of the control of animals.

I am very grateful to all of the authors for their time and effort in making their contributions and to the many others who have made helpful suggestions as to the changes that were required. I should particularly like to thank the President of the Chartered Institute of Environmental Health, Brian Hanna CBE, for his Foreword and for his earlier comments on the construction of the revisions.

As in UK law, the male pronoun is sometimes used alone for ease of reading. It should be taken to refer to both males and females.

W.H. Bassett
Exeter, February 2004

Extract from the preface to the first edition

Under the ever-widening scope of public health administration, and in view of the high standard demanded of students presenting themselves for the examination of the Royal Sanitary Institute and Sanitary Inspectors' Examination Joint Board, the need has arisen for a handbook, more comprehensive and more in accord with the progressive demands of modern times, than has been available hitherto. At the request of the publishers, this volume has been prepared to meet that need. Despite the number and variety of the subjects necessarily included in a book of this character, the author has endeavoured, as far as is practicable within the limits of a single volume of convenient size, to cover the whole range of the duties of a sanitary inspector, observing due proportion in the attention given to each subject, and while dealing fully and in detail with some subjects, omitting none that is of real importance.

The author has always felt that the subjects to be studied by sanitary inspectors are by no means of an uninteresting character. It has been his endeavour to make this book 'readable' and such as to rob systematic study of much of its inherent drudgery.

Few things are of more practical value to executive public health officers, either in examination or in practice, than the ability to illustrate answers or suggestions by sketches. In order to assist readers in this direction, the liberal illustration of technical matters has been made a special feature of the book; every illustration being from an original line drawing in which sectional details are shown, and which the reader may reproduce or develop for himself.

H.H.C.
London, April 1933.

Part One
Environmental Health – Definition and Organization

1 Introduction to environmental health

Rachel J. Flowers,
Ian S. Gray and
Ian D. MacArthur

We are at the threshold of change; in many ways this new millennium provides a defining moment for fundamentally rethinking the role and function of environmental health. Environmental health has in fact been here before; a similar moment in the UK occurred in the eighteenth and nineteenth centuries, when urbanization and industrialization brought millions of people together in crowded, sprawling and insanitary settlements. In response, society underwent a period of rapid change. New models for the delivery of public services were introduced – the municipal corporations of the 1830s and the new Boards of Health, largely conceived to deal with outbreaks of communicable disease and to afford basic measures of health protection. Subsequently, a new concept of health arose: the idea of public health. New philosophies for the delivery of education and other public services began to take shape, and a new form of democracy evolved – extended suffrage, which became universal in the early twentieth century.

At the beginning of the twenty-first century, we are at a comparable watershed. In the West, urbanization and industrialization are largely behind us. Our horizons are increasingly global, not merely national. We deal, typically, in symbols and abstractions, in information rather than in manufactured objects. Engineered solutions are no longer the answer in themselves and where they still have a role to play they need to be managed and discussed with those they affect. The problems themselves have assumed less tractable, more qualitative dimensions to do with lifestyles, quality of life and psychological health. More importantly, perhaps, the manageable, predominantly urban concept of public health that was developed in the nineteenth century has been replaced by a much bigger concept, that of the global biosphere. As a result, we are grappling with the much bigger issue of the future of life and health on earth, a question that rarely occurred to the Victorians. Environmental health, having invented itself in the nineteenth century, therefore needs to reinvent itself at the start of the twenty-first. And although many of the new threats are global, the problem, and the response, will often be regional or local, even individual.

The worst case scenario for the future takes as its social backdrop a picture of sharpening inequalities, continuing environmental degradation, a decaying family and community fabric, increasing stress and the deterioration of the state into a kind of reactive, coping mechanism. Against such a background, the emphasis within environmental health would be on cure rather than prevention, hard technological solutions rather than a softer people-based response, and probably professional compartmentalization rather than integration, since the former offers more certain career gains to individuals.

If prevention does remain a low priority and the integrated planning and delivery of services is drowned out by calls for quick fixes, then many diseases associated with modern lifestyles, such as cardiovascular disease and cancer, will take much longer to conquer. For example, the research needed to understand them will be slower to materialize and might well face a powerful rearguard action by well-organized industrial lobbies. There is, however, a further complexity to consider, when prevention does become a higher priority it is often in the context of finite or limited resources. This is where some hard decisions may have to be made about the shifting of resources from treatment and care to prevention and all the ethical issues that this entails. A review of National Health Service expenditure [1] has highlighted the economic benefits of upstream investment and the potential contribution of public health measures in reducing the burden of disease. However, there is a lack of solid evidence of which measures to use and success will require 'full engagement', that is, increased spending and high public engagement to achieve improved health status and rising healthy life expectancy.

The environmental impact on health is an inevitable by-product of human activity, and it is therefore the nature of that activity, and the attitudes that go with it, that hold the key. So it will be to the soft technologies – the technologies of mind, reason and social organization – that we must increasingly look for solutions. Hard 'engineered' technologies are at best a partial answer, at worst a diversion. The answers offered are much more difficult. They require vision, ambition and leadership; they also require debate, consensus and agreement. In the worst case scenarios described, these solutions will be more, not less, difficult to realize and they will need a new kind of environmental health practitioner to tackle them.

The World Health Organization (WHO) has also contectualized environmental health into human rights:

Human rights cannot be secured in a degraded or polluted environment. The fundamental right to life is threatened by soil degradation and deforestation and by exposures to toxic chemicals, hazardous wastes and contaminated drinking water. . . . Environmental conditions clearly help to determine the extent to which people enjoy their basic rights to life, health, adequate food and housing, and traditional livelihood and culture. It is time to recognize that those who pollute or destroy the natural environment are not just committing a crime against nature, but are violating human rights as well. (Klaus Toepfer, Executive Director of the United Nations Environment Programme at the 57th Session of the Commission on Human Rights, Geneva, 2001)

To fulfil their future role environmental health practitioners need to have an understanding of the following:

- The **definitions and principles** of environmental health.
- The **agenda** with which they need to be engaged.
- The **skills and expertise** required of their professional practice.
- The **objective** of their environmental health activities.

The following sections of this chapter provide accepted guidance on these subjects.

DEFINITIONS AND PRINCIPLES OF ENVIRONMENTAL HEALTH

The term environmental health is quite different from other definitions in that its meaning can be so wide, and in many respects it is this catch-all nature that creates unease and misunderstanding. By separating the two dimensions of **environment** and **health** we can illustrate the all-encompassing nature of the combined term: Albert Einstein noted that environment was 'everything that's not me', and WHO considers health to be a state of complete physical, mental and social well-being.

Despite the broad challenge of these concepts, several definitions exist for environmental health. In 1989, the WHO defined it 'as comprising of those aspects of human health and disease that are determined by factors in the environment. It also refers to the theory and practice of assessing and controlling factors in the environment that can potentially effect health' [2]. A more recent attempt at defining the term emerged from a meeting of WHO European member states in 1993. Their proposed definition was:

Environmental health comprises of those aspects of human health, including quality of life, that are determined by physical, biological, social and psycho-social factors in the environment. It also refers to the theory and practice of assessing, correcting and preventing those factors in the environment that can potentially affect adversely the health of present and future generations. [3]

This definition concludes that environmental health is in fact two things: first, certain aspects of human health; and second, a means by which to address these issues. Figure 1.1 demonstrates the interface of environmental health. It shows that it is a discipline concerned primarily with human health and it notes that humans operate in different environments and it gives the examples of the living environment, the home environment, the work environment and the recreational environment. The figure also sets out the various stresses that can impact on any of these environments. It is from this

perspective that a holistic view can be taken of human health and the environments in which people live. As a concept and as a means of delivering practical solutions, environmental health provides a strong basis upon which decision-makers can work towards sustainable development.

It has been clearly recognized that environmental health is a wide-ranging discipline that relies upon intersectoral co-operation and action. It is therefore essential that all the potential contributors to the development and implementation of environmental health programmes can recognize their role. Environmental Health Officers (EHOs) and other professionals cannot deal with all aspects of environmental health. Environmental health as a concept is far larger than the roles traditionally played by EHOs employed in local authorities. Many professionals in the public and private sectors make invaluable contributions towards environmental health; they may not perceive themselves as environmental health professionals, but they nevertheless perform tasks that contribute to the whole. In particular, there are many people, for example, public health specialists, health visitors, practice nurses, general practitioners, occupational health and safety staff and technical specialists who work within the broader public health movement and who can, and do, work with EHOs in a way that maximizes the human resources available.

Professionals whose work is on the periphery of environmental health work, for example urban planners, would also greatly benefit the system if they also understood the role of environmental health and could relate its principles to their work. It is important that they understand the principles of environmental health and how their diagnosis and subsequent action can assist in developing effective environmental health interventions. Furthermore, building capacities in environmental health should not be restricted to the public sector workforce, as there are many people employed in the private sector, and not just those in mainline environmental management positions, who recognize and build the principal precautions of environmental health management into everyday business practices.

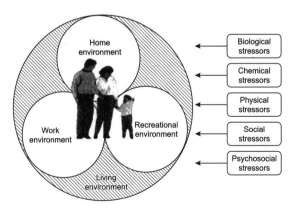

Fig. 1.1 The interface of environment health. (Source: MacArthur, I. D. and Bonnefoy, X. (1998).)

THE PRINCIPLES OF ENVIRONMENTAL HEALTH

The development of the environmental health approach has grown organically rather than by design. However, it has demonstrated that it can bring rhetoric to life and adds considerable value to the process of improving human health and quality of life.

In a world that is subject to constant and turbulent change, it is surely important to retain some sense of core values or principles as touchstones for our work. Environmental health and the mechanisms to deliver it are founded upon such fundamental principles. They do not apply simply to environmental health at the community level, where the main focus is local government. These principles apply to all levels of government and all sectors that contribute to environmental health. They relate to many government issues and depend upon the way in which governments at all levels relate and interact among themselves and with communities. This is a major challenge as many governments still fail to understand the true nature of environmental health, and therefore the significance and value of its approach.

Environmental health is relevant in three time phases. It must work to repair past damage, to control present risk and to prevent future problems. The emphasis given to each phase is determined by a complex formula of factors, depending largely on an assessment of the risks and resources available. It is of course important and correct to address the most pressing issues relating to environmental health urgently, but emphasis should also be given to addressing, and so avoiding, future problems. This is the basis of the precautionary principle, which is becoming widely accepted in all policies and programmes and ensures that environmental health action remains at the leading edge of improving the quality of life.

Principle 1
The maintenance and improvement of the human condition is at the centre of all environmental health action. A similar principle is contained in *Agenda 21* [4]. The principle is a recognition that the main target of environmental action is the well-being of the human race and those factors in the environment, however wide, that may affect it. This principle is fundemental and all others link into it. It reflects both the complexity and the potential that environmental health exhibits in the local and global picture.

Principle 2
The disadvantaged groups within society are often those that must live in the worst housing with poor environmental conditions, work in the most dangerous occupations, and that have limited access to a wholesome and varied food supply.

> The poor bear a disproportionate share of the global burden of ill-health and suffering. They often live in unsafe and overcrowded housing, in underserved rural areas or periurban slums. They are more likely than the well-off to be exposed to pollution and other health risks at home, at work and in their communities. They are also more likely to consume insufficient food, and food of poor quality, to smoke tobacco, and to be exposed to other risks harmful to health. This undermines their ability to lead socially and economically productive lives. [5]

The disadvantaged do not form a single homogeneous group: different people are at a disadvantage in different contexts. For example, low income households in northern European countries may be at risk of poor health because of damp and cold housing conditions, fuel poverty and/or inadequate nutrition. This phenomenon has been clearly recognized by the WHO, which acknowledges that access to the appropriate medical technology cannot in itself offset the adverse effects of environmental derogation and that good health will remain unobtainable unless the environments in which people live are health promoting. A reduction in inequality requires equal access to environmental health services and an uptake of services that relates to need. The provision of high technology services should not be restricted to certain sections of the population because of social or economic disadvantage in the others, and services should be sensitive to the needs of minority groups. To achieve this, the disadvantaged within the population will require special assistance and attention. Equity is therefore a core and primary element that underpins any action on environmental health.

Principle 3

A range of governance issues that can be described as the conditions for civic engagement must be in place. The adoption of the democratic principles of government is the cornerstone to the effective management of environmental health. For example, the European Charter of Environment and Health [2] sets out the basic entitlements of individuals including the rights of full information, active consultation and genuine participation in environmental health decisions.

Environmental health protection is based on a model of democracy in which experts and elected politicians make decisions on behalf of the general public. This sometimes paternalistic and frequently closed style of decision-making is a legacy of the democratic system developed in the eighteenth and nineteenth centuries and strengthened throughout most of the twentieth. But times are changing and demands have increased for greater public participation in all aspects of society. Democratic principles also require a two-way exchange and the involvement of non-governmental organizations and an informed public in the decision-making process is both necessary and practicable. It is therefore important that decision-makers are not only held accountable, but that they owe their accountability to the public.

Modernization of local services requires that people are not merely represented but actively participate in the development and delivery of local policies, services and projects and in doing so they can ensure that these meet their priorities and needs. This power sharing agenda can feel challenging and there are a variety of skills that people working in environmental health need to develop to ensure that a broad cross-section of the community are engaged and not only the 'usual suspects'. This approach to work can take more time and will not happen overnight and it must be remembered that it will also be undertaken within other service areas. It is therefore important to join up and to communicate with others within a locality and ensure that partner agencies work together to plan for and focus on community engagement.

Many of the problems that currently face environmental health will only be solved by communities acting as a whole, rather than as individuals. This is one of the key areas to challenge our traditional approach to solving environmental health problems, and it is certainly an area that requires research.

Principle 4

Co-operation and partnership. Isolated decisions and actions cannot normally solve problems in environmental health: an intersectoral approach is needed. The practice of co-operation and partnership in pursuit of improvements in environmental health, not only between the health and environment sectors, but also with economic sectors and with all social partners, is a crucial element, the origins of which can be traced back to the days of Edwin Chadwick. This is the principle that lies at the heart of effective environmental health management.

Intersectoral activity is at the core of good environmental health practice. Its usefulness, however, depends on how broadly it is interpreted and implemented. When intersectoral co-operation is understood to pertain only to the support of the programme of a particular sector, it fails. If properly interpreted and applied, intersectoral co-operation and co-ordination means that

- the problems tackled are common ones in which all participants have a stake;
- not only governmental agencies but also all the public and private sector organizations and interests active in the sector are involved;
- policy-makers, technical and service staff and volunteers at both national and local levels have actual or potential functions to perform;
- various participants may play leading and supporting roles with respect to specific issues;

- co-operation consists not only of ratifying proposals, but also of participation in defining issues, prioritizing needs, collecting and interpreting information, shaping and evaluating alternatives and building the capabilities necessary for implementation;
- stable co-operative mechanisms are established, nurtured and revised according to experience.

Principle 5
Sustainable development or sustainability. In a similar way to the term environmental health, this concept does not just encompass certain issues, but also requires particular ways of managing them. In the policy-making process relating to environmental health there are three particularly important threads that serve to confirm the almost overlapping nature of environmental health and sustainable development. They are:

- **policy integration**: the bringing of environmental health considerations into all other areas of policy and the tying together of different policy fields and different levels of government;
- **partnership**: consultation with and participation by all groups in society in the planning and implementation of sustainable development policies;
- **appropriate scale**: the handling of policy at the level of government (from local to international) at which each environmental health issue itself occurs, with a bias or emphasis towards the subsidiarity principle.

Principle 6
Environmental health issues are truly international in their character. International communication and travel is making the world an ever smaller place. Environmental health professionals have long recognized the fragility and proportions of the planet, and that the contaminants in our environment do not respect national boundaries.

The world of environmental health is also small. The worldwide community of professionals who dedicate their working lives to improving and protecting the places we live in for the common good are but a speck compared with those who work to exploit and deplete the world's resources in pursuit of wealth creation. However, the diminutive character of the world's environmental health community brings great advantages. We can and must communicate with ease. Although our languages may be different and our heritage and culture places us in different systems, our problems and approaches are mutual. Our commonwealth of knowledge can provide us with an irresistible resource for solving many of the perplexing dichotomies we are faced with in our daily work. International co-operation and collaboration is therefore a key principle for environmental health and is one that should not be overlooked, despite the distraction of our immediate surroundings and problems.

THE AGENDA FOR ENVIRONMENTAL HEALTH

One of the most significant influences on the subject of environmental health in recent times has been the work carried out by the UK Commission on Environmental Health and its subsequent report *Agendas for Change* [6]. This Commission was established by the Chartered Institute of Environmental Health (CIEH) in 1996 in recognition that the famous names of public health – such people as the Chadwicks, the Pastors and the Rowntrees – all reacted to the conditions of the industrial revolution, a revolution that created physical (and ultimately social) changes in our environment. It was those people who created and founded the institutional structures and approaches to environmental health problems that are still remembered today throughout the world. Yet, why should a system established in the 1800s be suitable for today's society, a society that still bears the physical hallmarks of the Victorian age but operates in a completely different and more complex manner than any before it?

It was against such a background that the CIEH took the imaginative decision to establish a wide-ranging Commission on Environmental Health with the following terms of reference:

- to consider the principles of environmental health and their application to the health of individuals and the pursuit of sustainable development of communities;
- to examine the relationship between environmental health and relevant socioeconomic factors;
- to recommend a framework for action in the United Kingdom to reinforce and take forward the principles of environmental health with the involvement of the whole community.

Under the chairmanship of Dr Barbara McGibbon, the Commission, consisting of individuals from fields related to environmental health and environmental health management, published their report, *Agendas for Change*, in 1997 [6]. This set out a new vision for environmental health that recognized the social, economic and environmental aspects of human activity and called for different approaches that move away from regulation and control and towards co-operation, partnership and management as the tools for lasting solutions. Figure 1.2 illustrates the new approach, which recognizes that many of the environmental health problems that face society today are created by society and can only be tackled by the whole of society taking action. It is relatively easy to get individuals to take action when it is a matter of individual health. However, the new challenge is how to get whole communities to act, not for individual benefit but for the advantage of society as a whole. The example of road transport and air quality illustrates this point well, as it is only when a complete modal shift in transport patterns occurs that air quality has a chance of improving. Individuals acting in isolation and not using their cars will not make a significant difference to air quality and will ultimately simply inconvenience those people. It will only be through working in partnership, in co-operation with the various stakeholders, that we can expect positive results. Managed solutions will be more sustainable. *Agendas for Change* is a wide-ranging report that

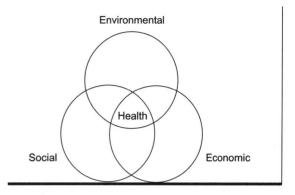

Fig. 1.2   The social, economic and environmental axis. (Reproduced from WHO EURO (9).)

contains many themes and 21 recommendations specific to the UK situation. The commission recognized, in particular, that

1. there is a strong need for the integration of public and environmental health;
2. there is a need for clear local accountability in service delivery;
3. many positive contributions are being made to environmental health under the heading of 'Agenda 21' and 'Local Agenda 21' (now termed 'Local Action 21');
4. sustainable development processes provide many of the elements of good environmental health and these elements need to be nurtured to ensure sustainable change.

The Commission's approach

In considering its approach, the Commission identified several ways to carry out a root and branch assessment of what environmental health might mean towards the year 2020. It considered: looking at the characteristics of 'environmental health'; examining the agreed schedule of environmental health interests; or examining illnesses related to environmental deficiencies or pollution. However, the lack of a simple definition of environmental health, and the complicated interactions between environment, behaviour and biology, make the whole topic extremely complex.

It has become clear that the majority of the forces that will shape and form how environmental health must work in the new century come from outside the worlds of the environmental or health professions. It was with this in mind that the Commission concentrated on the 'agendas' that are driving change in environmental health. These forces include: a new interest in quality of life, rather than length of life; social inequity; new attitudes to health, safety and risk; sustainable development; globalization; economic management; the information revolution; and balancing subsidiarity and the desire of direct public participation in policy-making and implementation.

These and other similar issues form the complex web that environmental health as a discipline must recognize and about which it must negotiate so that it continues to have added value in the years to come.

The Commission's view for 2020 assumes first and most importantly that healthy people and healthy environments go together. This may sound a truism, but if so it is not widely acknowledged. Science and technology, together with our capacity to create artificial environments, have contrived to obscure that message and to replace it with another one: that we are somehow above nature, that we don't need it and if it doesn't suit us, we can change it.

Five major principles were identified by the Commission as key components of environmental health and these informed the commission's approach:

- precautionary approach
- intersectoral collaboration
- addressing inequalities and inequities
- community participation
- sustainable development.

In addition, two themes feature in the 2020 vision and these are inequality and sustainability. First, urgent attention needs to be paid to reducing the disparities in wealth between social classes that relate so closely to disparities in environmental quality and health. The 2020 vision is therefore of a society where goods, whether they are measured in terms of monetary income or quality of life and physical and mental well-being, are much more equitably distributed than at present. The Commission recognized that equality and efficiency go together. An unequal and divided society is a malfunctioning society, and malfunction carries a heavy and measurable penalty, in ill-health, in deteriorating environment and in the resulting social and financial burden. Second, is sustainability which simply relies on the assumption that we inject the health dimension, along with the green dimension, into our planning for sustainable development, even when this involves managing our lifestyles better than we do now.

AGENDAS FOR CHANGE [6]

A healthy environment

The fundamental message of the Commission's report is that human beings can only be healthy in a healthy environment. It recognizes that we cannot insulate ourselves from our surroundings – the air we breathe, the water we drink, the food we eat, the buildings and the landscapes we inhabit. Directly or indirectly, they will affect our health and well-being. These may seem unexceptional statements yet they carry implications for policy-making that are far from understood. They mean, for example, that we need a far more integrated and comprehensive approach to policy formulation, one that embraces health impacts even where they may not be obvious. The problem – and one of the reasons why this integrated approach does not yet exist – is that the target seems to be constantly shifting.

The relationship between environment and health is a complex one, but only recently has this complexity been recognized. In the space of a generation our attitudes to the environment, and to the illness or improved health it may generate, have undergone a revolution. Our mindset – our expectations and attitudes – have changed fundamentally. We think of the planet as our home and conceive of it as a unity. Hence the new global agenda, one that deals in both economic opportunities – the global market – and in environmental limits, and the costs that breaching these limits may entail. We are aware of human rootedness in the environment, that the health of the biosphere, of air and water and soil, is a vital determinant of human health and well-being. We also realize that human well-being is a complex, variable and vulnerable state, encompassing physical, social and psychological factors that are not easily separated.

The holistic approach

To express such thinking, we have begun to evolve a new terminology: speaking of interdependence and interconnectedness, of the importance of a holistic perspective. From professionals, this demands a flexibility of response that may run counter to specialisms. For the public, it can mean that patience with established procedures is lower and expectations much higher. People not only expect that their environment should not damage their health, also increasingly expect that it should promote their health and well-being [7].

All these changes have occurred within the career-span of an individual, roughly the time it would have taken for a person to progress from professional education and training in environmental health to a senior position in the field. But this is not merely an issue for environmental health specialists, since the very nature of the changing agenda – the broadening of concerns, the perception of new links between previously separate areas – means that everybody working in the fields of environment and health is affected.

Most of us recognize these points in principle. Many professionals and policy-makers no doubt try to place their work in a wider context, but at some point they come up against obstacles. More often than not this is institutional, in the broadest sense: the fact that the framework for defining and delivering environmental health has not kept pace with the revolution in knowledge or attitudes. Where changes in the mechanism of delivery have occurred, they have often been piecemeal, inconsistent or driven by forces that have intrinsically little to do with the concept of a genuinely health-promoting environment: cost-effectiveness theories, for example, or laissez faire governance.

We are, in effect, driving an old vehicle into a new era. The question is therefore: do we leave it to creak along, attempting the odd roadside repair, or send it back to the design shop for a complete overhaul?

The agendas

The agendas for change identified by the Commission start from the premise that quality of life is fundamental to health. Both health and quality of life are subject to serious inequalities, but both also raise questions of lifestyle, which carry important implications for the environment. In addressing the two sets of issues – individual health and the integrity of the environment – the argument starts in the particular and the personal, moves on to the environmental and the global, and ends in the social, economic, professional and organizational – the areas where solutions must ultimately be found. The agendas that draw from the principles of environmental health are listed as Table 1.1. Although presented as a list, any of these agendas could be the starting point: these are cross-cutting themes and each provides an entry point into a larger agenda as outlined below.

Table 1.1 The agendas for change

- Quality of life
- Inequality
- Lifestyle
- Globalization
- Democracy
- Information
- Integration
- Sustainability

Quality of life

Discomfort, dysfunction and dissatisfaction are primary indicators of poor life quality and it has been long established that health relies upon quality of life as well as the absence of disease.

Quality of life is frequently seen only in terms of economic wealth, but a different, though related, approach is the attempt to reformulate human and social goals in terms other than the purely economic. The emphasis on economic growth is largely a product of the later twentieth century. However, we have learnt that whatever its positive effects, a higher economic standard of living also carries a number of negative effects on quality of life. It does not necessarily bring greater equality; it tends to degrade environments; and it does not necessarily lead to greater contentment. In fact, an exclusive concentration on economic objectives may destroy the very things – family, community, the social fabric – on which our physical and mental health relies.

A number of studies over the past two decades have shown that people feel better – less stressed, more content, less aggressive – in the presence of natural landscape features: greenery, trees, water. Their mental performance improves, they feel refreshed, relaxed and reinvigorated, and with these feelings comes improved physical health. These elusive, qualitative issues probably underpin the movement of people from the cities in search of better health and quality of life. Of course, many more people do not have such choices.

Inequality

By and large, the poorest people live in the worst housing, suffer the most degraded living environments, work in the worst jobs, have the lowest level of educational attainment, and eat the least wholesome food (and pay proportionally more for it). To say that poor environments and unhealthy lifestyles go together is not to say that they always go together and that individuals, through effort or initiative, cannot transcend them; but people born into such circumstances have the dice loaded against them.

The equality issue – the gap between the haves and the have nots – has emerged as one of the dominant themes of the 1990s. This is a gap that matters in human terms, and numerous examples exist of real health differentials, for example, *An Independent Inquiry into Inequalities in Health in England* (Stationery Office, London, 1998, ISBN 0-11-322173-8). Growing social and environmental inequality exact an enormous toll on health and life prospects. Resolving these issues will require a political commitment. Ultimately, the only way to remedy environmental and health indicators of relative deprivation is a conscious and sustained drive to reduce social inequalities in all policy areas. To prioritize it in health alone condemns the initiative to futility.

Lifestyle

There are two main components to the lifestyles agenda, both of which depend on the idea that there is such a thing as an identifiable Western lifestyle and that this carries implications for the health of individuals, society and the global environment. The first is that such lifestyles are unhealthy, and that they tend to reinforce social and environment inequality. The second is that they are unsustainable – that the Western lifestyle cannot be transferred wholesale to the rest of the world without enormous damage to the biosphere.

To many people, consumerism represents a lifestyle that must be curbed if the environment is to be protected. Western lifestyles make heavy demands on energy and resources and thus carry far reaching implications for environmental health. Green or ethical consumerism has proved a powerful force for change since the late 1980s. As a result, many people in commerce have been forced to reconsider how they do business and have injected the green dimension into purchasing and sales policies. Yet, although the marketplace has proved capable of reform, local and national governments have also been weakened by the power of the markets. What we have gained as consumers, we appear to have lost as citizens. A major theme of the Commission's report is that to protect our health we have to protect the environment, and this will not happen unless people are prepared to become more actively involved. Healthy and sustainable lifestyles, like green consumerism, require knowledge and empowerment.

Globalization

At the same time as new information technology has compressed distances, the market economy is extending its reach. The new global economy is dominated by transnational corporations and brand names that are recognized throughout the world. Its prevailing ideology is liberalization and deregulation: free trade.

Contrasting strongly with the apparently unstoppable spread of globalism has been the rise in concern for the local that takes many forms. For example, there is a new interest in a sense of place and local distinctiveness. Environmentalism has bred a radical, often passionate, attachment to particular landscapes. The global and the local are increasingly at odds with each other, sometimes in open conflict. If there are sensible alternatives to the process of globalization, they need to be explored urgently.

Democracy

Several factors have prompted the search for a new democratic model. They include the loss of local control to the new global economy; the difficulty of reconciling long-term issues of sustainable development with short-term political preoccupations; and the development of local action around specifically local environmental or health issues [8,9]. A widespread cynicism also attaches to politics and politicians. Local government has been emasculated: its powers and competencies have been undermined. At the same time there is an impatience with established procedures, particularly among younger people, and a new vogue for direct action. People are also, by and large, more educated, better informed and less deferential. Over the past three decades there has been a rapid expansion of civil society – the network of charities, voluntary organizations, and campaigning and pressure groups through which much political activity is focused and directed.

Cumulatively, such pressures point towards a new model for devolving power and involving people: a new model of community ownership. Although its likely shape remains unclear, there are many clues. There is demand for greater participation, although not necessarily in existing forms of government. There is a search for what might be called appropriate decision-making technology, for decisions to be taken at the most appropriate level. This often turns out to be the lowest level – the one nearest to those whose lives will be affected – compatible with efficiency and a strategic view. Hence, in part, the growth of interest in subsidiarity. There is talk of stakeholders and empowerment and there is an emphasis being placed on community involvement. Yet, there are also powerful ideological, bureaucratic and economic tendencies pushing in the opposite direction, towards centralization.

All of these issues bear on environmental health, yet environmental health and environmental protection are still based on a model inherited from the Municipal Corporations and Boards of Health of the nineteenth century in which experts and elected politicians make decisions on behalf of the public. This Victorian model, often distant and lacking transparency, sometimes paternalistic, is in urgent need of an overhaul.

Information

One of the key features of modern society is the explosion in the volume and complexity of information and the difficulty we have in handling it. Developments in information technology – the Internet, computer databases and so on – while creating new opportunities for participation, are adding to this problem of management and interpretation that applies across the range of environmental health issues.

To make their own judgements, people need good quality information that is independently and expertly validated and clearly communicated. Education in environmental health is central to understanding the information and encouraging informed individual and community action. Without that education and good quality information, the range of problems identified will grow more acute.

Integration

To meet the challenges of the outlines in the commission's vision will require new approaches to policy-making and implementation. These will stress flexibility, teamwork, partnership and co-operation. Within and between agencies, we will require new forms of co-operation. We need multidisciplinary and intersectoral partnerships and the involvement of the public. We need to develop local models of integration between different bodies which can then be disseminated by national government. Such local initiatives will need underpinning by enabling legislation, funding and resources.

Professional staff will need to be flexible in their approach and multidisciplinary in their skills: able to see the whole picture, to see how their own function dovetails with those of others and vice versa. They must also have the knowledge and the skills to be able to involve the public in the new forms of participatory structures suggested above. This means a move away from the 'statute-led' nature of much of environmental health work.

Sustainable development (also see Chapter 9)

Of all the issues that have emerged in recent years, the challenge of sustainability is probably the most crucial, cutting across traditional professional boundaries, requiring new forms of management and posing questions that are fundamental to human health. Several key ideas underpin the sustainability agenda. There is the attempt to place local behaviour in a global context and to see how global concerns can be expressed locally. There is the precautionary principle, acting where proof of harm may be lacking. There is the need to think as nature 'thinks', in other words like a living system, in cycles and not straight lines. And there is the need to operate across organizational or specialist boundaries because it is the impact of an activity on the environment – not the successful prosecution of a specialist function – that matters.

The challenge of sustainable development is enormous. We cannot talk about sustainable development without also talking about environmental health. In recent years we have seen several international and national initiatives that try to bring together sustainability and health, for example, the WHO Health For All strategy and Healthy Cities project, the UK government's several publications aiming to provide a national public health strategy and the Local Agenda 21 process [7]. One of the priorities in environmental health must be to recognize the links between these programmes and ensure that they are managed in an integrated way.

THE SKILLS AND EXPERTISE OF ENVIRONMENTAL HEALTH PRACTICE

In September 2002, the CIEH and the Health Development Agency for England published a vision statement for the development of environmental health over the next ten years [10]. The report describes the work of a project which explored the projected growth of the role of environmental health in improving the public's health and reducing inequalities. The aim of the project was to support the environmental health profession in developing a strategic vision for its contribution to health development and well-being.

The need for this project arose because of the increasing importance that is being placed on the multidisciplinary delivery of public health and the recognition that environmental health has, for a variety of reasons, become disconnected from the public health agenda and the public health organizations. The Environmental Health 2012 vision captured the challenges, constraints and ideas from environmental health, public health and health improvement professionals from the various sectors and it incorporated the advice of key practitioners, professionals and academics from environmental health, local authorities, health services, voluntary and community groups.

Participants in the project reported that the mainstream practice of environmental health had become fixed on the delivery of a narrow agenda, and that a number of factors were preventing it from achieving its traditional involvement in addressing the wider determinants of health. They expressed concern about the fragmentation of environmental health services, and lack of clarity on the nature of future environmental health roles and their contribution to health improvement and tackling health inequalities. There was a reported lack of available resources to deliver the new approaches and initiatives called for by the modern public health agenda, or to participate fully in the new organizational structures for public health. In particular, the culture of performance management of specific enforcement targets has resulted in environmental health officers having to take on predominantly technical and enforcement roles. This trend has been at the expense of the effective practice of the wider principles of environmental health protection, and has had the effect of deskilling many in the profession, leading to both dissatisfaction among existing environmental health officers and a diminishing number of applicants for student training. The project was carried out in England, but the findings have been relevent to all the countries within the United Kingdom and may have wider international application because similar issues have also been reported in other parts of the world, from America [11] to Australia [12].

The role of environmental health in public health

The Environmental Health 2012 strategic vision reported that by addressing the wider determinants of health, including food, housing standards, occupational health and safety, air quality, noise and environmental issues generally, environmental health makes a fundamental contribution to the maintenance and improvement of public health and improving quality of life and well-being.

In order to demonstrate the importance of these contributions it was necessary to describe the sphere of environmental health. Figure 1.3 takes as its starting point the interface of environmental health shown previously in Fig. 1.1 [13]. The original interface has been developed and extended so as to include the main areas of activity of environmental health in food, water, air, land and buildings. The stressors have also been extended and specified (see panel on stressors).

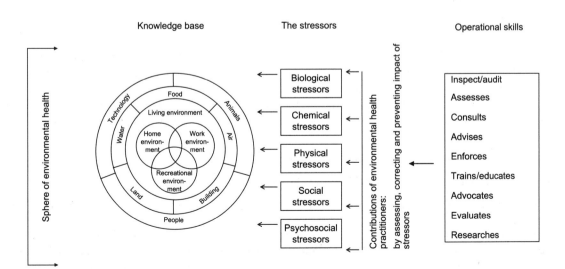

Fig. 1.3 The 2012 vision model of environmental health. (Developed from: Burke *et al.* (10).)

Environmental health stressors

Environmental health stressors are features of the environment that may induce harm in or damaging responses to a living system or organism.

Biological stressor

Those biological elements of the natural and man-made world which present a direct risk to human health through ingestion, inhalation, inoculation or physical contact, and those miscellaneous elements that may influence biological systems to the detriment of humans and their environments.

Chemical stressor

Those chemical entities (or their intermediates) which, through their presence in a particular environment, expose the human to risk through ingestion, inhalation, inoculation or absorption and/or which interfere in biological systems to the detriment of humans and their environments.

Physical stressor

Those measurable physical manifestations induced naturally or through human activity that may impact unfavourably on human health through their damaging effects on cells, tissues, organs and homeostatic systems, as well as their impact on mental and social well-being.

Social stressor

Those behaviours associated with human life that are a consequence of settlement in communities and habitation and which have impacts on health and well-being.

Psychosocial stressor

Those attitudes of mind and mental processes that may have an adverse impact on the health of a person or community.

Environmental health practice

The contributions of the environmental health practitioner have been defined as assessing, correcting and preventing the impact of the stressors on the living environment. The operational skills required by the environmental health practitioner have been identified as assessing, consulting, advising, enforcing, training/educating, advocating, evaluating and researching and each of these has again been described (see panel on operational skills).

Operational skills of environmental health practitioners

Assess – be aware of, and able to practice, those analytical skills that form the basis of professional judgement.
Consult – be aware of, and able to practice, the full range of techniques for giving and receiving information.
Advise – be able to communicate technically correct information for the purpose of informing colleagues, clients and others of the most appropriate course of action to be taken in a wide range of circumstances.

Enforce – be aware of, and be able to use, the full range of mechanisms available for securing compliance with legislative provisions, statutory requirements and standards, commensurate with the perceived level of risk.

Train/educate – be aware of and able to use a range of practical skills associated with education in an environmental health context for the purpose of:

- acquiring knowledge;
- raising awareness; and
- modifying behaviour.

Advocate – be able to support, promote and campaign on a range of issues.

Evaluate – be able to consider all aspects of an environmental health issue and be able to apportion values which can be supported and defended.

Research – be able to:

- discover, identify and use appropriate information sources;
- critically assess options in day-to-day practice;
- undertake a research exercise from planning to report stages.

The vision for the contribution of environmental health to public health in 2012 [10]

The strategic vision for environmental health 2012 identifies that environmental health officers

have a unique contribution to make to public health throughout their prime focus of maintaining health rather than curing illness. They use problem solving skills, supported by legal powers, to intervene in the causes of ill health in the home, workplace and community. Their actions directly influence health determinants and maintain healthy environments for the benefit of both individuals and communities, while also extending to the protection of the environment for future generations.

The strategic vision envisages that environmental health practitioners will

- play lead roles in local government community health and well-being stategies and actively contribute to the public health agenda;
- be key partners in protecting and improving the health and quality of life of individuals

and communities and reducing health inequalities;
- tackle the wider determinants of population health by identifying, controlling and preventing current and future risks.

Curriculum for environmental health education and training

The work of this project has profoundly influenced thinking and planning about the role of environmental health at a high level in the nations within the United Kingdom and Northern Ireland. In particular, the education and training for future environmental health practitioners is now being based on a curriculum which will properly equip our future generations to fulfill their role.

The curriculum encompasses the study of the physical, social and human worlds and their interface with the environment; the study of the 'stressors' on the environment; the public health impacts of the stressors and the identification and implementation of appropriate interventions for the purpose of eliminating, controlling or mitigating the various public health impacts and for identifying opportunities to promote health and well-being.

Key elements of the curriculum

The curriculum of the CIEH for courses leading to the registration of environmental health practitioners (endorsed by the Council on 3rd April 2003) contains the following key elements.

The physical, social and human worlds

- The biosphere and its processes
- The principles of ecology and their relationship to aspects of environmental health
- The principles of sociology and their importance to aspects of environmental health
- Psychological aspects of group function and its application in the practice of environmental health
- The principles of micro and macro economics and their importance in society
- The principles involved in the good governance of communities
- The concept of health, ill-health and disability, well-being and quality-of-life; the cost/benefit of health; the principles of personal choice/social responsibility and citizenship
- The concept of disease and ill-health arising from stressors acting on the sphere of environmental health, together with the roles and responsibilities of health practitioners
- The concept of risk and principles surrounding risk assessment
- The value and practice of good communications to environmental health
- The principles underpinning research and the importance of evidence to the good practice of environmental health.

The stressors and their implications for health

- The range of biological, chemical, physical, social and psychosocial stressors that may act upon the physical, social and human worlds
- The points of possible impact of the stressors
- The mechanisms of impact of the stressors
- The potential and actual public health implications that arise from the impact of stressors on the physical, social and human worlds.

The environmental health interventions

- Identifying the points at which the environmental health practitioner may intervene to prevent, control or mitigate the impact of stressors on the physical, social and human worlds
- Identifying partner organizations, fellow professionals and others with whom appropriate intervention strategies might need to be formulated
- Determining the most appropriate intervention to exercise, having regard to the factors (be they political, financial or technological) that are likely to influence the process of decision-making
- Implementing appropriate intervention strategies
- Monitoring and reviewing the effectiveness of the intervention strategy; altering or adapting it, where necessary, according to the actual or predicted outcome.

An international profile of the environmental health officer

A consultation by the WHO on the Role, Functions and Training Requirements of Environmental Health Officers (Sanitarians) in Europe in 1978, concluded: 'Experience of those countries in the Region which have officers with specialized training in environmental health who are recognized as constituting a specific profession clearly demonstrates their value. Thus it would be to the advantage of all Member states to introduce into their environmental health service staff of this kind whom the consultation called in English **environmental health officers**'. That conclusion was followed by a further series of considerations by the WHO, one of which looked at the development of environmental health manpower [14], and contained a professional profile of the environmental health officer, which may still be considered to provide a valid description:

A professional profile of the environmental health officer [14]

1. The environmental health officer is concerned with administration, inspection, education and regulations in respect of environmental health.

2. The numbers of environmental health officers should be sufficient to exercise adequate surveillance over health-related environmental conditions; this surveillance should include necessary monitoring activities. They must have a close association with the people in their area and be readily accessible to them, providing professional advice and guidance, and thereby gaining the community's confidence and encouraging its participation in improving environmental health. They should be members of multidisciplinary primary health care teams delivering comprehensive health care at community level.

3. Environmental health officers act as public arbiters of environmental health standards, maintaining close contact with the community. They must at all times be aware of the general environmental circumstances in their districts, and must know what industrial hazards to health may arise there and what resources are available in the event of an emergency.

4. They are professional officers capable of developing professional standards and applying them to their own work in relation to that of non-professionals involved in environmental health. Also, their role will obviously touch on aspects with which physicians, veterinarians, toxicologists, engineers, nurses and others deal in a more specialized manner.

5. One of their vital functions is to maintain effective liaison with other professional officers who have a contribution to make in the promotion of environmental health, for example, with regard to water resources management, waste management, housing, rodent, insect and other pest control and protection of the recreational environment.

6. Environmental health officers would carry out the well-established duties of sanitarians/sanitary inspectors, including inspection of housing and food hygiene, and also monitor and control the new hazards arising from intensive industrialization, for example, pollution by chemical and physical agents, which could harm the health of a community.

7. Even greater emphasis than in the past should be placed on the preventive role of environmental health officers in relation to environmental hazards to health.

8. They will obviously not possess the expertise of physicians in personal health, of veterinarians in animal health, of microbiologists in microbiology or of sanitary engineers in the provision of water supplies. However, they will have sufficient background and practical knowledge of these areas to understand the principles involved, and may develop some specialist expertise. They will be able to work easily with the other professionals. Their wider experience will enable them to formulate an approach on a broader base, to contribute to the decisions to be made, or to make these decisions alone in cases where they have the necessary authority. They must understand the environmental aspects of the problems that are the concern of other professionals, so that they may contribute to their solution.

9. They must be able to plan and co-ordinate activities between different professional disciplines, official agencies and authorities. They need to have continuing links with other professionals involved in environmental and health-related work. The other professionals with whom liaison will be appropriate include physicians, physicists, microbiologists, chemists, civil/building/sanitary engineers, veterinarians and lawyers.

10. In some countries and situations, the environmental health officers will initiate the collaboration; in others, they will provide the information and advice that are sought. This liaison role will

extend beyond the other professionals to technicians and a range of other specialists, including those concerned with the public health laboratory services. While being able to act independently in both advisory and enforcement capacities, exercising self-reliance and initiative, they should also be able to function as members of a team with other professionals in implementing environmental health programmes.

11. In industry and commerce, environmental control specialists interpret legislation, promote and maintain standards, and solve the problems that may come to light through, for example, a system of internal control or 'self-inspection'.

12. An important part of their functions must be to acquaint themselves with actual or potential environmental hazards and to ensure that appropriate action is taken to deal with them, for example, to safeguard the public from the hazards associated with microbiological contamination of food and with chemical residues in food substances, and to monitor and control potential and existing environmental hazards, with the backing of strong legislation.

13. A combination of training in public health and toxicology should enable them to cope with such problems as soil pollution due to degradation-resistant agricultural pesticides, leachates from industrial wastes, fallout from the plumes of chemical works, liquid radioactive wastes from industry and research; chemical pollution of the work environment from solvents and from dust arising from processes using silica, asbestos and lead; pollution of the home environment due to such products as cosmetics, detergents, paints, pesticides and gas used as fuel; heavy contamination of water resources by mercury, antimony, barium, cobalt and other metals due to industrial wastes, pesticides used in agriculture, etc.; and new problems in food safety such as the irradiation of food.

14. Environmental health officers in the public service should have the following basic functions:

    (a) improving human health and protecting it from environmental hazards;
    (b) enforcing environmental legislation;
    (c) developing liaison between the inhabitants and the local authority, and between the local and higher levels of administration;
    (d) acting independently to provide advice on environmental matters;
    (e) initiating and implementing health education programmes to promote an understanding of environmental principles.

15. Because of the range of functions, environmental health officers will operate in a managerial capacity and in collaboration with other environmental agencies and services.

The increasing complexities of environmental health problems require a continuing development of expertise and an updating of knowledge. To maintain a leading role in dealing with environmentally related health problems, environmental health officers must be in a position to respond to challenges presented by new hazards in the environment, and exercise influence in promoting and regulating environmental health activities. Their training should equip them with the necessary expertise to act at any level – national, provincial (or intermediate) or local – or within any sector, private or public.

Thus environmental health officers are trained as generalists across the range of the 15 basic environmental health activities identified above and therefore occupy a key position in the environmental health service. An important part of their functions must be to acquaint themselves with actual or potential environmental hazards, and to ensure that appropriate action is taken to deal with them. In some cases, they may have the required authority and expertise. In others, they may need to press particular agencies to take appropriate action. There will also be occasions when they will need to consult other professional officers, and to make use of laboratory and other expert scientific services. A vital part of their functions will, therefore, be to maintain effective liaison with other relevant professional officers who have a contribution to make in the promotion of environmental health. Environmental health is very much a team concept, and this must be recognized in any organizational arrangement.

AN OBJECTIVE FOR ENVIRONMENTAL HEALTH

The pupose of this introductory chapter has been to help to define the purpose and practice of environmental health. It is clearly the case that environmental health, whilst firmly rooted in the history of public health, needs to adapt and modernize in order to meet the demands and expectations of new generations. Most importantly, the modern practice of environmental health must be embedded within sustainable development in order to deliver the long-term and long lasting improvements in both our public health and our environment.

The acceptance of this and the incorporation of previous definitions would offer us a challenging objective for future environmental health practice:

To achieve mediating strategies between people and their environments, synthesizing personal choice and social responsibilities in health to prevent premature death and reduce disease, disability, dysfunction and discomfort and create supportive and sustainable environments to enable people to live healthy lives.

It is a common motive of environmental health practitioners that we wish to make a difference in people's lives and the authors of this chapter extend their best wishes to all who accept this challenge.

REFERENCES

1. HM Treasury (2002) *Securing Our Future Health: Taking a Long-Term View – the Wanless Report*, HM Treasury, London.
2. WHO (1989) *Environment and Health: A European Charter and Commentary*, WHO, Copenhagen.
3. CIEH (1995) *The UK Environmental Health Action Plan – A Response by the CIEH*, Chartered Institute of Environmental Health, London.
4. UN (1992) *Earth Summit. Agenda 21: The United Nations Programme of Action from Rio*, United Nations Department of Information, New York.
5. WHO (1998) *Health for all in the Twenty-first Century*, WHO, Geneva.
6. CIEH (1997) *Agendas for Change – Report of the Environmental Health Commission*, Chartered Institute of Environmental Health, London.
7. LGMB (1996) *Health and Sustainable Development – Local Agenda 21 – Roundtable Guidance*, Local Government Management Board, Luton.
8. WHO (1997) *Health and Environment in Sustainable Development – Five Years After the Earth Summit*, WHO, Geneva.
9. WHO EURO (1997) *Sustainable Development and Health: Concepts, Principles and Framework for Action for European Cities and Towns*, World Health Organization Regional Office for Europe, Copenhagen.
10. Burke, S., Gray, I., Paterson, K. and Meyrick, J. (2002) *Environmental Health 2012 – A Key Partner in Delivering the Public Health Agenda*, Health Development Agency, London.
11. CDC (2002) *A Strategy to Revitalize Environmental Health Services in the United States*, Centres for Disease Control and Prevention, Department of Health and Human Services, Atlanta Georgia. Unpublished working draft.
12. Bell, S. (2002) Environmental health: Victorian anachronism or dynamic discipline? *Journal of the Australian Institute of Environmental Health*, 2 (4).
13. MacArthur, I.D. and Bonnefoy, X. (1998) *Environmental Health Services in Europe – Policy Options*, WHO, Copenhagen.
14. WHO (1987) *Development of Environmental Health Manpower*, Environmental Health Series No. 18, World Health Organization Regional Office for Europe, Copenhagen.

OTHER USEFUL READING

Department of Health (1998) *Our Healthier Nation: A Contract for Health*. Cm.3852, The Stationery Office, London.
WHO (1998) *Environmental Health Services in Europe – An Overview of Practice in the 1990s*. London World Health Organization Stationery Office.

2 Historical development of environmental health in the United Kingdom

The late Eric W. Foskett
updated for this edition by
William H. Bassett

INTRODUCTION

The history of the development of environmental health control in the United Kingdom is long and complex. Any brief version cannot contain the wealth of detail that is available, and so the course traced here is that of streams of developments and important landmarks.

Environmental health is concerned with any effect by any environmental factor on human or animal health. Initially, concern was limited to those factors that were easily discernible as affecting human health, but as the environment became better understood, it was recognized that the role of environmental health should be expanded.

It is not possible to assign a specific date from which problems relating to environmental health emerged. While this study has to have a beginning in time, it must be understood that the need to control the environment in the interests of health has been evolving for a long time, and it continues to evolve as a consequence of man's occupation of the planet.

For the purpose of this study, it is possible to divide the evolution of environmental health control into four time zones to which some dates can be assigned. These are no more than convenient, relatively imprecise, time boundaries because many important events straddled those boundaries and, in many cases, there were parallel streams of different activities in the same time scale.

The first of these periods may be regarded as being about 1750–1850; the second 1850–1900; the third 1900–45; and the fourth 1945 to the present day.

A historical perspective is important. A knowledge of what has happened in the past, and how and why changes came about, will make it easier to understand the present position. If the present position can be understood, it may be possible to make some intelligent forecasts of future developments.

In attempting to understand what has happened, and why it occurred, it is important not to make value judgements of past events based on present day knowledge and attitudes, for, in the last two centuries, social philosophies and scientific and technological knowledge have undergone many changes. What is common knowledge or practice today was largely unknown or unacceptable 150 years ago. Indeed, it may well be that the fourth boundary proves to be inaccurate, and that boundary may, in the future, be understood to have been drawn at some time in the 1980s, meaning that this study was written at the beginning of the fifth period.

The conditions in the United Kingdom that have led to a system of environmental health control may be thought to be exaggerated or overemphasized, but the entire period is very well documented with numerous literary sources to be consulted. These sources are to be found in the reports of numerous inquiries, in political writings, in newspaper reports and in contemporary literature.

THE AGRICULTURAL REVOLUTION (1700–75)

It is common to associate the increasing need for intervention in the environment in the interests of health with the onset of industrialization. There is truth in this except that what is usually termed 'the Industrial Revolution' was preceded by an 'Agricultural Revolution' that contributed greatly to the movement of population from the country to the towns.

Agricultural improvements introduced by Charles Townshend, Thomas Coke, Jethro Tull, Robert Bakewell and others in husbandry, the breeding of better livestock, the introduction of farm machinery and the enclosure of land reduced the demand for agricultural labour, and the dispossessed workers migrated, mainly to the towns.

For those who remained as farm workers, their lot was usually a very low living standard characterized by housing that was damp, overcrowded and with few amenities. The greater productivity resulting from the Agricultural Revolution improved workers' diet a little, and made some contribution to a slow decline in the rural death rate. Nevertheless, work on the land, often in bad weather, was unremitting toil for scant wages. Disease ravaged the family. The children were especially vulnerable but, for the adults, poor nutrition, excessive child-bearing, a poor domestic environment and exposure to the elements while working contributed to a low life expectancy.

As towns expanded and became more remote from their supply hinterland, it became increasingly difficult to provide a satisfactory food supply. The situation was ripe for fraudulent substitution, and for food to be adulterated. Doubtless, most of the rural population would have preferred to remain in the hamlets and villages, but when they had to move they had few options. They could go to the small, often remote, villages in the hills where the presence of water power had induced the establishment of a textile mill, or to other villages that grew up around a mill, or they could retreat to the growing towns.

THE INDUSTRIAL REVOLUTION

The documentary evidence of environmental conditions is graphic. Accounts written in the first third of the nineteenth century relate to conditions created to cope with urban population increases due to the very high birth rates and the influx of people into the towns during the preceding 50 or so years.

The pace of industrialization quickened with the development of the steam engine, which enabled factories, and especially textile mills, to be set up in more favourable locations than villages on hillsides adjoining swift running water. With this transition came many environmental evils.

Machinery enabled production to be vastly increased with economies of scale to be grasped. As factories increased in size and complexity, they needed more workers, who had to be housed nearby because, in the absence of public transport, they had to be able to walk to work.

The need to house workers close to the factories resulted in street after street of small, ill-constructed houses. There was little provision for drainage or refuse disposal; water supplies were inadequate and usually grossly polluted. To add to these unfavourable conditions, the houses were overshadowed by the tall factories and they were polluted by smoke, grit and dust from both the factory furnaces and domestic fires emitting low level smoke.

The bad environmental conditions were matched by the poor social conditions. Even though the working hours were very long, wages were low. As a result, workers suffered from malnutrition, and their poor physical condition was made worse by the ill-ventilated factories and, in some textile mills, the high temperatures and humidity. In addition to the adverse working conditions, there were few sanitary amenities within the factories.

For many workers, factory practices involved working with dangerous machinery, and fatal or disabling accidents to workers were common.

If the adult workers suffered from bad working conditions the lot of the children was even worse. Child labour was often considered an essential factor in production. There was always work a child could do and there was a place for them in the mines and in the mills. Indeed, children in textile mills and pottery factories often worked longer hours than the adults as they stayed to clean the machinery at the end of the day, or arrived earlier than the adult workers to ensure that the latter could start full production as soon as possible.

Children started to work at an early age and, in textiles especially, they formed part of a team. Adults who could not take a child with them might find work hard to get and, where the shortage of children was acute, mill owners imported children from poor law institutions as apprentices.

The general social and economic conditions in the early nineteenth century have to be viewed against a background of, initially, an unreformed parliament, an inadequate local government system, and a political philosophy that permitted, in the name of individual liberty, all manner of what would now be considered to be social abuses. Because of the political and economic views current at the time, there was no political will to make deliberate attempts to ameliorate the conditions endured by most urban workers and their families.

In the first decades of the nineteenth century there was considerable political ferment urging change, but almost always in vain. The people whose lot warranted improvement had little individual political influence, and the ability to organize to bring political pressure to bear was a skill still to be learned. Nevertheless, their plight was noticed, and many politicians and philanthropists adopted causes with which they sympathized, and fought for

changes using their social and political connections. In the more populous areas, groups of people emerged who became interested in the welfare of their fellows, and they supported inquiries and investigations. Physicians, such as Thomas Perceval of Manchester and James Currie of Liverpool, visited slum areas and reported what they had found.

Environmental disadvantage was not uniformly spread. There were many districts that were comparatively salubrious, even though few adequate sanitary amenities were available. Some of the more prosperous boroughs spent money improving parts of their areas but, in many instances, improvements were carried out by improvement commissioners appointed to implement, in a defined locality, environmental improvements specifically authorized by a private act of parliament, the cost being defrayed by a 'rate' levied on the householders in that area.

In 1795, a contagious disease swept through children working at a mill near Manchester, and the local justices of the peace took what action they could to prevent a recurrence.

Early in the nineteenth century there was the first move towards improving the conditions of some juvenile workers. Sir Robert Peel, himself a mill owner, introduced a bill that became the **Health and Morals of Apprentices Act 1802.** This:

passed with little or no opposition. Its chief provisions may be summarized as follows. The working hours of apprentices were limited to 12 a day. Night work (by apprentices) was to be gradually discontinued and to cease entirely by June 1804. Apprentices were to be instructed in reading, writing and arithmetic, and a suit of clothing was to be given yearly to each apprentice. Factories were to be whitewashed twice a year, and at all times properly ventilated; separate sleeping apartments were to be provided for apprentices of different sexes, and not more than two were to share a bed. Apprentices were to attend church at least once a month . . . All mills and factories were to be registered annually with the Clerk of the Peace. The justices had power to inflict fines of £2 to £5 for neglect to observe the above regulations. [1]

That summary indicates clearly the nature of the environmental conditions in textile mills. To enforce the Act, the local bench of magistrates had to appoint two of their number, one of whom had to be a clergyman, to inspect mills. The Act was of little value. Strictly, it only applied to 'apprentices' and not to children whose parents consented to their working in the mills, and, in effect, the Act was an extension of the still existing Elizabethan poor law system.

Sadly, at that time, the forces pressing for change realized that they could not hope to improve working conditions, especially hours of work, for adult workers, and their best chance lay in an improvement in the terms of child labour. At least Peel's Bill showed that some of the worst environmental conditions had been recognized.

Although the first census was taken in 1801, the statistical picture of the times could only be achieved by analysing data derived from bills of mortality and the register of baptisms, neither of which were very good sources of data. While the compulsory registration of births, marriages and deaths was started after 1836, it was to take 50 more years to secure the notification of some communicable diseases. Using the above sources, an attempt was made in about 1835 to compare infant mortality. Over a century, there appears to have been some improvement in the mortality rate, but in 1829 disease was still killing children under five years of age at a rate equivalent to about 30% of the number of children born that year. The 'diseases of infancy' were, for many years, to take a great toll of young people. The social conditions, especially of poorer people, left most of the population vulnerable to the ravages of communicable diseases, especially tuberculosis, scarlet fever and smallpox.

By the beginning of the 1830s, there was a great deal of social unrest, especially over long working hours, the insanitary and dangerous conditions in work places, poverty and bad living conditions. The pressures that had led to the **Health and Morals of Apprentices Act 1802** were supplemented by other forces but, in the economic and political climate then prevailing, rapid change was unlikely. In 1819, the Factory Act was amended, but in a short period of time there were moves to change that too. The agitations for the reform of parliament and local government were close to bearing fruit. **The Reform Act 1832** changed the composition of parliament, and this was followed by the reform of the existing municipal corporations in 1835. Unfortunately, there was no perceived need to provide a comprehensive local government system.

In late 1831, cholera was introduced to Great Britain, the disease being imported from mainland Europe through the port of Sunderland. By the spring of 1832, cholera had spread to many towns and cities. In London, 5275 people died, with a mortality rate of almost 50%. Overall, a total of about 22 000 people died out of a population of 14 million.

Cholera is usually thought of as a tropical disease, but here the disease was manifesting itself in the cooler months of a temperate climate:

> Conditions in all the large towns were, at this time, very favourable to the spread of cholera. Much of the drinking water came from wells in the towns themselves in close proximity to cesspools. Even when piped water supplies existed, the water was often taken from rivers grossly polluted by human sewage . . . . No one realized that the drinking of polluted water was dangerous, and indeed even at the end of the century there were still a few die hards . . . who refused to believe that drinking water was an important vehicle of infection. [2, p. 67]

While this comment describes the state of the country's towns and the standard of scientific knowledge, it can also be emphasized that there was no adequate government machinery to deal with such an outbreak.

THE ROYAL COMMISSION ON CHILD LABOUR

The major social issue of the 1830s related to factory work, and the crucial aspects were hours of work and safety. There was general support for the reduction of children's hours and for a 10-hour day for adults. When a parliamentary bill to secure this was narrowly defeated, a 'Royal Commission on Child Labour' was set up to enquire quickly into the national position so as to assuage the anger of the working people.

Edwin Chadwick, recently appointed to public office as an assistant poor law commissioner, and rapidly promoted to full commissioner by virtue of the excellence of his work, was seconded temporarily in 1833 to be chief commissioner for this Royal Commission. Chadwick and a small number of close colleagues steered this inquiry, but the fieldwork was done inadequately by assistant commissioners appointed through political patronage.

Because the government demanded early action, Chadwick wrote the report personally. The report's recommendations pointed to fundamental changes in the industrial health and safety scene, and laid down important principles that were followed in later legislation.

Central to the problem was reducing the hours that children could and adults did work. In promoting a solution that enabled adults to work long periods of time while reducing the hours allowed for a juvenile, Chadwick argued:

> Why not reduce the hours of children even more drastically than the 10 hours' limitation suggested by the operatives – so drastically that two sets of children could be used to work against the normal adult day. In this way child labour would be used but not overworked: at the same time no reduction in the adult hours need take place. [3, p. 80]

Chadwick proposed that the Act should be enforced by a centralized inspectorate, the members of which would have wide powers, be salaried, and would act as a board and, by meeting at intervals, would produce some degree of uniformity of action. The **Factory Act 1833** empowered the inspectors to enter premises and to require the fencing etc. of dangerous machinery and the provision of sanitary conveniences. The inspectors had powers to require a satisfactory system of schooling for all children employed in the mills, and this was the first compulsory school system.

Chadwick's report was generally accepted, but he was never consulted about its implementation.

EDWIN CHADWICK

Born in Manchester in 1800, Chadwick went to London at an early age, there trained as a lawyer, and became involved with the followers of Jeremy Bentham who was a radical thinker and the father of the 'utilitarian' school of philosophy. Chadwick became a major exponent of Bentham's theories, and came to know others in Bentham's circle, some of whom became close working colleagues.

The work done on the Factory Act 1833 was a diversion from the task that was to result in Chadwick being a major influence in the development of environmental health administration.

THE ROYAL COMMISSION ON THE POOR LAWS

The poor law administration in Great Britain had not been significantly changed since the reign of Elizabeth I. By the 1830s, its administration was uneven and amateur, and it was costing the poor ratepayer too much. Hence it had to be reformed, and in February 1832 a 'Royal Commission for Enquiring into the Administration and Practical Operation of the Poor Laws' was appointed. Through the persuasion of his friend, Nassau Senior, who had been appointed a commissioner, Chadwick became an assistant commissioner with a remit to do a field study and to report. Assistants were supposed to submit a summary of their findings; Chadwick, typically, submitted a report on his research, his recommendations and indicated the philosophy on which those recommendations were based.

> It [Chadwick's contribution] was no Selection of extract but a complete Report, very long, (one-third of the entire volume was his) brilliantly executed; and working up to six clearly formulated and practical conclusions. [3, p. 103]

The Commission was prepared to ask Chadwick to draft proposals to be incorporated into the commission's report, but it was felt that it would be unfair to take advantage of his work without appointing him a full commissioner. This was done, and Chadwick was launched on a long, distinguished career as a reformer.

Although Chadwick had made a name in connection with the poor law reform, his lasting reputation came out of what was to be his life's best work. After the initial report on the poor law, progress on the reform was hindered by political procrastination and personal animosity, with the result that Chadwick was virtually excluded from the continuing discussions.

Employing his painstaking methods, Chadwick had looked at the causes of poverty, and from his investigations into the living conditions of the working population in England and Wales he could hardly have failed to understand the connection between poverty and ill health and an insanitary environment. This inquiry was initiated by the government following a resolution in the House of Lords, the Poor Law Commission being instructed to carry out that investigation.

By 1841, the report on the enquiries and the conclusions that Chadwick was drawing were almost complete but, because of political pressures, he was instructed to drop the report. However, political control changed shortly afterwards and he was required to complete the report. When it was completed, there was again political interference and, on the grounds that it would cause offence, the government refused to publish it as a government report, but Chadwick was allowed to print it under his own name as his personal view (*The Sanitary Conditions of the Labouring Population of Great Britain*). It was a brilliant success and the King's printer produced many more copies than was usual for such a publication, with 10 000 being distributed free of charge.

There had been previous reports with which Chadwick had been associated, and which had centred on the improvement of houses. Before he completed the 1842 report, Chadwick had changed his views, and the report advocated a shift from the improvement of houses themselves to the improvement of their external sanitation and drainage.

Furthermore, it proposed a system; house drainage, main drainage, paving and street cleansing were now to be considered as integral parts of a single process mechanically motivated by the constant supply of water at high pressure. [3, p. 211]

THE ROYAL COMMISSION ON THE HEALTH OF TOWNS

Following further political pressures for legislation to make sanitary reform possible, and in order to prepare the way, a 'Royal Commission on the Health of Towns' was set up.

The general principles of his [Chadwick's] Report were not to be questioned by the Royal Commission. Instead the Commission was to demonstrate the various means of applying these principles. [3, p. 123]

The Commission was instructed to follow Chadwick's plan of action and it heard reports about the sanitary conditions in 50 of England's largest towns in which about 18% of the total population lived. The Royal Commission submitted reports to parliament in 1844 and 1845, but when a bill was introduced it was so heavily criticized that it was abandoned. Much of the opposition to the proposed legislation came from the Health of Towns Associations and similar bodies.

London's Association, founded in 1844 and created by Chadwick's friend Dr Thomas Southwood Smith, was supported by many of the day's most eminent politicians. It had as its objectives the dissemination of knowledge of the evils arising from the existing insanitary environment.

THE CLAUSES ACTS AND PRIVATE LEGISLATION

By 1847, similar associations had been created in many of the more populous cities and towns. Many of these towns had already established the practice of promoting private legislation to enable them to carry out improvement to their areas. Because similar powers were being sought, parliament enacted **The Town Improvement Clauses Act 1847**. This Act set out clauses that local authorities could incorporate into their own legislation and, as such clauses had already been approved, opposition to them was unlikely when new Bills containing them were introduced into parliament.

The city of Liverpool had pioneered environmental and public health legislation in its **1846 Liverpool Sanitary Act**. This Act took powers to abolish some local bodies, such as the Commissioners of Sewers, and gave Liverpool's council sole powers to deal with drainage, paving and cleansing, etc. It also gave the council powers to appoint an officer of health, an inspector of nuisances, and a surveyor. Dr William H. Duncan, a distinguished local physician who knew Chadwick, Southwood Smith and William Farr was appointed as the officer of health, and Thomas Fresh was given the post of inspector of nuisances.

> The duties of the Medical Officer of Health are set out . . . and they are drafted as widely as possible. Under a later section [of the Act] the duties of the Inspector of Nuisances are defined, and it is clear that he was to be an officer of the Council independent of the Medical Officer of Health but compelled, by the nature of his functions, to co-operate closely with him. [4, p. 36]

Although Chadwick was very much aware of the non-medical skills required in providing a sanitary environment, his 1842 report contained a recommendation that it would be good economy to appoint an independent medical officer to 'initiate sanitary measures and reclaim the execution of the law' [4, p. 37].

The Liverpool Act became law in late 1846, and Dr Duncan took up his appointment as officer of health in January 1847.

THE PUBLIC HEALTH ACT 1848

There were further delays in making statutory provision for sanitary reform and, by 1848, there was no positive sign of progress until a cholera epidemic intervened – the country had experienced previous epidemics of cholera and the memory of

the death toll was fresh. The fear of cholera made the need for sanitary reform more urgent.

The measure placed before parliament in 1848 was far reaching as it had to create a code of sanitary legislation and an administrative machine to carry it into effect.

As the Bill went through its parliamentary stages, it was pruned and watered down. Wide opposition was generated and on many grounds, some of which reflected a fear of losing a vested interest or the reduction of personal rights. So as to get some measure on the statute book, the government appears to have surrendered to almost all the opposition. Even Chadwick argued for dropping the Lord's smoke clauses to placate manufacturers who might, otherwise, have jeopardized the whole Act [3, pp. 324–5]. The Bill, passed against a background of fear and the turbulence of the Chartist movement, was the first major public health legislation. The Act gave mainly permissive powers for the formation of local boards of health. It empowered the appointment of paid officials and there was a statutory duty requiring local boards of health to appoint a surveyor, inspector of nuisances and an officer of health. This latter post, statutorily protected, was available only to qualified medical staff.

The 1848 Act vested responsibility for all sewers in the local boards of health. It became illegal to build a house without drains, a sanitary convenience and an ashpit. All streets in a board's district had to be cleansed. Except for private streets, all streets had to be paved and drained. Slaughterhouses and common lodging houses had to be registered with the board, and cellars with ceilings less than 7 ft high were prohibited from use as dwellings. Powers were also given for cleansing filthy houses. Because of the complex pattern of authorities existing in London, the Act, which was to last for five years, did not extend to the city or the metropolitan area of London. It established a **General Board of Health**, which was to have oversight of local boards of health, but this board suffered one particular defect in that its president was not necessarily a minister accountable to parliament for its activities. When the General Board of Health was set up, Lord Morpeth, who had been the parliamentary pilot, became its president and

Chadwick the paid commissioner. The third commissioner was Lord Ashley who was a well-respected philanthropist. Southwood Smith, who had been a mainspring of the Healthy Towns movement, was made the chief medical inspector under the **Nuisances Removal and Diseases Prevention Act of 1848**, which was enacted to give operational powers not provided in the Public Health Act 1848 and was a response to the threat of cholera.

The 1848 Act was permissive and could be adopted by the then local authorities, which could create a local board of health. Previously, those authorities that had wished to improve environmental conditions in their town had had no central authority to advise them; they now had the General Board of Health. The board could compel the establishment of a local board of health in exceptional local circumstances, such as if the death rate exceeded 23 per 1000 in seven successive years or if, on a petition by 10% of the ratepayers, an enquiry by a superintending inspector reported that a local board was desirable. The cholera epidemic, which coincided with the 1848 Act and the establishment of the General Board of Health and local boards of health, was protracted and widespread. It started in Scotland in October 1848 after ravaging Europe, and by June the following year it was in full force; it was very severe in London, the Midlands and parts of Wales. There were 53 293 deaths in a population of about 15 million [2, p. 68].

THE ADULTERATION OF FOOD

Mention is made before about the adulteration of food partly as a consequence of the enlargement of the cities and towns and their isolation from what had been their 'door step' sources of food. Even before 1800 there had been an increasing tendency to adulterate food. Much of this was a fraud on people rather than a means of injuring their health but, while the adulteration of, say, bread with alum fell into this category, there were some forms of adulteration that were positively dangerous.

The analytical skills needed to detect such adulteration were scarce, although some chemists practised in this field. One of these was Frederick Accum, a skilled and well-known scientist. He undertook analyses over a long period of time, and in 1820 he published *A Treatise on Adulterations of Food and Culinary Poisons*. This was the first time that the issue had been discussed openly and objectively [5, p. 101]. Sadly, Accum later left to live abroad to avoid humiliation for a trivial offence.

Further publications of a less authoritative nature appeared from time to time to keep the issue alive, but in 1848 a book by John Mitchell (quoted in [5] and see also [5], pp. 106–8) largely resumed where Accum had left off. There was little public awareness of food adulteration, and no public body had power to deal systematically with it.

ASHLEY'S HOUSING ACTS

The various enquiries carried out into environmental conditions in the first half of the nineteenth century had all shown dramatically the bad conditions under which people lived. The survey that resulted in Chadwick's 1842 report *The Sanitary Conditions of the Labouring Population in Great Britain* was matched by Engels' *Condition of the Working Class in 1844* [6]. Engels is thought to have derived much information from earlier surveys.

Housing conditions *per se* were bad, and there was much overcrowding; many people shared houses and many lived in cellars. At the time that Liverpool secured its Sanitary Act 1846, 7668 cellars were inhabited by about 30 000 people [6] out of a population of 370 000.

Because of the low income of working families, many were unable to rent separate accommodation, and common lodging houses and houses in multiple occupation were common [4, p. 230]. In 1851, despite the difficulties associated with such shared dwellings, two Acts, promoted by Lord Ashley, gave powers in respect of lodging houses. The first of these, the **Common Lodging Houses Act 1851**, gave controls over common lodging houses which laid the foundations for later forms of control; the second Act, the **Labouring Classes Lodging Houses Act 1851**, gave powers to the then local authorities to create common lodging houses as a means of reducing homelessness.

Later, having become Lord Shaftesbury, Ashley admitted that his legislation had had little practical support.

THE GENERAL BOARD OF HEALTH

Strictly, the Ashley Housing Acts fell into the second time phase, and it is necessary to look again at the General Board of Health. The board was created by the Public Health Act 1848 which, itself, had a life of only five years. Under its three members, the board was very active in attempting to establish a public health system.

Inevitably, it created opposition as it conflicted with vested interests and, under Chadwick's influence, it had centralizing tendencies. Many of the Board's alleged faults were commonly ascribed to Chadwick, who became regarded as an evil genius guiding it. Because of the Board's unpopularity, the government failed to secure its renewal at the end of its five-year life. The Public Health Act 1848, however, survived.

Chadwick and Southwood Smith both retired from paid employment in 1856, but although Chadwick continued to be an influence for many more years, he never again held public office.

A new General Board of Health was set up in 1853 under the presidency of Sir Benjamin Hall. He appointed an advisory body which included Dr Neil Arnott, John Simon and William Farr. Simon was later appointed as the Board's salaried medical officer and, when the Board was disbanded in 1858 and the duties transferred to the Privy Council, John Simon became the Privy Council's medical officer, and began a career of immense importance in the development of the public health service.

COMMUNICABLE DISEASES TO 1900

A significant feature of nineteenth-century social life was the impact of communicable diseases such as cholera, diphtheria and scarlet fever. There was an interrelationship between the incidence of such diseases, poor domestic environments and the inadequacy of the sanitary infrastructure. The incidence of communicable diseases was great, the mortality rate very high, and there was little understanding of the causes of such diseases and the mechanics by which they were spread. Indeed, the actions proposed by Chadwick and others were based on the 'miasmic' (noxious vapours) theory of the spread of disease. The Public Health Act 1848 was followed by a massive cholera epidemic, and there was a further major visitation by the disease in 1853–54, when the first General Board of Health had fallen into disfavour. Nevertheless, those two outbreaks saw the beginning of the better understanding of the disease and how it was spread. John Simon formed views on this but the major advances were the work of John Snow.

> It is difficult . . . to appreciate the full measure of Snow's achievement, since what he proved seems so obvious, but . . . the climate of opinion before the days of bacteriology was completely different from that of the present time. Then, and for many years afterwards, the miasmic theory of the origin of infectious diseases was the one which was most generally believed. The theory was that soil polluted with excrement or refuse of any kind gave off an atmospheric 'miasma' which was the cause of certain epidemic diseases. There was nothing specific about it, and indeed it was commonly believed that one disease could change into another; for example even after the clinical distinction between typhoid and typhus had been made, it was commonly believed that one could change into the other. The 'miasmatic' theory was that on which the early movement for sanitary reform was based. [2, p. 68]

The results of Snow's first enquiries into the transmission of cholera appeared in a pamphlet in 1849 after the second epidemic. He continued his work in the later outbreak, also making use of the statistical material prepared by William Farr of the Registrar General's office. Out of these investigations came further publications recording his findings and conclusions, the practical importance of which he demonstrated in the celebrated **Broad**

Street pump. In this 1854 incident, the cases of cholera in a population served by contaminated water drawn by that pump dramatically declined when its use was discontinued by the removal of the pump handle.

Although Snow was not able to demonstrate the causative organism (that was left to a German bacteriologist, Robert Koch, in 1883), he rightly argued that a primary source of infection was consuming water polluted by infected faeces. Snow further demonstrated that propagation through the consumption of infected water was not the only method of spread, and he described the classical methods of disseminating such diseases through bad personal and culinary hygiene.

Simon's and Chadwick's belief in the miasmatic theory was understandable in the light of contemporary medical knowledge, but the reforms they promoted were appropriate in laying the foundations for the improvement of public health.

Cholera returned in 1866 as an epidemic disease, but other infectious diseases, especially those related to social conditions, were of continuing importance, especially during the last three decades of the nineteenth century. Diphtheria was important, and scarlet fever frequently left severe and permanent complications.

Records show that the two diseases were often confused in the years following the onset of industrialization, possibly because of the common symptom of a sore throat. After about 1830, scarlet fever became virulent again, having manifested itself in a milder form for several years.

Up to 1856, returns of scarlet fever and diphtheria were combined, and Simon said that diphtheria was unknown to the vast majority of British doctors until 1855. In 1860, Duncan, writing of the Liverpool experience, said that diphtheria was of minor importance, and that if it was a new disease it had only appeared about three years earlier.

In 1863, the annual death rate from scarlet fever was about 4000 per million children under the age of 15 (a note on John Simon, see [4], p. 42). After the two diseases were distinguished, scarlet fever remained the more significant until about 1885, when diphtheria became more common. It remained a major cause of infantile deaths for 60 years.

Although the mortality rate from scarlet fever, and to a lesser extent diphtheria, declined, the incidence declined at a slower rate. The virulence of haemolytic bacteria of the genus *Streptococcus* seems to have declined, but other factors may also have been important. The decline antedated modern chemotherapy, but slight improvements in housing, including a reduction in overcrowding, and slightly better dietary regimes may well have been contributory factors.

The incidence of most communicable diseases was greatest in the sections of the community that were poorest in terms of housing, nutrition, wealth and leisure. Tuberculosis was a disease that affected all classes. It was endemic, but its incidence was greatest in poorest areas where housing was bad, nutrition poor because of poverty, ignorance or social habits, and where workers laboured in damp, dusty or hot conditions and the work necessitated long, excessive hours of strenuous toil.

After about 1850, the incidence of tuberculosis declined continuously. There was no useful drug to combat the disease but rest, reduced stress, improved diet and fresh air were held to be curative and, from the last quarter of the century, sanatoriums providing treatment on these lines were built.

THE ADULTERATION OF FOOD TO 1900

While chemistry was more advanced in the mid-nineteenth century than medical science, little was done to follow the work of Accum in 1820 until almost the middle of the century, when John Mitchell showed that there had been a continued increase in food adulteration.

Food at that time was seriously adulterated, and it was discovered that the public had become accustomed to the flavour of such foods and liked neither the flavour nor appearance of unadulterated food. Some pioneers, such as the emerging co-operative societies, which aimed to provide a fair service (including pure food) for their members, met strong consumer opposition, and one society experienced such difficulty in selling unadulterated tea that it employed a lecturer to tell its members what good tea was really like!

Mitchell's work induced an active response, and much publicity, especially through articles published in the *Lancet*. Following public pressure, parliament appointed a select committee to enquire into food adulteration. The facts that were unearthed made parliamentary action inevitable, and resulted in the **Adulteration of Food Act 1860**.

The Act disappointed radical reformers, and evidence suggests that it was a failure. It gave no sampling powers to local authorities, but allowed the appointment of public analysts to deal with suspected food presented by private citizens prepared to pay for the analysis.

The position created by the 1860 Act was clearly unsatisfactory, and it was unlikely to achieve the reforms that were desired. In 1860, it was exceptional for there to be legislative interference in the free working of the economy, but the 1860 Act was a breach in that dike.

In the decade that followed, there was a persistent stream of criticism, and demands were made for more effective safeguards. In 1868, proposals were made to amend the 1860 Act. Those intentions met with obstruction and delay, and not until 1872 was the law amended by the **Adulteration of Food, Drink, and Drugs Act 1872**.

Progress towards a satisfactory code was being made slowly. The 1872 Act made it an offence to sell food, drink, or drugs that were not of the 'nature, substance or quality' demanded by the purchaser, and this has been the basis of all later food control legislation. The Act gave limited powers to appoint public analysts, but its most important provision was to permit inspectors of nuisances, as well as private individuals, to acquire samples of food for analysis. This power resulted in systematic and increased sampling of foods, and a marked increase in the number of cases of food adulteration being detected.

Despite the improved legislation, the 1872 Act still had some deficiencies, and a select committee was appointed in 1874 to examine how the 1872 Act was working. Out of its findings, the **Sale of Food and Drugs Act 1875** was passed.

The 1875 Act, although amended and extended later, formed the basic legislation followed in later revisions of the law. Some important issues had to be decided on appeal to the High Court, including the reversal of the judgement that an inspector purchasing a sample for analysis could not be prejudiced. Much of the adulteration at that time, such as the addition of water to milk and spirits, was fraudulent rather than a danger to health – provided that the water was pure!

The public analysts had a crucial role. The Society of Public Analysts was formed in 1874, and most practitioners were members. It published its proceedings, gave a stimulus to analytical chemistry, and developed new tests and standards of purity. In particular, the society played an important role in establishing the limits beyond which an article of food would be regarded as being adulterated, and many of its recommended standards achieved statutory recognition.

MILK AND MEAT TO 1900

For urban populations, the supply of milk and meat has always presented problems. For many years, the main source of milk for town dwellers was cows kept in sheds within the built-up area. For the supply of meat, the practice was to buy animals at country sales, drive them into the towns and slaughter them in small, back street slaughterhouses. Back street cowsheds and slaughterhouses were conducive to the transmission of disease and the creation of serious nuisance.

Cattle closely confined in insanitary sheds were very susceptible to disease. Tuberculosis, in particular, could be transmitted to those who drank the milk. Furthermore, in an age when food adulteration was extensively practised, milk was adulterated by the fraudulent addition of water. For residents near to the cowsheds there were nuisances from odours, flies and the disposal of manure. The coming of the railways made possible the rapid carriage into the towns of rurally produced milk, and this was the prime cause of the decline of the town-kept cow population.

Slaughtering animals in the small, urban slaughterhouses created nuisance from noise, odours, flies and the disposal of the by-products of slaughter. For many people the trade itself was repugnant.

The recognition of the potential for nuisance and the sale of diseased meat led to the early regulation of the trade by some of the growing towns. **The Manchester Police Act 1844** gave powers to control the slaughtering of animals to prevent nuisance, for the licensing of new and the registration of existing slaughterhouses, and for the appointment of slaughterhouse inspectors who had power to inspect not only slaughterhouses, but also meat and other foodstuffs.

The 1844 Act also showed that, at that time, there was no concept of an impartial and uncorrupt administration, as inspectors appointed under the Act were required to make a statutory declaration that they would act honestly.

A similar local Act in 1846 gave Manchester Borough Council powers to license butcher's and fishmonger's shops.

Although the Sale of Food and Drugs Act had been passed in 1875, its provisions were augmented by the **Public Health Act 1875**, which authorized medical officers of health and inspectors of nuisances to inspect food exposed or deposited for sale; it gave powers to seize unfit food; and it made provision for the named officers to enter slaughterhouses and premises used for the sale of meat, so that animals slaughtered could be inspected.

HOUSING TO 1900

Reference has already been made to Lord Shaftesbury's 1851 Lodging Houses Acts. These Acts had little impact. Despite the poor conditions under which most of the population lived, it was not until 1868 that there was further significant legislation. That Act and a subsequent housing statute in 1879 (both designated as the **Torrens Acts**) made changes that enabled local authorities to deal with individual insanitary houses. While these were useful powers, they did not permit local authorities to deal with areas of bad housing.

Some areas of bad housing were demolished under commercial pressures. With no security of tenure, possession of a tenanted house could be obtained easily by the owner, who would sell if he had a favourable offer from a developer or a railway company wanting access to a town centre. The first statutory powers to deal with areas of unfit houses were obtained by the city of Manchester in private legislation in 1867.

Further social pressures resulted in legislation in 1875 and 1879, which permitted local authorities to deal with areas of insanitary houses by clearing them and redeveloping the sites. Of the two Acts, collectively known as the **Cross Acts**, the first, the **Artisans and Labourers Dwelling Improvement Act 1875**, allowed councils to deal with unhealthy houses by buying the land and buildings for the purpose of improvement. Councils were allowed to build houses or let the land for building subject to schemes having special regard to providing accommodation for the working classes. Where owners refused to sell land, provision was made for compulsory purchase.

There was an interesting connection between the Cross Act 1875 and the **Public Health Act 1875**, for the latter gave powers to councils to make building by-laws. Progressive local authorities adopted such by-laws, and were thus able to have some control over the building standards for houses built to replace the insanitary dwellings that had been demolished.

Although some legislative provision had been made, the operation of the Torrens Acts and Cross Acts was expensive, and many local authorities showed little initiative in tackling insanitary housing conditions in their areas. Because of public pressure, parliament appointed two select committees of the House of Commons in 1881 and 1882, and the **Artisans Dwelling Act 1882** resulted.

Despite this and the efforts of the Local Government Board, progress was slow, and public opinion reflected this. In 1884, a Royal Commission was appointed to inquire into the housing of the working classes. A number of very distinguished people served on that Commission, and it heard evidence from such prominent public figures as Lord Shaftesbury and Chadwick, who appeared as president for the Association of Sanitary Inspectors.

Out of the work of the Royal Commission came the **Housing of the Working Classes Act 1885**. The Act required local authorities to use the powers they already had with regard to insanitary housing in order to achieve proper sanitary conditions of all dwellings in their district; it gave powers to make by-laws to deal with houses let in lodgings, and required the supervision of tents and vans used for dwellings.

With tenure pattern for housing being heavily dominated by rented dwellings, it is important to note that some charitable trusts both provided 'working class' houses and managed their properties with some enlightenment. In this regard, Octavia Hill (granddaughter of Southwood Smith) pioneered better housing management and, in particular, worked with the tenants of houses to improve their living standards.

In 1890, housing legislation was consolidated in the **Housing of the Working Classes Act 1890**. In terms of practical housing administration, the Act had three main parts dealing with unhealthy areas and improvement schemes, unfit dwelling houses, and powers for local authorities to provide lodging houses. This Act, both consolidating and pioneering, provided the administrative procedures and concepts that were followed in subsequent revisionary housing measures.

Although the housing legislation noted was enacted with good intentions, it was flawed to the extent that to implement it required local authorities to spend considerable sums of money which they had to raise themselves. At this time, local authorities were traditionally careful spenders of their revenue, and only in rare cases was there to be found a zeal for improving the local housing. Some of the large northern industrial towns engaged in longer term plans and were enterprising in finding legal authority for carrying out their plans of improvement.

AIR POLLUTION TO 1900

By the early 1850s, air pollution was still a major problem in London and the large cities. There was an increasing number of factories that burned coal, and the population continued to grow, which resulted in a rise in air pollution from coal-burning fires. The growth of mining, metal and chemical industries in the rural areas created air pollution in the countryside.

There was little effective control of air pollution in terms of either legislation or field enforcement. However, the situation was changing gradually, and failure to secure legislation focused attention on the problem.

Simon tried to secure some provision for smoke control in the city of London and eventually succeeded in 1851. While this measure applied only to the city, there was strong pressure for the powers to be extended to the whole of London. In 1852, a petition to parliament resulted in the promotion of an appropriate Bill to control smoke in London. The opposed Bill was pruned to meet the demands of critics, but it emerged as the **Smoke Nuisance (Metropolis) Act 1853**.

There was a bonus for this success in that the prime minister, Lord Palmerston, insisted that the Act should be properly enforced. Out of the defence arguments arose an interesting concept.

the danger now was whether or not the defendant had used best practicable means . . . . This formula, so common it became abbreviated to b.p.m., persists to this day . . . it began as a great obstacle to the enforcement of clean air laws, it evolved into an indispensible prescription for their effective enforcement. [7, p. 74]

There were further improvements to smoke control in the **Local Government Act 1858**, and the **Sanitary Act 1866**, but despite the power in the 1866 Act, local sanitary authorities were unable to reduce the smoke burden. There were three reasons for this failure. Some local powers under the Act were flawed by faulty procedural requirements; the fines imposed by magistrates were often derisory; and any improvement in industrial air pollution was offset by the increased pollution from the additional dwellings built to house the growing population.

THE EMERGENCE OF A CENTRAL POLLUTION INSPECTORATE

After about 1830, the alkali industry grew rapidly and large quantities of acid were discharged, despite the fact that by 1836 a remedy for this was known. Many plants in the chemical industry produced not only visible smoke, but also invisible fumes, which were dangerous to health and damaged the fabric of buildings and plants. Such emissions were fairly localized and usually came from plants situated in rural areas.

In 1862, this issue was brought before the House of Lords and a 'Select Committee was appointed to enquire into the injury resulting from noxious vapours evolved in certain manufacturing processes and into the law relating there to' [7, p. 21].

The prime mover in this, Lord Derby, was anxious not to make the issue one of landowners versus factory owners, but he wanted to discover whether legislative control was possible.

The select committee, advised by more skilled experts than had testified to earlier inquiries, reported very quickly. It recommended that there should be legislation enforced by independent officers free of local control and influence.

A Central inspectorate could ensure consistency, exchange information in control technique, and acquire an expertise that would not be possible amongst officials acting for a score or more Local Boards of Health, Boards of Guardians and the like. [7, p. 22]

The bill, introduced in 1863, fell short of the recommendations of the select committee, but it did create an inspectorate within the Board of Trade, and thus gave victory to the centralists. The Bill was opposed but was enacted as the **Alkali Act 1863**, and this made provision for the first **Alkali Inspectorate**.

The Act was a landmark. It confirmed the view that central government should take action to protect the public from noxious vapours; it allowed, for the first time, for inspectors to enter factories to protect not the workers, but people and property outside; the inspectors were to be experts from the beginning, would serve a central department and would be insulated from local interference.

The 1863 Act was to last for only five years; but it was renewed, the inspectorate became established and its role was gradually expanded. It devised methods of working that became

characteristic of its operation – patience and few prosecutions. In the 1880s, attention was being focused on industrial smoke in urban areas, and on domestic smoke in London. The struggle for clean air involved combating the argument that control action was not justified because smoke had not been proved to be injurious to health, although previous work had shown a correlation between smokey fog and mortality: in 1880, the Hon. F.A.R. Russell had published a book on London fogs in which he showed that they increased mortality; but because the effect was slow and diffused throughout the population, it received little attention.

OCCUPATIONAL HEALTH TO 1900

Much attention had been focused on the lot of the factory worker, and especially the juvenile employee; less note had been taken of the conditions under which shop workers laboured. Many shop premises were ill-ventilated, overcrowded with stock and had very limited welfare facilities. Shop assistants worked very long hours and had limited facilities for refreshment or rest. In many larger shops, the assistants lived on the premises, in attics over the shops.

In 1873, an attempt to limit hours of shop work failed but, because of pressure, a House of Commons committee looked at the position in 1886, and declared that the long hours worked normally were ruinous to health, especially to young women. Three hundred London doctors petitioned parliament to support the Bill before it. The **Shop Hours Act 1886** limited the working hours of young people to 74 per week including meal times. This Act was short-lived and a further parliamentary committee examined the position. Although the 1886 Act had failed, similar provisions were made in the **Shops Act 1892**, but even that Act was flawed and had to be amended in 1893 and 1895.

One group of workers whose practices were changed as a consequence of industrial development were the farm workers because of the introduction of mechanical aids. These increased productivity could be extremely hazardous, and specific legislation had to be introduced to give protection against some of the more dangerous machines.

As central government became aware of its inability to supervise the expanding number of premises that should be visited, intervention in industrial activities had to be increased. At first, the difficulty was met by increasing staff, but the **Workshop Regulation Act 1867** brought most manual workers under supervision and, experimentally, local authorities were involved.

The experiment lasted for four years, and was then abandoned. The reasons for the failure were that the powers given to local authorities were impracticable to work, and local authorities, on the whole, failed to administer the Act satisfactorily.

In some cases, there was deliberate inaction, and in others local influence adversely affected the local inspectorate. Thus, the **Factories Act 1871** enacted simply 'it shall cease to be the duty of the local authority to enforce the provisions of the Workshops Acts 1867–71 and it shall be the duties of the inspectors and sub-inspectors of factories to enforce the provisions of these Acts' [1, p. 230].

For some years, the enforcement of occupational health law was unsatisfactory. This was partly the consequence of changes in the legislation, which failed to provide either comprehensive powers or an efficient administrative system. Even the changes made by the **Public Health Act 1875** caused confusion.

By 1891, further factory legislation was required. Local authorities were again involved in occupational health issues, and in the **Factory Act 1891** they were made responsible for the sanitary conveniences in factories and for cleanliness, ventilation, overcrowding and limewashing in workshops. Local authority inspectors were given the same powers for their duties as factory inspectors under the principal Act.

In making this change, central government rehearsed all the arguments for not using local authorities but decided, on balance, that they had the staff and the local knowledge that was required. Local authorities were required to notify the factory inspectorate of workshops found within their area.

From the evidence available, it seems clear that local authorities performed rather better under the new Act, but still not well enough, and an 1895 Act required them to report back to the factory inspector on action taken to deal with complaints from him. This and other changes in local government attitudes led to a more vigorous enforcement and the employment of specialist officers to do the work.

CONSTITUTIONAL AND INSTITUTIONAL DEVELOPMENTS

A notable year for legislation that involved environmental health issues was 1875. In many ways, the **Public Health Act 1875** was one of the most important statutes of the century, for it contained specific provisions to improve public health, and it created an administrative framework within which local government was to develop in the following century or so.

Following the **Public Health Act 1848**, the public health service experienced many difficulties in establishing itself and, in particular, it never enjoyed the political protection of a high-ranking minister. Even when John Simon was responsible to the Privy Council for the public health service, the situation was little improved despite his enthusiasm, administrative skills and incisive reports.

After considerable pressure, agreement was reached in 1868 for the appointment of a **Royal Commission to examine the problems of sanitary administration**. There was a change in government, and the incoming Liberal administration appointed a Royal Commission in 1869. It was to examine the sanitary circumstances in England and Wales but excluded London, and it was to look at central as well as local organization.

Simon had strong views about what was needed – a strong central administration overseen by a minister, a system of local, all-purpose sanitary/public health authorities, and comprehensive new public health legislation building on experience and responding to contemporary need. Simon was not to see all of these achieved, despite the recommendations of the Royal Commission's report in 1871.

That report was accepted fully, and out of it sprang three Acts of Parliament.

In many ways, the most important of those three Acts was the **Local Government Board Act 1871**. This established the **Local Government Board**, which became responsible for poor law and public health functions, and other relevant ancillary activities. Under the **Public Health Act 1872**, an attempt was made to simplify the system of local authorities by mapping out areas for which a local authority would be created to exercise within its area all the Sanitary Acts, thus creating urban and rural sanitary authorities. The Local Government Board was also empowered to create sanitary authorities for ports.

While the Local Government Board was an important step in the creation of a comprehensive local government system that would enjoy specific parliamentary support, there were other factors to note. It was significant that the proposal for the 1869 Royal Commission was acceptable to the government, and to the opposition, which had initiated the proposal. Thus it was probable that future legislation would be easier to achieve because of an informal concensus of view.

Perhaps of greater significance was the change in informed public opinion towards issues of social importance, and with the weakening of *laissez-faire* influences interventionist policies became much easier to introduce and defend. Political power was shifting from centre and rightist bases, towards a somewhat more radical stance which underpinned the desire in influential quarters to extend the franchise into the lower middle classes.

In practical terms, the third Act, the **Public Health Act 1875**, made the greatest impact. This Act provided a comprehensive range of environmental health powers dealing with local authority areas and powers, sewage, drainage, water supplies, nuisances, offensive trades, the protection of food, infectious diseases, highways, street markets, slaughterhouses and the making of by-laws.

The Act of 1875 was drafted with some vision. Its provisions gave substance to an administrative system that was flexible and amenable to extension, and it made possible the very great progress made in the last quarter of the century in improving the sanitary circumstances of the country.

The Act was consolidated with amendments into **the Public Health Act 1936** and remained one of the principal pieces of legislation dealing with environmental health until well into the 1980s. Indeed, although now substantially repealed because of the environmental legislation of the last 25 years of the twentieth century, for example, **The Environmental Protection Act of 1990**, it remains in force and still contains important measures dealing with port health administration, sewerage, drainage and sanitary conveniences.

In 1876, the **Sanitary Institute of Great Britain** was founded. This organization was initiated by a distinguished group of people, including Chadwick, and was dedicated to the exchange of knowledge and to the examination of candidates for professional qualifications in surveying and as inspectors of nuisances. It later became the **Royal Sanitary Institute**, and is now the **Royal Society for the Promotion of Health**. For many years, it continued to examine sanitary inspectors, but later formed part of the examining body known as the Royal Sanitary Institute and the Sanitary Inspectors Examination Joint Board. The Royal Society of Health continues to have a diminishing interest in the overseas qualifications of environmental health officers.

In 1883, the **Sanitary Inspectors Association** was founded as an amalgamation of smaller local bodies, and the organization has continued, using various designations, to expand in membership and influence. It currently operates under a Royal Charter as the **Chartered Institute of Environmental Health**. Its first president was Chadwick, and while formerly the members of the Institute were all qualified environmental health officers, other grades of membership are now open to people holding some approved qualification appropriate to environmental health practice.

THE NOTIFICATION OF COMMUNICABLE DISEASES

A flaw in the Public Health Act 1875 was its failure to create a system for the notification of cases of infectious diseases. This was a long perceived need and medical officers for areas where communicable diseases were most prevalent – the densely populated inner cities – had recognized the impediment to epidemiological enquiries caused by the lack of the information that notification would provide.

Some local authorities took private act powers that enabled them to require notification of some diseases. Fourteen years were to elapse before a general notification requirement was attempted. In 1889, the **Infectious Diseases Notification Act** was passed which allowed local authorities to adopt powers to require notification.

This was an inadequate measure, but it remained for a decade until supplemented by the **Infectious Diseases Act (Extension Act) 1899**, which required all sanitary authorities to adopt the 1889 Act. This listed eleven diseases that had to be notified to the sanitary authority and, furthermore, it empowered the Local Government Board to list other diseases by order.

Although the notification of communicable diseases was not compulsory until near the end of the century, various powers existed that allowed local authorities to act to prevent such diseases, but the lack of knowledge of **all** cases clearly hampered satisfactory control. In this context, Frazer wrote, 'The system of notification . . . has proved to be of the greatest possible value to medical officers of health in dealing with epidemics because early notification is vital to prevent further cases arising' [4, p. 181].

The compulsory notification of specified communicable diseases was consolidated in the **Public Health Act 1936**. Subsequently, further diseases were added to the list either by regulations or, notably, by the **Food and Drugs Act 1938**, which required the notification of food poisoning. There was a further consolidation in the **Public Health (Prevention of Disease) Act 1984**.

The control of communicable diseases at ports of entry was a feature of the early development of public health (see Chapter 4).

REGISTRATION OF BIRTHS AND DEATHS

Apart from the 10-yearly census, public authorities had few authoritative statistics to guide them. A select committee of the House of Commons was appointed in 1833 to investigate the position with regard to births and deaths, and from their recommendations came the **Births and Deaths Registration Act 1836**.

The objectives of this Act were to establish the General Register office, to appoint a **Registrar General**, and to institute a system of registering all births and deaths. Before that date, the only statistics available were to be culled from parish registers.

The reputation achieved by the Registrar General's office in the next 40 years was mainly due to Dr William Farr, who was appointed as compiler of medical statistics. His work was pioneering and his reports were fundamental to the development of epidemiology and demographic studies. Farr's appointment is believed to have been due to Chadwick, and Farr is numbered along with Chadwick, among the great pioneers of public health.

MINISTRY OF HEALTH

Noted above was the formation of the Local Government Board. Because of subsequent developments and extensions in the public health and poor law services, the Board was identified as being a less than satisfactory administrative machine.

The formation of a Ministry of Health had been advocated for many years by various people for a variety of reasons. By 1918, it became obvious that changes had to be made in public health administration and, in 1919, the Ministry of Health was established to take over the functions of the Local Government Board and other duties. It was made responsible for some social services and most activities that had a health connotation, although a notable exception was the retention of the occupational health service by the Factory Inspectorate. The first Minister of Health was Dr Christopher Addison, later associated with housing legislation.

COMMUNICABLE DISEASES SINCE 1900

Although there were no further epidemics of cholera, other communicable diseases continued to be important in the first half of the twentieth century. The virulence of scarlet fever appeared to decline after about 1883, but it is difficult to assign a reason.

The disease with which scarlet fever was most associated, diphtheria, was, however, to remain a serious cause of infant mortality until the 1940s. There was a slow decline to around 300 deaths per million children under 15 per annum in the period 1921–5, and that remained a plateau until 1940 when immunization virtually eliminated the disease.

Although antitoxin was first used to treat the disease in 1895, and a satisfactory immunization regime was available in 1923, it was not utilized until 1940. Subsequently, a continuous immunization campaign has kept the disease under control with only the occasional case arising.

One communicable disease that was a potent cause of infant death was **summer diarrhoea**. The incidence and mortality rate had already been declining for 20 years, but in 1911 there were still 31 000 deaths of infants under 1 year due to diarrhoea; by 1931 this had been reduced to 11 705. There was medical controversy about its cause, and the state of bacteriological science at the turn of the century was such that it could throw no positive light on the causes or the mode of transmission.

The best opinion took the view that it was fly-borne, and this was an opinion well supported by the fact that the disease was prevalent in summer when flies bred in back street stable middens, and that the disease declined when the horse ceased to be a major form of transport. There were other factors in reducing the disease's incidence, such as higher standards of domestic hygiene and baby care.

For **tuberculosis**, a 'social' disease, both the incidence and the mortality rate continued to fall, although there were temporary distortions in the pattern in both the time and place of cases. The environmental conditions imposed by two world wars and economic depression encouraged continued decline. Factors that have contributed to the improved circumstances include better nutritional standards, improved domestic hygiene, better occupational hygiene, reduced working hours

with less physical stress, reduced overcrowding and better housing conditions.

Immunization against tuberculosis has played a part, as has the more recent (post-1945) availability of chemotherapy. Nevertheless, there have been some disturbing trends in urban communities with a high proportion of ethnic populations, or where there is poverty. Immigrants may be at especial risk where they arrive from areas where there has been no chance to acquire natural immunity and where, on arrival, dietary changes and poor housing may be contributory factors.

The elimination of **smallpox** in 1980, following the WHO campaign, has been a relief to public health workers. Two varieties of smallpox have been the cause of epidemics in the United Kingdom.

Although for many years smallpox was epidemic in Britain it is probably true that since 1914 major smallpox has always been exotic though it has not always been possible to prove this as source tracing has not always been successful. [2, p. 56]

The public health control of smallpox developed because it was a serious disease with a high mortality rate. The control encompassed vaccination, notification, isolation, and effective disinfection regimes. After 1906, the incidence of Asiatic smallpox – variola major – dropped to negligible proportions, but variola minor continued to be a serious problem, with 70 000 cases in the United Kingdom between 1925 and 1931. The elimination of smallpox as a human contagious disease represents a triumph for international control and co-operation. It is ironic that probably the last cases of smallpox occurred in the United Kingdom, the home of vaccination, through the accidental release of laboratory specimen viruses.

The incidence of communicable disease does vary and illnesses have different importance at different times. In the 1980s and 1990s, sexually transmitted disease was emphasized with the spread of HIV (Human Immunodeficiency Virus) and AIDS (Acquired Immune Deficiency Syndrome). Cholera became more widespread internationally, and the risk of importing cases through air travel was significant. Domestically, meningococcal meningitis became increasingly common, and there was a high degree of awareness of the risk of importing rabies. Also important have been outbreaks of Legionnaire's disease and food-borne disease due to *Listeria and E. Coli.*

In January 2002, the Chief Medical Officer of Health's report 'Getting Ahead of the Curve: A strategy for combating infectious diseases' was published [8]. The main recommendation was to combine the existing functions of four organizations and a range of professions into one agency; the Health Protection Agency (see Chapter 12).

TREATMENT OF COMMUNICABLE DISEASES

The high incidence and mortality of communicable diseases in the nineteenth century reflects the poor domestic environment of most people, a low standard of medical and nursing care because medical science had yet to make its major advances and the lack of drugs to treat the diseases. That the position gradually improved was due to the provision of drainage, safer water supplies, better refuse disposal, improved housing and better diet.

Significant discoveries concerning the causative organisms of many serious diseases were not made until the last quarter of the nineteenth century.

Modern chemotherapy may be said to have started in 1910 when Paul Ehrlich discovered Salvarsan (arsphenamine) and made the treatment of syphilis possible. Perhaps the next important step was the introduction in the 1930s of the sulphonamide drugs, which were a product of the dyestuffs industry. These transformed the treatment of many diseases, as did the introduction of penicillin, originally discovered by Sir Alexander Fleming in 1928, and later isolated and developed by Ernst Chain and Howard Florey during the Second World War. In the post-1945 period, the development of new drugs and antibiotics has proceeded apace.

Perhaps beginning with vaccination against smallpox, introduced by Edward Jenner in the late eighteenth century, there have subsequently been major advances in dealing with communicable diseases by the development of immunization regimes. Vaccination gives protection against infection, and modern thought is still that 'prevention is better than cure'.

HOUSING SINCE 1900

At the beginning of the twentieth century, the basic housing legislation was still the **Housing of the Working Classes Act 1890**, although amending Acts were passed in 1900 and 1903. Despite this legislation, the standard of housing for poorer people left much to be desired, although there had been some slum clearance and house building in London and northern cities.

A feature of the growth of the built environment was that not until 1909 were there any provisions for town planning. The planning powers in the **Housing and Town Planning Act 1909** were permissive for local authorities, which showed little enthusiasm to exploit its provisions, although some did. The provisions, permitting the control of land use, were intended to act as a brake on unrestricted development and had little regard to amenity and community interests. House construction was subject to little control because of the inadequacies of building by-laws.

As far as housing was concerned, the 1909 Act amended the **Housing of the Working Classes Act 1890**, and made it mandatory in all districts. It prohibited for the first time the erection of back-to-back houses, which were still being constructed in some northern industrial cities. The Act also recognized that public utility societies, such as building societies, could be responsible for house building.

The relationship between planning and a satisfactory built environment had long been recognized. In 1898, Ebenezer Howard published his seminal work, *Tomorrow, a Peaceful Path to Real Reform*, which advocated the building of new garden suburbs that were self-sufficient and isolated from other suburbs by a green belt, but with easy means of communication between them. Howard, a practical person as well as a philosophical theorist, was the driving force of the Garden Cities Association, and he witnessed the successful founding of Letchworth Garden city in 1902. The 'new town' movement developed largely from Howard's initiatives.

There was a further Housing and Town Planning Act in 1919, but the housing provisions of the 1909 and 1919 Acts were repealed by and re-enacted in the **Housing Act 1925**.

Because of the shortage of houses and promises made to servicemen returning from the First World War, attempts were made to stimulate house building, and powers were taken in the **Housing Act 1919**. The shortage of houses is indicated by the fact that nationally in 1911, 9.1% of the population lived at a density of more than two per room, and by 1921 the census showed that this had risen to 9.6%. With the aid of government subsidies, 176 000 houses were built under the 1919 Act by local authorities at considerable cost.

Further Housing Acts in 1923 and 1924 were designed to continue the stimulation of house building, but at a less extravagant rate of subsidy. This legislation was successful, and by 1927 an annual output exceeding 270 000 dwellings was achieved.

These 1920s Housing Acts were designed to stimulate house building, but they did little to address other housing problems, and the much amended and extended 1890 Act was no longer able to meet current needs. In view of this, the existing legislation concerning the repair, maintenance and sanitary condition of houses was consolidated by the **Housing Act 1925**.

By the late 1920s, it became apparent that much remained to be done to deal with the unsatisfactory housing situation. Building houses for sale rather than for renting was more profitable and the output of rentable houses declined. It was estimated that there were several million insanitary houses:

The outside estimate . . . was 4 000 000 but . . . that would depend upon the definition of a slum. In other words, a slum is what the Medical Officer of Health of the District believes to be a slum. What was beyond doubt,

however, was that the slum evil in many of the industrial towns, especially London, Liverpool, and Glasgow was of vast dimensions, and that much harm to the Public Health was being caused by the delay – inexplicable to many people – in dealing with this problem as a matter of the greatest possible urgency. [9, p. 60]

The **Housing Act 1930** was the government's response. It was not a fully comprehensive measure as it omitted to deal with overcrowding, but it made provision for the clearance of insanitary housing, and it prescribed procedures to be followed to make clearance and improvement areas. It required housing accommodation to be provided in advance for those to be rehoused, and created the principle that compensation should not be payable in respect of unfit houses.

With regard to the proposed improvement areas, which were envisaged as being large areas of houses and other properties, local authorities were empowered to demolish or repair unfit dwellings, buy land and demolish buildings so as to leave a developable site. Local authorities had to deal with overcrowding, which parliament failed to define.

The concept of the improvement area was directed against urban decay. While local authorities used the clearance powers widely, they were reluctant to experiment with improvement areas.

The **Housing Act 1935** made changes in the procedures for dealing with redevelopment, although this was little used. The Act did, however, tackle the problem of overcrowding by defining it. It required local authorities to survey all houses below a given rateable value and to certify the number of people allowed to dwell in such houses. Local authorities were also empowered to deal with overcrowding cases of which they became aware. The **Housing Act 1936** consolidated previous practical, as opposed to financial, housing legislation, and considerable slum clearance and repair activity was stimulated. The Second World War brought virtually all housing activity, slum clearance and new building to a stop. What repair resources were available were diverted to dealing with war damage, so that there was little routine maintenance of houses.

At the end of the war, housing was in extreme stress. Physically, the housing stock had been depleted by air raids; many houses had drifted into unfitness because of lack of maintenance; few of the formerly unfit houses had been demolished; and very few new houses had been built. Sociologically, the position was exacerbated by the effective demand for higher housing standards; by the abandonment of houses perceived as being unfit for habitation; and by the increased rate of household formation.

The housing question became important politically and sociologically, as well as from the aspect of public health. It became apparent that future action would have to encompass the removal of unfit housing, the elimination of overcrowding, and the repair and improvement of the housing stock. In the years after 1948, there was a succession of Housing Acts designed to promote competing political programmes for house building, slum clearance and improvement. The **Housing Act 1949** made the first provision for the improvement of houses through grant aid although, in the light of subsequent developments, this was an experimental and meagre approach.

The **Housing Act 1954** required local authorities to survey all the houses in their district to determine which dwellings were unfit for habitation. These surveys revealed the magnitude of the task. In the big cities, the rate of unfitness was staggering. In Birmingham, 16% of all houses were judged to be unfit, in Manchester 33%, and in Liverpool 43%.

The **Housing Act 1957** was a fundamental restatement of housing law that made significant changes to the law relating to repair and clearance of unfit properties. For the first time, England and Wales were given a statutory definition of what constituted a fit house, and that definition was wide enough to permit a liberal interpretation of what constituted unfitness. Prior to that the standard of unfitness had to be related to the general local standard of housing; the new definition applied nationally and allowed districts where housing standards were low to bring their housing gradually up to the national level.

The **Housing Act 1964** contained important provisions relating to the supply of water to houses and the possibility of grant aid in that regard. Attention began to focus on other aspects of housing that could affect human health, such as the improvement of houses, the universal provision of standard amenities, and the upgrading of whole areas by the repair and improvement of the houses and softening of the external environment.

The **Housing Act 1969** was largely the legislative outcome of the report of the **Denington Committee** (*Our Older Homes – a call for action*). It suggested the need for a fresh approach to house improvement, and emphasized the impact a bad external environment had on housing. Out of this legislation came the concept of the general improvement area.

It was, of course, impossible to discuss housing except by understanding what constituted a satisfactory house. While the Housing Act 1957 had set out a standard of fitness, it did not address such questions as space and amenity.

In 1953, a working party was set up under Parker Morris to inquire into, and make recommendations for, standards of accommodation in houses. The report from that working party was well received, and the standards it promoted were widely adopted by local authorities and better house builders. However, at a later date, in the interests of financial economy, these standards were less well observed.

In the 1970s and 1980s, there were further Housing Acts. Most of these were designed to liberalize and stimulate the repair, renovation and, especially, the improvement of houses individually and in areas. With the **Housing Act 1974** came the concept of the **housing action area**, where the intention was to provide minimum repair and improvement to houses that were of poor standard and likely to be included in clearance proposals within 10 years. This was an imaginative approach to the challenge to provide early improvement of living conditions in houses that were approaching the end of their useful life. Sadly, the impact of inflation on building and improvement costs made the schemes impracticable except to a standard of repair and improvement to ensure a 30-year life.

This Act also made radical, and enhancing, changes to the improvement grants schemes, and initiated a decade of great activity in the rehabilitation of houses.

The **Housing Act 1980** was designed to tidy up the provisions for house improvements, but it also made radical changes to tenure patterns, including the power of tenants of local authority-owned dwellings to purchase the houses at substantial discounts. This was a right denied to tenants of rented houses in other ownership.

By the early 1980s, it was realized that the post-war housing legislation needed consolidation, and this was achieved in the **Housing Act 1985**, which drew the threads together but made few changes. The main object of the **Housing Act 1988** was to alter, again, the pattern of tenures. The main thrust of the previous decade had been towards increased owner-occupation of houses, and houses sold out of the rented stock for owner-occupation obviously diminished the rented sector pool, which caused difficulties. By introducing new forms of tenancy, it was hoped to induce more owners to put houses into the pool for renting, but this did not occur.

The Act also introduced a scheme to create housing action trusts in which it was hoped that large, run-down council housing estates would opt for management by such trusts, which were promised liberal support denied to local authorities to enable them to deal with their own houses.

It had also become apparent that the liberalization of the house improvement scheme had led to much grant aid being given for properties not in the greatest need of repair and improvement, and the **Local Government and Housing Act 1989** aimed to redress that balance. There was also a great need to make the improvement grant system less cumbersome to operate. The 1989 Act addressed these problems, making radical changes in housing law, including a new concept of the fitness of a house, the nature of grants available (but retaining the concept of a mandatory grant), and provided for the testing of an applicant's ability to contribute to the cost. These new provisions and procedures made most local authorities revise their private housing sector activities.

There was a continuing plethora of housing legislation in the 1990s including the **Housing Grants, Construction and Regeneration Act 1996**, which dealt with grants and other assistance for housing purposes, and the **Housing Act 1996** which included provisions relating to houses in multiple occupation. The former has been superceded by a new regime of financial assistance for improving housing which was implemented through **the Regulatory Reform (Housing Assistance) (England and Wales) Order 2002** and is described in Chapter 17.

In the 1980s, there was a significant shift in political opinion in respect of public involvement in housing policies, from considerations of the health and welfare of house occupiers, to financial and sociological aspects. Local authority tenants were empowered to buy their houses, thus introducing enclaves of private property into monolithic, council-owned estates. It gradually became very difficult for councils to build new houses to enable them to continue to provide rented accommodation. Increasingly, the private sector and housing associations were entrusted with the provision of rented housing, while the drive towards increased owner-occupation encouraged private house building.

The statutory overcrowding provisions remained, but were largely ignored as being irrelevant in the postwar era, with most local authorities working to their own higher standards.

Throughout the industrial period, dwellings used for multiple or common occupation have been a problem. Multiple occupation continued as a serious issue in many towns and, although grant aid became available for improving such properties, many continued to be grossly unsatisfactory. There was much pressure by interested groups to secure legislation that would permit more effective control. The worst housing provision has always been considered to be the **common lodging house**. In the interwar years, the privately owned common lodging house run for profit virtually disappeared; those that remained were frequently owned by charitable organizations and were developed on almost institutional lines.

In the 1970s and 1980s, the situation changed as a number of smaller properties were adapted. This was usually done by a charitable body to provide sheltered accommodation for small groups of socially vulnerable people. In addition, former commercial premises were often adapted as night shelters for people who would otherwise sleep in the open. The common lodging house remains a social and public health hazard, but one that often needs a compassionate approach, although there is now no separate housing legislation dealing with them.

FOOD SINCE 1900

Although local authorities had had powers to inspect food for many years, it was becoming apparent that some control over the premises in which food was handled was required if clean and safe food was to be available. The **Public Health Act 1925** introduced the first powers to deal with food premises and, although the requirements were modest, they extended to water and washing facilities and cleanliness in certain food premises. A start had been made on the road to better food hygiene.

By 1920, large scale adulteration of food had ceased, although there were foods that were easy to adulterate, a typical example being milk. The basic law with regard to food adulteration had remained unchanged since 1875, and clearly needed to be consolidated. This was done in the **Food and Drugs (Adulteration) Act 1928**, which was to remain in force for a decade. The Act neither re-enacted the food safety provisions of the 1925 Public Health Act, nor included supplementary food safety powers.

The Ministry of Health advised through Memo 36/Foods how the Act was to be operated. A similar ministry memo (Memo 3/Foods) laid down inspection standards with regard to meat.

Some local authorities understood that their powers to deal with food-borne diseases and insanitary conditions in food premises were inadequate and, especially in the 1930s, through private act legislation, they took some regulatory powers of registration of some types of food premises and their occupiers.

The **Food and Drugs Act 1938** was both consolidatory and innovatory. It had three main parts: the protection of food supplies, the sampling of food and drugs and the control of certain food

premises. It made provision for the inspection of food and for the control of some premises in which food was handled. It gave the ministers powers to make regulations in respect of the registration of some food premises and some practices of food handlers. It reaffirmed sampling procedures and the provisions with regard to warranties, and it made possible the control of such premises as slaughterhouses, knacker's yards, cold stores and markets.

Regrettably, the new Act did not come into force until after the outbreak of the Second World War, but although its full implementation was delayed, especially with regard to the potential advances in food hygiene legislation, the war period was particularly active and productive in terms of food control. Faced with food shortages, a rationing and food control system ensured that the diet of the public was better than it had ever been. To ensure that food was the most nutritious available, a considerable number of regulations were promulgated, and most of these later found a place in permanent food law.

The 1938 Act made provision for ministers to introduce regulations on a variety of topics related to food safety, and in 1955 the first regulations designed to provide a food hygiene code were introduced. These were for general application to food premises, and were followed by comparable regulations for docks and warehouses, market stalls and delivery vehicles, and slaughterhouses. These were timely because in 1954 the meat trade ceased to be subject to control. The slaughterhouses that had been closed by the policy of concentrating slaughtering facilities were able to apply for the renewal of their licences, but very many failed to achieve the standards required and remained closed.

After the 1938 Act had been passed there were numerous changes in the food industry and its practices, especially in the replacement of small factories and shops and the development of larger scale production, distribution and retail units. There was a need for the modernization of the core legislation. This was achieved in the **Food and Drugs Act 1955**. Much food control law was contained in regulations that could be easily changed but, nevertheless, that law continued to lag behind developments in food technology and retailing, areas that were perpetually responding to market forces to meet changes in taste and lifestyle.

After the 1955 Act, there were many updating amendments to existing food law, particularly to the regulations, as attempts were made to meet current needs. Other developments were to come. After the experiences of the period between 1920 and 1945, there was a view that food adulteration was a solved problem. After about 1950 the industry changed to the large scale processing of food, which resulted in a lowering of compositional standards.

The post-1945 period saw a major increase in the number of cases of food-borne diseases. While notified cases of food poisoning had increased significantly, much attention was focused on unnotified incidents. Enquiries into causes and effects have suggested that changes in diet and dietary and social habits were major causes. Also implicated was poor food hygiene, which resulted in a call for more public information and training of food handlers.

The increasing importance of good food hygiene brought the realization that it was difficult to act swiftly and decisively to deal with food premises that were so insanitary as to pose a serious risk to health. In 1971, the city of Manchester took powers in the **Manchester Corporation Act 1971** to enable its environmental health officers to close insanitary food premises by direct application to the courts. The value of this power was clear and the **Food and Drugs (Amendment) Act 1976** gave similar, but less rigorous, powers to local authorities generally. The Manchester powers, slightly modified, were applied to the other nine metropolitan district councils in Greater Manchester when, countywide, local act legislation was consolidated.

There was pressure from all sides to review the food law and in 1977 the ministers asked for the views of interested bodies on how the food law should be amended or recast. Many organizations, including the Institution of Environmental Health Officers, responded. However, no action took place except to find parliamentary time to produce the **Food Act 1984**, which consolidated the existing law but made few changes of significance.

From 1973, the whole issue of food legislation was complicated by legislation from the European Economic Community (EEC), as it was then called. By the time the United Kingdom joined the EEC there had been a great deal of legislation, including laws relating to food, that had to be accepted as it was.

After 1973, the United Kingdom was able to participate in the discussions that preceded further food legislation, but was always at a disadvantage because Continental systems of food control were very different from those in the United Kingdom. There followed a considerable flow of statutory instruments converting EEC directives into UK law.

From the 1970s into the 1980s, the many organizations that had an interest in food legislation continued to press for amending legislation. This was met with little success until, after a series of food safety incidents, central government conducted a survey of views and issued a white paper on food safety. In due course, the **Food Safety Act 1990** was passed.

This legislation provided some satisfaction for those who had been pressing for change. However, although the Act conceded many principles that had been urged upon the government, the actual fulfilment of those principles was to be met largely by regulations. Control through statutory instruments has important advantages as they are easier to amend or extend, but they are also normally the means of writing in the detail to clothe the principles in the legislation and, as such, may fall short of real expectations or needs.

OCCUPATIONAL HEALTH AND SAFETY SINCE 1900

Factory law was reviewed, extended and consolidated by the important **Factory and Workshops Act 1901**, which was divided into 10 parts covering many aspects of industrial employment.

For many years, the work of the Factory Inspectorate had been increasing, and it had been difficult to do enough inspections. Despite the indifferent performance of local government in the health and safety duties assigned to it, it was considered imperative that its role should be extended. As a result, local authorities were made responsible for sanitary accommodation in all factories, and for creating a more effective system for dealing with 'homework'.

When the Shops Bill was before parliament, there was opposition to bad working environments and excessive hours. In 1910, Hallworth and Davies said that a large number of shop assistants in London worked all their waking hours on most days of the week [9, p. 64]. At a drapery store employing about 30 assistants, the staff worked until at least 10 p.m. on five nights of the week: 'They were herded at night into a dilapidated house in one of the neighbouring mean and dirty streets and, although they seldom left the shop before 10 o'clock at night, they had to be in the dormitory by 11 o'clock' [9, p. 64].

There was an unsuccessful attempt in 1909 to give shop workers some of the health protection that factory workers had, and there was a wish that the legislation, not then usually an environmental health function, would be enforced by the Factory Inspectorate.

However, the **Shops Acts 1912** and **1913** brought consolidation of the law and some improvement. But the law was still insufficient and inadequately enforced, and this demonstrated that shop workers' conditions, like those of the factory workers, were being improved in a piecemeal fashion.

The legislative position remained unchanged until the **Shops Act 1934**, which introduced important changes. It controlled the hours of work for those under 18 years of age, and for the first time it introduced health and welfare provisions for workers in shops and those in offices attached to shops.

Shop working hours were to be enforced by Shops Act authorities (county and county borough councils). The welfare provisions, which included sanitary and washing facilities, heating and ventilation, facilities for taking meals and the provision of seats for female workers, became the responsibility of sanitary authorities (rural and urban districts, municipal boroughs and county boroughs).

Up to this time, environmental health officers had had a declining interest in work involving workshops as such places decreased in number

and became factories in which, as officers, they had much more limited responsibilities. The Shops Act 1934 changed this, and there was a renewed interest in occupational health. On the whole, the work of the local authorities under the welfare provisions of the 1934 Act was so well done that it was later to lead to a greater involvement in occupational health and safety.

Although better provision was made for shop workers, there were still many non-industrial workers who had no occupational health protection at all, and this was to remain the case for some time. In 1946, the government set up an interdepartmental committee, chaired by Sir Ernest Gowers, to look at the needs of non-industrial workers. **The Gowers Committee** reported in 1949, and its report had a delayed, but important, impact on occupational health and safety legislation. The consolidating **Shops Act 1950** owed little to the influence of the Gowers Report. The **Factory Act 1951** was a consolidation of previous legislation, and it met the perceived need of the industrial worker. However, it gave little further protection to land workers or those in non-industrial employment.

Postwar developments in agricultural practices, machinery and the chemical control of weeds and pests posed a substantial threat to the health and safety of the workers involved. The first occupational health legislation that implemented recommendations of the Gowers Committee was the **Agriculture (Safety, Health and Welfare Provisions) Act 1956**. The enforcement of this Act was entrusted mainly to the Agricultural Inspectorate and not to local authorities, whose role under this Act was equated with their, then, limited duties under factory legislation.

The **Offices Act 1960**, a private member's bill, followed Gowers' recommendations. The Secretary of State could make regulations, after consulting interested bodies, prescribing standards for offices. However, such regulations were never made, probably because other measures were being prepared, and the Offices Act 1960 was never put into operation.

In 1963, the **Offices Shops and Railway Premises Act** was passed. It excluded many non-industrial workers, but it did cover offices and wholesale and retail premises. Most enforcement duties were assigned to local authorities, and records of local authority activities were subsumed into national statistics. Factory inspectors had some concurrent powers, and Superintending Inspectors of Factories had oversight over local authority work to ensure uniformity of practice and standards of enforcement. Under this Act, regulations were made prescribing health and safety standards. Environmental health officers, to whom the local authority role was largely assigned, were experienced in the hygiene of buildings, and the administration associated with legislation, but they needed to learn new skills in ergonomics and the safe operation of powered machinery.

When the Bill was before parliament, there was opposition to it being made a largely local authority function. In theory it is easier for a central authority to achieve common standards of performance and enforcement, and there can be little opportunity for local influence to minimize the scale of activity. But central authority is never likely to have the local knowledge of the pattern of the problem; or to be subject to the constant probing of the elected members of councils responsible for enforcement; or to have an inspectorate that is as sensitive to local cultural patterns and traditions as it should be. In contrast, centrally organized inspectorates can invest heavily in specialist skills and supporting scientific staff.

The 1963 Act was important to local authorities in respect of their involvement in the enforcement of occupational health and safety, and it gave them an opportunity to make a mark.

Although this Act was a large step forward, it still did not protect all non-industrial workers, and the law needed further extension. Legislation to protect workers had been developed on an *ad hoc* basis with separate inspectorates for different working environments, and there were gaps and overlaps in jurisdiction. It was obvious that the system needed a radical overhaul to enable it to deal with changes in technology, patterns of trade and commerce, lifestyles and public expectations.

It was clear that conditions within the workplace could affect the public outside it. Processes were often hazardous to the worker on the shop floor and to others in and outside the factory. Many thought that in order to promote safety, healthy working conditions and to avoid prosecutions for infractions of the law, improved education and training were needed. Others urged co-operative efforts to reduce hazards, improve health and replace coercion and confrontation.

In 1970, a **Committee of Inquiry** was set up under the chairmanship of **Sir Alfred Robens** to review the system of securing health and safety at work. It reported in 1972 [10] and it recommended the following:

1. A more united and integrated system to increase the effectiveness of the contribution to the health and safety at work made by the state.
2. Conditions for more effective self-regulation by employers and employees.
3. A Health and Safety Commission with an executive arm to be called the Health and Safety Executive.
4. A continued enforcement role for local authorities which would ultimately be responsible for most of the non-industrial sector.

The committee also laid down two other principles: no self-inspections or dual inspections were to be undertaken.

The recommendations of the committee found expression in the **Health and Safety at Work, Etc. Act 1974**, which brought six significant changes for environmental health authorities. Their role was to be expanded and the officers they appointed to administer the Act would have concurrent powers with Health and Safety Executive officers. Third, the subordinate role given under the 1963 Act was removed, and the local authority inspectorates were given equality of status with those of the Health and Safety Executive. Additionally, the changes required local authority inspectorates to recognize the Health and Safety Executive as the commission's administrative machine, and to support and co-operate with the Health and Safety Executive field staff.

The sixth change was in the administrative procedures. Although these brought new concepts to environmental health departments, they were readily and advantageously adopted.

The new creation was not perfect. Environmental health officers felt aggrieved at the time, because continued Crown immunity meant that they were excluded from inspecting civil service establishments (except those of the Health and Safety Executive), and they could not deal with premises occupied by other local authorities that were located in their area. It was also recognized that environmental health officers would need additional skills. Factory inspectors saw local authority officers as a challenge, an attitude later exacerbated by civil service retrenchments.

There was an unsubstantiated fear that local authority elected members would interfere with their officers' operations. The civil servants were ill-informed about the local government system, its traditions, training, and capacities and the practical achievements of the environmental health service. These were matched by a similar misunderstanding in local government circles of the civil service role, qualifications and training.

If the two major arms of enforcement were to work harmoniously to produce an effective national system with uniform standards of performance and enforcement a system of liaison had to be introduced. Nationally, a liaison committee, Health and Safety Executive/Local Authority Committee (**HELA**), representing the Local Authority Associations and the Health and Safety Executive was established as a forum to discuss problems, especially in the field of common standards, information exchange, statistics and developments, and to make recommendations for changes perceived to be necessary.

The HELA set up subgroups to have regard to training, standards of enforcement and statistics and, in 1984, a Local Authority Unit was established. This unit, designed to give local authority views open access to the system and to harmonize practices and approaches, was staffed, in part, in technical aspects, by environmental health officers and factory inspectors. It has been responsible for developing uniform standards and raising the local authority profile in this important work.

Locally, a senior factory inspector is appointed in each Health and Safety Executive area to act as the local Enforcement Liaison Officer, whose remit is to assist in the solution of local

demarcation and technical problems, and to call periodic meetings of local authority counterparts for discussions and exchanges of experience.

The local authority role was defined in the **Health and Safety (Enforcing Authority) Regulations 1977**. It was always the intention to extend the local authority role and, in 1983, discussions started with the object of revising these regulations. But despite agreement between the officers engaged in this task, there was considerable delay in new regulations (**Health and Safety (Enforcing Authority) Regulations 1989**) being promulgated, and the changes made were less extensive than had been originally envisaged. **The Health and Safety (Enforcing Authority) Regulations 1998** have now replaced those of 1989 and widen the local authority role (see Chapter 19).

AIR POLLUTION CONTROL SINCE 1900

This period is characterized more by a better understanding of the problems of air pollution, the emergence of air pollution monitoring and the growth of organizations committed to securing improved air quality, than by improvements in pollution control legislation. The emergence of pressure groups arguing for powers to deal with polluted air began at the turn of the century.

By the 1880s, the relationship between air pollution, fog formation and increased mortality had been accepted. In London and other big cities, choking winter fogs were commonplace. A **Fog and Smoke Committee**, an organization that had the practical help of many distinguished people, was established in London in 1880. However, the precise causal relationship between air pollution and ill health had still to be defined.

The political difficulty of controlling domestic smoke was also understood, and the Fog and Smoke Committee was determined to try educational means to overcome the problem.

In 1882, the Fog and Smoke Committee became the **National Smoke Abatement Institution**, which developed into a powerful lobby but was replaced in 1896 by the **Coal Smoke Abatement Society**. While this organization was primarily a London body, which for some time employed its own smoke inspectors, similar organizations were also formed in northern industrial towns.

Manchester, in many ways the cradle of the clean air movement, had a special subcommittee to research into air pollution. In 1909, it became the headquarters of the **Smoke Abatement League of Great Britain**, which was an amalgamation of a number of societies including Manchester's own **Noxious Vapour Abatement Committee**. It is important to emphasize the interest that these local societies had generated, and how they recognized that the strength of a unified organization was more likely to be an effective pressure group. Although there were now two national organizations, the League sought the co-operation of provincial local authorities leaving London to the Coal Smoke Abatement Society.

The First World War restricted the activities of the two major groups, but they resumed their campaigns after 1919. The Coal Smoke Abatement Society continued its pressure on politicians, and this led to a Ministry of Health Departmental Committee in 1926 and, later, the **Public Health (Smoke Abatement) Act 1926**. This Act was of minor importance in dealing with the smoke pollution burden, but it marked the increased interest in air pollution, and led to changes in the attitudes of central government.

There were some interesting developments in the 1920s. In conjunction with a London environmental health officer, E.D. Simon, a leading industrialist, politician and former Lord Mayor of Manchester wrote a book entitled *The Smokeless City* [11]. Apart from showing that the significance of domestic smoke was understood, the book makes clear that fuel and appliance technology that could have reduced air pollution from domestic fires was already available.

Although the positive relationship between air pollution and ill health had still to be established, in 1923 Dr James Niven drew attention to the air pollution in Manchester, the diminished sunlight and the high incidence of pulmonary and other diseases.

The increasing interest in air pollution during the mid-1920s induced central government to consider providing local authorities with advice. When the Alkali Inspectorate declined to be involved, it was decided that the task of collecting information should be given to one of the three Ministry of Health officials. Ministry minutes of the day recorded resistance to the idea of promoting further smoke control measures. It was considered that the public was not ready for measures as no causal connection had been established between polluted air and ill health. Whoever wrote that was unaware of the documented circumstantial evidence that showed a clear correlation between smoke, fog and increased mortality.

J.C. Dawes was then selected to be the advisory officer. He was a refuse disposal expert, an environmental health officer by profession who later became president of the Institution of Environmental Health Officers from 1938 to 1952.

In 1929, the Smoke Abatement League and the Coal Smoke Abatement Society amalgamated to form the **National Smoke Abatement Society** (NSAS). The League, based in Manchester, had appointed a full-time secretary, and the amalgamation was a positive move forward. Fortunately, the new body was able to retain the services of many of its prominent lay members who worked for the cause of clean air, and the support of environmental health professionals. It was recognized that smoke control was mainly the province of environmental health officers and their support for the movement for clean air was crucial.

It was, however, a lay member of the NSAS who produced the seminal concept of the smokeless zone. Charles Gandy, a Manchester barrister, conceived the idea. Manchester City Council, long committed to clean air, supported the concept.

From about 1928, some local authorities began measuring air pollution. Although the equipment then available was relatively crude, it did at least yield measurements that could be used as evidence and be compared over different time periods and geographical areas. Despite this and other research, practical action had to be delayed until after the Second World War.

In 1946, the city of Manchester was able to secure, in a private Act of Parliament, powers to establish a smokeless zone. Other authorities, encouraged by the NSAS, took similar powers. The new powers took some time to implement, and it was not until 1954 that Manchester had its first smokeless zone.

Local authorities became more interested in air pollution control and appreciated the transboundary nature of smoke pollution. This recognition made it easier to establish local committees to co-ordinate local smoke abatement, and some of these organizations were successful and long-lived.

Despite the increasing pressure, little had been done by central government to give adequate powers to local authorities to deal with smoke, especially domestic smoke.

In December 1952, London, as a consequence of persistent temperature inversion, suffered a particularly dense and persistent fog that had disastrous consequences in terms of human mortality. Later statistical analyses suggested that the mortality in excess of the normal death rate was in the order of 4000 people.

Under pressure, the government appointed a committee under a prominent industrialist, Sir Hugh Beaver, to investigate the problem and make recommendations.

The committee made rapid progress, but there was more pressure, and a private member, Gerald Nabarro, approached the NSAS for a draft Bill and other assistance. That Bill failed, but the pressure on government was so great that it adopted a similar Bill and, during its passage, Nabarro acted as the principal spokesman on amendments suggested by the NSAS.

The Beaver Committee was aware that much needed to be done, but there could be no quick solution. The aim would be to secure an 80% reduction in air pollution in the heavily populated areas over 10–15 years.

The new legislation in the **Clean Air Act 1956** (later consolidated with the subsequent **Clean Air Act 1968** into **the Clean Air Act 1993**) embodied principles that had been proposed for years by bodies pressing for clean air and by the authorities responsible for controlling air pollution.

Air pollution from industrial coal-burning plants was less of a problem because the economies achievable by the proper combustion of the fuel had made companies install adequate plant and operate it with skilled engineers.

The new legislation proposed to deal with domestic smoke by adapting the smokeless zone principle into **smoke control areas**, in which all

premises would be subject to some control. The control over industrial premises was already available, but domestic premises would be required to be able to burn smokeless fuel. Although the grant was only to be 70% of the cost of the change, that level was chosen to reflect the fact that the new appliances would be much more efficient and burn less fuel. No grant was available for commercial premises, which nevertheless had to comply.

The new interest in air pollution control caused a re-examination of the powers of the Alkali Inspectorate and local authorities. The Beaver Committee had urged that the central and local inspectorates should work closely together, and the subsequent closer contact brought better co-operation and improved working arrangements.

Many of the larger authorities argued that they could be responsible for all air pollution control, but in the end they were left with responsibility for a few installations while some of those they had previously been responsible for were transferred to the Alkali Inspectorate. Three large local authorities were allocated control over three small-sized power stations.

One effect of the Clean Air Act 1956 was a major increase in the monitoring of air pollution by local authorities, and the development of a national survey assisted by these local authority statistics. When the United Kingdom became a member of the EEC, it was required to meet the air pollution standards set down by it, and this necessitated a rearrangement of the monitoring activities.

OTHER FORMS OF POLLUTION

Originally, the investigation into air pollution was directed against the products of the combustion of coal, but after about 1970 there was increased interest in other gaseous pollutants, especially those emanating from internal combustion engines. This has its most recent manifestation in the system of air quality management operated by local authorities under regulations made through **the Environment Act 1995** but having a delayed implementaion in 2000.

A keen public interest began to develop in all forms of pollution, and this found expression in the control over the deposition of hazardous wastes. The first control was exercised in the early 1970s by the then equivalent to the district council through a licensing system.

Pressures for more effective law persisted, and the **Control of Pollution Act 1974** was passed as a consequence. This Act was designed to give better control over various forms of pollution, including air pollution and the pollution of land. This coincided with the reorganization of local government, and the duty of dealing with hazardous wastes was given to 'waste disposal authorities' which, in England, were the county councils.

The anti-pollution movement grew increasingly strong, and there was much public pressure for protection from environmental pollution, including noise pollution. As a result, the **Water Act 1989** was passed bringing in new measures to control water pollution.

The **Environmental Protection Act 1990** gave measures for the control of other forms of pollution. The objectives of the 1990 Act included the creation of an integrated national pollution inspectorate, which would involve local authorities in the control of certain scheduled industrial processes through a process of authorizations. This scheme is being gradually replaced by a new system of Integrated Pollution Prevention and Control through **the Pollution Prevention and Control Act 1999** and regulations made under it.

The pressures for more comprehensive environmental control included very considerable international pressure for action to be taken to deal with pollution problems that had transboundary effects, such as the pollution of international water courses, and the acidification of rain by the emission of sulphur gases from the combustion of fossil fuels. There was great interest in global warming, the rise in carbon dioxide levels and the use of chlorofluorocarbon compounds.

THE ROYAL COMMISSION ON ENVIRONMENTAL POLLUTION

Royal Commissions of Inquiry are one of the prestigious ways by which the UK government institutes a formal inquiry into a particular topic. It is carried out by a body of distinguished people who, while not necessarily experts in the particular topic under review, are capable of mounting a sustained and penetrating inquiry by means of examining papers and listening to oral evidence from expert witnesses.

Most Royal Commissions are appointed for a single task, but the Royal Commission on Environmental Pollution is a standing commission that has undertaken a series of investigations into issues of pollution, publishing an authoritative report after each inquiry.

PROFESSIONAL RELATIONSHIPS

The **Public Health Act 1848** required local boards of health to appoint an 'officer of health' and an inspector of nuisances. Both were separate appointments and, in this the Act followed the **Liverpool Sanitary Act 1846**. The duties of neither office were prescribed, but those of the **medical officer of health** were influenced by advice from the local government board.

Following the **Public Health Act 1872**, the local government board made a series of orders prescribing qualifications, appointment, salaries, and tenure of office of medical officers of health and **inspectors of nuisances**. The duties of the medical officers of health in London were defined in 1891.

In 1910, the **Sanitary Officers (Outside London) Order** defined medical officers' duties and it applied also to inspectors of nuisances. The conditions of office and tenure of medical officers and sanitary inspectors were prescribed by the **Public Health (Officers) Act 1921**. That Act required that their dismissal could only be with the consent of the minister, and it also required that the term 'inspector of nuisances' be replaced with the designation '**sanitary inspector**'.

Between that date and 1974, various other orders of a similar kind were promulgated, and these had the effect of requiring public health inspectors to work under the general direction of the medical officer of health.

In 1955, the designation was changed to public health inspector by the **Sanitary Inspectors (Change of Designation) Act 1955**. The current designation of environmental health officer began to be adopted about 1970, and became accepted as the designation after local government reorganization in 1974.

THE DEVELOPMENT OF LOCAL GOVERNMENT

Environmental health control has been associated with local government for many years; indeed, it would be true to say that local authorities grew out of the original environmental health authorities. The history of local government is a specialist topic, and all that is recorded here are the steps that created the local government bodies to which reference is made in past, and some existing, environmental health legislation.

The **Public Health Act 1848** established **local health boards**. The **Public Health Act 1872** created the **sanitary authorities**, which would be responsible for all sanitary functions within their areas. The **Public Health Act 1875** created **urban and rural sanitary districts**, with the former comprising either municipal boroughs, local government districts, or an Improvement Act district. Rural sanitary districts were the responsibility of the Board of Guardians.

The **Municipal Corporations Act 1882** created a modernized form of **municipal borough**, and the **Local Government Act 1888** set up **county councils** and **county boroughs**. This Act also created the **London County Council** and, with 32 **metropolitan boroughs**, replaced a wide miscellany of former authorities.

Urban District and Rural District Councils emerged from the **1894 Local Government Act** to replace the urban and rural sanitary districts.

After the **Public Health Act 1872** the local government board could approve the establishment of **Port Sanitary Authorities** and assign functions to them. Furthermore, under the **Public Health (Ports) Act 1896**, the local government board was empowered to invest Port Sanitary Authorities with duties in respect of communicable diseases.

The **Local Government Act 1929** reformed the poor law system. It abolished the Boards of Guardians and vested their social and medical work in **county councils and county boroughs**.

There were major changes in the pattern of local government in London and the Midlands in the 1960s. In the Greater London area, 34 London boroughs were set up to take over the functions of a large number of smaller authorities, and London boroughs had powers similar to those of county boroughs except that some did not become education authorities.

After a long deliberation by the Redcliffe–Maude Committee, proposals to change local government to a series of single-purpose authorities were not accepted by the government, and the **Local Government Act 1972** was used to create a new system. The 1300+ councils in England and Wales were reduced to about 400 **shire districts**.

Metropolitan counties were created to cover London and six other conurbations. Additionally, 36 **metropolitan district councils** were established by amalgamating a number of authorities. Constitutionally, these councils were very similar to the former county boroughs and, indeed, in most cases a former county borough formed the core of a new metropolitan district. The shire districts had limited functions but were responsible for housing and environmental health, but the metropolitan districts were 'all purpose' authorities. A comparable system of regional and district councils was created for Scotland.

One consequence of the 1974 local government reorganization was that conventional environmental health administration continued in two main streams. In many cases environmental health sections of multidisciplinary health departments became autonomous departments dealing with environmental health and peripheral issues. In others the environmental health function was combined with other activities, such as housing, in which the new combined department was often headed by an environmental health professional. In this type of development the environmental health function was sometimes subsumed into a technical services department. One of the features of the development of environmental health law in the 100 years prior to the 1974 reorganization of local government was that local authorities promoted private act legislation in which enhanced environmental health powers were incorporated. When central government promoted public health legislation the environmental health provisions in private Acts of Parliament, which had been tested in practice and proved to be practical, were often made the basis of general public health legislation.

The Local Government Act 1972, the basis of the reorganization, provided that the former private act legislation was to be consolidated on a countywide basis. The result of this was that the residents of many local authorities that had been absorbed into new authorities enjoyed a wider protection through the new consolidating acts than had previously been available to them.

In 1986, the Greater London Council and the six metropolitan counties were abolished. Functions that could not be returned to district councils were assigned to 'residuary bodies'. The Greater London Council and the metropolitan counties had no environmental health functions other than the disposal of refuse.

Further measures to reorganize local government took place in the 1990s. Some former county boroughs resumed their 'all purpose' status, and this was also accorded to some large district councils that had resulted from amalgamation. Reorganization in Wales and Scotland made the provision for local government to be the responsibility of 'unitary' authorities, each of which was an 'all purpose' authority. In all cases the environmental health function remained with local government at the lower tier.

CHANGES AND CHALLENGES, 1974–2000

The year 1974 was the beginning of almost continuous change, some of which affected environmental health practice. The factors that brought these changes were numerous. There was increasing public awareness of, and involvement in, environmental issues, and although environmental health might claim primacy because of its concern for human health, other streams of activity also claimed a place in the protection of the environment. Because of the economic policies of central

government, local government activities were constrained by financial and legislative moves. The conventional environmental health function remained, on the whole, with local authorities, but financial constraints on councils led to structural changes and resource reductions that often resulted in further amalgamations of the environmental health team with other activities. Government policies were directed to greater use of private sector resources and to centralization.

Local authorities were compelled to allow private companies to tender for such activities as refuse collection and removal. The duty to oversee this primary environmental health activity remained the function of local government. In many authorities, environmental health professionals had the responsibility for the day-to-day organization of this service but tended to be confined to specifying the level of service and monitoring the overall performance of the contractor.

Simultaneous with the local government reorganization in 1974 came the allocation to local authorities of extensive functions under the provisions of the Health and Safety at Work, Etc. Act 1974. Central government did this in the face of fierce opposition because local authorities had done well in administering the welfare provisions of the Shops Act 1934 and the Offices, Shops and Railway Premises Act 1964. Further, the expenses that local government would incur in such enforcement activities would be a charge on local budgets and not on that of a central government department. This was really the first time that environmental health professionals had had to work in an arena dominated by a centralized organization and with the expectations that enforcement standards equivalent to those of the central inspectorate would be achieved. At last local government was able to participate in high level consultations on standards.

Other considerations led to the creation of other centralized organizations that operate in fields that were either partly or wholly the province of environmental health professionals. While the movement had some clear political advantages in a climate that looked unfavourably on local government, there were other forces at work. Environmental control in the round was being influenced by advancing knowledge and the need for scientifically knowledgeable bases: field observation needed the support of science and was needed to furnish information to fuel further scientific advances. Such a system could conform better to a national pattern than it could to local needs.

Being a member of the now renamed European Union (EU) also placed pressures on government, for as knowledge advanced there was a considerable interest in achieving uniformity of both reaction and monitoring.

The Environment Act 1995 led to the creation of the Environment Agency, a centralized non-departmental public body. In it are vested the National Rivers Authority, H.M. Inspectorate of Pollution, Waste Regulation Authorities and some functions formerly in the domain of the then Department of the Environment.

A number of environmental health professionals were employed by the merging bodies and continued to serve as this new organization could properly be regarded as having high level specialist environmental health functions. The Environment Agency is now responsible for flood defences, water resources, fisheries, navigable waters (rivers and canals), recreation and conservation, and in respect of pollution control for water quality and water pollution, radioactive substances, standards and licensing for controlled wastes, mine water pollution and reporting on the management of contaminated land, although this latter is still a local authority task.

Air pollution control remains within the local authority domain and there are three main areas of action:

- traditional nuisance control (dark smoke from chimneys), approval of boiler plant and chimney heights;
- control of pollution from designated processes;
- air quality management, which requires local authorities to establish whether any part of their area is subject to air pollution such that specific standards would be exceeded by 2005, the Secretary of State having reserve powers to act in default. The stages required as these are part of the National Air Quality Management

Strategy, which is both following and leading EU legislation, are

- carrying out of assessments;
- using data obtained to model and implement counter measures, if necessary.

In some cases, centralization was seen as the answer to the perceived failings of local government and pressures to imitate European solutions. For example, the meat inspection service was integrated into a national service because of a perception of local government inadequacies despite the very long service rendered by local government in this field. A long-standing environmental health function was thus lost from local control to a centralized agency which, at least in its early days, did not meet the expectations aroused by centralization.

In the field of food control there was, after much delay, a revision of the legislation, still based on **the Food and Drugs Act 1955**, which reflected the increased importance of food safety. **The Food Safety Act 1990** demonstrated this principle and made provision for its application by giving ministers extensive powers to make regulations.

For many years it was perceived that the central government approach to food control and safety and hygiene standards lacked a balance between the needs of the producer and the consumer. These shortcomings, together with a serious incidence of animal disease and a rising rate of food-borne disease, led to the establishment of the Food Standards Agency, which oversees food production and distribution in the widest sense, provides a scientific foundation for the food industries and monitors the enforcement performance of those agencies appointed to ensure that the prescribed regulatory provisions are carried out, including that of local authorities.

It seems clear that there needs to be a balance struck between the need in many environmental health activities to have a central point of scientific reference, the need for some activities to be carried out by specialists in that field, and the need for very local environmental control; such a balance is extremely difficult for a centralized body to furnish economically. Environmental health professionals are trained to take a holistic approach to environmental health problems as the interplay of environmental health situations means that only such an approach is appropriate; this then allows decisions to be taken about which specialist professional help is required.

INTERNATIONAL INFLUENCES

Most of the legislation discussed in the preceding pages has been the consequence of the internal recognition that there was a need to deal with specific problems, but it is important to understand that because environmental health problems are no respecters of national boundaries there are some issues that need an international dimension.

One of the first problems to be faced internationally was the control of some important communicable diseases. As we have seen earlier, to some extent public health control in the United Kingdom was the direct result of a series of epidemics of cholera, a disease with a high mortality rate that is not indigenous to the British Isles and that affected all social classes. Most of the diseases that attracted attention were also characterized by their relationship to the movement of people: there is no recorded case of a communicable disease passing from one community to another faster than a person could travel.

Between 1851 and the late 1930s, a series of International Sanitary Conferences took place in an attempt to prevent the transmission of exotic diseases. (Note that the term 'sanitary' was then used to denote something appertaining to health rather than its modern connotation, which relates to the removal of body wastes.) The usual procedure was for the conference to produce a convention (a form of treaty) after negotiations, and while that convention was signed by all the countries attending the conference, it had to be formally ratified by each government before it became binding.

The first conferences took place because there was a genuine desire to make some progress but they were hampered by a lack of epidemiological knowledge. Later conferences were apprised of the latest advances in medical knowledge, but usually disregarded them.

The importance of a public health system that could deal with the problems of imported disease was recognized in the United Kingdom and appropriate powers were enacted in the Public Health Acts 1872 and 1875. Port Sanitary Authorities were established and they were assigned the powers necessary for their work by the Local Government Board.

By the mid-1870s, the old concept of quarantine was being challenged and medical inspection was recommended in its place. This view was adopted by the 1874 International Sanitary Conference in Vienna. The Seventh International Sanitary Conference was held in 1892, also in Vienna, and produced the 1892 Sanitary Convention, which focused attention on cholera. This formed a useful starting point for effective international public health control, although a further conference (in Dresden in 1894) was needed to ensure that countries that had signed the convention notified each other of any cases of cholera they had.

The British response was to enact the Public Health (Ports) Act 1896, which enabled the sanitary conventions to be observed. Further, the Local Government Board gave Port Sanitary Authorities powers under the provisions of the Infectious Diseases Prevention Act 1889. Regulations were made dealing specifically with cholera, yellow fever and plague; quarantine was abandoned and medical inspection procedures were prescribed. An Eleventh conference was held in Paris in 1903, and this resulted in the addition of plague to the list of 'convention diseases', but also produced the concept of a permanent international health organization. Thus, in 1907, the **Office International d'Hygiène Publique** was set up in Paris, creating a worldwide system of reporting infectious disease.

The new organization laid the foundations for a further conference in 1911/12. The sanitary convention that resulted from this conference added yellow fever to the list of convention diseases, and in addition to refining the processes for dealing with plague and cholera replaced the provisions of conventions signed in 1892, 1893, 1894, 1897 and 1903.

In 1923, the League of Nations set up an office in Geneva to complement the work of the Paris office, and within two years it was producing weekly epidemiological reports of the world position relative to the incidence of the convention diseases. A further conference in Paris in 1926 added typhus and smallpox to the convention diseases, and the reaction of the United Kingdom was to produce the comprehensive Port Sanitary Regulations 1933.

A major international conference in Paris in 1938 was notable because for the first time serious attention was paid to contemporary medical knowledge and the problems of travel by road, air and sea as it was now known that it was possible to import diseases during their incubation period. The Second World War precluded further consideration of the issues raised by the 1938 conference.

After operating through a series of health-related international organizations, the United Nations (UN) formed the World Health Organization (WHO) in 1946 to signal effect. It did away with the need for every country to ratify the conventions, which tended to make their implementation very slow, because the WHO was invested with the power to make regulations binding on every state that was a member. Wide consultations were required, but in 1951 the WHO produced the International Sanitary Regulations, to be effective from October 1952. The United Kingdom revised its law relating to domestic control through the Public Health (Ships) Regulations 1979, the Public Health (Aircraft) Regulations 1979 and the Public Health (Control of Diseases) Act 1984.

In dealing with world health problems, the WHO does not confine itself to publishing measures that member countries should follow but embarks on practical health programmes, one of which, through international co-ordination and co-operation succeeded in eliminating smallpox.

One of the more recent manifestations of the global nature of environmental health problems was the establishment of the International Federation of Environmental Health in 1985.

THE EUROPEAN UNION

Earlier in this chapter, reference was made to the possibility of tracing the development of environmental health law, and hence practice, in a series of phases, but it was also speculated that the four periods noted might, with hindsight, be increased to five; without question, a fifth period would include the influence of the EU on British environmental law and practice.

It should be remembered that the present EU has developed from an organization that was primarily centred not on health issues but on economic welfare, and it might be wondered why an organization driven by economic considerations should have such an impact on environmental health. The reason for this is that there are environmental health considerations in many commercial activities, and because the EU is activated by commercial considerations it is bound to involve environmental health. A simple case will illustrate this. One major concern of the EU is for the economic environment of every member to be equal. Thus, if the particular issue is metal smelting, common enough in industrial communities, then every enterprise that is engaged in it must comply with the same standards of fume emission. Now, while metal smelting may be a commercial activity, the effluent fumes are an environmental health matter, and the EU, rightly, takes steps to ensure that all smelting plants operate to the same rules both as far as production is concerned and also in respect of the protection of the workers and the general public.

POSTSCRIPT

Although this chapter is long, it affords only a brief introduction to the way environmental health control has evolved. Most of the topics covered here are treated in more detail elsewhere.

From the account given it should be apparent that, while there has been a continual stream of environmental health legislation, there have been some periods (1875, the 1930s, 1974 and the 1980s) when developments took place simultaneously in several streams of activity. Experience also suggests that changes only come as a consequence of pressures on government, and such pressures now tend to arise from the public rather than from official sources.

Environmental health has always been in the lead in 'social' legislation and has thus tended to reflect contemporary political and economic thinking. Because of this, environmental health law has been the subject of the impact of deregulation and provides good illustrations of the conflict between the need to provide adequate protection and a perceived need to reduce the burdens of compliance.

REFERENCES

1. Hutchins, B.L. and Harrison, L. (1911) *History of Factory Legislation*, P.S. King & Son, London.
2. Gale, A.H. (1955) *Epidemic Diseases*, Pelican, London.
3. Finer, S.H. (1956) *The Life and Times of Sir Edwin Chadwick*, Methuen, London.
4. Frazer, W.M. (1950) *A History of Public Health*, Ballière, Tindall & Cox, London.
5. Burnett, J. (1960) *Plenty and Want*, Pelican, London.
6. Engels, F. (1968) *Condition of the Working Class in 1844*, Allen & Unwin, London.
7. Ashby, E. and Anderson, M. (1979) *Politics of Clean Air*, Clarendon Press, Oxford.
8. *Getting Ahead of the Curve* (January 2002) Report of the Chief Medical Officer. DoH, London.
9. Hallworth, J. and Davies R.J. (1910) *The Working Life of Shop Assistants*, National Labour Press, Manchester.
10. Robens Committee (1972) *Safety and Health at Work*, Cmnd 5034, HMSO, London.
11. Simon, E.D. and Fitzgerald, M. (1922) *The Smokeless City*, Longmans, London.

FURTHER READING

Brockinton, W.F. (1952) *The World Health Organization*, Pelican, London.
Frazer, W.M. (1950) *A History of Public Health*, Baillière, Tindall & Cox, London.
Haig, N. (1987) *EEC Environmental Policy in Britain*, Longmans, London.
150 Years of Public Health (2000) CIEH, London.

3 The organization of environmental health in the United Kingdom

Janet Higgitt

THE OVERALL FRAMEWORK

The organization of environmental health is distributed across a number of organizations, and, whilst service delivery is effected at the local, community level, overall policy and legislation is driven by central government. Knowledge of the structure and function of all involved is fundamental to understanding the organization of environmental health in the United Kingdom.

This chapter deals mainly with the position in England and Wales although there are references to Scotland and to Northern Ireland. The three appendices at the end of this chapter deal in more detail with the environmental health service in Northern Ireland, Scotland and Wales.

LOCAL GOVERNMENT IN THE UNITED KINGDOM

History

The English shires are one of the most ancient units of local government to have survived anywhere in the Western world. The first mention of Hampshire appears in AD 757 and *The Doomsday Book* shows that the shire system south of the Tees was almost completely established as it was to remain until the major revisions of 1888 and 1974. Throughout the last two centuries the structure of local government has evolved and has been reformed a number of times (see also Chapter 2). Following the major changes that took place in 1974, a 1983 White Paper 'Streamlining the Cities' proposed the abolition of the Greater London Council (GLC) and six metropolitan county councils. This was implemented with the enactment of the 1985 Local Government Act, which *inter alia* abolished the GLC and established a directly elected Inner London Education Authority (ILEA), which in turn was later abolished in 1988, with the functions passing to the London boroughs.

In 1992 the Local Government Act supported further structural reorganization to create new unitary Councils. By the end of the process in 1998, 46 new unitary authorities had been created throughout England. A similar process took place in Wales with the 1994 Local Government (Wales) Act to establish a system that gave Wales 22 unitary authorities, which came into existence on 1 April 1996. Further development took place in Wales with the Government of Wales Act 1998 establishing the National Assembly for Wales. The Act also established a unique statutory Partnership Council between Welsh local government and the National Assembly.

Whilst the structure of local government has remained unchanged since, some key policy developments have taken place. In 1999 the Local Government Act brought Best Value and the Local Government Act 2000 introduced the concept of a directly elected mayor and gave options for the control system of local councils; the principle of directly electing mayors having been introduced for London by the Greater London Authority Act 1999 [1].

Local government in the United Kingdom grew out of a need to provide services for local people many of which are fundamental to environmental health such as sanitation, clean water supply and good housing.

Structure

Local government today is a complex creature. Over 21 000 elected councillors represent local communities and local people and serve on over 400 local authorities in England and Wales. Employing over two million people, local councils undertake an estimated 700 different functions [2]. Local government expenditure (currently £70 billion per annum) accounts for 25% of all public expenditure in the United Kingdom [3].

Currently local government is structured in two ways:

- A single tier all purpose organization delivering all local government functions. These include Scottish, Welsh and English unitaries, metropolitans and the London boroughs.
- A two-tier framework in the remainder of England with local government functions being split between them, with county councils occupying a larger geographical area and being split into a number of district councils (see Table 3.1).

The Greater London Authority (GLA) also has a number of functions: environment, working with the boroughs on air quality, waste; health, promoting the improvement of the health of Londoners; and also duties in relation to transport, economic development, planning, fire and culture.

The locus for environmental health is in the single-tier authorities and the district level of the two-tier framework. An understanding of where environmental health fits is best gained by a consideration of the totality of local government provision (see Table 3.2). This arrangement can have a variable effect on the delivery of environmental health as in the larger single-tier authorities environmental health will be competing to get funding with functions such as education and social services, but equally in the smaller district tier, resources may be scarce because there is no facility for economies of scale afforded in larger organizations. However there is general recognition in the United Kingdom that there are strong advantages in the core environmental health service operating at the community level where it will be in close contact with the population it serves and the environmental problems it faces. This approach is supported by WHO [4].

The link with the public health origins of local government and environmental health are being refuelled by the Local Government Act 2000, which gives local authorities the power to promote the economic, social and environmental

Table 3.1 Structure of UK local government

Single-tier authorities	
Unitary authorities (Wales)	22
Metropolitan authorities (England)	36
London boroughs	33 (London also has the Greater London Authority)
English shire unitary authorities	47 (including Isles of Scilly)
Two-tier authorities in England	
County councils	34
District councils	238

Table 3.2 Arrangement of UK local government functions

	Met/London authorities	Shire/unitary authorities			
	Met councils	London boroughs	District councils	Unitary authorities (England and Wales)	County councils
Education	×	×		×	×
Housing	×	×	×	×	
Planning applications	×	×	×	×	
Strategic planning	×	×		×	×
Transport planning	×			×	×
Passenger transport				×	×
Highways	×	×		×	×
Fire				×	×
Social services	×	×		×	×
Libraries	×	×		×	×
Leisure and recreation	×	×	×	×	
Waste collection	×	×	×	×	
Waste disposal	×	×		×	×
Environmental health	×	×	×	×	
Revenue collection	×	×	×	×	

Source: Adapted from LGA Fact Sheet on local government structure 26/06/2001.

well-being of their area. The act also makes a duty to review and make new arrangements, separating executive and scrutiny functions (see later).

Revenue

Local authorities raise their income in a number of different ways, with the council tax in 1999/2000 only raising 25% of total local authority revenue. The rest is made up of central government grants, which, at around 48%, form the majority of local government revenue. The Non-domestic Rate is a charge to businesses that is set by central government and this raises about 25% of local authority revenue with the remainder being made up by charges for services and reserves [5].

CENTRAL GOVERNMENT, DEPARTMENTS AND AGENCIES

Central government comprises a number of departments of state, which in most instances are headed by a Secretary of State, assisted by Ministers. Since environmental health is so broad there is no single central government department or agency to which it relates; rather a variety of departments and agencies play a role to a greater or lesser extent including:

• Office of the Deputy Prime Minister (ODPM)
  – local government
  – housing
  – neighbourhood renewal

- Department of Trade and Industry (DTI)
  - consumer issues
  - energy
- Department for Transport (DfT)
  - sustainable transport planning
- Department for Environment, Food and Rural Affairs (DEFRA)
  - safe food supply chain
  - sustainable development
  - environmental protection
  - drinking water via the Drinking Water Inspectorate
- Department of Health (DoH)
  - public health
- Department of Work and Pensions (DWP)
  - health and safety at work
- Food Standards Agency (FSA)
  - food standards and hygiene
- Health and Safety Executive (HSE)
  - health and safety at work
- Health and Safety Commission (HSC)
  - health and safety at work
- Health Protection Agency (HPA)
  - communicable disease
  - chemical hazard management
- Health Development Agency (HDA)
  - public health
- Environment Agency (EA)
  - environmental protection
  - waste management.

The number and responsibilities of government departments can be changed by the Prime Minister, for example in June 2001 the Department for Work and Pensions and Department for Environment, Food and Rural Affairs were established. In addition to the main departments there are hundreds of executive agencies and non-departmental public bodies (NDPB).

Executive agencies are established to undertake the executive functions of governments, as distinct from policy advice. Each agency has a framework document, which sets out its relationship with Ministers and accountability arrangements, for example Health Protection Agency (sponsored by DoH), Health and Safety Executive(sponsored by DWP).

Non-departmental public bodies (NDPB) have a role in the processes of national government but are not government departments or part of them, and therefore operate to a greater or lesser extent at arm's length from Ministers. There are three main types of NDPB: executive, advisory and tribunals.

(i) *Executive NDPBs.* These bodies carry out a wide variety of administrative, regulatory and commercial functions. They generally operate under statutory provisions, employ their own staff and have responsibility for their own budgets, for example the Environment Agency and the National Radiological Protection Board. A key feature of such organizations is that they have a national remit (i.e. their functions cover England, England and Wales, Great Britain or the United Kingdom); they are legally incorporated and have their own legal identity. Most executive NDPBs are financed mainly by grant-in-aid from their sponsor department; others may be self-financing.

It is important to note however that there are bodies which match all of the characteristics of executive agencies but are not classified as NDPBs – for example NHS Trusts. Such bodies have their own classification, that is NHS bodies. They also have their own accountability arrangements.

(ii) *Advisory NDPBs.* These are generally set up administratively by Ministers to advise them and their departments on matters within their sphere of interest. Some Royal Commissions are classified as advisory NDPBs, but departmental committees of officials are not. Advisory NDPBs are normally supported by staff from within the sponsor department, and do not incur expenditure on their own account; for example OFWAT National Customer Council, Expert Panel on Air Quality Standards (sponsored by DEFRA), Advisory Committee on Dangerous Pathogens (sponsored by DoH).

(iii) *Tribunal NDPBs.* This category of NDPB covers bodies with jurisdiction in a specialized field of law. Tribunals generally operate under statutory provisions and, independently of the executive, decide the rights and obligations of private citizens towards a government department or other public authority or towards each other. In general, tribunals are serviced by staff from the sponsor department. Tribunals vary widely in the kind and amount of work they do, for example the Meat Hygiene Appeals Tribunal for England and Wales, which considers appeals against a decision to refuse, revoke or suspend a licence or to impose conditions on a licence, for premises to engage in the slaughter, cutting and storage of red meat, white meat and game meat (both farmed and wild) and the Employment and Employment Appeal Tribunal, which amongst other things hear appeals against Health & Safety at Work Act notices.

Not all bodies fit neatly into a single category, for example some bodies, such as English Nature, have both executive and expert advisory functions. All NDPBs are subject to external audit, either by the Comptroller and Auditor General or by an auditor appointed by the Secretary of State.

Whilst the main establishment of government is located within the capital, over recent years there has been a move to decentralize with the founding of the Regional Co-ordination Unit (RCU) and Government Office Network. They promote the improved delivery of services that have cross-cutting outcomes to make a difference on the ground to local people.

The nine government offices (GOs) were set up in 1994 and now bring together the English regional services for the following departments [6]:

- Office of the Deputy Prime Minister
- Department of Trade and Industry
- Department for Education and Skills
- Department for Transport
- Department for Environment, Food and Rural Affairs
- Home Office
- Department for Culture, Media & Sport
- Department of Health
- Department for Work and Pensions.

FUNCTIONS OF CENTRAL GOVERNMENT

Legislative

The government formulates policy and introduces legislation in parliament. The drafting and processing of statutes, subordinate legislation such as regulations and orders relating to environmental health, is undertaken by the relevant central government department or agency. Once the government has decided that there is a need for a change in the current legal provision, it may well issue a **consultation paper** (sometimes called a Green Paper), which will seek the views of interested people and agencies. Having considered the results of such consultation, the government may issue a **command paper**, commonly known as a White Paper, which outlines the content of legislation that the government intends to introduce. It is thus a statement of intent to introduce legislation.

The issue of consultation and command papers is not always required in the process of legislation, but is often used by government as a way of gauging support for a proposed measure. In some cases however there is a statutory requirement for government to consult before introducing legislation, for example Section 50, Health and Safety at Work Act 1974. The legislative process within parliament is described in Chapter 6.

Administrative

Each central department or agency provides advice to local authority environmental health departments about the implementation of legislation and policy. This is done in a variety of ways:

- through the issue of ministerial circulars and guidance;
- through the local authority associations;
- through explicit liaison arrangements, for example Health and Safety Executive/Local Authority Enforcement Liaison Committee (HELA);

- though explicit agreements, for example Food Standards Agency Framework Agreement Monitoring.

Whilst each local authority is an independent organization delivering services to local people; increasingly local government has become more and more accountable to central government for its actions and its expenditure. Central government therefore needs to have assurances that monies are being spent appropriately and that the law is being enforced consistently. Consequently a range of monitoring mechanisms are in place. In some circumstance a central body such as the Audit Commission (see the following section) does this across the breadth of local authority service or across a single service area such as environmental health. For the specialized areas of environmental health such as food and health and safety there are some particular arrangements, for example the FSA and the HSE both have arrangements in place to audit local authority practice and require regular statistical returns.

Returns about local housing activity also are required by ODPM, as is the submission of an annual housing strategy review, which is linked to bids for capital spending. Further, returns to DEFRA are required about activities under environmental protection legislation.

The Audit Commission

The Audit Commission is an independent body responsible for ensuring that public money is used economically, efficiently and effectively. It is an NDPB sponsored by the ODPM together with the DoH and the National Assembly for Wales. It works within an agreed framework document with these sponsoring departments. The Commission has a chairman, deputy chair and up to 15 other members drawn from – but not representing – a wide range of interests including industry, local government, health and the accountancy profession. The sponsoring departments jointly appoint members. The day-to-day operations of the organization are managed by a team of directors led by the Controller of Audit The Commission undertakes its work in a variety of ways:

- By appointing auditors to audit accounts and producing value for money (VFM) reports on public service providers.
- Through inspections of local services. Local authorities are required by government to assess their own performance and put in place measures to ensure continual improvements in these services. The Audit Commission both audits local authority plans for improvement and also inspects local services to assess their quality and cost effectiveness and how likely they are to improve in the future. The Commission undertakes this (see the following section).
- By collecting and publishing performance information on local authority services; this enables monitoring and comparison of service performance. In addition to national indicators determined by central government, local authorities are encouraged to develop and use local performance indicators. Further, along with the 'well being' agenda, voluntary indicators for local authorities are being developed. They also identify key VFM indicators based on the recommendations from selected national studies and associated audits. These VFM indicators are then used to see what changes have taken place since the original study or audit, and the findings are published.
- By conducting national VFM studies of local services from the user's perspective, comparing performance and identifying and promoting good practice. The Audit Commission undertakes national research designed to promote economy, efficiency and effectiveness in local public services. These services include health and social care, police, education, housing, fire and other local government responsibilities. The aim is to promote improvement in public services by changing policy locally and nationally and improving public services management; assure the public of the proper stewardship of public finances and help local people contribute to improving public

services; and assist public service organizations, auditors and inspectors to deliver local assurance and improvement. Topics for the studies are chosen every year following extensive consultation with people who use services, run them and study them. Research programmes include surveys, site visits and desk research, often asking people who use the services what they think.

In preparation for Comprehensive Performance Assessment (CPA), they are compiling local authority corporate assessments, combining performance information data, audit inspection and other service assessments. (See section on CPA later.)

Best value

The principle of the compulsory exposure of local authority services to competition by tender was first put into practice by the Local Government, Planning and Land Act 1980 as a first step by the newly elected Conservative government to improve the efficiency and effectiveness of local government. The 1980 Act required competition with the private sector for various activities and also required the establishment of a separate direct labour organization (DLO) where the tender was won inhouse. This was followed by the Local Government Act 1988, which defined further activities to be subjected to a process of competition.

The incoming Labour government of 1997 indicated its intention to replace the compulsory competitive tendering (CCT) regime with a duty on local authorities to achieve Best Value. The government's White Paper on local government, 'Modern local government: In touch with the people' [7], and the 'Local voices' White Paper, published by the Welsh Office in July 1998 [8], announced an extensive reform agenda for the local government sector. At the centre of this change were the following key concepts:

- the need to put people first (i.e. consult with them on service provision and to discover their needs);

- the need for local authorities to provide services that compare with the best available in both the public and private sectors;
- the requirement to improve the quality and cost of services on a continual basis.

The Local Government Act 1999 introduced the concept of 'Best Value,' which requires councils to review, over a period of five years, the way in which they carry out all their functions, in consultation with local people, local businesses and the people who use their services. Each local authority in England and Wales must publish an annual Best Value performance plan (BVPP); and review all their services every five years, to ensure they are applying principles of continual improvement. Each service review must show that the local authority has applied the four C's of continual improvement to services, and show that it is:

- **challenging** why and how the service is provided;
- **comparing** performance with others (including non-local government providers);
- **competing** the authority must show that it has embraced the principles of fair competition in deciding who should deliver the service;
- **consulting** local service users and residents on their expectations about the service.

Authorities' reviews are inspected by the Audit Commission's Inspection Service, to ensure that services are being well run and have the potential to improve. Inspections look at the degree to which services meet the authorities' corporate aims and objectives, cost effectiveness, quality and customer focus. Once authorities have produced their BVPPs and reviewed a number of services, they agree with the Commission on which services will be inspected when. To ensure the level of inspection is proportionate to risk, the Commission decides the level of inspection required. Inspectors judge the service against the following criteria:

- How good are the services?

   1. Are the authority's aims clear and challenging?
   2. Does the service meet these aims?

3. How does its performance compare with its aims?

- Are the services going to improve?

1. Does the best value review drive improvements?
2. How good is the improvement plan?
3. Will the authority deliver the improvements?

About four weeks before the inspection the inspectors review all relevant documents – the authority's review documents, strategic plans, policy documents, and the like – to understand the context of the inspection. They then request any additional documents they would like to receive from the authority. Having understood the context, the inspectors plan reality checks, which will be aimed at examining particular points of concern or interest to have emerged from the review of the documentation and the inspectors tell the authority what they intend to inspect. The authority may suggest amendments to the programme, which the inspectors may choose to accept. The inspectors then examine the service from a customer/user's point of view (reality checks). It may involve interviewing managers, frontline staff and customers. It involves unannounced visits to service delivery points, so that practice can be compared with claims made for the service. These reality checks are then used to make informed comments on how effectively the service is being delivered.

Inspectors then tell the authority (the interim challenge) how they view the service and indicate how they rate it on a 0–3 star scale (poor–excellent), and how likely they think it is that the service will improve (from very unlikely to very likely). Inspectors take feedback from the authority, which may result in an amendment to their final verdict. Having considered the authority's response to the interim challenge, the inspectors issue a final report, which is available to the public. The final report will have clear recommendations, which will assist the authority in improving its services so that it is within the top quarter of performers nation-wide.

The inspectors conduct a follow-up to ensure that the points in the authority's review are being implemented and to note progress in adopting the inspectors' recommendations. In cases where the inspectors have concerns about the way in which a service is being delivered, they may propose a further inspection, or refer the authority and service to the appropriate Secretary of State, to consider an intervention.

Comprehensive Performance Assessment

Comprehensive Performance Assessment (CPA) is about helping local councils improve local services for their communities. As well as assessing the current state of key council services, their managerial effectiveness and ability to improve will also be reflected in the results. The assessment brings together evidence from a range of sources to assess the quality of most of each council's core services, how it uses its resources and its capacity to improve. These assessments are combined into an overall result placing each council in one of five categories: excellent, good, fair, weak or poor. CPA was proposed in the government's White Paper 'Strong local leadership – Quality public services' [9] published in December 2001. It is linked with other proposals to provide greater freedom to councils, including additional freedoms for well-performing councils, and support for those that need help to improve. CPA brings together judgements about:

- core service performance, drawing on inspection reports, performance indicators and plan assessments by government departments;
- use of resources, including an audit judgement; and
- the council's overall ability measured through a corporate assessment.

The Audit Commission's role in CPA was to form a judgement of the performance and proven capacity of every local authority in England by late autumn 2002 and continue this to deliver a similar judgement for district councils by autumn 2003. Once judgements have been formed, the Commission will produce a 'balanced scorecard' for every authority, available to the public.

INDIVIDUAL KEY GOVERNMENT DEPARTMENTS /AGENCIES AND THEIR RELATIONSHIP WITH ENVIRONMENTAL HEALTH

The Department of Health

The aim of the Department of Health (DoH) is to improve the health and well-being of people in England although it should be understood that the NHS is not part of the DoH. The department has a wide and varied role in relation to environmental health:

- Public health
- Accident prevention
- Health effects of air pollution
- Promoting immunization
- Health effects of noise pollution
- Sponsoring research on issues such as the social amplification of risk.

The Health Protection Agency

The Health Protection Agency (HPA) is an executive agency of the DoH and was established on 1 April 2003 by the amalgamation of a number of previous agencies and NDPBs:

- The Public Health Laboratory Service, including the Communicable Disease Surveillance Centre and Central Public Health Laboratory
- The Centre for Applied Microbiology and Research
- The National Focus for Chemical Incidents
- The Regional Service Provider Units that support the management of chemical incidents
- The National Poisons Information Service.

It is described in detail in Chapter 12.

The Health Development Agency

The Health Development Agency (HDA) is a special health authority established to identify what works to improve people's health and reduce health inequalities. It works alongside professionals to get evidence into practice, advising and

supporting policy makers and practitioners (see Chapter 9). Its establishment was announced in the White Paper, 'Saving Lives, Our Healthier Nation' in the summer of 1999 [10]. It works with key statutory and non-statutory organizations at national, regional and local levels to develop and maintain:

- an accessible evidence base;
- guidance on how to translate evidence into practice;
- resources to help those working locally.

The HDA is working with a range of organizations and agencies whose remit is health improvement – not just in the NHS but within national and local government, and the voluntary, academic and private sectors. Within local authorities they support elected members and staff from a range of functions including health strategy, environmental health, housing, transport, anti-poverty, education, healthy cities, social services, urban regeneration, health promotion and community safety.

It is focusing on the development of the public health workforce through a variety of means in order to ensure the delivery of health improvement. This is being achieved through the development of the evidence base on public health and the translation of that evidence into both policy and practice. Evidence is channelled into policymaking through direct policy comment and facilitation and support of the implementation of public health policy.

The role of environmental health is seen as fundamental to success in public health as stated by Steve Brown in *Health Development Today*, a bimonthly magazine published by the HDA, in May 2002 [11].

Environmental health officers play a key role in public health within local government. Other departments do have a part in protecting and promoting the public's health (housing and social services, for instance) but no other professionals have such wide ranging involvement in so many health issues. As such, they seem ideally placed to lead local government in its new role to promote the economic, social and environmental well-being of local communities – and to work with colleagues in the NHS and elsewhere on health improvement and modernisation plans (HImPs) (previously known as health improvement programmes – HImPs).

The HDA has been working with the Chartered Institute of Environmental Health (CIEH) to develop a vision of the role that environmental health professionals could play in government plans to improve the public's health and reduce inequalities in health. Indeed a joint post has been appointed to support the project, which will ensure that the broad-based skills of environmental health officers are fully utilized. Recently it published 'Environmental health 2012. *A key partner in delivering the public health agenda*' [14], the report describes the work of a joint CIEH/HDA project to support the environmental health profession in developing a strategic vision for its contribution to health development and well-being. It explores the projected growth of the profession's role in improving the public's health and reducing health inequalities over the next 10 years (see Chapter 9).

The Department for Environment, Food and Rural Affairs

The Department for Environment, Food and Rural Affairs (DEFRA) was formed on 8 June 2001 from the former Ministry of Agriculture, Fisheries and Food (MAFF). DEFRA took over most of MAFF's responsibilities, and also took on some responsibilities from the Department for the Environment, which was also disbanded at the same time. The Departments' aim is one of sustainable development, including a better environment at home and internationally, and sustainable use of natural resources.

A review of DEFRA's objectives demonstrate clear links to environmental health

- To protect and improve the rural, urban, marine and global environment and to lead integration of these with other policies across government and internationally.

- To enhance opportunity and tackle social exclusion in rural areas.
- To promote a sustainable, competitive and safe food supply chain, which meets consumers' requirements.
- To promote sustainable, diverse, modern and adaptable farming through domestic and international actions.
- To promote sustainable management and prudent use of natural resources domestically and internationally.
- To protect the public's interest in relation to environmental impacts and health, and ensure high standards of animal health and welfare.

DEFRA works through a number of executive agencies, for example the Pesticides Safety Directorate, the Veterinary Medicines Directorate and NDPBs such as Environment Agency and Central Science Laboratory

The Environment Agency

The work of the Environment Agency (EA) is executed through:

- a head office, split between Bristol and London;
- a combination of national centres (which provide technical and scientific expertise to support key areas of work, e.g. National Flood Warning Centre and the National Water Demand Management Centre) and National Services (e.g. the National Laboratory Service and the National Library and Information Service); and
- a series of Regional and Area Offices, which are responsible for the day-to-day management of the area and for making sure that the needs of the local community are met.

Enforcement for environmental protection is shared between the EA and local authorities (also see Chapter 34). Part I of the Environmental Protection Act 1990 established two pollution control systems: local air pollution control (LAPC) system enforced by local authorities in England and Wales and by the Scottish Environment Protection Agency in Scotland (referred to as 'local enforcing authorities'), and an integrated pollution control (IPC) system enforced by the EA in England and Wales and the Scottish Environment Protection Agency in Scotland. This system is gradually being replaced by new requirements under the Pollution Prevention and Control Act 1999 and associated regulations, the Pollution Prevention and Control Regulations 2000, which implement the European Union Directive 96/61 on integrated pollution prevention and control (IPPC). The new regime has three arms:

Regime A1, an integrated permitting regime. Emissions to the air, land and water of potentially more polluting processes are regulated. The EA is the regulator.

Regime A2, an integrated permitting regime. Emissions to the air, land and water of processes with a lesser potential to pollute are regulated. The local authority is the regulator.

Regime B, the permitting of processes with a lesser potential to pollute. Only emissions to the air are regulated. The local authority is the regulator.

Liaison is effected through an Industrial Pollution Liaison Committee (IPLC), a group of local authority representatives, DEFRA officers, and the Local Authority Unit technical officers, meeting twice yearly to discuss LAPC developments, and ensure government thinking is in tandem with local government.

The Food Standards Agency

The Food Standards Agency (FSA) is an independent food safety watchdog established by the Food Standards Act 1999 to protect the public's health and consumer interests in relation to food (also see Chapter 27). Although the FSA is a government agency, it works at 'arm's length' from government and unlike other government agencies involved in environmental health, according to its website, 'it doesn't report to a specific minister and is free to publish any advice it issues' [13].

The local authority Enforcement Division carries out a monitoring and audit role over local authority enforcement and the arrangements for

implementation of this is set out in the FSA Framework Agreement. This includes the arrangements for audit and details the information that the local authorities must supply to the FSA on an annual basis including:

- Compliance with inspection programme
- Visit activity
- Enforcement action
- Inspection outcomes
- Complaints
- Sampling
- Imported foods.

Authorities are chosen for audit on the basis of the monitoring information that they provide to the FSA, information from Best Value or peer review to include low and high performers and also a random selection. Once chosen local authorities are given a period of notice to prepare for the audit, which will be undertaken by trained auditors, who may include EHOs seconded from local government.

The on-site audit will include:

- an initial meeting to address the scope, objectives and arrangements for the audit including an opportunity for the local authority to raise any local issues that may affect the food service;
- an assessment of the implementation of the documented policies and procedures drawn up to meet the requirements of the standard following an audit protocol to ensure consistency;
- a meeting with relevant authority representatives where the auditors provide a summary of their initial findings, in particular the areas of apparent non-conformance including an opportunity to clarify any points of misunderstanding.

The auditors then prepare a draft report within 20 working days to which the local authority have to make a response with factual corrections and details of their proposed action plan, again within 20 days. Ten days later a final report, including the agreed action plan, is issued. Follow-up action will depend on the level and type of non-conformance identified and the action plan produced by the authority. The arrangements will, in some circumstances, include re-visits to local authorities. Where these arrangements identify a local authority failing to implement all or part of their action plan, subsequent agency action will be considered on a case-by-case basis. A mechanism is in place to deal with local authorities disputing the outcome of audits. Reports of audits are in the public domain. Although the enforcement of food law is primarily the responsibility of local authorities and more specifically Environmental Health Officers (EHOs) and Trading Standards Officers (TSOs); the FSA is also an enforcement authority in its own right. Its executive agency, the Meat Hygiene Service in England, Scotland and Wales, and through the Department of Agriculture and Rural Development Veterinary Service in Northern Ireland, is the enforcement authority for around 1700 licensed premises in the United Kingdom producing meat for sale for human consumption, including slaughterhouses, cutting plants and cold stores.

A number of EHOs work in the FSA; indeed currently the Director of Enforcement and Food Standards is a former director of environmental health in local government. There is an Enforcement Liaison Group, including representatives from central and local government and consumers. Its terms of reference include contributing to the development of agency strategies to improve the effectiveness and consistency of food law enforcement.

The Office of the Deputy Prime Minister

In May 2002 the Prime Minister announced that the Office of the Deputy Prime Minister (ODPM) was to be separated from the Cabinet Office and established as a central department in its own right, with new responsibilities covering a range of cross-cutting regional and local government issues. In addition to the Social Exclusion Unit, the Regional Co-ordination Unit and the government offices for the regions, the Office has now been expanded to include regional policy, local government, local government finance, planning, housing, urban policy, the Neighbourhood Renewal Unit and the Fire Service. Some of these responsibilities were previously the remit of the

former Department of Transport, Local Government and the Regions (DTLR) and the Department of the Environment. At this time a new Department for Transport (DfT) was created to focus solely on transport issues.

The Department for Work and Pensions

In July 2002 the Prime Minister announced that responsibility for the Health and Safety Commission (HSC) and Health and Safety Executive (HSE) would transfer to the Department for Work and Pensions (DWP) [14]. Departmental responsibility for HSC/E previously lay with the DTLR, however with the dismantling of this department to establish the DfT opportunities were taken to build on the existing co-operative arrangements between DWP and HSE to enhance work with respect to preventing disability and promoting rehabilitation.

The Health and Safety Executive and Health and Safety Commission

Both these organizations are executive agencies of DWP. The Health and Safety Commission's (HSC) function is to make arrangements for the health safety and welfare of people at work and the public; including proposing new laws and standards, conducting research and providing information and advice. The executive assists the Commission in its functions and has a statutory role in the enforcement of health and safety law.

The Local Authority Unit (LAU) is a freestanding unit within the Health and Safety Executive (HSE) and works with the HSE/Local Authority Enforcement Liaison Committee (HELA) to give national advice, information and guidance to local authorities. The Unit is the central focus for the development of local authority enforcement policy and also provides enforcement officers with training and support. Further it has close links with the local authority associations.

The HSE/HELA mission is 'to achieve, by leadership, a consistent approach to enforcement, among local authorities and between local authorities and HSE, to enable business to comply with health and safety law'.

The impact of these organizations on the environmental health service is described in Chapters 18 and 19.

THE EUROPEAN AND INTERNATIONAL FRAMEWORK

Most UK legislation on environmental health is now the result of European Union (EU) requirements laid down in various directives, particularly those relating to environmental protection, food safety and health and safety.

There is also an increasing trend for an administrative involvement of the EU in the work of UK environmental health departments, for example reports on activity under the Food Safety Act 1990 have to be submitted by each department through the UK government.

Understandably the greatest overlap internationally is where there is need for co-operation, for example:

- pollution does not respect national boundaries, neither do infectious diseases;
- the importation of food; or
- where to satisfy the 'level playing field' of free trade in the EU working standards need to be harmonized.

The FSA represents the UK government on food safety and standards issues in the EU, which also enables the Agency to share knowledge and expertise with other member states. The FSA aims to ensure that food imported from all countries meets the standards required in the United Kingdom resulting in the Agency playing an increasingly important role internationally, representing the UK government on joint international bodies and making food-safety information available to other countries and organizations [15].

The agency has an interest in the work of several international organizations:

- Joint FAO/WHO Codex Alimentarius Commission;
- World Health Organization (WHO);
- Food and Agriculture Organization of the United Nations;
- World Trade Organization.

In particular, the FSA represents the UK government in the joint FAO/WHO body, Codex Alimentarius, which is responsible for developing food standards, guidelines and related texts such as codes of practice under the joint FAO/WHO Food Standards Programme. The main purposes of this programme are protecting the health of consumers, ensuring fair trading practices and promoting co-ordination of all food standards work undertaken by international government and non-governmental organizations.

The HSE too are engaged with a number of international institutions in negotiating, developing and applying international standards and law, codes and guides on occupational health and safety. In the EU, these include the Directorates General of the Commission (Employment and Social Affairs, Environment, Energy and Transport, Internal Market, etc.) and their advisory committees and working groups, the European Agency for Safety and Health at Work, Eurostat and the European Committee for Standardization [16].

In addition, there are dealings with the Organization for Economic Co-operation and Development (OECD), the International Labour Organization (ILO), WHO and the International Atomic Energy Agency (IAEA). HSE also works with other regulators in fora such as the Senior Labour Inspectors' Committee, the International Association of Labour Inspectors, the International Liaison Group of Government Railway Inspectors, and the Western European and International Nuclear Regulators' Associations.

The task of negotiating specific directives, standards, conventions, and the like is carried out by the staff in dedicated teams within HSE's policy, technical and operational divisions who are responsible, via domestic law, for implementing their requirements in Great Britain.

The EA feeds information from its Pollution Inventory into the National Atmospheric Emissions Inventory (NAEI), which is compiled and annually updated by the National Environmental Technology Centre on behalf of the DEFRA. This information is also fed into the European Pollutant Emission Register (EPER), which provides useful information in establishing compliance with international treaties and conventions. The United Kingdom contributes to a number of international treaties and conventions that involve pollution inventories. The three main bodies through which we report emissions data are the EU, the OECD and the UN. Every two years the OECD publishes a compendium of environmental data to monitor environmental conditions and trends in OECD countries and to promote international harmonization of data. The compendium is a key reference for international environmental data [17].

A number of UN bodies handle issues related to environmental protection. In particular, the United Nations Economic Commission for Europe (UNECE) acts as a secretariat for a number of programmes specific to pollution inventories. The UNECE also facilitates the Aarhus Convention, which links human rights and environmental rights. Another UN body, the World Meteorological Organization, deals with issues specific to climate change, and facilitates conventions such as the United Nations Framework Convention on Climate Change (UNFCCC).

DEFRA has a direct interest in a number of international environmental research and demonstration programmes and initiatives including the European Union Framework Programmes and LIFE (Financial Instrument for the Environment) Programme, COST (European Co-operation in Science and Technology) and research activities of the Organization for Economic Co-operation and Development (OECD).The EU and the International Science Branch in the Europe Environment Division (EED) co-ordinate (where appropriate) and manage the Department's interests in the EU, OECD, COST and LIFE environment research and demonstration programmes ensuring that DEFRA's and the UK's policy objectives are met in the strategic formulation, coverage and content of these programmes [18].

The international branch of DoH is responsible for the co-ordination, monitoring and the development of health policies coming out of the EU and is responsible for the UK's interests in several international organizations concerned with health. The Chief Medical Officer represents the UK on the Executive Board (the governing body) of the WHO, which promotes technical co-operation for health among nations, carries out programmes to control and eradicate disease and strives to improve the quality of human life. Its four main functions are to give worldwide guidance in the field of health; to set global standards for health; to co-operate with governments in strengthening national health programmes; and to develop and transfer appropriate health technology, information and standards [19].

OTHER INFLUENCES UPON LOCAL GOVERNMENT AND ENVIRONMENTAL HEALTH PRACTICE

In addition to the central government, their agencies, organizations such as the Audit Commission and local politics there are a number of other organizations that influence environmental health within local government and lobby on its behalf.

The Local Government Associations

The Local Government Association (LGA), the Convention of Scottish Local Authorities (COSLA) and the Association of Local Authorities in Northern Ireland (ALANI) are the principal local authority associations in the United Kingdom. The LGA represents district, county, metropolitan and unitary local authorities in England and Wales. The LGA also represents police authorities, through the Association of Police Authorities (APA), fire authorities and passenger transport authorities. A key feature of the LGA regional structure is the Welsh Local Government Association (WLGA), which is a constituent part of the LGA but retains full autonomy in dealing with Welsh affairs. A separate association, the Association of London Government (ALG), exists to represent the interests of local authorities in the capital.

The local authority associations provide a national voice for local government. They aim, through lobbying, to be influential on behalf of local authorities and their communities. Their work is funded by subscriptions from their members. Each of the associations operates on broadly similar lines. They are staffed by full-time officials organized into six divisions in the LGA, of which one is the Economic and Environmental Policy Division, accountable to elected members nominated by the constituent member local authorities. It is organized in much the same way that local authorities operate, with a number of committees called 'executives' that determine policy [20].

The LGA's strategic objectives are:

- to achieve an equal partnership between central and local government;
- to secure the powers needed for local government to achieve their vision;
- to win greater freedom and flexibility in the exercise of these powers;
- to secure financial arrangements that enable local government to exercise its powers and duties effectively;
- to raise awareness of the role and achievements of local government; and
- to secure access for local authorities to the information, advice, services and support they need.

The executives are the driving force behind the development of LGA policy and the management of its business. Comprising small bodies of members, they enable the LGA to set the agenda for local government and to respond rapidly to events. The chairs of the executives are shared between the political groups to reflect the political balance of the organization. The central LGA executive focuses on board strategy, co-ordinating policy and local government finance and the management executive is responsible for the management of the LGA. There are further 17 executives each taking care of specific remits, for example Environment and Regeneration, Housing and Social Inclusion amongst others. The onus is on the executives to fully involve the wider local authority membership in their work, primarily through meetings of policy forums. The LGA's work is also supplemented by local authority secondees.

Local Authorities Co-Ordinators of Regulatory Services

Known for many years as LACOTS (Local Authority Co-ordinating Organization for Trading Standards) the organization was set up in 1978 to co-ordinate the enforcement activities of trading standards services. Later it took on the additional responsibility of co-ordinating food enforcement. Its new name reflects a widening regulatory brief that, since April 2002, has included the Registration Service for Births, Deaths and Marriages, public entertainment licensing and liquor licensing. It is a local government centralized body created by the UK LGAs to support local authority regulatory and related services. As an organization it is accountable to its own Board of Directors and a Management Committee made up of senior elected members nominated by the UK LGA. It is assisted in its work by a network of regionally nominated local authority advisers and recognized experts.

The vision of Local Authorities Co-Ordinators of Regulatory Services (LACORS) is: 'To make a major contribution to the development of high quality, consistent and co-ordinated local authority regulatory and related services across the UK' [21] and it is principally funded from monies 'top sliced' from the Revenue Support Grant. It provides and disseminates comprehensive advice, guidance, good practice and information to local authority enforcers. It makes clear however, that any advice it offers, especially on legal matters, is 'opinion'; as only the courts can interpret the law. However its opinions do enjoy the voluntary support of authorities because of:

- the shared desire for consistency;
- the advice represents the consensus views of local authority enforcement practitioners and experts; and
- its national-level consideration to views of other interested parties such as government and business.

Key aspects of its work relate to the promotion of good enforcement, consistency, disseminating good practice facilitating liaison between authorities and collaborating with government and the EU. LACORS supports the Home Authority Principle, a vital mechanism for liaison and co-ordination between local authorities dealing with businesses that have outlets in more than one local authority area and/or distribute goods and services beyond the boundaries of one local authority area.

LACORS administers statutory arrangements as the United Kingdom Single Liaison Body for trans-border food problems within the EU. It is a member of a number of pan-European co-operative bodies including Product Safety Enforcement Forum of Europe (PROSAFE), which deals with the safety of consumer products and the European Forum of Food Law Enforcement Practitioners (FLEP), which allows representatives of European food control authorities to meet, exchange information and address inconsistencies and practical enforcement difficulties.

The Chartered Institute of Environmental Health

The Chartered Institute of Environmental Health (CIEH) is the body that represents the professional interests of those working in the field of environmental health. The mission of the CEIH is to maintain, enhance and promote improvements in public and environmental health.

The organization that was, in due course, to become the CIEH was founded in London in 1883. Just over a hundred years later, having proved its ability to establish and maintain professional standards, it was granted a Royal Charter in 1984 as the Institution of EHOs; and in 1994, as a further mark of recognition, Her Majesty the Queen granted the body a new title – the Chartered Institute of Environmental Health.

The primary objective of the CIEH is the promotion of environmental health and the dissemination of knowledge about environmental health issues for the benefit of the public. It represents the views of its members, 9800 worldwide, the majority being located in England, Wales and Northern Ireland, on environmental and public health issues and is independent of central and local government. Whilst in the United Kingdom most members are employed in local government a significant number now work

in the private sector either for individual companies or as private environmental health consultants. There are also environmental health professionals working in central government and non-commercial organizations in addition to those teaching environmental health in universities and colleges throughout the country.

For organizational purposes the membership is divided into 17 centres; 15 of these are geographical and two are functional (Port Health and Commercial & Industrial). A centre council elected from the membership in the designated area runs the centres. A centre elects representatives to the general council, the governing body of the CIEH, in proportion to the number of members in the centre. At the time of publication this structure is under review, together with a consultation on widening the membership.

The charter governing the operation of the CIEH makes it clear that 'The object for which the Chartered Institute is established is to promote for the public benefit the theory and science of environmental health in all its aspects and the dissemination of knowledge about environmental health.'

To fulfil this objective, the activities of the CIEH include:

1. The holding of branch and centre meetings for the conduct of business and the discussion and development of good practice.
2. The provision of seminars, study weekends and other opportunities for professional development at both local and national levels.
3. The holding of an annual environmental health conference and an annual general meeting.
4. The accreditation and monitoring of academic programmes of environmental health, the operation of professional examinations and the assessment of work-based learning – all leading to qualification as an environmental health practitioner (see Chapter 4).
5. The operation of an assessment of professional competence (APC) scheme and a continuing professional development (CPD) scheme for its graduate and voting members (see Chapter 4).
6. The convening of expert advisory panels and task groups at the national level to develop policy and recommendations for good practice

and to respond to consultation documents published by government departments and other organizations with concerns for environmental health.

7. The distribution of a weekly newspaper and a monthly journal, and the publication of texts, practice notes and policy documents on environmental health issues. Recently this has extended to include the publication of *The Journal of Environmental Health Research*, which is intended to be a biannual publication.

8. The provision of advice to members on educational, constitutional and technical issues.

As well as providing services and information to its members, the CIEH also advises government departments and agencies on environmental health and is consulted by them on proposed legislation relevant to the work of environmental health professionals. Increasingly it is reaching out internationally, to campaign for improvements in public health and to facilitate debate about the global environmental challenges that threaten our health. Most of this work involves communicating the views of the profession to governments and other professional and international organizations including the EU and the UN. The CIEH became the WHO/EURO Collaborating Centre for Environmental Health Management in 1993.

The Improvement and Development Agency

The Improvement and Development Agency (IDEA) is based in central London and was established by and for local government in April 1999 with a mission to support self-sustaining improvement from within local government. Understandably therefore current priorities reflect those of local government, e-government, leadership, capacity building, improving council services and community well-being. It is a non-profit organization and is wholly owned by the LGA, and is funded via the Revenue Support Grant (RSG). The Office of the Deputy Prime Minister makes decisions on the recommendations by the LGA on the level of RSG paid to the IDEA. The Agency's board is responsible for the overall strategic direction and performance of the IDEA. It comprises members of all party groups from the LGA, and representatives of stakeholders including trade unions, the private sector, the Welsh LGA and central government. The IDEA aims to help local authorities engage effectively with communities, build local partnerships and integrate sustainable development within their decision making processes and delivery of services. The IDEA's work comprises a mix of services free to all councils, and more tailored support services to individual authorities [22].

The ombudsman

When a local authority customer has a complaint about local services their final recourse is to the local government ombudsman (LGO). He investigates complaints of 'maladministration'. Unfortunately there is no legal definition of this, however comments made by the LGO in reports suggest that it could include administrative shortcomings, bias, neglect, inattention, delay, incompetence or ineptitude. Some things are specifically excluded from his remit, for example personnel matters, certain education matters, complaints about action that affects all or most of the inhabitants of the authority's area, the commencement or conduct of civil or criminal proceedings before a court and action taken related to any commercial transaction. Complaints can only be made when the maladministration has led to injustice such as financial loss, being deprived of an opportunity to object to a decision or being deprived of an amenity, for example closure of a museum or something that causes a person concern, confusion or inconvenience. However injustice is not suffered by someone/group who think that others have suffered injustice. In fact, the main test of whether there has been maladministration is whether an authority has acted reasonably in accordance with the law, its own policies and generally accepted standards of local administration. 'Reasonableness' crops up again and again in letters and reports explaining the LGO's conclusions. The LGO is concerned with the way a decision is reached, not with the **merits** of decisions. He can investigate complaints about how the council has done something, but they cannot question what a council has done simply because someone does not agree with it.

The complainant must first give the council concerned an opportunity to deal with any complaint against it. Ideally using the council's own complaints procedure, if it has one. Then if the complainant is not satisfied with the action taken by the council, he or she can send a written complaint to the LGO or ask a councillor to do so on their behalf.

The role of ombudsman was created by the Parliamentary Commissioner Act 1967 and was rolled out to local authorities in 1974. The LGO's jurisdiction covers all local authorities (excluding town and parish councils); police authorities; education appeal panels; and a range of other bodies providing local services. The vast majority of the complaints received concern the actions of local authorities and that is why they have become known as the Local Government Ombudsmen. Under the 1974 Act however a complaint could only be made in writing to a member of the authority concerned [i.e. not directly to the LGO] and then the member had to refer the complaint to the LGO. However the 1988 Local Government Act now permits a complaint to be made directly to the LGO as an alternative to the 1974 procedure. Ninety per cent of complaints are direct and all must be made within 12 months from the day the aggrieved person first had notice of it. However the LGO may investigate out of time complaints if he thinks it is reasonable.

In 2000/2001 the top ten causes of maladministration were:

- delay in taking action;
- taking incorrect action;
- failure to provide information;
- failure to compile and maintain adequate records;
- failure to take action;
- failure to take relevant considerations into account in making a decision;
- failure to investigate;
- failure to deal with letters or other enquiries;
- failure to comply with legal requirements; and
- making misleading or inaccurate statements.

The outcome of an investigation is in the form of a report that recommends a remedy, for example setting right the maladministration or paying compensation to the complainant. Unfortunately this judgement is not enforceable; however if the LGO is dissatisfied with the local authority's response he can publish a second report thus hopefully embarrassing the local authority into action.

There are three LGOs in England and they each deal with complaints from different parts of the country, however each is assisted by a Deputy and between 35 and 40 investigative staff to whom they can delegate powers.

In practice environmental health receives very little attention from the LGO as can be seen from the following data gathered during the period 2002/03, when 17 600 new complaints were received. The categorization of complaints by subject was as follows:

- housing 38%
- planning 20%
- education 9%
- highways 8%
- social services 7%
- local taxation 6%
- environmental health 3%
- land 2%
- other 7%.

For the purposes of an investigation, an Ombudsman has the same powers as the High Court in respect of the attendance and examination of witnesses and the production of documents.

The Ombudsman may require anyone – whether a councillor, an officer or anyone else – to provide information and documents. But nobody is compelled to give any evidence that he or she could not be compelled to give in civil proceedings before the High Court. Anyone who, without lawful excuse, obstructs an Ombudsman in the performance of his or her functions is guilty of an act or omission that, if the investigation were a proceeding in the High Court, would constitute contempt of court (there has not, so far, ever been need to resort to proceedings for contempt). The Ombudsman is given unqualified discretion whether to initiate, continue or discontinue an investigation. Investigations are conducted in private [23].

Since 1996 a digest of cases referred to the Ombudsman has been published with the aim of promoting good local administration and increasing understanding of how and why decisions are reached. The most recent digest in 2002/03 included two cases listed as environmental health – both about noise nuisance [24]. The Local Government Ombudsman also produces an annual report, which includes details of complaints dealt with in the year [25].

THE LOCAL FRAMEWORK

Local government has to work within the powers laid down under various acts of Parliament and if they exceed their statutory powers they can be challenged in the courts. Some functions are mandatory, which means that the authority must do what is required by law; others are discretionary, allowing an authority to provide services if it wishes. Whilst the way in which they organize themselves internally, in terms of the numbers and type of departments, (see section on environmental health units) is to a large degree flexible, much of the constitutional framework in which they work is laid down.

As opposed to the traditional committee system, authorities and their local population can now choose from three options on the constitution (and therefore executive) they would like to implement for their council. This change was driven by the desire for more public involvement and better, more accountable decision making. It was implemented by the Local Government Act 2000 and the process for deciding on any of the three options is one of local choice as the electorate can call for a referendum on the way a council is managed.

Whilst the full council will continue to:

- decide new constitution;
- decide policy framework;
- decide budget;
- appoint Chief Officers;

there are three options that can be chosen for the executive:

- *A directly elected mayor with a cabinet.* The mayor is elected by the whole electorate and he then selects a cabinet from among the councillors. The cabinet can be drawn from a single party or a coalition. These cabinet members have portfolios for which they take executive decisions acting alone. The mayor is the political leader for the community, proposing policy for approval by the council and steering implementation by the cabinet through council officers. The chief executive and chief officers are appointed by the full council. The chief executive has particular responsibility for ensuring that both executive and backbench councillors receive all the facilities and officer support necessary to fulfil their respective roles. The office of directly elected mayor is separate from the traditional ceremonial mayor.

- *Leader and cabinet.* Under this option a leader is elected by the council and the cabinet is made up of councillors, either appointed by the leader or elected by the council. As with a directly elected mayor model, the cabinet can be drawn from a single party or a coalition. The model is very similar to the mayor and cabinet system except that the leader relies on the support of members of the council rather than the electorate for his or her authority and can be replaced by the council. While the leader could have similar executive powers to a directly elected mayor, in practice the leader's powers are less likely to be as broad as there is no direct mandate from the electorate for the leader's programme.

- *Mayor and Council Manager.* The mayor is directly elected to give a political lead to an officer or 'manager' to whom both strategic policy and day-to-day decisionmaking are delegated. The mayor's role is primarily one of influence, guidance and leadership rather than direct decisiontaking. Using a private sector analogy, the mayor might resemble a non-executive chairman of a company and the council manager its powerful chief executive. Again this can be separate from the traditional ceremonial mayor.

For those authorities with a population under 85 000 or have had a referendum for a mayor that has been rejected, the local electorate can have alternative arrangements whereby the full council has an enhanced policymaking role under alternative arrangements with all councillors acting together as the full council. The local authority will be able to delegate implementation of its policy to streamlined committees. Under alternative arrangements councillors will have the following roles:

- adopting the new constitution and any subsequent changes to it;
- adopting the local authority's code of conduct;
- conforming the local authority's policy framework and budget;
- making appointments to committees; and
- making or confirming appointment of the chief executive.

Alternative arrangements must involve effective overview and scrutiny.

The LGA reported in October 2002 [26] that the vast majority of local councils have gone for the second option – Leader and cabinet (316) with only 10 and 1 going for the Mayor and cabinet and Mayor and Council Manager respectively; 59 have taken up the Alternative Arrangements option.

Irrespective of the option chosen all councils must have a scruting committee. They are made up of councillors who are not members of the executive/cabinet and reflect the political balance of the authority. Scrutiny takes in a varied range of activities. These are:

- Review and development of the council's policies.
- Make policy and budget proposals to the council.
- Review of proposed executive decisions.
- Call in or review of decisions before they are implemented.
- Performance monitoring and review.
- Scrutiny of other local organizations, including health services.

Councils can choose whether to have scrutiny committees for particular services or themes, such as education or the environment, or whether to have only one scrutiny body. But there must be a scrutiny committee with a responsibility to have an overview of each service the council does. There will also be a scrutiny body that looks at local health service issues, and possibly other public services not run by the council itself. Scrutiny bodies can be called panels or commissions, or some other title, rather than committees. Scrutiny committees will investigate issues and report to the executive/cabinet or the full council. Other compulsory requirements are Regulatory Committees, reflecting the political make up of the authority overall for quasi-judicial decisions such as planning and licensing decisions, and include granting planning permission, licensing certain premises, licensing taxis, and so on, and Standards Committees that must have at least one independent or lay member to promote and maintain high standards of conduct in the authority.

The Local Government Act 2000 aimed to create new ways of working for authorities so that decisionmaking became more efficient, transparent and accountable. However feedback suggested that councillors felt excluded from the decision-making process. A review has been undertaken and its findings reported by the Transport, Local Government and Regional Affairs Select Committee [27] in 'How the Local Government Act 2000 is working' in 2002. Whilst the report acknowledged some potential failings in the new system they largely concluded that it was too soon to tell. The report also highlighted the Committee's concerns that local authorities are now subject to too much scrutiny by too many external organizations. The CPA is intended to be key to the achievement of a proportionate and co-ordinated inspection programme. The various inspectorates are working together to develop a programme that will significantly reduce the volume of inspection and make it more efficient. In addition it is proposed that those councils identified as 'excellent' will get a break in the cycle of inspections.

THE ENVIRONMENTAL HEALTH UNIT TODAY

The exact internal organization of a local council is largely up to the council itself. Historically there would have been dedicated 'environmental health departments', however internal reorganizations over the years, for example to reduce senior management positions has resulted in the amalgamation of departments in variety of ways with the inevitable 'swallowing up' of environmental health. Also the allocation of functions and 'names' given to such departments is a decision for the council itself. Some authorities still use the word department to describe an operational component while others use division or service. Today this means that environmental heath as a function could be found for example:

- in a dedicated environmental health department (rarely);
- in a housing and health department/division;
- in a planning and health department/division;
- in the environmental health and consumer service;

to name but a few of the variations now found. According to an unpublished survey by the Society of Environmental Health Officers (SHEO) in 1993, of those authorities responding, about [28]:

- 37% of units had formed separate departments;
- 58% were part of a broader directorate;
- 5% had structures where the environmental health function had been split.

It is possible that since 1993 there has been a continued trend towards the environmental health unit being part of a wider directorate, but no further surveys have been carried out. Given that the 'environmental health department' is a thing of the past this chapter uses the term 'environmental health unit' to describe the administrative base from which typical environmental health functions are executed.

Also the arrangements within the environmental health unit itself can differ. Traditionally EHOs looked after a geographical patch undertaking all environmental health duties therein. Nowadays however, with increasing technology, legislation, competency requirements, and the like, questions are asked as to whether this is the best use of resources. A more common arrangement therefore is to provide services on a specialist basis where a typical arrangement might mean:

- A commercial section – dealing with food and health and safety.
- A public health section – dealing with environmental protection and all other environmental health areas.

The Employers' Organization for local government and the Society of Environmental Health Officers conducted an Environmental Health Workforce Survey of all authorities in England, Wales and Northern Ireland during the summer of 2002 [29]. The survey achieved a response of 52.7% (212) of the 402 authorities with the environmental health function. The survey recorded a total of 11 728 staff employed in environmental health departments as of 1 January 2002, including:

- 5226 Environmental Health Officers (44.6%);
- 3949 Environmental Health Technicians (33.7%);
- 316 Student Environmental Health Officers (2.7%); and
- 2237 administrative staff (19.1%).

Seventy three per cent of non-administrative staff were in specialist roles. There were 7126 specialist posts in environmental health departments, 26.0% of which were managerial/supervisory and 74.0% non-managerial. The primary functions of specialist posts with one main function were pollution (24.7%), food (18.2%), housing (16.8%), and a further 12.2% combined food with health and safety.

Other key findings were:

- 87.4% of staff were employed full-time and 12.6% part-time;
- 40.4% of staff were EHO and 14.6% of administrative staff;
- 37.0% of non-administrative staff were female.

Tasks

Since the environmental health unit is the focal point of a much broader system of control, it follows that the tasks to be performed within the unit will be diverse. The following groupings indicate the range of activities.

Inspection and enforcement

One of the core functions of the environmental health unit is to ensure compliance with standards, including the enforcement of legislation, primarily statute law and secondary legislation, but also local legislation including by-laws.

The inspection of premises in order to enforce legislation is one of the main tasks of the environmental health unit, and the way in which this is approached is crucial to its effectiveness. The inspection activity must be structured so as to take full account of the risks inherent in any particular premises or type of premises. Thus more attention and a greater frequency of inspection should be given to those premises whose operations and/or past record involve a greater risk to public health. Indeed in some areas of environmental health, for example food and health and safety at work this approach has been stressed by the FSA and HSE respectively and both issue schemes by which priority ratings can be applied to individual premises. Thus risk assessment and an inspection programme based on it are essential ingredients of the enforcement function.

The last government's main vehicle for applying good enforcement procedures was Section 5 of the Deregulation and Contracting Out Act 1994 (DOCA). The procedures were applied in some areas of food safety, housing standards and health and safety. In December 1996 the Regulatory Impact Unit (then known as the Deregulation Unit) issued a consultation document on proposals for the extension of Section 5 enforcement procedures to trading standards, care services and environmental health. In their responses local authorities were strongly opposed to the proposals to extend Section 5, but they made it clear that they were not opposed to the principles behind the Section, but to the incorporation of them in legislation as being bureaucratic and inflexible.

In 1997 the new government decided to build on the agreement on the principles of good enforcement, while avoiding the inflexibility of the legislative route and in 1998 a concordat setting out principles and procedures of good enforcement was developed in consultation with business, local and central government, consumer groups and other interested parties. The Enforcement Concordat was signed in 1998 to bridge shortfalls in the DCOA.

Adoption of the Concordat is voluntary and to date 96% of all central and local government organizations with an enforcement function have adopted the Enforcement Concordat [30].

Other organizations have specific advice to give on enforcement policy, for example the HSC in 'Health and Safety at Work etc Act 1974: Enforcement Policy Statement 2001' [31], which requires that enforcement is carried out with regard to the principle of proportionality, consistency, transparency and targeting.

Surveillance

An essential task for the unit is to undertake ongoing and total surveillance of those environmental factors that affect health. Sampling/surveillance might arise in response to a specific problem, or as part of a wider proactive initiative. This involves the establishment of a wide range of direct activities within the unit, and of collaborative/monitoring arrangements with other organizations involved in environmental monitoring, both statutory and non-statutory. Examples of sampling/surveillance activity might include:

- The microbiological and chemical sampling of food as part of a local strategy as well as part of ongoing programmes on behalf of the FSA.
- Environmental sampling as part of the Air Quality Management Strategy or monitoring emissions from authorized or permitted processes.
- Monitoring private water supplies.
- Measurement of background noise levels as part of an assessment of proposals for a new development, or monitoring levels as part of a nuisance investigation.

In addition to the sampling/measurement of environment factors there will also be a need to collect and analyse data on, for example, enforcement activity, accident statistics, food poisoning and communicable diseases.

Provision of information, public involvement and education

The Environmental Information Regulations 1992, as amended in 1998, require a wide range of bodies, including local authorities, to make available to the public environmental information that they hold relating to:

1. The state of any water or air, the state of any flora or fauna, the state of any soil or of any natural site or other land.
2. Activities or measures (including noise or other nuisance) that adversely affect anything in (1) above.
3. Any measures (including environmental management programmes) that are designed to protect against these concerns.

There are some exceptions to these requirements, for example commercial confidentiality and information subject to legal proceedings.

In some circumstances there are other statutory requirements relating to environmental information, for example the keeping of registers for authorizations/permits of processes under part I of the Environmental Protection Act 1990 and Pollution Prevention and Control Act 1999 respectively. Under the Environment and Safety Information Act 1988, registers of information on enforcement notices, for requirements that affect the public, served under the Health and Safety at Work, Etc. Act 1974 must be kept available for public inspection.

Increasingly though there is a move to more openness and the Freedom of Information Act 2000 gives a general right of access to all types of 'recorded' information held by public authorities. The Act reflects a national policy shift in public administration from a culture of confidentiality to one of openness. The underlying principle is that all information held by a public authority should be freely available except for a small number of tightly defined exempt items. It seeks to balance three rights:

- The right to information
- The right to confidentiality
- The right to effective public administration.

All public bodies are covered by the Act; these include government departments, local authorities, NHS bodies (such as hospitals, as well as doctors, dentists, pharmacists and opticians), schools, colleges and universities, the Police, the House of Commons and the House of Lords, the Northern Ireland Assembly and the National Assembly for Wales. There is a provision in the Act for other authorities to be named later and for organizations to be designated by the Secretary of State as public authorities because they exercise functions of a public nature or provide a service under a contract, which is a function of that authority.

The Act will be brought into force fully by January 2005 and public authorities will have two main responsibilities under the Act. First they will have to produce a 'publication scheme' (effectively a guide to the information they hold, which is publicly available) and second they will have to deal with individual requests for information. The duty to adopt a publication scheme is being phased in and local authorities had to comply with this by 28 February 2003. All public authorities will be required to deal with individual requests for information from 1 January 2005 when the general right of access to information held by public authorities comes into force. The Act will be enforced by the Information Commissioner (previously the Data Protection Registrar).

In addition to the increasing amount of attention being given to the provision of environmental health information to the community on a full range of issues, for example food hazards, heart disease, environmental pollutants, there is also a need to inform the community of the services available and more importantly to involve the public in the decisions about the content of those services and the ways in which they are made available. Best Value requires consultation to be made with not only the public at large but also with stakeholders such as businesses.

Over the past few years, increasing attention has been given to identifying the customers for public services and a need to relate to them closely in terms of service standards and delivery. At government level this identified itself in the form of the Citizen's Charter [32]. Customer relations are now firmly placed as an issue within the environmental health service. One facet of this is the setting of performance indicators, another is the Charter Mark. At the heart of Charter Mark criteria is the fundamental question: 'What does the customer expect or hope for from the services you offer?' Charter Mark holders are able to demonstrate that they offer choice to their customers so that a wide range of needs are catered for; that the benefits of new technology are maximized; users and staff are consulted on where choices can be made; and communities have a say in the design and delivery of local services. This is an award for excellence in the provision of services, and several environmental health units have successfully applied for it.

As part of the response to Agenda 21 [38], there is an increasing tendency to establish widely based collaborative groups at local level to consider the environmental issues relating to that area. These groups include representatives of statutory and voluntary bodies, and may involve the Chamber of Commerce, Friends of the Earth and local consumer groups. The central, co-ordinating role played by the environmental health unit in these groups is very important. Changes in the public perception of environmental health issues demand a much closer relationship between the environmental health unit and its community.

Gauging public opinion has been embraced by central government too using the People's Panel a representative sample of 5000 adults based on age, gender, region and a wide range of other demographic indicators. It is used by government departments and agencies – and other publicly funded bodies – to test reaction to a range of policies and initiatives.

The provisions of the Food Safety Act 1990 and the Health and Safety at Work, Etc. Act 1974 provide for training to be given to staff employed in activities covered by the legislation – in effect the whole range of employment. Many employers look to the environmental health department for assistance, and the CIEH validates such courses through those departments.

Service provision

Local authorities provide a wide range of environmental health services directly to the community, some being statutorily required and others being discretionary. These include:

- refuse collection and street cleansing;
- pest control;
- dog warden;
- food hygiene training;
- health promotion.

These services may be provided either directly by the environmental health unit or by others on a contract from the unit. Some services may be provided by the environmental health units as the result of an agency agreement with the body carrying the primary responsibility for that service. One example is the arrangement between a district and a county council for the operation of the food standards function. In this case, the primary authority retains control over the standards of the service and resource allocation, while the district council bears responsibility for the operation of the service within those parameters. The exact mix of service delivery will vary according to how that council has decided to structure itself.

Investigative

Investigative work results from both community complaints, for example nuisances and also from notifications of for example accidents in the workplace and incidences of food poisoning. Many will be investigations within a legal framework, which may lead to the institution of legal proceedings, for example statutory nuisances (see Chapter 6), while others will have no statutory remedy and will require persuasion if the problem is to be eliminated.

Innovative

One of the essential tasks of the unit is to try new things, and to experiment with solutions to problems. Historically it has been the efforts of individual or groups of local authorities that have led to major legislative changes on issues such as clean air, food hygiene, dog control and health promotion.

A legitimate function of the unit is to stimulate change to the approach and operation of the environmental health service through experimentation. Examples are the creation of 'community contracts' for the street cleaning function in York, a scheme for the retrieval of supermarket trolleys in Exeter (now embodied in national legislation through the Environmental Protection Act 1990) and arrangements for health promotion in Oxford. Both the CIEH and the local authority associations have a vital role to play in promoting such innovation locally, and in pressing for national change based upon the success of local initiatives.

There is a developing role for environmental health officers in undertaking sponsored research as an aspect of environmental health as part of studying for a higher degree, and some of this work can lead to new ideas for environmental health policy and practice.

Agency services

The past 10 years have seen a growth of agency and advisory services to the health departments operating agency services for the undertaking of work related to renovation grants, and also for 'stay-put' schemes to enable elderly or disabled people to live more comfortably in their own homes as part of the Supporting People framework. These agency services include surveys, the identification of defects, the production of specifications, the letting and supervision of contracts and the identification of funding arrangements, including grant aid.

Delegation of authority

The Local Government Act 1972 allows a council to delegate decisionmaking to committees, subcommittees and officers, this is further developed to take account of the new administrative arrangements brought in by the Local Government Act 2000 whereby Section 15 permits the Council to arrange for the discharge of any of its executive functions by the cabinet, a committee of the cabinet, an individual cabinet member or an officer. The correct use of these powers are critical to the effective operation of services. Any delegation needs to be undertaken as a formal process by clear definition through resolution of the appropriate part of the administration. Because the environmental health service contains a high level of legislative enforcement, delegation to its chief officer is usually greater than to any other chief officer, and will need to indicate each legal process delegated, for example service of notices and institution of proceedings under each piece of legislation. In addition, to expedite actions powers are frequently delegated to individual officers. EHOs and associated technical staff should familiarize themselves with powers so delegated and ensure that they do not act outside them.

NEW INITIATIVES

Local decision making and public scrutiny

Decision making takes place at a variety of levels with some decisions such as the approval of the budget and the adoption of policies resting with the full council. Others, particularly in relation to implementation of policies, are for the executive to make in line with any delegation arrangements. All decision making however must be subject to rigorous review and scrutiny. Accordingly it must be made clear to the public who in the council is responsible for making decisions, how they can have their input and they must have access to meetings when key decisions are being taken. Information about and access to agendas, reports and minutes of meetings and certain background papers must also be available (see section on information).

By law local authorities are required to establish a Scrutiny Committee (known in some authorities as the Overview and Scrutiny Committee) to monitor decisions of the cabinet and, where appropriate, to advise the Council on matters of policy or service delivery. The idea of scrutiny in local government is based on the Westminster parliamentary system of Select Committees, which were set up as a means by which policies and their impact could be examined in detail by backbench MPs, with the power to call for evidence from ministers, civil servants and independent experts. Scrutiny looks in detail at the decisions of the cabinet to make sure the powers are used wisely. The scrutiny process is significantly different from the old committee system. Scrutiny is led by elected Councillors who are not members of the cabinet.

This requirement for overview and scrutiny is being rolled out to allow local authorities to set up Committees to scrutinize decisions made by NHS bodies The new power will enhance authorities' role as local community leaders and could prove an important opportunity for EHOs to brief their councillors as to what they should be looking for in terms of the public health/health improvement agenda.

Beacon Councils

Co-ordinated by the Improvement and Development Agency, the Beacon Council Scheme was set up to disseminate best practice in service delivery across local government. Each year, the government selects themes for the beacon scheme, announced one, two or more years in advance with some themes being repeated in future rounds.

Any local authority can apply to become a Beacon. The application process involves a written submission in the first instance followed by an assessment visit for shortlisted councils, the final decision is made by government ministers based on recommendations made by an independent advisory panel. The status is awarded for a year; and during this period IDEA works with them to facilitate the sharing of good practice through a series of learning exchanges, open days and other learning activities. Beacon status is granted to those authorities who can demonstrate a clear vision, excellent services and a willingness to innovate within a specific theme. However, to obtain

beacon status applicants must also demonstrate that they have good overall performance, and not just in the service area for which beacon status is awarded.

Six rounds of themes have been announced thus far with Round 4 just beginning at the time of publication. Effective environmental health is listed in Round 6 although previous themes such as 'neighbourhood renewal in urban and rural areas' in Round 3 and 'local environmental quality' in Round 2 have had an environmental health focus.

A beacon council should be able to demonstrate the following 10 clear attributes.

1. A clear vision of where they are going and a strategy to get them there.
2. Strong leadership (political and managerial).
3. A commitment to partnership working.
4. A learning culture and a willingness to innovate.
5. Decisionmaking based on relevant information.
6. Appropriate and effective processes for informing and involving the public and/or service users.
7. Clear, achievable outcome-driven performance targets, effective use of benchmarking comparisons including effective monitoring processes.
8. Effective, regular measurement of public satisfaction with services.
9. On the basis of the foregoing a record of sustained improvement or of maintaining a high standard of service provision.
10. Demonstrable readiness to exchange share practice and learning with others.

REFERENCES

1. Local Government Association (2002) Fact Sheet – *Key Dates in English and Welsh Local Government History*, Local Government Association Publications.
2. Local Government Association (2003) *Local Matters – Councils 2003/04 a snapshot*, Local Government Association Publications.

3. ODPM (2003) *Local Government Finance – Key Facts.*

4. WHO (1978) *Role, Functions and Training Requirements of Environmental Health Officers (Sanitarians) in Europe*, WHO Regional Office for Europe, Copenhagen.

5. Local Government Association (2001) *Fact Sheet – Local Government Structure*, Local Government Association Publications.

6. Office of the Deputy Prime Minister (2003) *Government offices*, http://www.odpm.gov.uk/go/index.htm (accessed 13 May 2003).

7. Department for Local Government Transport and the Regions (1998) *Modern local government: in touch with the people*, White Paper.

8. Welsh Office (1998) *Local voices – Modernising local government in Wales*, White Paper.

9. Office of the Deputy Prime Minister (2001) *Strong local leadership – Quality public services*, White Paper.

10. Department of Health (1999) *Saving Lives, Our Healthier Nation*, Department of Health Cm 4386.

11. Brown Steve (2002) EHOs countless opportunities *Health Development Today*, Health Development Agency, 8 May 2002.

12. Chartered Institute of Environmental Health and Health Development Agency (2002) *Environmental health 2012. A Key Partner in Delivering the Public Health Agenda.*

13. Food Standards Agency (2003) About us, http://www.foodstandards.gov.uk/aboutus/ (accessed 13 May 2003).

14. Health Safety Executive (2002) *Responsibility For Health And Safety Moves To Department For Work And Pensions*, Press release, 29 July 2002.

15. Food Standards Agency (2003) Developing beneficial international relations, http://www.foodstandards.gov.uk/aboutus/how_we_work/intdevelopment (accessed 13 May 2003).

16. Health Safety Executive (2003) International branch, http://www.hse.gov.uk/policy/puhome.htm#1 (accessed 13 May 2003).

17. Environment Agency (2003) 'Business', http://www.environment-agency.gov.uk/business/ (accessed 13 May 2003).

18. DEFRA (2003) EU and International Environment Research & Demonstration Programmes, http://www.defra.gov.uk/environment/internat/research/index.htm (accessed 13 May 2003).

19. World Health Organisation (2003) 'International health', http://www.doh.gov.uk/international/who.htm (accessed 13 May 2003).

20. Local Government Association (2003) 'LGA member structure', http://www.lga.gov.uk/ (accessed 13 May 2003).

21. Local Authorities Coordinators of Regulatory Services (2003) 'About Lacors', http://www.lacots.com/pages/trade/lacors.asp (accessed 13 May 2003).

22. Improvement and Development Agency (2003) http://www.idea.gov.uk/ (accessed 13 May 2003).

23. Commission for Local Administration in England (2003) The work of the Local Government Ombudsman, http://www.lgo.org.uk/index.htm (accessed 14 January 2004).

24. Commission for Local Administration in England (2003) Digest of cases, http://www.lgo.org.uk/digest.htm (accessed 14 January 2004).

25. Commission for Local Administration in England (2003) Annual report of Local Government Ombudsman, http://www.lgo.org.uk/annual.htm (accessed 14 January 2004).

26. Local Government Association (2002) *Fact Sheet on Local Government Act 2000: Political Management of English Local Authorities*, Local Government Association.

27. Office of the Deputy Prime Minister (2002) *The Government's Response to the Transport, Local Government and Regional Affairs Select Committee's Fourteenth Report on How the Local Government Act 2000 is Working* Cm 5687.

28. Society of Environmental Health Officers (SHEO) (1993).

29. Employers Organization for Local Government (2002) *Environmental Health Workforce Survey 2002*.

30. Cabinet Office (2003) Enforcement Concordat, http://www.cabinet-office.gov.uk/regulation/publicsector/enforcement/enforcement.htm (accessed 14 January 2004).

31. Health and Safety Commission (2001) Health and Safety at Work Etc. Act 1974: Enforcement Policy Statement 2001.

32. Her Majesty's Government (1991) *The Citizen's Charter – Raising the Standard* Cm 1599, Stationery Office, London, July.

33. UN (1992) Earth Summit, Agenda 21, The United Nations Programme of Action from Rio, UN Department of Information, New York.

APPENDIX A: THE ORGANIZATION OF ENVIRONMENTAL HEALTH IN NORTHERN IRELAND

SAM KNOX

The present system of local government in Northern Ireland was established in 1973. There are 26 single-tier local authorities, 5 city councils, 13 borough councils and 8 district councils. The population of the councils varies considerably, from Belfast with a population of 2 97 300, to Moyle with 15 000.

Local authority functions

The powers of local authorities in Northern Ireland are limited compared with local government elsewhere in the United Kingdom. Councils are not responsible for health services, social services, education, housing, water supply, sewerage or planning. These services are provided by government departments, agencies or nominated statutory boards as shown in Table 3.3. There is a statutory requirement for councils to be consulted by the planning service of the Department of the Environment on planning proposals.

Environmental health functions

With the exception of the environmental functions listed in Table 3.3, councils have responsibility for waste collection and disposal, street cleaning and a wide range of environmental health functions including food safety, food composition and labelling, food complaint investigation, health and safety at work, safety of consumer goods and air and noise pollution control. They also have responsibility for licensing and by-law controls in important areas of trade and business.

Not all these functions are allocated to the environmental health department of the councils for enforcement, but the majority are. Generally, where any of these functions are the concern of other departments, Environmental Health Officers (EHOs) have a significant input in terms of the provision of information and support.

In the majority of councils, environmental health still operates as a separate department but in 12 councils, the function has been merged with other services. However, in the majority of cases, the Directorate is under the control of an EHO.

Table 3.3 Responsibility for services in Northern Ireland

Service	Agency responsible
Health and social services	Health and Social Services Boards
Education	Education and Library Boards
Housing	Northern Ireland Housing Executive
Water supply and sewerage	Department of the Environment of Northern Ireland
Planning	Department of the Environment of Northern Ireland

The group environmental health system

A rather unique aspect of the environmental health service in Northern Ireland is the group system, whereby the 25 councils outside Belfast city council are statutorily joined together in groups for specific purposes while remaining independent enforcement authorities. A group system was first established in 1948 based on county council and county borough boundaries. When local government was reorganized in 1973, county councils were abolished and new groups were established with boundaries that were coterminous with those of the Health and Social Services Board areas. These arrangements were made under powers contained in the Local Government Act (Northern Ireland) 1972 by the Local Government (Employment of Environmental Health Officers) (Northern Ireland) Order 1973.

The purpose of the initial group system was to make available the services of specialist officers to even the smallest district council and to government bodies, and to ensure effective co-ordination and consistency of the service in district councils by having all EHOs working in the districts employed by the group and working under the general direction of the Group Environmental Health Committee's chief EHO.

As a result of objections from a number of district councils over the years, the system was subjected to scrutiny and to a 'Value for Money Study' which resulted in the introduction of a Local Government (Employment of Group Environmental Health Staff) Order (Northern Ireland) 1994. This radically changed the way the groups operated by returning employment of EHOs to each district council and by removing the general direction powers of the group's chief EHO.

The group committee still has its own staff of specialist officers who, with the committee's chief officer, the officer responsible for environmental health, provide specialist services to the constituent district councils, co-ordinate the environmental health service within and between groups, assist district councils to draw up annual environmental health plans and monitor district councils' provision of environmental health services. The group committees also employ a number of EHOs and water quality inspectors for control of water pollution, who are made available to the Environment and Heritage Service (see later). The group committees are still responsible for student and professional training and provide a pest control service to meet district council needs.

Environmental health at government level

In 1996, the restructuring of the former Environment Service led to the establishment within the Department of the Environment for Northern Ireland of the Environment and Heritage Service (EHS), a next steps agency that has the overall responsibility for the protection of the environment from pollution of air, water and land.

These changes led to a review of the role of the environmental health unit within the department, and as a result on 1 October 1996 there was a transfer of the government focus for environmental health and public health issues in Northern Ireland to the Department of Health and Social Services where the posts of chief and deputy chief EHOs were established.

The other EHOs remaining from the environmental health unit were charged with the responsibility for establishing an air and environmental quality unit and a waste and contaminated land unit within the EHS.

The air and environmental quality unit majors on the technical and policy advice aspects of ambient air quality and noise control. The waste and contaminated land unit exercises the enforcement role in these fields, previously exercised by district councils.

The chief EHO is responsible for maintaining a body of environmental health expertise within the Northern Ireland civil service and for providing policy advice on the full range of environmental health issues to the Northern Ireland ministers and departments. Additionally, he maintains an oversight of the environmental health service provided by the 26 district councils, for promoting good practice and advising on professional matters.

Liaison arrangements

Liaison in environmental health matters is principally effected through regular meetings of the Northern Ireland chief environmental health officers group (NICEHOG), which consists of the chief EHO, Department of Health, Social Services and Public Safety, the director of health and environmental services, Belfast, the four group chief EHOs and the directors of environmental health or heads of service of the other 25 district councils.

NICEHOG's vision is that environmental health shall be a key and valued service for the protection and promotion of health and well-being. To achieve this vision, NICEHOG is striving to be:

- Trusted experts with broad credibility
- Recognized advocates of healthy public policy
- And a quality high-profile service.

It is currently operating administratively through an executive committee and is supported by two task and finish sub-groups, namely information technology, Best Value and performance indicators and by three other continuing sub-groups, namely, Strategy, Marketing and Training.

There is also a number of sub-groups consisting of specialist EHOs who meet regularly to discuss health and safety, food safety and standards, consumer protection, public health and regulatory services, pollution and radiation. Each sub-group has a planned programme of work, which includes liaison with appropriate government departments and agencies, commenting on new and proposed legislation, organizing sampling and promotion programmes and developing appropriate standard procedures to ensure consistency of enforcement.

Legislation

Legislation on environmental health matters in Northern Ireland closely follows the legislation for England and Wales, but there is often a considerable time delay before legislation in Northern Ireland is brought into line with that in Great Britain. Guidance on this aspect of environmental health law is outlined briefly in the following.

Public health

Nuisances, drainage, sanitary conveniences, offensive trades and burial grounds are dealt with under the Public Health (Ireland) Acts 1878–1907. These Acts were amended regularly, and the provisions corresponded fairly closely to British legislation until the introduction of the Environmental Protection Act 1990.

Food

The principal food legislation is contained in the Food Safety (Northern Ireland) Order 1991, which closely follows the Food Safety Act 1990. Food hygiene regulations and food composition and labelling regulations follow British legislation. The enforcement of food composition and labelling law is the responsibility of district councils, and is carried out by EHOs. A few exceptions to this are specified in the Food Safety (Enforcement) Order (Northern Ireland) 1997, for example dairy farms, liquid milk plants and certain meat plants.

There is a well-developed system of co-ordination and liaison through the Northern Ireland Food Liaison group (NIFLG), which is made up of group food specialist officers, a representative of district council food specialist officers, a representative of the FSA and a representative of the chief EHOs group. The NIFLG has formed links with the specialist panels of LACORS, which deal with food safety and food labelling and composition, and participates fully in co-ordinated national sampling and enforcement programmes. The NIFLG is also represented on the LACORS Food Policy Forum and the FSA's Enforcement Liaison group.

The FSA operations began in Northern Ireland in April 2000. It is responsible for providing advice and draft legislation on issues across the food chain to the Minister responsible for Health, Social Services and Public Safety. Its role in Northern Ireland is augmented by advice from the Northern Ireland Advisory Committee. The FSA liaises closely with the Food Safety Promotion Board (a cross-border body) and the Food Safety Authority of Ireland in relation to food issues that have an all-island dimension.

An officer of the FSA for Northern Ireland chairs a Northern Ireland Food Law Enforcement group, which comprises members of the NIFLG and officers of Veterinary and Quality Assurance Divisions of the Department of Agriculture and Rural Development. Its remit is to discuss food safety and hygiene issues in respect of meat and dairy products.

Consumer protection

Legislation regarding the safety of consumer goods can be found in Parts II and IV of the Consumer Protection Act 1987 and a considerable number of safety regulations made under Section 11 of the Act. In addition, there are a small number of safety regulations made under the European Communities Act 1972, of which the General Products Safety Regulations 1994 are the most significant.

The responsibility for enforcement of Consumer Safety legislation in Northern Ireland rests with the environmental health departments of district councils. There is also a Northern Ireland Consumer Protection Group, which is a sub-group of the chief EHOs group and its purpose is to provide a forum of professional expertise as regards consumer protection matters in order to supply policy advice and assist in function co-ordination.

Occupational health and safety

General provisions for the control of occupational health and safety are contained in the Health and Safety at Work (Northern Ireland) Order 1978.

The Health and Safety at Work (Enforcing Authority) Regulations (Northern Ireland) 1999 allocates enforcement responsibility by using main activity criteria, as in the legislation for England and Wales. The regulations identify work activities, some of which are allocated to district councils while others are allocated to the HSE for Northern Ireland. The HSE has been established as a non-departmental public body under the Health and Safety at Work (Amendment) (Northern Ireland) Order 1998.

A transfer of enforcement responsibility by agreement between enforcement bodies has been introduced by the Health and Safety (Enforcing Authority) Regulations (Northern Ireland) 1999 or by the Department of Enterprise Trade and Industry.

There is no HSC in Northern Ireland, its role falling to the HSE for Northern Ireland.

Under Article 20(4)(b) of the Health and Safety at Work (Northern Ireland) Order 1978, as amended, the HSE for Northern Ireland is empowered to issue mandatory guidance on adequate arrangements for the enforcement of health and safety to relevant enforcement authorities. Currently, guidance is awaited on enforcement policies and service plans to encompass the principles that the HSE for Northern Ireland wish to see adopted when enforcing health and safety.

Liaison on health and safety matters is effected regionally through the HSE for Northern Ireland Local Authority Enforcement Liaison Committee (HELANI). HELANI ensures a consistent and proportionate response to the enforcement of health and safety legislation across Northern Ireland.

The Northern Ireland Health and Safety Liaison Group is formed of EHOs from the group committees, a representative from the district councils and the Head of the Local Authority Unit. The Local Authority Unit resides with the Health and Safety Executive for Northern Ireland and provides support to HELANI and the district councils.

HELANI, in 2002, produced a three-year strategic plan, which provides a framework within which district councils will work to undertake work in strategic themes and focus on certain key priority areas of work. This approach is in line with the HSC's role in revitalizing health and safety strategically in England and Wales.

Waste management

Waste collection and disposal is the duty of district councils, which are also responsible for street cleaning, litter and abandoned motor vehicles. Allocation of responsibility for these functions varies within councils, but even where EHOs do not have direct responsibility they currently have to undertake site licensing activities. However with the introduction of the new Waste Management Licensing Regulations (Northern Ireland) 2003, this function will transfer to the EHS.

The Northern Ireland Waste Management Strategy was launched in March 2000, which encouraged district councils to group together to prepare a waste management plan as mandated under Article 23 of the Waste and Contaminated Land (Northern Ireland) Order 1997. Three groups have been formed namely, Arc 21 (Antrim, Ards, Ballymena, Belfast, Carrickfergus, Castlereagh, Down, Larne, Lisburn, Newtownabbey and North Down), The North West Group (Ballymoney, Coleraine, Derry, Donegal, Limavady, Magherafelt and Moyle) and The Southern Waste Management Partnership (Armagh, Banbridge, Cookstown, Craigavon, Dungannon, Fermanagh, Newry and Mourne and Omagh). Each Group has submitted its plan to the Department of the Environment and is currently in the implementation phase.

Other articles of the Waste and Contaminated Land (Northern Ireland) Order 1997 and in particular those relating to contaminated land have yet to be enabled by commencement orders. It is envisaged that environmental health departments will have a significant role in the enforcement of the contaminated land provisions.

The Controlled Waste (Duty of Care) Regulations (Northern Ireland) 2002 require waste producers to ensure that their waste is properly disposed off and require transfer notes to be completed. This legislation is enforced by environmental health departments of district councils and EHS.

Noise control

Controls over noise nuisance, noise on construction sites, loudspeakers in the street, and powers to issue and approve codes of practice for minimizing noise are contained in the Pollution Control and Local Government (Northern Ireland) Order 1978. The legislative controls are similar to those contained in the Control of Pollution Act 1974 and are enforced by district councils in Northern Ireland. (The part dealing with noise abatement zones has not been implemented.)

The Integrated Pollution Prevention and Control Regulations control noise from IPPC installations. Noise from Part A processes will be controlled by the Industrial Pollution and Radiochemical Inspectorate (IPRI) of the EHS. Local authorities will continue to deal with noise issues arising from non-IPPC installations.

The Environmental Noise Directive 2002/49/EC relating to the management and assessment of environmental noise is to be transposed into Northern Ireland statute in July 2004. The Directive aims to establish common approaches to avoid, prevent or reduce exposure to environmental noise. This Directive will require governments to determine exposure through noise mapping and produce action plans if necessary. The timetable for completion of mapping and action plans is 2007/2008 and 2012/2013.

Atmospheric pollution

The principal legislation enforced by district councils at present is contained in the Clean Air (Northern Ireland) Order 1981, the Pollution Control and Local Government (Northern Ireland) Order 1978 and the nuisance provisions of the Public Health (Ireland) Act 1878.

This legislation is broadly similar to that contained in the Clean Air Act of 1993. Non-scheduled processes and domestic air pollution are controlled by EHOs of district councils.

Provisions similar to those in Section 5 of the Health and Safety at Work, Etc. Act 1974 are not contained in the Health and Safety at Work (Northern Ireland) Order 1978. Emissions to the Atmosphere Regulations have not been made.

The Alkali Etc. Works (Northern Ireland) Order 1991 as amended by the Alkali Amendment Order 1994 lists both the noxious or offensive gases and the registrable (scheduled) processes or works.

The Industrial Pollution and Radiochemical Inspectorate of the EHS of the Department of the Environment (Northern Ireland) enforces the Alkali Etc. Works Regulations Act 1906 as amended by the Orders of 1991 and 1994.

Similar controls to those in Part I of the Environmental Protection Act 1990 will be gradually introduced by the Pollution Control (Northern Ireland) Order 1998.

The Environment (Northern Ireland) Order 2002, which took effect from 17 January 2003, has introduced the new Pollution Prevention Control (PPC) regime through regulations effective from 1 April 2003. The new Pollution Prevention and Control Regulations (Northern Ireland) 2003 will co-exist along with the existing IPC regime, with the PPC regulations setting out the timetable for PPC to be phased in and IPC to be phased out.

Whilst the new legislation will by far have the greatest impact on Part A premises controlled by the Industrial Pollution and Radiochemical Inspectorate, Part C processes controlled by district councils will be affected. The first block of Part A installations permitted by the IPRI are due to make applications early in 2004.

Existing council-regulated IPC Part C processes will have to undergo an administrative change under the new regime, that is IPC authorizations must be changed to IPPC permits. Councils are required to determine deemed applications within 12 months of 1 April 2004, 2005 and 2006. Councils will be consulted on all Part A applications and may comment on any issues including local air quality and noise issues. Recommendations may be made on noise control and IPRI shall consider these within the context of Best Available Technology. Councils will continue to control noise activities from sources not considered to be part of the IPPC installation.

New installations proposing to come into operation after December 1999 or in the six months prior to the relevant date are required to seek a permit under the new regime.

New operations proposing the start up prior to six months before the relevant date will require an authorization, as at present under the IPC (Northern Ireland) Order 1997.

The Air Quality Standards Regulations (Northern Ireland) 1990 implement the EU directive requirements in relation to sulphur dioxide, suspended particulates, lead in the air and nitrogen dioxide.

The Air Quality Regulations (Northern Ireland) 2003 incorporate as one single piece of legislation, the relevant authority regulations and the air quality regulations that prescribed the air quality objectives.

The first edition Air Quality Strategy for England, Scotland, Wales and Northern Ireland was received in 1997 with further editions and amendments in 2000, 2001 with an addendum issued in 2003. The objectives are for eight main air pollutants to protect health, with two objectives to protect vegetation and ecosystems.

The Air Quality Strategy addendum sets new targets for PM_{10}, CO and Benzene. The Air Quality Legislation in Northern Ireland is the Environment (Northern Ireland) Order 2002, Air Quality Regulations (Northern Ireland) 2003, Air Quality Limit Value Regulations (Northern Ireland) 2002 and the Air Quality Ozone Regulations (Northern Ireland) 2003.

To date (2003) all 26 councils in Northern Ireland have completed Stage 1 of the Air Quality Review and Assessments, which was due in June 2000. Stages 2 and 3 Review and Assessments are due for completion by the end of 2003. Full reviews are proposed in 2006 and 2009 with progress reports in the intervening years.

The Local Air Quality Management Policy Guidance is issued under Article 16 of the Environment (Northern Ireland) Order 2002 and district councils and other public bodies must have regard to this guidance when carrying out their Local Air Quality Management duties under Part III of the Order. The guidance sets out the principles behind review and assessments of air quality up to 2010 and the recommended steps that relevant authorities should take. Contained within is a timetable for reviews and assessments up to 2010. Guidance is given on how the district councils should handle the designation of air quality management areas and the taking forward of the local and regional air quality strategies.

The Radioactive Substances Act 1993 extends to Northern Ireland.

Water pollution

Under the Water Act (Northern Ireland) 1972, the EHS, an agency within the Department of the Environment for Northern Ireland (DOENI), has a duty to promote the conservation of the water resources of Northern Ireland and the cleanliness of water in waterways and underground strata. In performing this duty, the EHS is required to have regard to the needs of industry and agriculture, the protection of fisheries, the protection of public health, the preservation of amenity and the conservation of flora and fauna. The EHS protects the aquatic environment by preparing water quality management plans, controlling effluent discharges, taking action to combat or minimize the effects of pollution and by monitoring water quality.

The provisions of the Water Act (Northern Ireland) 1972 are broadly similar to those formerly contained in the Control of Pollution Act 1974 in that a 'consent' is required to make a discharge of trade or sewage effluent or of any polluting matter to a waterway or underground stratum. The provisions of the Food and Environment Protection Act 1985 cover discharges and dumping into the sea.

EHOs (water pollution) employed by Belfast city council and the group committees carry out the work in connection with 'consents' and the investigation of pollution incidents on behalf of the EHS, reporting directly to the officers of that service.

Housing

The Northern Ireland Housing Executive, not the district councils, is the enforcing authority for the parts of housing orders that deal with housing conditions, individual unfit houses and houses in multiple occupation.

The main legislation is contained in the Housing (Northern Ireland) Order 1981, amended in 1983 and the Rent (Northern Ireland) Order 1978.

The Housing (Northern Ireland) Order introduced in 1991 brought many of the provisions of these housing orders into line with corresponding parts of the legislation for England and Wales contained in the Housing Act 1988 and the Local Government and Housing Act 1989.

A considerable amount of survey and inspection work is carried out on behalf of the Northern Ireland Housing Executive by the EHOs of district councils.

District councils have important duties under the Rent (Northern Ireland) Order 1978 where in the Regulated Tenancy Sector they ensure compliance by landlords with their repairing obligations through service of Certificates of Disrepair and follow-up enforcement action to ensure that repair work specified in the certificates is completed. District councils also have responsibility for issuing regulated rent certificates for the change from restricted tenancies, in which rents are tied to pre-1978 levels, to regulated tenancies.

The Housing (Northern Ireland) Order 1992 effected changes to the Rent (Northern Ireland) Order 1978, which gave district councils specific authority to investigate and prosecute offences of illegal eviction or tenant harassment.

The Housing (Northern Ireland) Order 2003 gives the district councils power to enforce the provisions under the Rent Order (Northern Ireland) 1978 in relation to the provision of rent books.

Communicable disease

The legislation for controlling communicable disease is contained in the Public Health Act (Northern Ireland) 1967 and the Health and Personal Social Services (Northern Ireland) Order 1972. Health and Social Services Boards through their Directors of Public Health and their Consultants in Communicable Disease Control enforce the legislation, which includes provision for notification of specified diseases and for prohibition from work of carriers, contacts, and the like.

There are four health boards, each of which is coterminous with a local environmental health group. Belfast, with the eastern group environmental health committee, forms the eastern health and social services board area.

A considerable amount of fieldwork on behalf of the health and social services boards is carried out by EHOs, some of whom are authorized under the aforementioned legislation. They are involved in the investigation of food poisoning outbreaks and in the sampling of food and water. Bacteriological examination of a wide range of samples is carried out by the laboratory at Belfast City Hospital, which acts as a regional laboratory for routine programmes of microbiological sampling and in investigations of food, water and environmental specimens in food poisoning inquiries.

The directors of public health in the health and social services boards liaise closely with the district directors of environmental health and the group chief EHOs. As a result written protocols and standard procedures exist to cover the various functions in which the boards and environmental health departments have a joint interest.

Port health

In relation to port health, district councils act as agents of the health and social services boards, which are the enforcing authorities under the relevant legislation: the Public Health (Aircraft) Regulations (Northern Ireland) 1971 and the Public Health (Ships) Regulations (Northern Ireland) 1971.

The EHOs at the ports carry out the full range of port health duties associated with hygiene and infectious disease.

District councils enforce the general provisions of the Imported Food (Northern Ireland) Regulations 1991, which were updated with regard to third country imports by the Imported Food Regulations (Northern Ireland) 1997.

District councils act jointly with the Department of Agriculture in enforcement of the Products of Animal Origin (Import and Export) Regulations (Northern Ireland) 1998.

Drinking water

At present the Water Quality Regulations (Northern Ireland) 1994 set out sampling and other regulatory requirements to demonstrate the wholesomeness of drinking water supplies for public supply and use in food production. A directive (98/83/EEC) on water intended for human consumption was adopted as European law in November 1998. As a consequence, the Water Quality Regulations have been replaced by the Water Supply (Water Quality) Regulations (Northern Ireland) 2002 and these come fully into effect from 25 December 2003.

Private Water Supplies Regulations (Northern Ireland) 1994 set standards for private water supplies and requires the Department of the Environment to monitor private supplies according to the classification category. The sampling of private water supplies is carried out on behalf of the Northern Ireland Drinking Water Inspectorate by environmental health departments. The sampling programme for dairy farms is carried out by the Quality Assurance Division of the Department of Agriculture and Rural Development.

Dogs and other animals

The Dogs (Northern Ireland) Order 1983 provides a range of dog control measures that are enforced by district councils. Requirements include an annual licensing fee of £5, with a discount for owners aged over 65 years living alone and exemptions for guide dogs for the blind and hearing dogs for the deaf.

The order provides penalties for the offences of allowing dogs to stray and for attacks on livestock or persons. Provision is made within the order for the issue of fixed penalty notices for the offences of keeping a dog without a licence, allowing a dog to stray, failing to display identification and allowing the fouling of footpaths contrary to by-laws.

The Dangerous Dogs (Northern Ireland) Order 1991 is enforced by district councils and introduced to Northern Ireland similar powers for the control of dogs as exist in this legislation in England and Wales.

The Welfare of Animals Act (Northern Ireland) 1972, which requires licensing of pet shops, animal boarding, riding and zoological establishments, is enforced by the Department of Agriculture and Rural Development.

Pest control

The Prevention of Damage by Pests Act 1949 (see p. 876) does not apply in Northern Ireland, where the relevant legislation is the Rats and Mice Destruction Act 1919.

Caravans and camping

The relevant legislation is the Caravans Act (Northern Ireland) 1963, the provisions of which are similar to those in the Caravan Sites (Control of Development) Act 1960. Model Licence Conditions were introduced by the Department of the Environment for residential sites in 1992 (holiday sites) and 1994 (permanent sites). The Housing (Northern Ireland) Order 2003 transfers control of travellers sites from district councils to the Northern Ireland Housing Executive.

Swimming, boating, etc.

Controls on swimming and boating are exercised under public health and health and safety at work legislation supplemented by local by-laws.

Entertainment licensing

District councils are responsible for the licensing of places of entertainment under the Local Government (Miscellaneous Provisions) (Northern Ireland) Order 1985, the provisions of which closely follow those in the Local Government (Miscellaneous Provisions) Act 1982.

Acupuncture, tattooing, ear piercing and electrolysis

Acupuncture, tattooing, ear piercing and electrolysis are subject to registration by the district council where the council has applied the Local Government (Miscellaneous Provisions) (Northern Ireland) Order 1985, which also contains the power to make by-laws. Similar provisions for England and Wales are contained in the Local Government (Miscellaneous Provisions) Act 1982.

Hairdressing

Registration of premises by district councils is required under the Hairdressers Act (Northern Ireland) 1939 and SR&O No. 86 of 1939. Provisions similar to those in the Public Health Act 1961 for the making of by-laws are contained in the legislation.

Street trading

The Street Trading Act (Northern Ireland) 2001 enables district councils in Northern Ireland to regulate street trading in their districts. Street trading is prohibited unless a trader has a licence or a temporary licence granted by the district council for the area in which trading is to take place. Streets in which trading takes place from a stationary position must be designated by the council and the council can only pass a resolution prescribing a street as designated after consulting with the public and other named agencies. The council may also decide to designate the classes of goods or services in which trading in a designated street will be permitted and each trading licence will be subject to a range of standard conditions. Less stringent requirements relate to the granting of temporary and mobile trading licences.

The future

Northern Ireland's present system of local government dates from the 1970s. The Local Government Act (Northern Ireland) 1972 produced a significant change in the roles and responsibilities of councils most notably the removal of their control of education and health care. Housing provision was transferred to a newly appointed body, the Northern Ireland Housing Executive. In recent years, there has been an expansion of local government with their role extending to embrace economic development, tourism, community development/relations and the arts. In addition, the Northern Ireland Office has requested councils to develop Community Safety Strategies to enhance the safety of local communities. Councils also are playing a role in the new police partnership boards and a central role in dispensing EU

Special Peace and Reconciliation Funds. Local Strategy Partnerships were formed and their role and remit is developing under the auspices of Peace II. In some areas, EHOs are working with the Local Strategy Partnerships on environmental and health and well-being issues and with their councils and communities on Community Safety Strategies.

A review of public administration was launched on 12 February 2002 by the Office of the First and Deputy First Minister and, although the Northern Ireland Assembly and its executive committeee is currently in suspension, the consultation process is continuing and a consultation document on the reform of public services is expected to issue in the autumn of 2003.

The wider agenda of environmental health is presently being considered and two documents in particular are helping to prepare the ground for EHOs with respect to their role in the new public health agenda for Northern Ireland. The first document is the Group Environmental Health: Vision 2005 [1], which examines how EHOs can deliver, in partnership with other professionals, improvements in communities' and individual's health and well-being and quality of life.[1] The second document 'Environmental Health 2012 – A Key Partner in Delivering the Public Health Agenda' [2] has issued from CIEH in consultation format, has been endorsed by NICEHOG and currently ownership for this Vision is being encouraged at all levels of the profession through a series of Open Members' Fora throughout Northern Ireland. These fora are being facilitated by the new Director of CIEH in Northern Ireland, Gary McFarlane. The delivery of both these Visions will provide local managers in environmental health both considerable challenges and opportunities. Lets hope we are all up to the challenge and the opportunities will be there for the taking.

References

1. *Group Environmental Health Vision* (2005).
2. *Environmental Health 2012 – A Key Partner in Delivering the Public Health Agenda* (2002) CIEH, London.

Further reading

Dickson, B. (1994) *The Legal System of Northern Ireland – The Law in Action*, SLS Publications, Belfast.

DoENI (1996) *A Corporate Framework for the Department of the Environment for Northern Ireland*, Department of the Environment (Northern Ireland), Belfast.

DoENI (1996) *Your Guide to the Department of the Environment for Northern Ireland and its Agencies*, Department of the Environment (Northern Ireland), Belfast.

Northern Ireland Year Book 2003: A Comprehensive Guide to the Political, Economic, and Social Life of Northern Ireland.

APPENDIX B: THE ORGANIZATION OF ENVIRONMENTAL HEALTH IN SCOTLAND

MICHAEL HALLS

There are several differences between the system of environmental health that operates in Scotland, and that in other parts of the United Kingdom. These differences embrace both the structure of the local authorities and the functions that are undertaken by environmental health departments. In this appendix, if there is no reference to a Scottish way of working it can be assumed that the arrangements are similar to those for England and Wales.

Local government system

As is generally the case in the rest of the United Kingdom, the delivery of environmental health services takes place at the local level and in Scotland this means that the 32 unitary authorities provide these services.

Until the last reorganization of local government in Scotland in 1996 most councils employed a Director of Environmental Health or a Chief Environmental Health Officer. Today, however, that situation has changed dramatically and now only a small number of authorities have made such appointments. Although some environmental

---

[1]Also see Department of Health, Social Services and Public Safety (March 2002) *Investing for Health*, DHSSPSNI, Belfost.

health departments remain under the control of an Environmental Health Officer (EHO), the majority of such officers work in multidisciplinary departments headed by a non-EHO.

Central government framework

When the Scottish people voted for the devolution of some powers from central control to the control of the Scottish Parliament in Edinburgh, the shape of the central government changed significantly.

What was at one time the Scottish Office is now known as the Scottish Executive and this body contains two departments with an interest in environmental health.

The public health role undertaken by local government is carried out in partnership with the National Health Service and comes under the jurisdiction of the Scottish Executive Health Department.

The Scottish Office Environment & Rural Affairs Department has control over other aspects of environmental health such as air quality, climate change, GMOs, food safety, water supplies, sustainable development, building standards, and so on.

The Scottish Executive enjoys a considerable degree of autonomy, and as a result some local government functions are discharged somewhat differently in Scotland than in England and Wales. Examples of the differences can be seen in food law enforcement, the law relating to milk and dairies and in waste disposal.

The Convention of Scottish Local Authorities (COSLA)

COSLA is the representative voice of Scottish local government. It also acts on behalf of its member councils as their employers' association, negotiating salaries, wages and conditions of service for local government employees with the relevant trade unions.

COSLA is funded by a levy paid by its member councils, calculated on the basis of population.

COSLA's objectives are to:

- develop effective working relationships with the Scottish Parliament, the Scottish Executive, European institutions and other bodies, so as to

promote the role of councils and ensure that local government has greater control over its own affairs;
- support councils in providing leadership for the communities they serve, strengthening local democracy and enhancing the public's awareness of and support for local government;
- support councils in the continuous improvement of service delivery and providing value for money.

The Food Standards Agency

The UK FSA has a national office in Aberdeen, which deals with the agency's functions north of the border.

The Meat Hygiene Service

The Meat Hygiene Service (MHS) is an executive agency of the FSA, just as it is in the rest of the United Kingdom.

The Scottish Environment Protection Agency

The Scottish Environment Protection Agency (SEPA) was set up in 1996 and has as its main aim the provision of an efficient and integrated environmental protection system for Scotland that will both improve the environment and contribute to the goal of sustainable development.

The outcomes that it strives to achieve are:

- Provision of an excellent environmental service for the people of Scotland;
- Improvement in the sustainable use of natural resources by minimizing waste, recovering value and ensuring best management of disposal;
- Maintenance and restoration of all water environments;
- Ensuring good air quality;
- Promotion of respect for the Scottish environment; and
- Creation of an environmental framework for the economic well-being of Scotland.

SEPA is now among the largest employers of EHOs in Scotland.

The Royal Environmental Health Institute of Scotland

The Royal Environmental Health Institute of Scotland (REHIS) is the professional body that represents the interests of the environmental health profession in Scotland. It was granted its Royal Charter in 2001. Its membership includes representatives of all officers engaged in the various aspects of environmental health work. EHOs account for the largest proportion of its membership, but other officers may also be given full membership provided they are suitably qualified.

Membership includes a number of consultants in public health medicine, veterinarians, meat inspectors, food safety officers and people involved in various aspects of environmental health education. Representatives of each of these groupings may be elected to the Council of REHIS. Elected members of local authorities and health boards and people engaged in commercial activities associated with environmental health, are eligible for associate membership.

In addition to representing the professional interests of environmental health officers, REHIS's main aims are to stimulate interest in, disseminate knowledge of, and promote education and training in matters relating to environmental health.

This is achieved by overseeing the professional training of, and by examining, environmental health officers, food safety officers and meat inspectors, by organizing regular training courses and by holding an annual national conference.

A journal that covers a wide range of topics of interest to the membership is published regularly. An annual report on environmental health in Scotland and a compendium of conference papers are also published each year.

REHIS requires that all members who are qualified EHOs must comply with its scheme of Continuing Professional Development (CPD). The purpose of this scheme is to maintain, improve and broaden the knowledge, skills and expertise of EHOs and to develop the personal qualities necessary to undertake professional tasks and duties. The scheme also assists in the development of the managerial skills necessary to supplement professional knowledge, and activities are undertaken regularly so as to ensure that the learning process is continuous throughout the officer's working life.

A similar voluntary scheme for other qualified officers who are members of REHIS is available for non-EHOs.

REHIS is also responsible for the organization of courses in meat inspection. The syllabus for these courses is approved by the Scottish Executive. Courses and examinations, which lead to the relevant diploma being awarded, are organized by REHIS.

REHIS is consulted by government departments and COSLA on proposed legislation and environmental health issues, and advises its members and individual local authorities on a wide variety of topics. The office that deals with the business and administrative affairs of REHIS is in Edinburgh at the following address:

The Royal Environmental Health Institute of Scotland
3 Manor Place
Edinburgh
EH3 7DH
Tel: 0131-225 6999
Fax: 0131-225 3993.
Email: contact@rehis.com

Training and Qualification of Environmental Health Officers

In Scotland, the route leading to full professional qualification in environmental health entails obtaining a BSc (Hons) degree in Environmental Health, and undertaking a period of practical training and assessment leading to the Diploma in Environmental Health of the Royal Environmental Health Institute of Scotland.

A four-year honours degree course following a syllabus accredited by REHIS has been established at the University of Strathclyde. The off-campus element requires students to complete 48 weeks of practical training with a local authority following a programme prescribed by REHIS. This can be undertaken either in separate blocks during university vacations, or 'end-on' as one period after graduating. The final stage is a two-day assessment of professional competence involving the submission of written work and interviews. If this is completed satisfactorily, the Diploma in Environmental Health is granted.

National co-ordinating groups for environmental health

As Scotland is a small country (with a population of just over five million), it has been possible to establish national co-ordinating bodies to provide a forum at which major environmental health issues can be discussed. Examples of these are described in the following.

The Scottish Centre for Infection and Environmental Health

The Scottish Centre for Infection and Environmental Health (SCIEH) was established in 1993 by the amalgamation of the Communicable Diseases (Scotland) Unit and the Environmental Health (Scotland) Unit. It is based at Clifton House, Clifton Place, Glasgow G3 7LN.

The SCIEH is a multidisciplinary organization comprising expertise in environmental health, clinical infectious diseases, information sciences, public health medicine, nursing, social sciences and veterinary public health.

The SCIEH is responsible for the surveillance, on a national basis, of environmental health hazards and the provision of advice, expertise and support to local authorities and Health Boards throughout Scotland.

Its mission statement is 'to improve the health of the Scottish population by providing the best possible information and expert support to practitioners, policy makers and others on infectious and environmental hazards'.

The SCIEH is mainly funded by central government but also earns income by providing advice on a consultancy basis to interested organizations. Further it has a role in education and training and carries out extensive research.

Scottish Food Enforcement Liaison Committee

The FSA Scotland set up this Committee as a non-statutory advisory body to assist it in its work in Scotland. The Committee's mission statement is 'to contribute to the development and maintenance of the Agency's strategies and policies by providing the best possible information and expert support in terms of food law enforcement in Scotland'.

Its objectives include working collectively to influence food policy and improving the efficiency and consistency of the enforcement of food law in Scotland. The Committee will endeavour to do so by taking a proactive approach to policy development and consideration of Scottish enforcement issues.

Its work will to an extent mirror the liaison arrangements for the rest of the United Kingdom and it will utilize working groups and sub-committees to assist it to achieve its aims.

The Scottish Food Safety Officer's Registration Board

The Scottish Food Safety Officer's Registration Board (SFSORB) was set up by REHIS to provide a means whereby food officers other than EHOs could obtain qualifications to enable them to carry out inspections of food premises or food standards. The SFSORB is approved by the Scottish Executive and is authorized by the Food Safety Act's Code of Practice to award the appropriate certificates.

The SFSORB consists of representatives of REHIS, the Scottish Executive, the Association of Meat Inspectors, the Institute of Food Science and Technology and the Scottish Food Safety Officers Association. It has developed syllabuses for the Ordinary and Higher Certificates in Food Premises Inspection and the Higher Certificate in Food Standards Inspection.

Applicants have to satisfy the board that their educational qualifications are acceptable or they have to sit for a written food examination set by the board. In addition, all candidates have to submit a case study or series of reports and are subsequently interviewed to assess their competence in carrying out food inspections.

The Board is now a standing committee of REHIS.

Environmental health in local government

The reorganization of local government that took place in 1996 led to one of the largest and most comprehensive changes to affect environmental health in the second half of the twentieth century. Whereas the authorities in existence until then had, more or less, adopted similar systems for delivering an environmental health service based on a discrete department headed by a director of environmental health, the new councils have adopted almost as many systems for delivering the service as there are councils!

Very few councils have a stand-alone environmental health department, and the most common location for the service is in a large department with other enforcement services, such as consumer protection or trading standards. Other authorities have adopted even greater amalgamations, and the titles of the departments charged with the environmental health function give a clue to the groupings established. Examples include: Environmental and Consumer Protection Services; Lifelong Care; Community Services; Environment; Protective Services; Technical and Leisure Services; Planning, Roads and Environment; and Environmental and Protective Services.

Having been redrawn, the local government map of Scotland still shows in graphic detail the differences of scale that have always existed with regard to both the size and population of local government areas. The council with the highest population is the city of Glasgow, with 609 000 people and that with the lowest is Orkney with 19 000. Discounting the island authorities, the council with the lowest population is Clackmannan with 48 500.

When areas are examined, huge differences become apparent, reflecting the sparsity of the more rural areas. Highland Council administers an area that, at 2.6 million hectares, is one-third the size of the whole of Scotland, and at the other end of the spectrum is the city of Dundee, which occupies only 5500 hectares.

The functions carried out by the departments responsible for providing the environmental health service in Scotland are very similar to those in other parts of the United Kingdom, although there are some differences due to the fact that some aspects of the law in Scotland are not the same as that in England and Wales. In general, however, the basic environmental services of food safety, occupational health and safety, aspects of pollution control, housing standards, communicable disease control and waste management fall on the department charged with providing the environmental health service.

EHOs act as both advisers and enforcers and, probably because their training is essentially practical and based on a holistic approach, they tend also to be given responsibility for a range of other local government functions, such as pest control, control of dogs, animal welfare provisions and, increasingly, matters relating to sustainable development.

Legislation applying to Scotland

A proportion of the legislation used on a daily basis by EHOs in Scotland applies to the whole of the United Kingdom. Examples include enactments relating to the environment and its protection, occupational health and safety and food safety. Other laws apply only to Scotland, although they generally resemble similar powers available in the rest of the United Kingdom. Peculiarly, Scottish legislation includes the Civic Government (Scotland) Act 1982, which gives powers to councils to deal with, *inter alia*, repairs to buildings, licensing of street traders, late catering licences, control of dogs and the making of by-laws – all of which are used by environmental health departments in the discharge of their duties.

A second example is the Licensing (Scotland) Act 1976. Although primarily concerned with the conditions attached to the issue of liquor licences, it also contains provisions that relate to food hygiene standards in licensed premises, and can also enable the licensing authority to impose noise control standards where music or entertainment is provided in licensed premises.

With regard to housing standards, there is a difference between Scotland, where the 'tolerable standard' has existed since the 1970s, and England and Wales, where a standard based on 'fitness' is used.

Financial accountability

The Commission for Local Authority Accounts in Scotland (the Accounts Commission) has responsibility for ensuring that all local authority accounts are externally audited. The Accounts Commission, through the office of Controller of Audit, exercises this function either directly using its own staff, or by the use of private firms of approved auditors. The accounts of environmental health departments are subject to this process, which requires the auditor to ensure that all statutory requirements have been met and proper practices have been observed in the preparation of the accounts. The Accounts Commission is also empowered under the Local Government Act 1988 to undertake value-for-money studies in areas that include environmental health functions. These studies can lead to the making of recommendations aimed at improving economy, efficiency and effectiveness, and at highlighting areas where management can be improved.

Useful websites

Convention of Scottish Local Authorities – www.cosla.gov.uk

Food Standards Agency – www.foodstandards. gov.uk

Royal Environmental Health Institute of Scotland – www.rehis.org

Scottish Centre for Infection and Environmental Health – www.show.scot.nhs.uk/scieh

Scottish Environment Protection Agency – www.sepa.org.uk

Scottish Executive – www.scotland.gov.uk

APPENDIX C: THE ORGANIZATION OF ENVIRONMENTAL HEALTH IN WALES

RONNIE ALEXANDER

The National Assembly for Wales

In July 1997, the government published a White Paper, 'A voice for Wales', which outlined proposals for devolution in Wales. These proposals were endorsed in the Referendum of 18 September 1997. Subsequently, Parliament passed the Government of Wales Act 1998, which established the National Assembly for Wales. The National Assembly for Wales (Transfer of Functions) Order 1999 made under the 1998 Act transferred a large number of powers and responsibilities from the Secretary of State for Wales to the Assembly with effect from 1 July 1999. Since then other transfer of functions orders have been made and many Acts of Parliament since 1999 have conferred new powers and duties on the Assembly.

The Assembly decides on its priorities and allocates the funds made available to it from the Treasury. Within its functions, the Assembly develops and implements policies that reflect the particular needs of the people of Wales. Decisions about these issues are made by politicians who are accountable, through the ballot box, to voters in Wales. Wales remains part of the United Kingdom and MPs from Welsh constituencies continue to have seats in Westminster. Most laws passed by Parliament in Westminster still apply to Wales.

The Elections for the First Assembly were held on 6 May 1999 and the Second Assembly was elected on 1 May 2003. Future elections will be held every four years. The Assembly has 60 elected members and each voter has 2 votes. The first vote is used to elect a local or constituency member in the same way as MPs are elected to the House of Commons. Forty Assembly members are elected on this 'first past the post' basis, one from each constituency in Wales. The second vote is used to elect 20 additional members on a regional basis, to ensure the overall number of seats for each political party reflects the share of the vote they receive. This is known as the Additional Member System (AMS), a form of proportional representation. There are five electoral regions: North Wales, Mid and West Wales, South Wales West, South Wales Central and South Wales East, and each region returns four Members to the Assembly.

The Assembly has considerable power to develop and implement policy within a range of areas, including agriculture, the environment, economic development, education, health and health services, housing, transport, planning and local government. Examples of important decisions the Assembly can make include administering funding for local authorities, establishing schemes for sustainable development and equality of opportunity and developing housing policy, including tackling homelessness.

The central structures and procedures for the Assembly are laid down in the Government of Wales Act 1998. The more detailed processes are set out in the Assembly Standing Orders. The first Plenary meeting of the Assembly took place on Wednesday 12 May 1999, when Members elected the First Presiding Officer, Deputy Presiding Officer and First Minister (then known as First Secretary) of the Assembly. When Assembly members meet in plenary session, the Assembly is chaired by the Presiding Officer, who has a similar role to the Speaker of the House of Commons and who is elected by the whole Assembly. Once elected, the Presiding Officer serves the Assembly impartially. There is also a Deputy Presiding Officer who is elected in the same way.

The 60 Assembly members agree to delegate most of the Assembly's functions (the making and implementing of decisions and secondary legislation) to the First Minister, who is elected by the whole Assembly and therefore usually represents the largest political party. The First Minister in turn delegates responsibility for delivering the executive functions to Assembly Ministers who form the cabinet. The cabinet makes many of the Assembly's day-to-day decisions, and its Ministers are responsible for individual subject areas such as health and education. The cabinet is accountable to the rest of the Assembly, which scrutinizes all its decisions and actions.

Members from all parties can voice their opinions on how the Assembly operates through Subject Committees such as Health and Social Services, which develop policies and examine what the Assembly does. Members are elected to serve on these Committees to reflect the balance of political groups within the Assembly.

Regional Committees represent the needs and interests of their localities and convey issues of local concern to the full Assembly and to the Subject Committees. There are four Regional Committees, which are made up of members of

the relevant constituency and electoral region. Most Committee meetings take place in public and a number are broadcast.

Plenary meetings of the Assembly take place in public and are broadcast on television. The Presiding Officer circulates an agenda for each Plenary session in advance and business is dealt with in the order in which it appears on the agenda. For each week the Assembly meets in Plenary Session, at least 30 minutes are allocated to oral questions to the First Minister. Each Minister also responds to oral questions at least once every four weeks, while any other member can propose a motion or topic for a short debate before the end of every other Plenary session. With the Presiding Officer's prior approval, any member can propose that the Assembly should immediately consider a matter of urgent public importance at any Plenary Session.

Time is also allocated within the annual Plenary cycle for certain categories of Assembly business, such as allocation of the Assembly budget and subordinate legislation – regulations, orders and other instruments, which are made under powers set out in acts of Parliament (or primary legislation).

The Secretary of State for Wales ensures that the interests and needs of Wales are fully considered in policy formulation within the UK government, is responsible for relations between the UK government and the Assembly in general terms, and for taking to Parliament primary legislation, which relates particularly to Wales. Under the Government of Wales Act 1998, the Secretary of State for Wales has the duty to transfer the Welsh budget from the Treasury to the Assembly and to consult and debate the government's legislative programme with the Assembly. He may participate in Plenary sessions of the Assembly but not to vote.

Legislation and guidance

The Government of Wales Act 1998 largely defines the constitution, powers and duties of the National Assembly. The basis of the Assembly's powers to form policy and take decisions for particular areas of responsibility is found in the various transfer of functions orders and in primary legislation. The Assembly's relationship with the UK government and Whitehall departments is set out in an overarching Memorandum of Understanding and a series of concordats agreed with individual departments. The Assembly can make subordinate legislation, which is scrutinized under the procedures contained in the Assembly's Standing Orders. Such legislation may have either general or local application. It is possible for the Assembly to make legislation other than by formal statutory instruments, for example in directions, determinations or statutory codes of practice or guidance.

Assembly staff and structure

Most Assembly staff support the Assembly cabinet. They help to formulate and implement policies on behalf of the cabinet and administer the public services for which the Assembly is responsible. The remainder of the staff support the Presiding Officer.

The Welsh Assembly government, headed by the First Minister and supported by the cabinet, includes the Minister for Health and Social Services. The Health and Social Services Committee of the National Assembly for Wales scrutinizes the work of the Minister, debates relevant matters and makes recommendations for action.

Within the civil service arm of the Assembly, there is a Health and Social Care Department. This comprises:

- A health and social care policy unit
- A strategy implementation unit
- Delivery based on Regional Office teams
- A corporate support team
- A unified finance and resources unit.

The CMO has overall accountability for the co-ordination and provision of independent professional advice with respect to public health, health services and social care, and for the organization of the development and training needs of the professional staff. The CMO is responsible for public health policy across the full range of ministerial portfolios and she is supported by advisers who head five Health Professional Groupings – the Medical Sub Group; Dental Division; Pharmaceutical Division; Environmental Health Division; and the Scientific Division.

The Environmental Health Division is small and consists of a Chief Environmental Health Adviser and a Deputy but with a direct reporting line to the Chief Medical Officer. The main areas of responsibility for the Division are to:

- Provide independent and consistently high quality professional public health and environmental health advice, both predictive and reactive, to the Welsh Assembly government to assist in the development of policy to achieve better health and well-being for the people of Wales.
- To collate, co-ordinate and formulate advice on issues relating to environmental health activity and to liaise with Directors of Public Protection in local authorities in Wales to provide advice and interpretation of National Assembly guidance.

The National Public Health Service for Wales

The abolition of the five health authorities in Wales from 1 April 2003 and their replacement by 22 Local Health Boards, necessitated the need for a home for the Public Health Departments of the five health authorities, and to provide the necessary public health support to the fledgling local health boards. This led to the development of the National Public Health Service (NPHS) for Wales.

The NPHS will provide public health expertise, primarily to service the needs of local health boards and local government, but also to incorporate the needs of NHS Trusts, government agencies and the voluntary sector in Wales. It will also incorporate the work of the Public Health Laboratory Service (PHLS) and the Communicable Disease Surveillance Centre (CDSC) in Wales, along with the All Wales NHS Child Protection Service

The Director of the NPHS is professionally accountable to the CMO and managerially accountable to the Chief Executive of Velindre NHS Trust, which hosts the service. The NPHS will have three regional directors of public health. They will work closely with the corresponding three regional teams of the NHS Wales Department. They will also have the functional national lead (on behalf of the Service) for specific areas of public health work.

The NPHS will work closely with the Health Protection Agency. It will maintain a strong commitment to public health specialist training for people from a medical and non-medical background. While this is the particular concern of the Wales Centre for Health, the NPHS will provide support in terms of training capacity.

The Public Health Laboratory Service

The Public Health Laboratory Service (PHLS) in Wales will function as a business unit of the NPHS and will be headed by a Director. Each Local Health Board will name a public health director from within the NPHS who will be a Faculty Accredited public health specialist. The health authority libraries, independent professional advisers and the medical advisory and audit groups will also be within this service.

The Wales Centre for Health

The Wales Centre for Health will play a unique and important role in Wales as an independent corporate partnership centre for public health advice, information, research and multi-professional development. It will:

- take responsibility for training, advice and research related to health protection and improvement;
- tackle the legacy of high levels of ill-health in Wales and the considerable health inequalities that exist between our communities;
- monitor health and disease trends, looking ahead to give early warning of future public health problems;
- identify gaps in information and carry out projects to highlight particular health issues. It will promote the development of public health skills and multidisciplinary training;

- work with the Assembly, public, voluntary and academic sectors to help inform the development of new health protection and improvement policies.

It is still in shadow form, hosted by the Velindre NHS Trust and headed by a Director, but when formally established under the Health (Wales) Act 2003 it will be an independent statutory body.

Local government structure in Wales

Local government reorganization in 1996 generally brought together the environmental health and trading standards services into new Public Protection departments within the Unitary Authorities. The professional representation of each prior to reorganization was through separate Chief Officer groups. These groups came together to form the Society of Directors of Public Protection with representation from each of the 22 councils in Wales.

The purpose of the society is to provide advice and guidance to each local authority, assist with the production of uniform policies and procedures, to be a reservoir of expertise, to respond to consultation documents and to advise the local government association in Wales. The society is also heavily committed to the public health agenda and partnership working. A clear example of this is the Collaboration for Health and Environment, a partnership with other key players in the health field.

The Society of Directors of Public Protection meets a number of times each year and the Chief Environmental Health Adviser of the National Assembly for Wales has Observer status at such meetings. In addition, Directors of Public Protection come to the offices of Welsh Assembly Government three times each year for liaison meetings with senior civil servants chaired by the Chief Environmental Health Adviser of the National Assembly.

Two sub-groups exist beneath the Society of Directors of Public Protection. These are the Wales Heads of Trading Standards Group and the Wales Heads of Environmental Health Group. There are a number of technical panels that operate in support of the Wales Heads of Environmental Health Group, for example licensing, training, food safety and communicable disease. Several strategic groups fall within the remit of the Wales Heads of Environmental Health Group, for example communication and improvement and performance management. In addition, a number of multi-agency fora exist, for example the Welsh Air Quality Forum and the Welsh Food Microbiological Forum.

4 Training and professional standards

Paul Robinson

TRAINING OF ENVIRONMENTAL
HEALTH PRACTITIONERS IN ENGLAND,
WALES AND NORTHERN IRELAND

The routes to qualification as an Environmental Health Practitioner (EHP) are many and varied, but all involve the following common elements:

- complete a BSc (Hons) degree or MSc in Environmental Health which must be accredited by the Chartered Institute of Environmental Health (CIEH);
- undertake a period of work-based learning and submit a satisfactory portfolio of evidence;
- pass the CIEH professional examinations.

Completion of all three of these elements leads to the graduate being registered by the Environmental Health Registration Board (EHRB) and the issuing of a Certificate of Registration, which is the formal qualification recognized for practice as an EHP within the United Kingdom.

A new curriculum[1] for accredited courses was developed in 2003 and all courses will be based on this from 2004. The new curriculum has a focus on the identification and implementation of appropriate interventions to protect the health and safety of individuals and communities.

All courses include the study of science, technology, social science and law to the levels necessary to be able to understand and appreciate the biological, chemical, physical, social and physchosocial stressors on human health produced by the impacts of urbanization, industrialization, transport, farming and food production on the human environment, be it living, working or leisure. Specific interventions are studied in relation to Food Safety, Health and Safety, Environmental Protection, Housing and Public Health with some interventions being studied in more depth towards the end of the course. The emphasis on particular aspects of environmental health work differs between courses, so students with an interest in a particular aspect of environmental health may well choose a course which best suits their particular needs.

Laboratory work, case studies, visits, group work and tutorials make the courses varied and interesting. Group work is encouraged in all aspects of the course, and students are also expected to do a lot of independent study. This is good preparation for a career that can involve working both alone and as part of a team.

The majority of accredited courses offer a part-time route that enables people already working in environmental health to obtain the requisite professional qualification while still remaining in employment. However, work-based learning remains an important element of the qualification, and a balance has to be struck between

[1]Also see p. 18.

obtaining the necessary work-based learning experience and carrying out day-to-day duties. Increasing use of distance learning and web-based learning is making courses of study more accessible to potential students who live or work some distance from an accredited academic course.

Most students pursue an integrated course of study where work-based learning occurs either in the third year of a four-year degree course or throughout the course. However, for those students who are not able to access a work-based learning placement during their course, the majority of accredited courses offer a non-integrated route, which allows the academic BSc or MSc to be completed prior to obtaining the necessary work-based learning experience.

Work-based learning can be undertaken in any organization that can provide access to the environmental health interventions that are needed to be accessed to facilitate the production of the portfolio of evidence necessary to pass the work-based learning assessment. Private sector companies, consultancies, government departments and agencies, voluntary bodies, primary care trusts and housing associations, as well as local authority placements are all possibilities for accessing work-based learning opportunities.

There are accredited courses throughout England, Wales, Scotland and Northern Ireland. The new curriculum has generated increased interest in course provision with a number of universities looking to develop courses for the first time. A list of accredited courses is provided in the appendix at the end of this chapter.

The work-based learning logbook provides guidance on the experiential learning to be obtained during the work placement. Each student must complete this which, with an accompanying portfolio of evidence produced by the student while undertaking the work-based learning, must be assessed by the CIEH as satisfactory.

The professional examinations, which can only be taken once the first two elements have been obtained, assess the eligibility of the graduate to integrate theory and practice and operate as an EHP in any aspect of environmental health work.

TRAINING OF ENVIRONMENTAL HEALTH TECHNICIANS

Environmental health technical support staff are drawn from a variety of backgrounds and are essential to local authorities to support their delivery of environmental health enforcement responsibilities. Some technicians have a construction background, others come from nursing or the food industry, and yet others have experience in applied sciences.

There is no specific prescription for, or definition of, the work of a technician, and it depends upon the needs of each employer.

Technical support staff working in the inspection and auditing of food premises must possess a Higher or Ordinary Certificate in Food Premises Inspection, depending on the level of risk associated with the premises being inspected. These certificates are issued by the EHRB, the Institute of Food Science and Technology (IFST) and the Royal Environmental Health Institute of Scotland (REHIS). There are different routes to obtaining these certificates, and it is advisable to contact the relevant body for details. Generally, it involves the completion of an appropriate course of study, an assessed period of practical training and some form of examination or assessment.

With the issuing of Section 18 guidance in 1996, under the Health and Safety at Work, Etc. Act 1974, specifying the competence requirements of health and safety enforcement officers, technical support staff working in the occupational health and safety field need a recognized qualification. The EHRB offers a Diploma in Health and Safety Enforcement which demonstrates attainment of the appropriate and necessary knowledge and skills to undertake this work. The EHRB also offers a Diploma in Environmental Protection for those dealing with air quality, noise, waste management, contaminated land and water management. Diplomas in Housing and Public Health are also being developed by the EHRB for support staff working in these areas. Assessments of all these qualifications will be similar to that of the Food Certificates, that is, by the completion of an appropriate course of study, an assessed period of practical training and some form of examination or assessment.

A list of those Academic Institutions currently accredited by the EHRB to deliver these courses can be found in the appendix at the end of this chapter.

ASSESSMENT OF PROFESSIONAL COMPETENCE

After qualifying to practise, EHPs are eligible for voting membership of the CIEH. After at least five years of professional practice, members who have passed the Assessment of Professional Competence (APC) can apply to become Chartered EHPs.

The assessment requires candidates to submit a work experience log, which summarizes the work they have undertaken during their professional practice. A case study must also be prepared and submitted, documenting how the EHP has resolved a particular problem that they have had to deal with which demonstrates use of their professional skills. Candidates must also attend a professional interview. Via the case study submitted and the interview, EHPs need to demonstrate to the assessor that they have developed their professional skills to a level that is acceptable for access to Chartered status. The skills assessed are:

1. investigative
2. analytical
3. interpretive
4. communicative
5. educative
6. organizational
7. attitudinal.

The APC scheme is due to be reviewed in 2004 and a revised scheme is expected to be in operation from 1 January 2006.

CONTINUING PROFESSIONAL DEVELOPMENT

The CIEH's scheme of Continuing Professional Development (CPD) was introduced in July 1992 on a voluntary basis and made mandatory in 1996. All members of the CIEH have a personal responsibility to maintain their professional competence throughout their professional practice.

The scheme requires Graduate and Voting members to undertake and record at least 20 hours of relevant CPD activity every year. Activities such as training courses, seminars, branch and centre meetings, research and the writing/presentation of papers are all considered to be relevant, as is preparing for and taking the APC. Chartered EHPs are required to complete at least 30 hours of CPD activity each year.

The CIEH calls in CPD records from a random sample of members each year.

The address of the CIEH headquarters is:

> The Chartered Institute of Environmental Health (CIEH)
> Chadwick Court
> 15 Hatfields
> London SE1 8DJ
> Tel: 0207–928 6006

Up to date information on qualification as an EHP or Technician can be found in the CIEH web site at: www.cieh.org or at www.ehocareers.org.

The Royal Environmental Health Institute of Scotland operate a similar qualification process which leads to the issue of a REHIS Diploma which is, similarly, recognized as a qualification for practice as an EHP within the United Kingdom (also see appendix B to Chapter 3).

The address of the REHIS headquarters is:

> The Royal Environmental Health Insititute of Scotland
> 3 Manor Place
> Edinburgh
> EH3 7DH

APPENDIX: UNIVERSITIES OFFERING COURSES IN ENVIRONMENTAL HEALTH

BSc (Hons) degree in Environmental Health/Science

Four-year integrated courses, the majority of which offer part-time and non-integrated routes – contact the course leader for details of options available:

Bristol

University of the West of England
Faculty of Applied Sciences
Coldharbour Lane
Frenchay, Bristol
BS16 1QY
Tel: 0117–965 6261
Ms Melanie Grey

Cardiff

University of Wales Institute
Cardiff School of Environmental Sciences
Western Avenue
Cardiff, CF5 2YB
Tel: 01222–551111
Mr Andrew Curnin

Glasgow

University of Strathclyde
John Anderson Building
107 Rotten Row
Glasgow G4 0NG
Tel: 0141–548 3539
Dr Tony Grimason

Leeds

Leeds Metropolitan University
Calverley Street
Leeds, LS1 3HE
Tel: 0113–283 2600
Ms Catherine Gairn

London

King's College London
School of Life Sciences
Franklin-Wilkins Building
150 Stamford Street
London, SE1 9NN
Tel: 020–7848 4109
Mr M Howard

Middlesex University
School of Applied Sciences
Bounds Green Road

London, N11 2NQ
Tel: 020–8368 1299
Mr Alan Page

Manchester

Manchester Metropolitan University
Department of Food and Consumer Technology
The Hollings Faculty
Old Hall Lane
Manchester, M14 6HR
Tel: 0161–247 2000
Mr Steve Turner

Nottingham

Nottingham Trent University
Burton Street
Nottingham, NG1 4BU
Tel: 0115–941 8418
Mr Paul Paddock

Salford

University of Salford
Dept of Environmental Management
Salford, M5 4WT
Tel: 0161–295 5000
Mr Chris Miller

Ulster

University of Ulster
Shore Road, Newtonabbey
Co. Antrim
N. Ireland, BT37 0QB
Tel: 01232–365131
Mr Oliver Hetherington

MSc in Environmental Health

For graduates with appropriate science degrees:

Birmingham

University of Birmingham
School of Biological Studies
Edgbaston
Birmingham, B15 2TT
Tel: 0121–414 7180
Mr Maurice Brennan

Bristol

University of the West of England
Faculty of Applied Sciences
Coldharbour Lane
Frenchay, Bristol
BS16 1QY
Tel: 0117–965 6261
Ms Melanie Grey

Derby

University of Derby
School of Environment & Applied Sciences
Kedleston Road
Derby, DE22 1GB
Tel: 01332–591703
Dr Derek Walton

Courses are also being considered for development at Northumbria University, Newcastle and at the University of Brighton. Please contact the Chartered Institute of Environmental Health for further details.

APPENDIX: UNIVERSITIES/COLLEGES
OFFERING COURSES FOR
TECHNICAL OFFICERS

Higher/Ordinary Certificate in Food Premises Inspection

Birmingham

University of Birmingham
School of Biological Studies
Edgbaston
Birmingham, B15 2TT
Tel: 0121–414 7180

Ms Madeleine Smith

London

Middlesex University
School of Applied Sciences
Bounds Green Road
London, N11 2NQ
Tel: 0181–368 1299
Mr Alan Page

Salford

University of Salford
Dept of Environmental Management
Salford, M5 4WT
Tel: 0161–295 5000
Ms Denise Rennie

Diploma in Health and Safety at Work Enforcement

Nottingham

Nottingham Trent University
Burton Street
Nottingham, NG1 4BU
Tel: 0115–941 8418
Mr Paul Paddock

Leeds

Leeds Metropolitan University
Calverley Street
Leeds, LS1 3HE
Tel: 0113–283 2600
Ms Catherine Gairn

As other courses are developed and accredited they will be advertised in the Environmental Health News and by letter to all EH Departments.

5 Port health

Peter Rotheram

DEVELOPMENT OF PORT HEALTH

Systematic quarantining of ships arriving from Levantine Turkey was adopted by Venice in the aftermath of the Black Death when it became apparent that the Levant had become a permanent reservoir of plague. The Venetian ships collected valuable cargoes of silk and spices that had been transported overland to the Levant, and they were isolated in quarantine for 40 days on returning to Venice to establish that they were not infected.

When plague spread to the Baltic and the Low Countries, ships arriving in England from infected areas were detained by an Order of Council, but it became increasingly difficult to restrict the illicit landing of people [1]. English common law did not oblige townspeople to maintain a watch between sunrise and sunset or to detain people unless they had committed a felony. The first Quarantine Act of 1710 removed these impediments to enforcing quarantine, but it was only after the Levant Company lost its monopoly of trade with Turkey in 1753 that legislation was introduced – an Act for Enlarging and Regulating Trade into the Levant Sea 1753 and an Act to Oblige Ships to Perform Quarantine 1753 [1] – requiring all ships loading in the Levant to undergo quarantine in the Mediterranean.

But the Dutch did not apply such rigorous quarantine [2], and with more than half the cotton used in England transhipped through Holland, British shipping was at a disadvantage. As a consequence, the law was subsequently amended.

The Act to Encourage Trade into the Levant Sea 1799 provided a more convenient mode of performing quarantine. Ships from the Levant were permitted to perform their quarantine in the Medway at Stangate Creek, hulks were provided for cargoes to be aired and Parliament voted funds for the building of a lazaret at Chetney Hill.

The relaxation of quarantine was achieved amid controversy about whether plague was 'epidemic' or 'contagious'. Charles Maclean, a ship's surgeon who had served with the East India Company, propounded an elegant hypothesis suggesting that diseases that were seasonal and could infect people more than once were 'epidemic', while those that occurred independently of season or the state of the air and only infected a person once were 'contagious'. His hypothesis [3] was based on the observations of ships' surgeons that fevers such as typhus and yellow fever were dependent on the season and the state of the air. The experience of Southwood Smith at the London Fever Hospitals brought him to the same conclusions as Maclean, and he developed his theme on the high economic costs of disease that was later adopted by Edwin Chadwick [4]. The spread of cholera to Great Britain prompted the General Board of Health to produce a report on quarantine [5]. Largely the work of Southwood Smith, it was decidedly anti-contagionist and led to the first International Sanitary Conference in 1851. The resulting convention was ratified by only France and Sardinia, however.

The British view was that the enforced detention of ships in quarantine was only necessary when the disease was actually on board, and that if the ship was in a foul condition it should be cleansed and disinfected before pratique was granted. Orders assigning to the Poor Law Authorities at endangered ports the power to deal with shipping arrivals suspected of having cholera on board were issued by the General Board of Health in 1849. Section 32 of the Sanitary Act 1866 subsequently made ships subject to the jurisdiction of the Nuisance Authority of the district in which they were moored. Regulations were introduced requiring ships to be inspected and dealt with as if they were a house. This arrangement was not entirely satisfactory when there was more than one Nuisance Authority in a port, and Section 20 of the Public Health Act 1872 empowered the Local Government Board to constitute one Port Sanitary Authority with jurisdiction over the district on any port established by the commissioner of customs.

These statutory powers were consolidated in 1875 and re-enacted in the Public Health Act 1936, which changed the designation Port Sanitary Authorities to Port Health Authorities. The statutory provisions relating to port health were subsequently incorporated into the Public Health (Control of Disease) Act 1984. In Scotland, port health functions are exercised by port local authorities, designated under the Public Health (Scotland) Act 1897. A Joint Board constituted as a port health authority precepts the constituent local authorities for its expenditure.

JURISDICTION

The functions of a port health authority are assigned directly by statute or in a statutory instrument. The area of jurisdiction of a port health authority is specified in a statutory order. It may comprise the whole or part of a customs port, including the whole of any wharf and the area within the dock gates. A port health authority may, within its district, exercise any of the functions of a local authority relating to public health, waste disposal or the control of pollution assigned to them in the order. The Prevention of Damage by Pests Act

1949 provides that the local authority for any port health district is the port health authority. It is also designated as a local authority under Parts I and III of the Environmental Protection Act 1990 and as a food authority under the Food Safety Act 1990. Any local authority having jurisdiction in any part of a port health district is excluded from exercising any functions assigned to the port health authority.

ORGANIZATION

Most of the work of a port health authority is undertaken by environmental health officers; where a district council is the port health authority, the work may devolve around a few specialists. The whole environmental health department may be involved, particularly where the port or airport is handling traffic outside normal business hours. When the port health authority is constituted as a joint board, it will employ its own environmental health officers and administration staff and precept the constituent local authorities for any expenditure incurred. In either case, the port health authority will appoint one or more port medical officers; district health authorities have a statutory obligation to provide these services free of charge to the port health authorities. The port medical officer may also be appointed as a medical inspector of aliens, to undertake medical examinations on behalf of the immigration service. Where a port or airport is approved by the European Union (EU) for the importation of produce of animal origin from third countries, an Official Veterinary Surgeon (OVS) has to be appointed by the competent authority.

At any port or airport the port health officers liaise with customs officers and traffic controllers to maintain surveillance of shipping and aircraft arrivals. Customs officers alert the port health officer whenever they become aware of any apparent contravention of the Public Health (Ships) Regulations 1979 and Public Health (Aircraft) Regulations 1979, or the Imported Food Regulations 1997. When the port health authority is not an enforcing authority under the Animal Health Act 1981 the port health officer will advise the responsible authority of any animals that are imported contrary to the rabies control legislation. Details of the cargo imported on a ship or aircraft can be obtained from a manifest, which can be obtained from the carrier. This information may be stored as computerized inventory control, which involves the importer entering details of each consignment on the computer system to obtain clearance from customs and port health. A case study of the London Port Health Authority is included as an appendix at the end of this chapter.

INTERNATIONAL HEALTH CONTROL

Successive International Sanitary Conferences eventually led to the elimination of quarantine [6], and under the auspices of the World Health Organization (WHO) the International Health Regulations 1969 specified the maximum restrictions that could be applied to ships and aircraft involved in international commerce. They are based on the assumption that the only effective protection against the spread of epidemics is the provision of wholesome supplies of food and water, the effective disposal of waste and the elimination of vectors of disease on ships and aircraft.

Following the increasing emphasis on epidemiological surveillance for communicable disease recognition and control, the WHO agreed to strengthen the use of epidemiological principles as applied internationally, to detect, reduce or eliminate the sources from which infection spreads, to improve sanitation in and around ports and airports, to prevent the dissemination of vectors and, in general, to encourage epidemiological activities on the national level so that there is little risk of outside infection establishing itself.

As a result of a worldwide campaign by the WHO, smallpox was eradicated and vaccination against the disease is no longer required. Efforts to control cholera by vaccination failed to prevent the spread of the disease, and people arriving from infected areas are no longer required to have cholera vaccinations. Yellow fever and plague are controlled by measures to control the vectors of the disease. A yellow fever vaccination is required for people arriving in countries where the mosquito vector of the disease is prevalent, but this does not include the United Kingdom.

Following the change in the approach to dealing with cholera and the elimination of smallpox, the International Health Regulations were amended in 1973 and 1981. The diseases subject to the Regulations are plague, yellow fever and cholera, and provision is made in the United Kingdom for implementing the government's obligations under these regulations in the Public Health (Ships) Regulations 1979 and Public Health (Aircraft) Regulations 1979.

The effect of global warming is causing concern, as warmer weather in the British Isles will facilitate the spread of species of mosquito that would otherwise not breed there, increasing the risk of the introduction of exotic disease, particularly by aircraft. It is hoped that the WHO will address this problem when it brings forward its proposals to review international health regulations.

THE PUBLIC HEALTH (SHIPS) AND (AIRCRAFT) REGULATIONS 1979

These provide for people arriving on infected or suspected ships or aircraft to be placed under surveillance; the periods are calculated from the time of leaving the infected area. Measures to control the spread of Lassa fever, rabies, viral haemorrhagic fever and Marburg disease are also specified. Requirements for the notification and prevention of the spread of other infectious diseases from arrivals at ports and airports are included in the Public Health (Infectious Disease) Regulations 1984 and 1988.

An **infected** ship or aircraft is:

1. A ship or aircraft that on arrival has on board a case of a disease subject to the International Health Regulations, or a case of Lassa fever, rabies, viral haemorrhagic fever or Marburg disease.
2. A ship or aircraft on which a plague infected rodent is found on arrival.
3. A ship or aircraft that has had on board during its journey:

   (a) a case of human plague developed by the person more than six days after his or her embarkation;

(b) a case of cholera within five days of arrival;
(c) a case of yellow fever or smallpox.

A **suspected** ship or aircraft is:

1. A ship or aircraft that, not having on arrival a case of human plague, has had on board during the journey a case of that disease developed by a person within six days of his or her embarkation.
2. A ship or aircraft on which there is evidence of abnormal mortality among rodents, the cause of which is unknown on arrival.
3. A ship or aircraft that has had on board during its journey a case of cholera more than five days before arrival.
4. A ship that left an area infected with yellow fever within six days of arrival.
5. A ship that on arrival has on board a person whom the medical officer considers may have been exposed to infection from Lassa fever, rabies, viral haemorrhagic fever or Marburg disease.

HEALTH CLEARANCE OF ARRIVALS

The master of a ship or the commander of an aircraft arriving from a foreign country must report to the port health authority not more than 12 hours and not less than 4 hours before arrival:

1. The occurrence on board during the passage (or the last four weeks if longer) of:

   (a) death other than by accident;
   (b) illness where the person concerned has or had a temperature of 38°C or greater that was accompanied by a rash, glandular swelling or jaundice, or persisted for more than 48 hours; or has or had diarrhoea severe enough to interfere with work or normal activities.

2. The presence on board of:

   (a) a person suffering from an infectious disease or who has symptoms that may indicate the presence of infectious disease;
   (b) any animal or captive bird of any species including rodents and poultry, or mortality or illness among such animals or birds.

3. Any other circumstances that are likely to cause the spread of infectious disease.

If none of these circumstances prevails, the ship or aircraft does not require health clearance (**free pratique**) unless otherwise directed by an authorized officer of the port or airport health authority. Otherwise the specified signals (Table 5.1) should be shown on arrival and no person may board or leave the ship or aircraft other than a customs officer, immigration officer, the pilot of a ship or an authorized officer until health clearance has been granted. These restrictions on boarding or disembarking do not apply when only the presence on board of any animal or captive bird requires reporting, unless plague has occurred or is suspected among rats and mice or there has been abnormal mortality among them.

When health clearance is required, a **maritime or aircraft declaration of health** has to be completed by the master or commander, except in the case of arrivals from Belgium, France, Greece, Italy, the Netherlands, Spain or the Irish Republic. Control measures to prevent the spread of infectious disease from ships and aircraft that are 'suspected' or 'infected' or which have arrived from an infected area, may include medical examination, or the disinfecting of any person suffering from or exposed to infectious disease, disinfection or disinsectation of ships, aircraft or clothing, measures to prevent the escape of plague-infected rodents and the removal of any contaminated food or water other than the cargo.

Table 5.1 International code of signals health clearance messages

Signal	Meaning
Q	My vessel is 'healthy' and I request free pratique
QQ	I require health clearance*
ZU	My maritime declaration of health has a positive answer to question(s) (indicated by appropriate number(s))
ZW	I require a port medical officer
ZY	You have health clearance

Note:
*By night, a red light over a white light may be shown, where it can best be seen, by ships repairing health clearance. These lights should only be about 2 m apart, should be exhibited within the precincts of a port and should be visible all round the horizon as nearly as possible.

AIRCRAFT DISINFESTATION

In order to minimize the risk of aircraft spreading vectors of disease, the commander of an aircraft that has landed in a risk area may be required to produce details of the application of residual insecticides or inflight disinsectation. The International Maritime Organization (IMO) has recommended [7] an initial treatment with permethrin to produce an even deposit of 0.5 g/m^2 on carpets and 0.2 g/m^2 on other surfaces (these rates may be halved for subsequent treatments). For the purposes of inflight disinsectation, an approved aerosol formulation is required to be dispensed uniformly in the enclosed space at the rate of 35 g/100 m^3. The serial numbers of the aerosol dispensers used in the treatment should be entered on the aircraft's declaration of health. In no circumstances should insecticide be used on an aircraft unless it has been specifically approved for that purpose.

SHIP DISINFESTATION

A ship arriving from a foreign port must have a **deratting certificate** issued in a designated approved port showing that the ship has been deratted within the previous six months, or a **deratting exemption certificate** issued in either an approved port or a designated approved port showing that the ship was inspected and found free from rodents and the plague vector. The list of approved and designated approved ports is published by the WHO, and amendments are notified in the *Weekly Epidemiological Record* [8].

If the certificate was issued more than six months previously, the ship must be inspected at an approved port and, if necessary, be deratted at a designated approved port. Where the ship is proceeding immediately to a designated/approved port, the validity of the certificate may be extended by one month. This provision would be applied where the amount of cargo remaining on board prevents the inspection of all the spaces for the issue of a deratting exemption certificate. If evidence of rodents is found during the inspection of a ship, the port health officer is authorized to require the master to apply the appropriate control measures. These may be either trapping, poisoning or fumigation of the infested spaces, together with the elimination of any harbourage.

Disinfestation methods on ships

The various types of ship construction (Fig. 5.1) together with the limited time available to disinfest ships limit the options for treatment. Consequently, acute poisons such as sodium fluoracetate or fumigation with hydrogen cyanide or methyl bromide may be required. These will usually be undertaken by a contractor, but it is the responsibility of the port health officer in charge to specify the treatment required and to have regard for the safety aspects. Acute rodenticides should only be used in spaces that can be secured to prevent access. While fumigation is taking place, the crew must be accommodated ashore. Care is required if it is intended to fumigate food with hydrogen cyanide or leave food exposed to hydrogen cyanide during fumigation.

If a ship, aircraft or cargo infested with rats arrives from an area in which plague is present, or if there is undue mortality among them, then fumigation should always be required. As methyl bromide is three times heavier than air, its use is particularly appropriate for the fumigation of foodstuffs, or a fully laden ship. Methyl bromide and hydrogen phosphide are also used for the elimination of arthropod infestations of cereals, nuts, dried fruit and herbs and spices. Such fumigation may take place prior to shipment or after loading on board the ship.

With regard to in-transit fumigation, the IMO recommends that the ship's master should obtain approval from his national administration. At least two members of the crew, including one officer, should be trained with particular reference to the behaviour and hazardous properties of the fumigant in air, the symptoms of poisoning and emergency medical treatment. They should brief the crew before the fumigation takes place and the ship should carry gas detection equipment, at least four sets of protective breathing apparatus and appropriate medicine and first-aid equipment. It is the responsibility of the fumigator to ensure that all spaces treated are gas tight and that warning notices have been posted at the entrances to any spaces considered unsafe. Adjacent spaces, accommodation

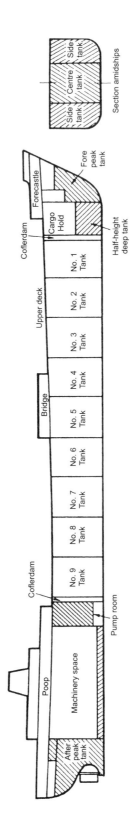

Tanker 15BH WB20500T Incl. 14500T in clean ballast tanks DTf3300T

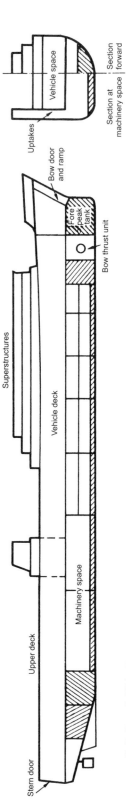

Ro Ro Cargo Coll BH to U dk 10 to 2nd dk WB350T incl. DTf90T

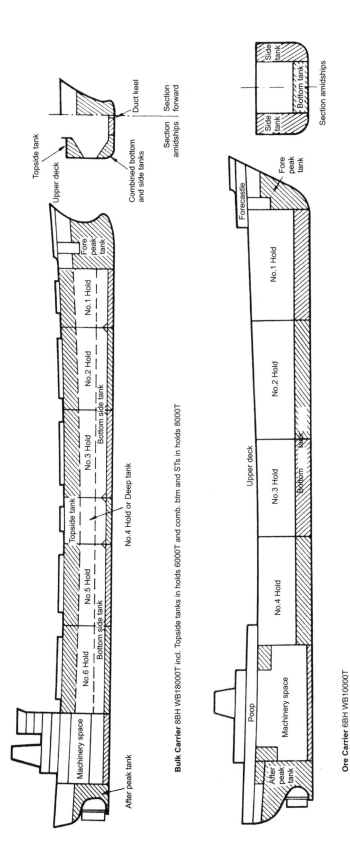

Bulk Carrier 8BH WB18000T incl. Topside tanks in holds 6000T and comb. btm and STs in holds 8000T

Ore Carrier 6BH WB10000T

Fig. 5.1 Profiles illustrating decks, superstructures, spaces and tanks in different types of ships. (Reproduced by kind permission of Lloyds Register of Shipping, London.)

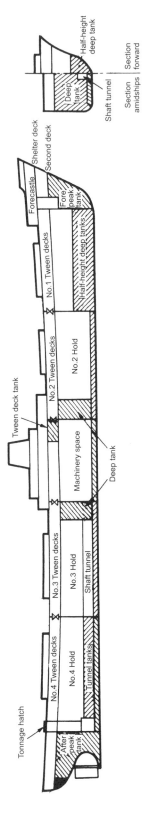

General Cargo Ship Coll BH & APBH to S dk 7 to 2nd dk WB4200T incl. DTma890T DTmf890T Tunnel tanks 400T UnDk a20T f20T

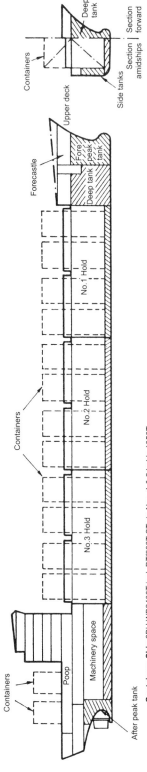

Containers Ship 6BH WB2400T incl. DTf300T STs in Nos. 1 & 2 holds 1350T

Fig. 5.1 Continued.

and working spaces should be checked for the level of gas concentration at least at eight-hour intervals, and the readings recorded in the ship's log book [7].

SHIPBOARD SANITATION

The provisions of the Public Health Acts relating to filthy, unwholesome and verminous premises or articles and verminous persons (see Chapter 6) apply to ships as if the vessel were a house, building or premises within the district, and the master or other officer in charge of the occupier. With regard to statutory nuisances (see Chapter 6), the Environmental Protection Act 1990 defines 'premises' to include any vessel. An abatement notice may be served on the master of a vessel as if he or she were the occupier of premises on which a statutory nuisance exists or is likely to occur. Except for a ship belonging to the Queen or visiting forces, the person in charge of a ship may be ordered to remedy any conditions on board that are prejudicial to health, and statutory nuisances may be dealt with summarily. The standardization of sanitary measures to be taken on ships is specified by the WHO in the *Guide to Ship Sanitation* [9]. It provides practical recommendations for protecting food and potable water, the disposal of waste and elimination of pests, all of which are necessary to prevent the spread of disease.

INTERNATIONAL TRAINS

An international train is a shuttle train or through train operating through the Channel Tunnel, and the person designated as the **train manager** on a journey terminating in the United Kingdom is responsible for advising the enforcement authority of the presence of any animal or sick person aboard the train. The Public Health (International Trains) Regulations 1994 define the enforcing authority as the port health or local authority at any place designated as a control area, freight depot or terminal control point. In relation to the records to be kept by an international train operator and in the event of a health alert, the Secretary of State is designated as the enforcing authority.

Where a stowaway animal that is, or was at the time of its death, capable of carrying rabies, plague or viral haemorrhagic fever is suspected aboard an international train, the enforcement authority at a stopping place may require the train and its contents to be deratted, disinfected or decontaminated. There are also powers for the disinfection or decontamination of any rolling stock or any article on board where a sick person has been identified on the train.

Sick traveller means a person who has a serious epidemic, endemic or infectious disease, or whom there are reasonable grounds to suspect has such a disease. It does not mean venereal disease or infection with human immunodeficiency virus (HIV). See also Chapter 14.

WATER SUPPLIES[1]

Every port and airport is required by the international health regulations to have a supply of pure drinking water. Potable water for ships and water boats must be obtained only from those water points approved by the port health authority. Water boats should have independent tanks and pumping systems for potable water. Hydrants for the supply of water from ashore should be located so as to prevent contamination. Supply pipes should be above the high water level of the port, and drainage openings for pipes need to be above any water surge from passing ships. Water supply hoses require a smooth, impervious lining and should be used exclusively for the delivery of potable water. When not in use the ends should be capped.

Ships' storage tanks for potable water should be independent of the hull (Fig. 5.2) unless the tank bottom is at least 45 cm above the deepest load line. The bottom of the tank should not be in contact with the top of any double bottom tank. Pipe lines carrying non-potable liquids should be routed so they cannot contaminate the potable water supply. A manhole on the tank top should have a coaming to keep the opening clear of the deck, and overflow pipes should have the open end pointing downwards. Where tanks do not meet these recommendations, the water may be used only for domestic purposes aboard ship, or should be chlorinated before use as potable water.

---

[1]Also see Health Protection Agency (2003) *Guidelines for water quality on board merchant ships including passenger vessels.* HPA, London.

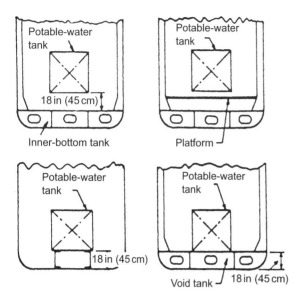

Fig. 5.2 Acceptable location of potable water in relation to bilge space of inner-bottom tank. (Source: [9], reprinted with amendments in 1987.)

Chlorine should preferably be applied as a hypochlorite solution with equipment that will produce a free chlorine residue of 0.2 mg/l. Whenever any potable water tanks or any part of the supply system has been contaminated, serviced or repaired, they should be cleansed, disinfected and flushed before being put back into operation. The chlorine solution for disinfection should be 50 mg/l for 24 hours, or in an emergency 100 mg/l for 1 hour. The heavily chlorinated water should be discharged and the system flushed with potable water before the system is used to supply potable water. Automatic chlorinators on an ultraviolet sterilization system should be fitted where low pressure evaporators are used for desalination. Sea water obtained when within 32 km of the land should never be used to produce potable water.

IMPORTED FOOD

Products of plant origin

The Imported Food Regulations 1997 (IFRs) prohibit the importation of food not of animal origin intended for sale for human consumption from a third country (a country or territory that is not part of the EU), which fails to meet food safety requirements or is unsound or unwholesome. Food is deemed to be from a third country if it originates in that country and has not been released into free circulation in another member state, or if it arrives from either the Channel Islands or the Isle of Man, having been under customs charge and sent for examination at a place in Great Britain or Northern Ireland.

Each food authority is required to enforce the IFRs where food has entered into the British Islands at a place that is in its area, unless that place is situated within the district of a port health authority where the latter is responsible. A food authority (other than a county council in England) can enforce the IFRs when the imported food arrives in its district, provided that it receives notice from the food authority at the place of landing that the food has not been inspected together with a copy of importers undertaking that the container containing the food has been sealed and will not be opened until it reaches its specified place of destination.

An authorized officer of a food authority may examine imported food subject to the IFRs, and the importer shall provide all such facilities as may reasonably be required for such examination. If the authorized officer considers that the food should be sampled he may require that once he has obtained his sample, the food shall not be removed from the place specified for six days excluding Saturdays, Sundays and public holidays. If the food is not clear of customs charge, an officer of customs may give written permission to a person to remove the food, but he shall notify the authorized officer of the food authority of his intention to do so.

If, after inspection or examination, it is found that any consignment, lot or batch of food that has been imported is unsound or unwholesome or fails to comply with food safety requirements, the authorized officer, after consulting the person importing the food, may give notice that the food may be used for purposes other than human consumption, ordering its re-export outside the European Community (EC) or seize the food and have it dealt with by a justice of the peace, or in Scotland a sheriff or magistrate, who may condemn the food or order its destruction. An importer aggrieved by such a notice may, within six days, appeal to a magistrate court or in Scotland to the Sheriff, who may cancel or affirm the notice.

The IFRs amend the Food Safety (General Food Hygiene) Regulations 1995 with regard to the carriage in bulk of liquid oils and fats in seagoing vessels. They allow the transport of such products that are intended for processing in tanks that are not exclusively used for foodstuffs, provided that the transportation is in a stainless steel tank or one lined with epoxy resin or its equivalent. The previous cargo transported in the tanks must have been an acceptable foodstuff. If the product is not carried in a stainless steel or epoxy-lined tank, the previous three cargoes should have been acceptable foodstuffs. If the oils or fats are not to be processed further and the tanks are not exclusively used for the transport of products intended for human consumption, they must be of stainless steel or epoxy lined, and the three previous cargoes transported in the tanks should have been foodstuffs.

The captain of the ship is required to keep documentary evidence of the three previous cargoes and the tank cleansing process used; he shall if required provide the food authority with documentary evidence.

Products of animal origin (POAO)

The Products of Animal Origin (Third Country Imports) (England) Regulations 2003 disapply the IFRs to POAO and apply EC requirements for veterinary checks on products entering the EC from third countries. All such products of animal origin must be imported through a Border Inspection Post (BIP), which is approved by the EC for veterinary checks on that product, where they must undergo specific checks to ensure that they comply with the requirements of Community legislation. The competent authority for enforcement of the regulations at BIPs where food for human consumption is landed is the port health authority, at airports the local authority. The BIP must employ an Official Veterinary Surgeon (OVS) to undertake veterinary checks of POAO, an environmental health officer appointed as an Official Fish Inspector (OFI) can carry out the regulatory checks on fishery products.

Importers must give the BIP 24 hours notice of the arrival of any consignment of POAO at a seaport or 6 hours at an airport, and submit a certificate of veterinary checks together with any requisite health certificate. Every consignment of POAO is subject to a veterinary check to verify the origin of the product and establish that any requisite health marks are present. Physical checks must be undertaken on all consignments to verify that the conditions of transport have maintained the wholesomeness of the products and that any temperature specifications have been complied with. An organoleptic examination must be carried out on a percentage of items in each consignment to check for any abnormalities. Additional physical and laboratory checks are carried out in the event of an adverse outcome, or when information from another member state or the result of checks on a previous consignment have been unfavourable. If all the checks are satisfactory, the CVC can be signed and the consignment released into free circulation, the person responsible for the consignment is required to pay all the prescribed charges for the veterinary checks carried out on the consignment.

Where products fail veterinary checks at a BIP, or are found by an authorized officer away from a BIP to be non-conforming products, the OVS or authorized officer shall serve notice on the person having charge of the products either to redispatch it to a third country within 60 days or destroy the product without delay.

Personal imports of animal and plant products

New rules, effective from 1 January 2003, prohibit personal imports into the EU of meat, meat products, milk or milk products, unless they comply with the requirements of the EC veterinary checks regime. Products not subject to EC veterinary checks include chocolate, confectionery and cakes, provided they do not contain fresh cream.

Exemptions are made for personal imports, from any country, of powdered infant milk, infant food and special foods required for medical reasons. These exemptions only apply to products that are packaged proprietary brands for direct sale to the final consumer, the packaging of which is unbroken and do not require refrigeration before opening. Up to 1 kg of animal products other than meat, milk, meat and milk products is permitted from countries permitted to export such products to the EU.

Meat, milk, meat products and milk products in small quantities for personal consumption are permitted from any State in the EU and from:

Greenland	Faroe Islands	Iceland
Andorra	San Marino	Liechtenstein
Switzerland	Estonia	Lithuania
Latvia	Poland	Norway
Hungary	Slovenia	Slovakia
Romania	Bulgaria	Malta
Cyprus	Czech Republic	

Fresh, smoked, cooked or dried meat and cheese may only be brought from another State in the EU, or from any of the countries mentioned here.

Small quantities of plants and plant products for personal consumption may be brought into the United Kingdom without a plant health certificate, provided they are contained in personal baggage, are free from signs of pests and disease and they are not intended for use in the course of trade or business.

Individuals may bring back from Algeria, Canary Islands, Cyprus, Egypt, Israel, Jordan, Lebanon, Libya, Malta, Morocco, Norway, Switzerland, Syria, Tunisia or Turkey, 5 plants and 2 kg bulbs, corms, tubers (other than seed or ware potatoes), rhizomes, 2 kg fruit, a bouquet of cut flowers and foliage and 5 retail packs of seed. From countries other than those listed here, individuals may bring 2 kg fruit, 1 bouquet of cut flowers and foliage and up to 5 retail packets of seed (but not the seeds of potatoes).

There are no restrictions on any animal products from countries within the EU, provided products are from approved establishments and that they are not from endangered species. And there are no restrictions on plant material brought into the United Kingdom from countries in the EU, provided it was grown in the EU.

HM Customs and Excise have been assigned responsibility for anti-smuggling controls for meat and animal products imported from non-EU countries.

Food labelling

The Food Labelling Regulations 1996 (as amended 1998) require that all food that is for delivery to the ultimate consumer or to a catering establishment

should, subject to specified exceptions, be marked or labelled with:

- the name of the food;
- a list of ingredients (see amendments relating to ingredients in the Food Labelling (Amendment) Regulations 1998/1398);
- indications of durability;
- any special storage conditions or conditions of use;
- the name and address or registered office of the manufacturer, packer or of a seller established within the EU;
- the place of origin if failure to give it would mislead;
- instructions for use if not apparent.

Organic food

Any food product of organic origin imported into the United Kingdom must have a Certificate of Origin (CoO) issued by the body certifying the goods as organic in the exporting country. If the consignment is to be marketed in the Community as an organic product, the CoO must be checked and endorsed by the port health authority.

Food safety

As a result of a number of incidents of food contamination that have come to light in the last few years, the EC is to introduce Regulations on Official Feed and Food Controls. A European Food Safety Authority will be responsible for official controls on imported feed and food, with particular reference to hygiene, zoonoses, residues and contaminants, pesticides and additives. The new Regulations will replace the Directives that currently apply to food control in the Community and extend checks on products of plant origin and animal feeds.

Where imported food is found to be contaminated, the current procedure is for the EC to impose special conditions on products from their countries. These can involve a prohibition on imports of certain products from specified regions, require testing and certification and place of origin and or compulsory sampling at the point of entry into the Community. Regulations have also been made to control mycotoxin levels in food, limits are also placed on the level of arsenic, chloroform, colours, extraction solvents, flavourings, lead, mineral hydrocarbons, pesticides, sweeteners, tin and various miscellaneous additives and constituents of materials used in contact with food, including plasters. Specific rules for fresh meat, fish and fishery products require packaging to be strong enough to provide effective protection during transport and not to alter the organoleptic character of the product.

Emergency arrangements

The Food and Environment Protection Act 1985 empowers ministers to make emergency orders where circumstances exist that are likely to create a hazard to human health through the consumption of contaminated food. Such orders may prohibit the distribution of affected produce (e.g. shellfish from layings affected by oil pollution) where foodstuffs may have been contaminated. If imported food is suspect, ministers may authorize port health officers to act on their behalf and sample and detain food landed in their district and prevent its distribution.

FOOD HYGIENE AND SAFETY

An order made under the Food Safety Act 1990 defines 'premises' as including any ship or aircraft specified in the order made by ministers.[2] Powers of entry conferred by the Act include the right to enter any ship or aircraft for the purpose of ascertaining whether there is any imported food present as part of the cargo. Provision is also made in the Act (Section 58) for treating any offshore installation in the territorial waters of the United Kingdom as if it were situated in the adjacent part of Great Britain for the purposes of food safety legislation.

Fishing vessels, including fish processing factories, are subject to the hygiene requirements of the Food Safety (Fishery Products on Fishing Vessels) Regulations 1998, which are intended to ensure that fish is handled and stored under hygienic conditions and protected from risk of contamination. Fish caught by a fishing vessel flying a third-country flag landed in a UK port is subject to the veterinary checks regimen; such direct landings do not have to be accompanied by a health certificate.

AIR POLLUTION

The prohibition of dark smoke from chimneys (see Chapter 36) applies to vessels, as references to a furnace include references to the engine of a vessel, so long as the vessel is in navigable waters contained within any port, river estuary, etc., for which charges other than light dues can be made in respect of vessels entering or using facilities therein.

The permitted periods in which vessels are allowed to emit dark smoke without committing an offence are contained in the Dark Smoke (Permitted Periods) (Vessels) Regulations 1958. But the provisions of Part III of the Environmental Protection Act 1990 relating to nuisances do not apply to a vessel powered by steam reciprocating machinery.

In relation to the release of other pollutants into the air, the Act can be used to abate or prevent nuisances arising from the handling of bulk cargoes [10]. The area of a port health authority includes the territorial waters to seaward of the district in respect of any nuisance to which Part III of the Act applies.

REFERENCES

1. Salisbury Manuscripts 1602/1603, Calender State Papers Domestic 1635/1636, Calender Treasury Papers 1708 and 1714.
2. Howard, J. (1789) *Lord Liverpool's Papers 1786* (Account of the Lazarettos Privy Council: Levant Company to Pitt 1792).
3. Maclean, C. (1796) *The Source of Epidemic and Pestilential Diseases*, Calcutta.

---

[2]See the Food Safety (Ships and Aircraft) (England and Scotland) order 2003 – p. 131/2.

4. Southwood Smith, T. (1866) *The Common Nature of Epidemics*, London.
5. General Board of Health (1849) *First Report on Quarantine*.
6. Howard-Jones, N. (1975) *The Scientific Background of the International Sanitary Conference 1851–1938*, World Health Organization, Geneva.
7. International Maritime Organization (1984) *Supplement to the Recommendation on the Safe Use of Pesticides in Ships*, IMO, London.
8. WHO (1985) *Weekly Epidemiological Record*, Vol. 60, World Health Organization, Geneva.
9. Lamoureux, V.B. (1967) *Guide to Ship Sanitation*, World Health Organization, Geneva.
10. Schofield, C. and Shillito, D. (1990) *Guide to the Handling of Dusty Materials in Ports*, 2nd edn, British Materials Handling Board, Ascot.

APPENDIX A: A CASE STUDY OF A PORT HEALTH AUTHORITY – LONDON PORT HEALTH AUTHORITY

Jon Averns

This appendix describes how the London Port Health Authority (LPHA), as the largest Port Health Authority (PHA) in the United Kingdom, is organized and outlines how it tackles its principle functions. Current challenges that face the Authority are examined and the impact of proposed changes to legislation and the organization of port health functions are also assessed.

The LPHA is part of the Corporation of London, the Local Authority for the 'Square Mile' or the financial district of London. Although a legally constituted PHA, it forms part of the Corporation's Department of Environmental Services, the structure of which is shown in Fig. 5A.1. The Service is the responsibility of the Port Health and Environmental Services Committee.

History

The LPHA was originally constituted under Section 20 of the Public Health Act 1872 as the Port of London Sanitary Authority, but the term 'Port Sanitary' was changed to 'Port Health' in 1936. The current London Port Health Authority Order, which assigns the jurisdiction, powers, duties and functions of the Authority, was made in 1965, but it is now being updated to meet modern-day needs and to define the area of jurisdiction more accurately.

History of the Corporation of London

The Corporation was established in 1189 and appointed as the Conservator for the Thames in 1193, although the Port of London Authority was established as the administrative body for the Port of London in 1909. The cost of administration of the LPHA was originally met from the Corporation's private funds but eventually became rate (and grant) aided.

In 2002 and 2003 the Corporation was as subject to an audit under the government CPA scheme that rated it as excellent. The Port Health Service was cited as being very high performing. It has recently changed its franchise to include the business vote as part of its modernizing agenda. However, its links with the maritime community remain strong, via Lloyd's, shipping brokers and insurance companies.

The port of London

London has been a commercial port since Roman times, but the upper river London docks were built in the eighteenth and nineteenth centuries. These are no longer used for merchant shipping, although training vessels and visiting navy ships still berth there. A floating hotel is also located in one of the docks and St Katherine's dock is now a marina. There were many riverside wharves, but most of these have been converted for residential use. However, the government is committed to retaining those that remain in commercial use for this purpose.

LPHA area of jurisdiction

The LPHA is responsible for port health functions throughout the whole of the tidal Thames, from

Teddington in the west, 94 miles down river to the 300 square miles of the estuary in the east. It encompasses the lower Medway in Kent and the rivers Crouch and Roach in Essex. Consequently, it covers not only a vast geographical area, but also virtually all the traditional port health functions. It is the food authority for the area and in common with other local authorities has a Food Law Enforcement Service Plan that details its functions and priorities.

Office locations

Although the Authority's headquarters are located in the City of London, in order to discharge its functions effectively the Authority has offices at five key locations: Charlton, Tilbury, Denton, Thamesport and Sheerness. The offices at Tilbury at Thamesport are open from 0800 to 2100 during weekdays and from 0800 to 1400 at weekends, whilst the other sites are staffed during normal working hours with shifts worked as required. Offices are located at:

- *Charlton.* This office covers the upper reaches of the Thames and central London as well as London City Airport. The staff at this office undertake all the food hygiene and food standards inspections in registered premises as well as co-ordinating and leading other staff on the inspection of cruise vessels.
- *Tilbury.* The port of Tilbury is the only remaining commercially active enclosed dock in the port of London. Vessels need to lock into the dock basin, but the larger container ships moor on the recently extended outside quay on the Thames. It is the largest Border Inspection Post in the United Kingdom for meat and meat products with some 13 000 consignments being imported per annum.
- *Denton.* A former quarantine ward as part of an isolation hospital, this now serves as the base for the down-stream part of the Thames and the estuary. It has its own jetty and pontoon, a workshop and a training room.
- *Thamesport (Isle of Grain, Kent).* Built on the former oil refinery, Thamesport is part of the same organization that owns the port

of Felixstowe, the largest container port in the United Kingdom. It is primarily a purpose-built modern container terminal, although bulk products were handled at the port until recently.
- *Sheerness.* The Port of Sheerness, which until the early 1960s was a Royal Navy base, is now the premier fruit importing port of the United Kingdom, with produce arriving from South Africa, Israel, South America and New Zealand.
- *Launch service.* An integral part of the River Division is the Launch Service, which is needed to undertake the full range of duties performed by the Authority. Currently the 'fleet' consists of two patrol launches – M.L. Lady Aileen and M.L. Londinium III, two Rigid Inflatable Boats (RIBs) – 'Hygiea' and an Avon RIB, as well as a Sea Otter Work Boat that is also used for access to the larger launches.

Structure and staffing

The structure of the Service is show in Fig. 5A.1. The management team consists of the Port Health Service Director, Assistant Port Health Services Director, the Divisional Port Health Inspectors and the Senior Navigator. The number and type of staff have been determined by an activity analysis that was also used to calculate the proportion of staff time spent on administering the veterinary checks regime. All costs associated with this function can be charged to importers so it is important that these are accurately assessed. Staff are encouraged to progress through the different levels and the grading system is intended to offer a career structure to all employees.

The designations of operational staff are as follow.

Port Health Inspectors

All the Port Health Inspectors (PHIs) are qualified Environmental Health Officers (EHOs). Newly qualified staff are expected to have undertaken 200 hours of meat inspection as part of their training as this forms a large part of their work at the Tilbury Border Inspection Post.

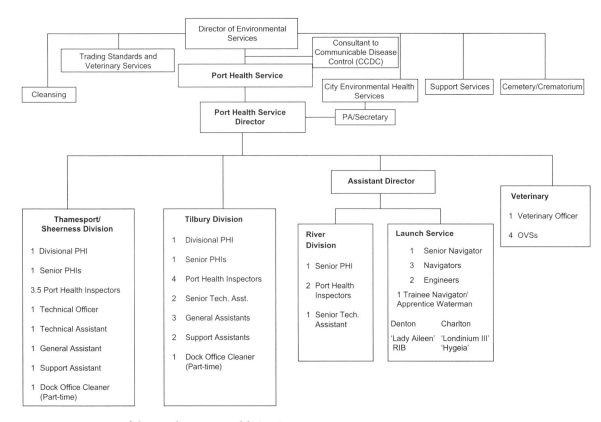

Fig. 5A.1 Structure of the London Port Health Service.

Most of the EHOs currently in post have considerable port health experience, but new recruits are expected to have a background in food safety or environmental protection. All the PHIs are designated as OFIs under the POAO Regulations and as such are responsible for inspections of fish and fishery products and certifying compliance with veterinary legislation.

Post-entry training is encouraged, particularly in food safety and an in-house Continual Professional Development (CPD) programme assists officers in achieving 10 hours of food related CPD per annum.

Official Veterinary Surgeons

European legislation stipulates that OVSs must oversee and certify the importation of products of animal origin (with the exception of fish and fishery products). It is a requirement that OVSs are present during the opening hours of the BIPs, and although they are assisted by other staff, the final decision regarding whether a consignment of POAO can be imported rests with the OVS.

Technical Assistants

This level of staff was originally employed to undertake inspections for deratting/deratting inspection certificates. They still undertake initial boarding of vessels, water sampling and rodent control, but are now more actively involved in assisting with some of the checks on imported food. All Technical Assistants (TAs) have a basic rodent control qualification, and are also encouraged to undertake relevant food safety training.

General Assistants

General Assistants (GAs) are the link between the administrative staff and the PHIs/OVSs, and as such undertake a range of documentary and practical tasks such as initial documentary checks on POAO, verification of the seal numbers on shipping containers with relevant veterinary certificates (Identity Checks) and assist with the sampling of foodstuffs that are subject to rigorous sampling plans, for example, for aflatoxin in nuts.

Support Assistants

The Support Assistants are employed to undertake a variety of administrative tasks, including data entry and recording, telephone duties and filing.

Launch crew

The launch crews consist of licensed watermen from the Company of Watermen and Lightermen and marine engineers. These staff are also trained to undertake inspections of vessels for deratting/deratting exemption certificates, water sampling and shellfish sampling.

Head office support staff

The Service is supported by the Departmental Support Services teams. These comprise a central administration unit, finance, personnel and data resources.

Functions of the LPHA

Imported food control

As indicated earlier, there are three main ports through which foodstuffs are imported. The basic methods of control are:

- Perusal of manifests – usually electronic but some still in paper form – to identify all food consignments and to initiate relevant controls.
- Performance of further documentary checks into the type of consignment and origin, confirm that appropriate certification is available and to determine whether examination is required.
- Examination of selected consignments in accordance with food sampling policy and protocols.

The procedures are outlined in Fig. 5A.2.

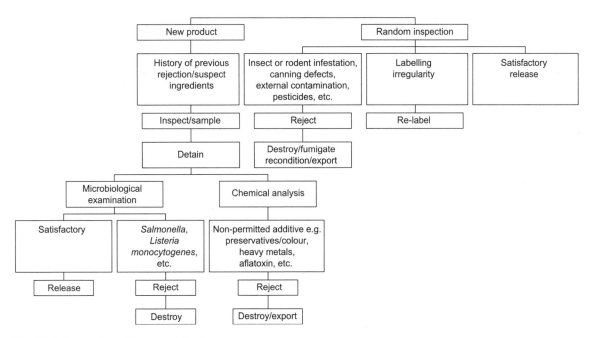

Fig. 5A.2 Inspection of imported foods.

Food sampling policy

The sampling policy prioritized the following foodstuffs:

1. Samples for analysis ('chemical' samples to be sent to the Public Analyst)

   - Spices (dried/ground) for mycotoxin contamination
   - Fresh fruit for pesticide residues
   - Confectionery for non-permitted/excess preservatives and excess colourings
   - Preserved vegetables for non-permitted/excess preservatives
   - Alcoholic Spirits for methanol content
   - Canned foodstuffs for heavy metal contamination
   - Soy sauce and soy sauce-related products for excess levels of 3-mcpd (3-monochlor-propane-1,2-diol)

2. Samples for examination ('bacteriological' samples to be sent to the Health Protection Agency laboratories)

   - Prepared/ready to eat salads
   - Spices (dried/ground) (for *Salmonella*)
   - Wet bean curd (for *Bacillus cereus*)
   - Bottled drinking water
   - Dried fruit

Although sampling activities are concentrated on the products listed here, the list is not exclusive; routine monitoring (and sampling), particularly of new food products is also undertaken.

The monitoring and control of imported foods and POAO is now the principal function of the Authority. These functions have received external accreditation under ISO 9000: 2001, and a programme is in place to achieve the standard for Food Hygiene and Shellfish controls.

Veterinary checks regime

The Products of Animal Origin (Third Country Imports) (England) Regulations 2003 are enforced at the EC approved BIPs of Tilbury and Thamesport. These Regulations implement Directive EC 97/78 and much of the work is based directly on EC regulations and decisions. The EC Food and Veterinary Office undertakes rigorous audits of BIPs and failure to address non-compliances can result in the BIP being delisted.

The BIP inspection facilities need to comply with EC Decision 2001/812/EC; BIPs are due to be upgraded to meet these standards and EC audit requirements.

The procedure for veterinary checks is as follows (see Figs 5A.3 and 5A.4):

- Perusal of manifests to identify all POAO consignments and to initiate relevant controls.
- The importer or agent for the importer submits relevant documents and the initial documentary check is undertaken by general assistants.
- Depending on the product and country of origin an identity check is carried out.
- Physical checks – a detailed examination upon the statutory percentage of consignments.
- Decision to release, detain or reject consignment depending on the outcome of checks.

Veterinary residue checks

Veterinary residues monitoring is undertaken on behalf of the Veterinary Medicines Directorate, an Executive Agency of the Department for Environment, Food and Rural Affairs protecting public health, animal health, the environment and promoting animal welfare by assuring the safety, quality and efficacy of veterinary medicines in the United Kingdom.

Infectious disease controls

The Authority provides a 24-hour stand-by service to ensure that it can meet the reporting requirements of the Public Health (Ships) Regulations and the Public Health (Aircraft) Regulations. In order to answer disease notifications at the seaports and airport as well as for major emergencies phones are provided at all sites and a member of the management team may be contacted around the clock via a single contact telephone number, which is redirected to the duty officer.

The creation of the Health Protection Agency (HPA, see Chapter 12) has changed the way in

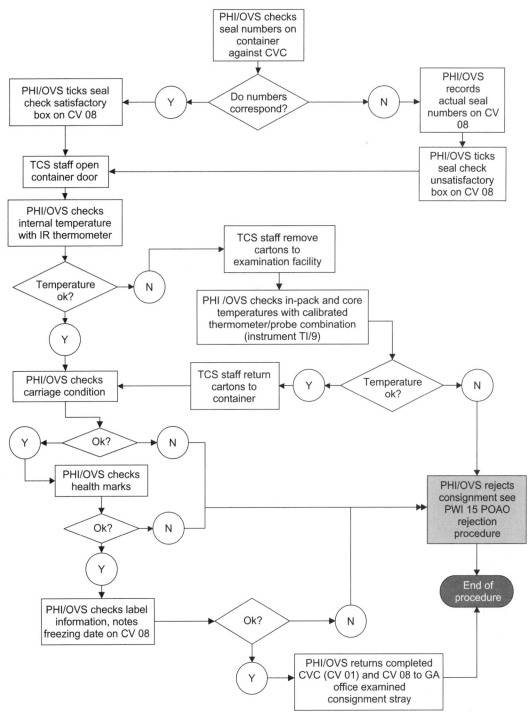

TCS, Tilbury Container Services – the berth operator
CV 08, Inspection form
CV 01, CVC – Certificate for Veterinary Checks

Fig. 5A.3 Identity check procedure.

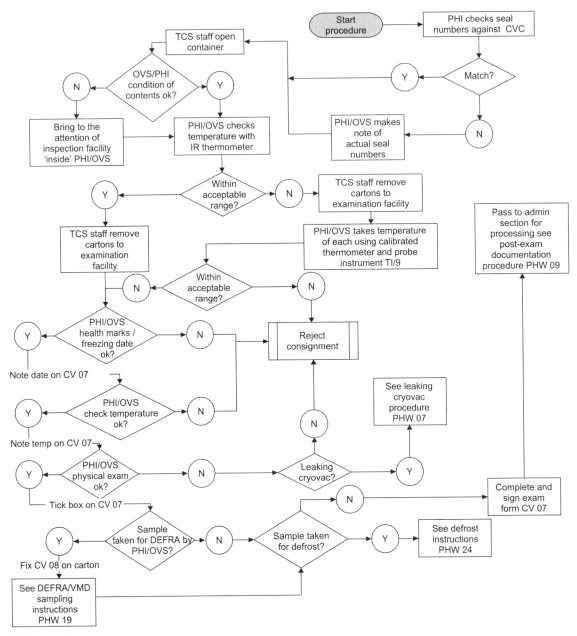

CV 07, Inspection document

Fig. 5A.4 Physical examination of POAO consignments.

which medical support is provided, as far as the LPHA is concerned. The North East London Health Authority Sector of the HPA provides the proper officer/medical officer for the port. However, due to the extended area of the LPHA,

a local GP practice is sub-contracted to undertake the boarding medical officer role for the lower river.

There are protocols in place for infectious diseases. In addition, exercises are held to verify the validity of the arrangements and a training programme for both medical officers and port health staff at the seaports and airport is in place.

The water supplies to, and on board, vessels are subject to a sampling programme in accordance with the HPA 'Guidelines for Water Quality on Board Merchant Ships Including Passenger Vessels'. There is a risk of Legionnaires Disease aboard vessels, particularly cruise ships, and inspection of vessels takes this into account.

Environmental protection

The Authority has 13 Prescribed Processes under the Environmental Protection Act 1990 for which authorizations have been issued. These include coal discharge and storage, cement discharge, paint spraying and timber treatment.

Uniquely for a PHA, LPHA has powers to take action under the Environmental Protection Act where there is evidence of a statutory nuisance caused by noise. Noise complaints are often received concerning 'disco boats', that is from those vessels that ply the river playing amplified music, which frequently operate in the small hours and disturb riverside residents during the summer months. Consequently the Authority's vessels are deployed to undertake noise patrols during the night and notices are served as appropriate.

Shellfish controls

The Thames Estuary has a large shellfish industry, notably cockles but also mussels and oysters. There are eight classified areas, all of which require ten samples to be taken per year to enable the classification group to be determined and maintained. In addition, samples of shellfish flesh and seawater are obtained for shellfish biotoxin checks (72 water samples and 12 flesh samples are taken annually). The Authority also has a contract with a neighbouring riparian borough to obtain its shellfish samples.

The Thames Estuary Shellfish Liaison group comprises representatives from all relevant local authorities, laboratories and the shellfish industry. This forum is used to discuss relevant enforcement issues and practical matters.

There have been considerable problems in recent years with an atypical form of Diarrhetic Shellfish Poisoning (DSP) being isolated in cockles and mussels. This has resulted in many of the shellfish beds being closed by Temporary Prohibition Orders issued under the Food Safety (Fishery Products and Live Shellfish) (Hygiene) Regulations 1998. Harbour Authorities, riparian local authorities, the Kent and Essex sea fisheries committee, the county police forces and the Ministry of Defence co-operate with the LPHA to ensure that the these orders are not transgressed.

Food premises

The Authority has 136 registered food premises, many of which are afloat, some moored but also pleasure craft that serve food. Some of these are A/B classification. All floating premises have particular problems associated with water supply, ventilation, refuse storage and disposal and sewage disposal.

Cruise ships

Tilbury was once one of the busiest cruise terminals in the country with vessels carrying both emigrants and immigrants calling at the port. It is still a port of call for some cruise ships today, but many more call in at the Pool of London near Tower Bridge or alongside H.M.S. 'Belfast'.

It is LPHA policy to board all cruise vessels upon arrival and if there is no documentation to demonstrate that a full inspection has been carried out recently an in-depth survey is undertaken. An example of the current passenger/cruise ship sanitation and hygiene certificate is reproduced in Fig. 5A.5. In addition the vessel and its owners are supplied with a detailed written report that covers all aspects of the inspection. As with merchant ships, where there are conditions that would warrant closure of a food premises, the Maritime and Coastguard Agency (MCA) is contacted. The MCA has powers to detain vessels until remedial work has been carried out to the satisfaction of its survey, which relies on the expertise of the PHI with regard to food hygiene.

PASSENGER/CRUISE SHIP SANITATION AND HYGIENE CERTIFICATE

To: .... Being the Master, Officer or person in charge of: -

Vessel: .... Port of Registry: ....

Official No: .... Tonnes gross: ....

Address of Owner or Agent: ....

....

TAKE NOTICE that this certificate records the inspection of the above-named vessel on .... by an officer of this Authority in accordance with Article 14 of the International Health Regulations, Section 79 of the Environmental Protection Act 1990, provisions contained in the Public Health (Ships) Regulations 1979 and other relevant legislation. The officer has also considered recommendations contained in the WHO Guide to Ship Sanitation.

RECORD OF INSPECTION

S = Satisfactory	**X = Unsatisfactory**	**O = Not Examined**

(1) POTABLE WATER

a) Filling lines
b) Storage tanks
c) Sampling results

(2) WASTE DISPOSAL

a) Water closets/Urinals
b) Wash hand basins
c) Waste disposal system
d) Vents and traps
e) Sewage treatment system
f) Refuse disposal

(3) ACCOMMODATION

a) Overcrowding
b) Cleanliness
c) Air conditioning
d) Heating and ventilation
e) Lighting
f) Vermin

(4) MISCELLANEOUS

a) Noise/Smoke/Dust
b) Animals
c) De-rat. Certificates
d) Vermin/Insect proofing

(5) SWIMMING POOLS/JACUZZIS

a) Water supply
b) Cleanliness/Chlorination
c) Circulation/Filtration
d) Sampling

(6) HEALTH, FITNESS AND BEAUTY

a) Sauna/Solarium
b) Gymnasium
c) Hairdressing salon

(7) FOOD SANITATION

a) HACCP/Safe food system
b) Quality and protection of food
c) Food handlers
d) Storage of food
e) Preparation and service
f) Water supply and drainage
g) Lighting
h) Ventilation
i) Vermin and other pests
j) Storage equipment and other utensils
k) Washing equipment
l) Toilet and hand washing facilities
m) Food sampling results
n) Equipment swab results

Details of defects found, if any, appear overleaf

Fig. 5A.5 Passenger/cruise ship sanitation and hygiene certificate.

Rabies controls

The Corporation undertakes the animal health function for most London boroughs. As far as LPHA is concerned, it undertakes rabies controls at the docks, wharves and marinas within Greater London, as well as London City Airport, and on an agency basis for Kent and Essex Authorities. This involves the monitoring of ships and aircraft for animals and birds whose presence should be declared under the Public Health (Ships/Aircraft) Regulations.

Houseboats

All along the River Thames there are moored vessels that are used for residential purposes. The Authority is responsible for enforcing the Houseboat Bylaws. It is intended that these will be updated to give greater powers to secure improvements.

Additional functions

Student training

The LPHA is the principal PHA in the United Kingdom for training of student EHOs. It offers a $3^1/_2$-day theoretical and practical course that covers all the main port health functions, and is currently free of charge to participants.

Rodent control contract

At the Port of Tilbury the LPHA has a contract with the port to monitor and control rodents within much of the port. The work is carried out by a TA under the supervision of a PHI.

Projects work

The Corporation is prepared to fund research projects by its departments and the Port Health Service is currently committed to two projects:

1. An epidemiological study into the potential effects of river water on recreational users. The Authority collaborated with the Environment Agency to undertake a survey into the microbiological quality of the River Thames. The new study is to determine whether the levels of contamination can lead to increased risks of illness amongst rowers on the river.
2. A project to isolate and identify human lung fluke metacercaria in Chinese Mitten Crabs collected from the Thames. This crab has colonized the Thames extensively, and as it is the intermediate host of the human lung fluke and the crab may be eaten raw, the project will determine whether there is a public health risk to consumers.

Powerboat handling courses

The Launch Service offers a Royal Yachting Association (RYA) approved Training School for courses on small powerboats handling and safety in boats. The navigators are qualified trainers and an annual programme of courses is available to all river users on a commercial basis.

Trends and pending changes

Government proposals regarding import controls

An outcome of the Foot and Mouth Disease epidemic was a study into import controls carried out by the Cabinet Office Machinery for Government Secretariat. Its recommendations were carried into the government's response to the Foot and Mouth Disease inquiries. The most significant requirement as far as PHAs are concerned is to bring about a 'step change' to improve the co-ordination and delivery of imported food (including POAO) controls within one year. Government will then take a hard look at the case for bringing these functions into a central agency, or delivering them from other routes.

This project to secure the step change is being led by the Food Standards Agency, which has its own Imported Food Division. The success is fundamental to the future of PHAs and within the LPHA an action plan has been approved by its committee to address all areas of concern highlighted by the Cabinet Office.

The Food Safety (Ships and Aircraft) (England and Scotland) Order 2003

The main effect of this Order is to extend the definition of 'premises' in the Food Safety Act 1990, to certain ships and aircraft in relation to enforcement of food hygiene and specified temperature control requirements. Exemptions are sovereign immune ships or aircraft or any ship of a state other than the United Kingdom. Thus the Order gives authorized officers the power of entry to ships and aircraft, to carry out food hygiene inspections, as required by the Food Safety (General Food Hygiene) Regulations 1995, (SI 1995 No. 1763), as amended, and certain duties in terms of the Food Safety (Temperature Control) Regulations 1995, (SI 1995 No. 2200), as amended.

The guidance issued under a Statutory Code of Practice will determine the impact on the LPHA in terms of resources required to meet its statutory obligations. The inspection frequency for different types of vessels and aircraft has yet to be determined, but it is inevitable that there will be an increase in the number of visits required.

European Food and Feed Regulations

The draft EC Food and Feed Regulations have specific requirements for import controls, but the final form that these will take is not yet clear. However, it is likely that for high-risk foods, there could be a system that is similar in principle to the veterinary checks regime. This will enable PHAs to recover their costs and to ensure that all high-risk foods are properly monitored in a consistent manner.

Shellfish controls

Also under consideration by the EC and the Food Standards Agency is the method of classification of shellfish harvesting areas – the number and frequency of sampling could be changed. If weekly samples were required instead of monthly this would clearly have resource implications.

Electronic service delivery

The LPHA seeks to meet government requirements regarding electronic service delivery. It is reviewing how it can offer more Internet-based services, as well as providing full details regarding legislative requirements and import procedures via its website.

Collaboration with other agencies

In the interests of 'joined-up-government', and improved service delivery, the LPHA collaborates with many other external bodies. Examples of these include the Environment Agency, the Port of London Authority, the Maritime and Coastguard Agency, riparian local authorities, the different police forces on the river, government departments and the Association of Port Health Authorities.

The stated aim of the Authority is to achieve continuous improvement, and it is constantly reviewing its policies, service delivery mechanisms and protocols to achieve this aim.

Part Two
Environmental Health Law and Administration

6 Environmental health law

Frank B. Wright

SOURCES

This chapter is concerned with the law of the United Kingdom. The United Kingdom comprises Great Britain and Northern Ireland, with Great Britain being made up of England, Wales and Scotland. The Channel Islands and the Isle of Man are Crown dependencies and are not part of the United Kingdom. The UK constitution is not contained in any single document but has evolved over the course of time. It was formed originally by customary law and later by the common law, then by statute and convention. The United Kingdom is a constitutional monarchy. It is governed by the Sovereign, who is both the Head of State and Head of Government. The organs of government are the legislature (Parliament), the executive and the judiciary. Although the powers of the monarchy are now very limited, being restricted primarily to advisory and ceremonial functions, some important duties that are reserved for the Sovereign remain. These include the summoning, the proroguing and the dissolving of Parliament, the appointment of ministers, including the Prime Minister, and the appointment of judges and certain senior officials.

International treaties, entered into on behalf of the United Kingdom, are concluded by ministers acting under the royal prerogative. Ministers are responsible to Parliament. International treaties must be incorporated into UK law by an Act of Parliament.

The international and European dimension

There is a presumption of interpretation that in incorporating an international treaty the United Kingdom intends to fulfil its international obligations. However, where the words of an Act of Parliament are clear and unambiguous, they must be given effect even if the decision results in a breach of international law. The primary sources of European Union (EU) law consist of the three founding treaties (European Coal and Steel Community, European Atomic Energy Community and European Economic Community) with their annexes and protocols, which supplement the treaties; the Convention on Certain Institutions Common to the European Communities (1957); the Merger Treaty (1965); certain treaties on budgetary matters; treaties on accession and their annexes; the Act of Council concerning direct elections for the European Parliament (1976); the Single European Act (1986); the Treaty on European Union (1991); the Treaty of Amsterdam (1997) and the Treaty of Nice (2001). The Treaties of the European Communities and European Community law were incorporated into UK law by the European Communities Act 1972 and the European Communities (Amendment) Act 1993. The courts of the United Kingdom have not demonstrated any reticence in applying EU law and construe UK legislation so far as possible in conformity with the purpose of EU legislation. Where it appears to the courts that there is inadvertently conflicting UK legislation, they will endeavour to give effect to EU law.

Common law

The earliest laws of which there is documentary evidence date from the Anglo-Saxon period of history before the Norman Conquest. These laws related to particular areas such as Kent, Wessex and Mercia. As a centralized system of law developed, customary law gave way to national law. The national law came to be known as common law. Common law was developed by the King's judges and is derived entirely from case law. In addition to settling principles of law, which were to be observed nationally, the courts began to establish formal rules of procedure for those who wished to bring cases before them. Court rules established that actions had to be commenced by royal writ and set out in an accepted form. Difficulties emerged in relation to the use of writs and forms of action. Litigants who were unable to get satisfaction petitioned the monarch, who in turn handed the petitions to the Lord Chancellor. The Lord Chancellor set up the Court of Chancery, which was guided by principles of fairness or equity in coming to a decision. The law of equity became established in the fifteenth century. Both common law and equity came to operate as parallel systems, each bound by its own judicial precedents. The Supreme Court of Judicature Acts 1873–75 reorganized the existing court structures and brought together the common law courts and the Courts of Chancery. This legislation has now been consolidated. Where common law and equity conflict, the law of equity prevails. All courts administer the principles of common law and equity, and grant the remedies of both as the case demands.

Statute law including the parliamentary process

An Act of Parliament that has been given royal assent and placed on the parliamentary roll is the law and must be given effect by the courts (see *British Railways Board* v. *Pickin* [1974] A.C. 765). Parliament is made up of three elements: the

Sovereign, the House of Lords and the House of Commons. In order for a legislative measure to become an Act of Parliament and be recognized as such by the courts it has to undergo one of several procedures. The measure is drafted, usually by parliamentary counsel to the Treasury, and presented to the House of Commons or House of Lords, more usually the House of Commons, as a Bill. Before a Public Bill becomes an Act of Parliament it must undergo five stages in each house: first reading; second reading; committee stage; report stage; and third reading. The first reading is a formality. There is no debate at this point. Nowadays this stage constitutes an order to print the Bill. At the second reading there is a debate on the principles of the Bill, while the committee stage sees the Bill examined in detail, clause by clause. In most cases the committee stage takes place in a standing committee, or the whole house may act as a committee. At the report stage there is a detailed review of the Bill as amended in the committee stage. On the third reading the Bill is finally debated. These stages can be taken quickly, although the process normally takes a number of months. Much depends on the Bill's political importance. A Bill may be considered first in either house but it has to pass through both houses to become law. Both houses must agree to the same text of a Bill, so that the amendments made by the second house are then considered in the originating house, and if not agreed, sent back or themselves amended, until agreement is reached. Under the Parliament Acts 1911 and 1949, if the House of Lords rejects any Public Bill (except one to prolong the life of a Parliament) that has been passed by the Commons in two successive sessions, then the Bill will become law without the consent of the Lords, unless the Commons directs to the contrary. The royal assent is signified by letters patent to such Bills and measures as have passed both Houses of Parliament (or Bills that have been passed under the Parliament Acts 1911 and 1949). The Sovereign has not given royal assent in person since 1854. The power to withhold assent resides with the Sovereign but has not been exercised in the United Kingdom since 1707, in the reign of Queen Anne. Public Bills promoted by a Member of Parliament who is not a member of the Government are known as Private Members' Bills.

Delegated legislation

This is law made by subordinate authorities acting under law-making powers delegated by Parliament or the Sovereign. Many statutes empower ministers to make delegated legislation. Such legislation includes the following.

Orders in Council

The Sovereign in Council, or Privy Council, was the chief source of executive power until cabinet government developed in the eighteenth century. Membership of the Privy Council is automatic upon appointment to certain government and judicial positions in the United Kingdom. Membership is also accorded by the Queen to eminent people in the United Kingdom and independent countries of the Commonwealth of which Her Majesty is Queen, on the recommendation of the British Prime Minister. The Privy Council Office is headed by the Lord President of the Council, who is a cabinet minister. Orders in Council are approved by the Queen in the presence of three Privy Councillors (enough to constitute a quorum) after which it is announced that the Queen held a Privy Council. The matters considered, however, will have previously been recommended by the responsible departments of government. Orders in Council may be made by ministers acting under the royal prerogative. See, for example, Section 84 of the Health and Safety at Work, Etc. Act 1974, which provides that Her Majesty may by Order in Council extend the provisions of the Act outside Great Britain.

Statutory Instruments

Many thousands of Statutory Instruments are issued annually by Ministers of the Crown acting under delegated powers provided by Acts of Parliament. Statutory Instruments are of considerable importance as a source of environmental health law because of the detailed technical nature of the law enforcement role. Each one

published is allocated a number for the year, see, for example, the Reporting of Injuries, Diseases and Dangerous Occurrences Regulations 1995 (SI 1995 No. 3163). Statutory Instruments fall into four broad categories: **Affirmative Instruments,** which are subject to the approval of both Houses of Parliament before they can come into or remain in force; **Negative Instruments,** which are subject to annulment by resolution of either House; **General Instruments,** which include those not required to be laid before Parliament and those that are required to be so laid but are not subject to approval or annulment; and **Special Procedure Orders,** against which parties outside may lodge petitions.

By-laws

Local authorities (i.e. district and London borough councils) have long had power, delegated by Parliament and subject to confirmation by ministers, to make by-laws, see, for example, by-laws made for the good rule and government of local authority areas and by-laws for the prevention and suppression of nuisances made under the Local Government Act 1972, Section 235. By-laws must be consistent with the common law and statute and must not be made if provision for that purpose has already been made, or is or may be made by any other enactment.

EU secondary legislation

The EU has policies on environmental conservation and protection, health, agriculture, fisheries and food, social policy including employment protection and health and safety at work, consumer protection, transport, energy and industry, amongst others. Many of these policies affect the work of the environmental health officer. The law-making powers of the EU institutions are to be found in Article 249 of the EC Treaty. It is there provided that: 'In order to carry out their task the Council and the Commission shall, in accordance with the provisions of this Treaty, make regulations, issue directives, take decisions, make recommendations or deliver opinions'. These measures, described as 'acts' are defined as follows.

- A **regulation** shall have general application. It shall be binding in its entirety and directly applicable in all member states.
- A **directive** shall be binding in its entirety, as to the result to be achieved, upon each member state to which it is addressed, but shall leave to the national authorities the choice of form and methods.
- A **decision** shall be binding in its entirety upon those to whom it is addressed.
- **Recommendations and opinions** shall have no binding force.

Regulations must be recognized as legal instruments and do not need national implementation – indeed it is impermissible to do so (*Commission* v. *Haly* Case 39/72 [1971] E.C.R. 1039). Most of the EU legislation affecting the work of the environmental health officer takes the form of directives. Examples of such directives are listed below.

1. 89/391/EEC Council Directive on the introduction of measures to encourage improvements in the safety and health of workers at work, 12 June 1989.
2. 89/397/EEC Council Directive on the official control of foodstuffs, 14 June 1989.
3. 89/437/EEC Council Directive on hygiene and health problems affecting the production and the placing on the market of egg products, 20 June 1989.
4. 89/428/EEC Council Directive on procedures for harmonizing the programmes for the reduction and eventual elimination of pollution caused by waste from the titanium dioxide industry, 21 June 1989.

The Court of Justice has held that in certain circumstances at least, directives and decisions might contain directly effective provisions (*Grad* v. *Finanzamt Traunstein* Case 9/70 [1970] E.C.R. 825). For this to occur two conditions must be fulfilled. First, there must be a clear and precise obligation, and second it must not require the intervention of any act on the part of institutions of the EU or the member states. Individuals

may invoke before their national courts, as against the state, such provisions despite the fact that they are contained in directives or decisions rather than regulations. Governments of member states to which directives have been addressed are obliged to implement them within the period of time specified. Failure to do so could lead to an action in the European Court of Justice by another member state or, more likely, by the Commission. (See Articles 226 and 227 EC Treaty.)

Actions are also possible in the domestic courts of the member state. In two cases (joined), *Francovich* v. *Italian State* and *Bonifaci* v. *Italian State* [1992] I.R.L.R. 84, it was held that where a member state fails to enact the legislation required in order to achieve the objective prescribed by a directive it may be possible for an aggrieved individual to take remedial action directly against his member state government. One possible route might lie in the context of the principles established by the European Court of Justice in *Francovich*.

Francovich implied that where damage is suffered through the acts or omissions of a private party which are incompatible with directly applicable provisions of a directive which has not been implemented by appropriate legislation, an action in damages will lie against the state. However, it would appear that three conditions must be fulfilled before such liability can be created. In the first place, the objective sought by the directive must include the creation of rights for individuals. The second is that the content of those rights must be ascertainable from the provisions of the directive itself. The third condition is the existence of a causal link between violation by the state of its duty to implement the directive and the loss sustained by the individual. Where these three conditions are met, EU law directly confers on individuals the right to obtain compensation against the state.

Recommendations and opinions have no binding force but may be indicative of important policy orientations, and member states may take such measures into account when enacting national legislation or when making administrative directions.

THE COURTS AND TRIBUNALS

European Court of Justice

The European Court of Justice is composed of 15 judges and eight advocates-general appointed for renewable six-year terms by the governments of the member states. The Court exists to safeguard the law in the interpretation and application of the Community Treaties, to decide on the legality of decisions made by the Council of Ministers or the Commission, and to determine violations of the treaties. Cases may be brought before it by member states, EU institutions, firms or individuals. Its decisions are directly binding in all member states. A Court of First Instance consisting of 15 judges appointed by common accord by the governments of the member states hears staff cases, cases involving EU competition law and certain applications under the European Coal and Steel Community Treaty. There is a right of appeal from this court, on matters of law, to the European Court of Justice.

The jurisdictions of the United Kingdom

There is one legislature for the United Kingdom but three separate legal jurisdictions: England and Wales; Scotland; and Northern Ireland. These jurisdictions have separate law, judicial procedure and court structure, although there is a common distinction between civil and criminal law.

The judicature of England and Wales

The supreme judicial authority for England and Wales is the **House of Lords**. It is staffed by the Lord Chancellor and 12 Lords of Appeal in Ordinary (Law Lords) who are members of the Upper House of the legislature. Cases are normally heard by a panel of five Law Lords. Each Law Lord expresses his own opinion in the form of a speech. In recent years a number of environmental health cases have been heard in this court. (See, for example, *Alphacell* v. *Woodward* [1972] A.C. 824, *Salford City Council* v. *McNally* [1976] A.C. 379, and *Austin Rover Group Limited* v. *Her Majesty's Inspector of Factories* [1990] A.C. 619.)

The **Supreme Court of Judicature** comprises the **Court of Appeal**, the **High Court of Justice** and the **Crown Court**. The Court of Appeal has two divisions: a Criminal Division and a Civil Division. The jurisdiction of the Court of Appeal includes civil and criminal appeals from the three divisions of the High Court, including divisional courts, from the county courts, from the Employment Appeal Tribunal, from the Lands Tribunal and the Transport Tribunal. For civil cases, the Master of the Rolls is the most senior judge. The President of the Family Division and the Vice Chancellor sit occasionally. The Criminal Division is presided over by the Lord Chief Justice. There are 37 Lords Justices to carry out the work of the two divisions. They are assisted on occasion by High Court judges. Three Appeal Court judges will normally hear a case and a majority is sufficient for a decision.

The High Court of Justice is the superior civil court. Its work is carried on by 108 High Court judges in three divisions: Queen's Bench Division; the Chancery Division; and the Family Division. The Queen's Bench Division deals with commercial and maritime law and with civil cases not assigned to other courts. Within the Queen's Bench Division is the Divisional Court, which reviews decisions of governmental and other public bodies and hears appeals from lower courts (see, for example, *R.* v. *Health and Safety Commission ex parte Spelthorne Borough Council* (1983), *The Times*, 18 July 1983). The Chancery Division is concerned mainly with equity, bankruptcy and contentious probate business, and the Family Division deals with matters relating to family law. Sittings are held at the Royal Courts of Justice in London or at Crown Court centres throughout England and Wales. High Court judges sit alone to hear cases at first instance. Appeals from lower courts are heard by two or three judges, or by a single judge of the appropriate division. Under the provisions of Section 81 of the Environmental Protection Act 1990, for example, a local authority may take action in the High Court for the purpose of securing the abatement, prohibition or restriction of any statutory nuisance where they are of the opinion that proceedings for an offence of contravening an abatement notice would not provide a sufficient remedy. (See, for example, *Hammersmith London Borough Council v. Magnum Automated Forecourts Ltd.* [1978] 1 W.L.R. 50.)

Most minor civil cases, including most cases under the Housing Acts, are dealt with by the 218 **county courts** throughout England and Wales. These courts are staffed by county court judges (who also sit as circuit judges in criminal cases), and district judges for smaller claims. Magistrates' courts can hear certain classes of civil cases, including family matters and debt collection, while committees of magistrates currently license public houses and restaurants (subject to the changes in the Licensing Act 2003), clubs and betting shops.

The Crown Court, brought into being by the Courts Act 1971, has an exclusively criminal jurisdiction. It sits in some 78 centres, divided into six circuits, and is presided over by High Court judges, full-time circuit judges, and part-time recorders and assistant recorders, sitting with a jury of 12 lay persons in all trials that are contested. It deals with a wide range of serious criminal offences, including those relating to environmental pollution, health and safety at work, food hygiene, the sentencing of offenders committed for sentence by magistrates' courts and appeals from lower courts. Magistrates usually sit with a circuit judge or recorder to deal with appeals and committals for sentence.

Minor criminal offences (summary offences) are dealt with by some 30 000 justices sitting in 620 **magistrates' courts**, which usually consist of three lay magistrates, sitting without a jury. The magistrates are advised on law and procedure by a fully qualified clerk to the justices. In the busier courts, a full-time, salaried and legally qualified district judge presides alone. There are currently around 104 district judges in England and Wales.

The Scottish judicature

Scotland has a legal system that differs substantially from that of England and Wales. Scotland is divided into six **sheriffdoms**, each with a full-time Sheriff Principal. The sheriffdoms are further divided into sheriff court districts, each of which has a legally qualified, resident sheriff or sheriffs, who are the judges of the court.

The **sheriff court** has a wide civil jurisdiction. Appeals against decisions of the sheriff may be made to the Sheriff Principal and thence to the **Court of Session** or directly to the Court of Session, which sits only in Edinburgh, and from there to the House of Lords in London. There will normally be at least two Scottish judges hearing appeals from the Scottish courts in the House of Lords. In criminal cases sheriffs principal and sheriffs have similar powers; sitting with a jury of 15 members they may try more serious cases on indictment, or sitting alone they may try lesser cases under summary procedure. Judges in the Sheriff Court may not impose a sentence of more than three years' imprisonment. Cases will be committed to the **High Court of Justiciary** if the sheriff forms the view on reading the papers that on conviction a court may wish to impose a more severe sentence.

The High Court of Justiciary consists of the same judges who sit in the Court of Session. It has jurisdiction over all of Scotland in respect of crimes committed unless a statute provides otherwise. It is both a trial and an appeal court. As a court of first instance it comprises a single judge sitting with a jury of 15. As a court of appeal it sits only in Edinburgh and then comprises at least three judges. In recent years a number of environmental health cases have been heard in this court. (See, for example, *Strathclyde Regional Council v. Tudhope* (1982) SCCR 286, *Docherty v. Stakis Hotels Ltd.* (1991) SCCR 7, *Kvaerner Govan Ltd. v. Her Majesty's Advocate* (1992) SCCR 10 and *Lockhart v. Kevin Oliphant Ltd.* (1993) SLT 179.) There is no appeal to the House of Lords in criminal cases.

Minor summary offences are dealt with in **district courts**, which are administered by the district and the island's local authorities and presided over by lay justices of the peace (of whom there are about 4000) and, in Glasgow only, by stipendiary magistrates. (See also Chapter 3, Appendix B.)

The Northern Ireland judicature

In Northern Ireland the legal system and the courts' structure closely resemble those of England and Wales; there are, however, differences in enacted law. (See also Chapter 3, Appendix A.)

Employment tribunals

Members of employment tribunals in England and Wales, Scotland and Northern Ireland, amongst their other duties, hear appeals against the service of improvement and prohibition notices issued by inspectors acting under powers granted under the Health and Safety at Work, Etc. Act 1974. Employment tribunal chairmen are legally qualified and are appointed by the Lord Chancellor. Lay members serving on tribunals in Great Britain are appointed by the Secretary of State for Trade and Industry following their nomination by specified employer and employee groups. Lay members serving on tribunals in Northern Ireland are appointed by the Department of Economic Development, Northern Ireland. Appeals may be made from a tribunal on a point of law only. The appeal is to the High Court (or the Court of Session in Scotland) and must be made within 42 days of the date of the entry of the decision in the register.

THE LAW RELATING TO STATUTORY NUISANCES

Since Edwin Chadwick, one of the founding leaders of the public health movement, sent poor law medical investigators into the London slums in 1838 and issued in 1842 his *Report on an Enquiry into the Sanitary Condition of the Labouring Population of Great Britain* (see Chapter 2), which led to the nineteenth-century Public Health Acts, the suppression and abatement of nuisances has been an important local authority responsibility. It is true that remedies to deal with nuisances were available at the time of the Chadwick Report, but resort to the law in the nineteenth century was the prerogative of the rich. In addition, strong social pressures were at work in militating against such court actions and the procedures were both unwieldy and time consuming. Consequently, local authorities had, and continue today to have, an important role to play in this area of environmental protection, both because of the clear public interest in maintaining public health standards and a pollution-free environment, and because bringing a nuisance action can be prohibitively expensive.

Public and private nuisances

The law of nuisance can be divided into public and private nuisance. **Public nuisances** are crimes, although they can be tortious in some circumstances. An act or omission that materially affects the reasonable comfort and convenience of a class of Her Majesty's subjects is a public nuisance and a criminal act. It is possible to obtain an injunction to restrain a public nuisance through the Attorney General on a relator action. Under the Local Government Act 1972, Section 222, local authorities are entitled to take **injunction proceedings** in the High Court in order to prevent harm to inhabitants of their area. Aside from these two avenues, the right to take action is only available if special damage has been suffered.

Private nuisances are always tortious and the principal remedies are damages and an injunction. A private nuisance is 'the unlawful interference with a person's use and enjoyment of land, or of some right over or in connection with it'. It is thus a property right. The basis for a claim in private nuisance is founded on a balancing exercise centred around the question of reasonableness, and in assessing this balance the court will take into account the locality of the nuisance, the duration of the nuisance and any hypersensitivity on the part of the plaintiff.

The law of statutory nuisance was consolidated in the Public Health Act 1936 after previous Acts in 1848, 1855, 1860 and 1875. The law was updated in a piecemeal fashion and again consolidated in Part III of the Environmental Protection Act 1990. The Environment Act 1995 extended the statutory nuisance provisions to Scotland, repealing the statutory nuisance provisions in the Public Health (Scotland) Act 1897.

The control of statutory nuisances

It is the duty of every local authority to inspect its area to detect statutory nuisances and investigate complaints. If, having done so, it is satisfied that a statutory nuisance exists and is likely to recur, it should serve an abatement notice. A statutory nuisance is a nuisance at common law. This is a private nuisance of public health significance, that is one created by unlawful interference with a person's use or enjoyment of land. This can take the form not only of physical damage to land but also of causing discomfort to the owner or the occupier. An element of repetition is required because a one-off incident will rarely constitute a nuisance. It is also necessary to put the alleged nuisance in the context of the locality, as something that may be nuisance in a residential area may not be in a purely industrial location.

Statutory nuisances with which the legislation is concerned are those created by any premises in a state deemed prejudicial to health generally or which, owing to the emission of smoke, fumes, gas or noise or, in the case of industrial, trade or business premises, dust, steam and smells, are prejudicial to health or a nuisance.

Section 79 of the Environmental Protection Act 1990 places a duty on every local authority to cause its area to be inspected from time to time to detect whether a nuisance is likely to occur or recur. Thus, if an individual within an area makes a complaint, the local authority is obliged to investigate it; if there is a failure to do so the remedy of judicial review is available to an aggrieved applicant (see *R. v. Crown Court at Liverpool and another, ex parte Cooke* [1996] All ER 589). Section 80 provides that where a local authority is satisfied that a nuisance exists, or is likely to occur or recur, it must serve an abatement notice on the person by whose act, default or sufferance the nuisance is attributable, or if that person cannot be found or the nuisance has not yet occurred, on the owner or occupier of the premises from which the nuisance arises or continues. The following categories of statutory nuisance are listed in the Environmental Protection Act 1990, Section 79.

(a) Any premises in such a state as to be prejudicial to health or a nuisance.

Note: 'Premises' includes land and, subject to subsection (12) and Section 81A(9), any vessel. A vessel powered by steam reciprocating machinery is not a vessel to which this part of this Act applies.

(b) Smoke emitted from premises so as to be prejudicial to health or a nuisance.

Note: This provision does not apply to smoke emitted from a chimney of a private dwelling within a smoke control area, dark smoke emitted from a chimney of a building or a chimney serving the furnace of a boiler or industrial plant attached to a building or for the time being fixed to or installed on any land, smoke emitted from a railway locomotive steam engine, or dark smoke emitted otherwise than as mentioned above from industrial or trade premises. This provision also does not apply to premises occupied on behalf of the Crown for naval, military or air force purposes or for the purposes of the department of the Secretary of State having responsibility for defence, or occupied by or for the purposes of a visiting force.

(c) Fumes or gases emitted from premises, which are private dwellings, so as to be prejudicial to health or a nuisance.

(d) Any dust, steam, smell or other effluvia arising on industrial, trade or business premises and being prejudicial to health or a nuisance.

Note: This provision does not apply to steam emitted from a railway locomotive engine.

(e) Any accumulation or deposit which is prejudicial to health or a nuisance.

(f) Any animal kept in such a place or manner as to be prejudicial to health or a nuisance.

(g) Noise, including vibration, emitted from premises so as to be prejudicial to health or a nuisance.

Note: This provision does not apply to noise caused by aircraft other than model aircraft and not to premises occupied on behalf of the Crown for naval, military or air force purposes or for the purposes of the department of the Secretary of State having responsibility for defence, or occupied by or for the purposes of a visiting force.

(ga) Noise, including vibration, that is prejudicial to health or a nuisance and is emitted from or caused by a vehicle, machinery or equipment in a street.

Note: This provision does not apply to noise made by traffic, by any naval, military or air force of the Crown or by a visiting force, or by a political demonstration or a demonstration supporting or opposing a cause or campaign (see also the provision of the Noise Act 1996, p. 660).

(h) Any other matter declared by any enactment to be a statutory nuisance.

Other matters declared by other enactments to be a statutory nuisance are:

Public Health Act 1936, Section 141

Any well, tank, cistern, or water-butt used for the supply of water for domestic purposes which is so placed, constructed or kept as to render the water therein liable to contamination prejudicial to health.

Public Health Act 1936, Section 259

1. Any pond, pool, ditch, gutter or watercourse which is so foul or in such a state as to be prejudicial to health or a nuisance.
2. Any part of a watercourse, not being a part ordinarily navigated by vessels employed in the carriage of goods by water, which is so choked or silted up as to obstruct or impede the proper flow of water and thereby to cause a nuisance, or give rise to conditions prejudicial to health.

Public Health Act 1936, Section 268

A tent, van, shed or similar structure used for human habitation:

(a) which is in such a state, or so overcrowded, as to be prejudicial to the health of the inmates; or
(b) the use of which, by reason of the absence of proper sanitary accommodation or otherwise, gives rise, whether on the site or on other

land, to a nuisance or to conditions prejudicial to health.

Mines and Quarries Act 1954, Section 151

A shaft or outlet of certain abandoned and disused mines where:

1. (a) it is not provided with a properly maintained device designed and constructed as to prevent persons from accidentally falling down the shaft or accidentally entering the outlet, or
   (b) by reason of its accessibility from a highway, or a place of public resort, it constitutes a danger to members of the public.

2. A quarry which is not provided with an efficient and properly maintained barrier so designed and constructed as to prevent persons from accidentally falling into it and which by reason of its accessibility from a highway, or a place of public resort, constitutes a danger to members of the public.

Though not a statutory nuisance under the Environmental Protection Act 1990, Section 59(1), the Building Act 1984 provides that if a local authority considers that certain features of a building, for example, a cesspool, private sewer, drain, soilpipe or rainwater pipe, is in such a condition as to be prejudicial to health or a nuisance, then it must serve a notice on the owner or occupier of the building requiring such work as may be necessary to be done.

Remedies for statutory nuisances

The procedures for the remedy of statutory nuisances are set out in Part III of the Environmental Protection Act 1990. An expedited procedure is set out in Section 76 of the Building Act 1984 where the need for a remedy is perceived to be urgent. (See p. 145.)

Abatement notice procedure

Section 80(1) of the Environmental Protection Act 1990 provides that where a local authority is satisfied that a statutory nuisance exists, or is likely to occur or recur, in the area of the authority, the local authority shall serve a notice ('**an abatement notice**') imposing all or any of the following requirements:

1. the abatement of the nuisance or prohibiting or restricting its occurrence or recurrence.
2. the execution of such works, and the taking of such other steps, as may be necessary for any of those purposes.

The abatement notice must specify the time or times within which the requirements of the notice are to be complied with.

The abatement notice must be served on the person responsible for the nuisance except where:

1. the nuisance arises from any defect of a structural character, when it must be served on the owner of the premises, or
2. where the person responsible for the nuisance cannot be found or the nuisance has not yet occurred, when it must be served on the owner or occupier of the premises.

The notice must specify the time or times within which the requirements of the notice are to be complied with.

If a person on whom an abatement notice is served, without reasonable excuse, contravenes or fails to comply with any requirement or prohibition imposed by the notice, he will be guilty of an offence.

Defences

By Section 80(7) of the Environmental Protection Act there is a defence of best practicable means for operators of industrial, business or trade premises. If a person on whom a notice is served without reasonable excuse contravenes any requirements of the notice he will be guilty of an offence under Part III of the Environmental Protection Act 1990.

Appeals procedure

A person served with an abatement notice has 21 days from the service of the notice in which to appeal against the notice to a magistrates' court. The grounds of appeal are set out in regulations made under Schedule 3, Environmental Protection Act, the Statutory Nuisance (Appeals) Regulations 1990 (SI 1990 No. 2276) and the Statutory Nuisance (Appeals) (Amendment) Regulations 1990 (SI 1990 No. 2483). These include:

(a) that the abatement notice is not justified in terms of Section 80;
(b) that there has been a substantive or procedural error in the service of the notice;
(c) that the authority has unreasonably refused to accept compliance with alternative requirements or that their requirements are unreasonable or unnecessary;
(d) that the period for compliance is unreasonable;
(e) that the best practicable means were used to counteract the effect of nuisance from trade or business premises.

The regulations further provide for the suspension of an abatement notice pending the court's decision, unless the local authority overrides the suspension in the abatement notice with a statement to the effect that the notice is to have effect regardless, and that:

(a) the nuisance is prejudicial to health;
(b) suspension would render the notice of no practical effect; or
(c) any expenditure incurred before an appeal would be disproportionate to the public benefit.

Sentencing powers for contravention of an abatement notice

A person who commits an offence on industrial, trade or business premises is liable on summary conviction to a fine not exceeding £20 000. Other persons are liable to a fine not exceeding level 5 on the standard scale (currently £5000), with a further fine of one-tenth of that level for each day on which the offence continues after conviction. Furthermore, if the person served fails to execute all or any of the works in accordance with the abatement notice, the local authority may execute those works and may recover from the person in default their costs incurred in so doing except such of the costs as that person shows were unnecessary in the circumstances (whether or not proceedings have been instituted). In proceedings brought by the local authority to recover such costs, the person in default cannot raise any question which he could have raised on appeal against the notice. The expenses become a charge on the premises and expenses carry interest at a reasonable rate – see Section 81A and 81B, Environmental Protection Act.

Expedited action for defective premises (Building Act 1984, Section 76) (does not extend to Scotland)

Where it appears to a local authority that any premises are in such a state as to be prejudicial to health or a nuisance, and that unreasonable delay in remedying the defective state would be occasioned by following the somewhat lengthy procedure set out in Part III of the Environmental Protection Act 1990, the local authority may serve on the person on whom it would have been appropriate to serve an abatement notice under the Environmental Protection Act 1990, a notice stating that the local authority intends to remedy the defective state and specifying the defects that it intends to remedy. The local authority may then, after the expiration of nine days after service of such a notice, execute the necessary works to remedy the premises' defective state, and recover reasonable expenses from the person on whom the notice was served.

However, if, within seven days after the service of the notice, the recipient of the notice serves a counter-notice that he intends to remedy the defects specified in the first-mentioned notice, the local authority must take no action unless the person who served the counter-notice fails within what seems to the local authority to be a reasonable period of time to begin to execute works to remedy the defects, or having begun to execute such works fails to make reasonable progress towards their completion. In proceedings to recover expenses for work carried out in default and initiated by the local authority, the court must inquire whether the local authority was justified in concluding that the premises were in a defective state, or that unreasonable delay in remedying the defective state would have been occasioned by following the procedure prescribed, and if the defendant proves that he served a counter-notice shall inquire whether the defendant failed to begin the works to remedy the defects within a reasonable time, or failed to make reasonable progress towards their completion, and if the court determines that the local authority was not justified in using this provision the local authority shall not recover the expenses or any part of them.

A local authority must not serve a notice under this provision or proceed with the execution of works in accordance with a notice so served if the execution of the works would, to its knowledge, be in contravention of a building preservation order under Section 29 of the Town and Country Planning Act 1947.

Proceedings in the High Court

Where a local authority considers that proceedings for an offence under Section 80 would afford an inadequate remedy in the case of any statutory nuisance, it may take proceedings in the High Court to secure the abatement, prohibition or restriction of the nuisance (usually by way of injunction.) See further *Hammersmith London Borough Council* v. *Magnum Automated Forecourts Ltd* [1978] 1 W.L.R. 50. In such proceedings it will be a defence to prove that noise was authorized by a construction site consent under Section 61 of the Control of Pollution Act 1974.

Individual action by persons aggrieved by statutory nuisances

Nowadays environmental health departments of local authorities are often understaffed and may be unable to deal fully with every statutory nuisance in their district. Section 82 of the Environmental Protection Act enables any person, after giving 21 days' notice to the defendant (three days in case of noise) to make a complaint to a magistrates' court on the ground that he is aggrieved by the existence of a statutory nuisance. This is a more expeditious and economic form than an action for private nuisance in the county court. If the magistrates' court is satisfied that the alleged nuisance exists, or that although abated it is likely to recur on the same premises or, in the case of a noise nuisance, is caused by noise emitted from or caused by an unattended vehicle or unattended machinery or equipment and is in the same street as before, the court must make an order for either or both of the following purposes:

(a) requiring the defendant to abate the nuisance, within a time specified in the order, and to execute works necessary for that purpose;
(b) prohibiting a recurrence of the nuisance, and requiring the defendant, within a time specified in the order, to execute any works necessary to prevent the recurrence.

The court may also impose on the defendant a fine not exceeding level 5 on the standard scale (currently £5000) together with a further fine of an amount equal to one-tenth of that level for each day on which the offence continues after the conviction.

If the magistrates' court is satisfied that the alleged nuisance exists and is such as, in the opinion of the court, to render premises unfit for human habitation, an order may be made to prohibit the use of the premises for human habitation until the premises are, to the satisfaction of the court, rendered fit for that purpose. Before instituting proceedings for such an order against any person, the person aggrieved by the nuisance shall give to that person such notice in writing of his intention to bring such proceedings and the notice must specify the matter complained of.

If a person is convicted of an offence, a magistrates' court may, after giving the local authority in whose area the nuisance has occurred an opportunity of being heard, direct the authority to do anything which the person convicted was required to do by the order to which the conviction relates. It is most important for individual complainants to note that if they wish to make complaint to the local authority or to a magistrates' court, sufficient and proper evidence (dates, times, severity and length of the nuisance if it has occurred and the number of people affected), is gathered or

there is strong evidence to show that the statutory nuisance is about to occur.

Power of entry

By schedule 3, para. 2 of the Environmental Protection Act 1990, any person authorized by a local authority may, on production, if so required, of his authority, enter any premises at any reasonable time for the purpose of ascertaining whether or not a statutory nuisance exists; or for the purpose of taking any action, or executing any work, authorized or required by Part III of the Environmental Protection Act 1990. It should be noted, however, that admission to any premises used wholly or mainly for residential purposes shall not, except in an emergency, be demanded as of right unless 24 hours' notice of the intended entry has been given to the occupier. If it is shown to the satisfaction of a justice of the peace on sworn information in writing that admission to any premises has been refused, or that refusal is apprehended, or that the premises are unoccupied or the occupier is temporarily absent, or that the case is one of emergency, or that an application for admission would defeat the object of the entry, and that there is reasonable ground for entry into the premises for the purpose for which entry is required, the justice may by warrant authorize the local authority, by any authorized person, to enter the premises, if need be by force. The warrant will continue in force until the purpose for which entry is required has been satisfied. An authorized person entering any premises has wide powers. He may take with him such other persons and such equipment as may be necessary; carry out such inspections, measurements and tests as he considers necessary for the discharge of any of the local authority's functions under Part III; and take away such samples or articles as he considers necessary for that purpose. The authorized person must secure any unoccupied premises on leaving.

For the purpose of taking any action, or executing any work, authorized by or required under Part III in relation to a statutory nuisance within Section 79(1)(ga) of the Environmental Protection Act 1990 caused by noise emitted from or caused by the vehicle, machinery or equipment, any person authorized by a local authority may after notifying the police and on production (if so required) of his authority enter or open a vehicle, machinery or equipment, if necessary by force, or remove a vehicle, machinery or equipment from a street to a secure place.

If entry has been gained to any unattended vehicle, machinery or equipment the authorized person shall leave it secured against interference or theft in such manner and as effectually as he found it. If the unattended vehicle, machinery or equipment cannot be left secured the authorized person must immobilize it by such means as he considers expedient, or remove it from the street to a secure place taking care not to cause more damage than is necessary. The local authority must then notify the police of its removal and current location.

Recovery of expenses

A local authority may recover its reasonable expenses in executing the powers set out above.

Offences relating to entry

A person who wilfully obstructs any person acting in the exercise of any powers set out above will be liable, on summary conviction, to a fine not exceeding level 3 on the standard scale.

If a person discloses any information relating to any trade secret obtained in the exercise of any powers set out above he shall, unless the disclosure was made in the performance of his duty or with the consent of the person having the right to disclose the information, be liable, on summary conviction, to a fine not exceeding level 5 on the standard scale.

FURTHER READING

Bassett, W.H. (2002) *Environmental Health Procedures*, 6th edn, Spon Press, London.

Burnett-Hall, R. (1995) *Environmental Law*, Sweet & Maxwell, London.

Hutter, B.M. (1988) *The Reasonable Arm of the Law*, Clarendon Press, Oxford.

Kramer, L. (1997) *E.C. Treaty and Environmental Law*, Sweet & Maxwell, London.

Neal, A.C. and Wright, F.B. (1992) *The European Communities' Health and Safety Legislation*, Chapman & Hall, London.

Wright, F.B. (1997) *Law of Health and Safety at Work*, Sweet & Maxwell, London.

7 Enforcement of environmental health law

Terence Moran J.P.

WHAT IS ENFORCEMENT?

First and foremost, enforcement should never simply be taken to mean prosecution. The Health and Safety Commission, in its policy statement on enforcement (HSC150) says that the term 'enforcement' has a wide meaning and applies to all dealings between enforcing authorities and those on whom the law places duties. It was Hawkins [1] who recognized that law may be enforced by compulsion and coercion, or by conciliation and compromise, and, in the words of Hutter [2] the term 'enforcement' should be used to accommodate a 'much wider concept, defining enforcement as the whole process of compelling observance with some broadly perceived objectives of the law'. It is this definition that is accepted for the purposes of this chapter and it is this process of compulsion, and the mechanisms to be employed to achieve this end that will be explored.

Adopting Hutter's definition, we may then define enforcement activity as 'the decisions environmental health officers make about how to dispose of individual cases and how, as members of a department, they interpret the law in terms of policies and strategy. Thus enforcement constitutes the bridge between the government's decision to intervene and protect the environment, and the impact of this intervention upon both the environment and the regulated' [2].

Historically, the enforcement of environmental legislation has been characterized by a relatively low number of prosecutions [3]. This has been attributed to an informal regulatory style, which, in the words of Vogel, is typified by: an absence of statutory standards; minimal use of prosecution; a flexible enforcement strategy; considerable administrative discretion; decentralized implementation; close co-operation between regulators and the regulated; and restrictions on the ability of non-industry constituents to participate in the regulatory process [4]. It is a style that has not always been seen as entirely satisfactory.

Those involved in the enforcement of environmental health law often find themselves uncertain of their role; are they to be police officers or friendly advisers? This ambivalence of attitude resulted in the development of what may be termed a dual-track approach, often using informal co-operation in preference to coercion. Hutter identified that many regulatory agencies (not just environmental health officers), had adopted this co-operative approach, which relied on negotiation, bargaining, education and advice to secure compliance [2]. Carson found:

[factory] inspectors do not see themselves as members of an industrial police force primarily concerned with the apprehension and subsequent punishment of offenders. Rather they perceive

their function to be that of securing compliance with the standards of safety, health and welfare required and thereby achieving the ends at which the legislation is directed. [5]

This attraction to informal techniques was similarly found to be displayed by water authorities. Quoting one officer, Hawkins records:

> the objective of the job is not to maximize the income of the exchequer by getting fines. The job is to make the best use we can of the water for the country . . . we get more co-operation if we use prosecution as a last resort. [1]

As Hutter [2] points out, the law in books is rarely implemented in a clear-cut fashion. It is in this area that environmental health professionals (EHPs) have to determine matters before them: what the law means; whether matters before them are covered by the legislation; and what the most appropriate method of resolution is. It is not enough simply to claim that it is the technical complexity of environmental health law that causes the difficulty in enforcement. The situation is altogether more complex, involving considerations of history, policy and jurisprudence.

A number of interacting factors have been identified, including:

- the range of penalties available and the perceived low level of penalties actually imposed;
- the exercise of discretion in decision-making regarding enforcement;
- the accountability of EHPs;
- limited resources;
- EHPs' perception of themselves as educators and advisers rather than police officers;
- the uncertainty of the criminality of the prohibited conduct in this area.

ENFORCEMENT POLICY

In this area there have been many recent and welcome changes. In 1991, an Audit Commission report [6] suggested that only a minority of local authorities had fully developed enforcement policies, and that even fewer had been published. For example, the Audit Commission found that only 43% of local authorities had a departmental policy for food hygiene and safety law enforcement.

Furthermore, those authorities that did have policies appeared to do little to publicize them. Thus, over 90% of respondents to a Department of Trade and Industry (DTI) business survey in 1994 did not know whether their local authority had an enforcement policy [7]. However, we now have the welcome sight of Best Value performance indicator 166 posing the question to Local Authorities 'Does the authority have written and published enforcement policy/policies, formally endorsed by its Members that cover all aspects of both Environmental Health and Trading Standards enforcement?'

The former lack of visibility of such policies inevitably raised questions of transparency, accountability and effectiveness. Yet, it was always relatively easy to formulate policies that incorporate the means to measure effectiveness. Rowan-Robinson and Ross [8] developed a framework that allowed some measurement of effectiveness. They suggested a nine-point approach as an aid to the measurement of the effectiveness of enforcement:

1. the clarity of the objectives of the legislation
2. the measurement of unlawful conduct
3. the character of the enforcement agency
4. the resources devoted to enforcement
5. the objectives of enforcement
6. the character of the deviant population
7. the organization of the enforcement agency
8. external dependency relationships
9. sanctions for a breach of control [8].

Rowan-Robinson and Ross suggested that the most important institutional pressure that structures the way that enforcement agencies seek to obtain legislative goals is the agency's own policy on enforcement [8]. They identify that 'the development of a formal "top-down" policy compels an agency to address the way in which enforcement practice may accommodate whatever constraints are imposed and still contribute in the most effective way to policy implementation' [8]. In the absence of a 'top-down' approach, enforcement policy will emerge from the 'bottom up' as the sum of day-to-day practice by officers. Policy is then formulated, rather than mediated, by officers on a case by case basis. This is the least desirable scenario, creating as it does a policy that is opaque, untargeted, and unlikely to be commensurate or, in the words of the Health and Safety Executive Local Authority Unit, 'proportionate to risk' [9]. A policy that is not commensurate must inevitably lack credibility.

Taking a lead from the Crown Prosecution Service, and again this is a key indicator when considering Best Value, a prosecution policy can be a public declaration of the principles upon which EHPs will exercise their functions. The purpose of the policy is then the promotion of efficient and consistent decision-making so as to develop, and thereafter maintain, public confidence in EHPs' performance of their duties. Such a policy can acknowledge the range of options available and how, depending on circumstances, EHPs may use those options to ensure protection of the public. These options extend from education and advice via warning letters and statutory notices to prosecutions and perhaps beyond.

The key to understanding the operation of such a policy is first to accept the value of the use of discretion. EHPs are given considerable discretion when applying the law. Hawkins saw this discretion as 'operationally efficient', as it is only the enforcing officer who really knows the target groups, their problems and their negotiating styles' [1]. However, Richardson *et al.* [10] noted that the exercising of discretion must be correctly guided and managed. The misuse of discretionary powers can carry severe consequences, not only for those suspected of criminal activity but also for the public at large and, importantly, for the reputation of the system of justice.

For all enforcing authorities it is not, nor has it ever been, simply a matter of whether or not to prosecute. Even the imposition of strict liability should not to be taken to require an automatic prosecution. In *Smedleys Ltd* v. *Breed* [1974], a case relating to contaminated food, Viscount Dilhorne asked:

In [the] circumstances, what useful purpose was served by the prosecution of the appellants . . . It

may have been the view that in every case where an offence was known or suspected, it was the duty of a food and drugs authority to institute a prosecution, that if evidence sufficed a prosecution should automatically be started . . . I do not find anything in the Act imposing on [an authority] a duty to prosecute automatically whenever an offence was known or suspected and I cannot believe that they should not consider whether the general interests of consumers were likely to be affected when deciding whether or not to institute proceedings . . . No duty is imposed on them to prosecute in every single case and although this Act imposes on the food and drugs authorities the duty of prosecuting for offences . . . it does not say – and I would find it surprising if it had – that they must prosecute in every case without regard to whether the public interest will be served by a prosecution.

A prosecution where no useful purpose was served was, and therefore remains, unnecessary.

The Health and Safety Executive/Local Authority Liaison Committee (HELA) recognized this in their guidance to local authorities [9] and suggested that local authorities should generally reserve prosecutions in the area of health and safety at work for the more serious offences, that is,

(a) where there is a blatant disregard for the law, particularly where the economic advantages of breaking the law are substantial and the law abiding are placed at a disadvantage to those who disregard it;
(b) when there appears to have been reckless disregard for the health and safety of workpeople or others;
(c) where there have been repeated breaches of legal requirements in an establishment, or in various branches of a multiple concern, and it appears that management is neither willing nor structured to deal adequately with these. An examination of the company safety policy, if any, would be particularly useful in such a situation;
(d) where a particular type of offence is prevalent in an activity or an area;

(e) where, as a result of a substantial legal contravention, there has been a serious accident or a case of ill health;

(f) where a particular contravention has caused serious public alarm;

(g) where there are persistent poor standards for control of health hazards.

There is, in fact, little evidence to suggest that local authorities routinely embark on unnecessary prosecutions. The DTI Interdepartmental Review Team report [7] found that fewer than 10% of those who responded showed a marked preference for formal action, the preferred approach being altogether more flexible.

If legal proceedings are to be instituted, it is important that such action is undertaken only after reference to appropriate policy guidance. Proceedings should then be instigated on a fair and consistent basis, a basis derived from a mature enforcement policy. Codes of Practice made under the Food Safety Act 1990 offer one model for the form such guidance may take. The codes have long recommended that, before deciding whether a prosecution should be made, food authorities should consider a number of factors, which may include:

- the seriousness of the alleged offence;
- the previous history of the party concerned;
- the likelihood of the defendant being able to establish a due diligence defence;
- the ability of any important witnesses and their willingness to co-operate;
- the willingness of the party to prevent a recurrence of the problem;
- the probable public benefit of a prosecution and the importance of the case;
- whether other action, such as issuing a formal caution, would be more appropriate or effective;
- any explanation offered by the affected company.

This is not, however, the only guidance that EHPs can draw on. Both the Health and Safety Executive (HSE) and the Environment Agency have published guidance on these matters.

The Health and Safety Commission (HSC) has developed this approach into an enforcement policy [11], which has embraced the principles of **proportionality, consistency, transparency** and **targeting**.[1] The HSC defines proportionality as meaning that enforcement action taken by an enforcing authority is to be proportionate to the seriousness of the breach or the risks to health and safety. Consistency here relates to the enforcement practice, that is, adopting a similar approach to the options of enforcement rather than ensuring uniformity. Transparency of the arrangements is the extent to which duty holders are clear about what is expected of them and what they can, in turn, expect of the enforcing agency. Targeting means enforcement action should be properly targeted at those who are responsible for the risk and at those whose activities give rise to the risks that are the most serious or least well controlled.

The approach identified by the HSE is echoed by that adopted by the Environment Agency, which has published the principles that it will adopt for its enforcement activities in its *Enforcement and Prosecution Policy*.

The code, as is typical, sets out four basic principles that should inform the enforcement of environmental protection law. These are: proportionality, consistency, targeting of action and transparency. By proportionality, the Environment Agency means relating enforcement action to the risks and costs. For the Environment Agency, consistency of approach does not mean merely 'uniformity', but instead means taking a similar approach in similar circumstances to achieve similar ends. Targeting for the Environment Agency is a matter of ensuring that regulatory effort is primarily directed towards those whose activities give rise to or create a risk of most serious environmental damage.

Consistent with published studies, it is mindful that other approaches to enforcement may prove more effective than prosecution. Whatever enforcement option officers may ultimately elect to take, their decision should be informed by a full appreciation of all the relevant facts. This is vital as not only will the officers then be in a position to make a proportionate enforcement decision, but they also will be able to meet certain statutory requirements, such as those relating to disclosure.

---

[1]See also the HSE's Enforcement home page at http://www.hse.gov.uk/enforce/index.htm

Officers considering any form of enforcement action must therefore be competent to undertake an appropriate level of investigation. This means that they must be aware of all of the statutory powers, options and obligations they are given as an aid to both investigation and enforcement.

Enforcement Concordat

In 1997, the Secretary of State for the Environment, Transport and the Regions signed, with the chairman of the Local Government Association (LGA), a framework document for partnership. The framework commits central and local government to working together to stengthen and sustain locally elected government in England and sets out arrangements for the conduct of central/local government relations.

Within this framework, in March 1998, the LGA and the government launched an Enforcement Concordat for adoption by local authorities that is relevant to all environmental health law enforcement.

In the concordat, the term 'enforcement' is deemed to include advisory visits and assisting with compliance as well as licensing and formal enforcement action. Adopting the concordat means that a local authority is committed to the following policies and procedures (which contribute to best value), and will provide information to show that they are observing them.

Principles of good enforcement policy

STANDARDS   In consultation with business and other relevant interested parties, including technical experts where appropriate, local authorities must draw up clear standards setting out the level of service and performance the public and business people can expect to receive. They must publish these standards and their annual performance against them. The standards will be made available to businesses and others who are regulated.

OPENNESS   Local authorities must provide information and advice in plain language on the rules that they apply and must disseminate this as widely as possible. They must be open about how they set about their work, including any charges that they set, consulting business, voluntary organizations, charities, consumers and workforce representatives. They must discuss general issues, specific compliance failures or problems with anyone experiencing difficulties.

HELPFULNESS   Local authorities should believe that prevention is better than cure and that their role therefore involves actively working with businesses, especially small and medium-sized businesses, to advise on and assist with compliance. They should provide a courteous and efficient service, and their staff should identify themselves by name. They should provide a contact point and telephone number for further dealings with them and encourage business to seek advice/information. Applications for approval of establishments, licenses, registrations, etc. should be dealt with efficiently and promptly. They should ensure that, wherever practicable, enforcement services are effectively co-ordinated to minimize unnecessary overlaps and time delays.

COMPLAINTS ABOUT SERVICE   Local authorities should provide well-publicized, effective and timely complaints procedures that are easily accessible to business, the public, employees and consumer groups. In cases where disputes cannot be resolved, any right of complaint or appeal must be explained, with details of the process and the likely timescales involved.

PROPORTIONALITY   Local authorities should minimize the costs of compliance for business by ensuring that any action required is proportionate to the risks. As far as the law allows, they should take account of the circumstances and attitude of the operator when considering action.

They should take particular care to work with small businesses, and voluntary and community organizations so that they can meet their legal obligations without unnecessary expense.

CONSISTENCY   Local authorities should carry out duties in a fair, equitable and consistent manner. While inspectors are expected to exercise judgement in individual cases, arrangements should be in place to promote consistency, including effective arrangements for liaison with other authorities and enforcement bodies through schemes such as those operated by the Local Authorities Co-ordinators of Regulatory Services.

HUMAN RIGHTS

One major recent piece of legislation that has brought about changes that are yet to be fully identified and understood is the Human Rights Act of 1998. The European Convention on Human Rights (ECHR), adopted in 1950, was ratified by the United Kingdom in 1951. The aim of the convention was to give effect to the UN Declaration on Human Rights, adopted in December 1948.

The Act, though it received Royal Assent in November 1998, did not come into force until October 2000. However, since that date the Act has had the effect of making rights derived from the European Convention on Human Rights enforceable in UK courts. Moreover the Act requires that all UK legislation, so far as it is possible to do so, be interpreted and applied in a way which is compatible with the Convention rights.

For the EHP the effect of the act is to require all public bodies, and thus their servants, to ensure that everything they do, and this of course includes enforcement activities, is compatible with Convention rights. The only exception to this being unless such compatibility is rendered impossible by an Act of Parliament.

When undertaking their activities EHPs are obliged to ensure that all legislation is interpreted and applied as far as possible in a manner that is consistent with the Convention.

This legislation thus creates a new landscape within which enforcement must operate and gives an ever-present backcloth to the content of this chapter. HELA LAC 63/5 (February 2002) gives useful guidance on the act that is applicable beyond simply the area of health and safety.

The Convention Rights

These are to be found in the 'Articles'.

Article 2 – Right to life

1. Everyone's right to life shall be protected by law. No one shall be deprived of his life intentionally save in the execution of a sentence of a court following his conviction of a crime for which this penalty is provided by law.

2. Deprivation of life shall not be regarded as inflicted in contravention of this article when it results from the use of force which is no more than absolutely necessary:

   (a) in defence of any person from unlawful violence;
   (b) in order to effect a lawful arrest or to prevent the escape of a person lawfully detained;
   (c) in action lawfully taken for the purpose of quelling a riot or insurrection.

Article 3 – Prohibition of torture
No one shall be subjected to torture or to inhuman or degrading treatment or punishment.

Article 4 – Prohibition of slavery and forced labour

1. No one shall be held in slavery or servitude.
2. No one shall be required to perform forced or compulsory labour.
3. For the purpose of this article the term 'forced or compulsory labour' shall not include:

   (a) any work required to be done in the ordinary course of detention imposed according to the provisions of Article 5 of this Convention or during conditional release from such detention;
   (b) any service of a military character or, in case of conscientious objectors in countries where they are recognized, service exacted instead of compulsory military service;
   (c) any service exacted in case of an emergency or calamity threatening the life or well-being of the community;
   (d) any work or service which forms part of normal civic obligations.

Article 5 – Right to liberty and security

1. Everyone has the right to liberty and security of person. No one shall be deprived of his liberty save in the following cases and in accordance with a procedure prescribed by law:

   (a) the lawful detention of a person after conviction by a competent court;
   (b) the lawful arrest or detention of a person for non-compliance with the lawful order

of a court or in order to secure the fulfilment of any obligation prescribed by law;

(c) the lawful arrest or detention of a person effected for the purpose of bringing him before the competent legal authority on reasonable suspicion of having committed an offence or when it is reasonably considered necessary to prevent his committing an offence or fleeing after having done so;

(d) the detention of a minor by lawful order for the purpose of educational supervision or his lawful detention for the purpose of bringing him before the competent legal authority;

(e) the lawful detention of persons for the prevention of the spreading of infectious diseases, of persons of unsound minds, alcoholics or drug addicts or vagrants;

(f) the lawful arrest or detention of a person to prevent his effecting an unauthorized entry into the country or of a person against whom action is being taken with a view to deportation or extradition.

2. Everyone who is arrested shall be informed promptly, in a language which he understands, of the reasons for his arrest and of any charge against him.

3. Everyone arrested or detained in accordance with the provisions of paragraph 1(c) of this Article shall be brought promptly before a judge or other officer authorized by law to exercise judicial power and shall be entitled to trial within a reasonable time or to release pending trial. Release may be conditioned by guarantees to appear for trial.

4. Everyone who is deprived of his liberty by arrest or detention shall be entitled to take proceedings by which the lawfulness of his detention shall be decided speedily by a court and his release ordered if the detention is not lawful.

5. Everyone who has been the victim of arrest or detention in contravention of the provisions of this Article shall have an enforceable right to compensation.

Article 6 – Right to a fair trial

1. In the determination of his civil rights and obligations or of any criminal charge against him, everyone is entitled to a fair and public hearing within a reasonable time by an independent and impartial tribunal established by law. Judgment shall be pronounced publicly but the press and public may be excluded from all or part of the trial in the interests of morals, public order or national security in a democratic society, where the interests of juveniles or the protection of the private life of the parties so require, or to the extent strictly necessary in the opinion of the court in special circumstances where publicity would prejudice the interests of justice.

2. Everyone charged with a criminal offence shall be presumed innocent until proved guilty according to law.

3. Everyone charged with a criminal offence has the following minimum rights:

(a) to be informed promptly, in a language which he understands and in detail, of the nature and cause of the accusation against him;

(b) to have adequate time and facilities for the preparation of his defence;

(c) to defend himself in person or through legal assistance of his own choosing or, if he has not sufficient means to pay for legal assistance, to be given it free when the interests of justice so require;

(d) to examine or have examined witnesses against him and to obtain the attendance and examination of witnesses on his behalf under the same conditions as witnesses against him;

(e) to have the free assistance of an interpreter if he cannot understand or speak the language used in court.

Article 7 – No punishment without law

1. No one shall be held guilty of any criminal offence on account of any act or omission which did not constitute a criminal offence under national or international law at the time when it was committed. Nor shall a heavier penalty be imposed than the one that was applicable at the time the criminal offence was committed.

2. This Article shall not prejudice the trial and punishment of any person for any act or omission

which, at the time when it was committed, was criminal according to the general principles of law recognized by civilized nations.

Article 8 – Right to respect for private and family life

1. Everyone has the right to respect for his private and family life, his home and his correspondence.
2. There shall be no interference by a public authority with the exercise of this right except such as is in accordance with the law and is necessary in a democratic society in the interests of national security, public safety or the economic well-being of the country, for the prevention of disorder or crime, for the protection of health or morals, or for the protection of the rights and freedoms of others.

Article 9 – Freedom of thought, conscience and religion

1. Everyone has the right to freedom of thought, conscience and religion; this right includes freedom to change his religion or belief and freedom, either alone or in community with others and in public or private, to manifest his religion or belief, in worship, teaching, practice and observance.
2. Freedom to manifest one's religion or beliefs shall be subject only to such limitations as are prescribed by law and are necessary in a democratic society in the interests of public safety, for the protection of public order, health or morals, or for the protection of the rights and freedoms of others.

Article 10 – Freedom of expression

1. Everyone has the right to freedom of expression. This right shall include freedom to hold opinions and to receive and impart information and ideas without interference by public authority and regardless of frontiers. This article shall not prevent States from requiring the licensing of broadcasting, television or cinema enterprises.
2. The exercise of these freedoms, since it carries with it duties and responsibilities, may be subject to such formalities, conditions, restrictions or penalties as are prescribed by law and are necessary in a democratic society, in the interests of national security, territorial integrity or public safety, for the prevention of disorder or crime, for the protection of health or morals, for the protection of the reputation or rights of others, for preventing the disclosure of information received in confidence, or for maintaining the authority and impartiality of the judiciary.

Article 11 – Freedom of assembly and association

1. Everyone has the right to freedom of peaceful assembly and to freedom of association with others, including the right to form and to join trade unions for the protection of his interests.
2. No restrictions shall be placed on the exercise of these rights other than such as are prescribed by law and are necessary in a democratic society in the interests of national security or public safety, for the prevention of disorder or crime, for the protection of health or morals or for the protection of the rights and freedoms of others. This Article shall not prevent the imposition of lawful restrictions on the exercise of these rights by members of the armed forces, of the police or of the administration of the State.

Article 12 – Right to marry

Men and women of marriageable age have the right to marry and to found a family, according to the national laws governing the exercise of this right.

Article 14 – Prohibition of discrimination

The enjoyment of the rights and freedoms set forth in this Convention shall be secured without discrimination on any ground such as sex, race, colour, language, religion, political or other opinion, national or social origin, association with a national minority, property, birth or other status.

Article 16 – Restrictions on political activity of aliens

Nothing in Articles 10, 11 and 14 shall be regarded as preventing the High Contracting Parties from imposing restrictions on the political activity of aliens.

Article 17 – Prohibition of abuse of rights
Nothing in this Convention may be interpreted as implying for any State, group or person any right to engage in any activity or perform any act aimed at the destruction of any of the rights and freedoms set forth herein or at their limitation to a greater extent than is provided for in the Convention.

Article 18 – Limitation on use of restrictions on rights
The restrictions permitted under this Convention to the said rights and freedoms shall not be applied for any purpose other than those for which they have been prescribed.

The First Protocol

Article 1 – Protection of property
Every natural or legal person is entitled to the peaceful enjoyment of his possessions. No one shall be deprived of his possessions except in the public interest and subject to the conditions provided for by law and by the general principles of international law.

The preceding provisions shall not, however, in any way impair the right of a State to enforce such laws as it deems necessary to control the use of property in accordance with the general interest or to secure the payment of taxes or other contributions or penalties.

Article 2 – Right to education
No person shall be denied the right to education. In the exercise of any functions which it assumes in relation to education and to teaching, the State shall respect the right of parents to ensure such education and teaching in conformity with their own religious and philosophical convictions.

Article 3 – Right to free elections
The High Contracting Parties undertake to hold free elections at reasonable intervals by secret ballot, under conditions which will ensure the free expression of the opinion of the people in the choice of the legislature.

The Sixth Protocol

Article 1 – Abolition of the death penalty
The death penalty shall be abolished. No one shall be condemned to such penalty or executed.

Article 2 – Death penalty in time of war
A State may make provision in its law for the death penalty in respect of acts committed in time of war or imminent threat of war; such penalty shall be applied only in the instances laid down in the law and in accordance with its provisions.

Not all the Convention rights are applied in the same way.

The types of rights can be said to be:

- absolute rights, for example, the Article 3 right to protection from torture, inhuman and degrading treatment and punishment;
- limited rights, for example, the Article 5 right to liberty and Article 6 right to a fair trial. These are limited under specific circumstances, set out in the Convention;
- qualified rights, for example, the Article 8 right to respect for private and family life. For these interference is permissible only if what is done contrary to the right:

  a. has its basis in law; and
  b. is done to secure a permissible aim set out in the relevant Article, for example, for the prevention of crime, or for the protection of public order or health; and
  c. is necessary in a democratic society, which means it must fulfil a pressing social need, pursue a legitimate aim and be proportionate to the aims being pursued.

Where a court finds that a public authority has acted (or proposes to act) unlawfully under the Act it may grant such remedy within its powers as it considers just and appropriate [12].

The full impact of this legislation is yet to be appreciated. However it has already been addressed in certain areas relevant to the work of the EHP. In *R* v. *Hertfordshire County Council, ex parte Green Environmental Industries Limited* [2000] 2 WLR 373 it was held that someone obliged by s.71 [2] Environmental Protection Act 1990 to provide information about waste-dumping activities could not claim the protection of Article 6 (the ground of self-incrimination) as this did not apply to extra-judicial enquiries.

In *Donoghue* v. *Poplar Housing and Regeneration Community Association Ltd* (2001) it was held that Housing associations are public authorities for the purposes of the Human Rights Act and in *Barnfather* v. *London Borough of Islington Education Authority and the Secretary of State for Education and Skills* (2003) the applicability of Article 6 to strict liability offences was considered, it being found that Article 6 does not entitle the courts to question the justification for strict liability offences.

Perhaps the most obvious and, so far as enforcement is concerned, arguably one of the most important changes the Human Rights Act (HRA) has stimulated has been the need to revise the Police and Criminal Evidence Act 1984 (PACE) codes of practice.

PACE, and particularly the Codes of Practice associated with it, have always been good guides to EHPs in terms of compliance with the ECHR.

Section 66 of the Act requires the Secretary of State to issue codes of practice in connection with

(a) the exercise by police officers of statutory powers

  (i) to search a person without first arresting him; or
  (ii) to search a vehicle without making an arrest;

(b) the detention, treatment, questioning and identification of persons by police officers;
(c) searches by police officers on persons or premises.

Since the passing of the Act there have been a number of versions of the codes.

The most recent codes issued under Section 66 of the Act, came into effect in April 2003 and regulate activity in the following areas:

A. The exercise by police officers of statutory powers of stop and search
B. The searching of premises and the seizure of property
C. The detention, treatment and questioning
D. The identification of people
E. Audio recording of Interviews
F. Visual recording of interviews – currently of limited geographical application.

It is clear that these codes have been changed, not least, as a result of the impact of the HRA. Code B for example explicitly states that

> The right to privacy and respect for personal property are key principles of the Human Rights Act 1998. Powers of entry, search and seizure should be fully and clearly justified before use because they may significantly interfere with the occupier's privacy. Officers should consider if the necessary objectives can be met by less intrusive means.

And that 'In all cases, police should exercise their powers courteously and with respect for persons and property and only use reasonable force when this is considered necessary and proportionate to the circumstances.'

It is established in the ECHR that everyone has the right to respect for their private and family life, home and correspondence. In recognition of this the code emphasizes that when undertaking a search with consent of tenanted property that 'every reasonable effort should be made to obtain the consent of the tenant, lodger or occupier. A search should not be made solely on the basis of the landlord's consent unless the tenant, lodger or occupier is unavailable and the matter is urgent.' Similarly the code advises that any officer undertaking a search of premises should identify him or herself, show their authorization, state the purpose of and grounds for the search and identify and introduce any person who is accompanying them on the search and describe that person's role in the process.

Code C has also been subject to change. At Section 10 a significant change has been to the nature of the caution that may need to be given to those suspected of a criminal offence. These changes arise from the European Court judgement in *John Murray* v. *United Kingdom* ([1996]

22 EHRR 29). This case concerned a man arrested in connection with terrorism charges. He remained silent throughout questioning, despite being cautioned that his refusal to answer could be held against him. Murray was questioned for a long period without being allowed access to legal advice and was ultimately convicted of conspiracy to murder and sentenced to eight years.

He appealed to the European Commission on Human Rights on the grounds that the drawing of adverse inferences from his refusal to speak at his trial and the denial of access to legal advice was a contravention of Article 6(1) of the Convention (right to a fair trial).

It was found that Article 6(1) did not give an absolute right to silence. However, when taken with a refusal to allow him access to a solicitor, this was held to breach Article 6(1).

Consequently, the caution to be administered is now in two forms. One which must be used if there is a restriction on drawing adverse inferences from silence and one to be used where there is not. For the latter the caution remains: 'You do not have to say anything. But it may harm your defence if you do not mention when questioned something which you later rely on in Court. Anything you do say may be given in evidence.' However, if the restriction on drawing adverse inferences does apply then the caution becomes 'You do not have to say anything, but anything you do say may be given in evidence.'

This restriction on drawing adverse inferences applies:

(a) to any detainee at a police station, who, before being interviewed, or being charged or informed they may be prosecuted

  (i) asked for legal advice;
  (ii) had not been allowed an opportunity to consult a solicitor; and
  (iii) had not changed their mind about wanting legal advice;

(b) to any person charged with, or informed they may be prosecuted for, an offence who:

  (i) has had brought to their notice a written statement made by another person or the content of an interview with another person which relates to that offence;

  (ii) is interviewed about that offence; or
  (iii) makes a written statement about that offence.

The restriction on drawing inferences from silence will not apply to anyone who has not been detained and who therefore cannot be prevented from seeking legal advice if they wish.

REGULATION OF INVESTIGATORY POWERS ACT 2000

The Regulation of Investigatory Powers Act came into force on 24th October 2000, its creation driven, in part, by the need to respond to the requirements of the Human Rights Act, the effect of which was to require the government to institute some form of statutory framework to regulate the techniques used by law enforcement agencies.

The Act is in five parts.

I. Interception of Communications and the Acquisition and Disclosure of Communications Data
II. Surveillance and Covert Human Intelligence Sources
III. Investigation of Electronic Data Protected by Encryption etc.
IV. Scrutiny of Investigatory Powers and Codes of Practice
V. Miscellaneous and Supplemental.

The main purpose of the legislation is to ensure that the relevant investigatory powers are used in accordance with human rights requirements. For the purposes of the Act relevant investigatory powers are:

- the interception of communications;
- the acquisition of communications data (e.g. billing data);
- intrusive surveillance (on residential premises/ in private vehicles);
- covert surveillance in the course of specific operations;
- the use of covert human intelligence sources (agents, informants, undercover officers);
- access to encrypted data.

The Act seeks to legislate to ensure that the law clearly covers:

- the purposes for which the powers may be used;
- which authorities can use the powers;
- who should authorize each use of the power;
- the use that can be made of the material gained;
- independent judicial oversight;
- a means of redress for the individual.

Part II of the Act, dealing with Surveillance and Covert Human Intelligence Sources, creates a framework of authorizations for various types of surveillance and use of human intelligence sources. The sort of areas where a local authority might use surveillance are, for example, when dealing with nuisance cases, anti-social behaviour, breach of planning control etc.

Section 26 of the Act details the conduct which can be authorized under Part II. Specifically, it regulates three types of activity: directed surveillance, intrusive surveillance and the conduct and use of covert human intelligence sources.

Directed surveillance is defined as:

> Covert surveillance that is undertaken in relation to a specific investigation or a specific operation which is likely to result in the obtaining of private information about a person (whether or not one specifically identified for the purposes of the investigation or operation); and otherwise than by way of an immediate response to events or circumstances the nature of which is such that it would not be reasonably practicable for an authorisation under this Part to be sought for the carrying out of the surveillance. Surveillance will be covert where it is carried out in a manner calculated to ensure that the person or persons subject to the surveillance are unaware that it is or may be taking place.

Intrusive surveillance is defined as:

> Covert surveillance carried out in relation to anything taking place on residential premises or in any private vehicle. This kind of surveillance may take place by means either of a person or device located inside residential premises or a private vehicle of the person who is subject to

the surveillance or by means of a device placed outside which consistently provides a product of equivalent quality and detail as a product which would be obtained from a device located inside.

Covert surveillance

This is governed by a Code of Practice Pursuant to Section 71 of the Regulation of Investigatory Powers Act 2000. The Regulation of Investigatory Powers (Covert Surveillance: Code of Practice) Order 2002. The code applies to every authorization of covert surveillance or of entry on, or interference with, property or with wireless telegraphy carried out under Part II of the Regulation of Investigatory Powers Act 2000 by public authorities.

The code makes it clear that general observation forms part of the duties of many law enforcement officers and other public authorities and is not usually regulated by the 2000 Act. The code notes that such observation may involve the use of equipment, such as binoculars, or the use of cameras, where this does not involve systematic surveillance of an individual. The proper authorization of surveillance should ensure the admissibility of evidence so obtained under the common law, Section 78 of the Police and Criminal Evidence Act 1984 and the Human Rights Act 1998.

Covert human intelligence source

A person is a covert human intelligence source if

> he establishes or maintains a personal or other relationship with a person for the covert purpose of using such a relationship to obtain information or to provide access to any information to another person or he covertly discloses information obtained by the use of such a relationship, or as a consequence of the existence of such a relationship.

Surveillance is covert if, and only if, it is carried out in a manner that is calculated to ensure that persons who are subject to the surveillance are unaware that it is or may be taking place; a purpose is covert, in relation to the establishment or maintenance of a personal or other relationship, if, and only if, the relationship is conducted in a manner that is calculated to ensure that one of the parties to the relationship is unaware of the purpose; and private information, in relation to a person, includes any information relating to his private or family life.

It is necessary for local authorities to demonstrate that using such techniques in a particular case is both necessary and proportionate to meet specified law enforcement objectives such as the prevention or detection of crime.

Section 27 – Lawful surveillance etc.

An authorization under Part II of the 2000 Act will provide lawful authority for a public authority to carry out surveillance.

All conduct defined in Section 26 will be lawful if

(a) an authorization under this Part confers an entitlement to engage in that conduct on the person whose conduct it is; and
(b) his conduct is in accordance with the authorization.

Authorization of directed surveillance

Directed surveillance is conducted where it involves the observation of a person or persons with the intention of gathering private information to produce a detailed picture of a person's life, activities and associations. However, it does not include covert surveillance carried out by way of an immediate response to events or circumstances which, by their very nature, could not have been foreseen. For example, a plain-clothes police officer would not require an authorization to conceal himself and observe a suspicious person who he comes across in the course of a patrol.

Directed surveillance must be authorized in each case by a designated person. Section 30 provides that the persons entitled to grant such authorizations will be such persons within the relevant public authorities that are designated by order of the Secretary of State.

Here the relevant order is the Regulation of Investigatory Powers (Prescription of Offices,

Ranks and Positions) Order 2000 and the Regulation of Investigatory Powers (Prescription of Offices, Ranks and Positions) (Amendment) Order 2002.

For a Local Authority (within the meaning of Section 1 of the Local Government Act 1999) the relevant 'rank' of officer is the Assistant Chief Officer responsible for the management of an investigation. Even if so authorized no one should grant an authorization for the carrying out of directed surveillance unless he believes

1. the authorization is necessary on specific grounds; and
2. the authorized activity is proportionate to what is sought to be achieved by it.

The specific grounds are that the authorization is necessary:

- in the interests of national security;
- for the purpose of preventing or detecting crime or preventing disorder;
- in the interests of the economic well-being of the United Kingdom;
- in the interests of public safety;
- for the purpose of protecting public health;
- for the purpose of assessing or collecting any tax, duty, levy or other imposition, contribution or charge payable to a government department; or
- for other purposes which may be specified by order of the Secretary of State.

The authorizing officer must give authorizations in writing, except that in urgent cases the authorizing officer may give them orally. In such cases, a statement that the authorizing officer has expressly authorized the action should be recorded in writing, by the applicant, as soon as is reasonably practicable. A case is not normally to be regarded as urgent unless the time that would elapse before the authorizing officer was available to grant the authorization would, in the judgement of the person giving the authorization, be likely to endanger life or jeopardize the investigation or operation for which the authorization was being given. Guidance suggests that an authorization is not to be regarded as urgent where the need for an authorization has been neglected or the urgency is of the authorizing officer's own making. Also authorizing officers should not be responsible for authorizing investigations or operations in which they are directly involved, although it is recognized in guidance that may sometimes be unavoidable, especially in the case of small organizations, or where it is necessary to act urgently. Where an authorizing officer authorizes such an investigation or operation the central record of authorizations should highlight this.

The legislation and associated guidance is alive to the possibility of 'Collateral Intrusion', that is, the possibility of intrusion into the privacy of persons other than those who are directly the subjects of the investigation.

Before authorizing surveillance the authorizing officer should also take into account such a risk. And measures should be taken, wherever practicable, to then avoid or minimize unnecessary intrusion into the lives of those not directly connected with the investigation. Any officer seeking an authorization should thus include an assessment of the risk of any collateral intrusion and the authorizing officer should take this into account, when considering the proportionality of the surveillance.

Information to be provided in applications for authorization

A written application for authorization for directed surveillance should describe any conduct to be authorized and the purpose of the investigation or operation. The application should also include:

- the reasons why the authorization is necessary in the particular case and on the grounds (e.g. for the purpose of preventing or detecting crime) listed in Section 28(3) of the 2000 Act;
- the reasons why the surveillance is considered proportionate to what it seeks to achieve;
- the nature of the surveillance;
- the identities, where known, of those to be the subject of the surveillance;
- an explanation of the information which it is desired to obtain as a result of the surveillance;

- the details of any potential collateral intrusion and why the intrusion is justified;
- the details of any confidential information that is likely to be obtained as a consequence of the surveillance;
- the level of authority required (or recommended where that is different) for the surveillance; and
- a subsequent record of whether authority was given or refused, by whom and the time and date.

Additionally, in urgent cases, the authorization should record (as the case may be):

- the reasons why the authorizing officer or the officer entitled to act in urgent cases considered the case so urgent that an oral instead of a written authorization was given; and/or
- the reasons why it was not reasonably practicable for the application to be considered by the authorizing officer.

Where the authorization is oral, the detail referred to above should be recorded in writing by the applicant as soon as reasonably practicable.

Duration of authorizations

A written authorization will cease to have effect (unless renewed) at the end of a period of three months beginning with the day on which it took effect. If at any time before an authorization would cease to have effect, the authorizing officer considers it necessary for the authorization to continue for the purpose for which it was given, he/she may renew it in writing for a further period, beginning with the day when the authorization would have expired but for the renewal.

Central record of all authorizations

A centrally retrievable record of all authorizations should be held by each public authority and regularly updated whenever an authorization is granted, renewed or cancelled. The record should be made available to the relevant Commissioner or an Inspector from the Office of Surveillance

Commissioners, upon request. These records should be retained for a period of at least three years from the ending of the authorization and should contain the following information:

- the type of authorization;
- the date the authorization was given;
- name and rank/grade of the authorizing officer;
- the unique reference number (URN) of the investigation or operation;
- the title of the investigation or operation, including a brief description and names of subjects, if known;
- whether the urgency provisions were used, and if so why;
- if the authorization is renewed, when it was renewed and who authorized the renewal, including the name and rank/grade of the authorizing officer;
- whether the investigation or operation is likely to result in obtaining confidential information as defined in the code of practice;
- the date the authorization was cancelled.

Intrusive surveillance

Section 32: Authorization of intrusive surveillance: Such authorizations may only be granted by the Secretary of State (see Sections 41 and 42) and by senior authorizing officers as listed in subsection (6). Local authorities are not included in the list and thus are not authorized under the terms of the act to undertake intrusive surveillance.

INSPECTION

Programming of inspections

The 'trigger' event for enforcement activity may take many forms. There may be a major food poisoning outbreak or a major pollution incident, there may be a workplace accident or a member of the public may make a formal complaint. In the sense that these events are unpredictable they are therefore beyond the control of EHPs. However, not all such 'triggers' are unanticipated or unplanned. There is, for example, the preplanned

inspection. Enforcement officers find themselves, now more than ever, involved in '**risk based approaches**' to enforcement. Whether that be in the area of health and safety at work using risk assessment, or in food safety using HACCP (hazard analysis of critical control points), it is a factor that influences enforcement activity.

Take, for example, the area of food safety. Code of Practice 9 on food hygiene inspections gives guidance on the priority planning and programming of inspections. Food authorities are obliged to adopt a programme of inspections and ensure, so far as is practicable, that inspection visits are carried out in accordance with that programme. Such a programme should be constructed to recognize that certain premises will pose a higher risk than others and are therefore to be inspected more frequently than other, lower risk, premises. The code in Annexe 1 gives an inspection rating scheme for food premises, and obliges a food authority to adopt this or a similar scheme to determine minimum inspection frequencies. It takes no great thought to see how frequency of inspection is an enforcement tool. Regular inspection is likely to make those subject to that inspection more aware of the need to comply with the law and more inclined to do so.

Inspections, or the threat or anticipation of them, may themselves secure compliance with the law or may give rise to other enforcement activity aimed at securing the same outcome. However, for such inspections to be productive an officer must first gain entry to the premises concerned.

Entry

One key statutory power given to EHPs is the **power to enter premises**. Because of the wide range of duties undertaken, the range of powers of entry, search and seizure is equally extensive. Self-evidently, routine entry into premises will be with permission. This '**express permission**' ensures that the officer concerned does not enter the property as a trespasser and thus enters and remains on the property as a lawful visitor.

Any unauthorized entry by any person, including an EHP, into private premises is an actionable trespass. Therefore, EHPs who enter a building with permission, but are subsequently told to go must do so or leave themselves, and perhaps their employer, open to an action for trespass. Entry on to land, however, may be lawful without there having been any express permission given, for example, there may be a right of way across land that makes it lawful for a stranger to be on that land.

However, for the EHP the power of entry is statutory. Here a specific statutory provision provides for officers to enter private premises for the execution of specified activities. Such statutory provisions do not provide an unlimited right of access for all persons to all premises at all times. Any empowering statute will inevitably stipulate the conditions precedent for the exercise of a power of entry, by whom the power is exercisable, and the limits of its application.

Every officer seeking to use any power of entry under any statutory provision must first be **authorized** by the relevant authority for that purpose and have the authorization document available. There is no national format for these documents and therefore their precise form and layout vary across the country.

Officers must be clear about the extent of their authority and it must never be assumed that an officer has been authorized. Being employed to do a job is not the same as being authorized for the purposes of an act; to confuse the two is to invite disaster.

Because of the range of provisions, there is no single, uniform procedure for gaining entry to premises. Each one of the statutory provisions giving the power to enter displays common features, but there is not sufficient uniformity to permit EHPs to assume that familiarity with one provides an adequate knowledge of all. Each is required to be understood in its own right and it is up to the EHPs to ensure that they know the relevant requirements of the power of entry they are seeking to use. For example, where the power to enter requires the officer to have reasonable grounds, just what constitutes reasonable grounds varies from case to case and will, in the final analysis, remain a question for an officer's professional judgement. What is clear, however, is that officers must understand that the test is an objective one.

The use of the words 'reasonable grounds' imposes the condition that such reasonable grounds must in fact exist and be known to the officer before the conditional power in question can validly be exercised. EHPs must understand that the issue is not whether they believed there to be reasonable grounds but, when looked at objectively, whether there were reasonable grounds. The credibility in terms of the quantity and, particularly, quality of information required to establish 'reasonable grounds' must necessarily be less than that that would be required to institute proceedings or take other statutory action. Nevertheless, EHPs must ensure that they can satisfy the central objective test. This means that all officers should record clearly those facts that lead them to conclude that they have reasonable grounds for their belief. This has a twofold benefit. First, it ensures that officers have approached the issue correctly and satisfactorily addressed this threshold test for entry. Second, it ensures that, if subsequently questioned, officers have clear and defensible reasons for coming to the conclusion they did.

Other conditions, such as those relating to the need to give prior notice, must similarly be acknowledged and accommodated by EHPs. For example, any prior notice of entry must be specific about its purpose, should identify the relevant authority, and should be specific about the relevant statutory provision concerned. A failure to comply with the stipulated requirements for prior notice will render any dependent entry invalid and any subsequent application for a warrant impossible.

Recording of information

Once inside any premises it is important that officers ensure that all information acquired is accurately and thoroughly recorded. A common method for this is the notebook. Often referred to as 'PACE' (Police and Criminal Evidence Act 1984) **notebooks**, these books can prove vital to the professional functioning of EHPs. Their significance, in terms of the preparation of reports, statements and memory refreshing in the witness box, is major. The record of events, if it is substantially contemporaneous, may, by the application of the 'memory refreshing rule', be used by EHPs to refresh their memory when later giving evidence to a court. If EHPs are to rely on their notebooks in this way, they must be kept in a disciplined fashion, particularly as the defence in any criminal case will be entitled to examine the notebooks to establish that entries are consistent with the evidence being given. Guidance previously given to police officers on the completion of notebooks has proved to be entirely appropriate for EHPs as well.

An alternative or an addition to the notebook as a means of recording information is the pre-prepared **inspection sheet**. As for a notebook, it is essential that accuracy is observed at all times, for it is from the observations recorded in a notebook or inspection proforma that any subsequent statements may be composed and thus prosecution instigated. It is therefore vital that inspection proformas are completed with the same discipline as for a notebook.

The aim of any officer inspecting premises should be to record all relevant information in such a manner that it

- will inform any subsequent enforcement decision;
- will be permissible under the memory refreshing rule;
- will allow a colleague examining the inspection findings, but not having previously visited the premises, to be able to understand and make a reasonable assessment of the conditions.

It is therefore important that EHPs take note of all possible sources of information. However, it is unwise to take everything at face value. For example, if told the property being inspected was rewired two years before, this should not be recorded as a fact in the inspection findings. To record that 'the property was rewired 2 years ago' without further investigation is unsustainable and evidentially is hearsay.

Generally, **a systematic approach to inspection** will pay dividends, both in savings of time and effort and in ensuring that a higher quality of information is recorded. Take, for example, the inspection of a house. The **accurate recording of findings** is critical. It is never wise to trust to one's memory. An inspection checklist or sheet can help as it can ensure that the inspecting officer keeps to the discipline of a properly planned schedule of inspection while providing an immediate and obvious way of ensuring that no important element is overlooked.

When recording defects, the use of words such as 'broken' or 'defective' should be avoided. Without further elaboration these words convey little information about the exact nature of the problem identified. Words such as 'cracked', 'missing', 'holed' and 'rotten' all convey a clearer picture of the conditions found. Equally, it is appropriate to attempt to quantify the extent of the defect noted.

For the same reasons as those applying to the keeping of a notebook, EHPs should aim to avoid leaving blank spaces on any inspection form used. If an element of the structure is sound, record its presence and mark it as such; a blank space in any inspection report might be taken to indicate that no defect was present or that the inspector forgot to look at that particular thing.

For a house inspection, the following should be considered as the minimum information to be recorded:

- the address and location of the property;
- the name of the inspecting officer;
- the date of inspection;
- the time of day;
- the weather conditions;
- the approximate age of the house;
- the type of property: terraced, back-to-back, semi, etc.;
- the type of construction: traditional, timber frame, etc.;
- the occupiers of the property: these details should include: names, ages, sexes, relationship, how long they have lived there, if tenanted in whose name is the tenancy, what rent is paid, to whom is the rent paid, is there a rent book, and was the rent book inspected;
- the type of occupancy of the house: house in multiple occupation (HMO), single household;
- the name and address of any managing agent and the name and address of the owner;

- details of any mortgagee;
- accommodation: number of rooms, type of rooms;
- method of orientation for the purpose of recording inspection findings.

With appropriate amendments, the same approach could be adopted for all inspections by EHPs.

In addition to relying on information obtained directly, EHPs might also have to rely on information obtained by others. Thus when dealing with members of the public it is vital to make clear all instructions about how they should complete any record they might make. First, the potential legal significance of what the person is doing should be made clear. It should be explained that he or she may be asked to give evidence in court based on any notes made. It is therefore vital that matters about memory refreshing are fully explained. Clarity and precision must be emphasized from the outset. Potential witnesses must have explained to them both the type and significance of the information they are to record. For example, they must make sure of the source of any alleged nuisance rather than assume it.

Investigative powers

It is clear that to be able to select an enforcement option EHPs must be prepared to use all investigative alternatives to inform their decision to take action and what action to take. For example, if that action is to serve a notice they must then decide what to include in the notice specification. To do otherwise is to run a number of risks: the risk of selecting the wrong notice, or of serving a notice improperly, the risk of failing to specify with sufficient particularity, or the risk of imposing conditions that are more onerous than are truly required.

EHPs are given extensive investigative powers under a wide range of statutory provisions. For example, consider Section 20 of the Health and Safety at Work, Etc. Act 1974 relating to the powers of inspectors. This section of the Act gives inspectors an extensive list of investigative tools or powers backed up by criminal sanctions. Anyone seeking to restrict the correct exercise of these powers is committing a criminal offence.

1. An inspector has the power to enter any premises at any reasonable time for the purposes of carrying into effect any of the relevant statutory provisions. If he is of the opinion that the situation is or may be dangerous, he may enter the premises at any time, and if the inspector has reasonable cause to apprehend serious obstruction in the execution of his duty he may take with him a constable.

2. An environmental health professional authorized under Section 20 may be accompanied by any other person duly authorized by the inspector's enforcing authority. He is also entitled to take on to the premises any equipment or materials which may be required for any purpose for which the power of entry is being exercised. The powers allow the environmental health professional to make such examination and investigation as may be necessary to carry into effect any of the relevant statutory provisions. An officer using these powers may also direct that any premises that he has entered or any part thereof is to be left undisturbed for so long as is reasonably necessary for the purpose of any examination or investigation.

3. The environmental health professional may take measurements, photographs and recordings and also samples of any articles or substances found on the premises. In the case of any article or substance which has caused or is likely to cause a danger to health or safety, an authorized officer may have it dismantled or subjected to any process or test. Further, the officer may take possession of the article and detain it for so long as is necessary to examine it, to ensure that it has not been tampered with and to ensure that it will be available as evidence in any proceedings.

4. Under this section the inspector acquires powers to question any person whom he believes to be able to give information relevant to an investigation. The inspector is able to require the production of documents required to be kept by virtue of any of the statutory provisions and any other book or document necessary for him to see as part of his enquiries.

5. The environmental health professional may require any person to afford him such facilities

and assistance with respect to any matter or thing within that person's control as is necessary to enable the inspector to exercise any of the powers conferred on him.

6. Finally, Section 20 gives a residual or 'catch-all formula' that gives the authorized officer any power necessary for the purpose of carrying into effect any of the relevant statutory provisions.

Officers should not view the use of these powers and similar ones contained in other statutes as optional. The question ought not to be 'which one of these shall I use?' but 'which of these is it not appropriate to use?' Information so gained may be the key to the correct selection of an enforcement option. It may also reveal some defence that makes the proposed enforcement action, such as the service of a notice, of questionable validity. The service of a notice under Section 80 of the Environmental Protection Act 1990 on a business is inadvisable when a best practicable means defence exists.

The only way to anticipate such matters properly is to use all appropriate investigative powers to the extent necessary to act in an informed manner. It is only in this way that the decision to prosecute or to serve notice can hope to withstand a challenge of this nature.

NOTICES

There is little doubt that one of the most useful enforcement options available to EHPs is the **statutory notice**. A wide range of statutes provides for the service of legally enforceable notices under many differing circumstances and there exist very sound reasons for this particular form of enforcement mechanism. The Robens Committee report [13] recognized the weakness of the pre-existing law relating to the enforcement of health and safety at work. It stated: 'The criminal courts are inevitably concerned more with events which have happened than with curing the underlying weaknesses which have caused them.'

The report noted that the penalty of imprisonment was hardly ever sought and that the then maximum level of fines was derisory. Examining this enforcement pattern the committee decided

not to recommend changes that would involve an increase in prosecutions, preferring instead to recommend a 'constructive means of ensuring that practical improvements are made and preventative measures adopted'. These were to be 'non-judicial administrative techniques for ensuring compliance with minimum standards of safety and health at work'.

New powers were recommended to give inspectors power to issue notices on their own authority, thereby achieving the stated aim. Although the Robens report was concerned with safety in the workplace, its findings offer an insight into the nature of many other forms of statutory notice. The use of a notice allows the option of enforcement without reference to a court, with the advantages that it is speedier, more flexible and less costly than the cumbersome operation associated with a criminal prosecution, while still achieving the aim of the relevant legislation.

The statutory notice represents one in a number of points in a continuum of options available to EHPs to deal with unsatisfactory situations. It offers an intermediate step between inactivity (here meaning no formal action based on the properly guided use of discretion) and prosecution.

Although the use of the enforcement notice does represent a position short of prosecution, it would be wrong not to recognize the legal significance of the notice and the need to approach the selection, drafting and service of such documents with complete discipline and thoroughness.

Statutory notices are basically of two types: those whose service is dependent on there having been a breach of a specific statutory provision; and those that are triggered by the existence of a 'substandard' state of affairs. Under the former category would come notices under provisions such as: Section 21 of the Health and Safety at Work, Etc. Act 1974 or Section 10 of the Food Safety Act 1990. The latter category would embrace notices such as Section 80 of the Environmental Protection Act 1990 and Section 189 of the Housing Act 1985. All categories would generally carry a penalty for non-compliance with the notice itself, but only in the former is the trigger event also a breach of the law. So, for example, there is no offence of causing a statutory nuisance but there is the offence of failing to comply with the terms of an abatement notice under the Environmental Protection Act 1990. Clearly, there may be one or two notices that do not sit easily in either of these categories, but as a general categorization the classification is sound.

A statutory notice is not a prosecution and its service does not invite the same standard of proof. For many statutory notices there is therefore no need to have obtained evidence 'beyond a reasonable doubt' about the state of affairs alleged. However, that is not to say that an officer can base the service of a notice on just any evidence. To be able to draft a valid and defensible notice, EHPs must be prepared to use all investigative options to inform their decision both to serve the notice and what to include in it. To do otherwise is to run the risk of serving a notice improperly. EHPs are given extensive investigative powers under a wide range of statutory provisions. Officers should not view these as optional. Again, the question ought not to be 'which of these shall I use?' but 'which of these is it not appropriate to use?' By considering all investigative options, and then positively rejecting those that are inappropriate, an officer can be sure that he will not overlook valuable information. This information may be the key to the correct drafting of the notice, as it may identify what the central issue in the problem is, and thus, when viewed within a developed enforcement policy, help the officer to decide on the appropriate specific remedial works. It may also reveal a defence, raising a question mark against the service of the notice in the first place.

Power to charge for enforcement action

Perhaps the need for more transparent enforcement is greater now that, in certain areas of work, provision is starting to be made to allow an authority to charge when a notice is served. The Housing Grants, Construction and Regeneration Act 1996 states that a local housing authority may now make such reasonable charges as it considers appropriate as a means of recovering administrative and other expenses incurred in taking action of the following kinds:

- serving a deferred action notice or deciding to renew such a notice;
- serving a repairs notice;

- making a closing order;
- making a demolition order.

In the Housing (Maximum Charge for Enforcement Action) Order 1996, the maximum charge was specified as £300. The expenses that may be recovered are those incurred in:

- determining whether to serve/renew the notice or make the order;
- identifying the works to be specified in the notice;
- serving the notice or order.

It is evident that arbitrary or unstructured enforcement policies are inconsistent with defensible charging regimes.

THE CAUTIONING OF OFFENDERS

As already stated, an EHP has a number of options in any given situation, ranging from taking no formal action, through the service of a notice, to prosecution and beyond. A further option on this enforcement continuum is the administration of a **formal caution** under the guidance to be found in Home Office Circular 18/1994. The purpose of the circular is to provide guidance on the cautioning of offenders, and in particular to:

- discourage the use of cautions in inappropriate cases, such as for offences that are triable on indictment only;
- seek greater consistency;
- promote the better recording of cautions.

Though obviously aimed at police officers, the circular is nevertheless appropriate for those involved in environmental health enforcement. Formal cautioning can be successfully incorporated into an enforcement policy and is relevant across the whole range of the work of the EHP. However, to be most effective this enforcement option must be deployed as part of an enforcement policy and not considered in isolation. Circular 18/94 recognizes the difficult interface that exists between informal warnings and formal cautions, and between formal cautions and prosecutions. To attempt to utilize the formal caution outside of a framework of guidance would be to run the risk of misusing a valuable enforcement tool. However, properly applied cautioning can be regarded as an effective form of disposal.

The decision to caution is always one for an enforcing authority, and should never be merely a matter of routine. The proper use of discretion, as has been said, is a matter to be guided by a thorough and refined enforcement policy, having regard to whether the circumstances are such that the caution is likely to be effective, and appropriate to the offence. Circular 18/94 identifies that the accurate recording of cautions is essential in order both to avoid multiple cautioning and to achieve greater consistency. A formal caution so recorded will then be expected to influence an authority in deciding whether to institute proceedings if the person should subsequently offend again. Additionally, and advantageously, an earlier caution may be cited in subsequent court proceedings for other offences if the person is found guilty.

Before a caution is given, departmental or other records should be checked to ascertain if the offender has received any such warnings previously. It is both possible and permissible for a person or organization to be cautioned on more than one occasion, although the policy implications of this should be acknowledged. The authors of the circular, aware that multiple cautioning may bring this option into disrepute, advise that cautions should not be administered to an offender in circumstances where there can be no reasonable expectation that this will curb his offending. Guidance suggests that it is only in the following circumstances that more than one caution should be considered:

- where the subsequent offence is trivial;
- where there has been a sufficient lapse of time since the first caution to suggest that it had some effect.

This is consistent with the identified purpose of formal cautioning, which is to deal quickly and simply with less serious offenders, to divert them from the criminal courts and to reduce the likelihood of their reoffending. Recognizing that there will be, in any enforcement policy, the option of simply giving a oral warning, the circular is clear that there is no intention of inhibiting this practice. Care should be taken, however, to ensure that this informal warning is not recorded as a formal caution. Unlike such a caution, a verbal warning may not be cited in subsequent court proceedings. It should therefore be clearly understood that a formal caution is not a sentence of the court and cannot be made conditional on the completion of a specific task.

Because of the seriousness of the decision to caution, it should not be taken lightly. In recognizing this, the circular stipulates certain conditions that must be met before a caution should be given:

- there must be evidence of the offender's guilt sufficient to give a realistic prospect of conviction;
- the offender must admit the offence;
- the offender must understand the significance of the caution and give informed consent to being cautioned.

Those proposing to use this method of disposal are warned to ensure that consent to the caution is not to be sought until it has been decided that cautioning is the correct course. The significance of the caution must be explained to the offender. The offender must understand that a record will be kept of the caution, that the fact of a previous caution may influence the decision of whether to prosecute if the person or company should offend again, and that it may be cited if the person or company should subsequently be found guilty of an offence. Where the evidence does not meet the required standard, the circular is clear that a caution cannot be administered. Nor will it be appropriate where a person does not make a clear and reliable admission of the offence. Cautioning is an alternative to a prosecution and therefore ought not to be administered where a prosecution could not be commenced.

Public interest principles

In the giving of a caution, as in any prosecution decision, public interest considerations apply that will inform the decision whether to administer a caution. Described in the *Code for Crown Prosecutors* (see later), these public interest principles are:

- the nature of the offence;
- the likely penalty if the offender were to be convicted by a court;
- the offender's age and state of health;
- previous criminal history;
- attitude of offender towards the offence, including practical expressions of regret.

The guidance in the circular was clearly formulated with the more mainstream criminal activities in mind, particularly when it considers the role and views of the victim when administering a formal caution. Yet, there is no reason why this guidance should not also be viewed as applicable to EHPs. For example, it advises that before a caution is administered it is desirable that the victims be contacted to establish their view of the offence and the nature and extent of any harm or loss suffered. These views should then be assessed, relative to the victims' circumstances, and consideration should be given to whether the offender has made any form of reparation or paid compensation. In some, though obviously not all, environmental health offences this would be sound guidance to follow. If a caution is being considered, it is little more than common sense and courtesy that its significance should be explained to the victim.

The code suggests that police officers administering cautions should be of a certain rank, and considers that it may be appropriate to nominate suitable cautioning officers. The concept of an officer of 'rank' is inappropriate for EHPs, although seniority is certainly a more accessible concept.

It is important that the giving or receiving of a caution is not seen as a soft option. This is why all formal cautions must be recorded and thorough records kept. As stated, formal cautions may be cited in court in subsequent proceedings if they are relevant to the offence then under consideration. This will occur after a guilty verdict, and although the existence of a prior caution cannot be relevant to any finding of guilt or innocence, it may have direct bearing on any sentence imposed. Care must be taken to distinguish previous cautions from previous convictions. Although the circular clearly envisages face to face formal cautioning, there appears to be no bar on the use of a formal cautioning system that relies on acceptance of the caution by letter.

PROSECUTION

Sayre [14] identified that the criminal law, which from early times had been used to punish those threatening the public health, was seized upon in the last century as a convenient instrument for the enforcement of a number of new regulations. This had little to do with moral disapproval or a heightened sense of the immorality of, for example, pollution, it was simply a question of convenience rather than legislative policy: using the criminal law to enforce social regulation in new fields of activity.

The criminal law continues to be used in this way, and has its ultimate manifestation in a criminal prosecution. Section 1 of the Prosecution of Offences Act 1985 created the **Crown Prosecution Service** (CPS). The CPS is a national prosecution agency for England and Wales and is responsible for the conduct of all proceedings instituted on behalf of a police force. No such similar agency exists for prosecutions brought by other public agencies. However, Section 6 of the Prosecution of Offences Act 1985 did preserve the right for private individuals and certain statutory bodies to commence criminal proceedings, thus allowing local authorities to be able to prosecute on relevant criminal matters independently of the CPS. In this respect, local authorities are very much their own masters when it comes to decisions relating to prosecution, and thus there are many occasions when a local authority will find itself a party to criminal proceedings. These proceedings will most commonly arise when an authority is undertaking one or more of its many enforcement activities under numerous statutory provisions. The general power of a local authority to institute proceedings is to be found in Section 222 of the Local

Government Act 1972, which states that where a local authority considers it expedient for the promotion or protection of the interests of the inhabitants of its area, it may:

- prosecute or defend or appear in any legal proceedings, and in the case of civil proceedings may institute them in its own name;
- in its own name, make representations in the interests of the inhabitants at any public inquiry held by or on behalf of any minister or public body under any enactment.

The prosecution of an offender is, at one level, merely another facet of the work of the EHP. It has its role and its purpose, both for the matter in hand and for the more general protection of health and the environment. However, it is clearly a serious step and one that should be undertaken only after appropriate consideration guided by a well-thought-out policy. The Environment Agency, for example, will consider prosecution where:

- it is appropriate in the circumstances as a way of drawing attention to the need for compliance and the maintenance of standards, especially where prosecution would be a normal expectation or where deterrence may be a consideration;
- there has been potential for considerable environmental harm arising from the breach;
- the gravity of the offence, taking into consideration the offender's record, warrants it.

Currently, the HSE may seek prosecution if the breach carries significant potential for harm, regardless of whether it caused an injury. In deciding whether to prosecute, the HSE and those local authorities adopting the HELA guidance, will also consider:

- the gravity of the offence;
- the general record and approach of the offender;
- whether it is desirable to be seen to produce some public effect, including the need to ensure remedial action and, through the punishment of offenders, to deter others from similar failures to comply with the law;

- whether the evidence available provides a realistic prospect of conviction.

In these respects the HSE and the Environment Agency are clearly guided by the *Code for Crown Prosecutors* published by the CPS.

In its *Code for Crown Prosecutors*, the CPS offers useful guidance on when proceedings are appropriate. The code, which is issued under Section 10 of the Prosecution of Offences Act 1985 and is a public document, is based on principles that have hitherto guided all who prosecute on behalf of the public. These principles obviously apply with no less vigour to EHPs; indeed the code makes it clear that it contains information that is important to all who work in the criminal justice system. The code stipulates two main tests to be borne in mind when considering a prosecution: the evidential sufficiency tests and the public interest test.

Evidential sufficiency

The *Code for Crown Prosecutors* states that a prosecution should be neither started nor continued unless the prosecutor is satisfied that there is admissible, substantial and reliable evidence that a criminal offence known to the law has been committed by an identifiable person or persons. The test to be applied is whether there is a realistic prospect of a conviction. A realistic prospect of conviction is identified as an objective test, and describes those circumstances when a jury or magistrates, when properly directed in accordance with the law and aware of all relevant facts, are more likely than not to convict the defendant of the charge alleged.

There are therefore certain matters that those considering a prosecution are expected to consider when evaluating evidence. These include:

- the requirements of the Police and Criminal Evidence Act 1984;
- any doubt on any admissions by the accused due to age, intelligence or apparent understanding of the accused;
- the reliability of witnesses; has a witness a motive for telling less than the whole truth?; might the defence attack his credibility?; are all

the necessary witnesses available and competent to give evidence?

The public interest criteria

If the evidential requirements are met, the CPS proposes that a prosecutor must then consider whether the public interest requires a prosecution. In this the CPS is guided by the view expressed by Sir Hartley (later Lord) Shawcross when he was Attorney General:

It has never been the rule in this country – I hope it never will be – that suspected criminal offences must automatically be the subject of prosecution. Indeed, the very first regulations under which the Director of Public Prosecutions worked provided that he should . . . prosecute 'wherever it appears that the offence or the circumstances of its commission is or are of such a character that a prosecution in respect thereof is required in the public interest'. That is still the dominant consideration . . . the effect which the prosecution, successful or unsuccessful as the case may be, would have upon public morale and order, and with any other considerations affecting public policy.

The factors that may lead to a decision not to prosecute will of course vary from case to case. However, the CPS broadly recognizes that the graver the offence, the less likelihood there will be that the public interest will allow a disposal other than prosecution. The code gives guidance on a number of matters in the decision which make a prosecution more or less likely. The code says that a prosecution is likely to be needed if:

(a) a conviction is likely to result in a significant sentence;
(b) a weapon was used or violence was threatened during the commission of the offence;
(c) the offence was committed against a person serving the public (e.g. a police or prison officer, or a nurse);
(d) the defendant was in a position of authority or trust;
(e) the evidence shows that the defendant was a ringleader or an organizer of the offence;

(f) there is evidence that the offence was premeditated;
(g) there is evidence that the offence was carried out by a group;
(h) the victim of the offence was vulnerable, has been put in considerable fear, or suffered personal attack, damage or disturbance;
(i) the offence was motivated by any form of discrimination against the victim's ethnic or national origin, sex, religious beliefs, political views or sexual orientation, or the suspect demonstrated hostility towards the victim based on any of those characteristics;
(j) there is a marked difference between the actual or mental ages of the defendant and the victim, or if there is any element of corruption;
(k) the defendant's previous convictions or cautions are relevant to the present offence;
(l) the defendant is alleged to have committed the offence whilst under an order of the court;
(m) there are grounds for believing that the offence is likely to be continued or repeated, for example, by a history of recurring conduct; or
(n) the offence, although not serious in itself, is widespread in the area where it was committed.

Equally, it goes onto to say that a prosecution is less likely to be needed if:

(a) the court is likely to impose a nominal penalty;
(b) the defendant has already been made the subject of a sentence and any further conviction would be unlikely to result in the imposition of an additional sentence or order, unless the nature of the particular offence requires a prosecution;
(c) the offence was committed as a result of a genuine mistake or misunderstanding (these factors must be balanced against the seriousness of the offence);
(d) the loss or harm can be described as minor and was the result of a single incident, particularly if it was caused by a misjudgement;
(e) there has been a long delay between the offence taking place and the date of the trial, unless:

- the offence is serious;
- the delay has been caused in part by the defendant;

- the offence has only recently come to light; or
- the complexity of the offence has meant that there has been a long investigation;

(f) a prosecution is likely to have a bad effect on the victim's physical or mental health, always bearing in mind the seriousness of the offence;

(g) the defendant is elderly or is, or was at the time of the offence, suffering from significant mental or physical ill health, unless the offence is serious or there is a real possibility that it may be repeated. The Crown Prosecution Service, where necessary, applies Home Office guidelines about how to deal with mentally disordered offenders. Crown Prosecutors must balance the desirability of diverting a defendant who is suffering from significant mental or physical ill health with the need to safeguard the general public;

(h) the defendant has put right the loss or harm that was caused (but defendants must not avoid prosecution solely because they pay compensation); or

(i) details may be made public that could harm sources of information, international relations or national security.

Each case will be different and, as the CPS points out, it should never become simply a matter of adding up the number of factors on each side of the argument. Although each of the foregoing factors is certainly relevant to the EHPs when formulating a policy and deciding upon a course of action, EHPs must decide on the importance of each factor on a case by case basis.

The approach so far would facilitate a more uniform pattern of prosecution but does not consider all factors at play. One reason for the non-use of prosecution as a means of enforcement is that there may be a belief that the available sanctions are neither appropriate not effective. This is not simply a case of fines being too small [2]. Sometimes the fine may be, in the opinion of the officers, too large, thereby removing the possibility of money being spent on preventative measures [10]. In more marginal cases where EHPs cannot readily decide whether to recommend prosecution or not, their view of the court as an 'awful place' [1] may even be decisive.

The reason for this opinion of the court as an awful place may be complex. Croal [15] notes that EHPs and other officers assume that offenders are dealt with sympathetically and leniently in court. This treatment is attributed to the fact that, in addition to the common use of strict liability in the framing of offences, there exists a 'cultural homogeneity' between the offenders and the personnel of the court. This, Croal records, has, arguably, led to the marginalization of offences through a denial of moral blameworthiness. Hawkins [1] observes that officers of environmental enforcement agencies often regard the courts as remote from and unsympathetic to the real problems of enforcing the law. Where pollution control is concerned, and arguably elsewhere, magistrates are often regarded as ignorant laymen, possessing neither the knowledge nor experience of field staff, ignorant of the causes and treatment of pollution, and lacking the technical and scientific awareness to make informed decisions. Hawkins [1] notes that there are situations where the courts are perceived by the regulating authority as not 'understanding pollution'. Hawkins puts this down to the existence of two discrepant views: that of the magistrates is the product of matters raised in court; that of the regulators is the product of the experience of the regulator with this case and more generally. Hawkins claims that magistrates see a cross-section of reality, whereas enforcement agencies have a longitudinal view of career. All too often enforcement officers can feel that when a case does come before the court the true issues are, in practice, obscured by the rules of evidence and procedure applicable to a criminal court.

EHPs must recognize this; to fail to do so is to view the system of prosecution through the courts not as the sum of its parts, but as an end point. Here, any or all perceived deficiencies are focused into a single event (a hearing or trial), and it is this single event and those who participate in it who are to blame if any deficiencies surface. This, for a professional, is too narrow a vision. It fails to recognize the complexity of the issue. The court is but one part in a complex and interrelated mechanism, affected by and in turn affecting the other parts.

Proceeding to a prosecution

If the option of a prosecution is taken, then the pathway of a case to a court is relatively straightforward. Individuals or companies appear before a court either because they have been charged with an offence or because a summons has been issued. The charging of a person will follow an arrest; this is not a power available to EHPs and therefore it is the latter process only that will be discussed below. The means by which a person is called before the magistrates' court to answer criminal matters is by a summons, and this is obtained by the laying of an information before the court.

Laying information

An information is essentially the means by which the court is informed of an alleged offence. It will ultimately form the charge that the accused will be obliged to answer at any subsequent trial. An information is therefore required to describe the offence in ordinary language. The **Magistrates' Courts Rules 1981** set the detail of what an information must contain. They require that every information and summons is sufficient if it describes the specific offence with which the accused is charged, in ordinary language, without necessarily stating all the elements of the offence.

Although it is unlikely that an EHP would lay the information, there is no prohibition against this. The Magistrates' Court Act 1980 provides that any person authorized by the prosecutor may lay an information before a Justice of the Peace (JP). The Act also stipulates that, upon an information being laid, the JP may issue a summons requiring the person accused to appear before a magistrates' court to answer the information, or issue a warrant to arrest that person and bring him before the court.

An information may be laid either orally or in writing. As a matter of practice all informations laid in connection with an environmental health offence will be in writing. The information is accepted as being laid when it is received at the office of the clerk to the justices. It is not necessary for the matter to be put before the clerk of the justices personally. Any EHP laying an information would be well advised to record the date and time that the information was handed to the office. There have certainly been cases where this has proved vital in establishing that the subsequent proceedings had not become time-barred under the Magistrates' Court Act 1980, Section 127.

Time period for proceedings

There is no time limit for proceedings for an indictable offence, although Section 127(1) of the Magistrates' Courts Act 1980 requires that, in the absence of an express statutory provision to the contrary, proceedings for summary matters must be commenced, that is, the information must be laid, within six months of the offence having been committed. For a summary offence this six-month period runs to the date the information is laid, not to the date when the accused actually appears in court. Therefore where an offence is triable only summarily, this is the relevant period and is an important limitation. The interdepartmental inertia that can arise when a prosecution file is passed from one local authority department to another may result in the unwary falling into the trap of attempting to prosecute on time-expired matters. In calculating the relevant time period the date on which the offence was committed is to be excluded (*Marren v. Dawson Bentley & Co.* [1961]). Thus for an offence committed on 1 January, the last date for laying an information is 1 July. Were the offence to be committed on 31 March, the deadline would be 30 September. For continuing offences the time period runs from each day the offence is committed. It is important to note that this six-month time limit does not apply to an indictable offence and therefore not to summary proceedings where the offence is triable either way.

Duplicity

It is essential that an information and summons does not charge more than one offence. The Magistrates' Court Rules 1981 states that magistrates' courts shall not proceed to the trial of an information that charges more than one offence.

Such an information is bad for duplicity and the court is prohibited from convicting on the matters alleged. If the duplicitous matters are identified during the trial, the court is obliged to request that the prosecutor elect which offence to proceed upon. Once identified, other matters are struck out and the trial proceeds on the amended information.

It is not duplicitous to set out common factual and legal matters in a preamble to the summons and then set out the various offences charged in numbered paragraphs if the offences are contrary to the same statutory provisions. In *Shah* v. *Swallow* [1984], a shop owner was convicted under the Food Hygiene (General) Regulations 1970 of four out of five offences contained in an information. The five offences were contained in a single information. The first paragraph of the document contained allegations of fact in relation to each of the offences; and the second paragraph identified the regulations that created the offences. The House of Lords found that the document was a single document that contained five informations. The fact that the preamble contained details that were common to a number of separate allegations did not tie them so as to make the information duplicitous.

Summons

As already indicated, Section 1 of the Magistrates' Court Act 1980 states that upon an information being laid before a JP that any person has, or is suspected of having, committed an offence, the JP may issue a summons requiring the person concerned to appear to answer the information. The summons should contain the nature of the information and the time and place at which the accused is required to appear. Although containing essentially the same information whatever the offence, the precise format of a summons may vary.

Complaint

Part II of the Magistrates' Court Act 1980 deals with the civil jurisdiction of the magistrates' court. It indicates how it is possible to obtain a summons from the court by the making of a complaint. A complaint is a written or verbal reference to a JP

or justice's clerk that a person or a corporation has committed a breach of the law that is not criminal. As for an information, once the complaint is received at the office of the clerk the complaint is accepted as having been duly made. Section 52 of the Magistrates' Court Act 1980 stipulates that where a complaint is made, a JP may issue a summons directed to that person requiring him to appear before a magistrates' court acting for that area to answer the complaint. This is the means for commencing a civil action in a magistrates' court where, according to Section 52 of the Magistrates' Court Act 1980, the court has jurisdiction.

It is not very common for EHPs to be party to such proceedings but it is possible. For example, Section 82 of the Environmental Protection Act 1990 allows an individual aggrieved by the existence of a statutory nuisance to bring an action in a magistrates' court. The statute provides that a magistrates' court may act under this section on a complaint made by an aggrieved person. In fact, in a number of cases it has been found that the procedure contained in Section 82 of the Environmental Protection Act 1990, although commenced by complaint, is in fact criminal in nature (*Herbert* v. *Lambeth LBC* [1991]; *Botross* v. *London Borough of Hammersmith and Fulham* [1995]).

THE FUTURE

It is an interesting speculation to pose the question 'if we were setting up a system now to protect public health would we start from here?'

There is, without doubt, a patchwork of agencies and responsibilities which may not always work together to best effect.

The nature of local authority Environmental Health enforcement work has inevitably focused on 'local' matters though certainly grounded on national agendas. This focus was always a limitation capable of exploitation by the more criminally enterprising or could lead to exclusion from national stages. Certainly evidence would suggest that there are always enterprising criminal elements waiting to exploit any profitable loophole in the law.

So it should perhaps thus come as no surprise that we have now seen this 'underside' of environmental health surface in complex trials. Trials where people have been convicted of offences related to wide ranging conspiracies to defraud the public, and of course seriously jeopardize their health, by selling meat for human consumption when it was unfit. Such conspiracies, coming as part of the seemingly endless string of food scares would not unnaturally prove to be a stimulus to action.

The Waste Food Task Force was set up by the Food Standards Agency and made up of EHOs, a trading standards officer and industry and consumer representatives as part of an action plan to prevent animal by-products finding their way into the human food chain.

In its exploration of the subject the task force found a patchwork or tapestry of enforcement agencies working to what was described a 'silo culture'. Between these silos a vacuum of supervision and enforcement was found to be possible. Criminals, unlike nature, do not abhor a vacuum. Instead they see an opportunity. Groups operating in Rotherham, Amber Valley and South Norfolk saw such an opportunity. Here criminal gangs were found to have 'filled' this vacuum by their criminal efforts. The task force made 23 recommendations, the full effect of which will take time to be realized. Though it could not quantify the scale of the problem the task force noted that 'what may appear to be a small-scale local problem may have much wider implications'. It advocated the need for effective co-ordination and central support for enforcement officers working in these 'silo' agencies.

The Food Standards Authority was seen as the appropriate agency to provide the necessary co-ordination and leadership. It was to work with Local Authorities that have experience of these activities to create a 'task force' of Local Authority staff who could devote some of their time to aiding fellow authorities experiencing these issues for the first time. To help ensure a consistent approach is taken on these issues, the Agency provides comprehensive training for task force members on legal matters, investigatory techniques and problems specific to this type of

fraud. In addition, the Agency is to produce guidance on investigations that will be issued to all local authorities in England. FSA Scotland, Wales and Northern Ireland will be issuing guidance to local authorities in those countries.

Food safety is not the only area, however, where the need for inter-agency co-operation when enforcing the law is prominent.

At present Health and Safety Enforcement authorities refer any evidence relating to the offence of manslaughter to the police, as they have no responsibility to conduct investigations into manslaughter. Again this 'silo' approach to thinking could have caused unnecessary problems. To address these there was, in 1998, published a protocol of liaison for investigating work-related deaths.

Under the 1998 Protocol there was established principles for effective liaison between the HSE, the Police and the CPS in relation to work-related deaths in England and Wales.

The stated aim, from the outset, was to create a mechanism that provided a framework that encouraged a partnership approach to the investigation of events leading to a work-related death. The signatories were to work together to ensure effective communication, co-ordinate information and where appropriate secure joint decision-making.

In 2003, the Health and Safety Executive (HSE), Crown Prosecution Service (CPS), Association of Chief Police Officers (ACPO), Local Government Association (LGA) and British Transport Police (BTP) published a revised protocol – Work-Related Deaths: A Protocol For Liaison. Now EHPs should follow this protocol taking guidance on joint working with police when investigating a workplace death. Meeting twice yearly the National Liaison Committee, made up of representatives from ACPO, BTP, CPS, HSE and LGA, makes suggestions for improvements as necessary aimed at ensuring the continuing effective operation of the protocol.

The de-compartmentalization of thinking addressed by the two initiatives discussed here represent welcome, if overdue changes. The activity of enforcement, when viewed as a process rather than merely an outcome, reveals the role of EHPs to be pivotal. Few public officials have vested in them the powers available to EHPs.

If public confidence is to be maintained, these powers, and especially the powers of enforcement, must be used wisely. The new initiatives explicitly invite Environmental Health professionals to dive deeper into the waters of enforcement. In doing so it is expected that those taking the plunge will be more accomplished swimmers. It is incumbent on all to ensure that they are up to the task. 'A little learning is a dangerous thing; drink deep, or taste not the Pierian spring: there shallow draughts intoxicate the brain, and drinking largely sobers us again.' (Alexander Pope (1688–1744) – An Essay on Criticism.)

REFERENCES

1. Hawkins, K. (1984) *Environment and Enforcement: Regulation and the Social Definition of Pollution*, Clarendon Press, Oxford.
2. Hutter, B.M. (1988) *The Reasonable Arm of the Law: The Law Enforcement Procedures of Environmental Health Officers*, Oxford Socio-Legal Studies, Clarendon Press, Oxford.
3. Carter, H. (1992) The criminal law as a tool for environmental protection. *Environmental Law*, 6(1), 187–96.
4. Vogel, D. (1986) *National Styles of Regulation*, Cornell University Press, USA.
5. Carson, W.G. (1970) White collar crimes and the enforcement of factory legislation. *British Journal of Criminology*, 10, 383–98.
6. Audit Commission (1991) *Towards a Healthier Environment, Managing Environmental Health Services*, Audit Commission, HMSO, London.
7. DTI (1994) *Local Government Enforcement, Report of the Interdepartmental Review Team*, Department of Trade and Industry, HMSO, London, p. 35.
8. Rowan-Robinson, J. and Ross, A. (1994) Enforcement of environmental regulation in Britain: strengthening the link. *Journal of Planning and Environmental Law*, March, 200–10.
9. HSE (2000) *Choice Of Appropriate Enforcement Procedure*, LAC 22/1, Health and Safety Executive. Local Authority Unit, London.
10. Richardson, G., Ogus, A. and Burrows, P. (1982) *Policing Pollution*, Oxford Socio-Legal Studies, Clarendon Press, Oxford.
11. *Health and Safety at Work, Etc. Act 1974: Enforcement Policy Statement*, HSC15, 1 January 2002.
12. Study Guide (2nd edn) *Human Rights Act 1998*, Lord Chancellor's Dept, October 2002.
13. Robens Committee (1972) *The Report of the Commission on Safety and Health at Work*, Cmnd 5034, HMSO, London.
14. Sayre, F.B. (1933) Public welfare offences. *Columbia Law Review*, 33, 55.
15. Croal, H. (1988) Mistakes, accidents and someone else's fault: the trading offender in court. *Journal of Law and Society*, 15(3), 293–315.

FURTHER READING

Bassett, W.H. (2002) *Environmental Health Procedures*, 6th edn, Spon Press, London.
Moran, T. (1997) *Legal Competence in Environmental Health*, E & FN Spon, London.

8 Fundamentals of information technology and its application in environmental health

Paul B. Paddock

INTRODUCTION

Over the past 50 years, organizations have been developing computer-based information systems. Throughout the 1960s, 1970s and early 1980s the vast majority of issues were associated with how to 'supply' information systems to organizations. As these issues have become better understood, and with many of the basic organizational systems having been automated, attention has turned to more imaginative and fruitful applications for information technology (IT) and the ascertaining of the 'demand' for information systems in organizations. In essence, organizations have witnessed a revolution in the way people do their jobs and in the development of new systems. The computer keyboard and screen are now familiar desktop tools in many offices. This may be a computer terminal used to access a remote mainframe computer or a personal computer used for word-processing and spreadsheet calculations or, increasingly these days, a combination of both.

With this 'explosion' in IT has come a plethora of technical terms, many of which are unknown to, or poorly understood by, the **end-users** – the people receiving the printouts and manning the computer workstations, as well as those who order and pay for the technology. An understanding of the fundamentals and terminology of IT will assist end-users to utilize the power of the computerized information systems introduced into an organization. In general terms, the objectives of IT in any organization are to make the organization more efficient, to make managers more effective and to achieve a competitive advantage for business (or to avoid being disadvantaged).

DATA AND INFORMATION

The terms 'data' and 'information' are used interchangeably in everyday speech to mean the same thing. However, for managers and information specialists these terms have specific meanings. **Data** are facts, events and transactions that have been recorded; they are the raw input materials from which information is produced. **Information** is data that have been processed in such a way as to be understood by and useful to the recipient.

The mere act of processing data does not itself produce information. Figure 8.1 outlines a model that is applicable to all information systems, whether manual or computerized, and that illustrates the important distinction between these two terms. The characteristics of good information are identified in Table 8.1. For a user to value

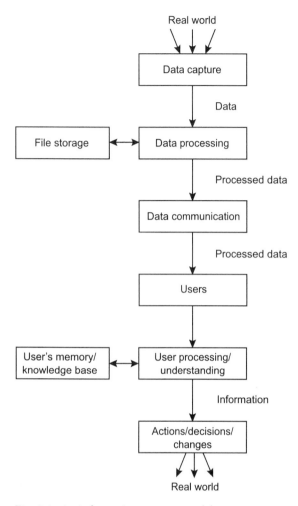

Fig. 8.1 An information systems model.

information, it must meet these characteristics because they enhance the usefulness of such information in decisionmaking.

Managers at all levels within an organization need relevant information to assist them to plan, control and make decisions. Relevant information increases knowledge, reduces uncertainty and is usable for the intended purpose. A suitable management information system should provide managers with relevant, complete, accurate, clear and timely information.

INFORMATION AS A RESOURCE

Information is a resource that needs to be managed just like any other resource and 'information resource management' (IRM) has become a widely accepted term. IRM brings together an amalgam of skills and attitudes from information science, data processing, management science and records management. Information must be regarded as a valuable asset or resource that deserves and needs the same kind of management disciplines given to other resources – financial, physical, material and natural. There are some similarities between information and other 'traditional' managed resources as listed in Table 8.2, indicating that information should be treated as a tangible entity. However, due to the very unique, even paradoxical, qualities of information summarized in Table 8.2, it needs to be treated very carefully, separate from other facets of resources management.

MANAGEMENT INFORMATION SYSTEMS AND INFORMATION TECHNOLOGY

Having established that information is a resource that needs to be managed, it is essential for any business, whether in the private or public sector, to have a strategy for its information and systems that is driven by the requirements of the business. If a business does not develop an information systems (IS) strategy, then this could result in the business being seriously disadvantaged within its business environment and/or incurring significant expenditure on IT investments but achieving few

Table 8.1 The qualities of good information

1. Relevant for its purpose
2. Sufficiently accurate for its purpose
3. Complete enough for the problem
4. From a source in which the user has confidence
5. Communicated to the right person
6. Communicated in time for its purpose
7. Contains the right level of detail
8. Communicated by an appropriate channel of communication
9. Understandable by the user

Table 8.2 Information as a resource

Similarities between information and other 'traditional' managed resources

1. Information is acquired at a definite measurable cost
2. Information possesses a definite value
3. Information consumption can be quantified
4. Cost accounting techniques can be applied to help control the costs of information
5. Information has identifiable and measurable characteristics
6. Information has a clear life-cycle: definition of requirements, collection, transmission, processing, storage, dissemination, use, disposal
7. Information may be processed and refined
8. choices are available to management in making trade-offs between different grades, types and prices for information

Unique qualities of information

1. Information is expandable, it increases with use
2. Information is compressible, able to be summarized, integrated, etc.
3. Information can substitute for other resources
4. Information is transportable virtually instantaneously
5. Information is diffusive, tending to leak from the straightjacket of secrecy and control
6. Information is sharable, not exchangeable, it can be given away and retained at the same time

---

business benefits. The principal components of such a strategy and the relationships between them are summarized in Fig. 8.2.

The term 'management information system' (MIS) has become almost synonymous with computer-based data processing. However, a more open interpretation is 'a system to convert data from internal and external sources into information and to communicate that information, in an appropriate form, to managers at all levels in all functions to enable them to make timely and effective decisions for planning, directing and controlling the activities for which they are responsible'. The emphasis of this definition is on the use of the information and not on how it is produced. However, it is recognized that the vast majority of organizations use IT to generate the information required by managers and decision makers.

An analysis of the present IT environment identifies three definite interlinking strands:

1. A more competitive hardware and software market, including the development of consistent standards such as the International Standards Organization's Open Systems Interconnection (OSI) standards as a basis for all future IT products.
2. Changing methods of delivering management information to the users, such as by the combination of information from different systems and by the development of geographical information systems (GIS).
3. Changing working methods in the office through the extensive use of office automation, the analysis of information flows and the possible consideration of business process redesign/ re-engineering (BPR).

With the rapid developments in microelectronics and telecommunications currently taking place, office work is being transformed, which is, in turn, influencing the availability and type of information that managers use.

RECENT MAJOR DEVELOPMENTS

Several major IT and communications developments have taken place over the past few years:

- the Internet and the World Wide Web(WWW)
- electronic mail (email)
- video conferencing
- broadband
- networks and wireless networking.

The Internet and the World Wide Web

The Internet is a global network of computer networks. It started at the end of the 1960s when the US military began to appreciate the vulnerability of its communication systems. The US Defense Department developed the idea of a computer network that would be able to withstand an attack on the system. If part of the network was disabled, information would still be able to find its way

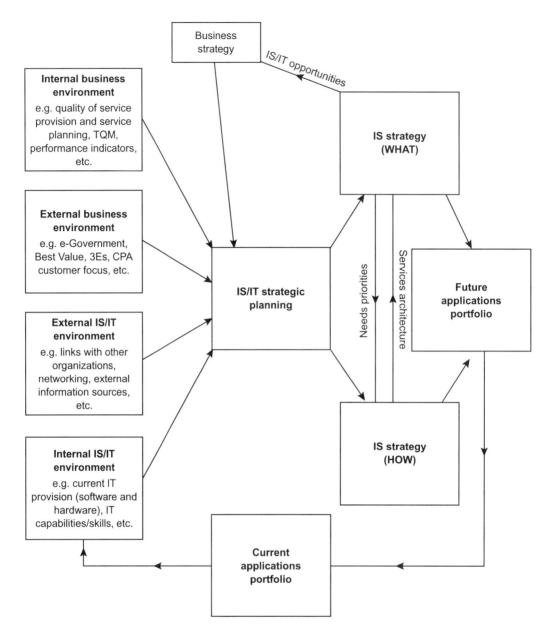

Fig. 8.2 Inputs and outputs to IS/IT strategy. TQM, total quality management; 3 Es = efficiency, economy and effectiveness; CPA, comprehensive performance assessment (adapted from [10]).

around the damaged part and communication lines would be secure. This development became known as ARPANET (Advanced Research Projects Agency Network). The computers that were developed for this system were, and still are, responsible for addressing and checking the communications, rather than the network itself. The computer data that was sent over the ARPANET was bundled into standard packages called Internet Protocol (IP) packets, each of which contained information about the addresses of the originating and destination computers.

During the 1970s and 1980s many academic institutions linked into ARPANET. However, it was in the late 1980s that the US Government's National Science Foundation (NSF) set up five supercomputer centres, which users were able to access from anywhere in the country. NSF created its own network, NSFNET, which used the same IP technology as ARPANET. Soon after, the system expanded rapidly, and with hardware upgrading of both computers and telephone lines, the network was opened up to commercial traffic in the early 1990s. This network of computer networks uses a language called TCP/IP (transmission control protocol/Internet protocol) which enables different computers to communicate with each other more easily.

The WWW is one of the main reasons for the considerable increase in the use of the Internet. It was developed at the European Laboratory for Particle Physics in Switzerland in 1990 and during the early part of 1995 it became the most popular means of accessing information on the Internet. The WWW allows users to 'browse the net' by the use of hyperlinks. All web pages are created using HyperText Markup Language (HTML), which started as a simple way of formatting text but has now expanded to include commands for integrating pictures, video, sound and even forms that can be filled in online. HTML is a standard language that all parties on the web have implemented, allowing the presentation of formatted text with images that look good across different computer platforms, and providing quick access to other Internet sites through hyperlinks ('hot' areas of text that open other HTML documents, irrespective of their location on the net).

A program known as a web browser is needed to access the WWW. Two of the most popular are Netscape Navigator® and Microsoft Network®. These programs allow access to addresses, also called Uniform Resource Locators (URLs), on the web. All web addresses start with http://, which stands for HyperText Transfer Protocol. There is also a newer version called https://, which incorporates data encryption enabling the secure transmission of data over the web. Examples of useful Internet sites with an environmental health perspective are given in Table 8.3.

Table 8.3 Some useful sites on the Internet for environmental health

Location (URL)*	Description
http://www.ukonline.gov.uk	The UK Government Information Service makes available a large number of government services, as well as usefully providing a daily update of government business
http://europa.eu.int	The European Union's Internet site
http://www.cieh.org.uk	The home page for the Chartered Institute of Environmental Health
http://www.who.ch	The home page for the World Health Organization
http://www.bsi.org.uk	The home page for the British Standards Institution

Note:
*URL, uniform resource locator; a method of addressing resources available on the Internet in a standard way.
All of these sites provide hot links to other related Internet addresses.

The final essential element required to gain access to the Internet is known as an Internet Service Provider (ISP), companies that sell connections to the net. Examples of ISPs include CompuServe™, AOL™ and Microsoft Network™. The choice of an ISP depends on the level of service required and the costs involved.

Electronic mail

Electronic mail (email) is a rapidly expanding development of the Internet. Any computer connected to the telephone network via a modem can leave messages at a central computer. These messages may be collected by others. All subscribers have their own 'pigeon hole'/'mail box' and can dial up the central computer to check to see if a message is waiting. It is possible to email many people at one time and it is fast and cheap. Email messages can have two or three components. The first is the header, which contains the addressing information, a description of the contents and a list of the recipients. The second component is the text of the message, and the third is an optional attachment. It is possible to attach virtually anything to a message – a word-processed file, an image and so on.

Email addresses are easy to understand and comprise a number of components. An example of an address would be another@hse.gov.uk The part after the @ is known as the domain name, uk means the domain is on a computer in the United Kingdom (every country has a two-letter code, the top-level domain code), gov means the domain belongs to a government organization, in this instance the Health and Safety Executive (HSE), another is the username. It is important to get the email address correct, otherwise the mail will be returned with a message from a mail server that the address was not recognized.

Video conferencing

Video conferencing is a method of communication that allows people to have face-to-face interaction over long distances and is set to change from an isolated function to a core element of most business functions. It is important to note that video conferencing does not negate the need for face-to-face meetings as it is important for people to meet on a regular basis. However, it does allow people to collaborate more often than they otherwise could, and to do so in a more satisfactory manner than just using the telephone. It is also useful for *ad hoc* communications, such as technical support tasks or to allow professionals to give more comprehensive advice.

This technology is developing at a considerable rate. In video conferencing, both data and video are interactively shared between various locations. The signal can be carried by a variety of means: the local area network (LAN); the Integrated Services Digital Network (ISDN), a digital mobile telephone (GSM) or the Public Switched Telephone Network (PSTN). With the advent of broadband technologies, networks are able to cope with large video signals to provide a real motion-video connection.

Broadband

Broadband refers to telecommunication in which a wide band of frequencies is available to transmit

information. Subsequently, information can be sent concurrently on many different frequencies or channels within the band, allowing more information to be transmitted in a given amount of time. Broadband is distinct from Narrowband – a term used to define traditional modes of transmission such as an analogue telephone line. Definers of Broadband have assigned a minimum data rate to the term, and these vary between 256 Kbps and 1 MB.

It is generally agreed that Digital Subscriber Line (DSL), Wireless Local Loop (WLL) and T1/E1 are modes of Broadband 'access network' – the physical infrastructure that connects homes and businesses to the service provider network, and the Internet. The provision of Broadband services to a local population depends on the availability of appropriate infrastructure reaching out to all premises. The UK broadband network falls under the government's remit and, in particular, Ofcom.

Networks and wireless networking

Networks are communication systems that link computers, storage devices, word-processors, printers and even the telephone system. There are a number of different types of network with names that describe the geographical area over which their different components are spread:

- **wide area networks** (WANs) span separate locations that may be many miles apart and use the general telecommunications network;
- **metropolitan area networks** (MANs) span a single city – they require special cables that are currently being laid in some cities to provide high speed communications in order to cope with the increasing graphical content of IT and the expected growth in the transmission of video and voice data;
- **local area networks** (LANs) are usually restricted to a single or a few buildings and are linked by direct cables rather than by general telecommunication lines.

Each type of network may comprise one or more types of computer system with geographical links through circuits provided by the national postal, telegraph and telephone authority (PTT). These circuits could be ordinary dial-up telephone lines or dedicated circuits (leased lines) used exclusively for computer communications.

Within organizations, the simplest type of network is one where it is possible to share disks and printers between several computers. **Peer-to-peer networks** allow any computer to make its disks and printers available to any other computer on the network. The advantage of this type of network is its simplicity and its low cost. However, as performance is likely to suffer, many organizations have installed networks using **dedicated servers**, which supply the requested services. Each computer on the LAN needs to be connected to the network through an appropriate interfacing card, which is dependent on the type of network installed, typically Ethernet® and Token Ring® networks. A further development in networking involves wireless connections. Wireless networking has many advantages over the wired alternative in that users can be fully mobile yet still have network connectivity for email, web and database applications. However, speed of data transfer over wireless networks still does not match that possible across cables. It should also be noted that there are several wireless networking standards, such as 802.11a, 802.11b, 802.11b+ and 802.11i, operating from 1 to 54 Mbits/s.

For LANs to work efficiently, network management programs are required. Examples of these are Novell NetWare®, LAN Server® and LAN Manager®. Critical to network operation is an effective server-based backup strategy, usually provided by tape drives. It is essential to institute daily backups in order to avoid disasters.

ELECTRONIC GOVERNMENT (e-GOVERNMENT)

As a result of the foregoing, and other, technological developments, government published a paper entitled 'Our competitive future; building the knowledge driven economy' in December 1998 [1]. This paper was an attempt to foster a new entrepreneurial spirit in order to achieve commercial success and prosperity and it also recognized

the key role of government in acting as a catalyst, investor and regulator. At the same time, it was recognized that new technology offered opportunities and choice and could give access by customers to public services 24 hours a day, 7 days a week. Government firmly stated its intention to be at the head of these developments and give effect to the vision that was about to be published in the 'Modernising government' White Paper [2].

'Modernising government' White Paper

In March 1999, the government published the 'Modernising government' White Paper in which it was stated that 'Government matters. We all want it to deliver policies, programmes and services that will make us more healthy, more secure and better equiped to tackle the challenges we face. Government should improve the quality of our lives. . . . Modernisation is vital if government is to achieve that ambition' [2]. The proposed reforms included a new target of all dealings with government being deliverable electronically by 2008. Chapter 5 of the White Paper was concerned with 'information age government' – the use of new technology to meet the needs of citizens and business, and not trail behind technological developments. At this time it was already appreciated that a corporate IT strategy for government needed to be developed as different public service agencies had developed their IT systems separately and that encouragement was needed to get them to converge and interconnect. The importance of e-commerce was also recognized along with the need to develop electronic services for citizens and business. Within the White Paper reference was made to driving up technology standards in the public sector, with the establishment of a central/local information age government concordat, which was meant to encourage innovation and co-operation between central and local service providers. Of particular significance was reference to a range of new frameworks across government to cover data standards, digital signatures, call centres, smartcards, digital TVwebsites, government gateways and better online services for businesses. However, the White Paper specifically set down new targets for electronic delivery – by 2002, 25%

of dealings with government should be capable of being done by the public electronically, with 50% of dealings by 2005 and 100% by 2008. Local authorities were expected to set and publish their own targets for electronic delivery and to work with the Local Government Association in order to bring this about. Whilst it was appreciated that there would be a range of technological developments and breakthroughs, a range of drivers for information age government were identified:

- Household access to electronic services;
- User-friendly, inexpensive and multifunctional technology;
- Less dependence on keyboard skills;
- Dramatic increases in computing power;
- Multi-purpose smartcards;
- Government forms and other processes that are interactive;
- Smarter knowledge management;
- Use of government websites;
- Repackaging of government services or functions;
- Flexible 'invest to save' approaches.

In September 1999, government set up the Local Government Modernisation Team to work with local authorities throughout England in taking forward the government's agenda for local government modernization. This team has been working on information age government, in particular electronic service delivery (ESD) and how it fits within the modernization agenda.

After the publication of this White Paper, the Prime Minister announced on 30 March 2000 that the target date of 2008 announced in the 'Modernising government' White Paper for all services to be available electronically, was to be advanced to 2005. This announcement came a few weeks after he promoted plans to ensure that everyone who wants it will have access to the Internet by 2005.

e-Government strategic framework

A strategic framework for public services in the information age was published in April 2000 [3]. e-Government was an initiative launched to ensure that government played a full part in the radical transformation of our society as a result of the information age revolution. In fact e-government fulfilled the government's commitment (in the 'Modernising government' White Paper) to publish a strategy for information age government and it was centred on four guiding principles:

- Building services around citizens' choices;
- Making government and its services more accessible;
- Social inclusion;
- Using information better.

The strategy was also meant to encourage innovators in government to identify new ways of working in partnership with the private sector along with a strong lead and effective support from the centre. To this end, the e-Envoy was set up with a view to owning the e-government strategy and to lead its implementation and to identify new opportunities for cross-cutting initiatives. In addition the Central IT Unit in the cabinet office was created to assist in the support of citizen-focused integration, to lead implementation of framework policies, standards and guidelines, to promote shared infrastructure and applications and to establish a government portal.

The vision within this document relies on services being available through many delivery channels as well as a drive towards better integration of services. It was envisaged that there would be government, sectoral and local portals with many of these accessible through private sector sites thereby removing the monopoly of entry from the public sector. The development of such portals has created a wide range of e-businesses, many in the private sector, operating within a regulatory framework comprising key building blocks for e-government.

This strategic framework promotes an architectural model outlining how individual departmental and sectoral initiatives relate to the framework and its associated standards. This model is shown in Fig. 8.3 and has three principal elements:

- access – services will be accessed by multiple technologies thus framework policies need to be developed;

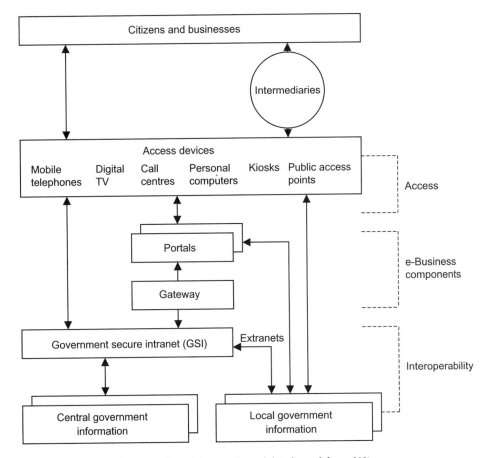

Fig. 8.3 e-Government strategic framework architectural model (adapted from [3]).

- e-business components – framework policies need to be developed for third-party service delivery channels, security of transactions and information, authentication and smart cards;
- interoperability – common standards and infrastructure need to be developed to enable electronic communication between government departments and the wider public sector.

Building blocks of e-government

Framework policies for access technologies

Services will be accessed by multiple technologies, including websites accessible from PCs, kiosks, mobile phone and digital TV, and call and contact centres and intermediaries may use these technologies to provide better face-to-face services. Framework policies on websites, call centres and digital TV (including interactive DiTV) have already been published.

e-Business components

Framework policies and standards have been published for third-party service delivery channels, security of transactions and information, authentication and smart cards. This will save work for service providers and it should in time create a familiar experience for users. It is recognized by government that engendering trust and confidence is vital to increase the uptake of e-government services. Therefore, government is working with various partners in other government departments and in the private sector to develop security policy that will allow citizens and businesses to use the right level of security and authentication and ensure the integrity of IT systems.

There are four principles that are essential for safe electronic transactions.

- **Confidentiality**: Keeping information private.
- **Integrity**: Ensuring information has not been changed or manipulated.
- **Non-Repudiation**: The individual who undertook the transaction cannot subsequently deny it.
- **Authentication**: Confirming the identity of the individual who undertook the transaction.

In addition, all services need to be compliant with the Data Protection Act 1998.

Interoperability

Interoperability enables the exchange of information between different computer systems. The e-Government Interoperability Framework (e-GIF) is an essential component of e-government strategy and sets out the policy and standards for interoperability across the public sector. It sets the architecture for joined-up and web-enabled government, for the UK online portal and Gateway, and for Electronic Service Delivery (ESD). The e-GIF:

- Adopts Internet and WWW standards for all government systems.
- Adopts XML (the data language of the internet) as the key standard for data interchange.
- Adopts standards that are well supported by the market.
- Aims to reduce the cost and risk for government systems.
- Makes the browser the key interface for access and manipulation of all information.
- Provides an implementation strategy using the UK GovTalk™ website as the mechanism for consultation and communication.

The Government Gateway allows secure authenticated transactions and joined-up government services to take place via the web. The Gateway is an authentication and routing engine built on open standards, allowing different systems in different government departments to communicate with the Gateway and with each other. This means that in the future, electronic transactions involving many different departments at once will be possible, ensuring a truly joined-up electronic public service.

'Modern councils, modern services – Access for all' White Paper

In August 2001, the government published a further White Paper entitled 'Modern councils, modern services – Access for all' [4]. This paper marked the next significant step for achieving e.revolution in English local government. In particular, it developed the vision of the modern council and provided a route-map for this vision. The e.revolution is all about making the most of new technologies and making them deliver, within the resources available, better quality and more accessible public services. As a consequence every council will need commitment at the highest level to achieve the vision of an e.council, in terms of a clear strategy (policy, procurement and technical aspects) whilst fostering a culture of innovation and learning. A modern council that has embraced the e.revolution will be striving to deliver services to meet user's needs, providing services in modern, convenient ways, empowering citizens to get involved, offering access to Information Communication Technology (ICT) to all and supporting new ways of working. Such a council will have re-engineered its services in accordance with the model in Fig. 8.4. It would appear that many councils have adopted this model for delivering their ESD target. The route-map envisaged in this White Paper consisted of 'three routes' – leadership and commitment, action and achieving a change in culture and is summarized in Fig. 8.5. Progress along route 1 is being monitored through the Best Value Performance Indicator (BVPI) 157 along with internal targets within councils' Implementing Electronic Government (IEG) statements, Best Value performance plans and community strategies. Route 2 action is more complex requiring the development of a range of access channels (telephones and effective call management, one-stop shops, websites, digital TV and smart cards), developing integrated back-office processes and commissioning/procuring effective delivery vehicles involving the development of local strategic partnerships (LSPs). This White Paper advocated that for the achievement of a vision of modern service delivery all councils should take into account the features indicated in Fig. 8.6. In addition, it also invited expressions of interest from councils and partnerships to act as local government online (LGOL)

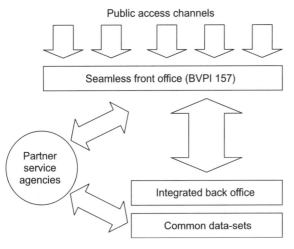

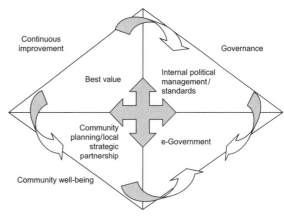

Fig. 8.6 Joining up modernization (adapted from [4]).

Fig. 8.4 Modern service delivery (adapted from [4]).

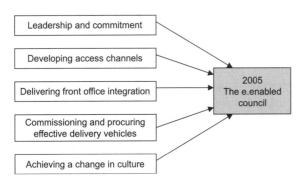

Fig. 8.5 The route-map to the e.revolution (adapted from [4]).

pathfinders, which would act as a focus of learning to enable all councils to meet the 2005 ESD target, to further develop products and disseminate their learning and good practice more widely, and to develop products for 'national rollout'. As a consequence, a substantial number of local authorities are now participating in the pathfinder scheme.

Office of the e-Envoy and e-Minister in cabinet

The Office of the e-Envoy (OeE) is part of the Prime Minister's Delivery and Reform team based in the cabinet office.[1] The creation of this office in September 1999 followed a central recommendation of the Performance and Innovation Unit's report entitled 'e-commerce@its.best.uk' [5]. Recommendation 14.2 of this report stated 'the Government should appoint an "e-envoy" with a wider remit than originally proposed. Covering both e-commerce and the IT elements of the Modernising Government White Paper. The e-envoy should be a high-level champion for Information Age issues across Government, based in the Cabinet Office with a direct link to the Prime Minister.' The OeE has responsibilities across the whole e-agenda and is organized into four principal work areas – e-Policy, Service Transformation, e-Delivery and e-Communications. Thus the primary focus of the OeE is to improve the delivery of public services and achieve long-term cost savings by joining-up online government services around the needs of customers. The e-Envoy is responsible for ensuring that all government services are available electronically by 2005 with key services achieving high levels of use. In addition to the OeE, the UK online strategy is overseen by the e-Minister, curently the Right Honourable Patricia Hewitt.[2] The post of e-Minister was created as a result of recommendation 14.1 of the

---

[1] A Head of E-government is to be appointed in 2004, which represents an evolution in the e-Envoy role.
[2] The e-Minister has announced new plans to support a private sector-led Digital Inclusion Panel in December 2003.

'e-commerce@ its.best.uk' report. Each government department has identified a senior (Board level) official to act as an 'e-champion' (formerly Information Age Government Champions). As a group, the e-champions support the e-Minister and e-Envoy in driving forward the UK online strategy.

The OeE has published three sets of guidelines for UK government websites as of May 2003:

1. 2001 – framework for senior managers;
2. May 2002 – illustrated handbook for web management teams (revised March 2003);
3. July 2002 – consultation draft framework for local government.

egov@local consultation paper

This paper [6] was published in April 2002 by the Office of the Deputy Prime Minister (ODPM) in order to work towards the development of a national strategy for local e-government. This paper was concerned with the realization of the e-government vision at the local level, at the point where the vast majority of services are delivered. The model offered for consultation aimed to develop previous guidance, such as the 'route-map' offered to local authorities in the 'Modern councils, modern services, access for all' White Paper

and to provide common tools and a framework for local e-government. The model itself contained three key components – the e-organization, joining it up (Fig. 8.7), priority outcomes and the national framework (Fig. 8.8). This draft strategy promoted considerable discussion, with 224 formal responses, and was broadly welcomed by all interested parties.

The national strategy for local e-government

In November 2002, the Office of the Deputy Prime Minister published the national strategy for local e-government [7] as a result of the aforementioned consultation paper. This strategy sets out the issues that councillors and chief executives will need to consider and the questions they will want to ask of their councils. In addition, the document explains the roles of government, the Local Government Association and various partner organizations in relation to delivering the national framework that local government will operate within.

Local e-government is about transforming services, renewing local democracy and promoting local economic vitality. However, it is recognized that councils will not be able to deliver local e-government alone but will need to work with

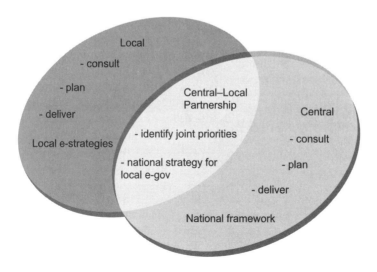

Fig. 8.7 Central-local relations [6].

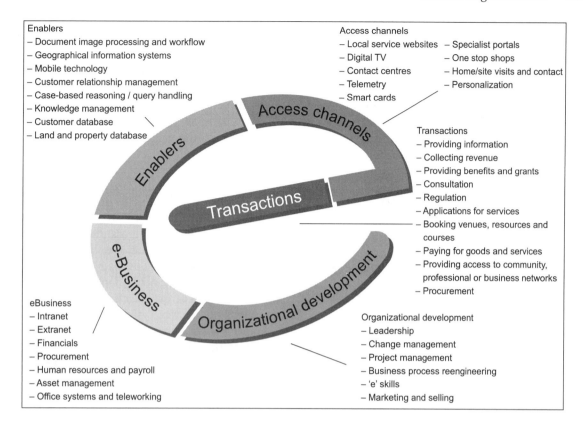

Enablers
– Document image processing and workflow
– Geographical information systems
– Mobile technology
– Customer relationship management
– Case-based reasoning / query handling
– Knowledge management
– Customer database
– Land and property database

Access channels
– Local service websites   – Specialist portals
– Digital TV   – One stop shops
– Contact centres   – Home/site visits and contact
– Telemetry   – Personalization
– Smart cards

Transactions
– Providing information
– Collecting revenue
– Providing benefits and grants
– Consultation
– Regulation
– Applications for services
– Booking venues, resources and
  courses
– Paying for goods and services
– Providing access to community,
  professional or business networks
– Procurement

eBusiness
– Intranet
– Extranet
– Financials
– Procurement
– Human resources and payroll
– Asset management
– Office systems and teleworking

Organizational development
– Leadership
– Change management
– Project management
– Business process reengineering
– 'e' skills
– Marketing and selling

Fig. 8.8 Local e-organization [6].

central government departments and agencies within a common framework and be able to share information over secure common networks.

October 2002 saw local authorities completing their second Implementing e-Government (IEG) statements. However, implementation of these statements of plans and priorities for e-government will require consideration of:

- how local priorities and targets support Community Strategy and Best Value Performance plans;
- identification of senior councillors and managers who will lead the delivery;
- an analysis of the skills possessed and the support required;
- the investment needed along with the expected returns;
- identification of local and regional partners in both the private and public sectors; and

- identification of what can be learnt and obtained from others (Pathfinders, etc.)

The success of local e-government will be reflected ultimately by the strength of local democracy and the satisfaction of local communities with their public services. Three key issues will be measured in order to assess the success or otherwise of local strategies:

- the availability of electronic services;
- their take up (reflecting ease of access and quality of services offered); and
- the value for money provided by local e-government.

Progress will be monitored through a combination of the BVPI (BVPI157 – a corporate health indicator), the IEG statements and the broader audit, inspection and assessment regime for local government. BVPI157 provides local authorities with a nationally consistent context within which to measure progress as it is concerned with the number of types of transactions enabled for electronic delivery as a percentage of the types of interactions that are legally permissible for electronic delivery. At present this indicator does not take into account the volume of individual types of interaction nor does it apply to all government departments and agencies. It is anticipated that these elements will be incorporated into a modified indicator.

It needs to be borne in mind that this strategy promotes objectives beyond availability, take up and value for money. Other objectives include:

- services that are joined-up around customer needs;
- councils that are open and accountable;
- communities that are engaged and active in shaping the views and services of councils; and
- a thriving, modern local community.

Future IEG statements will be required from local authorities – IEG 3 statements are required in October 2003. However, the government expects future statements to set and monitor local targets that reflect local strategies and priorities with the expectation that through the process, local authorities will be delivering modern, successful and effective local government.

The Improvement and Development Agency

The Improvement and Development Agency (IDeA) was founded in April 1999, by local government, to work with it and for it and help it do better. Acting at the national level, IDeA aims to provide a focus for the implementation of local e-government and to enable local authorities to co-ordinate and share progress. They are involved in a range of projects including:

- Marketplace – a national electronic procurement systems for local government;
- National Land and Property Gazatteer (NLPG);
- National Land Information Service (NLIS);
- National Land Use Database (NLUD);

- Local Authority Secure Electoral Register (LASER);
- Electronic Service Delivery toolkit (ESD-toolkit);
- National Street Gazetteer (NSG).

Gazetteers

Seamless services and better information to local people require councils to take a more holistic approach to information on addresses and assets. In the past the incremental development of information systems has meant that address, asset and other land and property information is frequently dispersed and duplicated across the IT infrastructure. Local authorities are aware that various government intiatives such as the NLPG and NLIS, as well as the requirement to provide seamless, joined-up services across their entire operation utilizing electronic service delivery, make it essential to implement a corporate land and property gazetteer. Commitment is required at the highest level ensuring co-operation across the local authority in providing the property database information from all departmental systems. At the same time, commitment is required at all levels of the organization in developing a central index of addresses, streets, land and properties.

The NLPG is a means by which local authority data can be consistently referenced to enable it to be identified, retrieved and integrated with other data. It is the foundation of the data infrastructure for joined-up government and is the key to unlocking data and making it compatible.

To facilitate the creation of the NLPG, British Standard 7666 (spatial datasets for geographical referencing) has been created [8]. This standard specifies a format for holding details on every property and street.

There are several parts to BS7666:

- a street gazetteer (BS7666 Part 1);
- a Land and Property Gazetteer (BS7666 Part 2);
- addresses (BS7666 Part 3);
- a data-set for recording Public Rights of Way (BS7666 Part 4).

BS7666 is based on the concept of a land parcel unit known as a Basic Land and Property Unit

(BLPU). Each BLPU has a unique reference number (UPRN), a spatial reference (grid co-ordinate) and one or more Land and Property Identifiers (LPI). The LPI is the address of the BLPU that uniquely identifies it in relation to a street as defined and held in the NSG. In order to achieve Best Value in any of an authority's services the adoption of BS7666 is an essential requirement.

As a consequence of standardizing address-based information on BS7666, Councils are now able to create fully corporate gazetteers with which they can merge property databases maintained in various departments, such as Council Tax, NNDR and electoral registration. Such data, coupled with the Ordnance Survey's 'Address-Point' will ensure that where possible properties are assigned a reference and geocode. This should provide the means to access the data from within a GIS. This more holistic approach, recognizing that departmental applications should share a common database of addresses, land and property and be spatially enabled by GIS technology, is currently being pursued by most local authorities.

A range of companies are now providing complete sets of modules for various local government functions within different departments that share a core BS7666-compliant database, covering such professional areas as development control, environmental health, local land charges, residential premises and electoral registration.

THE USE AND APPLICATION OF INFORMATION TECHNOLOGY IN THE DELIVERY OF ENVIRONMENTAL HEALTH SERVICES

Environmental health software application packages

Application packages are available that are designed to assist local authorities in the management of the statutory functions associated with environmental health and to address the e-government agenda. These packages are normally independent of operating system, database management system, network and graphical user interface. Such integrated systems are intended to deliver increased productivity and responsiveness to busy departments, improving efficiency and helping to deliver a consistently high level of service. Whilst such software application packages require substantial storage media, the hardware costs involved have considerably reduced over the last few years.

The software packages currently available have been designed with input from environmental health professionals and administrative staff and are structured to reflect the routine working patterns of all officers, enabling the input of all types of work through standard layouts. Environmental health modules comprise a range of facilities including:

- Complaints and service enquiries – quick input screens provide a rapid source of information and provide for response time monitoring and prioritization along with progress details. Users can define groups, categories and types of information and can set performance indicator targets for automatic calculation, which, when combined with the user-created sets of pre-defined activities and response targets (based on the local authority's peformance indicators), assist in generating performance indicator reports both for internal use and for external consumption. The database built up in relation to complaints can also be used for other activities such as pest control requests, refuse collection, noise complaints and various general enquiries.
- Inspections/audits – full details of all inspections and audits are recorded and information relating to risk scoring/rating can also be included. This allows for the programming of food and health and safety visits or EPA monitoring with visit dates automatically calculated on the basis of the risk score. Visit details, accident information, and a full risk and activity history can be generated, which is accessible at any time, or archived/closed.
- Premises records – provides a complete record of all premises with details of licences, occupancy, equipment, and a full history of activities. Addresses are verified with BS7666/NLPG-compatible gazetteers thereby ensuring

that addresses are accurate and consistent. Details of all letters and notices served are recorded, with modules often allowing the creation of the actual letter/notice or other documentation thereby assisting with the development of a 'paperless office'.

- Authorized processes – information on such processes can be easily retrieved with details of companies, processes and permitted prescribed substances, as well as assisting with the sorting of data into public domain and restricted information.
- Residential – can include such areas as stray/ lost dogs, private water supplies and burglar alarms register as well as details regarding houses in multiple occupation and housing health and safety risk findings.
- Prosecutions – if any activity results in a prosecution, this is recorded accordingly and this information links to all aspects of the application software, including the relevant premises record, related actions and time recording. Detailed information will include the legislation under which the prosecution was carried out, together with the pleas, results and fines levied.

A wide range of standard reports can also be generated dealing with such aspects as general workload and complaints, performance monitoring and analysis and statistical returns for CIPFA, FSA, HSC (LAU) and CIEH. In addition, powerful diary functions are normally included in order to plan total workload and to allow officers to prioritize their schedules of visits in line with defined risk categories. Further, it is very important that all such application packages must relate to and integrate with other office support systems within an 'open systems' environment.

Telecommunications

The information requirements demanded by environmental health departments cannot always be provided internally. Frequently, information must be sought from external sources such as **online databases**. For environmental health departments to search online databases, the necessary connections need to be made through a WAN, with

appropriate hardware in the department, and with access being controlled by a hierarchy of passwords.

Over the past few years there has been substantial growth in the offering of online databases by many organizations, such as Technical Indexes, Medline and Toxline. The costs of searching such databases varies from host to host and is made up of telecommunication charges, time spent searching and charges for displaying and printing records. The advantages of online searching are considerable due to the extensive range of databases and resources available.

Further sources of information pursued by many departments are provided by specialized CD-ROMs, such as Justis and OSH-UK. The range of titles is rapidly growing and they have the advantage of being straightforward and easy to use after a little tuition. These services are also regularly updated so as to incorporate, for example, new legislation, approved codes of practice and guidance. Many of the titles will allow 'downloading', the transference of some of the text from the CD-ROM into a word-processor or on to a disk to be saved for use at a future date.

Groupware

This is a term used to cover software aimed at supporting groups of people and includes email, diaries, scheduling, conferencing and bulletin boards.

Many environmental health departments have now instituted email systems across their LAN, allowing messages to be communicated by electronic means rather than by paper. Messages are displayed on a desktop terminal and incoming and outgoing messages are filed electronically, if required.

The facilities of a WAN are required in order to gain access to the **Telecom Gold** service. This is an electronic mail network widely used by public authorities and private businesses alike. Within the local government sector major users include trading standards departments, social services departments and environmental health departments. Most users of Telecom Gold are organized in private networks so that information distributed between them can be controlled and managed. This also ensures confidentiality. Within environmental health, Telecom Gold is regularly used for exchanging technical information and for more urgent matters such as food hazard warnings. It can also be used as a research tool to access technical reference libraries or to request information from colleagues. Examples of databases that are accessible directly via Telecom Gold are **Profile** (which includes the full text of most major newspapers), **Justis** (which contains the complete text of UK and European Union (EU) law reports and other related legal information) and **HSELine** (the HSE's abstract database of information on occupational health and safety). Another database accessible through Telecom Gold on a subscription basis is **Environmental Health Briefing**. This offers subscribers the chance to set up their own inhouse environmental health database that is updated each week. It also offers ready prepared reports on topical issues and an awareness service on current legal and technical topics.

A further development has been the introduction of **EHCnet**, the Environmental Health Communications Network, which aims to support environmental health officers with a fast communications link and to give them access to a wealth of professional information.

Another example of an electronic mail network is **Epinet** (Epidemiological Network) controlled by the Public Health Laboratory Service (PHLS) at Colindale in London. The network provides links for various bodies involved in communicable disease control including hospitals, the PHLS, environmental health departments and others.

End-user systems

End-user computing is a large and continually growing field. This situation is true of many environmental health departments where managers, professional and office staff have adopted a direct, hands-on approach to computers. Such usage can assist people to develop systems that help them to perform their functions more effectively. Staff are using IT to support their provision of what is, after all, a people-centred service – one in which experience, local knowledge and professional

judgement is supported, and not replaced, by technology.

Management, particularly senior management, has an important part to play in encouraging a positive attitude towards technology, especially as the value of informal learning is often underestimated. One of the most effective methods for individuals to gain confidence and to overcome any feelings of anxiety is to allow them to 'play' on the computer. Although this may involve investing extra time in such people, this investment often pays off as these individuals may bring fresh ideas and new ways of looking at problems. The allocation of time for training is also an important issue to be borne in mind by managers. Both informal learning and training will increase technical awareness and will promote the tendency for staff to use the information systems and technology.

When implementing any computer system in a department, it is necessary to consider the implications and impact of computer hardware. Who is to be given a terminal? Where should terminals be positioned? The current trend is to attempt to give everyone a terminal on their desk as this enables the system to be used more efficiently.

Hand-held technology

The use of hand-held technology has been introduced in a number of local authority departments in order to maximize the use of IT. Hand-held computers (i.e. laptops and palmtops), coupled with portable printers, enable officers to automate letter and notice production on site, improving the quality, quantity and speed of service achieved. Files generated out on the district can then be downloaded on to the office system through simple cable links once an officer returns to base.

CONCLUDING REMARKS

The Audit Commission report 'Towards a healthier environment: Managing environmental health services' [9] identified the need for the computerization of environmental health records and the creation of property registers relating to the main function areas. The problems of implementing such a process were also highlighted in the report. However, as indicated throughout this chapter, the present use of computers in the delivery of environmental health services, within the framework of the e-government strategic initiative, has grown considerably and now meets the need identified in the Audit Commission's report.

IT use within environmental health departments can be viewed on two distinct levels.

1. **Operational systems**: the management of departments can be assisted by the computerization of certain operational functions, such as registers of commercial premises and records of inspections.
2. **Strategic systems**: where the use of computers can assist in the setting of or achievement of the department's strategic objectives making a direct impact on the service delivered to customers.

The development of strategic IT systems should relate to and support the department's objectives, as detailed in service plans, and should identify several subsystems: the core systems (finance, payroll), office systems (word-processing, email, etc.) and strategic systems (complaints registers, staff monitoring, etc.). A properly designed computer system should assist departments to operate at levels of increased efficiency, while effectively reducing the workload on staff by removing a vast amount of work that is duplicated with manual systems. However, a suitably designed system requires all users to be trained as appropriate if its capabilities are to be fully realized and exploited.

Managing IT and realizing its potential are tasks made more difficult by the increasing pace of change. IT in all organizations, including local government, has progressed from a highly centralized activity to one of a mixture of central and departmental computer systems. The adoption of 'open systems' is enabling a number of other important developments:

- suppliers to provide components to a number of different hardware manufacturers;
- the same software to be run on a variety of hardware;

- information held on different machines to be communicated to other users on the same network;
- databases to be 'interrogated' and data abstracted to provide managers with timely information;
- staff using powerful PCs and user-friendly software to analyse and present information in an accessible manner, to integrate data and text, and to store, retrieve and communicate the information with relative ease [1, p. 75].

Some of the ways in which departments that provide environmental health services are using IT have been briefly discussed in this chapter. IT is used for a multitude of purposes at both operational and strategic levels. Increasingly, the monitoring and control of costs is vital when operating in an increasingly competitive environment. With the current and future changes departments face in the business/operating environment, such as Value for Money (VFM), it is important that systems are in place to cope with, control and anticipate future developments. The increasing need to provide services that represent best value, the requirement for performance indicators, increased legislative provisions, the pressure for statistical returns, and the continuing need to work towards quality service provision are all issues to be considered in a climate of diminishing resources. IT and computerization can assist environmental health departments in facing up to these challenges.

> Local e-government has much to offer, but it is no quick fix. If it is truly to support our twin objectives of strong local leadership and quality public services, then it must be built into everything local government does. This will take time, effort and money. Much good work has aleady been done and much progress made. But there is a long way to go if we are to meet all of our aspirations to improve services and to build the strong, vibrant and responsive councils that our communities want and deserve. (Foreword to The National Strategy for Local e-Government, 2002) [7].

REFERENCES

1. DTI (1998) *Our Competitive Future; Building the Knowledge Driven Economy*, DTI, London.
2. Cabinet Office (1999) Modernising government, White Paper, Cm 4310, The Stationery Office, London.
3. Cabinet Office (2000) *e-Government: a strategic Framework for Public Services in the Information Age*, Central IT Unit, London.
4. ODPM (2001) *Modern Councils, Modern Services: Access for All*, DTLR, London.
5. Cabinet Office (1999) *e-commerce@its.best.uk*, Performance and Innovation Unit Report, London.
6. ODPM (2002) egov@local; *Towards a National Strategy for Local e-Government*, DTLR, Wetherby.
7. ODPM (2002) *www.localegov.gov.uk*; *The National Strategy for Local e-Government*, London.
8. BSI (2000 and 2002) BS7666 *Spatial Datasets for Geographical Referencing*, Parts 1–4, BSI, London.
9. Audit Commission (1991) *Towards a Healthier Environment: Managing Environmental Health Services*, HMSO, London.
10. Edwards, C., Ward, J. and Bytheway, A. (1995) *The Essence of Information Systems*, 2nd edn, Prentice Hall Europe, London.

Part Three
Public Health and Safety

9 Public health, environmental health and sustainable development

Chris Church

INTRODUCTION – TEN YEARS OF CHANGE AND REORGANIZATION

The links between health and environmental quality are nothing new. Growing awareness of those links led to the development of the environmental health professions and indirectly to the environmental movement. To date, many of the working relationships between the environmental and the health sectors have not been fully explored or developed. This is, unsurprisingly, an important work area for environmental health professionals.

Over the last decade there has been a renewed focus on these links, and a growing body of work shows the value of cross-sectoral working, which has been fuelled in part by the integrated approach that is 'sustainable development'.

Sustainable development was originally defined as 'development that meets the needs of the present without compromising the ability of future generations to meet their own needs' [1]. The 1992 Earth Summit (see later) put this on the political agenda. This event produced the comprehensive global action programme 'Agenda 21' (see later). One of the key sections of this plan – chapter 6 – specifically addressed health issues (see section 2), and set out an ambitious programme of actions for every nation.

One of the primary needs of any individual or community is good health, and the sustainable development approach is that to meet that and other needs we have to have a safe environment, a secure economy for all, and long-term social equity. It is worth noting that the UK 1998 Acheson Report [16], which started from the perspective of health inequalities, arrived at much the same conclusion.

Sustainable development has been taken forward in many ways although progress has been slow. In the United Kingdom the Government's 'Quality of Life' strategy (10), has set out a national framework but much of this practice has developed at the local level and is often unrecognized.

The last decade has also been a time of major change for both the environment and health sectors and this has worked against developing long-term strategies, but there have been many positive initiatives, notably the new focus on health inequalities, while the new restructuring around Primary Care Trusts certainly offers the opportunity for more 'joined-up' working between the disciplines. How far the pressures of time and lack of resources mean these opportunities are neglected remains to be seen. To be successful any organization seeking to set up a joint initiative should understand the recent background and the changes that have taken place.

The health perspective

The opening paragraph of Agenda 21 [3] states:

> Health ultimately depends on the ability to manage successfully the interaction between the physical, spiritual, biological and economic/ social environment. Sound development is not possible without a healthy population; yet most development activities affect the environment to some degree, which in turn causes or exacerbates many health problems.

One reason why the United Kingdom has been slow to react to this has been the almost constant reorganization of the health service. This change started with the greatest change to the National Health Service (NHS) in 1991 with the implementation of the NHS and Community Care Act, which introduced an internal market and structural re-organization. In 1992, the *Health of the Nation* report [4] aimed to shift the focus from the delivery of clinical services to health. It encouraged health authorities to take on a more strategic role, namely that of maintaining and improving the health of the local population. This was the first time a strategic planning approach had been adopted in the United Kingdom and was also the first attempt to put in place a national health strategy based on the WHO Health for All (see later). Although a considerable step forward, *Health of the Nation* was criticized for placing too much emphasis on individual behaviour as an explanation for poor health and for disregarding important social determinants, particularly inequalities and poverty.

In 1997, further change came with the election of the Labour Government. Their first major policy document on the NHS, *The New NHS: Modern, Dependable* [5] set out to replace competition within the internal market with a system based upon collaboration and partnership. A new post of Minister of State for Public Health was also introduced. *The New NHS* proposed a strategic role for health authorities in partnership with local authorities, to identify and take action on social, environmental and economic issues which impact on the health of people.

This was followed in 1999 by a White Paper, *Saving Lives: Our Healthier Nation* (OHN) [6]. New responsibilities were also placed upon health authorities, involving drawing up health improvement programmes (HImPs) for their areas. In addition, some areas of extreme deprivation were designated health action zones (HAZs) and received special assistance to work on raising health standards among deprived groups.

In July 2001, plans [7] were published to further restructure the NHS at regional and local level. More responsibility was given to Primary Care Trusts/Groups (PCT/Gs) and health authorities reduced by about a third. A regional director of public health has been appointed in every regional office of government. At a local level a public health function will be delivered to PCTs via new strategic health authorities and public health networks. Each PCT is to have its own director of public health. It is still too early to see how this restructuring impacts on local environment and health issues.

Finally, in November 2002, the Department of Health produced a 'Cross-Cutting Review on Health Inequalities' [8].[1] This specifically identifies environmental improvement as one of four areas where interventions are likely to bring health benefits. This opens new opportunities for environmental health programmes.

The environmental perspective

Environmental issues have also changed significantly since the 1992 UN Earth Summit took place at the end of a period when environmental awareness had risen very rapidly. The 1986 Chernobyl disaster, CFCs and the ozone layer, tropical rain forest destruction and climate change had put global issues in the media, although little of this elated to the immediate concerns of environmental health professionals. In the mid-1990s the focus shifted to action in the United Kingdom, especially concerns about roads and transport with strong arguments over the impact of vehicle emissions on health. In 1995, the Committee On the Medical Effects of Air Pollution (COMEAP) issued its report on 'Quantification of the medical

---

[1]Also see Department of Health (2 July 2003) *Tackling Health Inequalities – A Programme for Action*. DoH, London.

effects of air pollution in the United Kingdom' [9] which concluded that between 12 000 and 24 000 people might die prematurely every year as a result of air pollution. Other reports suggested that, in urban areas, road transport was the major cause of such air pollution.

The 1997 change of government produced a new strategy for sustainable development, *A Better Quality of Life* [10]. It includes a section 'Better Health for All' and also refers at relevant points to the Health strategy '*Our Healthier Nation*' [6].

The other major UK environment and health issue has been the food we eat. A succession of 'food scares' pushed food and farming up the political agenda. The arrival of the new Department of Environment, Food and Rural Affairs was a significant development in this context. It seems certain that the overall focus on food is having at least limited health benefits, while reductions in pesticide use lower the public health risks from misuse and overuse.

The risks linked to the most dangerous pesticides have been the subject of intense activity over the last decade. The Stockholm Convention finally agreed to an international phase-out of the 'dirty dozen' persistent organic pollutants (POPs), most of which are pesticides, while locally many local councils have reviewed and reduced pesticide use. The EU is now working on a new Chemical Regulation that will limit the production and use of persistent endocrine-disrupting chemicals, and a range of UK NGOs are seeking to turn this 'limiting of production' into a phase-out programme.

The other major concern is global climate change. The nature and extent of this change remains open to debate, but there is an overwhelming consensus that change is taking place. The impact of all this on health remain unclear and much research is under way.

One new concern, which is likely to have increasing influence in the next decade is the concept of Environmental Justice – the idea that poorer communities suffer the additional impact of living in the worst environments. This started out in the United States of America but many UK groups are now looking at this approach and the Environment Agency has set up an Environmental Equality steering group.

The local agenda

The 1990s have also seen a steady expansion of local work on environmental issues. Much of this has resulted from or been helped by local council-run Local Agenda 21 programmes (also see p. 204) which have seen local government became more active on environmental issues, and to explore the sustainability agenda including issues such as health, poverty and social equity. Voluntary sector networks on issues such as food and waste have also expanded, aided in part by funding from National Lottery programmes.

This has further implications for change. Whereas in the 1980s most 'environmental' professionals were employed by national organizations, the majority are now spread throughout thousands of local projects, local councils and other organizations. This means that environmental health professionals may find more local stakeholders for new projects or programmes.

DRIVERS FOR CHANGE

Work to link environment, health and sustainable development is being influenced by a number of significant documents and processes which have emerged over the last decade and which provide much of the context for change. These include:

Saving Lives: Our Healthier Nation [6]

Saving Lives: Our Healthier Nation is a comprehensive Government wide public health strategy for England. It was published as a White Paper in July 1999 with twin goals:

- to improve health;
- to reduce the health gap (health inequalities).

It claimed to be 'the first comprehensive Government plan' and aims to prevent up to 300 000 untimely and unnecessary deaths by the year 2010. OHN expresses a commitment to setting goals for improving population health with more emphasis placed upon the social and environmental determinants of health. An important feature is an emphasis on tackling health inequalities: this will involve work on poverty, unemployment, poor housing and environmental pollution.

A Better Quality of Life, The Government's 'Quality of Life' sustainable development strategy [10]

The new Labour Government introduced a new strategy for sustainable development, *A Better Quality of Life* (DETR, 1999, see 1.3.C). While this remains something of an aspirational document, it is as good an example of 'joined-up thinking' as can be found anywhere in government. It includes fourteen 'headline indicators', one of which is 'expected years of healthy life'; two others (on housing and air pollution) link strongly to the health agenda.

Agendas for Change [11]

In 1996, the Chartered Institute of Environmental Health (CIEH) established a Commission on Environmental Health to undertake a broad review and to make recommendations for its future development in the United Kingdom. *Agendas for Change* was the final report, which suggested that, while health and the environment have always been closely related, in recent years the subjects have tended to drift apart. The Commission concluded that this drifting apart needs to be reversed and hoped that 'Agendas For Change' should provide 'a starting point for a wide ranging debate within the profession and beyond, leading to substantial reform and long-term improvements in environmental health'.

The Acheson report – 'The Independent Inquiry into Inequalities in Health' [16]

The 'Acheson Report' Inquiry, chaired by Sir Donald Acheson, Chairman of the International Centre for Health and Society at University College, London, was set up by the incoming Labour Government in 1997 and goes further than any other report to link poverty, health, environment and inequity. It was asked 'to review and summarise

inequalities in health in England' and to identify 'priority areas for future policy development . . . likely to offer opportunities for Government to develop beneficial, cost effective and affordable interventions to reduce health inequalities'.

The report suggests that 'the economic and social benefits of greater equality seem to go hand in hand. The quality of the social environment is worst where financial deprivation is greatest, such as in the inner cities.' It makes many recommendations, the first of which is 'that as part of health impact assessment, all policies likely to have a direct or indirect effect on health should be evaluated in terms of their impact on health inequalities, and should be formulated in such a way that by favouring the less well off they will, wherever possible, reduce such inequalities'.

Health and Environmental Impact Assessment

Typically, a government will spend less than 10% of its budget on the health sector, but it is the other 90% of its expenditure that arguably has the most impact on human health. Assessing the impact of such activities clearly makes sense in health terms. The WHO defines HIA as 'a combination of procedures or methods by which a policy, program or project may be judged as to the effects it may have on the health of a population'.[2] Health and Environmental Impact Assessment (HIA) is a tool designed to bring public health issues into the foreground of policy and decision-making. Key principles of HIA include an explicit focus on social and environmental justice, a multidisciplinary, participatory approach, and openness to public scrutiny.

HIA is being developed alongside Environmental Impact Assessment (EIA) procedures which are now established in the United Kingdom. They come under the responsibility of the Environment Agency and are required to aid decision-making on land use issues by planning authorities. Most Environmental Statements (the end product of an EIA) are provided by land developers seeking planning permission for building projects. An important part of EIA is the social impact assessment (SIA).

---

[2]See Health Development Agency (January 2004) *Clarifying Health Impact Assessment, Integrated Impact Assessment and Health Need Assessment*. HDA, London.

Although, under current legislation, EIAs address certain aspects of human health, for example, standards of water and air quality, noise levels, etc., the developer is not expected to comment explicitly in the Environmental Statement on how healthy (or not) their proposed development will be. The first formal example of prospective HIA in the United Kingdom was the assessment of the health impact of the proposed second runway at Manchester Airport undertaken by Manchester and Stockport Health Commissions and submitted to the Manchester Airport public enquiry in 1994.

In spite of recommendations by, among others, the Chartered Institute of Environmental Health (CIEH) and the British Medical Association (BMA) to the contrary, HIA work is still being developed in parallel to the EIA framework. This reflects the tradition over the past 30 years in the United Kingdom of addressing separately the protection of human health and the environment through the provision of health care services and public health legislation on the one hand, and through environmental protection measures on the other.

The World Health Organisation (WHO) and the 'Environment and Health for Europe' process

The WHO European regional office runs the 'Environment and Health for Europe' process as well as its work on 'Healthy Cities' (see section 3.x1 of WHO's Healthy Cities Programme) [12]. This was intended to promote closer interdisciplinary working and also to improve practice in the Central and Eastern Europe region and the Newly Independent States (former Soviet Union). To support this work in 1995 the WHO set up the European Environment and Health Committee (EEHC) to help implement Environmental Health Action Plan for Europe (EHAPE) adopted in 1994.

In June 1999, WHO Europe ran the Third European Ministerial Conference on Environment and Health in London, bringing together health and environment ministers from all parts of Europe. A parallel event for NGOs and other sectoral groups was also organized. The conference focused on issues including

- Transport Environment and Health
- Water and Health

- Children's health and the environment
- Human health effects of climate change
- Economics, environment and health
- Local Environment and Health Implementation
- Access to information and justice in environment and health matters.

There have been fewer follow-ups on many of the key London issues than was hoped at the time of the conference. This has been for several reasons, including lack of resources but it does seem that the underlying problem is a lack of commitment to joint working.

Agenda 21, Chapter 6: an action plan for health and sustainability? [3]

One of the drivers for the integration of environment and health within the overarching of these sustainable development has been the UN Agenda 21 action plan, agreed by governments at the 1992 UN Earth Summit. Chapter 6 specifically addresses health issues, with five key target areas:

(a) Meeting primary health care needs, particularly in rural areas
(b) Control of communicable diseases
(c) Protecting vulnerable groups
(d) Meeting the urban health challenge
(e) Reducing health risks from environmental pollution.

All these issues relate directly to environmental health priorities. Given that the United Kingdom has a fully developed National Health Service it is not surprising that the majority of the issues in Chapter 6 have been comprehensively tackled. The introduction of Primary Care Trusts brings the United Kingdom very much in line with the underlying thrust of Agenda 21, as does the recent work on Health Inequalities. However, there is still work to be done.

a. Meeting primary health care needs, particularly in rural areas

The prime objective, of 'meeting the basic health needs' has been met, with the proviso that there are worrying differences in accessibility to health services in rural areas, and that the public is very concerned about many aspects of health service delivery. However, health indicators show clearly that people living in rural areas tend to slightly better health than urban dwellers. This is linked to overall poverty figures for the United Kingdom, and conceals pockets of rural poverty and ill-health.

b. Control of communicable diseases

The anthrax-related events in the wake of 11 September 2001 have given a new focus to work on communicable diseases in the United Kingdom and one (bio-terrorism) that was not covered in Agenda 21. The Government published *Getting Ahead of the Curve – A strategy for combating infectious diseases* in March 2002 [13]. This set out a blueprint for a new Health Protection Agency (see Chapter 12) which will, by combining existing bodies, enable improved co-ordination and surveillance and detection of disease as well as rapid response to threats and epidemics.

Even though the United Kingdom is free of many of the diseases highlighted in Agenda 21, tackling diseases such as the AIDS/HIV pandemic, malaria and childhood diarrhoea require global support. UK development assistance has risen since Rio, but is still a very long way from the international target of 0.7% of GDP called for at Rio. The United Kingdom has also significantly failed to 'contain the resurgence of tuberculosis'. The continuing rise in the number of reported cases in London and other major cities highlights the need for increased spending to improve environmental health and to continue to tackle poverty.

c. Protecting vulnerable groups

This section stresses that 'specific emphasis has to be given to protecting and educating vulnerable groups, particularly infants, youth, women, indigenous people and the very poor' as a prerequisite for sustainable development. This can again be seen as primarily an issue for developing countries. However, the United Kingdom has poverty,

albeit not as absolute; there are vulnerable elderly people (and numbers are rising), and the health impacts are clear in areas such as fuel poverty and early deaths every winter among old people.

d. Meeting the urban health challenge

The 'urban health challenge' set out in Chapter 6 was very optimistic, calling for an improvement of '10 to 40%' in health indicators by the year 2000. The United Kingdom has tackled this issue, albeit in a piecemeal way. Many urban regeneration programmes have specifically tackled health issues (see section 3). Other work has been supported by the WHO European Healthy Cities programme. A number of UK cities are actively involved in this programme (see below).

e. Reducing health risks from environmental pollution

This section of Agenda 21 identifies 13 key pollution issues, all of which are tackled to some extent in the United Kingdom, some through health and safety work, others through the Environment Agency and through ongoing research.

Local Agenda 21 (LA21)

One of the most positive outcomes of Agenda 21 came from the single sentence which called on local authorities to 'consult with their communities' to produce a 'local Agenda 21'. Over seven thousand programmes have been developed across the world and this was taken up enthusiastically in the United Kingdom: a Regional Government Office survey shows that by 2001 over 93% of UK councils had produced a document, with widely varying results and approaches. Environmental health staff played a key role in this work; in some cases LA21 programmes were and are being run by EHPs.

Most LA21 programmes remained rooted in an environmental perspective, with some positive results including targets for waste minimization, energy saving (often linked to fuel poverty work), and work on biodiversity and transport. However, few LA21s made working links with the health sector although this is perhaps mostly due to the non-involvement of health sector organizations. LA21 is now being superseded in many ways by the new Community Strategies programme and the development of Local Strategic partnerships (see later).

THE KEY PLAYERS

Work on environment and health issues is now being done through a range of agencies. This means there are more stakeholders and also more potential partners for new programmes.

The National Health Service

The NHS is the largest employer and indeed the largest organization in the country. In 2001, over one million people were employed in NHS hospital and community health services. It is one of the largest users of resources and producers of waste in the United Kingdom. The NHS Purchasing & Supply Agency (PASA) influences about half of the annual NHS expenditure of £7 billion on goods and services: about £500 million is spent in the NHS every year on food, providing over 300 million meals for patients, staff and visitors. The overall priorities for the service are set out in the *NHS Plan* (July 2000) DoH, London and there is little consideration of environmental issues within these priorities.

Given the size of the organization, it is perhaps not surprising that the NHS has been slow to embrace sustainability. However, the benefits of environmental interventions on public health are well charted, and it is therefore disappointing that so little has been done to maximize the mutual benefits of improving internal environmental performance and public health. On top of the issues that any large organization faces the NHS has specific problems: NHS trusts produce over 100 000 tonnes of clinical wastes each year and GP's surgeries, dentists and others produce more than this again. The costs of disposing of such wastes are between £180 and £320 per tonne, so investment in a waste minimization strategy is likely to pay real dividends.

Change is taking place. Guidance by NHS Estates, an executive agency of the Department of Health, illustrates how sustainable development concepts and practices can help the NHS to gain most value from its estate and contribute to improving the quality of life in the United Kingdom.

External calls for change also came in a recent report by the King's Fund, 'Claiming the Health Dividend', [14] which suggests that the NHS could save millions of pounds and improve thousands of people's lives every year by investing in local communities and more sustainable practices. The report calls on the NHS to use its unparalleled purchasing power more effectively to promote better health (see later), and also to invest in tackling unemployment, boosting local business, reducing the amount of waste they produce and making their buildings more energy-efficient. It suggests that if they do, they will reap multiple benefits later on.

In April 2002, following 'Our Healthier Nation' and the government's sustainable development strategy, NHS Estates issued its' 'environmental pack' – two new documents and a CD ROM. One document was 'The new environmental strategy for the National Health Service', the other was on 'Sustainable Development in the NHS', and the CD ROM includes 'NEAT', the NHS Environmental Assessment Tool.

The Environmental Strategy points out that the NHS 'supports a healthier nation and environment' but focuses mostly on cost-saving measures for energy, waste and water. It sets out some actions for the NHS:

Environmentally sound policies should be reflected in NHS's core value. This will be achieved through the introduction and achievement of:

- Environmental appraisal
- Environmental management systems
- Environmental performance management

At the launch of these documents every Trust was asked to produce their own local strategies for energy, waste, water, transport and procurement by the end of October 2002. It is not clear how far this has progressed.

The Guide provides useful advice on environmental management and lists a few case studies of good practice. Some NHS operations do cause pollution and the NHS is covered by the Environmental Protection Act 1990. Some Health Trusts have already fallen foul of this, notably in regard to waste management, and are now being advised to recognize both the Polluter Pays Principle and the Precautionary Principle (see Chapter 34), which seems appropriate, given that it is at the core of public health work.

The Chartered Institute of Environmental Health

The Chartered Institute of Environmental Health (CIEH) has been at the centre of many debates about better linkages between the two disciplines of health and environment. Nationally and internationally the CIEH has worked to ensure health and sustainability issues were considered together, notably through the 'Agendas for Change' report [11], and had an active involvement in the UN 'Habitat II' international summit on human settlement issues in 1996.

However, the environmental health profession has also been under pressure to tackle difficult issues on limited and even declining resources. This means officers have tended to limit themselves to an agenda of enforcement and regulation and have had to ignore opportunities to take a more strategic role in sustainable development and to develop cross-disciplinary working. These points are looked at in the joint CIEH/Health Development Agency project on 'Environmental Health 2012' (report published 2002. CIEH/HDA, London) which sought to develop and set out a new vision.

The Environment Agency

The Environment Agency (EA) was set up as a non-departmental public body in 1996. It took over the functions of Her Majesty's Inspectorate of Pollution, the National Rivers Authority and the local waste regulation authorities, bringing together the many agencies responsible for environmental regulation in England and Wales. Equivalent agencies exist in Scotland and Northern Ireland. Most of the Agency's work with regard to the protection of the environment also has an impact on health and it is this impact that determines the setting of standards and actions required with regard to the environment. The Agency is also now developing work on Environmental Equality.

The European Union

Health policy is so high on national political agendas that most governments have generally resisted the EU 'interfering' with it and therefore in the past the EU did not develop public health policy. Until the Treaty of Maastricht was ratified in 1993 there was no legal basis for EU health policy. The Treaty states that 'The community shall contribute towards ensuring a high level of human health protection by encouraging co-operation between member states and, if necessary lending support to their action.'

There is now a new programme on public health which came into effect in January 2003.[3] The new programme has three 'Strands of Action':

1. improving information for the development of public health;
2. reacting rapidly to health threats;
3. tackling health determinants through health promotion and disease prevention.

THE KEY PROCESSES

Any work on environment and health must, if it is to be successful and recognized, fit in with and link to other ongoing processes. Several of these offer opportunities for joint working; indeed many actively require such partnership.

Health and Local Strategic Partnerships

Just as the Health Service has been reorganized, so local government has changed. The Local Government Act 2000 introduced the concept of Community Strategies for every local council to

---

[3]Also see a Communication adopted by the EC on 11 June 2003 on a Strategy for Health and the Environment COM/ 2003/0338 final.

improve local service delivery through a Local Strategic Partnership (LSP) which includes all sectors of society.

LSPs are central to delivering integrated local services in ways that meet local needs and most have brought the health sector in as active participants through PCTs or Health Authorities. It is envisaged (in the NHS plan) that PCTs will in time take a lead role:

- 'in ensuring alignment between Health Improvement Programmes, Primary Care Trust plans and Community Strategies and Neighbourhood Renewal Strategies;
- in ensuring that health input to local strategic partnerships from all key stakeholders in the local health community is properly co-ordinated.'

It is further expected that Health Action Zone integration will roll out alongside LSP development.

A 2002 report, 'Community Strategies and Health Improvement' by the HDA [15] sets out a range of ways to develop practice and shows how selected local authorities and health trusts are going forward together. There is however evidence emerging elsewhere that suggests that some PCTs that are joining their LSPs are not playing a very active role.

Health and regeneration

The need to regenerate poorer communities has been a central part of government policy for over twenty years. Following the introduction of the Single Regeneration Budget (SRB) programme it became easier to run regeneration programmes aimed at 'non-housing' outcomes. One theme for some of that work has been to improve the health of the people living in those areas. Relatively little work exists to show the health benefits of regeneration work. The groundbreaking Stepney Health Gain Project reported major health improvements from a major housing redevelopment: the overall rate of illness days fell from 37 per hundred (over one in three) to five per hundred (one in twenty) – a sevenfold improvement. To date little other work exists: the lack of resources for the necessary baseline studies may be one problem.

The Government now has a National Strategy for Neighbourhood Renewal. This goes much further than previous government work in making the case for integrated regeneration. It highlights the role of the Department of Health in narrowing the health gap between socio-economic groups, and between the most deprived areas and the rest of the country. Targets for health outcomes from regeneration are now being set. Environmental health programmes operating within regeneration areas may well contribute to such targets and may find ways of mobilizing regeneration resources to improve environmental health quality in the area.

Environmental epidemiology (see also Chapter 15)

One of the major issues concerning links between quality of the environment and impacts on human health is simply evidence. The nature or lack of causal relationships has been at the root of many arguments over transport, chemicals, nuclear power etc. The science of exploring and researching these relationships is known as environmental epidemiology.

Over the last ten years knowledge in this field has expanded substantially. Universities such as Southampton, Imperial College and the London School of Hygiene and Tropical Medicine have developed substantial experience in exploring these links, while more and more studies have refined methodologies and ways of working. The COM-EAP report on air pollution and transport in London was a defining moment in the transport debate in 1998.

Much environmental epidemiology is not concerned with what are usually seen by the public as 'environmental issues' but broader issues about human beings and their surroundings. However, matters such as food and access to facilities are increasingly central to the development of local strategies for sustainable development, and it may well be that a reassessment of what is and is not 'environmental' is under way as a result of the focus on sustainability. It is certainly the case that environmental epidemiological studies have provided the core evidence base for developing work in the United Kingdom on Environmental Justice (see later). It also provides new opportunities for links between researchers and professional staff with environmental health departments.

Healthy Living Centres

The shape and style of voluntary initiatives has changed markedly during the last decade, due in no small part to the National Lottery. This has funded many health-focused schemes through the Community Fund. This Fund was supplemented more recently by the New Opportunities Fund (NOF) which funds local projects in line with national guidelines. One innovative approach to public health that has emerged from this process is the Healthy Living Centre initiative. It was launched in January 1999 with a budget of £300 million to develop a network of Healthy Living Centres across the United Kingdom.

The WHO Healthy Cities Programme

The 'urban health challenge' set out in Chapter 6 of Agenda 21 called on local authorities to 'develop and implement municipal and local health plans'. This work was boosted in 1997 by Habitat II, the major international conference on sustainable cities and human settlements run by the United Nations Centre for Human Settlements (UNCHS). The action plan arising from this event makes many links between urban development and health.

Many of the advances in this field have been due to the WHO's Healthy Cities Programme [12]. This is a long-term international development project that aims to 'place health high on the agenda of decision makers in the cities of Europe, and to promote comprehensive local strategies for health and sustainable development based on the principles and objectives of the strategy for health for all for the twenty-first century and Local Agenda 21'.

It defines the qualities of a healthy city:

'A city should strive to provide

1. a clean, safe physical environment of high quality (including housing quality)
2. an ecosystem that is stable now and sustainable in the long term
3. a strong, mutually supportive and non-exploitative community
4. a high degree of participation and control by the public over decisions affecting their lives

5. the meeting of basic needs (food, water, shelter, income, safety and work) to all people
6. access to a wide variety of experiences and resources, for a wide variety of interaction
7. a diverse, vital and innovative city economy
8. the encouragement of connectedness with the past, and heritage of city-dwellers and others
9. a form that is compatible with the past, and enhances the preceding characteristics
10. an optimum level of appropriate public health and sick care services accessible to all
11. high health status (high levels of positive health and low levels of disease).'

BRIDGING THE DIVIDE

The arrival of PCTs and the DoH review of Health Inequalities suggest a 'new public health agenda'. Some have seen this as moving towards creating better links between public health and sustainable development and any analysis of the goals of sustainable development, of environmental health and of public health show an overwhelming similarity. Sustainable development will never be achieved without good public health, while good public health requires healthy and safe environments, an end to poverty and strong civil society.

Despite this operational links remain weak and fragmented. One obvious reason is that people in both fields are under-resourced and under pressure to carry out core tasks and to run basic regulatory systems. Yet there is more to be tackled. Public health can still be massively improved (the United Kingdom is well outside the top ten of countries with highest life expectancy) and much of the work, especially in poorer areas continues to be environmental.

The Stepney Health Gain project shows the undoubted health benefits of well-planned regeneration. The impact of traffic-generated air pollution has become clearer during this last decade planning and the recent work on health improvement from traffic reduction in Oxford suggests that environmental planning will have a role to play in tackling that, while less car use will certainly have a role to play in increasing walking and cycling and thus cardio-vascular health.

It seems to be clearly in the interests of both disciplines to improve the linkages between them. The question is 'how'? There may be six issues which environmental health professional may wish to consider when planning cross-disciplinary work.

(1) Better joint evaluation

Environmental or regeneration projects rarely take the time (or have the resources or capacity) to evaluate their work for health gains. In any case many of these may be long-term outcomes: very few have clear health-related outputs. But while evaluating the health benefits of environmental work may need extra funding, the methodologies exist. Ways to evaluate the environmental benefits of health and social development are often less clear. The direct outputs may often be limited and the longer-term outcomes may be beneficial but unclear.

There is a need for better cross-disciplinary work on identifying and agreeing outputs, outcomes and relevant evaluation criteria. Including criteria that will show the broader impacts of work in one discipline may help show practitioners the other side of the benefits of co-operation.

(2) Differences in professional perspective

There is concern amongst professionals about joint working. Most workers have more than enough within their work area to keep them busy. People seeking to work in both disciplines may find it hard to build working relationships in either. There are also differences in approach and language between health and environment staff: EHPs could play a role in building mutual understanding and working links.

(3) Greening the NHS

As the Kings' Fund 'Health Dividend' report [14] states, there 'has never been a better time to build a more sustainable health service. The NHS will receive unprecedented extra funds in the next five years. It is already committed to investing in childcare, energy efficiency and big staffing increases.'

Good public and environmental health programmes can make a real contribution to the NHS simply by preventing illness and freeing up beds in hospitals. The NHS itself could play an active part in making this happen. There is of course heavy pressure from short-term targets and reorganization, but the benefits of such a strategy will pay off in many different ways.

(4) The new local agenda – public health and sustainability

The requirement that each PCT should have a 'public health network' under the Director of Public Health, provides a new structure within which public health can be developed. At the same time the 'new agenda' for local government of integrated and improving local services, delivered under the guidance of an LSP offers new opportunities to deliver and integrate public health with work on local sustainability. There is a need to share ideas and experiences as this work develops.

(5) Integrating spatial planning and health development

'Unhealthy places' don't just happen: poor design has played its part along with pollution, economic deprivation etc. There is a need for a safe and healthy local environment with well-designed public and green space, for good quality local public services, including education and training opportunities, health care and community facilities, especially for leisure; and for a sense of place. This opens the way for joint working between environmental health and planning professionals.

(6) Linking 'Environmental Justice' and health inequalities

Recent years have seen a new focus on 'environmental justice' issues in the United Kingdom. This phrase originated in the United States where it developed out of recognition that communities suffering the worst pollution and most disproportionate environmental and health impacts were overwhelmingly poor black and Hispanic communities. This work has led to a greatly increased involvement of those ethnic groups in environmental action, often with a very local focus, and also led to action at a national level with the signing of a 'Presidential Order on Environmental Justice' designed to ensure that communities did not suffer from excessive pollution on grounds of race or ethnicity.

Environmental justice work in the United Kingdom has developed with a broader focus on all aspects of poverty and the environment. Informal link between a few practitioners in health, environment and academic networks has developed an initial base of case studies and a UK Environmental Justice Network is now under development.

The implications of this for work on environment and health are likely to be significant. A new focus on enabling and empowering local communities is likely to bring environmental issues into communities with health problems, environmental epidemiology studies may become the basis for more legal action, and public participation in environmental action may be extended well beyond the 'usual suspects' both locally and nationally. All this will of course link to the 2002 Health Inequalities review.

CONCLUSION

Perhaps the most important lesson of the last ten years is that joint working is feasible and that it delivers. The information and understanding have developed steadily; both the obstacles and the opportunities are that much clearer; the potential benefits have been set out in detail, both in terms of better public health and sustainable development. All that is needed now is the will, commitment and resources to develop pilot projects into common practice.

REFERENCES

1. World Commission on Environment and Development (1987) *Our Common Future*, Oxford University Press, Oxford.
2. The Report of the Committee of Inquiry into the Future Development of the Public Health Function (1988) *Public Health in England*, HMSO, London.
3. The UN Conference on Environment and Development April 1993, *Agenda 21 – Programme of Action for Sustainable Development*, UN Department of Public Information, New York.
4. Department of Health (July 1992) *The Health of the Nation – A Strategy for Health in England*, HMSO, London.
5. Department of Health (1997) *The New NHS: Modern, dependable*, DoH, London.
6. Department of Health (1999) *Saving lives: Our Healthier Nation*, HMSO, London.
7. Department of Health (2001) *Shifting the Balance of Power*, DoH, London.
8. Department of Health (2002) *Cross Cutting Review on Health Inequalities*, DoH, London.
9. Department of Health (1995) Committee on the Medical Effects of Air Pollution, *Quantification of the Medical Effects of Air Pollution*, HMSO, London.
10. DETR (1999) *A Better Quality of Life – a Strategy for Sustainable Development in the UK*, HMSO, London.
11. CIEH (1997) *Agendas for Change – Report of the Environmental Health Commission*, CIEH, London.
12. WHO Europe/Healthy Cities Network (1998) *City Planning for Health and Sustainable Development*, WHO, Copenhagen.
13. Department of Health (2002) *Getting Ahead of the Curve – A Strategy for Combating Infectious Diseases*, DoH, London.
14. Kings Fund (2001) *Claiming the Health Dividend*, Kings Fund, London.
15. Health Development Agency (2002) *Commuity Strategies and Health Improvement*, HDA, London.
16. Department of Health, NHS Executive (1998) *The Independent Inquiry into Inequalities in Health*, NHS Executive, Leeds.

FURTHER READING

Ambrose P. (1993) *Cost-effectiveness in Housing Investment*, University of Sussex.

Chartered Institute of Environmental Health/
WHO (2000) *A Source Book for Implementing
Local Environment and Health Projects*, CIEH,
London.

Chartered Institute of Environmental Health
(2002) *Environmental Health 2012*, CIEH/
Health Development Agency, London.

Church, C. and Elster, J. (2002) *Thinking Locally,
Acting Nationally*, Community Development
Foundation/Joseph Rowntree Foundation,
New York.

Church, C. (2003) *Healthy People, Healthy
Planet: Ten years of Progress on Health
and Environment Practice and Policy*, CIEH,
London.

DEFRA (2002) *Farming and Food – A Sustainable
Future*, London.

Department of Health (2001) *Health Effects of
Climate Change in the UK*, HMSO, London.

Friends of the Earth (1995) *Prescription for Change:
Health and the Environment*, FOE, London.

Hamer, L. and Easton, N. (2002) *Community
Strategies and Health Improvement: a Review
of Policy and Practice*, Health Development
Agency/IdeA/LGA/DTLR 2002, London.

Hamer, L. (1999) *Making T.H.E. Links –
Integrating Sustainable Transport, Health and
Environment Policies*, Health Education
Authority for the DoH and DETR, London.

Lafferty *et al.* (2001) *Sustainable Communities in
Europe*, Earthscan.

McArthur, I. and Bonnefoy, X. (1997)
*Environmental Health Services in Europe:
An Overview of Practice in the 1990s*, WHO,
Europe.

NHS Estates (2002) *Sustainable Development in
the NHS*, Leeds.

Regional Integrated Monitoring Centre (2001)
*Grass Roots and Common Ground: Guidelines
for Community-Based Environmental Health
Action*, University of Western Sydney,
Australia.

Seymour, J. (2000) *Poverty in Plenty: A Human
Development report for the UK*, Earthscan,
London.

Transport 2002 (1998) *The Healthy Transport
Toolkit*, London.

UKPHA (2002) *Why Health is the Key to the
Future of Food and Farming – A Report on the
Future of Farming and Food*, London.

UNEP (1992) *Agenda 21*, Available from
www.unep.org; published in the UK by Regency
Press, Sevenoaks, Kent.

WHO Europe/Healthy Cities Network/EU DG XI/
ESCTC (1998) *City Planning for Health and
Sustainable Development.*

WHO (2001) *Climate Change and Health* WHO
fact sheet 266 (www.who.int).

10 Risk management for environmental health

Norma J. Ford, Denise M. Rennie with Liz Falconer and Kath Ray

INTRODUCTION

Environmental health practitioners are professionally concerned with the risks that arise from the adverse effects of environmental stressors, such as pollution, conditions in the living and working environments and contaminated food. Progress in reducing these risks has been made with consequent improvements in health and quality of life but these improvements have been accompanied by increased expectations for a society free from involuntary risks and the imperative for state intervention to reduce risks. In addition government departments are subject to increasing demands for more openness and transparency in their decision-making processes on handling societal risks [1].

It is suggested that several factors are influencing society's expectations in relation to health, safety and environmental risks and the role of the state in managing them, namely:

- People are becoming more cautious regarding developments in science and technology. Disasters such as those at Chernobyl and Bhopal and mounting concerns about environmental threats like global warming have resulted in society being more circumspect in relation to industry's exploitation of technological advances. The public's readiness to embrace new technologies is now tempered with a need to be reassured that risks are properly controlled.
- Information and communication technology is having an increasing impact in shaping the perceptions of the public to risks. Information about and images from industrial disasters or scares occurring anywhere in the world are quickly and graphically communicated to other countries. In addition the Internet enables interest groups to be more effective in disseminating information.
- Globalization of economic activity is changing people's attitude to risk. There is an increasing realization that the United Kingdom is in competition with the rest of the world and that being uncompetitive will result in job losses. Thus there is support for the need for a measured response to risk, which is embedded within an effective policy for sustainable development.
- Ambivalence as to what government and regulators can or should do and the need to guard against over reaction in the face of transient shifts in public opinion [2].
- Mounting recognition that the concept of risk needs to be broadened to take account of society's responses to the nature of the risk (which may be varied and disparate) in addition to

judgements about the probability of harm [3]. The concept of risk is strongly shaped by individual and cultural influences and has evolved to include values that cannot be readily verified by traditional scientific methods [1].

In the face of these incongruent pressures the control of risks is a complex task. Whilst society accepts the need for the benefits of scientific and technological advances to be exploited there is also a corresponding belief that those who create the risks should ensure they are properly controlled and the state should be proactive in ensuring risks are mitigated. Simultaneously there is an underlying presumption, in a free market economy, that industry should be able to take advantage of these advances without unnecessary state intervention [1]. Thus the state must endeavour to reconcile the conflicting pressures, within the constraints imposed by finite resources and an incomplete understanding of many of the risks affecting health. Risk assessment is regarded as a powerful tool for informing, but not dictating, decisions on the management of risks so as to enable consistent, reliable, transparent judgements, and therefore, well-founded decisions to be made concerning risks. There are sound scientific, economic, political and sociological reasons for using risk assessment techniques, not least of which is their potential to facilitate accountability and transparency [4]. The value of risk assessment as a tool to guide decisionmaking for those whose task it is to balance safety, public concerns and expenditure may explain why risk assessment underpins much national and international policymaking and legislation.

The use of risk assessment as a tool to guide decisionmaking is evident at a number of levels. It is used to inform decisionmaking by government in relation to the strategies and policies to be adopted to manage and regulate risk [3,5]. The UK government requires a risk assessment to assist Ministers in reaching their decisions on any regulatory proposal affecting business [5]. At operational level it is used to guide the distribution of enforcement resources through processes such as inspection rating schemes. Moreover risk assessment is increasingly a legal requirement placing duties on those creating risks to introduce appropriate measures to eliminate or control them, for example, the Management of Health and Safety at Work Regulations and Food Safety (General Food Hygiene) Regulations. In addition it is a tool used by enforcement officers in assessing compliance with regulations.

The aim of this chapter is to enable environmental health students and practitioners to appreciate and understand both the advantages and limitations inherent in risk management. It addresses the development of the understanding of the concept of risk. Consideration is also given to the role of risk assessment and examples of risk assessment frameworks that can be applied. The components of risk assessment techniques are identified and the significance of risk perception and approaches to risk communication explored.

HAZARD AND RISK

The Royal Society in their seminal work on risk analysis, perception and management [6] define the terms hazard, risk, risk assessment and risk management as follows:

Hazard is a property or situation that in particular circumstances could lead to harm.

Risk is the probability that a particular adverse event occurs during a stated period of time, or results from a particular challenge (where an adverse event is an occurrence that produces harm).

Risk assessment is the term used to describe the study of decisions subject to uncertain consequences. It has two components: risk estimation and risk evaluation.

Risk estimation includes:

- the identification of outcomes;
- the estimation of the magnitude of the associated consequences of these outcomes; and
- the estimation of the probabilities of these outcomes.

Risk evaluation is the complex process of determining the significance of the estimated risks for those affected. It therefore includes the study of risk perception and the trade-off between perceived risks and perceived benefits.

Risk management flows from risk estimation and risk evaluation, and is the making of decisions concerning risks and their subsequent implementation.

These definitions have been used as the basis for policy development in a variety of applications but the increased focus on risk as a basis for the policy development and regulatory control in a range of settings since the 1980s has resulted in a proliferation of terms and risk assessment techniques. In the United Kingdom the government's Inter-departmental Liaison Group on Risk Assessment (ILGRA), an informal committee of senior policy-makers on risk issues, was established in 1996 and disbanded in 2002. It found that the use of risk assessment had not developed systematically in government, but had simply evolved within departments [5]. The lack of a common terminology was identified as a major obstacle to the promotion and understanding of the concept of risk assessment. The ILGRA was therefore tasked with considering how greater coherence could be achieved and to explore the possibility of developing risk-related protocols for integrating the various inputs into policy decisions and the regulatory framework. In their third and final report in 2002, ILGRA [7] noted that government departments had prepared and published framework documents explaining the procedures, protocols and criteria they applied in making risk-based decisions.

RISK ASSESSMENT

The value and validity of the methods available to assess, manage and communicate risk are subject to ongoing scrutiny and there have been disagreements about the role that risk assessment should play in the regulation of risk. The benefit of a properly conducted risk assessment is that its results provide essential information for reaching decisions on the management of risk.

The underlying concept of risk assessment is that of seeking to identify in some quantitative or otherwise comparable way the connection between hazards and actual exposure to harm and the significance of the estimated risk to those affected. It depends on an identification of hazards and damages, and consists of an estimation of the risks arising from them with a view to their control, avoidance or to a comparison of risk. The detail, scope and complexity of a given assessment will vary depending on factors such as the availability of time (relative to the urgency of action), the perceived seriousness of the hazard, knowledge of the risks, available resources and the outcomes and possible consequences of any decisions [4]. Uncertainty can affect all stages of risk assessment and management processes. For example, the science underpinning the assessment may be complex, ambiguous or incomplete and/or essential data may be unavailable. Analysing the nature and extent of the uncertainties can assist in highlighting the knowledge gaps and focus attention upon critical issues during the decisionmaking processes, such as the need to adopt a precautionary approach in the face of an incomplete understanding of the risks. In environmental health practice it will not always be possible to conduct a full risk assessment because the preliminary information and time available are inadequate. Nevertheless, decisions that have been informed even by a short, well-focused assessment, using the information that is available, will usually be better (and more defensible) than uninformed judgements. Risk assessment, comprising risk estimation and risk evaluation, results in identification and prioritization of risks, with a view to their reduction where this is reasonably practicable. A scheme of risk management measures tailored to priorities can then be introduced.

Those charged with the responsibility of carrying out risk assessments need a framework to approach the task and various sequential models exist. The environmental health examples summarized here share common steps of hazard identification, consideration of the consequences, evaluation, decisionmaking about appropriate controls and monitoring of both risk and the effectiveness of the controls. A European Directive on Hygiene of Foodstuffs following Hazard Analysis and Critical Control Point (HACCP) principles requires the following to be incorporated within management of food safety:

1. Analysis of the potential food hazards in a food business operation.
2. Identification of the points in those operations where food hazards may occur.
3. Deciding which of the points identifiable are critical to ensuring food safety.
4. Identification and implementation of effective control and monitoring procedures at those critical points.
5. Periodic review of the analysis of food hazards, the critical points and the control and monitoring procedures.[1]

The Department of the Environment, Food and Rural Affairs (DEFRA) Guidelines for Environmental Risk Assessment and Management 2000 [8] propose the framework, shown in Fig. 10.1 with a tiered approach in which the level of effort put into assessing each risk is proportionate to its priority (in relation to other risks) and its complexity (in relation to an understanding of its likely impacts). The guidelines identify a number of stages to be worked through during the risk screening and prioritization elements of the framework namely:

- Hazard identification;
- Identification of consequences;
- Magnitude of consequences;
- Probability of consequences; and
- Significance of the risk.

The Health and Safety Executive in the Management of Health and Safety at Work Regulations 1999 Approved Code of Practice make the point that there are no fixed rules about how a risk assessment should be carried out because factors such as the type of business and the nature and extent of hazards and risks differ and will have an influence upon the most appropriate course of action. It does recommend that risk assessment should be an inclusive process with active involvement of managers and employees. Guidance entitled Five Steps to Risk Assessment [9] has been issued by the HSE to help employers and the self-employed in the commercial, service and light industrial sectors to undertake risk

assessment. The initial stages are to look for the hazards and then to decide who might be harmed and how. Once the hazards and their effects have been identified, the risks are evaluated and decisions made as to whether existing precautions are adequate or more should be done. The final two stages are to record the findings and then to undertake periodic review and revisions where necessary.

A health and safety risk-rating system for housing has been under development in the United Kingdom since 2000. The planned scheme provides an evidence-based risk assessment approach for use in the assessment of housing conditions. The hazards considered are those that can be controlled or rectified by the owner or landlord of a dwelling. The Housing Health and Safety Rating System (HHSRS) (also see Chapter 16) addresses a combination of likelihood of exposure to the hazard and severity of exposure outcome weighted to reflect the degree of potential harm [10]. The technique enables the determination of a quantitative risk ranking reflecting the seriousness of hazards present such that this can be used to inform decisions on appropriate enforcement action.

Following any of these schemes risk evaluation may be undertaken using a combination of qualitative and quantitative approaches, depending upon the circumstances. Even if a purely quantified approach is taken to risk estimation, the total assessment of the risk would also involve qualitative issues such as the tolerability of the risk. This general approach is equally applicable to risks in the areas of food safety, environmental protection, health and safety at work or residential accommodation.

Quantitative risk assessment

There are several mathematical and engineering approaches to hazard identification and risk analysis that apply qualitative techniques to the identification of hazards and then apply quantitative methods to estimate the risk. They represent an attempt to develop reliable risk assessment techniques and can be used to demonstrate compliance with legislation and minimize adverse impact on health, safety and the environment.

---

[1]Proposed EU Food Hygienic Regulations will require documented food safety management systems which also include the establishment of corrective actions and verification of the measures applied (also see p. 513).

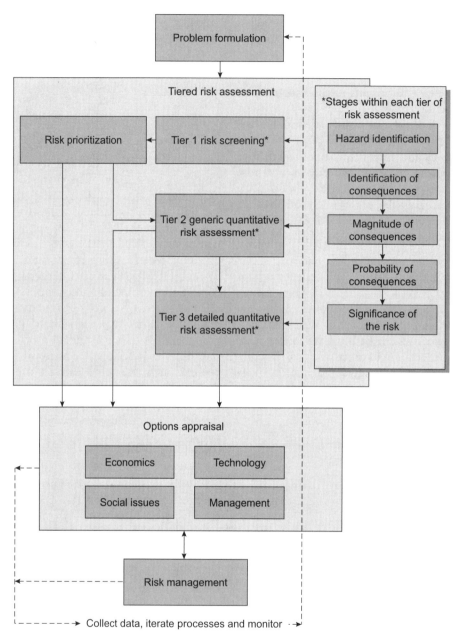

Fig. 10.1 DEFRA tiered approach to risk assessment and risk management [8]. (HMSO Click-Use Licence Ref C02W0003108.)

An example of a hazard identification technique is Hazard and Operability Study (HAZOP).

Like HACCP systems and Environmental Impact Assessment (EIA), HAZOP is a bottom-up approach that moves from potential cause to likely effect. It should be conducted by a team of people with a suitable range of expertise. The group should have an independent chair who ensures the provision of a forum for the individuals to use their imagination to determine possible hazards and operability problems in a process or system.

HAZOP is a qualitative procedure, which systematically examines the safety of processes by evaluating what could go wrong. HAZOP is often combined with risk ranking enabling the quantification of the hazards identified through the subjective views of the experts in the group. They have to judge the acceptability of risk or determine appropriate risk reduction measures. The quantitative scores produced can be used to inform managers about where risk reduction strategies are needed but it is important to remember that all the inputs to the system are qualitative.

Quantification of risks can be estimated through risk analysis techniques such as Failure Mode and Effect Analysis (FMEA), Failure Modes and (Criticality) Analysis (FMECA), Event Tree Analysis (ETA) and Fault Tree Analysis (FTA) The intention is to obtain an objective measure of the risk associated with a faulty piece of equipment or inadequately controlled process.

FMEA is mainly used to judge the impact of failures in mechanical or electrical systems. It was developed to identify critical areas and enable planners and analysts to modify designs to reduce the probability of failure [11]. The analyst considers what effects on the output will occur when there is a failure of input or components. Alternatively, each individual component of a system might be considered for each of its failure modes. It is then possible to assess the probability of failure rates for each component and hence for each system failure mode. FMECA extends the FMEA approach by ranking failures according to probability and severity of adverse effect.

Event Tree Analysis takes an inductive approach by starting from the occurrence of an initial failure and sequencing the possible follow on events and generating a range of possible outcomes. The paths or tree branches can include effective application of controls to prevent any adverse outcome. An example is the accidental release of LPG, which could result in safe dispersion or fire depending on whether or not there is a source of ignition and on the topographic circumstances.

Fault Tree Analysis, is a top-down technique that could, for example, be applied to an outbreak of food borne disease, a release of a damaging substance to the environment or an industrial accident. In constructing a fault tree, the top event is first identified (i.e. the outbreak, release or accident) and then questions are asked to identify the likely or actual causative factors. The findings could be in terms of either component failure or inappropriate operator action. Fault Tree Analysis is conducted by systematically describing the impacts and outcomes of series and combinations of possible occurrences. This is done through deductive analysis, that is, working backwards from the potential outbreak, release or accident and deciding on possible means of controlling the failures that might lead to the event. In developing fault trees there is a need to make assumptions and it is important that these are recognized and recorded in any plan to control the potential failures.

The accuracies of any of the available approaches are dependent upon the accuracy of the probability data used, and also on the questions which are asked. Groeneweg [12] found that, on presenting the same data and scenario to several groups of specialist assessors, their final assessments varied considerably. Similar techniques have been applied to the assessment of human reliability (e.g. [13]), particularly in potentially hazardous processes such as the nuclear and chemical industries. Whilst these techniques have been used with some success in these industries, the paucity of probability data on certain types of human error can make quantification problematic.

Human judgement plays a fundamental role in all risk estimation consequently it is argued that to suggest risk exists in a finite, measurable form is much too simple a view [14]. This is made more complex by the increasingly accepted view that it is simplistic to believe that there is a derivable quantifiable physical reality that most people will agree represents the 'true' risk from the hazard. Lewis [15] has argued that quantitative techniques that use scientific methods are fundamentally sound, being an implementation of logic. However, Groeneweg [12] contests this by suggesting that the inevitable inadequacy of information available to analysts, results in logic methods of risk estimation as simply identifying, '. . . *islands of knowledge in a sea of unknown events*' (p. 138).

With respect to scientific uncertainty, Adams [16] argues that in the majority of cases assessments of risk are not based upon conclusive scientific evidence since the statistical evidence for accurately judging the probability of harm occurring is usually absent or unreliable. In these circumstances it is essential to rely upon assumptions that do not derive directly from the evidence but from human judgement. Quantitative calculations of risk are based upon particular theoretical premises and mathematical extrapolations that are just as subjective (in the sense that they rely on human judgement) as the judgements of individual, non-expert risk-takers [17,18]. The difference is that the theoretical models used by experts are evaluated and legitimized by the scientific community. For instance, animal experiments are usually used to assess the risks from toxic chemicals and in the safety assessment of food additives. In these experiments, animals are subjected to high doses of a compound over a short period of time, and then the results are used to extrapolate to the effects on humans, who are generally subjected to very low doses of a chemical or food additive over long periods of time [18]. Different results can be obtained from these experiments depending upon which assumptions are used in extrapolating from the low-dosage to the high-dosage cases. A 'supra-linear model' assumes that responses will remain high even as the dosage is reduced, whilst a 'threshold model' assumes that under a certain dose there is no identifiable effect.

Mathematical and engineering approaches to risk assessment have generated a multiplicity of risk assessments by different rational participants using different techniques. Engineers and scientists have tended to see risk as a quantifiable entity whereas social scientists (particularly psychologists and sociologists) have argued that risk cannot be viewed as a one-dimensional objective concept. More recently it has been recognized that engineers who are familiar with a process have a tendency to underestimate errors that are made and thus not recognize the seriousness of the risk that can be presented [19]. It is generally agreed that the physical consequences of hazards such as deaths, injuries and environmental harm are objective facts, however, the assessment of

risk depends upon human judgement [20]. This has been contested by some technocrats and natural scientists [21] but there is evidence of the adoption by government of an understanding of risk in which social, cultural and political variables play a role [1,5]. Three short reports from the Royal Academy of Engineering also emphasize the importance of both scientific and societal measures of risk. These reports recognize that scientists and engineers have tended to blame society for not understanding or accepting the benefits of new technologies. There is an expressed desire for the development of a common language and understanding of risk issues by all stakeholders to facilitate participation in dialogue about existing and future technological applications [22].

RISK PERCEPTION

The Royal Society (5) devotes extensive coverage to the advances in social scientific theories of risk. In 1983, the previous Royal Society report: *Risk Assessment*, maintained a distinction between 'objective risk' as a quantifiable measure calculated by experts, and 'subjective risk' as the (inaccurate) perceptions of risk held by lay persons. Subsequent developments have resulted in a partial displacement of the quantitative definition of risk (the probability of adverse events occurring) as the sole one, and have instead emphasised risk as a multifaceted concept with a number of different qualitative dimensions. In its 1992 Report the Royal Society Study Group recognized this broader perspective and argued that the separation between objective risk and subjective risk could no longer be sustained [23].

Much of the early influential work on risk perception was undertaken by psychologists. Later during the 1980s, social anthropologists and sociologists investigated the importance of social and cultural influences on perceptions of risk. Distinct boundaries were apparent between the disciplines and paradigms until more recently when several attempts have been made to integrate the two approaches. Significant ideological and methodological differences still remain but a growing consensus has emerged that in order to understand people's judgement about and responses to risk it is necessary to know something about the context in which they were formed.

Psychological theories of risk perception

Psychological work on risk perception has a long and well-established history based upon considerable amounts of supporting empirical data. Early work comprised studies comparing lay assessments of quantitative risk, for example, the number of fatalities arising from a variety of stated hazards within a specified time period, with the actual reported number of fatalities (e.g. [24,25]). The risks from those activities with the highest numbers of fatalities were generally underestimated by the respondents and those with the lowest numbers of fatalities were generally overestimated. This has been explained by psychologists through the notion of 'cognitive heuristics', or 'mental rules of thumb', which are used by respondents to arrive at judgements of probability. A strategy called the 'availability heuristic' was considered to be particularly important in determining respondent's judgements of fatality frequencies. This suggests that the more information or images that are available for recall about an event the more likely respondents are to judge it likely to happen. Events that are particularly imaginable are those of which one has personal experience, and also those that have been given widespread or particularly vivid media coverage. An example of its effect is the considerable level of consumer resistance to the proposals to lift the UK prohibition on food irradiation in the late 1980s following the 1986 Chernobyl disaster [26]. Resistance to consumption of genetically modified food might similarly be heightened by media attention on the potential for cloning of animals and human beings. Further 'heuristics' that can influence the perception of risks are the desire for certainty, which causes people to estimate the probability of a certain course of events, rather than face uncertainty [17] and anchoring biases. Anchoring biases occur when people use a natural starting point as a first approximation to a judgement – an anchor. This anchor is then adjusted to accommodate the implications of new information.

Typically the adjustments are crude and imprecise and fail to take full account of the importance of the new information [25].

Investigations of why lay estimations of the frequency of adverse events occurring differed from the actual frequencies of these occurrences retained the quantitative concept of risk. In an alternative approach, Slovic *et al.* [27], identified a number of characteristics, which had been hypothesized to influence perceptions of actual or acceptable risk. Survey methods were used to ask respondents about their perceptions of risk and to rate the qualitative characteristics that they perceived various hazards to have.

The numerous dimensions or characteristics of risk expressed by the respondents were examined by the researchers and found to be highly intercorrelated. It was therefore possible to group the different qualitative characteristics into two factors that had a particular impact upon the risk perceptions elicited. The two factors 'dread risk' and 'unknown risk' are shown in Fig. 10.2 [27]. The dread risk factor includes dimensions of risk such as controllability, potential of fatality, equity of risk, catastrophic potential, risk to future generations and voluntariness of exposure. Hazards that rated high upon this scale included crime, warfare and terrorism and nuclear power. The unknown risk factor included knowledge about the risk, latency of effect, observability and the novelty of the risk. Hazards that rated high on this scale included space exploration, DNA research and food additives. Motor vehicles, alcoholic beverages, mountain climbing and downhill skiing, all of which are perceived by experts to be relatively high risk in terms of the probability of fatalities or injuries occurring, were perceived by respondents to be low on both the scales of dread risk and of unknown risk. The results showed that risks which rated low on the dread and unknown scales, that is, risks that were familiar, voluntary and had well-known and immediate consequences, were more tolerable to respondents than risks that rated high on these scales [18]. More recent studies have identified cross-cultural differences in people's perceptions of risk, which suggests that factors such as gender, race, political worldviews and affiliations impact upon perceptions of risk [25]. This

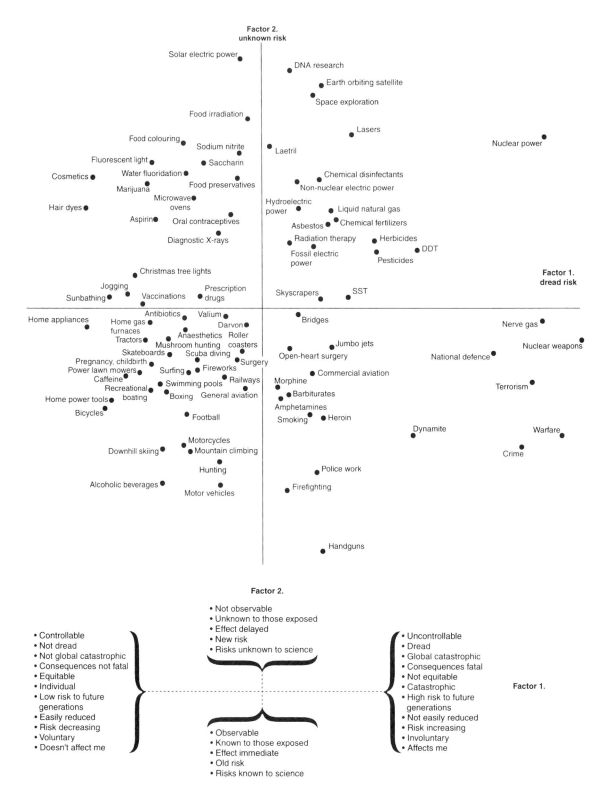

Fig. 10.2 Locations of 90 hazards on Factor 1 (dread risk) and Factor 2 (unknown risk) of the three-dimensional factor space derived from the interrelationships among 18 risk characteristics. Factor 3 (not shown) reflects the number of people exposed to the hazard and the degree of personal exposure. (Source: [27].)

finding has been regarded by some as the most significant contribution of the psychometric work [23]. Nevertheless there remains criticisms that because the work is centred on the psychology of the individual it has failed to broaden its scope to make a significant contribution to the understanding of the importance of social processes on perceptions of risk.

Drawbacks associated with the psychometric tradition of risk perception research include the fact that there are a variety of hazard sets, rating scales and multivariate statistical techniques that can be used, and these vary according to the particular study in question [28]. There is also the problem, common to all social scientific empirical research, that the way in which questions are presented can have a significant bearing upon the responses people give regarding their perceptions of risk. The use of standard hazard sets, in order to compare across studies, does not allow respondents to identify for themselves which risks are the most important to them. Despite these methodological drawbacks, it is clear that the identification of a range of qualitative dimensions of hazards, which impinge upon risk perception, has contributed greatly to understanding of the complexity of risk perception and tolerability. They also provide an opportunity to capture lay conceptions of risk, and provide an insight into how the public reacts to hazards and hazard management.

The findings have important consequences for those responsible for assessing and managing risk in society since they suggest that the results of quantitative risk assessments using, for example, accident probabilities, annual mortality rates or mean losses of life expectancy, are not the only, nor even the most important factor, to be taken into account when assessing and managing risk.

CULTURAL THEORIES OF RISK

Cultural theories of risk are often cited as an alternative to the psychometric tradition of risk perception research, because of their focus upon the individual as a social being rather than as an isolated individual. Cultural theories of risk emphasize that not only is risk a multifaceted concept with numerous qualitative dimensions but that the perception and assessment of the characteristics of risk will vary according to the culture of the perceiver. The general principle of cultural theories of risk is that individuals within societies participate in numerous social relationships within large- and small-scale social and institutional arrangements. It is through such relationships that people's attitudes, beliefs and values, and thus their 'worldviews', are constructed and maintained. These 'worldviews' are important determinants of risk perception, causing people selectively to emphasize certain elements of a risk [28]. The emphasis is on studying the individual, located within particular social networks and contexts, as an active receiver and interpreter of information about hazards.

Anthropological work by Douglas and Wildavsky [29] has been influential in the development of research into socio-cultural influences as explanations of bias in the perceptions of risk. Central to cultural interpretation is the concept of the 'grid group' approach, a typology of social structures that are claimed to reflect a range of identifiable worldviews or social orientations, which impact upon and account for differences in people's perceptions of risk in a predispositional sense. Worldviews are general social, cultural and political attitudes that appear to have an influence over people's judgements about complex issues [30]. The cultural theory of risk postulates that people's 'worldviews' can be classified into four types as shown in Table 10.1. The types reflect beliefs about human nature, the physical environment and equality and competition [29].

The implications for management of risks in society are fundamental, in that they suggest that disagreements over risk acceptability cannot necessarily be resolved by recourse to scientific evidence. People's 'worldviews' will also determine whether the evidence presented to them is regarded as trustworthy or unreliable. Indeed the selection of certain risks for attention is one means by which individuals defend their preferred way of life and place blame upon other groups.

Cultural theory of risk emphasizes the issue of trust in the institutions that are responsible for managing risk, as an integral element of people's

Table 10.1 Categorization of people's worldviews [29]

Type	Characteristics
Hierarchists	Respect scientific and administrative authority
	Believe in state regulation of risks for common good
Individualists	Believe in market forces and individual responsibility
	Oppose state regulation
Egalitarians	Favour regulation to protect the environment
	Assert the precautionary principle
Fatalists	Resigned to their fate and see no point in attempting to change or control things
	Little control over things that affect their own lives

'worldviews'. This issue was also highlighted as an important element in risk perception in Wynne's 1982 study of the Windscale inquiry in the United Kingdom [31]. Wynne suggested that the expert and public perceptions of the risks involved differed due to their different frames of reference. The expert point of view focused only upon the technical aspects of risk management, thus taking the trustworthiness of the institutions concerned for granted, whilst it was the lack of trust in the institutional arrangements for risk management at the plant that framed the opponents' perceptions of the risks concerned [28]. Trust in risk management, like risk perception, has been found to correlate with gender, race, worldviews and affect. Slovic [25] argues that much of the contention, which has been evident in risk management decisions, could be attributed to a climate of distrust that exists between the public, industry and risk management professionals.

Power differentials between different social groupings may explain varying levels of trust in the institutions engaged in managing risk, as marginalized groups have less control over and are less likely to receive the benefits of a variety of risk-creating activities. These issues are significant in relation to the relative power held by managers and workers, government and the population, enforcement agencies and businesses, etc.

Cultural theorists recognize that people's interpretative 'worldviews' or frameworks are not stable and coherent but contain inconsistencies and ambiguities and can thus be expressed quite differently according to context. Responses to hazards may be determined by the behaviours of significant others such as friends, family, colleagues and public figures. Alternatively risk perceptions, which are identified by research studies, may in fact be *post hoc* rationalizations of behaviour [17].

One of the key distinctions between psychologists and the cultural theorists is the status ascribed to lay knowledge and the extent to which subjectivity is accepted as being present in all conceptions of risk. The cultural theorists argue that whilst research by cognitive psychologists in the 1970s and 1980s identified differences between expert and lay perceptions of risk, its focus upon bounded rationality as an explanation, implicitly underplayed the potential for cultural bias amongst members of the scientific community. They contend that social and cultural influences impact upon expert's perceptions of risk as they do on lay persons perceptions [23,32,33].

THE SOCIAL AMPLIFICATION OF RISK

An attempt to integrate ideas from both psychological and cultural theories of risk exists in the social amplification of risk theory [34]. This focuses upon the means by which risks are communicated to people and how level of risk is intensified or attenuated through a variety of psychological, social and cultural processes [18, 28]. Behaviour patterns that flow from this process produce secondary social or economic consequences, which may also act to increase or decrease the physical risk itself. Secondary effects trigger demands for additional measures to be taken or alternatively (in the case of risk attenuation) constrain needed protective actions. The framework emphasizes the importance of understanding the impact of the way in which information about risk is communicated to individuals, through scientific communications, government agencies, politicians, community activists, 'significant others', the mass media, etc. to the risk management process. Each of these channels can act as a 'social amplification station', either emphasizing or downplaying certain elements of a hazard depending upon a variety of factors. Within this conceptualization there is no single true (absolute) picture of risk as compared to the distorted (socially determined) risk. Instead the information systems and characteristics of public responses that compose social amplification are essential elements in determining the nature and magnitude of risk.

The social amplification of risk framework (SARF) was conceived and empirically tested in North America. More recently the UK ILGRA has sought, through a programme of research, to explore the impact of the framework on risk communication in the United Kingdom. A component of this programme, an international workshop, explored SARF in the current risk decision and communication context and reviewed its perceived weaknesses. Its shortcomings include its characterization of the social amplification process as linear (although iteration is acknowledged), its focus upon individual rather than social or group processes, the over simplification of the public's responses to mediated information and its focus on the amplification of risk and the underplaying of the power and use of power by institutions, corporations and governments in the process [35].

Overall there was agreement that the social processes which influence risk interpretation beliefs and policy responses are too complex to be represented in any framework. Petts *et al.* [35] explored the role of the media in the amplification of risk issues and concluded that the SAR framework was overly simplistic and did not provide a coherent and comprehensive explanation of the impact of plural media and their relationship with the consumer. The findings suggested that the public were not passive recipients of media messages but sophisticated and 'media savvy' users.

In summary, there are several, clear theories of risk perception within which there are areas of consensus and contention. No single paradigm has proved to be sufficiently inclusive and comprehensive, in terms of the insights offered to supersede the alternative approaches. Nevertheless there is general agreement that social, political and cultural variables play an important role in shaping an individual's attitude towards risk. Expert calculations of risk, based upon the quantitative assessment of the probability of adverse events occurring, constitute one approach to risk assessment and to understand the public's reactions to risk, some account must be taken of the social and cultural context in which hazards arise and the way in which these variables shape people's attitudes, beliefs and behaviour [23].

These developments in understanding about risk perception have important implications for the processes of risk management and risk communication, since risk perception necessarily affects how people will respond in the face of new hazards and what societal and personal risks they will consider to be acceptable or tolerable.

The need for the public's viewpoint to be considered has been increasingly recognized. Reports on tolerability of risk following the Layfield Inquiry into the Sizewell B nuclear power station emphasized that the 'opinions of the public should underline the evaluation of risk' [36]. Subsequently the underlying philosophy, that risks can be classified as unacceptable, tolerable or broadly acceptable, has gained considerable acceptance by regulators and industry and as having wider application beyond nuclear power [1].

Environmental health practitioners must therefore recognize that risk is a multifaceted concept and risk judgements depend not only on the physical characteristics of the hazard itself but are also determined by broader psychological and sociological considerations. There is a need to acknowledge the differing perceptions of risk and the way in which they impinge upon their professional judgements. They are also an important factor in determining the responses of stakeholders with whom they interface in their work.

RISK MANAGEMENT

Environmental health risk management has been defined as the process of identifying, evaluating, selecting and implementing actions to reduce risks to human health and the ecosystem [37]. Its goal is to implement scientifically sound, cost effective, integrated actions that reduce or prevent risks while taking into account social, cultural, ethical, political and legal considerations.

Social and cultural theories of risk draw attention to the variety of social, political and ethical dimensions of decisionmaking, which impinge on risk management, and suggest that decisions made on the basis of quantitative assessments of the likelihood of deaths or injuries are incomplete because they ignore society's responses. Debates over the acceptability of risk accruing from the activities of industry in relation to workplace conditions, emissions from industrial premises or foods products, will be affected by the role and credibility of the institutions charged with the management and communication of risk. Other political and ethical dimensions of risk acceptability might include conflicting value judgements over the benefits of risk-creating activities, such as improved health care, improved economic growth, increased power of businesses and enhanced national independence. For example, there has been more support for the use of genetic modification techniques to improve medical treatment of diseases such as cystic fibrosis than there is for its use in food production [38]. Moreover, the perception of risk is multidimensional, with particular hazards meaning different things to different people depending upon their value and belief systems and the context of the risk [8].

Risk management at the institutional level is subject to a complex division of labour; the management strategies that are proposed and adopted reflect a wide range of influences. These range from the 'worldviews' of the civil servants responsible for drafting policy documents, to the views of the politicians commissioning the reports, to the views of the public who make up the constituencies of the politicians and the lobby groups that exert pressure upon them. It is now widely accepted at state level that economic, political, legal and social concerns play important roles throughout the assessment, evaluation and decisionmaking stages of risk management [8].

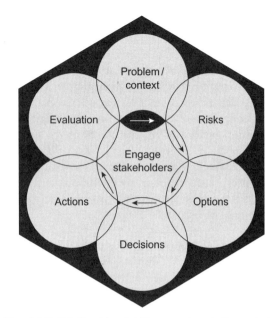

Fig. 10.3 US Presidential/Congressional Commission on Risk Assessment and Risk Management – framework [37].

The Presidential/Congressional Commission on Risk Assessment and Risk Management [37] shown in Fig. 10.3, proposed a risk-management framework primarily designed for risk decisions related to setting standards, controlling pollution, protecting health and cleaning up the environment. It encapsulates principles for making good risk-management decisions and for actively engaging stakeholders whilst enabling the level of effort and resources utilized to be scaled to the importance of the problem, potential severity and economic impact of the risk, level of controversy and resources constraints. The framework has six stages namely:

- Define the problem and put it in context;
- Analyse the risks associated with the problems in context;
- Examine the options for addressing the risks;
- Make decisions about the options to implement;
- Take actions to implement the decisions;
- Conduct an evaluation of the action's results.

These stages are conducted in collaboration with stakeholders and designed to be iterative should new relevant information emerge.

Despite the problems, organizations must still make risk decisions. Cox and Tait [39] offer an objective setting model, which itemizes the steps an organization might take in managing the safety of work systems as shown in Fig. 10.4. They use the term safety broadly, in the recognition that this model can be used to manage risks in work systems generally. The decision sequence diagram shows how assessment of risks includes reference not only to the system itself, but also to the wider environment. Risks therefore do not exist in isolation, but affect and are affected by the context in which they are situated and decisions regarding risks encompass the benefits that may come from taking the risk.

The increasing requirements in many areas of environmental health legislation for organizations to manage risks effectively have given rise to much advice for managers. For example, in the field of occupational safety and health, British Standard 8800:1996 offers two models of safety management, one of which is based upon the Health and Safety Executive's booklet HS(G)65 'Successful Health and Safety Management' [9], the other based upon BS EN ISO 14001: *Environmental Management Systems* (1996). The rationale of BS 8800 is to apply similar management practices in health and safety as other corporate matters and thus enable organizations to integrate the management of occupational risks into existing management systems, or indeed to use the occupational risk management systems to stimulate other management systems within the organization. Essentially the advice is the same whatever system is used, that is, that an effective risk management system includes the steps of planning, implementation and review.

Risk management decisions inevitably involve an aspect of cost. Recognizing that a zero-risk society is impossible to achieve, decisions in risk management revolve around questions of tolerability. The ALARP/ALARA (as low as reasonably practicable/achievable) principle in environmental protection accepts the need for a trade-off between risks and benefits. The Health and Safety at Work Act 1974, Section 2, places duties on employers to do certain things to reduce risks to their employees so far as is reasonably practicable. Case law has defined this as a balancing act between the quantum of risk on the one hand and the cost of remedying the risk on the other. The Food Safety (General Food Hygiene) Regulations place requirements on the operators of food businesses where these are appropriate and/or necessary for the purposes of ensuring the safety and wholesomeness of food. Similarly the selection of the Best Practicable Environmental Option (BPEO) acknowledges the impact of costs on the feasibility of environmental protection measures.

Risk decisions of course take place on a daily basis by all members of society, not just by professionals, and not all decisionmaking about risk involves extensive public debate. In situations where scientific uncertainty is low and the consequences of taking a wrong decision about risk are also low, then debates over risk management usually occur on a purely technical level. Where there is more scientific uncertainty, or the consequences of potential mistakes are higher, then the debate may also move into the realm of managerial competence. Finally, where scientific uncertainty and the consequences of being wrong are both very high, then the question of social values and 'worldviews' comes into play in debates about risk management.

RISK COMMUNICATION

Fischhoff [40] identifies a number of developmental stages in the study of risk communication beginning with a concern over conveying 'accurate' messages about risk to the public and moving on to a concern with the format of risk messages and with the trust that people hold in the institutions promoting messages about risk. It is important to acknowledge that it is not only the content, but also the format of communications, which has an influence.

The way in which the potential outcomes of hazards are expressed can affect people's perceptions of activities significantly. Simply reiterating that adverse events will not occur because of

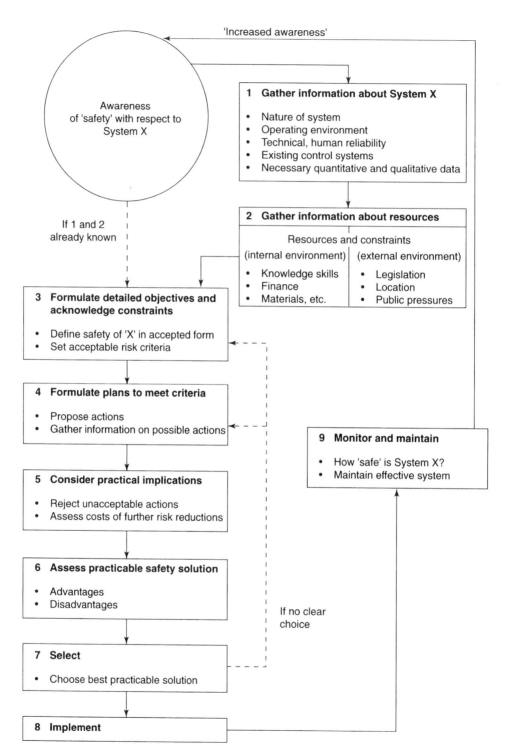

Fig. 10.4 An objective setting model for the 'safety' of System X. (Source: [39].)

safety features, that is, conveying quantitative risk 'accurately' can actually have the counter-productive effect of causing alarm rather than reassurance. This is because the communication may actually increase the imaginability of the risk to the receivers of the message and they will thus judge it more likely to occur.

The principle of using comparative figures in risk communication strategies has been found to be more meaningful to recipients than using absolute numbers about risk probabilities. However making comparisons across risks is very difficult given the multidimensional nature of risk perception. The use of mortality rates as a point of comparison does not take into account either the differential benefits of 'risky' activities or the perceived qualitative dimensions of various hazards. The qualitative characteristics of hazards render each hazard unique, and thus comparisons of hazards may not be at all meaningful to lay individuals. Such presentations provide only a small element of the total information that is used by the public in their risk-related decisionmaking [17].

Sandman *et al.* [41] examined means of conveying risk magnitudes to the public and found that the presentational format of a 'risk ladder' was most effective at explaining the relative magnitudes of risk. Respondents received information about differing levels of radon and asbestos in the home, smoking and other risks in a variety of formats. The response to the location of the hazard on the risk ladder is clearer and more logical than reaction to absolute numbers or to other comparative figures.

Risk communication must also focus upon the wider social and cultural contexts in which individual messages about risk are formulated and embedded. Suitable approaches acknowledge the complex web through which messages are communicated to the public as well as the cultural 'worldviews', which come into play in the interpretation of messages. Cultural theories of risk suggest that beliefs will determine the way in which evidence about risk is received and interpreted. Evidence that is congruent with beliefs is likely to be accepted, whilst that which conflicts with beliefs is more likely to be rejected as unreliable, unrepresentative or irrelevant [18].

It is impossible to present risk information in a neutral manner and thus selecting between different presentational strategies necessarily invokes a goal. This may be either the allaying or the enhancing of people's concerns over the hazards in question depending upon the 'worldviews' and political values of those communicating the message. Accordingly, this raises ethical concerns, and as Sandman *et al.* [41] note, 'the boundary between respectful communication and manipulation' is ambiguous and contested.

Government and regulatory agencies have unique responsibilities for regulating risks, setting guidelines, informing citizens and protecting the public. Regulators and those they regulate can hold divergent views because each group samples from a different world. Scientific uncertainties make risk communication difficult and responsible authorities sometimes deliberately withhold data or release limited information causing confusion and suspicion of their motives. Communication is not a one-way process nor is it merely the provision of facts or information. Risk communication is most effective when it is responsive and sustained. There are clear challenges in achieving this when communicating with the public where there are many different views and levels of understanding.

Greater risk perception by citizens can be due to uncertainty generated by a lack of openness in a regulatory process. The Food Standards Agency was established in 2000 with a remit to protect public health and restore public confidence in the way food safety decisions are made. Its core values are to put consumers first, to be open and accessible and to be an independent voice. There is an expressed policy to be open about uncertainty at the same time as recognizing best scientific advice and encouraging participation from all stakeholders.

In order to extend and advance participatory processes there is a need for mutually comprehensible data, public access to audit information or other evidence, explicit recognition of risk transfer and clear separation of scientific assessment from risk analysis. Scientific and consumer issues should be dealt with at the same time rather than separately. A precautionary approach and widespread public debate can be used to prevent escalation of public concerns and generation of controversies about risk.

SUMMARY

There is growing acceptance that there is a need to integrate qualitative and quantitative approaches in modelling and managing risk. Quantitative risk assessment is based upon numerical estimates of the probability of occurrences that have the potential to cause harm or injury. These may be extensive and complex and may involve an integration of different risks to provide a single measure of risk from a hazardous agent. They are usually based on predicting the statistical probability of a particular event occurring. In making predictions evidence from the past is analysed and preventive or protective actions introduced where the potential damage to environment or health is considered unacceptable.

Qualitative approaches have been criticized as being less precise and based on subjective rather than objective judgements. This criticism is unjustified given that quantitative measures of risk are used as the basis for policy decisions made by individuals or groups applying personal or professional judgements after interpreting the data available.

Public perceptions of risks are neither uniform nor static, rather they vary between cultures and in the light of prevailing economic conditions. There is suspicion of authorities, which have taken inadequate action to control risks in the past. Industrial and military operations in particular are seen to have direct interest in under assessing or under reporting the risk of activities in order to achieve their primary objectives at least economic cost. The parameters measured are also contentious. It is clearly difficult to equate death rates to less severe social impacts. The importance and significance of lay person's perceptions of risk have increasingly been recognized and the need for decisions in relation to risk to take these into account is now widely accepted by government. Thus the assessment and management of risk is no longer viewed as being a technical subject but is now regarded as a management approach. The aim of risk management is

therefore usually to control hazards to a tolerable level in the context of the benefit associated with a particular activity or process.

As knowledge about risk continues to expand, techniques of risk assessment and for reviewing the effectiveness of risk management practices are likely to take on even greater importance in environmental health. Current developments in approaches for the determination and regulation of housing fitness in the United Kingdom and the proposed mandatory extension of requirements for HACCP or similar systems in all food businesses exemplify this. Policymakers and enforcement agents increasingly recognize the influences on public risk perception, particularly the impacts of levels of trust in institutions, the relative ability to exert control over sources of risk and the significance of effective risk communication messages.

REFERENCES

1. HSE (1999) *Reducing Risks, Protecting People Discussion Document DDE11*, Sudbury, HSE Books, UK.
2. Cabinet Office (Strategy Unit) (2002) *Risk: Improving Government's Capability to Handle Risk and Uncertainty*, Summary report.
3. ILGRA (1998) *Risk Assessment and Risk Management: Improving Policy and Practice within Government Departments*, Sudbury, HSE Books, UK.
4. Advisory Committee on Dangerous Pathogens (1996) *Microbiological Risk Assessment: An Interim Report*, HMSO, London.
5. ILGRA (1996) *Use of Risk Assessment within Government Departments*, Sudbury, HSE Books, UK.
6. Royal Society (1992) *Risk Analysis, Perception and Management. Report of a Royal Society Study Group*, The Royal Society, London.
7. *Interdepartmental Liaison Group on Risk Assessment* (2002) Third report prepared by ILGRA.
8. DEFRA *et al.* (2000) *Guidelines for Environmental Risk Assessment and Management*, Revised Departmental Guidance, August 2000.

9. HSE (1993) *Successful Health and Safety Management HS(G)65*, HSE Books, UK.
10. Court, R. (2003) The UK's Housing Health and Safety Rating System: where we are now and where are we going? Presented at *un*Healthy Housing; promoting good health conference, University of Warwick, UK, March 2003.
11. Andrews, J.D. and Moss, T.R. (1993) *Reliability and Risk Assessment*, Longman Scientific and Technical.
12. Groeneweg, J. (1996) *Controlling the Controllable*, 3rd edn, DSWO Press, Leiden.
13. Kirwan, B. (1994) *A Guide to Practical Human Reliability Assessment*, Taylor & Francis, London.
14. Watson, S.R. (1981) On risks and acceptability. *Journal of the Society for Radiological Protection*, 1(4), 21–5.
15. Lewis, H.W. (1978) *Risk assessment review group report to the United States Nuclear Regulatory Commission*, NUREG CR 0400, US Nuclear Regulatory Commission.
16. Adams, J. (1995) *Risk*, UCL Press, London.
17. Slovic, P. (1987) Perception of risk. *Science Wash*, **236**, 280–5.
18. Soby, B.A., Simpson, A.C.D. and Ives, D.P. (1994) Managing food-related risks: integrating public and scientific judgements. *Food Control*, 5(1), 9–19.
19. Fisk, D. (1999) Perception of risk – is the public probably right? in *Risk Communication and Public Health* (eds P. Bennett and K. Calman), University Press, Oxford.
20. Hurst, N.W. (1998) *Risk Assessment: The Human Dimension*, The Royal Society of Chemistry, Cambridge.
21. Newby, H. (1997) Risk analysis and risk perception: the social limits of technological change. *Process Safety and Environmental Protection*, 75 (August), no. B3.
22. Royal Academy of Engineering Working Groups (2003) *Common Methodologies for Risk Assessment and Management*, The *Societal Aspects of Risk* and *Risks Posed by Humans in the Control Loop*, The Royal Academy of Engineering, London.
23. Weyman, A.K. and Kelly, C.J. (1999) *Risk Perception and Risk Communication: A Review of the Literature*, CRR 248/1999, HSE Books, Sudbury, UK.
24. Lichtenstein, S., Slovic, P., Fischoff, B., Layman, M. and Combs, B. (1978) Judged frequency of lethal events. *Journal of Experimental Psychology (Human Learning and Memory)*, **4**, 551–78.
25. Slovic, P. (2000) *The Perception of Risk*, Earthscan Publications Ltd, London.
26. Ford, N.J. and Rennie, D.M. (1987) Consumer understanding of food irradiation. *Journal of Consumer Studies and Home Economics*, **11**, 305–20.
27. Slovic, P., Fischoff, B. and Lichtenstein, S. (1980) Facts and fears: understanding perceived risk, in *Societal Risk Assessment: How Safe is Safe Enough* (eds R.C. Schwing and W.A. Albers), Plenum, New York.
28. Pidgeon, N., Hood, C., Jones, D., Turner, B. and Gibson, R. (1992) *Risk Perception in Risk Analysis, Perception and Management. Report of a Royal Society Study Group*, The Royal Society, London.
29. Douglas, M. and Wildavsky, A. (1982) *Risk and Culture: An Essay on the Selection of Technological and Environmental Dangers*, Berkeley University of California Press.
30. Dake, K. (1991) Orienting dispositions in the perception of risk – an analysis of contemporary worldviews and cultural biases. *Journal of Cross Cultural Psychology*, **22**, 61–82.
31. Wynne, B. (1982) *Rationality and Ritual: The Windscale Inquiry and Nuclear Decisions in Britain*, British Society for the History of Science, Chalfont St Giles.
32. Covello, V.T. and Johnson, B.B. (eds) (1987) *The Social and Cultural Construction of Risk: Issues, Methods and Case Studies*, Reidel Press, New York.
33. Jasanoff, S. (1986) *Risk, Uncertainty and the Legal Process*, Cornell University.
34. Kasperson, R., Renn, O., Slovic, P., Brown, H., Emel, J., Goble, R., Kasperson, J. and Ratick, S. (1988) The social amplification of risk: a conceptual framework. *Risk Analysis*, 8, 177–87.
35. Petts, J., Horlick-Jones, T. and Murdock, G. (2001) *Social Amplification of Risk: The*

Media and the Public, CRR332/2001, HSE Books, UK.

36. HSE (1998) *The Tolerability of Risk from Nuclear Power Stations*, London, HMSO.

37. The Presidential/Congressional Commission on Risk Assessment and Risk Management Framework for Environmental Health Risk Management (1997) Final Report, Volume 1, 529, 14th Street, Washington, DC.

38. Tait, J. (1988) Nimby and Niaby: public perception of biotechnology. *International Industrial Biotechnology*, 8, 5–9.

39. Cox, S.J. and Tait, N.R.S. (1991) *Reliability, Safety and Risk Management, an Integrated Approach*, Butterworth Heinemann, Oxford.

40. Fischhoff, B. (1995) Risk perception and communication unplugged: twenty years of process. *Risk Analysis*, **15**(2), 137–44.

41. Sandman, P.M., Weinstein, N.D. and Miller, P. (1994) High Risk or Low – How Location on a Risk Ladder Affects Perceived Risk. *Risk Analysis*, **14**(1), 35–45.

FURTHER READING

HSE (1995) *Generic Terms and Concepts in the Assessment and Regulation of Industrial Risks*, Discussion Document, HSE Books, London.

11 Specific public safety and health issues

Richard J. Palfrey

INTRODUCTION

There is something that is curiously beguiling about most of the attractions that make up the present day leisure industry. They have elements of surprise, excitement and mystery together with their often unique transient nature to be sampled today, or gone tomorrow. They are intended to transport their participants from the familiar to new sensations and excitement. Ensuring the safety of these events is a whole range of environmental health controls that mainly go unnoticed until some serious mishap occurs.

FAIRGROUNDS

Fairground apparatus has become increasingly sophisticated and the potential for serious personal injury is high if safe working practices are not established and followed rigidly. Fairgrounds have features that are difficult and complex, ensuing from the wide variety of largely non-standard devices, the diversity of sites and the method of operation. In the travelling section of the industry, there are problems of repeated 'build-up' and 'pull-down'.

Fairgrounds and amusement parks are generally considered to be relatively safe, although there have been a number of serious incidents involving the public and employees showing that high standards of control have not always been maintained throughout the industry. These prompted a review of fairground safety by the Health and Safety Executive (HSE) [1].

Guidance on safe practice

Useful advice can be found in *Fairgrounds and Amusement Parks: Guidance on Safe Practice [2]* prepared by the HSE through the Joint Advisory Committee on Fairground and Amusement Parks. The guidance deals with the principles associated with the overall safety management of attractions and places great emphasis on risk assessment, management of safety and assessment of conformity to design.

The guidance gives detailed advice on the key points of the system of steps and checks agreed with the industry as being appropriate measures for working safely and complying with the law and indicates the main responsibilities of duty holders. It also develops good practice and places increased emphasis on risk assessment, management of safety and the inspection stages known as design review, assessment of conformity to design, initial test and thorough examination. It introduces the industry's scheme for the registration of inspection bodies. The design review is an appraisal of the design by an inspection body to check the adequacy of a design specification and the assumptions on which it is based.

Each attraction is required to be checked by an inspection body to ensure that the ride is constructed to the design specification. An initial inspection must be carried out to ensure that the equipment is capable of meeting functional design requirements when operated. In addition a daily visual and functional check by operators and attendants must be undertaken to assess the condition of the equipment. Thorough examinations are also required by an appointed body to decide whether an amusement device may continue to be operated for a specified period of time.

Recognizing the importance of inspection in the steps and checks required to ensure safety, the Amusement Device Inspection Procedures Scheme (ADIPS) has been introduced by the industry with HSE support. The scheme covers the:

- four types of inspection required for amusement devices;
- documentation required by amusement device operators;
- registration and administrative control of appropriately qualified inspection bodies;
- transitional arrangements for introducing the scheme; and
- inspections required for coin-operated children's amusement devices.

Whilst the term 'thorough examination' is used in HSG 175 it should be borne in mind that the Provision and Use of Work Equipment Regulations 1998 require annual inspections of equipment.

Definition and enforcement

Fairgrounds are defined in the Health and Safety (Enforcing Authority) Regulations 1998 as meaning 'such part of premises as is for the time being used wholly or mainly for the operation of any fairground equipment (see below), other than a coin-operated ride, non-powered children's playground equipment, swimming pool slide, go-kart, or plant designed to be used by members of the public for entertainment purposes for bouncing upon'.

Enforcement of the provisions of the Health and Safety at Work, Etc. Act 1974 is the responsibility of the HSE by virtue of schedule 2 of the regulations. Fairgrounds at premises otherwise allocated to local authorities, for example, in holiday camps, will also fall to the HSE for inspection.

Fairground equipment

The definition of fairground equipment is not included in the regulations, but it is to be found in the Health and Safety at Work, Etc. Act 1974 as amended by the Consumer Protection Act 1987. Fairground equipment is here defined as any fairground ride or any similar plant that is designed to be in motion for entertainment purposes with members of the public on or inside it, or any plant that is designed to be used by members of the public for entertainment purposes, either as a slide or for bouncing upon. In this definition, the reference to plant that is designed to be in motion with members of the public on or inside it includes a reference to swings, dodgems and other plants that are designed to be in motion wholly or partly under the control of, or to be put in motion by, a member of the public.

Although playground equipment is excluded, the guide does cover the use of amusement devices in other premises, making it of use to local authority inspectors who come across equipment, the use of which may constitute only a minor activity on the premises.

A series of guidance notes has also been produced by the HSE in the Plant and Machinery Series. These describe various factors that contribute to accidents on fairground apparatus. The guidelines are based on HSE reports of incidents, visits to fairgrounds by inspectors and the considerable experience of fairground operators. The range of passenger-carrying amusement devices so far covered by these guidance notes include the waltzer (PM47), the octopus (PM48), the cyclone twist (PM49), the big wheel (PM57), the paratrooper (PM59), the chair-o-plane (PM61), roller coasters (PM68), ark/speedway devices (PM70), water chutes (PM71) and the trabant (PM72). Another source of information is the *Fairground User's Safety Code* published by the Royal Society for the Prevention of Accidents (RoSPA) [3].

Playground equipment

Playground equipment and rides are also to be found in hotels and restaurants, where Section 3 of the Health and Safety at Work, Etc. Act 1974 will apply (see p. 425). While the guidance on safe practice is principally aimed at fairground activities, it contains useful guidance on standards that could be required for children's rides. A useful guide on playground equipment standards is the booklet *Playground Management for Local Councils*, issued by the National Playing Fields Association [4]. 'Coin-operated kiddie rides: a safety guide' produced by the British Amusement Catering Trades Association helps to explain the requirements of HSG 175.

The Entertainment Services National Industry Group (NIG) of the HSE is also prepared to give advice on standards through the HSE Area Enforcement Liaison Officer Service.

HAIRDRESSING

Hazards

The potential occupational hazards are numerous. The industry is made up of very small units, usually employing a high proportion of young people, and there is always a large number of trainees and others waiting to enter the industry.

The most frequent occupational problems encountered are those of dermatitis of the hands and the ergonomic problems resulting from long periods of standing in tiring postures. There has been a growing awareness of the possible long-term hazards associated with the chemical dyes and sprays that are frequently found in these premises, and the risks from customers infected with transmissible blood diseases.

However, although there has been concern about the occupational hazards of hairdressing, it is not generally held to be a high-risk activity in respect of the transmission of serious infections. Nevertheless some of the practices employed in hairdressing may result in infection passing from customer to customer if hairdressing implements are not sterilized. Therefore the promotion of hygiene in the salon is important.

The two major infections about which there has been most concern, Human Immunodeficiency Virus (HIV) and hepatitis B (see Chapter 14), are both capable of transfer by small amounts of blood and serum (from an infected hairdresser or customer) to breaks in the skin. Although there is no danger if these diseases are not present (it is unlikely that HIV can survive for long periods on equipment anyway), it is poor practice to rely on this being the situation, and so high standards of hygiene and positive methods to ensure the destruction of likely organisms must be employed.

Less serious infections, including spots, boils, abscesses, impetigo (both streptococcal and staphylococcal), herpes, ringworm, headlice and warts, may also be passed from person to person if hygienic practices are not employed.

Legislation

There is specific legislation covering hairdressing salons. Local authorities may make by-laws that relate to hairdressers and barbers under Section 77 of the Public Health Act 1961 for the purpose of ensuring the cleanliness of the premises, equipment and staff. All hairdressing businesses must also comply with the requirements of the Health and Safety at Work, Etc. Act 1974, and local authorities are the enforcing authorities for these premises.

Under the Control of Substances Hazardous to Health Regulations 2002 (COSHH), there is an obligation to carry out an assessment of the risks of substances used at work. Exposure must then be adequately controlled (see Chapter 23). These provisions are important in the hairdressing trade.

Hygiene

The main vehicles of bacterial transmission are razors, scissors, clippers and styptic, with brushes, combs, massagers, rollers, towels and hands presenting only an occasional risk. To eliminate risk, it is wise to avoid the use of open razors, replacing them with disposable razors or disposable blade razors, which can be discarded after use. Electric razors can be difficult to sterilize and should also be avoided.

The use of scissors cannot be avoided, but if skin is punctured, wounds should be immediately treated, for example, with a prepacked spirit swab, and left to dry. The scissors should not be used again until they have been sterilized by autoclaving, boiling or soaking them in 70% alcohol chlorhexidine for 30 minutes. Because of the time required for treatment, hairdressers often find it helpful to keep two or more pairs available for use. Scissors should be washed regularly in hot water containing detergent, and then dried or wiped with an alcohol wipe before being allowed to dry.

Manual clippers with non-detachable blades should not be used, and care should be taken to ensure that the blades of electrical clippers are correctly aligned to avoid cuts to the skin. When this happens, the blades should be removed and treated. Properly adjusted clippers only need a regular wipe-over with an alcohol wipe.

Styptic, used to stop bleeding, should not be applied directly on to broken skin. It should either be applied on gauze or cotton wool or applied in aerosol form. It is recommended that it is not used at all and that bleeding is controlled by wiping with gauze or cotton wool, or by waiting for the bleeding to stop naturally.

Combs, brushes, massagers, etc. may be cleaned by washing with hot water and detergent after use for each customer, drying and using an alcohol and chlorhexidine wipe.Towels, capes and gowns require no special precautions but should be laundered regularly. Disposable paper items are more hygienic and are especially recommended for customers with skin problems. Apart from alcoholic disinfectants and bleach for blood spills, chemical disinfectants are not generally recommended in hairdressing salons, as they may regularly become contaminated and their concentration may vary. They are also often toxic and corrosive.

Automatic autoclaves are recommended as being the most effective means of sterilizing hairdressing equipment, and these should be used wherever possible. Glass bead sterilizers use the dry heat method of sterilization. These instruments need time to heat up and may be difficult to use for some items as they can only sterilize the parts in contact with the hot beads. It is also often argued that these instruments may also blunt pieces of equipment with sharp cutting edges.

If sterilization is not possible by either of the methods mentioned here, disinfection may be achieved by boiling or steaming for at least 10 minutes in equipment specially designed for hairdressing instruments. The ultraviolet light apparatus often found in hairdressing salons does not sterilize equipment and is therefore not as efficient as the use of autoclaves, etc.

Useful guidelines on satisfactory methods of sterilization and disinfection have been produced by the Public Health Laboratory Service [5]. The Department of Health has also produced an HIV/AIDS (Auto Immune Deficiency Syndrome) information leaflet for hairdressers [6].

Premises and equipment should be kept clean and staff should follow good standards of personal hygiene. The salon should be kept clean using proprietary cleaners, but alcohol-based disinfectant is specifically recommended for surfaces that need to be wiped three or four times a day. Staff should wash their hands before and after each customer. If staff suffer from dermatitis, disposable gloves should be worn.

Product safety

Considerable advice is now available on the composition and safety of products supplied to hairdressing businesses. These products will not present a risk to health and safety if they are used sensibly and in accordance with the instructions supplied by manufacturers. A useful source of information in this respect is the HSE publication *How to Use Hair Preparations Safely in the Salon* [7]. This booklet gives advice on the storage and sensible use of hair preparations.

Hair preparations are governed by the Cosmetic Products Regulations 1978, which, among other things, requires that cosmetic products shall not be liable to cause damage to human health when applied under normal conditions of use. These regulations lay down safety standards for all cosmetic products, including appropriate labelling requirements.

ACUPUNCTURE, TATTOOING, SKIN-PIERCING AND ELECTROLYSIS

During the 1970s, considerable concern was expressed about the possible transmission of blood diseases as a result of skin-piercing activities. Although this concern was initially associated with the spread of the hepatitis virus, the growing awareness of the dangers of HIV and AIDS have since made the control of these potentially dangerous activities essential.

Health and safety aspects

General duties are placed on operators of these businesses by the Health and Safety at Work, Etc. Act 1974 to conduct their operations in ways that are, as far as is reasonably practicable, safe and healthy, and do not put staff or customers at risk. Cosmetic and therapeutic skin-piercing, when not carried out under medical control and supervision, is allocated to local authorities for enforcement of the Act. Where a peripatetic practioner carries out work in a client's private home, this is the responsibility of the HSE.

Adoptive powers

All local authorities have specific powers to regulate businesses providing cosmetic body-piercing, semi-permanent make-up (or micropigmentation) and temporary tattooing.

Local authorities still rely on the detailed duties in the codes of practice and by-laws made under Part VIII of the Local Government (Miscellaneous Provisions) Act 1982 which have been beneficial, and use these specific powers. The powers are adoptive, and district and London borough councils are able to choose the provisions they wish to apply within their areas.

Acupuncture and tattooing, skin-piercing and electrolysis are treated separately for the purposes of making a resolution to adopt the powers, and local authorities may resolve that any of these activities be controlled, or that different ones be controlled, from different dates. Provision is made for adequate publicity before a resolution takes effect.

Acupuncture is not defined in the Act but is generally taken as meaning 'the insertion of needles into living tissue for remedial purposes'. Tattooing is referred to in the Tattooing of Minors Act 1969 as 'the insertion into the skin of any colouring material designed to leave a permanent mark'.

Registration

The effect of passing a resolution is to require the registration of persons undertaking skin-piercing activities, unless the activities are undertaken by a registered medical practitioner or a dentist. Premises must also be registered. Where a person travels offering skin-piercing services, that person's home must be registered.

There are no transitional periods for the benefit of existing traders. Existing traders must therefore ensure that they register early, and local authorities must ensure that they are able to process these applications before their resolution takes effect.

Registration cannot be refused, but where a previous registration has been cancelled by a magistrate court as a result of a conviction for an offence under the local authority's by-laws, the court's consent must be obtained before a person can be re-registered. The registration certificate issued and a copy of the by-laws must be displayed prominently on the premises, and a reasonable fee is payable to the local authority for registration. Failure to display the registration certificate or by-laws is an offence.

Offences

It is an offence punishable by a fine up to level 3 to carry out any skin-piercing activities unregistered. A similar fine is possible for a contravention of by-laws. A court can also suspend or cancel a registration by order instead of, or in addition to, levying a fine.

By-laws

Model by-laws produced by the then Department of the Environment and the Welsh Office contain provisions to secure the cleanliness of premises, sterilization of instruments and hygiene of the practitioners. Codes of practice are also drawn up by most councils to assist practitioners in complying with their by-laws. Most codes are derived from *A Guide to Hygienic Skin Piercing* [8].

The British Acupuncture Association, the Traditional Acupuncture Society and the Register of Oriental Medicine have also produced codes of practice.

Guidance on the risk of infection from skin-piercing activities has been given to local authorities by the HSE in an advisory circular, *Risk of Infection from Skin Piercing Activities* [9]. (See also Scottish Centre for Infection and Environmental Health, *Body and Skin Piercing – Guidance for Local Authorities*, SCIEH, Glasgow, 1998.)

In 1997, the Department of Health carried out a review of the legislation that gives local authorities the power to register/licence skin-piercing activities. Although the results of the review are not yet available, it is likely that any conclusion will be that on public health grounds there is a continuing need for regulation; but there may be scope for some deregulation where factors such as advances in technology or membership of a professional body provide equivalent protection to public health or where the current regime places unnecessary burdens on business. An extension of regulation to include the new fashion of cosmetic piercing and semi-permanent make-up has been introduced by changes to the 1982 Act made by section 120 of the Local Government Act 2003.

As most tattooists or ear piercers also carry out body-piercing, local authorities have an opportunity to work with businesses to promote safe and hygienic practices. Voluntary registration schemes and the production of good practice guidelines are being introduced by many local authorities and are advising the public of potential health risks and how to choose a reputable business. One worthy of note is that produced by the Bury and Rochdale MDCs.

The HSE has published guidance for local authorities on health and safety issues related to skin-piercing [10] and the CIEH publication 'Body art, Cosmetic therapies and other special treatments' [11] contains valuable information on techniques, infection control and legal matters.

PUBLIC ENTERTAINMENT LICENCES

Purpose of licensing

The objective of this regime is to ensure that such events are adequately controlled, that there is proper hygiene and safety and that nuisance is avoided. It is intended that this procedure will be replaced commencing in 2004 through the implementation of the Licensing Act 2003 – see later.

Licences for music and dancing

The Local Government (Miscellaneous Provisions) Act 1982 provides that public dancing, music or other similar public entertainment can only be provided under the terms of a licence. Any person providing such an entertainment without a licence is liable to a fine of £20 000 and/or imprisonment for up to six months. There are punitive measures for breaches in the terms, conditions and restrictions of a licence of a fine of up to level 5 and/or imprisonment of up to three months (the Entertainments (Increased Penalties) Act 1990).

There are exceptions to the licensing requirement in respect of music in places of worship, at religious meetings, at pleasure fairs and at entertainments held in the open air, unless the local authority has adopted the provisions relating to outdoor entertainments.

Licences can be granted for one or more occasions and relate to the entertainment, rather than the premises. Licences are required for either live or recorded music (although in the latter case there is no requirement for a licence in premises licensed for the sale of intoxicating liquor), and regardless of whether the entertainment is by the public or by a performer.

Private entertainment

Care must be made to distinguish between public entertainment, which requires a licence, and private entertainment, which does not under these provisions. It will be a matter of fact and degree in each case as to whether entertainments are private or public, and this is not always an easy distinction to make. There is, however, guidance in Home Office

Circular 95/84 and from a number of decided cases. The test set out in *Allen* v. *Emerson* [1944] KB 362 [1944] 1 A11 ER 344 DC gives guidance 'where public entertainment will be provided in a place open to members of the public without discrimination who desire to be entertained and where means of entertainment are provided'. There is no reference to payment, as payment for admission is immaterial to a person's status as a member of the public. Where a charge is made, the above judgement can be considered in the light of the test in *Gardner* v. *Morris*: 'it is not whether one or two (or any particular number) members of the public are present but whether, on the evidence, any reputable member of the public, on paying for admission could come in' (1961 59 LGR 1987 and *Frailing* v. *Messenger* 1867 31 JP 423).

Bona fide guests of members of clubs are not regarded as members of the public (*Severn View Social Club and Institute Ltd* v. *Chepstow Licensing JJ* (1968) 1 WLR 1512).

However, the device of becoming a member of a 'club' merely on the immediate payment of a fee and completion of an application form has been discredited (see *Panama (Picadilly) Ltd* v. *Newberry* (1962 1.WLR. 610).

The Home Office view is that it is likely that events will be required to be licensed if large numbers of people are able to gain admission simply upon payment of the required fee. This view was upheld in the High Court in the case of *Lunn* v. *Colston-Hayter* (*Times* Law Reports 28 Feb. 1991).

Premises available to the general public, such as a function room in a public house, do not need to be licensed when used for a private purpose, for example, a wedding reception.

For information on the licensing of private places of entertainment see later.

Licences for sporting events

The second category of entertainment for which a licence is required consists of, or includes, any public contest, exhibition or display of boxing, wrestling, judo, karate or similar sport. There are exceptions to this requirement for these entertainments when held at pleasure fairs, and when held in premises licensed for music and dancing

taking place wholly or mainly in the open air. The Fire Safety and Safety at Places of Sport Act 1987 requires a licence for indoor sporting events that the public attends. Sports entertainment at sports complexes also requires a licence.

Musical entertainment in the open air

The third category applies to any public musical entertainment held wholly, or mainly, in the open air and on private land. There are exemptions for events such as fêtes etc. This requirement relates to open-air pop festivals (see later) and other open-air entertainments in which music is a substantial ingredient. This control is adoptive and the procedure requires publicity following the appropriate resolution, with the provisions becoming effective from the date specified in the council resolution.

Application for licences

Schedule 1 of the 1982 Act contains detailed provisions on the procedure for applying for licences and the powers of local authorities to impose conditions.

The procedure for making an application is found in paragraph 6 of schedule 1 of the Act. Applications for the grant, renewal or transfer require the applicant to give 28 days notice to the local authority, chief police officer and fire authority. Where the required notice has not been given, the local authority has some discretion to consider it, but a prerequisite to this is that consultation takes place with the fire authority and chief officer of police. Applicants must furnish such particulars and give notice as the local authority may prescribe by regulations.

Subject to limited exceptions, a reasonable fee must be paid and the amount is at the discretion of the local authority. However, the fee must not be arbitrary, unreasonable and improper and should not exceed the cost of administration of the licensing system. The exceptions relate to licences for buildings occupied in connection with places of worship and for village, parish or community halls and similar buildings. The local authority may also remit the fee where the entertainment is of an educational, charitable or similar purpose.

Licence conditions

Licences are issued subject to standard conditions prescribed by regulations made by the licensing authority. Every licence then granted, renewed or transferred is presumed to have been issued subject to these conditions unless it has been expressly excluded or varied. It is open to the holder of a licence to apply for conditions to be varied. When attaching conditions, it should be borne in mind that these must relate to safety, health and the prevention of nuisance.

One way of ensuring that licence conditions are adhered to is to attach a licence condition requiring door supervisors to be registered and trained. Many local authorities do this as a means of raising the quality of door staff and reducing criminal activity in clubs. It is claimed that community safety in and around late-night venues is enhanced by such schemes. From 2004 a National Registration Scheme will be introduced by the Security Industry Authority.

The hours of opening are best controlled by special conditions as hours of opening are linked closely with liquor laws. There is a limit on the conditions that may be attached to a licence for an outdoor entertainment. These relate to the safety of those present, provision of access for emergency vehicles, the provision of adequate sanitary appliances and the prevention of noise nuisance (HSE Guidance Notes GS50 and IND(G)102L).

Provisional licences

A good feature of the 1982 Act is that it permits the issue of provisional licences, subject to confirmation, for premises about to be constructed, or under construction or alteration. This is useful for those engaged in elaborate or costly proposals, as it gives an early indication of whether the project is likely to receive a licence.

Consultation

When considering an application, the licensing authority is required to consider the observations submitted by the chief officer of police and the fire authority. Conditions adopted by many local authorities also require public notice of application to be given. It appears, however, that no other person has a right of objection, although any observations made would be considered.

Although no guidance is given in the 1982 Act about the hearing of objections when they are made, the applicant should be informed of the nature of the objections so that he or she can respond. In these situations, procedures for dealing with applications must be considered carefully or they will lead to appeals against licensing-authority decisions. All applications must be dealt with in accordance with the terms of natural justice, though in many cases an oral hearing will not be necessary and can be dealt with by written representation only. If an oral hearing is arranged it is wise for objectors to state their case first.

Offences

Where entertainments are provided without the necessary licence or in contravention to the terms, conditions or restrictions of a licence, an offence is committed. The only statutory defences are to prove that due diligence had been exercised, that all reasonable precautions had been taken or that a special order of exemption is in force under Section 74(4) of the Licensing Act 1964. These orders automatically override any conditions on permitted hours. An investigation of a breach of hours in licensed premises should therefore include a check on whether a special order of exemption is in force.

Public conveniences at places of entertainment

Certain local authorities are empowered to require sanitary facilities to be made available for public use in 'relevant' places (see the following paragraph) by Section 20 of the Local Government (Miscellaneous Provisions) Act 1976. The requirement by written notice can be occasional – for such occasion specified in the notice – or provided for continuing use. There is only a right of appeal against the latter, although unreasonable requirements under the former may be challenged in any prosecution for non-compliance.

'Relevant place' means any of the following:

1. A place that is normally used or is proposed to be normally used for any of the following purposes, namely:

   (a) the holding of any entertainment, exhibition or sporting event to which members of the public are admitted, either as spectators or otherwise;
   (b) the sale of food or drink to members of the public for consumption at the place.

2. A place that is used on some occasion or occasions or proposed to be used on some occasion or occasions for any of the purposes aforesaid.
3. A betting office.

A duty is imposed to have regard to the needs of disabled persons when complying with a notice. There may be some overlap with the requirements of some licensing procedures.

Pay parties

In view of the dangers to the public and the nuisance caused by so-called 'pay parties', the Entertainments (Increased Penalties) Act 1990 raised the penalties for the use of premises for which no licence was in force, or for contravention of the terms and conditions of a licence imposing a limit on the number of people who may be present at the entertainment to a £20 000 fine or six months imprisonment. The Criminal Justices Act 1988 (Confiscation Orders) Order 1990 gives magistrates the power to order the confiscation of the proceeds, where these exceed £10 000, made by people convicted of these offences.

'Pay parties' is the most commonly used generic name given to these events, the different names given to the various types of pay party reflect both the size and venue of the event:

- acid house party, dance parties, raves and warehouse parties,
- the smallest pay parties often held in domestic premises are called 'blues parties or 'she-beens'.

(See Joint Home Office and Department of the Environment Guidance booklet *Control of Noisy Parties*, September 1992.)

Sections 63–67 of the Criminal Justice and Public Order Act 1994 give the police powers to deal with 'raves' (a gathering of 100 or more people at which amplified music is played at night and which is likely to cause serious distress to inhabitants). These powers include the removal of people attending and seizure of sound equipment.

Powers of entry

There is a provision in the 1982 Act for the entry of places of entertainment by the police, or by authorized officers of the licensing authority and the fire authority, for enforcement of the licence. Proceedings for alleged breaches of the schedule may be instituted by any person and there is no limitation in the Act.

Revocation

If convicted of an offence under the Act, the licence may be revoked. The principles of natural justice apply, and the holder of a licence should, even though convicted of an offence, have an opportunity to state his or her case. There is a right of appeal against any adverse decision of a licensing authority, but the holder of a licence is not bound to implement the decision until either the 21-day appeal period has passed or the appeal has been determined. The appeal period cannot be extended. The procedure for an appeal is given in Section 34 of the Magistrates' Courts Rules of 1981. Appeals are by way of re-hearing and the appellate court is entitled to substitute its own opinion about the facts or merits of the case. (It would no doubt give some consideration to the fact that the original decision has been made by an elected representative body.) As the magistrate court can only summon those against whom it can make an order, only the licensing authority can be summoned to appear. If observations or objections were considered in arriving at a decision, it is up to the licensing authority to bring evidence of these before the magistrate court.

An appeal against the decision of the magistrate court may be brought to the crown court within 21 days, but in this case there is provision to extend this time for giving notice of appeal.

The Public Entertainment Licences (Drugs Misuse) Act 1997, which came into force on 1 May 1998, will make it easier for local authorities to close night clubs where there is a serious drugs problem.

Under the Act, an authority is able to revoke a public entertainment licence if informed by the police of serious problems relating to the supply or use of controlled drugs at or near the licensed premises and if it is satisfied that such action will assist in dealing with the problem.

Apart from refusing to renew or transfer a licence, local authorities may be able to impose conditions on a licence. Where there is an appeal against a decision to revoke or not renew a licence the licence is suspended until after the appeal.

The Private Places of Entertainment (Licensing) Act 1967

In areas where this Act has been adopted, private dancing, music or other entertainments of a like kind, which are promoted for private gain, must also be licensed. A definition of 'private gain' is found in the London Authorities Act 1991.

The Act is an adoptive one and therefore only applies if the appropriate authority – district councils, the London boroughs or the council for the Isles of Scilly – so resolves. The procedure for adopting the Act, together with the requirements for publicity, etc., are laid down in Part II of the schedule to the Act, and the powers available enable control of events involving music and dancing including pay parties.

When licences are granted, they can be made subject to terms, conditions or restrictions imposed by the licensing authority. This may include conditions providing power of entry to private premises for which generally there is no right of entry. Those also applicable to public entertainments will usually be equally relevant.

Enforcement provisions are identical to the provisions found in the Local Government (Miscellaneous Provisions) Act 1982, with respect to public entertainment and, as with that Act, any person may prosecute for the breaches defined in Section 4. Investigations into alleged contraventions of this Act need care and the procedures of the Police and Criminal Evidence Act 1984 must be followed.

When licences are refused, appeals to a magistrate court are available, and the appeal is in the form of a re-hearing of the application.

Private parties not for gain

In the case of private parties that are not held for gain and do not come within the scope of licensing legislation, the powers available to local authorities are restricted solely to the noise nuisance abatement powers of Part III of the Environmental Protection Act 1990. This includes the power under Section 81[5] to seek a High Court injunction if the authority considers that summary proceedings would provide an inadequate remedy.

THE LICENSING ACT 2003[1]

Enacted in 2003 and to become operative through Commencement Orders from 2004, this legislation will radically change the UK liquor and public entertainment licensing system. The powers currently with the magistrate courts to control the sale of liquor are to be transferred to local authorities who will also retain control of what will then be called 'regulated entertainment'. The change seeks to streamline the more than fifty statutes impacting on the licensing systems for alcohol, public entertainment and late-night refreshment houses into a single system.

The change is being brought about because the government recognizes that the issues of public disorder and anti-social behaviour are strongly associated with the consumption of alcohol, which is sold by many of the businesses to be controlled though the new Act. It is said that modern laws are required to ensure that people may enjoy their leisure whilst being adequately protected and without fear of violence, intimidation and disorder.

---

[1] See Department of Culture, Media and Sport (2003) Draft Guidance on the Licensing Act 2003 available on www.culture.gov.uk

Licensed activities

The legislation will provide for the licensing of the sale of alcohol, the supply of alcohol in certain clubs, the provision of regulated entertainment and the provision of late-night refreshment. The licensing authorities discretion will not be as wide as in current law and hearings and process will be strictly regulated, fees set centrally and guidance issued by the Secretary of State will be given sufficient influence to promote best practice and ensure consistent application of the legislation.

Four objectives will underpin the new system:

- the prevention of crime and disorder;
- public safety;
- the prevention of public nuisance;
- the protection of children from harm.

Informed by these objectives, by the licensing authorities own statements of licensing policy and by the Secretary of State's guidance, it is intended that a cultural change will be brought about in the nation's drinking and entertainment habits. It will also become easier for businesses to develop family-friendly facilities.

Licences

A **premises licence** will be required for premises holding regulated entertainment. Expert bodies (including police, fire, health and safety and environmental health) will be statutory consultees on every application and others (local residents, businesses, etc.) will be given the opportunity to make representations. If no representaions are made, licenses will be granted on the conditions and hours requested by the applicant. To avoid duplication with other legislation, licensing authorities will only be able to impose conditions that are necessary to promote the licensing objectives.

All decisions of the licensing authority will be subject to appeal to a magistrate court and premises licenses will not be time limited as is the current practice. Similar systems will apply to what are currently called 'registered members clubs'.

Temporary events (previously occasional licenses) will not require licences provided they last no longer than 96 hours and involve less than 500 people. The police will have the chance to object or to modify by agreement with the premises-user the conditions under which the event will take place.

Personal licences to sell alcohol is a new concept to be introduced by the Act. In most cases the issue of these will be a quick administrative process provided that applicants are at least 18, have not been convicted of relevant criminal offences, possess an adequate licensing qualification and have not forfeited a previous licence.

The Act will contain safeguard measures to deal with crime and disorder and anti-social behaviour. Police powers will be extended to close down offending premises or temporary events.

POP FESTIVALS

Background

These often accommodate in excess of 100 000 people and employ many hundreds of staff. They comprise a major sector of the leisure industry.

Guide to Health, Safety and Welfare at Pop Festivals and Similar Events

This 1993 guide [12] indicates the key points in the planning and arrangement of a pop concert. Advice is given on health and safety, fire and emergency planning and venue facilities.

Legislative controls

Under the Local Government (Miscellaneous Provisions) Act 1982, local authorities may adopt the controls for public musical entertainments taking place wholly, or mainly, in the open air and on private land. These were included specifically to control pop festivals. Whether the particular festival has the requisite degree of music, giving rise to a need for a licence, and whether it is a public or private entertainment, must be determined for these controls to be relevant.

Licence conditions can be imposed in the interest of health and safety and are limited to securing:

- the safety of performers and other persons present;
- adequate access for emergency vehicles;
- the provision of adequate sanitary accommodation;
- preventing neighbourhood disturbance by noise.

The law of nuisance (see Chapter 6) also provides a broad and powerful restriction on any kind of potentially intrusive activity. With the approval of the attorney general, local authorities or individuals who have reason to feel unhappy at the prospect of a pop festival (or indeed of any large gathering) can take the matter to the High Court and, if they can prove likelihood of substantial and unreasonable interference to the community, they will obtain an injunction. This will effectively put the promoter at risk of proceedings for contempt of court if nuisance is caused as a result of the pop festival. (Such action has been taken successfully by Windsor and Maidenhead borough council and Newbury district council in relation to pay parties.) Although the law of nuisance is not a very flexible means of control, it can give local authorities a very effective negotiating weapon.

The essence of a nuisance is a condition or activity that unduly interferes with the use or enjoyment of land. This has been applied to a number of cases involving the congregation of crowds, and in the case of *A.G. v. Great Western Festivals Ltd* (unreported) to pop festivals in particular. The features of a pop festival that might amount to a nuisance include noise, trespass and damage to adjoining property, traffic congestion and pollution by litter. The normal remedy for a nuisance is an action for an injunction. For temporary events, such as pop festivals, it will usually have to be a *quia timet* injunction – one granted before an event on the basis of evidence showing a strong possibility that a nuisance will occur.

The Noise at Work Regulations 1989 will also apply to persons at work at the pop festival. The general duties of the Health and Safety at Work, Etc. Act 1974 will also apply, including the management of Health and Safety at Work Regulations 1992, and these powers may prove useful in dealing with the many other hazards, such as laser equipment, disco lighting and pyrotechnics encountered on the site.

Other legislation such as the Theatres Act 1968, the Cinemas Act 1985[2] and the Building Regulations may also be relevant.

Forward planning

To ensure the success of any large pop festival, it is essential that planning starts at least six months before the event, that sufficient funds are available to invest in the event and that an efficient back-up organization and management policy is established. It will, therefore, be necessary to set up a working party composed of the promoters, local authority, police and fire officers and representatives of voluntary organizations at an early stage.

Many of the faults and failures of festivals in relation to matters that are the concern of the local authorities have arisen not only through a lack of co-operation between promoters and local authorities, but also through not allowing the time for such co-operation to be really effective. The prime responsibility rests with the promoters. It remains, however, a major responsibility of a local authority to create conditions under which such co-operation is possible.

In the initial stages, the local authority should be concerned with the location and suitability of the site, estimated attendance figures, legal and financial implications and all of the possible public health problems. Standards must be identified and met and the necessary safeguards observed.

If local legislation requires an application to be made for a public entertainment licence, this should be made clear to the promoters, who may then take the appropriate action in good time.

Sites vary from fields to theatres to sports stadiums – and even on occasions lakes and rivers.

---

[2]To be replaced by the Licensing Act 2003 during 2004.

The choice of site therefore has a direct bearing on the standards that will need to be applied.

Prediction of numbers

It is not possible to plan for a pop festival without a reasonable idea of the numbers expected to attend. Prediction of numbers is undeniably a difficult task, particularly for those with little experience of pop festivals.

Standards

Attention needs to be paid to the following issues, all of which are dealt with in the guide:

- crowd safety
- structural stability of stages, etc.
- protection of water sources
- refuse and litter
- food safety
- washing facilities (one for every five sanitary conveniences)
- drainage
- pest control
- noise
- sanitary accommodation (one closet per 100 females and three closets for every 500 males plus 1.5 m run of urinal per 500 males)
- access and signs for disabled people
- fire prevention and fighting (also see [13])
- power supply
- medical services
- security
- management of site and stage.

Other outdoor events

A *Code of Practice for Outdoor Events* [14] provides guidance on safety management at outdoor shows and meetings. It indicates standards for crowd control and site operations at events ranging from national athletic meetings to agricultural festivals and car-boot sales.

Managing Crowd Safety (IND(6)1426) published by the HSE gives guidance on management responsibilities for such events.

SAFETY AT SPORTS GROUNDS

A sports ground is any place where sports or other competitive activities take place in the open air, and where accommodation has been provided for spectators, consisting of artificial structures or of natural structures artificially modified for the purpose (Safety at Sports Grounds Act 1975).

The Wheatly Report, commissioned in 1972 as a result of the Ibrox Park disaster, resulted in the passing of the Safety at Sports Grounds Act 1975. The Act requires all designated sports grounds (those for which a designation order is in operation) with a capacity of over 10 000 people to be issued with safety certificates by the local authority. This capacity may be changed by order, and can be different for different classes of sports grounds. The Act was considerably extended by provisions contained in the Fire Safety and Safety of Places of Sports Act 1987 giving certifying authorities similar powers with respect to regulated stands, that is, covered stands with a capacity of 500 or more spectators.

The 1975 and 1987 Acts are administered in London and conurbations by the London borough councils and metropolitan authorities, respectively, and in the rest of England and Wales by unitary and county councils. In Scotland, the work is carried out by unitary councils.

It is the duty of every local authority to enforce the Act and its regulations, and arrange for periodic inspections of designated sports grounds. Powers of entry to, and inspections of, any sports ground for this purpose are provided by Section 11 of the 1975 Act.

The Football Spectators Act 1989 was introduced to control admission to designated matches by a membership scheme. It also provides for the safety of spectators by means of licences and safety certificates.

Safety certificates

Safety certificates may only contain conditions to secure safety at the sports ground, and may include a requirement to keep records of attendance and maintenance of safety measures. Before a certificate is issued, the local authority is required to consult the building authority and the chief officer of police. When determining an application for a safety certificate, it is the local authority's duty to determine whether the applicant is a qualified person, that is one who is likely to be in a position to prevent contravention of the terms and conditions of any certificate issued. The form of application is contained in the Safety of Sports Grounds Regulations 1987, which, among other things, lay down the procedure for making an application under the Act. Safety certificates may be amended either with or without the application of the holder, but amendments must be limited to safety measures. Transfers of certificates are provided for, but the local authority's duty to determine whether the person is a qualified person remains. In both cases, consultation must take place between the chief officer of police, the fire authority and the building authority before the local authority amends or transfers the certificate. Any alteration to a sports ground that may affect safety must be notified to the local authority in advance. If a person is judged by a local authority not to be a qualified person, there is a right of appeal to the magistrate court as there is against the inclusion or omission of anything from a certificate or a refusal to amend it.

Where a general safety certificate is in force, its provisions take precedence over certain other pieces of legislation that may impose terms and conditions. This may well affect the licence conditions issued in respect of any public entertainments licence.

Prohibition notices

A special procedure is detailed in the Act for dealing with situations where serious risk is posed to spectators. In that situation, local authorities have power to serve prohibition notices that specify the matters giving rise for concern, and restrict or prohibit admission to, or to parts of, a sports ground until matters have been remedied. These notices take effect immediately. A person aggrieved by a prohibition notice may appeal to the magistrates' court against the notice, but in view of the overriding requirement for safety, the bringing of an appeal does not have the effect of suspending it.

Offences

Various offences are detailed in the Act, including contravention of the certificates, its terms or conditions, etc. These are punishable by fine or summary conviction, or fine and/or imprisonment for not more than two years on indictment.

Guidance

Useful guidance on sports ground safety is included in the voluntary code, *Guide to Safety at Sports Grounds*, issued jointly by the Home Office and the Scottish Home Office and Health Department [15]. Advice is given in this document (which is known as the 'Green Guide') on the construction and layout of grounds, including details of access and egress and ground capacity estimation, terracing, barriers, stands, etc., as well as for other matters covered by a certificate. The third edition in 1990 incorporated lessons from the Hillsborough, Sheffield tragedy in 1989 and the Bradford City fire in 1985. Its recommendations may be put into statutory form by inclusion in safety certificates. The information contained in the guide is especially applicable to football grounds, but the advice is also of use in dealing with a variety of sporting events at grounds where the gathering of crowds may present a safety problem.

Toilet facilities at stadiums

In 1994, the Sports Council in association with the Football Trust published a guide [16] that offers advice on evaluating existing toilet facilities, improving standards and choice of equipment for new installations. It applies to rugby and hockey stadiums as well as to football grounds.

Specific guidance is given on planning, location and access, design of toilet areas, fittings and materials, provision of toilets for family areas, disabled spectators, non-spectator use and portable toilets. The section on design recommends that before finalizing any designs that omit doors or lobbies to toilet areas, the local environmental health officers should be consulted.

A chapter is devoted to the subject of the ratio of toilets to the number of male and female spectators. Table 11.1 shows minimum recommendations made by the guide for newly constructed or refurbished stadiums and stands, per accessible area.

Indoor sporting events

Safety at indoor sporting events is controlled by the public entertainments licence procedure detailed in Schedule 1 of the Local Government (Miscellaneous Provisions) Act 1982 (see p. 236). These are not covered by the Safety at Sports Ground Act 1975.

Health and safety enforcement

Responsibility for enforcing the Health and Safety at Work, Etc. Act 1974 in sports grounds falls to local authorities (unless they are sports grounds under the control of local authorities). However, because there are overlapping responsibilities relating to safety of sporting events at sports grounds, there is a need for liaison between the various authorities involved. It is the Health and Safety Commission's (HSC) policy that the provisions of the 1974 Act should not generally be enforced if public safety is adequately covered by enforcement of the specific legislation in the 1975 Act. In cases of urgent threat to life or injury, however, the use of the Act's powers would not be precluded if they could eliminate the risk or reduce it.

CINEMAS

Legislation

The main piece of legislation concerning the exhibition of films is the Cinemas Act 1985, which consolidates the Cinematograph Acts of 1905 and 1952, the Cinematographic (Amendment) Act 1982 and related enactments. This Act made no change to the previous law except that, on the recommendation of the Law Commission, it amended provisions derived from the Sunday Entertainments Act 1937 so as to extend the exemption from the Sunday Observance Act 1780 to cover exhibitions produced by means of videos.[3]

Licences

Subject to certain exemptions, film exhibitions may only be given in premises that have been licensed by the local authority in which they are situated. Before the consolidating legislation was passed, licensing control related to exhibitions of 'moving pictures' produced on a screen by means that included the projection of light. This definition was not flexible enough to encompass changes in modern technology, and did not include video exhibitions, which are transmitted by signal. While control now extends to videos shown in clubs and pubs, it does not extend to exhibitions of moving pictures arising out of the

Table 11.1 Minimum number of toilets and washing facilities for newly constructed or refurbished stadiums and stands

	Urinals	*WCs*	*Hand-wash basins*
Male	1 per 70 males	1 per 600 males, but minimum of 2 per toilet area	1 per 300 males, but minimum of 2 per toilet area
Female	—	1 per 35 females, but minimum of 2 per toilet area	1 per 70 females, but minimum of 2 per toilet area

Source: [16].

---

[3]This legislation is to be replaced during 2004 through the implementation of the Licensing Act 2004.

playing of video games, and places such as amusement arcades and public houses therefore do not require to be licensed under this legislation.

Licence conditions

The granting of a licence is discretionary and, subject to regulations made under the Act, licensing authorities may impose terms and conditions on the licence. The terms and conditions that may be imposed are not limited to those for securing safety but, when imposed, the test to be applied is that they should be reasonable and in the public interest. Subject to these restrictions, there is no fetter upon the power of the licensing authority. Model licensing conditions are detailed in Home Office Circulars No. 150/1955 and 63/1990.

The London Inter-borough Entertainments Working Party has produced its own set of rules relating to the showing of films for inclusion in licences issued under the Cinemas Act 1985.

Film exhibitions for children

When granting a licence, the licensing authority has a duty to impose conditions or restrictions prohibiting the admission of children to film exhibitions involving the showing of works designated unsuitable for them. The familiar classification of films is not based on statute, but is undertaken by a body known as the British Board of Film Classification (BBFC). Its system of classification, the object of which is principally to indicate which films are considered suitable for viewing by children, has been adopted by local authorities. For the classification of films by the BBFC see Home Office Circular No. 98/1982 and 63/1990.

Any film exhibition organized wholly or mainly for children requires the consent of the licensing authority, and in these circumstances they can impose special conditions or restrictions. A statutory obligation to provide for the safety of children's entertainments is contained in the Children and Young Person's Act 1933.

Applications for licences

When applying for a licence, or for its renewal or transfer, the applicant must give 28 days clear notice of intention to the licensing authority, the fire authority and the chief officer of police. When the requisite notice has not been given, the licensing authority can still grant a licence, but only after consultation with the other two authorities. (The mere sending of a letter does not constitute consultation.) Licences are granted for periods of up to 12 months. Fees for licences may be fixed by the licensing authority, but they must not exceed the sum stipulated in the Fees for Cinema Licence (Variation) Order 1991. These are currently set at £173 in the case of a grant or renewal for one year; in the case of a grant or renewal for any lesser period the fee is £200 for each month for which a licence is granted or renewed, but in this case the aggregate of the fees paid in any year is not permitted to exceed £600. The maximum fee for a transfer of licence is £120. These figures can be amended by order of the Secretary of State. (Useful advice on fees is contained in Association of District Councils' Circular 1986/119. The Association of District Councils is now part of the Local Government Association.)

Regulations

Section 4 of the Act requires film exhibitions to comply with regulations made by the Secretary of State. To date no new regulations have been made but, by virtue of the Interpretation Act 1978, the Cinematograph (Safety) Regulations 1955 as amended and the Cinematograph (Children) No. 2 Regulations 1955 have effect.

Exemptions

Certain exhibitions are exempted from the requirement to obtain a licence. These include exhibitions in private dwellinghouses where the public are not admitted and where there is no private gain, or where the sole or main purpose is to demonstrate or advertise products, goods or services or to provide information, education or instruction.

Where the public are not admitted, or admitted without payment, or the exhibitions are given by an exempted organization and conditions regarding private gain etc., as aforementioned, are fulfilled, no licence is required. Exempted organizations are defined in relation to a certificate given by the Secretary of State. This exemption does not apply to certain exhibitions for children as members of a club, the principal object of which is attendance at film exhibitions, unless in a private house or as part of the activities of an educational or religious institution, or to exempted organizations in cases where the premises were used for an exhibition for more than three days out of the previous seven days.

It is not necessary to obtain a licence when premises are not used for more than six days in a year, and where film exhibitions are held occasionally and exceptionally, and the occupier has given at least seven days notice in writing to the fire authority and the chief officer of police, and he or she complies with any regulations of conditions imposed by the licensing authority and notified to him or her in writing. Strictly speaking, licences are required where film exhibitions take place in premises on a regular basis, even if fewer than six times per year, as the use would not be exceptional, but many authorities do not require a licence in these circumstances.

Film exhibitions that take place in buildings or structures of a movable character only need to be licensed (by the licensing authority where the owner normally resides) where the owner has given at least two days notice to the fire authority and the chief officer of police, and he or she complies with any conditions imposed by the authority in writing.

Sunday opening

The Sunday Observance Act 1780 is not contravened by staging film exhibitions on a Sunday, but a licensing authority is entitled to impose conditions, including those aimed at preventing employment, where a person has been employed for the previous six days. There are exceptions to this restriction in cases of emergency notified to the licensing authority where a rest day is given in lieu and where an employer has, on making due enquiry, reasonable grounds for believing that a person has not been employed for the earlier six days.

Offences

Where premises are used without a licence or consent (in respect of children) or where terms, conditions or restrictions are contravened, those responsible for the organization or management, as well as the licence holder, are guilty of an offence. A maximum fine of £20 000 is stipulated for operating without a licence and in other cases level 5 on the standard scale. In addition, a court can order the forfeiture of anything produced to the court relating to the offence, as long as the owner has been given an opportunity of appearing to show cause why it should not be.

If the owner of a licence is convicted of an offence as stipulated in Section 10 of the Act, or failed to provide for the safety of children, the licensing authority may revoke his or her licence.

Power of entry

Right of entry to inspect premises to see whether the relevant provisions are being complied with is given to police constables and authorized officers of licensing and fire authorities. Inspections by the fire authority to check on fire precautions, however, require 24 hours notice. When authorized by warrant, constables or authorized officers of the licensing authority can enter and search premises when they have cause to believe that an offence has been, is being or is about to be committed. This power is subject to the restrictions of Section 9 of the Police and Criminal Evidence Act 1984, and authorized officers must produce authority when requested. Any person who intentionally obstructs an officer is liable on summary conviction to a fine not exceeding level 3 of the standard scale.

A constable or authorized officer who enters and searches any premises under the authority of a warrant issued under Section 13 of the Act, may seize and remove apparatus or equipment, etc., which he or she believes may be forfeited under Section 11.

Appeals

A person may appeal to the crown court against a refusal or revocation of a licence or terms, conditions or restrictions subject to which a licence is granted as he or she may against the refusal to renew or transfer a licence. Appeals in Scotland are to the sheriff's court. Refusals in England and Wales in relation to Sunday opening and any conditions in that respect are to the crown court. Where a licence has been revoked, it remains in force until the determination or the abandonment of the appeal or, if successful, it is renewed or transferred.

Local authorities in Greater London have powers to vary licences and grant provisional licences, but other authorities have no power to grant a licence except in respect of premises actually in existence.

Where introductory music is played at any premises, or is featured in an interval or at the conclusion of a show, it is considered to be part of the exhibition provided that the total time taken amounts to less than one-quarter of the time taken by the film exhibition. A public entertainment licence is therefore not required.

VIDEO JUKEBOXES

Commercial premises that promote cinematographic exhibitions (video jukeboxes) as a means of attracting custom come within the Act's control. It is not necessary for a charge to be made for admission for a licence to be required if the exhibitions were advertised and the sums paid for the facilities or services are for private gain. Additional requirements in respect of television exhibitions in Part 4 of the Cinematographic (Safety) Regulations 1955 have to be complied with.

Health and safety enforcement

Responsibility for enforcement of the provisions of the Health and Safety at Work, Etc. Act 1974 in cinemas lies with local authorities. However, where the main purpose of a cinema premises is for educational or vocational training, similar to that provided in the mainstream education system, such premises will be the responsibility of the HSE.

THEATRES

Generally speaking, any premises used for the public performance of plays is required to be licensed. The only exceptions to this arise with respect to buildings under the control of the armed forces or in buildings known as 'patent theatres' where the performance may take place by virtue of 'letters patent'. The licensing authorities are the London boroughs, unitary authorities and district councils.

The law relating to theatres was formerly embodied in the Theatres Act 1843, which at the time gave the Lord Chamberlain powers of absolute censorship over the presentation of any stage play. To assist in this measure, copies of every new stage play were required to be submitted to him. In these early days, theatre-going was something of a hazardous affair and the records show that in 1884 alone it was calculated that, worldwide, 41 theatres were burned down involving the death of over 1200 people. From these tragedies, lessons were learned and there followed important decisions and action regarding public safety, fire precautions and fire-fighting. The Public Health (Amendment) Act of 1890 enforced stricter fire regulations, but within a few years there followed demands that all theatre planning should be subject to municipal or state control. It was not until the Theatres Act 1968, however, that this legislation was repealed and the role of the Lord Chamberlain in respect of censorship was abolished. Copies of scripts of all new plays are now sent to the British Library. The censorship measures were replaced by provisions for the prevention of obscene performances vested in the courts, rather than by administrative or executive action.

Licences

When considering the need for a licence, the term 'play' is usually taken to mean any dramatic piece – whether improvised or not – by one or more persons who are actually present and performing. What the performers do, whether it consists of speech, singing or action, must constitute the whole or a major part of a performance and involve the playing of a role. For example a dialogue between persons in costume or action without words may constitute a dramatic piece. Ballets, whether they fall within these definitions or not, do require a theatre licence, but theatre licences do not cover public music or dancing events. Where, however, the music is incidental to a play, or it takes place in the interval, or the music and dancing forms part of a musical comedy, a public entertainment licence is not required. It is not unusual for a building to hold both licences.

Full licences are granted for periods of up to one year, but there is also provision to grant a licence in respect of one or more occasions (an occasional stage play licence).

When a full licence is applied for, the licensing authority and chief officer of police in whose area the premises are situated, must be given at least 21 days notice of intention to make an application. The information to be given with an application must be in accordance with regulations prescribed by a licensing authority. An application for a renewal of a licence must give at least 28 days notice of intention. Where, however, the application is for one or more particular occasions, only 14 days notice is required to be given, and there is, in this instance, no obligation to inform the chief officer of police.

Appeals

Where a licensing authority refuses a licence, there is a right of appeal to a magistrate court by way of complaint for an order, and the licensing authority will be a defendant. Although the chief officer of police has to be notified of an application, he or she is not party to appeal proceedings, even if the decision to refuse a licence was made after considering his or her recommendations to do so. There is no right of appeal against a licensing-authority decision to issue a licence. Persons aggrieved by the decision of a magistrate court may appeal to the crown court. When granting a licence, a licensing authority may impose conditions and restrictions, but they must act judicially in doing so, and the restrictions that the licensing authority is empowered to make are strictly controlled. They relate in the main to matters of health and safety, and do not extend to restrictions on the nature of plays or the manner of performance.

Licence conditions

It is worth emphasizing that the conditions that can be imposed by a licensing authority when granting, renewing or transferring a licence are strictly limited to protecting physical safety and health. This is usually achieved by adopting local conditions that are principally aimed at providing a safe means of escape. This is achieved by providing an escape route that allows normal people to get out from a theatre after an outbreak of fire to a place of safety by their own efforts, without being placed in jeopardy while doing so. In addition, the conditions will require a sufficient number of well-located exits with adequate lighting and direction signs throughout. Equipment and areas of potential hazard are required to be protected and there must be an efficient exit drill procedure to ensure orderly exit.

In addition to these physical requirements, control of psychological factors conducive to panic need to be addressed and alarm procedures should reflect this. Fire escape drills are helpful, but it should be borne in mind that in a building whose occupants are transient they may be of limited value unless permanent staff are trained to help the temporary occupants. This is usually reflected in a requirement for the provision of adequate stewards or attendants. These requirements are even more vital in situations when an audience consists mainly of children, and it is not unusual for theatrical performances to which Section 12 of the Children and Young Persons Act 1933 applies to provide an increased number of stewards. It is essential to liaise with the fire and rescue services when considering these aspects of a theatre licence. As to the provision of exits, entrances, etc., see also the Public Health Act 1936, Section 59 for further powers.

Conditions related to health include the provision of adequate ventilation, the prevention of overcrowding and the provision of adequate sanitary accommodation and washing facilities, including facilities for the use of disabled people.

Guidance on this may be found in the British Standard 6465: Part 1: 1984 Sanitary Installations, and powers for requiring it are contained in Section 20 of the Local Government (Miscellaneous Provisions) Act 1976.

In older, purpose-built theatres it is not unusual to find large spans of unsupported ceilings, often decorated with ornate plaster mouldings. In these circumstances, conditions often require the regular inspection of these features and the provision of certificates of safety. For further advice on the safety of ceilings and the responsibility of licensing authorities, see Home Office Circular No. 264/1947.

Relationship to liquor licences

When a theatre licence is granted, the licensee acquires the right to sell intoxicating liquor if he or she notifies the clerk to the licensing justices of his or her intention to do so, unless the licensing authority has issued the licence subject to restrictions prohibiting the sale of liquor. When imposing such restrictions it is important that the local authority considers each application on its merits, for it is not permitted to attach a restriction in pursuance of a general rule.

Where liquor is permitted to be sold by virtue of a theatre licence, the provisions of the Licensing Act 1964 apply, and it is therefore only permitted to be sold during the ordinary permitted hours applicable to licensed premises. A condition can be imposed by the licensing authority, however, limiting the sale of liquor to the times when the premises are used as a theatre. If such a condition is not imposed, the theatre can also sell liquor when it is not in use as a theatre, including days such as Sundays, Christmas Day and Good Friday, when no performances take place. As in other licensed premises, extension of the permitted hours is not permitted.

These provisions will be amended by the Licensing Act 2003 (see p. 240).

Fees

The person applying for the grant, renewal or transfer of a licence must pay the licensing authority a reasonable fee. This requirement is waived in respect of occasional licences if the performance is of an educational or like character, or is to be performed for a charitable or other like purpose.

Provisional licences

As with many other forms of licensing, it is possible to grant a provisional licence where premises are under construction or alteration. In such circumstances, a provisional licence may be issued if the licensing authority is satisfied that the completed premises would be in accordance with its requirements. A licence will be granted subject to the condition that it will be of no effect until confirmed.

Sunday performances

By virtue of Section 1 of the Sunday Theatre Act 1972, licensed theatres are permitted to open on Sundays despite the prohibition contained in the Sunday Observance Act 1780. Theatres are, however, required to be closed by 2 am (3 am in Inner London) after Saturday night performances and must remain closed until 2 pm.

Offences

It is an offence to use unlicensed premises and any person concerned in the organization or management of a performance is liable, on summary conviction, to a fine of up to level 4 on the standard scale, or imprisonment for three months, or both. A similar penalty is available for the breach of the licence conditions, and in this instance a licence can be revoked. Fines of up to £1000 and up to six months imprisonment can be imposed for presenting or directing obscene plays and those that incite racial hatred or provoke a breach of the peace by the use of threatening, abusive or insulting words. In the case of the latter offence, proceedings can only be authorized with the consent of the attorney general. Authorized officers of licensing authorities have certain powers of entry, and wilful obstruction of an officer is an offence.

It should be noted that, in many instances, the law concerning the licensing of theatres is also contained in local Acts that may supplement, modify or supersede the general law.

Health and safety enforcement

Enforcement of the provisions of the Health and Safety at Work, Etc. Act 1974 in theatre premises is allocated to local authorities under the Health and Safety (Enforcing Authority) Regulations 1998.

DEALING WITH LICENSING APPLICATIONS

Although most routine matters relating to licensing administration are delegated to officers of a council, decisions on whether or not to grant licences are usually dealt with by a committee. It is important that sound administrative procedures in accordance with the rules of 'natural justice' are followed to avoid procedural difficulties at later stages. It is proposed that, through the Licensing Act 2003, such processes will be subject to a greater degree of legislative requirement and prescription (see p. 240). In the meantime the following comments remain appropriate.

Committee hearings of licensing applications are generally more informal than court hearings as the rules on evidence do not apply and it is not given under oath. Nonetheless, the committee must endeavour to ensure that fair and orderly hearings take place and that applicants are given the opportunity of being heard before applications are refused, even if not expressly required by the law. Applicants should be permitted to be accompanied by a legal or other representative if desired, and as much notice of the hearing as is practicable should be given in order to enable them to prepare their case adequately. Similarly, any body or person wishing to make representations in respect of an application should also be given the opportunity of appearing before the committee.

It is important that committee members are given copies of every document. It is also desirable for applicants to be informed of the nature of any objections to their application so that they can respond to them. While local variations to procedure will exist, there are certain elements that should always be followed:

1. Those present should identify themselves and the chairman of the meeting should ascertain whether the applicant, if unaccompanied, was aware that he could be represented.
2. The chairman or appropriate officer of the council should open the hearing with an outline of the relevant details of the application.
3. The applicant should be invited to present his case, following which he may be questioned by members of the committee. Persons who have made representations may also be afforded this opportunity.
4. Comments are then invited from the technical officers, including the police and fire authority where present, following which the applicant should be permitted to ask questions of the officers.
5. The applicant should then to allowed to make a final statement. It is vital that the case of any party is prosecuted in the presence of the other, and it is essential that all the committee members remain present throughout the hearing.
6. Committee members should confine themselves to asking questions and must not indulge in any discussion of the merits of the case.
7. Any request for adjournment should be granted if refusal would prejudice a fair hearing and deny the applicant natural justice.
8. The applicant, third parties and officers of the council may be asked to withdraw at the end of the hearing to allow the committee to consider the matter. The committee's legal adviser and minutes secretary will remain, and if it is necessary to seek clarification and further advice parties may be recalled.
9. When a decision is reached the parties will be recalled and the decision announced to the applicant, together with an explanation of any conditions that are to be attached or reasons for a refusal. The information is then given in writing as soon as practicable together with details of any rights of appeal.

LIAISON WITH THE LICENSING JUSTICES

Although the administration of liquor licensing legislation in England and Wales is undertaken by the licensing justices, environmental health departments should play a significant part in the process (but see proposals to make local authorities responsible for this function through the Licensing Act 2003, see p. 240).

The formal system of notification of applications includes the requirement for the proper officer of the district council to be notified. The proper officer is responsible for returning comment on behalf of the council, and directors of environmental health, when designated as proper officers, are in an excellent position to coordinate the response and achieve improvements relating to food hygiene and health and safety.

A procedure must be set up to consult each relevant department of the council and coordinate the replies directly to the licensing justices. Where matters are raised it is often the case that an applicant will deal promptly with any deficiencies, or at least give a written undertaking to deal with them within a reasonable timescale, in order to ensure that a liquor licence is granted. If facilities are lacking or substandard and no informal agreement can be obtained, it is open to the proper officer to object to the issue of the licence. This will entail a personal appearance at the hearing, and so a system of early notification of this course of action to the council's legal representatives should be in place in the internal administrative procedures.

In order to keep the licensing justices informed of environmental health officers' continuing interest in premises that they license, it is useful to send them copies of any notices of requirements served. It is also useful to keep the licensing justices informed of public entertainment licence decisions, as these often also relate to premises having liquor licences in some shape or form.

SWIMMING AND LEISURE POOLS

To obtain the benefits and pleasure that swimming can give, the water of a swimming pool must be fresh and crystal clear, attractive in appearance and free from harmful and unpleasant bacteria. To achieve these characteristics, the water must be in a state of chemical balance. Only minimum amounts of chemicals should be used if they are not to cause discomfort to the delicate membranes of the bathers' eyes, nose, throat and skin.

The water supplied to pools will often be of varying quality, and therefore it can be seen that, apart from knowledge of the delicate adjustments necessary to maintain the correct balance, some knowledge of the quality of water making up each pool is necessary.

Pool pollution may arise from a variety of sources: dust, hair, body grease and excretions from the nose and throat, for example, collect on the surface of the water (the top 16 cm of water contains 75% of the bacterial pollution). Many of the insoluble pollutants, such as dirt, sand from filters and precipitated chemicals, may find their way to the bottom of the pool. In addition, there may be forms of dissolved pollution, such as urine, perspiration and cosmetics, and chemical pollution produced by reaction in the water treatment. While many of these factors are unpleasant and merely a nuisance rather than a risk to health, they must all be considered when designing a pool and selecting a water treatment plant. However, the most serious pollution comes from the living organisms introduced by the bathers themselves, and it is this form of pollution that gives rise to a number of unpleasant conditions and diseases, and poses the most serious risk unless the pool is properly controlled and facilities such as showers and footbaths are provided to reduce pollution loads (see Fig. 11.1).

Standards of operation

Primary responsibility for pool water quality obviously lies with the pool operator, although environmental health officers have a key role through their enforcement of public health and safety legislation to ensure pools are maintained in a clean and safe condition. It is therefore essential that operators receive adequate training and knowledge to ensure the correct balance between treatment and pool usage.

Sampling techniques

The use of correct sampling techniques is essential to provide reliable information to determine

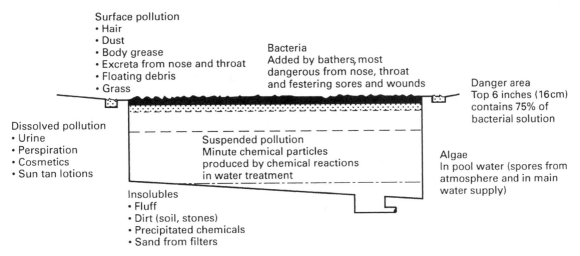

Surface pollution
• Hair
• Dust
• Body grease
• Excreta from nose and throat
• Floating debris
• Grass

Bacteria
Added by bathers, most
dangerous from nose, throat
and festering sores and wounds

Danger area
Top 6 inches (16 cm)
contains 75% of
bacterial solution

Dissolved pollution
• Urine
• Perspiration
• Cosmetics
• Sun tan lotions

Suspended pollution
Minute chemical particles
produced by chemical reactions
in water treatment

Algae
In pool water (spores from
atmosphere and in main
water supply)

Insolubles
• Fluff
• Dirt (soil, stones)
• Precipitated chemicals
• Sand from filters

Fig. 11.1 Physical, chemical and biological pollution of swimming pools.

whether disinfection is being carried out properly or not:

1. Although many pools will now have automatic monitoring equipment, these are not a substitute for routine testing by the operator, although they may allow a reduced programme of tests.
2. Training of pool operators should include advice on how to act effectively on the results and, as many of the tests require matching of colours, operators should be examined to ensure that they have no difficulty in reading results correctly.
3. While almost any sample of water from the distribution system may be typical of the whole this is not the case with swimming pools where pollution and disinfectant levels vary. The bottom at the deep end may receive little or no pollution over a period of time while continuing to be disinfected, and yet the shallow end could contain pollution added only a short time before the sample was taken. The state of the water therefore depends a great deal on the concentration of disinfectant and its speed of action.
4. Sampling normally is done at the shallow end of a pool when bathers are present and active,

but instructions on the proper place and time of sampling cannot be rigidly stipulated.
5. Frequent residual determinations of deep-end samples should be taken, but samples for bacteriological analysis at the deep end need not be done so frequently. For special pools (paddling, diving and remedial pools) variations on sampling technique are a matter of common sense.
6. A record should always be kept of the place, time, use of the pool, the pH and the amount of chlorine (residual and free) in the water. These do not need to be kept permanently, but are useful evidence of the state of a pool over the previous few weeks.
7. The frequency of tests depends upon the use of the pool, its equipment and past performance, and heavy use, changes in disinfectant and recent failures will require more frequent bacteriological checks. No pool, however, should go without its occasional bacteriological check, with the samples being taken unannounced on a day and time when the pool is in use.
8. Particular care must be taken to avoid contamination of samples and to neutralize the disinfectant in samples taken for bacteriological examination.

9. When samples are taken for bacteriological analysis a chlorine determination should be made on another sample at the same time and place as the first.
10. As results of bacteriological examinations are not immediately available, regular and frequent determinations of free residual (if chlorine is used) are an important check on the condition of the water and a guide to any action required. For no pool should this be less frequent than two to three times a day.
11. Pool attendants should be trained to do the tests, but officers should also do a test when they visit. Test kits must be properly maintained and kept clean at all times with any glassware being thoroughly washed in clean water to remove test reagents before making subsequent tests.
12. Reagents have a 'shelf life' and require replacing at regular intervals. They should be properly stored when not in use.

Purification

All swimming pools should be equipped with adequate purification plants to remove sources of pollution. The most usual method is to recirculate pool water after chlorination, filtration and aeration. By maintaining sufficient levels of a chosen disinfectant, usually chlorine, in the water, the rapid and immediate destruction of bacteria is assured. In any pool, water is also constantly being discharged, as part of the filtration process, and being replaced with fresh water. Pollution can also be progressively diluted. However, although a significant amount of pollution may be removed at the first turnover, the proportion falls with successive turnovers; even after the water has been through the filter system several times, a small proportion of pollution remains.

The **turnover rate** of a pool is determined by the number of hours it takes to pass the total pool water volume through the filter. This influences the choice of filter and its operation during each 24-hour period. However, when choosing a filter it should be borne in mind that not all pool water will actually pass through a filter in a single turnover due to remixing of the filtered water

when it is returned to the pool. It has been estimated that even in a pool with good circulation, something like seven 'turnovers' are required in order to filter 99% of the pool water.

The following rates of turnover are suggested as being minimum requirements:

1. **Private outdoor pools** with a small bathing load would normally need a filter with an eight-hour turnover period.
2. **Private indoor heated pools** would require a six-hour turnover period.
3. **School pools**, which are often heavily used, require a turnover of less than two hours.

(These figures will only be relevant if the plant is well maintained and operated in a good condition.)

Filtration

From the foregoing it will be seen that the selection of a filter should be undertaken with great care as it has an important bearing on the ultimate clarity, appearance and safety of the pool water. The filter is the heart of any pool installation, and without an effective one problems will arise in maintaining pool water in an acceptable condition for bathing. It is the correct use of a filter, in association with correct chemical disinfection, that maintains pool water bright and clear.

For organisms such as *Cryptosporidium*, good filtration is the only means of protection against infection because it is resistant to the residual disinfectants in normal use.

Filters are generally of the pressure type using sand, diatomaceous earth or cartridges as the filtering medium.

Sand filters

These are basically of three categories, depending on the rate of flow of water through the sand bed. The system employed by each type is similar. Water, under pressure, flows through a bed of graded sand enclosed in a container. These filters ultimately reach a point when they are unable to deal with further particulate matter, and they are then cleaned by 'back washing': the reversal of the water flow disturbing the sand particles and allowing the filtered-out debris to be backwashed to waste. Such filters, although expensive to install, are simple to operate and can be inexpensive in use, as the filter medium is reused time and time again.

Diatomaceous earth filters

Although of varying designs, these filters tend to work on the same filtration principle as sand filters. This involves the use of a fine fabric supported within the filter vessel through which the water flows. Diatomaceous earth is introduced to the vessel as a filter powder, and forms a fine coating over the fabric. This layer removes the dirt from the water until, when fully loaded as shown by a rise in pressure, the water flow is reversed, and the filter powder and dirt are discharged to waste. These filters can be very efficient and give good quality water when operated correctly. They do tend to be more expensive to use, however, as the diatomaceous earth has to be discharged to waste.

Cartridge filters

These filters are normally used in small pools, and rely on the passage of water through pads of bonded fibre and/or foam. The dirt-coated pads are removed and replaced at intervals because it is uneconomic to try to clean dirty pads. This system can be relatively inexpensive to purchase and operate, but the efficiency of filtration, while adequate, may not be as high as for those types already described.

Sterilization

While filters are an essential part of the water treatment process, a sterilant or disinfecting chemical has to be used to ensure that water is kept free from harmful bacteria and other organisms. While many disinfectants, such as ozone, bromine and iodine, are available, to date only chlorine has satisfied all the major requirements. Its effectiveness, low cost, speed of kill and relative ease of control, makes it the most likely disinfectant to be encountered.

Chlorine

In many pools chlorine is injected automatically in controlled quantities. In most large pools this method of sterilization is very sophisticated, and samples of water are tested and dosed to maintain continuous control of the levels of free and combined chlorine. As a result, many of the problems associated with high combined chlorine levels – chlorinous odours, stinging eyes and skin irritation – can be eliminated. Usually, the dose of chlorine is maintained at a sufficiently high level to produce a free residual of 1.5–3 ppm, but where there is heavy use, levels of up to 5 ppm free chlorine should not give rise to complaint, provided the pH is carefully controlled (pH 7.4–7.6 being considered ideal). There are, at present, four main chlorine donors used for swimming pool water treatment, and the chlorine content varies according to the donor chosen. The use of chlorine gas as a disinfectant for swimming pool water has caused several serious incidents and is no longer recommended.

Electrochlorination

This is an on-site process for the production of sodium hypochlorite. These units are basically small scale versions of the type of electrochlorination plant already widely used for water treatment at power stations and offshore drilling installations. The plant normally consists of a water softener and sodium chloride brine make-up system; an electrolytic cell, rectifier and associated electrical equipment; an intermediate bulk storage tank for sodium hypochlorite (day tank), although some installations use direct injection of sodium hypochlorite into the pool water recirculation system and, therefore, do not utilize a day tank; and water monitoring instruments, associated pumps, sampling lines and interconnecting piping.

Diluted brine is measured into the electrolytic cell, which converts the salt solution into sodium hypochlorite with hydrogen gas produced as a byproduct. The hypochlorite from the cell is either injected directly into the pool recirculation system or, more commonly, is stored ready for use in a day tank. Where hypochlorite is injected directly into the recirculation system rather than stored, the addition is controlled by the pool water monitoring system and the cell operation, which depends upon hypochlorite demand, is therefore intermittent.

Information on such plants is contained in the Department of the Environment booklet 'Swimming pool disinfection systems using ozone with residual free chlorine or electrolytic generation of hypochlorite – guidelines for design and operation'.

Sodium hypochlorite

This form of chlorine liquid is widely used and is an accepted method of chlorination. It is marketed commercially under a number of different names, and contains 10–15% chlorine. Because this product is very caustic and has a very high pH (pH 12), great care must be taken in its handling. In use it can cause the pH of the pool water to increase, will tend to increase total dissolved solids and can cause scale and corrosion of metal pipes and valves. As this material can decompose quickly if stored incorrectly – in heat, light or in metal containers – it is inadvisable to order more than one month's supply at a time. Figure 11.2 shows a full sodium hypochlorite treatment system.

Isocyanurates

These materials offer the advantage of being stable in sunlight, making them suitable for use in outdoor pools. They are available in powder or granule form, are safe to handle and have a long shelf life. The granules are added directly to the pool water or in solution through a feeder. They dissolve to leave no sedimentation and, being slightly acid, have little effect on pH levels. Depending upon their form, they have relatively high levels of available chlorine: sodium dichloroisocyanurate 56%, and trichloroisocyanuric acid 91%. When using these materials, cyanuric acid levels should be maintained at between 25 and 50 ppm; if it rises above 150 ppm, 'chlorine-lock' can occur.

Calcium hypochlorite

This is available in tablet or granular form, and is usually made into a 6% solution to be pumped into the circulation. The material contains 65% minimum of available chlorine that is stable if stored in cool, dark conditions prior to dilution. In use, this material has the advantage over sodium hypochlorite of not causing the pH level to rise by as much, and it does not add nearly as much total dissolved solids (TDS). A disadvantage is that a much larger dosing pump is required, and it needs constant agitation to keep it soluble when made up in solution.

Determination of residual chlorine

For low concentrations of chlorine in pool waters, use is made of the Palin-DPD (diethyl-phenylene-diamine) test for chlorine, and phenol red tablets (containing a reagent to counter chlorine bleaching) for the pH test. Reliable and simple to use, these kits use tablets and comparator discs to monitor chlorine and pH levels.

Chemistry of chlorine in pool water

When chlorine and water are mixed, the free chlorine killing agent hypochlorous acid is formed, together with hypochlorite ion and hydrochloric acid. The hypochlorous acid combines with organic pollutants (body fluids, dead skin and ammonium compounds) to form chloramines – combined chlorine. The hypochlorous acid (free chlorine) is rapid acting, whereas hypochlorite ion is slow acting, and the percentage of these constituents is influenced by the pH, for example, the efficiency of free chlorine is approximately 100% higher at pH 7.5 than it is at pH 8.0, and the former figure should be aimed for. However, in addition to adding bacteria to a pool, bathers constantly add nitrogenous matter, and chlorine is also needed to deal with these pollutants and remove chloramines in the pool water. Both hydrochloric acid, which reduces the pH of the water, and nitrogen trichloride can be formed unless adequate hypochlorous acid is added. Control of the amount of chlorine added is

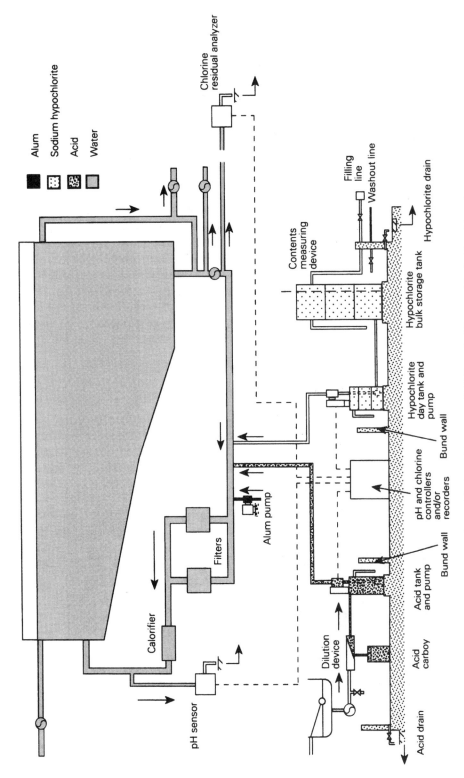

Fig. 11.2 Pool water treatment system using hypochlorite. (Courtesy of Wallace & Tiernan, Tonbridge, Kent.)

therefore essential to ensure a satisfactory free chlorine level, and that the proportion of combined chlorine is maintained at about one-third of the total chlorine present.

Breakpoint chlorination

If insufficient chlorine is available in a pool, the combined chlorine level will rise, the water may lose colour and clarity and nitrogen trichloride may be produced resulting in complaints of smarting eyes and skin irritation among bathers. If sufficient chlorine is added, it will combine with the pollutants and break them down to harmless substances. Further additions of chlorine will then be present in the water as **free chlorine**, available to deal with pollution as it arises.

A 'breakpoint' occurs and the residual falls, and from this point free chlorine can exist together with combined chlorine. This is the system adopted in most pools in the United Kingdom.

Ozone

The use of ozone in the purification of swimming pool waters is practised extensively throughout mainland Europe, and appears to be finding some favour in the United Kingdom. The initial cost of an ozone installation is high, but it is claimed that

if used in a new pool complex, the cost difference may be largely offset by capital savings in a heating and ventilation plant.

It is also claimed that ozone has several advantages over chlorine in that it is an effective bactericide, and is also effective against viruses. It oxidizes organic matter and the resultant pool water has excellent clarity. However, it has a short life in solution and is toxic and aggressive. As ozone does not remain in water as 'residual' ozone, it is necessary to chlorinate to provide residual protection in the water. The amount of chlorine necessary is, however, greatly reduced, and the process still results in an improved atmosphere in the pool hall, with a noticeable lack of characteristic monochloramine odour. Figure 11.3 shows a diagrammatic layout of an ozonization plant. The dose of ozone should be about 2 mg/l.

'Slipstream' ozonization is now available. This treats only 10% of the water with ozone, the rest being treated by a normal filter and chlorine dosing system.

Bromide

Treatment of swimming pool water with bromide is sometimes encountered as it is claimed to counteract some of the disadvantages of using chlorine as a disinfectant. It has been widely used in the

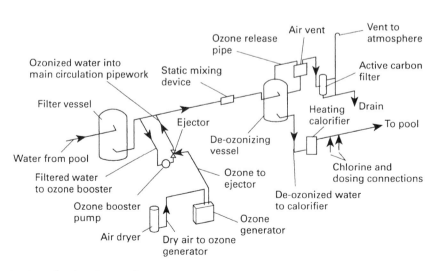

Fig. 11.3 Ozonization of swimming pool water.

United States since the 1940s, and is commonly used in its solid form under the name Dihalo, which combines bromine (66%) and chlorine (28%). It has been predicted that the use of this disinfectant may increase in the future, although its relatively high cost may restrict this.

In France, liquid bromine has been used in swimming pool water treatment, and its innocuousness to bathers has been confirmed in a number of toxicological studies. Its bactericidal, viricidal and algicidal properties are comparable with chlorine, and it is claimed that because the connection between its bactericidal activity and the pH is slight, this factor can be kept at a level where the risk of irritation is minimal.

The bromine content of water can be readily measured by chemical methods using colorimetry or by electrochemical methods. Reagents used are orthotolidine or DPD, the former being the most sensitive.

Other disinfectants

Iodine chlorine dioxide, ultraviolet light and metal ions (electro katadyn system) have been used, mainly outside the United Kingdom, but their use is rare as all have disadvantages.

The determination of pH values

The expression pH is used to indicate the degree of acidity or alkalinity of water. Pure neutral distilled water has a pH of 7.0; a figure below 7.0 will indicate that the water is acidic, above 7.0 that the water is alkaline. Normal waters contain a variety of substances in solution. Calcium and magnesium salts give the water its characteristic hardness, for example, and traces of many other compounds are naturally present and are affected by changes in pH.

Alkaline conditions can give rise to irritation if the pH rises much in excess of pH 8.0, and the effectiveness of chlorine as a sterilant is also greatly reduced.

In 'hard' water areas, another problem can arise with a high pH: the precipitation of calcium salts, which results in a cloudy pool and blocked filters. Should this occur, the filter must be thoroughly backwashed as the accumulation of the deposit may set hard and possibly ruin the filter.

If the pool water is allowed to go acidic, even very slightly, two major difficulties arise. First, corrosive conditions will exist, and any metal parts of filter systems etc. in the pool will be attacked, and any alkaline floor and wall finishes, such as the proprietary 'Marbelite', may also suffer. Second, in acidic conditions, unpleasant compounds are formed by the interaction of chlorine and polluting substances, which cause irritation to mucous membranes and eyes, and give rise to the so-called 'pungent odour of chlorine'. This is the reason for most of the complaints of 'over-chlorination', and the cause is not an excess of chlorine, but the formation of these irritant and smelly compounds due to the presence of too little chlorine in an acid water.

This is the reason for maintaining pool water at around pH 7.4–7.6, that is slightly alkaline. It will give some protection against the accidental production of acid conditions, while still allowing effective performance by the chlorine.

In measuring pH values, advantage is taken of the fact that certain dyes, known as indicators, change their colour in a definite and reproducible manner and degree, according to the pH value of the solution with which they are mixed. Phenol red or diphenol purple are the most suitable for pool water, their ranges being pH 6.8–8.4 and 7.0–8.6, respectively. Both are available in tablet form. For an accurate measurement of pH, a meter should be used.

Algal control

Algae are tiny aquatic plants that often first appear in pools as brown or green slimes on steps, walls and floors. If they are allowed to establish themselves, they can be a considerable nuisance. They grow rapidly, increase chlorine demand, block filters and are the main cause of discoloured water. In correctly maintained pools, algal problems are rare. Weekly additions of algicide will establish a buffer against temporary or inadvertent loss of chlorine, which may create the conditions that favour rapid algal growth.

Care must be taken when using algicides of the quaternary type, as over dosing may lead to excessive foaming or the formation of a foam layer on the surface of the pool.

Total dissolved solids

The addition of water treatment chemicals to pool water results in an increase in the amount of solids dissolved in the pool water. When these increase to undesirable levels the colour, clarity, appearance and taste of the water can be adversely affected, to the discomfort of bathers. A maximum of 1500 ppm should not be exceeded. The choice of disinfectant can have a significant effect on the total dissolved solids (TDS) content of pool water, and control can be obtained by accurate control of pH value (7.3–7.5), the use of minimum quantities of chemicals, regular backwashing of filters and dilution of pool water with fresh water.

Alkalinity

Alkalinity is the quantitative analysis of the amount of alkali present in the water as bicarbonate, which will act as a 'buffer' and be readily available to react with fluctuating pH conditions and maintain the clarity and comfort conditions in the pool. A level of around 100 mg/l is desirable. It must not be allowed to rise above 220 mg/l, otherwise corrosion can result. High levels of alkalinity can be lowered by the addition of hydrochloric acid to the pool water.

Water quality guidelines

It is recommended that where water quality values are set, they are seen as objectives for operators to follow rather than as rigid or inflexible standards. Colony counts and the test for *Escherichia coli* and other coliform organisms provide simple tests for checking the bacteriological quality of swimming pool water.

If coliform organisms are absent and pH and disinfectant residual levels are maintained at correct levels, the risk to bathers will be minimal. Colony count tests, which determine the number of organisms capable of living in the pool water, although

regularly carried out, are not essential for assessing bacteriological safety. They are, however, useful as indicators of the general quality of the water and that filtration and disinfection processes are operating correctly. A range of values can be established for each pool, depending on the local circumstances and disinfectant used. These will, after a time, establish a norm for the pool, and this can be used to identify significant changes in water quality. This is more useful to the enforcement officer than the actual numerical results themselves.

Guidelines on appropriate bacteriological standards are to be found in the Department of the Environment's booklet *The Treatment and Quality of Swimming Pool Water* [17]. The absence of coliform organisms with colony counts not greater than 10 (and always fewer than 100 organisms per millilitre) at 24 hours gives a good indication that the quality of the pool water is satisfactory. Counts above 100 organisms per millilitre require investigation. Occasional counts of up to 10 coliform organisms, in the absence of *E. coli*, are acceptable provided that they do not occur in consecutive samples, that pH and disinfectant levels are satisfactory and that the colony count levels are within the established norm for the pool.

Where there are persistently poor results, which may indicate disinfection and filtration failures, a full investigation and sampling programme must be implemented. In these circumstances, it may also be necessary to look for the presence of *Pseudomonas aeruginosa* and possibly *Staphylococcus aureus*, as extra indicators of water quality.

Procedures for taking samples are described in detail in *The Bacteriological Examination of Drinking Water Supplies* [18].

Whirlpool spas

These have become very popular forms of water recreation, and are now to be found in many hotels, clubs, leisure centres, etc. In theory, their disinfection should be relatively simple, but their popularity and the heavy loading they are subjected to make this a difficult task unless a strict regime is utilized.

Spa pools are commonly disinfected with Dihalo (bromine), although more are changing to hypochlorite disinfection for easier and more efficient control.

Most pools run at a temperature of around body heat (37°C), and this may induce bathers to sweat heavily. In leisure and fitness centres, pools are often used after heavy exercise often without showering before immersion, and because the pool volume is small, pollution levels will be high. The combination of pollution, temperature and bathers being close together in a confined space, coupled with inadequate surface water removal, insufficient filtration and unbalanced water can lead to discomfort and danger. The water temperature is ideal for skin rash inducement and respiratory infection, unless disinfection is efficient.

Pseudomonas aeruginosa folliculitis has been the most commonly identified infection associated with whirlpool spas, with the characteristic rash developing anywhere between eight hours and five days after bathing (mean an incubation period of 48 hours). When investigating complaints, it is important to differentiate this rash from others of an eczematous nature, sometimes associated with solid bromine treated pools. The folliculitis may, on occasions, be associated with mastitis and otitis externa in the bathers. Circumstantial evidence has also linked whirlpool spas to occasional urinary tract infections, and they may serve as sources of other infections including legionellosis and Pontiac fever. Defective maintenance or operation are usually common in reports of outbreaks.

The Swimming Pools and Allied Trades Association (SPATA) has produced standards for the installation and operation of commercial spa pools, and there is mention of SPATA in the Department of the Environment's booklet *The Treatment and Quality of Swimming Pool Water* [17]. These guidelines should be followed to ensure trouble-free operation. It is also recommended that, in addition to emptying and refilling pools at least weekly, the pH value, disinfectant concentration and temperature be tested regularly during the day, and records be kept in a log book. From time to time, tests for alkalinity, TDS and surface and calcium hardness should be undertaken. Control of bathing load and enforced intervals

between bathing sessions can assist in maintaining water quality.

Biological standards may be found in SPATA's *Standards for Spa Pools* [19], which states that the biological conditions of the water shall be at the judgement of the environmental health officer for the area. Notwithstanding this, the standards recommended for spa water are:

- total plate count at 37°C should not exceed 100 colonies for 1 ml of sample water
- no *E. coli* in 100 ml of sample water.

See also *Hygiene for Spa Pools* [20].

The health and safety guidelines for public spas and hot tubs issued by the US Department of Health and Human Services suggests: 'The presence of organisms of the coliform group, or a standard plate count of more than 200 bacteria per millilitre or both in two consecutive samples or in more than 10% of the samples in a series shall be deemed as unacceptable water quality'.

The health and safety aspects of pools

Guidance on safety in swimming pools, the risks associated with their operation and the precautions to be taken to achieve a safe environment for the public that uses them and the employees who work in them is to be found in the HSE publication *Managing Health and Safety in Swimming Pools* [21]. This booklet outlines the standards of good practice as a basis for decisionmaking by managers on what arrangements are best for their pools.

The risk of drowning in a swimming pool is not high in comparison with the overall national drowning problem, with only 3–5% of drownings occurring in pools. Considering the estimated 150 million visits to swimming pools each year, the figures are small. However, any death that can be avoided, especially in such a controlled environment where all risk should be removed, is inexcusable. The problem is therefore primarily one of management.

Existing legislation, which controls safety in swimming pools, is to be found in the Health and Safety at Work, Etc. Act 1974 and the Public Health Act 1936.

The Health and Safety at Work, Etc. Act 1974 places certain general obligations on all pool operators, and these responsibilities extend to protecting the public that may use the pool and any contractor working at pool premises. The HSE is the enforcing authority at local authority-run pools, including school pools. At pools that form part of residential accommodation (hotels, holiday camps, etc.) and leisure complexes, enforcement is the responsibility of the local authority.

The Public Health Act 1936 contains powers for local authorities to make by-laws for the regulation of swimming pools under their management. Model by-laws are available from the Department of the Environment, Transport and the Regions (DETR), and cover aspects such as water purity, hygiene, behaviour and the prevention of accidents.

Even though the legislation exists, effective safety relies on the general acceptance and adoption of recognized standards, such as the following.

Safe design of the pool structure, systems and equipment

Good design is essential for a safe pool environment, and safety is one of the important factors looked at by the Sports Council when considering schemes submitted for grant aid. Designers and sponsoring authorities are encouraged to meet the standards recommended in the booklet [21].

Details on the structure and finish of the pools and buildings are outlined to avoid dangerous situations, such as abrupt changes in floor level in wet areas, being built into a pool. Precautions to prevent people, particularly young children, having accidents or falling through open ledges of stairways and landings are included. Good planning and circulation layout ease management problems and enhance safety. Floor and wall surfaces and features next to wet circulation areas should not present a hazard to bathers, and slip-resistant flooring and well-designed walls that avoid sharp edges, projections or abrasive finishes are recommended to minimize these hazards. Suitably toughened glazing, as specified in BS6206, should be used in areas adjacent to wet circulation routes to reduce the risk of injury and damage.

Entry to and exit from a pool should not only be safe but easy, and the design and sighting of these facilities must be suitable for the pool.

Pool edges should be clearly visible – colour contrasted – so that bathers can avoid hitting the edges when diving or jumping in. This may not be feasible where the pool takes the form of a gently sloping 'beach', but in such leisure pools this is not so critical.

The profile of the pool bottom should not be a hazard to swimmers in the pool or those jumping in, and for rectangular pools the Sports Council has recommended a number of possible pool profiles, with water depths ranging from 2 m deep water to 90 cm shallow water with gradients of not more than 1:15. In irregular-shaped leisure pools, profiles will depend on pool layout and any features present.

Wave machine openings, sumps and inlets or outlets of the pool water circulation system should have suitable protective grilles or covers designed to prevent trapping. Undue suction should not be created at openings, which could otherwise result in a body being held against the grilles.

Any safety signs used should comply with the Safety Signs Regulations 1980, and the content and location of signs needs careful consideration as part of the overall safe pool environment. Clear signs, showing depths of water, areas where it is safe to swim or dive and those giving instruction on the safe use of diving or other equipment are particularly important. Examples are given in *Managing Health and Safety in Swimming Pools* [21].

The installation of heating, ventilation and air conditioning systems in pools needs careful consideration, as these factors can indirectly affect pool safety. They also promote rapid corrosion of pool-side structures if out of balance by permitting excess humidity.

A comfortable temperature should be maintained in the swimming pool hall and changing areas. A maximum temperature of around 27°C in the water with the air temperature about 1°C higher, to avoid excess condensation, is suitable.

Effective, draught-free ventilation should be provided, and humidity and air movement should be balanced to achieve comfortable conditions.

Adequate lighting, either natural or artificial, should be provided to avoid excessive glare or reflections from the pool water, to avoid solar gain and to ensure that the whole of the pool and its base are easily visible to lifeguards and bathers.

Wet and corrosive conditions in pools can compound the risks from electricity. Designers should be aware of the various risks of shock, burns, fire or explosion, and take these dangers into account.

Maintenance requirements and safe working practices

The correct planned maintenance of buildings and the plant is essential in ensuring the health and safety of pool users and employees. Arrangements should therefore be made for their thorough inspection and examination, either by utilizing manufacturers' instructions or by pool operators devising them as part of the pool operating procedure. Maintenance should take place at the specified intervals and records should be kept of any remedial work carried out.

Buildings should generally be inspected annually, but where high humidity levels increase the risk of corrosion, and chemicals in the atmosphere may increase the risk, some structures may require more frequent inspection, for example, every six months.

Any steam boilers and the plant should be maintained to the standards required by Sections 32–35 of the Factories Act 1961, including a regular, thorough examination by a competent person. After each examination, a certificate should be obtained and kept available for inspection.

Staff should be adequately trained for their pool duties and useful courses to provide this training are organized by the Institute of Baths and Recreation Management (IBRM). These include courses for supervisors, attendants and plant operators. Courses designed for the latter include technical aspects of pool water treatment.

Asbestos may be found in swimming pool premises since at one time it was widely used for insulation and fire protection. All work with asbestos is now subject to the Control of Asbestos at Work Regulations 1987 as amended and its associated codes of practice. Generally speaking, only persons licensed by the HSE under the Asbestos (Licensing) Regulations 1983 as amended may work on asbestos.

Access to exterior windows for cleaning can pose special problems in swimming pools. Suitable guidance on this aspect of maintenance can be found in *Prevention of Falls to Window Cleaners* [22]. Fixed electrical installations should be inspected and tested to the standards in the current edition of *Regulations for Electrical Installations* published by the Institution of Electrical Engineers [23]. Because of the adverse conditions of a pool environment, tests should be done at least annually.

The pool water treatment system

Whichever system of disinfection, filtration and circulation is used, it must be operated safely. The main risks associated with treatment systems include risks to bathers from unclear water, and risks to bathers and employees from the chemicals used in disinfection systems. There is often the added danger to employees of having to work on these items in confined spaces inside the plant. Written health and safety policy statements should include an assessment of all of the hazards associated with all aspects of the plant and the precautions to control the risks. Adequate staff training should be provided.

Advice on delivery, storage and handling of chemicals is given in a series of Department of the Environment booklets (see the section Further reading) giving guidelines for the design and operation of plants using different disinfectants. All chemical containers should be clearly labelled with their contents, and the packaging and labelling should comply with the Chemicals (Hazard Information and Packaging) Regulations 2002.

Storage facilities should be secure, dry, well ventilated, clearly marked and sited well away from public entrances and ventilation intakes. Safe systems of work should always be followed to safeguard employees from harmful materials; these may include the provision of protective clothing and, in some cases, respiratory protection. First-aid provision should be adequate to deal with the consequences of chemical splashes and the like.

Where any major, uncontrolled release of toxic gas is possible, written emergency procedures for dealing with such an incident should be prepared, and should include evacuation procedures and the notification and coordination of emergency services.

While the Department of the Environment booklets give good advice on the safe design and operation of the common disinfection systems, the more important hazards associated with these systems are listed in the following.

SODIUM HYPOCHLORITE AND ACID SYSTEMS   Used with automatic dosing systems these have been known to release chlorine gas when water pumps have failed. Correct siting of pumps to avoid them losing their prime and the provision of interlocks to prevent incorrect dosing, as well as additional sampling points and correct maintenance procedures, are measures to be adopted to eliminate this hazard.

ELECTROLYTIC GENERATION OF SODIUM HYPOCHLORITE   This can sometimes produce hydrogen and, occasionally, chlorine gas. In view of the flammable nature of hydrogen the selection, siting and maintenance of electrical equipment is likely to be a specialist job. More detailed guidance on the hazards arising from this type of plant and the precautions to be taken is given in the HSE/LA Enforcement Liason Committee circular 89/ 1 'Electrochlorination plant for use in the treatment of swimming pool water'.

OZONE SYSTEMS   These present hazards from the chemicals used and from the electrical ozone generating process. Guidance on the health hazards associated with ozone is to be found in *Ozone Health Hazards and Precautionary Measures* [24]. Ozonators should be provided with automatic shutdown devices to cope with any abnormal operation.

Where ozone devices are installed to remove ambient odours in changing rooms, ozone levels must not be allowed to exceed the recommended occupation health limits set out in the HSE guidance note *Occupational Exposure Limits* [25].

CHLORINE GAS SYSTEMS   Chlorine gas is particularly hazardous and, in view of the advice contained in

the statement on the use of chlorine gas in the treatment of water of swimming pools issued by Department of the Environment Circular 72/78 recommending that this use should cease by 1985, it will only rarely be found. Where it is found, the advice contained in the HSE booklet *Chlorine from Drums and Cylinders* (26) and the Department of the Environment booklet *Swimming Pool Disinfection Systems Using Chlorine Gas – Guidelines for Design and Operation* (27) should be followed.

ELEMENTAL LIQUID BROMINE SYSTEMS   Being less hazardous than other pressurized gas systems, the main problems of these systems relate to spillage. Adequate supplies of neutralizing material should be provided to deal with such emergencies.

CALCIUM HYPOCHLORITE, CHLOROISOCYANURATE, HALOGENATED DIMETHYLHYDANTOIN AND SOLID ANCILLARY SYSTEMS   The main risks relate to general chemical-handling and the generation of chlorine gas if chemicals are mixed or stored incorrectly.

PH ADJUSTMENT BY THE USE OF CARBON DIOXIDE   The system of metering carbon dioxide gas into the water circulation system is becoming more popular as it eliminates the risk of chlorine gas generation – unlike acid and hypochlorite systems. However, because of the risk of asphyxiation, carbon dioxide should be stored outside buildings.

SAND FILTERS   When it is necessary to enter filter vessels the advice published by the HSE in *Confined Spaces* [28] should be followed.

Supervision arrangements to safeguard pool users

All pools require supervision if they are to be operated safely. Pool operators should therefore consider carefully the main hazards associated with their pool, and make detailed arrangements to deal with them. The precautions taken by the pool operator must include a written operating procedure, which sets out the organization and arrangements for user safety, including details of staff training requirements. This is particularly important when a pool may be used without constant poolside supervision. These should be constantly reviewed and updated so as to take account of incidents experienced at the pool, thus keeping the procedures relevant.

By displaying suitable signs and posters such as those based on the *Swimming Pool User's Safety Code* published by RoSPA [29], bathers can be made aware of potential hazards and encouraged to act responsibly.

Generally, the DETR recommends that a minimum water area of 2 m^2 per bather be allowed for physical safety, but this is only a guideline and operators must assess the maximum number that can be safely admitted to their pool, taking account of bathers' behaviour and also the capacity of the pool water treatment system.

Where it is deemed necessary to provide constant poolside supervision, sufficient adequately trained lifeguards should be provided and effectively organized and supervised. Because of the variety of pool facilities and users, it is not feasible to make specific recommendations for lifeguard numbers. These must be arrived at by taking account of all relevant local factors, although as a starting point some advice on minimum numbers is set out in the Department of the Environment publication *The Treatment and Quality of Swimming Pool Water* [17].

REFERENCES

1. *The Review of Fairground Safety* available on HSE website www.hse.gov.uk/spd/noframes/spdleis.
2. Health and Safety Executive (1997) *Fairgrounds and Amusement Parks: Guidance on Safe Practice*, HSG 175, HSE/ Joint Advisory Committee on Fairground and Amusement Parks, HSE Books, Sudbury, Suffolk.
3. Royal Society for the Prevention of Accidents (undated) *Fairground User's Safety Code*, WS 53, RoSPA, Birmingham.
4. National Playing Fields Association (1983) *Playground Management for Local Councils*, NPFA, London.
5. Noah, N.D. (1987) *Guidelines for Hygienic Hairdressing*, Public Health Laboratory Service, Colindale.
6. Department of Health and Social Security (1987) *AIDS: Guidance for Hairdressers and Barbers*, HMSO, London.
7. Health and Safety Executive (1986) *How to Use Hair Preparations Safely in the Salon*, HSE Books, Sudbury, Suffolk.
8. Noah, N.D. (1983) *A Guide to Hygienic Skin Piercing*, PHLS Communicable Disease Surveillance Centre, Colindale.
9. Health and Safety Executive (1985) *Risk of Infection from Skin Piercing Activities*, LAC(T) 5.6.1, HSE Books, Sudbury, Suffolk.
10. HSE/LA Enforcement Liason Committee (2001) LA circular 76/2, *Enforcement of Skin Piercing Activities* and www.hse.gov.lau.lans176-2.
11. CIEH London (2001) *Body Art, Cosmetic Therapies and Other Special Treatments* ISBN 1-902423-80-1
12. Health and Safety Executive/Home Office/ Scottish Office (1993) *Guide to Health, Safety and Welfare at Pop Festivals and Similar Events*, HMSO, London.
13. Home Office (1990) *Guide to Fire Precautions in Existing Places of Entertainment and Like Premises*, HMSO, London.
14. National Outdoor Events Association (1997) *Code of Practice for Outdoor Events*, NOEA, Wallington, Surrey.
15. Home Office/Scottish Home Office and Health Department (1990) *Guide to Safety at Sports Grounds*, HMSO, London.
16. Sports Council/Football Trust (1994) *The Guide to Safety at Sports Grounds*, Sports Council, London.
17. Department of the Environment (1984) *The Treatment and Quality of Swimming Pool Water*, HMSO, London.
18. Department of the Environment/Department of Health/Public Health Laboratory Service (1983) *The Bacteriological Examination of Drinking Water Supplies*, Report 71, HMSO, London.
19. Swimming Pools and Allied Trades Association (1983) *Standards for Spa Pools*, SPATA, Croydon.

20. Public Health Laboratory Service (1994) *Hygiene for Spa Pools*, PHLS, Colindale.
21. Health and Safety Executive/Sports Council (1988) *Safety in Swimming Pools*, Sports Council, London.
22. Health and Safety Executive (1983) *Prevention of Falls to Window Cleaners*, Guidance Note GS25, HSE Books, Sudbury, Suffolk.
23. Institution of Electrical Engineers (1997) *Requirements for Electrical Installations*, IEE, Stevenage.
24. Health and Safety Executive (1983) *Ozone Health Hazards and Precautionary Measures*, Guidance Note EH38, HSE Books, Sudbury, Suffolk.
25. Health and Safety Executive (1998) *Occupational Exposure Limits*, Guidance Note EH40, HSE Books, Sudbury, Suffolk.
26. Health and Safety Executive (1987) *Chlorine from Drums and Cylinders*, Guidance Booklet HS(G)40, HMSO, London.
27. Department of the Environment (1980) *Swimming Pool Disinfection Systems Using Chlorine Gas – Guidelines for Design and Operation*, HMSO, London.
28. Health and Safety Executive (1998) *Entry into Confined Spaces*, 101, HSE Books, Sudbury, Suffolk.
29. Royal Society for the Prevention of Accidents (1992) *Swimming Pool User's Safety Code*, RoSPA, Birmingham.

FURTHER READING

Anon (1993) *Pool Manager's Handbook*, Olin (UK), Droitwich.
Association of Professional Piercers (1994) *Procedure Manual*, APP, San Francisco.
Chamberlain, M. (1985) Swimming pools – established safety standards. *The Safety Practitioner*, March.
Department of the Environment (1975, reprinted 1980) *The Purification of Swimming Pool Water*, HMSO, London.
Department of the Environment (1981) *Swimming Pool Disinfection Systems Using Calcium Hypochlorite, Chloroisocyanurates, Halogenated Dimethylhydantoins and Solid Ancillary Chemicals – Guidelines for Design and Operation*, HMSO, London.
Department of the Environment (1982) *Swimming Pool Disinfection Systems Using Chloroisocyanurates – A Survey of the Efficacy of Disinfection*, HMSO, London.
Department of the Environment (1982) *Swimming Pool Disinfection Systems Using Ozone with Residual Chlorination – Monitoring the Efficacy of Disinfection*, HMSO, London.
Department of the Environment (1982) *Swimming Pool Disinfection Systems Using Ozone with Residual Free Chlorine and Electrolytic Generation of Hypochlorite – Guidelines for Design and Operation*, HMSO, London.
Department of the Environment (1982) *Swimming Pool Disinfection Systems Using Sodium Hypochlorite and Calcium Hypochlorite – A Survey of the Efficacy of Disinfection*, HMSO, London.
Department of the Environment (1982) *Swimming Pool Disinfection Systems Using Sodium Hypochlorite – Guidelines for Design and Operation*, HMSO, London.
Department of the Environment (1983) *Swimming Pool Disinfection Systems Using Electrolytically Generated Sodium Hypochlorite – Monitoring the Efficacy of Disinfection*, HMSO, London.
Noise Council (1995) *Code of Practice for the Control of Noise from Outdoor Pop Concerts*, Noise Council, London.
Pool Water Treatment Advisory Group (1995) *Pool Water Guide*, PWTAG, Diss, Norfolk.
Swimming Pools and Allied Trades Association (1980) *Standards for Swimming Pools – Water and Chemical*, SPATA, Croydon.
Department of Health, London (2000) *Hepatitis C: Strategy for England*, Consultation paper.
Health and Safety Executive (1998) *The safe use of work equipment – the Provision and Use of Work Equipment Regulations 1998*, Approved Code of Practice and Guidance L22, HSE Books, Sudbury, Suffolk.
British Amusement Catering Trades Association (BACTA) (1998) *Coin Operated Kiddies Rides: A Safety Guide*, BACTA Regents Wharf, 6 All Saints Street, London, N1 9RQ.

Part Four
Epidemiology

12 Communicable disease – administration and law

Richard Elson

INTRODUCTION

Communicable diseases and their control remain high on the agenda of many countries worldwide, the optimism of the post-Second World War era being replaced with concerns over emerging and resurgent infections, the spread of antimicrobial resistance and the threat of bioterrorism. A robust legal framework underpinning the surveillance, investigation and control of communicable disease is required. However, the legislative and administrative arrangements currently in place in the United Kingdom have been criticized by many in their adequacy to fulfil this obligation. This chapter discusses the history of communicable disease law and administration in the United Kingdom, recent developments in international, regional and national communicable disease legislation, the formation of the Health Protection Agency and the need for a review of the present legislative and administrative arrangements.

BACKGROUND

The responsibility for the control of infectious diseases has traditionally been divided between the local authority (London boroughs, metropolitan boroughs and district councils) and the health authorities of the National Health Service (NHS). Local authorities (LAs) were responsible for the investigation and control of outbreaks of infectious disease and, through their appointed 'proper officer', for the enforcement of the various legislative provisions that exist for this purpose. Health authorities (HAs) were responsible for a wide range of services contributing to the prevention, control and treatment of communicable disease and infection.

The respective roles of each agency and that of the proper officer were never fully clarified and this situation led to calls for reform and a committee of inquiry was established which reported in January 1988 [1]. The report suggested three areas in which an improvement could be effected.

- Better and continuing collaboration between health and local authorities.
- Those responsible should be able to react quickly and decisively to problems.
- Clear recognition of the responsibilities of health authorities for the treatment, prevention and control of most communicable diseases and infections, whilst retaining a continuing role for local authorities in the prevention and control of notifiable diseases, particularly those that are food and water-borne.

The report also identified particular actions that could be taken to improve matters on these three issues including the need for one officer in each area to be made responsible for communicable disease and infection, the establishment of District

Control of Infection Committees and a review of the Public Health (Control of Disease) Act 1984.

With the exception of the legislative review, which was started but not completed, the recommendations of the Acheson Report were accepted by the government, and were implemented by administrative action in the issue by the Minister of Health of Circular 88/64 (now incorporated in NHS circular HSG (93) 56) setting out a framework for the administrative arrangements which were as follows.

1. Regional health authorities had to ensure that proper arrangements had been made between district health authorities and local authorities for the prevention of infection and communicable disease and the control of outbreaks.
2. District health authorities, either individually or on a joint basis where appropriate, had to appoint a consultant in communicable disease control (CCDC) with responsibility for those arrangements, taking necessary action and arranging appropriate co-ordination. This appointee was also to be designated as the proper officer of the local authority for the exercise of its control functions.
3. District Control of Infection Committees were to be established to provide advice and assist the CCDC in producing a written policy relating to the monitoring and surveillance of communicable diseases, outbreak investigation and collaborative arrangements including channels of liaison. The membership of the committee comprised a cross-section of those involved and included environmental health officers from the local authorities within the health district.

The essence of these arrangements was to retain the legal responsibilities of both the local and health authorities. In addition, the communicable disease function of the health authority and the duties of the proper officer of the LA were brought together under the post of the CCDC. A further review of these recommendations was undertaken by a joint Department of the Environment/Department of Health (DoE/DoH) committee as part of a wider study of the operation of the public health function following further changes in the organization and management of the NHS in 1990, that is, the establishment of NHS Trusts and the implementation of a purchaser/provider relationship. Revised guidance on communicable disease control was issued in EL(91)123 (included as Annexe B to NHS Management Executive Guideline HSG(93)56) which forms the basis of the current administrative arrangements.

'Getting Ahead of the Curve' and the Health Protection Agency in England and Wales

In early 1999, the Minister for Public Health established a Communicable Disease Strategy Group to look at reducing the amount of illness and premature death caused by communicable disease. This took place against a background of NHS restructuring[1] and culminated in the publication of the strategy 'Getting Ahead of the Curve' in January 2002 [2]. 'Getting Ahead of the Curve' was the Chief Medical Officers' (CMO) strategy for combating infectious diseases including other aspects of health protection. The strategy proposed 'a series of actions to prevent, investigate and control the infectious diseases threat and address health protection more widely'. Amongst these actions were:

- the establishment of a national Health Protection Agency and local health protection service including a strengthened and expanded system of infectious disease surveillance;
- the creation of a national expert panel on infectious diseases;

---

[1]In April 2002, District Health Authorities in England were replaced by Primary Care Trusts (PCTs) and Health Authorities were replaced by Strategic Health Authorities (SHAs). The role of PCTs is to plan and secure health services (including the control of infectious diseases) for their population. SHAs are responsible for the performance management of PCTs and developing strategies for the local health services, ensuring high quality performance and to ensure that national priorities are integrated into local plans.

- the appointment of an Inspector of Microbiology and rationalization of microbiology laboratories and the introduction of standards, and;
- a review of the law on infection control.

The Health Protection Agency was established, initially as a strategic health authority, on 1st April 2003, subsuming most of the existing functions of the Public Health Laboratory Service (including the Communicable Disease Surveillance Centre (CDSC)), the Microbiological Research Authority (including the Centre for Applied Microbiology and Research), the National Focus for Chemical Incidents and the National Radiological Protection Board.

In Wales, the National Public Health Service – Wales (NPHS-W) was formed in April 2003 and provides general and specialist public health functions, including Consultants in Communicable Disease Control CsCDC, laboratories and the Welsh CDSC. Local and regional health protection needs are met by the NPHS-W which works closely with the Health Protection Agency to ensure action on an England and Wales basis when needed, that is, emergency planning, counter measures, new policies and in specific work involving reference laboratories. The Health Protection Agency offers high-level advice and guidance and is answerable to the Welsh Assembly Government for national specialist functions [3].

Following changes to primary legislation,[2] the Health Protection Agency became a non-executive departmental body in April 2004. For the purposes of this chapter, the communicable disease function of the Health Protection Agency only will be discussed.

Health protection in Scotland and Northern Ireland

In contrast to England and Wales, the Scottish Executive issued a consultation document in 2002 on its own plans for health protection proposing several models concerning the functions provided by NHS Health Boards, the Scottish Centre for Infection and Environmental Health,

The National Focus for Chemical Incidents, the Scottish Poisons Information Bureau, the National Radiological Protection Board, Scottish National Reference laboratories and the Information and Statistics Division [4]. A review of the public health function in Northern Ireland, including how this will link to the Health Protection Agency, was carried out by the CMO in 2003 [5].

Structure of the communicable disease function of the Health Protection Agency

National level

Communicable disease surveillance

The communicable disease surveillance function of the Health Protection Agency (Fig. 12.1) is fulfilled by the CDSC which incorporates the surveillance of respiratory, gastro-intestinal, blood-borne and nosocomial infections. Surveillance of food and water is also a Health Protection Agency function. The collation of national data on communicable disease enables the Health Protection Agency to monitor the long-term trends in disease patterns, detect outbreaks and evaluate the application of control measures as well as informing the development of national policies for communicable disease prevention and control. The CDSC has strong links with its counterpart organizations in Scotland and Northern Ireland as well as internationally.

The CDSC provides expert advice and support in the investigation and control of outbreaks of communicable disease by regional and local health protection teams (see Section 'Division of Local and Regional Services'), Primary Care Trusts (PCTs), local authorities and infection control teams in hospitals. It gives priority to the field investigation of newly recognized infections and diseases of increasing or sustained high incidence and co-ordinates and collaborates in the investigation of national and international outbreaks. CDSC provides essential information to the Board and other Divisions of the Health Protection Agency, the Department of Health, other government departments, agencies and relevant advisory

---

[2]See the Health Protection Agency Act 2004.

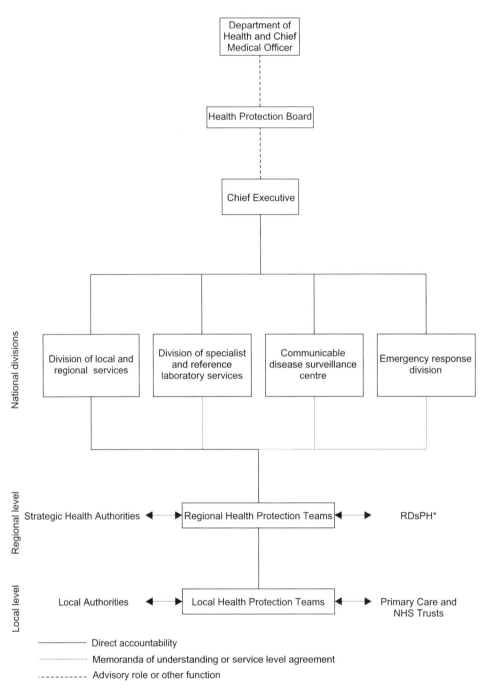

Fig. 12.1 Schematic representation of communicable disease function within the Health Protection Agency. (RDsPH = Regional Directors of Public Health.)

committees to inform the development of national policies for disease prevention and control.

Division of specialist and reference laboratory services

This Division is the national reference centre for medical microbiology in the United Kingdom and provides specialist expertise and advice to regional Health Protection Agency laboratories, NHS laboratories, local and regional health protection teams, medical staff, local authorities, other Health Protection Agency Divisions and other government departments. The Division carries out specialist tests not routinely performed in regional laboratories including traditional and molecular typing methods and works closely with the CDSC during outbreak investigations and surveys of novel or existing human pathogens.

Emergency response division

Increased concern surrounding terrorism was one of the driving forces behind the establishment of the Health Protection Agency. The broad remit of the emergency response division is to prepare and co-ordinate the emergency response capabilities of the Health Protection Agency and to provide support to the NHS and local and regional units of the Health Protection Agency in any emergency response including the deliberate release of biological agents. It also supports the DoH in developing policy in this area.

Division of local and regional services

This Division is responsible for the co-ordination and management of regional and local health protection services.

Regional health protection services

There are nine Regional Health Protection Teams in England covering the same geographical areas as regional government offices. Each Regional Health Protection Team has a regional epidemiologist, a regional public health microbiologist, emergency planners and information staff. The key functions of the Regional Health Protection Teams are to:

- provide the Regional Director of Public Health (RDPH) with operational support;
- co-ordinate the surveillance of communicable diseases at regional level;
- commission and co-ordinate regional public health microbiology services (including food, water and environmental microbiology) through the Regional Public Health Microbiologist, and;
- generally co-ordinate the Local Health Protection Teams including the input of the Health Protection Agency to major incidents which cross two or more Local Health Protection Teams.

The RDPH is responsible for health protection and for ensuring that accountability and operational systems for health protection are sufficient and functioning in each government region. The Regional Health Protection Teams are directly accountable to the Director of Local and Regional Services but are also accountable to their RDPH through memoranda of understanding or service level agreements.

Local health protection services

Each Health Protection Agency region contains a number of Local Health Protection Teams of which there are 42 in England. The key functions of the Local Health Protection Teams are to:

- Investigate and manage communicable disease incidents;
- Provide surveillance of communicable diseases and infections;
- Implement national strategies to prevent and control specific biological hazards, contribute to health emergency planning;
- Appoint medical inspectors,[3] and;

---

[3]Medical Inspectors are appointed under the Immigration Act 1971 and may disclose information to organizations prescribed in Section 133 of the Nationality, Immigration and Asylum Act 2002.

- Support local authorities (including port health authorities) in their responsibilities under the Public Health (Control of Disease) Act 1984 and associated regulations.

Local Health Protection Teams are staffed by consultants specializing in communicable diseases, nurses and other staff with specialist health protection skills and have access to expert advice on health emergency planning and communications. PCTs retain the overall responsibility for the public health protection of their populations; however, Local Health Protection Teams are accountable to PCTs through local health protection agreements and joint business plans drawn up with the involvement of the relevant local authorities and other bodies. All Local Health Protection Teams have common functions, however, the exact make-up and organizational setting of these teams reflects the arrangements deemed most appropriate locally.

ADMINISTRATIVE ARRANGEMENTS

At the time of writing, detailed working arrangements concerning communicable disease control between local, regional and national health protection teams and their counterparts were not finalized. However, it is envisaged that the existing arrangements detailed in Annexe B to NHS Management Executive Guideline HSG(93)56 (see Background) will continue unchanged until such time that a national framework or guidance is issued by the Health Protection Agency.

The communicable disease control plan – local arrangements

The communicable disease control plan is a written policy relating to the monitoring and surveillance of communicable diseases, outbreak investigation and collaborative arrangements between all the agencies involved in these areas. These arrangements should be in the form of a joint plan between the PCT and the Local Health Protection Team through consultation with the LA, provider units, NHS Trusts, general practitioner (GP)

fund holders, water companies, port health authorities and other agencies involved. The plan should include:

- a clear description of the role and the extent of the responsibilities of each of the organizations and individuals involved on a day to day basis or that may be involved when an outbreak occurs;
- the arrangements for informing and consulting the Communicable Disease Surveillance Centre, the Regional Health Protection Team and the Regional Director of Public Health;
- the arrangements that have been agreed with neighbouring LAs, PCTs, Local Health Protection Teams and others, both for managing individual cases of infectious disease that may have implications for those authorities, and for dealing with outbreaks of infection that cross boundaries;
- the arrangements that have been agreed among all the LAs and PCTs and Regional Health Protection Teams involved in areas where the boundaries between these authorities overlap;
- the arrangements for creating a control team to manage a significant outbreak of disease or other incident, the support that will be available to the team and what its duties will be. The plans for dealing with outbreaks should be sufficiently detailed and flexible to allow them to be implemented rapidly by all concerned in a situation requiring urgent action;
- the arrangements that have been made to provide the necessary staff and facilities outside normal working hours should this be required to manage an outbreak of disease, and;
- the arrangements for dealing with district immunization programmes, both on a day-to-day basis and in the event of an outbreak of disease.

These joint plans are to be reviewed annually, copied to all involved and to the Regional Director of Public Health and tested from time to time by audit and simulation exercises.

The proper officer

CsCDC should be formally appointed by each local authority as the proper officer under the

Public Health (Control of Disease) Act 1984, and the appropriate control powers should be formally delegated to that person. In certain circumstances, CsCDC may believe it is necessary or desirable for a Chief Environmental Health Officer (CEHO) to assume the lead in co-ordinating or managing a particular problem or task, for example, an outbreak of food or water-borne disease. In such cases, the CEHO should continue to seek and have regard to the advice provided by CsCDC. CsCDC are now employed by the Health Protection Agency and it is unclear how the new arrangements and the planned review of communicable disease legislation will affect their status as proper officer.

Local Authority support for the proper officer and Local Health Protection Service

LAs should provide professional and other staff to support the proper officer and the Local Health Protection Team and ensure that satisfactory arrangements are made for the receipt of notification of disease.

Liaison between agencies

A number of authorities and bodies are involved in the control of infectious disease. Each has its own area of responsibility but some of these overlap and will benefit from definition by the Health Protection Agency. Despite this, local communicable disease control has been largely successful in the past ten years. However, the issue of clarity concerning the roles and responsibilities of those involved in communicable disease control at local level was alluded to in the Pennington report on the circumstances leading to the 1996 outbreak of *E. coli* O157 infection in central Scotland [6]. This report commented that designated medical officers should contribute to the public health activities of local authorities and should report, at least once a year, to relevant council committees. Although this report was made in the Scottish context, it is in line with a more general recommendation made by the Environmental Health Commission that the public health function (including the control of infectious diseases) should be firmly placed as a local authority function [7], a notion more recently amplified by the Health Development Agency [8]. The Phillips Report following the BSE/CJD crisis highlighted a similar lack of clarity over who is responsible for public health at national level [9]. It is therefore clear that there is scope for overlap and in order for the system to work effectively, it is essential for there to be close liaison and co-ordination of effort between the authorities, agencies and individuals involved at all levels, both in respect of specific outbreaks and more generally. This involves the establishment of a good working relationship between those involved through regular contact, the sharing of information and the appreciation of the area of expertise of the other members of the team.

COMMUNICABLE DISEASE LEGISLATION

General

The dynamic nature of trade and migration make communicable diseases an important topic in international law. Infectious diseases do not observe political or national boundaries and, as such, require attention not only at national and regional levels but also at a global level.

International Health Regulations

The International Health Regulations (IHR) are a set of regulations set under the auspices of the World Health Organization (WHO) for the control and exchange of epidemiological information of specified diseases (plague, cholera and yellow fever) or of other diseases under surveillance by WHO. The IHR aim to provide security against the spread of infection with minimal interference to world traffic. The regulations list maximum public health measures applicable during outbreaks and provide rules concerning international traffic and travel, including the certification of travellers from areas infected by the three diseases covered to non-infected areas and the deratting, disinfecting and disinsecting of ships and aircraft. These areas are covered in more detail in Chapter 5.

The World Health Assembly (WHA) has adopted two resolutions aimed at linking the IHR with health activities at the global, regional and national levels. In 1995, resolution WHA48.7 requested that the IHR be revised to take more effective account of the threat posed by the international spread of new and re-emerging diseases. In 2001, resolution WHA54.14 expressly linked the revision of the IHR to WHO's activities to support its member states in identifying, verifying and responding to health emergencies of international concern.

Key focus areas of revision of IHR

- Global Health Security (epidemic alert and response);
- Public health emergencies of international concern;
- Routine prevention measures;
- National IHR focal points, and;
- The need for synergy between the IHR and other related international regimes, for example, Codex Alimentarius.

It is expected that a final version of the revised IHR will be available for adoption by the WHA in May 2004 [10].

Communicable disease in the European Community

In 1999, the European Parliament and Council decision 2119/98/EC established the Community Network for the Epidemiological Surveillance and Control of Communicable Diseases. The overall aim of this network is to prevent and control communicable diseases within the European Community. The network comprises an early warning and response system, co-ordination of existing epidemiological surveillance systems and allows surveillance information to be made available to the public. This network has been strengthened by decision 2000/96/EC that lists the communicable diseases or health issues (e.g. antimicrobial resistance) which should be subject to surveillance throughout the EU.

The European Union has had a public health competence since the Maastricht Treaty 1993, further strengthened by the Amsterdam Treaty 1997. The influence of European Law over national law can ensure that homogenous measures concerning communicable disease legislation can be applied across member states and devolved administrations in the United Kingdom. Europe's Commissioner for Health and Consumer Protection has also indicated a commitment to the creation of a European Centre for Disease Control by 2005 [11].

Communicable disease legislation in the United Kingdom

The core legislative tools concerning communicable disease in England and Wales are the Public Health (Control of Disease) Act 1984 and the Public Health (Infectious Diseases) Regulations 1988.

The notifiable diseases

The diseases notifiable under the provisions of the Public Health (Control of Disease) Act 1984 are:

1. cholera[4]
2. plague[4]
3. relapsing fever
4. smallpox[4]
5. typhus
6. food poisoning.

There are also 24 diseases notifiable under the provisions of the Public Health (Infectious Diseases) Regulations 1988. These are:

1. acute encephalitis
2. acute poliomyelitis
3. meningitis
4. meningococcal septicaemia
5. anthrax
6. diphtheria
7. dysentery (amoebic or bacillary)
8. paratyphoid fever

---

[4]Indicates a requirement to notify under the IHR as defined for the purposes of the Public Health (Infectious Diseases) Regulations 1988 irrespective of any changes to the International Health Regulations themselves.

9. typhoid fever
10. viral hepatitis
11. leprosy
12. leptospirosis
13. measles
14. mumps
15. rubella
16. whooping cough
17. malaria
18. tetanus
19. yellow fever[5]
20. ophthalmia neonatorum
21. rabies
22. scarlet fever
23. tuberculosis
24. viral haemorrhagic fever.

Similar provisions exist in Scotland under the Public Health (Notification of Infectious Diseases) (Scotland) Regulations 1988 and in Northern Ireland under the Public Health Notifiable Diseases Order (Northern Ireland) 1989 both of which also require notification of chickenpox and Legionellosis. Acquired immune deficiency syndrome (AIDS) is not notifiable, although certain regulations relating to disease control are applied to it. Other legislation contains the requirement that local authorities and the Health and Safety Executive should be notified of certain communicable diseases, for example, food legislation (Chapter 28) and the Reporting of Injuries, Diseases and Dangerous Occurrences Regulations 1995 (Chapter 24).

A local authority may make any disease notifiable within its area subject to an order approved by the Secretary of State for Health. An order must state the public health powers being adopted in respect of the disease and must be advertised with a copy of the order being sent to every registered medical practitioner within the district. In an emergency, the local authority may make an order without the prior approval of the Secretary of State, although the Secretary of State may subsequently approve or revoke the order. If he or she does not respond, the order ceases to have effect after one month.[6]

The notification process

A registered medical practitioner who becomes aware, or suspects, that a patient is suffering from a notifiable disease must notify the proper officer of the local authority in writing and provide the following information:

1. the name, age, sex and address of the patient;
2. the disease or particulars of the poisoning from which the patient is, or is suspected to be, suffering, and the date, as nearly as can be determined, of its onset;
3. if the premises are a hospital, the day on which the patient was admitted, the address from which the patient came, and whether or not the disease or poisoning leading to the notification was contracted in hospital.

There is a standard form for notification and the medical practitioner receives a fee from the health authority for each notification made.

Upon receipt of a notification the proper officer of the local authority must, within 48 hours, send a copy to:

1. the PCT within which the medical practice submitting the notification is situated;
2. the proper officer of the local authority from which the patient came, the PCT from which the patient came and, if relevant, the appropriate port health authority if the certificate is for a patient in hospital who came from outside the local authority or health authority in which the hospital is situated, and that patient did not contract the disease in hospital.

The proper officer is also required to make weekly and quarterly returns to the Registrar General of the notifications received. The CDSC of the Health Protection Agency collates the weekly returns and publishes analyses of local

---

[5]See footnote p. 276.
[6]In July 2003 the HPA published a discussion paper on making SARS (see p. 325), and certain other diseases and pathogenic, subject to notification.

and national trends in the weekly Communicable Disease Report. The proper officer is also required to notify, by telephone, the Chief Medical Officer (CMO) of the DoH of any serious outbreak of disease or food poisoning, as well as of any of the following diseases:

- a disease subject to the International Health Regulations as defined for the purposes of the Public Health (Infectious Diseases) Regulations 1988, that is, cholera, plague, smallpox or yellow fever;
- leprosy;
- malaria or rabies contracted in Great Britain;
- viral haemorrhagic fever.

Where a local authority or port health authority has reason to believe that rats in its district are infected with plague or are dying in unusual numbers, it is required to notify the CMO. At the time of writing, no changes to the notification procedure had been made, however, it is likely that notifications to the CMO will be through the Health Protection Agency.

It is recognized that notifiable diseases, particularly of 'food poisoning' or infectious intestinal disease, are under-reported. A number of steps must be taken before a confirmed case is recorded by national surveillance. Laboratory confirmed cases represent patients who have consulted a doctor, where the doctor has ordered a laboratory investigation, obtained a positive result from the laboratory and the laboratory or doctor have reported the result to their national communicable disease surveillance centre [12]. It has been estimated that for every 136 cases of infectious intestinal disease in the community, only one will be reported to national surveillance [13]. In addition, notifications of 'food poisoning' are not always accompanied by information indicating the cause of the infection and can be based on clinical suspicion rather than laboratory diagnosis.

Powers of investigation

Amongst the provisions of The Public Health (Control of Diseases) Act 1984 are a number of powers to enable officers to obtain information when investigating cases or outbreaks of communicable disease. These are powers of entry, provision of information and medical examination.

Powers of entry

An authorized officer of a local authority has the right to enter any premises at all reasonable hours in connection with the enforcement of his or her powers under this act. In certain circumstances, a warrant may be applied for from a justice of the peace.

Provision of information

The proper officer of a local authority may require information to enable measures to be taken to prevent the spread of the disease or, in the case of food poisoning, to trace the source. Information may be requested from occupiers of premises, including employers and head teachers. Powers granted to authorized officers under the Food Safety Act 1990 may provide a more pragmatic approach to outbreaks of food-borne illness.

Medical examination

If deemed necessary, the proper officer of a local authority can issue a written certificate seeking and justifying a mandatory medical examination. This examination can only be ordered by a justice of the peace where he or she is satisfied that there is reason to believe that:

- a person is or has been suffering from a notifiable disease or is carrying an organism that is capable of causing one;
- it is in the interests of that person, their family or the public interest for a medical examination to be carried out, and;
- where applicable, the person's doctor consents to such an order being made.

This procedure may be combined with a warrant of entry described earlier and allows a registered medical practitioner to enter any premises. Medical examination includes submitting the subject to bacteriological and radiological tests and similar investigations. Groups of people can also be examined under similar circumstances.

Powers of control

Amongst the range of infection or outbreak control measures available to the proper officer under the Public Health (Control of Disease) Act 1984 are:

1. *Control of infectious persons*

(a) Any person who knows he is suffering from a notifiable disease or has care of a child so affected must avoid exposing other people to the risk of infection, either directly from his or her presence in a public place, or at school, or indirectly by exposing others to the risk of handling infected bedding, personal clothing or articles that could carry infection.

(b) In order to prevent the spread of infection, restrictions may be put on the movement and disposal of bodies of persons who have died while suffering from a notifiable disease.

(c) A person suffering from a notifiable disease must not carry out any trade, business or occupation when there is a risk of spreading the infection. It may be necessary for the proper officer to require patients or carriers of notifiable disease not to attend work when there is a risk of their spreading infection. In the case of food workers, the obligation to ensure that infected people do not continue to work where they are liable to infect food is placed on the proprietor of the food business through various regulations, for example, the Food Safety (General Food) Hygiene Regulations 1995. However, LAs do have reserve powers of exclusion. Advice on the matter is given by the DoH [14].

(d) If it is considered that a person suffering from a notifiable disease should be admitted to hospital to prevent the spread of infection but the person refuses, a warrant may be obtained from a justice of the peace requiring that person to enter hospital and be detained.

2. *Control of infected premises, articles, etc.*

(a) When required, the local authority may cleanse and disinfect any premises or, when necessary, destroy any articles inside. The local authority may pay compensation for any damage resulting from its action and provide temporary accommodation while the disinfection is carried out.

(b) The local authority may provide a disinfecting station and may remove or disinfect articles free of charge.

(c) A person may not let any accommodation, whether it is a house or hotel that has been occupied by persons who are known to have suffered from a notifiable disease unless the accommodation has been satisfactorily disinfected.

(d) A local authority may prohibit home work on a premises where there has been a case of notifiable disease until such time as satisfactory disinfection has been undertaken.

3. *Food control*

The range of controls over food, food premises and food workers in order to prevent and control outbreaks of food-borne disease are dealt with in Chapter 28.

PORT HEALTH

In districts that have been designated as port health districts, communicable disease control is the responsibility of the port health authority. The powers of port health authorities in respect of the control of infectious disease are contained in the Public Health (Aircraft) and (Ships) Regulations 1979 (Chapter 5).

Channel tunnel

The Public Health (International Trains) Regulations 1994 were brought in to safeguard public health in relation to the opening of the Channel Tunnel and were made under the Public Health (Control of Disease) Act 1984. The regulations are broadly constructed so as to ensure the effective safeguarding of public health while minimizing disruption of this international service.

So far as communicable disease is concerned, the regulations provide for the following.

1. Notification by the train manager to the enforcement authority when there is a sick traveller (i.e. one who has a serious epidemic, endemic or infectious disease or where there are reasonable grounds for suspecting such disease) on board a train.
2. The enforcement authority may question train passengers whom they believe to be a significant danger to public health, either because they are sick or because they may have been exposed to similar infections, in order to ascertain their current state of health, contact with infection and previous and intended destinations.
3. The enforcement authority may require the disinfection of rolling stock or articles on board the train.

The regulations apply not only to the Channel Tunnel system but also to control areas designated by the Secretary of State, for example, terminal control points for passenger services. The enforcement authority is the local authority in whose district the control area is situated and, for situations on board a train, the Secretary of State. The diseases that are generally defined as being subject to the controls are those that are serious epidemic, endemic or infectious diseases. These are not specified, but venereal disease and human immunodeficiency virus (HIV) infection are specifically excluded. All the notifiable diseases are included in these definitions, including tuberculosis. This is interesting as tuberculosis is excluded from the control provisions for dealing with ships and aircraft. The growing concern about the increasing prevalence of tuberculosis is the reason for its inclusion in the Channel Tunnel controls. Some wider legislative controls still remain in relation to the use of the Channel Tunnel. Of particular note are those dealing with:

• the disinfestation of verminous persons, articles, etc. (Public Health Act 1936);

• the carriage of persons suffering from notifiable diseases on public conveyances, which includes trains (Public Health (Infectious Diseases) Regulations 1988);
• the powers of detention and removal to hospital of infectious persons, who may also be sick travellers (Public Health (Control of Disease) Act 1984).

Reference should also be made to the Report Concerning Frontier Controls and Policing, Co-operation in Criminal Justice, Public Safety and Mutual Assistance Relating to the Channel Fixed Line [15] which sets out the arrangements between the British and French governments on these issues.

LEGISLATIVE REVIEW AND REFORM

As described earlier, one of the recommendations of the Acheson Committee [1] was that the Public Health (Control of Disease) Act 1984 should be revised. A review of this legislation was undertaken by the DoH, and a consultation paper outlining the options for change was issued in 1989 [16]. A further review of the legislation was started by the DoH in 1997, however, no substantive progress was made and this task is now absorbed by the remit of the Health Protection Agency.

The law of communicable disease control in the United Kingdom, although relevant and practical in some parts, is considered by many to be outdated, fragmented and generally unclear to those who may be required to act within its remit. Contemporary legislation is effectively a conglomeration of old acts and regulations developed to deal with specific situations concerning specific pathogens. A recent assessment of health practitioners' concerns regarding communicable disease indicates that most concern revolves around antimicrobial resistant infections, nosocomial, respiratory, and sexually transmitted infections and infectious intestinal disease [17]. In addition, communicable disease legislation has been bypassed by huge changes in world trade, population movement, knowledge of communicable disease and the development of human rights and other legislation.

Although a number of legislative reviews have been started, none have been pursued to a conclusion. In simple terms, any future review and reform of legislation ultimately needs to define who is responsible for what at local, regional and national levels in terms of notification, surveillance, investigation and control of communicable diseases. The Nuffield Trust reviewed the current state of communicable disease law as part of a wider project and, as a contribution to Getting Ahead of the Curve, proposed a new legal framework for consideration in future legislative reviews [18]. The report concludes that:

- The current legal framework is not based on a modern understanding of communicable disease control nor does it adequately address the main communicable disease problems of today.
- There is no one body overseeing the application of and the development of the law, both primary and subordinate, relating to public health across the United Kingdom and there is genuine ambiguity about where leadership and responsibility for communicable disease control lie.
- Public health law, advice and guidance have been affected by the process of devolution with resultant duplication and inconsistencies adding complexity for the future.

The legal framework proposed by the report recognizes that relevant public health agencies should have adequate regulatory powers to safeguard public health whilst advocating restrictions on the use of these powers through a duty to demonstrate a sound scientific and economic justification if, and when, coercive powers are to be used. Such powers should also be compliant with relevant human rights legislation.[7] As a further pointer for legislative reform, this report suggests that new legislation should move the legal authority from local authorities to the Health Protection Agency, removing the need for a defined local authority 'Proper Officer'.

It is now clear that the Health Protection Agency will be responsible for leading the forthcoming review and modernization of legislation 'to support the effective prevention, investigation and control of infectious diseases' [2]. Specific areas of attention will be to address the current requirements for notification for surveillance purposes, the need to ensure adequate data protection whilst meeting health protection imperatives, remove old and irrelevant legislation and address the roles and responsibilities of health bodies and local authorities [2]. At the time of writing, there was no indication of when the review will commence.

ACKNOWLEDGEMENT

The author would like to thank Mr Ian Fisher of the CDSC Division of the Health Protection Agency for his assistance in preparing the section on Communicable Disease Surveillance in the European Community.

REFERENCES

1. The Acheson Report (1988) *The Report of the Committee of Inquiry into the Future Development of the Public Health Function: Public Health in England*, HMSO, London.
2. Department of Health (2002) *Getting Ahead of the Curve. A Strategy for Combating Infectious Diseases* (including other aspects of health protection), London.
3. National Public Health Service for Wales (2003) Details available at www.wales.nhs.uk
4. Scottish Executive (2003) *Health Protection In Scotland – A Consultation*, available at www.scotland.gov.uk
5. The NHS Confederation (2003) *Getting Ahead of the Curve and the Development of a National Health Protection Agency*. Briefing.
6. The Pennington Group (1997) *Report on the Circumstances Leading to the 1996 Outbreak of Infection with E. coli O157 in Central Scotland*, Scottish Office, Edinburgh.
7. Chartered Institute of Environmental Health (1997) *Agendas for Change: Report of the Environmental Health Commission*, CIEH, London.

---

[7]The Civil Contingencies Bui 2004 seeks to introduce wide ranging emergency powers including those associated with infectious diseases or their agents.

8. Burke, S., Gray, I., Paterson, K. and Meyrick, J. (2003) *Environmental Health 2012: A Key Partner in Delivering the Public Health Agenda*, Health Development Agency.

9. Lord Philips of Worth Matravers, Mrs June Bridgeman CB, and Professor Malcolm Ferguson-Smith FRS (2000) Report, evidence and supporting papers of the Inquiry into the emergence and identification of Bovine Spongifom Encephalopathy (BSE) and variant Creutzfeldt-Jacob Disease (vCJD) and the action taken in response to it up to 20 March 1996.

10. Aginam, O. (2002) International law and communicable diseases. *Bull World Health Organisation*, 80, 946–51.

11. Anonymous (2002) European Commissioner again pledges European centre for disease control by 2005. *Eurosurveillance Weekly*, 6.

12. Department for Environment Food and Rural Affairs (DEFRA) (2001) Zoonoses Report, United Kingdom 8. 2003. DEFRA.

13. Food Standards Agency (2000) *A Report of the Study of Infectious Intestinal Disease in England*, The Stationery Office, London.

14. Department of Health (1995) Food Handlers: Fitness for Work. Guidance for food businesses, enforcement officers and health professionals. Prepared by an expert working party convened by the Department of Health, HMSO, London.

15. Anonymous (1993) Report Concerning Frontier Controls and Policing, Co-operation in Criminal Justice, Public Safety and Mutual Assistance Relating to the Channel Fixed Line, HMSO, London.

16. Department of Health (1989) *Review of the Law of Infectious Disease Control*, HMSO, London.

17. Horby, P., Rushdy, A., Graham, C., O'Mahony, M. (2001) PHLS overview of communicable diseases 1999. *Commun Dis Public Health*, 4, 8–17.

18. Monaghan, S. (2002) *The State of Communicable Disease Law*, The Nuffield Trust, London.

13 Food-borne disease

Stephen R. Palmer

INTRODUCTION

The ingestion of food may give rise to disease under a variety of circumstances: some substances are inherently unsafe, for example, certain fungi; food may be contaminated with chemicals such as organophosphates; food stuffs may be handled or stored incorrectly such that toxic substances may form; and food may become contaminated by infectious pathogenic microorganisms such as *Salmonella* spp. or *Campylobacter* spp. Diseases resulting from such events are collectively known as food poisoning or food-borne disease.

The World Health Organization (WHO) definition of food-borne disease is 'any disease of an infectious or toxic nature caused by or thought to be caused by the consumption of food or water'.

Some predominately food-borne infections, for example, salmonellosis, can also be transmitted by the faecal–oral route, and some infections predominently spread by the faecal–oral route, for example, hepatitis A and *Shigella* spp., can occasionally be food-borne.

Cases of food poisoning occur singly (sporadic cases) or in outbreaks where two or more cases are epidemiologically related.

In recent years, outbreaks of food-borne disease have caused increasing public alarm, leading to government inquiries and considerable media attention.

NON-MICROBIAL FOOD-BORNE DISEASE

Fungi

Mushroom poisoning

Clinical mushroom poisoning is rarely seen. Most cases in Europe are caused by the *Amanita* genus of fungi. They cause one of two clinical syndromes depending upon the toxins they produce. *A. pantherina* (false blusher) and *A. muscaria* (fly agaric) ingestion results in the rapid onset of gastrointestinal symptoms followed by incoordination and other signs of neurological disorder due to the muscarine that they harbour. Recovery is usually within 24 hours. In contrast, the toxins in *A. phalloides* (death cap) and *A. virosa* (destroying angel) – amanitine and phalloidin – cause an initial enteritis followed a few days later by renal and hepatic failure, carrying a mortality rate of up to 90%.

Mycotoxicosis

Foods affected by toxin-producing moulds may cause illness in humans or animals, sometimes on an epidemic scale. Manifestations are as diverse as gangrene and convulsions (ergotism), renal disease (Balkan nephropathy) and liver cancer (aflatoxicosis). Low levels of aflatoxins are frequently found in peanuts and may play a part in the development of liver diseases in tropical countries, where food storage is conducive to fungal growth.

'Special' foods

A number of 'health foods' such as ginseng and liquorice may, if taken in large quantities over long periods, cause hallucinations, nausea, vertigo and other central nervous system (CNS) effects.

Red whelks

The salivary glands of this shellfish, which is readily distinguishable from the edible whelk, secrete tetramine. This toxin produces alarming symptoms, including muscle weakness and vertigo. They disappear within a few hours, however.

Red kidney beans

Though recognized as a cause of severe gastroenteritis since a large outbreak in 1948 was attributed to 'flaked beans', food poisoning from the consumption of raw or undercooked red kidney beans only came into prominence during the late 1970s and early 1980s, when a series of incidents were reported. A number of toxic substances can be extracted from the beans, but current evidence suggests that the haemagglutinin component is probably responsible for the diarrhoea and vomiting. Much of this substance is leached out by soaking the beans for several hours, and thorough cooking will render them safe. The onset of symptoms is usually within an hour or two of ingestion and rapid recovery is the rule.

Food contaminated with chemical substances and toxins

Pesticides

These substances are used to control pests of various kinds on wheat seed, fruit trees and vegetables. If ingested they are absorbed and particularly affect the CNS. A mortality of around 8% was recorded in a large outbreak in Iraq caused by bread made from seed treated with organomercurial compounds.

Metals

Mercurials discharged into the sea may be taken up by fish and have caused nephritis and incapacitating CNS damage to people who ate them. Minamata disease in Japan, resulting from wholesale discharges of industrial pollutants containing mercury into a bay from which the local population caught fish that formed the major part of their diet, left many people with severe neurological damage.

Zinc leached from galvanized pans when acid materials, such as fruit, are boiled in them is toxic and causes acute abdominal symptoms. A variety of foods including apples, rhubarb, chicken, spinach and alcoholic punch have caused outbreaks. Vomiting and/or diarrhoea appear rapidly after ingestion, often within a few minutes. Recovery is generally quick. The presence of zinc levels in suspected foods will confirm the diagnosis.

Dinoflagellates

'Blooms' or 'red tides' of proliferating dinoflagellate organisms (*Gonyaulax* spp.) periodically infest coastal waters where they are ingested by bivalve molluscs, particularly mussels. The toxin produced by the organism is concentrated in the flesh of shellfish, but causes no harm to them. In humans the toxin affects the CNS, causing paraesthesia of the mouth, lips, face and limbs, which may progress to a paralytic state depending upon the species of *Gonyaulax* involved. Regular monitoring of toxin concentrations in shellfish is carried out on susceptible coasts, and public warnings are given when potentially dangerous levels are found.

Additives

Monosodium glutamate, used extensively in Chinese cookery, may induce temporary burning sensations over the trunk, face and arms, headaches, tightness in the chest and, occasionally, abdominal pain and nausea.

Ciguatera

Small tropical fish may feed on the dinoflagellate *Gambierdiscus*, which will render their flesh toxic. They in turn are eaten by large edible fish, for example, groupers, in which the toxin accumulates in high concentration. Rare cases occur in the United Kingdom when fish from tropical waters are consumed. The symptoms, which develop within a few hours of ingestion, are similar to those with dinoflagellates.

Food stored incorrectly

Scombrotoxic fish poisoning

Scombroid fish, tuna, skipjack and mackerel and, occasionally, herring and sardine, may become contaminated with spoilage organisms, which can convert the amino acid histidine in the tissue to histamine. This substance, together with other as yet poorly characterized toxins, causes acute facial flushing, rapid pulse, headaches and mild gastroenteritis within a few hours of ingestion. The symptoms are alarming but short-lived. Proper refrigeration of fish will prevent the proliferation of the organisms responsible. Histamine levels of 20 mg or more per 100 g or more of fish are diagnostic, but it should be noted that suspect fish must be kept refrigerated during transport to the laboratory for testing.

Solanine

Potatoes that are left to sprout or that are exposed to sunlight such that the skin surface becomes green, will accumulate the alkaloid solanine in the skin and just below the surface. Peeling and washing will render them safe, but jacket potatoes have caused some cases of poisoning. A large outbreak at a school in London was associated with potatoes stored in sacks over a holiday period. Symptoms are those of gastroenteritis together with varying degrees of confusion and other neurological signs. Recovery within 24 hours is usual, although prolonged indisposition may occur and the occasional fatality has been recorded. It is surprising that the condition, relating as it does to one of our commonest foods, is so uncommon.

FOOD-BORNE DISEASE CAUSED BY PATHOGENIC MICRO-ORGANISMS OR THEIR TOXINS

Incidence

The true incidence of food-borne infection is difficult to determine since asymptomatic infection is common, because only a minority of people with symptomatic infection will seek medical treatment and because only a minority of patients will be investigated microbiologically. Suspected food poisoning is statutorily notifiable in most countries, and many countries also conduct surveillance by collating laboratory results [1]. Food-borne infection appears to be one of the most common infectious diseases, and the incidence is still increasing in many countries. For a review of the factors that have contributed to the changing patterns of food-borne diseases, see WHO Surveillance Programme for Control of Foodborne Infections and Intoxicants in Europe (October 1998, *Newsletter 57*, WHO, Copenhagen).

In many countries, reference laboratories are able to subtype food-poisoning organisms. For example, in England and Wales isolations of *Salmonella* spp. and certain other enteric pathogenic organisms are referred to the Health Protection Agency Specialist and Reference Microbiology Service for confirmation of identity. This information is combined and correlated with that obtained by the Health Protection Agency Communicable Disease Surveillance Centre (CDSC), and that from statutory notifications to the Office of National Statistics. Parallel systems operate in Scotland and Northern Ireland. The figures are published at regular intervals and can be regarded as a reasonably reliable guide to the trends in incidence of the various forms of microbial food-borne disease.

Trends in the incidence of food poisoning in England and Wales are illustrated in Table 13.1. The number of cases has risen steadily since the mid-1960s, but the unprecedented increase since 1985 became the cause of grave concern. Since 1999 the trend has reversed probably due to strict controls applied to egg production.

Organisms responsible – microbial characteristics

Six groups of organisms have traditionally been responsible for most cases and outbreaks of food-borne disease. These are *Salmonella* spp., *Campylobacter* spp., *Clostridium perfringens*, *Staphylococcus aureus*, *Bacillus* spp. and Small Round-Structured Viruses (SRSVs). Of these, the

Table 13.1 Food poisoning notifications – annual totals for England and Wales, 1982–2002

Year	Total	Formally notified	Otherwise ascertained
1982	14 253	9964	4289
1983	17 735	12 273	5462
1984	20 702	13 247	7455
1985	19 242	13 143	6099
1986	23 948	16 502	7446
1987	29 331	20 363	8968
1988	39 713	27 826	11 887
1989	52 557	38 086	14 471
1990	52 145	36 945	15 200
1991	52 543	35 291	17 252
1992	63 347	42 551	20 796
1993	68 587	44 271	24 316
1994	81 833	50 412	31 421
1995	82 041	50 761	31 280
1996	83 233	50 718	32 515
1997	93 901	54 233	39 668
1998	93 932	53 764	40 168
1999	86 316	48 454	37 862
2000	86 528	46 481	40 047
2001	85 468	46 768	38 700
2002	72 649	38 541	34 108

Source: Statutory Notifications of Infectious Diseases. Last updated on 11 August 2003.

campylobacters and salmonellas far outnumber all the others. More recently, *E. coli O157* and *Listeria* spp. have emerged as important food-borne pathogens. Occasionally, parasites, including *Giardia lamblia* and *Cryptosporidium parvum*, can be food-borne. Table 13.2 summarizes the organisms according to whether their common reservoirs are human, animal or environmental.

Pathogenesis

The effects of food poisoning bacteria on the gastrointestinal tract are either mediated through the presence of the organisms themselves or through the production of powerful exotoxins, which may exert their influence in the absence of the organisms. The former are known as the infection type, and the latter the toxic (Table 13.3). Generally, the

incubation period of the toxic type illnesses is shorter than the infective, as the toxins may be preformed in the food. Proliferation in the intestinal tract is necessary with the infective type before a sufficiently large number of organisms has accumulated to cause damage and initiate symptoms. The toxin of *Clostridium botulinum* is one of the most toxic substances known to man. There may be difficulties in the detection of organisms of the toxic type in foods. Tests may be required to demonstrate the toxin, which will remain after the organisms that produced it have died out. Some of the organisms causing food-borne disease may have both toxic and invasive potential. Technical details on the biological and cultural characteristics of the organisms can be found in standard microbiology texts. The following sections provide information that is particularly relevant to their ability to cause food poisoning.

Salmonella *spp.*

The two major groups are those causing enteric fevers (typhoid and paratyphoid), and those causing food poisoning (the non-typhoid salmonellas). The enteric fever group consists of *Salmonella enterica serovar typhi* and *Salmonella enterica serovar paratyphi* A and B. With rare exceptions these organisms infect humans only. The non-typhoid group consists of over 2000 species, all of which have their primary reservoirs in animals – both wild and domestic.

The classification into serotypes is made on the basis of antigens on their surfaces (O antigens) and on their flagellae (H antigens). Kaufman and White [2] developed a scheme using these antigens in combination to enable identifications to be made. Certain more common species can be more precisely characterized using special techniques such as phage typing, plasmid analysis and fingerprinting. The organisms can grow in many foods at a wide range of temperatures from 10°C to over 40°C. They are destroyed by pasteurization temperatures and are killed in 20 minutes at 60°C. Salmonellas survive well outside a person or an animal in faeces, on vegetables, in animal feeds and in many foods for long periods. While

Table 13.2 Food poisoning organisms by origin

	Human only	*Animal*	*Environment*
Common	*Staphylococcus aureus** Viruses	*Salmonella* spp. (non-typhoid) *Campylobacter* spp.	*Clostridium perfringens*
Less common	*Salmonella typhi* *Salmonella paratyphi B* *Escherichia coli* (enteropathogenic, enteroinvasive, enterotoxigenic) *Shigella* spp.	*Listeria monocytogenes* *Yersinia enterocolitica* *Escherichia coli* (enterohaemorrhagic)	*Bacillus* spp. *Clostridium botulinum* *Vibrio parahaemolyticus* *Aeromonas* spp. *Pleisomonas* spp. *Listeria* spp.

*Rare phage type 42D is of bovine origin.

Table 13.3 Organisms causing food poisoning – infectious and toxic

Infectious	*Incubation period*	*Toxic*	*Incubation period*
Salmonella spp.	12–72 hours	*Clostridium perfringens*	8–18 hours
Campylobacter spp.	2–5 days	*Clostridium botulinum*	12–36 hours
Escherichia coli	1–2 days	*Staphylococcus aureus*	2–6 hours
Vibrio parahaemolyticus	12–24 hours	*Bacillus* spp.	1–4 hours
Yersinia enterocolitica	3–7 days	*Escherichia coli*	1–6 days
Listeria monocytogenes	1–10 weeks	(enterohaemorrhagic)	
Aeromonas spp.	Variable		
Pleisomonas spp.	Variable		

10 serotypes account for 90% of all human isolates, two serotypes, *Salmonella enterica serovar enteritidis* and *Salmonella enterica serovar typhimurium*, constitute 80% of the total in the United Kingdom.

Campylobacter *spp.*

The species causing food-borne disease are *C. jejuni* and *C. coli*. They are found as part of the normal gut flora of many animals including mammals, birds and reptiles. Their growth requirements are more exacting than those for salmonellas, and the organisms do not grow on food under normal circumstances. Subtyping can be undertaken but is less precise than with salmonellas. Growth occurs up to 42°C and the organism survives well in water and raw milk.

Clostridium perfringens

This organism is found in animal faeces, soil, dust, vegetation and elsewhere in the environment. It will only grow anaerobically, and produces a variety of exotoxins, some of which act upon the gastrointestinal tract. Spores may be formed in adverse conditions, which enables the organisms to survive circumstances that would normally kill vegetative bacteria. Five types are recognized on the basis of toxins and enzymes produced. Type A strains are responsible for most food poisoning outbreaks.

Staphylococcus aureus

Many people carry *Staph. aureus* in their noses and some carry it on their skin. Many strains produce enterotoxins, of which there are at least five: A, B, C, D and E. Growth occurs on many foods, particularly those with a high protein content, and the toxin is readily produced at ambient temperatures. The organisms are killed at 60°C in 30 minutes, but the toxin will survive boiling for the same period of time. More precise identification of staphylococci is achieved by phage typing. Because of the commonness and ubiquity of this organism, such methods must be used when attempting to identify the sources of outbreaks.

Norovirus

These agents once known as SRSVs were first identified in the faeces of patients in the town of Norwalk in the United States during an outbreak of gastroenteritis. The organisms are routinely detectable only by electron microscopy. Other strains are named either by their appearance, for example, *Calicivirus*, Astrovirus, Norwalk-like or by their geographical location of origin, for example, Snow River agent. They appear to be entirely human in distribution and are found in sewage effluent. They can be concentrated in shellfish such as mussels and oysters. The SRSVs will survive well on surfaces and food, and are relatively heat-resistant. A sudden onset of vomiting can produce aerosols of virus that can widely contaminate the environment, leading to food-borne and person to person outbreaks.

Bacillus cereus

This free-living organism is widespread in the environment and is particularly, but not exclusively, associated with foods involving grains and cereals. Strains produce two distinct exotoxins, one of which causes vomiting and the other diarrhoea. Spores are formed that resist boiling and frying for short periods. There are numerous serotypes, some of which are more commonly found as pathogens than others. There is good evidence that other *Bacillus* species, for example, *B. subtilis* and *B. licheniformis*, may also be responsible for cases and outbreaks of food poisoning.

Escherichia coli

Strains characterized by certain biological properties and known as enteropathogenic (EPEC), enteroinvasive (EIEC) and enterotoxigenic *E. coli* (ETEC) can cause gastroenteritis, and may be defined serologically according to their somatic (O) and flagellar (H) antigens.

Another group of increasing importance consists of around 50 serotypes known as enterohaemorrhagic *E. coli* (EHEC). These are now definitely associated with haemorrhagic colitis and the haemolytic uraemic syndrome. All produce a characteristic toxin known as verotoxin (VT) and strains may be referred to as VTEC in some texts. (In the United States and Canada the same toxin has been referred to as Shiga-like toxin, SLT.)

Escherichia coli type O157 in the United Kingdom is by far the most frequently reported strain in human cases and appears to be largely of beef or milk origin. Further typing systems have been developed and can be used for epidemiological tracing.

Shigella *spp.*

The dysentery group of organisms are pathogens of man only and are usually transmitted by the faecal–oral route in circumstances of poor general hygiene. Food contamination readily occurs, although in the United Kingdom *Shigella* spp. are rarely associated with food poisoning. There are four major species, *S. sonnei*, *S. flexneri*, *S. boydii* and *S. dysenteriae*, and numerous subtypes.

Vibrio parahaemolyticus

This marine organism is found in coastal and brackish waters. It is halophilic (salt-loving) and grows best in media with a high salt content. Infection is associated with seafoods and is particularly prevalent where raw fish is eaten, for example, in Japan. There are many antigenic subtypes.

Clostridium botulinum

This anaerobic, spore-bearing organism is extensively distributed in the environment and produces a powerful exotoxin affecting the CNS. The organism will grow particularly well on low acid (pH > 4.5) foods such as fish, fruit and vegetables. There are seven antigenic types, A–G, each with their own distinct toxin. Human intoxications are usually with types A, B and E, though a recent Swiss outbreak was caused by type G. The toxins, powerful as they are, are readily destroyed by heat. The spores, if type A, resist boiling for several hours; those of other types are slightly less resistant.

Listeria monocytogenes

Because these organisms are very widely distributed in the gastrointestinal tracts of wild and domestic animals, they contaminate the environment generally. They can grow over a wide range of temperatures from as low as +4°C , are comparatively resistant to disinfectants and changes in pH and can survive extreme environmental conditions. Strains can be distinguished by serotyping, types 1/2a and 4b being most commonly found in strains isolated from pathological sources.

Yersinia enterocolitica

Like *L. monocytogenes*, these organisms are widely distributed in animals, including mammals, birds, flies and fish, and in the environment, particularly water, soil and vegetation. They may grow at refrigeration temperatures, and at pHs from 5.0 to 9.0. Many antigenic types occur, and 0:3, 0:8 and 0:9 are most frequently associated with human disease. Most clinically significant strains produce an enterotoxin, but only at temperatures up to 30°C . The part played by this toxin in mediating disease is not yet clear.

Other organisms

Evidence that *Aeromonas* (see The risk to public health of *Aeromonas* in ready-to-eat salad products, *Communicable Disease and Public Health*,

1 (4) December 1998) and *Plesiomonas* spp. can cause food poisoning comes largely from studies in countries in the Far East where seafoods, raw and cooked, form a large part of the diet of the population. *A. sobria* is the most commonly isolated strain in the United Kingdom.

Cases caused by *Brucella* species are reported from time to time due mainly to the consumption of imported dairy products, usually cheese. Indigenous cases are now very rare.

The association between the protozoal parasites *G. lamblia* and *Cryptosporidium* spp., and food-borne disease is tenuous, though one school party is thought to have been infected by the latter through tasting silage.

'HIGH-RISK' FOODS

While food-borne disease has been linked to almost every kind of food from peppercorns to chocolate bars, certain foods are far more prone to significant contamination than others. Most outbreaks are associated with the following foods.

Foods of animal origin

Meats

Salmonellas and campylobacters are found in the gastrointestinal tracts of cattle and poultry and less often in other farm animals. Contamination of meats during slaughter and processing is common, particularly so with poultry. Raw carcasses are therefore potentially hazardous and nearly all 'oven ready' chickens will yield one or other of the organisms on culture. Undercooking of red meat and poultry may therefore leave residual pathogens. Salmonellas but not campylobacters are liable to proliferate if cooked-meat storage conditions are inadequate.

Dairy products

Milk always contains organisms that are derived mainly from the gastrointestinal tract of the cow, and pathogens may be part of their flora. Raw and imperfectly pasteurized milk may be contaminated with salmonellas, campylobacters, listerias, yersinias or enterotoxigenic *E. coli*. Dairy products like creams and soft cheeses (rarely hard cheese) may consequently also be infected.

Eggs

Salmonellas may contaminate eggs through cracks in their surfaces or, with certain strains, may infect the eggs within the oviduct. Where such eggs are lightly cooked, temperatures within the albumen of the yolk may well not be sufficient to destroy the organisms. Mayonnaise and other dishes in which raw eggs are used may also transmit salmonellas.

Shellfish

Bivalve molluscs, for example, mussels, oysters and, to a lesser extent, gastropods such as cockles and whelks, are grown commercially in river estuaries, many of which are chronically and heavily polluted with human sewage. Purification processes can reliably remove bacteria but viruses become fixed in their flesh where their concentration may reach five times that in the water from which they are derived. Decontamination procedures are often ineffective and raw or undercooked shellfish are frequently the source of viral gastroenteritis outbreaks.

CLINICAL AND EPIDEMIOLOGICAL FEATURES

Most notified cases are sporadic. Since most households carry several 'high risk' foods, and since most such food is used within a short time after purchase, it is not generally possible to trace with any certainty the foods responsible for these cases.

Outbreaks may occur within families, in institutions such as hospitals or within the community, usually at functions such as receptions, weddings, barbecues and dinners. Sometimes clusters of cases occur in the community in which no connection in time or place is readily apparent. Imaginative and assiduous investigation may establish a common link such as the purchase of cooked ham from outlets supplied by one butcher.

Salmonellosis

Enteric fever

Though not generally considered to be 'food poisoning', typhoid and paratyphoid can be transmitted on food from a patient or carrier to further patients or in water soiled by urine or faeces. The disease presents after an incubation period of usually 10–14 days with fever, malaise, anorexia, aches and pains and constipation. Diarrhoea is uncommon before the tenth day after onset. The organisms are hardy and the source, always a case or a carrier, may be remote from the vehicle of transmission. In the Aberdeen outbreak of 1963, the vehicle was corned beef that had been infected in the Argentine. Outbreaks are rare in the United Kingdom but a number have been recorded in mainland Europe, the United States and Canada within the past 20 years in which such vehicles as water, shellfish and dairy products were implicated. Carriers with good hygiene seem to pose little risk in domestic situations. Proven cases working in the commercial food industry, however, must undergo complex clearance procedures before returning to work. Details are provided in a PHLS document [1].

Non-typhoid salmonellosis

Presentation and course   The incubation period is between 12 and 48 hours, extending sometimes to 72 hours. Onset is rapid with abdominal pain, diarrhoea and often vomiting in the early stages. The stool is liquid and commonly contains little or no blood or mucus. Spontaneous uncomplicated recovery over five to seven days is the usual course, only supportive treatment being needed. In the very young, frail and elderly, significant dehydration can develop within a few hours and hospital treatment with intravenous fluids may be required. Rarely the organisms become invasive and a serious septicaemic illness ensues for which appropriate treatment must be given. Mortality in outbreaks is in the region of 1–4%, almost entirely in the particularly susceptible individuals mentioned here.

Epidemiology   Transmission of salmonellosis follows two basic patterns. In the first, food or, more rarely, water contaminated with the organisms is the vehicle; in the second, the organism is passed from person to person by way of direct contact with infected excreta or indirectly through handling objects, for example, clothes, bedding or toys, contaminated with infected excreta. In effect, spread is by the faecal–oral route, and this is a serious problem when cases occur in institutions. An outbreak can therefore initially occur as a result of the consumption of infected food, and further people may subsequently be infected through contact with one of the original cases, having themselves no connection with the original causative circumstances. The former are known as primary cases, the latter as secondary.

Because of the wide distribution of salmonellas in animals and the environment, a great variety of foodstuffs has been associated with outbreaks, from the exotic, for example, bean sprouts and peppercorns, to the more mundane, for example, eggs. The organisms grow well in meats, milk, eggs and dairy products, and the majority of cases and outbreaks, where traced, ultimately lead back to one or other of these sources. Currently, poultry meat, eggs and egg products are of particular importance in the United Kingdom, and are responsible for the unprecedented rise in cases in the past few years (Table 13.4).

Strains of *Salmonella enterica serovar enteritidis* are the cause of a high proportion of cases and outbreaks in the present epidemic, which extends to many countries throughout the world. Fresh milk, raw or imperfectly pasteurized, and dried milks are also responsible for numerous outbreaks, while red meats, whole and processed, are rather less common as causes of salmonellosis than they were.

The way in which these foods become contaminated in the first place takes in modern systems of feeding of commercial animals, husbandry methods and the techniques employed in food processing from slaughter to distribution. Infection may be acquired directly by eating contaminated raw foods such as unpasteurized milk, undercooked meats and poultry or eggs or egg products in which the temperatures reached in cooking were insufficient to destroy the organisms. Thus raw eggs used in home-made products

Table 13.4 Salmonella in humans: Faecal and unknown reports excluding *S. typhi* and *S. para.* in England and Wales, 1981–2001

Year	S. typhimurium	S. enteritidis	*Other serotypes*
1981	3922	1087	5172
1982	6089	1101	5132
1983	7785	1774	5596
1984	7264	2071	5392
1985	5478	3095	4757
1986	7094	4771	5111
1987	7660	6858	6014
1988	6444	15 427	5607
1989	7306	15 773	6919
1990	5451	18 840	5821
1991	5331	17 460	4902
1992	5401	20 094	5860
1993	4778	20 254	5618
1994	5522	17 371	7518
1995	6743	16 044	6527
1996	5542	18 256	5185
1997	4778	23 008	4810
1998	3039	16 397	4292
1999	2424	10 775	4333
2000	2651	8 468	3725
2001*	2085	10 755	3625

*Provisional data.
Source: PHLS Laboratory of Enteric Pathogens 1981–1991; PHLS Salmonella dataset 1992 onwards; Last updated on 12 March 2002.

such as mayonnaise, meringue and glaze are regularly shown to be the source of outbreaks, and the use of pasteurized liquid egg is recommended instead where possible. The Chief Medical Officer has issued guidelines to the public advising on simple procedures to prevent egg-borne salmonellosis [3].

Other cases arise from contamination or recontamination of foods during home or institutional catering procedures and are caused by failure to observe basic food hygiene rules. Incorrectly prepared food may allow the survival of contaminating pathogens, and incorrectly stored food may allow proliferation of pathogens originally present in insignificant numbers. Contamination of food at source is, however, the initial event in any food-borne *Salmonella* incident, and control of the initial contamination is as important as food hygiene is in the long term. An important example is the success of the Lion Brand scheme for preventing salmonella in eggs. A further source of salmonellas is imported food and, in recent years, the organisms have been isolated from foods as various as frog legs, pasta, cuttlefish, pâté and herbal tea.

Patients convalescing from salmonellosis may continue to excrete the organisms in faeces asymptomatically for substantial periods. Only about 50% of patients will have ceased to excrete the organism after five weeks and 90% by nine weeks. This may pose problems with those involved in catering and food production.

Although person to person spread is well recognized in salmonellosis, the role of the commercial food handler excreter in the cause of outbreaks is unclear. Most of those found to be excreting salmonellas during an outbreak have also eaten the suspect food, and may actually be victims rather than the source of infection. A food handler is defined for this purpose as 'one who handles food which is either to be eaten raw or which is not to be further cooked before consumption' on the basis that adequate cooking will destroy any organisms allowed to contaminate food. The real risk from food handlers is when they are symptomatic, and therefore no commercial food handler with diarrhoea should be permitted to handle any food at all. Details of exclusions of personnel with salmonellosis and other food-borne diseases, as well as clearance policies, are given in a Department of Health document [4].

The prospect of all food becoming free from salmonellas is remote. Because of their distribution, resistance to unfavourable environments, ability to grow in many foods at a wide range of temperatures, propensity to give rise to secondary cases and prolonged asymptomatic convalescent excretion, they are unique in the problems posed in investigation and both short- and long-term control.

Gamma irradiation of foodstuffs can eliminate pathogenic organisms including salmonellas from the high-risk foods. Problems still remain in the general acceptance of the safety of such techniques in the public perception.

Illustrative outbreaks of salmonellosis

Outbreaks of food-borne salmonellosis are generally present either as point source incidents, or extended common source incidents. In the former, most of the patients involved are infected at roughly the same time, usually at the same place or event, for example, a wedding reception. In the latter, the organism may continue to infect people over a period of time, either because the vehicle of transmission is widely distributed both in time and place, or because it is not identified for weeks or months and cases present apparently sporadically rather than in epidemic form.

An example of the first type was an outbreak at a large psychogeriatric hospital in which 358 patients and 50 staff developed *Salmonella* gastroenteritis, mostly over a period of three to four days. Nineteen patients died. Studies suggested that cold roast beef, probably contaminated while in a refrigerator, was the cause. More than 80% of the patients became ill within the usual limits of the incubation period, and the inference was that they were infected at the same meal. A small number of secondary cases occurred later.

In contrast, imported chocolate bars contaminated with *S. napoli* were distributed to many parts of the country, and cases were reported with no clear relationship either in time or place until careful epidemiology suggested the likely means of transmission. Similar outbreaks have occurred with processed sausage (*S. typhimurium*), cochineal used to measure intestinal transit times in a hospital in the United States (*S. cubana*), and contaminated baby foods in the United Kingdom (*S. ealing*). In all these incidents, the cases seemed to be sporadic and unrelated to each other, but continued to occur in spite of measures taken to prevent spread. The appearance of numbers of unusual strains, however scattered they may be in distribution, should raise suspicion that they may be related epidemiologically. Where common strains, for example *S. enteritidis*, are the cause of widespread infection, recognition of relatedness can be slow to develop.

Campylobacter spp.

Presentation and course

The incubation period is between two and five days, rarely longer. Abdominal pain and cramps may be quite severe. Diarrhoea and vomiting is of acute onset, and lasts from four to seven days without specific therapy. The stool frequently contains both blood and mucus. Arthritis occurs in 1–2% of cases, but invasive disease and other complications are rare. Erythromycin is effective in severe cases and will also reduce the duration of carriage.

Epidemiology

Although *Campylobacter* spp. can be found in the intestines of most animals, only in sheep do they cause any ill effect. Otherwise they behave as part of the normal flora. Transmission to humans, however, occurs readily, either through food or from contamination of the environment.

The infection in humans is common worldwide, and now accounts for more reported cases of gastroenteritis than does salmonellosis. All age groups in all climates are affected, although the incidence in temperate areas is seasonal, with a rise in the number of cases in late spring and summer. As in salmonellosis, the sources of sporadic cases are usually impossible to define. Large outbreaks have been traced to the contamination of water or milk with animal excreta. A surprising form of contamination is the transfer of *Campylobacter* from the beaks of birds, mainly magpies, to bottled milk left on the doorstep when they peck through the top. Undercooked or cross-contaminated poultry is also a source, and is the only food regularly associated with the disease. In families, especially where there are young children, kittens and puppies with diarrhoea have been responsible. Secondary cases are, in contrast to salmonellosis, uncommon, although infants with liquid stools may disseminate the organism widely within a household. Carriage may occur over some weeks, but its significance is much less than in salmonellosis.

Illustrative outbreak

Three to four days after an extended family gathered for a reunion dinner cooked at the home of the matriarch, 7 of the 10 people present developed severe abdominal cramps and diarrhoea. One child was admitted to hospital for observation with suspected appendicitis. *Campylobacter jejuni* was grown from all five stool samples submitted. Food history analysis suggested that turkey, which formed the major part of the main course, was responsible. Two of the three guests that did not become ill were vegetarian. No food from the meal was available for analysis, but a number of the patients had noticed that the meat near the bone was pink and seemed undercooked. The turkey had been cooked on the morning of the dinner after overnight thawing at room temperature. Campylobacters were isolated from other birds from the same farm that supplied the family. All strains were of the same biotype; serotyping was not done. Recovery took up to nine days. In spite of there being close contact between cases and children, no secondary cases occurred.

This outbreak illustrates the severe clinical symptoms, the long incubation period and the lack of person to person spread characteristic of *Campylobacter* infection. It is likely that the thawing time was insufficient and that parts of the turkey did not reach temperatures adequate to destroy the organisms.

Clostridium perfringens

Presentation and course

After an incubation period of 8–18 hours, nausea and colicky abdominal pain is followed by diarrhoea and, less often, vomiting. The course is characteristically milder and shorter (one to two days) than with *Salmonella* or *Campylobacter* infections, and complications rarely occur.

Epidemiology

This ubiquitous organism, which is found in human and animal excreta and in soil and dust, can readily contaminate food. Transmission through flies and other insects may also occur. The ability to cause disease, however, is dose-dependent, and fairly exacting conditions for growth must be satisfied if the number of organisms sufficient to cause disease is to be reached. Cooked meats, both red and white, stews and gravies provide suitable anaerobic environments. Spores of the organism survive cooking heat, and unrefrigerated storage will provide optimum temperatures for germination of the spores at some stage during the slow period of cooling. Proliferation will occur with the subsequent production of the toxin. Toxin is only formed by actively sporulating organisms, and thus the temperature conditions allowing the spores to develop into vegetative forms are critical. The toxin is heat-resistant and is not generally destroyed when the food is reheated.

Outbreaks are associated with relatively large catering concerns, where bulk cooking of sizeable cuts of meats will result in slower cooling than with the usual, much smaller, domestic-sized joints and poultry. Classically, meat dishes are prepared well in advance, set out to cool for some hours or overnight at ambient temperatures, and then reheated for a short time before serving. Stock pots can provide excellent conditions for the sporulation of *Cl. perfringens* and should be discouraged. Cooling in a refrigerator will reduce temperatures sufficiently quickly that the foods will pass rapidly enough through the critical range to prevent germination taking place.

Illustrative outbreak

Fifty-seven patients and 12 staff members from all eight wards of a small, 200-bed mental institution developed diarrhoea over an eight-hour period. Preliminary enquiries suggested that all the patients had shared a meal at lunch time the previous day, about 12 hours before the onset of the first cases. Food histories strongly pointed to a chicken broth as the most likely source. *Cl. perfringens* type 71 was isolated from the faeces of 24 of the cases and the remnants of the broth. In the latter, the same organisms were present at a concentration of over 10^4 per millilitre. Enquiries into the food preparation revealed that the broth had been prepared the evening before, transferred to a number of large bowls and placed in a refrigerator overnight.

In the morning, the bowls were noticed to be warmer than they should have been, and it was suspected that the refrigerator had not been working properly. The broth was put back into a vat, brought to the boil and served over the following two hours. The patients and staff all recovered within three days.

The failure to recognize that both the nature of the food and the conditions of storage provided near perfect conditions for clostridial sporulation and germination was the prime cause of this outbreak. Refrigeration temperatures should be checked and recorded at least daily, and the widely held notion that heating will render any food safe must be dispelled.

Staphylococcus aureus

Presentation and course

The onset is usually about two to six hours after ingestion of food containing enterotoxin, and may be dramatically severe, particularly when a large number of people are involved in an outbreak. Abdominal pain, nausea and violent vomiting may cause rapid exhaustion, prostration and collapse. Diarrhoea may follow some hours later. Dehydration requiring intravenous fluid and electrolyte restoration is not uncommon. Recovery, however, generally follows quickly after the acute phase.

Epidemiology

Food is almost always contaminated through contact with a staphylococcal lesion on the skin of a food handler. The hand is most often the source, but other exposed parts such as the eye, ear and nose may be the infected sites. Some outbreaks in the United States have been attributed to nasal carriers without overt lesions, but these are thought to be very unusual as large concentrations of organisms, 10^6 per gram of food and 10^7 per millilitre of milk, are probably required to produce enough enterotoxin to cause symptoms.

Protein-rich foods provide particularly good conditions for the growth of organisms. Most frequently implicated are meats, sliced and processed, pies, cured hams and dairy products such as cream, mayonnaise, pastries and custards. Staphylococci can grow in the presence of high concentrations of salt and other food preservatives and thus in cured and pickled meats. Once inoculated, the bacteria multiply rapidly at ambient temperatures and within two to six hours, depending on the temperature and initial contaminating 'dose', sufficient toxin is produced to initiate illness. The toxin is moderately heat stable and mild reheating or cooking, that is, boiling for 30 minutes, may not render affected food safe.

The bovine strain of *Staph. aureus* phage type 42D may be found in raw milk from cows with mastitis and, before pasteurization was almost universally adopted, caused outbreaks from time to time. However, temperatures above 20–25°C are necessary for growth and toxin production to occur.

Illustrative outbreak

Seventy-two out of 123 people who attended a wedding reception fell ill with acute, persistent and severe vomiting between the time of the reception and when the evening's celebrations were due to begin. Many were temporarily prostrated, including the bride and groom, and a few elderly cases required hospital admission for parenteral treatment. The reception was held in a marquee on a very warm, dry day in mid-summer. The caterers prepared the food over the previous 24 hours and it was laid out in the tent at about 10 am. The, buffet, which began at 3 pm, consisted of meats, poultry, salads, mayonnaise, sweets with cream, salmon and pastries. Staphylococci of the same phage type were isolated from patients and from a number of the foods. Food analysis did not clearly implicate one food, but it was thought that the mayonnaise, which had been used on many items, may have been the major source. The catering establishment was found to be in breach of some basic food hygiene regulations, and one food handler had an eye infection from which the same phage type of staphylococcus was isolated. This person presumably transferred the organism from his eye to the food, which was subsequently kept at a high ambient temperature in the tent during

which time bacterial proliferation and toxin production occurred. The two major errors were permitting an employee with an obviously infected lesion to handle food, and failure to recognize the hazard of leaving foods at high ambient temperatures for prolonged periods.

Viral food-borne disease

Presentation and course

The incubation period is usually between one and two days, and the disease presents as a gastroenteritis of sudden onset often with precipitate violent vomiting. Resolution is rapid, generally within 24 to 48 hours. Complications are rare but secondary cases occur frequently.

Epidemiology

Many, possibly most, cases of viral gastroenteritis, whether sporadic or epidemic, are of a person to person type and are of unknown origin without any demonstrable link to food.

Food-borne outbreaks, which are far less common, are particularly associated with infected food handlers working while symptomatic or with seafoods that are normally eaten raw, for example, oysters, or lightly cooked, for example mussels. The molluscs become contaminated during cultivation in sewage-polluted waters, and depuration techniques, which can be relied upon to clear the shellfish of bacteria, are not yet shown to be reliable for viral clearance. Secondary faecal–oral spread following food-borne outbreaks is a characteristic feature. Food that is handled prior to consumption can be a vehicle of SRSV infection.

There is evidence from a number of outbreaks that excretion of the virus may continue in sufficient quantity to transmit the infection for some time after apparent clinical recovery, and convalescent catering staff should avoid handling raw or not to be reheated foods for 48 hours after symptoms have resolved.

The inability to confirm cases other than by electron microscopy, which requires very high concentrations of organisms in specimens, renders the study of outbreaks other than by epidemiological methods difficult. The viruses are no longer detectable in faecal specimens 24 to 48 hours after the onset of illness, and are not yet routinely detectable in foodstuffs. It is anticipated that modern technology such as the use of DNA techniques will be applied to these problems in the foreseeable future.

Illustrative outbreak

A reception was held at a well-known institution to provide an opportunity for the producers of unusual and exotic foods to display their wares to invited guests representing a wide range of interests in the food trade and its regulation. The following day a number of the guests reported sudden attacks of vomiting and diarrhoea with mild abdominal pain. Investigations revealed that about 40 of the 200 people attending had been unwell. Electron microscopy of stool specimens of eight of the cases revealed SRSV particles. Analysis of questionnaires completed by all those attending, whether ill or not, demonstrated a striking relationship between the eating of raw oysters and lightly cooked mussels, and the development of symptoms. The shellfish, which had been satisfactorily depurated of bacteria, had been grown in an estuary known to be heavily polluted with human sewage.

Bacillus cereus

Presentation and course

There are two clinical forms of the disease depending on which of two toxins is produced by the strain involved in the incident. One causes vomiting one to five hours after ingestion, and the other diarrhoea after 8 to 16 hours. Symptoms are generally short-lived and person to person spread does not occur.

Epidemiology

The organism is widely distributed in the environment and faecal carriage among the clinically normal population in one study was 14%. It is therefore not surprising that isolations have been made from many foods. Small numbers of organisms are clearly of no concern, and it is only under circumstances when spore germination can take place that sufficient organisms and toxins are produced to cause disease. The spores of different strains vary in their resistance to heat, but most will withstand boiling and quick frying.

Meat, vegetable and cereal dishes are most often implicated. The toxins are produced when contaminated food is kept warm over prolonged periods, but at temperatures insufficient to kill vegetative forms of the bacteria or to prevent the germination of the spore forms. Dishes where rice has been pre-cooked in quantity and then kept at ambient temperatures until being mildly reheated just before consumption, have been responsible for a number of outbreaks.

Cereal-based outbreaks are usually of the short-incubation vomiting type, while the diarrhoeal syndrome is more often associated with other foods, but this is by no means always the case.

Illustrative case

Three groups of people complained to an environmental health department in the course of one evening that they had felt unwell and had moderately severe vomiting some four hours after eating take-away Chinese meals from one particular outlet. Though the main dishes varied considerably, all had had portions of fried rice. The rice had been prepared the previous day by boiling and allowed to cool overnight. When customers ordered fried rice, portions were dipped in simmering deep fat for about a minute and then placed in containers. *Bacillus cereus* was isolated from the uncooked rice in small quantities, and from the boiled rice and fried rice in large numbers. As in the case of *Cl. perfringens*, the *B. cereus*, which can grow either aerobically or anaerobically, was able to sporulate and germinate as the slow cooling process passed through optimal temperatures for these processes to occur.

Escherichia coli

EPEC, EIEC and ETEC are particularly associated with infantile diarrhoea in developing countries, and with diarrhoea in travellers to countries with poor hygiene standards. Human carriers are the reservoir for the organisms, and sewage contamination of food and water is responsible for most cases. Outbreaks in developed countries caused by water pollution have been reported, but are rare events.

Infections associated with enterohaemorrhagic *E. coli* (EHEC, VTEC), most commonly *E. coli O157*, are generally sporadic, but clusters of cases and outbreaks are reported from time to time. Clinical presentation is of diarrhoea, commonly with heavily blood-stained stools and abdominal cramps. About 5% of cases subsequently develop the haemolytic uraemic syndrome, particularly young children, and renal failure may occur in a few patients.

The incubation period is around one to six days, and the very young are the most frequently and severely affected. Person to person spread is reported relatively frequently.

The incidence of VTEC appears to be increasing in both the United Kingdom and the United States. In Scotland in 1996 a major outbreak causing hundreds of cases resulted from cross-contamination of cooked meats in one butcher's shop. The resulting inquiry [5] led to measures to enforce the separation of cooked and raw meats in butcher's shops [6]. Direct contact with farm animals is also a risk factor for VTEC.

For a review of risk factors for and prevention of sporadic infections with EHEC, including VTEC, see [7–9].

Illustrative outbreak

Over a two-month period a cluster of more than 20 cases of *E. coli O157* were identified in one region, 16 of which were all of one phage type. The onset of most of the latter cases clustered over a two-week period. There was considerable geographical scatter, but eight cases were from a single town.

A food questionnaire was administered to all the cases and to 39 others in a case–control study. A strong association with a locally produced yoghurt became apparent. Inspection of the production unit showed a basically good hygiene practice, but there were areas where cross-contamination of the milk might have occurred after pasteurization,

although no direct microbiological evidence was obtained. Modifications were recommended and no further cases have occurred.

Clostridium botulinum

Presentation and course

A few cases may present with gastroenteritis. Most, however, show CNS involvement from the outset. Symptoms such as dizziness, double vision and inability to open the eyes fully may progress alarmingly quickly to rapidly developing paralysis. The incubation period is 12 to 36 hours, but may be longer. Mortality varies from 15% to over 90%, partly depending on the treatment available.

Epidemiology

The organism is extensively distributed in soil and mud and is therefore also found in animals that tend to forage in or around water, and fish. In societies where raw fish or seal meats are allowed to ferment, the disease is relatively more common because such conditions will encourage the growth and toxin production of *Cl. botulinum*. Elsewhere, home preservation by canning of many types of food, particularly vegetables, which are quite likely to be contaminated with spores, has been responsible for outbreaks. Boiling will not destroy the spores, though the use of a pressure cooker may enable lethal temperatures to be reached. The toxin is thermolabile and is destroyed by cooking. Thus it is only food not subject to further heating just before consumption that constitutes a risk.

Botulism is rare in the United Kingdom – only 52 cases having been reported since the first in 1922. Twenty-seven cases were from a single outbreak in 1989 in which commercially produced hazelnut yoghurt was the vehicle of transmission. Other incidents were associated with duck, rabbit, hare, pigeon, nut brawn, meat pie, macaroni cheese and fish.

Illustrative outbreak

Four patients were admitted in hospital in quick succession with CNS symptoms suggestive of botulism.

The diagnosis was rapidly confirmed by mouse inoculation. All had eaten tinned salmon from a single tin that had come from a canning factory in the northwest of Canada. *Cl. botulinum* was isolated from the can opener and from the tin, and was shown to produce type E toxin, the same as that in the serum of the patients, two of whom died. Investigations showed a small defect in the can sealing, which probably occurred at the factory, allowing access of the organism at a late stage in the processing.

Listeriosis

Presentation and course

As many cases present only a mild, flu-like illness, it is likely that very few are diagnosed clinically and that even fewer are confirmed bacteriologically. Severe disease is generally confined to the pregnant, the neonate and the immuno-compromised. Infection in pregnancy may result in abortion or in overwhelming sepsis with brain involvement in the foetus or newborn. Meningitis is the usual manifestation in the compromised. Since so many foods contain *Listeria monocytogenes*, the incubation period is difficult to estimate. Published figures vary from 1 to 70 days. Overall mortality is about 30% in the United Kingdom for severe disease.

Epidemiology

The disease, at least in its severe form, remains rare in the United Kingdom. Outbreaks presenting as clusters of abortions or neonatal infection in North America and Switzerland were traced to contaminated coleslaw, milk and soft or Mexican cheeses. In the United Kingdom, a nationwide outbreak occurred from imported pâté. *L. monocytogenes* is present in varying quantities in many foodstuffs, with a particularly high incidence in chicken, soft cheeses and certain processed meats such as pâtés. However, the factors determining whether clinical disease will result from ingesting contaminated food remain obstinately obscure.

Yersinia enterocolitica

While most cases are probably acquired by the faecal–oral route and are not food-borne, in Scandinavia and some continental and transatlantic countries, epidemics with patterns suggesting food origin occur regularly. Ground or raw pork has been identified as an important source in Belgium, and in the United States large outbreaks involving milk or milk products have been recorded.

Vibrio parahaemolyticus

Most cases seen in the United Kingdom have been acquired abroad, particularly in Southeast Asia and Japan. Cross-contamination between cooked and uncooked seafood or the use of sea water rinse have been responsible for a number of outbreaks. Imported, frozen, cooked shellfish, particularly prawns, are another fairly common source of infection. Sporadic cases of presumed local origin are seen occasionally. Other members of the *Vibrio* group of organisms, *Aeromonas* spp. and *Plesiomonas* spp., are also linked to seafood and water.

Shigellosis

The *Shigella* organisms are classically transmitted person to person by the faecal–oral route, and epidemics are associated with institutions where personal hygiene may be unreliable. In the past, outbreaks in which contaminated foods of many kinds were implicated were commonplace. Nowadays, food-borne shigellosis is very rare in the United Kingdom. However, see 'Shigella Outbreak in a School Associated with Eating Canteen Food and Person-to-person Spread' (1998, *Communicable Disease and Public Health*, 1 (4) December).

THE PRINCIPLES OF INVESTIGATION AND CONTROL OF FOOD-BORNE DISEASES

Detailed consideration of the investigation, analysis and control of outbreaks of communicable diseases in general is presented in Chapter 14.

The following section provides a brief account of the application of the principles and practices described therein to food-borne disease.

Outbreaks may come to the attention of health authorities through a variety of sources. Notification as required by law may be received from one or more general practitioners or from hospital control of infection officers. Patients may present at accident and emergency departments. Complaints may be made directly to environmental health departments. Most outbreaks of food-borne disease declare themselves by virtue of the numbers involved over a short period of time among a well-defined group of people, for example, coach party, wedding reception or works outing. However, in circumstances where the victims may be widely dispersed, as in an outbreak caused by *S. ealing* in baby milk powder, where symptoms other than gastroenteritis prevail (botulism and listeriosis), or where other forms of transmission (person to person) might have occurred, only careful investigation will provide the basis for control.

The purpose of investigation is to stop any further spread of the outbreak as quickly as possible, and to provide information to prevent recurrences. To achieve this, the organism responsible must be identified, the food concerned defined and the means whereby contamination took place discovered. The steps to be taken are: first, the collection of data; second, the analysis of that data; and third, the implementation of evidence-based control measures.

Recommendations on the investigation and control of food poisoning are issued by the Department of Health and the Welsh Office [10]. Similar advice is given by the relevant departments in Scotland and Northern Ireland.

Investigations

General

A quick history taken from a number of cases may reveal one or more possible sources common to them all. Investigations designed around a theory of the cause of an outbreak are often more fruitful than a wide approach with no clear target in mind. Unfortunately, this is not always possible.

Laboratory

Laboratory studies to determine the microbiological cause of the outbreak should be initiated at the same time as the epidemiological data are collected. Early identification of the organism responsible will influence the depth and extent of the investigation required.

Faecal samples

These should be obtained from the clinically ill, others exposed to the same possible sources and, if relevant, any food handlers involved. Suspect food should be submitted as soon as possible with proper documentation of its source, handling and storage.

Inspections

Where premises from which suspect food may have come are identified, inspection for the state of hygiene of the kitchens and for the food-handling practices must be carried out. Environmental swabs may be taken to determine the extent and nature of any contamination of both fixed and movable fittings.

Epidemiology

In any outbreak in which there is uncertainty about the source and/or the pattern of transmission, further investigations will need to be carried out.

The purpose is to define:

- the timing of the outbreak
- the place or geographical situation of people affected at or around the time they were infected
- certain personal characteristics of those involved.

This information should be obtained through questionnaires designed according to the principles given in Chapter 14.

Timing

By plotting the dates and times of the onset of symptoms against the number of cases, an epidemic curve of the outbreak can be constructed. The shape of such a curve will provide information about the type of outbreak. A curve with a single high peak where nearly all cases fall within the range of a single incubation period is characteristic of a point source outbreak, for example, a company dinner party, when all cases are infected within a short time of each other (Fig. 13.1). A broad, lower curve suggests person to person spread or an ongoing source, for example, from a dried or frozen food which is stored for varying times before it is used (Fig. 13.2).

Place

Details of the geographical locations of individuals at or around the times of the onset of symptoms may, in circumstances where there is no obvious occasion linking them, be useful. Apparently randomly distributed cases may be related by such things as delivery rounds, supplies of unusual foods, water sources, movement of personnel or the attendance at particular institutions or functions.

Persons

Personal characteristics that may be relevant include age, sex, occupation, travel, medical history, eating habits and relationships to other sufferers. The extent of the food histories required will depend upon whether a particular occasion can, with reasonable confidence, be identified. Details of food eaten and not eaten from a defined menu can often be obtained without difficulty. It may be necessary, however, to ask patients to recall food eaten three or more days before becoming ill.

Analysis

On the basis of the information obtained it may be reasonably clear what the likely food source was. It is, however, only too easy to arrive at doubtful or even incorrect conclusions unless great care is taken in the assessment of the data. For example, asymptomatic food handlers are often found to be excreting the same organism as the victims of a food poisoning episode, and come under suspicion as the sources of the organism, but they rarely, if

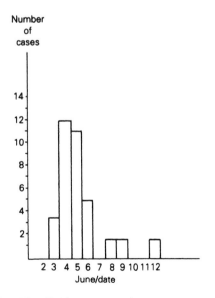

Fig. 13.1 Epidemic curve from a point source hospital outbreak caused by *Salmonella* bacteria in a chicken dish. Note the small number of secondary cases. (Data supplied by CDSC.)

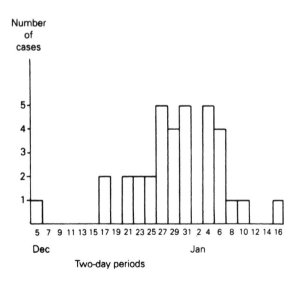

Fig. 13.2 Epidemic curve from an extended common food source caused by commercially produced and widely distributed spiced sausage. (Data supplied by CDSC.)

ever, are. Food handlers handle, and not infrequently nibble at, the food they are dealing with and thus become victims themselves.

The two analytical epidemiological methods used in the investigation of outbreaks are the case–control study and the cohort study. Both essentially compare data from infected people with that on uninfected people (controls). The principles and applications of the methods are given in Chapter 15.

Control

Personnel (also see p. 524)

Patients with skin lesions or in the acute diarrhoeal phase of gastroenteritis will be excreting large numbers of organisms and must be regarded as potentially infectious. Secondary spread is, however, unlikely with organisms other than SRSVs, *Salmonella* and, to a lesser extent, *Campylobacter*.

Patients involved in catering and food handling should be excluded from work during this phase. Patients with good personal hygiene need not be restricted after clinical recovery, and will only require clearance at follow-up if involved in commercial food-handling activities. Details of recommended exclusions for all food-borne infections, which should be applied to patients, their contacts and food handlers, are given in a PHLS document [1] and a DoH guidance note [4].

Food and food hygiene

Once suspect foods have been identified, the most appropriate action will depend upon the way in which contamination has taken place. Some foods, particularly red and white meats in the raw

Table 13.5 Factors contributing to 1479 outbreaks of food poisoning, England and Wales, 1970–82

Contributing factors	Number of outbreaks in which factors recorded (%)					
	Salmonella	C. perfringens	Staph. aureus	B. cereus	Other	Total
(i) Preparation too far in advance	240 (42)	464 (88)	80 (48)	54 (86)	6 (4)	844 (57)
(ii) Storage at ambient temperature	172 (30)	276 (53)	75 (45)	39 (62)	4 (3)	566 (38)
(iii) Inadequate cooling	125 (22)	313 (60)	12 (7)	17 (27)	1 (<1)	468 (32)
(iv) Inadequate reheating	76 (13)	275 (52)	5 (3)	33 (52)	2 (1)	391 (26)
(v) Contaminated processed food	100 (18)	19 (4)	27 (16)	4 (6)	86 (54)	236 (16)
(vi) Undercooking	139 (25)	74 (14)	2 (1)	1 (2)	7 (4)	223 (15)
(vii) Contaminated canned food	2 (<1)	4 (<1)	42 (25)	1 (2)	55 (35)	104 (7)
(viii) Inadequate thawing	61 (11)	34 (6)				95 (6)
(ix) Cross-contamination	84 (15)	8 (2)	2 (1)			94 (6)
(x) Raw food consumed	84 (15)		1 (< 1)		8 (5)	93 (6)
(xi) Improper warm holding	15 (3)	52 (10)		8 (13)	2 (1)	77 (5)
(xii) Infected food handlers	13 (2)		50 (30)		2 (1)	65 (4)
(xiii) Use of leftovers	25 (4)	25 (5)	11 (7)	1 (2)		62 (4)
(xiv) Extra large quantities prepared	29 (5)	17 (3)	2 (1)			48 (3)
Total	566	525	166	63	159	1479

Source: [11]; reproduced with the permission of the publishers and the author.

Note: The total at the foot of the table represents the total number of outbreaks for each type of organism being examined; some outbreaks have more than one cause. The percentages given are for each causal factor in the total number of outbreaks due to each organism.

state, must be assumed to harbour pathogens. Where milk or tinned foods are suspect, failures in pasteurization or manufacturing may require investigation. Food that is adequately cooked and, if not eaten immediately, stored correctly, that is, cooked and raw foods kept separately at appropriate temperatures, is safe. Breaches in food hygiene standards may take many forms (Table 13.5) [11]. The Food Standards Agency in the United Kingdom monitors food safety and may decide to issue food hazard warnings and initiate food withdrawals.

Veterinary aspects

Meat, poultry and other foods derived from animals may be contaminated with organisms known to cause food poisoning in humans. Frequently, those organisms may neither cause any illness in the animal nor interfere with growth or weight gain. Thus those involved in the husbandry of domestic animals are more often than not unaware that any particular animal is carrying a pathogen. Moreover, so extensive is the contamination of the environment with both human and animal excreta that many vehicles, from flies to feed, are available whereby organisms can be transferred to and between animals. During the lairage, slaughter and processing of cattle and poultry, further opportunities for cross-contamination arise. Control at this level is clearly difficult and is largely the province of DEFRA and the National Meat Hygiene Service.

CONCLUSION

The global increase in food-borne infection, particularly salmonellosis, and the emergence of new food-borne diseases, for example, *E. coli O157* and Noroviruses, has made food safety an important public health and political problem. Control of food-borne infection requires a full understanding of the microbiology, epidemiology, veterinary and food industry factors. Control measures need to address the whole food chain, from the contamination of food at source to cross-contamination at lairage and abattoir, hazard analysis and critical

control points (HACCP) approaches to food processing and awareness of consumers. The role of environmental health officers is crucial in this area, both in identifying the causes of food-borne infection and in implementing control measures.

REFERENCES

1. Public Health Laboratory Service *Salmonella* Committee Working Party (1995) The prevention of human transmission of infections, infestations and bacterial intoxications. *Communicable Disease Report Review*, 5, R158–172.
2. Old, D.C. and Threfall, E.G. (1998) *Salmonellae*, in *Topley and Wilson's Microbiology and Microbial Infections*, Volume 2, 9th edn (eds L. Collier, A. Balaw and M. Sussman), Oxford University Press, Oxford.
3. Chief Medical Officer (1988) *Advice on Raw Eggs Consumption*, Press Release 88/445.5, December, Department of Health, London.
4. Department of Health (1995) *Food Handlers Fitness to Work. Guidance for Food Businesses, Enforcement Officers and Health Professionals. Prepared by an Expert Working Party convened by the Department of Health*, DoH, London.
5. The Pennington Group (1997) *Report on the Circumstances Leading to the 1996 Outbreak of Infection with* E. coli O157 *in Central Scotland: The Implications for Food Safety and the Lessons to be Learned*, Stationery Office, Edinburgh.
6. Ministry of Agriculture, Fisheries and Food (1997) Government's response to the Pennington final report. *MAFF Food Information Bulletin*, No. 84, May.
7. Anon (date unavailable) EHEC/VTEC. *Lancet*, 351(9108).
8. Parry S. and Palmer S (2002) E. coli *Environmental Health Issues of VTEC 0157*, Spon Press, London and New York.
9. Anon (1997) *World Health Organization Workshop*, the Stationery Office, London.

10. Department of Health (1994) *Management of Outbreaks of Foodborne Illness. Guidance Produced by a Department of Health Working Group*, HMSO, London.

11. Roberts, D. (1993) Factors contributing to outbreaks of food poisoning, in *Food Poisoning and Food Hygiene*, 6th edn (eds C.B. Hobb and D. Roberts), Edward Arnold, London.

FURTHER READING

Advisory Committee on the Microbiological Safety of Food (2001) *2nd Report on Salmonella in Eggs*, The Stationery Office, London.

Advisory Committee on the Microbiological Safety of Food (1999) *Report on Microbial Antibiotic Resistance In Relation to Food Safety*, The Stationery Office, London.

Advisory Committee on the Microbiological Safety of Food (1998) *Report on Foodborne Viral Infections*, The Stationery Office, London.

Advisory Committee on the Microbiological Safety of Food (1995) *Veromytotoxin Producing* Escherichia Coli, HMSO, London.

Department of Health and Social Security (1986) *Report of the Committee of Inquiry into an Outbreak of Food Poisoning at Stanley Royd Hospital*, HMSO, London.

Palmer, S.R., Lord Soulsby and Simpson, D.I.H. (1998) *Zoonoses*, Oxford University Press, Oxford.

Wall, P.G., de Louvois, J., Gilbert, R.J. and Rowe, B. (1996) Food poisoning notifications, laboratory reports and outbreaks – where do the statistics come from and what do they mean? *Communicable Disease Review*, 6, R93–100.

14 Communicable disease control

Stephen R. Palmer

INTRODUCTION

Improvements in the sanitation, housing, nutrition and other social conditions of the British population over the past century have resulted in a dramatic decline in such epidemic diseases as cholera, typhoid, typhus, dysentery, tuberculosis and diphtheria. Medical advances in antibiotic therapy and vaccine production after the Second World War brought communicable disease under control still further. Consequently, a body of opinion gained ground, which saw communicable disease as a rapidly diminishing area of concern.

In the past 30 years or so, however, communicable diseases have re-emerged as a major public health issue [1]. The growth in **international travel** has brought a significant problem from imported infections, such as typhoid and malaria, and the fear that newly discovered infections, such as Lassa fever, may become epidemic in the United Kingdom has led to the development of national policies for control. Discoveries of new diseases such as Legionnaires' disease have revealed new hazards from **modern living conditions**. Changes in the **food industry**, with mass production and national and international distribution of foods, have provided the opportunity for widespread outbreaks of food-borne disease. Diseases such as hepatitis B and Acquired Immune Deficiency Syndrome (AIDS) have highlighted the influence of **personal behaviour and lifestyles** on the risk of infection. The AIDS pandemic has ensured that communicable diseases now have a very high priority in public health.

More recently, there has been concern over **global warming** and predictions that the range of vector-borne diseases such as dengue and malaria will expand. **War** and **civil unrest** remain major causes of epidemic disease, and recent conflicts have seen the re-emergence of polio among the Kurds and in Chechnya, typhoid in Bosnia and cholera in Rwanda.

Many of the emerging infections of global concern are zoonoses (diseases transmitted from animals to humans [2]), and changes in **animal husbandry** and **wildlife distribution** can have major effects on communicable disease epidemiology. For example, new variant Creutzfeldt-Jakob Disease (CJD) has emerged following the Bovine Spongiform Encephalopathy (BSE) epidemic in cattle; Lyme disease is epidemic in parts of the United States because of the increasing number of deer living in proximity to human dwellings. Further, there are implications of infectious agents in chronic diseases previously attributed to lifestyle factors, for example, heart disease and stomach ulcers (H. pylori), various cancers, etc.

The SARS outbreak in South East Asia and Canada is the most recent in a spate of newly recognized infections that provide new challenges for public health services throughout the world. Increased public awareness and expectations have ensured a high profile for infectious diseases in the political arena.

As the disease pattern has changed, so the need has grown for public health professionals to re-establish public health infrastructures and re-examine the methods of control to keep them appropriate to the new circumstances. Modern public health now has to focus much more on national and international surveillance to identify and assess communicable disease problems and, increasingly, epidemiological methods are being applied to their investigation and control.

A communicable disease is the result of a complex interaction among an infecting agent, a host and environmental factors. Consequently, successful investigation of infectious disease incidents and their control depends upon understanding not only the microbiology of the organism, but also the environmental and host factors that result in exposure to the agent and the development of disease. The study of the interaction of these factors in a population is the basis of epidemiology, which may be defined as 'the study of the distribution and determinants of disease in populations and its application to control'.

Accurate laboratory diagnosis is usually a major factor in successful control, but epidemiological methods alone may be sufficient to introduce interim control measures. For example, the means by which AIDS spread were identified by epidemiological methods two years before the infecting agent, Human Immunodeficiency Virus (HIV), was discovered.

ORGANISMS THAT CAUSE DISEASE

Agents vary in their ability to cause disease (pathogenicity) and different strains of the same agent may cause more or less severe disease (virulence). Some agents cause disease only in certain animals. *Salmonella enterica serovar typhi* is a pathogen of humans, whereas other agents can cause disease in several species, for example, *Salmonella enterica serovar enteritidis*. The following is a classification of the infective agents.

Viruses

These consist of a single nucleic acid, RNA or DNA, surrounded by a protein envelope. They measure less than 300 nm, and cannot multiply outside of living cells.

Changes may take place in the molecular structure of the DNA or RNA during passage of the virus through a host to another, allowing it to bypass the immunity that has been acquired by past exposure to the virus. A slow alteration in the make-up of the virus is called **antigenic drift**, and a sudden change is called **antigenic shift**. Hosts previously exposed to new strains have little or no acquired immunity, and large-scale epidemics can result.

Viruses can be identified by the following methods.

- **Electron microscopy (EM)** This is used for pox viruses, for example, chickenpox and orf, and for gastroenteritis viruses, for example, rotavirus and Norovirus (formerly SRSV and Norwalk-like viruses).
- **Culture** A virus may be grown from body fluids, skin, throat swabs, etc. on culture media such as cell cultures, in embryonated eggs, and in suckling mice. Most viruses can be cultured, although highly specialized techniques may be required and the yield is usually not good. Dangerous viruses should only be cultured in high containment reference laboratories.
- **Serology** Virus particles (antigens) may be detected by mixing the sample of blood or faeces (e.g. hepatitis B surface antigen in blood and the rotavirus in faeces) with antibodies against the virus. Alternatively, specific antibodies against the virus can be detected. Following infection, the patient produces antibodies in the blood. Two blood samples are taken, the first as soon as possible after onset of illness, and the second after about two to three weeks. An increase in antibody level may be detected by a variety of techniques, including complement fixation tests, immunofluorescence, radioimmunoassay and enzyme-linked immunosorbent assays. These serological methods can be applied to population surveys for

epidemiological purposes. Newer techniques are being developed in which saliva and urine are used instead of blood.

Bacteria

These are unicellular organisms without nuclei that are classified by shape, for example, spherical cocci, cylindrical bacilli, helical spirochaetes; by their ability to stain with different chemicals, for example, Gram's staining: Gram-positive or Gram-negative bacteria; by culture characteristic, for example, their dependence on O_2 for growth and their appearance on culture plates; by biochemical reactions (particularly their ability to ferment different sugars); and by antigenic structure. Some bacteria produce spores that resist heat and humidity, allowing the organism to survive for prolonged periods in the environment (e.g. *Bacillus anthracis*). Other organisms are delicate and survive for only a very short time outside the body, for example, gonococcus.

The differences between bacteria in growth requirements are used to selectively culture pathogens for diagnostic purposes. All bacteria require water for growth, but they differ in their oxygen requirements. Some organisms, for example, *Campylobacter*, require additional carbon dioxide. Most bacterial pathogens prefer a temperature of about 37°C for growth, but some will grow at refrigeration temperatures, for example, *Yersinia* and *Listeria*, and this can be used to culture them selectively (cold enrichment). The differences in the nutritional requirements of bacteria are used to make culture media, which suppress some organisms and encourage the growth of others.

Bacteria contain a single chromosome but may have additional genetic material within the cell, for example, plasmids. Plasmids may contain genetic information coding for resistance to antibiotics. The pattern of plasmids within a bacteria can be used to type organisms. Mutation of genetic material takes place and strains with altered antibiotic resistance may emerge. These may be distinguished by their resistance pattern to a range of antibiotics.

Bacteria cause disease by two means: invasion of tissues and production of toxins. Toxins may be liberated outside the bacteria (exotoxins), which can then circulate via the bloodstream to cause tissue damage away from the site of infection (e.g. diphtheria). These toxins may be pre-formed in food (e.g. botulism). Toxins that are part of the structure of the bacteria (endotoxins) cause damage at the site of the infection or, when the cell dies, they can circulate around the body. Detection of toxins in blood or faeces is used to confirm certain infections, for example, botulism.

Pathogenic bacteria may be detected by the following means.

- **Microscopy**   Light microscopy of body fluids, faeces, etc. using staining techniques.
- **Culture**   Bacteria may be cultured from tissues or the environment. The material is put onto plates of culture media containing nutrients, for example, meat extract, blood or agar, and incubated at 37°C, usually for 24–48 hours.
- **Serology**   As with viruses, antibody detection can sometimes be used to confirm infection, for example, Legionnaires' disease, but this is not useful for most enteric infections.

Chlamydias, Coxiellae and *Rickettsiae*

These organisms, like viruses, will only multiply in living cells but are considered to be bacteria. They cause a variety of diseases. *Chlamydias* cause psittacosis and *Coxiellae* cause Q fever. *Rickettsiae* cause a wide range of serious illnesses, such as typhus and Rocky Mountain spotted fever, which are very rare in the United Kingdom.

Yeasts and fungi

These are forms of plant life that obtain energy by parasitism. They may live in the environment, or they may be normal inhabitants of animals and humans, in which case they cause disease only when the host's defences are depleted, as in AIDS patients. Environmental fungi may be widespread, such as *Aspergillus*, which can cause lung disease when inhaled by immunosuppressed patients or have a limited geographical distribution and cause disease in unusual circumstances, like *Histoplasma*.

Dermatophytoses, such as athlete's foot, and nail infections are transmitted by direct and indirect person to person contact. Zoonotic fungi such as ringworm are transmitted from animals to humans by direct contact. Other fungi are not communicable from person to person. Diagnosis is by microscopy of lesions and sometimes culture.

Protozoa

These are single-celled nucleated organisms that may have complex life-cycles involving sexual and asexual reproduction. Examples include organisms of the genera *Plasmodium* (which cause malaria), *Toxoplasma*, *Cryptosporidium* and *Giardia*. The latter three occur commonly in the United Kingdom.

Helminths

Tapeworms (cestodes), flukes (trematodes) and roundworms (nematodes) are all helminths (worms). They have complex life-cycles that must be understood before the diseases they cause can be controlled. Most are rare causes of disease in the United Kingdom.

THE HOST

Host defences

The body has a general resistance to the invasion and multiplication of organisms. The skin and mucous membranes are natural barriers to infection, while allowing organisms to live as commensals in their surface without causing disease. The acidity of the stomach kills most organisms that are ingested. If an organism does penetrate these barriers, circulating cells called macrophages may attack and kill them.

In addition to this general protection, the immune system provides more specific defences. Foreign material such as the surface of an infecting organism is recognized by the immune system as 'foreign' (antigen). As a result, proteins called

antibodies are produced by the cells of the immune system, which bind with the antigen to inactivate it and bring about its destruction. The first time the immune system meets a particular antigen, the response may be relatively inefficient. If the immune response overcomes the infection, bringing about recovery from the infection, the immune system remembers the encounter and the next occasion the antigen is encountered a more rapid response is mounted, which usually prevents the disease from developing at all. This explains why second bouts of measles or chickenpox do not occur. However, if the organism changes (like influenza) through antigenic drift (see earlier) it can evade the immune response and cause another episode of illness.

Host factors and disease

Certain factors may reduce immunity and place a person at greater risk of developing an infection. The elderly are at greater risk because of declining natural resistance and waning immunity; the very young are also at increased risk because of the immaturity of their immune systems. Poor nutrition also increases susceptibility, leading to, for example, a high mortality from measles in developing countries and a high tuberculosis rate in alcoholics and vagrants. Natural barriers may be compromised. Thus smokers are at greater risk of respiratory infections including Legionnaires' disease; those taking antacid medication are at greater risk from gastrotintestinal pathogens.

THE ENVIRONMENT

Particular occupations may place workers at increased risk from certain diseases (e.g. psittacosis in poultry processors). Poor housing with overcrowding and lack of hygienic facilities increases the risk of disease such as tuberculosis and dysentery. Climatic conditions also influence the incidence of disease. Food poisoning is commoner in the summer, partly because many organisms multiply faster in food that is at higher ambient temperatures. Respiratory infection is commoner in the winter, probably because of colder temperatures and more time spent indoors with poorer ventilation. Air pollution and smoking may increase susceptibility to respiratory infections. Disruption of populations because of war, famine or migration results in epidemics from poor sanitation, contamination of water supplies, increase in vermin and lack of personal hygiene.

BASIC CONCEPTS IN INFECTIOUS DISEASE EPIDEMIOLOGY

Reservoir of infection

This is where the agent normally lives and multiplies and what it mainly depends on for survival. This may be humans, for example, chickenpox; animals, for example, brucellosis; or the environment, for example, tetanus. It is not necessarily the same as the source of infection in a particular incident.

Source of infections

Infection may arise from the organisms normally living in a person, or those from another human being, an animal (zoonoses) or the environment. The source of an infection may sometimes be different from its reservoir. For example, in an outbreak of listeriosis in Canada in 1981, the reservoir of infection was a flock of sheep, from which manure was used as fertilizer on a cabbage field. Contaminated cabbages from the field were used to make coleslaw, which became the source of infection for humans. When the source of infection is inanimate, for example, food, water or fomites, it is termed the **vehicle** of infection.

Methods of spread

The routes by which an infectious agent passes from source to host can be classified as follows.

1. Food-, drink- or water-borne infection (e.g. typhoid and cholera). The term 'food poisoning' is often used for incidents of acute disease in which the agent has multiplied in the food vehicle before ingestion (e.g. *Salmonella* food poisoning), and where it may have formed toxins, for example, botulism. Other agents such as viral gastroenteritis agents may be carried on the food but do not multiply in it. This subject is dealt with fully in Chapter 13.

2. Direct or indirect contact. This includes spread from cases or carriers, animals or the environment to other people, who are 'contacts'. (A carrier is someone who is excreting the organism but who is not ill.) Within this category possible routes include:

   (a) faeces to hand to mouth spread (e.g. shigellosis)
   (b) sexual transmission (e.g. syphilis)
   (c) skin contact (e.g. wound infection and cutaneous anthrax).

3. Percutaneous infection. This includes:

   (a) insect-borne transmission via the bite of an infected insect, either directly from saliva (e.g. malaria), or indirectly from insect faeces contaminating the bite wound (e.g. typhus);
   (b) inoculation of contaminated blood or a blood product, either by transfusion, by sharing intravenous needles or by contaminated tattoo or acupuncture needles (e.g. hepatitis B);
   (c) the agent passing directly through intact skin (e.g. schistosomiasis) or through broken skin (e.g. leptospirosis).

4. Airborne infectious organisms may be inhaled as:

   (a) droplets and droplet nuclei (e.g. tuberculosis, SARS);
   (b) aerosols (e.g. Legionnaires' disease);
   (c) dust (e.g. ornithosis) (Table 14.1).

5. Mother to foetus. Organisms may pass from the mother across the placenta to the foetus before birth (e.g. rubella), or via blood at the time of birth (e.g. hepatitis B).

Occurrence

An infection that is always present in a population is said to be **endemic**. An increase in incidence

Table 14.1 Childhood infections transmitted by close contact with saliva, respiratory secretions or airborne droplets

Disease	Agent	Clinical features	Incubation period	Period of communicability	Control
Chickenpox	Virus	Fever, rash	Usually 13–17 days	5 days before rash to 5 days after first crop of vesicles	Specific immunoglobulin to high-risk contacts; antiviral agents
Measles	Virus	Fever, rash, cough	Usually 10–14 days	Onset to 7 days after rash	Routine vaccination of all children; specific immunoglobulin to contacts in some circumstances
Mumps	Virus	Fever, parotitis	Usually 18–21 days	1 week before to 10 days after parotitis	Routine vaccination of all children
Rubella	Virus	Fever, rash	Usually 17–18 days	1 week before to 4 days after rash	Routine vaccination of all children, and some women of childbearing age
Scarlet fever	*Streptococcus*	Fever, rash, pharyngitis, tonsillitis	1–3 days	Several weeks if not treated	Penicillin to cases and possibly to contacts
Whooping cough	*Bordetella pertussis*	Fever, paroxysmal cough	Usually 10–14 days	3 weeks from onset	Routine vaccination of all infants; antibiotic treatment of cases and possibly contacts

above the endemic level is described as an **epidemic**, or **pandemic** when the epidemic is worldwide. Cases may be **sporadic** when they are not known to be linked to other cases, or clustered in **outbreaks** when two or more linked cases or infections occur, suggesting that there was a common source or there has been spread from person to person. Two commonly used measures of occurrence of disease or infection are the **incidence rate**, the number of new cases occurring in a defined population over a specific time period expressed as a proportion of the total population, for example, 10 cases per 100 000 people per year; and **prevalence**, the proportion of a defined population with the disease at a point in time.

In infectious diseases propagated from person to person, for example, measles, an epidemic occurs only when a sufficiently large proportion of the population is susceptible to infection. The resistance of a population to the epidemic, because a sufficient proportion of the population is immune, is called **herd immunity**.

The **attack rate** during an outbreak is the proportion of the population at risk who were ill during the period of the outbreak. The **secondary attack rate** is the attack rate in the contacts of primary cases due to person to person spread.

Incubation period

This is the time from infection to the onset of symptoms. For each organism there is a characteristic range within which **the infecting dose** and the **portal of entry**, as well as other **host factors** (e.g. age and other illness) give rise to individual variability. For example, in rabies the period is shorter the closer the bite wound is to the head. The virus travels up the nerves to the brain and has less far to go the closer the bite is to the head.

Communicability

The infectious agent may be present in the host and passed to others over a long period of time. This is known as **the period of communicability**. Some infections can be passed on even when the host is well. These people are then known as temporary or chronic **carriers**, for example, typhoid carriers. In some diseases, transmission from person to person occurs before symptoms develop. For example, people with hepatitis A are most infectious to others just before they become ill.

Variables

Epidemiology involves measuring attributes or factors that vary in character or quantity. Some variables are fixed, that is, they are either present or absent (e.g. sex, occupation and nationality); or they may be discrete (e.g. the number of people in a household); or they may be continuous (e.g. age, height and weight). Analysis of the distribution of fixed variables in a population will usually be by calculating the proportion of people who fall within certain categories, or the rates of occurrence of disease within subgroups of the population (e.g. death rates by residence or occupational group). Analysis of continuous variables is more complicated since values obtained from a population will lie along a range, and these values are usually summarized by an average.

DETECTING PROBLEMS

The process of detecting trends in the occurrence of disease and infection in a population and reporting information to those responsible for public health action is called **epidemiological** or **population surveillance**. Langmuir defines it as: 'the continued watchfulness over the distribution and trends of incidence through the systematic collection, consolidation and evaluation of morbidity and mortality reports and other relevant data' [3]. Epidemiological surveillance has become increasingly important in identifying outbreaks due to nationally and internationally distributed contaminated foodstuffs, and it may be the only way to detect outbreaks when the victims have travelled during the incubation period to many different destinations.

The stages of surveillance are:

1. systematic collection of data;
2. analysis of the data to produce statistics;
3. interpretation of the statistics to provide information;
4. distribution of this information to all those who require it so that action can be taken;
5. continuing surveillance to evaluate the action.

Data may be collected especially for surveillance purposes (**active systems**) or use may be made of routine data (**passive systems**). Most active data-collecting systems are based on a carefully designed standard case definition, such as the clinical reporting system set up in 1982 to monitor the AIDS epidemic. An internationally agreed case definition was essential if data from different countries were to be compared.

Passive data collection systems are usually based on a microbiological or clinical diagnosis that is not precisely defined, and this may lead to problems of interpretation. For example, for the notifiable diseases a doctor only has to suspect the diagnosis in order to report a case. If all these were followed up, not all would be true cases.

Nevertheless, such data are invaluable for monitoring trends and for detecting episodes or cases for further investigation.

The main sources of surveillance data for communicable diseases in the United Kingdom are outlined in the following. Possible weaknesses in the accuracy and completeness of the data should always be borne in mind.

Death certification and registration

Every week, copies of death entries in the local death register for the preceding week are sent to the Office of National Statistics (ONS) by all registrars in England and Wales. The underlying cause of death is coded in accordance with World Health Organization (WHO) manuals, and statistics are published weekly, monthly, quarterly and annually in varying detail.

Death certification and registration is virtually 100% complete, but errors in the data can occur at any stage from diagnosis through certification and coding, to processing and analysis. The death entry is a public document and this may sometimes deter the doctor from entering the correct diagnosis (e.g. in cases of syphilis and AIDS), although it is possible for the doctor to provide further information about the death that is not entered on the public record.

The present system depends upon identifying a single cause of death for analysis. This may be unrealistic, particularly in the elderly, and limits the usefulness of published statistics.

Infection that contributed to death but was not considered to be the underlying cause of death (e.g. pneumonia complicating chronic bronchitis) will not be coded under the present routine system, and much important data on infectious diseases therefore does not appear in published statistics. Furthermore, most infectious diseases in England and Wales do not result in death, so that in these cases mortality data are not useful in monitoring trends.

Mortality data are analyzed weekly and published within seven days of collection so that they can be used to identify increases in mortality rate quickly at the beginning of, for example, influenza epidemics.

Statutory notifications of infectious disease

In England and Wales, the clinician making or suspecting the diagnosis is required to notify the proper officer appointed by the local authority for the control of infectious disease who, in turn, sends a weekly return, and these are reported weekly by the Health Protection Agency's (HPA) Communicable Disease Surveillance Centre (CDSC) on behalf of the ONS. These data are corrected quarterly and the analyses are published quarterly and annually. Similar systems operate in Scotland and Northern Ireland. The data are available quickly and are related to defined populations, so that rates by age and sex can be calculated. For some diseases that are not often confirmed in the laboratory (e.g. measles and whooping cough), notifications provide an invaluable means of monitoring trends and are available over many decades. However, the clinical diagnosis may not always be correct, and most infections are considerably undernotified. (See also Chapter 12.)

Laboratory reporting and mircobiological data

Laboratory reporting of infections forms the core of communicable disease surveillance in the United Kingdom. Medical microbiologists report specified infections each week to the directors of the HPA's CDSC and the Scottish Centre for Infection and Environmental Health (SCIEH) via electronic reporting systems. These data are analysed, and within a week of receipt the resultant information is published in the weekly *Communicable Disease Report* (CDR) and *SCIEH Report*. Information is also made available via electronic systems such as Epi-Net.

The data are limited to infections in which there is a suitable laboratory test; those that are easy to diagnose clinically are poorly covered. Not all microbiology laboratories report. However, the laboratory-based data have proved invaluable in national and international surveillance of communicable diseases.

General practice reporting of clinical data

The Royal College of General Practitioners (RCGP) set up a clinical data collecting system in 1966 in a small number of volunteer practices. The data are now published by the HPA CDSC. Similar systems exist in Wales and Scotland, and district-based systems have also been established. The data cover diseases not usually needing hospital admissions or laboratory investigations, and can be related to a defined population. The GP surveillance data have been especially useful in influenza surveillance.

INVESTIGATING PROBLEMS

There are three complementary approaches: the epidemiological, microbiological and environmental measurement and inspection methods. In an outbreak of food poisoning, for example, the **microbiological approach** relies upon culturing the causative organism from food sources. The **environmental approach** would be to document how the food was prepared, identify faults in kitchen practices and measure cooking and refrigeration temperatures. The additional need for the **epidemiological approach** is not always appreciated but can be shown as follows.

In an outbreak of *Salmonella* food poisoning, the reservoir of infection may be commercially reared chickens, but utensils and surfaces may have been cross-contaminated and bacteria transferred to other foods that, when eaten, become the source of infection. Isolation of the causative organism from food, or surfaces, cannot alone distinguish the order of events leading to the outbreak. Very often an outbreak investigation begins after all foods served have been consumed or discarded. Foods that remain may have been contaminated after the event. Sometimes the significance of the isolation of organisms from food is unknown, for example, *Listeria* from sandwiches bought in a shop, and only by showing an association between being ill and eating the food can the risk be clarified. The epidemiological approach looks for evidence of association between eating the food and illness.

Stages in the investigation of an outbreak

Preliminary enquiry

The purpose of the preliminary enquiry is to confirm that the outbreak is genuine; confirm the diagnosis; agree on a case definition; formulate ideas about the source and spread of the disease; start immediate control measures if necessary; and decide on the management of the incident.

Confirming the outbreak

An increase in the reported number of cases of a disease may be due to misinterpretation of data. There may be increased recognition of the disease because a new or more sensitive diagnostic technique has been introduced. If a doctor has a special interest in a disease, it may lead to increased investigation and more frequent recognition of that disease in a particular locality. Occasionally, a laboratory error causes a 'pseudo-outbreak'.

Confirming the diagnosis and case definition

The clinical diagnosis is usually established by a study of the case histories of a few affected people. Laboratory tests are essential to confirm the diagnosis in most infections, but epidemiological investigation should begin immediately and should not usually await laboratory results. A clear case definition is essential for case searching to be carried out. This case definition should be agreed by all involved in the investigation and used consistently throughout the investigation by all investigators; this is especially important in a previously unrecognized disease, or in one in which there are no satisfactory confirmatory laboratory tests.

Tentative hypotheses and immediate control

The preliminary enquiry should include detailed interviews with a few affected people so that common features may be identified, such as an attendance at a function. Symptoms, dates of onset and possible exposures should be documented. Ideas can then be developed about the source and spread of infection, and a questionnaire can be designed to test these hypotheses in subsequent analytical studies. However, it may be necessary to take immediate control measures before confirmation so that further cases may be prevented. When a common vehicle or source of infection is suspected, appropriate action should be taken to interrupt the spread and control the source.

Management of an incident

If the preliminary enquiry confirms that the incident is real, a decision should be taken on its management. Small outbreaks will usually be managed informally. In serious outbreaks, an outbreak control team should be set up. In addition to the environmental health officer, this should include: the consultant for communicable disease control, the local microbiologist, a PHLS consultant microbiologist and possibly a consultant epidemiologist from the CDSC or SCIEH. The control team may require an administrator or epidemiologist to manage an 'incident room' where information on the outbreak should be collated and made available to those who require it. Each local authority must have an incident management plan that details the responsibilities and duties of each member of the team (see Chapter 12).

Identification of cases, collection and analysis of data

The cases first reported in an outbreak are usually only a small proportion of all the cases and may not be representative. Focusing only on these cases can be misleading. The exposed population should be identified so that thorough case finding can be carried out.

School or hotel registers, lists of institutional residents, pay-rolls and other occupational records and lists of people attending functions associated with the disease are useful ways of identifying cases.

The aim of the enquiry will be to collect data from those affected, and those who were at risk but were not affected. The data routinely sought from cases include name, date of birth or age, sex, address, occupation, recent travel, immunization history, date of onset of symptoms, description of the illness and the names and addresses of the medical attendants. Other details will depend on the nature of the infection and possible methods of spread.

To ensure accurate and comparable records of everyone included in the enquiry, and to help analysis, the data should be collected on a carefully designed standard form or questionnaire. Administration of the questionnaire will often be by face-to-face interview by a single investigator or group of investigators trained to administer the questionnaire. Interview by telephone may be useful in obtaining data quickly. When numbers are large and the enquiry is straightforward, a self-administered postal questionnaire is cheaper and quicker to administer, but the response rate and accuracy may be worse. Errors in recall can be reduced by providing background details of events, and making use of other sources to check data such as diaries, menus, discussion with relatives, etc.

The data from the cases should be analysed by time, place and person to determine the mode of spread, source of infection and people who may have been exposed.

Time

The time of importance is the time of onset of the disease, since from this and a knowledge of the incubation period of the infection, the period of possible exposure can be determined. These data are presented graphically, usually in the form of a histogram (Fig. 14.1). In point-source outbreaks, all cases are exposed at a given time, and the onsets of symptoms of all primary cases cluster within the range of the incubation period. An epidemic that extends beyond a single incubation period range suggests either a continuing or recurring source of infection, or the possibility of secondary transmission. In outbreaks spread from person to person, cases will be spread over a longer period, with peaks at intervals in the incubation period.

Place

The place of residence, work or other exposure of cases should be plotted on a map to show the geographical spread of the outbreak. Cases that do not follow the general time or geographic distribution may provide invaluable evidence of the source of infection. Cases clustering in a particular place

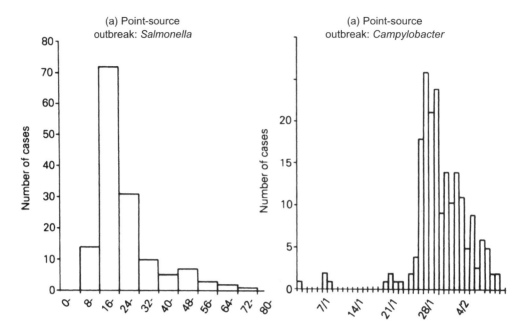

Fig. 14.1 Epidemic curves of two outbreaks. (a) Point-source outbreak of *Salmonella* food poisoning in people eating a buffer meal. Almost all cases occurred within the usual range of the incubation period (8–72 hours). (b) A community common-source outbreak of *Campylobacter* from raw milk. The contaminated milk was delivered over a period of several days, so the dates of onset cover a period greater than the incubation period for *Campylobacter* (2–10 days). A diminishing level of contamination of milk during the week was likely.

of work or neighbourhood may indicate the existence of a point-source of infection or of person to person spread.

Testing hypotheses

The data analysed so far may indicate that most cases ate a particular food or worked near a particular cooling tower. Great care, however, should be taken when interpreting such data. It is almost always necessary to have a control group to find out if, say, all the population are fond of a particular food or visit premises near the cooling tower. This is where analytical epidemiology is necessary [4].

Cohort and case–control studies

The analytical cohort study attempts to investigate causes of disease by using a natural experiment in which only a proportion of a population is exposed to a factor such as a food,

and it compares attack rates in the exposed and unexposed. For example, when investigating a food poisoning outbreak in an institution or hotel, it is usually possible to identify retrospectively most of those exposed and to calculate attack rates in people who did and did not eat particular foods.

A case–control study approaches the question from the opposite direction and begins by identifying people with and without the infection, and then tries to identify factors associated with disease. A group of cases is compared with a group of people who were not ill but had equal access to the likely source of infection. Controls can be taken from electoral registers, hospital admissions lists, the telephone directory, general practitioner (GP) age/sex registers, hotel and reception guest lists, family members of cases, neighbours of cases acquaintances nominated by cases and people who were investigated by the laboratory but were negative for the disease in question. The statistical power of the study can be increased by increasing the number of controls per case.

There are many possible pitfalls in conducting analytical studies and careful design is essential to minimize bias. An important possible bias may arise from the loss of cases or controls from the study because of refusal to be interviewed or failure to trace patients. Patients' recall may be biased by their own preconceptions or by press and media speculations. Cases will often have been interviewed on many occasions before an analytical study is carried out, and this may have introduced bias from suggestions made by interviewers, as well as prompting a more detailed recall. Bias may result from the interviewer knowing the disease status of the person and having his or her own suspicion or prejudice about the source.

Incomplete histories may be taken in which, for example, the patient with *Salmonella* poisoning is asked in detail about only one food. Training and experience in the technique of interviewing, and use of a structured questionnaire are safeguards.

In both cohort and case–control studies, the basic analysis is by a comparison of proportions. The date can be presented in a contingency table (Table 14.2) and analysed as follows.

In cohort studies, $a/a + b$ is the attack rate in the exposed and $c/c + d$ is the attack rate in the unexposed. The ratio $(a/a + b)/(c/c + d)$, which is the ratio of attack rates in the exposed and unexposed, is called the **relative risk**. The size of the relative risk is an indication of the causative role of the factor concerned. In case–control studies, the ratio $a : a + b$ is not meaningful, since b is usually an unknown fraction of the total of the well population that was exposed. However, a statistic called the odds ratio or cross-product ratio –

$ad : bc$ – is a useful measure of association between disease and exposure.

CONTROL

Control measures may be directed towards the source, the method of spread, the people at risk or a combination of these.

Control of source

Some infections that are spread from a human source can be controlled by putting the case or carrier in isolation (e.g. diphtheria and typhoid fever). When animals are the source of an infection, it is sometimes possible to control an outbreak by eradication (e.g. rodent control for leptospirosis). Rabies may be controlled by the destruction of rabid animals and wild or stray animals, and by the muzzling of domestic dogs. Outbreaks of food-borne zoonoses are usually controlled by removing the vehicle of infection. Eradication of animal reservoirs has played a major part in the long-term control of zoonoses such as bovine tuberculosis and brucellosis (Fig. 14.2).

Environmental sources for diseases like Legionnaires' disease (water cooling systems of air-conditioning plants, domestic hot water systems in large buildings or whirlpool spas) may be controlled by cleansing and disinfection. Other pathogens, which contaminate worksurfaces, utensils and equipment in kitchens, and clothing can also be controlled by the cleansing, disinfection or sterilization of these environmental sources.

Sterilization is defined as a process used to render an object free from viable micro-organisms, including bacterial spores and viruses. **Disinfection** is a process that reduces the number of viable micro-organisms but does not necessarily inactivate some viruses and bacterial spores.

Steam sterilization

Steam under pressure can be heated above 100°C and direct contact will kill vegetative

Table 14.2 Contingency table

	Ill	*Well*	*Total*
Exposed	a	b	a + b
Not exposed	c	d	c + d
Total	a + c	b + d	a + b + c + d

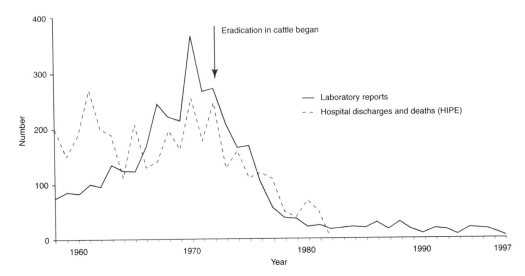

Fig. 14.2 Human brucellosis in England and Wales, 1958–1997. HIPE, hospital inpatient enquiry. (Source: PHLS and ONS, prepared by the CDSC.)

micro-organisms and their heat-resistant spores and viruses. The temperature and contact time must be precisely controlled for guaranteed results.

Hot-air sterilization

Dry heat at 160°C for two hours, 170°C for one hour or 180°C for 30 minutes will kill all micro-organisms. Hot-air ovens are used for materials that will not be damaged by high temperatures.

Ethylene oxide sterilization

This will kill most bacteria, spores and viruses, and is usually used at subatmospheric pressures with an inert diluent gas. It is toxic, potentially carcinogenic and flammable. It is used for heat-sensitive equipment.

Low temperature steam and formaldehyde

This combination of dry saturated steam and formaldehyde kills vegetative bacteria, spores and most viruses.

Objects are exposed to dry saturated steam at 73°C at subatmospheric pressure into which formaldehyde is introduced. It is used for items

that are not damaged by this process but that are unsuitable for steam or dry heat sterilization.

Sterilization by irradiation

Irradiation by gamma rays or accelerated electrons in excess of 25 k gray (Gy) provides adequate sterility. This method is used widely for single-use medical devices.

Low temperature steam disinfection

This is a disinfection or pasteurization process that kills most vegetative micro-organisms and viruses. The process usually involves exposure to dry saturated steam at 73°C for more than 10 minutes at below atmospheric pressures.

Disinfection with washers

Washer disinfectors use physical cleaning and heat to disinfect contaminated re-usable items. Items have to be able to withstand wet heat at 80°C.

Decontamination by manual cleaning

Physical removal of contamination itself is important. It reduces the initial load for disinfection and is a necessary preparatory step before sterilization or disinfection is carried out.

Chemical disinfection

There must be good contact between the item and the disinfectant for a predetermined minimum time period. The precise choice of disinfectant will be determined by the particular task at hand.

Boiling water disinfection

Water at 100°C for more than five minutes will kill most micro-organisms. Items have to be cleansed before immersion. This is a commonly used method for disinfecting small items of medical equipment.

Disposal by incineration

This is applicable to all micro-organisms where temperatures are in excess of 850°C, or 1000°C if cytotoxic drugs are in the waste stream.

Control of spread of infection

Disease spread by food, milk and water is prevented by the withdrawal or treatment of the contaminated product. Diseases spread by direct contact may be controlled by avoiding contact. Those spread by indirect contact may be prevented by hand-washing. Spread by the faecal–oral route, for example, dysentery, is prevented by hand-washing and the disinfection of surfaces in lavatories.

Diseases due to insect bites, such as malaria, typhus and yellow fever, are controlled by vector destruction, protective clothing and insect repellents.

Hepatitis B and HIV infections, which can be spread by blood, are controlled by preventing accidental inoculation or contamination of broken skin or mucous membranes by infected blood or tissue fluids. The spread of airborne infections may be limited by ventilation in buildings. In special circumstances, physical isolation of infectious cases (e.g. Lassa fever) is required, and highly susceptible people, for example, children with leukaemia, may be placed in protective isolation.

Control of people at risk

Some diseases can be controlled by immunization or by giving antibiotics to people at risk. Human immunoglobulins are commonly used prophylactically (e.g. when immunoglobulin is given to travellers to prevent hepatitis A). Rabies immunoglobulin and vaccine are given to people following the bite of a rabid, or possibly rabid, animal. Antibiotics are given to contacts of cases of meningococcal infection to eradicate carriage of the organism in the nose and throat of contacts, thereby preventing them from passing on the infection to others.

THE CONTROL OF IMPORTANT DISEASES IN THE UNITED KINGDOM

Acute encephalitis

Infection of the brain can be caused by several organisms, the commonest being viruses such as mumps. The incubation period varies with the agent. Specific treatment may be available for certain agents.

Acute poliomyelitis

Polio has been a worldwide problem, especially in children in developing countries. The virus infection is often asymptomatic, but it can cause paralysis and death or permanent disability. The virus enters the body orally, and invades the central nervous system. Diagnosis is by clinical presentation and isolation of the polio virus from faeces.

Source and spread

Humans are the reservoir. The virus is present in pharyngeal secretions and is excreted in faeces. Transmission occurs by faecal–oral spread and by direct contact.

Incubation period

Three to five days.

Control

Routine immunization of all infants. Cases should be nursed with enteric precautions. Contacts should be immunized. Outbreaks can be controlled by emergency mass vaccination. Travellers to endemic areas should be immunized.

AIDS

AIDS is the result of infection with HIV, which attacks the body's white blood cells thereby reducing resistance to other infectious agents. The pandemic of AIDS began in the 1970s, and all countries of the world have cases. The highest incidence at present is in Central Africa and North America. Cases present with a variety of unusual opportunist infections. Diagnosis is based on fulfilling strict clinical/microbiological criteria. At present, there is no curative treatment, although new antiviral therapies are prolonging survival.

Source and spread

The origin of HIV is uncertain, but humans are the reservoir, and the virus is present in blood and body fluids. Transmission occurs as a result of contact with the blood or body fluids of an infected person in three main ways:

- anal or vaginal sexual intercourse
- perinatal transmission from mother to foetus
- blood transfusions and the use of blood products, and the sharing of contaminated needles by intravenous drug users.

Health workers exposed to inoculation accidents with HIV infected blood have contracted the virus, although the risk is very small; no risk from intimate, but non-sexual, contact has been shown.

Incubation period

Several months to many years. A median of 10 years has been suggested.

Control

All blood donations must be tested for HIV antibodies and contaminated blood removed. Only sterile needles should be used. Intravenous drug users should not share needles. Health education should alert people to the risks from unprotected sexual intercourse with multiple partners and sharing needles, and to the lack of risk from casual contact. Confidential HIV counselling and testing should be promoted.

Treatment of other sexually transmitted infections will reduce risk.

Anthrax

Anthrax is an acute bacterial infection of humans and animals caused by *Bacillus anthracis*, which may be fatal. Infection is usually of the skin with an ulcer and scab formed. It may be followed within a few days by septicaemia and meningitis.

The disease is enzootic in certain African and Asian countries, but is very rare in the United Kingdom. It is an occupational hazard of people such as wool-sorters, fellmongers, knackermen, farm workers and veterinarians in contact with infected animals or their products, for example, blood, wool, hides and bones. Diagnosis is by culture of wounds, and treatment is with penicillin.

Source and spread

All domestic, zoo and wild animals are potentially at risk of infection. Anthrax bacilli are released from infected carcasses, and form resistant spores on exposure to air. These spores contaminate soil for many years. Humans are usually infected by inoculation into cuts and abrasions from direct contact with infected animals, carcasses or animal products and contaminated soil. Inhalation or ingestion of spores may occur. Animals are infected from contaminated feed, forage, water or carcasses.

Incubation period

Three to ten days.

Control

Prohibit contact with infected animals and their products. Establish environmental and personal hygiene, for example, ventilation and protective clothing, where a special risk exists. Disinfect imports of hairs and wool. Vaccination may protect those occupationally exposed to risk. Carcasses suspected of infection should not be opened, but should be disposed of safely.

Campylobacteriosis

This occurs worldwide and is the commonest diagnosed diarrhoeal disease in the United Kingdom. It is usually caused by *Campylobacter jejuni*. Diagnosis is by isolation of the organism from faeces. Treatment is usually confined to fluid replacement.

Source and spread

There is widespread intestinal carriage in most mammals and birds. Poultry and cattle are believed to be the main reservoirs for human infection, which may be acquired by ingesting contaminated raw milk, undercooked chicken or other food contaminated in the kitchen. Direct faecal–oral spread from animals may occur, especially from puppies and, exceptionally, there is person to person spread. Large water-borne outbreaks have occurred.

Milk from bottles pecked by magpies has been shown to cause a significant number of cases in the United Kingdom.

Incubation period

One to ten days.

Control

Pasteurize milk, chlorinate drinking water supplies, thoroughly cook meat (especially poultry), practice good kitchen hygiene and protect doorstep-delivered milk from bird attack.

Cholera

Cholera is a bacterial infection that is endemic in many developing countries, with recent epidemics in South America. It may cause epidemics of profuse watery diarrhoea, particularly among refugees and people living in overcrowded and poor sanitary conditions. Diagnosis is by faecal culture. Death due to dehydration occurs, but can be prevented by prompt fluid replacement.

Source and spread

Humans are the reservoir and transmission occurs when water or food that have been contaminated by infected faeces is consumed. Sewage contamination of water supplies causes epidemic disease. Outbreaks due to raw or undercooked seafood have occurred.

Incubation period

A few hours to 5 days.

Control

The separation of water supply from sewage disposal effectively eradicates the disease. Infected people should not handle food. Unchlorinated drinking water should be boiled.

Cryptosporidiosis

A recently recognized, but common, protozoal cause of gastroenteritis, especially in children, worldwide. It can cause life-threatening disease in immunosuppressed patients. Diagnosis is by identification of oocysts in faeces under light microscopy. There is no specific therapy.

Source and spread

Cryptosporidia have been identified in the faeces of most animal species. Human infection results from person to person faecal–oral spread, especially in children, and from raw milk or direct contact with farm animals, especially calves. Water-borne outbreaks have been reported, especially when there has been failure of water filtration.

Incubation period

One to ten days.

Control

Ensure personal hygiene. Pasteurize milk and ensure adequate filtration of drinking water and maintenance of swimming pools.

Diphtheria

This is caused by the bacterium *Corynebacterium diphtheriae*, which produces a toxin leading to sloughing of epithelial tissue in the throat and the formation of a membrane, sometimes with cardiac and nerve involvement. The fatality rate is high if not treated early. Mass immunization has made this a rare disease in temperate climates, where it was once common. In tropical countries, skin and wound infections are more common. Diagnosis is by culture of the organism from the throat, nose or wound swabs. Treatment is with antibiotics and, possibly, specific antitoxin.

Source and spread

Humans are the reservoir, and transmission occurs via droplets and nasal and skin discharges to close contacts of cases or carriers. Rarely transmission occurs via contaminated food such as milk.

Incubation period

Two to five days.

Control

Routine immunization of all infants. Isolate infected cases. Search for carriers in close contacts and eradicate carriage with antibiotics. Contacts should be quarantined, put under daily surveillance for a week and given antibiotics if not previously immunized. All previously immunized contacts should be given a booster dose.

Escherichia coli – enterohaemorrhagic strains

This is a relatively newly recognized (1982) cause of bloody diarrhoea (haemorrhagic colitis) and kidney disease (haemolytic uraemic syndrome), and has a significant mortality. It is caused by strains of *E. coli* that produce verocytotoxin (VTEC strains) most commonly in the O157 group. Diagnosis is by isolation of the organism from faeces. The value of antibiotic treatment is unclear.

Source and spread

Cattle are the probable major reservoir. Food-borne outbreaks, particularly associated with undercooked hamburgers and raw milk, have occurred. Person to person faecal–oral spread, especially in nurseries, has been well documented. Water-borne outbreaks have been reported.

Incubation period

Twelve to sixty hours.

Control

Thorough cooking of raw meat, pasteurize milk, chlorinate water supplies and good kitchen and personal hygiene.

Giardiasis

This is a common cause of diarrhoea due to infection with the protozoa *Giardia lamblia*. Symptoms tend to be prolonged (approximately 3 weeks) or intermittent abdominal pain, and loose stools. Diagnosis is by identification of the organism in faeces by light microscopy. Treatment is with metronidazole.

Source and spread

Humans are the usual reservoir, with person to person faecal–oral spread, especially between young children in nurseries. Zoonotic infection in North America is relatively common with beaver and muskrats contaminating water supplies.

Incubation period

Five to twenty-five days.

Control

Good personal hygiene. Exclude children with diarrhoea from nursery or school. Filter and chlorinate water supplies.

Legionnaires' disease

A relatively uncommon form of pneumonia caused by the bacterium *Legionella pneumophila*, which has a fatality rate of about 20%. The immunosuppressed, aged, heavy smokers and those with pre-existing heart and lung disease are at greatest risk. Point-source outbreaks associated with hotels and hospitals are reported. Diagnosis is by isolation of the organism from sputum or biopsy, and by serological identification of antibodies. Treatment is with antibiotics.

Source and spread

Legionellae are free-living organisms that are ubiquitous in standing waters and soil, and colonize plumbing systems, cooling towers, etc. Most environmental isolations are not related to cases of infection. Human infection occurs when aerosols of contaminated water are created and inhaled. The risk of infection will depend upon the dose of the organism present in the aerosol, and the susceptibility of the person exposed. Aerosols can be created by air-conditioning cooling towers, domestic showers and taps. Person to person spread does not occur.

Incubation period

Two to ten days.

Control

Domestic water should be routinely chlorinated. All buildings with wet cooling tower systems should be identified, and cooling towers regularly cleaned and disinfected. Where possible, dry cooling systems should replace them. Domestic water supplies should be maintained at a hot water temperature of more than 50°C and a cold water temperature of below 20°C. In an outbreak, higher temperatures and hyperchlorination may be necessary to render water systems safe.

Leprosy

This is a chronic bacterial infection caused by *Mycobacterium leprae*, which causes skin and nerve damage, and may lead to traumatic injury and deformation. Several forms are recognized, some of which may heal spontaneously. It is still common in the tropics and subtropics. Diagnosis is by clinical presentation and identification of the organism in skin scraping or biopsy material. Treatment is with long-term antibiotics. Physiotherapy and surgery may help deformities.

Source and spread

Humans are the reservoir. Transmission does not occur very readily, and usually only in household settings, probably from contact with nasal discharge in children.

Incubation period

Several years.

Control

Early identification and treatment of cases. Infectious cases should be isolated until treatment is established. Examine all household contacts. Treatment may need to be given to children in the household to prevent illness.

Leptospirosis

A sporadic bacterial disease of varying severity transmitted by contact with infected animal urine. The causative agent is the spirochaete *Leptospira*, with over 170 serotypes. It occurs worldwide, with areas in which host-adapted serotypes predominate, for example, *L. hardjo* in cattle in the United Kingdom. Weil's disease, a particularly severe form with liver and kidney involvement, is caused by infection with *L. icterohaemorrhagiae*. Diagnosis is by serology. Treatment is with antibiotics.

Source and spread

Most animal species may be hosts of leptospires, but the main natural reservoirs for human infection vary with the serotype: *L. canicola* in dogs, *L. hardjo* in cattle and *L. icterohaemorrhagiae* in rats. Leptospires are excreted in urine, which contaminates the environment, especially watercourses. Humans are infected by direct contact with the animal or contaminated environment, and leptospires enter the body through abrasions, wounds or mucous membranes. Person to person spread does not occur.

Incubation period

Three to twenty days.

Control

Control rodents, avoid exposure to contaminated waters and protective clothing is needed for workers at special risk.

Listeriosis

A relatively uncommon, but increasingly recognized, bacterial disease in humans caused by *Listeria monocytogenes*. It occurs worldwide. Cases are almost exclusively in pregnant women, neonates, immunosuppressed patients and the elderly. It can cause fatal meningoencephalitis and abortion. Diagnosis is by culture of the organism. Treatment is with antibiotics.

Source and spread

The agent is widely distributed in animals, birds, humans and soil. The main reservoir for human infection is not clear. The organism is excreted in animals faeces. Outbreaks of food- and milk-borne infection have occurred in humans. Soft cheeses and pâté have been identified as high-risk foods. *Listeria* grow slowly at normal refrigeration temperatures. Cross-infection in hospitals has been reported.

Incubation period

Uncertain, but probably a few days.

Control

Ensure good personal hygiene and care in the storage and preparation of food. Heat-treat dairy products. Ensure safe handling of infected animals, and avoid contact with possibly infected materials during pregnancy.

Lyme disease

This infection, with the spirochaete *Borrelia burgdorferi*, causes a characteristic skin rash, erythema migrans, followed by variable cardiac, nerve and joint manifestations. It appears to be common in certain parts of North America and Europe, but it is still rare in the United Kingdom. Diagnosis is based on clinical features and laboratory tests. Early treatment with antibiotics may prevent the more serious sequelae.

Source and spread

Deer and mice are known to be reservoirs in the United States. Transmission to humans is via the *Ixodes* tick, which feeds on infected animals, but many patients do not remember being bitten. Person to person spread does not occur.

Incubation period

Days to several weeks.

Control

Avoid tick-infested areas. Remove ticks from the skin quickly. Cover exposed parts of the body when in tick-infested areas. Early treatment with antibiotics.

Malaria

Worldwide, malaria is one of the most important infections and a common cause of death. Malaria is caused by one of the four species of the protozoan parasite *Plasmodium falciparum, P. ovale, P. malariae* and *P. vivax*. It occurs in the tropics and subtropics. Diagnosis is by clinical presentation and identification of organisms in the blood. Treatment is with chloroquine, quinine and other antimalarials. Chloroquine resistance of the parasite is a major problem throughout the world.

Source and spread

Humans are the reservoir and transmission occurs via the bite of the *Anopheles* mosquito.

Incubation period

Usually 10–15 days. May be much longer with *P. vivax*.

Control

Protect against mosquito bites by staying inside after dark, covering arms and legs and using a repellent spray. Eradicate mosquito breeding sites. Insecticide spraying of dwellings. Prompt treatment of cases. Travellers to endemic areas require chemoprophylaxis for which medical advice must be sought. (For mosquito control generally, see Chapter 41.)

Meningococcal meningitis and septicaemia

Meningitis is an infection of the brain lining by various bacteria or viruses, causing severe illness; it has a high mortality if untreated. Viral meningitis is usually less severe than bacterial meningitis.

The latter may be associated with a generalized (blood) infection (septicaemia), which has a worse prognosis. Infection caused by the meningococcus receives the greatest public health attention in the United Kingdom, where it is endemic with cyclical epidemic waves.

Diagnosis of bacterial meningitis is by the isolation of the organism from blood or cerebrospinal fluid. Viral meningitis is more usually diagnosed serologically. Treatment is by appropriate antibiotics.

Source and spread

The natural reservoir of the meningococcus is the human nasopharynx. The carriage rate for all meningococci in the normal healthy population is about 10%. Transmission occurs by droplet spread between people who are close contacts. Only a small proportion of people who acquire the organism ever develop the disease.

Incubation period

Two to ten days.

Control

All cases of meningococcal meningitis and septicaemia should be notified promptly to public health authorities. Household and kissing contacts should be traced and offered antibiotics to clear pharyngeal carriage in an attempt to prevent transmission to other susceptible contacts. A vaccine against groups A and C is available and trials of group B vaccine are underway. Group B is the commonest form of the disease in the United Kingdom. In Africa and Asia, group A disease is more common, and a vaccine is offered to travellers going to highly endemic areas.

Plague

A highly dangerous bacterial infection with *Yersinia pestis*, which historically caused plague in large epidemics. Its main geographical distribution today is in Asia. Diagnosis is by direct microscopy of lesions and culture of the organism. Treatment is with antibiotics.

Source and spread

The natural source of bubonic plague is the brown rat and rat flea. The flea ingests infected blood, *Y. pestis* multiplies in the flea's stomach and is excreted in its faeces or is regurgitated. Rats are infected by flea bites. Humans are infected by rat flea bites, or by handling infected rats, or by person to person droplet spread (pneumonic plague). The death of infected rats causes fleas to seek other hosts, such as the black rat, which carries fleas into close contact with humans.

Incubation period

Two to six days.

Control

Isolated cases. Travellers to endemic areas can be immunized. Antibiotics may be given to contacts to prevent disease. Control rodents.

Psittacosis, ornithosis

A febrile bacterial infection with *Chlamydia psittaci* causing fever and pneumonia. It occurs worldwide. Diagnosis is by serology and treatment is with tetracyclines and erythromycin.

Source and spread

Psittacine and other birds, including ducks, turkeys and pigeons, are the usually identified source of human infections. Sheep strains may infect pregnant women. Infection is via inhalation of aerosols or of infected dust contaminated by bird faeces, nasal discharges or sheep products of gestation or abortion. *C. psittaci* may survive in dust for many months. Person to person transmission of avian or ovine strains is rare. Outbreaks occur among aviary and quarantine station workers and poultry processing workers.

Incubation period

Usually 4–15 days.

Control

Quarantine infected birds. Provide good ventilation of poultry processing plants and heat-treat feathers. Pregnant women should avoid contact with flocks during lambing in enzootic areas.

Q fever

A disease caused by *Coxiella burnetti*, which presents with fever, pneumonia and sometimes endocarditis. It occurs worldwide. Diagnosis is by serology and treatment is with antibiotics.

Source and spread

Many animals, as well as ticks, are natural hosts. The reservoir for human infection is usually sheep and cattle. The organisms are abundant in placentae and birth fluids, and remain viable in dust and litter for months. Infection results from inhalation of contaminated dust, handling infected carcasses or by consumption of contaminated milk, or possibly, from tick bites.

Incubation period

Two to four weeks.

Control

Pasteurize milk. Take hygienic precautions in abattoirs. Prevent contamination of urban areas with infected straw, etc.

Rabies

A viral infection of the central nervous system that is invariably fatal in the non-immunized. Rabies occurs in all continents except Australia and Antarctica. The British Isles are rabies-free. Diagnosis is by visualization of the virus in the brains of animals, and by serology in humans. There is no definitive treatment available.

Source and spread

The virus can infect all warm-blooded animals and birds. Two cycles of transmission are recognized: urban dog rabies, which is now largely confined to the less developed countries; and sylvatic or wildlife rabies, which is the main type in the United States and much of Europe, with various reservoir hosts, for example, skunks, raccoons and foxes in the United States and Canada; Arctic foxes in the Arctic; mongooses and jackals in Africa; foxes in Europe; blood-feeding bats in South America; and other bats in the Americas and Europe. Rabies is fatal in carnivores, and it is the population density that determines the maintenance and spread of infection by biting. Transmission among bats by contrast is by aerosol inhalation. Most human infections are from bites of domestic carnivores or, in South America, vampire bats. Person to person transmission has resulted from infected corneal transplant grafts.

Incubation period

This can be from 10 days to a year or longer. The incubation period is shorter the nearer the bite is to the head.

Control

In enzootic areas, avoid contact with wild animals and promptly cleanse any bite wounds. Specific immunoglobulin is vital as soon as possible after exposure. Pre- and post-exposure vaccination can be given. In enzootic areas, vaccinate dogs, cats and cattle. Quarantine all carnivores on importation (six months in the United Kingdom for carnivores). Some countries require vaccination of dogs and cats before importation. Vaccinate dogs at frontiers between enzootic and rabies-free areas. (For the control of dogs generally see Chapter 42.)

Relapsing fever

A widely distributed bacterial infection caused by *Borrelia* species with high fatality. The endemic form is tick-borne and occurs in Africa, the Americas, Asia and, possibly, parts of Europe. Epidemic relapsing fever is louse-borne, and is limited to parts of Asia, Africa and South America. Diagnosis is by microscopy of blood or culture of the organism. Treatment is with tetracycline.

Source and spread

Humans are the reservoir of epidemic louse-borne infection. Infection occurs when the louse is crushed on the bite wound by scratching. Endemic tick-borne relapsing fever is transmitted from the natural wild rodent reservoir by tick bites to humans.

Incubation period

Five to fifteen days.

Control

Treat cases, clothing and bedding with insecticide. Control ticks.

Severe Acute Respiratory Syndrome

On 14 March 2003 the World Health Organisation (WHO) issued a global health alert to be aware of a new atypical pneumonia called Severe Acute Repiratory Syndrome (SARS) reported in several countries in South East Asia.

Source and spread

The main symptoms of SARS are high fever (>38°C), cough, shortness of breath or breathing difficulties. Changes in chest X-rays indicative of pneumonia also occur. SARS appears to be less infectious than influenza and based on currently available evidence, close contact with an infected person poses the highest risk of the infective agent to spread from one person to another.

Incubation period

Two to seven days.

Control

Antibiotics and antiviral drugs are being used to treat the illness, but there is no specific treatment.

Shigella dysentery

Of the three main types, *Shigella sonnei* is the commonest in the United Kingdom, usually causing transient mild diarrhoea in children. *S. flexneri* and *S. dysenteriae* are usually imported infections, and more often cause severe bloody diarrhoea.

Diagnosis is by isolation of the organism from the faeces. Treatment is fluid replacement, and antibiotic therapy in severe cases.

Source and spread

Humans are the reservoir and the organism is excreted in faeces. Prolonged carriage is possible. Transmission is usually by the direct faecal–oral route, especially in families with small children, and in nurseries and schools. Food-borne outbreaks have occurred, and in developing countries water-borne outbreaks occur due to sewage contamination of drinking water.

Incubation period

One to seven days.

Control

Strict personal hygiene. Sanitary disposal of faeces and protection of water supplies. Infected people should not handle food. Children with diarrhoea should be excluded from school until they are symptom-free and their stools are formed. In school outbreaks supervision by staff of hand-washing in young children and adequate provision of hot water and clean towels in toilets is imperative.

Tetanus

This occurs worldwide, but is rare in the United Kingdom. Painful contraction of muscles is caused by a toxin produced by *Clostridium tetani*. Diagnosis is usually based on clinical presentation. Treatment is with specific immunoglobulin and penicillin.

Source and spread

Clostridia are normal intestinal flora, and also survive as spores in the soil. Infection may be by contamination of deep, penetrating wounds. Neonatal tetanus occurs in developing countries, and is caused by contamination of the umbilicus.

Incubation period

Usually three to twenty-one days.

Control

Clean wounds thoroughly. Immunize children routinely and following laceration injuries if the immune status of the child is in doubt or if the wound is particularly dirty. Booster doses of tetanus toxoid are also necessary if 10 years have elapsed since the last dose.

Toxocariasis

A common roundworm infection (*Toxocara canis* and *T. cati*) of dogs and cats. It is acquired by children worldwide. Symptomatic disease is rare, but does occur, particularly in children subject to pica. Diagnosis is by serology and biopsy.

Source and spread

Natural hosts are dogs and cats. Eggs that are excreted in faeces require a maturation period in soil. The eggs hatch in the intestine and larvae penetrate the intestinal wall to enter the blood vessels. In puppies younger than five weeks old, larvae migrate to the intestines via the lungs and complete their maturation. Dormant larvae in adult bitches reactivate during their pregnancy, and migrate to cross the placenta or may infect puppies via milk. Larvae excreted in faeces by puppies may mature in the bitch once ingested. Humans are infected by ingesting eggs from contaminated soil and grass.

Incubation period

Weeks or months.

Control

Teach children good hygiene and prevent access to dog faeces. Keep dogs away from children's play areas. Cover sand pits when they are not in use. Clean up faeces when exercising dogs in public parks. Worm all dogs regularly. (For the control of dogs generally see Chapter 42.)

Toxoplasmosis/congenital toxoplasmosis

A common and usually asymptomatic protozoal infection of humans caused by *Toxoplasma gondii*. It occurs worldwide. Infection in the womb in humans can lead to serious brain lesions. Diagnosis is by serology. Antibiotic therapy may be effective.

Source and spread

Definitive hosts are cats, which are infected by eating raw meat, birds or mice that contain parasite cysts. Humans may be infected by eating raw or inadequately cooked meat (mainly sheep, pigs, cattle or goats), and unwashed salad vegetables, or by ingestion of faecal oocysts from cats. Congenital infection of the foetus occurs when the human mother acquires a primary infection in pregnancy.

Incubation period

Uncertain, but possibly one to three weeks.

Control

Pregnant women should avoid handling cat litter, and should wash their hands after handling raw meat. Freezing may kill cysts in meat, but thorough cooking is strongly recommended. Pregnant women should also avoid contact with lambing ewes.

Tuberculosis

A potentially severe chronic bacterial disease caused by *Mycobacterium tuberculosis* and *M. bovis* that occurs worldwide. *M. bovis* has been almost eradicated from the cattle of several developed countries, including the United Kingdom. *M. tuberculosis* is of increasing concern in association with the AIDS epidemic. Diagnosis is by sputum-staining and culture. Treatment is with long-term antibiotics.

Source and spread

Humans are the reservoir for *M. tuberculosis* and transmission is by direct contact and airborne spread. Cattle (and possibly badgers in the United Kingdom) are the natural reservoir of infection with *M. bovis*, and transmission to humans is via the consumption of raw milk.

Incubation period

Four weeks to several years.

Control

Identify and treat cases promptly. Screen contacts for infection and treat early. Heat-treat all milk. BCG vaccination in children.

Typhoid and paratyphoid

A severe infection with *Salmonella typhi* or *S. paratyphi* A or B, with a high mortality if untreated. Most UK cases are imported. Diagnosis is by the isolation of the organism from blood, faeces or urine. Treatment is with antibiotics.

Source and spread

Humans are the definitive host for these species of *Salmonella*. After infection, a person may remain a symptomless faecal or urinary carrier for months or years. Transmission occurs usually by contamination of food or water by faeces or urine. Outbreaks may result from infected food handlers and from sewage pollution of water supplies. Cases may also occur by person to person spread within households.

Incubation period

One to three weeks.

Control

Separation of water supply and sewage disposal. Infected food handlers should be excluded from work until at least six consecutive monthly faecal and urine samples are negative. Household contacts of cases and carriers should be screened, and food handlers excluded from work until two negative faecal and urine samples, at least 48 hours apart, have been obtained. Travellers to endemic areas should be immunized.

Viral haemorrhagic fevers

A variety of viruses may cause severe, often fatal, infection. These viruses are not endemic in the United Kingdom and are only a problem for people visiting exotic areas. For example, Lassa fever is endemic in rural parts of West Africa. Diagnosis is based on serology and isolation of the virus from body secretions. Antiviral agents may be used in the treatment of the infection.

Source and spread

A variety of animals and birds may harbour the viruses. Wild rodents are the reservoir for Lassa fever. Transmission of Lassa fever occurs by direct or indirect contact with rodent urine, which may contaminate food. Person to person spread does occur in some circumstances, such as in hospitals by direct contact with blood, secretions and droplets of infected patients. Other viral haemorrhagic fevers may be transmitted by mosquito and tick bites.

Incubation period

Usually a few days to three weeks.

Control

Avoid endemic areas. Isolate suspected cases and take precautions to avoid respiratory spread. Disinfect contaminated objects, clothing, etc. Specific immunoglobulin may be used to prevent infection in contacts. Ribavirin is used to prevent the spread of Lassa fever. Close contacts of proven cases should be quarantined for three weeks after the last contact.

Viral gastroenteritis

Probably the commonest form of infective gastroenteritis, although not the most commonly reported. It is caused by a variety of viruses including Noroviruses (Norwalk-like) and rotavirus. Worldwide it is a major cause of mortality due to dehydration in young children. Diagnosis is by electron microscopy of faeces obtained within 48 hours of onset, or by serological identification of virus antigen in faeces.

Source and spread

Humans are the reservoir and the virus is excreted in huge quantities in faeces and vomitus during the first few days of illness. Transmission occurs by direct contact; faecal–oral spread and from infected food handlers up to two days after recovery from illness; and indirectly when virus particles contaminating surfaces are transferred to food or fingers and then to the mouth. Secondary household transmission following food-borne outbreaks is common.

Incubation period

Eighteen to seventy-two hours.

Control

Strict personal hygiene. Food handlers should exclude themselves from work immediately if they suffer from gastroenteritis, and for at least 48 hours after symptoms subside. Disinfection of contaminated areas and surfaces.

Viral hepatitis – hepatitis A

This is a common, enterically acquired infection of the liver, commonest in children, who may often not have jaundice. It occurs worldwide. Diagnosis is by serological methods to identify specific antibodies against the virus. There is no specific treatment, but passive immunization with immunoglobulin confers protection for a limited period.

Source and spread

The reservoir of infection is humans. The virus is excreted in faeces before the onset of symptoms. Transmission occurs commonly in families by direct contact, via the faecal–oral route and sometimes by food- and water-borne spread. Outbreaks may occur in nurseries and schools.

Incubation period

Two to six weeks.

Control

Strict personal hygiene and sanitary disposal of faeces. Schools should provide adequate hand-washing facilities. A food handler who is not immune and who has contact with cases may be advised not to handle food for four weeks after his or her last contact with an infectious person, and may be given immunoglobulin. Family contacts and travellers to endemic areas may be given normal human immunoglobulin. Immunoglobulin is effective after exposure if given within two weeks.

Viral hepatitis – hepatitis B and C

These are less common than hepatitis A in the United Kingdom, but worldwide they are very common infections of the liver and carry an increased risk of chronic liver disease and hepatic carcinoma. Diagnosis is by serological identification of viral antigens and antibodies. There is no specific treatment.

Source and spread

Humans are the reservoir. Hepatitis B and C are carried in the blood and body secretions. Infected people may remain carriers for months or years after an acute infection. Transmission occurs by sexual intercourse; from mother to baby at birth; blood contamination of shared needles, as in intravenous drug users; unhygienic tattooing; and from blood transfusions.

Incubation period

Two to six months.

Source and spread

Screen blood donors and exclude contaminated blood. Sterilize intravenous needles and avoid sharing needles. Avoid unprotected sexual intercourse with carriers. Maternal to infant transmission of

hepatitis B can be prevented by giving specific immunoglobulin and vaccine to the infant immediately after birth. A safe and effective vaccine for hepatitis B is now available and is recommended for all health care workers who may be exposed to blood, and to household and sexual contacts of cases.

Yellow fever

A severe, febrile, mosquito-borne viral disease leading to liver and kidney failure and death in many cases. It occurs as a zoonosis mainly in forest dwellers in the rainforest areas of northern South Africa, and Central and South America.

Source and spread

The natural reservoirs of infection are forest monkeys, marmosets and humans. Transmission among monkeys is by the bite of various species of mosquito in forests, and to people who enter or live near infected forests. There may also be an independent human/mosquito cycle in urban areas. The urban cycle has been almost eliminated by eradication of the mosquito vector.

Incubation period

Three to six days.

Control

Control mosquitoes and avoid their bites. Immunize the exposed population and travellers to endemic areas. Immunization confers lifelong protection and is obtained from yellow fever vaccination centres. (For mosquito control generally, see Chapter 41.)

REFERENCES

1. Committee of Inquiry into the future development of the Public Health Function (1988) *Public Health in England*, Cmnd 289, HMSO, London.
2. Palmer, S.R., Lord Soulsby and Simpson, D.I.H. (1998) *Zoonoses*, Oxford Medical Publications, Oxford.

3. Langmuir, A.D. (1963) The surveillance of communicable diseases of national importance. *New England Journal of Medicine*, **268**, 182–92.

4. Palmer, S.R. (1990) Review article: epidemiology in search of infectious diseases: methods in outbreak investigation. *Journal of Epidemiology and Community Health*, **43**, 311–14.

FURTHER READING

Chin J. (ed.) (2000) *Control of Communicable Diseases Manual*, 17th edn, American Public Health Association, Washington.

Detels, R., McEwen, J., Beaglehole, R. and Tanaka, H. (eds) (1997) *Oxford Textbook of Public Health*, 4th edn, Oxford Medical Publications, Oxford.

Giesecke, J. (2002) *Modern Infectious Diseases Epidemiology*, 2nd edn, Edward Arnold, London.

Noah, N. and O'Mahony, M. (eds) (1998) *Communicable Disease Epidemiology and Control*, John Wiley, Chichester.

Russell, A.D., Hugo, W.B. and Ayliffe, G.A.J. (1992) *Principles and Practice of Disinfection, Preservation and Sterilization*, Blackwell Scientific, Oxford.

15 Environmental epidemiology and non-infectious diseases

Richard T. Mayon-White

INTRODUCTION

Historically, environmental health professionals have concentrated upon the health effects of infectious diseases and accidental injuries, and their control, investigation and regulation. Protocols for these established areas of intervention are well developed, with roots well into the nineteenth century. Success in these traditional areas has been followed by an increasing interest in non-infectious diseases, especially cancers and ischaemic heart disease. Non-infectious conditions are often called chronic diseases, referring to the fact that they occur mostly in older adults as a consequence of factors that have slow or delayed effects. Another difference from the traditional infectious diseases and accidental injuries is that chronic diseases often have more than one essential causative factor, and genetic inheritance may play as large a part as external factors in determining who gets which disease. Amongst the external factors, lifestyle is likely to be as important as what is in the environment, not only in the causation of disease but also in affecting the exposure to environmental factors. This complicates the understanding of the causes and prevention of non-infectious diseases.

EXPOSURE

Exposure to a factor in the environment is a fundamental concept, and needs to be measured if risks are to be properly understood. The timing of exposure, both of populations and individuals, may take three basic forms: acute, subacute and chronic.

Acute exposures are usually due to some accidental release of material, perhaps from an industrial accident or fire. Other than in anecdotal or publicly reported form, no long-term records are currently kept of incidents leading to exposure of populations, and little accurate information about the precise nature of the materials, the exposure levels or the people exposed. Clearly, there are a number of potential victims of a sudden chemical release in this way: employees and others in the immediate vicinity when the release takes place; the population within the exposure area; and the emergency services and others who have to deal with the immediate problem and effects of the release.

Subacute exposure may take place where a population lives in houses built on contaminated land or near an old industrial waste tip. There may be a continuous low-level exposure that is episodically increased by development or climatic causes. In such environments, children are particularly at risk because of their play activities.

Chronic exposure may occur where a well-controlled industrial process makes low-level releases into the atmosphere, or where there is a continuous dietary or occupational exposure, for example, drinking water supplies containing chloroform, increased mercury dosage due to an over-large consumption of tuna, or benzpyrenes accumulating in cooking oil that has been used at too high a temperature and not changed regularly.

It is important to remember that, particularly in the case of non-biodegradable materials and those that cause cumulative damage, it is the total burden of exposure from all sources that is likely to have an influence on the health of an individual. The amount of exposure can be assessed in six ways:

1. Interviews, questions and diaries to record what people observe of their own exposure, both in time and degree
2. Physical, chemical, or microbiological measurements on water, air or soil
3. Sampling the personal environment, for example, foods eaten or air breathed
4. Measurement of dose of exposure (individual radiation monitors are one example)
5. Concentration of chemicals in human tissues (e.g. lead in blood or teeth)
6. Markers of the physiological effects of exposure (e.g. carboxyhaemoglobin as an indicator of exposure to tobacco smoke).

THE EPIDEMIOLOGICAL APPROACH

The complexity of environmental factors in the causation of chronic diseases has meant that epidemiological methods used to investigate them must be rigorous. Epidemiology can show which environmental factors cause disease or injury, even when the mechanism or precise factor is unknown. When the nature of the harm is understood, epidemiology can determine if the effects can be prevented or reduced to an acceptable level. If hazards have been identified and

controlled, epidemiology can be used to monitor the process. In short, environmental epidemiology is the measurement and evaluation of the health effects of environmental hazards. In this chapter, the main epidemiological methods are described and illustrated with examples drawn from non-infectious (chronic) diseases.

One of the characteristics of epidemiology is that it enables us to detect unexpected or uncommon effects. This may lead some people to distrust the findings of epidemiological studies because they know of many people who have been exposed to the hazard without falling ill. They dismiss the evidence as circumstantial because the damage cannot be shown to be 'cause and effect' in laboratory experiments. This is pertinent in legal proceedings when lawyers and juries are unfamiliar with statistical arguments. We must therefore expect to have to argue carefully and persistently to get public action to control the environmental hazards detected by epidemiology.

DESCRIPTIVE EPIDEMIOLOGY AND ECOLOGICAL STUDIES

The straightforward enumeration of cases is a familiar method in public health. For more than a century, public health records have enumerated the number of deaths from certain causes, of notifiable infections, of cancers and of respiratory conditions. The fact that these reports were regular (annual) and related to well-defined areas (countries, regions or counties, towns or rural districts) led to obvious ways of looking for environmental factors. The familiarity of routine statistics disguises the very real difficulties in collecting reliable data. The value of data that are routinely collected lies in the large number of people who can be studied for long periods of time. If large differences in the incidence of disease are observed in different places, or at different times, then it is obvious to look for factors that vary in place or time and might explain the different rates of disease. As the proportion of children and older people in populations will also vary with time and place, the death rates have to be standardized, that is, calculated to refer to a standard population, producing an standardized mortality ratio (SMR). A population with an SMR above 100 means that the population has a higher death rate than other populations in the comparison. This is illustrated in Table 15.1, which shows the variation in mortality rates in different parts of the United Kingdom, with higher rates in the north than in the south.

This approach of comparing disease rates in different places and times is sometimes called an **ecological study**. An example to illustrate ecological studies is provided by **malignant melanoma**, a cancer of the pigment-forming cells of the skin. The cancer may start in a mole (a benign pigmented skin tumour) or in apparently normal skin. Once the melanoma becomes cancerous, it tends to metastasize (spread to other parts of the body via the bloodstream) early, so that surgical treatment is not always effective. Consequently, the case-fatality rate (the proportion of patients who die) for malignant melanoma is high, and the death rate from this condition is a good indication of its incidence. The diagnosis of malignant melanoma is usually accurate in developed countries because skin tumours removed by surgeons are checked by pathologists. This can lead to the diagnosis being recorded in a cancer register, so that all cases,

Table 15.1 Standardized Mortality Ratios in the United Kingdom in 2000 [1]

Country and English Region	SMR
United Kingdom (the standard population in these calculations)	100
Scotland	118
Northern Ireland	106
Wales	102
England	98
North East Region	110
North West Region	107
West Midlands	102
Yorkshire and Humberside	101
East Midlands	98
London	94
East Anglia	93
South West	92
South East	92

including survivors, can be counted. As malignant melanomas occur in young and middle-aged adults, the diagnosis is likely to be made with particular care. For these various reasons, the high rate of melanomas in Queensland, Australia was an important observation. The best explanation is that subtropical levels of sunshine stimulate the pigment cells of fair-skinned people to the point that some cells become malignant.

The incidence of melanomas is increasing in the United Kingdom by more than 6% per year, which reflects the popularity of holidays with lots of sunbathing. Melanomas are uncommon in black people, so there is clearly a genetic factor affording more tolerance of sunshine. But the environmental factor of this skin cancer is evident from the description of the geographical variation in incidence. The important part of the ultraviolet light spectrum is the B band (UV-B) from 280 to 320 nm. The carcinogenic potential of UV-B is not new information, but the link to a severe cancer like melanoma has raised public concern, leading to advice on prevention, both personal (care in sunbathing) and global (by the protection of the ozone layer in the lower stratosphere).

Changes in incidence may also lead to the discovery of the cause. The change usually has to be an increase rather than a decrease in incidence in order to stimulate research. **Lung cancer** became more common in the United Kingdom in the first half of the twentieth century. By 1950, there were a number of theories about the cause of this increase, including the suggestion that environmental factors like the exhaust fumes of cars and lorries were the reason. The attraction of this theory was that the rise in the use and number of vehicles propelled by internal combustion engines ran in parallel with the rise in lung cancer mortality. However, as is now well known and widely accepted, the cause was in fact part of the social environment: the increase in cigarette smoking. The association of lung cancer with tobacco was determined by a **case-control study** of the type described later in this chapter. The annual average consumption of cigarettes had risen from 0.5 lb to 4.5 lb (0.23–2.04 kg) per person between 1900 and 1947, followed by a 10-fold rise in deaths from lung cancer in the same period.

Another illness and cause of death that increased markedly during the twentieth century is **ischaemic heart disease**. There is a link to smoking, but other factors are also important. The study of ischaemic heart disease is complicated by the fact that it is part of a larger process of arterial disease, by inaccuracies in the diagnosis, and by the effects that medical care has on where the diagnosis is made and when death occurs. Despite these complications, there is consistent evidence of **geographical and temporal differences** in incidence. High rates of coronary heart disease correlate with high dietary consumption of saturated fats when the data from different countries are compared. The disease is distinctly more common in men than in women, something that is not simply explained by theories about diet. So it is clear that coronary heart disease has multiple causes. Which causes are the most important cannot be resolved by descriptive epidemiology. Theoretically, the importance of different possible factors can be calculated as relative risks in **cohort studies** (described later). In practice, people are likely to be convinced only by studies in which the incidence of disease is reduced by removing one or more of the risk factors, that is, by conducting controlled trials of prevention, 'intervention studies'.

The descriptive approach to environmental epidemiology can be summarized as follows.

1. Define who is to be counted as a case of the disease in question.
2. Find the cases from various sources of information (mortality statistics, hospital records, occupational health records, special surveys).
3. For each case, find the age, sex, place of residence, date when the disease started or was diagnosed or caused death, and any other items that are deemed to be relevant.
4. Find the size of the population in which the cases have arisen.
5. Examine the data for patterns, calculate rates and compare with information about other populations.
6. See if there is correlation between the disease and the attributes of different times and different places.

7. Think of the possible explanations of what has been found, but **be cautious about making any firm conclusions**.

The reason for the note of caution about firm conclusions is that descriptive studies are not very powerful tests of theories on causation, although they are useful in the formulation of ideas. In modern times there are large computer banks of data on disease incidence and environmental factors. Remarkable coincidences are bound to be found when the computers are programmed to search through hundreds of combinations of disease and environmental factor. Not all the data are accurate and precise, particularly when they have been collected for purposes other than epidemiological research.

CLUSTERS OF DISEASE

Coincidences are especially difficult to interpret when the number of cases is small but significantly clustered in one place. The cluster of childhood leukaemias in south-west Cumbria, near to and possibly associated with the nuclear processing plant at Sellafield, is one well-known example. But in other parts of the country, doctors can point to villages or small areas where there have been five or six cases of leukaemia when only one or two should be expected from the regional or national averages. In some cases, this has added to the focus on nuclear installations, including the military establishments at Dounreay in Scotland and Aldermaston in Berkshire. There is another theory to explain these clusters of leukaemia, based on the influx of the workers and their families into areas that were previously much less populated. Viruses are a possible cause of leukaemia (at least in cattle and mice), and viruses spread more than usual when new populations are formed, in this case by industrial development. In such new populations, children would be likely to encounter more viral infections, and so might have an increased chance of getting leukaemia. This theory explains why leukaemia cases near Sellafield have not continued to be higher than the average, although improved radiological protection is an alternative explanation.

Clusters of severe diseases like leukaemia and other cancers naturally excite local interest, resulting in all manner of hypotheses and rumours. It is important to be careful and methodical when investigating such clusters. The Leukaemia Research Fund has published a useful handbook on the subject [2] and the Centers for Disease Control in the USA have written detailed guidance [3]. Essentially, the study of clusters of disease in a defined area is the comparison between the number of observed cases and the number that would be expected if the rates of disease in a 'reference' population applied in the study area. The reference population is usually the population of a larger area, for example, the whole of the county or the region. To make a valid comparison, the rates of disease need to be specific for age and sex. Other causative factors must be controlled, for example, smoking which plays a role in the development of several types of cancer. A single study with small numbers is unlikely to prove or disprove a causative hypotheses, but it should promote and sustain large well-designed studies that take several years to complete. Suggestions of local environmental causes for clusters of diseases should not be made without careful thought about the fear that may arise, fear that is not easily dispelled by later research. An epidemiological approach is to begin by thinking about the various causes of bias that could have led to the observation of an unusual cluster. Has a single label, for example, 'cancer', been attached to a mixed collection of diseases that are most unlikely to have a single common environmental cause? How reliable is the diagnosis, because doctors and patients are influenced by their knowledge of what is common in the area? Have the cases been brought together by medical facilities, either for diagnosis (and hence registration) or for treatment, where the cluster has nothing to do with the cause? Has the cluster been formed by finding one or two cases, looking for more and stopping when, and only when, the incidence appears to be abnormal? Is the ascertainment of cases more complete in the area of the cluster than elsewhere? Has publicity affected the evidence submitted by people involved?

These critical questions may give local people the impression that the epidemiologist does not understand their fears of a local cause. This impression is, of course, strengthened by statements about the unreliability of small numbers and the necessity for further studies. There is no easy answer to avoid this tension between understandable anxieties and the scientific approach, so caution is needed in reporting clusters of disease.

Once the cause of a disease is known, clusters of patients with the condition take on an entirely different significance. As with outbreaks of infectious disease, clusters of a non-infectious disease point to the need for more public health action. **Lead poisoning** causes anaemia and neuropathy. The clinical disease is well defined and readily confirmed by blood tests, which can show not only highly abnormal levels of lead but also the metabolic disturbances caused by lead. Lead is an accumulative poison, so prolonged exposure (e.g. in houses fitted with lead pipes in soft water areas; children whose furniture and toys are painted with lead-based paints; and in places where the streets are filled with cars emitting lead from petrol) increases the risk of disease, in contrast to an infectious agent, which would normally stimulate immunity or tolerance. There are two epidemiological methods to apply to this problem. The first is to set up surveillance to monitor the incidence of lead poisoning seen in clinical practice, so that every detected case is followed up with checks on the home, and the place of work of adult patients. The other method is by a **prevalence survey**, aimed at detecting overexposure before clinical disease has reached the stage of frank poisoning.

PREVALENCE (CROSS-SECTIONAL) SURVEYS

The **prevalence** of a condition is the proportion of people in a given population who have the condition; it is not the same as the **incidence**, which is the proportion who develop the condition in a given period of time, although the two terms are often confused. A prevalence survey is an obvious method for counting diseases or conditions that are chronic. In a chronic condition, some people may not seek medical advice until late, while others may stop attending doctors when they find that treatment does no more than palliate the condition. For both these reasons, medical records will underestimate the number of people with the condition. Yet other people will have subclinical effects, that is, will have no symptoms despite detectable pathology.

Following on with the example of lead poisoning started above, a prevalence survey could be the collection of symptoms and blood samples from a sample of the population in a town with a soft water supply. The purpose of the survey would be to find parts of the town, types of houses, age groups and social factors that were associated with abnormal lead levels and related illness. It follows that the resources to find and treat cases, and to repair the defects in the environment, should be concentrated on the areas of greatest need. It should be stressed that it would be unethical to make such a survey, with the intrusion into the lives of the people in the sample, unless it was expected that the survey could find a health problem and that the problem could be corrected.

In the example of a whole town being the population under consideration, it is likely to be more efficient to study a sample instead of every resident for a prevalence survey. A study of road traffic pollution and chest diseases in Munich, Germany, looked at children who were about 10 years old. The sample was children who were in one particular grade in the local schools and who had lived in the area for five years. The study found higher rates of respiratory illness and reduced lung function at the schools in districts with more car traffic. In other cases, such as the workforce of a factory making lead batteries, there may be strong arguments for studying all the staff to ensure that everyone is screened and that there are no unexpected pockets of exposure or disease. The decision about whether to survey a sample or the whole population must be made with statistical advice on the appropriate number of people to study. If a sample will give sufficient numbers, the statistical advice should be extended to the method of selecting a sample that properly represents the population under consideration.

CASE-CONTROL STUDIES

In a case-control study, all the cases of the disease in question are taken and their histories of exposure to the possible environmental factor compared with the rate of exposure in controls. The controls are selected to be people without the disease in question, who are a comparison group for the cases. The controls should have had the same chance of exposure as the cases before the start of the enquiry. The history of exposure is often found by asking cases and their controls the same questions, either by interviews or by written questionnaires. The early cases whose medical histories suggested a link between the factor and the disease in the first place should not be included in a case-control study to test the hypothesis.

As mentioned above, tobacco smoking was suspected as a cause of lung cancer in the United Kingdom and the United States in 1947. Since the original case-control study that demonstrated the association between tobacco smoking and lung cancer 40 years ago, there have been questions about the cause of lung cancer in people who do not themselves smoke. One possibility has been that at least some of these non-smoker cases have been at risk from exposure to other people's tobacco smoke, the so-called '**passive smoking**'. Lung cancer in non-smokers is rare, but case-control studies are suitable in this situation. In the case-control studies on exposure to environmental tobacco smoke, the groups of cases have been formed by taking patients with a proven diagnosis of lung cancer but who have never smoked. In some studies, controls were chosen from among patients in hospital with other diseases, to be of about the same age and to have the same proportion of men and women. In other studies, the controls were drawn from a sample of the population, and were not confined to hospital patients. Exposure to an environment of tobacco smoke was defined as living with a smoker (usually having a spouse who smoked). More of the cases with lung cancer had lived with cigarette smokers than the controls. The difference is not very large, as living with a smoker increases the risk of lung cancer by about 30%. Nevertheless, this level of risk is sufficient to cause more than 200 deaths a year from lung cancer in non-smokers in the United Kingdom. If any domestic product or industrial air pollutant had this effect, there would be urgent action to control it. What has occurred is a slow but determined public health movement to reduce exposure to other people's tobacco smoke.

A single case-control study finding this level of increased risk may not be very convincing on its own. Whenever possible, a second case-control study should be conducted on a different population by another team of investigators. This is exactly the same process as applies in the physical sciences, when experiments are repeated before their results are accepted. In the example of passive smoking, there have been at least 37 case-control studies, and their combined results were used to make the estimate of increased risk given above. The pooling of the results of several studies is called **meta-analysis**. Meta-analysis is being used more and more in epidemiology because it helps to estimate the risk of an environment factor more accurately than can be achieved by a single study.

The difficulty in convincing people of risks on the basis of case-control studies should not be underestimated. The method is not familiar to the general public and politicians. Case-control studies have shown the causes of disease before the pathological mechanism is known and before the causative agent is precisely identified. The task is even more difficult because epidemiological evidence is unlikely to be available before a chemical, a product or a physical process is in widespread use. So there will be an established lobby of producers and users to resist the allegations of a health hazard. The case-control method has advantages in determining the environmental causes of disease. Because it depends on cases that have already occurred, it can produce results relatively quickly. By concentrating on cases and a similar number of controls, it is relatively inexpensive. There is no need to depend on other estimates of the normal incidence of disease, the uncertainty of which can bedevil descriptive studies.

In conducting case-control studies, the control group must be chosen carefully to avoid **bias**. If the controls or the people who interview them know what the suspected factor is and that they are controls, their answers are likely to reflect this knowledge. This may happen when the control group is formed and studied after the cases. The controls should be of the same age, sex and social background as the cases. Controls of this sort may be obtained by asking patients to name friends and neighbours. If there are several ways of choosing controls, for example, other hospital patients and neighbours, the option of using more than one group of controls should be considered carefully. Using more than one group may increase the work, but it may also improve the reliability of the result.

One problem that has to be guarded against is **confounding**. Confounding happens when the link between a disease and a supposed factor is not a direct cause, but is due to something else to which both disease and factor are linked. In some British cities, mothers belonging to ethnic minorities are more likely than other mothers to have babies of low birth weight. Ethnicity is a confounder, because low birth weight is caused by other factors. The cause is environmental and lies in the quality of antenatal care, poverty, diet and home conditions. Once confounding is suspected, it can be taken into account in the analysis and prevented from confusing the results of a study.

DOSES OF ENVIRONMENTAL FACTORS

Another aspect of environmental epidemiology has been demonstrated by the case-control studies on lung cancer and tobacco smoke. This is the effect of **increasing dosage**. The importance of considering the extent of exposure has been described above. Logically, one would expect that the greater the exposure, the greater the chance of disease if it is true that the factor under study is the cause of the disease. If one found the inverse (higher dosage, less risk of disease), one could doubt that the association between the preceding factor, for example smoking, and the disease was a cause and effect relationship. With both smokers and non-smokers who live with smokers, the risk of lung cancer increases with the number of cigarettes smoked and the number of years exposed to tobacco smoke. So a **dose effect** is present.

Another situation in which the dose effect of an environmental factor is of practical importance in epidemiology is when there has been exposure to a single source of an airborne toxic material. This was illustrated by the gas leak of methyl isocyanate and other gases at the **Union Carbide plant at Bhopal**, India, in 1984. The immediate toxic effects were well known as soon as the accident happened. What was less certain was whether **chronic lung disease** would result from the lower concentrations experienced over a wider area. An international commission made a cross-sectional study on a sample of the population 10 years after the incident. They found that there was a gradient of effect, with respiratory symptoms and diminished lung function (an objective measurement) being more common the closer the people were to the factory at the time of the explosion. These findings were controlled for height, age, smoking and literacy. The effects were strongest on people who lived within 2 km of the factory, and were less at distances of over 6 km.

Working in the other direction, one might ask if there is a dose below which there is no ill-effect. This appears to be so with some toxins and some infectious agents. When it is possible to collect information that measures the amount of exposure, this should be done because of the difficulties of removing substances that have been in common use. With saturated fats in the diet and the risk of ischaemic heart disease, the advice is to modify the diet without a total abstention from saturated fats. With carcinogens, the present view is that there are no thresholds of safety. The decision on 'safe levels' becomes a balance of risks between the use of carcinogenic substances, like tobacco or asbestos, and radiation on the one hand, and the use of alternatives or doing without. In this decision, the calculation of risk needs to be accurate. One disadvantage of the case-control method is that the calculation of risk has to be made indirectly, in contrast to the next method described.

COHORT STUDIES

The limitations of the case-control method may be overcome, at least in part, by applying the **cohort method**. With this method, a group or a sample of a population is observed over a period of time to see who becomes exposed to various environmental factors and who develops disease. This method is sometimes called a **prospective study** (in the sense that one starts with a group of well people and looks forward to future events, like the onset of disease) or a **longitudinal study** (because it is conducted over a long period of time). The immediate advantages of this method are that one can plan to generate the required information instead of relying on existing records and the memories of the subjects. The level of exposure can be measured instead of estimated. The possibility of bias in the choice of controls is avoided. The price to be paid for these advantages includes the greater cost and organization employed in following up a large group of people, and the delay until the group has been exposed and the disease has had time to develop.

One example of a cohort study is the follow-up of adults who lived in the US town of **Framingham, Massachusetts** in 1950. The purpose of this was to learn more about the causes and development of **ischaemic heart disease and hypertension**. The study was run for more than 30 years. At two-year intervals, the adult subjects in the study were medically examined, and blood pressure and blood cholesterol levels, among other things, were measured. Admissions to hospital for, and deaths from, heart disease were recorded as outcomes. With the passage of time it became clear that a high blood cholesterol at the start predicted an increased risk of a heart attack, especially in men. The increase in risk was shown to depend on the level of cholesterol. From this, doctors could calculate how much heart disease could be prevented by reducing cholesterol levels to normal in those most at risk. This result was only the end of the first stage in preventing heart disease, because the methods of reducing cholesterol by modifying diets still needed to be proved.

One problem that can arise in a cohort study, and which affected the Framingham study, was that a proportion of the subjects recruited at the beginning moved away, or stopped being followed up for other reasons. This could seriously weaken the conclusions of a cohort study, partly because the size, and therefore the power, of the study falls, and partly because the people lost to follow-up may be significantly different from those who stay. There are therefore advantages in using cohorts which are more likely than a sample of the general population to stay in the study. An example of such a cohort is the employees in a large organization. Their qualifications, their pensions and the occupational health services all help to keep the subjects known to the researchers. Civil servants in Whitehall, London have been followed up like the citizens of Framingham, and have demonstrated the benefits of exercise in preventing heart disease. Like other well-conducted cohort studies, it provided a large amount of interesting observations, not the least of which being the effects of socioeconomic status on health. The people in the better-paid posts were generally healthier than those in the lower-paid posts, despite the initial idea that the senior posts were desk jobs with little physical activity. The difference remained after adjusting for the healthier lifestyles (fewer smokers or reduced consumption, better diets, more exercise) of the better-paid staff, and could not be explained by the possibility that fitter people get better jobs. This observation opens more questions for research into environmental factors that are linked to economic status.

Workers in nuclear installations provide another example of a group of people followed in a cohort study. They are monitored for their exposure to **radiation** (including measurements of dosage and frequency) and for their incidence of cancer. At Sellafield, there have been slightly more deaths in workers from myeloma than expected (7 instead of 4) and slightly more prostatic cancer deaths (19 instead of 16), but other types of cancer, including leukaemia, were rather less common. Staff exposed to radiation were not more likely to die early or get cancer than those who were unexposed. This helps to show that careful environmental practices can control hazards to acceptable levels of risk.

Ischaemic heart disease has been sufficiently common among non-smokers for cohort studies of the risk from **environmental tobacco smoke** (passive smoking) to be assessed. In a meta-analysis of 19 studies, including one that followed up a generation of British doctors, a non-smoker who lives with a smoker has a risk of ischaemic heart disease that is 23% higher than other non-smokers, because small concentrations of tobacco smoke increase the clotting of blood in the circulation.

A cohort study may seem to be a slow, deliberate task, but there are occasions when this method is used quickly in an emergency. An example is the investigation into the health effects of the oil spill from the *Braer* **tanker disaster** in Shetland in 1993. In this investigation, all the islanders who lived within 4.5 km of the tanker's grounding were followed up to see what symptoms they experienced and what illnesses emerged. The study was very useful in recognizing the symptoms of headache, throat irritation and itchy eyes, and in reassuring the local population that no severe diseases were caused by the exposure to airborne oil droplets.

INTERVENTION STUDIES

The epidemiological methods discussed so far could be considered to provide circumstantial evidence rather than direct proof. Direct proof would be given by intervening to remove a suspected environmental cause of disease. The logic is that if factor X is truly a cause of disease Y, then intervening to remove X will reduce the incidence of Y. If Y does not fall with such an intervention, then X is not a cause. In the late eighteenth century, it was recognized that fresh fruits and vegetables, particularly lemons, prevented scurvy in sailors in the British Navy, long before the exact cause of the disease, vitamin C deficiency, was identified.

In medical practice, there has been a revolution of thinking in the past 50 years in how doctors choose between different drugs and other treatments. In the past, empirical choices were made: the treatments that appeared to work were used again, and passed on in conventional teaching. Perhaps because the treatments were not very powerful, and perhaps because there was greater respect for the 'experience of the master', the traditional system was used for centuries. Nowadays, new drugs are tested in **randomized controlled trials**, designed to remove any bias. These trials come close to the standard of biological experiments in their objectivity, and they have set the pace for preventive medicine.

In intervention trials, it is essential to make the groups of patients given different treatments as equal as possible in all respects before treatment starts. This is best done by allocating patients to the different groups by randomization, which gives all patients an equal chance of receiving any of the treatments being tested. This system of randomized allocation can be used in trials of vaccines or other prophylaxes that are given to individuals, although randomized trials of prevention require a formidable number of subjects.

It is much more difficult to use randomization in designing intervention studies of environmental change. In most circumstances, the nature of the intervention is such that there has to be a deliberate choice about which population is treated. It is tempting to assume that the incidence of disease in the treated population before and after the change will show whether the intervention works. But it is wise to have a contemporary untreated population as a control. A project in Finland to prevent heart disease is a good example. **North Karelia** was a Finnish community with a high incidence of **ischaemic heart disease**. The preventive strategy was aimed at the multiple factors of heart disease, including smoking, blood pressure and diet. A neighbouring and similar community, **Kuopio**, was selected as a control population. This proved essential to the interpretation of the effects. Although smoking was reduced and blood pressure was treated, similar improvements were observed throughout Finland. Mortality from ischaemic heart disease fell in both Karelia and the neighbouring Kuopio.

Fluoridation of the water supply has been restricted by pressure groups to a minority of households in the United Kingdom. This has created a 'natural' experiment whereby the intervention could be tested. The populations of Birmingham and other places supplied with water that has additional fluoride added have experienced far less dental caries than people living elsewhere. There will be other opportunities for natural experiments when local authorities take effective action to reduce road traffic in cities. If banning traffic from a city centre increases the traffic density in the suburbs, there could be a health loss, because the change could increase the air pollution in residential areas, where young children are most exposed. If the change does not increase air pollution, but traffic flow and speeds rise, there may be more injuries due to accidents. A good study of the health of a sample of the population in both the city centre and the suburbs before and after the change in traffic will be enlightening. The important public health point is that environmental change has unpredictable effects, and we must anticipate and encourage epidemiological research whenever and wherever opportunities arise. Another point is that the health effects of environment factors cannot be adequately assessed by laboratory tests alone, particularly not by tests on laboratory animals.

ASSESSMENT OF EPIDEMIOLOGICAL STUDIES

Epidemiology is the first line of investigation of environmental hazards. The results of epidemiological studies may therefore be the first evidence of the risks involved. It is not difficult for the significance of the early evidence to be exaggerated by journalists, or to be underrated by those whose business or lifestyle is so challenged. To assess early evidence, the following checklist of questions may help.

1. What do you know about the authors of the report? Have they been trained to be objective and scientific?
2. Is the method used in the study adequately described? The possible weaknesses in the methodology should be discussed by the authors.
3. How many subjects recruited into the study left it before it was completed? Studies with drop-out rates of more than 30% should be treated with caution.

4. If there is a control group, has it been chosen in a way that offers a fair comparison with the cases (case-control study) or treated group (intervention trial)?

5. If there is no control group, is there a reasonable calculation of what might have been found without the factor or treatment under study?

6. Is there a dosage effect, and do the conclusions make biological sense? There should be a plausible explanation of how the environmental factor causes disease, even if the detail is not yet known.

7. Is the association between the environmental factor and the disease a strong one? The association should be tested by conventional statistical methods, but a lay person's guide is to ask how many of the cases are explained by the alleged causative factor.

8. Is there supporting evidence from independent studies? This is a very important question.

9. Does the study lead on to further research, by showing how it can be repeated elsewhere either to gain independent confirmation or to explore ideas and problems thrown up by the study itself?

10. Does the study lead to some practical means of preventing disease?

A summary comparison of epidemiological methods is shown in Table 15.2.

CONTROL OF ENVIRONMENTAL CAUSES OF CHRONIC DISEASES

The control of environmental causes of chronic diseases is simplest when a non-infectious disease has a known cause, usually from a toxic polluting substance, which can be analysed epidemiologically. An example is the prevention of mesothelioma (a cancer of the chest) by the control of asbestos dust. Once the hazards of asbestos were recognized, alternatives to the substance were used as far as possible, especially in heat insulation. When asbestos has to be encountered (e.g. when removing it from a building), exposure is minimized by dust control and protective clothing and respirators. When the cause is not so easily identified and removed, as is often the case, a precautionary approach is required, trying to reduce exposure to possible or suspected hazards. This has led to the adoption of standards set for the maximum exposure that people should encounter. While statutory controls on many polluting substances are applied under environmental and occupational legislation, they are based on technical standards rather than on an evaluation of the health risks involved.

In considering exposure risk to environmental pollutants that may cause non-infectious diseases, it is important to recognize that most are **multisource in origin** and that many are **cumulative in effect**. It is only by considering the total potential

Table 15.2 Comparison of epidemiological methods

Useful for	Descriptive	Prevalence	Case-control	Cohort	Intervention
Rare disease	Yes	No	Yes	No	No
Rare cause	Yes	No	No	Yes	Yes
Time relations	Yes	No	No	Yes	Yes
Clustering	Yes	Yes	No	Yes	No
Risk assessment	No	No	Yes	Yes	No
Quick results	Yes	Yes	Yes	No	No
Low resources	Decreasing from descriptive to intervention				
Scientific rigour	Increasing from descriptive to intervention				
Avoiding bias	Increasing from descriptive to intervention				
People who may move	Yes	Yes	Yes	No	No

exposure, in a holistic way, that the risk to a particular individual may be quantified. It is the sum of occupational exposure, environmental exposure and dietary exposure, plus domestic, social and leisure exposure that determines the total biological impact of a substance on a particular individual or group. For that reason, it is vital that environmental health professionals should be constantly aware that the particular situation they are assessing is likely to be but part of the exposure pattern.

Generally, it is difficult to make allowances for individual sensitivity to a substance or predisposition to disease, and much greater research is required to even begin to estimate dose–response rates. However, children generally have a greater sensitivity to toxic materials and have a smaller body mass, which, taken with their lifestyle, particularly in the case of younger children, is likely to expose them to higher levels of contaminants, for example, from dust and soil. Behaviour patterns such as pica may also add to this exposure. The risk from the effect of toxic materials is thus generally much greater for children, and they merit special consideration and higher levels of protection.

CONTROL OF MATERIALS

In the United Kingdom, environmental pollutants are controlled almost entirely by reference to technical or statutory standards, without necessarily any direct relevance to their local health impacts. Rather, what are often the determining local factors are the existing background levels and the standards that are achievable against this background. Generally, standards are adopted after a review of the existing literature and established industrial practice, plus a degree of harmonization with other European countries. They are often carried forward without updating or further research into their relevance. Table 15.3 outlines existing environmental health controls on potentially harmful materials, although the relevant chapters should be read for detailed information. While these standards lack a holistic interrelationship and are not directly related to overall health effects, they do still provide a threshold in each

case and have undoubtedly provided a level of protection. A large amount of monitoring of these standards is undertaken by local authorities and government agencies, but there is no obvious link to the work of other professionals on the health consequences. A collaborative approach would clearly be beneficial, with possible medium-term health gains available.

In the United States, the approach is somewhat different. Almost all states have established procedures for investigating clusters of health events, and the federal Agency for Toxic Substances and Disease Registry (ATSDR) is required by Congress to make an annual priority list of toxic substances and, more importantly, include a public health statement on each. The ATSDR's top 20 toxic substances for 1995 are listed in Table 15.4 [4]. The ATSDR listed a total of 275 substances in its annual statement to Congress in 1995. British readers may find some of the highest priority substances surprising, as they include some of the oldest toxic substances known and used by humans. Perhaps it is because of their ubiquitous presence that they have gained an acceptance as being part of the urban environment, but it is this very widespread nature of their existence and consequent multisource availability that makes for greater risk and difficulty in assessment; and time has not reduced their toxicity. Some substances, such as lead, were at one time thought no longer to pose a public health risk, until research commencing in the 1950s demonstrated more subtle toxic effects at subclinical dose rates.

In the ATSDR's public health statements, most (but not all) of the top 20 are regarded as either having been established as being carcinogenic, or being reasonably anticipated to be so. In addition, they all cause a variety of long-term, chronic health damage, and in some cases are capable of producing severe acute effects at relatively low dosage rates. In the case of most substances there is an established need to carry out considerable further research before the limits of their chronic health effects, including cancers, can be clarified. Much of the current information is based on animal studies rather than on epidemiological surveys of human populations.

Table 15.3 Current environmental health controls on potentially harmful substances

Sector	Regulations applying/substances affected
Food	Contaminants (mostly metals) Additives (colour, flavouring, preservatives, sophisticants, etc.) Packaging materials
Water	WHO guidelines EU standards Bathing waters (18 physiochemical parameters) Abstracted waters (21 physiochemical parameters) Drinking waters (51 physiochemical parameters)
Occupations[1]	Occupational health, COSHH regulations Maximum exposure limits, 42 substances Occupational exposure standards, 450 substances
Housing	Radon Indoor pollutants (nitrous oxides, sulphur dioxides, etc.) Leaching of methane Possible electromagnetic radiation
Air and land[2]	WHO air quality guidelines EU directive standards and UK regulations (lead, nitrogen dioxide, sulphur dioxide and smoke) Recommendations of the DoE Expert Panel VOCs: benzene

Notes:
1 Responsibilities are placed on employers to carry out monitoring and assessments, including combined effects.
2 There are no action levels or guide levels for land; standards are based on large-scale land surveys and practical, achievable levels in a particular area.
WHO, World Health Organization; COSHH, Control of Substances Hazardous to Health Regulations; EU, European Union; DoE, former Department of the Environment; VOCs, volatile organic compounds.

CONTROL PROGRAMMES

If a local control programme is to be introduced there should be an adequate organizational structure and sufficient resources to allow for long-term evaluation and intervention initiatives. A collaborative approach is clearly essential, partly because of the widespread interest in health events once they enter the public domain, and also because of the very technical and professional complexity of the issues involved. It is also important to remember that pollution and disease are no respecters of geographical boundaries, and that interauthority working is as important as intersectoral cooperation. Consideration must be given to how the effect of the control programme is to be monitored by health statistics, and this may be difficult where local authority and health service boundaries are not co-terminus (as in England). It may take a long time for the intervention to produce an change in a chronic disease, and multiple factors may enhance or hinder the planned intervention. This may be illustrated by the example of an integrated transport strategy to reduce motor vehicles in a city: the health benefits would come from increasing the physical activity of the citizens, and hence improving their cardiovascular fitness and bone density, while at the same time reducing air pollution, respiratory symptoms, and decreasing the motor vehicle accidents. But such a strategy might meet opposition from businesses and car owners in the city, and there would have to be concerted efforts to make such a strategy work. The strategy would have to be monitored carefully to ensure that there

Table 15.4 The top 20 toxic substances listed by the ATSDR in 1995 [4]

Ranking	Toxic substance
1.	Lead
2.	Arsenic
3.	Mercury metallic
4.	Vinyl chloride
5.	Benzene
6.	Polychlorinated biphenyls (PCBs)
7.	Cadmium
8.	Benzo(a)pyrene
9.	Chloroform
10.	Benzo(b)fluoranthene
11.	DDT
12.	Arochlor 1260
13.	Trichloroethylene
14.	Arochlor 1254
15.	Chromium
16.	Chlordane
17.	Dibenz [a,h] anthracene
18.	Hexachlorobutadiene
19.	DDD
20.	Dieldren

are no perverse effects, for example, increasing traffic speeds without protecting cyclists and pedestrians from motor vehicles could lead to more fatal accidents. The cardiovascular and bone density benefits might take a generation to become apparent, so a substantial commitment would be needed from all parties.

An outline of the component parts of an organization for providing a medium- to long-term evaluation programme as well as a timely public health response to events, if necessary, is given below.

Programme director

This person is a prime necessity, although it is not a full-time position and the person is nominated rather than appointed. The nominee could be one of a number of local individuals, from the local authority, the health service or a local academic. Clearly, local circumstances will determine who best fits this post, but it is important that the individual should have sufficient stature, authority and reputation to command the respect of others (e.g. Director of Public Health, Director of Environmental Health, or the head of a University Department with health and/or environmental interests).

A professional advisory group

This is necessary to advise the programme director from time to time. Scientific, technical and professional issues are complex and require high-level contributions on clinical and epidemiological aspects of non-infectious diseases, chemical analysis and chemical engineering, sampling and survey systems, toxicology, public health medicine and environmental health as a minimum. This may be an *ad hoc* group formed for the specific programme, or may be a group that already exists, perhaps as an academic team within a medical or public health faculty.

A public liaison group

This is necessary to consider the wider interests of the community. It should include representatives of local communities, health services, pressure groups, the voluntary sector, central government agencies and, possibly, the media. Its purpose is to identify areas of concern, discuss issues and advise the director on priorities.

Operational arrangements and resources

Nothing will be achieved without the dedication of local resources to enable analysis and recording of information and epidemiological investigation of cases. As a minimum the following should be established:

- a local database of information on chronic, acute and subacute toxic exposures to local populations, bringing together information on public health surveillance and investigation of health events;
- accurate, corrected local mortality and disease statistics to assist with identification of potential problems;
- an established protocol for investigating health events;
- adequate, dedicated field staff to carry out investigations and conducting studies;
- availability of high quality epidemiological participation capable of evaluating the data accurately.

Administrative procedures for corrective intervention

In general, this should be by established routes rather than by making special arrangements, and will need to be put into action by the appropriate agency.

CONCLUSION

Non-infectious diseases with environmental causes have been identified as the main health problems and immediate challenge for the future. Steps have been taken, particularly in the field of health education and promotion, to intervene, but as yet no co-ordinated local action to examine and seek to control potential environmental causes has been established in the United Kingdom. While it will not be possible to control non-infectious diseases in the same way or over the same timescale as infectious diseases, they remain of greater importance and capable of offering a greater health gain in the longer term.

REFERENCES

1. National Statistics (2002) *Key Population and Vital Statistics 2000 VS no. 27 PPI no 23*, The Stationary Office, London.
2. Leukaemia Research Fund Centre (1997) *Handbook and Guide to the Investigation of Clusters of Diseases*, University of Leeds, Leeds.
3. CDC (1990) *Guidelines for Investigating Clusters of Health Events*, Centers for Disease Control, Atlanta.
4. ATSDR (1995) *Toxicological Profiles*, Agency for Toxic Substances and Disease Registry, Atlanta.

FURTHER READING

Bithell, J.F. and Stone, R.A. (1989) Statistical methods for analysing the geographical distribution of cancer cases near nuclear installations. *Journal of Epidemial Community Health*, **43**, 79–85.

British Medical Association (1990) *Guide to Living with Risk*, Penguin, London.

Campbell, D., Cox, D., Crum, J. *et al.* (1993) Initial effects of the grounding of the tanker *Braer* on health in Shetland. *British Medical Journal*, **307**, 1251–5.

CIEH (1993) *The Health of the Nation for Environmental Health*, Chartered Institute of Environmental Health, London.

Cullinan, P., Acquilla, S. and Ramana Dhara, V. (1997) Respiratory morbidity after the Union Carbide gas leak at Bhopal: a cross sectional survey. *British Medical Journal*, **314**, 338–43.

Detels, R., Holland, W.W., McEwen, J. and Omenn, G.S. (eds) (1997) *Oxford Textbook of Public Health*, volume 2, *Epidemiological and Biostatistical Approaches*, Oxford University Press, Oxford.

Hacksaw, A.K., Law, M.R. and Wald, N.J. (1997) Accumulated evidence on lung cancer and environmental tobacco smoke. *British Medical Journal*, **315**, 980–8.

Jacobs, R. (1992) *Polychlorinated Biphenyls, Dioxins and Furans: Human Health Effects and Population Studies*, Welsh Office, Cardiff.

Johnson, B.L., Andrews, J.S. Jr, Xintaras, C. and Mehlman, M.A. (1998) *Advances in Modern Environmental Toxicology*, volume 25, *Hazardous Waste: Toxicology and Health Effects*, Princeton Scientific Publishing, New Jersey.

Law, M.R., Morris, J.K. and Wald, N.J. (1997) Environmental tobacco smoke exposure and heart disease: an evaluation of the evidence. *British Medical Journal*, **315**, 973–80.

World Health Organisation (1991) *Investigating Environmental Disease Outbreaks – A Training Manual*, WHO/PEP/91.35, Geneva.

World Health Organisation (1992) *Our Planet, Our Health – Report of the WHO Commission on Health and the Environment*, WHO, Geneva.

WHO (1998) *Assessing the Health Consequences of Major Chemical Incidents – Epidemiological Approaches*, Stationery Office, London.

Part Five
Housing

16 Housing conditions and health

David Ormandy

INTRODUCTION

It is accepted that there is a direct relationship between the environment and our health. For human society, housing is a fundamental necessity. It is where we spend a major part of our lives, and is a complex environment in its own right. It provides a haven from the outside environment, gives privacy for the individual and the household; it is where we cook, eat, sleep and wash ourselves and our clothes; it is where children learn to crawl and then walk, and where they start to learn about their world; it is where we retreat when unwell; and it is where we retire.

It may be obvious that, as such a vital part of the human environment, housing will influence health, well-being and development. However, demonstrating the potential relevance of individual housing conditions and features, and of combinations of them, is not straightforward.

Housing is expected to meet the needs and aspirations of a wide spectrum of different households. What provides a safe and healthy environment for one household may be inappropriate for another.

There are several reasons why it is important to show how and to what extent housing conditions can and do influence health. As well as showing the impact on the individual, the household and on society, identifying the cause can indicate where action can be directed.

Physical injuries and ill health have a direct and indirect cost to society. There is the cost to the health service treating health conditions and injuries. Working days lost through illness and injuries cost both the economy and the individual through loss of earnings. Less easy to quantify is the cost in suffering and the effect of missed education.

Where the source of potential threats to health and safety is the design, construction or maintenance, these can be avoided or reduced through changes to minimum standards. Building regulations can help prevent identified design and construction problems, intervention standards can help deal with problems in older housing, and subsidies can encourage improvements. Home safety and health education can help avoid habits and practices which may increase or cause threats to health or safety in and around the house.

DEFINITIONS

To discuss the relationship between **housing** and **health** it is necessary to be clear about what constitutes each. (It is also important to note that definitions are given for particular purposes; the definitions given here for discussion purposes will differ from those given in legislation.)

First, a definition of **health** (see also the definition and discussion in Chapter 1). The World

Health Organization defined health as 'a state of complete physical, mental and social well-being and not merely the absence of disease or infirmity' [1]. This definition is clearly very wide, but is nonetheless a target. It should be noted that it includes mental and social well-being, and is not limited to soundness of body (a dictionary definition of health).

For discussion of its relationship with health, **housing** is made up of four interrelated components. These are the **dwelling**, the occupying **household**, the **neighbourhood** in which the dwelling is situated, and the human and social **community** in the neighbourhood.

Dwelling and household

The central component of a **dwelling** is a structure (or a part of one) which is used, or intended to be used, for human habitation. It also includes, however, any private or shared space and outbuildings associated with that structure. Thus, the term is wide enough to include both a house with its own private garden, and a bedsit and its associated shared use of a bathroom and kitchen. It also includes a caravan (or other moveable structure) and, where relevant, any communal personal washing facilities and sanitary accommodation provided in association with the standing or pitch.

At the very least, the dwelling should be capable of satisfying the basic and fundamental needs for human existence. It should provide shelter, space and facilities to fulfil the requirements for everyday domestic and natural life. A garden or yard associated with the structure will provide amenity or recreational space.

The **household** is the individual, or group of two or more individuals, occupying a dwelling. While the term 'home' is often used as a synonym for dwelling, more properly it is the social, cultural and economic structure established by a household. The characteristics of a home may vary considerably, dependant on the composition of the household, and on its priorities, needs and expectations. A household with young children will have very different priorities and needs compared to one composed of elderly and retired adults; similarly the home established by a household with reasonable financial resources will be very different to that of a household on low income or reliant on State benefits.

Some dwellings are designed (or adapted) to suit the perceived needs of a particular group – for example, dwellings for the frail elderly may have alarms connected to a warden. However, the majority of dwellings are somewhat of a compromise, designed to meet current social expectations for the area, a price range, and a general category of households. This means that a dwelling will be expected to meet the general needs of a spectrum of households who may occupy it.

Neighbourhood and community

The **neighbourhood** is the physical environment in which the dwelling is situated. It includes both the public areas, such as the roads, footpaths and public parks, and the commonly shared space separate from the public space. Internal commonly shared space includes staircases, halls and passages providing access within a building containing dwellings (such as a block of flats). External commonly shared space can be the greenspace, parking areas, footpaths and other shared facilities (refuse containers) associated and enjoyed with a group of dwellings.

The **community** is the residents in the neighbourhood and those who provide services and support to its inhabitants. This includes the shopkeepers, local religious leaders, local politicians, publicans, school staff and other community-based officials and representatives.

RELATIONSHIP BETWEEN HOUSING AND HEALTH

It should be a general principle that housing should provide an environment which is as safe and healthy as possible. However, it is impossible and undesirable to remove all potential dangers and hazards. Some are a necessary part of the housing environment, others are endured or even expected because the benefits gained make the risk acceptable.

In the United Kingdom, it is expected that there will be electricity for lighting and perhaps for cooking and heating. Supplies of gas are normal, and the use of solid fuel is not unusual. These are potentially dangerous, and require both adequate built-in safety precautions and also a recognition and appreciation of the dangers by the household. Similarly, there are design features and facilities expected and accepted in dwellings, such as stairs, windows, doors, supplies of hot water and sanitary accommodation. These also can be dangerous, increasing the possibility of falls, of trapped fingers, of scalds and of the spread of disease. Again, there is a need for safe design and for safe usage. There are similar and some greater risks, such as road traffic, in the immediate housing environment and the neighbourhood generally.

As it is not possible to remove all risks to health and safety from housing, the aim should be for all necessary and unavoidable dangers to be made as safe as possible and to make households and communities safety conscious.

The approach should be to acknowledge that there are some threats to health and safety which may be solely related to the design, construction and maintenance of the dwelling and the neighbourhood, some which may be solely related to careless or unreasonable human behaviour and others which are a combination of the two. This approach accepts that a dwelling should be capable of being occupied by a range of households with a spectrum of lifestyles, and that a neighbourhood should meet the needs of a wide range of households whose members are likely to include both the elderly and the very young.

Also, it is arguable that it is the most vulnerable sections of the population who spend the largest proportion of time at home and who make the greatest demands on it. This includes the very young, the elderly, pregnant women, the unemployed and those vulnerable because of sickness or other physical or mental conditions. Those making the least use of dwellings and making the least demands on them are the healthy, employed members of society.

The personal or human factors identified in the research on Sick Building Syndrome (SBS) can be transposed to housing. The term SBS is used to describe the phenomenon where users of a specific office building experience a range of symptoms which either disappear or considerably diminish when the victim is away from that building for a period. Research has found that SBS is a result of a combination of apparently unrelated and otherwise minor factors which can be divided into building factors and personal (or human) factors [2]. Among the human factors identified, the symptoms were found more likely to occur in women, those in routine jobs, those with a history of allergy, and those who felt they have little control over the indoor environment.

In housing terms, many of the tasks involved in housekeeping are routine – washing and drying clothes, dusting, vacuuming and bed-making; and many of those who are ill, whether through infection or allergic reactions, stay indoors at home and the policy of Care in the Community is ensuring that more of those who are vulnerable spend time in their home environment. Finally, tenants and low income owner-occupiers will feel that they have little control over their indoor environment, and being unemployed usually means low income which again will mean having little control over the indoor environment.

'The person-centred approach to "building safety" begins with "users" and focuses on **safety in use.** It first seeks to establish, through the use of accident and incident data, the nature and prevalence of hazards, hazardous situations and hazardous events.' [3]

DEMONSTRATING THE POTENTIAL HEALTH IMPACT OF HOUSING ON HEALTH

There is a considerable library of such studies establishing associations between housing conditions and their potential impact on health. The majority of these studies, to demonstrate the impact of a particular issue or aspect of housing, such as dampness, cold or accidents, have isolated or highlighted that particular issue. The aim being to investigate a proposed direct causal link or association between the issue and health status, and, if it exists, to quantify that association.

Such studies will often include taking account of, and attempting to compensate for, confounding factors such as social, economic and behavioural factors. Where they do so, and show a direct causal link, they provide solid and uncontroversial proof of the health impact. The reverse, however, is not true. Where they do not prove a link between isolated factors may be because the threat to health results from a combination of more than one factor, perhaps one a building factor and the other a human factor. For example, a dwelling which is difficult to heat because of limited thermal insulation may be satisfactory for a household on a reasonable income who can afford the cost of heating; but would be unsatisfactory for a household on low income. Similarly, a dwelling which seems satisfactory for a household consisting of two working adults, may be unable to cope with the moisture generated by a household which includes very young children spending most of the time indoors when it is cold and wet outside.

Studies are now acknowledging that housing is a complex environment and that the interaction between building factors and human factors should not be discounted; compensating for the so-called confounding factors can hide housing problems. Poverty can force people to live in unsatisfactory housing, and those with less control over their housing conditions, such as tenants and low income owner-occupiers, will feel stress and dissatisfaction if their housing is of poor quality. And social expectations, influenced by the media (especially television) can lead to dissatisfaction. These and other additional risk factors mean that a more holistic approach to housing–health research, while more complex, will provide more meaningful and useful results.

This recognition suggests that there is a need for an extended view of housing and health, integrating and acknowledging that the mental, social, cultural and economic aspects of the household and the community are fundamental to the housing–health relationship. From this perspective, what may seem to be confounding issues become central to understanding the association between housing and health and to informing housing policies.

Spot the research method

Studies to examine the relationship between housing and health, unlike laboratory experiments, cannot adopt simple strict controls. This means that there will be limitations on interpretation of the results, depending on the methodology adopted. It can be useful to identify the particular type of study, as this can give an indication of the strengths and weaknesses of the findings. Studies will usually fall into one of the following categories:

Descriptive or cross-sectional – These usually describe housing conditions and the health of the occupants at a particular time. Deductions will be made based on the association between housing and health, such as dampness and respiratory conditions. Individually, such studies give an indication of a possible link between housing conditions and health, rather than conclusive proof that the state of health is a direct result of the housing.

Case control – This type of study will start by identifying two groups of people, one with a particular health condition such as asthma, and one without that condition but otherwise similar. It will then compare the housing conditions of both groups to try to isolate potential links between housing and health. Where both a particular housing issue and health condition exist in one group and both are absent in the other, then such studies generally provide relatively strong support for the proposition that there is a link between the two.

Longitudinal – As with a case control study, two groups or cohorts of people are identified, one of which will be exposed to housing conditions considered to have some health impact. For this type of study, the health and housing conditions are monitored over a period of time or at several time intervals. The aim is to assess whether and how, by comparing the health of both groups, the housing conditions have an impact on health. Such studies provide strong evidence of the relationship.

Intervention – These assess the impact of some change on the health of people, for example, the effect of upgrading the heating system, thermal insulation or the immediate housing environment. For these to be useful there must be similar, detailed and comparable information of the situation before and after the intervention. Any relatively major housing intervention is likely to have negative affects through the disruption to the occupiers' daily life. The negative impacts may be short term, and should be clearly distinguished from any longer term effects.

Extrapolative – For this type of study, a health impact from a particular housing condition is presumed and calculated based on information from other studies, often in different situations, usually non-housing. This approach can be useful where the threat to health could be particularly serious; for example, it is this approach which has been used to assess the threat to health from Radon based on extrapolations from studies of the effects on miners.

Evidence on specific housing issues

While it is important to be aware of all the caveats on research methodology, the considerable library of research provides, on the balance of probability, powerful substantiation of the potential health impact of specific housing conditions. Several detailed reviews of the research studies provide summaries of the findings, particularly on the building-related factors. These include *Building Regulation and Health and Safety, Statistical Evidence to Support the Housing Health and Safety Rating System: Volume II – Summary of Findings* and *The Housing Health and Safety Rating System: Guidance (Version 2).* [4]

Some important issues which show the interrelationship between human and building factors are discussed next.

Dampness

Dampness has the potential to affect health in several ways. First, excess moisture affecting the structure will reduce the thermal insulation capabilities of the material, both through the increase in the thermal conductivity and to complete the evaporation process. This cools the dwelling. Second, the presence of moisture in structural timber increases the possibility of fungal attack. As well as the damage that this can lead to, as fungus develops it will release huge quantities of spores into the confined space of the dwelling. Fungal spores, such as those from *Serpula lacrymans* (dry rot) are potential respiratory allergens. Third, dampness in any form, but in particular, high levels of relative humidity within the dwelling, can increase the prevalence of house dust mites and of mould growth. Mites, mite debris and mould spores are all potent allergens. In addition, the presence of dampness, and of mould growth, can have a detrimental affect on the social and mental well-being. Occupants can feel ashamed, and reluctant to invite friends and relatives into their home.

Most of the considerable number of studies into the potential health effects of dampness have concentrated on the relationship between respiratory conditions and dampness rather than exploring the specific cause of any association. Those that have looked at specific mechanisms have confirmed that house dust mites and their debris and mould spores are allergens, and that the prevalence of mites and mould spores increases in damp conditions.

Studies have consistently shown an increase in respiratory problems such as cough, wheeze and rhinitis in children living in damp affected dwellings compared to those in dry ones. It also appears that a reduction in childhood asthma symptoms may be linked to the installation of central heating. Other studies have found increases in nausea, breathing difficulties and stress in adults in damp dwellings [5]. In addition, there are indications supporting the proposition that dampness has a deleterious affect on mental health and social well-being [6].

Home accidents

More accidents occur in and around the dwelling at work or on the roads (see Table 16.1).

There are many hazards which could cause physical injuries in and around housing, many of which society considers necessary, such as gas and electricity supplies and appliances, steps and

Table 16.1 Location of accidents

	Killed	Serious injury	Minor injury	Total
Road (1999)	3600	39 000	278 000	320 600
Work (1998/99)	620	132 300	500 000	632 920
Home and garden (1998)	4300	172 600	2 667 400	2 844 300

Based on Home Accident Surveillance System 22nd Annual Report, DTI, and data from Royal Society for the Prevention of Accidents.

stairs, and balconies. Most of these can be made relatively, but not completely, safe. There are, however, some structural features which may increase the likelihood of an accident. For example, horizontal bars to a balcony or landing guarding will provide a climbing frame for small children, and small changes in floor levels can be a trip hazard, and secondary hazards, such as non-safety glass, can increase the severity of an injury if there is a fall.

Normal human behaviour can be a major cause of accidents. Very young children lack the knowledge and experience to recognize danger, and are inquisitive by nature. In the elderly, mobility and sight may be impaired. Sometimes, a person may be distracted by something, such as an unexpected noise. In some cases, however, a person may take risks, while others may be maladroit or just careless. And in other cases the occupiers may introduce and create hazards by trailing electric cables to appliances, leaving obstacles on stairs, or leaving medicines and cleaning products readily accessible to small children.

Just less than half of all home accidents occur just outside the dwelling – in the garden, on paths and drives. Of those accidents occurring inside the home, most happen in the living (or dining) room, followed by the kitchen [7].

Excess winter deaths

Between December and March, it is estimated that there are around 40 000 more deaths than could be expected from the death rates in other months of the year [8]. This seasonal variation is larger in the United Kingdom than in many other countries of continental Europe and Scandinavia. While changes in the ambient outdoor temperature are a major contributing factor, indoor temperature, seasonal infections, changes in behavioural patterns and air quality levels are also implicated.

Indoor temperature is a function both of dwelling characteristics and the occupier, particularly the disposable income. It is, however, difficult to isolate dwelling factors, such as the energy efficiency and the effectiveness of the heating system, from human factors. Some households may prefer a relatively low indoor temperature, others must do so because of available finance, while for others inadequate provision for heating and insulation give no option.

Prolonged indoor temperatures below 18°C are not only uncomfortable, but increase the risk of respiratory infections, bronchitis, heart attacks and strokes. There is a risk of hypothermia, particularly for the elderly, where temperatures remain below 10°C for prolonged periods.

Housing perception

Housing is more than just a physically safe and healthy environment. We are affected by the aesthetics of the structure, by its location, and by the perception of the immediate environment and by the neighbourhood.

The demands and expectations for the neighbourhood will depend on the composition and socio-economic structure of the household. Access to schools, public transport, green space, recreation facilities and shops are all important factors affecting satisfaction and social well-being [9]. The housing neighbourhood should also allow for opportunities for social interaction without fear of crime.

Monitoring housing conditions and health impact

While a considerable amount of data is collected locally and nationally on housing conditions and on injuries and ill-health, currently there is little attempt to match data to evaluate the relationship between housing and health or to assess the effect of changes in housing conditions on health.

The majority of housing in the United Kingdom is in private ownership, only about 30% being provided by the public sector through housing associations and local authorities [10]. Nonetheless, housing is considered a national asset, subject to central and local policies which are directed to trying to ensure that the housing stock is maintained, that worn-out housing is replaced, and that additional housing is provided to satisfy demands. To inform policy decisions, a considerable amount of data is collected regularly on the physical state of the housing stock, together with some on the relationship between household and their dwellings and on household satisfaction.

Using this data and monitoring trends, resources and efforts are directed towards ensuring that dwellings are maintained in a reasonable state of repair, meet minimum standards, satisfy expectations, and that there is sufficient housing to meet the demand.

Policies informed by data on physical conditions are primarily directed to maintaining the housing stock. Such policies may also help to ensure that the health and safety of the occupants is protected. This, however, is an assumption and a fortuitous consequence. It is an assumption as it relies on the premise that minimum standards, both for new and existing dwellings, are health and safety based. It is fortuitous without sufficient data on the impact of housing policies on health is used to inform the decisions.

From a public health perspective, policy decisions should be informed not only by data on the physical state of the housing stock, but also by data on injuries and health which can be directly attributed to housing.

Sources of data

There is a considerable amount of data available on housing conditions and characteristics and on the health impact of housing.

Data on housing condition

Both locally and nationally, sample house condition surveys are generally accepted as the most satisfactory method of monitoring housing conditions to inform decisions on the priorities for action. Since 1967, national surveys have been carried out in England. The original concept for such surveys followed recommendations of the Dennington Committee [11]. The results from that first national survey indicated that around 1.8 million dwellings would fail what was then the statutory minimum housing standard [12]. This was more than double the number and showed a wider distribution than previous estimates provided by local authorities.

Since 1971, a national English House Condition Survey (EHCS) has been carried out every five years until 2001[1] [13]. The EHCS is a sample survey, which in 1996, included interviews with around 15 000 households, superficial physical inspection of some 12 000 dwellings, a valuation survey, and postal survey of both public and private sector landlords. It provides information on

- the composition of the housing, including characteristics, ownership, physical condition and the range and quality of facilities, and how this has changed since the previous survey;
- the profile of households and the housing in which they live; and
- the relationship between housing conditions and the circumstances of different household groups.

Some (limited) data is also collected on the health of occupants and on domestic accidents and fires in the home.

---

[1]See Chapter 17 pages 366 and 369 for information an the results of the 2001 survey.

Results from the EHCS are grossed to national averages and are presented in respect of all dwellings (including vacant), occupied dwellings only and households (these give different figures as some dwellings are occupied by more than one household). Although grossing samples introduces some uncertainty, most results given are outside the margins of error from the sampling and the measurement methods used.

Since April 2002, the EHCS has changed to a more continuous form, the fieldwork now being carried out four times a year, each lasting eight weeks. The aim is to try to provide early information on gradual changes, and allow for more detailed analyses of subsectors of the housing stock.

Local authorities are required to collect data to inform their policies and to provide a means to comparing the local and national conditions. This is generally through local stock condition surveys, which may be samples to reflect the whole or a part of the authority's district. Surveys may also concentrate on particular issues, such as multi-occupation. Advice and guidance on local surveys is given in *Collecting, Managing and Using Housing Stock Information* [14].

Data on health

The Home Accident Surveillance System (HASS) provides information on domestic accidents resulting in injuries requiring medical attention. The data has been collected since 1976 from a sample of Accident and Emergency Units in hospitals throughout England and Wales. It includes details of the dwelling feature involved and the type and seriousness of the injury caused. This data shows that in 1999, for example, there were 2.8 million home accidents, or over 7600 a day, which required medical treatment, and that this is estimated that the cost to the UK society is around £25 000 million per year [15].

As the data includes details of the dwelling feature involved in home accidents, it can be shown that falls associated with stairs and steps result in over 290 000 injuries each year. Analysis of this data shows that the likelihood of a fall on stairs is related to various design and construction

details such as steepness and the presence of handrails.

Records of fires attended by the Fire Brigade are collected by the Home Office. These include details of fatalities and injuries caused by these fires. The Home Office, through the British Crime Survey, also maintains data from sample surveys on burglary and on fires not attended by the Fire Brigade.

Other sources of data on housing and health

Other sources of housing data include the Survey of English Housing,[2] the Neighbourhood Statistics Dataset, ACORN and Residata. Complementing the EHCS data is that gathered through the Survey of English Housing (SEH). This is a national survey and concentrates on the opinion of the householder, and also provides information on household tenure, aspirations and costs through interviews with a representative sample of around 20 000 households in England. The Neighbourhood Statistics Dataset [16] gives population estimates at ward level by age group, and provides estimates of the economically active adults. ACORN and Residata [17] are two commercially developed datasets. ACORN gives dwelling characteristics and a classification based on key demographic variables. Residata provides information on house characteristics, including age, size and tenure.

Some data on health is available from national morbidity studies, statutory notification of diseases system and data on mortality. The national morbidity studies contain details of patient consultations at a sample of general practices in England and Wales, and includes socio-economic data. The Public Health Laboratory Service maintains data submitted on statutorily notifiable diseases, and the Office of National Statistics compiles details on numbers and causes of deaths.

Using data to inform decisions on policy and action

The various sources of housing and of health data are collected for their own specific reasons and are currently unco-ordinated. If some of them could

---

[2]See ODPM *Housing in England 2001/02*, TSO, London.

be matched and analysed, they may serve additional purposes. It was to this end that Indicators were developed.

An indicator is something that helps explain the current situation in relation to the target, and provide guidance on the appropriate action to move towards that target. Indicators provide a means to identify and use data in a way that will assist in making decisions on action and policy.

Indicators have been developed for many areas [18], and there is now an inter-nationally agreed methodology for the development of Indicators. Using this methodology, Environmental Indicators have been developed to illustrate conditions and trends in the environment, including the quality of the atmosphere, and of surface water. Similarly, Health Indicators have been developed to illustrate trends in health, such as life expectancy, or mortality rates from a specific cause.

Currently (2003), the World Health Organization, supported by the European Commission, is developing Environmental Health Indicators to relate the environment issues and health [19]. These are intended to illustrate a health outcome resulting from the exposure to an environmental hazard. The methodology requires that these Indicators fit into the **DPSEEA** framework:

Driving Force	the underlying dynamics
Pressure	the factors generated and influenced by the driving forces
State	the condition resulting from the pressure
Exposure	putting individuals in a situation where they may be affected by the condition
Effect	the health impact that can result from the exposure
Action	measures which may remove or reduce the impact.

Although housing has been included in the list of Environmental Health Indicators, it is recognized that it is a complex issue in its own right and it is intended that a separate set of Housing-Health Indicators will be developed.

A Housing-Health Indicator is an expression of the link between housing and health highlighting a specific policy issue in a form which assists

informed decisions on action. It should be based on an interpretable relationship between housing conditions and health, and that relationship should be known, consistent, unambiguous and largely unconfounded. To be of use it should satisfy certain criteria, for example, that:

- there is a demonstrated or hypothesized link between defined housing issues and a particular health outcome;
- there is a source of data, or the potential for gathering that data;
- it is robust, and based on sound logic;
- it provides an opportunity for action;
- it is capable of showing changes; and
- it is sensitive enough to monitoring changes.

Using the DPSEEA framework, it can be seen that there may be options for Action which may influence any of the five stages (see Fig. 16.1). This approach shows that identifying Actions which could remove or reduce the Exposure by manipulating the Driving Forces, the Pressure and/or the State, would both protect the occupiers and relieve the demands on the health services dealing with the Effect.

The Housing Health and Safety Rating System

In 1990, a revised minimum statutory standard for housing was introduced [20]. Responding to criticisms of this standard, central government commissioned a series of research projects to examine its application and to review the legal controls on minimum standards in existing housing generally [21]. At around the same time, other research reviewed the implications of building design and condition on health and on safety [22].

Together, these sets of research highlighted two main problems. First, that most serious housing hazards were either not covered, or covered inadequately, by legal standards applicable to existing housing. Consequently, there was no legal means of protecting occupiers from those hazards. Second, as the minimum standard was a pass/fail model, there was no indication of the degree of failure – it was not apparent whether conditions posed a relatively remote threat to health or safety

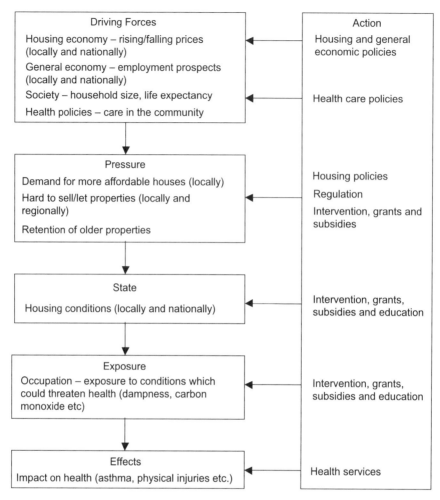

Fig. 16.1 Housing and Health in the DPSEEA framework.

or were an imminent danger of severe injury or death.

To overcome these and other problems, the concept of a system to rate or grade housing conditions was proposed and, in 1998, central government commissioned development work. The result was the Housing Health and Safety Rating System (HHSRS), Version 1 being released in July 2002. Version 2 of the System is scheduled for release in early 2004.[3]

There are two main aspects to the HHSRS. First, it provides a logical and transparent approach by requiring a two-stage approach – an assessment of the likelihood of an occurrence which could cause harm (an injury or other health outcome), and an assessment of the potential severity of any harm that could result from such an occurrence. Second, to inform judgements, a method was developed which, based on matching and analysing datasets, gave an indication of average likelihoods and severities of harm. This provides yardsticks against which conditions found in individual dwellings could be evaluated.

The analyses of housing and health data also gave information on the general order of hazards in the UK housing stock, ranked by occurrence

---

[3]The system is included in the Housing Bill which is subject to parliamentary procedures in early 2004.

and severity. Finally, the development provided an indication of where additional research is needed.

Principles and development [23]

The main principle behind the HHSRS is that a dwelling, including the structure and associated outbuildings and garden, yard and/or other amenity space, should provide a safe and healthy environment for the occupants and any visitors. As there will always be some hazards in a dwelling (such as gas and electricity supplies, stairs, etc.), these should be as safe as possible.

The Rating System concentrates on hazards – that is, it is concerned with the potential effect of a defect. This means that, for the assessment, it is not magnitude of the defect nor the cost of remedying it that is taken into account, only the potential threat to health and/or safety.

The development involved identifying and categorizing all the potential housing hazards (see Fig. 16.2). This relied on reported research which demonstrated and quantified a link between housing conditions or features and health. Using this evidence, analyses were made of datasets created by matching databases on housing conditions and characteristics with databases on reported illnesses, injuries and other health conditions. The categories of hazards were limited to those which were solely or primarily attributable to housing conditions – excluded were those which were solely a result of human behaviour (e.g. carelessness or negligence).

The hazards identified had different characteristics. In some cases, an occurrence would be relatively instantaneous – for example, fall hazards, which would involve a slip or trip resulting in a fall which caused a physical injury. Others required exposure to the hazard over a period to cause any harm – such as dampness or excess cold. In some cases, a fatal outcome is possible from a hazard, while in others the most serious outcome would be relatively minor and death is almost impossible. In addition, some hazards could result in a physical injury, while others could cause illness or a health condition. To be able to compare the severity of hazards at a dwelling required creating a method which would take account of these

differences. This was achieved by devising a formula using both the assessed likelihood and outcome to generate a Hazard Score.

The formula uses three sets of figures. As likelihood is often expressed as a ratio (e.g. 1 in 100, 1 in 250) this provided one figure – the likelihood of the hazard exposing a person to an occurrence which could cause harm as assessed by the surveyor. The different kinds of harm or health outcome which could result from hazards were classified according to the degree of incapacity caused and were given a weighting to reflect that incapacity (see Fig. 16.3). These weightings provided a second set of figures. Finally, while there may be a most likely outcome, there may also be other outcomes possible. To reflect this, the surveyor is required to judge the spread on a percentage basis (e.g. 60% chance of a Class III outcome, 30% chance of Class IV, and 10% chance of Class II). The Hazard Score is the sum of the products generated by multiplying the weighting for each Class of Harm by the likelihood expressed as a ratio and by the outcome for each Class of Harm expressed as a percentage (see Fig. 16.4).

The formula reflects the severity of the threat to health or safety and so allows comparison of very different hazards, such as those where there is a very high likelihood of a relatively minor outcome with ones where there is a very small chance of an extreme outcome (see Box 16.1).

The assessment of a dwelling under the HHSRS, as for any other form of assessment, is made on completion of a survey. This assessment relies on informed professional judgements – first, to determine to which hazard(s) faults identified in the dwelling could contribute; and second, to judge both the likelihood and the potential outcome of any hazard.

To promote consistency, assessment of the likelihood is based on two suppositions. Each hazard is assessed in relation to a member of the age group most vulnerable to that hazard. This approach means that a vacant dwelling can be assessed, and that the assessment of a dwelling relates to the dwelling and is not influenced by a change of occupier. In addition, it is the likelihood of an occurrence over the following twelve months that is to be assessed. This takes account of both the

A **PHYSIOLOGICAL REQUIREMENTS**
   Hygrothermal Conditions
   1 Damp and mould growth
   2 Excess cold
   3 Excess heat
   Pollutants (non-microbial)
   4 Asbestos (and MMFs)
   5 Biocides
   6 Carbon Monoxide and fuel combustion products
   7 Lead
   8 Radiation
   9 Uncombusted fuel gas
   10 Volatile Organic Compounds

B **PSYCHOLOGICAL REQUIREMENTS**
   Space, Security, Light and Noise
   11 Crowding and space
   12 Entry by intruders
   13 Lighting
   14 Noise

C **PROTECTION AGAINST INFECTION**
   Hygiene, Sanitation and Water Supply
   15 Domestic hygiene, Pests and Refuse
   16 Food safety
   17 Personal hygiene, Sanitation and Drainage
   18 Water supply

D **PROTECTION AGAINST ACCIDENTS**
   Falls
   19 Falls associated with baths etc
   20 Falls on the level
   21 Falls associated with stairs and steps
   22 Falls between levels
   Electric Shocks, Fires, Burns and Scalds
   23 Electrical hazards
   24 Fire
   25 Hot surfaces and materials
   Collisions, Cuts and Strains
   26 Collision and entrapment
   27 Explosions
   28 Ergonomics
   29 Structural collapse and falling elements

Fig. 16.2 Categories of Potential Housing Hazards, HHSRS Version 2 (2004).

difference in the rate of the effect (i.e. it covers both the insidious and the relatively instantaneous), and the influence of the seasons on some hazards.

Since the original development, the supporting statistical evidence has been refined and updated, Version 1 has been evaluated, and additional work has been carried out on the application of the System to multi-occupied buildings [24]. The results from these three projects have been taken into account to develop Version 2.

Class of harm	Examples	Weighting
Class I	Death Permanent paralysis below the neck Malignant lung cancer Permanent loss of consciousness 80% burns	10 000
Class II	Stroke Loss of hand or foot Serious fractures Serious burns Loss of consciousness for days	1000
Class III	Loss of a finger Malignant but treatable skin cancer Fractured skull Severe concussion Serious puncture wound Severe burns to hands	300
Class IV	Occasional severe discomfort Chronic skin irritation Some benign tumours Moderate cuts to face or body Severe bruising to body 10% burns	10

Fig. 16.3 Examples of Classes of Harm, HHSRS Version 1 (2000).

Class of harm	Weighting		Likelihood 1 in		Spread of harm (%)		
I	10 000	÷	100	×	0	=	0
II	1000	÷	100	×	10	=	100
III	300	÷	100	×	30	=	90
IV	10	÷	100	×	60	=	6
			Hazard score			=	196

Fig. 16.4 Example of Hazard Score Formula, HHSRS Version 1 (2000).

Box 16.1

Comparison of two hazards, one where an occurrence is likely but the outcome relatively minor, and one where an occurrence is 100 times less likely but the outcome serious.

Noise – because of poor sound insulation and close proximity to rail and road traffic, it is assessed that there is a 1 in 3 likelihood of an occurrence which could cause harm. The most likely outcome from such an occurrence is judged to be severe discomfort (Class IV), with a possibility of more serious stress (Class III). This would give a Hazard Score of over 1300.

Radon – measurements for radon are taken in a dwelling situated in a radon affected area with butt jointed timber ground floor; these give readings of over 3000 Bqm^{-3}. The likelihood of a harmful occurrence is assessed as 1 in 320. The most likely outcome is very serious, malignant lung cancer or death (Class I), with a possibility of the cancer being treatable (Class II). This would give a Hazard Score of over 2800.

Application and uses

It is important to note that the HHSRS is an approach to the assessment of housing conditions. It allows comparison of the severity of hazards within a dwelling, and the comparison of the different dwellings. It does not set standards, but is amenable to being used to so by specifying maximum or target scores.

Utilizing the potential for setting standards and assessing threats to health and safety, the HHSRS is to be incorporated into the legislation to replace the minimum standard originally introduced in 1954 [25]. For these purposes, guidance on the use and application of the HHSRS has been prepared [26]. However, it is important to note that the HHSRS is capable of wider application and that the principles and approach developed have been recognized as being international.

Keeping up-to-date

A major problem with any published work is the continuous development of the subjects discussed. This work is no exception, and it is important to be aware of sources of up-to-date information.

These days, as well as physical libraries, the internet provides access to detailed information and links to research reports and other published works. Government websites are useful sources of data and current government sponsored research, (e.g. http://www.housing.odpm.gov.uk/research/), and specialist bodies provide information on particular issues such as radiation (e.g. http://www.nrpb.org/radiation_topics/index.htm) and accidents (e.g. http://www.rospa.co.uk/CMS/) or on building and housing related issues (e.g. http://www.bre.co.uk/). Many universities, particularly those which have established centres or units devoted to research into aspects of housing, have sites providing details of their research, information on other projects and links to other sites (e.g. http://www.cf.ac.uk/cass/projects/housing_research.html, and http://www.cchr.net/cchr.php?b=1) and research funding bodies will give information on current and completed projects (e.g. http://www.jrf.org.uk/). Details of international research and developments can be found from networks and academic organizations (e.g. http:// www.enhr.ibf.uu.se/index.html, http://www.colorado.edu/plan/housing-info/menu0.html, and http://www.library.utoronto.ca/hnc/links.htm).

REFERENCES

1. Constitution of the World Health Organization, signed and adopted 1948.
2. *Sick Building Syndrome: A Review of the Evidence on Causes and Solutions*, HSE Contract Research Report No.42/1992 (1992) HMSO.
3. Cox, S.J. and O'Sullivan, E.F. (1995) *Building Regulation and Safety*, BRE, p. 4.
4. *Building Regulation and Health and Safety* (2000), CRC, *Statistical Evidence to Support the Housing Health and Safety Rating System: Volume II – Summary of Findings* (2003), ODPM; *The Housing Health and Safety Rating System: Guidance* (Version 2), (2004), ODPM.
5. See, for example, Strachan, D.P. and Elton, R.A. (1986) *Fam Pract*, **3**, 137–42; Martin, C.J., Platt, S.D. and Hunt, S.M. (1987) *British Medical Journal*, **294**, 1125–7; Platt, S., Martin, C.J., Hunt, S.M. and Lewis, C.W. (1989) *British Medical Journal*, **298**, 1673–8; Hyndman, S.J., (1990) *Social Science & Medicine*, **30**(1), 131–41; Billings, C.G. and Howard, P. (1998) *Monaldi Arch Chest Dis*, **53**(1), 43–9; Williamson, I.J. *et al.* (1997) *Thorax*, **52**(3), 229–34; Garrett, M.H. *et al.* (1998) *Clinical and Experimental Allergy*, **28**(4), 459–67; Peat, J.K., Dickerson, J. and Li, J. (1998) *Allergy*, **53**(2), 120–8; Somerville, M. *et al.* (2000) *Public Health*, **114**, 434–9; Dotterud, L.K., Korsgaard, J. and Falk, E.S. (1995) *Allergy*, **50**, 788–93; Mohamed, N. *et al.* (1995) *Thorax*, **50**, 74–8; Lindfors, A. *et al.* (1995) *Archives of Disease in Childhood*, **73**, 408–12.
6. Packer, C.N., Stewart-Brown, S. and Fowle, S.E. (1994) *Journal of Epidemiology and Community Health*, **48**, 555–9.
7. *Research on the Pattern and Trends in Home Accidents* (1999) DTI; *Accidental*

Falls: Fatalities and Injuries (1999) DTI/ University of Newcastle Upon Tyne; Cayless, S.M. (2001) *Applied Ergonomics*, **32**(2) 155–62.

8. Wilkinson, P., *et al.* (2001) *Cold Comfort: the Social and Environmental Determinants of Excess Winter Death in England, 1986–1996*, Policy Press.

9. Attwell, K. (2003) *The Housing Environment in a Greenscape Perspective – a Review of Danish Studies*; and Rex, D. *et al.* (2003) *Healthy Local Environments: Monitoring and Improving Access to Food*. Papers presented at Healthy Housing: Promoting Good Health, Warwick University.

10. Census *2001*.

11. *Our Older Homes: a Call for Action* (1966), Ministry of Housing and Local Government.

12. The Standard of Fitness given in s4 Housing Act 1957.

13. Separate surveys are carried out in Scotland, Wales and Northern Ireland.

14. DETR (2000), see also *Decent Homes: Capturing the Standard at a Local Level* (2002) DTLR.

15. Home Accident Surveillance System, 23rd Annual Report (2002) Department of Trade and Industry. The HASS ceased in 2003 and data archives are kept by the Royal Society for the Prevention of Accidents.

16. Developed by the Social Disadvantage Research Group of Oxford University and available from the National Statistics website.

17. Developed by CACI Ltd and Intermediary Systems Ltd respectively.

18. See http://www.sustainablemeasures.com/ Indicators/index.html for a comprehensive list of Indicators on sustainable environment.

19. See http://www.euro.who.int/EHindicators/ Indicators/.

20. Local Government and Housing Act 1989, HMSO, London.

21. Monitoring the New Housing Fitness Standard (1993) HMSO, and Controlling Minimum Standards in Existing Housing (1998), Legal Research Institute, Warwick Law School.

22. Cox, S.J. and O'Sullivan, E.F. (eds) *Building Regulation and Safety* (1995), Construction Research Communications; and Raw, G.J. and Hamilton, R.M. (eds) *Building Regulation and Health* (1995), Construction Research Communications, Peterborough.

23. For full details see Report on Development of the Housing Health & Safety Rating System (2000) DTLR.

24. Statistical Evidence to Support the Housing Health & Safety Rating System (2003) ODPM; Evaluation of Version 1 of the Housing Health & Safety Rating System (2003) ODPM; and The Application of the Housing Health & Safety Rating System to Houses in Multiple Occupation (2003) ODPM.

25. The Standard of Fitness for Human Habitation, originally s9 Housing Repairs and Rents Act 1954, subsequently, s4 Housing Act 1957 and then s604 Housing Act 1985.

26. Guidance on the Use of the HHSRS for Enforcement Purposes (2004) ODPM. This provides useful summaries of the health impact of housing conditions with comprehensive references.

17 Housing: standards and enforcement

Jill Stewart

INTRODUCTION

National housing strategy

Local authorities have responsibilities across all housing tenures although the environmental health officer's (EHO) role is more commonly associated with private housing, comprising owner occupation and the private rented sector, where an increasingly enabling and advisory role is administered.

Local authorities' housing capital programmes cover the provision and renovation of their own housing and support for the private sector (as 'Single Pot' housing finance), and by financing housing association developments. The government seeks in its capital resource allocation to enable authorities to carry out their statutory functions, particularly their duties towards homeless people in priority need and specifically to:

- ensure that housing in their ownership is renovated where necessary and efficiently managed and maintained;
- assist and enable private owners who could not otherwise afford necessary repair or improvement;
- support and supplement housing association and private sector investment in their areas where necessary to meet demand from people in need;
- ensure the maintenance of safety and fitness standards in the private rented sector, currently subject to fundamental change discussed later.

Resources for local authority capital expenditure are made up of:

- capital grants from central government in support of certain expenditures
- new borrowing authorized by credit approvals granted under the capital finance system
- authorities' own capital receipts (net of amount set aside to repay debt) and contribution from revenue
- capital grants from the European Union (EU) in support of specific programmes
- the housing partnership fund.

Funding to support local authority capital expenditure is available through a number of established programmes:

- the Housing Investment Programme (HIP)
- the Single Regeneration Budget (SRB)
- capital challenge programmes
- the European Regional Development Fund (ERDF)
- the New Deal for Communities programme (NDC).

All these funding routes involve a detailed bidding process, the nature of which varies according to whether funding is to be in support of ongoing programmes of repair and improvement (HIP) or specific targeted regeneration programmes (SRB, ERDF, capital challenge, NDC).

At the time of writing, the housing functions of EHOs in local government are subject to radical overhaul through direct legal changes as well as a wider reintegration of health and policy. Traditionally involved in the enforcement of satisfactory standards of provision and repair in private sector housing, the reduction and changes in capital resource allocations as well as the changes in the local authority housing role mean that EHOs have an important duty to perform in developing local housing and wider community strategy. Such strategy must now be closely allied to an increasingly partnership-based role as part of the wider public health agenda and focus around evidence-based need that reintegrates health, domestic safety and housing particularly when a tenure-neutral approach is encouraged as area regeneration is increasingly favoured.

Local authorities have been subject to the duty to ensure 'Best Value' across services since April 2000, requiring fundamental review over a 5-year period with year on year improvement leading to high-level performance agendas nationally in horizontal service integration, particularly in areas of health, commmunity safety, poverty and sustainability. Best Value applies to all local authority housing functions, and for private sector housing particularly incorporates renewal/development, which includes disabled facilities grants; area renewal activity; HMO works; house condition surveys; home improvement agencies; fitness and other enforcement; empty properties and energy efficiency [1].

The Housing Green Paper, the Community Plan and the Housing Bill 2003

The Housing Green Paper [2] and subsequent policy documents [3,4] set out the government's proposals for housing in the context of a wider strategic approach to address inequality, and momentum for improved standards in the social and private housing sectors has continued at a rapid rate. Objectives include a decent home for all, support for sustainable home ownership, a revival of the private rented sector and intervention to help prevent market collapse and to protect vulnerable households.

The Regulatory Reform (Housing Assistance) (England and Wales) Order 2002 subsumed earlier grant legislation and required that local authorities publish a policy for private sector housing by July 2003 consolidating wider strategic objectives and grant alternatives, and firmly setting objectives for private sector housing within a wider strategic context of furthering of personal responsibility for private housing conditions.

The action plan, *Sustainable Communities: Building for the Future* (the 'Community Plan') [5] lies at the heart of government proposals to create thriving, sustainable communities in all regions, and consolidates and extends many regimes across housing tenures. It seeks to ensure that social, economic and environmental community needs, and not just housing delivery, are sustainably addressed. The Community Plan proposes a continuation of partnership approaches as part of the public health agenda. It is wide-ranging and encompasses: providing more quality affordable housing, housing demand and changing social trends; meeting the housing need of key workers; designing attractive towns, cities and public places; making better use of brownfield sites and more efficient use of greenfield sites for development; regenerating declining communities; responding to regional, urban and rural differences in housing demand; tackling social exclusion and homelessness; tackling empty homes; bringing 'decent homes' and neighbourhoods for all to the core of policy; improving the planning system so that it is faster, fairer and more efficient; empowering local and regional government and improving performance standards.

Decent housing is defined as being wind and weather tight, warm and having modern facilities, being statutorily fit, in reasonable repair (with reference to property age), with reasonably modern facilities and services, having adequate insulation against external noise, having adequate size and layout of common areas in flats and having a reasonable degree of thermal comfort (heating and insulation). All social housing should meet the decency standard by 2010, with the worst stock being tackled first, with a separate target for private sector housing [6].

The Housing Bill [7] (see Fig. 17.1) was published on 31 March 2003 and sets out specific legislation – with new definitions for 'dwelling house' and 'house in multiple occupation' – in the following areas:

- Replacing the existing statutory standard of fitness with the housing health and safety rating system to enforce against unsatisfactory housing conditions (see later section on statutory fitness);
- Improving controls for houses in multiple occupation (HMO), including a mandatory licensing system to tackle poor physical conditions and management standards (see later section on HMOs);
- Granting local authorities powers to license all landlords in areas of low housing demand or similar areas where an extended and poorly managed private sector adversely impacts attempts to secure sustainable communities (see later section on licensing landlords);
- Requiring those marketing a home to put together a home information pack to help reduce uncertainty and unnecessary costs (enforceable by Weights and Measures Authorities, hence not covered in this chapter);
- Modernizing the Right to Buy scheme, to help prevent profiteering and emphasizing the purchaser's responsibilities; seen to contribute to affordable housing (not covered in this chapter).

Current housing position

There are 21.1 million dwellings in England compared to 20.3 million in 1996 [8A], although there are still serious imbalances in areas of high demand, mainly in London, the southeast and the larger cities. The amount of space per person is among the highest in Europe, and almost all houses now have basic amenities. There are, however, some major problems of disrepair in the rented sector, both public and private, and particularly acute ones in some urban areas.[1]

70% of houses in England and Wales are owner-occupied, one of the highest percentages in the world, and the private rented sector has declined from about 90% of the total stock in 1914 to only 10% in 2001. Council-owned housing has increased from 10% of all housing before the Second World War to about 13% [8A]. This percentage has declined in recent years due to right-to-buy sales and large-scale voluntary transfers to registered social landlords (RSLs). RSLs have expanded their role to the point that they own around 7% of the total stock. Figure 17.2 illustrates dwelling tenure changes in England between 1914 and 1996.

The situation in any area is also affected by the size, type, location and condition of the houses, and whether they are suitable to meet the needs of different sections of the community, including young people, single people, families, the elderly and the disabled. Demographic changes taking place in the population and the projected growth in the number of elderly people is of particular significance at present.

In those areas where there is an imbalance between housing need and availability, the shortage is demonstrated by:

- insufficient rented accommodation in the private and public sectors
- high occupation densities with congestion and overcrowding
- constant pressure for rented accommodation
- poor quality accommodation
- multi-occupied houses with periods of occupation well beyond what might be considered reasonable
- high land and house prices and, consequently, a depressed housing market and difficulties for first-time buyers.

Inevitably, this situation results in an increase in the number of homeless people, many of whom approach the local authority for assistance.

Local housing strategy

Each housing authority is required to submit annually to the Office of the Deputy Prime Minister (ODPM) a strategy that, among other things, should include a brief narrative description of its general housing policy, setting out

---

[1]Figure A3.7 of the English House Condition Survey 2001 [8A] shows the distribution of decent/non-decent homes by tenure for 1996 and 2001.

PART 1: HOUSING CONDITIONS
Chapter 1 – Introductory (Sections 1–4)

Standards for condition of residential accommodation	Repeal of fitness for human habitation; introduction of HHSRS
General enforcement of standards	Duty to take action for category 1 hazards

Chapter 2 – Improvement Notices (Sections 5–29)

Improvement Notices (IN)	Duty to serve IN in relation to category 1 hazard; power to serve IN in relation to category 2 hazard; service for dwelling house, HMO and common parts

Operation of improvement notices; appeals and reviews; main offences; powers of enforcement; supplementary provisions

Chapter 3 – Prohibition Orders (Sections 30–52)

Prohibition Orders (PO)	Duty to make PO in relation to category 1 hazard; power to make PO in relation to category 2 hazard; contents of PO

Service and operation of PO; appeals and reviews; main offences; other enforcement provisions; supplementary provisions

Chapter 4 – Other enforcement action (Sections 53–58)
Warning notices about housing conditions; Demolition Orders; Clearance Areas; other enforcement action

Chapter 5 – Additional provision about housing conditions (Sections 59–60)
Power to charge for certain enforcement action

PART 2: LICENSING OF HOUSES IN MULTIPLE OCCUPATION (Sections 61–81)

Introductory	Mandatory application where HMO is 3 storey with 5 occupants

Designation of additional licencing areas; licences; grant or refusal of licences; variation revocation and appeals; supplementary provisions

PART 3: SELECTIVE LICENSING OF OTHER RESIDENTIAL ACCOMMODATION (Sections 82–95)

Designation of selective licensing areas	Discretionary power

Confirmation of designations; duration, review and revocation of designations; licensing requirements; licence protocol; supplementary provisions

PART 4: ADDITIONAL CONTROL PROVISIONS IN RELATION TO RESIDENTIAL ACCOMMODATION (Sections 96–131)
Interim Management Orders; Final Management Orders; Management regulations; Overcrowding Notices; Supplementary provisions

PART 5: HOME INFORMATION PACKS (Sections 132–152)
Enforcement by Weights and Measures Authorities; requirement for Home Information Packs ('Seller's Pack'); Home Condition Reports, etc.

PART 6: OTHER PROVISIONS ABOUT HOUSING (Sections 153–157)
Amendments to 'Right to Buy' provisions; Social Housing Ombudsman for Wales

PART 7: SUPPLEMENTARY AND FINAL PROVISIONS (Sections 158–189)
Information provisions; meaning of HMO; other general interpretation, etc.

Fig. 17.1 Summary of the draft Housing Bill 2003. (Source: [7].)

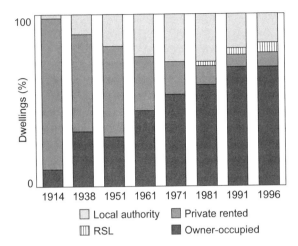

Fig. 17.2 Tenure change: 1914–96. Note: vacant dwellings are included within the tenure of their previous occupancy. (RSL, registered social landlords.) (Source: [8].)[2]

major aspects of the housing situation in its area, the overall approach to those circumstances and highlighting particular issues as appropriate. To achieve a consistent approach by local authorities, the Department of the Environment (DoE) as it was then produced a strategy guidance manual to be used in the 1995–96 HIP process. For the 1998–99 HIP, the DETR went a stage further and introduced a 190-point questionnaire for local authorities to complete. This document was submitted with a brief strategy statement or review of progress compared with the previous bidding year. The intention was to allocate HIP resources on a formula combining local housing need with local authority performance in delivering its HIP.

It is not possible for a local authority to make a once and for all housing policy. Changes in population, economic factors, social conditions and public expectation are all major considerations, as well as changing national political and financial strategies. An effective policy needs to be planned, programmed and regularly reviewed and monitored.

A local authority's policy for the private sector (see DoE Circular 17/96, *Private Sector*

Renewal: A Strategic Approach [9] and DETR (1998) *Private Sector Housing Renewal Strategies: A Code of Practice*) should refer to such matters as:

- an integrated, planned and programmed approach to renewal areas, clearance, assistance regimes;
- multi-occupied houses and tenanted properties;
- enforcement policies;
- vacant houses;
- housing aid and advice including harassment and eviction prevention;
- design standards and open space provision;
- the special needs of groups such as elderly and disabled people;
- the relationship with housing associations and voluntary organizations.

As a result of the decline in capital funding for local authority house-building programmes, it has been necessary for local authorities to develop new initiatives for reducing local housing registers. Much work has been undertaken through the government housing partnership fund to renovate unused private sector stock, providing financial assistance to the owners in exchange for the local authority having nomination rights for a period of years. EHOs, local authority housing officers and benefit officers work closely together on these schemes to ensure that appropriate affordable accommodation is provided and suitable tenants are nominated. Many of these schemes have proved so successful that funding has been obtained from other sources when housing partnership funding has run out.

More specific guidance on local authorities' policies in the private sector is given by the Audit Commission [10], which emphasizes that authorities, rather than simply reacting to requests for service, should have a proactive strategy that defines its priorities. Strategies should be based on information on the nature and scale of problems in the area and on the location of properties that need action. Councils should also pay close attention to the efficiency of their operation, setting

---

[2]Now also see ODPM (January 2004) Survey of English Housing 2001–02, TSO, London.

target response times and adopting inspection procedures that ensure accuracy and consistency.

THE CONTROL OF HOUSING STANDARDS

National house condition surveys

The English house condition survey is carried out every 5 years and is the government's main source of information on the condition of the housing stock in England. These surveys were first recommended by the Dennington Committee report [11] in 1966, and the first was carried out in 1967. A second was undertaken in 1971, and others have followed at 5-year intervals since. The survey comprises four elements: a physical inspection of dwellings, interviews with householders, a postal survey of local authorities and housing associations and a market value survey. The reports based on these surveys provide valuable information on the housing stock in terms of its composition, ownership and condition. The national housing policy can be based on this information and trends can be monitored. The first national Energy Report was carried out as part of the English house condition survey in 1991.

The 2001 English House Condition Survey [8A] found the following amongst other things:

1. At the end of 2001 there were 21.1 million dwellings in England containing 20.5 million households.
2. 39% of housing stock was built before 1945 and 21% before 1919.
3. 81% of dwellings were houses; 19% of dwellings were flats; 68% of the oldest (pre-1919) stock comprised terraced houses or flats.
4. There were 1.1 million houses in multiple occupation of which 82 000 were houses converted into bedsits, comprising 27% of households in multiple occupation.
5. Recently built dwellings are smaller in size: the average size for post 1980 homes is 83 cubic metres but 88 for pre-1980 homes.
6. 33% of dwellings are non-decent, a fall of a quarter since 1996. The most common reasons for being non-decent are their thermal comfort (26%), disrepair (9%), fitness (4%) and lack of modernization (2%).
7. Energy efficiency has generally improved but older dwellings are less energy efficient.
8. 4.2% of the housing stock was unfit, a reduction from 7.2% in 1996. This was mainly due to disrepair (46%) and was more prevalent in pre-1919 stock (10%) and in urban areas (8%).
9. 10% of the private rented sector was unfit; unfitness in the owner-occupied sector has reduced and disrepair has reduced generally since 1996.
10. Within the private sector there was a significant relationship between households' incomes and ethnicity and their housing conditions. There remains a 'hard core' of the worst unfit properties.

The ODPM is updating its website www.housing. odpm.gov.uk/research/ehcs/continuous/index.htm, to enable closer monitoring of its decent housing policy, more regular update of stock condition data to help improve resource allocation, as well as to create an enhanced database for *ad hoc* analysis.

The **Survey of English Housing** is a continuous ODPM survey that began in 1993 and reports on trends in tenure, the composition of households, their mobility and accommodation across all tenures. It also considers issues such as attitudes of tenants towards their landlord and mortgage payments and arrears, Housing Benefit receipt, rent payment and arrears and factors affecting rent levels. Results are analysed using the Index of Multiple Deprivation to identify households living in the most deprived wards (which are frequently over-represented by ethnic minority and lone parent households) and to enable comparisons elsewhere [12].

Housing assessments

Effective housing strategies cannot be developed without knowledge of the local housing conditions. Section 605 of the Housing Act 1985 (as substituted by paragraph 85 of Schedule 9 to the Local Government and Housing Act 1989) requires local housing authorities to consider, at least once in each year, the housing conditions in their district with a view to determining what action to take in performance of their duties in relation to repair notices, slum clearance, houses in multiple occupation, renewal areas and assistance. The Secretary of State is empowered to issue direction about the manner in which the duty to consider housing conditions should be undertaken.

Section 605 does not require a physical inspection of housing once a year, but local authorities are expected to keep up to date the information acquired on earlier comprehensive surveys. Local housing market conditions, the demand for housing in relation to the existing supply, renewal and repair policies will all have had some effect over a 12-month period.

Types of local survey

House condition surveys can be conveniently grouped under three headings according to the purpose they are intended to serve:

- strategy development
- action planning
- implementation.

Surveys falling into the first two categories aim to provide a statistical picture of housing characteristics that is sufficiently precise and reliable to be used for planning purposes. They do not however provide detailed specifications and costings of remedial work that could form the basis of a contract. This role is performed by the third category, the implementation survey.

Strategy development surveys

These surveys are used to identify housing problems within a local authority to determine:

- the nature and extent of the problems;
- the appropriate levels of investment to tackle them;

- priorities for action between tenures and between different locations within the district;
- the effectiveness of decisions already taken and expenditure already committed, by measuring changes in the situation over time.

They may cover the whole of the stock or focus on a particular sector; for example, separate surveys may be conducted on public and private sector housing. They are sample surveys that consider a representative selection of dwellings from the stock rather than all dwellings. They are useful for:

- the provision of information for the development of HIP strategy statements
- the provision of some of the data required in the HIP forms
- compliance with an authority's duty under Section 605 of the 1985 Housing Act (as amended by Section 85 of Schedule 9 of the 1989 Local Government and Housing Act).

Action planning surveys

Such surveys are needed when an authority has already identified specific housing problems and decided the priority in tackling them. The next step is to design appropriate courses of action against the problems. These surveys could focus on:

- the provision of information from which a range of costed options can be generated;
- determining the relationship between the housing problems and household circumstances, and establishing what solutions households would find appropriate;
- the detailed description and understanding of particular problems in a specific subsector of the stock, for example, HMOs.

They are normally confined to a relatively small geographical area or very specific subsection of the stock. Information is needed about each of the dwellings in the defined population but not necessarily at the same level of detail for each.

Examples of such surveys are those undertaken:

- as part of a neighbourhood renewal assessment;
- to develop an estate action programme;
- to pursue an action programme involving HMOs.

Implementation surveys

Before beginning a programme of work on a stock of dwellings it is usual to draw up a schedule of work to identify precisely what is to be done to each individual property. The information required is collected in an implementation survey. Normally all dwellings to be included in the programme are surveyed.

House condition surveys tend to be labour intensive and therefore costly. It is important that an authority is clear about why it wants to conduct a survey and how it plans to use the information before it embarks on the exercise.

The DoE provides general advice on conducting surveys (in a guidance manual [13]), as well as detailed methodology to enable local authorities to carry out surveys using a tried and tested approach. The manual is based on experience gained during the English house condition survey.

Inspection

An inspecting officer should have a means of identification stating the legislation, provisions and purposes for which the officer is authorized.

In all cases in which subsequent proceedings are contemplated, the name(s) of the owner or owners are required. Powers of local authorities to require information on ownership of premises, including houses, are set out in Section 16 of the Local Government (Miscellaneous Provisions) Act 1976.

Inspections under the Housing Acts should be carried out in the knowledge that they may be the subject of examination at a public inquiry, in the county court or by the ombudsman, in the course of which every detail may be contested. The inspector's procedure must be correct, and reports must be complete, accurate and quite unassailable in the impressions they convey.

The undermentioned information should be recorded for each premises inspected:

- address
- date
- brief description
- owner agent
- occupier(s)
- commencement date of tenancy
- rent.

It is important that inspections should be undertaken in a systematic way, taking the order of inspection in a logically progressive fashion. A systematic approach of this nature can be applied to the most complicated property and ensures that no important factors are missed. The introduction of the housing health and safety rating system (discussed later) is set to fundamentally alter the way in which houses are inspected and assessed in the future.

Standard of fitness

The standard of fitness is set out in Section 604 of the Housing Act 1985 as amended by paragraph 83 of Schedule 9 to the Housing Act 1989. Although advisory by nature, local housing authorities are asked to have regard to the guidance contained in Annex A to DoE Circular 17/96 [9] when determining whether a dwelling-house is fit or unfit.

The statutory standard for determining whether a dwelling is fit for human habitation is when, in the opinion of the local housing authority, it fails to meet one or more of the requirements listed here and by reason of that failure the dwelling is not reasonably suitable for occupation:

- it is structurally stable;
- it is free from serious disrepair;
- it is free from dampness prejudicial to the health of the occupants (if any);
- it has adequate provision for lighting, heating and ventilation;
- it has an adequate piped supply of wholesome water;
- there are satisfactory facilities in the dwelling-house for the preparation and cooking of food,

including a sink with a satisfactory supply of hot and cold water;

- it has a suitably located water closet for the exclusive use of the occupants (if any);
- it has for the exclusive use of the occupants (if any) a suitably located fixed bath or shower and wash-hand basin, each of which is provided with a satisfactory supply of hot and cold water;
- it has an effective system for the drainage of foul, waste and surface water.

Whether or not a dwelling-house that is a flat satisfies these requirements, it is unfit for human habitation if, in the opinion of the local authority, the building or part of the building outside the flat fails to meet one or more of the requirements listed in the following and, by reason of that failure, the flat is not reasonably suitable for occupation:

- the building or part is structurally stable;
- it is free from serious disrepair;
- it is free from dampness;
- it has adequate provision for ventilation;
- it has an effective system for the drainage of foul, waste and surface water.

In deciding whether a dwelling-house is fit or unfit for human habitation, discomfort, inconvenience and inefficiency may be relevant factors but the primary concern should lie in safeguarding the health and safety of any occupants.

The extent to which a building presents a risk to health and safety is governed by the nature of the defects present. However, the probability of accidents or damage to health may be increased by either the severity or extent of those defects. The location and duration of defects may also be a material factor that needs to be taken into consideration.

For the purposes of Section 604(1), a dwelling-house is defined as including 'any yard, garden, out-houses and appurtenances belonging to it or usually enjoyed with it'. The condition of out-buildings, boundary walls and the surfaces of yards and paths should be taken into account in assessing the standard of fitness, but it is not expected that the poor condition of these items would be sufficient to render the house unfit unless they formed part of other defects, particularly repair, stability and drainage.

The Section 604 standard may also apply to HMOs by considering the whole house as the dwelling-house irrespective of any subdivision that is not fully self-contained. Flats in multiple occupation may be similarly considered.

To encourage an objective interpretation of the standard, detailed guidance on its application is given in Annexe A of DoE Circular 17/96 [9]. In deciding whether a dwelling-house is unfit, an authority should determine for each of the statutory requirements in turn whether the dwelling-house is reasonably suitable for occupation because of a failure of that particular matter.

The Housing Bill 2003 and the Housing Health and Safety Rating System (also see Chapter 16)

There have been calls for review of the statutory standard of fitness for some time [2,14–17] and the Housing Bill 2003 [7] Part 1 sets out primary legislation to relace it with a completely new evidence-based regime known as the Housing Health and Safety Rating System (HHSRS). The HHSRS is a risk-based approach to assessing housing conditions and incorporates 24 domestic health and safety hazards, each of which would need to be individually assessed to determine a hazard rating score for comparison with the 'ideal', which comprises an equivalent annual risk of death. Conditions falling short of this are assessed for their potential to cause harm. This two-stage assessment considers the likelihood of occurrence over a 12-month period and the range of probable harm outcomes that may result. The person most 'vulnerable' to the identified hazard by age banding is taken into account. The combination gives a hazard weighting score, which may be acceptable or unacceptable, triggering appropriate action at a given threshold.

ODPM [7] reports that the benefits of the HHSRS are difficult to quantify but could be substantial, particularly in respect of domestic accident and poor health, to help reduce mortality rates in substandard housing, reduce isolation, fear of crime and improve involvement in community activities, improved mental health and lower rates of medical interventions. The HHSRS is argued to better target health-related intervention activity than the current statutory standard of fitness.

Central government and local authorities have been developing and implementing strategic action to be able to deliver the HHSRS. The (then) DETR presented national seminars to EHOs and requested feedback. Pilot authorities have tested the system and identified some issues of concern, particularly in relation to training, streamlined application, calculation of score, application to HMOs, enforcement and its relationship to area-based housing activity and (then) grant activity (now 'assistance'). Application of the HHSRS requires use of electronic survey equipment, which records faults. This is designed to be as consistent as possible, but continued training and advice is neccesary to help ensure a 'failsafe' approach, paticularly at a time of such fundamental change [18].

The Housing Bill provides that hazards will fall within category 1 or category 2, depending on the severity as calculated under the prescribed method. ODPM [7] has compared fitness enforcement with the HHSRS in determining intervention, and risk band categories scoring over 1000 (i.e. risk bands A, B and C) are selected as presenting an 'unacceptable risk', triggering mandatory local authority intervention. Further technical guidance on the HHSRS is expected by the end of 2003. There will be a new duty for local authorities to take action in relation to category 1 hazards for dwelling-houses and HMOs, with powers for category 2 hazards. New mandatory enforcement clauses (as the 'best mandatory course of action') will include improvement notices, prohibition orders, mandatory and voluntary warning notices, demolition orders and clearance areas, with new enforcement procedures and protocol, including the facility to issue notices electronically. The Minister has delegated powers to issue further orders to provide discretion as to enforcement in respect of category 1 hazards and in enforcement protocol, which is subject to continued consultation. The Bill also provides the power to charge for certain enforcement action.

DEALING WITH UNFIT PROPERTIES

The Secretary of State for the Environment has issued a code of guidance [9] that all local housing authorities are required to have regard to when deciding, for the purposes of the Housing Acts, whether the **'most satisfactory course of action'** in respect of premises that have been identified as unfit for human habitation is one of the following:

- **Repair:** the service of a repair notice in accordance with Section 189(1) or (1A) of the Housing Act 1985.
- **Deferred action:** the service of a deferred action notice in accordance with Section 81 of the Housing Act 1996 or the renewal of a deferred action notice in accordance with Section 84 of that Act.
- **Closure:** the making of a closing order in accordance with subsection (1) or subsection (2) of Section 264 of the Housing Act 1985.
- **Demolition:** the making of a demolition order in accordance with subsection (1) or subsection (2) of Section 265 of the Housing Act 1985.
- **Clearance:** the declaration of the area in which the premises are situated to be a clearance area in accordance with Section 289 of the Housing Act 1985.

In deciding the most satisfactory course of action an authority needs to have regard to a wide range of factors, which may include:

- whether there is imminent risk to the health and safety of the occupants;
- the characteristics of properties adjoining the unfit premises;
- the local authority's private sector renewal strategy;
- the impact on the local environment and community of any decision;
- the life expectancy of the premises if repaired.

Whatever the circumstances, the local authority must satisfy itself that any enforcement action taken represents the most satisfactory course of action. It must also be able to provide reasons for the decision and be able to demonstrate that it has had regard to the guidance code in reaching the decision.

The code also recommends that where local authorities do not have sufficient information to make a decision they should conduct a survey and assessment using the neighbourhood renewal assessment (NRA) techniques set out in Circular 17/96. This procedure will help to decide the most satisfactory course of action in the prevailing circumstances and explain the chosen course of action to those directly affected.

In reaching a decision on the most satisfactory course of action, a local authority should also have regard to the following.

Repair

In deciding whether to serve a repair notice under Section 189(1) or (1A), a local authority should:

- consider whether the authority should provide financial or other assistance;
- consider the circumstances and wishes of the owner and occupants, including the extent to which they are willing and able to carry out repairs; and the advice and assistance that might be available, or made available, locally to help with that;
- take into account the suitability of the premises for inclusion in a group repair scheme and the extent to which proposals for the preparation of such a scheme have been developed.

Deferred action

Section 81 of the 1996 Act enables a local authority to serve a deferred action notice on an unfit property where it is satisfied that this is the most satisfactory course of action. Section 81 also provides that a deferred action notice that has become operative is a local land charge so long as it remains operative.

A deferred action notice must:

- state that the premises are unfit for human habitation
- specify the works that, in the opinion of the local authority, are required to make the premises fit

- state the other courses of action available to the authority if the premises remain unfit.

The service of a deferred action notice does not prevent a local authority from taking other action in relation to the premises at any time. A deferred action notice must be reviewed not later than two years after the notice becomes operative and at intervals of not more than two years thereafter.

The deferred action notice has been introduced for the following reasons:

- to help authorities in the exercise of their fitness enforcement duties having regard to assistance;
- to provide authorities with additional flexibility to develop and implement strategies for tackling the worst private sector renewal problems identified in the areas, having regard to their finite resources;
- to enable authorities to respond more readily to the wishes of those who might not want to face the upheaval that making their homes fit might entail or who might not want to leave their home in cases where those wishes, when weighed with all other relevant factors, point to deferred action as being the most satisfactory course of action.

When serving or renewing a deferred action notice, local authorities should consider whether to provide the person on whom the notice has been served with advice and assistance, which may include the following:

- how to remedy the unfitness problems
- ways to finance the works
- how to employ a suitable builder
- agency services that might be able to assist.

In deciding whether to serve a deferred action notice under Section 81 of the 1996 Act, or renew a notice under Section 84 of that Act, a local authority should:

- consider the circumstances and wishes of the owner and occupants of the premises, the extent to which they are able to carry out repairs and any advice and assistance that may be available locally to help with the matter;

- consider the health and needs of the owner and occupants and the extent to which these might be adversely affected by a deferment of action;
- consider the physical condition of the premises, for example, whether it constitutes an immediate health and safety risk to the occupants;
- consider the cost and nature of the works required to make the premises fit and whether, in the case of those likely to qualify for assistance, the local authority strategy for assistance.

Closure

In order to close a property, a local authority must make a closing order under Section 264 of the 1985 Act and should:

- consider whether the premises are a listed building or a building protected by notice pending listing, where repair is not the most satisfactory course of action; serving a deferred action notice on or closure of a listed or protected building should always be considered in preference to demolition;
- take account of the position of the premises in relation to neighbouring buildings; where repair is not the most satisfactory course of action and demolition would have an adverse affect on the stability of neighbouring buildings, closure or the service of a deferred action notice may be the only realistic option;
- irrespective of any proposals the owner may have, consider the potential alternative uses of the premises;
- take into account the existence of a conservation or renewal area and proposals generally for the area in which the premises are situated; short-term closure may be an option if the long-term objective is revitalization of the area;
- consider the effect of closure on the cohesion and well-being of the local community and the appearance of the locality;
- consider the availability of local accommodation for rehousing any displaced occupants.

A closing order may be made in respect of a dwelling-house or a building containing flats, some or all of which are unfit.

Demolition

Under Section 265 of the 1985 Act, a local authority may make a demolition order and on doing so should:

- take into account the availability of local accommodation for rehousing the occupants
- consider the prospective use of the cleared site
- consider the local environment, the suitability of the area for continued residential occupation and the impact of a cleared site on the appearance and character of the neighbourhood.

A demolition order may be made in respect of a dwelling-house or a building containing flats, some or all of which are unfit.

Clearance

In deciding whether to declare an area in which unfit premises are situated to be a clearance area under Section 289 of the Housing Act 1985, a local authority should have regard to:

- the degree of concentration of unfit premises within the area;
- the density of the buildings and the street pattern around which they are arranged;
- the overall availability of housing accommodation in the wider neighbourhood in relation to housing needs;
- the proportion of fit premises and other, non-residential, premises in sound condition that would also need to be cleared to arrive at a suitable site;
- whether it would be necessary to acquire land surrounding or adjoining the proposed clearance area; and whether added lands can be acquired by agreement with the owners;
- the existence of any listed buildings or buildings protected by notice pending listing; listed and protected buildings should only be included in a clearance area in exceptional circumstances and only when building consent is given;
- the results of the statutory consultations;
- the arrangements necessary for rehousing the displaced occupants and the extent to which occupants are satisfied with those arrangements;

- the impact of clearance on, and the scope for relocating, commercial premises, for example, corner shops;
- the suitability of the proposed after-use(s) of the site having regard to its shape and size, the needs of the wider neighbourhood and the socioenvironmental benefits that the after-use(s) would bring, the degree of support by the local residents and the extent to which such uses would attract private investment to the area.

Repair of a house that is not unfit

Local authorities also have powers to require works to be carried out in premises that are in disrepair but that are not unfit. In such cases, an authority must be satisfied that:

- a dwelling-house is in such a state of disrepair that, although not unfit for human habitation, substantial repairs are necessary to bring it up to a reasonable standard having regard to its age, character and locality
- a dwelling-house is in such a state of disrepair that, although not unfit for human habitation, its condition is such as to interfere materially with the personal comfort of the occupying tenant.

If one or other of these criteria are met, the authority may serve a repair notice on the person having control (Section 190 on the Housing Act 1985). Section 190 notices may only be served in respect of an owner-occupied dwelling if the property concerned is in a renewal area.

The notice must specify a date by which work should be started and completed. The same rules apply regarding works in default, prosecution for non-compliance, land charges registration and recovery of costs as for notices under Section 189 of the Housing Act 1985. In assessing the repair of a flat, local authorities are enabled to have regard to the condition of any part of the building containing it, and notices can be served where it is such as to affect the condition of the flat or to interfere with the personal comfort of the occupying tenant of the flat.

Appeals against a notice should be made within 21 days of the service. No grounds for appeal are specified, except under Section 191A, that the works are the responsibility of someone else who is the owner (defined at Section 207). Similar to the situation with Section 189 notices, where a person served with a notice has difficulty in finding a builder able to complete the work in the required timescale, Section 191A allows the local authority to execute the works at the expense of the recipient and with his or her agreement. Local authority enforcement options to improve a defective property are shown in Fig. 17.3.

Statutory nuisance and housing conditions

The Environmental Protection Act 1990 Section 79(1)(a) is the statutory nuisance provision for poor living accommodation, which encompasses 'any premises in such as state as to be prejudicial to health or a nuisance'. A premise is defined to include all land, as well as buildings and vessels, and so has wider application than for a 'dwelling-house' under housing legislation, and whilst it can be applied to a dwelling-house, it could equally be applied to a caravan, houseboat or mobile home. The statutory nuisance may result from one defect or accumulation of defects, the effect of which (not the defect itself or its source) is likely to cause injury to health, although health is not defined. (Also see Chapter 6.)

Local authorities' enforcement officers have the right to enter and inspect for statutory nuisance and a duty to act where statutory nuisance exists. Enforcement is by service of an Abatement Notice on the person responsible for nuisance, subject to the right of appeal within 21 days. It is a criminal offence to not comply without reasonable excuse and the local authority can prosecute and/or instigate works in default.

Where statutory nuisance exists, the procedure can be accelerated by applying the Building Act 1984 Section 76 if unreasonable delay would result from the normal procedure. Using this provision, a local authority may serve notice specifying the defects and stating its intention to carry out remedial works within nine days, unless the person served serves a counter notice in seven days.

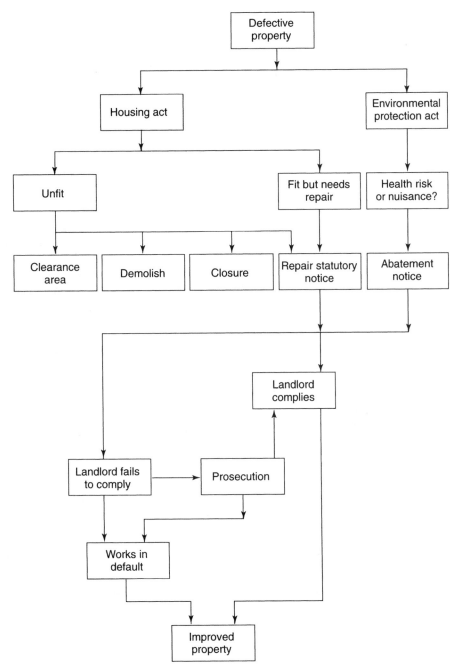

Fig. 17.3 Enforcement action to tackle poor condition housing: local authorities' options to improve a defective property. (Based on [10].)

Removal of obstructive buildings

An obstructive building is defined as 'a building which by reason only of its contact with, or proximity to, other buildings is dangerous, or injurious to health' (Section 283 of the Housing Act 1985). If a building appears to the authority to be an obstructive building, it may proceed to order and secure its demolition. The following steps must be taken.

1. Serve upon the owner or owners notice of a time and place at which the question of ordering the building to be demolished will be considered by the authority. (The owners are entitled to be heard when the matter is under consideration.)
2. If satisfied that the building is obstructive and it or any part of it ought to be demolished, make a demolition order accordingly.
3. Serve a copy of the order upon the owner or owners.
4. If requested to do so by the owners, purchase the building, the price to be assessed as prescribed by the Act, and as soon as possible after obtaining possession, carry out the demolition.
5. Alternatively, and in default of the owners themselves carrying out the demolition within a time prescribed in the order, the authority may enter and carry out the demolition and sell the materials rendered available thereby.

The expenses incurred by the authority and not covered by the sale of materials may be recovered as a simple contract debt. The local authority must compensate the owner or owners in respect of loss arising from the demolition, the amount to be assessed as prescribed by the Act.

Cleansing of buildings before demolition

In order to prevent the migration of vermin from a building about to be demolished by reason of a demolition order to other buildings adjoining it, and to prevent the sale of verminous building materials, local authorities are empowered by Section 273 of the Housing Act 1985 to secure the disinfestation of any building before demolition.

The local authority may at any time between the date on which the order is made and the date on which it becomes operative serve notice on the owner of the building that it intends to cleanse it before it is demolished; the authority may then enter and carry out such work as it thinks requisite for the purpose of destroying vermin. The owner who has received such notice may require the work to be done within 14 days; at the expiration of that time he or she may proceed with the demolition.

Power to charge for enforcement action

Section 87 of the 1996 Act provides local authorities with the power to make a reasonable charge as a means of recovering expenses incurred in:

- serving a repair notice under Sections 189 and 190 of the 1985 Act;
- serving or deciding to renew a deferred action notice;
- making a closing order under Section 264 of the 1985 Act;
- making a demolition order under Section 265 of the 1985 Act.

The Secretary of State has the power to specify by order the maximum amount of any charge. The Housing (Maximum Charge for Enforcement Action) Order 1996 specifies the maximum amount, which was set at £300 in 1996.

In deciding whether to exercise their powers to make a charge, and the level of any charge they do decide to make, local authorities should take into account the circumstances of the person or persons against whom enforcement action is being taken. The sum recoverable by the authority is, until recovered, a charge on the premises concerned.

HOUSE AND AREA REGENERATION

Private Sector Housing Renewal Policy and the Regulatory Reform (Housing Assistance) (England and Wales) Order 2002

Successive policy documents have recognized the substantial role of home owners in improving and adapting the housing stock through personal income, savings and loans. Government has increasingly emphasized home owner's responsibility to keep their home in good repair, but recognized that some lower income owners are not in a position to do so [2–4,19]. As a result, the Local Government and Housing Act 1989 introduced a new targeted form of grants, linking fitness to mandatory means tested grant assistance, which were later made discretionary under the Housing Grants, Construction and Regeneration Act 1996 (with the exception of Disabled Facilities Grants, which remain mandatory subject to certain criteria discussed in the following section) heralding a policy move away from grant provision, with increasing local authority discretion.

The emphasis on personal responsibility for private sector housing conditions has been furthered by the The Regulatory Reform (Housing Assistance) (England and Wales) Order 2002 and supporting *Housing Renewal Guidance* [20], which subsumed earlier grant provisions for renovation, and introduced a new wide-ranging power enabling local authorities to improve living conditions in their area by providing 'assistance' for housing renewal. The Order makes the following significant changes:

- Introduces a general power enabling local housing authorities to provide assistance for housing renewal.
- Repeals grant legislation under the Housing Grants, Construction and Regeneration Act 1996 for Renovation Grants, Common Parts Grants, HMO Grants, Group Repair and Home Repair Assistance.
- Repeals provisions under the Housing Act 1985 in respect of local housing authority loans for housing renewal, except for local authorities who are not housing authorities.
- Streamlines provisions relating to Renewal Areas.
- Makes changes to Disabled Facilities Grant provision, although mandatory DFGs remain largely unchanged.

This forms part of a wider change to support policy review, encourages partnership working,

local solutions to local housing issues and ends the dependency on grants. The new power therefore enables local authorities to provide assistance to any person for:

- The acquisition of living accommodation, where the authority wish to purchase a person's home or as an alternative to adapting, improving or repairing it.
- The adaptation or improvement of living accommodation (including alteration, coversion or enlargement, and by the installation of things or injection of substances).
- The repair of living accommodation.
- The demolition of buildings comprising or including living accommodation.
- The construction of living accommodation to replace living accommodation that has been demolished.

In doing so, local authorities are required to have regard to a person's ability to meet a contribution of repaying the assistance before imposing such a conditon and taking steps to enforce it, and government recognizes that some home owners, particularly the elderly and most vulnerable, do not have resources to keep their homes in good repair [20]. Additionally, local authorities have to provide a written statement of these conditions and place a charge on the property in order to secure the amount, gaining prior consent of the owner in respect of assisted work, and before varying or revoking any relevant condition. Local authorities must adopt and publish their policy, and are empowered to require information and evidence for this new power. The Order is seen to provide more flexibility in developing strategic action to deal with poor private sector housing through new policies and partnership arrangements as part of a wider move towards tackling poverty and social exclusion, health inequalities and neighbourhood decline [20].

Local authorities must have published its policy on private sector housing renewal by 19 July 2003, and in the light of wider strategic objectives, decided on the furture of grant assistance and consider alternatives to support home owners. *Housing Renewal Guidance* [20] provides advice on developing private sector renewal strategy and the policy tools available, together with procedures necessary to publish and continually review such policy, replacing much of DoE Circular 17/96, but it is emphasized that much of it remains in draft form initially and has non-statutory status.

Alternatives to grant assistance remain largely untested, although some 'not for profit' schemes are currently offering equity release products to some home owners to help finance repairs and improvements, often through Home Improvement Agencies. Other options include loan schemes, low-cost capital advances and home maintenance services and insurance-based products, but it remains too early to evaluate how such schemes might dovetail into sustainable local authority-led area regeneration strategies, when many products are essentially based on the individual and not the property, and subject to the workings of the market.

At the time of writing, the extent to which local authorities have been developing policy to respond to this new Order remain unclear. As a result, research has been commissioned to the Centre for Urban and Regional Studies, University of Birmingham to explore the current stage of published policy required under the Order, consultation, relationship to existing grant policies, proposed nature of strategic interventions, how activity is seen to address private sector housing conditions, access and likelihood of uptake of financial loans, etc., costs and benefits anticipated and challenges faced in the implementation of proposed policies.

Disabled facilities grants

The Housing Grants, Construction and Regeneration Act 1996 Section 23 provides for a 'disabled facilities grant' (DFG) specifically for works to provide facilities and to carry out adaptations to dwellings, or to the common parts of buildings containing flats, for the benefit of disabled people. Disabled facilities grants are mandatory and are primarily aimed at facilitating access for the disabled person into and around his or her home, including access to, or in some cases provision of, essential amenities such as toilets, bathroom and kitchen facilities, and adapting the controls of any heating, lighting or power supplies in order to make them suitable for use by the disabled occupants. Council tenants and housing association tenants are assessed for needs on the same basis as private owners and tenants and under the same means testing arrangements. Works eligible for mandatory DFG are set out in Section 23(1) of the 1996 Act and now include:

- Facilitating access and provision so that the disabled person is able have reasonable access into and around their home.
- Making a dwelling or building safe by providing certain adaptations required for the disabled person's need, for example, in the case of behavioural problems.
- Provision of a room suitable for sleeping, to maintain normal sleeping arrangements where no other room is appropriate.
- Provison of a lavatory, washing, bathing and showering facilities.
- Facilitating preparation and cooking of food, including rearrangement or enlargement of kitchen, or modified facilities to meet the disabled person's need independently.
- Provision or improvement of existing heating, lighting and power and full use of appropriate controls.
- Enabling works required to assist the disabled person to care for a dependant person who normally resides in the dwelling, for example, spouse, partner or child [21].

On receipt of an application, local housing authorities are required by Section 24 to consult the appropriate social services authority on whether the relevant works are 'necessary and appropriate' to meet the needs of the disabled person, but it is ultimately housing authorities who decide the level of adaptations for which grant is approved, whether works are reasonable and practicable in respect of both the disabled applicant's need and property condition as well as the opportunity to take into account good practice guidance jointly issued by ODPM and the Department of Health, *Responding to the Need for Adaptations*. The local authority must follow the procedural aspects of DFG application closely in respect of works eligible and certificates and conditions, including cases of successive applications where the disabled person's condition is degenerative. Local authorities have the option of recovering specialist equipment when it is no longer required to reassign it to someone else in need.

The Regulatory Reform (Housing Assistance) (England and Wales) Order 2002 and Annex D of recent Housing Renewal Guidance [21] amended earlier provisions governing mandatory DFGs. The Order makes two changes to DFGs from 19 July 2003:

- It extends mandatory DFG eligibility available for a broad range of essential adaptations to those occupying park homes and houseboats.
- It removes the power to give discretionary DFGs that were previously available to make the dwelling suitable for the accommodation, welfare or employment of the disabled person.

Additionally, the Order enables discretionary assistance in any form (e.g. grant, loan) for works such as small-scale adaptation or top-up mandatory DFG, or to assist in a move to alternative living accommodation, with no restriction on the amount that can be given, but this must be part of their published policy.

FUEL POVERTY

The first national Energy Report was carried out as part of the English House Condition Survey in 1991 [22] and revealed that 40% of privately rented homes, 20% of socially rented homes and 10% of owner-occupied homes were grossly energy inefficient. Additionally, some 30% (or 7 million) of UK households lived in fuel poverty, mainly due to poor insulation and heating systems [23].

Initiatives such as the Home Energy Conservation Act 1995 and Energy Conservation Act 1996 to reduce domestic energy emissions, and Home Energy Efficiency Scheme (HEES; and New HEES) grant assistance had limited impact but is now superceded by Warm Front Team grants managed by the EAGA partnership and TXU Warm Front Team under contract to the Department for Environment, Food and Rural Affairs (DEFRA).

The Fuel Poverty Strategy [24] sought to tackle the problem of some 40 000 excess winter deaths in Britain by lifting 800 000 vulnerable households in England out of fuel poverty by 2004 and to end fuel poverty by 2010. A partnership approach is seen as fundamental and Warm Zones have been piloted to deliver area-based sustainable change. These are currently funded by national government and energy utilities, working in partnership with energy installers, training organizations, primary care trusts (PCTs), local authorities and the business and voluntary sectors. Local authorities and their home improvement agencies remain well placed to identify fuel poverty and make appropriate referral for assistance, particularly in the private housing sector, which other partner organizations can find hard to reach.

RENEWAL AREAS

Renewal areas are not just about housing, but a comprehensive regeneration of an area, through improving homes, shops, commercial areas, the local environment, infrastructure and community opportunities and local authorities are increasingly encouraged to adopt area-based renewal strategies responding to local need with increased rigour and impetus. Renewal areas have several benefits that seek to encourage an increasingly partnership-based approach to engage communities and stimulate private sector investment alongside public resource, which encourages increased confidence in the area. They provide a local strategic framework for wider regeneration, demonstrate the local authority's commitment to the area and provide local authorities with additional powers, including to carry out environmental works and exercise acquisition of land and property.

The Regulatory Reform (Housing Assistance) (England and Wales) Order 2002 relaxed criteria under the Local Government and Housing Act 1989 in respect of declaration criteria for renewal areas to encourage further activity, allied to the new power to give assistance in renewal areas. Local authorities however retain earlier requirements to undertake a socio-economic assessment; to ensure that a renewal area is the most effective way of improving conditions in the area; to carry out consultation with local residents before, during and after declaration and in respect of any changes; to declare the time period of the renewal area; and to notify residents on excluding land, and time of local authority exit from, a renewal area. This is supported by earlier *Good Practice Guidance: Running and Sustaining Renewal Areas*, issued by the then DETR in 1999.

Local authorities must comply with several conditions prior to, during and at the end of the renewal area's life and must prepare a report according to the provision of the 1989 Act, which should contain details of:

- the living conditions in the area;
- ways in which those conditions may be improved;
- powers available and the local authority's detailed proposals to exercise these; and
- the cost and the financial resources available.

Neighbourhood renewal assessments

A neighbourhood renewal assessment (NRA) is a comprehensive appraisal of an area to ensure that the local authority has a thorough understanding of relevant issues, problems and stakeholders' views – as well as an appraisal of the housing market – to ensure a sustainable approach to forthcoming housing renewal activity. ODPM have issued detailed guidance on carrying out an NRA in Annex H of its *Housing Renewal Guidance* [20]. The NRA provides assurance that declaration of a renewal area is the most effective way of improving living conditions in the area. The NRA process involves:

- an understanding of present and future housing demand, the local population and economy, possibly working with adjacent local authorities;
- an awareness of the role of other services including education, transport infrastructure, health and crime in respect of the local market; and

- collation of stock condition information across tenures and the investment need over the proposed renewal area period.

The success of the NRA process requires a corporate approach from the local authority. Renewal assessments are not just about housing, and an authority will need to consider the full range of disciplines that should be included in the assessment team. The NRA process is based on a series of logical steps that, when taken together, provide a thorough and systematic appraisal for considering alternative courses of action in an area. One of the main reasons for undertaking an NRA is to decide whether or not it is likely to be the right way of tackling a neighbourhood's problems.

The development and justification of a preferred strategy forms the crux of the NRA methodology, and it is essential that an authority should know exactly what it wants to achieve in an area, and why and how it can go about it before it embarks on the process and commitment of declaring a statutory renewal area.

HOUSES IN MULTIPLE OCCUPATION

The majority of HMOs are in the private rented sector where they provide an important source of accommodation for low income households. Unfortunately, some of the worst housing conditions can be found in HMOs, including disrepair, inadequate means of escape from fire, lack of basic amenities and unsatisfactory management of such properties. See, for example, the article by Shaw *et al.* (1998, Health problems in houses in multiple occupation, *Environmental Health Journal*, **106** (10), 278–81).

The Housing Bill 2003 establishes provisions to introduce a national licencing scheme for higher risk HMOs, which will completely transform current enforcement protocol; this is discussed in more detail later in this section.

Definition

The legislation covering HMOs is principally found in Part XI of the Housing Act 1985, as amended by Parts VII and VIII and Schedules 9 and 11 of the Local Government and Housing Act 1989. The Housing Act 1985 defined a house in multiple occupation as 'a house which is occupied by persons who do not form a single household', but this was extended by paragraph 44 of Schedule 9 of the 1989 Act to include a flat in multiple occupation. The term now encompasses any purpose-built or converted flat whose occupants do not form a single household. The Chartered Institute of Environmental Health (CIEH) has defined six categories of HMOs:

1. Houses divided into flats or bedsitters where some facilities are shared.
2. Houses occupied on a shared basis where occupiers have rooms of their own.
3. Lodging accommodation where resident landlords let rooms.
4. Hostels, lodging houses and bed-and-breakfast hotels.
5. Registered residential homes.
6. Self-contained flats with common parts such as stairways.

Advice is also set out in DoE Circular 12/93 [25].

Control of HMOs

The provisions in Part II of the Housing Act 1996 are directed towards improving health and safety standards in HMOs by giving landlords clearer responsibilities and local authorities stronger powers to take action. The main provisions, described in the following, came into force during 1997 apart from paragraph 1, which is awaiting the issue of an HMO code of practice. The Act:

1. Places a duty on landlords of HMOs to ensure that their properties are safe, in accordance with advice in an HMO code of practice, or risk prosecution with a fine of up to £5000. This will commence when the Secretary of State issues a code of practice and is aimed at updating the standards of fitness for HMOs currently set out in DoE circular 12/92 [26].
2. Provides for a revised HMO registration scheme that will allow local authorities to adopt one of two model registration schemes. A local authority is able to refuse to register any property where conditions are substandard. The scheme may be altered with the Secretary of State's agreement, particularly to adopt special control provisions that will allow the closure of HMOs if they cause a nuisance or annoyance to the neighbourhood. Authorities will also be able to prevent new HMOs from opening if they would be detrimental to the area.
3. Allows authorities to charge landlords of HMOs registration fees and fees for reregistration after 5 years to help meet the costs of HMO enforcement activity. See the Houses in Multiple Occupation (Fees for Registration Schemes) Order 1997, amended in 1998. The maximum fine for not complying with a registration scheme containing control provisions will be increased to £5000.
4. Extends local authorities' mandatory duty to ensure there are adequate means of escape from fire in larger HMOs to include the provision of other fire safety precautions.
5. Protects landlords of HMOs from being forced to undertake works twice because standards have been revised. It will not be possible to serve a second HMO enforcement notice requiring works to be carried out to make a property fit for the number of occupants within a five-year period in respect of the same requirements unless there has been a change of circumstances in relation to the premises.
6. Allows authorities to recover from landlords of HMOs the reasonable administrative expenses, up to a maximum amount to be prescribed by order, they incur in serving an enforcement notice under Section 352 of the Housing Act 1985 to make an HMO fit for the number of occupants.

Earlier guidance from the DoE [25] provides comprehensive guidance on developing and implementing strategies specifically for HMOs, including the following:

- identifying and locating HMOs;
- the interface of planning policies with those for HMOs;

- risk assessment to identify priorities for scarce resources;
- enforcement action on HMOs.

Figure 17.4 indicates the various enforcement options available to local authorities to deal with HMOs.

Standards of management

Problems can occur in multi-occupied houses, particularly in common areas and in shared facilities where no individual tenant has overall responsibility. Local authorities have powers to deal with unsatisfactory standards of management in order to tackle this problem. Management regulations for HMOs made under Section 369 of the Housing Act 1985 apply automatically to all HMOs. The Housing (Management of Houses in Multiple Occupation) Regulations 1990 make provision for ensuring that the person managing a house in multiple occupation observes proper standards of management.

Regulation 2 varies the definition of 'person managing' in Section 398(6) of the Housing Act 1985, and that definition as so varied is used in these regulations. The 'person managing' is referred to in the regulations as 'the manager'.

The manager is required by the regulations to ensure the repair, maintenance, cleansing or, as the case may be, good order of:

- all means of water supply and drainage in the house (regulation 4);
- parts of the house and installations in common use (regulations 6 and 7);
- living accommodation (regulation 8);
- windows and other means of ventilation (regulation 9);
- means of escape from fire and apparatus, systems and other things provided by way of fire precautions (regulation 10);
- outbuildings, yards, etc. in common use (regulation 11).

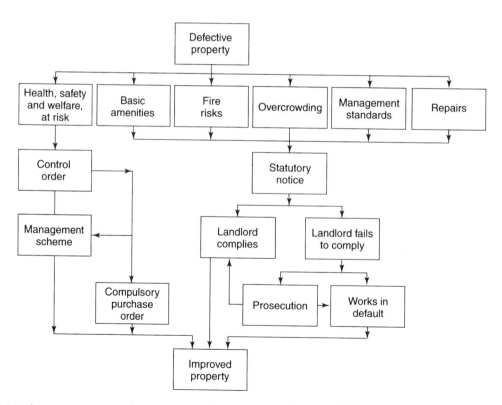

Fig. 17.4 Enforcement action on houses in multiple occupation. (Based on [10].)

The manager is also required to:

- make satisfactory arrangements for the disposal of refuse and litter from the house (regulation 12)
- ensure the taking of reasonable precautions for the general safety of residents (regulation 13)
- display in the house a notice of the name, address and telephone number, if any, of the manager (regulation 14)
- provide specified information to the local housing authority about the occupancy of the house where the authority gives him or her written notice to that effect (regulation 15).

Regulation 16 imposes duties on people who live in the house, for the purpose of ensuring that the manager can effectively carry out the duties imposed by the regulations.

Under Section 369(5) of the 1985 Act, knowingly to contravene, or to fail, without reasonable excuse, to comply with a notice[3] relating to any of these regulations is an offence punishable on summary conviction by a fine not exceeding level 3 on the standard scale.

Failure to comply with the regulations is a criminal offence, and it is important that local authorities ensure that they are given extensive continuing publicity to allow those affected to make any arrangements necessary to check the management of their properties. Authorities should stress that the regulations are binding not only on managers, but also on the occupants, and that either could be liable for prosecution in the event of a breach. Enforcing authorities should also ensure that care is taken to establish where poor standards are due to management neglect or to misuse of the property by the occupants. Advice on the implementation of the regulations is given in the DoE Circulars 12/92 [26] and 12/93 [25].

Where a house does not comply with the management regulations, the local authority may serve the manager with a notice requiring him to execute any necessary work. The notice must specify dates for commencement and completion of the works (Section 372 of the Housing Act 1985 as amended by Schedule 9 of the Housing and Local Government Act 1989). The commencement date must not be earlier than 21 days after the service of the notice. The time required for the work to be done must be reasonable, and the local authority may give written permission for this period to be extended. Any known owner or lessee must be informed of the notice.

As with repairs notices and Section 352 notices, local authorities can serve notice of intention to carry out the remedial works themselves if reasonable progress is not being made, and recover their expenses under Schedule 10.

Any person served with a notice may appeal to the magistrate court within 21 days on the grounds that:

- the condition of the house did not justify the service of a notice;
- there has been some material informality, defect or error in connection with the notice;
- the date specified for the beginning of the work is not reasonable;
- the time allowed for doing the work is inadequate;
- the works required are unreasonable in character or extent, or are unnecessary, or the local authority has refused the execution of alternative works;
- that some person, other than the appellant, is wholly or in part responsible for the state of affairs that necessitates the service of a notice, or will derive benefit from the work to be done as holder of an estate or interest in the premises.

HMOs – fitness standard for human habitation

The standard of fitness applies to HMOs by virtue of Sections 604(3) and (4) of the Housing Act 1985 as amended, and provides that an HMO is fit for human habitation unless, in the authority's opinion, it fails to meet one or more of the specified requirements in Section 604(1) and by reason of that failure is not reasonably suitable for occupation.

---

[3]The present requirement to serve a preliminary notice before implementing this procedure is currently under review.

Whether or not a HMO that is a flat in multiple occupation satisfies those requirements, it is unfit for human habitation if, in the opinion of the local housing authority, the building or a part of the building outside the flat fails to comply with the requirements of Section 604(2), and by reason of that failure the flat is not reasonably suitable for occupation.

Under Section 604A of the 1985 Act, local housing authorities are required to consider the most satisfactory course of action for dealing with such unfit HMOs in the context of Parts VI (repair notices) and IX (slum clearance), having regard to the statutory Code of Guidance for Dealing with Unfit Premises (Annexe B of DoE Circular 17/96) [9]. They are required to consider once a year the best means of discharging their functions under Part XI as under other parts of the 1985 Act.

HMOs – standard of fitness for number of occupants

Section 352 of the Housing Act 1985 as amended specifies the standard of fitness for the number of occupants of an HMO. (See also [27].) An HMO does not meet the standard where in the authority's opinion it fails to meet one or more of the requirements mentioned in the following list and, having regard to the number of occupants accommodated there, by reason of that failure it is not reasonably suitable for occupation by those occupants. The requirements are that the HMO has:

- satisfactory facilities for the storage, preparation and cooking of food including an adequate number of sinks with a satisfactory supply of hot and cold water;
- an adequate number of suitably located WCs for the exclusive use of occupants;
- an adequate number of suitably located fixed baths or showers and wash-hand basins, each provided with a satisfactory supply of hot and cold water for the exclusive use of the occupants;
- adequate means of escape from fire;
- other adequate fire precautions.

Guidance on this fitness standard for HMOs was issued in DoE Circular 12/92 [26].

Except in the instances referred to in the next paragraph, local authorities have the discretion to serve notice under Section 352 as amended, to render the property fit for the number of occupants or take alternative action under Section 354 (directions limiting the number of occupants).

Local authorities are able to recover reasonable administrative expenses up to a maximum of £300 incurred in the serving of a notice under Section 352 of the Housing (Recovery of Expenses for Section 352 Notices) Order 1997.

Fire precautions

Section 352 enables an authority to require the installation of adequate fire precautions, such as fire detection systems, fire warning systems and fire fighting equipment in appropriate circumstances, as well as means of escape. Where an HMO does not have adequate means of escape from fire, an authority may exercise its powers under Section 368 (as amended by paragraph 55 of Schedule 9) securing that part of an HMO that is not used for human habitation as an alternative to the provision of a means of escape. However, for some properties of at least three storeys in height, a local authority is obligated to exercise one of these powers by the Housing (Fire Safety in HMOs) Order 1997, which came into operation incrementally between 1997 and 2000.

Where the local authority is proposing action to deal with fire precautions under Section 352 or 368, it must consult with the fire authority.

Fire warning systems can make a valuable contribution to preserving life by giving people an early indication of a fire on the premises. Details of such systems for hostel-type accommodation and houses converted into self-contained dwelling units are contained in DoE Circular 12/92 [26]. The circular provides an aid to the process of consultation between housing and fire authority that is required by Section 365(3) as amended.

If a local authority considers that there is a serious risk to life in the event of fire and is unable to take suitable action under housing legislation, it can ask the fire authority to use its limited prohibition powers under Section 10 of the Fire Precautions Act 1971.

In the case of a multi-occupied flat, works – including means of escape – necessary under Section 352 may only be required within the flat itself, and cannot be specified in respect of parts of the building outside the flat. If a flat is not large enough to accommodate the necessary work, then local authorities should consider the use of a direction order under Section 354.

HMO notices

The single prescribed HMO notice can be used either where an authority decides that serving a repair notice under Section 189 (unfit for human habitation) is the most satisfactory course of action, where work is required under Section 352 (fit for number of occupants), or both. The notice can be served on the person having control, or on the person managing the house.

Section 352(4) empowers the authority to specify dates for commencement and completion of works. The grounds of appeal are similar to those following service of a notice under Section 372 of the Housing Act 1985 as amended.

The process for enforcing HMO notices is similar to that for repairs notices under Section 189. Section 352 notices are local land charges, thereby giving notice to subsequent owners that a notice requiring the carrying out of works has been served, and that new notices do not need to be served following a change of ownership.

The Housing Act 1985 generally requires an authority to inform owners and lessees of a property of any action that it is taking on that property. The Local Government and Housing Act 1989 imposed three new requirements in respect of Section 352 notices:

- all occupants of the house must be informed when notices are served;
- local authorities must keep a register of the notices served and make it available for public inspection;
- on request, local authorities must supply copies of notices to members of the public, and may charge a reasonable fee for doing so.

Where no appeal is lodged against the reasonableness of the start date specified in Section 352 or 372 notices, work should be completed no later than the designated final date and local authority can instigate works in default if it considers that reasonable progress is not being made towards compliance, giving the owner or recipient of the original notice not less than seven days warning of its intention. The provisions of Schedule 10 of the Housing Act 1985 as amended apply where such notice is served, including where the work is then carried out by people other than the authority, or by someone appointed by the authority. Paragraph 13(3) of Schedule 15 to the 1988 Act introduced a ground of appeal where an authority undertakes work because reasonable progress is not being made. The appellant can argue that expenses should not be recovered because reasonable progress was being made. The right of appeal is only exercisable after the works have been completed.

Expenses are recoverable from the person having control of, or managing, an HMO at the time the demand for expenses is made. Paragraph 70(4) of Schedule 9 of the Local Government and Housing Act 1989 enables authorities to sequester rents in order to recoup the cost of works in default.

After a Section 352 notice has been served and complied with, the local authority cannot serve a further notice for similar requirements for 5 years unless there has been a change in circumstances (Section 71 of the Housing Act 1996).

It is an offence for a person served with an HMO notice or a notice under Section 372 to maintain proper standards of management wilfully to fail to comply with the requirements of the notice. The maximum penalty on conviction is a fine not exceeding level 4 on the standards scale (Section 37 of the Criminal Justice Act 1982).

Direction to prevent or reduce overcrowding

As an alternative to or in conjunction with requiring work under Section 352 of the Housing Act 1985, a local authority can issue a direction under Section 354 limiting the number of occupants in an HMO. The direction may:

- specify a maximum number of individuals or households or both able to occupy the house in its existing condition;

- in conjunction with a notice requiring work to make the house fit for multi-occupation, specify the maximum number able to occupy the house after completion of the work.

The direction imposes an obligation on the occupier or any other person entitled to permit persons to take up residence:

- to ensure that no person takes up residence to increase the number living there above the limit set;
- or, if the number already exceeds the limit, not to permit the number to increase further.

At least seven days before giving a direction, a local authority must serve notice of its intention on the owner and all known lessees, and also post a notice in a part of the house accessible to all those living there. The person on whom the notice is served may make representation to the local authority, and the local authority must, within 7 days of giving the direction, serve a copy on the owner and lessees and post a copy in the house.

A local authority may revoke or vary a direction upon application of any person having an interest or estate in the premises after any changed circumstances. If a local authority refuses an application for revocation etc., or does not within 35 days of the making of the application notify the applicant of its decision, the applicant can appeal to the county court.

At any time while a direction is in force, a local authority may serve notice on the occupier of an HMO requiring him or her to furnish within seven days a statement showing the number of people living in the premises, their names and the rooms they occupy.

Overcrowding notices

Section 354 of the Housing Act 1985 deals with overcrowding in HMOs only as far as the number of occupants relates to the standard of fitness for multi-occupation, but further overcrowding provisions are included in Sections 358–364.

In this case, if it appears to a local authority that an excessive number of persons is, or is likely to be, accommodated having regard to the rooms available, it must serve a notice on the occupier and/or person having control. The notice must state the maximum number of persons permitted to occupy each room for sleeping accommodation at any one time, or indicate that a particular room cannot be used for this purpose. The numbers to be fixed here are at the discretion of each authority, and will depend upon that authority's own housing circumstances.

Having specified the permitted occupancy, the notice can then require that either:

- sleeping accommodation shall not be arranged other than in accordance with the overcrowding notice and in such a way that persons of opposite sexes over the age of 12 years (other than husband and wife) do not share the same sleeping accommodation
- or that new residents will not occupy sleeping accommodation other than in accordance with the notice and in such a way that persons of opposite sexes over the age of 12 years do not share the same room.

Before serving the notice, the local authority must give seven days notice to the occupier and person having control, and ensure as far as possible that each resident is informed of its intention, giving an opportunity for representation to be made. Appeals against an overcrowding notice may be made to the county court within 21 days by any person aggrieved, and the authority may at any time revoke or vary it so as to allow more people to be accommodated (but not fewer).

In order to determine the number of persons occupying the house at any time, the local authority may require the occupier to provide a written statement within seven days specifying the mode of occupation. The local authority may on the application of a person having an interest or estate in the premises, revoke or vary the notice so as to allow more people to be accommodated. If such an application is refused or no decision is made within 35 days, the applicant may appeal to the county court.

It is an offence to fail to comply with an overcrowding notice (maximum fine not exceeding level 4 on the standard scale), or to fail to reply to a requisition or to knowingly give false information (maximum fine level 2 on the standard scale).

Control orders (Sections 379–394 of the Housing Act 1985)

A local authority may make a control order in respect of a house in multiple occupation where it is satisfied that:

- a notice has been or could be served under Section 352 of the Housing Act 1985 as amended requiring works to make the house fit for the number of individuals or households in the house;
- or a notice has been or could be served under Section 372 requiring works to make good neglected standards of management;
- or a direction has been or could be given under Section 354 limiting the number of individuals or households to be accommodated;

and living conditions in the house are such that an order is necessary to protect the safety, health or welfare of the residents.

A control order comes into force when it is made and, as soon as possible afterwards, the local authority must do what is necessary to protect the health, welfare and safety of the occupants, and must maintain proper standards of management.

As soon as possible after making the order, a copy of it, together with notices setting out the grounds for making the order, the effect of it and indicating the right of appeal, must be posted in the house, and served on the person having control of the house and on any owner, lessee or mortgagee.

Compensation is payable quarterly to a dispossessed proprietor. After the order has been made, the local authority must prepare a management scheme and within eight weeks after the order comes into force serve a copy of it on the dispossessed proprietor, on the owner, lessee or mortgagee, and on any person who received a copy of the control order.

The scheme must give the following:

1. Particulars of works involving capital expenditure that the local authority would require

under the Housing Act 1985, or other housing or public health legislation.
2. An estimate of the cost of carrying out the works.
3. The highest number of people or households that should live in the house from time to time, as works progress.
4. An estimate of the balances that, from time to time, will accrue to the local authority from the rent received after deducting compensation payable to the dispossessed proprietor and all expenditure (other than capital expenditure) and establishment charges.

A local authority may at any time vary the payment of surpluses on revenue account so as to increase them.

While the control order is in force, the local authority must keep full accounts in respect of the house and afford the dispossessed proprietor facilities for examining the accounts and inspecting the house.

If a control order is made in respect of furnished accommodation, the right to possession of the furniture will vest in the local authority for the time during which the order remains in force, and the authority may, on written application by the owner of the furniture, renounce its right to possession by giving him two weeks notice. A local authority may fit out, furnish and supply a house subject to a control order with such furniture, fittings and conveniences as appear to be required.

At any time after the making of the order, but not less than 6 weeks from the date of the service of the copy of the scheme, a person having an estate or interest in the house or any other person who may be prejudiced by the making of the order may appeal to the county court against the order.

Within 6 weeks of the date on which a copy of the scheme is served, any person having an estate or interest in the house may appeal to the county court on the grounds set out in Section 386(2) and Schedule 13 paragraph 3(1).

A control order ceases to have effect at the expiration of 5 years beginning with the date it came into force, but the local authority may on an application or on its own initiative revoke the order. The local authority must give at least

21 days notice to occupiers, owners, lessees and mortgagees of its intention to revoke an order.

Within 28 days of the making of a control order, a local housing authority may make a compulsory purchase order for the acquisition of the premises under Part II of the 1985 Act. The duty to prepare and serve a management scheme is then suspended until the authority has been notified of the Secretary of State's decision.

Other statutory powers

In respect of HMOs, housing legislation will nearly always be the most appropriate. However, there are certain instances when other statutory powers should be used to effect repairs and improvements. These powers include:

- the statutory nuisance provisions contained in Sections 79–81 of the Environmental Protection Act 1990
- the Health and Safety at Work, Etc. Act 1974
- the dangerous building and drainage powers of the Building Act 1984 and Public Health Act 1961
- the Fire Precaution Act 1971
- Section 33 of the Local Government (Miscellaneous Provisions) Act 1976.

The Housing Bill 2003 and licensing of HMOs

Much legislation covering HMOs remains reactive and discretionary, leading to calls to introduce a mandatory licensing scheme. Delay with introducing such a system has been compounded by debate about which HMOs should be licensed, how authorities would prioritize and programme licensing work, whether landlords would withdraw from the sector from over-regulation, how the scheme could be administered, etc. However, the general proposal for HMO licensing is seen as a long-term way forward in controlling conditions. It could help to more accurately profile HMOs locally and nationally and add credibility to standards and enforcement through a new proactive and streamlined approach.

The Housing Bill [7] provides a new HMO definition as a house, hostel, self-contained flat or relevant building (specified by the Minister), which wholly or to a significant degree is occupied as relevant accommodation by persons who do not form a single household and in which one or more amenities (toilet, personal washing facilities or cooking facilities) are shared, subject to some exemptions.

Parts 2 and 4 of the Housing Bill establish provisions for a mandatory licensing scheme for high-risk HMOs including three storey houses occupied by five or more persons. The licence must authorize the maximum number of persons or households and may include provisions relating to the management, behaviour of occupants or visitors, condition and amenities, requirements to carry out works and timescales for compliance, limits on numbers of households or persons permitted in particular parts of the house, and can prevent parts of the house being occupied for particular purposes. Standards may be 'prescribed' under the legislation in respect of the number, type and quality of facilities and amenities such as bathroom, toilets, food preparation areas, etc., as well as for health and safety.

It will be an offence for the person managing or having control of the HMO to not hold a licence, and is it up to local housing authorities to take steps to ensure that licence applications are made as soon as practicable in accordance with the legislation and local authority specifications, and to have regard to assessing whether the proposed licence holder (or manager) is a fit and proper person with regard to evidence about offences (fraud, dishonesty, violence or drugs), unlawful discrimination or contravention of landlord or tenant law. Local housing authorities will have a duty to make an Interim Management Order (IMO) to protect the health, safety and welfare of persons occupying the house or persons occupying or having an interest in any premises in the vicinity, if it considers that an unlicensed HMO has no reasonable prospect of being licensed in the near future, but this would be terminated once the HMO were licensed or subject to a Final Management Order (FMO). An FMO must contain a management scheme specifying works to be carried out by the local housing authority, an estimate of costs and the maximum number of households or persons for whom the HMO is reasonably suitable for occupation and full records must be kept. The Minister may make regulations as to management in HMOs. The Bill also contains specific HMO overcrowding provisions where the HMO is not required to be licensed and is not subject to an IMO or FMO.

The Bill also provides discretionary powers to license other HMO categories and privately rented two storey houses as well as granting additional powers to deal with individual problematic HMOs outside of the mandatory licensing regime, including the facility to apply to the county court to make an IMP, possibly followed up by an FMO, but this remains subject to consultation.

The Housing Bill 2003 and selective licensing of other residential accommodation

Parts 3 and 4 of the Housing Bill [7] provide for selective licensing for areas of low demand or that are likely to fall into low demand in England. The proposed discretionary power would allow local housing authorities to introduce a licensing scheme for private landlords covering all or part of their area, through which landlords would be required to obtain a licence based on their personal fitness and management standards. The purpose is to help local housing authorities stop bad – even criminal – landlords and antisocial tenants from undermining attempts to secure sustainable communities as well as to help raise standards in the private rented sector. The designation needs to be part of a wider local housing authority and partnership strategy to contribute to the improvement of socio-economic conditions. The licence must include provisions requiring annual submission of documents relating to the safety of the house (gas, electric, furniture), smoke alarms, references from persons intending to occupy the house and a written statement as to terms on which they occupy the house. The licence may include provisions relating to the management of the house, behaviour of occupants or visitors, the requirement to carry out works and timescale and to prevent certain parts of the house from occupation.

OVERCROWDING

The present overcrowding standards were first set down in the Housing Act 1935, and were re-enacted in 1985 despite the Minister of Health's statement in 1935 that:

> It is relevant to point out that the standard does not represent any ideal standard of housing, but the minimum which is in the view of Parliament desirable while at the same time capable of early enforcement.

Nevertheless, according to recent figures [10], underoccupation is considerably more widespread, with overcrowding constituting only a small residual problem.

It should be noted that this section on overcrowding deals with the number of people who may sleep in the space available in a dwelling-house. Sections 354–357 of the Housing Act 1985 are referred to in the section on multi-occupation, where the number of occupants can be restricted having regard to the standard of fitness specified in Section 352 as amended.

Where it appears to a local authority that there is need to prepare a report on overcrowding in its district or part of its district, or if the Secretary of State so directs, a survey must be made and details submitted to the Secretary of State on any proposal for providing new dwellings.

Overcrowding standards

The 1985 Act contains two principal overcrowding standards, both of which must be satisfied by all dwellings. These are:

- proper separation of the sexes
- the number and floor areas of the rooms.

The Act stipulates that a dwelling is overcrowded when the number of people sleeping there:

- is such that any two of those persons being 10 years old or more of opposite sexes and not living together as husband and wife must sleep in the same room (Section 325);

- is in relation to the number and floor area of the rooms of which the house consists in excess of the permitted number of persons as defined in Section 326.

In determining the number of people sleeping in a house for the purpose of Section 326, no account is taken of a child less than 1 year old and a child aged between 1 and 10 years is reckoned as one-half of one unit.

The age limit of 10 years mentioned here is arbitrary and not based on a view of sexual maturity. The 1985 Act also deals with separation of the sexes at 8 years of age in the case of a local authority's own lodging houses (Section 23(3)) and 12 years of age in multi-occupied premises (Section 360(2)).

For the purpose of Sections 325 and 326, the following definitions apply:

- Section 343 'Dwelling' means any premises used or suitable for use as a separate dwelling. Under this definition any room – or part of a house – sublet to a tenant and occupied separately by him or her as a dwelling, constitutes a separate 'dwelling-house'.
- 'Room' does not include any room of a type not normally used in the locality, either as a living room or as a bedroom. Under this definition, in ordinary circumstances, rooms used as kitchens, sculleries, etc. must be excluded from calculations made under Section 326.

Section 326 states that for the purposes of Part IV of the Act, the expression 'the permitted number of persons' means, in relation to any dwelling-house, either:

1. the number specified in the second column of Table I in the Annexe that follows in relation to a house consisting of the number of rooms of which that house consists; or
2. the aggregate for all the rooms in the house obtained by reckoning, for each room therein of the floor area specified in the first column of Table II in the following Annexe, the number specified in the second column of that table in relation to that area, whichever is lesser;

ANNEXE

Table I	Table II
Where a house consists of:	Where the floor area of a room is:
1. one room 2	1. 110 sq ft or more 2
2. two rooms 3	2. 90 sq ft or more, but less than 110 sq ft $1\frac{1}{2}$
3. three rooms 5	3. 70 sq ft or more, but less than 90 sq ft 1
4. four rooms $7\frac{1}{2}$	
5. five rooms or more with an additional two in respect of each room in excess of five 10	4. 50 sq ft or more, but less than 70 sq ft $\frac{1}{2}$
	5. under 50 sq ft nil

provided that in computing for the purposes of the said Table I the number of rooms in a house, no regard shall be had to any room having a floor of area of less than 50 sqft. The maximum or 'permitted' number of persons who may occupy any dwelling-house is calculated from the two tables.

Taking as an example a house of four rooms, the floor areas of which are, say, approximately 120, 100, 80 and 60 sq ft respectively, the permitted number of occupants under Table I would be $7\frac{1}{2}$. Under Table II it would be 5; that is, the rooms 120 sq ft – 2; 100 sq ft – $1\frac{1}{2}$; 80 sq ft – 1; 60 sq ft – $\frac{1}{2}$. As the number computed under Table II is the lower, namely 5, this would be 'the permitted number of persons' who might occupy the house.

This standard is not satisfactory, and many authorities consider that a bedroom standard would be preferable to the prescribed general room standard, since under these present provisions all living rooms in addition to bedrooms are taken as being capable of providing sleeping accommodation.

Under Section 330, licences to exceed the permitted number of a dwelling may be granted by a local authority to take account of exceptional circumstances. A licence remains in force for a maximum period of 12 months.

Rent book information

Section 4 of the Landlord and Tenant Act 1985 requires a landlord to provide a rent book for tenants whose rent is payable on a weekly basis. The rent book is required to state:

- the name and address of the landlord (Section 5 of the Landlord and Tenant Act 1985)
- a summary of Sections 324–331 of the Housing Act 1985 in the prescribed format
- the permitted number of people that can live in the dwelling.

Occupiers must produce their rent books for inspection by any duly authorized officer. Local authorities must, upon application of the landlord or of the occupier, inform the applicant in writing of the permitted number of people that may occupy the house.

Overcrowding notices

It is the duty of the local authority to enforce the provisions of the Act relating to overcrowding in dwelling-houses in their districts (Section 339). In such cases, the local authority may serve upon the occupier a notice requiring that the overcrowding be abated within 14 days and, if at any time within 3 months from the end of that time the dwelling is occupied by the same occupier or a member of his or her family and is overcrowded, the local authority may apply to the county court for possession. On application, the court must order possession to be given to the landlord within a period of between 14 and 28 days.

The local authority may serve notice on the occupier of a dwelling requiring him or her to provide within 14 days a written statement of the number, ages and sexes of the people sleeping in the dwelling. The situations under which the occupier and the landlord are guilty of committing an offence by causing or permitting a dwelling to be overcrowded are set out in Sections 327 and 331, respectively, of the Housing Act 1985 (as amended).

Measurement of rooms

Rules for the measurement of rooms etc. are laid down in the Housing Act (Overcrowding and Miscellaneous Forms) Regulations 1937. They state:

1. measurements are to be taken at floor level and to the back of any projecting skirting boards;
2. any part of the floor over which the ceiling height is less than 5 ft must be excluded;
3. the area covered by fixed cupboards or chimney breasts and within the area of any bay is to be included.

At the time of writing, the Housing (Overcrowding) Bill is going through Parliament. The main effects of this Bill are to reform existing overcrowding provisions discussed here; to introduce a new bedroom standard to measure overcrowding; to modernize the space standard; to require local authorities to inspect their districts and prepare a report; and to require the Secretary of State to have regard to this report in considering financial allocations to local authorities.

Other overcrowding provisions

1. A tent, van or shed or similar structure used for human habitation that is so overcrowded as to be 'prejudicial to the health of the inmates' is a statutory nuisance for the purposes of Part III of the Public Health Act 1936, Section 268.
2. Hop, fruit or vegetable picking – power to make by-laws (Section 270 of the Public Health Act 1936).
3. Canal boats – duty of the Secretary of State to make regulations (Section 49 of the Public Health (Control of Disease) Act 1984).
4. Houses owned and managed by a local authority – power of an authority to make by-laws for management, use and regulation.

THE LAW OF LANDLORD AND TENANT

Action by local authorities to improve conditions in the private housing sector can cause friction and disagreement between landlord and tenant and local authorities should develop effective corporate policies to ensure that any action they take does not undermine a tenant's position, and that procedures are in place to deal promptly with harassment and threats of eviction.

Harassment

The law protects people living in residential accommodation from harassment and illegal eviction as a criminal offence whereby someone who is harassed or illegally evicted can claim damages through the civil court. The law against harassment applies to all people living in residential property as tenancies or licences, and it applies to the acts of anybody acting on behalf of a landlord.

The Protection from Eviction Act 1977 as amended by the Housing Act 1988 makes it an offence to:

- do acts likely to interfere with the peace or comfort of a tenant or anyone living with him or her;
- persistently withdraw or withhold services for which the tenant has a reasonable need to live in the premises as a home.

It is an offence to do either of these having reasonable cause to believe that they would cause the tenant to leave his home or stop using part of it, or stop doing the things a tenant should normally expect to be able to do. It is also an offence to take someone's home away from him unlawfully. A conviction can lead to a fine and/or imprisonment.

In housing terms, harassment can cover a wide range of activities. It can take many forms short of physical violence, and it may not always be obvious to others that particular sorts of activity are intended to drive the tenant from the property. Examples of harassment are:

- withdrawal of services like gas, electricity, water;
- withholding keys;

- antisocial behaviour by a landlord or his agent;
- removing a tenant's belongings;
- entering the tenant's accommodation without permission;
- threatening language or behaviour;
- failing to complete repairs.

Illegal eviction

A landlord cannot normally enforce his right to get his property back from a residential tenant or, in many cases, a licensee unless he does it through the courts. A landlord seeking possession from a periodic residential tenant (other than a tenant under the Housing Acts 1988 and 1996) or from a licensee must generally serve a written notice to quit giving at least four weeks notice. A landlord seeking possession from an assured tenant under the Housing Acts 1988 and 1996 must tell the tenant of his intention to start court proceedings by serving a notice on him. Depending on the grounds on which the landlord is seeking possession, the period of notice will be two months, except in a few cases where the tenancy agreement allows for longer notice. The tenant is not required to leave the property until the notice expires, and even then may not be evicted without an order of the court.

The Housing Act 1988 makes it a general requirement for a licensor to obtain a court order before he can evict a licensee. However, certain licences and tenancies are excluded from this requirement. They are broadly licences or tenancies granted on or after 15 January 1989:

- by resident landlords to people with whom they or a member of the landlord's family share accommodation;
- to trespassers;
- to those occupying a property for a holiday or rent-free;
- licences granted to the people living in certain publicly funded hostels.

However, although it is not necessary to get a court order to evict someone in the excluded categories, there is a common-law requirement for a landlord to serve a periodic tenant with notice equivalent to the period of the tenancy. This means, for example, that if the tenancy was from month to month, the landlord must give a month's notice. (In the case of yearly tenancies, he must give six months notice.) Under common law, a licensee must be given notice that is reasonable in all the circumstances.

Forms of tenure

The Housing Act 1988 created new forms of letting: the assured and assured shorthold tenancy, and the assured agricultural occupancy. These are now the standard forms of letting to new private tenants. However, tenants who were living in their present home before 15 January 1989, as well as tenants in unfurnished accommodation provided by a resident landlord under an agreement made before 14 August 1974, may very well have a regulated tenancy under the Rent Act 1977.

An **assured tenancy** was the standard form of tenancy for most new residential lettings made between private landlords and their tenants between 15 January 1989 and 28 February 1997. A tenant cannot have an assured tenancy of a property that is not let as a separate dwelling, nor his or her only or main home.

An assured tenancy may be written, or there may be no written agreement, just a spoken agreement between the people concerned. They will agree on the rent and other terms of the tenancy, such as whether the tenant may sublet, at the beginning. That rent will be whatever the landlord and tenant agree. The tenant will have long-term security of tenure; he does not have to leave unless he wishes to go, or unless the landlord has grounds for regaining possession that he can prove in court.

The tenancy may in the first place be for a fixed number of months or years – a **fixed-term tenancy**. Or it may be granted for an indefinite period with payment of rent being made on a weekly or monthly (or other periodic) basis: this is a **contractual periodic tenancy**. If the tenancy is a fixed-term tenancy, when the term ends, the landlord and the tenant may agree to another fixed term, or to a periodic tenancy. If they do not,

then a periodic tenancy arises automatically. This type of tenancy is known as a **statutory periodic tenancy**.

If the tenant has a fixed-term tenancy, the agreement will normally provide:

- either for the rent to be fixed for the length of the term;
- or for it to increase one or more times, at regular intervals on a specified basis during the term.

A landlord must deal with requests for rental increase in accordance with requirements under the relevant tenancy.

A **shorthold tenancy** is a kind of assured tenancy that offers the landlord a guaranteed right to repossess his property at the end of the tenancy. He does not have to prove any of the grounds described here. As a result of changes introduced by the Housing Act 1996, new tenancies are automatically shorthold tenancies unless the landlord gives written notice that it will not be a shorthold tenancy. The landlord must serve at least two months notice to gain possession of a property, although under some circumstances it may have to be longer. He may, however, bring it to an end at two weeks notice if he is relying on certain possession grounds.

The landlord and tenant can freely agree on the rent and the other terms of the tenancy between them. However, the tenant does not have the right during the first 6 months to refer the rent to the assessment committee, which is relevant if he is paying more than the market rent for the property concerned.

When the initial, fixed term of the tenancy comes to an end, if the landlord has not served notice that he must leave, the tenant may stay in the property. Unless a new tenancy is agreed, a statutory periodic shorthold tenancy can arise. The landlord may at any time after the fixed term serve two months notice (or longer in certain cases) to recover possession if he wishes; but the agreement can continue indefinitely if it suits both parties.

A tenant who is not able to afford the rent that he and a landlord have agreed between them for a property may be eligible for housing benefit towards that rent.

Most residential lettings by non-resident private landlords that began before 15 January 1989 will be **regulated tenancies** under the Rent Act 1977. It does not matter whether the letting is furnished or unfurnished. A few new lettings made after this date will also be regulated tenancies. A regulated tenant has certain important rights concerning the amount of rent he can be charged and security of tenure:

1. the landlord cannot evict the tenant unless he gets a possession order from the courts;
2. if the tenant dies, his spouse will normally take over the regulated tenancy – a family member who has been living in the home can take over an assured tenancy;
3. either the landlord or the tenant can apply to the rent officer for a fair rent to be registered;
4. once a rent is registered, it is the maximum the landlord can charge until it is reviewed or cancelled;
5. even if a rent is not registered, the landlord can only increase the rent in certain circumstances;
6. the landlord is usually responsible for major repairs;
7. the landlord, or in some cases the tenant, can ask the local authority for assistance towards certain repairs and improvements subject to the local authority's policy.

Some lettings will not be a tenancy but a '**licence to occupy**', for example, where all the accommodation is shared with someone occupying it under a separate agreement or where the occupier has to live there because of his job. Licensees generally have fewer rights than tenants. The distinction between licences and tenancies is not straightforward. The exact position depends on the term of the agreement.

Bringing tenancies to an end

A landlord can get possession of his property from an assured tenant by serving notice that he wishes to have his property back and by going to court. However, landlords seeking possession of properties on shorthold or assured lettings from tenants unlikely to have a defence do not need to go to court. The landlord must provide written evidence

to support his claim and this must be served on the tenant at the outset of the proceedings. A judge decides whether to order possession on the basis of the papers.

Licensees have very limited security of tenure, and normally four weeks notice from the landlord requiring posession is sufficient, but if the licensee does not move out, the landlord still usually needs a court order.

Local authority powers to take legal proceedings

Where a local authority considers it expedient for the promotion of protection of the interest of the inhabitants of its area, it may prosecute, defend or appear in any legal proceedings, and in the case of civil proceedings may institute them in its own name (Local Government Act 1972, Section 222). This power is used by district councils to take proceedings on behalf of tenants in relation to harassment and illegal eviction, etc.

Implied and expressed condition of tenancy

Section 8 of the Landlord and Tenant Act 1985 provides that in contracts for the letting of certain small houses for human habitation there are statutory implied terms to the effect that the house is, at the commencement of its tenancy and during its continuance, fit for human habitation. Subsection 1 of this section details the small houses to which the section applies, which depends on the date the contract was made, the location of the house, the local authority and the rent at which it is let.

The same implied condition of fitness and undertaking applies to the letting of houses to workmen engaged in agriculture as part of their remuneration or wages (Section 9). The term 'house' in this section includes part of a house.

Sections 11–17 impose burdens on landlords of dwellings subject to short tenancies, that is, fewer than seven years, where there are no specific repairing obligations on either side. The conditions implied as a landlord obligation are:

- to keep in repair the structure and exterior of the dwelling-house including drains, gutters and external pipes

- to keep in repair and proper working order the installations in the dwelling-house:

  (a) for the supply of water, gas and electricity and for sanitation but not for the fixtures, fittings and appliances for making use of the supply of water, gas or electricity
  (b) for space heating or heating water.

Section 14 limits the application of these provisions by excluding from them new leases granted to specified educational institutions, other specified bodies, housing associations, local authorities and the Crown, other than to the Crown Estate Commissioners.

Restoration of services to houses

Section 33 of the Local Government (Miscellaneous Provisions) Act 1976 (as amended) deals with the restoration or continuation of supply of water, gas and electricity, and allows a unitary, district or London borough council, at the written request of the occupier, to make such arrangements as it thinks fit with the utility suppliers for the supply to be either restored or continued where it has either been cut off as a result of the failure of the owner to pay for the supply or is likely to be cut off as a result of such failure. Sums incurred by a local authority in restoring or securing the continuation of a supply of water, gas or electricity where these have been cut off are a charge on the relevant property.

HOMELESSNESS

Local housing authorities were made statutorily responsible for the homeless by the Housing (Homeless Persons) Act 1977, which was consolidated as Part III of the Housing Act 1985, amended to form Part VII of the Housing Act 1996 and now superceded by the Homelessness Act 2002. Reasons for becoming homeless include:

- dispute with family, landlord or friends;
- landlord requiring property for own use;

- loss or expiry of tenancy;
- mortgage default or rent arrears;
- fire, flood, etc.

With the growth of homelessness there has been a still more rapid growth in the use of temporary housing, including bed-and-breakfast accommodation, and numbers of homeless people continues to increase [28]. The significance of this for EHOs is that houses generally referred to as 'hostels', 'guest houses', 'bed-and-breakfast accommodation', which provide shelter for people with no other permanent place of residence (as distinct from a hotel providing accommodation solely for visitors to an area who have another permanent home), are houses in multiple occupation. As such they should comply with both national statutory requirements and the amenity and space standards of the authority in whose area they are situated.

The Homelessness Act 2002 requires that every local authority develops a strategy to help prevent homelessness, but also to ensure accommodation and support for the homeless and those at risk of becoming homeless. It strengthens statutory protection available to homeless people and empowers local authorities to rehouse families with children who are homeless through no fault of their own until a more settled home becomes available. Increasing evidence-based information is available on the complexities of homelessness and the need for local strategy and practical activities to deal with acute health issues, notably for children in bed-and-breakfast accommodation and people sleeping on the streets [28].

HOME IMPROVEMENT AGENCIES

Home Improvement Agency (HIA) services have been established and funded by a wide range of organizations including local authorities, housing associations, independent or charitable bodies, private companies, professional firms, etc. They can provide an important part of partnership working between health, social services and housing departments (see e.g. [29], which gives useful strategic guidance based on a 12-point checklist designed to illustrate best practice in interdisciplinary working). The assistance provided by these agencies to help elderly, disabled and vulnerable people to remain independently in their own homes for as long as possible varies but can include:

- general advice on housing options;
- guidance on welfare rights;
- financial advice;
- assistance in obtaining grants, asssistance and loans from public or private sector organizations;
- technical advice on building problems;
- identification of necessary repairs and improvements;
- aid in finding suitable builders or contractors
- supervision of work;
- carrying out work.

Some HIAs target a particular client group, such as elderly or disabled people, while others assist people living in a particular area. The first HIAs were developed separately by some local authorities and voluntary sector organizations. Those organized by local authorities arose from recognition of the fact that some households found difficulty in coping with (then) grant procedures. The objective of the agency service was the provision of technical help with the application, identifying a builder and carrying out work [30]. Around the same time, organizations in the housing association and voluntary sectors were focusing on the housing needs of older people. Additional help was required in view of:

- the ageing of the population and, in particular, the anticipated growth in the number of frail elderly from 6% to 12% by 2000;
- the rise in the level of owner-occupation among older households;
- the number of older people in houses in poor physical condition;
- the general state of the housing stock.

The 2001 English House Condition Survey [8A] found that it is lone, older householders who are most likely to live in a non-decent home, and are increasingly likely to where they are low income private sector households aged over 85 years and have lived there for more than 30 years. The oldest housing presents problems of upkeep for all older households in the owner-occupied sector. Here, both lone and two-person older households are disproportionately represented in the worst condition of the pre-1919 stock, particularly where the homes are large.

In addition to the limited number of (then) grant-based schemes run by local authorities, both Shelter and the Anchor Housing Trust established agency projects around the country. These aimed to provide help to older home owners to carry out repairs of a minor nature, or to secure more significant repairs or renovations to enable them to remain in their own homes for as long as possible. The Shelter schemes were set up under an umbrella organization originally called Care and Repair, and Anchor's were known as Staying Put projects.

Some local authority schemes that were meant to be self-financing on the basis of fees, were financed to a large extent by (then) renovation grants. Few voluntary schemes charged fees, however, and most were reliant on funding from charitable sources, housing associations, the Housing Corporation or local authorities, so their financial position was extremely precarious. Research into HIA services has found that services offered met a wide range of needs and were generally effective, but that they lacked the resources to reach their full potential [31,32].

The government has continued to meet a proportion of the running costs of approved projects and it set up a new national coordinating organization, Care and Repair (now superceded by a national coordinating organization known as Foundations), with a remit to develop new projects and to provide support to and monitor the progress of existing projects. To ensure the continued financial stability of an agency, local authorities are required to ensure that a set proportion of income comes from charging fees for its work.

Under Section 4(3) of the Housing Association Act 1985, housing associations may also 'provide services of any description for owners or occupiers of houses in arranging or carrying out works of maintenance, repair or improvement or facilitating the carrying out of such works'. Local authorities can provide housing associations that have this as one of their objectives with financial assistance. They may also provide such assistance to any charity or other body specified by the Secretary of State. The authority must also encourage the association, charity or body to take such measures as are reasonably available to it to secure contributions from other people. However, if any authority wishes to give financial assistance to a body other than a housing association or charity in connection with the provision of a HIA for which no exchequer support is being paid, it will first be necessary to obtain the specific written consent of the Secretary of State (DoE Circular 5/91).

Local authorities considering whether to fund an agency, either independently or in conjunction with the government or a housing association, should take into account the:

1. housing needs of the residents
2. aims and objectives of the scheme
3. revenue and capital financial implications and the sources of available funds
4. level and type of support required from the local authority.

Most local authorities now operate with HIAs, many of which receive financial support from ODPM, although each provide differing support services. As responsibility increasingly shifts towards homeowners, there has been a renewed emphasis on helping people access resource for private sector housing renewal, helping people make use of other sources of funding and providing advice on loans, insurance and equity release with an emphasis on partnership arrangements [33,34].

The ODPM and DoH have jointly published a Consultation Paper outlining proposals to reform HIAs as part of the 'Supporting People' programme, strategic private sector regeneration and debt and poverty reduction through delivering client centred services [35]. This is set to introduce new arrangements to commission and fund housing-related support services from 2003/04, recognizing the growing interrelationship of heath and housing and partnership working, with improvements in the following areas:

- The need for structural and organizational reform to optimize the role of HIAs.
- Equity in geographic availability and standard of services.
- Clarification of HIA role(s).
- Delivering services that represent value for money.

SPECIALIZED FORMS OF HOUSING

Temporary and movable structures

The fitness standard, with its emphasis on the provision of fixed basic amenities, cannot be applied to temporary and movable structures, and the relevant sections of the 1985 Act have been repealed. The provisions of the Public Health Act 1936 continue to apply. In the case of caravans, site owners have powers under the Mobile Homes Act 1983 to require the removal from protected sites of residential mobile homes that have a detrimental effect on the amenity of the site.

Tents, vans, sheds, caravans, etc. used for human habitation

Tents, vans, sheds and similar structures used for human habitation, are, by Section 268 of the Public Health Act 1936, subject to the provisions of that Act relating to nuisances, filthy or verminous premises, and the prevention of disease, as if they were houses or buildings used for human habitation. (For nuisances in tents etc. see Chapter 6.)

A local authority may make by-laws for promoting cleanliness in, and the habitable condition of, tents, vans, sheds and similar structures used for human habitation, and for preventing the spread of infectious disease and generally for the prevention of nuisances in connection therewith.

The principal provisions of the Model Bye-law Series XVII 1956 refer to:

1. weatherproofing, cleansing, etc.;
2. precautions in the case of infectious disease;

3. site maintenance by the owner including clearance of ditches, space between tents, etc.;
4. provision of sanitary accommodation and water supply;
5. waste water and refuse disposal.

Under Section 269 (as amended by Section 30 of the Caravan Sites Control and Development Act 1960), a movable dwelling includes any tent, van or other conveyance whether on wheels or not, or any shed or similar structure used for habitation to which the building regulations do not apply. A local authority may grant licenses authorizing people to allow land occupied by them within the district to be used as sites for movable dwelling; and licenses authorizing people to erect or station, and use, such dwellings within the district. It may attach conditions concerning the number, type, space between, water supply, sanitary conditions, etc.

Consequently, a person shall not allow any land occupied by him or her to be used for camping purposes on more than 42 consecutive days or more than 60 days in any 12 consecutive months, unless he holds a licence from the local authority, or each person using the land as a site for a movable dwelling holds a licence.

A local authority shall be deemed to have granted a licence unconditionally unless within four weeks of the receipt of an application it gives notice to the applicant stating that it is refused, or stating the conditions subject to which a licence is granted, and if an applicant is aggrieved by the refusal, or by any condition attached, he may appeal to a court of summary jurisdiction.

The section does not apply to a movable dwelling that:

1. is kept by its owner on land occupied by him in connection with his dwelling-house and is used for habitation only by him or by members of his household;
2. is kept by its owner on agricultural land occupied by him and is used for habitation only at certain seasons and only by persons employed in farming operations on that land;
3. is not in use for human habitation and is being kept on premises on which the occupier permits

no movable dwelling to be kept except such as are for the time being not in use for human habitation.

If an organization satisfies the Minister that it takes reasonable steps for ensuring that camping sites belonging to or provided by it, or used by its members, are properly managed and kept in good sanitary condition, and that movable dwellings used by its members are so used as not to give rise to any nuisance, the Minister may grant that organization a certificate of exemption.

Caravan sites

Part I of the Caravan Sites and Control of Development Act 1960 (see also DoE Circulars 42/60, 75/77, 119/77 and 12/78) deals with the licensing conditions of caravan sites, their standards, registers of sites, consultation between planning and licensing authorities, power of entry and the matters that principally concern the environmental health officer. The licensing system is administered by district councils. The minister has specified a 'Model Standard, 1989', with respect to the layout, provision of facilities, services and equipment to which local authorities must have regard in determining any conditions they wish to attach to a caravan site licence. Part II relates to the general control, development, planning and enforcement. Part III deals with repeals, financial provisions, interpretation, etc. Part I does not apply to London but, with certain modifications, it does apply to Scotland.

In the interpretation of the Act:

1. **Caravan** means any structure designed or adapted for human habitation that is capable of being moved from one place to another, and any motor vehicle so designed or adapted, but does not include any railway rolling stock and any tent (Section 29, as amended by Section 13 of the Caravan Sites Act 1968). Due to the changing nature of modern 'caravans', they are increasingly referred to as *mobile homes* or *park homes* – see discussion in the following section.
2. **Caravan site** means land on which a caravan is stationed for the purposes of human habitation,

and land that is used in conjunction with land on which a caravan is so stationed.

3. **Site licence** means a licence under Part I of the Act authorizing the use of land as a caravan site (Section 1(1)).

4. **Occupier** means, in relation to any land, the person who, by virtue of an estate or interest therein held by him, is entitled to possession thereof or would be so entitled but for the rights of any other person under any licence granted in respect of the land (Section 1(3)).

Licensing of caravan sites

No occupier of land may cause or permit any part of the land to be used as a caravan site unless he is the holder of a site licence. A penalty is provided (Section 1). (See also [36].) However, no site licence shall be required for the use of land as a caravan site in any of the circumstances specified in the First Schedule, which include:

1. land for site incidental to its enjoyment as such of a dwelling-house within the curtilage of which the land is situated

2. land for a caravan for not more than two nights – provided no other caravan is situated there, and the land has not been used for stationing caravans for more than 28 days in the past 12 months

3. holdings of 5 acres (2.02 hectares) or more if the 28 days use has not been exceeded and not more than three caravans were so stationed at any one time

4. land that is occupied by an organization holding an exemption certificate

5. land for which there is in force a certificate issued by an exempted organization if not more that five caravans are at the time used for human habitation on such land

6. agricultural land for a caravan site for accommodation of persons employed in farming operations on land in the same occupation

7. land used similarly for accommodation of forestry workers

8. land for use as a caravan site that forms part of land on which building or engineering operations are being carried out, if that use is for the accommodation of persons employed in such operations

9. land used as a caravan site by a bona fide travelling showman

10. land for use as a caravan site occupied by the local authority in whose area the land is situated.

Local authorities may apply to the Minister for the withdrawal of any or all of the exemptions prescribed in the First Schedule. Generally, a site licence cannot be issued for a limited period. This is only possible where planning permission is for a limited period, and in these cases the site licence must expire at the same time (Section 4).

A prerequisite for a site licence is the entitlement by the applicant for the planning permission for use of the land as a caravan site (Section 3(3)). The conditions that may be attached to site licences are prescribed (Section 5). Provision is made for appeal to a magistrate court against conditions attached to a site licence or against alterations to such conditions. There is a time limit of 28 days after notification within which such appeals must be made.

Where an occupier of land fails within the time specified in a condition attached to a site licence held by him to complete to the satisfaction of the local authority any works required to be completed, the local authority may carry out those works, and may recover as a simple contract debt any expenses reasonably incurred by it (Section 9).

Provision is made whereby transfer of licences can take place, the surrender of licences for alteration can be demanded, and the rights of occupiers of land subject to a licence or special tenancy are protected (Sections 10–12). Exemptions from the requirements of Section I are provided for sites existing at the time the Act came into force. Restrictions are provided on any increase in the number of caravans on existing sites (Section 16).

Local authorities are required to keep registers of the site licences issued for caravan sites in their areas. Local authorities that have provided sites themselves must make available particulars of these sites to people inspecting the statutory register (Section 25).

Mobile homes, park homes and caravan site standards

The Secretary of State for the Environment has specified separate 'model standards' for mobile homes on permanent residential sites [37] and for static holiday caravan sites [38]. Where sites are mixed, the standards for residential sites apply. Local authorities should have regard to the standards in deciding what conditions to apply to a site licence.

The model standards refer to such matters as:

- density and space between caravans
- roads, gateways and footpaths
- hard-standings
- fire-fighting appliances
- storage of liquefied petroleum gas
- electrical installations
- water supply
- drainage, sanitation and washing facilities
- refuse disposal
- parking
- recreation space
- notices.

The model standards represent the standards normally to be expected as a matter of good practice on sites for residential mobile homes or static holiday caravans or both. They are not intended to apply to other types of caravan sites. They should be applied with due regard to the particular circumstances of each case, including the physical character of the site, any services that may already be available within convenient reach and any other local conditions.

Consideration should be given to a carefully phased introduction of any new standards after consultation with site owners, the caravan occupiers and the fire authority, as appropriate.

The Park Homes Working Party

There have been calls for several years for legislation to more accurately respond to the needs of permanently stationed residential caravans and/or mobile homes, now more commonly referred to as **Park Homes**. The Park Homes Working Party was established in 1998 to consider issues such as awareness and security of ownership rights and the scope and effectiveness of existing enforcement regimes, including alleged over-regulation, and to make proposals for change. The Working Party's report was published by the DETR in July 2000, and made several key recommendations, including providing guidance material, amendments to current Model Standards and the introduction of good practice codes on harassment and site licensing [39]. Some 1100 reponses were received during the consultation exercise from stakeholders.

The government's response to the Working Party's report was published in November 2001 [40] and focused on achieving a fair balance between the rights and responsibilities of residents and site owners. Some recommendations were implemented immediately, and further research and consultation was commissioned in respect of other issues, such as resource implications for possible legislative changes.

Travellers

The term 'traveller' is used to describe a way of life that has been adopted by certain groups of people. The two most commonly known groups are **new age travellers** and **gypsies**. Both groups reside for the most part in mobile homes, although of differing types. Gypsies use specially designed caravans, while new age travellers may use various types of vehicles that have been converted to meet their needs.

New age travellers

New age travellers are different from gypsies and travellers because they have a much shorter history. They came to prominence in 1992 due mainly to media focus on conflicts between the travellers and the police. New age travellers appear to be people who have left a settled community because of poor housing and unemployment. Under the Race Relations Act 1971, they are not defined as a distinct ethnic group.

Gypsies

A legal definition of a gypsy is contained within Section 24 of the Caravan Sites and Control of Development Act 1960, as amended by Section 80 of the Criminal Justice and Public Order Act 1994:

Gypsies means persons of nomadic habit of life, whatever race or origin, but does not include members of an organized group of travelling showmen, or of persons engaged in travelling circuses, travelling together as such.

This definition appears to include all travellers regardless of cultural background. However, the Court of Appeal, in *R. v. South Hams District Council and another, ex parte Gibb* [41] and other applications reported since, held that gypsies meant:

Persons who wandered or travelled for the purpose of making or seeking their livelihood, and did not include persons who moved from place to place without any connection between their movement and their means of livelihood.

The true gypsies are said to be the Romany people who left India about 1000 years ago and were first reported in England in 1505. Today, gypsies are part of distinctive English, Welsh, Scottish and Irish traveller communities, each having their own dialects.

A count of travellers and gypsies is carried out twice a year in order to provide information for planning, health and education purposes and also whether there is a need to provide more sites.

Legislation

The Caravan Sites Act 1968 provided local authorities with a mandatory duty to provide sites for travellers. Where authorized sites were provided, the powers of eviction from unauthorized sites were strengthened. Financial support was also made available to local authorities where new sites were created.

The lifestyle and customs of travellers and gypsies together with the growth in the use of unauthorized sites and festivals probably played a part in the introduction of the Criminal Justice and Public Order Act 1994. The Act changed the way that local authorities provide and manage sites and evict unauthorized campers. The relevant sections are contained in Part V of the Act and are supported by advice contained in DoE Circular 1/94 [42]. Some of the main provisions of the Act are as follows:

- removal of the mandatory duty placed on local authorities to provide sites;
- trespass made a criminal offence;
- power given to the police to remove trespassers on land;
- powers given to police officers to seize vehicles;
- powers in relation to raves;
- direct unauthorized campers to leave land;
- tolerance where unauthorized camping is not causing a nuisance;
- consideration of gypsy accommodation when preparing local authority structure and development plans.

DoE Circular 18/94 [43] recognizes that in some circumstances where gypsies are camped unlawfully on council land but are not causing a level of nuisance that cannot be effectively controlled, an immediate forced eviction might result in unauthorized camping elsewhere in the area, which could give rise to a greater nuisance. Accordingly it is suggested that it may be in the public interest for authorities to consider tolerating gypsies' presence on such land for short periods. The circular also indicates that local authorities should try to identify possible emergency stopping places as close as possible to the transit sites used by gypsies where gypsy families would be allowed to camp for short periods. Authorities should also consider providing basic amenities on these temporary sites; see also [44].

In January 2002, there were some 325 local authority-owned Gypsy sites in England providing pitches for 5005 caravans, just under 50% of all Gypsy caravans. Sites are also provided privately, and Gypsies prefer to provide their own sites. Up to 2000 residential pitches and up to 5000 additional transit sites will be needed over the next five years for the growing gypsy population [45,46].

Many local authorities make no provision at all for gypsies.

The government has encouraged local strategy to respond to this growing need in partnership with settled communities and relevant stakeholders (including the local authorities and the police) to help manage the growth in unauthorized camping in a more flexible and viable way, as evictions alone do not resolve the need for stopping places or access to wider services. Local authorities are encouraged, but not required, to make adequate provision for sites as well as management and maintenance arrangements, and to give consideration to providing emergency stopping places with facilities for refuse, water and temporary toilets [44,45].

In view of growing problems, some £17 million is to be made available to local authorities as part of the Spending Review between 2001/02 and 2003/04 for temporary sites and authorized encampments to discourage unauthorized camping and its associated problems [45–47]. However the emphasis is on a local response, not national policy.

Temporary buildings

The control exercised by local authorities over tents, vans, sheds and similar structures is strengthened by provisions for the control of buildings constructed of materials that are short-lived or otherwise unsuitable for use in permanent buildings. Section 19 of the Building Act 1984 allows local authorities to reject the plans for such buildings or fix a time within which the building must be removed.

Accommodation of hop pickers, fruit pickers, etc.

Under Section 270 of the Public Health Act 1936, a local authority may make by-laws for securing the decent lodging and accommodation of hop pickers and other people engaged temporarily in picking, gathering or lifting fruit, flowers, bulbs, roots or vegetables within its district.

The defects common to tents, vans and sheds used for human habitation are those most common in connection with structures provided for the accommodation of hop pickers, etc. Complaints generally concern improper or insufficient sanitary accommodation, or inadequate arrangements for the storage and removal of refuse.

The principal provisions of the Model Bye-laws, Series XX, 1956 are as follows. Any person who, for persons engaged in hop picking or in the picking of fruit and vegetables, provides any lodging not ordinarily occupied for human habitation must:

1. give the council **28 days** notice in writing of his or her intention to erect new lodging;
2. give **28 days** notice in writing of his or her intention to reoccupy the lodging in any year after the first year of usage;
3. maintain the lodging in a clean, dry and weatherproof condition at all times when in use;
4. new lodgings to be not less than **20 ft** (6 m) apart at the front, and not less than **15 ft** (4.5 m) apart elsewhere, and have an impervious path at least **30 in** (762 mm) wide along the whole front;
5. no obstruction to interfere with the access of air and light;
6. provide sufficient means of ventilation and natural light;
7. minimum floor space of **20 sq ft** (1.85 sq m) for each occupant (two children under **10 years** to be counted as one person);
8. separation of the sexes;
9. provide suitable accommodation for cooking and drying of clothes (one cooking house for every **16 persons**);
10. provide at all times a supply of good, wholesome water for domestic use in some suitable place (within **150 yd** (135 m));
11. provide clean, dry, suitable bedding;
12. cause walls and ceiling to be cleansed and/or limewashed **once in every year not more than two months before occupation;**
13. provide covered refuse bins, one for **16 persons;**
14. provide proper sanitary accommodation – separate for the sexes – not less than one for every **20 persons** (including children) lodged.

Rooms unfit for occupation

The occupation of certain rooms is, by reason of their situation and irrespective of their condition, prohibited.

Under Section 49(1) of the Public Health Act 1936, a room any part of which is immediately over a closet other than a water closet or earth closet or immediately over a cesspool, midden or ashpit, shall not be occupied as a living room, sleeping room or workroom. Any person who after notice from the local authority, occupies or permits the occupation of such a room is liable to a penalty.

Canal boats and residential houseboats

The Public Health (Control of Disease) Act 1984, Sections 49–53, and the Canal Boats Regulations 1878, 1925 and 1931 require the owner of a canal boat who wishes to use the boat as a dwelling to apply for registration to a local authority whose district abuts the canal on which the boat is accustomed or intended to ply. These provisions do not apply, however, to boats carrying a cargo of petroleum. A canal boat is defined as:

Any vessel, however propelled, which is used for the conveyance of goods along a canal not being:

1. a sailing barge which belongs to the class generally known as 'Thames Sailing Barge' and is registered under the Merchant Shipping Act 1874–1928 either in the Port of London or elsewhere; or
2. a sea-going ship so registered; or
3. vessel used for pleasure purposes only.

Before registration can be effected, the boat must comply with the structural requirements laid down in the regulations. Provisions are also made to deal with the lettering, marking and numbering of registered boats, cleanliness and habitable condition, and the spread of infectious diseases.

The Public Health (Control of Disease) Act 1984 provides the Secretary of State a duty to make regulations:

- Fixing standards for space and occupancy, ventilation, provision for separation of the

sexes, general healthiness and convenience of occupation
- Promoting cleanliness and habitable conditions
- To prevent the spread of infectious disease on canal boats.

The provisions of the Canal Boat Regulations 1878 (as amended in 1925 and 1931) still in force are now quite dated and may be difficult to enforce because of outmoded definitions relating to current boat use. The extent to which the regulations are enforced by used by local authorities is unclear, although local authorities are empowered to enter and inspect canal boats.

It is more likely that local authorities would use their powers under the Environmental Protection Act 1990 statutory nuisance provisions in conjunction with British Waterways, who also have enforcement responsibilities to 'residential houseboats' as they are more commonly defined, under the British Waterways Acts 1983 and 1995. The enforcement process is difficult, not least because each is unique in style, construction and occupation, but also because boats are mobile, and there are invariably additional mooring implications and controls to consider.

The British Waterways Act 1995 introduced the Boat Safety Certificate (BSC) to improve safety standards, including minimum standards for construction and equipment. Where conditions are met, a BSC is issued and is valid for four years. Failure to comply can lead to prosecution and/or removal of the boat from the water.

Environmental Health Officers in local authorities are able to apply the statutory nuisance provision of the Environmental Protection Act 1990. The whole boat should be considered in deciding whether conditions were prejudicial to the health of the occupants. This goes further than the BSC, which is only concerned with safety, and not amenities required for residential accommodation.

British Waterways have produced guidelines for residential boats that incorporate the requirements for:

- Water storage and supply
- Sanitary accommodation
- Facilities for the preparation and cooking of food

- Space heating
- Artificial lighting
- Ventilation
- Fire and safety precautions
- Life saving equipment
- Repair.

Ironically, a houseboat may pass the BSC but remain a statutory nuisance due to inadequate amenities and enforcement action may prove difficult. Some boats can be in such poor condition that they have to be removed from the water, leading to possible re-housing implications. An early partnership approach with a realistic way forward is therefore necessary.

REFERENCES

1. Department of the Environment, Transport and the Regions (2000) *Best Value in Housing Framework*, DETR, London. Online. Available at www.housing.dtlr.gov.uk/information/bvhf/4.htm (accessed 3 April 2002).
2. Department of the Environment, Transport and the Regions and the Department of Social Security (2000) *Quality and Choice: A Decent Home for All*, DETR, London. Online. Available at www.housing. detr.gov.uk/information/consult/homes/index.htm (accessed 20 August 2002).
3. Office of the Deputy Prime Minister (2002) *Housing Renewal Guidance (Consultative Document) Housing Research Summary 163, 2002*, ODPM, London. Online. Available at www.housing.odpm.gov.uk/hrs/hrs163/index.htm (acccessed 28 August 2002).
4. Office of the Deputy Prime Minister (2002) *Housing and Housing Policy: Housing Renewal in the Private Sector*. Online. Available at www.housing.odpm. gov.uk/information//index07.htm (accessed 28 August 2002).
5. Office of the Deputy Prime Minister (2003) *Sustainable Communities: Building for the Future*, ODPM, London. Online. Available at www.communities.odpm.gov.uk/plan/main/overview.htm (accessed 6 March 2003).
6. Office of the Deputy Prime Minister (2002) *A Decent Home: The Revised Definition and Guidance for Implementation. Section 1 and 2*, ODPM, London. Online. Available at www.housing.odpm.gov.uk/information/dhg/definition/02.htm (accessed 28 August 2002).
7. Office of the Deputy Prime Minister (2003) *Housing Bill – Consultation on Draft Legislation*, ODPM, London. Online. Available at www.odpm.gov.uk/information/consult/housingbill/01.htm (accessed 1 April 2003).
8. Department of the Environment (1998) *English House Condition Survey 1996*, HMSO, London.
8A. Office of the Deputy Prime Minister (2003) *English House Condition Survey 2001, Key Facts*, HMSO, London.
9. Department of the Environment (1996) *Private Sector Renewal: A Strategic Approach*, Circular 17/96, HMSO, London.
10. Audit Commission (1991) *Audit Commission Local Government Report No. 6: Healthy Housing: the Role of Environmental Health Services*, HMSO, London.
11. Dennington Committee (1966) *Our Older Houses – A Call for Action*, HMSO, London.
12. Office of the Deputy Prime Minister (2001) *Housing Statistics Summary: Report of the 1999/00 Survey of English Housing (Summary 009, August 2001)*, ODPM, London. Online. Available at www.housing.odpm.gov.uk/statistics/publicat/summaries/009/index.htm (accessed 11 August 2002).
13. Department of the Environment (1993) *Local House Condition Surveys*, HMSO, London.
14. Department of the Environment, Transport and the Regions (2000) *Housing Research Summary No. 122, 2000. Development of the Housing Health and Safety Rating System*, DETR, London. Online. Available at www.housing.detr.gov.uk/hrs/hrs122/htm (accessed 6 October 2000).
15. Department of the Environment, Transport and the Regions (2000) *Housing Research Summary No. 123, 2000. Housing Health and Safety Rating System Quick Guide*, DETR, London. Online. Available at

www.housing.detr.gov.uk/hrs/hrs123/htm (accessed 6 October 2000).

16. Department of the Environment, Transport and the Regions (2001) *Health and Safety in Housing: Replacement of the Housing Fitness Standard by the Housing Health and Safety Rating System: A Consultation Paper*, DETR, London. Online. Available at www.housing. dtlr.gov.uk/information/consult/hhsrs/pdf/ housinghealth.pdf (accessed 13 November 2001).

17. Department of Transport, Local Government and the Regions (2001) *Housing Research Summary No. 142, 2001. Worked Examples to Support the Housing Health and Safety Rating System*, DTLR, London. Online Available at www.housing.dtlr.gov.uk/hrs/hrs142/ pdf/hous142.pdf (accessed 1 November 2001).

18. Griffiths, A. (2001) 'How Does It Rate Now?' *Environmental Health Journal*, February 2001. Online. Available at www.ehj-online.com/ archive/2000/feb2001/february3.html (accessed 27 February 2003).

19. White Paper (1987) Housing: the government's proposals, Cmnd 214, HMSO, London.

20. Office of the Deputy Prime Minister (2002) *Housing Renewal Guidance*, ODPM, London. Online. Available at www.housing.odpm. gov.uk/information/consult/renewal/01.htm (accessed 20 February 2003).

21. Office of the Deputy Prime Minister (2002) *Housing Renewal Guidance, Annex D: Disabled Facilities Grant*, ODPM, London. Online. Available at www.housing.odpm. gov.uk/information/consult/renewal/annex04. htm (accessed 20 February 2003).

22. Department of the Environment (1996) *Housing Research Summary No. 2 (1996): English House Condition Survey 1991 Energy Report*, HMSO, London.

23. Boardman, B. (1991) *Fuel Poverty*, Belhaven, London.

24. Department of Environment, Food and Rural Affairs and Department of Trade and Industry (2001) *UK Fuel Poverty Strategy*, DEFRA and DTI, London. Online. Available at www.dti.gov.uk/energy/fuelpoverty/fuelpeng. html (accessed 12 January 2002).

25. Department of the Environment (1993) *HMOs: Guidance to Local Authorities on Managing the Stock in Their Area*, Circular 12/93, HMSO, London.

26. Department of the Environment (1992) *HMOs: Guidance on Standards of Fitness*, Circular 12/92, HMSO, London.

27. Chartered Institute of Environmental Health (1994) *Amenity Standards for HMOs*, Professional Practice Note, CIEH, London.

28. Office of the Deputy Prime Minister (2002) *More Than a Roof: A Report into Tackling Homelessness*, ODPM, London. Online. Available at www.housing.odpm.gov.uk/ information/homelessness/morethanaroof/ 04.htm (accessed 18 March 2003).

29. Department of Health and Department of the Environment (1997) *Housing and Community Care: Establishing a Strategic Framework*, HMSO, London.

30. Thomas, A.D. (1981) *Local Authority Agency Services: Their Role in Home Improvement*, Research Memo No. 88, University of Birmingham CURS, Birmingham.

31. Leather, P. and Reid, M. (1989) *Investing in older housing: A study of home improvements in Bristol*, SAUS, Bristol.

32. Leather, P. and MacIntosh, S. (1991) *Monitoring Assisted Agency Services*, Department of the Environment, HMSO, London.

33. Office of the Deputy Prime Minister (2002) Home Improvement Agencies: development and reform. *Housing Signpost: A Guide to Research and Statistics*, 14: 3.

34. Department of Transport, Local Government and the Regions (2001) *Housing and Housing Policy: Home Improvement Agencies*, DTLR, London. Online. Available at www.housing. dtlr.gov.uk/information/hia/index.htm (accessed 14 January 2002).

35. Department of Health and Office of the Deputy Prime Minister (2002) *Home Improvement Agencies: Development and Reform: A Consultation Paper*, ODPM, London. Online. Available at www.housing. odpm.gov.uk/information/consult/hia/index. htm (accessed 14 January 2003).

36. Chartered Institute of Environmental Health (1994) *Licensing of Permanent Residential Mobile Home Sites*, Professional Practice Note, CIEH, London.

37. Department of the Environment (1989) *Caravan Sites and Control of Development Act 1960, Section 5, Model Standards: Permanent Residential Mobile Home Sites*, HMSO, London.

38. Department of the Environment (1989) *Caravan Site and Control of Development Act 1960, Section 5, Model Standards: Holiday Caravan Sites*, HMSO, London.

39. Department of the Environment, Transport and the Regions (2000) *Housing Research Summary (No. 121, 2000) Report of the Park Homes Working Party*. Online. Available at www.housing.detr.gov.uk/hrs.hrs121.htm (accessed 30 August 2000).

40. Department of Transport, Local Government and the Regions (2001) *Government's Response to the Report of the Park Homes Working Party*. Online. Available at www.housing.odpm.gov.uk/information/park home/pdf/parkhome.pdf (accessed 6 March 2003).

41. *R. v. South Hams District Council and another, exparte Gibb* (1994) 4 A11 ER 1012.

42. Department of the Environment (1994) *Gypsy Sites Policy and Unauthorized Camping*, Circular 1/94, HMSO, London.

43. Department of the Environment (1994) *Gypsy Sites Policy and Unauthorized Camping*, Circular 18/94, HMSO, London.

44. Department of the Environment, Transport and the Regions and Home Office (1998) *Joint Good Practice Guidance for Local Authorities and the Police on Managing Unauthorised Camping*, HMSO, London.

45. Office of the Deputy Prime Minister (2002) *Housing and Housing Policy: Gypsy Sites Policy*, ODPM, London. Online. Available at www.housing.odpm.gov.uk/information/ index14.htm (accessed 28 October 2002).

46. Office of the Deputy Prime Minister (2002) Local authority gypsy/traveller sites in England: there is a need for more local authority residential and transit sites for gypsies and other travellers, *Housing Signpost: A Guide to Research and Statistics*, **14**: 4.

47. Office of the Deputy Prime Minister (2002) *Housing Research Summary 150: Monitoring the Good Practice Guidance on Managing Unauthorised Camping*, ODPM, London. Online. Available at www.housing.odpm. gov.uk/hrs/hrs150/index.htm (accessed 28 October 2002).

FURTHER READING

Audit Commission (1992) *Audit Commission Local Government Report No. 9: Developing Local Authority Housing Strategies*, HMSO, London.

Association of District Councils (1988) *The Challenge of Multiple Occupancy*, Local Government Association, London.

Chartered Institute of Environmental Health (1995) *Fire Safety for Houses in Multiple Occupation – An Illustrated Guide*, CIEH, London.

Chartered Institute of Environmental Health (1995) *Travellers and Gypsies: An Alternative Strategy*, CIEH, London.

Department of the Environment, Transport and the Regions (1997) *Private Sector Housing Renewal Strategies – A Good Practice Guide*, HMSO, London.

Leather, P. (2000) Crumbling Castles? *Helping owners to Repair and Maintain Their Homes*, Joseph Rowntree Foundation, York.

Moran, T. (1998) *Legal Competence in Environmental Health*, E. & F.N. Spon, London.

Stewart, J. (2001) *Environmental Health and Housing*, E. & F.N. Spon, London.

Stewart, J., Bushell, F. and Habgood, V. (forthcoming 2004) *Environmental Health as Public Health*, Chadwick House Group Ltd, London.

Part Six
Occupational Health and Safety

18 Introduction to occupational health and safety

Brian Etheridge and William H. Bassett

INTRODUCTION TO HEALTH AND SAFETY ENFORCEMENT

The Health and Safety at Work, Etc. Act [1] is 30 years old. Despite talk of a new Safety Bill at the time of the government's review of health and safety in 2000, there has been little activity to make this a reality. While the Act is likely, therefore, to remain the legislative foundation for the UK system of health and safety control for some time to come, the current approaches to enforcement, priorities and organizational arrangements are in a state of flux. The Health and Safety Commission (HSC) has been considering a new strategy for health and safety enforcement to focus on the government's priority of delivery, and to establish a new direction and programmes of work for the Health and Safety Executive (HSE) and local authorities.

The health and safety system, originally envisaged by Robens in his report on health and safety at work in 1972 [2], was largely implemented in 1975. The intervening years have, however, shaped and refined the system to meet the requirements of current work practices and safety and health concerns. The HSC wants to ensure that future change remains in pace and relevant to the changing world.

In 2003, in response to a year on year slowing of improvements in safety statistics and emerging occupational health issues, the HSC established a new vision for health and safety – to gain recognition of health and safety as a cornerstone of a civilized society, and with that, to achieve a record of workplace health and safety that leads the world. The HSC wants to ensure that the health and safety system continues to adapt to a changing society and workplace and to deliver continued improvements. It recognizes the need to raise the profile of health and safety, to address constantly evolving health and safety challenges in a dynamic economy, and to develop new strategies and tactics to augment and complement traditional approaches.

This introduction to health and safety enforcement looks at Robens' original vision for health and safety enforcement. It outlines the principles, systems and practices that have been developed to underpin that vision, the influences on them, and the way they have changed and will continue to change. It looks at the government's Revitalising Health and Safety Strategy Statement of 2000[3] and the HSC's most recent approach to defining a new strategy for 2010 and beyond. It considers emerging trends and what they might mean for health and safety enforcement in the future.

A knowledge and understanding of these is essential for enforcement officers and those engaged in the management of enforcement.

THE DEVELOPMENT OF THE MODERN HEALTH AND SAFETY FRAMEWORK

The Robens Committee on Safety and Health at Work

The enforcement of health and safety has been a feature of the regulatory system in the United Kingdom for more than 150 years. Its origins lie in the protection of children and young workers. In 1833, an Act to Regulate the Labour of Children and Young Persons in the Mills and Factories of the UK appointed four inspectors of factories, and set in place the framework for modern day regulation and enforcement. The development of the role of local authorities in public health is explained more fully in previous chapters, but the most notable landmark in the history of the regulation of health and safety, and the role of local authorities within it, was the report of the Committee on Safety and Health at Work chaired by Lord Robens in 1972 [2].

The Robens Committee of Inquiry was asked by the government of the day to consider whether any changes were needed to the scope of the law for occupational health and safety and the measures for protecting the public against hazards of industrial origin. His inquiry arose from the perceived health and safety difficulties of the time, a lack of confidence in the existing system, and a recognition that there had never been a comprehensive review of the system as a whole. His report makes stark reading, showing that every year in the region of 1000 people were killed at their work, 500 000 suffered injuries, and 23 million days were lost annually to industrial injury and disease. He hypothesized that this unhealthy state of affairs resulted from a single important cause: apathy. Notwithstanding the confidence that could be applied to the statistics at the time, Robens concluded that a major overhaul of the health and safety framework in the United Kingdom was needed.

Robens' diagnosis of the problem was:

- there was too much law, much of which was intrinsically unsatisfactory;
- health and safety could not be ensured by an ever expanding body of legal regulations enforced by an ever increasing army of inspectors;
- the primary responsibility for doing something about occupational accidents and disease should lie with those who created the risks and those who worked with them;
- the roles of regulatory law and government action should lie not in detailed prescription, but with influencing attitudes and creating a framework for better organized safety and health action by industry itself.

The outcome of Robens' deliberations was the enactment of the Health and Safety at Work, Etc. Act 1974, which sets out the general duties that employers have towards employees and members of the public, those that employees have to themselves and each other, and the duties of the self-employed. It also established the HSC and HSE, and confirmed a role for local authorities in health and safety enforcement.

The HSC currently comprises 10 people appointed by the Secretary of State for Work and Pensions. Its role is to propose new or updated health and safety legislation, to conduct research, and to provide information and advice. It is advised and assisted by the HSE, which has statutory responsibilities of its own, including the enforcement of the Health and Safety at Work, Etc. Act 1974, and other relevant statutory provisions, a function it shares with local authorities.

Since its inception, there have been significant changes in the environment in which it operates, including:

- changing patterns of employment;
- the development of European and international policies on health and safety enforcement and risk assessment;
- a changing perception of risk, predominantly by the public;
- new and emerging safety and health issues.

To remain flexible and to accommodate changes, much of modern health and safety law consists of goal-setting regulations made under the principal Act. Goal-setting legislation identifies what should be achieved but not necessarily how to achieve it. This may be set out in Approved Codes of Practice (ACOPs) or other guidance, including advice on good practice. Some legislation is, however, prescriptive and spells out absolutely what must be done or achieved. This legislative system is based on certain key principles, which require further explanation.

Self-regulation

Robens envisaged a more self-regulating system for health and safety at work, set out under a comprehensive and orderly set of revised provisions under an enabling Act. The Health and Safety at Work, Etc. Act 1974 carries through Robens' philosophy by establishing a general framework of responsibilities and also a quick and effective response to flagrant breaches of the law and a discriminating and efficient approach to other breaches.

Risk assessment

The concept of risk assessment (see also Chapter 10) is central to the health and safety system and Robens' concept of self-regulation. But it is not an easy concept, and elements of it such as hazard, risk and probability are not well understood. Predictably, individuals are more likely to accept higher risks when they are of their own making rather than when imposed upon them at a place of work. There is also a tendency to concentrate on the consequences of failure rather than the probability involved. The key to risk assessment and management in the workplace is a disciplined approach, which the legislation requires. Duty holders are required to judge for themselves the extent and nature of hazards in the workplace, the risks these pose and the appropriate control measures required to reduce them. Risk assessment is not a one-off, it must be continually revisited and reviewed in the light of changing circumstances and improved knowledge.

The Management of Health and Safety at Work Regulations 1999[4] make more explicit what employers are required to do to manage health and safety under the Act. The main requirement on employers is to carry out a risk assessment, to make arrangements for implementing measures identified as necessary by that risk assessment, and to provide clear information and training to employees. HSC aims to give employers and other duty holders guidance to assist them in these tasks, and plays a role in developing acceptable standards.

See also HSE (1998) Five Steps to Risk Assessment – Case Studies, HSG.183, HSE Books, Sudbury, Suffolk, and HSE (1998) Five Steps to Risk Assessment, INDG 163 (rev. 1), HSE Books, Sudbury, Suffolk.

So far as is reasonably practicable

The general provisions of health and safety law are qualified by the important test of reasonable practicability expressed as the proviso 'so far as is reasonably practicable' (SFAIRP). This concept, which has been derived largely from case law, implies a balance between the degree of risk associated with an activity that could cause harm, and the sacrifice in terms of time, trouble and money needed to overcome that risk. In practice, the application of SFAIRP should reduce any risks to as low as is reasonably practicable (ALARP) so that the activity (which is assumed to be desirable) can be undertaken without the assurance of absolute safety being incompatible with the continuation of the activity.

Although this concept is embedded in this and other regulatory systems in the United Kingdom, the European Union (EU) continues to have some difficulty with it. The EU is not convinced that it allows our system to ensure the health and safety of workers in their undertakings in the way that the EU's framework directive (89/391/EEC) on health and safety demands. It continues to consider formal infraction proceedings against the United Kingdom for a perceived breach of its treaty obligations.

Others have also argued that the application of risk assessment has led to some blurring of health and safety responsibilities. There is a growing demand, particularly among some small businesses, for a return to greater prescription to counter the perceived uncertainties and inconsistencies of risk assessment. The HSC is firmly wedded to the principle that action should be proportionate to risk. A clear understanding of the degree to which risks should be controlled is, however, essential if a self-regulatory regime is to work. It is also central to the effectiveness with which regulators can go about their business.

Action to deal with risks can take many forms, including: elimination or substitution; enclosure, guarding or segregation; safe systems and management; and personal protective equipment. At all levels in the system, duty holders are pressing regulators for an explanation of how they should address risk issues at both the general and detailed specific levels. The HSE defines this process as understanding the tolerability of risk so that the necessary steps can be taken to reduce risks such that they are regarded as tolerable to those who may be affected by them.

The tolerability of risk

Tolerability of risk accepts that there is a balancing act between risks and benefits. There is a gap between what is absolutely safe and absolutely unsafe, with what is unsafe with tolerable risks (what is safe enough) occupying the ground in between. In 1988, the HSE set out its philosophy for deciding the degree to which risks should be controlled in the nuclear power generating industry [5]. The philosophy has since been increasingly used within other industries to decide on the degree to which risks should be controlled. The tolerability of risk philosophy – that risks can be classified as intolerable, tolerable and broadly acceptable – is applicable across the full spectrum of risks whether they are quantifiable or not.

In recent years, it has been recognized that the generalized application of the philosophy needs to be made more explicit in order to:

- enable the regulators' approach to be more transparent and open to scrutiny;

- assist duty holders to manage risks in ways that are compatible with the expectations of regulators;
- reassure the public.

One of the main purposes of the enforcing authorities is to ensure that duty holders manage and control risks effectively to prevent harm. To enable the enforcement process to be more transparent, the HSE developed for use by it and local authorities the Enforcement Management Model (http://www.hse.gov.uk/enforce/emm.pdf). This model:

- provides a framework for consistent risk-based enforcement;
- enables others to understand enforcement decisions;
- assists in the training of less experienced inspectors.

The HSE has also produced a suite of guidance on the ALARP principle to identify what inspectors should expect to see in duty holders' demonstrations that the risk has been reduced to reasonable levels in unusual work activities where good practice has not yet been established or where there is a risk of disasters (http://www.hse.gov.uk/dst/alarp1.htm).

In 2001, HSE published Reducing Risks, Protecting People – HSE's decision-making process [6] (http://www.hse.gov.uk/dst/r2p2.pdf) to make clear the process of decision-making including risk assessment and risk management. It considers both individual risk and societal concerns in the assessment of risk, adopts a precautionary approach when addressing uncertainty and describes the tolerability of risk framework.

Consultation

Consultation is a further principle that underpins the work of the HSC and health and safety legislation. Again, the basis of the approach has its origins in Robens' work. He believed that, to be effective, health and safety should be a consideration at all levels of an organization from board room to shop floor, and that a wider involvement would secure higher standards.

The current law [7,8] requires employers to consult all their employees on health and safety matters. To be effective, employers must give information to employees and also listen to, and take account of, what they have to say when health and safety decisions are made. Where safety representatives exist, these must also be consulted on matters affecting the group or groups of employees they represent. Groups not so represented must also be consulted. Regulators also involve safety representatives during inspections and examinations.

Guidance

The HSC and HSE publish guidance in support of the statutory framework. The main purposes of guidance are:

- to interpret: helping people to understand what the law says, including, for example, how requirements based on EU directives fit with those under the Health and Safety at Work, Etc. Act 1974;
- to help people to comply with the law;
- to give technical advice.

The HSC also has a particular role in issuing guidance to local authorities in respect of their enforcement activities.

While compliance with guidance is not in itself an indication that statutory provisions are being complied with, it provides an indication of what is normally sufficient to comply with the law. ACOPs give further practical examples of good practice. They provide assistance in deciding what is reasonably practicable or suitable and sufficient. ACOPs have a special legal status and provide a benchmark for courts to decide whether enough has been done to comply with the law.

THE EU AND ENFORCEMENT

Since the 1980s, the main influence on the development of legislation in the United Kingdom has been the EU. UK regulations have been made to implement EU directives made under the Framework Directive on Health and Safety (see Chapter 19). Where possible, these regulations have been made in a way that is complementary to existing law and philosophy. This has not always been easy, and new regulations have been problematic for both enforcers and those on whom they are being enforced.

Although embracing the principles of risk assessment, the tendency of directives has been towards a greater level of prescription, which has not always been easy to accommodate within the general framework of the Health and Safety at Work, Etc. Act 1974 and its underpinning principle of SFAIRP. This has resulted in some misunderstandings by duty holders, which regulators have had to address through education, training and advice.

Enforcement in other member states

The arrangements for the administration of health and safety in the United Kingdom are unique among EU member states, although a number of parallels and comparisons can be drawn. The arrangements for the involvement of employers, employees and their representatives are common in varying guises in other European countries and reflect the arrangements incorporated in the make-up of the HSC to represent all relevant stakeholders. The acknowledgement of the role of safety representatives and safety committees is also a feature.

In most EU member states, the principal responsibilities rest with a centralized or government labour inspectorate that is responsible for other labour and employment matters in addition to pure health and safety. A common feature of continental European health and safety frameworks is the role of the insurance industry, which provides compensation payments for employees in the event of an accident. In some cases, because of their ability to issue guidance (to which members must adhere) and apply sanctions, they occupy an important place in the health and safety system.

The labour inspectorates throughout the EU coordinate their activities through the Senior Labour Inspectors Committee (SLIC), on which the UK government is represented by the HSE. SLIC collates information on activities and organizations throughout the EU, and further information about the organization of health and safety in other member states can be found in its publications [9].

The devolvement of enforcement responsibilities to a local or regional level is less common except in the case of fire safety, which is often locally administered.

THE CURRENT STRATEGIC FRAMEWORK

In recent years, the HSC has published a Strategic Plan normally to cover a three-year period. Each year, the HSC publishes a supporting business plan (a more detailed plan of work) to reflect in year priorities. The current strategic framework reflects the outcomes of the government's strategic appraisal of health and safety carried out in 1999. This aimed to:

- inject new impetus into the health and safety agenda;
- identify new approaches to reduce further the rates of accidents and ill health caused by work, especially approaches relevant to small firms;
- ensure that the approach to health and safety regulation remains relevant for the changing world of work over the next 25 years;
- gain maximum benefit from links between occupational health and safety and other government programmes.

The Revitalizing Health and Safety Strategy Statement, published in June 2000, set out for the first time targets for Great Britain's health and safety system, namely

- to reduce the number of working days lost per 100 000 workers from work-related injury and ill health by 30% by 2010;
- to reduce the incidence rate of fatal and major injury accidents by 10% by 2010;
- to reduce the incidence rate of cases of work-related ill health by 20% by 2010;

- to achieve half the improvement under each target by 2004.

A 10-point strategy statement set a framework for further action in the following areas:

- promoting a better working environment characterized by motivated workers and competent managers;
- recognizing and promoting the contribution of a workforce that is 'happy, healthy and here' to productivity and competitiveness;
- occupational health as a top priority;
- positive engagement of small firms;
- motivating employers through compensation, benefits and insurance systems;
- reinforcing the culture of self-regulation;
- partnership on health and safety issues;
- government leading by example;
- education in health and safety skills;
- designing in safety to processes and products.

This was supported by a 44-point action plan that included suggested measures to:

- motivate employers;
- engage small firms;
- put the government's own house in order;
- promote coverage of occupational health in local Health Improvement Programmes in England and co-ordinated government action on rehabilitation;
- secure greater coverage of risk concepts in education.

In July 2000, the HSC launched an occupational health strategy for Great Britain, Securing Health Together [11]:

- to reduce ill health, both in workers and the public, caused, or made worse, by work;
- to help people who have been made ill, whether caused by work or not, to return to work;
- to improve work opportunities for people currently not in employment due to ill health or disability;
- to use the work environment to help people to maintain and improve their health.

To achieve these targets, the strategy established five key programmes of work relating to compliance, continuous improvement, knowledge, skills and support mechanisms.

EMERGING TRENDS AND FUTURE STRATEGIES

A new strategic approach

In 2003, HSC recognised that a change of approach was needed. While building on a history of success and a record of health and safety which is the envy of the world, more of the same would not be sufficient to address new and emerging health issues or the demands of stakeholders. HSE and local authorities had to be clearer about their priorities, to concentrate on the things that they are best placed to do and to do them where they would have the greatest impact. This signalled a need for a greater understanding, reliance and trust in the contributions of others to health and safety improvements. The Commission believed that regulation should no longer be the automatic response to new issues and, in the future, we should aim for a state where the regulators were no longer the principal drivers for improvement.

The Commission highlighted some new areas for action including:

- more effective strategic partnerships, including with local authorities;
- new ways to advise, guide and support businesses with greater penetration;
- a focus on occupational health and rehabilitation;
- more worker involvement in local health and safety management;
- better communication of the health and safety message.

Most of all, the Commission recognised that, to remain relevant, it must be better placed to respond to the changing world of work. Key to this would be simplification of the concept of risk assessment to ensure a sensible approach to risk management.

Employment and accident trends

The Cabinet Office has promoted strategic futures work among government departments to improve strategic decision-making. It has set up a programme to improve the ability of government to think about, react and shape the future. Its website (www.cabinet-office.gov.uk) contains a collection of material from a wide range of sources. Along with the government's annual labour force projections and the Institute for Employment Research (IER, an independent research centre) a useful picture of future employment and labour force trends can be established. An understanding of these issues is essential for forward planning and policy-making in health and safety enforcement.

The roles of part-time working and self-employment have grown rapidly since 1987 and both are expected to continue to grow in importance. By 2011, part-time work is likely to account for 30% of total employment. The hugely diverse sector of self-employment already represents over 13% of total employment. The lack of standardization, the number of small businesses and the complex and numerous relationships that exist between client and contractor, present many challenges for health and safety enforcement.

By 2015, it is likely that there will be a higher percentage of workers aged 60 plus and an active and ageing workforce could present one of the important future challenges. An increased polarization of skilled and unskilled workers, with shortages continuing in trade and craft skills is also predicted. The total number of people in employment is predicted to remain relatively stable at around 27 million with parity between male and female workers.

Manufacturing and primary sectors have been declining steadily and manufacturing is predicted to account for less than 15% of employment by 2024. Service and knowledge based enterprises and small businesses continue to grow along with potentially profitable high technology. There is also likely to be greater fluidity in the labour market, more flexible working, virtual teams and organizations and less hierarchical management. Some predictions claim that by 2010, 50% of management jobs will be carried out from home but the high number of working hours will continue.

The allocation of enforcement responsibilities between the HSE and local authorities is set out in the Health and Safety (Enforcing Authority) Regulations 1998. These regulations maintain the broad allocation of service sector businesses to local authority enforcement, and manufacturing to the HSE. On current trends, it is possible to predict continued growth in the local authority enforced sector and continued decline in the HSE enforced sector.

The number of fatal injuries to workers in the local authority enforced sector fluctuates year on year but the long term trend in the last ten years is downward. The picture for non-fatal injuries is similar although survey results have indicated a recent rise. In the last five years there has been a rising trend in the number of reported fatal injuries to members of the public while non-fatal injuries has fluctuated considerably. At the same time, there have been year-on-year reductions in the number of local authority staff with consequent reductions in their rate of visiting and the number of premises visited. The number of full-time equivalent inspectors working in local authorities has reduced from 1590 in 1996/7 to 1060 in 2001/02 [12]. The data is updated annually and can be found in the series of HELA Annual reports available from HSE books.

Small firms

Among the most significant trends is the development of the small firms sector in the United Kingdom. Between 1979 and 1993, the number of businesses doubled from 1.8 million to 3.6 million. Of these 3.6 million firms in the United Kingdom, some 99% have fewer than 50 employees [13]. Local authorities are responsible for health and safety enforcement in approximately two-thirds of these. The picture is a very transient one: some 400 000 new firms are established every year, but approximately 300 000 others close in any one year. As part of its ongoing commitment to ensure a robust health and safety system, the HSC has adopted the need to address health and safety issues in small firms as a priority. This policy culminated in the launch of a strategy for small firms in February 1997 [13] and a HELA strategy for small firms in December 1997 [14].

The essence of the HSC's strategy is its wish to improve communication with small firms and its commitment to modernize and simplify the laws controlling health and safety at work backed by straightforward and easy to understand guidance. HELA's strategy is a response to the HSC's wishes. It commits local authorities to raising standards of compliance with health and safety law in small firms by identifying, assessing and disseminating information about local authority initiatives and good practice.

The emergence of occupational health as a priority

In 1997, the HSC agreed that the HSE should undertake work to produce a forward looking long-term strategy for occupational health with a wide-ranging remit to consider both the effect of work on health and the effect of health on work. This culminated in the publication of a long-term strategy mentioned above.

The importance in health and safety terms is not inconsiderable. In 1990, an estimated 2.2 million people reported that they had suffered ill-health caused or made worse by their work. In broad terms this equates to the loss of approximately 33 million working days per annum and costs of sickness absence of between £4000 million and £5000 million per year [15].

While the health and safety system has always dealt with occupational ill-health, recent health initiatives and the promotion of health in the workplace may indicate a shifting balance within the system between ill-health and accidents. Arguably, the systems for dealing with accident prevention, the management of risks and the control of hazards have reached maturity, whereas the approaches to health and lifestyle issues such as stress, to which the working environment may make a contribution, are still in their relative infancy.

Future strategies

In response to the HSC's recognition of the need to keep pace with a changing world, the HSE conducted a series of reviews during 2002/03. This culminated in a change programme designed to ensure that it made more of an effort to identify and address the constantly changing health and safety issues in the evolving workplace. These sentiments are now expressed in its mission which seeks to protect people's health and safety by ensuring risks in the changing workplace are properly controlled.

The HSE has also established a new suite of high level aims that are set out in terms of the things that the HSE does currently and intends to continue and those things that it intends to do in future.

What the HSE currently does and will continue to do:

- it protects people by providing information and advice; by promoting and assuring a goal-setting system of regulation; by undertaking research; and by enforcing the law where necessary;
- it influences organizations to embrace high standards of health and safety and to recognize the social and economic benefits;
- it works with business to prevent catastrophic failures in major hazard industries;
- it seeks to optimize the use of resources to deliver its mission and vision.

What the HSE aims to do:

- it will develop new ways to establish and maintain an effective health and safety culture in a changing economy, so that all employers take their responsibilities seriously, the workforce is fully involved and risks are properly managed;
- it will do more to address the new and emerging work-related health issues;
- it will achieve higher levels of recognition and respect for health and safety as an integral part of a modern, competitive business and public sector and as a contribution to social justice and inclusion;
- it will exemplify public sector best practice in managing resources.

These new aims and more details on HSC's new strategic approach are set out in A Strategy for Workplace Health and Safety in Great Britain to 2010 and beyond (10).

LOCAL AUTHORITY ENFORCEMENT

Uniquely, in the United Kingdom, both the HSE and the local authorities are responsible for the enforcement of the Health and Safety at Work, Etc. Act 1974 and other relevant statutory provisions (see Chapter 19 for local authorities' responsibilities). They do so under the general guidance and direction of the HSC. To ensure an appropriate balance between locally determined priorities and the needs of the health and safety system as a whole, the HSC has certain powers (see later) to direct local authority activity in pursuit of its aims and objectives.

The need for co-ordination and liaison between the HSE and local authorities to ensure a consistent approach has been a feature of the health and safety system since Robens first recommended a continued role for local authorities. Various mechanisms exist to cement the relationship between the HSE and local authorities and the partnership approach they have adopted to their common responsibilities. At the highest level, this includes a member of the HSC with a local authority background, usually nominated by the local authority associations.

Memorandum of understanding

The memorandum of understanding forms part of a wider picture of co-operation between central and local government. In 1997, the partnership arrangements between the HSE and local authorities were strengthened by the signing of a memorandum of understanding by the local authority associations and the HSE. The memorandum sets out the commitment of local authorities to enforce relevant statutory provisions in a spirit of partnership with the HSE. It commits them to working together in a co-ordinated and co-operative way that avoids duplication of effort, in which experience and expertise is shared, and where appropriate operational policy is developed and agreed. It also expresses specific support for the HELA.

The HELA

The Health and Safety Executive/Local Authority Enforcement Liaison Committee (HELA) exists to provide a fixed liaison forum to ensure consistency of enforcement between the HSE and local authorities and among local authorities. It is jointly chaired by a senior local authority manager, who has responsibility for health and safety enforcement, and the Deputy Director General of the HSE. Local authority representatives who serve on HELA are nominated by the local authority associations.

In recent years, HELA has developed an annual strategy that aims to influence the activity of local authorities in health and safety enforcement. In 1998, the nature and timing of this strategy was changed to establish a more direct link with the HSC's own priority objectives and continuing aims. The strategy ensures that local authorities, through HELA, are engaged in a genuine dialogue with the HSC about priorities in the local authority enforced sector, and local authority activity is oriented in pursuit of the HSC's objectives.

HELA is engaged in an ongoing dialogue with the HSC and makes an annual report to it on health and safety in the local authority enforced sector. HELA develops and issues advice and guidance for local authorities and conducts research and enquiry into local authority enforcement issues and health and safety in the local authority enforced sector.

The HSE Local Authority Unit

Support for the HELA is usually administered through the HSE's Local Authority Unit (LAU). The unit is headed by an environmental health officer who is supported by a team of administrative and technical staff, including HSE and local authority inspectors. The unit exists to support local authorities and to ensure that the HSC's objectives are achieved in the local authority enforced sector.

In practice this is achieved by:

- maintaining an effective channel of communication between local authorities and the HSC/HSE;
- promoting consistency, transparency, targeting and proportionality of enforcement;
- promoting and monitoring the effectiveness of HSC policies and objectives;
- enhancing the standing of local authority health and safety enforcement among local authorities, the HSC/HSE, other government departments, business, trade unions and consumers.

The Local Authority Unit's yearly work plan flows directly from HELA's strategy and the HSC's key objectives and ongoing aims.

Enforcement liaison officers

In addition to liaison arrangements at the strategic level, arrangements are also in place to secure liaison and co-operation at the operational level. Each HSE operational region has an enforcement liaison officer (ELO) who is responsible for engaging in an ongoing dialogue with, and providing operational support and in some instances training to, local authority enforcement staff. ELOs also act as the gateway to specialist technical expertise and services that the HSE makes available to local authorities.

Commission guidance

Section 18 of the Health and Safety at Work, Etc. Act 1974 places a duty on local authorities to make adequate arrangements for enforcing relevant statutory provisions. The same section enables the HSC to issue guidance to local authorities, to which they must have regard.

There have been very few occasions when the HSC has exercised its right to issue such guidance. Its general approach is to work by consensus, and this is no less vigorously applied in its dealing with local authorities. The HSC believes that the co-operation and goodwill achieved by this approach may be undermined if the HSC relies too heavily on its statutory guidance for the support of local authorities in pursuit of its objectives.

There are cases, however, where HSC guidance has been thought necessary and appropriate. In 1996, it issued guidance that clarified the general duty of local authorities to make adequate arrangements for the enforcement of health and safety within their areas of responsibility, and this was updated in 2002 [16]. At the same time, LAU, on behalf of HELA and HSC began a series of audits of LAs to review arrangements being made for enforcement.

In the view of the HSC, the following elements are essential for an LA to adequately discharge its duties as an enforcing authority:

- a clear published statement of enforcement policy and practice;
- a system of prioritized planned inspection activity according to hazard and risk, and consistent with any advice given by the HELA;
- a service plan detailing the LA's priorities and its aims and objectives for the enforcement of health and safety;
- the capacity to investigate workplace accidents and to respond to complaints by employees and others against allegations of health and safety failures;
- arrangements for benchmarking performance with peer LAs HELA has developed an inter-authority audit protocol to assist this process;
- provision of a trained and competent inspectorate;
- arrangements for liaison and cooperation in respect of the Lead Authority Partnership Schemes.

In essence, these requirements establish a framework for health and safety enforcement, and some of the principles that define them are therefore worthy of further mention.

Enforcement policy

The HSC believes that health and safety enforcement should be informed by the principles of proportionality, consistency, targeting, and transparency (see also Chapter 7).

Proportionality

Proportionality is necessary to ensure that enforcement action is related to risk. Clearly, some health and safety requirements are mandatory whereas others rely on the concept of proportionality built into the principle of SFAIRP, as mentioned above. These principles give discretion to both duty holders and enforcers.

Consistency

Consistency should not be interpreted as uniformity. The HSC believes that consistency of enforcement means taking a similar approach in similar circumstances to achieve similar ends. Both the HELA and local authorities have devised a number of feedback loops to ensure dialogue among and between enforcers. In addition, the HSE has conducted further research into the management of consistency within local authorities to enable local authorities to identify the processes that may lead to inconsistency, and the management tools that might assist in the control of the many variables that make the achievement of consistency so complex. The results of this research have yet to be published but are being used by the LAU to advise local authorities.

Transparency

The HSC believes that effective control of risk to health and safety is achieved if duty holders understand what is expected of them and what they should expect from enforcing authorities. Duty holders should understand what they are expected to do and not do by distinguishing between statutory requirements and advice or guidance.

Targeting

The HSC expects local authorities to have in place a system for prioritizing its activities according to risk. This, in essence, means ensuring that inspection activity is targeted at those activities that give rise to the most serious risks or where hazards are least well controlled. Similarly, action should be focused on the duty holders responsible for the risk. This may mean targeting action at employers, employees, manufacturers, suppliers, or others who are in the most appropriate position to control the risk. LAs are required also to take account of the HSC's strategic plan, the HELA strategy and HELA guidance.

Good enforcement practice

The principles of good enforcement policy have been recognized more widely within government. In March 1998, the Cabinet Office, the Department of Trade and Industry and the Local Government Association adopted a concordat on good enforcement (see p. 152). This committed relevant stakeholders to protecting the public while maintaining a fair and safe trading environment in which the national and local economies could thrive without unnecessary burdens upon them. The concordat acknowledges the importance of proportionality and consistency, but also recognizes the importance of:

- setting clear standards of enforcement performance and services;
- being open, consultative and clear in the giving of information and advice;
- being helpful, efficient and courteous on the basis that prevention is better than cure;
- providing a well-published, effective and timely complaints procedure that is easily accessible to business.

The Section 18 guidance requires that the approach of LAs is consistent with the policy set out in the HSC's most current statement on Enforcement Policy.

Training and competence of local authority staff

Competence of enforcement staff is an essential ingredient in the delivery of appropriate enforcement. The HSC considers that, as part of an enforcing authority's duty to make adequate arrangements for the enforcement of relevant statutory provisions, each local authority must satisfy itself that inspectors authorized to exercise some or all of their enforcement powers meet standards of competence set out in the HSC's guidance.

Requirement to undergo an audit and develop an action plan

This is a new requirement in the most recent guidance. All LAs should have in place arrangements to promote consistency in the exercise of their enforcement discretion, including effective arrangements for liaison with other LAs. This particular guidance ensures that these arrangements are monitored, reviewed and audited. HSC expects LAs to undergo such an audit at least once every five years.

REFERENCES

1. The Health and Safety at Work, Etc. Act 1974 (1974) *c.*37, HMSO, London.
2. The Committee on Safety and Health at Work (Robens Committee) (1972) Command 5034, HMSO, London.
3. Revitalising Health and Safety Strategy Statement (June 2000) *Health and Safety Commission and Department of the Environment, Transport and the Regions*, HMSO, London.
4. The Management of Health and Safety at Work Regulations 1999, HMSO, London.
5. Health and Safety Executive (1988, revised 1992) *The Tolerability of Risk from Nuclear Power Stations*, HSE Books, Sudbury, Suffolk.
6. Reducing Risks, Protecting People – HSE's decision making process (ref ISBN 0 7176 2151 0) (2001), HMSO, London.
7. The Health and Safety (Consultation with Employees) Regulations 1996, SI 1996. No. 1513.
8. The Safety Representatives and Safety Committees Regulations 1977, SI 1977. No. 500.
9. European Commission's Senior Labour Inspectors Committee (1997) *Labour Inspection (Health and Safety) in the European Union*, EU, Luxemburg.

10. A Strategy for Workplace Health and Safety in Great Britain for 2010 and beyond, HSE Books, Sudbury, Suffolk.

11. Securing Health Together 2000 – a long-term occupational health strategy for England, Scotland and Wales, Misc 225 07/00 C250, HSE Books, Sudbury, Suffolk.

12. HELA (2003) National Picture of health and safety in the local authority enforced sector, 10/03 C10, HSE Books, Sudbury, Suffolk.

13. Health and Safety Executive (1997) *Health and Safety in Small Firms*, MISC071, HSE Books, Sudbury, Suffolk.

14. HELA (1997) *HELA Strategy for Small Firms*, MISC100, HSE Books, Sudbury, Suffolk.

15. Hodgson, J.T., Jones, J.R., Elliot, R.C. and Osmant, J. (1993) *Self-Reported Work-Related Illness: Results from a Trailer Questionnaire on the 1990 Labour Force Survey in England and Wales*, Research Report 33, HSE Books, Sudbury, Suffolk.

16. Section 18 HSC guidance to Local Authorities (2002) C50, HSE Books, Sudbury, Suffolk.

19 Legislative and administrative framework

Paul C. Belcher and John D. Wildsmith

THE HEALTH AND SAFETY AT WORK, ETC. ACT 1974

This Act [1] broadly implemented the recommendations of the Robens Committee (see p. 412) [2] and was brought into force over a period of six months commencing on 1 October 1974.

General purposes

The objectives of the Act are set out in Section 1 and are:

1. securing the health, safety and welfare of persons at work;
2. protecting persons other than persons at work against risks to health and safety arising from work activities;
3. controlling explosive, highly flammable or dangerous substances;
4. controlling the emission of noxious or offensive substances from prescribed classes of premises;
5. to enable the previously existing health and safety legislation to be progressively replaced by a system of regulations and Approved Codes of Practice operating in combination.

Scope

All people at work, with the single exception of people employed as domestic servants in private households, are covered by the Act (Section 51). Section 52 makes it clear that the Act includes self-employed people. The effect of this legislation was to bring some eight million people who were not covered by previous health and safety legislation within the scope of this Act. Emphasis is now placed upon people and their activities rather than specified premises and processes. In addition to people at work, the general public are protected against risks to their health or safety caused by work activities.

General duties

Sections 2 to 7 inclusive of the Act set out a series of provisions placing upon both employers and employees general duties in relation to their activities; these provisions form the main thrust of the Act. All of the requirements are qualified by the provision that they are to be applied '**so far as is reasonably practicable**' (see p. 413).

Employers' duties to their employees (Section 2)

1. To ensure their health, safety and welfare at work.

2. Provide and maintain plant and systems of work that is safe and without risks to health.
3. Ensure safety and absence of risks to health in the use, handling, storage of articles and substances.
4. Provide necessary information, instruction, training and supervision.
5. Maintain workplaces under their control in a safe condition without risks to health, and provide and maintain safe means of access and egress.
6. Provide a safe working environment without risks to health together with adequate welfare arrangements and facilities.

Provisions are also contained in the Health and Safety (Information for Employees) Regulations 1989 as amended [3,4] and the Health and Safety (Consultation with Employees) Regulations 1996 [5]. It should be noted that the terms 'health', 'safety' and 'welfare' are not defined in the Act.

Duties of employers and the self-employed to persons other than their employees (Section 3)

1. Conduct the undertaking in such a way that persons not in their employment are not exposed to risks to their health and safety.
2. Self-employed persons must ensure that both themselves and other persons not employed by them are not exposed to risks to their health and safety.
3. If prescribed by regulations, to give persons not employed by them information about how their health and safety could be affected by their undertaking.

Duties of persons concerned with premises to persons other than their employees (Section 4)

Persons concerned with non-domestic premises must take measures to ensure that, all means of access and egress, and plant and substances provided for use in the premises, are safe and without risks to health.

Duties of persons in control of premises in relation to harmful emissions into the atmosphere (Section 5)

1. For classes of premises prescribed by regulations, to prevent the emission into the atmosphere of noxious or offensive substances.
2. Properly use, supervise and maintain plant provided to control these emissions.

(See also the Health and Safety (Emissions into the Atmosphere) Regulations 1983 (as amended) [6,7].)

Duties of manufacturers etc. with regard to articles and substances for use at work (Section 6)

During inspections, environmental health officers will become aware of matters that contravene the requirements of this section. In such cases, the case should be referred to the enforcement liaison officer of the Health and Safety Executive (see p. 420) for action.

1. Designers, manufacturers, importers and suppliers of any article for use at work must:

   (a) ensure that it is designed and constructed to be safe and without risks to health when properly used;
   (b) carry out such examination and testing as may be necessary;
   (c) ensure that adequate information is available about the use and testing of articles, and about any conditions necessary to ensure that it will be safe and without risks to health when it is put into use.

2. Designers and manufacturers must carry out research with a view to the discovery and elimination or minimization of any risks to health and safety.
3. Persons who erect or install articles for use at work must ensure that it is not done in a way which makes it unsafe or a risk to health when properly used.

4. Manufacturers, importers and suppliers of any substance for use at work must:

    (a) ensure that it is safe and without risks to health when properly used;

    (b) carry out such testing and examination as necessary;

    (c) make available adequate information about the results of relevant tests.

5. Manufacturers of substances for use at work must carry out necessary research with a view to the discovery and elimination or minimization of risks to health and safety which the substances may give rise to.

(See also the Health and Safety (Leasing Arrangements) Regulations 1992 [8].)

Section 6 was modified by the Noise at Work Regulations 1989 [9] (Regulation 12) so that the duty imposed under subsection (1) 'shall include a duty to ensure that where any such article is likely to cause any employee to be exposed to the "first action level" (i.e. a daily personal noise exposure of 85 dB(A) or above) or to the "peak action level" (i.e. a level of peak sound pressure of 200 Pa or above), then adequate information must be provided concerning the level of noise likely to be generated'. At the time of writing the current regulations are likely to be modified, to comply with updated standards created by the adoption of European Directive 2003/10/EC on the minimum health and safety requirements arising from physical agents (noise). Lower exposure action values of 80 and 85 dB(A) must be introduced by February 2006.

With regard to this area of work, reference must also be made to the Supply of Machinery (Safety) Regulations 1992 (as amended) [10,11], which implement Council Directive 89/392/EEC 'on the approximation of the laws of the member states relating to machinery' (the Machinery Directive) [12] as amended by Council Directive 91/368/EEC. The regulations mean that 'relevant machinery' (defined in Regulation 3) cannot be supplied unless it satisfies the relevant essential health and safety requirements laid down in the regulations. The appropriate conformity assessment procedure must also have been carried out.

The Health and Safety Executive is made responsible for these regulations (Part IV and Schedule 6) in relation to relevant machinery for use at work.

Duties of employees (Section 7)

Employees while at work must:

1. take reasonable care for the health and safety of themselves and other people who may be affected by their acts and omissions;
2. co-operate with their employers and others who may have legal responsibilities under the Act.

Self-regulation

The Act lays great emphasis on the principle of 'self-regulation' (see also p. 413) of the workplace by the employer and the employees, and the main thrust is set out in Section 2(3) in the requirement that each employer must:

> prepare and as often as may be appropriate revise a written statement of his general policy with respect to the health and safety at work of his employees and the organization and arrangements for the time being in force for carrying out that policy, and to bring the statement and any revision of it to the notice of his employees.

The Employer's Health and Safety Policy Statement (Exceptions) Regulations 1975 [13] release employers with fewer than five employees from this general obligation.

The objective of this provision is to ensure that each employer systematically evaluates the premises, operations, processes and practices, and produces systems of work and organizational and other arrangements to deal with them in such a way that hazards to health and safety are eliminated or minimized.

It is not desirable to issue a 'standard health and safety policy' document as this would negate the evaluation process. It is, however, possible to produce general guidance on the main elements, which should be as follows.

1. **General statement**: a declaration of the employer's intention to seek to provide, so far as reasonably practicable, safe and healthy working conditions. The statement should be signed by the person in the most senior executive position in the company, and should give the name of the person who is responsible for fulfilling the objectives in the policy.
2. **Organization**: following an evaluation of the business, this part of the statement should indicate in detail the degree of health and safety responsibility appropriate to the various levels of management. It should name the key posts in the organization, and define their responsibilities, including line management, supervisors and health and safety specialists. The arrangements for joint consultation on health and safety matters should be set out, including the names of any appointed union safety representatives and the terms of reference and arrangements made for any joint health and safety committees.
3. **Arrangements**: this part should follow a comprehensive study of the full range of work activities to include:

   (a) procedures for dealing with common and special hazards;
   (b) safe systems of work;
   (c) accident reporting and investigations;
   (d) provision and use of protective clothing and equipment;
   (e) procedures for introducing new machinery, processes and substances;
   (f) emergency procedures including fire and explosion 'drills';
   (g) arrangements for informing staff about health and safety issues;
   (h) health and safety training provisions;
   (i) safety Inspections, audits etc.

It should be noted that the involvement of employees in occupational health and safety in the workplace is encouraged. This is particularly stated in Section 2 of the Act, which allows the appointment of 'safety representatives' by recognized trade unions in accordance with regulations drawn up by the Secretary of State (the Safety Representatives and Safety Committee Regulations 1977) [14]. Employers have a duty under this section to consult with such safety representatives. In prescribed cases or when safety representatives request it, employers are obliged to formalize the relationships by establishing 'safety committees'.

More recently, the Safety Representatives and Safety Committees Regulations 1977 [14] required employees to allow representatives appointed by recognized trade unions to carry out various functions as highlighted above. There is therefore no similar duty placed upon an employer where there is no trade union recognition. However, under the European Health and Safety Framework Directive (89/391/EEC) which came into effect in Member States in January 1992, consultation with employees or their representatives is required in **all** work situations.

In England and Wales, the provisions of this Directive were implemented by the Health and Safety (Consultation with Employees) Regulations 1996 [15] whereby employers have a legal duty to consult either

(i) with employees directly; or
(ii) with representatives elected by the employees concerned (Representatives of Employee Safety) (ROES).

It must be noted that these ROES do not have the full range of duties and functions accorded to Safety Representatives. Where the employer chooses to consult with ROES he must inform the employees of the representative's name and identify the group represented.

Where the employer consults with ROES he/she must make available information relating to matters that are reportable under the Reporting of Injuries, Diseases and Dangerous Occurrence Regulations 1995. As in the case of Safety Representatives employers are not required to make available **all** information.

ADMINISTRATION

As suggested by the Robens Committee, the Act established both a **Health and Safety Commission**

(HSC) and the **Health and Safety Executive** (HSE) (see Chapter 6), detailed provisions being set out in Sections 10 to 14 inclusive and Schedule 2.

The main functions of the HSC include:

1. making general arrangements for the implementation of Part 1 of the Act and providing assistance to others concerned;
2. carrying out and encouraging research, publishing results and providing and encouraging training and information;
3. providing information and advisory services to all concerned with or affected by the Act;
4. preparing and submitting to the Secretary of State proposals for new regulations and approving codes of practice and guidance notes;
5. directing the Executive to investigate major accidents or occurrences or other relevant issues.

The HSC is serviced by the HSE, which is also its main operational arm. The HSE has direct duties in relation to the enforcement of the Act, but it also provides research and laboratory facilities.

It also provides the use of such services and other information to local authorities, to assist them with their enforcement of the Act.

Knowledge of the current guidance available on any topic is essential for any inspector enforcing this legislation.

The Employment Medical Advisory Service (**EMAS**) was continued by Part 2 of the Act, but now it operates as the 'medical arm' of the HSC. The services of EMAS can also be made available to local authorities.

ENFORCEMENT

Responsibilities for enforcing the Act are shared between the HSE and local authorities, and enforcement responsibilities are allocated on the basis of the 'main activity' carried out at each premises to which the Act applies. Currently, the arrangements are made according to the Health and Safety (Enforcing Authority) Regulations 1998 [16]. Schedule 1 (see below) lists the main activities that determine whether the local authority

is the enforcing authority, while Schedule 2 (see below) and Regulation 4 list activities for which, wherever they are carried out, the HSE is the enforcing authority. Regulation 4 refers to HSE premises, which include those occupied by local authorities.

The Health and Safety (Enforcing Authority) Regulations 1998, Schedule 1: Main activities which determine whether local authorities will be the enforcing authorities

1. The sale of goods, or the storage of goods for retail or wholesale distribution, except:

   (a) at container depots where the main activity is the storage of goods in the course of transit to or from dock premises, an airport or a railway station;
   (b) where the main activity is the sale or storage for wholesale distribution of any substance or preparation dangerous for supply;
   (c) where the main activity is the sale or storage of water or sewage or their by-products or natural or town gas;

   and for the purposes of this paragraph where the main activity carried on in premises is the sale and fitting of motor car tyres, exhausts, windscreens or sunroofs the main activity shall be deemed to be the sale of goods.
2. The display or demonstration of goods at an exhibition for the purposes of offer or advertisement for sale.
3. Office activities.
4. Catering services.
5. The provision of permanent or temporary residential accommodation including the provision of a site for caravans or campers.
6. Consumer services provided in a shop except dry cleaning or radio and television repairs, and in this paragraph 'consumer services' means services of a type ordinarily supplied to persons who receive them otherwise than in the course of a trade, business or other undertaking carried on by them (whether for profit or not).
7. Cleaning (wet or dry) in coin-operated units in launderettes and similar premises.

8. The use of a bath, sauna or solarium, massaging, hair transplanting, skin-piercing, manicuring or other cosmetic services and therapeutic treatments, except where they are carried out under the supervision or control of a registered medical practitioner, a dentist registered under the Dentists Act 1984, a physiotherapist, an osteopath or a chiropractor.

9. The practice or presentation of the arts, sports, games, entertainment or other cultural or recreational activities except where the main activity is the exhibition of a cave to the public.

10. The hiring out of pleasure craft for use on inland waters.

11. The care, treatment, accommodation or exhibition of animals, birds or other creatures, except where the main activity is horse-breeding or horse-training at a stable, or is an agricultural activity or veterinary surgery.

12. The activities of an undertaker, except where the main activity is embalming or the making of coffins.

13. Church worship or religious meetings.

14. The provision of car parking facilities within the perimeter of an airport.

15. The provision of childcare, or playgroup or nursery facilities.

Note: the 1998 regulations transferred responsibility for various activities from the HSE to local authorities including the following:

- pre-school childcare except where they are in domestic premises;
- storage of goods in retail or wholesale premises which are part of a transport undertaking;
- mobile vendors;
- horticultural activities in garden centres;
- theatres, art galleries and museums.

At the time of writing, further changes in the allocation of enforcement activities are under consideration.

Schedule 2: Activities in respect of which the Health and Safety Executive is the enforcing authority.

1. Any activity in a mine or quarry other than a quarry in respect of which notice of abandonment has been given under Section 139(2) of the Mines and Quarries Act 1954.

2. Any activity in a fairground.

3. Any activity in premises occupied by a radio, television or film undertaking in which the activity of broadcasting, recording or filming is carried on, and the activity of broadcasting, recording or filming wherever carried on, and for this purpose 'film' includes video.

4. The following activities carried on at any premises by persons who do not normally work in the premises:

   (a) construction work if:

      (i) Regulation 7(1) of the Construction (Design and Management) Regulations 1994 (which requires projects which include or are intended to include construction work to be notified to the Executive) applies to the project which includes the work;

      (ii) the whole or part of the work contracted to be undertaken by the contractor at the premises is to the external fabric or other external part of a building or structure;

      (iii) it is carried out in a physically segregated area of the premises, the activities normally carried out in that area have been suspended for the purpose of enabling the construction work to be carried out, the contractor has authority to exclude from that area persons who are not attending in connection with the carrying out of the work and the work is not the maintenance of insulation on pipes, boilers or other parts of heating or water systems or its removal from them;

   (b) the installation, maintenance or repair of any gas system, or any work in relation to a gas fitting;

   (c) the installation, maintenance or repair of electricity systems;

   (d) work with ionizing radiations except work in one or more of the categories set out in Schedule 3 to the Ionising Radiations Regulations 1985.

5. The use of ionizing radiations for medical exposure (within the meaning of Regulation 2(1) of the Ionising Radiations Regulations 1985).
6. Any activity in premises occupied by a radiography undertaking in which there is carried on any work with ionizing radiations.
7. Agricultural activities, and any activity at an agricultural show which involves the handling of livestock or the working of agricultural equipment.
8. Any activity on board a sea-going ship.
9. Any activity in relation to a ski slope, ski lift, ski tow or cable car.
10. Fish, maggot and game breeding except in a zoo.
11. Any activity in relation to a pipeline within the meaning of Regulation 3 of the Pipelines Safety Regulations 1996.
12. The operation of a railway.

The regulations also indicate responsibility for common parts of buildings where there is dual enforcement, identify arrangements for transfer of responsibility between the HSE and the local authorities (and vice versa) by agreement, and include the arrangements for enforcement to be assigned in cases of uncertainty.

In order to secure a degree of uniformity in the application of this legislation, both the HSC and the HSE issue a considerable amount of guidance to enforcement authorities and employers (see p. 419).

Liaison between local authority staff and HSE staff is achieved in several ways, such as:

- the establishment of a national Health and Safety Executive/Local Authority Liaison Committee (**HELA**) (see p. 420);
- local arrangements made through the appointment of an enforcement liaison officer (**ELO**) within each area office of the HSE (see p. 420).

Every enforcing authority may appoint such inspectors, having such qualifications as it thinks necessary (Section 19). Each appointment must be made in writing, specifying which of the powers conferred upon inspectors (Section 20) are conferred upon the person appointed.

It should be noted that it is **the inspector**, and not the enforcing authority, who has a number of powers conferred upon him, the principal ones being:

1. **Improvement notice (Section 21).** Where an inspector is of the opinion that a person is contravening one or more of the 'relevant statutory provisions' listed in Schedule 1 of the Act, or has contravened one or more of those conditions in circumstances that make it likely that the contravention will continue or be repeated, he may serve upon him an improvement notice, stating the particulars and his opinion, and requiring that the stated matters be remedied within a specified period (not less than the period allowed for appeals which is laid down in the Employment Tribunals (Constitution and Rules of Procedure) Regulations 2001) [17,18].
2. **Prohibition notice (Section 22).** This applies to any activities that are being or are about to be carried on by or under the control of any person, which involve or will involve the risk of serious personal injury. A prohibition notice must specify the matters that cause the inspector to form his opinion, and directs a person not to carry out the specified activities. Such a notice may come into effect immediately where the risk is imminent, or it may be deferred, to come into operation after a specified period where circumstances allow this course of action.

A number of supplementary provisions relating to improvement and prohibition notices are given in Section 23. These state, for example, that notices may include directions on the measures to be taken to remedy matters to which the notice relates, and also that such directions may be framed by reference to any approved code of practice.

Notices that do not take immediate effect may be withdrawn by an inspector at any time before any time period specified in the notice, and the period specified may be extended at any time when an appeal is not pending against the notice.

This allows considerable flexibility in ensuring compliance with notices.

Arrangements for **appeals** against improvement and prohibition notices are dealt with in Section 24 of the Act. Appeals are normally made to **industrial tribunals**, which are independent judicial bodies created to adjudicate in certain matters of dispute in the employment field in a speedy, informal and informed way. After its adjudication, the tribunal will issue written decisions that are binding upon the parties concerned. It should be noted that an appeal automatically suspends the operation of an improvement notice, but not a prohibition notice. However, this type of notice may be suspended by the tribunal after due consideration. The tribunal may cancel, modify or confirm any notice.

Where it is considered that there has been some administrative defect, a tribunal may be requested to review its decision. An appeal may be made to the High Court on a point of law.

Many additional powers are available under Section 20, although one important feature of this legislation is the power to deal with the cause of any imminent danger under Section 25, where the inspector may seize any article or substances and '**cause it to be rendered harmless whether by destruction or otherwise**'.

The procedures for dealing with improvement and prohibition notices and for the seizure of articles and substances are detailed in *Environmental Health Procedures* [19] by the use of flow charts.

Offences

Section 33 lists matters that are considered to be offences under the Act. These include:

1. failure to comply with the general duties under Sections 2 to 7;
2. contraventions of health and safety regulations;
3. contraventions relating to statutory enquiries;
4. contraventions of improvement and prohibition notices;
5. obstructing an inspector;
6. refusing to reveal information;
7. misusing or wrongfully revealing information obtained under statutory powers;
8. making false statements, entries in registers and forging documents;
9. failure to comply with court orders under Section 42.

Where an offence is committed by one person but is due to the act or default of another person, that other person may be charged with the offence either in addition to, or in substitution for, the original person (Section 36). Where offences are committed by corporate bodies, directors, managers, secretaries and others may be prosecuted if it can be proved that the offence was committed with their consent or connivance, or that any neglect can be attributed to them (Section 37). The combined effect of these two sections is to render liable to prosecution any person with designated or implied health and safety responsibilities within an organization.

EUROPEAN INITIATIVES

One of the subsidiary aims of the European Union (EU) was to improve the quality of the working environment. In the early years between 1957 and 1974, the newly formed European Economic Community (as it was then called) concentrated on the problems of occupational illness and disease. Following the adoption of the Social Action Programme in 1974, a number of directives were introduced to approximate the laws of member states on issues such as safety signs [20] and exposure to vinyl chloride monomer [21].

As with other aspects of the Union, 'action programmes' were developed and implemented in 1978 and 1984. Since 1988, however, the approach adopted to regulate health and safety at work has been via the enactment of the 'framework directive' on the introduction of measures to encourage improvements in the safety and health of workers (89/391/EEC) [22].

Under this 'framework directive', a number of 'daughter directives' have now been adopted by member states. Those currently of note include directives on:

- the minimum safety and health requirements for the workplace;
- the minimum safety and health requirements for the use of work equipment by workers at work;
- the minimum health and safety requirements for use by workers of personal protective equipment;
- the minimum health and safety requirements for the manual handling of loads where there is a risk, particularly of back injury, to workers;
- the minimum safety and health requirements for work with visual screen equipment.

The daughter directives are of real significance in the United Kingdom as they have led to the development of a variety of new regulations, commonly referred to as the 'six pack', which are listed below. Each of the provisions is dealt with in subsequent chapters. A useful reference text on the social legislative background of European health and safety legislation has been produced by Neal and Wright [23].

The 'six pack'

1. The Management of Health and Safety at Work Regulations 1999.
2. The Workplace (Health, Safety and Welfare) Regulations 1992.
3. The Provision and Use of Work Equipment Regulations 1998.
4. The Manual Handling Operations Regulations 1992.
5. The Health and Safety (Display Screen Equipment) Regulations 1992.
6. The Personal Protective Equipment Regulations 1992.

The report of a research study, *An Evaluation of the Six-pack Regulations 1992* was published by HSE in 1998. HSE Books, Sudbury, Suffolk.

REFERENCES

1. The Health and Safety at Work, Etc. Act 1974 (1974) *c.*37. HMSO, London.
2. The Committee on Safety and Health at Work (Robens Committee) (1972) Command 5034, HMSO, London.
3. The Health and Safety (Information for Employees) Regulations 1989. S.I. 1989. No. 682, HMSO, London.
4. The Health and Safety (Information for Employees) (Modifications and Repeals) Regulations 1995 S.I. 1995. No. 2923, HMSO, London.
5. The Health and Safety (Consultation with Employees) Regulations 1996. S.I. 1996. No. 1513, HMSO, London.
6. The Health and Safety (Emissions into the Atmosphere) Regulations 1983. S.I. 1983. No. 943, HMSO, London.
7. The Health and Safety (Emissions into the Atmosphere) Regulations 1989. S.I. 1989. No. 319, HMSO, London.
8. The Health and Safety (Leasing Arrangements) Regulations 1992. S.I. 1992. No. 1524, HMSO, London.
9. The Noise at Work Regulations 1989. S.I. 1989. No. 1790, HMSO, London.
10. Supply of Machinery (Safety) Regulations 1992. S.I. 1992. No. 3073, HMSO, London.
11. Supply of Machinery (Safety) (Amendment) Regulations 1994. S.I. 1994. No. 2063, HMSO, London.
12. Council Directive 89/392/EEC 'on the approximation of the laws of the Member States relating to machinery'. OJ No. L183, 29.6.89, p. 9.
13. The Employer's Health and Safety Policy Statement (Exceptions) Regulations 1975. S.I. 1975. No. 1584, HMSO, London.
14. The Safety Representatives and Safety Committee Regulations 1977. S.I. 1977. No. 500, HMSO, London.
15. Health and Safety (Consultation with Employees) Regulations 1996.
16. The Health and Safety (Enforcing Authority) Regulations 1998. S.I. 1998. No. 494, HMSO, London.

17. Employment Tribunals (Constitution and Rules of Procedure) Regulations 2001. S.I. 2001. No. 1171

18. Employment Tribunals (Constitution and Rules of Procedure) Regulations SCOTLAND 2001. S.I. 2001. No. 1170.

19. Bassett, W.H. (2002) *Environmental Health Procedures,* 6th edn, E. & F.N. Spon Press, London.

20. Council Directive on the approximation of the laws, regulations and administrative provisions of Member States relating to the provision of safety signs at places of work (77/576/EEC) as amended by Directive 79/640/EEC.

21. Council Directive on the approximation of the laws, regulations and administrative provisions of the Member States on the protection of the health of workers exposed to vinyl chloride monomer (78/610/EEC).

22. Council Directive on the introduction of measures to encourage improvements in the safety and health of workers (89/391/ EEC).

23. Neal, A.C. and Wright, F.B. (eds) (1998) *The European Communities Health and Safety Legislation,* Chapman & Hall, London.

20 The working environment

Paul C. Belcher and John D. Wildsmith

While one can lay great emphasis on the responsibility of individual workers to ensure that their health, safety and welfare is maintained, one must never forget the responsibility that rests upon the employer to provide a safe working environment. This chapter looks at some of the factors that help to ensure that workers have a workplace that does not present a risk to their health, safety and welfare, and that does not impose undue stress on them. These factors must be considered with reference to the identification of priority areas of work for enforcement officers, as highlighted in Chapter 18.

WORKPLACE STRESS

Stress can occur in a variety of forms. Some would argue that it is a desirable part of our everyday existence. However, prolonged exposure to stress can be extremely harmful, leading to coronary heart disease, hypertension and gastrointestinal disorders. The Health and Safety Executive (HSE) define stress as: '**The adverse reaction people have to excessive pressures or other types of demands placed on them**'.

At a social level, excess stress can result in the disruption of relationships. Within the workplace, people suffering from excessive stress may be more likely to lose concentration, possibly at vital moments, thereby increasing the risk of death by injury to themselves or to a fellow employee. Prolonged exposure to stressors can lead workers to the point of nervous breakdown or serious adverse impact on the health of individuals.

Given that stress can be extremely harmful to many people, it is important for both employers and enforcement officers to recognize those factors, including the physical factors discussed here, that may result in stress.

In response to the reported adverse effects of stress in the workplace, the HSE have developed a number of useful guides and policy documents for employers, managers and employees, which are available via the HSE's website (www.hse.gov.uk/stress/information.htm).

There is a growing body of case law on this topic, combined with examples where employees' claims have been settled out of court. Perhaps the most significant case is that reported in the Court of Appeal in the case of *Sutherland and others* v. *Hatton and others* (2002) [1]. The court set out a number of practical propositions for future claims concerning workplace stress.

(a) Employers are entitled to take what they are told by employees at face value unless they have good reason to think otherwise. They do not have a duty to make searching enquiries about employee's mental health.

(b) An employer will not be in breach of duty in allowing a willing employee to continue in a stressful job if the only alternative is to dismiss or demote them. The employee must decide whether to risk a breakdown in their health by staying in the job.

(c) Indications of impending harm to health at work must be clear enough to show an employer that action should be taken, in order for a duty on an employer to take action to arise.

(d) The employer is in breach of duty only if he fails to take steps, which are reasonable, bearing in mind the size of the risk, the gravity of harm, the cost of preventing it and any justification for taking the risk.

(e) No type of work may be regarded as intrinsically dangerous to mental health.

(f) Employers, who offer a confidential counselling advice service, with access to treatment, are unlikely to be found in breach of duty.

(g) Employees must show that illness has been caused by a breach of duty, not merely occupational stress.

(h) Compensation will be reduced to take account of pre-existing conditions or the chance that the claimant would have fallen ill in any event.

WORKPLACE HEALTH, SAFETY AND WELFARE

As far as local authority enforcing officers are concerned, control over workplace health and safety can now be thought of as being embodied in the Health and Safety at Work, Etc. Act 1974 and in the Workplace (Health, Safety and Welfare) Regulations 1992 [2]. These regulations implement most of the requirements of the European Union (EU) Workplace Directive (89/654/EC), which is concerned with minimum standards for workplace health and safety. In most cases, the regulations bring together the legislative controls that existed prior to the 1974 Act. Detailed guidance on the regulations is contained in the Approved Code of Practice [3] and in HSE publications [4,5].

The regulations contain a general requirement for every employer and others to ensure that any workplace under their control, and where their employees work, complies with any applicable requirement. This duty is extended to those who have control of a workplace in connection with any trade, business or other undertaking (whether for profit or not).

Maintenance of workplace, equipment, devices and systems

Regulation 5 requires that the workplace and certain equipment, devices and systems be regularly maintained in an efficient state, in efficient working order and in good repair. The equipment etc. covered by this regulation is any equipment etc. in which a fault is liable to result in a failure to comply with any of the regulations.

The Approved Code of Practice [3] suggests that any system of maintenance should ensure that:

- regular maintenance (including, as necessary, inspection, testing, adjustment, lubrication and cleaning) is carried out at suitable intervals
- any potentially dangerous defects are remedied, and that access to the defective equipment is prevented in the meantime
- regular maintenance and remedial work is carried out properly
- a suitable record is kept to ensure that the system is properly implemented and to assist in validating maintenance programmes.

The frequency of any maintenance should clearly depend on the nature of the equipment and its age. Regard should be had to advice from, for example, the manufacturers and the HSE.

Examples of equipment covered by this regulation include emergency lighting, fencing, fixed equipment used for window cleaning, anchorage points used for safety harnesses and escalators.

Ventilation

By virtue of Regulation 6, every enclosed workplace must be ventilated by a sufficient quantity of fresh or purified air. This requirement does not extend to work carried on in confined spaces where breathing apparatus may be necessary. In most cases, openable windows or other openings will provide satisfactory ventilation. However, if found to be necessary, mechanical ventilation should be provided. It is important that such systems be regularly maintained, particularly because of the possible risk of Legionnaires' disease.

Temperature in indoor workplaces

Many working environments can be uncomfortable because of excessive heat or cold. Under such conditions, it is possible for workers to suffer from heat or cold stress. Regulation 7 requires that, during working hours, the temperature in all workplaces inside buildings should be reasonable. The Approved Code of Practice [3] suggests a sedentary minimum temperature of 16°C, and 13°C where there is severe physical effort. The only exception to this rule is where processes such as cold storage require lower temperatures. In such circumstances the employer should take all reasonable steps to provide protective clothing and sufficient rest periods for employees.

The regulations also contain a general requirement for employers to provide a sufficient number of thermometers to enable employees to ascertain the temperature.

Gill [6] argues that to obtain a correct assessment of the thermal environment, four parameters need to be measured together:

- the air dry bulb temperature
- the air wet bulb temperature
- the radiant temperature
- the air velocity.

A whirling hygrometer can be used to measure wet and dry bulb temperature, while a simple globe thermometer can be use to measure the radiant temperature. To measure the air velocity either a kata thermometer or an airflow meter can be used. A detailed description of how to use these measurements to assess the thermal environment is found in Gill [7]. In addition, 'thermal comfort meters' are now available. These not only carry out many of these measurements automatically, but are also capable of being linked to recorders and computers to enable rapid analysis of findings.

Lighting

Adequate lighting is an important prerequisite for ensuring the safety and comfort of people at work. While natural light is the most desirous form of lighting, in many work situations this has to be either supplemented or completely replaced by some form of artificial lighting.

Suitable and sufficient lighting is required by Regulation 8. In addition, emergency lighting must be provided where failure of any artificial lighting could pose a risk to health and safety.

The amount and type of lighting provided in any work situation quite clearly depends upon the type of work being carried out there. The HSE has suggested levels of lighting for different work environments [8]. The Chartered Institute of Building Services Engineers has also produced a code for interior lighting [9].

Measurement of lighting levels can be simply carried out using a proprietary light meter. Where possible a light meter with a remote sensing device should be used, as this removes the risk of readings being affected by shadows from the operator.

Cleanliness

Every workplace, and the furniture, furnishings and fittings contained in it, are to be kept sufficiently clean (Regulation 9). In addition, the surfaces of the floors, walls and ceilings of all workplaces shall be capable of being kept sufficiently clean, and all waste material must be kept in suitable receptacles.

It is important to bear in mind that only an adequately planned cleaning programme will satisfy this general requirement.

Space

Regulation 10 requires that every room that people work in shall have sufficient floor area, height and unoccupied space for the purposes of health, safety and welfare.

The Approved Code of Practice [3] establishes that personal space should be at least 11 m^3. In calculating personal space it is assumed that there is a minimum notional ceiling height of 3 m.

Workstations and seating

Workstations must be arranged so as to be suitable for any person who is likely to work there, and for any work likely to be done there (Regulation 11). For workstations that are located outside, there must be protection from adverse weather conditions. These workstations must also be arranged so as to enable a swift means of escape in cases of emergency.

Where necessary, a suitable seat must be provided. Suitability is defined as being not only suitable for the operation being carried out, but also for the person it is provided for. Where necessary, a footrest must also be provided [10].

Workstations where visual display units, process control screens, microfiche readers and similar display units are used are covered by the Health and Safety (Display Screen Equipment) Regulations 1992 [11] (see later).

Conditions of floors and traffic routes

This covers another of the priority areas of work set for enforcement officers with regard to the prevention of slips, trips and falls. Floors and traffic routes should be of sound construction and must be constructed so as to be suitable for the purpose for which they are used (Regulation 12). A traffic route is defined in the regulations as: 'a route for pedestrian traffic, vehicles or both and includes any stairs, staircase, fixed ladder, doorway, gateway, loading bay or ramp'.

The regulations specify that there should be no holes, slope, unevenness or slipperiness that could pose a risk to health or safety. In addition, such routes should be kept free from obstruction, and, where necessary, be sufficiently drained.

Handrails should be provided where necessary. Open sides to staircases should be protected by an upper rail at 900 mm or higher, and a lower rail.

Falls or falling objects

In many workplace situations, for example, construction sites, factories or offices, work has to take place at height. Such work will involve activities such as maintenance operations, installation of plant and equipment or window cleaning.

The risk of falling, with often fatal results, is a common hazard of such operations and because of reported accident statistics, this has also been set as a priority topic for enforcement officers to pay attention to. Dewis and Stranks [8] recognize that there are two outstanding causes of this type of accident, relating first to the means of access to the work situation, and second to the system of work adopted once the working position is reached.

In many cases employers will have considered the risks to employees of working at height. Exercises such as risk assessments, safety surveys and safety audits will reveal hazards, and meetings of safety committees and discussion with safety representatives will also lead to the establishment of safe systems of work (see Chapter 22).

When considering the risks of working at height, factors such as the provision of safety harnesses and belts, including the necessary anchorage points for their safe use, and protective clothing such as suitable headcovering, gloves and safety footwear for employees need to be borne in mind.

The provisions of Regulation 13 extend this consideration not only to the prevention of falls but also to the protection of persons from being struck by a falling object likely to cause personal injury. In addition, every tank or pit shall be securely fenced, and every traffic route, where there is a risk of falling into a dangerous substance, shall also be securely fenced. The regulations define a dangerous substance as:

- any substance likely to scald or burn
- any poisonous substance
- any corrosive substance
- any fume, gas or vapour likely to overcome a person
- any granular or free-flowing solid substance, or any viscous substance that, in any case, is of a nature or quantity which is likely to cause danger to any person.

The Approved Code of Practice [3] suggests that fencing should be provided at any place where a person might fall 2 m or more. Such fencing, where provided for the first time, should be at least 1100 mm above the surface. The fencing should also prevent objects being knocked off the surface.

Where fixed ladders are provided they should be of sound construction. If they are at an angle of less than 15° to the vertical and more than 2.5 m high, they shall be fitted with suitable safety hoops or other fall arrest systems. The Code of Practice [3] gives guidance on such ladders.

Of particular concern are racking systems, particularly those found in warehouses. The Code of Practice suggests that certain precautions be taken to ensure the minimization of risk to health and safety. In addition, the HSE has published guidance on warehouse stacking [13].

Windows, transparent or translucent doors, gates and walls

All windows, doors, gates, walls and partitions glazed wholly or partially so that they are transparent or translucent must, where necessary for health and safety, be of safe material and be appropriately marked (Regulation 15). The Code of Practice [3] defines safe materials as:

- materials that are inherently robust, such as polycarbonates or glass blocks
- glass that, if it breaks, breaks safely
- ordinary annealed glass that meets the thickness criteria laid down in the Approved Code of Practice.

As an alternative to using safe materials, transparent or translucent surfaces may be protected by the use of screens or barriers.

Regulation 15 further requires that windows, skylights or ventilators that can be opened should not expose people opening them to a risk to their health and safety, and when opened such windows etc. should not expose people to a risk to their health or safety.

Finally, Regulation 16 deals with the cleaning of windows and skylights in workplaces. With regard to the advice that should be given to those employed in any work at height including the cleaning of windows, attention is drawn to the various sector-specific guidance given by the HSE and the contents of the relevant British Standard [14].

Organization of traffic routes etc.

This is a further priority area of work for enforcement officers due to the seriousness of accidents involving pedestrians and vehicular traffic. Regulation 17 deals with the safe circulation of pedestrians and vehicles in the workplace. There must be sufficient separation between vehicles and pedestrians. All routes must be adequately signed.

All doors and gates are to be suitably constructed to include the fitting of any necessary safety devices (Regulation 18). In particular, the regulations require that:

- any door or gate has a device to prevent it coming off its track during use
- any upward opening door or gate has a device to prevent it falling back
- any powered door or gate has suitable and effective features to prevent it causing injury by trapping any person
- where necessary for reasons of health or safety, any powered door or gate can be operated manually unless it opens automatically if the power fails
- any door or gate that is capable of opening by being pushed from either side is of such a construction as to provide, when closed, a clear view of the space close to both sides.

Regulation 19 requires that all escalators and moving walkways function safely, be equipped with any necessary safety device and be fitted with sufficient emergency stop controls. There is guidance issued by the HSE [15].

Welfare facilities

Regulations 20–25 deal with the provision of sanitary conveniences, washing facilities, adequate wholesome drinking water, suitable and sufficient facilities for changing clothing and suitable and sufficient facilities for rest and to eat meals. The Approved Code of Practice [3] gives details about the minimum number of sanitary conveniences etc. With regard to rest areas, both the regulations and the Code of Practice stress the need to have regard to the risks of passive smoking, and employers are required to ensure that suitable arrangements are made so as to ensure that no discomfort occurs [16].

Display screen equipment

Display screen equipment (DSE) is now covered by a specific set of regulations, the Health and Safety (Display Screen Equipment) Regulations 1992 [17]. The regulations implement a European Directive (90/270/EEC) on DSE.

There is some degree of overlap between this set of regulations and the Management of Health and Safety at Work Regulations 1999 [18], particularly with regard to issues such as risk assessment. The guidance on the regulations [19] points out that employers are required to comply with both the specific requirements of the DSE regulations, and the general requirements of the management regulations. However, carrying out a risk assessment on a workstation under Regulation 2 of the DSE regulations will also satisfy the requirements of the management regulations, but only for that workstation.

Health effects of display screen equipment

DSE has been blamed for a whole range of adverse health effects. One of the most significant is that of musculo-skeletal injury. This has also been established as a priority enforcement action topic for local authorities. There is also some epidemiological evidence to suggest that DSE can have some impact on other aspects of health status. These effects are linked to the visual system and working posture and include:

- repetitive strain injury, or work-related upper limb disorders;
- eye or eyesight defects;
- photosensitive epilepsy;
- fatigue and stress.

For a detailed consideration of the health effects of DSE, attention is drawn to a number of publications from the HSE [10,20–23], the International Labour Office [24] and the National Radiological Protection Board (NRPB).

The requirements of the display screen equipment regulations

DSE is defined as 'any alphanumeric or graphic display screen, regardless of the display process involved'. An operator or user is someone who 'habitually uses display screen equipment as a significant part of his or her work'. The regulations suggest that generally an individual will be classified as a user or operator if most or all of the following apply:

- the individual depends on the use of DSE to do the job as alternative means are not readily available for achieving the same results;
- the individual has no discretion about using or not using the DSE;
- the individual needs significant training and/or particular skills in the use of DSE to do the job;
- the individual normally uses DSE for continuous spells of an hour or more at a time;
- the individual uses DSE in this way more or less daily;
- fast transfer of information between the user and screen is an important requirement of the job;
- the performance requirements of the system demand high levels of attention and concentration by the user, for example, where the consequences of error may be critical.

The guidance document [19] gives the following examples of users:

- word-processing operators
- secretaries or typists
- data input operators
- journalists
- air traffic controllers.

Regulation 2 requires every employer to carry out a suitable and sufficient analysis of workstations used by users or operators, to assess the possible risks to health and safety. The assessment must be reviewed if there is a significant change or if the assessment is no longer valid. The employer, having identified the risks, is then required to reduce them to the lowest extent reasonably practicable.

Regulation 3 requires that all workstations meet the detailed requirements of the Schedule to the Regulations. The schedule deals with issues such as the display screen, the keyboard, the work desk or work surface, the work chair, space requirements, lighting, reflections and glare, noise, heat, radiation, humidity and the interface between the computer and the operator/user. The guidance document gives detailed advice on the interpretation of the schedule.

Regulation 4 requires employers to plan the activities of users so as to ensure that sufficient rest periods are provided. The guidance document suggests that short frequent breaks are more satisfactory than longer breaks, for example, 5–10 minutes for every 50–60 minutes of work.

There is a general provision made in Regulation 5 for eye and eyesight tests to be made available on request and at regular intervals. If any 'special corrective appliances' (generally speaking these are spectacles/lenses) are required, then currently these have to be paid for by the employer.

Regulation 6 makes it a requirement for the provision of health and safety training for DSE users. In addition, Regulation 7 requires employers to make available for users information about breaks, eye and eyesight tests and training, both initially and when the workstation is modified.

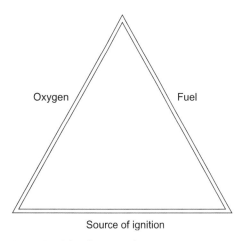

Fig. 20.1 The fire triangle.

FIRE

The 'fire triangle' is perhaps the most convenient way of studying the fire situation (Fig. 20.1). It is argued that fuel, oxygen and a source of ignition are needed in order for any fire to be sustained. As a result, in most cases, the removal of any one of these components of the triangle results in either the fire being extinguished, or fire prevention if no fire is occurring.

Classes of fire

Fires may be broken down into four main classifications.

1. *Class A*. These are fires involving solid materials of a cellulosic nature such as wood, paper, cardboard, coal and natural fibres. The risk of fire is at its greatest when the material is finely divided, for example, wood shavings.
2. *Class B*. Fires involving liquids or liquefiable solids. With regard to liquids, one can further divide the classification into:

   B.1 liquids that mix with water, for example, methanol, acetone and acetic acid;
   B.2 liquids that do not mix with water, for example, waxes, fats, petrol and solvents.

3. *Class C*. Fires involving gases or liquefied gases in the form of liquid gas spillage or gas leaks. Examples are methane, propane and butane.
4. *Class D*. Fires involving certain flammable metals such as aluminium and magnesium.

Electrical fires

This type of fire is often caused by the arcing or overheating of conductors. The problem with this type of fire is that the electrical apparatus often remains alive, causing a further risk to those involved in fighting the fire.

Dealing with fires

In order to extinguish a fire, one must have a knowledge of the combustion process. For the combustion reaction to proceed, the concentration of fuel to air must be in the flammable range, and the mixture must receive a certain amount of energy. The reaction will therefore cease if one or other of these criteria is not met.

One can employ this knowledge and deal with fires in one of three ways:

1. **Starvation:** reducing the concentration of the fuel. This can be achieved by, for example, switching off the gas supply or building fire-breaks.
2. **Smothering:** limiting the concentration of air. This can be achieved by allowing the fire to consume all the oxygen in the air while at the same time preventing any further airflow. Examples of this type of approach are the wrapping of a person whose clothes are on fire in a rug or covering fires with sand.
3. **Cooling:** reducing the energy input. This is by far the most common method of dealing with fires. By using water, the fire is put out because of the cooling effect of the water.

Fire-fighting agents

By applying the principles of extinction outlined in the previous section it is clear that, in the main, any fire-fighting agent will be used either to cool the fire or to starve the fire of oxygen. The most commonly used fire-fighting agents are listed in the following.

Water

As outlined earlier, water is an extremely efficient cooling agent. It is generally applied to the fire in the form of a fine spray or as a jet. Application by use of a spray is more effective than a jet, and more heat can be absorbed by the drops in the spray than by the bulk flow in the jet. However, with large fires, heavy jets are needed to provide the necessary quantity of water.

Foam

Foam consists of a mass of small bubbles that form an air-excluding and cooling blanket over the fire. There are two basic types of foam: chemical and mechanical. Chemical foam is produced by the mixing of two chemicals, usually sodium bicarbonate and aluminium sulphate. This can be represented by the following equation:

$$6NaHCO_3 + Al_2(SO_4)_3 \rightarrow 2Al(OH)_3 + 3Na_2SO_4 + 6CO_2$$

Sodium bicarbonate + aluminium sulphate gives aluminium hydroxide + sodium sulphate + carbon dioxide

'Mechanical foam' is produced by the release of a foam solution via a self-aspirating nozzle.

Carbon dioxide

Carbon dioxide acts by reducing the oxygen content of the atmosphere surrounding the fire. It is contained in a liquid form under pressure, and boils off to a gas when released.

Dry powders

Finely reduced sodium bicarbonate or ammonium phosphate has been found to be extremely effective fire-fighting agents when applied as a concentrated cloud.

Choice of portable extinguisher

The HSE in conjunction with the Home Office, give extensive advice on the choice of fire extinguishers as well as general advice on fire safety and fire risk assessment [25]. As a general rule the type of fire extinguisher to be used is determined by the class of fire (see earlier) that is being dealt with (Table 20.1).

Means of escape in case of fire

As mentioned earlier, it is imperative that all workplaces have a clearly defined procedure for ensuring that people know what to do and where to go in the case of a fire. To this end, all fire escapes should be clearly marked in accordance with the relevant British Standards, and all routes out of the workplace should also be protected from fire. In addition, staff, particularly new members of staff, should practise fire drills regularly.

Table 20.1 Classification of fire for portable extinguishers

Class of fire	Description	Extinguisher
A	Solid materials	Water, foam, dry powder, vapourizing liquid and CO_2
B	Liquids and liquefiable solids:	
	B.1 miscible with water	Water, foam, CO_2, dry powder, vaporizing liquid
	B.2 immiscible with water	Foam, dry powder, vapourizing liquid, CO_2
C	Gas or liquefied gas	—
D	Metals	—
—	Electrical equipment	Dry powder, vapourizing liquids, CO_2 (not computers)

REFERENCES

1. *Hatton and others* v *Sutherland and others* (2002) EWCA Civ76.
2. Workplace (Health, Safety and Welfare) Regulations (1992), S.I. 1992, No. 3004.
3. Health and Safety Executive (1992) *Workplace Health, Safety and Welfare – Approved Code of Practice to the Workplace (Health Safety and Welfare) Regulations 1992, L21*, HSE Books, Sudbury, Suffolk.
4. Health and Safety Executive (1994) *Workplace Health, Safety and Welfare IND(G)170L*, HSE Books, Sudbury, Suffolk.
5. Health and Safety Executive (1997) *Workplace Health, Safety and Welfare IND(G)244L*, HSE Books, Sudbury, Suffolk.
6. Gill, F.S. (2003) Workplace pollution, heat and ventilation, in *Safety at Work* (ed. J. Channing and J. Ridley), 6th edn, Butterworths, London.
7. Gill, F.S. (1995) The thermal environment, in *Occupational Hygiene* (eds M.J. Harrington and K. Gardiner), Blackwell, Oxford.
8. Health and Safety Executive (1998) *Lighting at Work HS(G)38*, HSE Books, Sudbury, Suffolk.
9. Chartered Institute of Building Services Engineers (2002) *Code for Lighting*, CIBSE, London.
10. Health and Safety Executive (1994) *A Pain in Your Workplace – Ergonomic Problems and Solutions HS(G)121*, HSE Books, Sudbury, Suffolk.
11. Health and Safety (Display ScreenEquipment) Regulations (1992), S.I. 1992, No. 2792.
12. Dewis, M. (2002) *Tolley's Health & Safety at Work Handbook*, Tolley Publishing Company, Croydon, Surrey.
13. Health and Safety Executive (1992) *Health and Safety in Retail and Wholesale Warehouses HS(G)76*, HSE Books, Sudbury, Suffolk.
14. British Standards Institution (1991) *BS 8213 Part 1 Windows, Doors and Rooflights: Code of Practice for Safety in Use and During Cleaning of Windows*, BSI, London.
15. Health and Safety Executive (1984) *Safety in the Use of Escalators PM34*, HSE Books, Sudbury, Suffolk.
16. Health and Safety Executive (2000) *Passive Smoking at Work INDG63 Revision 1*, HSE Books, Sudbury, Suffolk.
17. *Health and Safety (Display Screen Equipment) Regulations* (1992), S.I. 1992, No. 2792.
18. *Management of Health and Safety at Work Regulations* (1999), S.I. 1999, No. 3242.
19. Health and Safety Executive (1992) *Display Screen Equipment at Work, Guidance on Regulations L26*, HSE Books, Sudbury, Suffolk.
20. Health and Safety Executive (1992) *Working with VDUs IND(G)36L*, HSE Books, Sudbury, Suffolk.
21. Health and Safety Executive (1991) *Seating at Work HS(G)57*, HSE Books, Sudbury, Suffolk.
22. Health and Safety Executive (1990) *Work Related Upper Limb Disorders: A Guide to Prevention HS(G)60*, HSE Books, Sudbury, Suffolk.
23. Health and Safety Exectuive (1998) *DSE Health Problems: User-based Assessments of DSE Health Risks*, HSE Books, Sudbury, Suffolk.
24. International Labour Office (1994) *Visual Display Units: Radiation Protection Guidance*, ILO Publications, Geneva.
25. *Home Office, Fire Safety – An Employers' Guide* (1999) HMSO.

21 Protection of persons

Paul C. Belcher and John D. Wildsmith

THE PERSONAL PROTECTIVE EQUIPMENT AT WORK REGULATIONS 1992 [1]

These regulations replaced all prescriptive legislation made prior to the Health and Safety at Work, etc. Act 1974 and implemented the requirements of a European Directive. With the regulations came a set of Guidance Notes published by the Health and Safety Executive (HSE) [2].

The regulations place duties on employers and the self-employed, providing a framework for the provision of personal protective equipment (PPE) in circumstances where any assessment has shown a need for such protection. The regulations do not apply to the provision of most respiratory protective equipment, ear protectors and some other types of PPE because requirements are already laid down in regulations such as the Control of Substances Hazardous to Health Regulations 2002 (as amended) (COSHH) [3], the Noise at Work Regulations 1989 [4] and the Control of Asbestos at Work Regulations 2002 [5] covering the use of PPE in particular circumstances. The main aim of the 1992 regulations is to ensure the proper provision of PPE following a risk assessment.

PPE is defined in Regulation 2 as:

all equipment (including clothing affording protection against the weather) which is intended to be worn or held by a person at work and which protects him against one or more risks to his health or safety, and any addition or accessory designed to meet that objective.

As well as there being specific legal requirements, one must not forget the general duty of care owed to every employee by an employer. All employers must protect their workforce from the risk of reasonably foreseeable injury. Therefore as part and parcel of this general duty, it is reasonable to suggest that not only must such PPE be provided free of charge (see Section 9 of the Health and Safety at Work, Etc. Act 1974), but it must also be readily available at all times.

In the guidance to the regulations [2], PPE is seen as a last resort. Rather, employers should seek to introduce changes to the process or safe systems of work in an effort to reduce the risk to the health and safety of their employees. (See Chapter 10 for a fuller discussion of risk assessment techniques.)

Any PPE provided by the employer must be 'suitable'. The regulations state that PPE shall not be suitable unless:

1. it is appropriate for the risk or risks involved and the conditions at the place where exposure to the risk may occur;
2. it takes account of ergonomic requirements and the state of health of the person or persons who may wear it;
3. it is capable of fitting the wearer correctly, if necessary after adjustments within the range for which it is designed;

4. so far as is practicable, it is effective to prevent or adequately control the risk or risks involved without increasing the overall risk;
5. it complies with any enactment (whether in an Act or instrument) which implements in Great Britain any provision on design or manufacture with respect to health or safety in any relevant European Union (EU) Directive listed in Schedule 1 which is applicable to that item of personal protective equipment.

Before choosing any PPE, Regulation 6 states that any employer or self-employed person must carry out an assessment to ensure that the proposed PPE is suitable. This assessment follows on from, but does not duplicate, the risk assessment requirement of the Management of Health and Safety at Work Regulations 1999 [6], which are concerned with the whole range of hazards present in the workplace and the evaluation of the extent of any risks involved. The assessment required under Regulation 6 shall include:

1. an assessment of any risk or risks to health or safety which have not been avoided by other means;
2. the definition of the characteristics which personal protective equipment must have in order to be effective against the risks referred to in point 1 above, taking into account any risks which the equipment itself may create;
3. comparison of the characteristics of the personal protective equipment available with the characteristics referred to in point 2 above.

A specimen risk survey table is contained in Appendix 1 of the Guidance to the Regulations [2].
    The regulations further require that all PPE be maintained, and that those who use it be adequately trained in its use and the limits of the PPE involved. Furthermore, employers are under an obligation to ensure that employees use any PPE provided in an approved way. Finally, employees are under an obligation to notify their employer of any shortcomings with the PPE they have been provided with.

RISKS TO HEALTH AT WORK

It may be convenient to divide the hazards to health into three categories. These are physical, chemical and biological hazards.

Physical hazards

These include the well-documented hazards produced by, for example, light, heat, cold, noise and vibration. In addition, it may also be convenient to think of hazards such as ultraviolet, infra-red, and ionizing radiation under this heading.

Chemical hazards

This heading includes the whole range of substances and compounds found in organic and inorganic chemistry. From an Occupational Health perspective, these hazards should be controlled by the regime imposed by the COSHH Regulations [3]. In addition, substances such as asbestos would also be classified as a significant chemical hazard.

Biological hazards

Those at risk may be involved in the handling of bacteria, viruses, plants, animals or animal products. More recently, there has been concern expressed about people who come into contact with those infected with HIV (human immunodeficiency virus). In this case, attention is drawn to the Chartered Institute of Environmental Health guidance notes [7].

TOXICITY OF SUBSTANCES

Having made this distinction between various hazardous substances generally, it should be made clear that the toxicity of any substance depends upon a number of factors such as:

• the nature of the substance;
• the amount taken in, compared with the weight of the person;

- the physical condition and age of the person when exposure takes place;
- the sex of the person exposed to the substance.

Therefore, before one can attribute the term toxic to any substance and hence determine the need for protection of any worker against exposure to that substance, one must establish some of the factors listed above.

Modes of entry of harmful substances

Having established the nature and range of some of the harmful substances that can affect the human body, it is now necessary to look at the ways in which these substances can gain access to the body. It is convenient to think of these 'modes of entry' as falling under four main headings, although it should be borne in mind that some substances may have more than one mode of entry.

Inhalation

A wide range of substances is carried into the human body on the breath. While the nose, airways and lungs behave as fairly efficient filters against many substances, there is a critical size range where penetration can occur along the complete length of the respiratory tract.

Coates and Clarke [8] estimate that particles larger than 10 μm in diameter are filtered off by the nasal hairs. Particles that are between 5 μm and 10 μm tend to settle in the bronchi and bronchioles, are then moved upwards to the throat by ciliary hairs, and are then coughed out. Particles that are smaller than 5 μm are able to reach the lung tissue. Coates and Clarke point out that fibres, which predispose to disease, have a length to diameter ratio of at least 3:1, and a diameter of 3 μm or less. Therefore, the longer the fibre the more damaging it may be.

Ingestion

Many substances pass in to the body via the digestive system. If the substance is absorbed into the body, then it may be passed to the liver where it is rendered less toxic (detoxification) before being excreted. Some substances, such as bacteria and some chemicals, however, can cause harm without leaving the digestive tract.

Entry through the skin

The skin is a very substantial defence against many substances. However, some substances are able to pass directly through the skin and into the underlying tissue leaving the skin intact. In other cases, the substances pass into the epidermis but not through it, resulting in conditions such as dermatitis and some forms of cancer. Some substances, particularly solvents, are able to reduce the ability of the skin to protect the body against attack.

Irradiation

This is the term used to describe the exposure of the body to both ionizing and non-ionizing radiation. Exposure may result in body surface penetration. This topic is dealt with in detail in Chapters 25 and 26.

Defence against harmful substances

As well as the protective equipment available to protect the worker, the body also possesses its own defence mechanisms. These are:

- respiratory filtration
- cell defence mechanisms
- inflammatory response
- immune response
- thermoregulation
- metabolic transformation.

It is suggested that the reader refer to more specific texts dealing with human physiology for a comprehensive discussion of these factors.

PRINCIPLES OF PROTECTION

As has already been made clear, the use of PPE to give protection against a particular hazard should not be seen as a substitute for other methods of dealing with the danger. For example, at a drilling machine with an exposed rotating spindle where there is a risk of entanglement, the aim should be to eliminate the risk by proper guarding of the machine, and not to rely on the operator wearing suitable head covering.

Having said this, it should be borne in mind that personal protection is not an easy option, and it is important that the correct protection is given for a particular hazard. In addition, one must be satisfied that the equipment being used is of sufficient quality in order to afford the worker the required protection.

If one is to fulfil legal and moral obligations, it is essential that a programme looking at all aspects of the provision and use of personal protective equipment is in place. Else [9] gives three key elements of information required for a personal protection scheme. These are:

1. the nature of the danger;
2. performance data of personal protective equipment;
3. the acceptable level for exposure to danger.

Nature of the danger

It is important to know some details about the hazards to be faced. For example, with regard to a physical hazard such as noise, one would need to know the sound level and frequency characteristics of the noise. In addition, information could be gained from recorded accident/incident experiences, safety representatives, safety audits or surveys and medical records.

Performance data of personal protective equipment

The choice of equipment is extremely important. Its quality, durability, suitability, and lack of interference with the user's faculties and movements are important considerations affecting choice. In the United Kingdom, the British Standards Institution (BSI) has traditionally conducted assessment of equipment and produced British Standards. It should be noted, however, that over the next few years many of the existing British Standards will be replaced so as to ensure harmonization with European Standards or 'Norms' (ENs).

As a consequence of the Personal Protective Equipment Regulations 2002 [10], which came into force in May 2002, the implementation of the Personal Protective Equipment Product Directive (Council Directive 89/686/EEC) has been updated and extended. In effect, this requires that all PPE supplied for use at work must be independently assessed so as to ensure that it meets 'basic safety requirements'. Satisfactory testing results in a 'certificate of compliance', and entitles the manufacturer to display the 'CE' mark on its product. The other effect of these regulations is to make it illegal for any supplier to sell PPE unless it carries the 'CE' mark.

Acceptable level for exposure to danger

This factor is extremely important. Quite clearly for some dangers, such as exposure to potentially carcinogenic substances, the only acceptable level can be zero. Both the employer and the enforcing officer must have a sound working knowledge of fixed legal standards, such as 'hygiene limits', and also of those standards of protection that are merely advisory in nature.

These suggested considerations must be viewed in conjunction with the general requirement to engage in a risk assessment contained in the Management of Health and Safety at Work Regulations 1999 [6], and the more specific requirements with regard to PPE contained in the Personal Protective Equipment at Work Regulations 1992 [1].

SELECTION OF PERSONAL PROTECTIVE EQUIPMENT

The general discussion of the Personal Protective Equipment at Work Regulations 1992 above has highlighted the need for any PPE chosen to be both 'suitable' and capable of protecting against the risks to health and safety identified. Part 2 of the guidance [2] lays down some general principles about the way equipment should be chosen.

In addition, reference should be made to the information available from suppliers. Again, it should be remembered that all PPE must carry the 'CE' mark [10,11].

Types of personal protection

PPE may be divided into the following broad categories:

- hearing protection
- respiratory protection
- eye and face protection
- protective clothing
- skin protection.

Hearing protection

Hearing protection can be divided into two main types: (a) earplugs and (b) earmuffs. Earplugs are designed to be inserted into the ear canal. Those that are designed to be disposed of after use are usually made from either mineral down, which is an extremely fine glass down, or from polyurethane foam. Reusable earplugs are made of soft rubber or plastic. They must be thoroughly washed after use. Reusable plugs should be fitted in the first instance by a trained person who should provide advice to the wearer about the correct method of inserting the plugs. A British Standard [12] gives an appropriate specification for earplugs.

Earmuffs are designed to cover the external ear. They consist of rigid cups that fit over the ears and are sealed to the head with soft cushion seals. They have several advantages over earplugs. One size will usually fit a wide range of people, they tend to offer greater protection, and they are easy to remove and replace. However, they are not without their disadvantages. They have a tendency to make the ears hot and are rather bulky. A British Standard [13] deals with the required specification for earmuffs. It is imperative that any type of hearing protection is of sufficient quality to reduce the noise level at the wearer's ear to below any recommended limit. The use of data from octave band analysis should be compared with design data provided by the manufacturer in order to ensure maximum protection. In 1993, the BSI produced a standard method for the measurement of sound attenuation of hearing protectors [14].

The Noise at Work Regulations 1989 [4] lay down a requirement for employers to provide hearing protection in certain circumstances (Regulation 8). The advent of the regulations resulted in the HSE producing a useful practical guide, Reducing Noise at Work [15] which specifies five criteria for the selection of hearing protection. It states that attention should be paid to:

- the level and nature of the noise exposure;
- the job and working environment;
- compatibility with any other protective equipment or special clothing worn;
- the fit to the wearer;
- any difficulty or discomfort experienced by the wearer.

Respiratory protection

As has already been emphasized, personal protection is a form of last resort protection. In the case of respirable dusts and fumes, every effort must be made to try and enclose the process and provide exhaust ventilation. Where this is not possible, suitable protection must be provided.

Respiratory protective equipment may be divided into two broad categories: respirators that purify the air by drawing it through a filter, thereby removing the contamination, and breathing apparatus, which supplies clean air from an uncontaminated source.

Respirators

There are five basic types of respirators.

1. **Dust respirators** are the most commonly available form of respirators. They afford protection against solid particles of matter or aerosol sprays, but not against gases. They generally cover the nose and mouth of the wearer. BS 4275 [16] provides a specification for developing and implementing an effective programme for the use of respiratory protective equipment.

2. As well as dust respirators, there are light, simple **face masks,** which protect the wearer from the effect of nuisance dusts or non-toxic sprays. It should be remembered that while these are a very popular form of respirator, they do not offer the wearer any real protection.

3. **Cartridge type respirators** give protection against low concentrations of relatively non-toxic gases and vapours. This is achieved by the use of replacement filter cartridges. Care should be taken to change the cartridge regularly to ensure maximum protection.

4. **Canister type respirators** incorporate a full facepiece connected to a replaceable filter canister. This type of respirator offers far more protection than the cartridge type respirator. It is important that the correct canister is fitted to the equipment to ensure maximum protection for the wearer.

5. **Positive pressure-powered respirators** can either cover the nose and mouth or the whole of the face. Air is drawn in through a battery-powered suction unit, through filters and fed to the facepiece at a controlled flow. The excess of air escaping around the edges of the facepiece prevents leakage inwards. This type of protection is mainly used when working with disease-producing dusts, for example, asbestos.

Breathing apparatus

The choice of breathing apparatus is extremely complex and should be made by those well versed in the use and limitations of such equipment. BS 4275 [16] also gives advice on the selection, use and maintenance of such equipment. There are three basic types of breathing apparatus.

1. **Closed circuit systems,** which supply either oxygen or air from a cylinder carried by the wearer. BS EN 400 [17] gives a specification for this type of apparatus. The air is supplied via a demand valve. The system gets its name from the fact that the wearer breathes the same air over and over again. When the wearer exhales, the exhaled air is purified and passes into a breathing bag where it is enriched with fresh oxygen from the cylinder. This results in the apparatus being suitable for use over a longer period of time.

2. **Open circuit systems** provide compressed air or oxygen from a cylinder worn by the worker. However, there is no breathing bag, hence the wearing time is greatly reduced [18].

3. **Airline breathing apparatus** has a full facepiece connected to a source of uncontaminated air by a hose. The equipment allows the operative to work in most types of toxic atmospheres, but the trailing tube can limit the movement of the wearer [19].

It should be emphasized that although breathing apparatus provides the most effective protection against risks, because of its complexity its use requires specialized training and supervision. Detailed guidance has been published by the HSE [20] on the selection and use of respiratory protective equipment.

Eye and face protection

Eye injuries are extremely common. The main hazards are solid particles, dust, chemical splashes, molten metal, glare, radiation and laser beams. Injuries often result in severe pain and discomfort and, in many cases, long-term impairment of vision. It is extremely important that hazards are fully examined before any form of eye protection is chosen.

Industrial eye protectors, such as goggles, visors, spectacles and face screens, are required to satisfy the minimum requirements of BS EN 166 [21]. BS 7028 gives advice on the selection, use and maintenance of eye protection [22]. In addition, part 2 of the HSE's guidance [2] gives some advice on the choice of eye protection.

Protective clothing

There is currently a wide range of protective clothing available for use in most industrial situations. This ranges from well-known items such as safety footwear and overalls, to the more specialized gloves and aprons worn in certain work situations.

There is a whole range of British Standards and proposed European Norms dealing with the full range of protective clothing. Once again, the HSE's guidance [2] is useful in appreciating the criteria to be employed in selecting protective clothing.

It is important, as pointed out above, that those who wear protective clothing should be fully aware of the limitations of its use.

Skin protection

In some circumstances, it may not be possible or desirable to use gloves. In such situations, a proprietary barrier cream may be an alternative. Hartley [23] divides skin protection preparations into three broad groups.

1. **Water miscible** – protects against organic solvents, mineral oils and greases, but not metal-working oils mixed with water.
2. **Water repellent** – protects against aqueous solutions, acids, alkalis, salts, oils and cooling agents that contain water.
3. **Special group** – cannot be assigned to a group by their composition. They are formulated for specific applications.

It should be remembered that these creams are of only limited use as they are rapidly removed during the working day.

REFERENCES

1. The Personal Protection at Work Regulations 1992. S.I. 1992. No. 2966, HMSO, London.
2. Health and Safety Executive (1992) *Personal Protective Equipment at Work, Guidance on Regulations L25*, HSE Books, Sudbury, Suffolk.
3. Control of Substances Hazardous to Health Regulations 2002. S.I. 2002. No. 2677.
4. Noise at Work Regulations 1989. S.I. 1989. No. 1790, HMSO, London.
5. Control of Asbestos at Work Regulations 2002. S.I. No. 2675.
6. Management of Health and Safety at Work Regulations 1999. S.I. 1999. No. 3242. HMSO, London.
7. Chartered Institute of Environmental Health (1991) *Acquired Immune Deficiency Syndrome: Guidance Notes for Environmental Health Officers*, CIEH, London.
8. Coates, T. and Clarke, A.R.L. (2003) Occupational diseases, in *Risk Management – Safety at Work* (ed. Ridley J. & Channing J.), 6th edn, Butterworths, London.
9. Else, D. (1981) *Occupational Health Practice*, 2nd edn, Butterworths, London.
10. The Personal Protective Equipment Regulations 2002. S.I. 2002. No. 1144, HMSO, London.
11. Health and Safety Executive (1995) *A Short Guide to the Personal Protective Equipment at Work Regulations 1992, ING(G)174L*, HSE Books, Sudbury, Suffolk.
12. BS EN 352–2 (2002) *Hearing Protectors, Ear Plugs*, British Standards Institution, London.
13. BS EN 352–1 (2002) *Hearing Protectors, Ear Muffs*, British Standards Institution, London.
14. BS EN 24869–1 (1993) *Acoustic Hearing Protectors. Sound Attenuation of Hearing Protectors*, British Standards Institution, London.
15. Reducing Noise at Work (1998) L108, HSE Books, Sudbury, Suffolk.
16. BS 4275 (1997) *Guide to Implementing an Effective Respiratory Protective Device Programme*, British Standards Institution, London.
17. BS EN 400 (1993) *Respiratory Protective Devices for Self-rescue – Self-contained Closed Circuit Breathing Apparatus* BSI, London.
18. BS EN 1146 (1997) *Respiratory Protective Devices for Self-rescue – Self-contained Open Circuit Systems*, British Standards Institution, London.
19. BS EN 12419 (1999) *Respiratory Protective Devices – Airline Breathing Apparatus*, British Standards Institution, London.
20. Health and Safety Executive (1998) Selection, Use and maintenance of *Respiratory Protective Equipment*, HS(G)53, HSE Books, Sudbury, Suffolk.
21. BS EN 166 (2002) *Personal Eye Protection: Specifications*, British Standards Institution, London.
22. BS 7028 (1999) *Guidance for the Selection, Use and Maintenance of Eye Protection for Industrial and Other Uses*, British Standards Institution, London.
23. Hartley, C. (1999) Occupational Hygiene, in *Risk Management – Safety at Work* (ed. J. Ridley), 4th edn, Butterworths, London.

22 Plant and systems of work

Paul C. Belcher and John D. Wildsmith

DANGEROUS MACHINES

Many serious accidents at work involve some type of machinery and most machinery must therefore be regarded as being intrinsically dangerous. The Health and Safety Executive (HSE) has published a helpful guide that summarizes many of the key points that must be considered by employers, employees and enforcement officers [1].

TYPES OF INJURY

Injuries may result from any of the following:

1. (a) Trapping between moving parts and fixed parts, for example, guillotine blades, garment pressers and sliding tables.
   (b) Trapping between moving parts, for example, rollers, cogs, drive belts, food beaters and mixers. This category includes trapping by entanglement of hair or clothing, resulting in part of the body being brought into contact with the dangerous machine part.

Common sources of injuries of this type are 'nips' and 'pinches'. Of particular concern are 'in-running nips'. Examples of this form of trapping and other hazards posed by a wide variety of dangerous machinery can be found in BS EN 292-1 [2].

2. Contact with moving parts, for example, cutting blades, abrasive wheels, gear wheels – even slowly rotating parts must be considered to be potentially dangerous. This is particularly true in the case of gears or beaters such as those found in food processing machinery.
3. Burns, for example, from hot exhausts and deep fat fryers.
4. Electrocution, for example, from exposed electrical conductors.
5. Contact with moving workpieces, for example, lathe.
6. Striking by machine parts or workpieces thrown from the machine, for example, fractured abrasive wheels and non-secure cutter blades.

Other non-mechanical hazards and environmental considerations are dealt with in the following chapters.

PRINCIPLES OF MACHINE GUARDING

The principles of guarding machinery were established as a result of the introduction of the Factories Act 1961 [3]. The established standards were supported by a considerable body of case law. More recently, the Provision and Use of Work Equipment Regulations 1998 (PUWER) [4] (see the following section) have amplified the requirements for the guarding of equipment and applied them across all industrial, commercial and service sectors.

THE PROVISION AND USE OF WORK EQUIPMENT REGULATIONS 1998

These regulations [4] implement European Directive 89/655/EEC (minimum health and safety requirements for the use of work equipment by workers at work).

The regulations place duties on employers and others, including certain people who control work equipment, people who use, supervise or manage its use or how it is used, to the limit of their control.

The regulations address the following aspects of work equipment:

- suitability of work equipment
- maintenance
- inspection
- specific risks
- information and instructions
- training
- conformity with European Union (EU) requirements
- dangerous parts of machinery
- protection against specified hazards
- high or very low temperatures
- controls for starting or making a significant change in operating conditions
- stop controls
- emergency stop controls
- controls
- control systems
- isolation from sources of energy
- stability
- lighting
- maintenance operations
- markings
- warnings.

When selecting a control system, employers must ensure that they consider the failures and faults that might be expected to occur during its normal use. Where the safety of work equipment depends on its installation, employers must also ensure that it is inspected by a competent person after each installation, prior to being put into service for the first time, or after assembly at a new site/location. The results of such inspection must be recorded and kept until the next inspection. If any work equipment is used outside the business site, records of the last inspection must accompany it.

The regulations also cover mobile work equipment such as tractors and lift trucks (see also p. 454). The provisions relate to:

- employees carried on mobile work equipment
- rolling over of mobile work equipment
- overturning of fork-lift trucks
- self-propelled work equipment
- remote-controlled self-propelled work equipment
- drive shafts.

THE LIFTING OPERATIONS AND LIFTING EQUIPMENT REGULATIONS 1998

These regulations [5] lay down the health and safety requirements for lifting equipment. The regulations apply to employers and the self-employed but not to suppliers of lifting equipment. Lifting equipment is defined as: 'work equipment for lifting or lowering loads, including any attachments used for anchoring, fixing or supporting it'.

Employers must ensure that:

- lifting equipment is of suitable and sufficient strength and stability for each load, having consideration to the stress placed on mountings and fixing points;
- all parts of a load, anything attached to it and anything used in lifting is of adequate strength;
- lifting equipment for lifting people will prevent passengers from being crushed, trapped or struck, or from falling from the carrier;
- lifting equipment for lifting people, so far as is reasonably practicable, will prevent anyone who is using it to undertake activities from the carrier from being crushed, trapped or struck, or from falling from the carrier;
- lifting equipment for lifting people has suitable devices to prevent the carrier falling; if these risks cannot be prevented, the carrier must have an enhanced safety coefficient suspension rope or chain, which should be inspected daily by a competent person; if anyone becomes

trapped in a carrier, he should not be exposed to danger and should be able to be freed;

- lifting equipment must be positioned/installed so as to minimize the risk of the equipment/load striking a person, or of the load drifting, falling freely or being released unintentionally; there must also be suitable devices for preventing people from falling down shafts/hoist ways;
- machinery and accessories must be clearly marked to indicate their safe working loads (SWLs); separate SWLs must be given for different configurations; accessories must be marked so that the characteristics needed for their safe use may be identified; lifting equipment for lifting people must be clearly and correctly marked, and equipment that is not meant for lifting people but that could be used as such should be clearly marked to the effect that it should not be used for such a purpose;
- all lifting operations involving lifting equipment are properly planned, supervised and undertaken in a safe manner;
- before any lifting equipment is used for the first time it must be thoroughly examined; where safety depends on its installation, lifting equipment must be thoroughly examined after installation and prior to first use, after assembly and prior to use at a new site/location;
- lifting equipment and accessories used for lifting people are thoroughly examined at least every 6 months, and other lifting equipment at least every 12 months; it must also be examined when an incident has occurred that may adversely affect the safety of the equipment;
- records of the last thorough examination must accompany lifting equipment used outside the business.

The regulations outline the requirements regarding the making and keeping of reports and exemptions for the armed forces.

MACHINERY PROTECTION

There are three basic principles of machinery protection that should be taken into consideration when applying occupational safety legislation.

Intrinsic safety

Intrinsically safe machinery is that which is safe without any further additions or alterations. This means, for example, that when dangerous parts are in motion, or in use in some other way, they are not accessible to the operator or any other person.

This principle cannot always be applied. However, it is essential that potentially dangerous machinery should be made intrinsically safe wherever possible by virtue of design. This is an important aspect of BS EN 292-1 [2].

Safety by position

Great care should be taken concerning this aspect of machine safety. A machine should not be assumed to be safe because of its position; for example, just because it is only approached infrequently as it is 'out of the way'. In such cases, a judgement must be made of the particular circumstances involved, bearing in mind the requirements of the Health and Safety at Work, Etc. Act 1974.

Fencing or guarding of machinery

Ideally, guards should be designed as an integral part of the machine and not added on as an afterthought. Essentially, fixed guards must be robust, able to withstand severe treatment, adjustable and safe to use. They should also be capable of protecting operators or people in the vicinity against injury. This may mean that as well as forming a physical barrier, the guard may have additional functions, such as assisting the removal of toxic fumes or reducing noise to a safe level.

Many different types of guard are available but, in general, there is an order of preference, which is given in the following.

Fixed guard

This type of guard is usually preferable, as it is securely attached to the machine and permanently guards the dangerous part or parts. The guard should be sufficiently large to stop anyone reaching over it, and it should be sufficiently distant from the dangerous parts so that it does not form a trapping zone between itself and any moving parts.

This type of guard is not suitable in many circumstances, for example, where workpieces have to be inserted and removed regularly, or where regular maintenance is necessary.

It is often designed with a necessary opening in it, for example, for a workpiece to contact a cutting blade. In such circumstances, the opening must be as limited as possible. In some circumstances, the operator is removed from the dangerous part by the incorporation of a sleeve or tunnel guard. This topic is dealt with in greater detail in BS EN 292-1 [2].

When examining fixed guards, a number of points should be considered. The guard must be of sound and substantial construction, in good repair and trapping or entanglement should not be possible. Any openings should be acceptable only if the trapping area is not accessible through them.

Interlocked guard

There are two types of interlocked guard.

1. *Electrical interlock.* This consists of an electrical switch mechanism, which can be installed in a variety of ways. In one form, when the guard is moved, to give access to the dangerous parts, then the power to the machine is cut off, thus rendering the machine relatively safe. This is not acceptable if a dangerous part such as a cutting blade can still be contacted. By corollary, if such a guard is open, then the machine cannot be started up.

   It is important that if such systems fail, then they 'fail to safety', that is, the electrical interlock will ensure that the machine remains incapable of being operated until the interlock is replaced or repaired.
2. *Mechanical interlock.* This is a moveable guard where, when it is open, the machine cannot be operated, and when the machine is in motion, it is not possible to open the guard. An interlock guard should satisfy the tests for a fixed guard. In addition, the extent to which the guard can be opened can be examined. Also, as the guard is closed, the operation of any limit switches should be noted, that is, the extent to which the guard is in its correct

position, before the machine will operate. Because it is a moveable guard, signs of wear should be looked for.

Automatic guard

This type of guard is one where the dangerous parts are enclosed while the machine is prepared for operation. The guard may open to an extent where the work can be removed when a work cycle is completed. When the work is removed, the guard closes again. Normally, the guard is closely fitted around the workpiece.

Examination of this type of guard should extend to the tests for fixed guards. In addition, when the moving part of the guard is closed; it should not be possible to reach the dangerous parts. Finally, the extent of wear of any of the guard components should be assessed.

Trip guard

These may be moveable mechanical guards, which are operated when a person approaches a dangerous machine beyond a safe distance. A variation on this is where the guard incorporates either photoelectric cells, where a light beam can be disrupted, or an induction field detection coil, where the presence of a person will alter an electrical field. In both cases, when the sensor device is triggered the machine can be shut off, to render it safe. Trip guards must satisfy the tests outlined here for the other guards.

It should be remembered that machine safety does not stop with the presence of a guard. Any guards must be made from suitable materials, and must be maintained regularly. If maintenance can only be carried out by removing guards, consideration must be given to providing some other form of protection to ensure a safe system of work.

All machine controls must be located in such a position that they cannot be operated unintentionally. Also, emergency cut-off switches should be clearly identified, readily accessible and in working order. People operating machines should understand any risks involved, and be capable of rendering the machine safe in any emergency situation.

A common source of potentially dangerous machines is commercial kitchens. The hazards associated with a variety of machines, including food slicers, food processors, planetary mixers and waste-compactors, pie- and tart-making machines, incorporating some of the devices referred to here, can be found in two HSE publications [6,7].

Lift trucks

The legislative controls relating to these are now contained in the Lifting Operations and Lifting Equipment Regulations 1998 [5].

Forklift trucks represent a special type of mobile machine that has become commonplace in factories, warehouses and large shops. They are especially useful for moving and storing goods, but they do feature predominantly in accident statistics. Some injuries involving lift trucks are fatal. The key aspects involved in the safe use of these machines are dealt with in an HSE publication [8].

Frequently, forklift truck accidents are associated with a lack of suitable operator training. Therefore, the need for employers to provide suitable training in this field should be recognized in order to comply with their legal duties under the Health and Safety at Work, Etc. Act 1974.

It is important to appreciate the characteristics of forklift trucks, so that the hazards that arise during their operation can be understood. It is also important to ensure that the racks used to store and unload many of the materials where lift trucks are employed are capable of carrying imposed loads, are in a good state of repair and are fixed securely in position.

Forklift trucks are designed to lift relatively heavy loads, most commonly at the front of the vehicle, although there are some side-loading lift trucks. If the load is too heavy, the truck can be tipped over. Also, if the vehicle is unevenly loaded, driven on sloping or uneven ground or cornered at excessive speed, the stability of the vehicle can be affected, resulting in either shedding of the load or turning over of the truck.

The truck is more likely to turn over if the load is raised to a high level, as the centre of gravity of the vehicle is raised.

The construction and use of forklift trucks is controlled by a variety of British Standard specifications. Operators and inspectors should be familiar with the information that is required to be displayed on forklift trucks, which is as follows:

1. the manufacturer's name
2. the type of truck
3. the serial number
4. the unladen weight
5. the lifting capacity
6. the load centre distance
7. the maximum lift height.

These details must be recorded when investigating any accident or dangerous occurrence involving forklift trucks.

Lift trucks are mobile machines, powered either by electric batteries or internal combustion engines. There are risks associated with each of these types. Electric batteries have to be recharged, and this process involves the liberation of hydrogen gas. Battery charging should therefore only be carried out in a well-ventilated area where smoking is prohibited and sources of ignition are adequately protected or removed.

In buildings where forklift trucks are driven by internal combustion engines, effective ventilation is necessary in order to remove exhaust gases, which are toxic. In addition, areas used for refilling vehicles with petrol or diesel fuel should be located in the open air. Liquefied petroleum gas (LPG) cylinders should also be changed in a well-ventilated area. There must be no ignition sources in the vicinity of refuelling areas.

Because of the risk of causing an explosion, forklift trucks should not be used in premises where flammable vapours, gases or dusts are likely to be present, unless they have been specially protected for such use.

Operator training

Because of the extent and nature of accidents involving forklift trucks, the HSE has published an Approved Code of Practice [9]. This document essentially provides practical guidance for employers with regard to their duties under Section 2 of the

Health and Safety at Work, Etc. Act 1974: 'to provide such information, instruction, training and supervision as is necessary to ensure, so far as is reasonably practicable, the health and safety at work of his employees'.

One important point of this Approved Code of Practice is that operator training must only be carried out by instructors who have themselves undergone appropriate training in instructional techniques and skills assessment.

The training provided should be largely practical in nature, and should be provided 'off-the-job', so that trainees and instructors are not diverted by other considerations.

Testing of trainees should be carried out by continuous assessment as well as by a test or tests of the skills and knowledge required for safe lift-truck operation. The employer should keep records of each employee who has completed the basic training and testing procedure. In the case of an accident, the availability of this type of record could be helpful to the outcome of the investigation.

LEGIONNAIRES' DISEASE

Legionnaires' disease (see also p. 321) is a pneumonia-like condition caused by the bacterium *Legionella pneumophila*. The symptoms of this disease are similar to those of influenza: headache, nausea, vomiting, aching muscles and cough [10].

Legionnaires' disease was first identified after an outbreak of pneumonia among delegates attending an American Legion Convention for service veterans in Philadelphia in 1976, hence the name. Diagnostic tests were developed, and by testing stored specimens it was discovered that the disease could be traced back to the 1940s. The infection had escaped detection because the causative organism did not grow on conventional culture media.

Infection is caused by inhaling droplets that contain viable *L. pneumophila* organisms and are fine enough to penetrate deeply into the lungs. There is no evidence that the disease can be transmitted from person to person, nor is the dose required to infect a person known. Males are more likely to be affected than females, and most cases have occurred in the 40–70-year-old age group.

During the past few years, there has been a growing awareness of this condition, and some highly publicized cases have heightened public awareness. In England and Wales, about 200–250 cases of Legionnaires' disease have been reported each year, of which between 10% and 20% have been fatal.

This infection has caused much concern because of its association with hot water systems and the cooling towers of air conditioning and industrial cooling systems, where there is a generation of fine water droplets. The organism may therefore be present in a large number of workplaces and any risk of infection is clearly within an employer's duties to employees and non-employees under Sections 2 and 3 of the Health and Safety at Work, Etc. Act 1974.

It is therefore important to ensure that any high-risk areas in workplaces, such as cooling towers, are inspected on a regular basis to ensure that they present no risk to health and safety as far as is reasonably practicable.

The procedures for the identification and assessment of risk of legionnellosis from work activities and water sources as well as the measures to be taken to reduce the risk from exposure to the organism are explained in the Approved Code of Practice [10].

Procedures for the sampling of water services in all types of building, including domestic, commercial and industrial premises, are provided in BS 7592 [11]. This includes sampling a variety of water services such as hot water services, coldwater services, cooling systems and associated equipment. It should be noted that the British Standard is not applicable for sampling potable-water services or vending machines. Research work in this field continues, and it should be noted that in 1998 the British Standards Institution produced a culture method for the isolation of *Legionella* organisms and estimation of their numbers in water and related materials [12].

Where a potential risk exists, it is essential that cooling towers and associated water systems are subject to routine sampling as part of the general water-treatment programme. It is suggested that records of the analytical laboratory employed, the source of the sample and the results obtained should be kept readily available for inspection.

Many systems are colonized by *Legionella* bacteria without being associated with infections. The risk of infection can be minimized by good engineering practice in the design, construction, operation and maintenance of water installations.

In the case of systems that are liable to produce a spray or aerosol in operation, or where a spray could be generated incidentally during the cleaning or maintenance, additional precautions should be taken by the person who has responsibility for the plant under the Health and Safety at Work, Etc. Act 1974.

Cooling towers are of particular concern, and it is important that all parts of the system are thoroughly cleaned and disinfected, normally twice per year (in spring and autumn). In order to achieve adequate disinfection, sodium hypochlorite solution may be used to give a concentration of at least five parts per million (ppm) of free available chlorine. Water should be sampled periodically near the circulation return point. Chlorine levels can be determined using swimming pool testing kits (see Chapter 20).

Regular chlorination of 1–2 ppm is normally advocated, but this is not appropriate in all cases as it may react adversely with other water treatments. Some biocides are known to be effective against *Legionella* bacteria, although resistance has been known to develop. In such cases, shock dosing using chlorination of 25–50 ppm of free chlorine is effective. A level of at least 10 ppm should be maintained in the system for 12–24 hours. The use of biocides can be alternated with the use of chlorination to achieve a satisfactory level of disinfection.

In view of the risk of legionnellosis, the Notification of Cooling Towers and Evaporative Condensers Regulations 1992 [13] requires any person who has, to any extent, control of non-domestic premises to notify the local authority that such plant is situated on those premises, giving information set out in the schedule to the regulations. This assists environmental health practitioners in ensuring that appropriate, effective treatment and control systems are in place, as well as being vital in the case of any outbreak.

It is important that a safe system of work is devised in all premises where cleaning, maintenance and testing of the type of plant just discussed is carried out. This system of work, as well as any instruction and training necessary, must be made known to the personnel involved. For example, the HSE recommends that suitable respiratory equipment, such as a high-efficiency positive-pressure respiratory with either a full facepiece or a hood and blouse, be used. Also, such operations must be carried out in such a way as to avoid risks to other people working in the vicinity or members of the public.

In addition to its occurrence in cooling systems and water services, it should also be noted that spa baths (see p. 260) have been linked with various infections, including Legionnaires' disease. Spa baths are small pools where water is vigorously circulated by means of water and/or air jets. The water is usually quite warm (typically above 30°C), and is not changed between bathers as it is in the case of whirlpool baths. Careful water treatment is therefore needed for spa baths.

If an outbreak of Legionnaires' disease is suspected, past cases are a valuable source of reference. One useful publication is the *Broadcasting House Legionnaires' Disease*, which charts the investigation carried out by Westminster City Council in 1988 [14].

ELECTRICAL SAFETY

Every year about 50 people are killed by electric shocks in the workplace. In addition, poor electrical standards often lead to the occurrence of fires, for example, as a result of overheating electrical conductors. Also, electrical installations and appliances can act as ignition sources and lead to explosions. It must be noted that in flammable atmospheres, static electricity from movement of plant, materials or even clothing can be a great potential hazard.

It should also be noted that, with prompt action, employees with a basic training in first aid can save the life of a person who has received an electric shock. The HSE publishes a poster containing appropriate guidance for dealing with cases of electric shock [15]. This can be displayed in potentially dangerous areas and brought to the attention of employees.

In order to ensure that precautions are taken to prevent injury or fatality, the Electricity at Work Regulations 1989 [16] have been made under the Health and Safety at Work, Etc. Act 1974. In order to assist people with duties under the Act and regulations, the HSE has published guidance on them [17]. Although intended to assist engineers, technicians and managers to understand the nature of the precautions to be taken, the guidance is also of great use to those enforcing the provisions.

When dealing with electrical installations, reference is frequently made to the Institution of Electrical Engineers Regulations for Electrical Installations (IEE Regulations). These are non-statutory regulations evolved by the IEE that relate to the design, selection, erection, inspection and testing of electrical installations. They form a code of practice that has been widely recognized and accepted in the United Kingdom, and compliance with them is likely to achieve compliance with the statutory regulations. A useful commentary has been produced by Scadden (2002) [18].

In addition to the guidance memorandum [17], the Health and Safety Commission (HSC) and the HSE have published a number of guidance documents on various aspects of the supply of electricity and specific items of plant and equipment.

While carrying out an inspection of premises, the electrical installation and the electrical supply to both fixed and portable machines should be examined. The inspector may be able to recognize simple visual faults that may indicate that the electrical system requires examination by a competent electrician. It is stressed that such an examination must be carried out without risk to other employees or non-employees. Depending upon the circumstances, the employer may be required to provide an electrician's report, or the inspector may engage their own electrician (who must be properly authorized to carry out investigations on the premises).

Consideration must be given to ascertaining whether or not a particular business involves the use of portable electrical equipment. If such equipment is used, then its condition should be ascertained and enquiries made into how often the equipment is tested and the standards used. Any records kept should be examined. Where portable equipment is used in an external environment, then an isolating transformer or residual current device should be used to protect operators and others in case of an emergency. Some equipment can run on lower voltage supplies than mains voltage. In the case of some parts of electrical installations, for example, display lighting, supplies can be as low as 12 volts.

Some basic points that should be considered are listed in the following.

Flexible leads and cables

The correctly rated 'flex' or cable should be used for the application. Flexes and leads should be positioned and fixed correctly, and be in good repair. Common faults are incorrect fixing to plugs or equipment, and damaged insulation, for example, cuts, abrasions or heat damage. Joints should always be made using a suitable connector or coupler – joints made with insulating tape or 'block connectors' are normally dangerous. Care must be taken to ensure that cables are joined up with the correct polarities in each half of the connector.

Electrical socket outlets

Sockets should be sufficient in number and they must be positioned correctly to avoid tripping hazards from trailing flexes and overloading of the circuit due to the use of multipoint adapters.

Plugs

The flexible lead must be held firmly in the grip or clamp incorporated into the plug, and the individual wires must be connected firmly to the correct plug terminals. A suitably rated fuse must be installed. Some tools and lamps are 'double insulated', and these have only two wires in the flex. Such appliances are recognized by the box-in-box symbol marked on them.

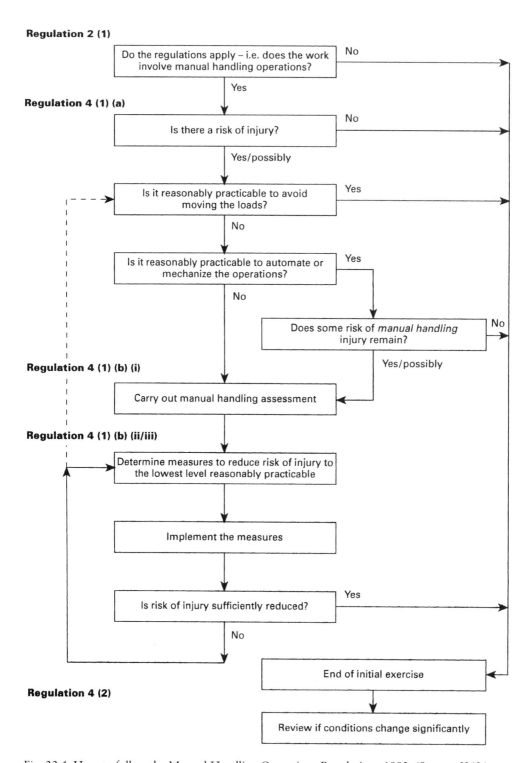

Regulation 2 (1)

Do the regulations apply – i.e. does the work involve manual handling operations? No

Yes

Regulation 4 (1) (a)

Is there a risk of injury? No

Yes/possibly

Is it reasonably practicable to avoid moving the loads? Yes

No

Is it reasonably practicable to automate or mechanize the operations? Yes

No

Does some risk of *manual handling* injury remain? No

Yes/possibly

Regulation 4 (1) (b) (i)

Carry out manual handling assessment

Regulation 4 (1) (b) (ii/iii)

Determine measures to reduce risk of injury to the lowest level reasonably practicable

Implement the measures

Is risk of injury sufficiently reduced? Yes

No

End of initial exercise

Regulation 4 (2)

Review if conditions change significantly

Fig. 22.1 How to follow the Manual Handling Operations Regulations 1992. (Source: [21].)

Switches

Every fixed machine must have a switch or isolator beside the machine that can switch off the power in an emergency. Power cables to machines must be armoured, heavily insulated or installed in a conduit.

Explosive atmospheres

Where the atmosphere of a workstation is either dusty or potentially flammable, then specially protected equipment must be used. Inspectors must consider this risk, for example, when they are taking monitoring equipment into premises. The HSE has produced a range of guidance documents for those working with electrical devices in explosive atmospheres.

MANUAL HANDLING

The Manual Handling Operations Regulations 1992 [19] came into operation as a result of the EU legislation mentioned earlier. In the guidance on the regulations issued by the HSE [20], it is claimed that more than one-quarter of the accidents reported to the enforcing authorities each year are associated with manual handling (i.e. the transporting or supporting of loads by hand or bodily force). Although manual handling accidents are rarely fatal, the majority are major injuries such as fractured arms.

Sprains and strains arise from the incorrect application or prolongation of force. Poor posture and excessive repetition of movement can be important factors in the development of such injuries. The main sites of injuries are the back (45%), finger and thumb (16%) and arm (13%). It should be noted that the injuries may occur over a considerable period of time.

The Regulations establish a clear hierarchy of measures which, in the first case, requires that employers should, as far as reasonably practicable, avoid the need for employees to undertake any manual handling operations at work that involve a risk of injury. This may require the redesigning, automation or mechanization of tasks. In order to comply with this requirement, a 'general assessment' carried out under the Management of Health and Safety at Work Regulations 1999 [21] must be carried out. However, under these regulations there is an additional, specific duty to carry out risk assessments of all manual handling operations where it is not reasonably practicable to avoid them.

Also, where manual handling operations cannot be avoided, the employer must take appropriate steps to reduce the risk of injury to the lowest level reasonably practicable.

The flowchart in Fig. 22.1 illustrates how to follow the regulations.

REFERENCES

1. Health and Safety Executive (1999) *Essentials of Health and Safety at Work*, Revised edn, HSE Books, Sudbury, Suffolk.
2. BS EN 292-1 (1991) *Safety of Machinery*, British Standards Institution, London.
3. *The Factories Act* (1961) 9 & 10 Elizabeth 2, c34, HMSO, London.
4. *The Provision and Use of Work Equipment Regulations 1998*, SI 1998, No. 2306, HMSO, London.
5. *The Lifting Operations and Lifting Equipment Regulations 1998*, SI 1998, No. 2307.
6. Health and Safety Executive (1990) *Health and Safety in Kitchens and Food Preparation Areas*, HS(G)55, HSE Books, Sudbury, Suffolk.
7. Health and Safety Executive (1986) *Pie and Tart Machines*, HS(G)31, HSE Books, Sudbury, Suffolk.
8. Health and Safety Executive (2002) *Safety in Working with Lift Trucks*, HSE Books, Sudbury, Suffolk.
9. Health and Safety Executive (1999) *Rider Operated Lift Trucks – Operator Training. Approved Code of Practice and Guidance*, HSE Books, Sudbury, Suffolk.
10. Health and Safety Commission (2000) *Legionnaire's Disease: The Control of Legionella bacteria in Water Systems*.

Approved Code of Practice and Guidance, HSE Books, Sudbury, Suffolk.

11. BS 7592 (1992) *Methods for Sampling for Legionella Organisms in Water and Related Materials*, British Standards Institution, London.

12. BS 6068-4.12 (1998) *Method for Detection and Enumeration of Legionella Organisms in Water and Related Materials*, British Standards Institution, London.

13. *Notification of Cooling Towers and Evaporative Condensers Regulations* (1992) SI No. 2225, HMSO, London.

14. Westminster City Council (1988) *Broadcasting House Legionnaire's Disease*, WCC, London.

15. Health and Safety Executive (2000) *Electric Shock – First Aid Guidance*, HSE Books, Sudbury, Suffolk.

16. *Electricity at Work Regulations* (1989) SI No. 635, HMSO, London.

17. Health and Safety Executive (1998) *Memorandum of Guidance on the Electricity at Work Regulations 1989*, HS(R)25, HSE Books, Sudbury, Suffolk.

18. Scadden, B. (2002) *Sixteenth Edition IEEE Wiring Regulations Explained and Illustrated*, 6th edn, Newnes, London.

19. *Manual Handling Operations Regulations* (1992) SI No. 2793, HMSO, London.

20. Health and Safety Executive (1992) *Manual Handling – Guidance on Regulations*, L23, HSE Books, Sudbury, Suffolk.

21. *The Management of Health and Safety at Work Regulations 1999*, SI 1999, No. 3242.

23 Toxic and dangerous substances

Paul C. Belcher and John D. Wildsmith

CLASSIFICATION OF HAZARDOUS SUBSTANCES

Many of the substances used in the workplace are obviously dangerous, but there are many more that are not obviously hazardous. These hazards may be biological, physical or chemical. Control of the hazards posed by dangerous substances has been developed to comply with the provisions of a number of European Union (EU) directives.

These directives are now implemented by the Chemicals (Hazard Information and Packaging for Supply) Regulations 2002 [1] (known as CHIP3), which impose requirements upon the suppliers of 'substances or preparations which are deemed to be dangerous for supply and carriage' with the exception of substances defined in Regulation 3.

The CHIP regulations provide a system designed to protect people (and the environment) from the ill effects of chemicals. Fundamentally, they require the suppliers and consigners of chemicals to label them clearly and package them safely, but in addition they require an initial 'classification' of the substances and the provision of safety data sheets for dangerous chemicals.

The fundamental requirement of CHIP3 is for a supplier to make an initial decision about whether the chemicals involved are dangerous. If they are, then a decision has to be made concerning the type of hazard, which must then be described using one of the appropriate 'risk phrases' (which must then be used on labels and safety data sheets). This process is referred to as 'classification'.

A substance is defined as '**dangerous**' if it is so classified in the '**Approved Supply List**' [2]. It is also 'dangerous for supply' if it has the properties described in Schedule 1 of the Regulations (Table 23.1).

Substances that are classified as dangerous must be marked with appropriate 'indications of danger', which include 'hazard warning signs'. Indications of danger and the associated hazard warning sign are given in Schedule 2 of the regulations (Table 23.2).

Substances and preparations are defined as 'dangerous for carriage' if they are so classified in the 'Approved Carriage List' [3], which is also approved by the HSC, or if they have the properties described in Schedule 1 (Table 23.1).

Safety data sheets must be provided by the supplier for dangerous substances, for the information of the recipient of the substance or preparation. The information provided must be given, using the following obligatory headings, as required by Schedule 4:

1. Identification of the substance or preparation and the company or undertaking.
2. Composition/information on ingredients.
3. Hazards identification.

Table 23.1 Schedule 1, Regulations 2(1) and 4(4), classification of dangerous substances and dangerous preparations, categories of danger

Category of danger	Property[a]	Symbol-letter
Physico-chemical properties		
Explosive	Solid, liquid, pasty or gelatinous substances and preparations that may react exothermically without atmospheric oxygen thereby quickly evolving gases, and which under defined test conditions detonate, quickly deflagrate or upon heating explode when partially confined.	E
Oxidizing	Substances and preparations that give rise to a highly exothermic reaction in contact with other substances, particularly flammable substances.	O
Extremely flammable	Liquid substances and preparations having an extremely low flash point and a low boiling point and gaseous substances and preparations that are flammable in contact with air at ambient temperature and pressure.	F+
Highly flammable	The following substances and preparations, namely: (a) substances and preparations that may become hot and finally catch fire in contact with air at ambient temperature without any application of energy (b) solid substances and preparations that may readily catch fire after brief contact with a source of ignition and that continue to burn or to be consumed after removal of the source of ignition (c) liquid substances and preparations having a very low flash point (d) substances and preparations that, in contact with water or damp air, evolve extremely flammable gases in dangerous quantities.	F
Flammable	Liquid substances and preparations having a low flash point.	None
Health effects		
Very toxic	Substances and preparations that in very low quantities cause death or acute or chronic damage to health when inhaled, swallowed or absorbed via the skin.	T+
Toxic	Substances and preparations that in low quantities cause death or acute or chronic damage to health when inhaled, swallowed or absorbed via the skin.	T
Harmful	Substances and preparations that may cause death or acute or chronic damage to health when inhaled, swallowed or absorbed via the skin.	Xn
Corrosive	Substances and preparations that may, on contact with living tissues, destroy them.	C
Irritant	Non-corrosive substances and preparations that, through immediate, prolonged or repeated contact with the skin or mucous membrane, may cause inflammation.	Xi
Sensitizing	Substances and preparations that, if they are inhaled or if they penetrate the skin, are capable of eliciting a reaction by hypersensitization such that on further exposure to the substance or preparation, characteristic	

Table 23.1 Continued

Category of danger	Property[a]	Symbol-letter
	adverse effects are produced.	
Sensitizing by inhalation		Xn
Sensitizing by skin contact		Xi
Carcinogenic[b]	Substances and preparations that, if they are inhaled or ingested or if they penetrate the skin, may induce cancer or increase its incidence	
Category 1		T
Category 2		T
Category 3		Xn
Mutagenic[b]	Substances and preparations that, if they are inhaled or ingested or if they penetrate the skin, may induce heritable genetic defects or increase their incidence.	
Category 1		T
Category 2		T
Category 3		Xn
Toxic for reproduction[b]	Substances and preparations that, if they are inhaled or ingested or if they penetrate the skin, may produce or increase the incidence of non-heritable adverse effects in the progeny and/or of male or female reproductive functions or capacity.	
Category 1		T
Category 2		T
Category 3		Xn
Environmental effects		
Dangerous for the environment[c]	Substances and preparations that, were they to enter into the environment, would present or might present an immediate or delayed danger for one or more components of the environment.	N

Notes:
a As further described in the approved classification and labelling guide.
b The categories are specified in the approved classification and labelling guide.
c In certain cases specified in the approved supply list and in the approved classification and labelling guide substances and preparations classified as dangerous for the environment do not require to be labelled with the symbol and indication of danger.

 4. First aid measures.
 5. Fire-fighting measures.
 6. Accidental release measures.
 7. Handling and storage.
 8. Exposure control/personal protection.
 9. Physical and chemical properties.
10. Stability and reactivity.
11. Toxicological information.
12. Ecological information.
13. Disposal considerations.
14. Transport information.
15. Regulatory information.
16. Other information.

The objective is to provide the recipient with sufficient, accurate data to enable them to take any necessary measures relating to the protection of health and safety at work or the protection of the environment. It should be noted that safety data sheets are not required to be provided

Table 23.2 Schedule 2, Regulations 2(1) and 10(6), indications of danger and symbols for dangerous substances and dangerous preparations

Indication of danger	Symbol-letter	Symbol
Explosive	E	
Oxidizing	O	
Extremely flammable	F+	
Highly flammable	F	
Very toxic	T+	
Toxic	T	
Harmful	Xn	
Corrosive	C	
Irritant	Xi	
Dangerous for the environment	N	

when these materials are sold to the public from shops.

Suppliers are required by CHIP3 to exercise all due diligence in complying with the requirements. Therefore, if a particular supplier wishes to use a classification assigned by a manufacturer or supplier higher up the supply chain, it is expected to have made appropriate enquiries to ensure that the classification is appropriate. Where reliable, reputable suppliers are used; the checks required would normally be quite simple, although enforcement officers may look for evidence that appropriate enquiries have indeed been carried out. The HSC have published guidance on the compilation of safety data sheets [4].

CHIP3 is constantly being updated, and information on the recent changes can be found on the HSE website www.hse.gov.uk/chip/changes/index.htm

THE CONTROL OF SUBSTANCES HAZARDOUS TO HEALTH REGULATIONS 2002 (AS AMENDED)

These regulations [5] referred to as COSHH, were made under the Health and Safety at Work, Etc. Act 1974. The COSHH regulations apply to substances that have already been classified as being very toxic, toxic, harmful, corrosive or irritant under the provisions of the CHIP3 regulations as outlined earlier. They also apply to:

1. substances that have an approved maximum exposure limits or occupational exposure standards;
2. substances that are a biological agent;
3. a substantial concentration of dust of any kind;
4. any other substance that is not listed in any of the aforementioned categories, but that creates a risk to health due to its chemical or toxicological properties, or the way in which it is used.

It must be noted that not all substances that can be a hazard to health are controlled by these regulations. Some examples of these 'exempted' substances include:

1. lead – so far as the Control of Lead at Work Regulations 2002 [7] apply;

2. asbestos – as far as the Control of Asbestos at Work Regulations 2002 [8] apply;
3. substances that are only hazardous to health, solely by virtue of any radioactive, explosive or flammable properties, or solely because it is at a high or low temperature or a high pressure.

There are a number of substances that are 'prohibited' under these regulations. The substances and the extent to which they are prohibited are contained in Schedule 2 of COSHH.

The critical point of these regulations is that an employer shall not carry on any work that is liable to expose any employees to any substance hazardous to health unless he has **made a suitable and sufficient assessment of the risks and of the steps that need to be taken to meet the requirements of the regulations.** Such assessments must be reviewed where a significant change in work occurs or the original assessment becomes invalid for any other reason.

The assessment itself must include:

1. an assessment of the risks to health
2. the steps that need to be taken to achieve adequate control of exposure to hazardous substances
3. identification of any other action necessary to comply with the regulations.

The assessment should consider what types of substances employees are liable to be exposed to, and this must include consideration of the consequences of the possible failure of measures provided to control exposure, the form in which the substances may be present and their effect upon the body. Of great importance is a consideration of the extent to which other workers or other people (including non-employees) are likely to come into contact with the hazardous substances being assessed.

The employer must make an estimate of exposure levels and compare these to any available, valid standards. If the assessment indicates that control is, or is likely to be, inadequate, then the employer must determine the steps that must be taken to obtain adequate control. It should be remembered that the COSHH regulations require that personal protective equipment should be used as a method of exposure control only after all other methods have been employed as far as reasonably practicable.

An assessment can be considered to be suitable and sufficient if the detail and expertise with which it is carried out are commensurate with the nature and degree of the risk involved with the work. In some circumstances, it will only be necessary to read manufacturers' or suppliers' safety information sheets to ensure that current working practices are satisfactory. In other cases, considerable atmospheric monitoring may be necessary before true exposure levels can be ascertained.

In addition to making an assessment of risks, employers are also required to ensure that the exposure of employees to substances hazardous to health is either prevented or, where this is not reasonably practicable, adequately controlled (by means other than personal protective equipment).

Where an employer provides any control measure to meet the requirements of these regulations, he is also required to ensure that those measures are maintained in an efficient state, in efficient working order and in good repair. Thorough examination and testing of engineering controls is required.

Where monitoring is carried out or is specifically required to be carried out, then a record of that monitoring must be kept, and the record itself or a summary must be kept available. Where the record is representative of the personal exposures of identifiable employees, this must be kept for at least 40 years. In any other case, the record must he kept for at least 5 years.

Where it is appropriate for the protection of the health of employees who are, or are liable to be, exposed to a substance hazardous to health, the employer must ensure that those employees are subject to suitable health surveillance, including biological monitoring where necessary.

An additional requirement of these regulations is that where any employee is exposed, or may be exposed, to substances hazardous to health, then the employer must provide the employee with such information instruction and training as is suitable and sufficient for that employee to know:

1. the risks to health created by such exposure;
2. the precautions that should be taken.

This would include the disclosure of the results of any monitoring of exposure in the workplace and information on any collective health surveillance (so presented that it cannot be related to any individual).

Any person carrying on a work activity must understand the nature of any hazards associated with the materials being used. It is essential that a hazard data sheet be obtained from the supplier when acquiring potentially hazardous materials. This data must be read and understood and kept in a readily accessible place. The contents of such data sheets will form the basis of instructions and training given to employees, as well as of the selection and provision of any necessary protective equipment.

Safe arrangements must be made for the reception, storage and use of such materials and, as far as reasonably practicable, emergency procedures should be designed in liaison with agencies such as the HSE, local authority and fire authority as necessary.

Finally, suitable arrangements should be made for any medical examinations (including health surveillance) required or deemed necessary, and first-aid facilities should be made available to the standard required by the Health and Safety at Work, Etc. Act 1974 enforcing authority.

It can therefore be seen that the COSHH Regulations 2002 provide a comprehensive set of measures that will act to improve the health and safety at work of employees and others affected by work activities. The HSC has published an Approved Code of Practice [6].

Occupational exposure limits

For some years now, the adequacy of the control of inhalation of substances hazardous to health has been assessed by reference to the concentration of those substances in air inhaled as a result of work activities. These concentrations of hazardous substances are referred to as occupational exposure limits (OELs). These OEL figures are normally reviewed each year by the HSE in the form of a guidance note [9]. The guidance also indicates where appropriate whether a substance is capable of causing

respiratory sensitization and/or can be absorbed through the skin.

It should be noted that this guidance does not extend to detailed considerations of asbestos and lead exposure levels, where specific legislation exists, nor does it cover situations where work is below ground or exposure to microorganisms. These OEL standards have evolved alongside the development of the COSHH regulations.

Maximum exposure limit

The maximum exposure limit (MEL) is described in HSE Guidance Note EH40 [9] as a limit set for substances that may cause the most serious health effects, such as cancer and occupational asthma, and for which 'safe' levels of exposure cannot be determined. A MEL is also set for substances which, although safe levels may exist, it is not reasonably practicable to control to those levels. The MEL is expressed as concentration in air of a substance, expressed in both parts per million (ppm) and milligrams per cubic metre (mg/m^3) averaged over one of two specified 'reference periods': 15 minutes or 8 hours (see later). Concentrations of mineral fibres are expressed as fibres per millilitre of air (fibres/ml).

The significance of the MEL is that where an exposure limit has been specified, the control of exposure by inhalation can only be considered to be adequate where the level of exposure is reduced so far as reasonably practicable and, in any circumstances, below the prescribed MEL.

Occupational exposure standards

An 'occupational exposure standard' (OES) is also a concentration of an airborne substance averaged over the same reference periods. Substances are assigned an OES after consideration of the available scientific data by the Working Group on the Assessment of Toxic Chemicals. In the case of these substances, concentrations are such that there is considered to be no evidence that the substance is likely to be injurious to employees if they are exposed to that concentration day after day. If exposure by inhalation exceeds the OES, then control is normally considered to be adequate as long as the employer has identified the reason that the OES has been exceeded, and that suitable steps are being taken to reduce exposure to achieve the OES as soon as is reasonably practicable.

Reference periods

Substances that are considered to be hazardous to health may cause adverse effects that range from irritation of the skin or mucous membranes through to death. The effect may be produced either over a very short exposure period or over a much longer period. It is, therefore, important to develop exposure limits that reflect these differences, and this has been done with regard to both MELs and OESs. These are listed either in the COSHH regulations or Guidance Note EH40 [9] as either 'long-term exposure limits' (8-hour time-weighted average reference period) or 'short-term exposure limits' (15-minute reference period).

Long-term reference period

The 8-hour reference period refers to a well-established procedure whereby the sum of individual occupational exposures over a 24-hour period are treated as being equivalent to a single, uniform exposure for an 8-hour period. This is known as the '8-hour time-weighted average' (TWA) for the exposure, and may be represented mathematically by:

$$\frac{C_1 \times T_1 + C_2 \times T_2 + C_3 \times T_3 + \cdots + C_n \times T_n}{8}$$

where C_1, C_2, C_n, etc. are the individual measured occupational exposures, and T_1, T_2, T_n, etc. are the corresponding times for each exposure period.

Example: A person works in a premises performing a variety of different jobs, in some of which he is exposed to substances hazardous to health. Monitoring is carried out to give the results shown in Table 23.3.

Exposure is known to be 0 during breaks or activities carried out in other parts of the premises; this exposure may also have to be measured in practice.

Table 23.3 Results of monitoring an employee's exposure to substances hazardous to health

Work period (time)	Exposure (mg/m³)	Sample
7.30–9.30	0.24	2.00
9.30–9.45	0	0.25
9.45–12.30	0.18	2.75
12.30–13.00	0	0.50
13.00–15.00	0.34	2.00
15.00–15.15	0	0.25
15.45–17.00	0.65	1.25

The 8-hour TWA is therefore:

$$\{0.24 \times 2 + 0.18 \times 2.75 + 0.34 \times 2.0 + 0.65$$
$$\times 1.25 + 0 \times [0.25 + 0.50 + 0.25]\}/8$$
$$= \{0.48 + 0.50 + 0.68 + 0.81 + 0\}/8$$
$$= 2.47 \text{ mg/m}^3.$$

Short-term reference period

Here exposure should normally be measured over the prescribed period for the hazardous substance concerned: this is normally 15 minutes. Measurements for periods greater than 15 minutes should not be used to calculate the short-term exposure.

Long-term exposure limits are thus designed to protect against the chronic effects of exposure, while short-term exposure limits are designed to avoid the acute effects of toxicants. Eight hours was selected to reflect the typical exposure during one working shift averaged over a 24-hour period, while the 15-minute period represents any 15-minute period during the working day. Where only short-term samples are taken, information such as episodic peak concentrations may be missed, whilst if a long-term sample is taken, significant high peak concentrations, which could be harmful, might not be detected as the result is averaged over that longer period. Monitoring systems and equipment must therefore be carefully selected and positioned in order to ensure that the data gathered can be usefully interpreted in the light of the available standards and guidance.

MONITORING STRATEGIES FOR TOXIC SUBSTANCES

Carrying out monitoring can be an expensive and time-consuming exercise, therefore some form of monitoring strategy must be applied in order to gain sufficient data for the purposes of the exercise in as cost-efficient a way as possible. The data must, however, be accurate and reliable so that the requirements imposed by any health and safety legislation can be fulfilled. Suitable advice is contained in the HSE's Guidance Note HSC173 [10]. This document discusses the factors that influence airborne concentrations of hazardous substances, and urges that a structured approach be taken. This procedure involves a number of distinct phrases for the monitoring work:

1. initial appraisal
2. basic survey
3. detailed survey
4. routine monitoring

With regard to these types of monitoring, a decision has to be made about the level of sophistication that the survey requires, both in terms of the quantity and quality of the data collected. The guidance note proposes a three-level approach to the carrying out of such an exercise. Finally, advice is given concerning the interpretation of results. It is essential that anyone carrying out or recommending an assessment under COSHH is familiar with this document.

Various other documents produced by the HSE are of significance when considering this topic, General Method for the Gravimetric Determination of Respirable and Total Inhalable Dust [11], and General Methods for Sampling Gases and Vapours [12].

Asbestos

Around 3000 people a year in Great Britain die from diseases caused by past exposure to asbestos, and the figure is set to rise to nearer 10 000 by 2010. The Control of Asbestos at Work Regulations 2002 [8] implement various European directives on the protection of workers from the risks related to exposure to asbestos at work. In particular they introduce a new 'duty to manage' regulation (Regulation 4). This requires employers to:

- manage the risk from asbestos by establishing the extent of asbestos on the premises;
- keeping accurate records of location and condition;
- assessing the risk from the material;
- prepare a plan showing how asbestos risks are to be managed;
- implement the plan;
- review and monitor the plan and the arrangements to put it into place;
- provide information to those who may work on or disturb it.

This requirement is a key part of the HSE's work to raise awareness of the risks and also reduce risk from asbestos exposure.

When considering the problems associated with exposure to asbestos, it is important to understand the concept of 'the fibre'. This term is defined in the 'European Reference Method for the determination of personal exposure, using the membrane filter method' (see later). The principle of this method is that a measured volume of air is drawn through a membrane filter, which is subsequently rendered transparent and mounted on a microscope slide. Fibres on a known fraction of the filter are counted under phase contrast microscopy, and the number of fibres in a unit volume of the air can be calculated and compared with the standards indicated in the following. Details of the method are given in an HSE publication [13].

In this document the definition of a 'fibre' is given as 'particles with a length greater than 5 micrometres, a maximum diameter of less than 3 micrometres and a length to diameter ratio of greater than three to one'. These are the particles that are counted, but it should be noted that, as it is not possible to identify the chemical nature of the fibres collected, all 'fibres' are counted as being asbestiform and used in determining whether control limits and action levels have been exceeded.

The key features of the regulations are as follows.

Assessments

Work with asbestos must be adequately assessed by an employer to determine the nature and degree of any exposure, and the steps that must be taken in order to prevent or reduce that risk. In the case of work that consists of the removal of asbestos, a suitable written 'plan of work' must be prepared. This must include details of:

1. the nature and probable duration of the work;
2. the location of the place of work;
3. the methods to be applied;
4. the characteristics of the equipment to be used for the protection and decontamination of those carrying out the work and the protection of other persons on or near the worksite.

Employers are required to identify the type of asbestos involved in the work, or to assume that it contains brown asbestos (amosite) or blue asbestos (crocidolite), which are the most hazardous types.

Employers must generally provide the enforcing authority with the particulars specified in Schedule 1 of the regulations at least 14 days before commencing the work, unless the enforcing authority agrees to a shorter period.

Prevention of exposure

Employers have a duty to prevent or reduce the spread of asbestos from a workplace to the lowest level reasonably practicable. They must also prevent or reduce employees' exposure by means other than the use of respiratory protective equipment. Other related duties include the need to monitor the air for asbestos fibres and to provide information and training related to the risks involved and the precautions that must be taken.

Control limits

These represent a level of asbestos in the air above which respiratory protective equipment must be worn [8].

The limits refer to one of the following concentrations of asbestos in the atmosphere when measured or calculated by a method approved by the HSC:

1. For chrysotile (white asbestos):
   (a) 0.3 fibres/ml of air averaged over any continuous period of 4 hours
   (b) 0.9 fibres/ml of air averaged over any continuous period of 10 minutes.

2. For any other form of asbestos, either alone or in mixtures with any other form of asbestos:
   (a) 0.2 fibres/ml of air averaged over any continuous period of 4 hours
   (b) 0.6 fibres/ml of air averaged over any continuous period of 10 minutes.

Action level

This represents a cumulative exposure level over a period of 12 weeks [8]. If an action level is likely to be exceeded, then the work must be reported to the HSE, 'asbestos areas' must be designated, medical surveillance of workers must take place and health records must be kept. The limits are:

1. where the exposure is solely to chrysotile, 72 fibre-hours/ml of air
2. where the exposure is to any other form of asbestos either alone or in mixtures including mixtures of chrysotile with any other form of asbestos, 48 fibre-hours/ml of air
3. where both types of exposure can occur separately during the 12-week period, a proportionate number of fibres per millilitre of air.

A number of Approved Codes of Practice have been issued in relation to working with asbestos. Details of these can be found at www.hse.gov.uk

A further useful publication is the HSE's Guidance Note EH10 [14]. This document gives good practical advice on the measurement of airborne asbestos and the associated calculations of fibre levels. Of particular importance is the methodology for '**clearance monitoring**', which must be carried out before a site is handed back to the control of its occupiers after asbestos removal operations. Using the specified method, the lowest fibre level that can be reliably detected above background levels in a 480 l sample is about 0.01 fibres/ml, and this level is therefore taken as the '**clearance indicator level**'. Sites should not be considered to be satisfactory until such a test has been successfully completed. Guidance Note EH10 [14] also gives further reference to guidance notes, videos and other material that will be useful in dealing with asbestos in a variety of work situations.

Reference should be made to the Control of Asbestos in the Air Regulations 1990 [15], which prescribe a limit value for the discharge of asbestos from outlets into the air during the use of asbestos. These regulations also provide for the regular measurement of asbestos emissions from specified types of premises. Provision is also made for the control of environmental pollution by asbestos emitted into the air as a result of the working of products from the demolition of buildings, structures or installations containing asbestos. Finally, the Asbestos (Prohibitions) Regulations 1992 (as amended) [16] implement EU directive 91/659/EEC 'on restrictions relating to the marketing and use of dangerous substances and preparations (asbestos)'.

REFERENCES

1. *Chemicals (Hazard Information and Packaging for Supply) Regulations 2002*, SI 2002, No. 1689, HMSO, London.
2. Health and Safety Commission (2002) *Approved Supply List: Information Approved for the Classification and Labelling of Substances and Preparations Dangerous for Supply*, 7th edn, HSE Books, Sudbury, Suffolk.
3. Health and Safety Commission (1999) *Approved Carriage List: Information Approved for the Carriage of Dangerous Goods by Road and Rail, Other than Explosives and Radioactive Material*, HSE Books, Sudbury, Suffolk.
4. Health and Safety Commission (2002) *The Compilation of Safety Data Sheets*, 3rd edn, HSE Books, Sudbury, Suffolk.
5. *Control of Substances Hazardous to Health Regulations 2002*, SI 2002, No. 2677, HMSO, London.
6. Health and Safety Commission (2002) *Approved Code of Practice: Control of Substances Hazardous to Health*, HSE Books, Sudbury, Suffolk.
7. *Control of Lead at Work Regulations 2002*, SI 2002, No. 2676, HMSO London.
8. *Control of Asbestos at Work Regulations 2002*, SI 2002, No. 2675, HMSO London.
9. Health and Safety Executive (2003) *Occupational Exposure Limits 2002 Plus 2003 Supplement*, Guidance Note EH40/2002, HSE Books, Sudbury, Suffolk.
10. Health and Safety Executive (1997) *Monitoring Strategies for Toxic Substances*, HSG173, HSE Books, Sudbury, Suffolk.
11. Health and Safety Executive (2000) *General Methods for Sampling and Gravimetric Analysis of Respirable and Inhalable Dust*, MDHS14/3, HSE Books, Sudbury, Suffolk.
12. Health and Safety Executive (1990) *General Methods for Sampling Gases and Vapours*, MDH70, HSE Books, Sudbury, Suffolk.
13. Health and Safety Executive (1995) *Asbestos Fibres in Air*, MDHS39/4, HSE Books, Sudbury, Suffolk.
14. Health and Safety Executive (2002) *Asbestos – Exposure Limits and Measurement of Airborne Dust Concentrations*, EH10, HSE Books, Sudbury, Suffolk.
15. *Control of Asbestos in the Air Regulations 1990*, SI 1990, No. 556, HMSO, London.
16. *Asbestos (Prohibitions) Regulations 1992 (as amended)*, SI 1992, No. 3067, HMSO, London.

24 Accident prevention and investigation

Paul C. Belcher and John D. Wildsmith

In 1956 the then Ministry of Labour postulated six principles of accident prevention [1]:

1. Accident prevention is an essential part of good management and good workmanship.
2. Management and workers must co-operate wholeheartedly in securing freedom from accidents.
3. Top management must take the lead in organizing safety in the works.
4. There must be a definite and known safety policy in the workplace.
5. The organization and resources necessary to carry out the policy must exist.
6. The best available knowledge and methods must be applied.

It is interesting that it was not until 18 years later, with the advent of the Health and Safety at Work, Etc. Act 1974 and its attendant regulations, that these important principles received legislative support. Before looking at the whole process of accident investigation and prevention, it is necessary to examine what is meant by the term accident. The Oxford English Dictionary's definition of accident is: 'an event without apparent cause, unexpected event, unintentional act, mishap'.

An early judicial definition was put forward by Lord MacNaughton in *Fenton* v. *Thorley & Co.*

Ltd [1903] AC 443 where 'accident' was defined as 'some concrete happening which intervenes or obtrudes itself upon the normal course of employment. It has the ordinary everyday meaning of an unlooked for mishap or an untoward event, which is not expected or designed by the victim.'

The thinking behind such a statement has greatly influenced the actions of those involved in accident prevention. However, this definition can be seen as being somewhat narrow in that it lays emphasis on the actions of the worker, and that it seems to infer that all accidents result in injury that is evident at the time of the incident.

Bamber [2] offers a wider definition of accident when he says that an accident is 'an unexpected, unplanned event in a sequence of events, that occurs through a combination of causes; it results in physical harm (injury or disease) to an individual, damage to property, a near-miss, a loss, or any combination of these effects'.

This is a far more satisfactory definition of the term accident in that it recognizes a wider view of the causes and effects of accidents.

CAUSES OF ACCIDENTS

Accidents don't just happen, they have a cause. Once this principle is accepted, then the whole notion of accident prevention and investigation takes on a new dimension. In addition, one must accept that the cause of the accident may not be immediately obvious, and that the employee or employees involved may not be primarily responsible for the accident's occurrence. The Health and Safety Executive (HSE) [3] emphasizes this point when it states 'the majority of accidents and incidents are not caused by "careless workers," but by failures in control (either within the organization or within the particular job), which are the responsibility of the management'.

Therefore, whilst on the surface the accident investigation may identify certain factors that led to the accident or incident occurring – the primary causes – they must be seen as symptoms of more fundamental underlying causes that allow these factors to exist or persist – the secondary causes.

Examples of primary causes are:

- working without authority
- using defective equipment
- horseplay
- failing to wear personal protective equipment
- inadequate or missing guards to machines.

Secondary causes may include:

- financial pressures
- lack of policy
- lack of commitment
- lack of knowledge.

As the HSE [3] makes clear, failure at this level can lead to:

unrealistic timescales for the implementation of plans which put pressure on people to cut corners and reduce supervision; . . . job and control systems which failed to recognize or allow for the fact that people were likely to make mistakes and might have difficulties communicating with each other.

ACCIDENT NOTIFICATION

There should be an efficient and effective system of accident notification in any industrialized society. The reasons for this include:

- to ensure compliance with current legislation;
- to assist in the monitoring and development of health and safety policies;
- to provide a feedback mechanism that will assist in the development of safe systems of work;
- to provide information to enforcing authorities so as to enable the refining of hazard analysis and inspection strategies.

The current legislation dealing with the notification of accidents can be found in the Reporting of Injuries, Diseases and Dangerous Occurrences Regulations [4] (RIDDOR), which came into force on 1 April 1996. Detailed guidance relating to the regulations is contained in a booklet published by the HSE [5].

The effect of this legislation is that it places a duty on all employers and other 'responsible persons' who have control over employees and work premises to report to the relevant enforcing authority accidents, both fatal and non-fatal, that occur in the workplace. In certain circumstances this notification must take place immediately by the quickest practicable means, and must be reported in writing within ten days to the relevant 'enforcing' authority. Regulation 10 details the exceptions to this requirement.

Regulation 3(1) lists those accidents that are reportable. As well as including accidents that result in physical injury, the regulations require the notification of other incidents that result in, for example, decompression sickness, loss of consciousness resulting from lack of oxygen, and acute illnesses that are the result of exposure to harmful substances, pathogens or infected material.

The regulations also require the notification of occupational diseases such as poisoning, skin diseases, lung diseases, infections and some cancers (see Schedule 3).

In addition to requiring the notification of accidents and occupational diseases, the regulations also require the notification of 'dangerous occurrences'. These include explosion of pressure vessels, the release of more than 1 tonne of flammable liquid, and the partial or total collapse of a building. The full list of defined dangerous occurrences can be found in Schedule 2 of the regulations.

A fuller description of the required notification procedures together with a flowchart can be found in Environmental Health Procedures [6].

INVESTIGATION OF ACCIDENTS

Having been informed of an accident or dangerous occurrence, it is incumbent upon the enforcing authority to investigate the incident as quickly as possible. In the case of serious or fatal injuries, an investigating officer should preferably be at the scene as soon as practicable.

Under the provisions of the Health and Safety at Work, Etc. Act 1974, an inspector has the power to ask questions, the answers to which may be written down in the form of a statement. The investigating officer must ensure that all such questioning and statement collection is carried out in accordance with the provisions of the Police and Criminal Evidence Act 1984 (see Chapter 7).

In addition to taking statements, the inspecting officer may also take away documents, for example, safety policies, and may take photographs at the scene of the accident. The object of the investigation is to establish as clearly as possible the sequence of events that led up to the accident, and subsequently to consider any recommendations that could be made to ensure that the incident does not re-occur.

As soon as possible after the investigation has taken place, the investigating officer should prepare a written report, which should be concise and, where necessary, should be supported by sketches, photographs and statements. It should be borne in mind that any report may not only be required in a criminal action, but may also be requested in any claim for damages taken through the civil courts.

As to any subsequent action that may be taken, this matter is often taken out of the hands of the investigating officer. However, in deciding what course of action should be taken regard should be had to:

- the seriousness of any contravention;
- the degree of risk present at the time;
- the means available to remedy the situation.

ACCIDENT PREVENTION

Quite clearly, any accident warrants full investigation, not only to establish the exact cause of the accident, but also to establish ways in which a similar incident can be prevented from happening again.

An important underlying principle of the Health and Safety at Work, Etc. Act 1974 is that any employer should provide a safe place of work and safe systems of work.

In its publication *Successful Health and Safety Management* [3], the HSE argues that 'all accidents, ill-health and incidents are preventable'.

If employers are to realize this ideal within their workplaces, then due consideration must be given to the systematic development of accident prevention strategies. In 1998, the HSE issued a discussion paper, which was subsequently followed by a consultation document from the Health and Safety Commission (HSC) in 2001, on the notion of placing a statutory duty on employers to investigate accidents. At the time of writing (2003), the HSC had indicated their intention to issue guidance, to help employers investigate incidents that cause injuries and ill health in the work place. The decision to issue the guidance, rather than to recommend legislation to require employers to investigate incidents, was taken by the HSC early in 2003, after taking views in response to its wide-ranging consultation exercise.

Strategies for accident prevention include: hazard analysis, safety inspections, accident investigations, checking of safety devices, publicity, training and education, and safe systems of work. Some of these principles are examined below.

Hazard analysis

Central to the whole notion of accident prevention is the identification of all potential hazards in the workplace. A hazard may be thought of as 'the potential to cause harm, including ill-health or injury; damage to property, plant, products or the environment; production losses or increased liabilities' [3].

To some extent this exercise will have already been carried out in many workplaces in order to ensure compliance with the Control of Substances Hazardous to Health Regulations 2002 (as amended) (COSHH) [7].

This principle has been extended by the Management of Health and Safety at Work Regulations 1999 [8]. These regulations require every employer and self-employed person to carry out a suitable and sufficient assessment of the health and safety risks to employees and others not in their employment arising out of or in connection with their operation (Regulation 3). There is also a requirement for those who employ five or more persons to record any 'significant findings' from such an assessment. The *Approved Code of Practice* [9] suggests that any such record should contain a statement of significant hazards identified, the control measures in place and the extent to which they control the risk, and the population exposed to the risk.

There are several ways in which the existence of hazards may be established. These include:

- workplace inspections
- analysis of production/work programmes
- analysis of individual job specifications.

Risk assessment

Once an analysis of potential hazards has been carried out, it may then be possible to evaluate the potential risk of the individual operations within the workplace, and also to identify those parts of the process that are particularly hazardous (see also Chapter 10). Risk may be thought of as 'the likelihood that a specified undesired event will occur due to the realization of a hazard by, or during, work activities or by the products and services created by work activities' [3].

Within the *Approved Code of Practice* [9], a more concise definition of risk is presented as 'the likelihood that the harm from a particular hazard is realized'.

Any evaluation carried out will ultimately lead to the establishment of a hazard rating for that particular premises and/or the particular operation(s) carried out at the premises. Such a hazard rating could be used to prioritize the inspection regime to be used by the enforcing authority.

There are a number of ways in which such a hazard rating can be developed. Most are based on the simple formula:

$$\text{Risk} = \text{severity estimate} \times \text{likelihood of occurrence}$$

Examples of such estimating techniques can be found in *Successful Health and Safety Management* [3] and *Five Steps to Risk Assessment* [10].

Risk control

Having established the potential hazards within a workplace, the next step is to establish systems to eliminate or reduce the potential risk to the worker. Bamber [11] suggests a gradation of measures from long-term (permanent) to short-term (temporary). He presents the grading as:

- Long-term

  - eliminate hazard at source
  - reduce hazard at source
  - remove employee from hazard
  - contain hazard by enclosure
  - reduce employee exposure to hazard

- Short-term

  - utilize protective equipment.

Such a system of grading is extremely useful, with regard to the reduction of physical as well as chemical hazards. It may also be regarded as an order of priority, in that the long-term aim must always be to eliminate the hazard at source, but while achieving this, more short-term actions may be employed. A similar hierarchy of control is proposed by the HSE [3].

There are various ways in which hazards may be controlled. For example, guarding of machines, efficient ventilation of the workplace, monitoring of the use of hazardous substances, and the use of protective equipment. However, as indicated, the use of protective equipment must always be seen as the short-term solution to the controlling of hazards within the workplace.

Having engaged in such activity it is vital to monitor and review both hazards and risks and any control measures introduced. To this end, the Management of Health and Safety at Work Regulations 1999 [8] not only require that employers must plan and implement such a review process (Regulation 4), but also impose on employees a duty to inform their employer of any shortcomings in health and safety arrangements (Regulation 12). This point is reinforced by the *Approved Code of Practice* [9], in which the HSE states:

the avoidance, prevention and reduction of risks at work needs to be an accepted part of the approach and attitudes at all levels of the organization and to apply to all its activities, i.e. **the existence of an active health and safety culture affecting the organization as a whole needs to be assured.**

Safe systems of work

Safe systems of work are fundamental to the whole notion of accident prevention. They have been defined by the HSE [12] as 'a formal procedure which results from systematic examination of a task in order to identify all the hazards. It defines safe methods to ensure that hazards are eliminated or risks minimised'.

Consideration of this definition reveals that any safe system of work that has been properly thought through will of necessity form part of the 'arrangements' section of any safety policy (see below). Any safe system of work will include consideration of some of the following issues:

- assessment of the task
- hazard identification and risk assessment
- the safe design, installation, and use of all plant, tools and equipment

- effective planned maintenance of plant and equipment
- a safe working environment
- suitable and sufficient training for all employees
- compliance with safety policy
- adequate consideration of vulnerable employees, for example, young persons, pregnant women and disabled persons
- regular reviews of all procedures employed within the workplace to ensure maintenance of high standards.

In certain circumstances it may be necessary to employ a **'permit to work'** system. This is, in essence, an extension of the principle of a safe system of work, and requires written permission before a particular job may commence. The permit is a written document identifying the plant to be worked on and details the precautions to be taken before work can begin.

Permit to work systems are required by legislation in certain circumstances, for example, entry into confined spaces. They may also be advisable in circumstances such as maintenance work on high-risk machinery, work in high-risk fire areas, working with asbestos and work with corrosive or toxic substances.

Safety training

Section 2 of the Health and Safety at Work, Etc. Act 1974 places a duty on an employer to provide such information, instruction, training and supervision as is necessary to ensure, so far as is reasonably practicable, the health and safety at work of employees.

It is essential that training be seen as being necessary for all employees, that is, from manager to manual worker. In addition, training courses should reflect the individual needs of the workforce, for example, driver training for fork-lift truck drivers, safe use of personal protective equipment, fire safety training.

The need for training presents the enforcing authority with an ideal opportunity to enter the workplace in a 'non-threatening' capacity, enabling it to win the confidence of both the management and the general workforce. A number of different options are available for meeting training needs. The introduction of National Vocational Qualifications (NVQs) allows for workplace assessment of competence in health and safety matters. In addition, there are the Basic and Advanced Certificates in Health and Safety at Work offered by the Chartered Institute of Environmental Health. The HSE lays down some guidelines for the assessment of training needs [3].

SAFETY POLICIES AS AN AID TO ACCIDENT PREVENTION

The Health and Safety at Work, Etc. Act 1974 places employers under a general duty to prepare a written statement of health and safety policy. The only exception to this general rule is where there are less than five persons employed on the premises. The HSE published advice on the preparation of safety policies for small businesses [13].

In essence there are three parts to any safety policy:

1. a general statement of intent
2. organization
3. arrangements.

These provisions have been expanded by the Management of Health and Safety at Work Regulations 1999 [8]. The risk assessments and significant findings discussed above will form an essential part of the safety policy as they form the basis of the control measures to be implemented and the responsibilities within the organization.

The general statement of intent outlines in broad terms the organization's overall philosophy in relation to the management of health and safety. Such statements should set the direction for the organization and ideally should identify ways in which health and safety can contribute to overall business performance. As the HSE [3] makes clear 'health and safety and quality are two sides of the same coin'.

Part two of the policy statement outlines the chain of command in terms of health and safety management. It indicates those people who have positions of responsibility or accountability, and

their sphere of operation. It will also contain details of how the implementation of the policy is to be monitored. This section will contain information on the role and function of safety committees and safety representatives or representatives of employee safety. A full discussion of this aspect of health and safety policies is contained in *Successful Health and Safety Management* [3].

The final section of the policy deals with the practical arrangements by which the policy will be effectively implemented. This will include details relating to safety training, safe systems of work, machine/area guarding, housekeeping, noise control, radiation safety, dust control, health checks, fire precautions, accident reporting and investigation systems, emergency procedures and workplace monitoring. Many of these will be identified by the risk assessments undertaken as a requirement of the Management of Health and Safety at Work Regulations 1999 [8]. It should be emphasized that this is only an indicative list. Each safety policy is a unique document that should be tailored to the circumstances of each workplace.

Quite clearly, the health and safety policy is an extremely important document. A well-written policy will enable the inspecting officer to be able to identify all potential risks and hazards within the workplace. It will also enable the officer to assess the organization's perception of the risks that exist, and enable him to tailor the inspection and subsequent recommendations accordingly.

An essential feature of the Robens Report [14] was that health and safety was everyone's business. This has received renewed emphasis from the HSE [3] and as a result of the introduction of the Management of Health and Safety at Work Regulations 1999 [8]. In order to hammer this point home, Robens suggested that all employees should participate in the making and monitoring of arrangements for safety and health in the workplace. It is therefore not surprising that one of the important features of the health and safety policy is the use of safety committees and safety representatives or representatives of employee safety.

The Safety Representatives and Safety Committees Regulations 1977 [15] gave statutory effect to this requirement for joint consultation. In order to comply with European legislation the requirement to consult employees was extended to those workplaces employing staff who were not members of a recognized trade union [16]. To complement these regulations, a Code of Practice has been prepared, supplemented by Guidance Notes [17].

REFERENCES

1. Ministry of Labour and National Service (1956) *Industrial Accident Prevention, Report of the Industrial Safety Sub-Committee of the National Joint Advisory Council*, HMSO, London.
2. Bamber, L. (2003) Principles of the Management of Risk, in *Safety at Work*, 6th edn (eds J. Ridley and J. Channing), Butterworths, London.
3. Health and Safety Executive (2003) *Successful Health and Safety Management*, HS(G) 65, HSE Books, Sudbury, Suffolk.
4. Reporting of Injuries, Disease and Dangerous Occurrences Regulations 1995. SI 1995. No. 3163, HMSO, London.
5. Health and Safety Executive (1999) *A Guide to the Reporting of Injuries, Diseases and Dangerous Occurrences Regulations 1995*, HSE Books, Sudbury, Suffolk.
6. Bassett, W.H. (2002) *Environmental Health Procedures*, 6th edn, Spon Press, London.
7. Control of Substances Hazardous to Health Regulations 2002. SI 2002. No. 2677.
8. Management of Health and Safety at Work Regulations 1999. SI 1999. No. 3242, HMSO, London.
9. Health and Safety Executive (2000) *The Management of Health and Safety at Work: Approved Code of Practice*, L21, HSE Books, Sudbury, Suffolk.
10. Health and Safety Executive (1998) *Five Steps to Risk Assessment*, IND G 163 (rev 1.), HSE Books, Sudbury, Suffolk.
11. Bamber, L. (2003) Risk Management Techniques and Practices, in *Safety at Work*, 6th edn (ed J. Ridley and J. Channing), Butterworths, London.
12. Health and Safety Executive, (1992) *Safe Systems of Work*. IND(G)76L HSE Books, Sudbury, Suffolk.
13. Health and Safety Executive (2000) *Stating Your Business: Guidance on Preparing a Health and Safety Policy Document for Small Firms*. INDG 324, HSE Books, Sudbury, Suffolk.
14. The Committee on Safety and Health at Work (Robens Committee) (1972) Cmnd 5034, HMSO, London.
15. The Safety Representatives and Safety Committees Regulations 1977. SI 1977. No. 500.
16. The Health and Safety (Consultation with Employees) Regulations 1996. SI 1996. No. 1513.
17. Health and Safety Commission (1996) *Safety Representatives and Safety Committees*, HSE Books, Sudbury, Suffolk.

25 Ionizing radiation

Paul C. Belcher and John D. Wildsmith

TYPES OF RADIATION

Environmental health officers may encounter problems with ionizing radiation when dealing with waste disposal matters or with the issue of radon gas and its radioactive 'daughter elements' in dwellings and, increasingly, in the workplace.

The atoms of some naturally occurring substances are found to be 'unstable', and therefore undergo a spontaneous transformation or decay process so as to achieve a more stable state. These substances are known as **radioisotopes**, and they reach their stable condition by emitting radiation, principally in the form of alpha particles, beta particles, gamma radiation and X-radiation. The transformation process is called **radioactivity**, and the emissions may be referred to collectively as **ionizing radiation**.

The time it takes for a radioisotope to decay to form a new element (which itself may be radioactive) varies between materials. The indicator used to describe the rate of decay is the **half-life** ($T_{1/2}$). This is the time taken for half of the radioactive atoms present to decay from the original element.

The various forms of ionizing radiation are emitted with different energies and, as they are different in form, they have different properties. Some of the principal properties are discussed below so that the nature and effects of ionizing radiation can be understood.

Alpha radiation is mainly emitted by the isotopes of heavier elements. It consists of helium nuclei, which are made up of two protons and two neutrons, and are tightly bound together to create a particle. This is a relatively large particle with a mass of 4 units, with 2 units of positive charge. Most of the transformations that result in the emission of alpha particles are accompanied by gamma radiation or X-radiation, which adjusts the energy balance of the emitter.

Beta radiation is mainly emitted by intermediate and lighter elements. It consists of high-speed electrons that originate in the atomic nucleus and have a mass of approximately 1/1840 mass units and carry 1 unit of negative charge. Most of the transformations that give rise to beta radiation also emit energy adjusting gamma radiation or X-radiation.

Gamma radiation is only emitted as a consequence of alpha or beta emissions. It is a form of electromagnetic radiation that consists of quanta of energy in the form of a wave motion. It is therefore non-particulate and uncharged.

X-radiation is electromagnetic radiation that differs from the other forms mentioned above in that it is non-nuclear in origin. It is normally generated electrically, although it can be generated when atomic electrons undergo a change in orbit, such as when beta particles react with other matter.

ENERGY TRANSFER

The energy contained in ionizing radiation is expressed in electron volts. One electron volt is the amount of energy gained by an electron when passing through an electrical potential of 1 volt. The energy of a particle basically depends upon its mass and its velocity and this system can therefore also be applied to the other forms of ionizing radiation. Because this is such a small unit it is usually expressed in kilo-electron volts (keV) or mega-electron volts (MeV).

Alpha particles typically have a charge of 2–8 MeV, while beta particles have 1 keV to 5 MeV. Gamma radiation is typically in the range 1 keV to 6 MeV, while X-radiation can be much higher [1].

Another fundamental characteristic of the principal types of radiation is the ability to penetrate materials. Heavy, slow alpha particles will not penetrate a sheet of paper and have a range of a few centimetres in air. Because the radiation is particulate and charged, this means that alpha radiation interacts both physically and electrically with the media through which it passes and, as it is effectively 'stopped', it must be considered as giving up all of its inherent energy by transferring it to the interacting material. Normally, this energy excites electrons of the absorbing medium, and this can cause the electrons to be released from their atomic orbits, which produces ions. Therefore, the effect of alpha particles can be to ionize material that they come into contact with, possibly with serious biological consequences.

Beta particles produce a similar ionization. However, they are much lighter and thus more penetrative than alpha particles. Beta particles will penetrate a sheet of paper and have a range in air of a few metres.

Gamma radiation and X-radiation have a very large range in air and are extremely penetrative. As they are non-particulate in nature, they interact with the electrons of matter by a series of different mechanisms. One mechanism is referred to as the 'photoelectric effect', where a gamma photon or X-photon is completely absorbed by an atomic electron, which is subsequently ejected from its atom. These electrons can then shed their energy to other atoms in the same way as beta particles.

Ionizing radiation is therefore of sufficiently high power to ionize the various atoms of which living tissues are composed, including water content.

The rate at which radioactive particles lose their energy during their passage through matter can be described by the term '**linear energy transfer**' (**LET**), which is the average energy deposited per unit length travelled. Heavy, slow particles have a high LET, whereas light, fast particles have a low LET.

EFFECTS OF IONIZING RADIATION

The potential for ionizing radiation to produce damage in living tissues is considerable. Essentially, in the case of humans, this can be considered in relation to two routes, that is, whether the radiation source is inside or outside the body [2].

Alpha radiation is stopped quite easily, therefore any source outside the body can be shielded and rendered safe. Because of its short range, even in air, external sources should normally present very little risk from alpha particles. However, if an alpha emitter is in contact with the skin or taken into the body, there will be considerable energy transfer to the living tissues, and this will present a significant risk to health. When ionization occurs, the tissue close to the track of the energy deposition will contain a large number of chemically active chemical species such as ions, free radicals and excited atoms. In these conditions, unconventional chemical reactions of an unpredictable nature take place. These unusual reaction products can produce adverse side-effects on the surrounding tissue. Also, DNA molecules are susceptible to alteration, with potentially serious consequences.

Some protection from the effects of beta radiation is afforded by ordinary clothing, therefore the risks from external sources are slightly lower than for alpha particles. Also, because of their penetrative properties, beta emissions from inside the body may penetrate to the outside and thus also pose less of a risk than alpha particles, although this may still be a significant risk.

Gamma radiation and X-radiation may pass through the human body depositing little energy on the way.

UNITS

There are a number of units used to describe the characteristics and effects of ionizing radiation. The more common ones are explained below [3].

Radioactivity

This represents the 'activity' of a radioisotope in terms of the number of disintegrations occurring each second. The unit is referred to as the '**becquerel**' (Bq), and is defined as:

$$1 \text{ Bq} = 1 \text{ atomic disintegration per second}$$

This is a rather small unit and is of dubious practical significance because it does not differentiate between the types of ionizing radiation and therefore does not indicate anything about the biological effect of the radiation concerned.

Radiation absorbed dose

This is the amount of energy absorbed by a receiving medium per unit mass from all sources. The unit of absorbed dose is referred to as the '**gray**' (Gy) and is defined as:

$$1 \text{ Gy} = 1 \text{ joule/kg}$$

Absorbed dose is quite a useful physical concept, but as it does not differentiate between the different types of ionizing radiation, it is not a good indicator of damage caused to biological systems.

Radiation dose equivalent

The difference in the 'radiobiological effectiveness' of different types of radiation is taken into consideration by the development of a unit of radiation dose equivalent. This unit is referred to as the '**sievert**' (Sv). The sievert is derived in the form of a weighted absorbed dose. Therefore, the absorbed dose (in grays) is multiplied by a so-called 'quality factor' (Q), which is itself derived to account for the distribution of the absorbed dose. This product is further multiplied by a factor designated 'N', which can be used to take into account further modifying factors. In fact, the value of 'N' has been assigned a value of unity, and so the overall equation for radiation dose equivalent can be expressed as:

$$\text{dose equivalent (Sv)} = \text{absorbed dose (Gy)} \times Q \times N$$

The values assigned to the radiation types mentioned above are 20 for alpha particles and 1 for gamma radiation, X-radiation and beta radiation, reflecting the potential of these radiations to produce biological damage.

Dose rate

Both the gray and sievert are units that relate to an amount of energy that is received without reference to any time period. Radiation hazards may be assessed or controls applied either in terms of a simple dose, or in terms of a dose rate. The dose rate is expressed in either grays per hour or sieverts per hour. The standards that must be applied may be derived, for example, from standards such as those contained in the Ionising Radiation Regulations 1999 [4].

CONTROL OF RADIATION EXPOSURE

Limits of radiation exposure are rather unusual in that they are agreed on an international level. The International Commission on Radiological Protection (ICRP) was formed in 1928 to perform this function. Since then, it has reviewed the available scientific evidence relating to the effects of exposure to ionizing radiation and modified exposure limits where necessary. In the United Kingdom, a great deal of advice and a large variety of published material, as well as monitoring and advisory services, are available from the National Radiological Protection Board (NRPB) [www.nrpb.org]. This organization is now part of the wider Health Protection Agency.

One aspect of protection against ionizing radiation is that the biological damage may be confined to an irradiated individual or, depending on the site of the damage, adverse effects may be manifested in subsequent offspring, indicating a hereditary consequence. These types of effect are referred to as '**somatic**' and '**hereditary**' effects, respectively. Occupational limits (discussed below) reflect this with particular regard to exposure of the abdomen of women [5].

There are two other expressions used by the ICRP in connection with the effects of radiation. An effect for which the probability of the event occurring, rather than its severity, is regarded as a function of received dose without a threshold or trigger level, is referred to as a '**stochastic**' effect. On the other hand, an effect for which the severity varies directly with the received dose, and therefore for which a threshold or trigger value may exist, is referred to as a '**non-stochastic**' effect.

IONIZING RADIATION REGULATIONS 1999

These regulations and the associated Approved Code of Practice [4,6] are made under the provisions of the Health and Safety at Work, Etc. Act 1974 to implement the basic requirements of various European Directives. They lay down basic standards for the protection of workers and the general public against the dangers arising from the use of ionizing radiation in work activities.

One of the requirements of the Regulations is that, with certain exceptions, employers and self-employed persons are required to notify the Health and Safety Executive (HSE) that they are working with ionizing radiation.

As with the Control of Substances Hazardous to Health Regulations 2002 (COSHH) [7], employers are required to take all necessary steps to restrict the exposure of employees and non-employees to ionizing radiation, so far as is reasonably practicable. Specific exposure limits are contained in Schedule 4 of the Ionising Radiation Regulations 1999 [4]. Various dose limits are listed for different groups of employees and different parts of the body.

Examples of these dose limits are:

- for employees aged 18 years and over, 20 mSv in any calendar year
- for trainees aged under 18 years, 6 mSv in any calendar year
- for any person under 16 years of age, 1 mSv in any calendar year.

Areas where people are more likely to receive more than the specified doses must be designated as 'controlled' or 'supervised' areas, and access must be restricted. Employees who receive more than the specified doses of radiation are required to be specially designated as 'classified persons'. The doses of radiation received by such persons must also be monitored and recorded by dosimetry services approved by the HSE. Also, certain employees must be subjected to medical surveillance. Radiation levels have to be monitored in controlled and supervised areas.

Finally, every employer who works with ionizing radiation is required to make an assessment of the hazards that are likely to arise from the work. Where more than the specified amounts of radioactive material are involved, the assessment must be sent to the HSE.

DETECTION AND MEASUREMENT OF IONIZING RADIATION

As with many hazardous environmental agents, the presence of ionizing radiation must be discovered by employing some form of detection device. As well as establishing the presence of ionizing radiation, it may also produce some form of measurement of radiation levels.

The instrumentation commonly available is based upon one or more of the following devices.

1. With the **ionization chamber**, a small voltage is applied between two electrodes situated in an air-filled chamber, and ions generated by the ionizing radiation flow towards either the anode or the cathode, depending upon their electrical charge. The ionization process therefore generates an electrical current that is in proportion

to the radiation intensity, and measuring the current allows the radiation intensity to be measured.

2. The **Geiger-Müller tube** also works on the principle of gas ionization. However, here a very high voltage is applied between the electrodes. This accelerates the electrons formed upon ionization, which themselves cause further ionizations and hence a multiplication of the current generated. Because of this form of amplification, a single ionization can generate a detectable pulse of current.

3. The action of the **scintillation counter** relies upon a solid material, the atoms of which can be excited by the ionizing radiation, thus storing energy. When the electrons return to their original energy state, the energy that was stored is lost, usually in the form of photons of visible light, hence the term 'scintillation'. The scintillations produced are detected and amplified by a photomultiplying device, and an electrical signal is produced that is proportional to the energy of the original ionizing radiation.

4. The **film badge** is the most common form of dosimeter, consisting of a small section of photographic film encased in a specially designed holder. There are many different types, but essentially the film contains coatings of different sensitivities. The film is blackened by impinging radiation, and the degree of blackening (when referred to a standard) can be related to an approximate radiation dose. The holder is designed to contain a variety of sections with different screening or filtering properties, and therefore by comparing the blackening that occurs between different density screens, the nature of the radiation type can be assessed.

These devices are used extensively for logging personal, whole-body exposure.

5. The **thermoluminescent dosimeter** is a device that uses the properties of certain materials to emit light photons in response to impinging ionizing radiation. These devices can give an electrical signal related to radiation dose, which can then be processed and stored on logging equipment.

Radon

Issues relating to the occupational exposure to radon are discussed in Chapter 38.

REFERENCES

1. Goldfinch, E.P. (ed.) (1989) *Radiation Protection – Theory and Practice*, Institute of Physics, London.
2. Martin, A. and Harbison, S.A. (1996) *An Introduction to Radiation Protection*, 4th edn, Arnold, London.
3. National Radiological Protection Board (1998) *Living with Radiation*, NRPB, 5th edn, Harwell.
4. Ionising Radiation Regulations (1999) S.I. 1999. No. 3232, HMSO, London.
5. United Nations Environment Programme (1991) *Radiation Doses, Effects and Risks*, Blackwell, London.
6. Health and Safety Executive (2000), *Work with Ionising Radiation: Approved Code of Practice and Guidance* L121. HSE Books, Sudbury, Suffolk.
7. Control of Substances Hazardous to Health Regulations 2002. S.I. 2002. No. 2677, HMSO, London.

26 Non-ionizing radiation

Paul C. Belcher and John D. Wildsmith

Non-ionizing radiation is generally regarded as being that part of the electromagnetic spectrum with wavelengths greater than 0.1 nm. Should radiation of this energy level impinge upon tissue, the energy imparted is not sufficient to produce excitation or ionization in the form described in Chapter 20. Non-ionizing radiation may still be hazardous, however, as it can cause tissue disruption by thermal damage, and there is a growing body of opinion about a number of less tangible effects upon the human body, which include a number of specific hazards in the workplace. There are many parts of the electromagnetic spectrum that should be considered under the umbrella of 'non-ionizing radiation'. These are identified under separate headings. It must be appreciated that the radiations consist of part or parts of a continuous spectrum and that the divisions employed are therefore somewhat artificial. They are, however, necessary to assist our understanding of this form of potential hazard.

The various types of radiation are described in an order ranging from the shorter wavelengths to the longer wavelengths. Devices such as lasers, which may emit more than one type of non-ionizing radiation, are then considered. The approximate wavelengths of the principal types of non-ionizing radiation are indicated in Fig. 26.1. It should be noted that most forms of non-ionizing radiation fall outside the visible part of the electromagnetic spectrum and that the hazard created may therefore not be detected until actual damage has been done. This means that the potential risk posed by all sources of non-ionizing radiation must be understood and appropriate action, including surveys using appropriate monitoring equipment, must be carried out wherever a risk exists.

ULTRAVIOLET RADIATION

This form of radiation is emitted naturally by the sun as well as a number of artificial sources such as insect killing devices, mercury discharge lamps, photocopiers and sun beds. There are three principal classes of ultraviolet radiation, which are broadly based on the physical properties and biological effect of the radiation. These three classes are: UV(A), with a wavelength of 315–400 nm; UV(B), with a wavelength of 280–315 nm; and UV(C), with a wavelength of 100–280 nm.

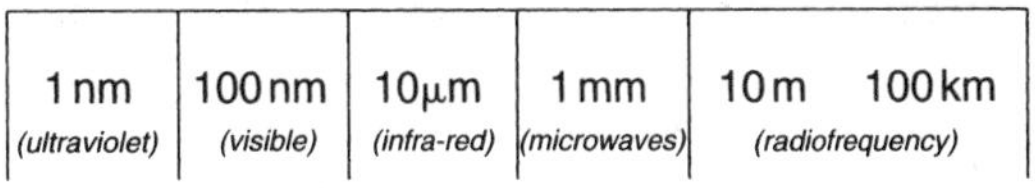

1 nm	100 nm	10 μm	1 mm	10 m 100 km
(ultraviolet)	(visible)	(infra-red)	(microwaves)	(radiofrequency)

Fig. 26.1 The electromagnetic spectrum.

The sun emits ultraviolet radiation in all of these classes, although the atmosphere will absorb most of the UV(A) and UV(B), particularly if pollution levels are high. Exposure to solar radiation is increasingly being associated with a growing incidence of skin cancer.

Biological effects of exposure

These depend on a number of factors, such as the energy of the radiation concerned, the dose received and the part of the body affected. This can produce both short- and long-term effects, which are generally well understood.

Skin effects

Short-term effects are erythema (reddening of the skin), which is similar to sunburn, as the blood vessels are dilated and the blood supply to the affected tissues increases. UV(A) induces rapid tanning of the skin as a result of the existing pigmentation (melanin) being stimulated. This does not give a long-term tan. UV(B) and UV(C), which are of higher energy than UV(A), tend to produce a long-term tanning effect by actually increasing the activity of the melanin-generating cells in the skin. It should also be noted that UV(A) has greater penetrative properties and can therefore produce adverse effects in deeper tissues.

A further long-term effect is that of premature ageing of the skin. The skin becomes dry, cracked and thickened. A more serious effect is the induction of a variety of skin cancers, although the class of ultraviolet radiation most likely to induce such cancers is unknown.

The main conclusion of a report funded by the Health and Safety Executive (HSE) [1] is that current recommendations on exposure limits to ultraviolet radiation (UVR) generally afford adequate protection against acute effects to the skin, but may not be adequate when considering eye data, particularly in the spectral region 310–340 nm.

Eye effects

Ultraviolet radiation affects the eye in a number of ways. Inflammation of the cornea and conjunctiva can be induced, which are painful conditions although there are usually no permanent effects. This irritation may, though, have other safety consequences, for example, when using dangerous machinery or handling hazardous chemicals in the workplace, or even driving motor vehicles or forklift trucks.

Ultraviolet radiation can also damage the retina if large doses are received, although this is not common as the lens and cornea tend to absorb the energy. Children are at particular risk as their eyes transmit much more UV(A) to the retina than do those of older people.

The principal long-term effect to the eye that has been identified is the production of opacity in the lens. This is known as cataract formation.

It should therefore be appreciated that any equipment considered to pose a risk to the health of workers or the public must be dealt with within the provisions of the Health and Safety at Work, Etc. Act 1974. Welders, for example, should wear suitable goggles or a mask, and they must also ensure that any passersby are not exposed to the ultraviolet radiation generated (in addition to other hazards).

Ultraviolet lamps must be of an appropriate type, for example, in insect killing devices. Equipment such as photocopiers must be properly enclosed, and staff who use the equipment must be properly instructed in the use of the machines and apprised of any risks involved.

One difficult area is commercial ultraviolet tanning equipment. This is still very widely used and here people deliberately expose themselves to a source of ultraviolet radiation. The HSE has recognized this risk and has produced a guidance note [2], which gives the scale and nature of the risks and offers advice on how these risks can be minimized. One extremely important area covered in this document is the fact that customers must be given information before using this type of equipment so that they can evaluate the risks involved, and have the freedom to choose whether to use the equipment. It is, however, important that the information provided is actually understood, and that the equipment is tested as far as is reasonably practicable, to ensure that it falls within the HSE guidance.

Further examination of the risk of skin cancer (cutaneous melanoma) arising from exposure to fluorescent lights and ultraviolet lamps can be found in the work of Swerdlow *et al.* [3]. Additional information can be found on the NRPB and HSE websites (www.nrpb.org and www.hse.gov.uk).

VISIBLE LIGHT

A lack of adequate lighting can render any workplace hazardous, and this is the main problem associated with this type of non-ionizing radiation. The requirement to provide adequate levels of illumination is dealt with in Chapter 20. It should, however, be noted that there are some additional hazards associated with visible light. One of the principal hazards is dealt with in the section on lasers. Unsuitable forms of lighting can introduce an element of hazard, for example, when fluorescent lights are used to illuminate moving equipment. It is possible that since the fluorescent light will 'flicker' at a high rate that cannot normally be detected by the human eye, it may produce a stroboscopic effect on moving parts of machinery, giving the illusion that a rapidly spinning fan, for example, is static. Clearly, this could prove to be hazardous and it is therefore important to ensure that the type of lighting used is suitable for the purpose.

Very bright lights, such as are sometimes found in welding operations and in places of entertainment, can induce temporary blindness. People working in such situations must wear suitable forms of eye protection.

INFRA-RED RADIATION

All objects that have a measurable temperature emit infrared radiation. Normally, the sensory mechanisms of the body will detect a build-up of heat before tissue disruption occurs. Infrared radiation can produce effects similar to sunburn, that is, a general irritation of exposed skin. It must be noted, however, that as with some other forms of non-ionizing radiation, the eye is particularly vulnerable to damage: parts such as the lens have no protective, cooling mechanism and therefore as the heat builds up, damage such as coagulation of the proteins in the lens can occur, giving rise to cataract formation.

Quite a wide variety of infrared sources are encountered in the workplace, such as lights (including some lasers) and heaters. The normal solution to the problem of exposure to infrared radiation is to wear eye protection. This means that employers must be able to select and provide suitable protective equipment and supply adequate information to staff who may be exposed to this risk. British Standard BS EN 175 [4] provides further guidance on this subject.

MICROWAVES AND RADIOFREQUENCY RADIATION

These two types of non-ionizing radiation are very similar and are often dealt with together. One way of defining each term is to call radiations with wavelengths of between 10 m and 10^4 m 'radiofrequency radiation', and those with wavelengths of between 1 mm and 10 m 'microwave radiation'. For the purposes of control of this type of radiation in the working environment, these figures should be regarded as a guide rather than a precise definition.

Microwave radiation

Microwave radiation is principally used in communications systems, heat-treatment processes in industry, ranging devices and devices for the heating and cooking of foodstuffs. The proliferation of such devices over a relatively short period of time has led to the need for an understanding of the biological effects of exposure to this type of radiation. Microwaves (as with other forms of non-ionizing radiation) can be generated as a continuous output or in very short bursts of energy. This factor can be important, for example, when using detection devices, since the average power delivered from a pulsed system will be less than the peak power. A high peak power delivered at a rapid pulse repetition rate could cause significant tissue damage, while a continuous output of the same mean power may be far less damaging. This aspect of non-ionizing radiation is also mentioned in connection with lasers.

The development of exposure standards for microwave radiation in the United Kingdom is based on the degree of thermal damage caused. This is not the same as in certain other countries where rather more subtle effects on the human body are attributed to this form of energy. The currently accepted UK upper exposure limit is 10 mW/cm^2 for continuous generators, with a lower limit of 5 mW/cm^2 [5].

The most common source of microwave energy is probably the microwave oven. BS EN 60335 [6] relates to the construction standards and testing of such appliances. Equipment should be in a good state of repair and particular attention should be paid to the condition of the door seals as the accumulation of food debris can allow radiation leakage.

Research into other effects caused by microwaves, such as hormonal changes and effects on the nervous system, continue. One useful source of reference on this issue is found in an NRPB report on electromagnetic radiation and risks of developing cancer [7].

Radiofrequency radiation

Radiofrequency radiation is principally used for communications. However, it is also employed quite extensively in induction heating units for processing a wide variety of materials such as wood, plastics, paper, ceramics and textiles. Such units must be designed and sited correctly to ensure that stray energy is not allowed to affect workers or passersby. This involves screening the radiofrequency generating unit efficiently, as at suitably high exposure levels, this form of non-ionizing radiation can produce deep-seated tissue damage.

There are more extensive problems in particular industries and the HSE has produced several guidance notes of some significance to environmental health officers. The advice contains useful guidance on the monitoring equipment and techniques that should be used in the detection and control of radiofrequency radiation in the workplace. The International Labour Office has also published a useful practical guide [8].

LASERS

The word laser is an acronym for the term 'light amplification by the stimulated emission of radiation'. These devices produce a unique kind of non-ionizing radiation, usually with a very small range of wavelengths in the form of a collimated beam, that is, a beam of energy that can be considered not to disperse as it moves through the environment, hence the energy is contained within the beam to a large extent, and can therefore be transmitted over great distances.

Laser radiation may be in the ultraviolet, infrared or the visible part of the electromagnetic spectrum. The most common forms are visible lasers, and thus the radiation is normally referred to as laser 'light' (typically with wavelengths of 0.4–0.7 μm).

There are several principal types of laser device named after the lasing material used: solid state (crystal) gas semi-conductor; and liquid dyestuffs.

Lasers are commonplace in work situations such as product launches, night clubs, theatres, outdoor and indoor displays, supermarket checkout systems, beauty therapy studios and the repair of devices such as compact disc players and DVD players.

General guidance concerning the assessment of safety is available in two significant documents: European Standard EN 60825 [9] and an HSE publication [10].

European Standard EN 60825

This standard [9], which also has the status of a British Standard, describes the biological effects of lasers and includes details of how to assess the risk posed to the skin and eyes in the form of a series of algorithms. It also classifies laser products into one of four basic types depending on the power output to which a person has access.

Class 1

These are inherently safe laser products that are either very low-power devices, or are high-power devices that are designed in such a way that the power to which access can be obtained does not exceed the maximum permissible exposure level, for example, by the use of interlocked enclosures.

Class 2

These are low-powered products emitting visible radiation. They are not inherently safe, although a suitable degree of protection is normally provided by the body's normal aversion responses, including blinking of the eye.

Class 3

This class is divided into 3A and 3B. Type 3A products have a power output of visible radiation of up to 5 mW for continuous lasers, and 5 mW peak power for pulsed and scanning lasers. The power classification is also limited to 2.5 mW/cm², and so protection is afforded to the eye by aversion responses. However, the use of common optical devices, such as binoculars, may prove to be hazardous unless they contain suitable filters.

Type 3B products are more powerful, and these lasers may be visible or invisible. Continuous lasers must have a maximum power of 0.5 W, while the limit for pulsed lasers is specified at 10^5 J/m². Direct viewing of laser beams of this intensity is hazardous, and reflection from mirrored surfaces (such as may readily be found in places of entertainment) is also deemed to be hazardous.

Class 4

This class is the highest laser class and contains all lasers with an output that exceeds that specified for class 3B. In addition to biological hazards, this class may pose a risk of fire as the beam is so powerful. Where such systems are used to provide entertainment, for example, a public display, it is advisable for the Health and Safety at Work, Etc. Act 1974 enforcing authority, the Public Entertainment Licensing Authority, and the fire authority to liase to ensure the safety of all associated with the event.

Health and Safety Executive guidance [10]

This document reiterates the European standard. It is, however, designed to cover the situation of indoor and outdoor laser displays, which are usually the situations where the most powerful laser products and the most dangerous systems in terms of additional hazards are encountered.

A helpful practical guide to the issues that must be considered by environmental health officers is included, and covers the following matters.

Prior to the public use of any display laser product, the operator of the system must supply the enforcing authority with sufficient information to assess the risks to employees and members of the public who may view the display. This information will include appropriate radiometric data shown on diagrams of the beam paths. Relevant calculations should also be provided before the display takes place.

Once the information is provided it can be used in conjunction with a series of tables that give the maximum permissible exposure levels (MPELs) for the eyes and skin and accessible emission limits (AELs) for the various classes of laser. The rather complex data in the various tables are converted into simple graphical representations that are easy to interpret. An example of the tabulated information along with the corresponding graphical data for the MPEL for direct ocular exposure of the cornea to laser radiation is shown.

With regard to the operation of the laser system, the display area should be demarcated using the calculations made by the operator to ensure that the audience or the general public (as well as the operators and any performers) are protected from the laser radiation within the guide limits. The level of exposure should not exceed the MPEL at any point where the public is permitted during the display and, in addition, unless effective means are employed to prevent access to the laser beam(s), the MPEL should not be exceeded at any point:

- less than 3 m above any surface upon which the audience or general public is permitted to stand

• less than 2.5 m in lateral separation from any position where a person in the audience or the general public is permitted during the display.

Where displays are outdoors, consideration must be given to people who are liable to view the beam directly within the normal optical hazard distance and also those who might view the beam or its reflections using optical aids. All publicity material, including tickets and posters, must contain appropriate warnings where relevant to ensure that viewing aids such as binoculars and telescopes are not used.

In the case of people who carry out maintenance work on all types of laser products, there may be a considerable risk. They must receive proper training and must be equipped with appropriate eye protection while carrying out maintenance operations. They must also ensure that safety interlocks are not overridden, and that passersby are not affected when systems are being tested.

Finally, the risks posed to health and safety enforcing officers must not be ignored. It is important that where such people are likely to be put at risk, they too are equipped with suitable forms of protection. They must also be supplied with, or should be able to obtain the use of, a suitable form of measuring equipment.

REFERENCES

1. National Radiological Protection Board (1994) *A Critique of Recommended Limits of Exposure to Ultraviolet Radiation with Particular Reference to Skin Cancer*, NRPB for the HSE, HSE Books, Sudbury, Suffolk.

2. Health and Safety Executive (2003) *Controlling Health Risks from the Use of Ultraviolet Tanning Equipment*, INDG209, HSE Books, Sudbury, Suffolk.

3. Swerdlow, A.J., English, J.F.C., Mackie, R.M., *et al.* (1988) Fluorescent lights, ultraviolet lamps and the risk of cutaneous melanoma. *British Medical Journal*, 297, 10 September, 647–50.

4. British Standards Institution (1997 as amended) *Personal Protection Equipment for Eye and Face Protection during Welding and Allied Processes*, BS EN 175, BSI, London.

5. Health and Safety Executive (2001) HSE/Local Authority Enforcement Liaison Committee (HELA) Local Authority Circular 60/3 Microwave Ovens – Exposure Control.

6. British Standards Institution (2003) *Safety of Household and Similar Electrical Appliances. Particular Requirements: Microwave Ovens including Combination Microwave Ovens*, BS EN 60335-2-25, BSI, London.

7. National Radiological Protection Board (2003) *ELF Electromagnetic Fields and the Risk of Cancer*, Report of an Advisory Group on Non-Ionising Radiation, NRPB, Harwell.

8. International Labour Office (1994) *The Protection of Workers from Power Frequency Electric and Magnetic Fields*, ILO Publications, Geneva.

9. British Standards Institution (1994) *Radiation Safety of Laser Products, Equipment Classification, Requirements and User's Guide*, European Standard EN 60825, BSI, London.

10. Health and Safety Executive (1996) *The Radiation Safety of Lasers Used for Display Purposes*, HS(G)95, HSE Books, Sudbury, Suffolk.

Part Seven
Food Safety and Hygiene

27 Introduction to food safety

Madeleine Smith

BACKGROUND

Production and consumption of food in the twenty-first century is a complex process. Food production may occur in small-scale premises following traditional methods, in massive factory units using engineering systems such as just-in-time deliveries and in all other combinations and scales in between. Food production and consumption is also an international business, with food being imported and exported in a complex web involving fresh produce, raw materials and finished product.

INFRASTRUCTURE

Control of the production and sale of food involves a number of agencies at the international, national and local level. Some of these agencies are funded and empowered on the expectation that their primary role will be in public protection. Other agencies, whose funding dictates a primary interest in particular areas such as surveillance, research or trade, still have a significant input into safe food production. Although technically independent of each other, the nature and scale of modern food production causes interplay between the major agencies described below, and the outcome of this interaction may have a major impact on the enforcement of food safety legislation at the ground level.

International

The European Union

As is appropriate for an organization whose roots are firmly based in matters of trade, one of the main objectives of food legislation so far implemented by the European Union (EU) in member states has been to ensure free movement of goods across the borders of member states. The Official Control of Foodstuffs Directive (89/397/ECC) [1] set down standards of hygiene and control that must be applied in food production in member states. The purpose of the Directive was to harmonize the inspection and control of production of foodstuffs across member states in such a way that consumers' interests were protected without erecting barriers to trade. Then a debate in the European Parliament in 1997 [2] (itself arising from concerns about what has become popularly known as the 'BSE Crisis') precipitated a reconsideration of the current regulatory system and the rationale behind it. More recent documents have highlighted the importance, not just of consumers' interests but specifically of health protection and food safety. The White Paper on Food Safety (COM (1999) 719 final) [3] states that a key policy of the Commission is assuring the highest standards of food safety within the EU. This policy is reflected in subsequent legislation, for example, the Regulation that establishes the European Food Safety Authority, Regulation (EC)

No 178/2002 [4]. Article 1 (the Aim and Scope) of the Regulation states: 'This Regulation provides the basis for the assurance of a high level of protection of human health … in relation to food … .'

Another important principle is that the requirements and standards supported by the EU should be based upon good scientific grounds. This, too, is stated in Regulation (EC) No 178/2002 [4], and is carried through the other food hygiene proposals currently under consideration by the Council, Parliament and Commission.

Directorate General Health and Consumer Protection

The Directorate General Health and Consumer Protection is part of the European Commission. It is the Directorate General that is responsible for ensuring that a high level of human health and consumer protection is attained throughout the EU. The subject areas under its direct jurisdiction include Public Health, Food Safety, Veterinary and Phytosanitary Standards and Controls including Animal Welfare, Scientific Advice and Consumer Protection.

The Directorate General Health and Consumer Protection drafts the regulations, directives and decisions relating to food safety in member states. It also carries out inspections of food premises in member states to determine if member states are complying with Community standards. As a consequence the DG Health and Consumer Protection has an advisory, legislative and enforcement role in food safety.

European Food Safety Authority

The Regulation that established the European Food Safety Authority, Regulation (EC) No 178/2002, [4] was adopted on 28 January 2002. The main purpose of the European Food Safety Authority is to provide information to the EC on matters of food safety. The primary focus of its activities will be in scientifically assessing food safety risks. The Authority will be expected to develop scientific and technical information related to food safety through a system of technical committees. These committees should be formed of independent experts appointed for their scientific expertise. Other Agency tasks will include safety evaluations of new processes or substances wishing to gain community-level approval, collection, assessment and collation of scientific data, and identification of emerging risks through surveillance and monitoring. The Agency will also set up networks that allow closer co-operation and liaison between food safety organizations (agencies) in different member states and be included in the Rapid Alert System managed by the Commission. Although the European Food Safety Authority may communicate with or respond to independent requests, the European Commission, Council of EU Ministers and EU Parliament will retain the responsibility for risk management and communication. The role of the European Food Safety Authority is mainly that of risk assessment. Further information may be found on the Authority's website http://www.efsa.eu.int

Food and Veterinary Office

The Food and Veterinary Office (FVO) is the agency responsible for monitoring the implementation of and compliance with food safety legislation within member states and elsewhere, for example in non-member states wishing to export food to the EU. The FVO is part of the Directorate General Health and Consumer Protection. The FVO carries out inspections in member states and in third countries exporting to member states to ensure that the EU legislation is being correctly enforced by the competent authority. The reports of the inspections are published and are available on the FVO's website http://europa.eu.int/comm/food/fs/inspections/index_en.html. The FVO designs an annual inspection programme (also available on the website) which it carries out using the FVO inspectors.

Codex Alimentarius Commission

The Codex Alimentarius Commission is an organization that was created in 1963 by the Food and Agriculture Organization (FAO) and World Health Organization (WHO). The organization's objectives are to protect the health of the consumers and ensure fair trade practices in the food trade. The main role of Codex Alimentarius is to develop guidance and codes of practice on food standards (including hygiene) which members may use. This is carried out primarily through a programme known as the Joint FAO/WHO Food Standards Programme. This programme promotes co-ordination of all food standards work undertaken by international governmental and non-governmental organizations and results in publications on hygiene, HACCP, microbiological standards and product specific standards. A list of these guidelines and codes of practice may be found on their website http://www.codexalimentarius.net/

All Member Nations and Associate Members of the FAO and WHO are eligible for membership of the Codex Alimentarius Commission. National delegations to meetings of the Codex Alimentarius Commission are led by appointed senior government officials. Delegations may also include representatives of industry, consumers' organizations and academic institutes if invited by the member government. Countries that are not yet members of the Commission sometimes attend meetings as observers.

The authority of the Codex Alimentarius Commission is advisory. It does not have enforcement powers. However, member governments and the European Commission take note of the advice offered, and may use the guidance and standards issued as a basis for international trade agreements, national legislation or importation standards.

National

Food Standards Agency

The Food Standards Agency (FSA) was established on 1 April 2000 as a result of the Food Standards Act 1999 [5]. It is a powerful independent Agency whose main purpose is to protect the public's health and consumer interests in relation to food. The FSA was established to address concerns about food safety enforcement that had arisen during the previous decades. According to the James Report published in 1997 [6], these concerns included:

- the potential for conflicts of interest within the Ministry of Agriculture Fisheries and Food (MAFF) arising from its dual responsibility for protecting public health and for sponsoring the agriculture and food industries;
- fragmentation and lack of co-ordination between the various government bodies involved in food safety;
- uneven enforcement of food law.

Consultation on the James report in the form of a White Paper in 1998 (The Food Standards Agency: A Force for Change) [7] indicated that there was broad support for the development of an independent agency such as the FSA and the Government initiated the process to establish it. This involved a number of interim steps involving those Ministers and Departments responsible for food safety at the time. In September 1997, a Joint Food Safety and Standards Group (JFSSG) had been set up comprising both MAFF and Department of Health (DoH) staff. The group was headed by a DoH official and handled matters of food safety as an interim measure while at the same time developing the process that would result in the new FSA. The JFSSG became part of the FSA when it was established in April 2000.

Although the FSA is a Government agency, it is said to be an 'arm's length' Agency in that it is independent from Government. The Agency is accountable to Parliament through Health Ministers, and to the devolved administrations in Scotland, Wales and Northern Ireland for its activities within their areas, but is free to publish any advice, research or guidance independently of the government position or opinion.

The Agency is led by a Board that has been appointed to act in the public interest and not to represent particular sectors. Board members have a wide range of relevant skills and experience. Officers are civil servants and/or seconded experts appointed through open competition or nominated.

The Agency has responsibility for drafting secondary legislation in food safety and food standards. It has overall responsibility for enforcement in food safety matters and liaises with the EC and Codex Alimentarius Commission on food safety and standards matters.

The FSA has no direct management responsibility for Local Authority food enforcement sections, but it does have a proactive role in setting and monitoring standards with respect to enforcement. The Agency has the power to

- set standards for enforcement;
- monitor local authority enforcement activity and performance;
- require information from enforcement authorities;
- audit local authorities;
- publish information on local authority performance.

In addition to the above mentioned powers under the Food Standards Act 1999, under the Food Safety Act 1990 (Section 42) [8] the Agency may also be empowered by the Secretary of State to discharge the duties of a food authority in place of that food authority if the food authority is deemed to be failing to discharge those duties. The Agency may also commission research. As a consequence, the FSA has legislative, enforcement, advisory and research authority.

The Agency is funded with public money from Central Government. The Agency is managed and run by an appointed board who are expected to act in the interests of the public rather than represent any particular sector. The board holds regular meetings and is unusual in that these meetings are held in public. The agendas, papers and decisions are also published. The main Headquarters of the FSA is based in London. There are, of course, other premises for the FSA Scotland, FSA Wales, and FSA Northern Ireland, but the London HQ houses the staff with agency wide responsibility and certain specialisms. The staff at the London Headquarters are divided into three groups covering the main roles and responsibilities of the Agency. These three groups are:

The Food Safety Policy Group
The Enforcement and Food Standards Group
The Corporate Resources & Strategy Group.

FOOD SAFETY POLICY GROUP   The main responsibility of the Food Safety Policy Group is food safety and nutrition. Within the group there are a number of specialist divisions responsible for particular areas of food safety and nutrition.

The Novel Foods Division develops and implements policy on novel foods, including genetically modified foods, functional foods and food irradiation.

The Chemical Contaminants & Animal Feed Division. As the name suggests this division is responsible for chemical contaminants in food and for animal feed. The chemical contaminant side of the division provides advice on such potential food contaminants as heavy metals, mycotoxins, organic chemicals and radio-nuclides, as well as process contaminants. The Animal Feed section negotiates and administers international standards for the composition and labelling of animal feeds and develops policy on these matters.

The Chemical Safety & Toxicology Division sets standards for food additives and food contact materials, and carries out research on food allergies. This division sets safety limits for chemicals in food, and considers the safety of natural toxicants, pesticides and veterinary medicines.

The Radiological Protection and Research Management Division works on methods of analysis and sampling, and advises on radioactive waste disposals and food irradiation. This division also co-ordinates the Agency's research programmes.

The Food Chain Strategy Division is responsible for investigation of food safety and standards throughout the food chain. It co-ordinates work on assurance schemes, organic foods and traceability.

The Microbiological Safety Division is responsible for developing the strategy for reducing foodborne illness. Other responsibilities include the promotion of food hygiene and the development of guidance for the food industry and the public. This division also deals with microbiological food hazards, liaising with the Public Health Laboratory Service and local public health officials as appropriate.

The Nutrition Division provides scientific data on the nutritional make-up of foods. It provides guidance and develops policy on the relationship between diet and health.

ENFORCEMENT AND FOOD STANDARDS GROUP   As the title implies this group co-ordinates enforcement of food legislation and deals with food standards and hygiene matters. There are several divisions within the group, specializing in various aspects of food hygiene and standards.

The **Food Labelling Standards & Consumer Protection Division** is responsible for developing policy on food labelling and authenticity; functional foods and dietary supplements. This division is also involved in the international aspects of food legislation, liaising with Codex Alimentarius and other international bodies on trade, food composition and quality standards; processing and transport.

The **Local Authority Enforcement Division** develops policy and guidance on enforcement, especially with regard to Local Authority food law enforcement and international food enforcement issues. It is responsible for monitoring and audit and also interacts with the EU on food legislation and other matters.

The **Meat Hygiene Division** is responsible for meat inspection and food safety in licensed meat plants (this includes fresh meat, poultry meat, farmed and wild game meat). Enforcement is carried out by qualified meat inspectors and veterinary surgeons. This division also developed the Clean Livestock Policy.

The **Meat Science & Strategy Division** is responsible for improving public health protection through the modernization of meat hygiene control. The Division manages the Agency's meat hygiene and BSE research contracts.

The **BSE Division** deals with the food safety implications of BSE and other transmissible spongiform encephalopathies. It also administers the Meat Hygiene Appeals Tribunal.

The **Veterinary Public Health Operations Division** has auditing and licensing roles as well as advisory and policy responsibility. It licenses meat plants and approves meat and combined meat products premises; it also audits the Meat Hygiene Service. Other responsibilities include ensuring animal welfare at slaughter and providing veterinary public health advice.

The **Communications Division** sets the strategy for the Agency's external communications. This division is responsible for electronic and printed communication and liaises with the media.

CORPORATE RESOURCES & STRATEGY GROUP   The Corporate Resources and Strategy Group plays a co-ordination and administrative role in the Agency. It comprises personnel, financial, statistical and secretariat sections and provides services in these areas to the Board, the Chief Executive and the staff of the Agency. It also provides strategic management advice and liaises with other international bodies.

Details about the structure and work of the FSA can be found on their website http://www.food-standards.gov.uk/[1]

Department of Health

Prior to the development of the FSA, the DoH was responsible for issues of food hygiene, food safety and nutrition. The majority of these duties have now been allocated to the FSA, including the role of liaising with the EC, drafting secondary legislation and providing enforcement guidance for food authorities. However, a number of useful documents were published for the DoH during the time it held responsibility for food safety. These are still in print and applicable (e.g. Department of Health Food Handlers Fitness to Work Guidance for Food Businesses [9]) and are published under the auspices of the DoH rather than the FSA.

Ministry of Agriculture, Fisheries and Food

The perceived conflict of interest that resulted from the MAFF's joint responsibility for the farming and fishery industries as well as for food safety was one of the issues that led directly to the development of the FSA in 2000. As with the DoH, the majority of MAFF's food safety responsibilities have now been taken over by the FSA, but some older documents published by the MAFF are applicable to food safety, for example the

---

[1]Now also see www.food.gov.uk/enforcement/ (which was established in late 2003 to assist food law enforcers) and www.food.gov.uk/imports (which holds useful information for consumers, food businesses and enforcers).

Guidance Notes on the Minced Meat and Meat Preparations (Hygiene) Regulations 1995 [10].

LACORS

LACORS (the Local Authorities Coordinators of Regulatory Services) is the new name for the organization formerly known as LACOTS. The acronym LACOTS stood for the Local Authorities Coordinating Body for Trading Standards and Food. This indicates its original interest in food standards and safety enforcement. LACORS is a central association developed by the UK Local Government Associations (the Local Government Association, Welsh Local Government Association, Convention of Scottish Local Authorities and Northern Ireland Local Government Association) to provide guidance and support to these organizations. LACORS has a Board of Directors and Management Committee made up of senior Elected Members. These Elected Members are nominated by the four UK Local Government Associations mentioned above. Its role in food safety is to support Local Authority Officers by giving advice on matters of food safety and standards and enforcement. It does this via circulars and other publications. Its role is advisory. Officers are appointed via open competition or nomination. The organization is paid for by Local Authorities, that is, by its members and further information can be obtained about its activities from its website http://www. lacors.gov.uk

Other

In addition to the main national organizations mentioned above, there are many other organizations with an interest in food and food safety that provide advice, training and/or guidance for members, the industry or the public. Examples of such organizations include the Institute of Food Science and Technology, the Chartered Institute of Environmental Health, The Royal Institute of Public Health and Industry associations such as the Bakery Alliance. These organizations can be excellent sources of guidance and information and may form powerful lobbies, although they do not generally have a direct regulatory role.

Local

Food Authority

The Food Safety Act 1990 [8] identifies the Food Authority as the organization with responsibility for enforcement of the Act at the local level. The Food Authority is defined as the council of each London Borough, district or non-metropolitan county, the Common Council in the City of London and the island or district councils in Scotland. Local Authorities with responsibility for food safety enforcement are required to produce a service plan for food enforcement that details

- the services offered by the Local Authority;
- the means by which those services will be offered;
- the way in which the standards required by the services will be achieved; and
- how performance of the service will be reviewed.

The FSA considers these service plans integral to its data gathering and audit process, and has issued guidance for local authorities on the matter in Framework Agreement on Local Authority Food Law Enforcement [11].

The Food Authority is required to authorize properly trained and qualified people to act on the authority's behalf to enforce the Act and associated Regulations. The Food Safety Act 1990 details the powers and duties of these authorized officers of the Food Authority. These 'authorized officers' are frequently Environmental Health Officers, although other officers with appropriate training and qualifications may also be authorized to discharge certain duties, for example, Trading Standards Officers, Veterinary surgeons, holders of the Higher or Ordinary Certificate in Food Premises Inspection. Information and guidance on proper qualifications may be found in Code of Practice No. 19 [12] associated with the Food Safety Act 1990 and the Framework Agreement on Local Authority Food Law Enforcement [11].

Although each Food Authority acts autonomously in the discharge of their duties under the Food Safety Act 1990, it is important that the legislation is enforced consistently and accurately across the country. There are several systems that assist Food Authorities in this matter.

CODES OF PRACTICE

Section 40 of the Food Safety Act 1990 [8] gives the Secretary of State the power to issue codes of recommended practice with regard to the execution and enforcement of this Act and of regulations and orders made under it. The FSA is responsible for consulting with organizations that are likely to be substantially affected by the code before issuing it. These codes are not in themselves legislation but are issued for the guidance of food authorities. Food authorities are expected to follow the advice given in them and may be obliged to do so by the FSA after consultation with the Secretary of State.

There have been a number of codes of practice issued under Section 40 of the Food Safety Act 1990, the first being issued shortly after the Food Safety Act 1990 was enacted. Some of the codes have been revised as changes have occurred, but the majority still exist in their original form. Table 27.1 lists the codes as they existed at the beginning of 2004. A major revision of the codes of practice has been undertaken and the final draft of this revision has been released for comment [13]. The proposed draft consolidates the codes of practice into two documents, the first entitled 'Food Safety Act 1990 Code of Practice', and the second entitled 'Food Safety Act 1990 Practice Guidance'. Once agreed, the 'Food Safety Act 1990 Code of Practice' would replace all previously made codes of practice concerning the execution and enforcement of the Food Safety Act 1990 and any regulations made under it.

Table 27.1 Codes of practice

Number	Title
1	Responsibility for Enforcement of the Food Safety Act 1990
2	Legal Matters
3	Inspection Procedures – General
4	Inspection, Detention and Seizure of Suspect Food
5	The Use of Improvement Notices (revised April 1994)
6	Prohibition Procedures
7	Sampling for Analysis or Examination
8	Food Standards Inspections (revised August 1996)
9	Food Hygiene Inspections (revised September 1995)
10	Enforcement of Temperature Control Requirements of Food Hygiene Regulations. Enforcement of Temperature Monitoring and Temperature Measurement (revised February 1994)
11	Enforcement of the Food Premises (Registration) Regulations
12	Quick Frozen Foodstuffs. Division of Enforcement Responsibilities; Enforcement of Temperature Monitoring and Temperature Measurement (revised February 1994)
13	Enforcement of the Food Safety Act 1990 in Relation to Crown Premises
14	Enforcement of the Food Safety (Live Bivalve Molluscs and Other Shellfish) Regulations 1992
15	Enforcement of the Food Safety (Fishery Products) Regulations 1992
16	Enforcement of the Food Safety Act 1990 in Relation to the Food Hazard Warning System
17	Enforcement of the Meat Products (Hygiene) Regulations 1994
18	Enforcement of the Dairy Products (Hygiene) Regulations 1995
19	Qualifications and Experience of Authorized Officers and Experts
20	Exchange of Information between Member States of the EU on Routine Food Control Matters

This Code of Practice will be issued under section 40 of the Food Safety Act 1990 and Food Authorities will therefore be required to follow the Code and implement its provisions. The second document, 'Food Safety Act 1990 Practice Guidance' is issued by the FSA to assist Food Authorities in their enforcement duties. As the title implies, the document contains guidance for Food Authorities on the interpretation of the Food Safety Act 1990 and its associated Regulations Copies of the draft may be accessed at www. consultation.gov.uk.

HOME AUTHORITY PRINCIPLE

The 'home authority principle' (HAP) is a system aimed at improving co-ordination and consistency of standards for companies with sites of operation in many different local authority areas. The local authority area in which the company's decision-making base (usually the head office) is located takes responsibility for ensuring that food safety standards are applied consistently. This includes legal compliance on food safety and hygiene issues. The home authority also acts as a liaison point for other local authorities in their dealings with the company. The system leaves the ultimate sanction of legal proceedings with the authority in which the contravention happened (the enforcing authority). This principle has proved to be an aid to effective enforcement based upon good practice and common sense. The main features for preventing infringements are offering advice at source and encouraging enforcement authorities and businesses to work in liaison with the home authority in order to minimize duplication and public expenditure. It is designed to assist businesses to comply with the law in a spirit of consultation rather than confrontation, although nothing can remove the onus of compliance from a business, or the primary duty of law enforcement by the authority responsible for an area in which a specific food contamination incident or breach of food hygiene law has taken place.

Under the HAP the following definitions are important.

- **The home authority**. This is the authority where the relevant decision-making base of an enterprise is located. It may be the place of the head office or factory, service centre or place of importation. In decentralized businesses, the role and location of the authority is a matter that may require discussion with other authorities, taking the views of the enterprise into consideration.
- **The originating authority**. This term is used to refer to the authority in which a decentralized enterprise produces or packages goods or services. The originating authority is one that will have a special responsibility for ensuring that goods and services produced within its area conform to legal requirements. Most enterprises in the United Kingdom operate from a single base, and in these circumstances the functions of home authority and originating authority are combined.
- **The enforcing authority**. This term refers to all other authorities undertaking inspections, receiving complaints, making enquiries or detecting infringements. It most commonly relates to enforcement in the 'marketplace', but can be extended to include enforcement throughout the distribution chain.

Under the HAP a home authority should be willing to offer counsel and guidance on policy issues, compliance with the law, adherence to recognized standards and codes of good practice and advice on remedial actions.

The home authority is the primary link between enforcement authorities, originating authorities, and enterprises that are based in the home authority area. The responsible officer within the home authority should be willing to advise colleagues and assist in the resolution of complaints and enquiries. Responsibility can be given to a named officer who should maintain a record of relevant incidents, company policies and diligence systems together with a record of significant advice offered to each business within the area.

Because of resource or other constraints, an authority may be unable to accept home authority responsibilities. It might, however, be able to suggest ways in which another authority or group of authorities might assume the responsibility.

Home authorities should be notified by originating authorities of significant findings regarding manufacturing, packaging, servicing and hygiene problems together with details of companies' diligence procedures and commitment to quality assurance. There are likely to be circumstances where an originating authority is able to undertake many of the functions of the home authority on issues that are straightforward.

Each local authority retains its statutory responsibility for the enforcement of the law. However, it may be appropriate when infringements are detected for the enforcing authority to liaise with the home authority or the originating authority before embarking upon detailed investigations or legal actions. This may help to avoid unnecessary duplication and provide the investigating officer with essential background information on company policies or long-standing problems of which the home authority is well aware.

The enforcing authority should ensure that decisions to prosecute and the results of legal proceedings are notified to the home authority as routine. Professional judgement will determine whether brief details of administrative actions such as written warnings, suspension notices or improvement notices need to be referred to the home authority incident record. Minor corporate matters should be referred to the home authority to take whatever action it judges to be appropriate. Further information and guidance on the HAP principle may be found on the LACORS website or in the published guidance on the matter [14,15].

The local authority associations have resolved that LACORS should receive reports of legal actions taken contrary to the advice of the home authority or the advice of a LACORS national panel.

LACORS has advocated the development of HAP throughout the EU and the European Free Trade Area. The European Forum of Food Law Practitioners has adopted the HAP, and in the initial stages LACORS is available to authorities in dealings between those authorities and food enforcement bodies in other member states.

SUMMARY

There are an extensive number of organizations with a stake in safe food production. The interaction of these stakeholders is an important issue in food safety as no action or decision is taken in isolation. Satisfactory understanding of the inter-relationships and the impact theses agencies have on each other is essential to appropriate enforcement of food safety legislation.

REFERENCES

1. Council Directive 93/43/ EEC on the Hygiene of Foodstuffs; OJ L 175; 19.7.93.
2. O'Rourke, R. (1999) *European Food Law*, Palladian Law Publishing, pp. 1–23; 135–46.
3. White Paper on Food Safety; Commission of the European Communities; COM (1999) 719 final. http://europa.eu.int/comm/dgs/ health_consumer/library/ pub/pub06_en.pdf
4. Regulation (EC) No 178/2002 of the European Parliament and of the Council.
5. Food Standards Act 1999; ISBN 0 10 542899 X.
6. James, P. (1997) *Food Standards Agency – An Interim Proposal by Professor Philip James*, the Cabinet Office, London.
7. Food Standards Agency (1998) *A Force for Change*, The Stationary Office, London.
8. Food Safety Act 1990; ISBN 0105416908.
9. Food Handlers; Fitness to Work; Guidelines for Food Business Managers, Department of Health, London, 1996.
10. Guidance Notes on the enforcement of Minced Meat and Meat Preparations (Hygiene) Regulations 1995; MAFF, the DOH, Welsh Office and Scottish Office.
11. Food Standards Agency (2001) *Framework Agreement on Local Authority Food Law Enforcement*, London.
12. Food Safety Act 1990 (Revised 2000) *Code of Practice No. 19: Qualifications and Experience*

of Authorised Officers, Department of Health, MAFF, the Welsh Office and Scottish Office, HMSO, London.

13. Food Safety Act 1990 (2003) draft Code of Practice and Practice Guidance, www.consultation.gov.uk

14. LACOTS (1997) *Guidelines for Home Authorities*, LACOTS, Croydon.

15. LACOTS (1998) *Guidance on Food Complaints*, 2nd edition, LACOTS, Croydon.

WEBSITES

http://europa.eu.int/comm/food/fs/inspections/index_en.html
http://www.codexalimentarius.net/
http://www.foodstandards.gov.uk/
http://www.efsa.eu.int
http://www.lacors.gov.uk
http://www.food.gov.uk/enforcement/
http://www.food.gov.uk/imports

28 Food safety law

Madeleine Smith

LEGISLATIVE PRINCIPLES

Food Safety legislation in the United Kingdom must cover several issues. It must

- set out objectives for food safety, consumer protection and standards of compliance;
- allow the authorities to address new or evolving situations;
- identify the powers and duties of relevant bodies;
- provide the necessary administrative (and other) support to achieve its objectives.

The Food Safety Act 1990 [1] is the main piece of primary legislation that addresses these requirements. The Food Safety Act 1990 maintains some of the traditional legal requirements dating back to the Food Act 1875 and also introduces new ones. Two principles that underpin the Act are:

1. The need to ensure food offered for human consumption is safe, meaning that it is free from any microbiological, toxic or other contamination that would render it unfit.
2. The need to ensure that consumers are not fraudulently treated when purchasing food, meaning that the food should be of the nature, substance and quality demanded by the consumer.

Specific sections of the Food Safety Act 1990 address these principles and use them to set duties and responsibilities on food proprietors that will maintain a high level of food safety and public protection. Details of the specific requirements and standards that must be applied to the production and sale of food are contained in a series of statutory instruments (Regulations) dealing with specific foods, premises or aspects of hygiene.

There are two situations in Britain that typically result in the need for new food safety legislation. One is a need to comply with European food safety law. The other is a response to a food safety incident that has identified a need for some action to further protect the public. The Food Safety Act 1990 allows an appropriate response to either of these two situations by delegating powers to ministers to make regulations. As a consequence some of the Regulations currently in force have originated as EU Directives, for example, Food Safety (General Food Hygiene) Regulations 1995 [2] while others have been purely domestic, for example, Food Safety (General Food Hygiene) (Butchers' Shops) Amendment Regulations 2000 [3].

The Food Safety Act 1990 also provides the powers and administrative support to ensure compliance with the food safety and standards set out. The Act itself contains a number of powers for authorized officers, for ministers and for the courts. Together with a series of associated Regulations the Act also provides the administrative framework for enforcement.

Some of the main provisions of the Act are summarized briefly below, followed by a consideration of the main hygiene regulations.

Food Safety Act 1990

Section 3: presumption on sale of food

Any food or its constituents, if sold from or found in food premises, are presumed to be intended for human consumption. The implication of this section is that all food found in a food premises must be fit to eat unless separated and clearly marked as unfit. This presumption is a great aid to food officers when investigating potential offences.

Section 7: rendering food injurious to health

A person is guilty of an offence if he renders any food injurious to health by means of any of the following operations:

- adding any article or substance to the food;
- using any article or substance as an ingredient in the preparation of food;
- abstracting any constituent from the food;
- subjecting the food to any other process or treatment with intent that it shall be sold for human consumption.

To decide whether any food is injurious to health, for the purposes of this section, any probable cumulative effect of consuming the food in ordinary quantities should be considered as well as any probable acute effect. 'Injury' in relation to health, can mean any impairment. This impairment can be permanent or temporary, and 'injurious to health' is to be construed accordingly.

This section therefore provides the power to deal with manufacturing processes or procedures where, either by negligence or design, the safety of the food is undermined and could cause injury when eaten either as a single item or in small quantities as part of normal dietary intake over a period.

Section 8: selling food not complying with food safety requirements

This is a wide-ranging provision that defines food as failing to comply with food safety requirements if:

- it has been rendered **injurious to health** by means of any of the operations mentioned in Section 7;

- it is **unfit for human consumption**;
- it is so **contaminated**, whether by extraneous matter or otherwise, that it would not be reasonable to expect it to be used for human consumption in that state.

Where any food that fails to comply with food safety requirements is part of a batch, lot or consignment of food of the same class or description, it shall be presumed for the purposes of this section, until the contrary is proved, that all the food in that batch, lot or consignment fails to comply with those requirements.

This presumption provides the power to deal with a larger consignment even where only a small proportion of it has been found to be unfit or contaminated. The question of fitness is a matter to be determined by the court on the basis of the circumstances of the particular case. Contamination could be such as to make the food unfit or unsound and unmarketable.

Section 9: inspection and seizure of suspected food

This provides power for an authorized officer of a food authority to inspect and seize suspected food. Where the authorized officer considers that food is likely to cause food poisoning or any disease communicable to humans, or otherwise not to comply with food safety requirements, he can order its detention in some specified place, or seize the food and remove it in order to have it dealt with by a justice of the peace. Where the food is detained, the officer must, within 21 days, determine whether he is satisfied that the food complies with food safety requirements, and either withdraw the detention or seize it within that period for it to be dealt with by a justice of the peace.

If a detention notice is withdrawn or a justice of the peace refuses to condemn it, the food authority will be liable for compensation to the owner of the food for any depreciation in its value arising from the activity of the officer.

The environmental health officer will want the food examined by a justice of the peace as soon as possible. This is crucial as the food may deteriorate further. It is important for a justice of the peace to see the food, especially if it is decomposing or infested, in a condition that is as close as possible to that which it was in when seized. This removes the defence of the argument that the food was satisfactory when seized and only deteriorated in the local authority's care. If a justice of the peace is satisfied, on the evidence presented, that the food is unfit, he will 'condemn' it. An order is then made to destroy the food or prevent it from being used for human consumption. After the order is made, any expenses incurred by the local authority in disposing of the food can be reclaimed from its owner.

Section 10: improvement notices

This provides the power for an authorized officer to serve an 'improvement notice' where he has reasonable grounds for believing that the proprietor of a food business is failing to comply with any regulations to which this section applies, for example, food hygiene regulations. Notices are subject to appeal procedure. (See also Code of Practice No. 5 Improvement Notices [4], and the Food Safety (Improvement and Prohibition – Prescribed Forms) Regulations 1991 [5].)

The improvement notice:

1. states the officer's grounds for believing that the proprietor is failing to comply with the regulations;
2. specifies the matters that constitute the proprietor's failure so to comply;
3. specifies the measures which, in the officer's opinion, the proprietor must take in order to secure compliance;
4. requires the proprietor to take those measures, or measures that are at least equivalent to them, within such period (being not less than 14 days) as may be specified in the notice;
5. indicates that any person who fails to comply with an improvement notice shall be guilty of an offence.

The service of an improvement notice was an important new power introduced by the Food Safety Act 1990, as no such notice was recognized or specified in previous Food Acts.

In accordance with Code of Practice No. 5[4] Improvement Notices (Revised April 1994):

- the use of improvement notices should not generally be considered as the first option when breaches are found on inspection – officers should use informal procedures where it is considered this will secure compliance within a reasonable timescale
- notices may only be signed by authorized officers who are properly trained and competent. According to Code of Practice No. 19 (Qualifications and Experience of Authorized Officers) as revised in October 2000 [6], these will include environmental health officers, official veterinary surgeons, holders of the Higher Certificate in Food Premises Inspection or in food premises of risk categories C–F, holders of the Ordinary Certificate in Food Premises Inspection. Authorized officers should not sign notices on behalf of other others, whether authorized or not. The officer signing the improvement notice should have witnessed the contravention.

Failure to comply with the improvement notice is itself an offence.

Prior to 1999, the Deregulation (Improvement of Enforcement Procedures) (Food Safety Act 1990) Order 1996 made it a requirement to serve a 'minded-to' notice prior to serving an improvement notice. These Regulations have now been repealed and such preliminary actions are no longer required before service of an improvement notice under Section 10 of the Food Safety Act 1990.

Section 11: prohibition notices

This section provides the courts with prohibition powers that can be used where there is risk of injury to health. The power can be used to prohibit the use of premises or of equipment, and can also be used to prohibit a person from managing any food business. The first step in a prohibition procedure is for the food authority to successfully prosecute the food proprietor for food safety offences. This may be for food hygiene infringements, processing contraventions or failure to comply with a notice or order. The courts then consider whether the premises, equipment or process puts the public at risk of injury to health. If so, the courts should impose a prohibition order. The courts may also issue a prohibition order on a person, prohibiting that person from managing any food business. If/when the health risk condition no longer applies to the business, the local authority is under an obligation to take action to lift the prohibition order if it applies to the premises, process of equipment. Only the courts may lift a prohibition order relating to a person. Code of Practice No. 6 Prohibition Procedures [7] and the Food Safety (Improvement and Prohibition – Prescribed Forms) Regulations 1991 [5] are relevant to this section.

Section 12: emergency prohibition notices and orders

These are important and substantial powers that allow enforcement officers to deal rapidly with serious public health risks. This section allows for immediate closure of the premises or stoppage of the process. By serving an emergency prohibition notice on the proprietor of the business, the officer imposes an immediate prohibition on the use of a process or equipment or on the premises themselves. According to the Code of Practice No. 19 (Qualifications and Experience of Authorised Officers) as revised in October 2000 [6] an emergency prohibition notice must be signed by an environmental health officer with two years post qualification experience in food safety matters and who is currently involved in food enforcement.

For this section to apply there must be an 'imminent risk of injury to health' present. Code of Practice No. 6 Prohibition Procedures [7] gives some examples of instances which could constitute an 'imminent risk of injury to health'. These include:

- A serious pest infestation that has resulted in actual contamination or which offers a real risk of contamination of the food;
- serious drainage defects leading to flooding and contamination from sewage;

- inadequate temperature control;
- serious hygiene contraventions in a premises that is associated with an outbreak;
- operation outside critical control criteria.

As this is such a draconian power, the food authority must follow up the service of an emergency prohibition notice by an application to the court for an emergency prohibition order. This confirms that the food authority has acted correctly and that imminent risk to health did exist. The food authority must serve notice on the proprietor of the business of its intention to apply for this order, as the proprietor has the right to attend the court hearing. The proprietor must be given notice at least one day before the date of such an application. A food authority must apply for the emergency prohibition order within a period of three days from service of the emergency prohibition notice. If it does not, or if it does but the court is not satisfied that the health risk condition was fulfilled, the food authority will be liable for compensation to the proprietor of the business for any loss suffered. Failure by the food authority to apply to the courts within three days also means that the Emergency Prohibition Notice ceases to be in effect.

Once the proprietor has taken sufficient steps to remove the 'imminent risk of injury to health', the enforcement authority must issue a certificate lifting the Emergency Prohibition Notice or Emergency Prohibition Order. Figure 28.1 summarizes the steps in the application of Section 12 to a food business.

Section 13: emergency control orders

This section provides the power for the Secretary of State to make an order prohibiting the carrying out of a commercial operation with respect to food, food sources or contact materials that may involve imminent risk of injury to health. The purpose of this power is to allow the Minister to deal with a widespread emergency, for example, one that encompasses more than one food authority. This section provides substantial power to back up food hazards or emergencies by preventing the consumption of food in national distribution that

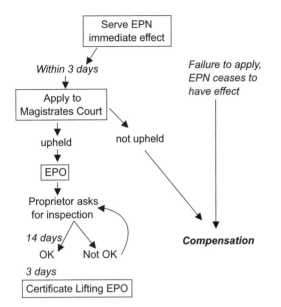

Fig. 28.1 Emergency prohibition powers, Section 12 Food Safety Act 1990.

is known to be a hazard to health. Normally, these hazards are dealt with by the voluntary co-operation of the parties involved, but if such co-operation is not forthcoming, the reserve power of an emergency control order is available.

Section 14: consumer protection

This section makes it an offence for any person to sell to the purchaser's prejudice any food that is not of the nature or substance or quality demanded by the purchaser. The purchaser of a food product has the right to expect that he will get what he asks for. Action would normally be taken by the food authority rather than the purchaser under this section.

This section creates a number of different legal offences and action should be taken under one of the options, that is, *either* nature *or* substance *or* quality. It is important that action is taken under the correct subsection. For example:

1 **Nature**, that is, different variety or because of substitution:

- margarine sold as butter
- cod sold as haddock
- monkfish tail sold as scampi.

2 **Substance**, that is, not containing expected ingredients because of adulteration or foreign matter:

- slug in a bottle of milk
- high lead levels in canned food
- glass in a cake.

3 **Quality**, that is, inferior product because of fraud or an error:

- excess water in ham
- sausages containing less than 65% pork
- stale bread.

Extensive case law exists relating to this section of the Food Safety Act 1990. Careful consideration of the relevant case law may help with the selection of the correct category when considering formal action.

Section 15: falsely describing or presenting food

This section provides offences of labelling or falsely describing food so as to mislead about the nature, substance or quality of the food.

Sections 16–19: delegated power to make regulations

These sections delegate the power to the Secretary of State to make regulations to address certain matters.

Section 16 allows Regulations to be drafted to deal with food safety and consumer protection, including

- substances and the composition of food;
- microbiological standards;
- processes or treatments in the preparation of food;
- the observance of hygienic conditions and practices in commercial operations;
- the labelling, marking, presentation and advertising of food;
- other general regulations for the purposes of ensuring that food complies with food safety requirements, or in the interest of public health, or for the purpose of protecting or promoting the interests of consumers;

- the hygiene of materials that come into contact with food intended for human consumption, and their labelling.

Section 17 allows ministers to make regulations to comply with EU obligations, and under **Section 18** ministers may make regulations regarding: novel foods; genetically modified food sources or foods derived from such sources; and special designations for milk and the licensing of milk.

Section 19 deals with registration and licensing of premises. Regulations made by the Secretary of State under this section deal with the registration by enforcement authorities of premises used or proposed to be used for the purpose of a food business, for prohibiting the use for those purposes of any premises that are not registered in accordance with regulations and for licensing of food businesses. The Food Premises (Registration) Regulations 1991 (as amended) [8] were made under this power, as were the Food Safety (General Food Hygiene)(Butchers' Shops) Amendment Regulations 2000 [3].

Section 20: offences due to the fault of another person

This section enables enforcement officers to 'bypass' the immediate seller or distributor of the food and to prosecute the person who caused the offence. For example, a food shop may sell some food containing a foreign body. Technically, they have committed an offence. However, the enforcement officer may find on investigation that it was the manufacturer of the food that allowed the foreign body to get into the food. This section allows prosecution of that manufacturer.

Section 21: defence of due diligence

Most modern food safety legislation does not require a guilty mind or intent for an offence to occur. The offences are therefore known as strict (or absolute) liability offences. A food proprietor may commit an offence unknowingly, unintentionally or in spite of his best attempts to avoid doing so. To accommodate this situation, in the interests of natural justice, the Food Safety Act

1990 provides a statutory defence of due diligence. In proceedings under the Food Safety Act 1990, a defendant can plead that he took all reasonable precautions and exercised all due diligence to avoid the commission of the offence by himself or by a person under his control and should therefore be considered not guilty.

The proprietor must prove both that he took all reasonable precaution AND that he showed all due diligence to avoid the commission of the offence, but the burden of proof on the defendant is on balance of probability. Put simply, taking all reasonable precautions means having a system to ensure things do not go wrong, while showing all due diligence means seeing the system works properly. As this is a one-limbed defence, the proprietor must prove both parts.

Most of the current legal precedent pertaining to due diligence decisions has resulted from trade descriptions and consumer protection cases. Consideration of these cases makes it possible to develop some principles that apply generally to due diligence defences. These principles include:

- The system must be under the directing will of the company. It is reasonable to expect that delegation of tasks will occur, but this must be formal and documented and there must be a system of training, checking and supervision.
- The precautions that need to be taken will relate to the size and resources of the company as well as the risks posed. What might be a reasonable precaution for a large multi-national company may not be reasonable for a village shop.
- Warranties may form part of a system but are not adequate as the only mechanism of assurance.
- The system for ensuring due diligence must be proactive and reactive, documented, checked and revised as necessary.
- The system must cover all aspects of the business that are covered by the Act and consequent regulations, not just some parts.

As the defence will be specifically related to the particular food premises, there may be other issues not listed above that need to be covered for a defence to be successful. It should be emphasized that whether a premises has a due diligence defence is a matter for the courts to decide. All officers can do is to assess the systems in place and evaluate whether any reasonable precautions have been omitted, or whether there are flaws in the systems that might undermine the diligence.

Section 23: *provision of training by food authorities*

This section gives the food authority permission to provide, either within or outside its area, training courses in food hygiene. These courses should be for people who are, or who intend to become, involved in food businesses in any capacity.

Section 29: *procurement of samples*

Section 29 gives an authorized officer of an enforcement authority power to purchase or take a sample of

- any food; or
- any such substance capable of being used in the preparation of food; or
- any such substance which appears to him to be intended for sale, or to have been sold, for human consumption; or is found by him on or in any premises which he is authorized to enter by or under section 32 below.

An authorized officer can also take a sample from any food source, or a sample of any contact material, which is found by him on or in any such premises. Section 29 (d) gives the authorized officer the power to take a sample of any article or substance which is found by him on or in any such premises and which he has reason to believe may be required as evidence in proceedings under any of the provisions of this Act or of regulations or orders made under it.

Section 32: *power of entry*

Section 32 sets out who may enter premises to enforce the Act and explains what they may do while on the premises. Officers can:

- enter all premises within the area of the authority at reasonable hours;

- enter all business premises in Britain to establish offence;
- enter domestic premises on 24 hours notice (or with a warrant).

Once inside the premises, authorized officers may inspect anything in order to establish if an offence has been committed or use any of the other powers allocated to them under the Act, for example, powers to seize or detain food. Authorized officers may take into the premises with them anyone they consider necessary.

This section also makes it an offence for an officer to make unauthorized disclosures of any information obtained when using the Act's powers. Code of Practice No. 2, part B gives further advice on this section.

Section 33: *obstruction*

Section 33 makes it an offence deliberately to obstruct a person enforcing the Act or to provide false or misleading information.

In addition to the main powers and duties under the sections of the Food Safety Act 1990, there are a number of statutory instruments that help food authorities and others implement the Act. Appendix A lists some of these, for example those detailing particular forms, powers or exemptions.

FOOD HYGIENE REGULATIONS

Many of the food hygiene regulations currently in force have resulted from EU directives. The directives are often referred to as being either vertical, that is, applying to specific processes or premises, or horizontal. Horizontal directives apply to all premises and processes unless that premises or process is exempted by a vertical directive. The hygiene regulations follow this lead.

Directive 93/43 on the Hygiene of Foodstuffs is a horizontal directive [9]. In addition to outlining certain standards and hygiene conditions that must be met by food businesses, this directive introduced a number of new ideas and provisions.

These include:

- Hazard analysis and risk assessment in food safety. Article 3 requires all proprietors (of food businesses) to identify any step in their activities that is critical to food safety and to ensure that these steps are adequately controlled, monitored and reviewed in line with the principles of HACCP (Hazard Analysis Critical Control Points).
- The introduction of Industry Guides on good hygiene practice. These guides are to be developed by industry and acknowledged by the competent authorities. Enforcement officers must take account of them when implementing the relevant legislation.
- The need for food hygiene training for food handlers. This is a general requirement that employees in a food business should be trained in food hygiene matters and/or supervised commensurate with their responsibilities.
- Risk-based enforcement and inspection. Article 8 requires the food authorities to assess the critical control points in a food business when carrying out inspections and to carry out these inspections with a frequency based on risk.

Two sets of domestic regulations enact this directive – the Food Safety (General Food Hygiene) Regulations 1995 [2] and the Food Safety (Temperature Control) Regulations 1995 [10].

FOOD SAFETY (GENERAL FOOD HYGIENE) REGULATIONS 1995

These regulations replaced all the existing regulations relating to food hygiene (general, markets, stalls and delivery vehicles, docks, wharves and on ships) in England, Scotland and Wales.

The regulations do not apply to any part of a food business that is subject to the requirements of other specified regulations that implement EU product-specific directives, for example, meat and poultry meat, meat products, egg products, etc. The main obligations on proprietors of food businesses are contained in Regulation 4 and its associated schedule. They may be summarized under three headings, that is, hygienic operations, hazard analysis and risk assessment, and staff illness.

Hygienic operations

There is a duty placed on the proprietor of a food business to ensure that all operations in the food business are carried out in a hygienic way. These operations include preparation, processing, manufacturing, packaging, storage, transportation, distribution, handling and offering for sale or supply, of food. Details of hygiene standards that apply to premises, food and food handlers are contained in Schedule 1. This schedule is divided into chapters that cover

(a) general requirements for food premises;
(b) specific requirements for rooms where foodstuffs are prepared, treated or processed;
(c) requirements for moveable and/or temporary premises (such as marquees, market stalls, mobile sales vehicles);
(d) requirements for premises used primarily as private dwelling-houses, premises used only occasionally for catering purposes and vending machines;
(e) transport;
(i) equipment for food premises;
(g) food waste;
(h) water supply;
(i) personal hygiene of employees;
(j) specific provisions regarding food, for example, foods known to be contaminated should not be accepted by food companies;
(k) food hygiene training – there is a general requirement that employees should be supervised and instructed and/or trained in food hygiene matters commensurate with their work activity.

The industry guides to compliance give extensive advice on how to comply with the requirements of the schedule. Various sectors of the food industry have (with government assistance) developed guides and such guides currently exist for the following sectors:

- Baking
- Catering
- Markets and fairs
- Retail with a separate supplement on Butchers' Shop Licensing
- Wholesale
- Vending machines.

The guides are published by the Chadwick House Group through the Chartered Institute of Environmental Health [11].

Hazard analysis and risk assessment

Regulation 4[3] enacts the requirements of Article 3 of the directive – that is, the need for the proprietor to identify critical control points in the process. The proprietor must

(a) consider the potential food hazards in the business operation;
(b) identify the points in the operation where food hazards may occur;
(c) decide which of the points identified are critical to ensuring food safety (the 'critical points');
(d) identify and put in place effective control and monitoring procedures at those critical points;
(e) periodically, particularly when there are changes in food operations, key personnel or structural changes, new purchase of equipment, etc., review the potential food hazards, the critical control points, and the control and monitoring procedures.

Staff illness

Regulation 5 requires that the proprietor must be notified by any staff member who is affected by a food-borne illness or infection or who has an infected wound, a skin infection, sores, diarrhoea or any analogous medical condition. What the food proprietor is supposed to do about the situation is not stated in the Regulations, but the Department of Health has developed

guidance which is available for food business managers [12].

FOOD AUTHORITIES

The Food Safety (General Food Hygiene) Regulations 1995 [2] also identifies duties and responsibilities of food authorities.

The main obligations on food authorities are to assess risks to food safety and to regard the guidance on good hygiene practice.

To assess risks to food safety

Food authorities are under a duty to:

(a) enforce the regulations;
(b) ensure that premises are inspected at a frequency commensurate with risk;
(c) include in their inspections a general assessment of the potential food safety hazards associated with the particular food business;
(d) pay attention to critical control points;
(e) consider if the proprietor has followed relevant guides to good practice.

When assessing risks to food safety, or wholesomeness of food, environmental health officers should have regard to the nature of the food, the manner in which it is handled and packed, any process to which the food is subjected, and the conditions under which it is displayed or stored. The Code of Practice No. 9 Food Hygiene Inspections (revised 2000) [13] gives advice on how the food authority can discharge some of these duties, including the frequency with which premises should be inspected. It should be noted, however, that in the draft code of practice currently under consultation [14] the process of risk rating for inspection frequency differs from that in the current code.

To regard the guidance on good hygiene practice

A food authority must give due consideration to any relevant guide to good hygiene practice recognized by the government, or any European guide, details of which have been published by the European Commission in the C series of the *Official Journal of the European Communities*. 'Relevant' means relevant to the food business in question and includes the official Industry guides mentioned above.

THE FOOD SAFETY (TEMPERATURE CONTROL) REGULATIONS 1995

The Food Safety (Temperature Control) Regulations 1995 [10] implement paragraphs 4 and 5 of Chapter IX of the Annex to Council Directive 93/43/EEC [9] on the hygiene of foodstuffs. The Regulations also contain certain national provisions relating to food temperature control.

Regulation 2 is an interpretation provision. Regulation 3 deals with the application of these regulations: they apply to all stages of food production except primary production, but – subject to an exception that relates to fishery products – they do not apply to the activities of food businesses that are regulated by or are under the regulations listed in Regulation 3[2], that is, the product specific regulations resulting from vertical directives.

Part II of the regulations contains the food temperature control requirements for England and Wales. Regulation 4 contains a requirement that food that needs to be kept chilled because it is likely to support the growth of pathogenic micro-organisms or the formation of toxins must be kept at or below 8°C, although this does not apply to mail order food, which is the subject of a separate offence. There are certain exemptions, for example, for some pastry products that are to be sold within 24 hours of production (Regulation 5, see Table 28.1) and a provision that allows for the upward variation of the standard temperature in appropriate circumstances (Regulation 6[1], see Table 28.2). Any such variation must, however, be based on a well-founded scientific assessment of the safety of the food at the new temperature (Regulation 6[2]). There are also defences that relate to the tolerance periods for

which food may be held outside temperature control (Regulation 7).

Regulation 8 contains a requirement that food that needs to be kept hot in order to control the growth of pathogenic micro-organisms or the formation of toxins must be kept at a minimum temperature of 63°C. There are defences that allow for downward variation of this minimum temperature in appropriate circumstances and for a tolerance period of two hours (Regulation 9). Regulation 10 adds a general temperature control requirement that prohibits keeping perishable foodstuffs at temperatures that would result in a risk to health, and Regulation 11 contains a further requirement in relation to the cooling of food. Regulation 12 deals with the evidential value, in certain circumstances, of guides to good hygiene practice. It should be noted that the Industry Guides will apply to the Food Safety (Temperature Control) Regulations 1995 as they do to the Food Safety (General Food Hygiene) Regulations 1995 because both sets of Regulations enact the Directive 43/93 on the Hygiene of Foodstuffs.

Part III of the guide describes the temperature control requirements in Scotland.

THE FOOD SAFETY (GENERAL FOOD HYGIENE) (BUTCHERS' SHOPS) AMENDMENT REGULATIONS 2000

These Regulations apply in England and introduce the requirement for retail butchers to be licensed. These Regulations amend the Food Safety (General Food Hygiene) Regulations 1995 and must be read in conjunction with them. The Food Safety (General Food Hygiene) (Butchers' Shops) Amendment Regulations 2000 [3] implement some of the recommendations of the Pennington Group report on the outbreak on *Escherichia coli* O157 in Lanarkshire in 1996 [15].

The Regulations apply to retail butchers, fixed and mobile, who handle both raw meat and ready

Table 28.1 Food Safety (Temperature Control) Regulations 1995: foods that are exempt from the 8°C limit

Exempt food	Examples
Foods to be sold within 24 hours of production	Uncut baked egg and milk pastry products such as custard tarts, cooked pies and pastries to which nothing has been added after baking and sausage rolls
Foods that can be kept at room temperature throughout their shelf-life without causing a health risk	Some cured or smoked products; certain bakery products that are to be sold quickly
Food that goes through a preservation process, for example, canning or dehydration	Most canned or dried foods are stable at room temperatures until the can is opened or the food rehydrated; some cans of ham or similar cooked meats may only have been pasteurized, and must therefore be kept chilled[1]
Food that must be opened or matured at room temperature	Soft or mould-ripened cheeses[2]
Raw food intended for further processing (including cooking) that will ensure that the food is fit for human consumption	Fresh meat and fish, except where they are intended to be eaten raw, for example, as steak tartare or sushi
Mail order food[3]	Smoked salmon by post

Source: [18,19].

Notes

1 Once the seal of a can is broken the food must be kept chilled if it contains any of the food types described earlier; for high acidity canned food, such as fruit and some vegetables, chilled storage is not essential.

2 Once fully ripened or matured, the food must be stored and/or displayed by food businesses at or below 8°C.

3 Although exempt from the 8°C control, mail order foods must be supplied at temperatures that will not present a health risk.

to eat foods. Such premises are required to be licensed by the food authority. Trading without such a licence is an offence, and the licence must be renewed annually. In order to be awarded a licence, the premises must

- comply with the Food Safety (General Food Hygiene) Regulations 1995 and The Food Safety (Temperature Control) Regulations 1995;
- have HACCP procedures in place;
- have all food handlers trained in food hygiene;
- have at least one person working in the shop trained in hygiene to a level that enables him to

supervise the activities and to ensure that the HACCP procedures are followed.

The food authority must be satisfied that the premises complies with requirements before the licence is issued, and they have 28 days from receipt of the butcher's application for a licence to determine it. The current cost to the proprietor for the licence is £100.

In Wales and Northern Ireland, The Food Safety (General Food Hygiene) (Butchers' Shops) (Amendment) (Wales) Regulations 2000 and The Food Safety (General Food Hygiene) (Amendment) Regulations (Northern Ireland) 2001 apply. The

Table 28.2 Food Safety (Temperature Control) Regulations 1995: allowable variations

The regulations recognize that it is impractical to keep foods at the suggested temperatures at all times. At certain times, they therefore allow a degree of flexibility, called tolerances, which are explained below.

Service or display
Food normally requiring temperature control may be kept above 8°C for a single period of up to four hours to allow it to be served or displayed. Only one tolerance period of service or display is allowed. After this period, any food remaining should be thrown away or chilled to 8°C or below until used. This would include self-service food, buffets and some foods displayed in restaurants and cafés, and food served or displayed in shops.

Food that will be served hot may be kept for service or on display for sale to consumers out of temperature control (63°C or above) for a period of two hours. After this time, the food should either be discarded or cooled as quickly as possible to a temperature of 8°C or below, before final reheating for sale.

Handling and unloading
Limited periods outside chill control are also allowed where:

- food is being loaded or unloaded from a refrigerated vehicle for transfer to or from food premises;
- there are unavoidable circumstances, for example when food has to be handled during and after processing and preparation, or if equipment is defrosted or temporarily breaks down.

Good management practice should ensure that food is kept under these conditions for the shortest time possible. The regulations do not give a specific time limit, but normally it should not be more than two hours, in order to avoid undue risk to the food.

Other tolerances
Fresh cream cakes can be kept at a maximum of 12°C for one continuous period of up to 16 hours. This upward variation is based on challenge testing of cream cakes with common pathogens. The research has been accepted by the DOH and LACORS.

Source: [18,19].

Butchers Shop Licensing Supplement to the Retail Guide applies in Wales and Northern Ireland as well.

PRODUCT-SPECIFIC REGULATIONS

As mentioned above, in addition to the Food Safety (General Food Hygiene) Regulations 1995 and The Food Safety (Temperature Control) Regulations 1995 that apply to the majority of food premises, there are also product or premises specific regulations. These resulted for EC Directives that laid down single European standards for specific products and the premises relating to them, for example, Council Directive 94/65/EC dealing with Minced Meat, Council Directive 91/493/EEC for Fishery Products. The Department of Health and Ministry of Agriculture, Fisheries and Food (MAFF) then drafted National Regulations to implement these Directives. Some of the main product specific hygiene regulations currently in force are:

- Dairy Products (Hygiene) Regulations 1995
- Minced Meat and Meat Preparations (Hygiene) Regulations 1995
- Meat Products (Hygiene) Regulations 1994
- Food Safety (Fishery Products and Live Shellfish) (Hygiene) Regulations 1998.

There are also a number relating to red and white meat, game and wild game. These are dealt with in Chapters 31–33.

These product specific regulations have some things in common. Generally, they apply to premises that are NOT selling to the final consumer and they normally require approval of the premises by the Food Authority. The approval process allocates a unique approval number to each premises. This number should be placed on foods produced in the premises to allow traceability. The individual regulations set standards for compliance with regard to premises and product, including sampling requirements for some foods. Even and consistent enforcement of the product specific regulations has sometimes been difficult. In reality premises may not conform to the categories identified in the legislation. This has been a particular problem with regard to meat products premises who sell both to the final consumer and to other food premises. At present most of the Regulations have Codes of Practice and MAFF/Department of Health Guidance associated with them. However, the Codes of Practice are currently under review by the Food Standards Agency and the product specific directives are also being re-considered by the European Commission, Parliament and Council (see below). The outcome of these reviews may mean changes in the product specific legislation and in the way premises are inspected and approved.

EU PROPOSALS (THE NEAR FUTURE)

There are currently five proposals under consideration by the European Council, Parliament and Commission (Four Regulations and one Directive) [16]. This draft legislation covers the following:

- Hygiene of foodstuffs
- Hygiene rules for food of animal origin
- Official controls for products of animal origin
- Animal health rules for products of animal origin
- Repeal and amendment of certain directives.

One of the main aims of the proposed regulations is to standardize hygiene rules between member states. A press release from the Health & Consumer Protection DG [17] states

> The proposal takes the form of a Council and Parliament Regulation rather than a Directive to ensure uniform application, better transparency and to facilitate rapid updating in the light of new technical and scientific developments. The Commission is particularly pleased that the Council acknowledges the need for regulations as opposed to the looser framework of Directives.

Other aids to standardization include definitions of some words and phrases that have caused particular difficulty with regard to even enforcement, for example, 'retail trade' and 'final consumer'.

Another aim of the review was to consolidate the different hygiene regimes implemented by various directives and to set out the obligations of proprietor throughout the food chain, thereby ensuring a high level of health protection. The proposed legislation consists of a Regulation that will apply to all food premises with specific additional chapters that will apply to certain product specific premises. For products of animal origin, the Hygiene of Foodstuffs Regulation would apply and also some specific regulations pertaining to the individual foods. Most parts of the proposal have already been agreed by the Institutions. The regulation for food of animal origin was expected to go to the parliament for a second reading by June 2003, so decisions on the drafts are expected in the near future. Progress can be checked by visiting Food Safety Press Releases at http://europa.eu.int/comm/food/index_en.html

Clearly, the implementation of Regulations and Directives from the European Council and Commission has a major impact on food safety legislation in member states such as the United Kingdom. The Food Safety Act 1990 is drafted in such a way that major changes may not be required to accommodate the draft legislation under consultation. However, it is likely that statutory instruments such as the Meat Products (Hygiene) Regulations 1994 or Food Safety (General Food Hygiene) Regulation 1995 will alter in the near future to comply with any new EEC Regulations and Directive.

APPENDIX A

Statutory instrument	Function
The Food Safety Act 1990 (Commencement no. 1 and no. 2) Orders 1990	This brought into force the main provisions of the Act and introduced new powers for ministers to use in emergencies.
The Food Safety (Enforcement Authority) (England and Wales) Order 1990	This sets out the division of responsibility of enforcement in the shire counties of England and Wales.
The Food (Sampling and Qualifications) Regulations 1990	These regulations set out the procedures to be followed by enforcement officers when taking samples for analysis or microbiological examination. They also set out qualification requirements for public analysts and food examiners. These regulations apply to Great Britain.
The Food Safety Act 1990 (Consequential Modifications) (England and Wales) Order 1990	This brings existing regulations into line with the updated provisions in the Food Safety Act and provides for the continuation of milk and dairies legislation by amending regulations relating to milk so that provisions contained in the Food Act 1984 are now contained in those regulations.
The Food Safety Act 1990 (Consequential Modifications) (No. 2) (Great Britain) Order 1990	This brings existing regulations into line with the updated provisions in the Food Safety Act.
Detention of Food (Prescribed Forms) Regulations 1990	These prescribe the forms of notice that may be issued in connection with the detention of food under Section 9 of the Act.
Food Safety (Improvement and Prohibition – Prescribed Forms) Regulations 1991	These set out the forms of notice that may be used in connection with the improvement notices under Section 10 of the Act, prohibition orders under Section 11 or emergency prohibition notices or orders under Section 12.
The Food Premises (Registration) Regulations 1991	These provide for the registration of food premises (including vehicles and other moveable structures) by food authorities.
The Food Premises (Registrations) (Amendment) Regulations 1993	These exempt childminders caring for no more than six children from the requirement to register their premises as a food business.
The Food Premises (Registrations) (Amendment) Regulations 1997	These exempt people who prepare food at home for sale in W.I. Country Markets Ltd from the requirement to register their premises as a food business.

REFERENCES

1. The Food Safety Act 1990; ISBN 0105416908.
2. Food Safety (General Food Hygiene) Regulations (1995) SI 1995/1763 as amended; HMSO, London.
3. The Food Safety (General Food Hygiene) (Butchers' Shops) Amendment Regulations (2000) SI 2000/930; HMSO, London.
4. Food Safety Act 1990; Code of Practice No 5; Improvement Notices; Department of Health, MAFF, the Welsh Office and Scottish Office, HMSO.
5. The Food Safety (Improvement and Prohibition – Prescribed Forms) Regulations (1991) SI 1991/100; HMSO, London.
6. Food Safety Act 1990 (Revised 2000) *Code of Practice No 19: Qualifications and Experience of Authorised Officers*, Department of Health, MAFF, the Welsh Office and Scottish Office, HMSO.
7. Food Safety Act 1990 (1990) *Code of Practice No 6: Prohibition Procedures*, Department of Health, MAFF, the Welsh Office and Scottish Office, HMSO.
8. The Food Premises (Registration) Regulations 1991 as amended; SI 1991/ 2825; HMSO, London.
9. Council Directive 93/43/ EEC on the Hygiene of Foodstuffs; OJ L 175; 19.7.93.
10. The Food Safety (Temperature Control) Regulations 1995; SI 1995/2200 as amended; HMSO, London.
11. *Industry Guide to Good Hygiene Practice: Catering Guide* (1997); Chadwick House Group Ltd, London.
12. Food Handlers: Fitness to Work, Guidelines for Food Business Managers; Department of Health, London.
13. Food Safety Act 1990 (1990 revised 2000) *Code of Practice No 9: Food Hygiene Inspections*, Department of Health, MAFF, the Welsh Office and Scottish Office, HMSO.
14. Food Safety Act 1990 (2003) draft *Code of Practice and Practice Guidance*, www.consultation.gov.uk
15. Pennington, T.H. (1997) *The Pennington Group Report*, The Stationary Office, Edinburgh.
16. Proposal for a Regulation of the European Parliament and of the Council on the Hygiene of Foodstuffs; Proposal for a Regulation of the European Parliament and of the Council laying down specific hygiene rules for food of animal origin; Proposal for a Regulation of the European Parliament and of the Council laying down detailed rules for the organization of official controls on the products of animal origin intended for human consumption; Proposal for a Council Regulation laying down the animal health rules governing the production, placing on the market and importation of products of animal origin intended for human consumption; Proposal for a Directive of the European Parliament and of the Council repealing certain directives on the hygiene of foodstuffs and the health conditions for the production and placing on the market of certain products of animal origin intended for human consumption, and amending Directives 89/662/EEC and 91/67/EEC; Commission of the European Communities; Brussels 14.07.2000 COM (2000) 438 final.
17. Health and Consumer Protection DG Press Release IP/02/948 27/06/2002, http://europa.eu.int/comm/food/index_en.html
18. *A Guide to the Food Safety (Temperature Control) Regulations 1995* (1996) Department of Health, HMSO, London.
19. *Industry Guide to Good Hygiene Practice: Baking Guide* (1997) Chadwick House Group Ltd, London.

29 Food safety controls

Madeleine Smith

The controls commonly used to try and ensure the production of safe food can be conveniently grouped into two sections for consideration. One grouping would comprise the management systems used by food producers to ensure the safety of their food. The other would include the control systems used by enforcement and other agencies to check that the food producers have been successful in producing safe food.

The overall objectives of both types of control systems are the same – to prevent unsafe food reaching the consumer. Although the approach may differ, both groups need to focus on the same type of issues:

- Eliminating anything injurious to health from the food
- Preventing contamination of safe food with anything that might be injurious to health
- Controlling or reducing to the absolute minimum anything that cannot be eliminated or prevented and that might be injurious to health.

HAZARDS

Anything that might be injurious to health is usually referred to as a hazard, and it is convenient to group hazards into three types:

1. Physical
2. Chemical
3. Biological.

Physical hazards generally mean inanimate objects found in the food that could cause some damage to the eater, or might render the food so contaminated that it would not be reasonable to expect someone to eat it in that state. Examples of physical hazards include pieces of equipment such as metal swarf or nuts and bolts, inherent contamination of raw materials such as stones, parts of the food handler (hair, finger nails) or their clothing or jewellery or parts of the premises such as glass from a broken light fitting.

Contamination by certain chemicals may also render the food injurious to health, or contaminate it so that it becomes inedible. Examples can include contamination associated with raw materials such as pesticides in fruit and vegetables or resulting from cross-contamination during storage or processing, such as may happen with cleaning fluids.

Biological hazards include the food poisoning bacteria and viruses that are food associated and cause gastrointestinal and other symptoms, fungi that produce poisonous mycotoxins in the food, algae that can cause seafood-associated poisoning and bacteria that may be non-pathogenic themselves but which cause changes in the food to render it injurious to health such as *Morganella* and the associated scombrotoxin food poisoning in fish.

The main difficulty with controlling physical, chemical and biological hazards in food is that, in the overwhelming majority of cases, the hazard is not easily discerned by the human senses. Contamination of foods with common pathogenic bacteria does not alter its taste, colour, smell or appearance in any way, and neither will the many chemical contaminants. Even physical contaminants, once incorporated into a food stuff, become very difficult to spot. All food safety control systems are an attempt to identify and control the interaction of these hazards with food destined for sale.

CONTROL SYSTEMS IN FOOD PRODUCTION

The systems used by food businesses to ensure safe food production can be considered in two categories – those systems that are common to all food premises and those that are product or premises specific. The first group are often referred to as prerequisite systems or underpinning systems. They comprise the basic rules of hygienic food handling and form the basis for the prescriptive sections of food hygiene regulations (e.g. the schedules in the Food Safety (General Food Hygiene) Regulations 1995). These systems address general hygiene issues and would include

- Hygiene policy
- Cleaning schedule
- Sickness policy
- Training policy
- Pest control system.

Although the details of these systems may differ from premises to premises, all food businesses expecting to produce safe food would need to have in place some management control procedures addressing these common issues. Further discussion of each of these systems is found below.

The second type of management system that is required is a product-specific system. This type of management system addresses the particular hazards associated with whatever food is being produced in the business. It would typically be some form of risk-based assessment of the process and hazards, for example, a Hazard Analysis Critical Control Points (HACCP) system.

UNDERPINNING OR PREREQUISITE SYSTEMS

Hygiene policy

A hygiene policy generally summarizes the standards and objectives of the company with regard to hygiene. It should be cross-referenced to other management systems typically the HACCP plan and training policy. The hygiene policy would be expected to describe the management structure of a food business and detail the hygiene responsibilities of the personnel. Such responsibilities would include, for example, the people responsible for training, for implementing the HACCP plan, for monitoring cleaning, for investigating customer complaints, for liaising with environmental health, etc. The policy should also indicate who is authorized to speak on behalf of the company should a formal interview under caution ever be required, with regard to food safety matters.

The hygiene policy may also indicate which, if any, standards are being applied in the premises. In addition to the *Industry Guides to Good Hygiene Practice* referred to in Chapter 28, there are guides to good manufacturing practice that may apply in particular premises. A cook-chill premises, for example, would be expected to have taken into account the Department of Health Guidelines on Cook Chill [1], while a bean sprout manufacturer might wish to follow the recommendations of the guidance published by Campden Food Research Association on the process [2]. It is reasonable to expect any such reference documents to be indicated in the company's hygiene policy. Similarly any standards set by customers would also be indicated. A hygiene policy is something that should describe the company in terms of its hygiene. It may include additional matters to those mentioned here. Its contents will depend to some extent on the other management systems and policies used by the company to manage food safety.

Cleaning schedules

The two objectives of carrying out cleaning in a food premises are

1. to kill pathogenic organisms, or reduce them to a safe level
2. to remove dirt and grease that might provide protective environments situations in which they might thrive.

The structure, equipment and working surfaces should be cleaned according to a planned programme, devised for the particular premises. The programme should be based on the 'clean as you go' principle. The cleaning programme should be in written form (the cleaning schedule), agreed with the staff and monitored by management. An example is shown in Table 29.1.

The cleaning schedule needs to be clearly written so that it is easy to understand and implement. The schedule should include the following:

- What is to be cleaned
- Who is responsible for cleaning it
- When it is to be cleaned (i.e. how often and at what time(s) throughout the work period)
- How it is to be cleaned, including the chemicals, equipment and other materials that are needed to clean it
- Any precautions or protective equipment needed to carry out the cleaning
- How to clean the cleaning equipment if necessary
- Who is responsible for checking that the cleaning has been carried out.

In some cases it may also be necessary to stipulate how long the cleaning should last – for example, if using a chemical disinfectant on a piece of equipment, how long must the disinfectant be in contact with the equipment.

Staff will need training to be able to clean properly. In particular they need to have the principles of cleaning and disinfection explained. The cleaning schedule should then be explained in terms of these principles.

There has been a tendency for the manufacturers of cleaning chemicals to persuade food proprietors to implement expensive regimes of variously coloured fluids in food premises. Because the manner in which cleaning chemicals work is often poorly understood, food proprietors may be wasting money on superfluous chemicals or, more dangerously, relying on chemicals to do a job that they were not intended to perform. In particular, the action of chemical disinfectants and their interplay with detergents may be unclear.

The terminology associated with cleaning chemicals is based on their chemistry, typically how the active ingredient ionizes in solution. These terms may be unfamiliar. Some terms commonly used in discussions of cleaning fluids are:

Detergents – fluids that reduce surface tension and remove grease.
Disinfectants – kill bacteria.
Anionic – describes the bulk of common liquid and powder household detergents. Anionic detergents are high foaming with good rinsing properties.
Cationic – complex chemicals that are capable of dissolving grease and may also be disinfectants. They may be mixed with non-ionic components to form what is known as a sanitizer.
Non-ionic – useful in hard-water areas, has less tendency to foam than the anionic detergents and can dissolve mineral oils very efficiently.
Amphoteric – act as either anionic or cationic depending on the pH of the solution.

Cleaning in a food premises requires the removal of dirt, grease and food debris **before** any consideration can be given to the need for disinfection. If the energetic stage of physically removing the debris and grease with a detergent is omitted, the debris and grease are likely to inhibit any disinfection step. The process of thorough cleaning using a detergent and power (e.g. 'elbow grease') can remove a significant number of contaminating bacteria [3] and also reduce the protective environment around any that remain.

Once this basic cleaning has been achieved the business should consider whether a disinfection stage is needed, and if so which would be the most suitable. Disinfection should reduce the number of vegetative bacterial cells to a safe level. Many pathogenic bacteria are susceptible to heat and/or desiccation and in some cases one or both of these

Table 29.1 Cleaning schedule – school kitchen

Description	Product	Concentration %	Method	Frequency
Floor – kitchen	Bactericidal detergent	0.5 (as specified by manu facturer)	Fill bucket or container from tap proportioner; wash or scrub manually with scrubbing brush, mop or squeegee; final rinse with clean cold water	At least daily – at end of day
			Supplement with sweeping	As often as necessary
Floor – pantry, larder, vegetable store			Sweep with brush; wash (as above)	Daily/weekly
Walls – tiled to 1.2 m	General detergent for general cleaning or as above	1.2	Apply manually with low-pressure spray; final rinse with clean water (heavy soiling areas); dust accessible fixtures	Once a week

Each working day |
Ceiling, light fittings, ventilation hoods (exterior), etc.			Brush or sweep manually; vacuum clean if possible	Once a week
Ventilation hood (interior)	General detergent	As specified	Wipe with cloth	Once a week
Drainage – open channel	Sodium hypochlorite solution for sterilization and stain removal	1.4	Pour solution down head of drains at end of cleansing procedure to disinfect and deodorize	Each working day
Utensil sinks	Bactericidal detergent	0.5	Fill bucket or container from tap proportioner; wash or scrub manually; final rinse with clean cold water	Daily
Wash-hand basins (if glazed stoneware)	Fine abrasive powder containing chlorine-based bleach, e.g. Vim, Glitto	Neat	Sprinkle on surface; wipe off with damp cloth; rinse off with clean water; wipe dry	Each working day
Refrigerators	Bactericidal detergent	0.5	Wash or scrub manually; final rinse with clean water	As required
Stainless-steel tables	Bactericidal detergent	0.5	Wipe with a clean cloth; re-rinse with water; apply with cloth, mop or brush; ideally, use a warm solution, leave for approx. 10 minutes; final rinse with clean water	After each use

Table 29.1 Continued

Description	Product	Concentration %	Method	Frequency
Framework, including underside of table			Wash or scrub manually	At least once month
Machinery – parts that come into contact with food	Bactericidal detergent	0.5	Wipe clean of all food particles and, where possible, detach and wash with cloth or brush in a hot solution; leave for approx. 10 minutes; final rinse with clean water	After each use
Mechanical parts that do not come into contact with food	Bactericidal detergent	0.5	Wipe clean of all food particles and apply detergent solution with a cloth; rinse cloth and wipe off	Daily
Structural or covering members	Bactericidal detergent	0.5	Dust with a brush; wash clean with a cloth; ideally use a hot solution; leave for 10 minutes and rinse with clean water; dry and polish with a clean dry cloth	Daily/once a week
Ovens and cookery equipment – tops, sides, shelves	Powdered detergent for cleaning all metal surfaces	0.4	Spray or brush on solution; ideally, use hot solution (60–70°C); leave for 10 minutes; scrub or wash thoroughly; final rinse with clean water	Daily
Insides	As above	As above	As above	Weekly

Safety instructions

It is important for safe handling of chemicals to follow any instructions specified in product literature.

Removal of gross soiling

As a prerequisite to cleaning, gross soiling matter should be removed first. This will ensure maximum effectiveness of the detergent/disinfectant.

General housekeeping

The necessity of 'cleaning as you go' is emphasized as this will minimize the extent and degree of soiling for the cleaning programme. Supplement daily cleaning with a thorough cleaning once a week and take the opportunity to remove any articles that have no business in the kitchen.

Temperature of detergent solution

Wherever possible, hot detergent solution should be used to achieve the quickest and best results.

Contact time

The efficient use of a detergent depends on the time allowed for it to be in contact with the soiling matter. It is essential that contact times are adequate.

Cleaning utensils

These should be kept clean. Cloths, dishcloths, etc. should be boiled in soap and water after each day's use. Scrubbing brushes must be washed clean in hot water and detergent and rinsed in clean water after each use. Scrubbing pails should be rinsed in hot water and stood upside down after use.

may be sufficient – for example, a dishwasher rinses and heats the plates so that they are hot and dry at the end of the cycle, typically achieving rinse temperatures of above 80°C. This is generally sufficient to destroy cells such as *E. coli* O157 on the plates, provided there is no greasy food residue that would protect the cells from the heat and desiccation. The Fresh Meat (Hygiene and Inspection) Regulations 1995 specify this method of disinfection for equipment (see Schedule 7, Part I, 3(d)), '. . . all equipment and implements which come into contact with fresh meat are cleansed and subsequently disinfected in water at a temperature of not less than +82°C' [4]. Heat and desiccation may not be appropriate for all equipment or surfaces in food businesses, and in such cases chemical disinfectants may be needed. The main types of chemical disinfectants are listed in the following.

Hypochlorites

Hypochlorites are inexpensive anionic disinfectants that are very effective and widely available. Hypochlorites possess a wide antimicrobial activity, being very effective against both Gram-positive and Gram-negative bacteria, including some spores. They can be fast acting and, depending on the strength of the solution, may need only short contact times (3–4 minutes) in order to inactivate vegetative cells. Their chief disadvantage is that they are easily inactivated by organic material. As a consequence, prior removal of food debris and grease is vital before using a hypochlorite as a disinfectant. Undiluted hypochlorites may have a strong smell and it is recommended that they are rinsed off any surface with which open food is likely to come into direct contact. Metallic materials or equipment may be corroded by prolonged immersion in hypochlorite solution. Hypochlorites should not be used with cationic detergents that could inactivate them. They should never be mixed with acidic cleaners either as the reaction can produce chlorine, which can act as an irritant to the person carrying out the cleaning. Named examples of hypochlorites are Chloros, Domestos and Milton, but the common generic name is bleach.

Iodophors

Iodophors are more expensive than hypochlorites. They are very similar in action, although they do tend to be less sporicidal. Generally, all iodophors incorporate a detergent. They work best in acidic conditions. Like Hypochlorites, Iodophors are inactivated by organic matter. As they are based on iodine, these disinfectants are a yellow-brown colour which may stain some plastics. Examples are Vanodine and Wescodyne.

Quaternary ammonium compounds (QAC)

Quaternary ammonium compounds (QAC) have a more limited range of antimicrobial activity than those just mentioned, but remain a popular cationic disinfectant for food premises. They are adequate in alkaline solutions, if used correctly, against Gram-positive bacteria and also against most moulds. They do, however, tend to be inactivated by organic material, soap, hard water, wood, cotton, nylon, cellulose sponge mops and a few plastics. Not only is the preliminary cleaning stage imperative when using QACs, but the type of cleaning materials, bacteria, water and surface are also limiting. Some bacteria will even grow in QAC solutions, and special care is therefore needed when preparing and using solutions. Named examples are Hytox, Dettox, Nonidet, Benalkonium chloride (Zephiran™) and Ceytlpyridinium chloride (Cepacol™).

Phenolic disinfectants

There are several types of phenolic disinfectants, but their main disadvantage is that many brands possess a very strong smell and taste so they are not always suitable for use in food businesses. They are also inactivated by rubber and some plastics. The white fluid phenolics and the clear soluble type have a range of antimicrobial activity similar to that possessed by the hypochlorites and iodophors, and they are less easily inactivated by materials of an organic nature. The chlorinated phenolic disinfectants are less effective as they show a reduced range of antimicrobial activity, and organic material will more readily inactivate them. Examples of white fluid phenolics are: Izal and White Cylelin; examples of clear soluble fluid phenolics are: Hycolin, Clearsol and Stericol; an example of a chlorinated phenol is Dettol.

Amphoteric surfactants

Amphoteric surfactants have some disinfectant ability in acid conditions, but they are fairly expensive. Their range of antimicrobial activity is narrow and many materials, such as hard water, organic material, wood, rubber, cotton, nylon, cellulose sponge and some plastics, will readily inactivate them. They are good detergents, but their high foaming characteristics mean they may be unsuitable for some systems. An example is Tego.

Anionic disinfectants, such as hypochlorites and phenolics, are compatible with anionic detergents, but deactivated by cationic detergent. Similarly, cationic disinfectants such as QAC are compatible with cationic or non-ionic detergents, but inactivated by anionic detergents. A sanitizer is a compound cleaning fluid that is a combination of detergent (often a non-ionic one) and a disinfectant. Such cleaning fluids are useful in that they combine degreasing and disinfection in one step. However they are not very effective at removing heavy soil and grease, and should only be used for lightly soiled areas. In order for them to be effective, it is essential that they are used exactly as directed.

An additional point with regard to chemical disinfectants is that they need a certain contact time in order to destroy bacteria. None are instantaneous. The instructions accompanying disinfectants will indicate the minimum contact time required for bactericidal action, often as long as 30 minutes. The efficacy of these chemical disinfectants is reliant on this contact time, and if this contact time is reduced, bacteria will not be inactivated. Unless the equipment or surface is allowed sufficient contact with the disinfectant, there is little point in using one and the disinfection power of the chemical cannot be relied upon to provide safe conditions.

In summary, when a chemical disinfectant step is incorporated into a cleaning schedule, it is very important to ensure that:

1. Dirt, grease and debris are removed before the chemical disinfection stage.
2. The disinfectant is compatible with the detergent **or** the detergent is thoroughly rinsed off prior to the disinfection stage.
3. The disinfectant is left in contact with the surface or equipment long enough for it to affect the bacteria.

Cleaning in a food business will need to be monitored. Generally this is done by visual inspection, but there may not be a good correlation between a high level of visual cleanliness and absence of microbial contamination [5]. For this reason some premises may swab equipment or use other monitoring techniques to help validate the cleaning schedule. Swabbing equipment on a regular basis and examining the swabs for the total number of bacteria present or for the numbers of Enterobacteriaceae can provide information about the consistency of cleaning. If counts increase, more attention needs to be paid to effective cleaning in the area swabbed. If the counts are low and remain low, cleaning is consistent. There are no agreed levels of contamination above which equipment can be said to be unacceptable and below which it is acceptable. The objective is to have levels that are as low as possible. In the case of the Enterobacteriaceae on surfaces or equipment that come into contact with ready to eat food, it is to be hoped that the levels would be below the limits of detection, as the main source of this group is the mammalian gut. Contamination of equipment or surfaces with the Enterobacteriaceae indicates a route of cross-contamination leading back ultimately to some mammal's faeces. The mammal could be either a food handler or a one used for meat. The Meat (Hazard Analysis and Critical Control Point) (England) Regulations 2002 [6] are one of the few sets of regulations to include a requirement for surface sampling as a monitoring mechanism for cleaning. Other product-specific legislation such as the Meat Products (Hygiene) Regulations 1994, refer to the option of using sampling to monitor cleaning but do not make it a legal requirement. Schedule 17C of the Meat (Hazard Analysis and Critical Control Point) (England) Regulations 2002 gives details of the microbiological checks and standards that must be achieved in licensed slaughter houses and cutting plants. The schedule specifies the sampling methods to be used (contact plate or swab) and establishes maximum acceptable values for contamination of cleaned surfaces. These values are very stringent. A level of Enterobacteriaceae above one colony forming unit per square centimetre is considered to be unacceptable in such premises and should precipitate an assessment of the causes of such a result.

There are also other monitoring methods that help to indicate whether cleaning is thorough. These methods are generally based on detecting protein or DNA that remains on the equipment or surfaces after cleaning. Positive results indicate that either food debris or bacteria (probably both) are still present. There are two types of system, one based on ATP bioluminescence, the other on direct detection of protein residue by a reagent that changes colour if protein is present.

ATP Bioluminescence

The systems based on ATP Bioluminescence use an enzyme to release light from any ATP that may be present in a swab or sample. The amount of light generated is proportional to the amount of ATP present, so the more light given off, the more ATP there was in the sample. As all living cells contain ATP, the amount of ATP in a sample and therefore the amount of light generated will be related to the original number of cells in the sample. In foods ATP is present in:

- animal cells such as meat, fish and poultry
- plant cells such as fruit, vegetables, herbs, spices
- microbial cells such as spoilage or pathogenic bacteria.

If a surface is sampled and tested using the ATP bioluminescence test, and it produces high light levels, it can be inferred that there is a residue of animal, plant or microbial cells on the surface. It has not, therefore, been adequately cleaned.

ATP bioluminescence is a useful tool to monitor cleaning in food premises. It should be used in the same way as swabbing – to give a series of readings that can be used to indicate whether cleaning is being carried out to a consistent standard. The advantage of bioluminescence over swabbing is its speed. Results can be available in 2–3 minutes rather than 2–3 days. However as with environmental swabbing, there is no universally accepted light emission level (or level of Relative Light Units, RLU) below which a surface can be said to be clean and above which it is said to be dirty. Before such a system can be used in an establishment, the management or proprietor must determine criteria for the sampling. This is usually done by monitoring critical sites when they are known to be clean and when they are known to be dirty, and then using these measurements to establish a level for pass/fail in terms of RLU values for that area of the premises. A number of commercial systems are available for use in the food industry. Named examples of such systems include Bio-Orbit, Hy-lite, Inspector/System SURE, Uni-Lite/Uni-Lite excel. The basic procedure is the same for all tests, although some are single-use disposable and some require reagents to be made up in batches. There is good agreement between the results of swabbing and the results of an ATP bioluminescence test in the majority of cases. However, where there is not good agreement it is usually because the surface is dirty by ATP assay but clean by plate count. This will be due to high levels of food residues with low microbial counts. However as good cleaning regimes also undertake to remove food debris as well as microbes, recleaning the area is appropriate in such circumstances.

Non-ATP based methods

The other methods for monitoring cleaning are generally based on detecting protein residues that have been left behind on surfaces by inadequate cleaning. The surfaces are sampled by swabbing or wiping, a reagent mixes with the sample and the presence of food residue is indicated by a colour change. A variety of systems on the market, for example, Pro-tect™ (Biotrace), Check PRO

(Unilever). The advantages of these systems are that they are quick, relatively cheap and easy to use – minimal training is required, unlike swabbing and ATP bioluminescence, which require some training. The main disadvantage is its lack of sensitivity.

A cleaning schedule that is properly designed, implemented and monitored is one of the most crucial systems for managing food safety in a food business. From this discussion it is evident that, although all food premises should have a cleaning schedule, the details of the schedule will vary between premises. The manner in which the cleaning is carried out, the chemicals used and the style of monitoring must not just fit the process but must also allow accurate implementation by the staff and monitoring by the management.

Sickness policy

People who have contracted certain food poisoning organisms such as *Salmonella* spp. are thought to be potential sources of food poisoning if they handle food in an unhygienic manner. With some *Salmonella* species and serovars it is possible to become an asymptomatic carrier, that is, someone who excretes the pathogen in their faeces but does not show any symptoms of the disease. Similarly food handlers with infected boils or cuts may transfer *Staphylococcus aureus* to foods. As a consequence people who are or have been suffering from certain types of illness should not handle open ready to eat food to minimize the chance of passing on their disease. A sickness policy should detail the processes to be followed if staff contract any disease that might be transmitted to others via the food, and any steps that might be in place to try and prevent the illness from occurring in the first place.

European Union (EU) directives on poultry, meat, fresh meat and meat products require staff to be medically examined before beginning food handling work. The Directive on the Hygiene of Foodstuffs (93/43/EEC) states that:

No person known or suspected to be suffering from, or to be a carrier of, a disease likely to be transmitted through food, or while afflicted for example with infected wounds, skin infections, sores or with diarrhoea, shall be permitted to work in any food handling area in any capacity in which there is any likelihood of directly or indirectly contaminating food with pathogenic micro-organisms. [7]

This provision clearly places the responsibility on proprietors or managers of food businesses and the employees themselves to ensure that food handling personnel are medically fit to handle food. Proprietors and managers may therefore require medical examinations of employees to ensure compliance with this requirement.

However, routine examinations may be of little value as they can only reveal the health status of the worker at the time of the examination, and cannot take into account later bouts of diarrhoea or other infectious conditions. Food handlers should be trained and instructed to report any illness immediately, particularly any gastrointestinal symptoms. They should be made aware of the potential hazard they represent if they are carrying food poisoning pathogens and handling food. In 1995, a Department of Health expert working group examined the issue of food handlers' standards of fitness for work. This Committee produced detailed guidance for food business managers to assist them [8]. It sets out the responsibilities of food managers including

- explaining good hygiene practices to employees;
- training, instructing and supervising employees with regard to the control of food borne illness;
- excluding infectious or potentially infectious food handlers from handling food.

The guidance outlines the risk factors and what action needs to be taken for food handlers with diarrhoea and vomiting. It also covers certain specific infections that require more stringent action, including typhoid, paratyphoid, hepatitis A and verotoxigenic *E. coli* (e.g. *E. coli* O157).

The guidance also sets out a pre-employment questionnaire, which proprietors may find useful to incorporate into their sickness policy.

Any sickness policy should take into account visitors to the premises such as maintenance staff, pest control officers and enforcement officers. Asking such visitors to complete questionnaires that confirm they are in a good state of health before they enter the food preparation area is a common method of trying to address this issue in a sickness policy.

Training policy

The training policy in a food business will need to cover all training related to the food safety management systems implemented by the business. In addition to training in the principles of food hygiene, the policy will need to address training on cleaning (see earlier), pest control (see later), Hazard Analysis/Risk Assessment (see later) and any monitoring and recording that might go with these issues.

The training policy should include details of

- the type of training (subject matter and style, e.g. a formal course, a seminar, self-administered, etc.);
- who is to receive it and when (during work time or outside work, and at what point during their time with the company);
- who is to deliver it and where (in house, external, etc.);
- the expected outcomes (e.g. must the operative pass any sort of assessment and if so how is success or failure linked to their job);
- any refresher or top-up training;
- records and certificates.

A well-constructed training policy that is properly implemented shows good management commitment to the staff. It may also help in sustaining a due diligence defence (see Chapter 28, Food Safety Act 1990 Section 21) should that be required by the proprietor.

With regard to training in the principles of food hygiene, all food handlers must be trained and/or supervised to a level that is commensurate with their responsibility. The Industry Guides to Good Hygiene Practice (see Chapter 28) give extensive advice to proprietors on how to comply with this requirement. All food handlers should be taught the rudiments of good hygiene practice. This should form part of the induction training and should cover the following instructions:

1. When to wash and dry hands:

   (a) before and after handling food
   (b) after going to the toilet.

2. To report illness to the management.
3. Not to work if suffering from diarrhoea and/or vomiting.
4. Not to handle food if affected with scaly, weeping or infected skin that cannot be totally covered during food handling.
5. To ensure cuts and abrasions on exposed areas are totally covered with a distinctively coloured waterproof dressing.
6. Not to spit in food handling areas.
7. Not to smoke in food handling areas.
8. Not to eat or chew gum in food handing areas.
9. To wear clean protective clothing, including appropriate hair covering.
10. To ensure work surfaces and utensils are clean.
11. How to do all of the above.

This basic instruction will need to be followed by formal training in most cases. Food handlers who are handling open, ready to eat food are considered to need a higher level of understanding of the hazards associated with food because of the increased risk of causing injury to health with such products. This training should have some type of formal assessment to ensure the food handler has an adequate grasp of the important control mechanisms for common food-borne hazards. A number of organizations offer food hygiene training courses. The Chartered Institute of Environmental Health (CIEH) and Royal Institute of Public Health (RIPH) have both established a three-level approach to food hygiene training.

The first level is to give food workers an appreciation of good food hygiene practice. The basic (or foundation) course lasts 6 hours plus a short examination. The syllabus includes: the causes and symptoms of food poisoning, including the names and sources of the common food-poisoning bacteria, the prevention of cross-contamination through good personal hygiene, temperature control, cleaning and disinfection; design and construction of premises and equipment; pest control; and a limited consideration of the legislation.

The second level is aimed at food business supervisors, and includes owners and managers of small to medium-sized businesses. The intermediate course lasts 18 hours over 3 or 4 days and has a 2-hour written examination. The intermediate course is a more detailed study of the subjects covered on the basic course. For instance, in the area of food poisoning and food-borne disease, there is a need for the student to be able to state the sources, types of food commonly involved, vehicles and routes of transmission, onset times, symptoms and control measures of:

* *Salmonella*
* *Staphylococcus aureus*
* *Bacillus cereus*
* *Clostridium perfringens* and *Clostridium botulinum*.

It can be seen from this that the student is required to have a considerably higher level of understanding and knowledge than on the basic course. The intermediate course also includes new areas of study, including supervisory management. This requires the student to understand how a supervisor should carry out his job in a food business and the influence he can have on food safety. Passing the course gives people in supervisory positions in the food industry a great awareness of their role and responsibilities.

The highest level is the advanced course, which is for people who are managerially accountable for a food business. The course requires detailed study of a wide range of food safety-related subjects. The expected outcome is that the student understands the relationship between food hygiene and food poisoning, the socio-economic costs of food poisoning, and their relationship to food safety. The course equips successful candidates with the skills and knowledge to ensure that their companies operate in a hygienic and efficient manner. The course is assessed by examination of all aspects of managing and supervising a food business, including hazard identification and management systems such as HACCP. This is via assignments and an unseen examination.

These hygiene courses are supplemented by additional specialist courses, for example, courses dealing specifically with HACCP. These courses are internationally recognized and delivered by accredited trainers. Details can be obtained from the websites of the both the CIEH and RIPH.

The successful completion of a course in food hygiene provides the theoretical knowledge needed to produce safe food. This should not be seen by a food proprietor as the end of his obligation. The knowledge must also be implemented properly to produce the desired outcome. This happens to a greater degree in food premises where there is a policy of systematically training staff than it appears to in premises where this culture is absent [9]. A good training policy should therefore include details of how the proprietor checks that the hygiene knowledge is being implemented. It should also make provision for refresher and update training even though these are not required by law.

Pest control system

There is no legal requirement for a food business to have an external pest control contract. The proprietor is responsible for ensuring that the premises are free form pests and that food is protected from the risk of contamination. Many proprietors do buy expertise and have a pest control system that includes routine inspection by a qualified person. This may not be necessary in all businesses, but all food businesses should have some sort of proactive system for ensuring the premises remain free from pests. This would include frequent, regular and routine inspections of the premises by someone who can recognize problems that may lead to pest infestations and who can take any necessary steps to prevent such a problem occurring. The responsible person should also be able to recognize the signs of an infestation and know what steps to take to contain and eliminate the infestation as appropriate.

PRODUCT-SPECIFIC MANAGEMENT SYSTEMS

Each foodstuff can be considered as having a unique range of hazards associated with it. By considering even the modest number of hazards listed in the first pages of this chapter, it is easy to see that a premises bottling apple juice, for example, will need to address hazards that differ from those found in a premises making sandwiches. The sandwich maker's hazards will differ again from those encountered by a business producing smoked mackerel or selling live mussels. Table 29.2 lists a few examples of biological hazards typically associated with common foodstuffs and which would need to be addressed by the management systems of a premises producing any of them.

End product testing

One traditional system for managing food safety was to sample a certain amount of the end product and to test it for whichever hazards were considered to be appropriate for that foodstuff. There are a number of inherent problems with such a system, the most serious being that it is not very accurate, especially for biological and some physical hazards. Microbial contamination is not evenly distributed throughout the foodstuff and neither, usually, is physical contamination. This means that food can be heavily contaminated but not detected unless one of the contaminated products happened to be selected for sampling. The testing process is destructive, so the outcome of a sampling system is that the only sample that can be guaranteed to be safe is the one that has just been destroyed by the testing. Another disadvantage of end product testing is that it is expensive and time consuming to carry out. When sampling a food for common enteric pathogens, for example, it will be several days before the results are available. This is because the process of isolating and identifying the bacteria is a multi-step process, which requires long incubation periods. If the food has been sold and a problem is identified from the sampling, some of the food may have been eaten and any that has not been eaten must be recalled. The former means the public's health has been put at risk. The latter is expensive and extremely undesirable. If, on the other hand, the food is held until the results are known (positive release), there will be the cost of storing the food during that time, and the loss of value if the food

Table 29.2 Examples of biological hazards that may be associated with particular foods

Food	Some biological hazards associated with each food
Wet fish	Scombrotoxin, parasite infection
Vacuum-packed fish	*Clostridium botulinum* and toxin, parasite infection, Scombrotoxin
Cream cakes	*Staphylococcus aureus*
Sliced, packaged bread	Moulds
Raw chicken	*Salmonella*, *Campylobacter*
Raw meat	*Salmonella*, *E. coli* O157, *Campylobacter*
Bivalve molluscs	*Vibrio parahaemolyticus*, viruses, toxins such as DSP, PSP, ASP, enteric pathogens such as *Salmonella*,
Soft unpasteurized cheese	*Listeria monocytogenes*, *Brucella*, *E. coli* O157
Mature hard cheese	Moulds
Apple juice	*E. coli* O157, patulin
Ready to eat Sandwiches	*Staphylococcus aureus*

has a short shelf-life. The system of positive release is also clearly inappropriate for the non-manufacturing sector, as restaurants and cafes serve food immediately to customers, thereby severely limiting the sampling potential. In an attempt to apply end product testing to the catering sector, some institutions such as hospitals and schools have, in the past, implemented a system of retaining small samples of the meals for a prescribed period of time. The theory behind such a system was that if any food-borne illness arose, these samples could then be examined for the relevant pathogens. Alternatively if no reports were received, the samples could be discarded after the incubation period had passed for any suspect organism. The reality of this system was that many catering institutions serving very high-risk individuals used significant portions of their chilled storage space to house tiny pots of slowly deteriorating food. Because of the non-homogeneity of microbial contamination, it is likely that such samples would have been entirely uninformative had any outbreaks arisen.

Some of the legislation implementing vertical directives retains a requirement for end product testing (e.g. The Dairy Products (Hygiene) Regulations 1995 (10) and Minced Meat and Meat Preparations (Hygiene) Regulations 1995 [11]). Generally, however, end product testing is now used mainly as a verification or validation procedure within a hazard analysis-type food safety management system.

The hazard analysis/risk assessment approach

A more favoured approach to food safety management is through the use of a proactive system that is based on the engineering system of 'failure mode and effects analyses'. The rationale behind such management systems is that:

- by identifying all possible flaws or hazards in the process;
- determining which are likely to pose a serious risk; and
- then implementing control measures to prevent them

the resultant product should be flawless or at least free from serious hazards. Current food safety regulations require this approach. Regulation 4(3) of The Food Safety (General Food Hygiene) Regulations 1995 [12] requires that:

A proprietor of a food business shall identify any step in the activities of the food business which is critical to ensuring food safety and ensure that adequate safety procedures are identified, implemented, maintained and reviewed on the basis of the following principles:

(a) analysis of the potential food hazards . . .
(b) identification of the points in those operations where food hazards may occur
(c) deciding which points are critical to food safety
(d) identification and implementation of effective control and monitoring at those critical points and
(e) review . . . periodically and whenever the operations change.

The product-specific regulations that enact the vertical directives also require assessment of critical control points (CCP), implementation of control measures and monitoring. These regulations also specify particular records and the time for keeping them available to inspectors.

There are several named systems that follow this preventative approach. Probably the most well known is HACCP. HACCP was originally developed in the United States in the early 1960s and has gradually become established throughout the world as a satisfactory system for managing food safety. HACCP as a system is supported by Codex Alimentarius, the EU, the Food Standards Agency and a number of stakeholders in food safety.

Hazard Analysis Critical Control Points

HACCP is a management system for producing safe food. The system ensures safe food by identifying potential hazards and developing appropriate controls. HACCP differs from other food safety management systems in that it also provides a framework to prove its efficacy. HACCP is structured and logical, proactive and preventative. It was originally designed by Pillsbury and NASA to ensure safe food during space travel, but its advantages mean that major food businesses have adopted it so as to be able move away from end product testing and positive release. HACCP is ideal for manufacturing companies with good technical expertise and factory production lines.

Codex Alimenarius has published guidance and recommendations on developing and using HACCP as a food safety management system [13]. There are seven principles of HACCP that should be implemented in sequence to provide a HACCP plan for a process/premises. These principles are:

(1) *Analyse the potential hazards in the process* To carry out the first step, the product must be accurately described and the process determined and characterized. The scope of the HACCP plan should be determined and a process flow diagram developed for the process. This needs to be verified on site. Then potential hazards should be identified at each step, and finally an assessment made as to which hazards need to be included. The justification for inclusion/exclusion should be based on sound scientific knowledge, surveillance, severity of the hazards and any other relevant data. The final result of the hazard analysis step should be:

- A list of steps which reflect the process
- A list of hazards at each step which are plausible and important

(2) *Identify the Critical Control Points (CCPs)* To help determine whether any step is a CCP, the following questions may be asked in sequence. Codex Alimentarius has organized these questions into a decision tree that leads the user through the questions to reach a 'yes/no' answer to the question 'is this step a CCP?' The questions are:

  (i) Is there a serious risk to the public if this step is not controlled? (If YES go onto (ii)).
  (ii) Is this step designed to eliminate or reduce the hazard to an acceptable level? (If Yes the step is a CCP. If No go to (iii)).

  (iii) Could the hazard increase to an unacceptable level? (If No, the step is not a CCP. If Yes go into (iv)).
  (iv) Will a subsequent step eliminate the hazard? (If Yes the step is not a CCP, if No the step is a CCP.)

It should be noted that the questions must be asked at **each** step for **each** hazard. It is quite possible for a step to be critical for one hazard, but not for another. For example the step of cooking may be critical to control the vegetative cells of *Salmonella* but not critical for controlling germination and growth of a spore-forming pathogen such as *Clostridium perfringens*. Similarly the step may not be critical for contamination by metal as a subsequent step (metal detection) may control this, but would be critical for physical contamination by glass, where there will be no subsequent step to control it. The decision that a step is or is not a CCP should always be stated with regard to a particular hazard.

It is worth considering the difference between a control point and a critical control point. A **control point** is defined as a step where something can be done to prevent or eliminate a food safety hazard (or reduce it to an acceptable level). A **critical control point** is a step where something **must** be done to ensure safe food (by preventing or eliminating a food safety hazard).

CCPs are required by law to be correctly identified and properly controlled to ensure safe food (Food Safety (General Food Hygiene) Regulations 1995 [12] and regulations enacting the vertical directives). Control points are places where the proprietor may implement controls, but is not legally required to do so. Many proprietors implement control measures at control points in addition to CCPs because it is good practice to do so. However, the legal requirement and the principles of HACCP differentiate between CCPs and control points and stipulate control at CCPs.

(3) *Determine controls (and critical limits) for these CCPs* A control measure is an action or measure that prevents or eliminates a food

safety hazard or reduces it to an acceptable level. Each hazard has own control. Control measures need to be monitored. In order to monitor whether the step is under control, it is necessary to identify levels at which the food will be safe. These levels will include a 'target' and 'critical limits'. A target value is a predetermined value for a CCP, which is guaranteed to eliminate the hazard. For example, if the hazard that is under consideration is the germination of *Bacillus cereus* spores and subsequent growth and toxin formation, an appropriate control measure might be to store the product in chilled conditions with a target of 4°C. Naturally some variation will occur from a target, so critical limits must also be set at any CCP. Critical limits are the acceptable range of deviation from the target that can occur before corrective action is taken. To continue with the example given earlier, the proprietor may find it acceptable if the food is at a higher temperature than 4°C provided it does not go above 8°C. This would make 8°C the critical limit. Once the storage temperature exceeds these limits, some corrective action must be taken.

(4) *Establish a system to monitor the controls* The system to monitor the control measures should be a series of planned observations and measurements. They should describe the situation at the CCP and determine whether the system is under control or deviating from the critical limits. Ideally monitoring should be continuous and indicate trends. A trend towards loss of control may then be identified before critical limits are breached and the situation can be retrieved before the food becomes unsafe. The responsibility for monitoring should be assigned and clearly indicated in the HACCP plan. Training will be required for the person carrying out the monitoring. This training must include how to monitor, what to do when deviation from target occurs and what to do when deviation from the critical limits occurs. The best parameters to monitor are physical or chemical (i.e. temperature, time, pH, moisture level) as measurements are straightforward and repeatable.

Microbiological measurements are usually best for verification rather than monitoring. Monitoring records need linking to day and time of production, to the process step and to the record keeper.

(5) *Establish corrective action for when the critical limits have been breached* If a step goes outside the critical limits, corrective action must be specified and taken. Corrective actions must be determined in advance. The action to be taken should be clearly indicated, as should the person responsible for taking the action. The corrective actions must also be monitored and recorded. The records should include the reason for the deviation and the steps taken to rectify the deviation. Most importantly, what happened to the affected food should also be recorded.

(6) *Verify the controls are working* It is essential to verify and validate the HACCP plan. This ensures that the HACCP plan effectively addresses the hazards in the system and confirms that the plan is operating effectively. As the plan is process/premises specific it is obviously essential to review the plan if and when any changes occur. Such a review process ensures the plan changes as required and continues to address the hazards correctly.

(7) *Document the controls* Although the main concern of the HACCP plan documentation is the monitoring of the control measures at the CCPs, other documentation is also important to be able to use the plan to prove that the production of the food is safe. Other documents that should be included in the HACCP plan include the following:

- Process flow diagram, hazards for each step and justification.
- CCPs.
- Indication of the critical limits, records of monitoring, list of corrective actions for each CCP.
- Verification process – how this was carried out and the results of any research that might have been carried out (e.g. challenge testing).
- Persons responsible for the foregoing.

In a well-managed business there may also be supplementary documentation relating to the prerequisite or underpinning systems mentioned here such as

- training records
- cleaning schedules
- hygiene policy
- pest control records.

These may be cross-referenced to control measures to streamline the HACCP plan.

The objective of developing a HACCP plan using these principles is to identify which points in the process constitute the major risks to food safety. Once identified, the proprietor can control these points to ensure the hazards are not realized. It allows the proprietor to target resources and training at points that matter the most. Provided the CCPs are correctly identified and always controlled correctly, the proprietor can be confident that the food produced in the premises will be safe for consumption.

Although HACCP as a named system is not required by law in Britain, a good HACCP plan will discharge the responsibility of a proprietor under regulation 4(3) of the Food Safety (General Food Hygiene) Regulations 1995 [12] or the equivalent requirements under the product-specific regulations. A further consideration for many food businesses is the way a HACCP plan can contribute to establishing a due diligence defence under Section 21 of the Food Safety Act 1990. In order to establish a defence the proprietor must take all reasonable precautions **and** show all due diligence. As discussed in Chapter 28, being able to prove such a defence requires a system that meets certain criteria. The system must be written down, it must be proactive and reactive – it must prevent faults and correct them when they occur, and the operation of the system must be checked and the results recorded. A good HACCP plan will address these requirements through principles 7 (*documentation*), 1 (*hazard analysis*), 2 (*CCPs*), 3 (*critical limits*), 4 (*corrective action*) and 6 (*verification*) respectively.

Developing a HACCP plan is resource intensive. It requires extensive technical knowledge, not just of the process and it associated hazards but also of the appropriate control mechanisms and the technology associated with them. Developing the plan also requires knowledge of HACCP itself and the way the principles must be used. A high level of understanding of the product specification is required, including parameters such as the pH (level of acidity), Aw (water activity), E_h (Redox potential) as well as the constituents and any changes that may result from processing. This level of expertise and resource is most often found in large businesses. This is cited as one of the reasons that HACCP has been mainly implemented in these businesses. The Food Standards Agency, in their strategy for the wider implementation on HACCP [14], report that in Britain 59% manufacturers have HACCP in place. HACCP is less favoured in other types of food business with 19% caterers and 16% retailers having HACCP plans in their premises. This disparity is not restricted to Britain. The Food and Veterinary Office inspectors reported that [their]

. . . visits [to member states] show that progress still needs to be made by firms in the food sector (particularly small & medium sized enterprises) with regard to setting up HACCP systems . . . this should be . . . a matter of priority by Member states. [15]

The need for a multidisciplinary team to address the development of a HACCP plan is a major stumbling block for small businesses, particularly the micro businesses with less than 10 employees. These constitute the majority of food businesses in Britain, numerically. Codex Alimentarius acknowledges that small companies may not have expertise on site, and in their discussion paper on implementation of HACCP in small and less developed businesses [16] makes the following recommendations to assist these businesses:

- governments and trade/industry associations should provide adequate technical support
- governments should ensure access to appropriate, current scientific support and to low-cost analytical services

- governments should develop sector-specific guides and generic HACCP plans for use in these businesses.

The industry guides on good hygiene practice developed under Article 5 of the Directive 93/43 on the Hygiene of Foodstuffs [7] give examples of generic HACCP plans that can be used as a basis for plans in small and medium-sized businesses. The major disadvantage with using generic HACCP plans in the way described by Codex is that if the proprietor has insufficient expertise and support to be able to implement the plan from scratch, he is unlikely to have sufficient to be able to revise it when necessary. It is also possible that he may have insufficient understanding to recognize when such review is needed. Using a generic HACCP plan without this understanding reduces it to another set of prescriptive rules and alters the system from a product/premises specific management system to another general hygiene system.

HACCP for butchers

One common generic system found in England is the HACCP system developed by the Meat and Livestock Commission (MLC) for retail butchers [17]. The MLC developed a course, known as the Accelerated HACCP for Butchers, together with supporting study material that presented HACCP in terms of the common processes found in the retail butchery trade. The advantage of the system as originally developed was that not only did each butcher attend an accredited course and receive the study material, but up to 8 hours of consultancy support were also provided. The butchers were able to use the generic HACCP plans as a base and then, with the help of the consultant if needed, to adapt and customize these plans to their premises and products. This system successfully addressed the main weakness of generic plans as mentioned earlier.

Assured Safe Catering

As discussed earlier, HACCP appears to suit factory-based manufacturing systems best and has not been

widely implemented in small and medium-sized businesses, especially caterers. Another named food safety management system was developed by the Department of Health to address this problem and is known as Assured Safe Catering [18]. Assured Safe Catering treats the common food treatments in catering as individual production lines (see Fig. 29.1) and identifies hazards that are common to the processes rather than to individual dishes. Extensive advice on planning and implementing the system is provided, but hazard identification is generic. Translating this system into a safe management system still requires a high level of technical expertise from the proprietor, particularly with regard to microbial pathogens.

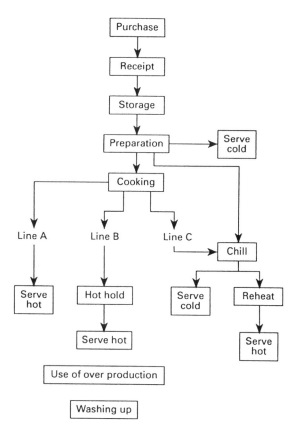

Fig. 29.1 Assured safe catering: flow diagram of any catering operation.

CONTROL SYSTEMS USED BY ENFORCEMENT AGENCIES

The control and management systems used by the authorities to manage food safety consist mainly of the following systems:

(1) Inspections

    (a) of food businesses by enforcement officers

    (b) of food authorities by the competent authority

    (c) of the competent authority and food authorities by the food and Veterinary Office.

(2) Food hazard warnings

(3) Surveillance

(4) Microbiological risk assessment.

Inspections of food businesses by enforcement officers

The main way enforcement officers control hygiene in food businesses is by inspecting the businesses and then taking action as may be required to address any problems identified.

Code of Practice 3 (Food Safety Act 1990) [19] defines an inspection as a visit to any food premises, which involves inspection of one or more of:

* premises
* equipment, including cleaning and maintenance
* process or procedure
* hygiene or practices of personnel
* labels or advertising material
* records.

An inspection is not a visit that:

* deals solely with a complaint
* is solely to obtain samples (although sampling may form part of an inspection)
* is solely to give advice
* is only to check compliance with improvement notices
* is only part of a survey.

Food premises inspections may be classified as food hygiene inspections or food standards inspections. Food hygiene inspections are inspections that consider the legal requirements according to the Food Safety (General Food Hygiene) Regulations 1995 [12] and the Food Safety (Temperature Control) Regulations 1995 [20] or product-specific regulations derived from the vertical directives. Food standards inspections cover quality, composition, labelling, presentation and advertising of materials or articles in contact with food.

According to Code of Practice 9 (Food Safety Act 1990) as amended in October 2000 [21] the objectives of an inspector carrying out food hygiene inspections are:

* To identify compliance with food safety legislation
* To identify potential risks arising from activities pursued at the premises.

To achieve these objectives an inspector must identify food safety hazards associated with the business, observe processes, check the structure and procedures of the premises against the schedules of the relevant legislation and inspect any records. The inspector will also need to discuss processes and procedures with the proprietor and/or staff, measure temperatures and possibly take samples.

Once the inspection process has finished, the inspector should discuss any breaches of legislation with the proprietor. The inspector should explain what will happen next (if anything), for example, whether the proprietor should expect a letter or notice. After programmed inspections a written report should be provided indicating:

* date, time and place of inspection
* what was examined
* any samples taken
* officers conclusions.

When conducting an inspection, officers should go at an appropriate time. Appropriate means at a time when the officer will be able to examine and inspect what is needed. The Food Safety Act 1990 gives authorized officers power to enter at all reasonable times (see Chapter 28). In this context, 'reasonable' should be taken to mean when the business would normally be functioning. It does not mean convenient for the proprietor. Code of Practice 3 (Food Safety Act 1990) [19] states that as a general rule, no prior warning would be given for inspections. However advance notice may be given at the officer's discretion. Whether prior warning is given or not, on arrival it is important to ensure the proprietor knows the purpose of the visit. When dealing with large premises Code of Practice No. 9 (Food Safety Act 1990) [21] suggests it is acceptable to divide the premises into subdivisions and make each subdivision the subject of a separate inspection. This approach allows for a more frequent inspection of high-risk sections as well as more manageable length inspections.

When planning an inspection, the officer should be properly prepared. This will include checking the background and property file for the premises, familiarizing himself with the relevant legislation and guidance, if necessary, and assembling any equipment that will be required. This will include clean protective clothing and head gear as appropriate. If the business being inspected offers its own protective clothing to the officer, Code of Practice 9 (Food Safety Act 1990) [21] states that this should be worn in preference to that provided by the food authority.

Inspections of enforcement organizations

The organizations carrying out inspections of food premises and enforcement of food safety legislation are themselves inspected and monitored. Food authorities are required by the Food Standards Agency to maintain a management system to monitor the quality and nature of inspections undertaken by officers. They must draw up a programme for food inspections. According to the Regulation 8 of Food Safety (General Food Hygiene) Regulations 1995 [12] food businesses posing a greater risk should be inspected more frequently than those with lesser risk. Officers should assess the risk of the premises inspected according to specific criteria contained in Code of Practice 9 (Food Safety Act 1990)

[21] and designate the premises to a hygiene category. The inspection frequency for premises is decided according to their category.

- Category A premises should be inspected at least every 6 months.
- Category B premises should be inspected at least every 12 months.
- Category C premises should be inspected at least every 18 months.
- Category D premises should be inspected at least every 2 years.
- Category E premises should be inspected at least every 3 years.
- Category F premises should be inspected at least every 5 years.

As stated in previous chapters, the Codes of Practice associated with the Food Safety Act 1990 are in the process of being extensively revised. The draft of the new code makes substantial changes to the risk rating scheme just described, and if enacted in its current draft form, would result in more frequent inspection of many high-risk premises such as meat products premises [22].

The Food Standards Agency requires returns on the ability of food authorities to meet their programmed inspection targets. These are published and failing local authorities can be relieved of their responsibilities (although none have so far). The Food Standards Agency also conducts audits of food authorities (see Chapter 27).

Inspections by the Food and Veterinary Office

As mentioned in Chapter 27, the Food and Veterinary Office (FVO) carries out inspections in member states to ensure that the EU legislation is being correctly transposed and enforced by the competent authority. The competent authority in the United Kingdom is the Food Standards Agency, although most food premises inspections are actually carried out by local authority inspectors. In January 2001, the United Kingdom had such an inspection to evaluate the controls in place with regard to the preparation of meat products, minced meat and meat preparations. FVO inspectors visited six premises altogether, four producing meat products (three were industrial meat products premises) and two producing minced meat or meat preparations, one of which was registered and the other approved. The final report of this inspection can be accessed at: http://europa.eu.int/comm/food/fs/inspections/vi/reports/united_kingdom/vi_rep_unik_3175-2001_en.pdf. The response by the Food Standards Agency may be found at www.foodstandards.gov.uk/ multimedia/pdfs/fvo_poultrytable and www.foodstandards.gov.uk/multimedia/webpage/fvoletter.

FOOD HAZARD WARNINGS

Food hazard warnings are a way of managing incidents that may compromise food safety on a national scale. They are issued by the Food Standards Agency to alert and instruct food authorities about problems associated with food. The Food Standards Agency issues these warnings based on information that comes to their attention from a number of sources. These sources will include the food industry, public complaints, consumer agencies and food authorities. Code of Practice No. 16 (Food Safety Act 1990) [23] covers this topic. The food hazard warnings are sent to food authorities with instructions for action and may be viewed by the public on the Food Standards Agency website. It is also possible to subscribe to an electronic paging system that will alert the subscriber to the warning. Prior to the development of the Food Standards Agency the food hazard warnings were issued by the Department of Health, Ministry of Agriculture, Forestry and Fisheries (MAFF) and the Scottish and Welsh offices.

There are four categories of food hazard warning:

A – for immediate action
B – for action
C – action as deemed necessary
D – for information only.

With categories A and B the action required of the food authority by the Food Standards Agency will be detailed and food authorities are expected to act accordingly. With categories C and D the food authority may be left to decide what action to take, if any.

When deciding to issue a food hazard warning, the Food Standards Agency must assess a number of parameters, including:

- The nature of the hazard.
- Its toxicity.
- The vulnerability of the target groups.
- The quantity and distribution – a small amount of locally produced food may not merit a nation-wide alert.
- The cooperation of the producer.
- The ability to identify affected batch – there is very little value in issuing a warning if the affected foodstuffs cannot be identified.
- The effectiveness of action.
- The significance for other products.
- Whether there was malicious intent.

In many cases the identification of a widespread problem that could warrant a food hazard warning will originate with a food authority. The food authority should contact the Food Standards Agency and provide sufficient details for the Agency to determine the appropriate course of action.

RISK ANALYSIS

Risk analysis (also see Chapter 10) is a system for assessing and managing food safety hazards. It is used by governments and non-governmental agencies as a way of assessing and advising on the risk posed by a hazard to a particular group. Such agencies want to identify what constitutes an acceptable level of risk and to use this information to set food safety objectives to manage that risk. Risk analysis separates the process of risk assessment, which is factual and scientific, from the process of risk management. Risk management will take into consideration additional points such as public perception.

The process of risk analysis consists of three interconnected concepts.

- Risk assessment or what is the risk?
- Risk management or what needs to be done about the risk?
- Risk communication or how can the risk be communicated?

The first stage, risk assessment, is a scientifically based process of four steps

(1a) hazard identification
(1b) hazard characterization
(1c) exposure assessment
(1d) risk characterization.

Microbiological risk assessment (**MRA**) is a particular form of risk assessment [24] that attempts to estimate the public health impact of a **microbiological** food hazard using existing scientific data. Typically a microbiological risk assessment will use published data (dose/response curve) to model the probability and severity of illness due to that hazard in a population (see e.g. Cassin *et al*. 1998 for an assessment of *E. coli* O157 in burgers [25] or Gerwen *et al*. 2000 for an assessment of cheese spread [26]). This process evolved from the approach used in toxicology and is based on measured data, models and estimates of disease incidence.

The first steps (hazard identification and characterization) describe the pathogen using clinical and epidemiological studies, surveillance, laboratory experiments on microbial genetics, transmission, virulence, infectivity and ecology of pathogens or related organisms and animal studies to arrive at an accurate description of the way the microbe behaves.

The third step in the risk assessment stage is an exposure assessment. This step tries to assess the extent to which the target group will be exposed to the hazard. This is a very difficult calculation. The assessment will be based on studies of dose/response and attack rates, as well as the levels on contamination in the product and the eating patterns associated with the product. Unfortunately, the data required for such an assessment is not always complete, consistent or easy to interpret. The

dose/response data available for pathogen-specific responses is very limited and very variable. Host susceptibility differs between patients. Not only can the attack rates of each pathogen vary, but the virulence of a species can also vary. In addition to the characteristics of the pathogen, scientists must also consider parameters such as the natural microbial ecology of the food, the way the processing and preparation will affect the microbial population, the initial contamination of raw materials, consumer characteristics and pattern of consumption. All of these parameters are also difficult to quantify accurately. As a result of all the confounding factors associated with the input data, the exposure assessment may be very broad and/or may only apply in very limited circumstances.

Step four in the risk assessment is risk characterization. This is the outcome of the hazard identification, hazard characterization and exposure assessment. It should determine how severe the health effects may be and how likely they are to happen. As with any estimate, the accuracy of this outcome depends on the validity of the data at the previous stages. How the risk characterization is described depends to some extent on the purpose of the risk assessment. The risk characterization may be described as a probability, for example, '*there is a risk of 1 person per X. becoming ill from a given pathogen from eating Y food*'. Alternately the risk characterization might be quite general, for example, '*this product is low risk because . . .*'. MRA typically results in a number or range of values, for example, the probability of illness due to *E. coli* O157 from a single burger in a single meal in United States is 5.7×10^{-7}–1.2×10^{-6} [25].

How is MRA used?

The risk characterization produced by an MRA can be used in several ways. For example, it can indicate the risk to a population of eating a food produced under particular circumstances. It can also be used to determine the initial contamination levels in raw materials and processes that are acceptable because that level will achieve a certain population level risk of disease. The advantages of MRA as a food safety management system are that the reasoning is transparent and based on published data. There may also be value in the process irrespective of the outcome because the process can indicate points in the food chain that constitute the major source of contamination or major control point.

The disadvantage of MRA really relates to problems with the initial data – it is always incomplete, frequently inconsistent and may be inaccurate. However, the process of risk assessment and MRA in particular is valuable in that it can be used to clarify the management of the risk and underpin the communication.

Risk assessment is the first stage of risk analysis. At the end of this stage there should be some quantitative assessment of the risk of the hazard to the target group. This will typically be in the form of a probability. The second stage of risk analysis is risk management. This stage addresses what needs to be done about the risk.

The process of risk management means selecting appropriate prevention and control options. While the stage of risk assessment is a scientific process relying on data calculation, the second stage is very much a management process.

Although the second stage will use the risk assessment data, it will not be the only consideration in the selection of the most appropriate control mechanism. Political, practical and financial issues will also need to be considered, for example. Typically this stage would involve some consultation with interested parties. The outcome would be the selection of an appropriate way to control the risk of the hazard to the target group. Risk management is used by governments and food agencies to set targets and food safety objectives. A food safety objective can be defined as the level of a hazard in food that is considered to be acceptable for consumer protection.

Examples of recently set food safety objectives include:

- Food Standards Agency – identified the reduction of *Salmonella* in raw materials as a target to reduce the number of notifications in humans.
- European commission (EC) – set maximum levels for aflatoxins in foodstuffs.

By using a risk analysis approach, the actual risk (probability × severity) will be one input to the production of food safety objectives, but political, commercial and consumer matters will also be taken into account. This is likely to produce a more workable objective, thereby maximizing its potential for success. It should also make the reasoning and choices that underpin the food safety objective transparent.

The third stage in the process of risk analysis is that of risk communication. This stage has to address the problem of how the risk can be communicated, particularly to the target group. This stage is especially problematic as the communication must accurately convey the risk of the hazard, without unduly frightening the target group. There are two major issues associated with accurate risk communication:

- Risk means different things to different people.
- The information to be conveyed is complex.

Risk is a combination of the probability of a hazard occurring and the severity if it does occur. To many people it may mean one or the other, or both but not equally. In other words, the definition of 'risk' varies between individuals. Severity is easier to understand (and explain) than probability. This part of the risk is often exploited by special interest groups to gain support for a particular view. Probability is more difficult to communicate and may be difficult for the target group to relate to. The response of a target group may not be based on the actual risk, even if that has been accurately conveyed and understood. Other issues such as familiarity, perceived control, that is, whether the exposure is optional, perceived benefit and whether the effect is catastrophic or by accretion all affect the way people respond to a risk. An unfamiliar hazard (e.g. dioxin contamination) may elicit a more extreme response than a familiar one (e.g. crossing a road) with a higher or equal risk. Similarly an action that has a very high risk, but that is optional with perceived benefit (e.g. smoking or eating oysters) may be actively engaged, irrespective of the risk. Another factor that can affect the target group response to a risk is whether there is equity associated with the risk, that is, whether the exposure is equal for all. The reputation of the organization providing the information also has a major impact.

The process of risk communication must account for the fact that the meaning of risk varies according to the individual and whether the recipient is a specialist or consumer, and that the response to the communication will depend on the context of the risk. Generally risk communication is more successful when the message is consistent and clear. The disadvantages need to be included as well as advantages, and unknowns should be clearly identified. Explaining a risk by comparison to other risks is not usually considered to be very effective. It can appear as though the organization is trying to trivialize the risk in question, and the two situations are unlikely to match exactly for issues such as benefits, familiarity, control, etc. Risk communication is a very difficult process to carry out successfully. It is much easier to find examples of where it has failed than where it has not, but the consequences of failure can prove catastrophic.

The WHO/Codex [27], the European Food Safety Authority and European Commission [28] and Food Standards Agency are in favour of using risk analysis to underpin food safety decisions for all food hazards. The aim of doing this is to be able to base trade and legislative decisions on sound science and risk-based food safety. Risk analysis incorporates human factors into the decisionmaking but requires transparency and valid science to support the outcome. This approach should result in a more consistent response from relevant agencies. The weaknesses in the risk analysis approach result from the flawed basic data sources and in the problems associated with risk communication. However, the agencies supporting the use of risk analysis are addressing some of these issues [29].

SUMMARY

The aim of all food safety control systems is to ensure that food reaching the consumer is safe to eat. Although the control systems or interventions may vary in detail depending on whether they are implemented by a food business to ensure the safety of the product, or by an enforcement, government or non-governmental agency to monitor the food producers, the principles of hazard analysis and risk assessment underpin them all. The move from a prescriptive set of rules to a process or premises-specific management system is international and supported by all the main agencies associated with food enforcement, including the Food Standards Agency, The European Commission and European Food Safety Authority and Codex Alimentarius. As a consequence, food proprietors and enforcement officers are obliged to implement and assess these systems.

REFERENCES

1. Department of Health, *Chilled and Frozen: Guidelines on Cook-Chill and Cook-Freeze Catering Systems*, HMSO, London.
2. Brown, K.L. and Oscroft, C.A. (1989) *Guidelines for the Hygienic Manufacture, Distribution and Retail Sale of Sprouted Seeds with Particular Reference to Mung Beans*, The Campden Food and Drink Research Association Technical Manual No. 25.
3. Tezcucano Molina, A.C. (2002) Assessment of the efficacy of disinfectant impreganated cloth and spray disinfecting on contaminated stainless steel surfaces, Unpublished MSc thesis, University of Birmingham.
4. *Fresh Meat (Hygiene and Inspection) Regulations 1995*, SI 1995/539 as amended, HMSO.
5. Tebbutt, G.M. (1991), Development of standardized inspections in restaurants using visual assessments and microbiological sampling to quantify the risks. *Epidemiology and Infection*, 107 (2), 393–404.
6. *The Meat (Hazard Analysis and Critical Control Point) (England) Regulations* 2002, SI 2002/889, HMSO.
7. Council Directive 93/43/ EEC on the Hygiene of Foodstuffs, OJ L 175, 19.7.93.
8. *Food Handlers: Fitness to Work, Guidelines for Food Business Managers*, Department of Health.

9. Kirby, M. and Gardiner, K. (1997) The effectiveness of hygiene training for food handlers. *International Journal of Environmental Health Research,* 7, 251–258.

10. *The Dairy Products (Hygiene) Regulations 1995,* SI 1995/1086 as amended, HMSO.

11. *Minced Meat and Meat Preparations (Hygiene) Regulations 1995,* SI 1995/3205, HMSO.

12. *Food Safety (General Food Hygiene) Regulations 1995,* SI 1995/1763 as amended, HMSO.

13. Codex Alimentarius Commission, Food Hygiene Basic Texts, Joint FAO/WHO Food Standards Programme (1997) ISBN 92-5-104021-4 (also available from the Codex Alimentarius website).

14. Food Standards Agency (2001) Agenda item 4, paper FSA 01/07/02, 14 November 2001.

15. Commission of the European Communities (1999) *Report from the Commission on the Second Series of Visits by Commission Representatives to Member States Pursuant to Article 5 of Council Directive 93/99/EEC with a View to Evaluating the National Systems for the Official Control Of Foodstuffs,* COM, 751.

16. Codex Alimentarius Commission (1999) Discussion paper on the implementation of HACCP in small and/or less developed businesses, Agenda item 9, CX/FH, 99/9.

17. Meat and Livestock Commission, *A Practical Guide to HACCP for Retail Butchers,* Milton Keynes.

18. Department of Health, *Assured Safe Catering: A Management System for Hazard Analysis.*

19. Food Safety Act 1990, Code of Practice No. 3, *Inspection Procedures – General,* Department of Health, MAFF, the Welsh Office and Scottish Office, HMSO.

20. The Food Safety (Temperature Control) Regulations 1995, SI 1995/2200 as amended, HMSO.

21. Food Safety Act 1990, Code of Practice No. 9, *Food Hygiene Inspections,* Second Revision October 2000, Department of Health, MAFF, the Welsh Office and Scottish Office, HMSO.

22. Food Safety Act 1990 draft Code of Practice and Practice Guidance (2003) www.consultation.gov.uk.

23. Food Safety Act 1990, Code of Practice No. 16, *Enforcement of the Food Safety Act 1990 in Relation to the Food Hazard Warning System,* Department of Health, MAFF, the Welsh Office and Scottish Office, HMSO.

24. Mitchell, R.T. (2000) *Practical Microbiological Risk Analysis,* Chandos Publishing, Oxford.

25. Cassin, M., Lammerding, A., Todd, E., Ross, W. and McColl R.S. (1998) Quantitative risk assessment for *Escherichia coli* O157 in Burgers. *International Journal of Food Microbiology,* **41**, 21–44.

26. Gerwen S., te Giffel, M.C., van 't Riet, K., Beumer, R. and Zwietering, M.H. (2000), Stepwise quantitative risk assessment as a tool for characterization of microbiological food safety. *Journal of Applied Microbiology,* **88**, 938–951.

27. WHO (1995) *Application of Risk Analysis to Food Standards Issues,* Report of the Joint FAO/WHO Expert Consultation, WHO/FNU/FOS/95.3.

28. Regulation (EC) No. 178/2002 of the European Parliament and of the Council of January 28 2002, OJ L 31/1 1.2.2002.

29. WHO (1999) *The Application of Risk Communication to Food Standards and Safety Matters,* FAO Food and Nutrition Paper No. 70, WHO/FAO, Rome.

30 The preservation of food

Philip W. Cox and Deborah Kendale

INTRODUCTION

Most foods intended for human consumption are of biological origin, apart from the obvious exceptions of salt, water, etc., and it is important that this basic fact is not forgotten. Biological systems are seldom inert; during the course of life, the metabolism of the living organism changes to create the end product of the adult animal, or continuously produces the leaves, tubers, buds and seeds of the plant. Pick an apple, net a herring or pod a pea, and you will halt the metabolic processes. The animal or plant will die and so cease to grow. During its life, it possessed a complicated and highly efficient mechanism of protection from bacteria, yeasts and moulds. With death, this mechanism no longer operates. The condition of the living plant or animal will change over time but, after death, changes occur that are likely to be cumulative, irreversible and largely deteriorative unless steps are taken to halt them. It has always been one of the primary objectives of research in the food industry to develop and perfect techniques for arresting these changes during the necessary periods of distribution and storage. Food can be preserved in a number of ways, for example, drying, canning and freezing. Other methods such as smoking, pickling and the use of salt, oils, sugar or syrups alter the characteristics of the food in order to preserve it.

MICROBIAL FOOD SPOILAGE

Foods are not stable commodities, and when they have to be imported or transported over long distances, as much of our food is, it may deteriorate very quickly. This deterioration is brought about by a combination of physical damage, chemical breakdown or contamination, insect or rodent activity, enzyme breakdown or spoilage caused by micro-organisms, whose access is aided by many of these changes. Changes in taste, smell, colour, texture and appearance are the main features of microbial spoilage. Some of these changes are very distinct, such as the 'sulphur stinker' spoilage of canned foods, or the production of 'ropiness' in bread. The main agents of spoilage are shown in Table 30.1.

Table 30.1 The main agents of food spoilage and the types of food they affect

Food type	Agent of spoilage
Canned foods	Bacteria or moulds
Dried goods	Moulds or yeasts
Modified atmosphere packed goods	Anaerobic/microaerophilic bacteria
Preserved foods	Resistant micro-organisms
Chilled foods	Bacteria, e.g. *Pseudomonas*
Frozen foods	Enzymes

The ability of all types of organisms to adapt in order to be able to grow on food should never be underestimated, although a differentiation needs to be made between those organisms that affect food quality by spoilage, and those that have public health consequences. In some cases growth of spoilage organisms may even be tolerated, for example, the growth of fungi on the surface of otherwise perfectly acceptable jam or cheese. In some cases, the growth of foreign organisms is encouraged or promoted in order to develop taste, texture or appearance, examples of this include Stilton or French cheeses. The pattern of spoilage will depend on the microbial load of the particular foodstuff, and is governed to a large degree by the type of food and its origin. Even so, the load will still be subject to change as a result of the storage, handling and processing of the food, thus allowing selected categories of organisms to become dominant. For example, during the process of pasteurization, heat-sensitive bacteria, yeasts and moulds may be destroyed, but heat-resistant spoilage organisms remain; and while storing food under chilled conditions inhibits the growth of mesophiles, it is still possible for psychrophiles to grow unhindered. Typically, most preservation methods rely on reducing the amount of water available for microbial growth to produce their effects. For instance, the addition of solutes to water reduces its availability for growth or with freezing where ice crystals are formed and so free water is removed. This may be expressed by equation 1 below

$$a_w = P/P_0 \qquad (1)$$

where P is the partial pressure of the solution and P_0 the partial pressure of pure water (a_w of pure water = 1).

So, as the concentration of solutes increase or more water is removed from a system, the value of a_w falls. Broadly speaking, the more traditional methods of food preservation typically affect the characteristics of the foods in this way, for instance, smoking, pickling or salting, whereas, more modern methods attempt to preserve the freshness and quality of the food, for example, freezing or chilling.

As well as spoilage, which renders the food unfit for consumption, preservation is also required to prevent the enumeration of pathogens that can cause a variety of, sometimes very severe, food poisonings, for example, Listerosis or *Campylobacter* infection. Food poisoning falls into two camps: first, there is infection from organisms present in the food, which enumerate in the gut and cause their symptoms, and second, if left to grow before preservation is carried out, toxins may resist the preserving treatment, which efficiently kills vegetative cells, and still enter the gut to cause poisoning, for example, botulism. Hence food preservation techniques need to be timely as well as effective.

Food spoilage organisms

The main groups of organisms involved in food spoilage are moulds, yeasts and bacteria (Table 30.1). Generally, **moulds** affect the surface of foodstuffs when their spores are able to germinate. They produce a mycelium, which gradually penetrates the food, producing the familiar fluffy whiskers and eventually the distinctly coloured spores. The food may also eventually become musty, softer and sticky or slimy. As well as spoiling the more common perishable foods, they are particularly capable of affecting foods with a high sugar and salt content, and 'dry' foods that have become damp as a result of bad storage, for example, the colonization of the surface of jam, cheese or dry bread.

Various strains of **yeast** will grow in anaerobic as well as aerobic conditions, that is, with or without oxygen (O_2), and also in acid and high sugar solutions. They may also be the fermentative type, which break down sugars to produce alcohols, carbon dioxide (CO_2) and acids, or the oxidative type, which oxidize sugars, organic acids and alcohol, raising the pH in the process. Some osmophilic yeasts (and xerophilic moulds) can spoil dried fruits and concentrated fruit juices, as they are able to tolerate conditions of low a_w (0.62), while others can tolerate a high salt content and may contribute to the spoilage of foods preserved by salting.

Although **bacteria** are simple, single-cell organisms that occur widely and cause spoilage under many conditions, they cannot grow with an a_w below 0.91 and do not feature in the spoilage of dried foods. However, in other foods that provide a suitable nutrient environment, they are able to bring about several types of spoilage. Bacteria readily grow on the moist surfaces of meat, fish and vegetables and will not only produce taints and odours, but can also bring about a degeneration of the food to produce slimes, which, in turn, may produce pigments capable of bringing about colour changes, for example, *Pseudomonas fluorescens* (fluorescent green) and *Serratia marcescens* (red).

A wide variety of organisms are capable of producing a viscous material (rope), which can affect milk, soft drinks, wine, vinegar and bread, for example, *Bacillus subtilis*. Many spoilage bacteria possess the ability to ferment carbohydrates. Homofermentative lactic acid bacteria primarily produce lactic acid, while heterofermentative bacteria produce butyric and propionic acid in addition to the gases such as CO_2 and other, frequently noxious gasses. Foods that have been improperly processed and packaged to give anaerobic conditions may be tainted with hydrogen sulphide, ammonia, amines or other foul-smelling products produced as a result of putrefaction by the anaerobic decomposition of protein. While aerobic hydrolysis of proteins may produce bitter flavours. However, in some cases these may enhance the palatability of certain foods and are therefore not always detrimental. Growth can be slowed or arrested by killing the bacteria (by sterilization as in canning), depriving the bacteria of the water that is essential for their metabolic activities (as in drying and freezing), altering the pH (as in acid pickling or fermentation), depriving the bacteria of available water by increasing the osmotic pressure of the surrounding medium (e.g. by the addition of sugar in jam making), by placing antimicrobial compounds at the food surface (e.g. smoking) or by sharply reducing the temperature below that at which bacteria can grow and multiply (as in chilling). The methods of food preservation mentioned here are discussed in greater detail in the following sections.

Although these treatments are designed to achieve a complete cessation of the biological activity, limited enzymatic activity is inevitable and, over long periods, will cause deteriorative changes. However, with efficient preservation it may take years before these deteriorative changes can be recognized by taste or appearance. But when dried foods become moist, canned foods are punctured, or quick-frozen foods (QFF) raised in temperature, the familiar pattern of spoilage reasserts itself.

DEHYDRATION

Drying deprives bacteria of the moisture necessary for their growth and reproduction, and also has the advantage of reducing the activity of the enzymes that cause foodstuffs to ripen and, ultimately, to rot.

The preservation of food by drying goes back thousands of years, and fruit, fish, meat, etc. are still dried in the sun and eaten by millions of people. However, sun-dried foods are subject to contamination by insects, birds and airborne infection. Because of the limitation of this traditional method, most foods are now dried by artificial methods.

Bed drying for solid foods

Today, 'solid' foods usually undergo some treatment prior to dehydration, for example, vegetables are usually blanched by steam or hot water to inactivate enzymes, hot alkaline dips are used to remove wax from the surface of fruits and meat, and fish and poultry are usually precooked, but this cannot be relied upon to remove pathogens.

An essential of the dehydration process is efficient contact between the product and the hot air, together with the quick removal of moisture. Various techniques are used.

The fluidized bed dryer is a common type of dryer that has proved useful for drying vegetables. It consists of a series of perforated beds with fans underneath. The food, for example, peas, is held in suspension by the upward blast of the hot air, which also provides the efficient mixing. As the product drops from one bed to another, the temperature is increased (from 40°C to a maximum of 55°C at the end) and the moisture content is reduced from 80% to 50%.

The peas are then subjected to a further form of treatment in another type of dryer – a hot air bed where moisture falls to approximately 20%. The hot air bed dryer is a chamber in which the product is placed on perforated trays, through which hot air is passed from below, but there is no fluidization. Peas then receive a further treatment in dryers until the moisture content falls to approximately 5%.

These hot air systems can cause undesirable effects as a result of the reactivation of enzymes. An alternative is vacuum drying.

Spray drying for liquid foods

Liquid foods are usually dried by one of the following methods.

1. Spraying a thin film of the liquid food on to heated rollers, for example, potato flakes, tomato paste, instant breakfast cereals and animal foods. The food is scraped from the rollers in flakes and is then powdered. Drum-dried foods are generally not as satisfactory as spray-dried foods as they possess a more cooked character and do not reconstitute as well.
2. Spray drying the liquid food through an atomizer into a heated chamber, for example, milk, egg products, tea and coffee powders. A development of this involves the injection of a gas into the feed line, which produces a foam that dries more rapidly.

Since these products are hydroscopic, it is essential that they are kept dry and correctly packaged to prevent contamination.

Freeze drying

This is the most advanced technique for eliminating the water contained in foodstuff. The product to be dried is first frozen and then exposed to a vacuum of not less than 1 torr (1 mmHg). The water contained in the product in the form of ice is directly sublimated into vapour, avoiding the liquid phase.

Condensers then eliminate the vapour. Freeze drying is also described as drying by sublimation, or the more microbiological term lypholization. The shape, colour and taste of the original product are preserved, and freeze-dried foodstuffs are easily re-hydrated, instantly regaining their original properties, even when adding cold water. The process is carried out in either steel cabinets or long, insulated tunnels fitted with vacuum locks and gates for continuous operation. It should be noted that a variety of micro-organisms would withstand lyophilization and may be capable of growth post re-hydration.

Due to the development of more efficient methods of heat transfer, accelerated freeze drying (AFD) was developed, which has given rise to the 'instant meal' consisting of pre-cooked freeze-dried foods to which hot water is added about 2–15 minutes before serving. Foods dried by this method do not distort but take the form of a spongy mass that can be quickly reconstituted. It is used to dry a variety of foods including meat, poultry, fish, shellfish, vegetables, noodles and fruit.

Freeze-dried foods are stable at room temperatures but should be refrigerated for long storage. They are usually vacuum packed or packed within an inert gas ('**gas flushing**') to prevent oxidation.

Infra-red drying

This is a system that uses high-powered sources of radiation (heat). It has been used in the dehydration of fruit, vegetables, meat, fish and cereals.

Use of solvents

Yet another method of dehydration is the use of a solvent such as ethyl acetate, which forms a low boiling point mixture with water that can be removed by distillation, the residual solvent being removed by vacuum drying. Liquid CO_2 is commonly replacing organic solvents in this process as the liquid CO_2 is easily distilled off and leaves no detectable residue. This process is also used to produce decaffeinated coffee, as caffeine is soluble in the CO_2.

With all drying methods, it is important to remember that the microbiological load in the product will depend upon the initial load before dehydration, and that when water is added to effect reconstitution, the organisms begin to grow again. Even with AFD methods, few bacteria are killed, and food must therefore be processed under strict hygienic conditions.

REFRIGERATION

Refrigeration has long been used for the preservation of food. There are three basic categories of refrigerated food: chilled food, frozen food and quick-frozen food (QFF). Chilled food is held at a temperature slightly above 0°C where no ice crystals form in the food. Examples of this include: wet fish; bacon, pies and sausages sold from a refrigerated display cabinet; vegetables are hydro-cooled by immersion in an ice-water slush and spray tank between picking and transport to market.

Initially, refrigeration concentrated on chilling food, not on the actual temperature of the foodstuffs. The increased sophistication of refrigeration equipment, along with improvements in monitoring technology, have extended storage times while maintaining food quality.

In a commercial or industrial food situation, it is important to differentiate between product chilling and maintenance of temperature-controlling systems. A hotel fridge may be capable of chilling small quantities of food and dealing with the consequent water vapour, but it cannot deal with larger joints of meat or other food without compromising operating temperatures and therefore food safety.

Rapid chilling of food to the holding temperature is vital for food safety. Bacteria may survive cooking as spores, so multiplication is possible if the transition between cooking and the chilled temperature is too long. Cooling of food from the cooking to the chilled temperature needs considerable care. This is particularly true of large quantities of food where core temperatures may take a considerable amount of time to reach 0–5°C. The equipment used in commercial or industrial situations must be capable of rapid chilling and dispersal of water vapour without excessively dehydrating the food. Blast chillers are capable of this, with a reduction

in food temperatures to below 3°C within 90 minutes or less depending on the loading of the equipment. This is achieved by blowing high velocity cold air over the food. It is important to remember that the formation of large ice crystals from slow freezing should be avoided or food quality will suffer.

After blast chilling, food must be transferred to temperature maintenance systems as soon as possible. These are designed to hold food at steady temperatures, usually with low energy inputs. This equipment has sophisticated monitoring equipment that checks either zones or layers of the holding area. Circulation systems ensure that the chilled air is evenly distributed without causing excessive dehydration of the foods.

Temperature control and enforcement

Enforcing the temperature requirements of food legislation (see pages 510 and 547) is made difficult because:

- there can be up to 7°C difference between incoming air and outgoing air in fridges and chill cabinets
- there is a 2 hour derogation on shop storage not in a cooling system
- there can be large variations in temperature across display cabinets and holding rooms
- food could have just been put in the fridge
- it can always be argued that the fridge is in a defrost cycle.

The enforcement officer must ensure that thermometers are properly and regularly calibrated. At least monthly, or more regularly if possible, thermometers should be checked against a reference thermometer held in the department. Records should be kept of each instrument so that drifts in calibration can be spotted rapidly and the instrument removed from service. The reference thermometer (and the officer's as necessary) should be recalibrated at least annually. While following this procedure, some departments also send their thermometers away for checking if officers have used them in investigations or inspections that will lead to legal proceedings.

Chilling

This is an imprecise term, the meaning of which in terms of temperature tends to vary with the product. Therefore, it can mean any reduction in normal temperature. The imprecise definition is used because different foods deteriorate at different temperatures, for example, fruit (apples will still brown if the temperature is low due to enzymatic activity).

Refrigeration temperature is usually taken to be below 5°C, and is the temperature that inhibits the growth of pathogens, although they still remain viable of course. *Clostridium botulinum* will grow and produce toxin between 3°C and 4°C, so temperatures are best kept at between 0°C and 3°C. A code of practice exists to govern the use of chilling and can be seen in 'Temperature Control Requirements' (no. 10 ISBN 0-11-321456-0) or advice can be gained from relevant research associations, for example, Campden and Chorleywood Food Research Association (CCFRA) in Chipping Campden Gloucestershire.

Cook-chill

A major trend in food processing has been the development of convenience foods. This has led to cook-chill ready meals and a huge growth in sandwich sales, together with continuing development of dairy products, particularly desserts and gâteauxs, and many other products that busy consumers would prefer to buy rather than make at home. The development of high-quality nutritious snacks has resulted in people eating fewer regular meals and a decrease in formal family eating.

The desire for 'additive free' or 'natural' foods has resulted in increased pressure on food manufacturers to reduce or eliminate preservatives, synthetic colours and anything else associated with 'E' numbers. Increased public knowledge about nutrition issues has raised the profile of food components such as salt, sugar, cholesterol, fats and fibre, and has resulted in the development of an increasing number of products with specific claims in these areas. As part of the increased nutritional awareness, low-fat products in the dairy sector have flourished. An increase in the popularity of

ethnic foods and different packing forms and cooking methods, such as microwave ovens, has widened the technological base of the convenience foods available for consumers.

The removal of preservatives from a wide range of products has increased the risk of premature yeast and mould problems for manufacturers, as well as increasing the potential for pathogen growth. An increasing emphasis on process control, pack integrity and chill chain control must form a part of food manufacturers' policy. The Chilled Food Association has issued guidelines for good hygienic practice in the manufacture of chilled foods other than those produced by catering operations [1]. Four categories of prepared chilled food are covered:

1. those prepared from raw components
2. those prepared from cooked and raw components processed to extend the safe shelf life
3. those prepared from only cooked components
4. those cooked in their own packaging prior to distribution.

The guidelines are intended to ensure that the cooking processes are sufficient to destroy vegetative pathogens, particularly *Listeria monocytogenes*.

The discussion of cook-chill foods is continued in the section detailing the thermal pasteurization of foods.

FREEZING

Frozen food is maintained at a temperature of at least −1°C, and this includes most of the meat sold in the United Kingdom. QFF pass through the zone of maximum crystal formation (0°C to −5°C) within a very short time, which is never more than 3 hours, and are held at a temperature of −18°C, although some sensitive foods, such as fish (if stored for a long time), should be maintained at −29°C or even lower if the texture is not to be damaged. Typically, the supply chain between manufacture, distributor, retailer and consumer will try to ensure that frozen food is maintained at −18°C.

Quick freezing

The conventional method of quick freezing is by plate frosters. Here the product is pre-packed in packages that are as slim as is practical, and is held in contact under pressure between the upper and lower freezing plates, through both of which a refrigerant is circulated. The plates are on a hydraulic ram that allows contact to be maintained during the inevitable expansion during freezing. At the end of the freezing time, the product is removed from contact with the plates and placed in cold store. This method is suitable for products that take a long time to freeze, for example, meat and fish.

A method of quick freezing that has gained acceptance, particularly for a free-flowing product such as peas, is that of air blast freezing. This method can be continuous, as opposed to the batch process of plate freezing, and consists of cascading the product on to a belt that moves through a freezing tunnel. Either air is circulated in the tunnel around and across the belt, or the belt is dispensed with and an upward flowing cushion of refrigerated air carries the product up and forward in a 'fluid bed' technique. This method is suitable for freezing vegetables including peas and beans.

Another method – **cryogenic freezing** – involves bringing the food into contact with a non-toxic refrigerant such as liquid nitrogen (N_2) or O_2. Freezing in this case can be accomplished in seconds rather than hours, and this can give quality advantages to delicately textured products. Most freezers used commercially for this method are spray freezers, which consist of an open-ended insulated freezing tunnel through which the food passes on a belt conveyor. At the exit of the tunnel are the liquid N_2 sprays. Passing through the tunnel, which takes about 6 minutes, brings the product close to its freezing point, and the sprays effect freezing in $1–1\frac{1}{2}$ minutes. Liquid N_2 boils at $-196°C$ and boiling occurs when the food is immersed as a result of rapid heat transfer. The method is used for foods such as strawberries and raspberries. In the pulse method of freezing, the wrapped food is placed in an insulated chest into which liquid N_2 is sprayed intermittently, the temperature of the food being allowed to equilibrate between each pulse. More recently, freon has been approved in the United States for direct contact freezing of food. Foods may also be quick frozen by immersing them in very cold fluids such as invert sugar solutions or brine.

In addition to the maintenance of the correct temperature ($-18°C$) of QFF, it is also important that the original packaging of the food is kept intact as a barrier to moisture vapour and O_2. Intense illumination should also be avoided for long periods where green vegetables in translucent wrappings are concerned, as this may lead to loss of colour.

In judging the fitness of QFF for human consumption, it is important to distinguish between a loss in quality and a hazard to health. QFF that have been allowed to thaw are not necessarily unfit for human consumption, and should therefore be judged on the same basis as any other perishable foods.

Although microbiological growth virtually ceases at temperatures below $-10°C$, and freezing may destroy a percentage of vegetative cells, it cannot be relied upon to prevent food poisoning, as once the food is thawed the micro-organisms will grow. Freezing has little effect on spores and viruses, and toxins are stable in this condition. There is even a species of yeast that will grow at temperatures as low as $-34°C$. Nutritional loss, by protein denaturation and rancidity, as well as dehydration by sublimation, can occur at lower temperatures. Figure 30.1, shows the rate of loss of vitamin C (ascorbic acid), the vitamin most sensitive to high storage temperatures. Figure 30.2 illustrates the temperature range of the growth of food poisoning and psychrotrophic organisms.

Both of these figures demonstrate the importance of maintaining QFF at $-18°C$ at all times. It follows that in assessing the fitness of QFF that have suffered temperature fluctuations, both the period and the degree of fluctuation are vital factors.

While the loss of quality may not necessarily be evident to the eye or by taste, the physical appearance of a quick-frozen product can sometimes indicate whether it has been subjected to temperature fluctuation. Loss of colour in green vegetables,

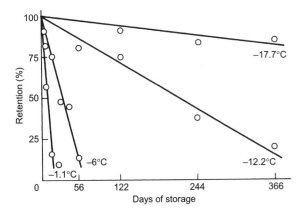

Fig. 30.1 Effects of temperature on ascorbic acid retention in peas.

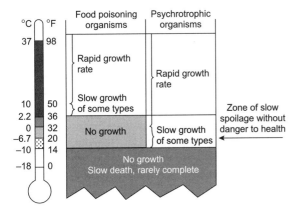

Fig. 30.2 The temperature range for growth of food poisoning organisms and psychotropic organisms.

a darkening of the surface area on poultry, and a lightening of the surface area of other meat products can all indicate that temperature fluctuations may have occurred. While these are generalizations, they can be useful indicators. It is important to realize that changes caused by temperature fluctuations are irreversible and cumulative. Successive large fluctuations lead to sublimation of moisture from the food and may well, over a long period, give rise to a loss of weight as well as a loss of quality, all of which will occur more rapidly where permeable or defective packaging materials are used.

The Quick Frozen Foodstuff Regulations 1990

These regulations implement European Union (EU) Directive 89/108/EEC (1998). QFF are perceived to be of a higher quality than ordinary frozen foods, hence the reason for the EU directive and the regulations. Environmental health officers and trading standards officers are responsible for enforcement. These regulations apply to food that has undergone 'quick freezing' whereby the zone of maximum crystallization is crossed as rapidly as possible. Products must be suitably pre-packaged so as to protect them from microbial and other contamination and dehydration, and must remain so packaged until sale.

A product available for sale must be labelled with:

1. an indication of the date of minimum durability
2. an indication of the maximum advisable storage time
3. an indication of the temperature at which and the equipment in which it is advisable to store it
4. a batch reference
5. a clear notice not to refreeze after defrosting.

QFF must be stored at or below −18°C. Provisions controlling the monitoring of temperature during transport, and for sampling and analysing the temperature of quick-frozen foodstuffs have been applied as a result of the Quick Frozen Foodstuffs Regulations (1990). *Code of Practice No. 12* (revised February 1994) [2] under the Food Safety Act 1990 provides advice on the division of enforcement responsibilities between environmental health officers and trading standards officers for these regulations. The code also makes it clear that the temperature monitoring of QFF should not be a high priority for food authorities since the principal objective of manufacturers is to produce high quality products, and temperature parameters are concerned with food quality not food safety. Most monitoring is likely to be carried out by the home authority on outgoing products at the factory, or at the distribution warehouses of haulage companies, and retail store. QFF temperature checks will mainly occur during the course of programmed visits or as a result of complaints. These checks are not only concerned with actual recorded temperatures, but with the positioning and operation of manufacturers' and distributors' own temperature-monitoring equipment and records.

Cabinet hygiene

The proper and efficient use of retail cabinets can only be achieved by following the principles laid down in the guidance issued by the UK Association of Frozen Food Producers [3]. Everyone concerned with the handling of QFF should study this code.

COLD CHAIN VEHICLE DISTRIBUTION

Temperature-controlled vehicles

The cold chain is the maintenance of an integrated temperature throughout the processes of production and distribution. Temperature-controlled vehicles are a vital part of this. There are several types of vehicles designed and built to **maintain** temperature, not to change the temperature of either chilled or frozen foods. In the United Kingdom, as in some other EU countries, there is no legislation to control the thermal efficiency of insulation other than on international journeys, where the carriage of foodstuffs comes under the International Carriage of Perishable Foodstuffs Regulations 1985 (as amended 1996 and 1997). These lay down minimum values of insulation efficiency and are only valid for a certain period, during which the container requires certification.

The various refrigeration systems employed in vehicles include the following.

1. **Water ice systems** with temperature parameters of 3–10°C (very few such systems are currently in operation).
2. **Total loss systems** use either CO_2 (dry ice) or liquid N_2. The former skips the liquid phase altogether, and by a process of sublimation vaporizes directly from the solid state.
3. **Liquid N_2**, carried in vacuum-insulated tanks on the vehicle, boils at −196°C. It is the gas which, via solenoid valves, is piped into the container through a number of spray heads. This system is silent, has a high degree of reliability and has the ability to pull down the container temperature very rapidly. However, local areas of under-cooling can still occur.
4. **Eutectic systems** use roof-mounted tubes or plates, which are fitted to the front bulkhead or sidewalls of the vehicle and are filled with a brine solution that is cooled by freon passing through pipes within the plates or tubes. Similar to the N_2 system, this system does not have forced air circulation. Cooling can be provided for a finite period only, that is, until all the eutectic has melted. The system is very inflexible, providing either frozen only or chill only.
5. **Vapour compression system.** Commonly known as mechanical refrigeration, this system is driven by hydraulic motors, electric motors or diesel engines. It is very flexible, and capable of maintaining temperatures at between +20°C and −30°C. Its flexible temperature capability coupled with the type of drive unit makes it the most common transport refrigeration unit. It can switch between heating and cooling automatically in response to the thermostat in order to maintain present load space temperature, irrespective of whether the outside is at a higher or lower temperature.

Vehicles are now being designed for the carriage of multi-temperature perishable foodstuffs. These have the ability to divide frozen products from chilled, and also to control the chilled product air temperatures and split certain chilled products into two different temperature bands. Position switches are accessible to the driver so that each temperature position can be preset and then locked. Temperatures can be preset to whatever parameters the retail customers require the foods to be delivered at.

Temperature monitoring equipment is now an essential component of such vehicles. Sensors are positioned in the warmest part of the container. Electronic data collection coupled with on-board printers and displays are now common [4]. These provide valuable information for comparison with the food officer's own temperature readings.

FERMENTATION

The pickling of vegetables has long been performed commercially and in the home. Here vinegar is added to the food in sealed containers, usually after salting to dehydrate the food, to lower the pH and so preserve the contents as well and producing the distinct taste. However, the process of anaerobic respiration, or fermentation, can be used to preserve several types of food and to produce many others. This is done by allowing the accumulation of natural organic acids as a result of the growth of organisms (typically bacteria) that produce lactic acid. Pickled cabbage (sauerkraut) and olives are examples of foods preserved in this way. Products such as tempeh (fermented soybeans) soy sauce and miso (fermented rice) rely on fungal fermentations within the high salt containing foodstuffs to perform the same role. Finally, alcoholic drinks have always been produced by the fermentation of yeast to convert sugar to alcohol.

BIOTECHNOLOGY AND GENETIC MODIFICATION

Biotechnology has been used in food production and processing for thousands of years. Almost every ingredient used in the production of food has as its source a living animal, plant or micro-organism. The food sources available to early humans, both plant and animal, evolved through the process of natural selection. Genetic diversity arising from spontaneous genetic changes was first exploited (i.e. the use of biotechnology) when farmers saved the seeds from their best crops for later sowings and used their best animals for breeding.

Genetic modification techniques have long been applied to the micro-organisms that produce fermented foods, including those involved in bread and beer production. For example, the production of a more efficient system for metabolizing maltose can reduce the time it takes to make bread. In the dairy industry, lactic acid producing bacteria can be genetically modified to produce strains with improved phage resistance or bacteriocin or flavour production. Some higher fungi, including some edible species, can be used to convert lignocellulose or other cheap carbon sources directly into fungal protein suitable for human consumption. Single-cell (bacterial or fungal) protein has been developed as a source of human and animal food, for example, the organism *Fusarium venenatum* is grown on hydrolysed starch to produce meat analogues for low-fat diets.

Individual microbial products may also be exploited. Rennet is the common milk clotting preparation used to produce cheese and other dairy products and is by far the largest single use of enzymes by the dairy industry. Extracellular fungal proteinases can serve as rennet substitutes and the production of 'vegetarian' cheese is now a marketed selling point for such products.

By exploiting micro-organisms, higher quality foods have been produced, which require less further processing for their preparation and storage. As bacteria, viruses, fungi, nematodes and insects can all be used as biological control agents in pest control (e.g. *Bacillus thuringiensis* as an insecticide). Many viruses that are pathogenic to insects are now being developed commercially and fungal insecticides have been used successfully since the beginning of the twentieth century. The application of biotechnology has also resulted in the development of rapid and sensitive analytical methods that may have many applications in food analysis. For example, the use of DNA probes and immunoassay methods is expected to prove useful in improving the detection of food contaminants and to detect the use of undeclared genetically modified organisms (GMO) foods in product formulations.

The Food and Agriculture Organization (FAO) and the World Health Organization (WHO) have examined the safety of foods produced by biotechnology. The prime consideration is to ensure that pathogens should not be introduced into foods by the process. In the evaluation of foods produced by biotechnology, databases are required on:

1. the nutrient and toxicant contents of food
2. the molecular analysis of organisms used in food production
3. the molecular nutritional and toxicant content of genetically modified organisms intended for use in food production.

FAO and WHO also recommend the need for consumers to be provided with sound scientifically based information on biotechnology in food production [5].

Recent advances in genetic engineering have allowed for new developments. As stated initially biotechnology has been a human activity for thousands of years (i.e. bread and beer making). However, the technology now exists to perform targeted mutagenisis on the organisms (microbe, plant or animal) which go on to become foodstuffs. In essence, this is the same as a selective breeding programme to enhance product traits, but without the trial and error associated with the traditional methods. Prime examples of this are the introduction of herbicide resistant soya beans or tomatoes with extended shelf lives. While these possibilities are now within easy reach of food manufacturers, consumer resistance to such introductions has been vehement. The use of genetically modified foods has been seen by many to be interference with the 'natural order'. Such consumer feeling follows close on the heels of the rejection of food irradiation, national concern over food poisoning out-breaks, Bovine Spongiform Encephalopathy (BSE) and foot and mouth diseases in meat and a general swing away from food additives and preservatives to minimally processed, nutritious foods. As a footnote it should be noted that the United Kingdom placed the most stringent legislative constraints upon GMO research of almost any country. Currently, GMO legislature in the United Kingdom is gradually relaxing to workable safe limits; whereas, other countries are tightening their controls.

Detailed below is the governing legislature for the use of genetically modified organisms and foods in the United Kingdom.

The Novel Foods and Novel Food Ingredients Regulations 1997

These regulations make provision for the enforcement of EU Council Regulation 258/97 (OJ L 43 of 14 February 1997) and designate the Food Standards Agency as the competent food assessment body. The EU regulation requires that before novel foods and novel food ingredients are placed on the market they be subjected to a pre-market safety assessment.

The Novel Foods and Novel Food Ingredients (Fees) Regulations 1997

These regulations provide for charging for assessment of novel foods and novel food ingredients.

The Genetically Modified Organisms (Deliberate Release) Regulations 1992 (as amended 1993, 1995 and 1996)

These regulations implement EU Council Regulation 90/220/EEC (OJ L 117 of 8 May 1990) on the deliberate release into the environment of GMOs. They provide the circumstances under which GMOs require consent for release into the environment and, in the case of putting crops on the market, marketing consent.

Domestic regulations to implement Council Regulation EC No. 1139/98 dealing with the labelling of certain foods produced from genetically modified organisms came into operation on 19 March 1999. There is a great deal of public, professional and media interest at present in issues surrounding genetically modified (GM) food. Some have called for a ban on GM foods, while others call for a moratorium on growing GM crops in the United Kingdom. The reader is referred to the following recent literature:

The Royal Society (February 2002) *Genetically Modified Plants for Food Use. An Update*, The Royal Society, London.

FOOD IRRADIATION

The process of food irradiation involves the exposure of food to ionizing radiations such as gamma rays, X-rays or electrons [6]. Gamma rays and X-rays are forms of radiant energy like sunlight or radio waves, but only the high energy forms of radiation are ionizing. Ionizing radiation interacts with the material being irradiated by transferring some of its energy to particles in the molecules of the material, producing electrically charged particles called ions. The energy transfer causes disruption of microbial DNA and so kills the organisms. These reactions destroy or prevent organisms such as species of *Salmonella* from multiplying. The technique of irradiating food has been patented in the United Kingdom since 1905, and is used in 21 countries including Belgium, France, the Netherlands and the United States. In the United Kingdom, irradiation has now been used for many decades to sterilize medical equipment, certain pharmaceuticals, cosmetics and packaging. Irradiation may also be used to stop the over-ripening of fruits, or the sprouting of vegetables or grains. Irradiation cannot make bad food good, and it cannot improve the appearance or taste of the food, or mask unpleasant odours.

The main use of irradiation is likely to remain the reduction in the number of harmful bacteria, such as species of *Salmonella, Listeria* and *Campylobacter* that can be present in certain high-risk foods, such as poultry, prawns or shrimps. This is particularly true for organisms such as *Listeria monocytogenes* which has a low infective dose (thought to be a low as a single cell; this is in contrast to Salmonella whose infective dose is around 10^6 cells per ml). For produce such as dried herbs and spices, irradiation could be used to destroy insects, pests and bacteria in place of the existing chemical fumigation method. However, irradiation is not suitable for use on all foods – some fruits, for example, are softened by irradiation, and some fatty foods develop a rancid flavour. Irradiation of food initially suffered consumer rejection but is likely to re-enter the market place as concerns over food safety have moved to newer areas (e.g. GMOs) and this forgotten technology may resurface.

All foods carry a low level of natural radioactivity, and there should be no measurable increase in the level of radioactivity after food is irradiated under approved conditions. Over the past 40 years, food irradiation has been studied more than any other process. The UK Advisory Committee on Irradiated and Novel Foods, WHO, the FAO Joint Expert Committee on Food Irradiation, and the EU Scientific Committee for Food have all concluded that there is no hazard associated with food that has been exposed to doses of ionizing radiation up to an overall average dose of 10 kilograys

(10 kGy is the amount of energy required to raise the temperature of 1 kg of water by 2.4°C), and irradiation introduces no significant nutritional or microbiological problems.

Following the government's decision to proceed with making irradiation an acceptable process of food treatment in the United Kingdom, the proposals for legislation follow those recommended by the CODEX Alimentarius Commission, a joint WHO/FAO body set up to prepare international food standards. In 1983, the commission adopted both a general standard for irradiated food and a code of practice for the operation of food irradiation facilities. This is the basis of the controls adopted in the United Kingdom. The important elements include:

1. a maximum limit on the irradiation dose that may be applied
2. licensing on a product-by-product and irradiation plant-by-plant basis
3. regular inspections of irradiation plants by expert central government staff
4. microbiological testing to ensure that only food of normal sound quality is irradiated
5. controls over imported irradiated food through official verification that the controls and standards achieved by exporting countries are equivalent to those in the United Kingdom
6. full labelling of irradiated foods so that consumers can make a choice about whether to buy irradiated or non-irradiated food.

These principles are embodied in the Food (Control of Irradiation) Regulations 1990 and the Food Labelling (Amendment) (Irradiated Food) Regulations 1990, which came into effect on 1 January 1991. The government has commissioned research into the development of suitable detection tests for irradiated food. WHO has recognized the benefits of irradiating food to reduce the number of bacteria in foods and has recommended [7] that consumers should select irradiated fresh or frozen poultry wherever possible.

The general conclusion is that food irradiation is a safe process, but considerable public concern on the matter remains despite official reassurances.

PROCESSED AND PRESERVED FOODS

The term 'processed food' is generally taken to mean foods that have been preserved (or their shelf life extended) by heat (thermal) treatment. Thermal treatment of food is probably the most common form of preservation used in commercial food manufacture. Not only does thermal processing extend shelf life but also produces desirable changes to most products. Equally some undesirable changes may also occur. Desirable changes usually mean the development of texture and taste, for example, the cooking of meats or pulses. Here the thermal process changes the structure of the food, that is, cooks the meat to give texture or softens the pulses by entry of water; as well as sterilizing or pasteurizing the food. However, thermal treatment may adversely affect the vitamin or organoleptic content of the food and so over processing might affect quality. In some notable cases over processing of food might be a desirable trait and to add to the quality of the food, for example, the more cooking tomato puree receives the stronger the taste, or the more processing baked beans or 'mushy' peas received the softer the beans become and so have a higher consumer value.

Thermal processing falls broadly into two camps: first, sterilized foods, for example, canned goods, that have received enough heat (usually performed at 121.1°C) for sufficient periods of time to ensure the total kill of all vegetative microbial cells and the spores of important food pathogens (usually based upon the death kinetics of *C. botulinum*). Second, pasteurization, a milder treatment regime, is used to kill vegetative cells so that foods have a longer shelf life, especially in cook-chill products where vegetative microbial cells are destroyed and the subsequent growth of spores is arrested (or at least slowed) by storage at refrigeration temperatures.

It should be noted that foods processed in anaerobic environments present an ideal medium for the growth of bacteria if the heat process is inadequate to destroy all pathogens and spoilage organisms within the pack. Also if the closure of the can or pack is not effective in stopping the entry of micro-organisms. Further hazards may arise from improper handling of products during processing, which causes damage and subsequent contamination of the internal contents of the can or pack.

STERILIZED FOODS

The most common form of sterilized food available is canned goods. Here it is the aim of canned food manufacturers to produce an encased 'sterile' product; one in which all micro-organisms and microbial spores are destroyed by the thermal treatment. The robust and impervious packaging with a double-folded mechanical seal then makes re-infection impossible and protects the food during handling and prolonged storage. However, in some instances it is not practical to achieve this and, in these situations, the storage temperature or the acidity of the contents are relied upon to be sufficiently adverse to the small number of heat-resistant spores that remain to prevent their growth. The food is then considered commercially sterile.

As micro-organisms that are capable of causing spoilage at a pH lower than 4.5 are generally of low heat resistance, acid and high-acid canned foods are subjected to a less severe heat treatment process than the medium- or low-acid canned foods. It is essential that the latter two categories be heated sufficiently to destroy the heat-resistant spores of *C. botulinum*. Where low-acid canned foods, for example, canned meats, incorporate curing salts, which are also microbial inhibitors, they can be given milder heat treatments. The heat treatment given is sufficient to destroy vegetative cells and limit the number of heat-resistant spores to no more than 10 000/g. In view of their limited treatment, such products should be stored under refrigeration. *C. botulinum* represents the major risk to such foods because of its heat resistance but the bacterium will not grow below a pH of 4.5. Due to *C. botulinum*'s inability to grow at reduced pH acid foods, such as fruit conserve, only receive a relatively low heat process. However, meats and vegetables stored without added salt, with a much higher pH, are given a process known as a botulinum cook to render them commercially sterile. Process time and temperature is determined for each pack, and will depend on such things as composition, weight, headspace, the viscosity of the contents, the presence of preservatives and the intended storage conditions. Scheduled processes for all low-acid products have been established and must be adhered to by canned food manufacturers [8]. They must be sufficient to ensure the destruction or inactivation of all pathogens. Certain heat-resistant spore-forming spoilage organisms that may be present, for example, *Bacillus stearothermophilus*, are more resistant than *C. botulinum*, the most heat-resistant toxin-producing pathogen; so by ensuring the destruction of spores of the former the latter is also destroyed and the food is made safe.

The food industry defines heat-treatment processes by the application of process values to thermally treated containers; the term used to describe this is 'integrated lethality'. The process values are defined by equation 2 below and relate the centre temperature of a product during heating and the holding time at a target temperature. This gives an estimate of sterility, that is, an integrated measurement of temperature/time of the process is produced.

$$F = \int 10^{(T-T_{ref})/z} dt \qquad (2)$$

where F is the process value (minutes), T the measured temperature of the food, T_{ref} the reference temperature for the process (in the case of canned goods this is 121.1°C) and z is the slope of the lethality curve (K) that is, the temperature required to achieve a 1 log reduction in spore numbers and t the process time. The z values are usually set to 10 K for most processes as this mimics the thermal death kinetics of *C. botulinum* spores. However, z might be set to any value that corresponds to a target spoilage organism. Or indeed z might be set to show the deterioration of any target attribute, for example, vitamin destruction.

In all thermal processes, a statistical approach to 'ensured sterility' is assumed. Treatments are based around a '12D' cook where the process ensures 12 decimal reductions in microbial numbers. Such

assumptions are to ensure that the majority (almost 100%) of cans is sterile and only the smallest probability that unsterile cans remain. In most cases, contaminated cans are generally found to be caused by poor processing and handling, rather than from the statistical probability of microbial growth.

After the heat-treatment process, great care is exercised to ensure that products are not recontaminated with pathogens or spoilage organisms. The entry of food poisoning organisms is prevented by ensuring, for example, that there is no human handling while seals are wet, rapid drying of containers, disinfection of all wet postprocessing handling equipment and that the disinfection of cooling water, by the addition of chlorine, is adequate. There have been outbreaks of food poisoning associated with canned meats and fish where contamination of the product has arisen during the cooling process by way of contaminants in cooling water and the canning environment finding a way through the seams during expansion/contraction. The bacteria involved have included *Staphylococcus aureus*, *Salmonella typhi* and *C. botulinum*. CCFRA publish a comprehensive range of manuals covering all aspects of the canning and heat treatment processes – see www.campden.co.uk

Spoilage of canned foods

In canned foods, spoilage is generally indicated by a 'blown' condition of the cans. The contents of cans in this condition are more or less decomposed and are unfit for human consumption. As stated earlier, the principles of sterilization are now so well understood that faulty cans, rather than under sterilization, are the most frequent cause of spoilage.

Cans of certain fruits, notably plums, cherries, raspberries, loganberries and blackcurrants, can have a blown appearance that makes them unmarketable, even though the fruit itself is both sound and sterile. The gas producing the pressure that blows the cans is found on analysis to consist largely of hydrogen. In consequence, this type of failure has come to be known as '**hydrogen swell**'. The cause of the trouble is principally the action of the fruit acids on the tin-plate. In the course of time, the action results in perforation of the can. The action occurs principally along the seams and other parts of the can that are subjected to severe strain during manufacture, and where the protective lacquer coating is defective. Cans lacquered after manufacture are rarely affected. Hydrogen swell occurs only in cans containing fruit, and is unlikely to occur in cans that have been sealed for periods of less than 12 months.

Taints and catty odours, so called because they resemble the odour of cats' urine, onion or garlic, have occasionally plagued the food industry in several countries. Such objectionable odours and flavours usually affect canned meat and vegetables and obviously make the food unacceptable to the consumer. Investigations carried out by British Food Manufacturing Industries Research Association demonstrated that contamination of the food by mesityl oxide, an unsaturated ketone, could give rise to these odours, particularly where it is able to react with hydrogen sulphide.

Examination of canned foods

Cans are an excellent and durable container. However, they are potentially dangerous if they have been processed incorrectly or the can fails in storage. Dangers arise from canned goods as a result of changes in the contents, defects in the canning, impurities from the can, or the use of preservatives and colouring agents. Superficially cans are inspected:

(a) by looking for dents, rust holes, blown and leaking cans; stained cases often demonstrate leaking cans;
(b) by palpation, that is, pressing the ends of a can with the fingers, will detect flippers, springers, and soft and hard swells (see later);
(c) percussion that is tapping with the finger should produce a dull note; a drum-like sound indicates the presence of gas; any can failing to give a dull note all over should be rejected; and
(d) shaking – meat that is in an advanced state of decomposition liquefies, allowing easy identification using this method.

Cans that are dented as a result of rough handling are always suspect, particularly if the damage is on a seam as this may result in its opening and air being admitted to the can. Occasionally, end seam defects such as false seams and irregularities such as wrinkles and pleats (where the metal folds back over itself during the first operation of seaming) will become apparent during inspection, Uniform, light-coloured rust spots on the labels seldom indicate leakage. However, if any of the spots have a darker inner area, the can may be perforated.

The inspection of canned good has generated its own vocabulary of descriptors with which to describe possible failures in the canning process. Phrases such as '**flipper**', '**springer**' and **swell** refer to gas generation inside the sealed cans; either by microbial or enzymatic action. '**Sours**' and '**stinkers**' describe unpleasant odours and flavours and inedible food, which have not produced gases; most notable of these is the '**sulphur stinker**' in which hydrogen sulphide has been formed to give the unmistakable 'rotten egg' or 'stink bomb' smell. Finally, cans might be stained, by the production of sulphides, or corroded to give leaking cans.

Cans of salmon and other seafoods are sometimes found to contain crystals that resemble glass. They are, in fact, crystals of magnesium ammonium phosphate, a normal product of the digestive system of the fish and are quite harmless. Crystals of calcium tartrate have been reported in canned cherries, and crystals of potassium hydrogen tartrate in canned grapes. All these crystals, with the exception of potassium hydrogen tartrate, will dissolve in vinegar.

OTHER METHODS

Liquid food, for example, milk may be sterilized in a continuous mode by using the HTST (high-temperature short time) or UHT (ultra-high temperature processes. Pasteurized milk is typically held at 63°C for not less than 30 minutes, or 72°C for not less than 16 seconds. However, with UHT or HTST the holding temperature is increased to 130–140°C with a few seconds holding time to ensure sterility while not significantly affecting quality. More recently, milk products have been produced which have received no heat treatment, but instead have been filtered to the exclusion of all microbes and therefore retain 'green milk' characteristics.

PASTEURIZATION

Convenience foods, such as cook-chill products and *sous-vide* packed foods (see next section) now form a major part of the modern diet. The main route for preservation and extended shelf life for these products is to pasteurize them before distribution. Here the thermal treatment destroys vegetative microbial cells and so heat-resistant spores might survive the process; it is therefore the maintenance of the chilled storage temperatures that prevents spore germination and enumeration. Typically, these treatments are performed by immersion in hot water baths or passing the foods through continuous tunnel ovens or pasteurizers. In the pasteurizers heated water rains down onto prepacked products as they pass through the tunnel on conveyer belts. Therefore, it is the rate of passage (i.e. belt speed) that defines the residence time in the oven; and the thermal treatment applied. Even high-acid foods (e.g. pickles) will routinely receive a pasteurization cycle during manufacture to ensure a minimal microbial load before distribution.

The milder thermal treatments encountered in pasteurization can equally be described by the integrated lethality equation shown in equation 2. Except that here the reference temperatures are generally lower and the F value has now become a P value or process value (still retaining the unit of minutes). As a base line P value, foods should be processed to the equivalent of 2 minutes at 70°C. However, this base limit is rarely approached as processors take great care in ensuring that all products are as safe (low microbial load) as possible and expend large amounts of energy and processing time ensuring all products are more that adequately pasteurized.

THE *SOUS-VIDE* SYSTEM FOR PREPARING CHILLED MEALS

This system involves a vacuum-packed product designed to retain the quality of the food while at the same time offering an improved shelf life for a chilled product. The system is particularly suitable for the hotel and restaurant sector of the catering industry, and was developed in the mid-1970s in France. It is now used in hundreds of restaurants in France and the rest of Europe and is typically used for reasonably high value products with complex formulations or expensive ingredients. In its simplest form, the system involves:

1. the preparation of high quality raw ingredients
2. pre-cooking, for example, browning, if necessary
3. placing the food into special heat-stable air and moisture barrier plastic bags or pouches
4. creating a vacuum and sealing the pouch
5. either steam cooking at specific times and temperatures to ensure pasteurization of the food, or cooking the food for immediate consumption
6. rapid chilling to reduce the temperature to 0–3°C within 90 minutes of the end of cooking;
7. labelling and controlled refrigerated storage within this temperature range until required for consumption within 5 days of date of production.

It is claimed that the aroma, flavour and texture of the food is preserved and maximized as a result of the construction of the pouch. No natural flavours are lost, but the product does rely on high quality original raw ingredients. *Sous-vide* packs are generally quite large and so care must be taken to ensure their proper pasteurization during processing.

Only small regeneration kitchens are necessary to reheat the food – minimal equipment and less skilled manpower are required compared with a conventional restaurant.

The problem with the system is that the creation of the anaerobic environment inside the pouch increases the risk of food poisoning from *C. botulinum*, *C. perfringens* and other anaerobic pathogens. The whole system depends heavily on the integrity of the process and the level of knowledge of all the operatives involved in ensuring safety. Internal temperature checks by thermocouples or thermistors can check the core temperatures of food before vacuumization. Temperature monitoring devices should be independent of the oven to enable the product sample to be monitored throughout its cooking and chilling cycle. Ideally, temperature printouts of cooking, chilling and storage cycles should be made [9].

The *Sous-Vide* Advisory Committee (SVAC) has produced a code of practice on the use of the system [10].

ANAEROBIC FOOD PACKS

As mentioned in the previous section anaerobic packs present an ideal growth environment for important food poisoning and spoilage organisms and so should receive special attention. Following the outbreak of *C. botulinum* from yoghurt in the north-west of England in the summer of 1989, the Department of Health sent a letter to all chief environmental health officers [11] asking for appropriate checks to be made on manufacturers of foods produced in anaerobic environments. The increasing use of vacuum packs in small manufacturing and retail premises to preserve the quality and shelf life of foods does represent a substantial risk factor, and the Department of Health letter requested chief environmental health officers to identify other food producers that may be using processes that have not been subjected to a hazard analysis. In the view of the Department of Health, it was likely that companies in this category would be the smaller manufacturers, particularly those that had begun food manufacture recently, or that had made changes to their products/processes without expert advice.

Developments in food processing that could introduce hazards are the increasing use of vacuum packing, a reduction in the use of preservatives (particularly salt and nitrites), a reduction in sugar content to produce a lower calorie product, and the tendency to use lower cooking temperatures to reduce loss of quality and improve the colour and texture of products.

Environmental health officers were asked as a matter of urgency to carry out an assessment of the safety of all types of food processes undertaken within their districts where the foods produced were packaged anaerobically. These included heat-treated, cured, smoked or fermented products in metal enclosures or vacuum packs, or other forms of packaging. The food manufacturers concerned were asked to supply evidence of the scientific and technical evaluation of the safety of their process. If there was any doubt, the companies were asked to obtain an expert evaluation of their process and to submit it without any delay to the local environmental health department. A list of organizations expert in food safety assessment that companies could be referred to was given in the annexe to the Department of Health letter.

Anaerobic process inspection

In carrying out an assessment of food processes, the following questions are important. The answers should enable food officers to judge whether potential hazards exist.

1. What is the range of products produced, the pack size and the number of different variations in each product?
2. Are technical personnel employed or retained by the company, and what is the training and background of those people?
3. What are the pH, salt and sugar levels of the products and their water activity? (If the processor is unable to provide satisfactory answers to these questions and seems unaware of their significance in relation to product safety, this should alert food officers to potential defects in the process.)
4. What is the total product life of all products? What are the storage temperatures required to achieve the prescribed product life and how has the life been technically determined?
5. How is the product packaged and what steps are being taken to ensure container/packaging integrity?
6. To whom are the products supplied and for what purpose (i.e. retail sales or as ingredients for other manufactured products)? What advice is given to customers about the storage and use of the products?
7. What evaluation has been made of raw materials and their suppliers? How is the safety of the raw materials determined?
8. What records are kept and what is the system for the tracing of products and recall procedures? What codes are applied?
9. What is the system for action in the event of process deviations from production schedules?

ADVISORY COMMITTEE ON THE MICROBIOLOGICAL SAFETY OF FOOD

The Advisory Committee on the Microbiological Safety of Food (ACMSF) held an initial review of the microbiological aspects of the safety of vacuum-packed and other hermetically sealed foods in June 1991. It concluded that a careful assessment was needed of the action that might be required to protect the public from any risk of botulism or other dangers. A working group was set up to carry out a detailed investigation. It presented its report to the ACMSF in January 1993 [12,13]. Among the important recommendations contained in the report were the following.

1. Food manufacturers should critically assess all new food process procedures to ensure elimination of the risk of botulism.
2. The commercial use of home preservation methods, for example, home canning or bottling of low-acid products such as vegetables and meats and home vacuum packaging (except for frozen products) should be actively discouraged.
3. In addition to chilled temperatures of 10°C, prepared chilled foods with an assigned shelf life of more than 10 days should contain one or more controlling factors at levels to prevent the growth of and toxin formation by strains of psychrotrophic *C. botulinum*.
4. The cooking or reheating of food should not alone be relied upon to destroy any botulinus toxin present. Other control factors should be used in foods susceptible to the growth of psychrotrophic *C. botulinum* in order to prevent its survival.
5. In addition to the maintenance of chilled temperatures throughout the chilled chain, the following control factors should be used singly or in combination to prevent the growth of and toxin production by psychrotrophic *C. botulinum* in prepared chilled foods with a shelf life of more than 10 days:

   (a) a heat treatment of 90°C for 10 minutes or equivalent lethality

   (b) a pH of 5 or less throughout the food and throughout all components of complex foods
   (c) a minimum salt level of 3.5% in the aqueous phase throughout the food and throughout all components of complex foods
   (d) an a_w of 0.97 or lower throughout the food and throughout all components of complex foods
   (e) a combination of heat preservative factors that can be consistently shown to prevent the growth of and toxin production by psychrotrophic *C. botulinum*.

6. The risk of *C. botulinum* hazard in chilled food must be addressed by manufacturers, caterers and retailers. Examples of at-risk foods include smoked salmon and trout cuisine, *sous-vide* products and modified atmosphere packaged sandwiches.
7. Food manufacturers and caterers should ensure that they have a thorough understanding of the operational capabilities of vacuum packaging machines, that the appropriate pouch or tray materials are used, and that the machinery used for establishing the vacuum function is maintained at the required specification.
8. Vacuum packaging machine manufacturers should alert users to food safety hazards and the risk from organisms such as *C. botulinum*.
9. The cooking of *sous-vide* products should not be undertaken unless operators have the technical expertise to ensure that pasteurization equipment and operating procedures are adequately designed and tested to give a uniform and known heat load to all containers during each cycle.
10. Information should be made available to consumers about the correct handling and cooking practices of vacuum modified atmosphere packaged foods.
11. Foods sent by mail order should be subject to control factors in addition to temperature to prevent the growth of pathogenic micro-organisms including pyschrotrophic *C. botulinum*.
12. There should be a comprehensive and authoritative code of practice for the manufacture of vacuum and modified atmosphere

packed chilled foods with particular regard to the risks of botulism. It should include guidance on:

(a) raw material specifications
(b) awareness and use of hazard analysis and critical control points (HACCP)
(c) process establishment and validation (including thermal process)
(d) packaging requirements
(e) temperature control throughout production, distribution and retail
(f) factory auditing and quality management systems (including awareness and use of HACCP)
(g) thorough understanding of the requirements to establish a safe shelf life
(h) thorough understanding of the control factors necessary to prevent the growth of and toxin production by psychrotrophic *C. botulinum* in chilled foods
(i) application of challenge testing
(j) equipment specifications, particularly with regard to heating and refrigeration
(k) training.

FOOD ADDITIVES

Additives are chemicals that do not form a natural part of the food. Many of these are now synthesized. They are used to:

• maintain or enhance the nutritional quality
• maintain or enhance stability
• improve appearance
• provide an aid to food processing.

See also the Miscellaneous Food Additives Regulations 2001 amended 2003.

Certain chemicals have been found to interfere with the principal agents that cause deterioration in food. They possess the ability to slow or arrest microbial growth by interfering with cell permeability, enzyme activity, or their genetic mechanism.

Substances such as salt or sugar, which have traditionally been in use for thousands of years, exert a preservative effect by dissolving in the water of the food and forming a concentrated solution that spoilage organisms are unable to live in. The concentrated aqueous solution is able to exert a strong osmotic pressure, and the cells are consequently deprived of water. Brines utilize salt to preserve meats, whereas sugar acts as a preservative in products such as jam, syrup and honey.

Many permitted preservatives, such as sulphur dioxide and proprionic, benzoic and sorbic acids, exert their effect by virtue of their acidity and are also important mould inhibitors. Some preservatives occur naturally, such as benzoic acid (found in cranberries), and these tend to be more effective against moulds and yeasts. In addition to exerting a preservative effect, some chemicals may also be of benefit in helping to maintain colour in ingredients that are going to be processed.

Labelling of additives

The Food Labelling Regulations 1996 (as amended 1998) require that a general description of the additives used as ingredients in pre-packed food should appear on the label. Additives performing certain functions must be declared by the appropriate name category followed by their specific name or serial number. The categories are:

• acid
• acidity regulator
• anticaking agent
• colour
• emulsifier
• emulsifying salts
• firming agent
• flavour enhancer
• flour treatment agent
• gelling agent
• glazing agent
• antifoaming agent
• antioxidant
• bulking agent
• humectant
• modified starch
• preservative
• propellant gas
• raising agent
• stabilizer
• sweetener
• thickener.

Flavourings may be declared simply as 'flavouring', or by a more specific name. If an additive serves more than one function, the category name that represents its principal function should be used. There are exemptions for non-pre-packed foods and foods packed on premises that are exempted by the regulations.

PACKAGING OF FOOD

If the methods of preservation described in the previous section are successful then the correct packaging of the microbially free foods is imperative to prevent re-infection. Packs may be simple barriers, such as paper, to encase stable foods, or they may provide a hermetic seal and barrier for susceptible products such as meat.

The Materials and Articles in Contact with Food Regulations 1987 implement requirements in EC Directive 83/229/EEC with regard to materials and articles that come into contact with food, including packaging.

The essential requirements of a packaging material are that it should:

- be inert, that is, it should not transfer its constituents to the food product
- protect the product from adverse environmental conditions
- present the product to the consumer in an appealing manner
- be easy for the consumer to use.

The Materials and Articles in Contact with Food (Amendment) Regulations 1994 deal with:

- purity standards for regenerated cellulose film
- residues migrating into foodstuffs
- administrative amendments to bring the 1987 regulations into line with the single European market.

The Plastic Materials and Articles in Contact with Food Regulations 1998 implement Commission Directive 97/48/EC. These regulations:

- set overall migration limits for all food contact plastics
- establish 'positive lists' of monomers and starting substances permitted for use in the manufacture of food contact plastics
- lay down rules for checking compliance with the regulations, for example, test times etc.

Choice of packaging

The type of packaging chosen for a particular foodstuff depends on a number of factors, not the least important of which is the nature of the goods to be packaged, especially if they are moist or greasy. Dry goods, such as flour or sugar, can be packed in simple materials such as paper, but foods such as coffee or biscuits, which are susceptible to humidity changes, must be packed in moisture-proof packages to keep them in a saleable condition. Greasy foods require greaseproof packaging, and moist goods need moisture-proof packages. According to the nature of the goods, a number of packaging materials may be suitable. The final choice will be based on the physical requirements of the pack in order for it to be stored and distributed in good condition.

When comparing the availability of materials, one of the main considerations (apart from cost) is that of compatibility. There are also subtle considerations such as the elimination of odour, dye or solvent contamination, and the matter of aesthetic appeal.

Paper

This is probably one of the oldest and most common packaging materials. Not only is it cheap and easily disposed of, but also it is easily printed on and is permitted to come into contact with food. However, it is clear that in its usual form it is only suitable for dry goods or as an outer wrapper for confectionery.

Waxed paper

This has been found suitable for wrapping bread and protecting sliced loaves from disintegration and loss of moisture. However, it is not possible to wrap bread in a completely waterproof material as this makes it liable to become soggy and mouldy. Bread is therefore cooled (to reduce condensation) and wrapped in waxed paper with the ends heat-sealed. The seal is 'loose' enough to allow some moisture to escape, gives the sliced loaf a more acceptable shelf life, and keeps it clean. Waxed paper or light gauge card is familiar as a material for forming disposable cartons for liquids such as milk. It is also extensively used for frozen products, where it gives mechanical protection to the contents.

Glass

Glass, in the form of jars or bottles with hermetically sealed covers, makes a most suitable food preservation container. The material possesses advantages over tinplate in that there is no risk of metallic poisoning, it is always possible to see the condition of the contents and the costly vacuum sealing machinery necessary for cans is not needed. On the other hand, it can easily be broken or splintered and is more expensive. Because of the danger of breakage, lower temperatures and longer exposures are necessary when treating the contents; care must also be taken when transferring glass containers from heating regimes to cooling baths as the thermal shock can shatter the glass. Typically, glass jars etc. will either be left to cool slowly at ambient temperatures or will undergo a two- or even three-stage cooling cycle as they are transferred from cooling baths of progressively lower temperature.

Clingfilms

Plastic packaging materials such as clingfilms offer many benefits. In particular, they help to prevent contamination of food by microorganisms. Clingfilms are also very convenient for consumers. However, in the 1980s it became known that, under certain circumstances, small amounts of chemicals could be transferred from tightly wrapped clingfilms on to foods. There is no evidence to suggest that these have caused any harm to health, and clingfilms remain suitable for most food uses (see following section). Nevertheless, as a result of this discovery, clingfilms have been the subject of much government research and advice.

In 1986, the government issued general advice about the use of clingfilms in cooking. Further work, discussed in a report issued in 1990 [14], showed that chemicals in clingfilm tend to transfer

most readily on to high-fat foods such as cheese. Government experts therefore advised that clingfilms should not be used to wrap food of this type. Their advice to consumers is as follows.

1. Do not use clingfilms in conventional ovens.
2. In a microwave oven, use clingfilms for defrosting or reheating foods; when cooking, use them to cover containers, but do not allow them to come into contact with food or use them to line dishes.
3. Do not use them in a way that makes them come into contact with high-fat food. If in doubt, use an alternative wrapping material – there are many to choose from.

Manufacturers are recommended to label packs along these lines, and retailers are recommended to follow the advice given in the final point earlier.

The Food Safety Directorate of the then Ministry of Agriculture, Fisheries and Food (MAFF, now DEFRA) issued further advice in November 1990 [15] following media interest and public concern over the use of clingfilms with food. The advice covered **all plastic** food wrapping films that have a 'cling' property, whatever their composition and whether used in the home or in shops. The advice does not cover the thicker, non-cling plastic wrapping often used by retailers and manufacturers.

The Tetra Pak

This is a familiar type of pack used to contain UHT milk and other liquids. It is formed from a paper base and polythene liner. Filling and sealing is a continuous and aseptic process, which can create a variety of package types and geometries.

Polythene

Being moisture-proof but not entirely vapour-proof, polythene is put to innumerable uses. Being a thermoplastic it is useful for heat sealing and may be overprinted to a degree. Its main use is as a cheap bulk package and shrink-wrap material for fruit and vegetables. Polythene packed meats, such as hams and large cuts of bacon, may be scalded after packing to shrink the packaging material into intimate contact with the surface. This will kill heat-sensitive surface organisms but, even with highly salted foods, bacterial growth may continue at a slow rate unless the products are refrigerated.

Polystyrene

Although brittle, polystyrene can be extruded and this makes it suitable for forming containers for cream, yoghurt, milk, etc. In its foamed form it is extensively used for meat trays, insulated cups and egg containers.

Pliofilm

This is a chlorine rubber product, similar in appearance to polythene. It is widely used for meat packaging as it is heat shrinkable and moisture-proof. It is also permeable to O_2, which allows the surface of packed meats to stay fresh and red.

Nylon and PVC

These are usually used in combination with Cellophane to produce vacuum-pack envelopes. The layers of the substances are bonded together with adhesives, and a thermoplastic layer is placed internally to allow the pack to be heat-sealed. These containers are suitable for conditions where it is essential that no air or moisture gets into the pack and no grease gets out. Crisps may be packed in this way with an inert atmosphere of dry N_2 to keep them crisp and prevent rancidity. N_2 packs are also used for products such as dried milk and coffee and dehydrated goods.

Cellulose

Cellulose used in combination with other materials is another form of transparent packaging, but may not be moisture- or gas-proof, although it can be reasonably greaseproof. It is ideal for packaging bread, confectionery, sausages, meats and cheese and, although not thermoplastic and therefore not heat sealable, it allows moisture to evaporate without condensing and encouraging mould growth.

Aluminium

This has the disadvantage of being expensive, opaque and not heat sealable, but as a metal it is both moisture- and gas-proof and has good heat-reflective qualities. It is an ideal packaging material for dairy goods, such as cheese, and being malleable makes ideal tops for milk bottles, jam jars, yoghurt containers, etc. Used as a laminate, it is useful for sacheting liquids and dehydrated foods and is used in N_2 packed goods as it gives great strength and long-term gas permanence to the package.

Tin plate and Lacquered plate

Tin plate cans have been discussed extensively earlier.

Retortable pouches

A new method of encasing food for sterilization and resale has become prevalent on the modern market. Retortable pouches have now become familiar for a very large variety of foodstuffs. The vacuum-sealed pouches are composed of metal foil laminated into plastic film, which are extremely durable and can be retorted. The reduction in the metal content and their increased handling properties make them ideal replacement for the traditional metal can.

Environmental responsibility

Once more, this is an area in which the consumer has made, reasonable, demands upon the food processing industry. Consumers have indicated that they wish to see less packaging used in food wrapping. Also the packaging that is used should be environmentally friendly, that is, recyclable, and should be easy to open.

Gas or controlled atmosphere packing

Many foodstuffs are so sensitive to O_2 that the only way to ensure an adequate shelf life is the complete removal or exclusion of the O_2. Ways of doing this are to:

- pack the container so tightly that there is little room for air

- evacuate the air (vacuum packing)
- remove the air and replace it with a non-oxidizing gas (gas flushing).

It is now possible to pack foods in atmospheres known to inhibit deteriorative changes. The technique has been used successfully for meats, vegetables and dairy products, and is also effective for fish. It is necessary to determine the optimum mixture of gases for each type of food. Thus, for meat it is necessary to increase the level of O_2 to maintain the desirable red colour, but for nuts, O_2 should be excluded to prevent oxidative rancidity. With fish, the rate of bacterial spoilage can be slowed down by increasing the level of CO_2.

The required atmosphere is carefully controlled at the time of packing, hence the name controlled atmosphere packaging (CAP). An alternative description for the same process is modified atmosphere packaging (MAP), as the atmosphere within the pack will change during storage. It is necessary to distinguish between this technique and the more complicated system of controlled atmosphere storage, in which the individual gases of the atmosphere are monitored continuously and maintained at predetermined levels throughout storage.

A number of machines providing different degrees of sophistication are now available for CAP. With the simplest machines, the product is placed in a plastic sleeve, flushed with a gas mixture and then sealed. More sophisticated machines provide an integrated packaging system. These machines will thermoform base trays from a roll of plastic film into which the product is placed. The loaded trays pass along the machine to be evacuated and are then filled with the required mixture of gases. A second roll of film is heat-sealed to the top edges of the trays, and the finished packs pass out of the machine.

The advantages of these packs include: an extended shelf life if stored under carefully chilled conditions; the containment of odour and drip of products such as fish; the packs can be attractively labelled; and a wide variety of products can be displayed by retailers without the need for specialist skills.

The gases CO_2, N_2 and O_2 are commonly used in CAP for fish, whose autolytic enzymes and subsequent bacterial infection can severely reduce shelf life. Extensive trials show that CO_2 inhibits bacterial activity, but autolytic changes proceed. N_2 is an inert gas that is included in the gas mixture to offset the adverse effects of high levels of CO_2 by maintaining the shape of the pack. Oxygen appears to reduce the amount of drip from white fish, but this gas may be beneficially omitted from packs of fatty fish to delay the onset of oxidative rancidity. Inclusion of O_2 does not overcome the potential hazard of toxin production by *C. botulinum*.

Thorough bacteriological testing has shown that toxin production is primarily dependent on temperature and that maintaining the fish at 4°C or below is the best preventive measure. Thus, packing fish in a controlled atmosphere presents no more danger than other forms of fish packaging with regard to a potential botulinogenic hazard, providing that chill temperature storage is maintained. To obtain the maximum benefit from packing in a controlled atmosphere, it is recommended that the temperature of packs is accurately maintained at 0–2°C throughout production, distribution and retailing [16].

Smart packaging technology

New packaging technologies are now entering the market place. These 'so-called' smart packs offer a range of enhancements beyond simple barriers to microbes, moisture, etc. and are designed to add quality as well as protection. The use of jar lids that indicate if a seal has been broken are now familiar to a wide range of product lines. However, beyond this a whole new vista of opportunity is available. Products that incorporate antimicrobial layers in their structure are now being produced. Carbonated drinks are commonly supplied in inexpensive plastic bottles that are gas impervious. Indicators of shelf life, beyond simple dating, are being applied to stored products. And metallic films are being incorporated into the packaging of microwaveable products so that the microwave energy can be used to brown food as well as just heat it.

Another familiar technology, which is used to label high value products, are anti-theft labels.

These tags are interrogated by radio frequencies and sound alarms when passed through detectors. Although of limited used for low unit cost foods the radio frequency identification (RFID) technology is expected to be used to produce smart pack that can (a) pass heating instructions directly to the microwave ovens, so that correct cooking protocols are used; and (b) be interrogated by 'intelligent' fridges so that 'use by dates' are flagged up by the fridge if a food is stored for excessive periods of time without use.

REFERENCES

1. Chilled Food Association (1998) *Guidelines for Good Hygienic Practice in the Manufacture of Chilled Foods*, 3rd edn, Chilled Foods Association, London.
2. Food Safety Act 1990 (1990, revised 1994) *Code of Practice No. 12: Quick Frozen Foodstuffs. Division of Enforcement Responsibilities: Enforcement of Temperature Monitoring and Temperature Measurement*, HMSO, London.
3. The UK Association of Frozen Food Producers (1994) *Guide to the Storage and Handling of Frozen Food*, UK AFFP, London.
4. Slocum, P. (1990) Cold chain distribution and other transport matters. Paper given at the Leatherhead Food RA Symposium on Food Safety Legislative Aspects, 7 November.
5. FAO/WHO (1992) *Strategies for Assessing the Safety of Foods Produced by Biotechnology: Report of a Joint FAO/WHO Consultation*, World Health Organization, Geneva.
6. Ministry of Agriculture, Fisheries and Food (1990) *Food Irradiation, Some Questions Answered*, MAFF Food Safety Directorate Information Issue No. 2, HMSO, London.
7. World Health Organization (undated) *Golden Rules for Safe Food Preparation*, WHO, Geneva.
8. Department of Health (1994) *Guidelines for the Safe Production of Heat Preserved Foods*, HMSO, London.
9. Schafheitle, J.M. (1990) The *sous-vide* system for preparing chilled meat. *British Food Journal*, **92**(5), 23–7.
10. *Sous-Vide* Advisory Committee (1991) *Code of Practice For* Sous-Vide *Catering Systems*, SVAC, Tetbury, Gloucestershire.
11. Department of Health (1989) *Guidance for Environmental Health Departments on the Prevention of Botulism*, letter to CEHOs and chief post health inspectors, EL (89), P145.
12. Advisory Committee On The Microbiological Safety Of Food (1993) *Report On Vacuum Packaging And Associated Processes*, HMSO, London.
13. Ministry of Agriculture, Fisheries and Food (1992) *Advisory Committee on the Microbiological Safety of Food Report on Vacuum Packaging And Associated Processes: Recommendations and Government's Response*, MAFF, London.
14. Ministry of Agriculture, Fisheries and Food (1990) *Plasticizers: Continuing Surveillance*, MAFF Food Surveillance Paper No. 30, HMSO, London.
15. Ministry of Agriculture, Fisheries and Food (1990) *Cling Film*, MAFF Food Safety Directorate Information, Issue No. 7.1, special edn, HMSO, London.
16. Seafish Industry Authority (1985) *Guidelines for the Handling of Fish Packed in a Controlled Atmosphere*, SFIA, Edinburgh.

31 Food safety: controls on milk, eggs and dairy products

Amanda Wheway

INTRODUCTION

Epidemiological information can be used to clearly show the link between the consumption of contaminated egg, milk and dairy products, and outbreaks of food-borne illness in the United Kingdom [1,2]. It has long been recognized that to prevent outbreaks of human illness associated with the consumption of egg, milk and dairy products, high standards of hygiene management must prevail over the production and processing of such products.

Milk and eggs are both highly nutritious foodstuffs with an ample supply of nutrients providing an excellent medium for microbial growth. In the 1920s and 1930s, 'milk was regarded as the most dangerous item in the diet' and consumption of contaminated milk was linked with thousands of cases of brucellosis, paratyphoid fever and bovine tuberculosis [3]. Heat-treatment of milk, in the United Kingdom, was initially introduced by the dairy industry around 1890 to prevent souring of milk and spread of milk-borne disease. Since that time, the dairy industry has developed rapidly, milk quality has improved and 'extensive adoption of pasteurisation of milk throughout the United Kingdom has reduced the number of outbreaks of illness' [4]. The introduction of the Food Safety Act 1990 [5] and the subsequent Dairy Products (Hygiene) Regulations 1995 [6]

placed extra regulatory pressure on the dairy trade.

The number of general outbreaks of food-borne illness in which milk and dairy products have been implicated as a source of infection has fallen over the past few decades [7]. Yet, in recent years, there have still been documented outbreaks of food-borne illness that have been traced back to the consumption of milk and dairy products. 'Salmonella and camplylobacter infections associated with unpasteurised or inadequately heat-treated milk have continued to be a public health problem' [8]. Causative organisms identified in 31 outbreaks of intestinal disease associated with milk, and dairy products, between 1994 and 1999, have included Salmonella *species, Camplylobacter, Escherichia coli O157, Small Round Structured Virus* and *Cryptosporidium* [9].

The fact that eggs remain the commonest vehicles of infection, in outbreaks associated with an identified food vehicle, reinforces the need for proper food safety control of this product from farm to fork [10]. In recent years, outbreaks of food-borne illness have been traced back to the consumption of eggs contaminated with *Salmonella* organisms, such as *Salmonella enteritidis* and *typhimurium* [11].

Food processing technology and equipment available to the dairy industry has become increasingly sophisticated and enforcement officers inspecting egg and dairy establishments require a sound understanding of relevant food processing controls. Food Authority officers involved in the inspection of on-farm dairies, and dairy and egg products establishments play an important role in safeguarding public health, by establishing whether the requirements of relevant food law are being met and taking appropriate action to ensure compliance with food law.

Officers carrying out inspections of establishments require to be competent and have a good understanding of the principles of the hazard analysis critical control point (HACCP) approach and risk assessment (see Chapter 29). The factors that contribute to food safety for particular products must be well understood by enforcement officers so that they can assess whether establishments are adequately controlling microbiological, chemical and physical food hazards. Enforcement officers must have a detailed knowledge of the relevant product-specific Regulations that relate to the establishments that they are responsible for inspecting. The Food Standards Agency requires that officers of Food Authorities, working in the United Kingdom, undertaking inspection of establishments manufacturing egg or dairy products 'should have received additional training and have demonstrated their competence to carry out such inspections' [12].

This chapter summarizes the necessary food safety control measures for production and processing of milk, egg and dairy products, and outlines the legislative framework.

LEGISLATIVE FRAMEWORK

Food hygiene legislation is almost completely harmonized throughout the European Union, under the Single Market. It consists of a number of product-specific hygiene Directives, covering products of animal origin including milk and milk products, shell eggs and egg products, meat hygiene, and fish and shellfish. These Directives cover their products through the food chain up to, but not including, retail and catering. The retail and catering sectors and the remaining food products not of animal origin are covered by a general food hygiene Directive.

Milk and dairy products

Council Directive 92/46/EEC (as amended) lays down health rules for all European Union (EU) member states regarding the production and placing on the market of raw milk, heat-treated milk and milk-based products from cows, ewes, goats and buffaloes, intended for human consumption. Member states were required to implement the requirements of this Directive from 1 January 1994.

The Directive contains detailed requirements for the conditions of raw milk prior to admission to treatment establishments; the hygiene of the production holding; the hygiene of the standardization centres; the microbiological standards to be met on admission to heat-treatment establishments; special requirements for the approval of treatment establishments; registration of collection and standardization centres; the hygiene requirements relating to the premises' equipment and staff of establishments; standards relating to heat-treatment; packaging and labelling of final heat-treated products; the storage of pasteurized milk; and transportation after heat-treatment. It also sets out the microbiological product standards and hygiene standards for production and processing plants, and the rules for packaging, transport, storage and labelling.

The Directive is implemented in England and Wales by the Dairy Products (Hygiene) Regulations 1995 (as amended). Regulations with similar intent apply in Northern Ireland and Scotland. The detailed requirements of the Dairy Products (Hygiene) Regulations 1995 include conditions for production holdings that produce raw milk and dairy establishments that carry out heat-treatment of milk. The Regulations are enforced at production holdings by the Dairy Hygiene Inspectorate, which is part of the Farming and Rural Conservation Agency. Food Authorities enforce the legislation at dairy establishments.

During 1999, a public consultation was carried out on draft proposals to amend, update and consolidate the Dairy Products (Hygiene) Regulations 1995. The new Regulations were due to be implemented in 2000, however, this did not happen. The European Commission is currently working on proposals to consolidate and simplify existing EU food hygiene legislation into one food hygiene directive. Any future changes to the Dairy Products (Hygiene) Regulations 1995 are likely to occur following the consolidation of food hygiene legislation by the European Commission.

The Food Labelling Regulations 1996 (as amended) require milk and milk products to be correctly labelled when pre-packaged and sold for human consumption.

Eggs

The EU Directive on Hygiene and Health Problems Affecting the Production and the Placing on the Market of Egg Products (89/437/EEC) (as amended) sets out detailed requirements for the processing of egg products. The Directive is implemented in the United Kingdom by the Egg Products Regulations 1993 [13] (as amended), which lays down health rules for the preparation and manufacture of egg products intended for sale for human consumption. The Regulations cover premises defined as 'egg products establishments'.

'Egg products' are defined as products obtained from eggs once the shell and outer membrane have been removed. These may be made from fresh whole egg, yolk, whole egg and yolk, or albumen. The definition includes liquid, concentrated, crystallized, frozen, quick frozen, coagulated and dried egg products to which some other foodstuff or additive has been added. Boiled eggs are excluded. 'Egg' includes eggs laid by chicken, duck, goose, turkey, guinea fowl or quail. Incubated, pulled (from the carcass) or broken eggs may not be used. The Regulations include requirements relating to preparation of egg products, pasteurization, microbiological criteria, storage, transport, general hygiene conditions, packaging of egg products and marking of egg products.

Code of practice

Food Authorities are required, when discharging their duties to enforce the Food Safety Act 1990 and the Regulations made under it, to have regard to the [draft] consolidated Food Safety Act 1990 Code of Practice [14] issued under Section 40 of the Act.[1] This new consolidated Code of Practice will replace the previous 20 individual Codes of Practice and sets out instructions and criteria to which the Food Authority should have regard to when engaged in the enforcement of food law.

Supplementary guidance for Food Authorities is also contained in the [draft] Food Safety Act 1990 Practice Guidance [15] that provides practical advice on the approach to be followed by food authorities in enforcing various aspects of the legislation. The Practice Guidance includes a chapter relating to approval of premises subject to product-specific food hygiene regulations. The Practice Guidance replaces almost all of the previous guidance issued by the Central Government on the Code of Practice and Regulations made under the Food Safety Act 1990.

LIQUID MILK

Outbreaks of food-borne illness that 'have been investigated thoroughly, combining epidemiological and laboratory data, suggest that many milk borne infections may be due to the consumption of unpasteurised milk' [16]. To ensure good microbiological quality of raw milk it is important to ensure that good animal husbandry practices and hygienic milking procedures are followed.

It is recognized that 'raw milk when received by a dairy may be contaminated with pathogens such as *Brucella*, *Camplylobacter*, *Salmonella*, *Listeria monocytogenes* and *Mycobacterium bovis* and psychrotrophic bacteria such as pseudomonads. In addition, the milk may be contaminated by antibiotics' [17]. Other pathogens including *Staphylococcus aureous*, *Bacillus cereus* and *Yersinia enterocolitica* may also be found in raw milk.

In Scotland, the sale of raw or thermized milk or cream for consumption was banned in 1983, on the grounds of public health, based on epidemiological

---

[1]Until this is issued see Code of Practice No. 18: Enforcement of the Dairy Products (Hygiene) Regulations 1995.

evidence. In January 1999, Ministers announced, following a period of public consultation, that sales of raw cows' drinking milk in England and Wales would not be banned. Although unpasteurized milk sold in bottles and cartons requires to be labelled with a health warning.

Historically, tuberculosis caused by *Mycobacterium bovis/tuberculosis* organisms and brucellosis caused by *Brucella* subspecies were considered to be the most significant pathogenic bacteria in milk. A programme of regular testing and culling of infected animals has helped to limit the incidence of these infections in the United Kingdom. The badger population is thought to be a reservoir of *Mycobacterium bovis* and in recent years there has been a resurgence of bovine tuberculosis in dairy herds mainly confined to the South West and West England and South Wales. These pathogens are destroyed by pasteurization, so the spread of infection through raw milk and dairy products made with unpasteurized milk is of concern. Guidance on Officially Tuberculosis Free Status and the Dairy Products (Hygiene) Regulations 1995 was issued in March 2000 [18].

Attention in recent years has focused on human exposure to *Mycobacterium Avium* subspecies *paratuberculosis* (MAP) from consumption of cow's milk. MAP is the bacterium that causes Johne's disease, a chronic gastrointestinal infection in cattle and other ruminants. There have been suggestions that there may be a possible link between MAP and Crohn's disease in humans. Crohn's disease is a chronic inflammatory bowel disease of humans. The Advisory Committee on Dangerous Pathogens, the European Union and the Food Standards Agency have examined the evidence for and against such a link. A similar conclusion has been reached by all of these pieces of work, 'that on the basis of the available information a link between MAP and Crohn's disease can neither be proved nor disproved' [19].

'The results of a Food Standards Agency commissioned survey found MAP in approximately 2% of samples of pasteurised milk' [20]. As a precautionary measure when information first showed the presence of MAP in milk some of the major dairies in the UK increased the pasteurization holding time of milk from 15 to 25 seconds. Research to determine whether increasing the holding time to 25 seconds would be effective in eliminating MAP has indicated that this does not guarantee the elimination of MAP. However, the work showed 'that increasing the pasteurisation temperature had much less effect than increasing the pasteurisation time' [21].

Heat-treatment of milk

Pasteurisation is intended to avoid public health hazards in the sense that, although it may not destroy all the pathogenic micro-organisms which may be present, it reduces the number of micro-organisms to a level at which they do not constitute a significant health hazard. Pasteurisation also extends the keeping quality of some products by reducing the number of spoilage microorganism in the product [22].

It is important to remember that 'various combinations of temperature and duration of heating have an equivalent and minimum bacterial effect necessary for the pasteurisation of milk and skimmed milk' [23].

Heat-treatment of milk is a critical control point and operators are required to have reliable means of monitoring its effectiveness. There is a range of heat-treatment equipment available. Operating methods vary considerably according to the type of plant, its design and the degree of automation of its control systems. Whatever operating method is used it is essential that during commissioning, the plant is set up correctly by a competent person and subject to the necessary checks. The plant must be maintained and regularly serviced. Thermometers must be calibrated to ensure accuracy.

During February 1996, the Milk Pasteurization Hygienic Quality Campaign was launched to advise on-farm pasteurizing operators on how to meet the requirements contained in the Dairy Products (Hygiene) Regulations 1995. The campaign was funded by the Ministry of Agriculture, Fisheries and Food (MAFF) and free ADAS (formerly the Agricultural Development Advisory Service)/ Environmental Health Officer advisory visits were carried out to milk processors as part of the campaign. The aim of the campaign was to promote a greater awareness of food safety issues and encourage the use of simple practical procedures to control food safety hazards.

Milk pasteurization

The continuous high temperature/short time (HTST) system using a plate heat exchanger is the most commonly used method in the United Kingdom and is well proven. The minimum heat-treatment combinations recommended for the pasteurization of milk are 71.6°C for 15 seconds by the HTST method. The flow of milk through an HTST plant is as follows. The raw milk is tipped into a weighing tank, weighed, released into a receiving tank, and then pumped into a storage tank. From the storage tank it flows to the balance tank (where the level is maintained by a float valve), and from this tank it is pumped, at a regulated rate, to the regenerative section of the heat exchanger. Here it is preheated by the pasteurized milk leaving the holder and then passes through the filter to the heating section, where its temperature is raised to just above the legal minimum by the circulation of hot water. The heated milk flows from the heat exchanger to the holder, during its passage through which it is held for the prescribed period at the required temperature. From the holder it is fed, via the flow diversion valve, back to the heat exchanger for cooling and, later, chilling.

After processing, the milk gravitates, or is pumped, to bottle filler that fills and caps the bottles fed to it by conveyor from a bottle-fed washing machine. Sometimes the milk is fed into cartons. Modern sophisticated milk pasteurization techniques are becoming more common.

Two examples of improved technologies are the Tetra Therm Lacta type CA automatic pasteurization line (Fig. 31.1) and the Tetra Alcope 97 pasteurizer control panel (Fig. 31.2).

Throughout the process the temperature of the milk and flow divert activity is recorded continuously on a thermograph chart, thus proving an invaluable record of the process results. It is important that the thermograph is set up properly and the recording pens are working. The thermograph records are required to be kept by the occupier of the dairy establishment for inspection by enforcement officers.

To ensure that milk is effectively pasteurized, the HTST process must be well managed. If the milk flow rate within the process is too great, the milk will pass through the holding tube too rapidly and this may result in the time/temperature combination required for pasteurization not being achieved. The gaskets and the heat-exchange plates in the regeneration section must be in good condition. If there are problems with the plate integrity it is possible for raw milk from one side of the section to contaminate milk in the pasteurized side.

The holder (batch) method is generally only used to pasteurize small quantities of milk, for example, up to 1000 l per day. The minimum heat-treatment combinations are 62.8°C for 30 minutes by the holder (batch) process. Typically, for this method milk is heated by hot water in a stainless steel vessel incorporating an insulated outer jacket. An agitator mixes the milk slowly during pasteurization. The temperature of the milk is held at a minimum temperature of 62.8°C for at least 30 minutes before the milk is cooled to 5°C or below either in the vessel or externally.

The prevention of post-pasteurization contamination of milk is critical to food safety. Results from the national study on the microbiological quality and heat processing of cows' milk that took place over a period of 18 months, between March 1999 and August 2000 highlighted that, in some cases, improvements could be made by dairy industry. 'Although the majority of samples of pasteurised milk were of satisfactory microbiological quality, a number were found to contain coliforms and *E. coli*.' [24] The presence of coliforms and *E. coli* in the pasteurized milk samples suggests that possible post-pasteurization contamination occurred.

Phosphatase test

Following pasteurization the phosphatase test is used to determine whether milk has been properly pasteurized. Phosphatase is an enzyme that is always present in raw milk. The time–temperature combination necessary for pasteurization, as specified in UK dairy legislation, ensures the

APPLICATION

Tetra Therm Lacta CA is designed for processing and pasteurization of market milk, cheese milk, yoghurt milk, cream, ice cream mix, etc.

WORKING PRINCIPLE

After the balance tank, the product is preheated in the regenerative section of the plate heat exchanger by the outgoing pasteurized product. In the heating section, the product is heated to pasteurization temperature by hot water generated in a steam heated hot water unit. Then the product passes through the holding tube, dimensioned for a specific holding time.

After the holding tube, the product is cooled by the incoming product in the regenerative section. If required, the product is further cooled by ice water in the final cooling section.

Right after the holding tube, the flow diversion valve automatically diverts insufficiently heated product to the balance tank.

Process control is fully automatic. The operator interface is used for process monitoring and selection of required functions and sequences. The process controller controls and supervises both the basic process and subordinate modules such as the deaerator, separator, standardization module and homogenizer. Main process sequences are sterilization, filling, production/circulation, emptying and cleaning.

Using the voltage-free contacts or the optional serial protocol, the process controller can communicate with other systems.

Pasteurization and sterilization/cooling temperatures, and the position of the flow diversion valve, are continuously recorded.

BASIC DESIGN

Tetra Therm Lacta CA is a pre-tested pasteurization module. All components are mounted on a stainless steel frame, expect for the PHE.

Scope of supply:

- Balance tank (BTD) with level control and automatic CIP.
- Centrifugal product feed pump.
- Flow control (mechanical flow controller, frequency controlled pump or homogenizer as timing device).
- Free-standing plate heat exchanger (PHE) with cooling, regenerative and heating sections.
- Holding tube, on top of the module.
- Hot water unit, incl. PHE, pump, steam valve and trap, expansion vessel, shut-off valves, etc.
- Control panel including process controller (PLC), operator interface (HMI), recorder, solenoid valves and motor starters.
- Automatic PLC operated sequences.
- Automatic pasteurization temperature control.
- Automatic flow diversion, interlocked with temperature failure before and after the holding tube.
- Automatic product circulation when fault occurs.
- Automatic process interaction with up- and down-stream equipment as tanks and CIP.
- Automatic fault supervision and action for pumps, temperatures and flow diversion valve.
- Registration of pasteurization and outlet temperatures, and position of flow diversion valve.
- CIP by external circulation.

INSTALLATION DATA

Typical footprint for basic module excluding PHE:

Capacity (l/h)	≤ 10 000	> 10 000
Length (mm)	2400	3000
Width (mm)	1400	1400
Height (mm)	2300*	2300*

* Higher if holding time more than 15 seconds.

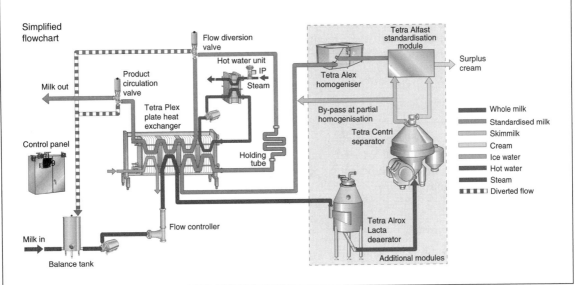

Fig. 31.1 Tetra Therm Lacta type CA – pasteurization module based on automatic processing and PLC-operated sequences. (Reproduced by kind permission of Tetra Processing UK Ltd.)

OPTIONS

- Design complies with relevant EU directives:

  - Higher pressure on pasteurized side
  - Differential pressure device for display, registration and alarm
  - Protection sheets on PHE.

- Cream circulation, interlocked with flow diversion and BTD low level switch.
- Increased/decreased heat regeneration.
- Automatic outlet and/or intermediate temperature control.
- CIP by internal circulation over BTD.
- Automatic dosing of cleaning detergents.
- Interface for serial communication.

ADDITIONAL MODULES

- Tetra Alrox Lacta I deaeration module
- Tetra Centri separation module
- Tetra Alfast automatic standardization module
- Tetra Alex homogenizer for full or partial homogenization
- Spiral holding cell for increased holding time
- Tetra Alcross M microfiltration module.

CAPACITY

Milk pasteurization   2000–35 000 l/h
Cream pasteurization   1000–10 000 l/h

STANDARD PROCESS DATA

	Market milk	Yoghurt milk	Cheese milk	Cream
Inlet temp. (°C)	4–8	4–8	4–8	8/50–60
Pasteurization temp. (°C)	72	90–95	72	85–95
Holding time (s)	15	300	15	3
Separation and homogenization temp. (°C)	. 55	—	—	—
Outlet temp. (°C)	4	40–45	28–32	6–8
Deaeration temp. (°C)	68–70	68–70	—	—
Heat regeneration (%)	90	90	90	80

Fig. 31.1 Continued.

destruction of phosphatase. Therefore, the test provides a simple method of determining whether milk has been adequately heat-treated to destroy pathogenic bacteria. A negative phosphatase test should demonstrate that the required time–temperature combinations of pasteurization have been achieved.

Coliforms

Coliforms do not survive pasteurization and should be absent from pasteurized milk. Presence of coliforms is often an indicator of post-pasteurization contamination.

Pathogenic micro-organisms

The regulations contain a requirement that pathogens must be absent in pasteurized milk. There is also a requirement that pathogenic micro-organisms and their toxins must not be present in raw milk sold directly to the final consumer.

Peroxidase test

The peroxide test (also known as the Storch test) is used to detect overheating of HTST pasteurized milk. The regulations specify a positive test for peroxidase.

Plate count at 21°C

The plate count is designed to measure the number of bacteria that have survived the pasteurization process and have gained access into the milk during the milk processing and packaging stages.

Sterilized milk

Sterilized milk is preheated to a temperature of 43–49°C, and is either filtered or clarified by passing it through a centrifugal clarifier. The milk is again heated to a temperature of 66–71°C, and then homogenized at about 122060 kg per m^2. This process ruptures the fat globules, which

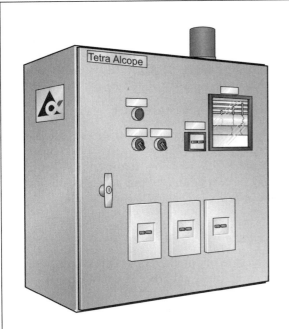

APPLICATION

Control of heat-treatment processes, such as pasteurization. The panel provides a limited degree of automation.

WORKING PRINCIPLE

The control panel features the following functions/controls:

- Operating power on/off switch.
- Temperature controller with digital display.
- Recorder for the temperature (0–150 °C) and event. The recorder draws a temperature graph and also marks the chart whenever an event occurs. In order to save paper, the recorder is only in operation when the feed pump is operating.
- Alarm beacon, flashing red when actual temperature dropping below set point causing the return valve to change over to recirculation.
- Alarm reset switch.
- Automatic supervision and control of the return valve.
- Indication lamp for return valve in forward position.
- Start/stop for 3 motors.

BASIC UNIT

Panel casing for wall mounting, with key-lock door. Temperature recorder, one solenoid valve for return valve. Two selector switches, one with indication lamp for forward flow, and alarm. Transmitter set including two Pt100 with bushing and 10 m of cable. Alarm beacon light. Three motor starters with on/off lamps. Pilot power transformer and rectifier. Two sets of documentation.

Material. Panel casing of stainless steel.

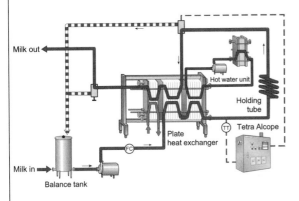

Simplified pasteurizer flow diagram including Tetra Alcope 97 connections.

Fig. 31.2 Tetra Alcope 97: Control Panel for Pasteurizer. (Reproduced by kind permission of Tetra Processing UK Ltd.)

allows them to remain in suspension in the milk. The hot homogenized milk is then fed into hot sterile bottles, which are then sealed with a 'crown' cap.

The bottles are then processed, in either a batch or continuous retort, where they are heated at approximately 104°C for 30 minutes. The bottles are then cooled, and are ready for sale. There is an alternative method in which the milk is sterilized by a continuous-flow method and then put into the aseptic containers in which it is to be supplied to the consumer.

Ultra-high temperature (UHT) treatment

The first UHT treatment was developed in Switzerland by Alpura Ltd, which named the process uperization. In this process, the milk is pumped into a preheater, which consists of a double pipe coil where the milk is preheated to 78°C; it is then transferred to the uperization unit, where it is heated to 150°C and held at this temperature in a timing tube for 2.4 seconds, which is considered necessary to ensure absolute sterilization. By this method, water is incorporated into the milk, and the water is removed by pumping the milk through an expansion nozzle at a reduced pressure, and through a centrifugal separator.

The water-free milk passes to a sterile homogenizer and then to a cooler, where it is cooled to a temperature of approximately 15.6°C, when it is aseptically poured into cartons. There are two other methods of UHT treatment: Tetra Therm Aseptic vacu-therm-instant-sterilizer (VTIS) and the APV process. In the Tetra Therm process, the milk is heated by direct steam injection, and in the AVP process by hot water and steam circulation. The method used by the Tetra Therm VTIS system is as follows. From the balance tank, the milk or other food product is pumped through a series of plate heat-exchanger sections where it is preheated by hot product flowing towards the filling machine. The preheating temperature is chosen to avoid any fouling. Instantaneous heating to UHT temperature by injection of steam directly into the product then takes place in the ring nozzle steam injector. Micro-organisms and spores are killed by holding the product at maximum temperatures for a few seconds as it passes through the holding tube.

The equivalent amount of injected steam is flashed off in a vacuum vessel, which causes the product to cool rapidly. It is even possible to adjust the pressure of the vacuum vessel to control the water/dry matter content of the end product. The built-in condenser simplifies plant sterilization and cleaning in place (CIP).

For optimal product texture and stability, aseptic homogenization is included as part of the processing. From the homogenizer, the product is re-circulated through plate heat-exchanger sections where it is cooled by heat transfer to incoming product, thereby achieving product-to-product heat regeneration. The process is illustrated in Fig. 31.3.

Sterilization of heat-treatment plant

This is an important aspect of the efficient processing of milk [25]. It consists of two operations: cleaning followed by sterilization. Sterilization may be carried out by hot water, steam or one of the approved chemical agents.

Cleaning

Subject to minor alterations depending on the type of plant, the usual procedure followed is:

1. preliminary rinse
2. washing or scrubbing with detergent
3. final rinse.

The preliminary rinse removes all loose residues, and the detergent is intended to remove all hardened residues that have not been removed by the preliminary rinse. The final rinse clears the plant of detergent and freed residues, and leaves the surfaces clean for sterilization. The preliminary rinse should be carried out as soon as possible after the day's work in order to prevent milk residues drying and hardening, and the detergent solution should then be used at the recommended concentration. In the case of HTST plant, it is usually circulated at a temperature of 72°C for 40 minutes, after which the plant is dismantled, scrubbed and rinsed.

Sterilization

Generally, when using hot water, the plant surfaces should be held at a temperature of 85°C for 10 minutes, and in the case of steam 99°C for 5 minutes. The sterilants permitted are approved brands of sodium hypochlorite and chlorine, or quarternary ammonium compounds. It is generally accepted that heat is more effective if residues remain on the plant surfaces, but if the plant has

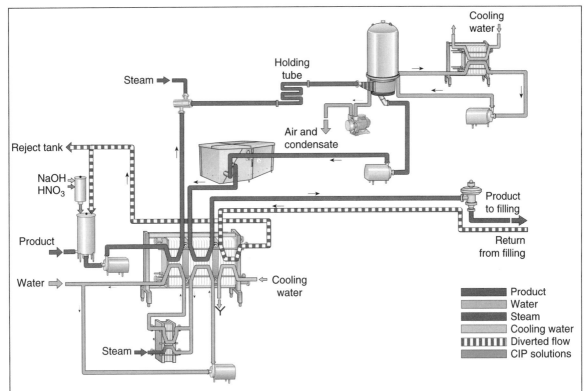

APPLICATION

Aseptic processing module for continuous UHT treatment with direct steam injection. Mainly for heat sensitive, low acid products, such as milk, enriched milk, cream, soy milk, formulated dairy products, ice cream mix, dairy desserts as well as (extended shelf life (ESL)) products.

WORKING PRINCIPLE

The module is fully automated to safeguard the aseptic status while in production. The operation can be divided into four steps:

- Pre-sterilization
- Production
- Aseptic Intermediate Cleaning (AIC)
- Cleaning in place, CIP

The Tetra Therm Aseptic VTIS 10 is sterilized by circulating hot water under pressure for 30 minutes. After sterilization the module is cooled down step by step to production temperatures. Finally, sterile water is circulated through the product circuit.

When a downstream tank or filling machine is ready, production can start by filling the module with product via the balance tank. The product displaces the water/product mix to drain or reject tank. A balance tank with product bowl minimizes the mixing phases.

If product supply fails or a stop at the aseptic tank or filling machine occurs, sterile water replaces the product and the module goes into sterile water circulation.

During production the product is preheated in a heat exchanger to about 80°C and the heat energy from the outgoing product is regeneratively recovered. Instant heating to sterilization temperature takes place in the steam injector by continuous injection of high pressure steam into the product.

The product enters a holding tube where it is held at sterilization temperature for the required period of time. The product then enters the flash vessel where the pressure and temperature drops instantly. The excess water added as steam is flashed off.

For optimal product stability the product passes an aseptic homogenizer before final cooling in the heat exchanger.

In order to prolong the production period between full CIP, an AIC can be performed. This is done under

Fig. 31.3 Tetra Therm Aseptic VTIS 10 – direct UHT treatment module. (Courtesy of Tetra Processing UK Ltd., Swan House, Peregrine Business Park, Gomm Road, High Wycombe, HP13 7DL.)

aseptic conditions and the module remains sterile. After each production run the module is cleaned (CIP) with both lye and acid.

PROCESSING PARAMETERS

Standard temperature programs for white milk
5–80–140/4 s–81–homogenization–20/25 °C
Optional temperature programme
5–80–150/6 s–81–homogenization–20/25 °C

CAPACITY

Variable production capacity with a maximum of 1:2 within the range 2000 up to 26 000 l/h.

BASIC UNIT

- Main module with:

  – Product balance tank BTD with level control and product bowl.
  – Centrifugal pump with frequency converter for product.
  – Centrifugal pump for water.
  – Flow meter for water.
  – Flow meter for product, display in control panel.
  – Brazed PHE for heating in the water circuit.
  – Batch header CIP dosing system.
  – Valves, pipes, fittings.
  – Frequency converters, mounted on the frame.
  – Pre-wired, signal/power cables (excl. homogenizer cable).

  – Control panel with Allen Bradley SLC-500, mounted on main module.
  – TPOP, Human Machine interface (HMI), mounted on control panel.
  – Recorder to register three temperatures and one event.

- Direct heating module with:

  – Steam injector for direct heating of the product.
  – Centrifugal pump with frequency converter as product pump after expansion vessel.
  – Vacuum pump.
  – Expansion vessel for flash cooling with built-in condenser.
  – Plate heat exchanger in stainless steel for cooling of recirculating water for the condenser in the expansion vessel.
  – Centrifugal pump for cooling water in the expansion vessel.
  – Valves, pipes, fittings.
  – PHE, Tetra Plex.
  – Pneumatic tools for the PHE.
  – Steam separator.
  – Pneumatic, remote controlled sanitary valves.
  – Product piping in AISI 316.
  – Set of pipes, bends, valves, internal signal wiring, pipes for signal wiring and fittings required for the pre-assembly of the UHT system.
  – Commissioning kit intended for the start-up.
  – Pre-assembly and water test before delivery.
  – Engineering, programming.
  – Technical documentation.

Fig. 31.3 Continued.

been effectively cleaned beforehand, chemical sterilization can be just as good.

'In place' cleaning

Because the cleaning and sterilization of dairy plant is time-consuming and arduous, and there is always the risk of recontamination during reassembly, the technique of 'in place' cleaning has been developed. The system involves cleaning and sterilization (it is sometimes possible to carry out both operations simultaneously) without the complete dismantling of the plant.

As the nature of the residue varies in different parts of the plant, for example, in the parts deal-ing with raw and heated milk, different detergents are necessary and, in order to avoid the wasteful application of chemicals and the risk of corrosion in parts of the plant, it is usual to divide the plant into circuits for the purposes of cleaning. In the case of larger plants, a further division into circuits enables cleaning to be carried out in the sections that have been cleared of milk while milk is still being processed or bottled. A typical circuit arrangement for an HTST plant would be:

1. raw milk reception and ancillary pipes
2. tanks
3. pasteurizing equipment.

A suitable cleaning procedure for the part of the plant containing raw milk would be:

1. cold water rinse: 5 minutes
2. hot detergent: 10–20 minutes
3. cold water rinse: 5 minutes
4. sterilization by hot water or chemicals: 15 minutes
5. cold rinse: 5 minutes (only after chemical sterilization).

For pasteurizing equipment, the method usually used is: a cold water rinse followed by the circulation of a detergent to soften the water and attack inorganic constituents of the deposits; a second detergent wash to remove the organic matter; sterilizing; and rinsing. Enclosed tanks are usually cleaned and sterilized by applying the various solutions to the inside of the tank using pressure sprays. The failure of heat-treated milk to satisfy the prescribed tests should be followed by repeat sampling and examination of the thermograph charts with, if necessary, the swabbing and test rinsing of the plant.

Swabbing

The apparatus consists of a piece of stainless steel wire about 350 mm long formed into a loop at one end and notched at the other to hold either a wad of cotton wool or some unmedicated ribbon gauze. Where possible, an area of 930 cm^2 should be examined. The sterile swab is received from the laboratory in a test tube containing quarter-strength Ringer's solution and plugged with cotton wool. Before use, the excess liquid is squeezed out of the swab by pressing it with a rolling motion against the inner surface of the tube; it is then rubbed with a fairly heavy 'to and fro' motion over the area being examined. The area should be covered twice, the swab being rotated to ensure all parts make contact with the surface being tested. After use, the swab is replaced in the tube, which is plugged with cotton wool and taken to the laboratory.

Rinsing

For utensils such as bottles etc., rinsing is more satisfactory, but care must be taken to see that the rinse solution comes into contact with all parts of the utensil. Quarter-strength Ringer's solution is used, the quantity depending upon the capacity of the vessel. For all sizes of bottles, 20 ml is used, the bottle being rotated about 12 times.

Guidance documents for milk processing

The MAFF produced three short guides for dairy farmers and dairy product processors [26–28]. These set out:

- the conditions to be met in the production premises;
- the standards for raw milk, heat-treated milk and milk products, including microbiological criteria, sampling protocols and analysis methods;
- the rules on wrapping, packaging, health marking, labelling and documentation;
- health checks.

British Standard Code of Practice for the pasteurization of milk on farms and in small dairies

Environmental health professionals involved in the inspection of on-farm dairy establishments pasteurizing milk and occupiers of dairies would be well advised to obtain a copy of British Standard BS 7771:1994 'Code of practice for Pasteurisation of milk on farms and in small dairies' [29]. This British Standard gives recommendations for the process principles, design features and operation of equipment used on individual farms by producer processors and in small dairies for the pasteurization of cows' milk by means of the holder (batch) and continuous flow (HTST) methods. Guidance is given on the requirements of UK and EEC legislation for dairies pasteurizing milk. Guidance is also given on the necessary hygienic and control measures required to achieve safety and high quality in pasteurized milk.

The Lancashire and Greater Manchester Food Officers Groups guidance notes for on-farm pasteurizers

The Lancashire and Greater Manchester Food Officers Groups have produced some very useful guidance notes intended for use by environmental health professionals involved in the approval, audit and inspection of on-farm pasteurization plants [30]. The guidance notes contain sections on commissioning of pasteurizers, cleaning and operating procedures, pasteurized bulk milk tank, cream separator pre-use disinfection and operation, homogenizer pre-use disinfection and operation, bottle filling machines, cleaning and disinfection of a bottle filler, bottle washing and guidelines for granting approval for milk and milk products heat-treatment and processing establishments. The guidance also incorporates a sample inspection form for use when inspecting dairy establishments.

CHEESE

In terms of food safety even though there has been an improvement in cheesemaking methods and process controls, there has still been a number of reported outbreaks of illness in the United Kingdom associated with the consumption of cheese made from both raw and pasteurized milk. Two important outbreaks of food-borne disease arose from the consumption of cheese made from unpasteurized milk in 1989. In one, 155 people were made ill (vomiting within 6–24 hours), and this was thought to be caused by an undetectable staphylococcal enterotoxin (no organism was found). In the other outbreak, 39 people fell ill from *Salmonella dublin* in a cheese imported from the Republic of Ireland. As a result of these incidents, both the cheese producers involved subsequently decided to introduce pasteurization of the milk during the manufacture of their cheese.

An outbreak of *E. coli* (VTEC) O157: H7 infection associated with the consumption of raw milk cheese, involving 22 cases was reported in Scotland in 1994 [31]. Another outbreak of *E. coli* (VTEC) O157 was linked to consumption of

Cotherstone cheese made from unpasteurized cows' milk during 1999 [32]. In 1996, an outbreak associated with cheese was caused by *Salmonella goldcoast*. Results of phosphatase tests and thermograph tracings at the cheese factory showed that pasteurization had failed on a number of occasions [33].

Review of epidemiological literature related to raw milk cheeses demonstrates that 'aged raw milk cheeses have a remarkable food safety record as evidenced by the infrequency of large cheese-associated outbreaks' [34]. Pasteurization of milk used for cheese is a critical control point and helps to ensure a safe product but sometimes at the expense of flavour. There are a number of factors other than pasteurization that act as contributors to cheese safety and these include: 'milk quality and management; lactic acid bacteria culture management; pH; salt; controlled ageing conditions; and natural inhibitory substances in raw milk' [35]. It is essential that cheesemakers identify the points that are critical to food safety for their particular cheese-making process and ensure that the critical points are adequately controlled and monitored.

The Institute of Food Science and Technology (IFST) have drawn attention to the potential public health hazard posed by pathogenic bacteria in soft and semi-soft cheeses made from raw milk. These cheeses have a higher pH and thus can permit the growth of *Listeria monocytogenes*. Production of soft and mould-ripened cheese 'is therefore a higher risk operation than production of hard cheese of higher acidity and salt content' [36].

Milk quality is the first critical control point of the process for those cheesemakers producing raw milk cheeses. The milk must be of good microbiological quality and not contain antibiotic and chemical residues that may interfere with the fermentation process. Some cheesemakers producing raw milk cheeses that have their own dairy herd, operate a 'closed herd policy' in which calves are bred from only within the herd to minimize the possibility of introducing infections from outside the farm. Some cheesemakers producing raw milk cheese may receive their milk from known farms with which they have agreed milk specifications. When the milk is off-loaded from the milk tanker it is important to ensure that the end of the tanker hose is disinfected, as otherwise the hose could be a source of potential microbiological contamination.

The words 'made with raw milk' must appear clearly in the labelling of milk-based products manufactured from raw milk whose manufacturing process does not include heat treatment. A large proportion of all imported Continental cheese of the soft and mould-ripened variety is made from untreated milk.

The development of controlled acidity is critical to the manufacture of cheese and to inhibit the proliferation of pathogenic and spoilage organisms. Failure of the starter culture to work properly is a hazard that must be controlled. Starters used during cheese production can be either purchased as a Direct to Vat Inoculation (DVI) or be sub-cultured in the dairy. DVI starter culture is usually purchased from a reputable supplier and a fresh culture is used for each batch of cheese. Problems can occur, however, with sub-cultured starters if they become contaminated with harmful bacteria and then are re-used over a period of time. If sub-cultured starters become infected with bacteriophage (a bacterial virus) this can inhibit the fermentation process and lead to an 'increased risk that pathogens might grow' [37].

Measurement of the acidity of whey throughout the process allows cheesemakers to develop a satisfactory acid development profile for their product and take corrective action if future deviations occur. Cheesemakers are strongly recommended to keep a cheesemaking log for each batch of cheese produced. The log forms part of their food safety management system records.

Maturation time and conditions during storage act as contributors to cheese safety. Pathogenic bacteria such as *Salmonella, Brucella, Campylobacter, Listeria* can survive in cheeses made from raw milk and may not be destroyed by some heat processes. However, 'some protection against the transmission of disease, especially brucellosis, is achieved by requiring cheese made from other than pasteurized milk to be aged for 60 days at a temperature of not less than 4.4°C'

[38]. In the United States regulations governing the use of raw, heat-treated and pasteurized milk advocate that cheesemakers choose either one of the two options to assure the safety of the cheese, that is either 'pasteurise milk destined for cheesemaking; or hold cheese at a temperature of not less than 2°C for a minimum of 60 days' [39].

In January 2002, the Food Standards Agency launched the Food Safety Management Awareness Initiative for Specialist Cheesemakers. The scheme was developed by ADAS in consultation with the Specialist Cheesemakers Association (SCA) and the Local Authority Co-cordinator of Regulatory Services (LACORS, formerly known as LACOTS). The aim of the scheme was to support consistent food safety management in specialist cheese-making premises. Cheesemakers in England, Wales and Northern Ireland were encouraged to volunteer for a free ADAS consultancy visit. The consultation was undertaken in partnership with local authority officers. A food safety workbook for the manufacture of specialist cheese was completed and any necessary actions agreed by all parties. A follow-up visit was carried out for the proprietor and local authority enforcement officer to meet the consultant on site and discuss progress.

Cheese production guidance documents

The Dairy Industry Federation has produced comprehensive guidelines to cover the manufacture of dairy-based products [40]. The Creamery Producers Association has produced guidelines for large cheese plants [41] and the Milk Marketing Board has produced similar guidance for the manufacture of cheese in small dairies and 'on-farm' plants, particularly in relation to *Listeria* risks [42]. The Specialist Cheese Makers' Association have produced a Code of Best Practice (1997) aimed at all those with an interest in specialist cheeses [43].

The Food Standards Agency (Scotland) in association with the Scottish Executive have produced a 'Report on Small-Scale Cheese Production in Scotland' written by the small-scale cheese production and Food Safety Expert

Working Group (2000). In the report the Group considered food poisoning incidents in Scotland associated with the production, distribution and retailing of cheese, by small-scale cheesemakers. The resulting good practice guideline document is for both enforcement officers and the trade and contains a guidance generic hazard analysis and critical control point system for cheese production.

SOURED AND FERMENTED MILKS

Soured and fermented milks include products such as buttermilk, yoghurt, kefir and crème fraiche. The production of these products differs from cheese in that rennet is not used to bring about coagulation and the whey is not removed. Starter culture is added to cause thickening produced by acidification of lactic acid bacteria. It is important that milk used to make yoghurt is free from antibiotic residues or cleaning residues as this could lead to inhibition of the starter culture, which could potentially allow pathogens to grow.

There have been few outbreaks of illness associated with yoghurt 'because of its high acidity and low pH (usually 3.8–4.2), yoghurt is an inhospitable medium for growth of pathogens which will not grow and survive well' [44]. Although in June 1989 there was an outbreak of botulism associated with yoghurt in which 27 people were affected. 'The yoghurt was flavoured with hazelnut conserve sweetened with aspartame rather than sugar. *Clostridium botulinum* type B was detected in a blown can of the conserve and opened and unopened cartons of yoghurt' [45].

BUTTER

Butter is an emulsion of water droplets in a continuous fat phase. It has a higher fat content than milk (80%) and uses pasteurized cream. There are different types of butter including sweet cream and ripened-cream butters.

Traditional homemade butter is made by beating the cream until it forms a solid mass. The liquid or buttermilk is poured off and the butter is pressed using two butter pats and then rinsed under cold running water until all the liquid comes out clear.

To make a conventional ripened cream butter, the starter culture is added after the cream has been chilled to allow the fat globules to harden and cluster. Starter culture is added and the cream is incubated at around 20°C to allow flavour production to take place. Once the butter has formed, the buttermilk is poured off, the butter grains are washed with water and salt may be added prior to the butter being worked.

To ensure good microbiological quality of butter it is essential that the cream has been effectively pasteurized and that the starter culture and the water used to wash the butter is of good microbiological quality. The process must be carried out in a hygienic manner and equipment must be capable of being effectively cleaned and sanitized. Good air quality and a low humidity within the production area helps to prevent the formation of yeasts and moulds. Finally, aluminium foil wrappers as opposed to oxygen-permeable parchment wrappers are preferred 'as they will help to discourage mould growth' [46].

ICE CREAM

Unlike much Continental ice cream, most of the ice cream produced in the United Kingdom has a vegetable rather than a dairy fat base. Dairy fat based ice cream is also popular in the United Kingdom, however, and its manufacture is covered by Directive 92/46/EEC and the Dairy Products (Hygiene) Regulations 1995.

The ingredients used in the manufacture of ice cream are required to be pasteurized by one of the three specified methods, or sterilized, and thereafter must be kept at a temperature below 7.2°C until the freezing process has begun. It is an offence to sell or offer to sell ice cream that has not been so treated (except in the case of certain water ices and ice lollies), or that has been allowed to reach a temperature exceeding −2.2°C without its being heat-treated again.

In the pasteurization process, after the ingredients have been mixed, the mixture must not be kept for more than 1 hour above 7.2°C before being raised to, and kept at, a temperature of pasteurization (normally 71.1°C for 10 minutes, or 79.4°C for at least 15 seconds). It must then, within 1 hour 30 minutes, be reduced to not more than 7.2°C and kept there until freezing has begun. Alternatively, processing of ingredients can also be safeguarded by sterilization; the mixture must be raised to and kept at a temperature of not less than 148.9°C for at least 2 seconds. A mixture so sterilized need not be cooled in the same way as pasteurized products, provided it is immediately canned under sterile conditions. If the temperature of any ice cream rises above −2.2°C, it must be subject to a reheating process prior to sale.

Ice cream and food-borne illness

In the past, in the United Kingdom, large outbreaks of food-borne illness were linked with the consumption of ice cream, but there have been very few since 1951. An outbreak of *S. enteritidis* PT4 infection, during 1992, was associated with eating ice cream. Illness was significantly associated with eating commercially produced ice cream. The ice cream was stored 'semi-frozen' in a refrigerator at too high a temperature and was probably cross-contaminated.

The Milk Marketing Board's Code of Practice on the hygienic manufacture of ice cream [47] gives excellent guidance and support and contains information on the following topics.

- Quality assurance: includes an example production flow diagram and a comprehensive HACCP framework for ice-cream production, with a full analysis of possible critical control points and control systems.
- Premises: includes guidance on the design and construction of suitable production facilities.
- Ingredients: sets out the important factors for raw materials and ingredients.
- Equipment: gives key design factors.
- Finished product: gives critical factors for satisfactory storage.
- Cleaning and disinfection: its essential role in the production of good quality ice cream is explained.
- Personnel: sets out the personal hygiene issues.

- Legislation: outlines the relevant legislation.
- Laboratory control: explains the importance of quality control in ensuring that ice cream meets the hygiene and legal requirements.

The guide is excellent for both producers and enforcers. It does not have the status of an industry guide under the Food Safety Act 1990 as yet, but it is up to the standards required. The Ice Cream Alliance and the Ice Cream Federation have also produced guidance on the hygienic retail of ice cream [48].

Soft ice cream

Soft ice cream is made from powder or liquid mixes that have previously been heat processed. The mix is tipped into a hopper in the dispensing machine where it is fed, by means of a drip feed or pump, into the freezing compartment. Air is introduced during freezing to improve texture and to reduce the shock to the tongue when the ice cream is eaten. It also increases the bulk (over-run). The mixture is then frozen in the freezer barrel and delivered to the service nozzle at a temperature of −5°C. 'Thick shakes' served in fast-food restaurants use similar equipment. The parts of the machinery that come into contact with food may be complex and somewhat difficult to clean; complete dismantling is necessary in order to clean and disinfect individual components. Where the machine is located in a mobile vehicle, cleaning and disinfection facilities may not always be readily at hand. Manufacturers' instructions on the use of machines must be followed, and the personal hygiene of operators is very important, during both the dispensing of soft ice cream and cleaning operations.

Self-pasteurization machines

After business is completed for the day, these machines employ a pasteurization cycle in which any liquid mix stored in the machine and all food contact parts are heated to pasteurization temperatures to eliminate any build up of contamination. Self-pasteurization temperatures (65.5–76°C for up to 35 minutes) are followed by the machine being kept at 4°C until freezing recommences. However, this pasteurization process at the end of working periods is an alternative to dismantling and cleaning the machine frequently; it does not serve to safeguard the ice cream itself.

If self-pasteurization cleaning takes place at the end of each work session, the manufacturers of these machines claim that thorough cleaning of the individual components and internal surfaces of the machine need only occur at intervals of two weeks to one month. This claim has been tested by the Public Health Laboratory Service (PHLS). Using sterile mixes, grade 1 results on methylene blue tests were produced for up to one month's operation without cleaning. In the summer of 1990, a self-pasteurizing ice-cream machine was tested for *Listeria* species by an independent laboratory, and was found to be satisfactory at eliminating the organism by the self-pasteurization process.

The manufacturers' claims for these machines would therefore appear to be justified, and, under normal commercial conditions, it would seem safe to operate for a period of two weeks using self-pasteurization at the end of each working period. After this period, however, the machines should be dismantled for thorough and effective cleaning as for other soft ice-cream machines.

Figure 31.4 shows a shake and soft ice-cream freezers using a self-pasteurization technique. Model PH 90 has an on-board computer and a refined monitoring system. It can disable or 'lock-out' a machine if a correct heat cycle has not been completed within the previous 24 hours. It will also 'lock-out' if a complete disassembly and brush clean is not carried out after the thirteenth heat cycle. A liquid crystal display shows the temperature in the hopper or freezing cylinder at any point of the operation. It can also display information on the previous 14 heat cycles. It has a fault diagnostic system with an audible alarm to alert the operator if a fault occurs.

EGGS

Eggs are an ubiquitous food. They are a vital ingredient of many recipes and products and are useful on their own. Their high nutritional value, which is of such benefit to humans, also makes them an ideal growth medium for a wide range of bacteria.

Eggs are sold in two main forms: fresh and preserved. Fresh eggs are eaten in large numbers: the UK consumption is estimated to be 10 thousand million per year [49]. The speed of modern distribution systems means that freshness is not usually a major problem, even for imported eggs. However, the key factors contributing to an egg's freshness are age, storage conditions, shell strength and the condition of the contents. The usual way of testing for freshness is to 'candle' the egg, that is to hold it in front of a strong light. In a fresh egg, candling will reveal a transparent albumen and a clearly visible yolk.

Eggs and salmonellosis

Eggs have long been recognized as an important vector of bacterial food poisoning, but the situation was particularly highlighted in the 1980s as a result of numerous outbreaks of *S. enteritidis* and some of *S. typhimurium* in both of which the epidemiological evidence showed a strong association with eggs. As a result, in August 1988, the chief medical officer issued a general warning to the public about the dangers of eating raw eggs, recommending that they be avoided [50].

In October 2002, following several outbreaks of *Salmonella* in the United Kingdom associated with shell eggs imported from Spain, the Food Standards Agency re-issued its advice to the catering trade to only use pasteurized egg in egg dishes that will not be cooked or only lightly cooked and to be careful about handling raw shell eggs to avoid cross-contamination.

S. enteritidis phage type 4 in particular has consistently been shown to be associated with eggs. Research has indicated that this bacterium had an especially invasive nature and that the transfer of infection was from the chicken to the egg via the transovarian route. The problem was compounded by the fact that, with the exception of *S. enteritidis* phage type 4, *Salmonella* infections

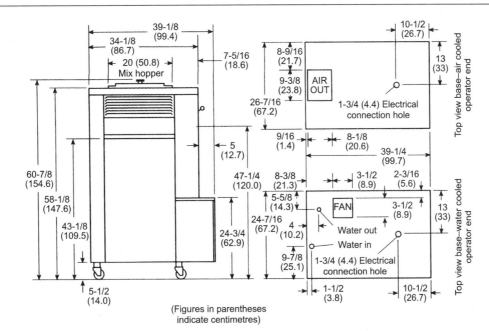

(Figures in parentheses indicate centimetres)

FEATURES

Freezing cylinder

Shake side, 7 quarts (6.6 l).
Soft serve side, 3.4 quarts (3.2 l).

Mix hopper

Two, 20 quarts (18.9 l) mix hopper capacity. The insulated reservoirs are refrigerated during normal and standby operation to maintain the mix below 408F (4°C).

Indicator lights

A warning flashes alerting operator of insufficient mix level. When the mix out light flashes, the unit automatically is placed in the 'Standby' mode, protecting the freezer from damage.

Electronic controls

The software in the universal control is programmed to monitor all modes of operation. Product viscosity is continually measured to assure consistent quality at all times.

Auto dispensing

Shake side: one dispensing spout provides four separate flavours. Flavours are selected from touch pads at the front of the unit. Solid state portion control automatically closes draw valve at proper fill level and prevents syrup carry over. Syrup calibration is microprocessor-controlled and aids the operator in accurately setting the proper syrup flow.

Power interrupt and lock-out

This computer-based safety feature is programmed to alert the operator if product reaches unsafe tempera-tures as a result of power loss. If heat treatment features are improperly used, the unit automatically locks all freezer functions until proper measures are taken to ensure health codes are met.

LCD readout

To assure proper performance, the LCD (liquid crystal display) readout will identify temperatures in the hopper or freezing cylinder at any point of operation. It can also provide a history of temperatures and times during the last 13 heat cycles.

Thorough agitation

Product is continually blended in the hopper during the heat treatment cycle, and the mix is maintained at or below 40°F (4°C) in 'Auto' or 'Standby'.

Air/mix pump

Coaxial air/mix pumps are located in the refrigerated hoppers providing consistent overrun and fast ejection.

Self-contained syrup compartment

Lower, front compartment contains four individually regulated 1 gallon (3.8 l) syrup tanks. (A self-contained air compressor is located in the freezer.) Use only single strength syrup that is free of pulp and seeds.

Syrup rail

Four compartment syrup rails is recessed in front of freezer: two heated and two room temperature. The 2 quart (1.9 l) reservoirs serve a wide variety of syrups and toppings.

Fig. 31.4 Shake and soft ice-cream freezer. (Courtesy of Taylor Freezer UK.)

had previously been of minimal economic significance to the poultry industry. However, many commercial flocks have now become symptomless carriers of *Salmonella*, which can be passed on through meat or eggs. *Salmonella* bacteria readily colonize the alimentary tract of chickens, and are therefore present in droppings. They are thus passed on as a consequence of intensive breeding methods.

The significance of *S. enteritidis* phage type 4 is that it infects the reproductive tract, not just the gut. Previously, contamination of eggs with food poisoning bacteria resulted from external contact of the eggs with faecal matter, but now eggs can become infected during formation in the oviduct.

In 1991, a PHLS survey demonstrated that 0.6% of intact eggs had been contaminated internally [51]. In a later study [52], out of 46 380 eggs purchased in boxes of six from high street retail outlets, 0.2% were found to be contaminated with *Salmonella*, nine isolations were made from the shell surface and eight from the contents. The percentage of infection, whether internal or external, rises significantly if the eggs are stored incorrectly. Inoculation of <10 organisms will rise to between 10^{10} and 10^{11} organisms when stored for two days at room temperature. This has important food safety implications because it shows that *Salmonella* bacteria can multiply within the egg contents.

Refrigeration has a marked effect on slowing the multiplication of bacteria, and testing by the PHLS has shown that it also makes the bacteria more heat-sensitive. However, in its report [53] on *Salmonella* in eggs in 1993, the Advisory Committee on the Microbiological Safety of Food (ACMSF) considered that while eggs must be regarded as an important source of human infection with *Salmonella*, the contribution they made to current levels of human salmonellosis could not be quantified precisely. The report found that only a small percentage of eggs were considered to be contaminated with *S. enteritidis*, and that studies indicated that the number of organisms they contained when laid was very low. The ACMSF recommended: that eggs should be consumed within three weeks of being laid, and that use by dates should be stamped on both egg packs and the eggs themselves; that industry and retailers should draw up a code of practice for the storage and handling of eggs; and that once purchased, eggs should be stored in the refrigerator. The government considered the recommendation on use by dates and has taken up the other recommendations [54]. In 1998, the ACMSF set up a working group to establish the factors which determine the presence of *Salmonella* contamination in or on eggs.

The deputy chief medical officer reiterated the Department of Health's earlier advice to the public to avoid eating raw eggs or uncooked foods made from them in 1998 [55]. The advice is particularly aimed at those who are most vulnerable to the ill effects of raw or partially cooked eggs, that is the elderly, babies and pregnant women. Such individuals are advised to eat only eggs that have been cooked until the white and yolk are solid and to avoid recipes that require the egg to be cooked only lightly, such as meringue.

It is emphasized that present evidence suggests that the consumption of cooked eggs does not pose a similar risk and that no increase in cooking times is therefore advised. The following simple hygiene measures are advised for the storage and use of eggs.

1. Eggs should be stored in a cool, dry place, preferably under refrigeration (at 8°C or below).
2. They should be stored away from potential contaminants such as raw meat.
3. Stocks of eggs should be rotated.
4. Hands should be washed before and after eating eggs.
5. Cracked eggs should not be used.
6. Preparation surfaces, utensils and containers should be cleaned regularly and always between the preparation of different dishes.
7. Egg dishes should be consumed as soon as possible after preparation or refrigerated if not for immediate use.
8. Egg dishes to be eaten cold should be refrigerated.

Commercially produced products are normally manufactured with bulk pasteurized egg and so are safe. The issue has continued to cause concern, however, and in February 1998, in evidence to the House of Commons Agriculture Committee, the Local Authorities Co-ordinating Body on Food and Trading Standards (LACOTS), now known as LACORS, and the Local Government Association called for a ban on the use of raw eggs by caterers.

Eggs and mayonnaise

Outbreaks of *Salmonella* food poisoning have been associated with contaminated mayonnaise. The low pH of mayonnaise and the presence of organic acids, particularly acetic acid, normally make it an unfavourable environment for the survival or growth of most bacteria. *Salmonellae*, however, may survive but not multiply in mayonnaise for at least 24 hours. The US Food and Drug Administration recommends that the pH in mayonnaise should be 4.1 or lower, and that the product should be held at 18–22°C for at least 72 hours before use unless pasteurized egg has been used in it. Recent outbreaks of salmonellosis associated with mayonnaise were probably due to contaminated raw egg being used, but mishandling of products in the early stages of preparation may also have been a contributing factor.

Pasteurization of eggs

The process of pasteurization is specified in the Egg Products Regulations 1993 as 2.5 minutes at 64.4°C for whole egg and yolk, but there is provision for food authority approval of alternative, equivalent heat processes. No time/temperature parameters are specified for the treatment of albumen – food authorities must be satisfied by the operators of businesses of the effectiveness of proposed processes and the degree of expertise involved. Process records are required to be kept and made available to the food authority.

Alpha-amylase test

To determine whether the pasteurization process has been effective the UK egg legislation contains a requirement that at least one sample of each batch of whole egg or yolk pasteurized should be tested using the alpha-amylase test.

Microbiological standards and other tests for lactic acid, butyric acid and succinic acid are contained in the regulations.

The Ungraded Eggs (Hygiene) Regulations 1990

These regulations prohibit the retail sale by producers of cracked eggs at the farm gate, in local markets and at doorsteps because of the hygiene risks they represent. The eggs must be visibly cracked such that the producer could reasonably have been expected to have identified and removed them from sale. It includes leaking eggs and eggs in which the shell, when viewed in the ordinary light by the naked eye, is visibly cracked. Enforcement responsibility for these regulations lies with food authorities.

REFERENCES

1. Ejidokun, O.O., Killalea, D., Cooper, M., Holmyard, S., Cross, A. and Kemp, C. (2000) Four linked outbreaks of Salmonella enteritidis phage type 4 infection – the continuing egg threat. *Communicable Disease and Public Health*, 3, 95–100.
2. Djuretic, T., Wall, P.G. and Nicholls, G. (1997) General outbreaks of infectious intestinal disease associated with milk and dairy products in England and Wales: 1992 to 1996. *Communicable Disease Report*, 7(3), R41–4.
3. Djuretic, T., Wall, P.G., and Nicholls, G. (1997) General outbreaks of infectious intestinal disease associated with milk and dairy products in England and Wales: 1992 to

1996. *Communicable Disease Report*, 7(3), R41–4.

4. Djuretic, T., Wall, P.G. and Nicholls, G. (1997) General outbreaks of infectious intestinal disease associated with milk and dairy products in England and Wales: 1992 to 1996. *Communicable Disease Report*, 7(3), R41–4.

5. The Food Safety Act 1990, HMSO, London.

6. The Dairy Products (Hygiene) Regulations 1995 Statutory Instrument No. 1086, HMSO, London.

7. Djuretic, T., Wall, P.G. and Nicholls, G. (1997) General outbreaks of infectious intestinal disease associated with milk and dairy products in England and Wales: 1992 to 1996. *Communicable Disease Report*, 7(3), R41–4.

8. Djuretic, T., Wall, P.G. and Nicholls, G. (1997) General outbreaks of infectious intestinal disease associated with milk and dairy products in England and Wales: 1992 to 1996. *Communicable Disease Report*, 7(3), R41–4.

9. Food and Veterinary Office (2000) 'Final report of a mission carried out in the United Kingdom from 19 to 20 January and 24 to 28 January 2000 to evaluate the application of council directive 92/46/EEC laying down the health rules for the production and placing on the market of milk and milk-based products', European Commission, Health and Consumer Protection Directorate-General, 18/05/2000, 24.

10. Evans, H.S., Madden, P., Douglas, C., Adak, G.K., O'Brien, S.J., Djuretic, T. *et al.* (1998) General outbreaks of intestinal disease in England and Wales: 1995 and 1996. *Communicable Disease and Public Health*, 1, 165–71.

11. Ejidokun, O.O., Killalea, D., Cooper, M., Holmyard, S., Cross, A. and Kemp, C. (2000) Four linked outbreaks of Salmonella enteritidis phage type 4 infection – the continuing egg threat: *Communicable Disease and Public Health*, 3, 95–100.

12. Food Safety Act 1990, Section 40: [draft] Code of Practice.[2]

13. The Egg Products Regulations 1993 Statutory Instrument No. 1520. HMSO, London.

14. Food Safety Act 1990, Section 40: [draft] Code of Practice.[2]

15. Food Safety Act 1990, Section 40: [draft] Practice Guidance.[2]

16. Djuretic, T., Wall, P.G. and Nicholls, G. (1997) General outbreaks of infectious intestinal disease associated with milk and dairy products in England and Wales: 1992 to 1996. *Communicable Disease Report*, 7(3), R41–4.

17. ICMSF (1998) HACCP in microbiological safety and quality. Blackwell Scientific Publications, Oxford, p. 48.

18. Food Standards Agency. Guidance on Officially Tuberculosis Free status and the Dairy Products (Hygiene) Regulations 1995, as circulated under cover of the CEHO letter Ref: CEHO/00/7 of 13 March 2000.

19. Food Standards Agency (2001) 'Control of Mycobacterium Avium subspecies Paratuberculosis (MAP) in cow's milk', Paper note 01/08/02.

20. Food Standards Agency (2001) 'Control of Mycobacterium Avium subspecies Paratuberculosis (MAP) in cow's milk', Paper note 01/08/02.

21. Food Standards Agency (2001) 'Control of Mycobacterium Avium subspecies Paratuberculosis (MAP) in cow's milk', Paper note 01/08/02.

22. IDF Document D 222:1991 Published by the International Dairy Federation. Brussels.

23. IDF Bulletin No. 292/1994. Recommendations for the hygienic manufacture of milk and milk-based products. Published by the International Dairy Federation. Brussels.

24. Food Standards Agency (2002) Report of the National Study on the Microbiological Quality and Heat Processing of Cows' Milk.

25. Ministry of Agriculture, Fisheries and Food (1975) *A Guide to Clean Milk Production*, BS 5226, HMSO, London.

26. Ministry of Agriculture, Fisheries and Food (1995) *Milk Hygiene: A Guide to the Dairy Products (Hygiene) Regulations for Dairy*

[2]Draft in January 2004 – to be published later in 2004.

Product Processors, MAFF Publications, London.

27. Ministry of Agriculture, Fisheries and Food (1995) *Milk Hygiene: A Guide to the Dairy Products (Hygiene) Regulations for Farmers Producing and Processing Milk from Goats and Sheep*, MAFF Publications, London.

28. Ministry of Agriculture, Fisheries and Food (1995) *Milk Hygiene: A Short Guide to the Dairy Products (Hygiene) Regulations for Dairy Farmers*, MAFF Publications, London.

29. British Standards Institution Code of Practice 7771:1994 (1994) Pasteurisation of milk on farms and in small dairies.

30. The Lancashire and Greater Manchester Food Officers Groups (2001) Guidance notes for on farm pasteurisers.

31. Anonymous (1997) Vero cytotoxin producing *Escherichia coli* O157. *Communicable Diseases Report*, 7, 409, 412.

32. Anonymous (1999) *Escherichia coli* O157 associated with eating unpasteurised cheese. *Communicable Diseases Report*, 7, 93, 96.

33. Anonymous (1997) Salmonella gold-coast and cheddar cheese: update. *Communicable Diseases Report*, 7, 409, 412.

34. Donnelly, C.W. (2001) Factors associated with hygienic control and quality of cheeses prepared from raw milk: a review. Cheese in all their aspects. Bulletin of the International Dairy Federation IDF 369/2001, pp. 35.

35. Donnelly, C.W. (2001) Factors associated with hygienic control and quality of cheeses prepared from raw milk: a review. Cheese in all their aspects. Bulletin of the International Dairy Federation IDF 369/2001.

36. Institute of Food Science and Technology (2000). Position statement of food safety and cheese. http://ifst.org/hottop15.htm.

37. Adams, M.R. and Moss, M.O. (1995) Food Microbiology, University of Surrey, Guildford. The Royal Society of Chemistry – Information Services.

38. ICMSF (1998) HACCP in microbiological safety and quality. Blackwell Scientific Publications.

39. Donnelly, C.W. (2001) Factors associated with hygienic control and quality of cheeses prepared from raw milk: a review. Cheese in all their aspects. Bulletin of the International Dairy Federation IDF 369/2001, pp. 16–27.

40. Dairy Industry Federation (1995) *Guidance for Good Hygiene Practice in the Manufacture of Dairy-Based Products*, DIF, London.

41. The Creamery Producers Association (1988) *Guidelines for Good Hygienic Practice in the Manufacture of Soft Fresh Cheeses*, CPA, London.

42. Milk Marketing Board (1989) *Guidelines for Good Hygienic Practice for the Manufacture of Soft and Fresh Cheeses in Small Farm Based Production Units*, MMB, Thames Ditton.

43. Specialist Cheesemakers' Association (SCA) (1997) The Specialist Cheesemakers' Code of Best Practice, PO Box 448, Newcastle-under-Lyme, Staffordshire, ST5 OBF.

44. Adams, M.R and Moss, M.O. (1995) Food Microbiology, University of Surrey, Guildford. The Royal Society of Chemistry – Information Services, pp. 265.

45. Sockett, P.N. (1991) Communicable Disease associated with milk and dairy products: England and Wales, 1987–1989. *Communicable Disease Review*, 1(1), R9–12.

46. Adams, M.R. and Moss, M.O. (1995) Food Microbiology, University of Surrey, Guildford. The Royal Society of Chemistry – Information Services, pp. 102.

47. Milk Marketing Board (1992) *Code of Practice for the Hygienic Manufacture of Ice Cream*, MMB, Thames Ditton.

48. Ice Cream Alliance/Ice Cream Federation (1997) *A Guide to the Safe Handling and Service of Ice Cream*, Ice Cream Alliance, London.

49. Egg Marketing Board (1998) *Annual Report 1997*, EMB, London.

50. Department of Health (1988) *Avoid Eating Raw Eggs*, press release, DoH, London.

51. Humphrey, T.J., Whitehead, A., Gawler, A.H.L. *et al.* (1991) Numbers of *Salmonella enteritidis* in the contents of naturally contaminated hens' eggs. *Epidemiological Infection*, **106**, 489–96.

52. de Louvois, J. (1993) *Salmonella* contamination of stored hens' eggs. *PHLS Microbiology Digest*, **11**(4), 203–5.

53. Advisory Committee on the Microbiological Safety of Food (1993) *Report on* Salmonella *in Eggs*, HMSO, London.

54. Advisory Committee on the Microbiological Safety of Food (1993) *Report on* Salmonella *in Eggs: Recommendations and Government's Response*, MAFF, London.

55. Deputy Chief Medical Officer (1998) *Raw Shell Eggs*, EL/98/p138, Department of Health, London.

32 Fish and shellfish

Chris Melville

The fish industry continues to endure turbulent influences which have in recent decades conspired to affect its viability. The traditional British meal of fish and chips was threatened by the burgeoning fast food industry in the 1960s. The 'cod wars' of that period also closed the traditional Icelandic fishing grounds to the distant water vessels of the home fleet. Quota restrictions, allegations of over fishing and environmental concerns further prejudice the success of the British fishing industry. Consumers are reportedly put off by the perceptions of the smell of fish, bones, cost, availability and lack of familiarity.

In contrast, the industry has been able to prosper from the increased consumption of fish in preference to meat products implicated in scares such as BSE in beef and *salmonella* in poultry. The industry has also been keen to promote the health-giving effects of increased fish consumption offering both a reduced saturated fat intake and the beneficial effects of the consumption of omega 3 oils, found particularly in the oily fish species including, herring, mackerel and tuna. TV and magazine chefs have striven to promote the range and versatility of fish dishes and the wide variety of alternative species to those in short supply.

The conflict between increased consumption and diminishing traditional supplies has forced the industry to become more imaginative, sourcing from different fishing grounds, promoting the consumption of novel species, for example, Hoki and squat lobster, promoting the use of less popular, underexploited species and sourcing raw materials and services from emerging markets.

Fish is traded as an international commodity. As an example, the ubiquitous fish finger may have been manufactured in the United Kingdom from fish caught in the Barents Sea by the Russian fleet, part processed and frozen on board, landed into the EU, exported to China for primary processing to be re-imported into the United Kingdom for final production, sale and potential export.

In contrast, the fresh fish industry now dominated by supermarkets, flies fresh fish into the United Kingdom not only from the more exotic locations but also traditional markets such as Iceland, in a desperate bid to secure the freshest product at the point of retail sale.

The greatest increase in sales has been seen in 'added value' products; principally, fish-based entrees and main courses offering the consumer a wide range of ready-to-eat products with minimal preparation, handling and waste.

Fish consumption is particularly seasonal. Some species are only available in reasonable quantities and quality at certain times of the year. With the exception of the seaside trade, traditional species such as cod and haddock will not sell in hot weather, consumers preferring barbecue products of salmon, monkfish, tuna and sardines. Supermarket chains will monitor weather forecasts in an effort to ensure suppliers can meet predicted demand.

LEGISLATION[1]

In common with all other 'vertical' legislation, that is, product specific, hygiene standards in the fish industry are controlled by regulations deriving from EU directives. Directives 91/492 and 493 respectively control the shellfish and fish industry from catching through to auction, processing, storage and transport. Both directives and several subsequent additions have become consolidated in the Food Safety (Fishery Products and Live Shellfish (Hygiene) Regulations 1998). Two amendments to these regulations have been made revising shellfish documentation, amending drafting errors and introducing a charging regime.

This legislation is also currently subject to the EU review on the consolidation of the vertical directives when it is anticipated that the various pieces of legislation covering diverse industries such as fish, meat products, dairy and eggs will be streamlined in an effort to ensure similar standards and removing inconsistencies.

Although the regulatory demands of the legislation have much in common for both the fish and shellfish industry, particularly in respect of structural hygiene, training and risk assessment, some aspects of the regulations have greater emphasis for the shellfish industry.

Shellfish are predominantly filter feeders, deriving their nutrients from the waters from which they are harvested. Pollutants, whether chemical, microbiological or natural, may be absorbed into the flesh of the fish. If the fish is consumed without further preparation, this contamination may have a direct effect on the consumer. The regulations seek to reduce these risks by, amongst others, categorizing the quality of harvesting areas based upon the trends of bacterial contamination. The categories of A, B and C reflect whether the fish is satisfactory for immediate consumption, consumption after relaying and purification or heat treatment.

Coastal Authorities must administer the enforcement of the harvesting area, in co-operation with the Food Standards Authority (FSA) and the Centre for Environment, Fisheries and Aquaculture Science (CEFAS) classifying and monitoring the production areas.

In an effort to control the movement of shellfish from one area to another, movement documents must be submitted to the Local Authority. A model movement document published by the FSA accommodates the movement of shellfish from gatherer, production area, and relaying and purification area. The gatherer and recipient must retain copies of the document for a period of 12 months.

Live bivalves must also be accompanied to the point of sale by a healthmark including the day and month of wrapping and the approval number of the dispatch centre. The label must also bear a warning that the animals must be alive when sold, or carry a date of durability. Because of the risk of food poisoning from these species, the certificates must be retained for 12 months after sale to assist any investigation into tracing the source of the outbreak.

All establishments involved in the fish industry must be approved by the Local Authority and the approval number must appear on packaging, however, unlike meat legislation, the size, shape and location of the mark are not specified.

One unique aspect of the legislation requires the industry to pay for the costs of enforcement. Charges are imposed on landings, vessel inspections and the throughput of product through an establishment. Charges are typically one Euro per tonne with a reduction of 55% for establishments carrying out their own checks regime. The value of the Euro to be used in calculating charges is published by the EU in the preceding September and is the same figure used in the imported foods regime.

The Seafish Industry Authority (SFIA) has published a number of excellent reference documents detailing standards, systems and good manufacturing practice throughout the industry, providing guidance to industry and enforcers (www.seafish.co.uk). CEFAS is also a valuable source of advice and guidance on the fish and shellfish industry, in addition to its role managing aspects of fisheries legislation and being

---

[1]Also see Bassett, W.H. (2002) *Environmental Health Procedures*, 6th edn, Spon Press, London at pages 358–75 dealing with shellfish and fisher products.

designated the Community and National Reference Laboratory (www.cefas.co.uk).

ASSESSMENT OF FITNESS

The assessment of fitness in fish can be highly subjective. One man's 'unfit' fish will be another's stock in trade. Freshness and quality is not the same thing. After spawning, fish will be exhausted and thin with little yield. It will be difficult to fillet and will not keep well. Paradoxically, fish from grounds with ample food stocks may also be poor, needing little effort to prosper and consequently having little musculature and texture. The merchant will be looking for fish with a good yield, that is, the proportion of flesh to bone, typically from 50% to as little as 20%.

The Fishery Products regulations specify maximum standards of TVB-N (Total Volatile Basic-Nitrogen), TMA (Trimethylamine-Nitrogen) for different species. Within the industry these standards are viewed with a degree of caution. TVB-N concentrations will increase as fish deteriorates; however a point is reached when the concentration starts to fall away again, giving rise to potential misinterpretation. The regulations also demand that fish of the Scombridae family undergo analysis to determine the presence of Histamine.

The assessment of freshness is generally easier in whole fish. Fresh fish will have clear, convex eyes, clean reddish gills, and bright shiny skin with a vibrant colour. Any visible blood in the body cavities will be bright red in colour. Very fresh fish will display rigor mortis, however this will dissipate rapidly and it is unlikely that all but the very freshest fish will display this feature.

On the other hand, unfit fish will have a combination of cloudy sunken eyes, brown sticky offensive gills, pallid, dull skin and discoloured offensive blood. The flesh will be soft and tear easily. In the case of the cartilaginous fish, dogs, rays, skate etc., a distinctive smell of ammonia will be present as urea which is naturally present within the flesh decomposes.

It is important to note that the characteristics of unfitness must be assessed in combination. The inspector is looking for most, if not all of them to be present to some considerable degree before considering condemnation. The inspection of fillets is not as straightforward as most of the indicators have been removed. Here, the inspector will be limited to the smell, texture and colour of the fillet to make the assessment.

In addition to these assessments the fish processing industry uses the 'Torry' scoring system to determine the quality of fish (see Table 32.1).

In this system samples of the product are boiled, generally in a microwave and tasted. The taste characteristics are then compared with benchmark descriptions. In order to enable the less experienced taster to use this system some assessors have enlarged the stilted vocabulary available in the original scheme. Tasters are invited to attach their own descriptors to samples of a given score and these phrases will be added to the text until a comprehensive vocabulary exists for each point in the table. Curiously, this became necessary with the increase in trade with China. Lactose intolerance in that population meant that the dairy descriptors, boiled milk, sour milk etc. were of little use to their producers and a revised set of phrases was developed.

Buyers and processors are able to communicate and negotiate over consignments in a quantifiable form. A supermarket may, for instance, specify 'fresh' fish supplies to be not less than Torry 8. Both the sender and the buyer will be able to monitor their stocks and intakes accordingly.

To retard deterioration in fish it must be maintained at chill temperature throughout the supply chain, from the vessel, through the market, processor, transport, retail and into the domestic environment. The legislation refers to fish 'at the temperature of melting ice'. Any break in this chain will accelerate decomposition. Fish handled well may have a shelf life, in extremes beyond two weeks whereas fish allowed out of the cold chain for even brief periods may last as little as one or two days. In warm weather, fish out of temperature control can 'turn' within hours.

In an effort to secure control over all parts of the supply chain, leading supermarket chains have now commissioned their own catching vessels whose catch is dedicated to their outlets via approved processors and hauliers. This consistency of ownership bypasses one of the fishing industry's problems, lack of continuous ownership

Table 32.1 Determination of the quality of fish

Odour	Flavour	Texture, mouth feel and appearance	Score
Initially weak odour of sweet, boiled milk, starchy, followed by strengthening of these odours	Watery, metallic, starchy, initially no sweetness but meaty flavours with slight sweetness may develop	Dry, crumbly with short, tough fibres	10
Shellfish, seaweed, boiled meat, raw green plant	Sweet, meaty, creamy, green plant characteristic		9
Loss of odour, neutral odour	Sweet and characteristic flavours but reduced in intensity	Succulent, fibrous; initially firm going softer with storage; Appearance originally white and opaque going yellowish and waxy on storage	8
Wood shavings, woodsap, vanillin	Neutral		7
Condensed milk, caramel, toffee-like	Insipid		6
Milk jug odours, boiled potato, boiled clothes-like	Slight sourness, trace of 'off' flavours		5
Lactic acid, sour milk, 'byre-like'	Slight bitterness, sour, 'off' flavours		4
Lower fatty acids (e.g. acetic or butyric acids), composted grass, soapy, turnipy, tallowy	Strong, bitter, rubber, slight sulphide		3

Source: Sensory Assessment of Fish Quality, Note 91 (1989) Torry Research Station, Aberdeen.

where the catching, auctioning, processing and transport companies take individual and often uncoordinated responsibility for the product.

FISH CURING

Smoking

The trade in fish smoking has origins in the ready potential for fish to spoil. Catches had to be processed soon after landing to minimize the effects of decomposition and smoking was one of the options available to the fishing community. In the original process the combined effects of salt and smoke served to arrest decomposition and to hide the taste of poor fish; however, the intensities of both would be too strong for the modern palate.

With the advent of better transport, refrigeration and quality control, the need to smoke for preservation has been superseded by the organoleptic qualities attributed to the smoking process.

The curing industry despairs of the often-quoted maxim elsewhere in the fish industry that smoking can salvage poor fish. The end product can only reflect the quality of the raw material; good quality cured fish must be manufactured from good quality raw material.

Fish can either be smoked traditionally in chimneys or mechanically in kilns. In both cases, the fish will be headed, skinned or filleted according to the specification of the end product. The fillets are brined in a saturated salt solution which may include a dye to impart a stronger colour to the finished product. Natural dyes have largely replaced the traditional azo dyes of tartrazine, E102 and sunset yellow E110 with turmeric, annatto and crocin. There is, however a move away from the use of any dyes in keeping with the health-giving properties of fish consumption and the absence of any additive. The end product may however appear pallid to the experienced consumer.

In traditional smoking the brined fish will be suspended over smouldering beds of hardwood or softwood sawdust or shavings and smoked for periods, from hours to days, depending upon the species. Cod, haddock etc. are supported on speights whereas kippers and Finnan haddock may be hung on tenterhooks.

The edible quality of the end product will be influenced by ambient temperature, wind speed and humidity and relies upon the skill and experience of the manufacturer to overcome these variables. Successful traditional fish smoking remains the province of the artisan.

In an effort to combat these outside influences the 'Torry' kiln or mechanical kiln was designed to provide a programmed smoking environment where temperature, time, humidity and smoke density can all be controlled. In modern units these parameters are controlled by computer programmes specific to species and even customer.

Fish fillets are loaded onto wire racks or speights on wheeled trolleys and pushed into a series of interlinked chambers where hardwood smoke from a small furnace is blown in horizontal streams. The closely controlled environment ensures a more consistent end product in a much faster time; however, the purist will insist that the traditional method delivers superior edible quality.

One area in which the mechanical kiln excels is in 'hot' smoking. In this method the fish will be cooked as well as smoked and although this can be achieved successfully by the traditional method, the finer controls of the mechanical kiln make the task easier and potentially safer. Often the kiln will be fitted with two sets of doors, raw fish entering from one side to be removed by different operatives on another designing out risks of cross contamination.

Once smoked the fish must be handled appropriately to minimize risk. Cold smoked fish has not undergone cooking and must still be considered raw. It should, however be kept away from wet fish to minimize taint. An exception to this rule includes species such as salmon and halibut designed to be eaten raw. In common with hot smoked fish it must be handled as a ready to eat high-risk product and kept entirely separate from raw fish, cold smoked fish and raw shellfish.

Salting

The origins of the salting industry have a lot in common with smoking, although in contrast, poor fish may be successfully salted without detriment to the finished product.

Layers of fish are interleaved with layers of salt and stacked for periods of several months. The action of the salt draws moisture from the fish and in concert with the bacteriostatic effect of the salt density prohibits the onset or progress of putrefaction. Before consumption, the fish will need thorough soaking to return the fish to a palatable condition. The majority of salt fish production in the United Kingdom is destined for export.

DISEASES AND CONDITIONS OF FISH

Bacterial conditions

Fish are by no means a common source of foodborne infections. However, the common types of food poisoning organisms will grow on seafoods if the conditions are suitable.

1. *Clostridum botulinum* occurs naturally in the marine environment, the type E strain usually being associated with fishery products. Several deaths per year are caused throughout the world, and are usually attributable to underprocessed fish or faulty canning techniques. Large numbers of this organism have been found in commercial trout farms, and so strict attention should therefore be paid to all stages

of production to ensure that the organism is not given a chance to produce its potent neurotoxin.

2. *Salmonellae* and *Staphylococcus aureus* are not regular inhabitants of either the fresh or salt-water environments, but fishery products can become contaminated during processing.

3. *Vibrio parahaemoliticus* occurs naturally in fish and is a major cause of gastroenteritis in Japan, where fish forms an important part of the diet. The organism is generally killed by cooking and most outbreaks result from under-processed or cross-contaminated products.

Heavy metals

Incidents of heavy metal poisoning arising from the consumption of exotic fish have been caused by the combined effects of industrial pollution and the scavenging habits of some species of fish. Consumers, particularly sensitive groups have been warned about the consumption of species including shark and tuna. Responsible traders and retailers will insist upon analysis and certification of consignments.

Parasitic infections

1. *Roundworms* are occasionally found in the flesh of fish. Commonly misnamed cod worms they may be found in the flesh of many other species. Infestations are proportional to seal activity in the fishing grounds as the seal is a co-host of the worm. The worms migrate from the gut of the fish and are generally found in the thicker fillet around the belly. Although harmless and killed by cooking and freezing, the worms are aesthetically undesirable and should be removed during filleting. 'Candling', passing the fillets over an illuminated table, facilitates detection but is labour intensive adding to the cost of production and is often only practised to satisfy a commercial contract.

2. *Anisakis simplex* is similar to the codworm (see earlier), which is white in colour and therefore difficult to detect. Judgement is as with the codworm.

3. *Chloromysciumthyrsites* is a protozoal infection that affects the hake in particular. The organism produces enzyme actions that severely soften the flesh within a few days of infection. Fish affected are often termed 'milky' hake.

4. *Dibothriocephalus latus* is a tapeworm that can infect humans who have eaten affected fish that has been undercooked. Pike, cod, grayling, perch and salmon have been found to harbour cysts of the adult worm, which are readily recognized as a white body in the flesh of fish. Affected fish are unfit for human consumption and may be seized.

5. *Grillitia ermaceus* is found in its adult stage in rays but the halibut is the host for the completion of the larval stage. It is harmless to humans.

6. *Sarcotases* is a parasite that penetrates the musculature or visceral cavity of fish and grows to a length of several centimetres. It is repulsive in appearance as its gut consists of a bag filled with black fluid (digested blood); it is often buried deep in the muscle of the fish.

Diseases

1. **Furunculosis** is an infectious condition caused by the bacillus *Salmonicida*, which affects many species of freshwater fish, including salmon and trout. Affected fish develop nodules beneath the skin which burst or discharge fluid. Affected fish should be regarded as unfit for human consumption as this is a highly contagious disease.

2. **Salmon disease** (salmon plague) affects fish of the salmon family and results in white patches that grow to cover the fish and kill it. This is another highly contagious disease, and it is recommended that affected fish are burnt to destroy the causative organisms.

3. **Tuberculosis**, in the form of nodules within the flesh or gut, is occasionally found in fish. The lesions result from infection by *Mycobacterium tuberculosis*, but the condition is not communicable to humans.

4. **Tumours**, particularly sarcomata, occur in fish and may be malignant or benign. They appear

as fibrous growths in almost any location, with osteomatas running down the radial bone of the fish. Halibut are often affected, and catfish and cod are affected to a lesser degree. Localized lesions may be trimmed, but badly affected fish should be rejected as unfit for consumption.

5. **Ulcerative dermal necrosis (UDN)** causes an ulcerative condition of the gills and head of salmon. It is a viral disease that is often associated with a secondary fungal infection, which gains access through damaged tissue.

Scombrotoxin food poisoning (also see Chapter 13)

Fish seldom presents risks to health, becoming unpalatable and organoleptically offensive fairly readily serving to deter the consumer. The trade occasionally quotes the maxim 'fish never killed anyone'!

Fish of the scombridae family, mackerel, tuna, bonito etc. are occasionally, however, associated with outbreaks of a toxin-based food poisoning. Previously associated with the traditional consumption of mackerel, the increased market for fresh tuna has seen a rise in cases within this species.

Symptoms include an extremely rapid onset of temperature, burning sensation, flushing, and diarrhoea although accompanied by an equally rapid recovery rate. The symptoms are caused by the ingestion of histamine produced by the denaturing of histadine naturally present in the flesh of these species. The formation of histamine in the fish is accelerated by increased temperature and consequently it is essential fish of these species are comprehensively iced immediately upon harvest and then throughout the supply chain.

Large fish, such as tuna, can have widely varying levels of the toxin throughout their flesh and it is thought this is caused when some part of the fish protrudes above the ice, particularly in the warm climates of the catching areas. Because the results of analysis may be inconclusive, not being representative of the whole fish, some major importers refuse to trade in tuna.

FISH BLOCKS

With the increasing market share enjoyed by 'added value' chilled and frozen fish products, pies, portions and coated items, it is generally this area with which the environmental health professional will have greater involvement. Almost without exception products of this nature will be manufactured from fish blocks.

The block is truly universal, sourced throughout the world and manufactured both at sea and in land-based establishments. Blocks weigh $16\frac{1}{2}$ pounds (7.48 kg) in weight and have a size of $20 \times 2 \times 10$ inches ($484 \times 256 \times 63$ mm).

A waxed cardboard liner is placed in a rectangular metal former. Mince or fillets are loaded into the liner until the correct weight has been achieved. The mince will be a combination of off cuts of the filleting industry and saw waste from the portioning industry. The integral waxed lid of the carton is folded over and the whole assembly is placed in a plate freezer. The combination of pressure and freezing forces the fish into the shape of the former producing a regular, dense block of frozen fish. The blocks are easy to pack and stack and will accurately fit the dedicated cutting equipment of the industry.

At the processing factory, the frozen blocks may be re-minced to make fish cakes and shapes, or cut and sawn to make fingers, portions and shapes. Intriguingly, manufacturers may even press fillet block rectangles back into fillet shapes for authenticity.

A further advancement of the block industry is the interleaved block. These blocks comprise individual fillets separated by polythene film. Also known as 'shatterpacks' these blocks will provide 'real' fillets for the processing industry as well as a resource for the frying and catering sector.

Although the purist may prefer 'fresh' fish in preference to frozen fish, frozen sea fish will have been processed within minutes of catching, almost eradicating any possibility of deterioration. The frozen blocks are compact, easy to store and give a consistency of quality and supply. They also minimize stock wastage and the difficulty of handling and storing wet fish.

SURIMI

Surimi is the technical name for synthesized products most commonly marketed in the United Kingdom as crabsticks. Derived from a traditional Japanese fish process, the products are generally manufactured from Alaskan pollack.

The highly complex manufacturing process involves water washing the minced flesh and the resulting combination of physical and chemical effects produces a product with the ability to be formed into strings. By aligning series of strings, adding colours and flavours, forming, pressing or extruding the mix and cooking, products such as crab, shrimp and other shellfish can be simulated.

FARMED FISH

The fish farming industry in Europe was initially dedicated to salmon and trout. The success of farms in Scotland, Ireland, Scandinavia and Iceland has resulted in a glut of cheap fish on the market. The industry was further damaged by allegations that the farms polluted the environment with excess feed and fish waste. Salmon lice infestations prosper in the overstocked pens and migrate into the natural environment damaging wild fish stocks. Pesticides used to combat the lice are alleged to be harming the environment and residues have been found in samples of the fish flesh. Concern has also been expressed that the dye, canthaxanthin,[2] used in feed to enhance the colour of the fish flesh is harmful to consumers. The fish itself is often criticized as soft, pallid and tasteless compared to the wild species.

In an effort to overcome these prejudices sectors of the industry have improved animal husbandry, practised rotational farming, used natural lice control methods and eradicated synthetic dyes. Farms can be located in fast flowing waters where the fish have to fight against the current, building up texture and musculature.

In addition to salmon and trout, other species are now being farmed, including expensive species such as turbot and bass and endangered species such as

cod. The international trade in farmed fish located in the Far East specializes in warm water prawns.

This industry has also suffered adverse publicity associated with residues. The unregulated use of antibiotics, including chloramphenicol, to minimize infections in the farms, resulted in residues being found consistently in samples of imported prawns. In the case of China, shortcomings in the regulation of veterinary medicine and non-observance of commitments and guarantees resulted in an EU-wide ban on all aspects of fish imports in 2002. Imports were progressively reintroduced following a highly restrictive import and sampling policy of positive release at Border Inspection Posts and favourable audits by EU inspection teams in China.

SHELLFISH

Distinguishing features

The term shellfish includes creatures of the order *Mollusca*, which can be subdivided into bivalves (mussels, oysters, clams, cockles, scallops), univalves (periwinkles and whelks) and crustacea (crabs, lobsters, crayfish, prawns, shrimps, etc.) (see Fig. 32.1).

Molluscs

1. **Cockle (*Cerastoderma edule*).** Roughly circular concave shells with ribs radiating from the hinge. The colour of the shell is usually cream or fawn, with some darker markings. Size: up to 4 cm in diameter. The flesh is partly grey, partly yellow, with occasional orange patches.
2. **Clam (*Venus merceneria*).** Harp-like symmetrical shells with clear concentric growth lines. The colour of the shells is a pale fawn to a yellowish brown. Size: approximately 15 cm long. The flesh varies from cream to a yellowish fawn in colour. The syphon is black and there is a small purple patch on the otherwise white interior of the shell.
3. **Scallop (*Pecten maximus*).** Two concave, tight-fitting, fan-shaped shells that possess distinctive

---

[2]It is intended to reduce the maximum level of canthaxanthin (E1619) in fish feed from 80 to 25 mg/kg by UK legislation, probably in 2004.

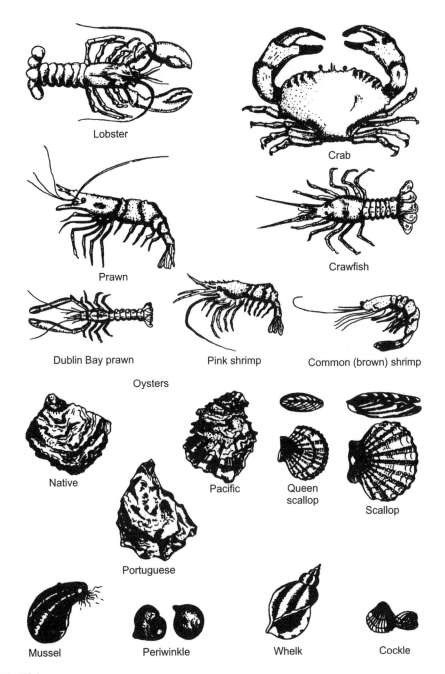

Fig. 32.1 Shellfish.

ribs. The upper shell is flat and the lower concave. The shell is pink or orangey brown. Size: approximately 10 cm. The flesh is white (adductor muscle) and the roe orange. When prepared for sale, the other organs are removed.

4. **Mussel (*Mytilis edulis*).** Two smooth, concave, symmetrical shells. The colour of the shells is blue-black. Size: approximately 7.5 cm long and 2.5 cm wide. The flesh is beige and orange in colour.

5. Oysters

(a) Native oyster (*Ostrea edulis*). A roughly circular, thin upper shell with a slight protrusion at the hinge or heel. The lower shell is deeper and heavier. Colour: varies from pale beige to greenish brown, but annual growth rings are always visible on the shell. Size: 6 cm in diameter. The adductor muscle is a pinkish or yellowish beige-white, the surrounding cillae and alimentary system is somewhat darker. Oysters from certain areas possess a greenish tinge.

(b) Portuguese oyster (*Crassostrea angulata*). The shell is in two parts, the upper flatter than the lower, which is concave. Both shells are more irregular and longer than the shells of the native oyster. The shell is flint-like and greenish grey in colour. Size: 12 cm long. The flesh is a pale greenish white.

(c) Pacific oyster (*Crossostrea gigas*). Two roughly triangular shells, the upper flatter than the lower, with a frilled appearance derived from the irregular annual shell growth, which protudes from below that of the previous year. Colour varies from bluish grey to dark green if covered with marine organisms. Size: 7 cm in diameter. The flesh is cream-beige.

6. **Periwinkle** (*Littorina littorea*). A small, compressed, thick spiral shell that is almost spherical. The entrance is bluish white and is guarded by a black foot, but the shell is black or grey-black. Size: approximately 2.5 cm. The flesh is dark grey to black in colour.

7. **Queen scallop** (*Chlamys opercularis*). Very similar to the scallop but both shells are concave and smaller than those of the scallop. The colour is usually pink, red-brown or yellow marked with white. Size: approximately 9 cm. The flesh is white with an orange roe that is smaller than that of the scallop. (Where the fish are mechanically processed the roe is removed with the other viscera.)

8. **Whelk** (*Buccinum undatum*) ('buckie'). A strong, thick spiral shell that is pale sandy-brown in colour with lighter and darker markings. The entrance may be closed by the extension of the 'foot', a round bony plate. Size: approximately 7.5–10 cm. The flesh is dull yellow with darker coloured viscera and is long and thin.

Crustacea

1. **Crab** (*Cancer pagurus*). A roughly oblong carapace with rounded corners. When alive, the upper surface is brownish red and the underside yellowish white. The first pair of legs have been modified into heavy black claws, each with a moveable pincer. The shell surface is minutely granulated and the front margin is divided into lobes. After boiling, the colouring lightens and becomes more red than brown. Size: across the back up to 30 cm. The white meat (muscle) is removed from the claws and the brown meat (liver and gonads) is found in the carapace. Sexual difference: the tail or apron of the female is heart shaped and broad, while in the male it is narrow.

2. **Crawfish** (*Palinurus elephas*) ('spiny lobster', 'langouste'). The carapace is reddish brown and spiny with two strong spines protecting the eyes which are large and globular. The tail is cylindrical with sharp-edged bony plates on the flanks, and ends in a bony fan. There are single acting claws on the first pair of legs, and the antennae are long with spiny bases. Size: up to 45 cm. The flesh of the tail is used for food and is similar to that of the lobster.

3. **Dublin Bay prawn** (*Nephrops norvegicus*) ('scampi', 'Norway lobster', 'langoustine'). The body and tail are narrow and cylindrical. The first pair of legs are slender and long with unequal claws, each having one moveable pincer. The eyes are large; the rostrum is long and has three teeth on each side. It has four short and two long antennae. The colour is pale flesh, darker in parts, which becomes a more opaque orange after boiling. Length: up to 20 cm. The tail meat is white with some pale pink bands.

4. **Lobster** (*Homarus gammarus*). The carapace is cylindrical with a well developed rostrum. The tail segments taper only slightly and terminate

in a broad fan of overlapping shell segments. The claws are very large and unequal in size, and each has a moveable pincer. The remaining legs are round and weak. It has four short and two long antennae. The colouring is blue-black on the back and flanks with a little creamy or orange spotted colouring on the underside. When boiled it becomes bright scarlet. Length: up to 45 cm. The male has five body segments and relatively heavier claws. The female has seven body segments. The tail meat is pinkish white; that of the claws is covered with a pinkish orange skin.

5. **Prawn** (*Pandalus borealis*). The carapace has a long, slightly upward curved rostrum; the antennae are long and pink and the eyes are large. It has a series of sharp points along the head. When fresh, the prawn is translucent grey, spotted, and lined with purplish grey. When boiled, prawns turn pinkish red. Length: up to 10 cm. The meat of the tail is pinkish white and segmented.

6. **Shrimps**

   (a) Brown shrimp (*Crangon crangen*) (common shrimp). The brown shrimp has no rostrum, the forward edge of the carapace having a series of serrations. The eyes are conspicuous and close together. It has two long and four short antennae, with scales on the base of the first pair. The first pair of legs are robust, smooth and strong and form small claws. Length: up to 6 cm. It is almost colourless but develops its colouring when boiled.

   (b) Pink shrimp (*Pandalus montagui*). The rostrum is long and almost straight with a slight curve upwards at the extremity. There are seven to eight teeth on the upper side, and three on the underside. It has large eyes and long antennae, and the first pair of legs are short. It is almost colourless but develops its colour when boiled. Length: up to 5 cm.

Shellfish toxins (also see p. 284 and 290)

Filter feeding shellfish will concentrate toxins from naturally occurring algal blooms. The incidence increases in warmer weather, more favourable to the growth of algae. ASP, DSP, PSP, NSP, respectively Amnesic, Diarrhetic, Paralytic, and Neurotoxic shellfish poisoning can cause a widely varying range of symptoms including nausea, parasthesia, disorientation, fatigue and in extreme cases, death. The onset can be as little as 15 minutes.

A recently discovered algal toxin, azaspiracid, is also thought to cause necrosis in the thymus, intestine and liver.

The toxins are heat stable and will not be destroyed by any form of preparation. Consequently, the only form of control is an extensive and well-regulated sampling regime implemented by coastal authorities and co-ordinated by the FSA and CEFAS. Elevated levels of toxins in samples will result in the harvesting areas being closed until the algal levels drop.

USEFUL WEBSITES

1. The Seafish Industry Association – www.seafish.co.uk
2. The Centre for Environment, Fisheries and Aquaculture Science – www.cefas.co.uk
3. www.globefish.org
4. Food and Agriculture Organisation of the United Nations – www.fao.org

33 Meat hygiene

Robert Butler

BACKGROUND

The role of the Environmental Health Officer (EHO) with regard to involvement in meat safety issues has undergone a major metamorphosis during the past decade. Prior to the introduction of the Meat Hygiene Service (MHS) in 1995, Local Authorities had a statutory duty to provide a meat inspection service. Although this duty has now been transferred to the MHS the role of the EHO in conjunction with other Agencies is now of paramount importance in minimizing the risk of illegal/unfit/diseased meat entering the food chain with the associated potential serious risk to public health.

During the past decade, the media has continuously identified and highlighted numerous 'food scares' and 'food scams' that have seriously affected public confidence especially in relation to the consumption of meat and meat products. These have included reports of illegally slaughtered animals, illegal imports and the large-scale criminal activities that were successfully investigated by Local Authorities such as Rotherham, Norfolk and Amber Valley whereby diseased unfit poultry meat was entering the human food chain. The Environmental Health News 'Stamp it out' campaign [1] has been consistently in the forefront in highlighting these criminal activities and reiterating the risks to both public and animal health from illegal imports [2] and illegal slaughter [3] and the necessity for all enforcement bodies to work closely together in policing the food industry.

In response to earlier consumer concerns the Food Standards Agency (FSA) Act [4] established the FSA with the prime objective of protecting public health and to act in the consumers' interest at any stage in the food production/supply chain. Stringent efforts have been taken by the FSA/Local Authorities/Port Health Authorities to close loopholes in legislation that has led to major changes in both the importation of meat and meat products. There is now a ban on personal imports of meat (possible causal factor of recent and devastating foot and mouth outbreak) and increased resources at border entry points with the use of detector dogs to identify miscreants. (It is therefore unfortunate that the excellent Port Health Hit Squads, who had powers to search passengers entering the United Kingdom together with the responsibility of tackling illegal meat imports, and who attained phenomenal success have now been disbanded and seen their powers shifted to Customs and Excise Officers.)

Changes have been made to the disposal of unfit meat and the introduction of widespread enforcement powers outside licensed premises has allowed EHO employed by Local Authorities to take appropriate action including the seizure of meat that has been produced illegally, i.e., 'smokies' (unskinned sheep/goats).

LEGISLATION

Although major changes in meat legislation have been incorporated in this chapter the reader must be aware that these requirements are subject to continual review from bodies such as the FSA, Department of the Environment and Rural Affairs (DEFRA), Department of Health (DoH) and specialist committees such as the Spongiform Encephalopathy Advisory Committee (SEAC) etc. that may culminate in new legal powers or amendments to present legislation in response to new scientific research/findings. The text includes where appropriate possible future changes of enforcement responsibilities between Local Authorities and the MHS in relation to 'catering butchers', wild game processing plants and cold stores.

RED MEAT

Fresh meat legislation

In April 1995, responsibility for meat inspection was transferred from the Local Authorities to the MHS who became responsible for the enforcement and execution of the regulations. The MHS was initially an executive Agency for the Ministry of Agriculture, Fisheries and Food (MAFF), which is now in part DEFRA. Finally, on 1 April 2000 the MHS moved over from MAFF to become part of the newly created FSA. The Fresh Meat (Hygiene and Inspection) Regulations 1995 [5] gave effect to Council Directive 91/497/EEC that updated and amended Directive 64/433/EEC on health problems affecting intra-community trade in fresh meat. It extended it to the production and marketing of fresh meat and in part to Council Directive 91/495/EEC on public health problems affecting the production and placing on the market of farmed game meat. The regulations require that slaughterhouses, cutting plants, cold stores, farmed game handling facilities and farmed game processing facilities must have a current licence issued by the minister. However, the regulations do not apply to premises where fresh meat is used for the production of meat products, meat preparations, minced meat and mechanically recovered meat. Neither do the regulations apply to premises where fresh meat is cut up for sale to the 'final consumer' or to the exemptions currently permitted for 'Farmers Markets' if certain criteria are met [6].

The main provisions of the regulations cover:

- the licensing of slaughterhouses, cutting premises, cold stores and farmed game handling and processing facilities;
- supervision and control of premises;
- conditions for the marketing of fresh meat;
- the admission to and detention of animals and carcasses in slaughterhouse and farmed game processing facilities;
- the construction and layout of and the equipment in licensed premises;
- hygiene requirements;
- ante-mortem inspection;
- slaughter and dressing practices;
- post-mortem inspection;
- health marking and certification;
- cutting practices, wrapping and packaging of fresh meat;
- storage, (chiller temperature may not exceed +4°C for poultry, +7°C for red meat and +3°C for offal) freezing (may not exceed −12°C) and transport of fresh meat.

The regulations apply to domestic species of animals such as bovines (including buffalo, bison), pigs, sheep/goats and horses, while the term farmed game has the meaning of wild game animals reared and slaughtered in captivity. In addition, every licensed slaughterhouse must be under the supervision of an Official Veterinary Surgeon (OVS) appointed by the minister and who is directly responsible for ensuring that meat is examined in accordance with legislative requirements and then suitably health marked. The Fresh Meat 1997 Amendment Regulations finally removed the previous loophole of 'private kill' that was being frequently abused so that now all meat animals killed in licensed slaughterhouses for human consumption must be treated as if they are intended for sale. They are subject to full inspection and health marking as specified in the legislation. The only exception to this rule is when the farmer kills the animal, processes it himself and consumes it himself. However, should he supply the meat to any other person even to a member of his own family then he is committing an offence. MHS is responsible for the enforcement of hygiene rules in all **licensed** slaughterhouses, cutting plants, re-wrapping centres, re-packaging centres and cold stores. Local authorities' EHOs continue to enforce legislation at unlicensed premises and have full responsibility once meat leaves unlicensed plants and moves to unlicensed meat plants. At present 'catering butchers' are exempt from the Fresh Meat Regulations as they supply the 'final consumer' and enforcement and supervision is the responsibility of the local authority. Due to a tightening of the definition of 'final consumer' it is proposed that catering butchers will come under veterinary supervision similar to that at present in wholesale cutting plants and therefore enforcement of these particular premises will move from local authorities to the MHS. In addition, it is proposed that cold stores handling meat would not require supervision by official veterinarians and supervision of these premises would return to local authorities.

Ante-mortem inspection of animals and Clean Livestock Policy

The ante-mortem inspection of food animals is an essential prerequisite in that it allows the OVS/Inspector to decide if the animal may be approved for slaughter/slaughtered under special precautions/slaughter delayed or condemned. When examining animals, notice must be taken of the movement of the animal, eyes, temperature, any swellings or enlargements especially of the lymph nodes, any discharges, respiration and signs of fever (i.e. shivering). It should allow for the detection of diseases such as anthrax, rabies, botulism, foot and mouth, swine fever and tetanus, orf, and provides an important role in protecting personnel from potential zoonotic disease. There is now a realization that the major threat to public health is to be found in the carrier animal that exhibits no symptoms of disease but excretes organisms that cannot be detected during either ante or post-mortem inspection (e.g. *Escherichia coli* O157).

However, the introduction of the Clean Livestock Policy [7] has prevented soiled and contaminated animals entering the slaughter hall thereby minimizing the consequential likelihood that the organisms present on the hide/fleece would contaminate plant/equipment/surfaces and the hands of personnel during dressing of the animal. These measures have been further augmented with the introduction of the Meat (Hazard and Critical Control Points) Regulations 2002 in licensed premises [8].

CARCASS AND ORGANS   A carcass of meat consists of the skeleton with its muscles and membranes and also the kidneys and fat surrounding them. The offal includes the head, tongue, feet, skin and all internal organs except the kidneys.

The carcass is separated internally by the diaphragm, a strong tendinous musculomembranous partition between the abdominal and thoracic cavities that extends across the body from above the end of the sternum (breast bone) to immediately below the kidneys. Known to butchers as the 'skirt', the diaphragm is pierced by the oesophagus (or gullet) and large blood vessels. The fore part of the carcass (the thorax, thoracic cavity or chest) is lined by a thin, transparent serous membrane called the parietal pleura. A similar membrane lines the hind part (the abdomen, abdominal cavity or belly) and is known as the parietal peritoneum. In healthy animals, these membranes are smooth and glistening, and free from inflammation or adhesions of any kind.

The comparative anatomy of animals presented for slaughter is given in Table 33.1.

THORAX   The thorax or chest contains the lungs, gullet (oesophagus), windpipe (trachea), heart, pericardium and visceral pleura.

Healthy **lungs** in properly bled animals are light red in colour, and the surface is smooth and glistening. On palpation, their substance should be spongy and homogeneous, and free from any solid parts, lesions, inflammation, tubercles or cysts. During life, the lungs of a healthy animal expand to occupy practically the whole of the chest cavity. After death, healthy lungs collapse. From the pharynx, the **oesophagus (gullet)** passes between the large lobes of the lungs and continues through

Table 33.1 Comparative anatomy

Teeth
Bovine — No incisors on upper jaw
Horse — 6 incisors on each jaw
Sheep — 4 pairs of incisors on lower jaw
Pig — 3 pairs of incisors and 1 pair of canines or tusks on lower jaw

Tongue
Bovine — Thick at base; pointed; firm in texture; upper surface rough and bristly; several circumvallate papillae on each side; epiglottis thick and broad, sometimes partly black
Horse — Flattened and broad at its free extremity – like a palette knife; has only two well-marked circumvallate papillae; the epiglottis is pointed; texture is soft; not firm like that of bovines
Sheep — Thick at its base, broad at its tip; has a deep middle furrow and a thick, strong sheath
Pig — Smooth with no dorsal ridge; one well defined circumvallate papilla on each side, tip is rounded

Ribs
Bovine — 13 pairs, flat, broad
Horse — 18 pairs, round rather than flat in section
Sheep — 13 pairs
Pig — 14 pairs

Lungs
Bovine — Left: 3 lobes
Right: 4 and sometimes 5 lobes
Horse — Left: 2 lobes
Right: 3 lobes
Sheep — Similar to bovines; texture very firm
Pig — Left: 2 or 3 lobes
Right: 3 or 4 lobes; tissue is very soft and breaks down under pressure

Liver
Bovine — 1 large and 1 small lobe (or thumb piece); gall-bladder is attached; weight up to 4.5 kg
Horse — Usually 4 lobes; 2 middle lobes shortened; dark; no gall-bladder
Sheep — 2 large lobes and a small triangular thumb piece; weight up to 1 kg

Liver cont.
Pig — 5 long thin lobes; fine interlobular or mottled appearance on its surface; weight up to 1.8 kg

Kidneys
Bovine — Distinctly lobulatted; right is bean-sheaped; left is three-sided (appears twisted)
Horse — Non-lobulated; right is heart-shaped left is long and narrow
Sheep — Bean-shaped, non-lobulated; shorter, rounder and thicker than those of the pig
Pig — Flat and haricot bean-shaped

Spleen
Bovine — Extended oval in shape; firm and convex when healthy
Horse — Long and scythe-shaped
Sheep — Oyster-shaped, reddish brown, elastic
Pig — Long, narrow bright red strip with one edge sharp and the other thick and round

Pluck
Bovine — Lungs, trachea, heart (heart usually sold separately)
Horse — Lungs, trachea, heart
Sheep — Lungs, trachea, heart, liver, spleen
Pig — Lungs, trachea, oesophagus and larynx (no spleen), liver and heart

Stomach
Bovine and ruminants — 4 compartments
Horse — Simple stomach
Pig — Simple stomach

Heart
Bovine — More pointed than in the horse; has 2 bones in the partition between the auricles; hard, white fat at its base; 3 grooves on its exterior
Horse — Less conical than in boviness; no bone; 2 grooves on exterior
Sheep — Similar to bovine but no bone; fat is white and very-firm
Pig — Round at its apex; fat is softer than in sheep

the diaphragm into the stomach. The **trachea** (**windpipe**) begins at the lower border of the larynx, and on reaching the lungs branches into the left and right bronchi. The **heart** lies somewhere to the front and rather to the left side of the middle line behind the sternum; it has a lobe of the lungs on each side of it. It is somewhat pear-shaped, its apex being the pointed end. Normally it is brownish-red in colour and its covering membrane – the epicardium – is smooth and glistening. In properly bled animals, the ventricles contain only a small quantity of coagulated blood. The **pericardium** is the sac or bag (serous membrane) in which the heart lies.

In healthy animals, the **visceral pleura** is a thin, transparent, glistening membrane that covers all the organs in the chest cavity.

ABDOMEN   The abdomen contains the stomach, liver, pancreas, spleen, omentum, mesentery, intestines, kidneys and, in female animals, the ovaries and uterus.

In ruminants, the **stomach** consists of four compartments: the rumen or paunch, the reticulum or honeycomb, the omasum or manifold, and the abomasum, reed or rennet – the true stomach.

The **liver** is situated towards the right side and touches the diaphragm. The gall bladder is attached to the liver and is close to the portal space. (The horse has no gall bladder.) After removal from the animal, the liver becomes reddish-brown in colour. The substance of the organ should be moderately firm; the blood vessels should not be congested with blood; the bile ducts should not be hard or prominent; and it should be free from abscesses, nodules, cysts or discoloured areas. The **pancreas** (gut bread) is an irregularly shaped glandular organ attached to the liver. The **spleen** (melt) is a flat, reddish-brown organ that lies alongside the stomach, to which it is attached. It has high blood content. Its external membrane should be free from inflammation. It should maintain its shape after removal from the animal. Failure to do this, or any abnormal softening of the substance, is indicative of disease.

The **omentum** (caul) envelops the abdominal organs. It consists of two layers of the visceral peritoneum that have a varying quantity of fat between them. The surfaces should be smooth and glistening.

The **mesentery** is a fold of the peritoneum containing fat that forms the medium of attachment between the intestines. The surface should be smooth and glistening.

The **intestines** lie posteriorly to the stomach. They are convoluted, and contained by the mesentery, from which they are separated after removal from the animal. Their surfaces should be smooth and glistening.

The **visceral peritoneum** is similar to the visceral pleura, except that the organs covered are those in the abdomen.

Lymph glands or nodes   Lymph glands are nodular in form, for which reason they are often referred to in meat inspection as 'nodes', and vary in size from a pin-head to a hen's egg. They may be round, flattened, oval or kidney-shaped. They are supplied with afferent and efferent vessels. Afferent vessels convey lymph directly from the various tissues to their respective lymph nodes. After traversing a node, the composition of the lymph is altered and it leaves the gland or node by the efferent vessels, which convey it back to certain circulatory ducts or veins. They are adjuncts to the vascular system. While the lymphatic system may protect the body from disease, it may also furnish a route of entry for disease. Each gland functions in a certain region and is named after that region. Lymphatic glands are of great importance in meat inspection. On section, they should be quite smooth, although they vary in colour. Glands may be darkened in colour by normal pigment or particles of carbon; such glands are not diseased. It should be noted that the other glands, such as the salivary and thyroid glands and the pancreas, are secreting glands only. These are irregular in shape, and may be distinguished from lymph glands by the fact that they are lobulated. The approximate positions of the glands to be found in the carcasses of bovines and swine when dressed, and which are important in meat inspection, are indicated in Figs 33.1 and 33.2. It is extremely important that the reader is fully acquainted with the lymphatic system and respective drainage areas in order that appropriate action can be instigated when

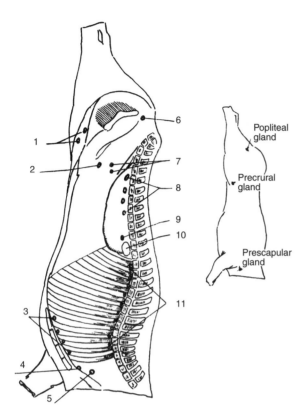

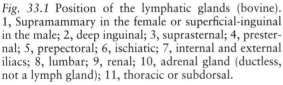

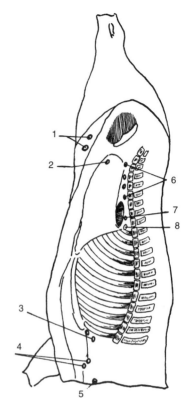

Fig. 33.1 Position of the lymphatic glands (bovine). 1, Supramammary in the female or superficial-inguinal in the male; 2, deep inguinal; 3, suprasternal; 4, presternal; 5, prepectoral; 6, ischiatic; 7, internal and external iliacs; 8, lumbar; 9, renal; 10, adrenal gland (ductless, not a lymph gland); 11, thoracic or subdorsal.

Fig. 33.2 Position of lymphatic glands (swine). 1, Supramammary in the female or superficial-inguinal in the male; 2, iliac; 3, prepectoral; 4, cervical; 5, retropharyngeal; 6, lumbar; 7, renal; 8, adrenal gland.

problems are encountered while examining meat in the food chain.

The **submaxillary glands** are placed superficially in the lower part of the jaw, one on each side between the maxillary (jaw) bone and the submaxillary salivary gland, that is, just in front of the jawbone, where it curves abruptly upwards. These glands are generally removed together with the tongue (drains head, nose and mouth into retropharyngeal).

The **retropharyngeal glands** are at the base of the tongue bones, one on each side of the posterior wall of the pharynx (drains tongue).

The **parotid glands** are situated one on each side of the cheek near the junction of the upper and lower jaws, partly embedded in the parotid salivary glands (drains muscle of head, eye, ear, tongue and cranium).

In cattle, the **gastrosplenic glands** are located in the folds between the second compartment (reticulum) and the fourth compartment or true stomach (abomasum). In swine, they are three or four in number and are situated under the pancreas. The splenic glands lie in a fissure of the spleen close to the stomach, and when the spleen is removed, the glands are very often left on the stomach.

The **mesenteric glands** lie between the folds of the mesentery adjacent to the intestines. They consist of a continuous chain situated along and near the serrated edge of the mesentery (drains small intestines ileum, jejenum).

The **hepatic** or **portal glands** are located on the posterior surface of the liver around the portal fissure. In swine, they are usually removed with the intestines (drains liver, pancreas and duodenum).

The **renal glands** are situated in the fatty tissue on the course of the renal artery close to the second lumbar vertebra (kidney and adrenal body).

There are two **adrenal** or **suprarenal capsules**, one above each kidney embedded in the fatty tissue (kidney fat) close to the first lumber vertebra. They are reddish-brown in colour. On section, each gland is found to be composed of two distinct portions; the outer cortex is darker in colour than the inner medulla. Note that the adrenals are ductless glands that are not connected to the lymphatic system and are therefore not lymph glands. They may, however, be affected by disease and their examination should form part of all routine inspections of carcasses.

The **bronchial glands** are located in the fat on each side of the trachea at the top of the lungs (drains lungs). The **mediastinal glands** (usually three or four in number) are located in the fat between the two large lobes of the lungs (drains lungs, diaphragm and via the diaphragm peritoneum).

The **superficial inguinal glands** are found in the male animal in the fat at the neck of the scrotum (drains genitals). The **supramammary gland** is found in the female animal behind and above the udder (drains udder).

The **lower cervical** and **prepectoral glands** are located in the fatty tissue on the lower side of the entrance to the thorax (drains efferents from middle cervical and prescapular). The **presternal gland** is located against the sternum and first rib (this gland is superficial, and is known to butchers as the inspector's gland) (drains diaphragm, abdominal muscles, intercostal muscles, parietal and visceral pleura and peritoneum). The **suprasternal glands** are located along the course of the internal thoracic vein and artery and are covered by muscular tissue (drains diaphragm, abdominal muscles, parietal and visceral pleura and peritoneum). They lie in the spaces between the ribs close to the junction with the sternal cartilages.

The **thoracic** or **subdorsal glands** are located in the spaces between the ribs, embedded in the intercostal muscles along each side of the dorsal vertebrae (Drains muscle of dorsal region, intercostal muscles, ribs and parietal pleura).

The **prescapular gland** is located above and inward from the shoulder joint, embedded in fat and covered by muscular tissue. (It is incised from the outside.) (Drains head, neck, shoulder and forelimbs.)

The **iliac** (**internal**) **gland** is a large gland found in the loin fat (suet) on each side of the pelvic region (drains muscle and pelvic viscera, genital organs and kidneys, and receives efferents from external iliac precrural, ischiatic and superficial inguinal/supermammary nodes). The **lumbar glands** are embedded in the fat bordering the large blood vessels beneath the lumbar vertebrae (drains lumbar region and peritoneum). The **precrural glands** are large glands, one on each side, in the flanks at a point where the butcher separates the thin and thick flanks. (They are incised from the outside.) (Drains skin and superficial muscles.)

The **popliteal gland** is a deep-seated gland situated in the fat in the silverside. (It is incised from the outside.) The **ischiatic gland** is a deep-seated gland adjacent to the external surface of the ischiatic bone and is frequently seen when the butcher separates the rump from the H-bone (drains lower part of leg and foot).

Diseases and conditions encountered in red meat animals

The diseases or conditions in the United Kingdom that may render a carcass, its organs or part of them unfit for human consumption include the following.

ABSCESSES   Abscesses are common, particularly in pigs. Modern intensive methods of husbandry are linked to an increase in fighting, tail-biting and septic infections. They are also common in bovine livers especially if the animal has been barley fed. The disease *pyaemia* is caused by pus producing organisms entering the bloodstream or lymphatic system and appearing as secondary/multiple abscess infection throughout the carcass or organs with evidence of systemic disturbance. However, multiple abcesses may be found especially in sheep that are due to contaminated needles used for vaccination purposes. In these instances if the abscess is encapsulated, chronic in nature and there is no evidence of systemic disturbance then local trimming is all that is necessary.

ACTINOBACILLOSIS AND ACTINOMYCOSIS These are two closely related diseases that affect cattle, pigs and man. Actinomycosis is usually confined in cattle to the head and is known as 'lumpy jaw' with a thickened bone that has a honeycomb appearance. While in pigs the udder tissue is affected with nodular appearance of the mammary tissue caused by suckling pigs. Actinobacillosis affects lymph nodes and the softer tissue such as the cheek, palate and tongue (wooden tongue) and the condition may spread throughout the carcass and organs and show itself as enlarged yellowish lymph nodes with a distinct nodular appearance embedded in a fibrous connective tissue.

ANTHRAX A disease that is communicable to humans and notifiable under the Anthrax Order 1991. Anthrax may occur in all food animals, sheep and bovines being most susceptible. The disease is caused by ingestion, inhalation or inoculation of the anthrax bacillus or its spores. The latter form only in the presence of oxygen, and may remain alive in air, for example on a pasture, for long periods, even years. The course of the disease is rapid: the first indication may be the finding of a dead animal. In such cases, the oozing of blood or bloodstained fluid from the natural openings should be regarded with suspicion. In a carcass that has been opened, indications of anthrax are a dark discoloration or tarry condition of the blood and, in particular, distinct enlargement of the spleen, the pulp being soft and dark, sometimes described as resembling blackcurrant jelly. The muscular tissues, glands, organs and membranes are dark and haemorrhagic. The carcass putrefies rapidly.

BRUISING In the majority of instances local bruising may be trimmed. However, in severe cases where the lymph nodes are enlarged and haemorrhagic it is frequently necessary to joint the carcass to discover extent of injury.

CASEOUS LYMPHADENITIS This is a chronic bacterial infection that is usually caused by innoculation, shearing or castration and is mostly found in the external lymph nodes namely, prescapular, precrural or popliteal although it can appear rarely in other sites including the offal. The signs of infection are enlarged lymph nodes, sometimes containing pus, while in older chronic conditions the node appears laminated and is surrounded by fibrous connective tissue.

CONTAMINATION Carcasses can become contaminated with animal faecal material, which may contain pathogenic organisms. The contamination usually results from contaminated hides or the leakage of faecal material from animal intestines during their removal at slaughter. The risk of contamination can be reduced by the adoption of clean livestock policies and good dressing practice when evisceration takes place.

The increased significance of contamination with, for example, the organism *E. coli* O157 highlights the need to ensure that meat is produced in as hygienically safe a manner as possible. O157 is a highly virulent strain of *E. coli*, and relatively few organisms are needed to cause illness in humans.

DECOMPOSITION Evidence of decomposition will be usually obvious in that the meat will have a distinctive and offensive odour and feel slimy to the touch with changes in colour from red to grey to green. In animals where there has been a delayed evisceration or where moribund animals have been slaughtered there is frequently noted a green colour on the peritoneal wall following the invasion of putrefactive bacteria from the intestine.

EMACIATION Emaciated animals have a lack of muscle together with the non-existence of fat that is especially noted in the loin and kidney region and the fat that remains will be jelly-like in consistency. The hindquarter will show marked wasting of the muscle and the meat along the loin will be sunken towards the bone.

FOOT AND MOUTH This is an acute contagious febrile disease affecting cloven-footed animals.

Vesicles appear on the feet and mucous membranes of the mouth. As the vesicles appear there is an increase in saliva flow, and it becomes thick and bubbly. Lameness, particularly on hard ground, is noticeable. A total of 4 077 000 animals were slaughtered for disease purposes while a further 2 573 000 were slaughtered for welfare purposes during the recent outbreak of foot-and-mouth disease and the total costs to farmers/tourism has been estimated at 5.8–6.3 billion pounds [10]. A significant causal factor in the recent outbreak has been associated with the major movements of animals throughout the country thereby allowing such a virulent organism to spread rapidly among the animal population.

JAUNDICE OR (ICTERUS)   Jaundice is a symptom of the presence of bilirubin (orange-yellow) and biliverdin (green) bile pigments circulating in the blood. The condition is caused by reabsorption of bile pigment into the circulatory system and may result from mechanical obstruction of the flow of bile, severe cirrhosis of the liver or the action of toxic substances on liver cells.

OEDEMA   This denotes the presence of abnormal amounts of fluid in the tissues and is closely associated with emaciation.

PLEURISY   Usually associated with pneumonia, pleurisy is characterized by the production of fibrous adhesions between the parietal pleura and the lung surface. The incidence can be extremely high in pigs. In post-mortem inspection pleurisy can appear in various stages namely chronic (fibrous), acute (fever), septic (pus) and diffuse (widespread).

SWINE ERYSIPELAS   This disease is characterized by dark red discoloration of the skin, in the early stages as spots, and when advanced as a generalized condition over the skin. The heart and liver show signs of deterioration, the spleen and kidneys are enlarged and dark, and the pleura and peritoneum are haemorrhagic. The carcass sets badly and putrefies rapidly. The disease can assume a milder form known as **urticaria**. An eruption on the skin of diamond-shaped haemorrhagic areas is characteristic. In chronic cases, there is arthritis

present and verrucose endocarditis (vegetative growths on mitral heart valve).

SWINE FEVER   This disease most frequently attacks young pigs and spreads rapidly. A red rash appears on the skin, especially at the base of the tail, under the belly, on the inside of the thighs and on the ears. Round ulcer-like projections develop in the intestine, particularly in the vicinity of the ileocaecal valve, a typical lesion being about the size of a one pence piece. The mucous membrane of the intestines is more or less inflamed. The lymphatic glands present a 'strawberry' appearance, and there may be haemorrhages in the kidneys.

SWINE VESICULAR DISEASE   This disease, which affects only pigs, is caused by *Enterovirus*, which may be carried by faeces and swill and on vehicles, boots and clothing. The signs are similar to those of foot-and-mouth disease, that is, anorexia, fever, lameness, and vesicles in the interdigital clefts and on the coronary bands. Smaller vesicles sometimes occur on the snout, hock and knees. When an outbreak occurs, control measures similar to those for foot-and-mouth disease are put into operation.

TUBERCULOSIS   This particular disease was once the scourge of the British beef industry and unfortunately, this disease is making an unwelcome return with increased incidence and spread. Figures for the three months of 2002 show a rise in new cases of about 25% more than the comparable figure in 2000. Following the reporting of these increased levels of tuberculosis the Advisory Committee in Microbiological Safety of Food (ACMSA) have recommended that people should not eat raw beef such as tartare steak. Lesions of tuberculosis are predominantly encountered in the lymph nodes and appear as enlarged, encapsulated abscesses that are grey in colour and appear gritty and cannot be shelled out. It is also characterized by the formation of tubercules in or on various parts of the carcass (principally the serous membranes and lymph glands) and in the organs. In section, the tubercules are purulent, caseous (cheesy) or calcareous. On the serous membranes (pleura and peritoneum), tubercules may be found clustered together and resembling bunches of grapes. If tuberculosis (TB) is found in the bovine then it must be notified to the appropriate authorities in order that an investigation can be undertaken under the provisions of the 1964 TB Order.

TUMOURS   A tumour is a growth of new cells that have become insensitive to normal growth control mechanisms and develop in an unrestrained and disorderly manner. All cases of tumours in bovines must be notified to the appropriate authorities under the provisions of the Bovine Leucosis Order 1977.

The parasitic diseases commonly found in red meat animals in the United Kingdom include the following

ASCARIDES   Ascarides are long cylindrical worms that infect the intestines of sheep, pigs and calves. Heavy infestation causes an unpleasant odour in the flesh. The intestines should be condemned in all cases where ascarides are present.

CESTODES (TAPEWORMS)   Tapeworms that can be observed in the cystic stage in intermediate hosts include the following.

1. *Cysticercus ovis* (the tapeworm is *Taenia ovis*) occurs in sheep and goats. The cysts are parasites of intermuscular connective tissue and are found in the heart, diaphragm and skeletal muscles. Affected organs and carcasses must be rejected as unfit for human consumption in accordance with legislation.
2. *Cysticercus bovis* (the tapeworm is *Taenia saginata*) is also known as measly beef. The cysts are found in the muscles of the cheeks, tongue, cervical region, thorax, heart, diaphragm and skeletal muscles. In cases of localized infection, carcasses may be made safe by prescribed cold storage namely at a temperature not exceeding $-7°C$ for not less than 3 weeks or at a temperature not exceeding $-10°C$ for not less than 2 weeks.
3. *Cysticercus cellulosae* (the tapeworm is *Taenia solium*) is also known as measly pork. This cystode has not been seen in the United Kingdom recently, but imported pork may contain the cysts. The cysts are found in the cheek muscles, tongue, heart, diaphragm and the cervical and skeletal muscles.
4. *Echinococcus* or hydatid cysts (the tapeworm is *Echinococcus granulosus*) are commonly found in the liver and can also be found in the lungs. They consist of a cuticular membrane and an internal germinal layer with a number of small papillae in its inner surface, which are developing brood capsules. These become bladder-like, and a number of tapeworm heads develop inside them.

NEMATODES   Trichinosis occurs in humans and other animals as a result of the ingestion of raw or inadequately cooked pork containing the encysted larvae of the parasite *Trichinella spiralis*. The muscles generally affected are the diaphragm, tongue and larynx. Sections of the muscles should be examined under the microscope. It is now extremely rare in the United Kingdom.

TREMATODES   The only trematodes that are of importance in meat inspection in the United Kingdom are 'flukes'. *Fasciola hepatica* is the common liver fluke. It is flattened and oval in shape, and pale brown in colour. Their habitat is the bile duct of the liver of cattle and sheep. Bile ducts affected are visibly distended and in the case of bovines thickened (pipey livers) and flukes may be easily extruded from them.

WARBLE FLY   These are two-winged insects that affect animals in the summer grazing season. The larvae have a parasitic existence in the animal's body. The larvae migrate to the animal's back, where they pierce a hole in the skin to obtain oxygen, which causes economic loss for the hides.

BOVINE SPONGIFORM ENCEPHALOPATHY   In November 1986, Great Britain became the first country to diagnose Bovine Spongiform Encephalopathy (BSE) in cattle. The disease is similar to scrapie, which has been recognized in sheep in western Europe for more than 250 years. When the consumption of beef from animals affected by BSE was linked to the tragic death of young persons from New Variant Creutzfeldt

Jakob Disease (vCJD) the initial consumer reaction was rejection of the product and stringent measures had to be adopted to protect human health and restore public confidence. Transmissible Spongiform Encephalopathy (TSE) or prion disease are groups of progressive neurogenerative diseases that affect both animals and man and can be transmitted both through inoculation and orally and is associated with abnormal accumulations in the brain of the host encoded glycoprotein (PrP) [9]. Although the source of vCJD is not known for certain it is now accepted that the most likely origin was human exposure to meat or meat products containing the infective organism especially the consumption of products containing mechanically recovered meat (MRM).

Although reported cases of BSE in cattle are continuing to fall, stringent measures are now in place to minimize potential risk to public health and maintain public confidence. At present, all bovine animals are subject at the slaughterhouse to passport and dentition check and if the bovine is over 30 months (42 months in the case of farm assurance animals) then it is removed from the food chain (currently under review by SEAC). In addition, the following Specified Risk Material (SRM) is removed at the point of slaughter, stained with a patent blue dye and then disposed of in accordance with Regulations.

In the case of bovines **of all ages** the intestine from the duodenum to the rectum and the mesentery is treated as SRM while in animals over 6 months the entire head (excluding the tongue if harvested, but including the brain, the eyes, trigeminal ganglia), intestines, thymus, spleen, spinal cord and tonsils. (It is also proposed that tonsils from animals under 6 months should also now be classed as SRM.) In addition, the Beef Bones (amendment) Regulations still retain the prohibition on the use of bovine bones and bone in beef in manufactured and processed foods. There is now increasing concern of a theoretical and potential risk that the BSE organism may be masked in sheep by 'scrapie' [10,11]. The current SRM controls for sheep/goats are the removal/staining and disposal of **spleens of all animals irrespective of age**. In the case of sheep/goats with one

permanent incisor erupted (i.e. about 12 months old) the skull (head including brain and eyes, tonsils) are removed/stained and disposed of as SRM and the carcass is required to be split and the spinal cord removed. Animals under 12 months do not yet require splitting (currently under review) but must be marked with an MHS YL stamp to indicate that the age was verified at the slaughterhouse by inspection staff undertaking a dentition check. Due to the potential theoretical risk of BSE in sheep, it is being recommended that the ileum of sheep/goats of all ages will be classed as SRM and that sheep intestines should also be added to the list of SRM.

Regulation 77, of the TSE Regulations 2002 [12] which for enforcement purposes has replaced the Specified Risk Material Regulations 1997 as amended allows for Section 9 of the Food Safety Act to be applied. This enables EHOs to inspect, seize and detain meat suspected of containing SRM. However, it must be noted that EHOs authorized under the Food Safety Act 1990 are not authorized to inspect and seize meat under the TSE Regulations 2002 and require further authorization under the European Communities Act 1972 [13]. Once seized the SRM or meat containing SRM must be detained and stored pending the outcome of the investigation and therefore chiller/freezer space will be required pending destruction.

POULTRY MEAT

Legislation

The Poultry Meat, Farmed Game Bird Meat and Rabbit Meat (Hygiene and Inspection) Regulations 1995 came into operation on 1 April 1995 [14] to implement Directives 71/118/EEC, 91/495/ EEC, 91/494/EEC and 93/121/EEC. The regulations do not apply to premises where fresh meat is cut up or stored or re-wrapped for sale direct to the final consumer. Poultry means domestic fowl, turkeys, guinea fowl, ducks and geese. Farmed game birds means birds other than poultry, including ratites (ostriches and emus), not generally considered to be domestic but which are bred, reared and slaughtered in captivity. Rabbit means domestic rabbit.

The main provisions of the regulations cover:

- the licensing of slaughterhouses, cutting premises, cold stores and re-wrapping centres;
- supervision and control of premises, including the authorization of plant inspection assistants (PIAs) working under the supervision of the OVS;
- the conditions for the marketing of fresh meat;
- slaughter conditions;
- construction, layout and equipment of licensed premises;
- hygienic requirements for staff, premises and equipment;
- preslaughter health inspection;
- post-mortem health inspection;
- health-marking and certification;
- cutting practices, wrapping and packaging, and transport.

Trade terms

Poussins

Young chickens weighing 222–454 g.

Spent hens

Hens that have laid eggs and are usually over 2 years old. They are used for poultry meat product manufacturing purposes.

Broilers

Young chickens or fowl reared intensively for the table and killed at 9–12 weeks with a live weight of 1.36–1.8 kg. This method of poultry production is the main source of table poultry in the United Kingdom. The process is outlined here.

- Birds are received and suspended by the legs ready for killing.
- After electrical stunning, the killing is usually done by cutting the jugular vein.
- The bird then passes through the bleeding tunnel and on to the plucking machine.

- In the wet plucking machine, the birds are immersed for 1–3 minutes in water at a temperature of about 52°C. Rubber strips on a rapidly revolving drum then strike the birds, removing the feathers. Stub and wing feathers are removed in another machine.
- After removal of the feet and head, the birds are eviscerated in a separate evisceration room.
- They are drained and dried, packed, weighed, packaged and placed in a cold store. Chilled birds must be held at a maximum temperature of 4°C, frozen birds at a minimum temperature of −12°C.

Inspection points for the processing operation should be established at: the reception area; after plucking for whole bird inspection; and at the evisceration point for full inspection.

The regulations have now been amended whereby the previous exemption from the requirements of poultry meat legislation for slaughterhouses killing less than 10 000 birds or rabbits per year does not apply now to slaughterhouses not on agricultural land (i.e. off-farm slaughterhouses). The FSA has also acted following concerns expressed by the ACMSF about the risk of food poisoning from contamination of other foods by leakage from intestines or inexperience in evisceration by banning the production of 'New York Dressed' (uneviscerated) poultry in 'off-farm' slaughterhouses. This exemption will now only apply to farmers who sell fresh meat from birds and rabbits reared and slaughtered on their holding. Now all 'off-farm' slaughterhouses will require a licence to operate and come under the supervision of the MHS. In addition, the other exemption from the requirement for fresh poultry meat to be accompanied by transport documentation has also been revoked. Poultry meat transported from licensed premises direct to the 'final consumer' will now have to be accompanied by an invoice or delivery note.

Inspection

Age

Young birds are usually identified by the pliability of the posterior end of the breast bone, smooth legs and smooth, glistening feet. In older birds, the meat is darker, breast muscle is sunken, abundance of fat in abdomen and the legs are rough and the feet are hornier.

Freshness

When fresh, the eyes are bright and prominent and the feet moist and pliable. The flesh gives easily when pressed. Later, the eyes become dark and sunken, the feet hard and stiff, and at a later stage, the carcass becomes greenish at the tail vent, followed successively by the greening of the abdomen, back, ribs and neck. The carcass gives off an offensive odour.

Diseases and conditions of poultry

The **abnormal conditions** most commonly found in the inspection of poultry are infection of the liver, emaciation, decomposition and oedema.

Abscesses

Avian abscesses differ from that of mammals in that the pus is normally dry, odourless and not encapsulated.

Avian leucosis complex

This affects adult poultry and can produce anaemia, emaciation and weakness. On post-mortem inspection, the liver is usually found to be much enlarged and friable. Fowl so affected are unfit for human consumption.

Blackhead

An infectious disease of turkeys, particularly young ones, and occasionally chickens, pheasants, partridges, etc. A post-mortem examination shows degenerate circular yellowish areas up to 2.5 cm in diameter on the liver and enlargement of the caecal tubes, which often contain a solid white core. Peritonitis may also be present.

Blood splashing

Small capillary haemorrhages throughout the muscle caused by stress, inexpert handling and delays between stunning and sticking.

Breast blisters

False bursa forms on sternal ridge and contains serous fluid that frequently becomes septic. Associated with birds with diseased joints and condition is exacerbated by excessive rubbing of the breast.

Egg peritonitis

Inflammation of the abdominal cavity and its contents owing to the presence of egg yolk or even shelled eggs floating free in the abdominal cavity.

Escherichia coli septicaemia

A disease of young chickens that is almost exclusively confined to the broiler house. It occurs either as a septicaemic disease in young birds or as granulomatous lesions in the intestines of old birds. A post-mortem examination reveals fibrinous pericarditis, and a mass of pus can usually be seen around the heart. The carcass may be emaciated. The strains of organism that cause the disease in poultry are unlikely to affect humans, but if the carcass is emaciated or if there is pus around the heart, the carcass should be condemned.

Infectious synovitis and arthritis

Infectious synovitis is characterized by swelling of the hock joints, footpads, sternal bursa and sometimes the wing joints. The swellings contain grey viscous fluid, often with spots of pus, and the liver and spleen are usually swollen and congested. Septic arthritis also causes an accumulation of pus around the hock joints.

Impaction of the oviduct

The oviduct is the tube that carries the developing egg from the ovary to the cloaca. In some cases, the oviduct is distended and completely impacted with large masses of cheesy egg material laid down in concentric rings.

Oregon disease

Degenerative myopathy or 'green muscle disease' that affects the deep pectoral muscle giving it a sunken appearance, the muscle appears green in colour, dry with a fibrous texture and can be unilateral or bilateral.

Parasitic conditions

None of these is transmissible to humans, and judgement of the carcass will depend on the degree of emaciation.

Salmonellosis

From the point of view of public health, this is probably one of the most important diseases. It is a disease of chicks, and the organisms of the *Salmonella* group that cause the greatest concern to public health officers are *S. enteritidis* and *S. typhimurium*. Salmonellosis is rarely a disease of mature chickens, although many do harbour the organism in the intestinal tract and gall bladder, so infection can be readily transferred.

Tuberculosis

Avian tuberculosis is a chronic infectious disease, but owing to its long incubation period, it is unlikely to be seen in broilers. The causal organism is the avian type of tubercle bacillus (*Mycobacterium avium*). Cattle, pigs and humans are susceptible to it. Emaciation is a characteristic feature of the disease, and the liver is usually enlarged and studded with white or yellow nodules that range in size from a pinpoint to 13 mm in diameter. Similar lesions may be found in the spleen, intestines, bone marrow (usually femur and tibia) and lungs.

Tumours

These are of various kinds and are usually seen as swellings in the flesh, skin, bones or organs. The most frequent sites are the breast, liver, skin, leg and wing bones, and ovaries.

ANIMAL WELFARE

The Welfare of Animals (Slaughter or Killing) Regulations 1995 [15] and the Amendment Regulations 2003 implement Directive 93/119/EEC on the protection of animals at the time of slaughter or killing. The regulations apply to the movement, lairaging, restraint, stunning, slaughter and killing of animals bred and kept for the production of meat, skin, fur and other products, and to the killing of animals for the purpose of disease control. They cover:

- the licensing of slaughtermen;
- the construction of equipment, and the maintenance of slaughterhouses and knacker's yards;
- requirements for animals awaiting slaughter;
- restraint of animals before stunning and slaughter;
- the stunning and killing of animals;
- bleeding of animals;
- killing of pigs and birds by exposure to gas mixtures;
- provisions for slaughter by a religious method.

WILD GAME

The Wild Game Meat (Hygiene and Inspection) Regulations 1995 [16] implement the provisions of the EU Wild Game Directive 92/45/EEC in respect of wild game processing facilities requiring to use the EU health mark. Wild game meat means meat derived from wild land mammals that are hunted and wild birds. The definition of wild game does not include wild land mammals, rabbits and game birds that are bred, reared and slaughtered in captivity. These are covered by the Fresh Meat (Hygiene and Inspection) Regulations 1995 and the Poultry Meat, Farmed Game Bird Meat and Rabbit Meat (Hygiene and Inspection) Regulations 1995.

The hygiene requirements for wild game meat premises follow principles that do not differ fundamentally from the operation of other meat premises. It is now proposed that all wild game processing plants including those currently supplying only the domestic market will require veterinary supervision and enforcement responsibility for those premises will transfer to the MHS. National rules will cover hunters supplying small quantities of game to the final consumer or retailers directly supplying the final consumer. Despite many years of experience with wild game meat production, no significant public health problem associated with the consumption of wild game meat has been established. The major problem that is associated with wild game is that decomposition following poor practices/processes at point of slaughter. Following slaughter animals are graloched (eviscerated) and if this is not undertaken in a hygienic way then there is serious risk of contamination from intestinal/stomach contents. Should the animal not be adequately protected following slaughter then there is also the risk of the meat becoming 'fly blown' with consequential maggot infestation.

MINCED MEAT AND MEAT PREPARATIONS

The Minced Meat and Meat Preparations (Hygiene) Regulations 1995 [17] apply to premises engaged in the production of these products. Production includes manufacturing, preparing, processing, packaging, wrapping and re-wrapping.

The regulations lay down the structural, hygiene and supervision standards; for the purposes of enforcement food authorities or the appropriate minister are the competent authorities. Food authorities approve, supervise and enforce all aspects of the regulations except in the following cases, when the appropriate minister is responsible (the MHS undertakes the work):

- combined slaughtering and minced meat/meat preparation premises;
- combined cutting premises with minced meat/meat preparation premises;
- combined licensed cold stores and minced meat/meat preparation premises.

Minced meat means meat that has been minced into fragments or passed through a spiral screw mincer. It includes meat with less that 1% salt added. Types of food that fall within the scope of the definition of meat preparation include raw sausages, burgers and flash fried products that are not fully cooked. Retail 'barbecue packs', which may include sausages, burgers, steaks and any pastry product that contains meat, such as uncooked sausage rolls or 'ready to cook' meals in which the meat has not been heat-treated or cooked prior to sale, should be regarded as meat preparations. Minced meat that contains any seasoning, more than 1% salt or other foodstuffs, for example, haggis, is considered to be meat preparations.

The definition of MRM has also been revised under the provisions of the Meat Products Regulations 2003 [18] as 'the product resulting from the mechanical separation of meat left on bones, or on poultry carcasses, so that the cellular structure of the meat is broken'. The description MRM applies to the product obtained under high pressure where little or fibrillar structure can be seen under a light microscope. It is not intended to apply to products obtained by low pressure or mechanical deboners, which leave the cellular structure substantially intact. These products may still be considered meat and can be counted towards the Quantitative Ingredient Declaration (QUID) declaration. MRM is used in the manufacture of meat products and must therefore be produced under hygienic conditions and strict temperature controls, which are both subject to producer and consumer abuse.

MEAT PRODUCTS

The Meat Products (Hygiene) Regulations 1994 (as amended) [19] implement Directive 77/99/EEC and specify structural, hygiene and supervision standards for the production of meat products. They are supported by *Code of Practice No. 17*. The MHS enforces the regulations in approved meat product premises that are combined with or are adjacent to a licensed slaughterhouse or a licensed cold store, and local authorities enforce the regulations in all other meat products premises including those combined with licensed cutting plants. For a detailed description of the local authority approval process, see [20].

Meat products are those that are prepared from or with meats (red, poultry, farmed or wild game) that have undergone treatments such that the cut surface does not have the characteristic appearance of fresh meat. Such treatments include heating, smoking, salting, marinating, curing, drying or a combination of these processes. Such products include cooked meat, meat pies, meat-based and meat-filled sandwiches, meat-based prepared meals, cured products such as bacon fermented sausages such as salami.

The Meat Products Regulations 2003 [18] amend the definition of 'meat' for **labelling** purposes and will have important ramifications for producers of meat products/meat preparations. Previously, the Meat Products and Spreadable Fish Products Regulations 1984 laid down compositional and labelling **national** requirements for meat products/meat preparations sold in the United Kingdom. The new regulations now amend the generic term of meat for the purpose of labelling as follows:

- Restricts the generic term 'meat' (as well as species names such as beef, pork, chicken etc.) to skeletal muscle with naturally included or adherent fat and connective tissue.
- Introduces maximum numerical limits for associated fat and connective tissue depending on species of meat. Any fat/connective tissue in excess of these limits cannot be counted towards the meat content and must be declared separately in the ingredients list.
- Excludes MRM which must already be declared separately in the ingredients list. MRM cannot be counted towards the meat content.
- Requires that other parts such as liver, kidney, heart etc. must be labelled as such. The generic term 'offal' cannot be used and neither can these parts be counted towards the QUID declaration for meat ingredient.

BEEF LABELLING (ENFORCEMENT) REGULATIONS 2000

These regulations [20] specify that any operator selling fresh beef/veal (including mince) anywhere in the food chain must comply with both the 'compulsory' and 'approved' labelling requirements. The key element in this system is that it enables beef/veal sold to be traced back to where it originated from. The following seven indicators are 'compulsory' and must be shown.

- Reference number or code – allows traceability.
- Slaughtered in (name of Member State or third country).
- Licence number of slaughterhouse.
- Cutting/cut in (name of Member State or third country).
- Licence number of cutting plant.
- Name of Member State or third country in which the animal or groups of animals were born.
- Name of Member State where animal or groups of animals were reared.

'Approval' is required by recognized independent verifiers if any claims are made relating to the product such as, breed/age/gender/method of production/method of slaughter (Halal/Kosher), and length of maturation.

ANIMAL BY-PRODUCTS

The Animal By-Products (Identification) (Amendment) Regulations 2002 make provision for the sterilization and staining of animal by-products and the control of movement of such products. An animal by-product is any carcass or part of a carcass that is not intended for direct human consumption that has come from an animal that has died or been killed in a knacker's yard or place other than a slaughterhouse, or has been slaughtered and has shown signs of disease communicable to humans, or has not been presented for post-mortem in accordance with the regulations. Following the recent investigations in Rotherham, Amber Valley and Norwich whereby unfit meat was diverted into the human food chain the new regulations now also require the staining and sterilization of high-risk poultry by-products, high-risk red meat and poultry meat in licensed cutting plants and in licensed cold stores. Animal by-products include animals that have been spoiled so that they present a risk to human or animal health, or that contain residues that pose a similar risk. They also include carcasses or parts of carcasses derived from bovine animals that are slaughtered for human consumption and were more than 30 months of age at the time of slaughter. The regulations require the immediate staining or sterilization of animal by-products (currently under review), prohibit their freezing, or storage in the same room as products intended for human consumption, and prohibit the movement of unstained or unsterilized animal by-products from slaughterhouses.

The Animal By-products Order 1999 came into force on 1 April 1999. The Order consolidates the rules on the processing and disposal of animal by-products and the processing of catering waste intended for feeding to pigs or poultry. It will be enforced by the local authority or the Meat Hygiene Service where appropriate.

HYGIENIC PRODUCTION OF MEAT

If not carried out correctly, the processes undertaken in a slaughterhouse can lead to the potential contamination of raw meat by pathogenic bacteria. These bacteria include *Salmonella* spp, *Clostridium* spp, *Campylobacter* spp and *E. coli* spp.

Raw meat can become contaminated by pathogenic bacteria from the hide, fleece or feathers during the evisceration process, from the spillage of gut contents, from poor handling and dressing by the slaughter hall staff, or from contact with unclean equipment. Correct policies and the principle of 'best practice' are all designed to reduce the level of potential contamination. It must be emphasized that there exists no stage in the processing of raw meat at which the prevention/ elimination of bacterial contamination can be guaranteed. However, significant measures are taken by the MHS to reduce the likelihood of bacterial contamination in raw meat at the slaughterhouses.

In August 1997, the MHS introduced a clean livestock policy, which details actions that should be taken on the farm, on the animal transporter, at the market, by the dealer/agent and in the slaughterhouse in order to prevent dirty livestock from entering the slaughterhouse. The dirty coat of an animal has the capacity to contaminate a clean carcass during dressing. Dirt from a dry coat can flake off the coat on to the carcass, and the capacity to contaminate increases when the dirty coat becomes wet. Water provides a ready means of transportation for any pathogenic organisms and also increases their ability to adhere to the surface of a carcass. Research has shown that there is a direct relationship between the visible contamination of the coat and the microbiological status of the finished carcass [21]. At the slaughterhouse, the OVS has the legal power to prohibit the slaughter of any animal if he thinks that it is so dirty that it would inhibit hygienic dressing.

Hazard Analysis and Critical Control Points

The report of the Pennington Group Report into the large-scale outbreak of verotoxin producing *E. coli* O157 in Lanarkshire that culminated in a total of 496 persons in Scotland becoming affected by the disease led to the successful introduction of licensing for butchers shops [22] – see section on Licensing of butchers. Conditions for granting the licence necessitated the implementation of all seven principles of Hazard Analysis and Critical Control Points (HACCP) and appropriate staff training for retail butchers selling both open raw and ready to eat foods. A further recommendation of the report was that HACCP system should be enshrined in the slaughterhouse legislation and this has now been implemented by the Meat (HACCP) Regulations 2002 covering slaughterhouses, cutting plants, cold stores, re-packaging centres and re-wrapping centres.

These regulations amend the Fresh Meat Regulations by implementing and maintaining a permanent procedure in accordance with the principles of HACCP as a prerequisite to the introduction of a 'risk based meat inspection' service. In addition to carrying out regular checks on the general hygiene of conditions of production in those premises they must also in the case of licensed fresh meat slaughterhouses and cutting plants carry out microbiological checks in accordance with set procedures and it is being recommended that poultry plants should also adopt microbiological procedures.

The HACCP food safety system is made up of seven principles:

1. Identify hazards to be prevented/eliminated or reduced.
2. Identify the Critical Control Points (CCPs).
3. Establish critical limits at CCPs.
4. Establish/implement effective monitoring procedures at CCPs.
5. Establish corrective action to be taken if CCPs not under control.
6. Establish procedures to verify if measures are working effectively.
7. Establish documents/records to demonstrate application of above measures.

CCPs are steps in the process which when controlled will prevent, eliminate or reduce a food hazard. This concept identifies all the CCPs that could lead to contamination during processing and recommends the use of 'best practice' to ensure that the risk of contamination is reduced. The FSA have identified the following as suggested CCPs in meat plants.

Cattle/sheep	*Poultry*	Despatch and transport (if under operators control)
Acceptance of animals for slaughter	Live bird supply (including microbiological health status)	
Hide fleece removal	Evisceration	*Offal*
Evisceration (including 'bunging/rodding')	Final carcass wash	Chill
Chill and storage (if applicable)	Chill and storage (if applicable)	Despatch and transport (if under operators control)
Despatch and transport (if under operator control)	Despatch and transport (if under operators control)	
Pigs	*Cutting plants/re-packing centres/re-wrapping centres*	*Cold stores*
Scald (if singeing is not practised/singe (if practiced)	Receipt of meat	Receipt of meat chilled storage
Evisceration (including 'bunging')	Pre-cut inspection (red meat only)	Despatch and ransport (if under operators control)
Chill and storage (if applicable)	Chilled storage	
Despatch and transport (if under operator control)		

THE MEAT (ENHANCED ENFORCEMENT POWERS) REGULATIONS 2000

These regulations [23] give wide-reaching powers to Local Authorities to stop an operation or prohibit a process if food hygiene is seriously compromised and to seize meat/meat products/meat preparations, which have not been produced in accordance with meat hygiene regulations. When it is identified that the meat/meat products/meat preparations have not been handled, stored or transported in accordance with meat regulations then it may be seized and treated for the purposes of Section 9 (Inspection and Seizure of Suspected Foods) of the Food Safety Act as failing to comply with food safety requirements and both the persons selling and supplying such meat may be prosecuted. In addition, the regulations now make it a statutory requirement to notify the names of plant operators, managers or directors, and of any subsequent changes and failure to do so is a breach of the regulations that could culminate in revocation or suspension of the approval.

LICENSING OF BUTCHERS (ALSO SEE P. 510)

The Food Safety (General Food Hygiene) (Butcher Shops) Amendment Regulations 2000 amended the Food Safety (General Food Hygiene) Regulations 1995 and now require that all butcher shops and other retail outlets that handle unwrapped raw meat and ready to eat foods from the same premises require annual licensing by Local Authority EHOs. The Pennington Group Report identified cross-contamination from raw meat to ready to eat foods through poor hygiene and handling practices in a butcher shop as the causal factor in the Lanarkshire outbreak of *E. coli* O157 in 1996.

In addition to the requirement to comply with general hygiene and temperature control legislative requirements there is the additional need for specific food hygiene training for both staff and supervisors together with a fully documented HACCP food safety management to be in place before a licence can be issued. This detailed examination of the premises/practices/processes in place will necessitate the EHO putting into place all their meat safety training and full understanding of HACCP prerequisites and requirements – see previous section.

UNFIT MEAT

The removal of meat from the food chain is a very important role for the EHO because of the fact that in the case of diseased animals or if meat is found in the later stages of decomposition then it could be and is likely to be accompanied by food poisoning pathogens. During the inspection of any butcher shop premises or catering premises, there is a need to identify all sources of raw meat/poultry meat/wild game meat/minced meat and meat preparations and meat products to ensure that it is sourced from premises where the meat has been produced/stored/handled and fully complies with legislative requirements. In **all** cases

when animals are legally slaughtered documentation should be available which would indicate where the animal was slaughtered and the destination of the carcasse with full contact details for both including the relevant licence numbers. Recent events have highlighted the necessity that all enforcement officers need to work together to identify sources of illegal/unfit meat entering the food chain and then to adopt all necessary procedures to ensure that those responsible are prosecuted and that the product is disposed of in accordance with legislative requirements. However, when examining meat it must be noted that the absence of clear health marks on its own does not constitute evidence of illegally produced meat as they may have been removed during trimming or lost when reduced into primal joints etc. but it may require further checks if there are grounds for suspicion such as detailed examination of the meat and any relevant documentation.

Suspicion would be aroused if the meat was obviously stained, badly bled, or there was evidence of poor slaughter, or contamination of the product from sources such as faeces, bile, grease etc. or evidence of decomposition on the peritoneum that may suggest that the animal was moribund at the time of slaughter. However, should there be any evidence that SRM is present then it is essential that the FSA/MHS are informed in order that a full investigation into the source can be undertaken. In the case of poultry there should be no missing parts/haemorrhages/bruising/broken appendages/abnormal swellings or lacerations to the skin and there must be no contamination from faeces/blood or bile. It is essential that all the enforcement authorities work together and liaise effectively in order to ensure that public health is protected.

REFERENCES

1. *Environmental Health News* (31 January 2003) **18**(4).
2. *Environmental Health News* (25 October 2002) **17**(41).
3. *Environmental Health News* (21 February 2003) **18**(7).

4. Food Standards Act (1999) HMSO, London.

5. Fresh Meat (Hygiene and Inspection) Regulations 1995 As amended HMSO, London.

6. Food Standards Agency (2002) *Cutting of Meat for Direct Sale by Farmers*, Guidance for Local Authorities.

7. *Clean Livestock Policy* (1997) MHS, York.

8. *Meat (Hazard Analysis and Critical Control Points) Regulations* (2003) HMSO, London.

9. Prusiner, S.B. (1991) *Molecular Biology of Prion Diseases*: Science, **252**, 1515–22.

10. Kao, R.R., Gravenor, M.B., Baylis, M., Bostock, C.J., Chihota, C.M., Evans, J.C., Goldman, W., Smith, A.J.A. and McLean, A.R. (2002) The potential size and duration of an epidemic of B.S.E. in British Sheep. *Science*, **295**, 332–5.

11. Ferguson, N.M., Ghani, J.C., Donnelly, C.A., Hagenaars, T.J. Anderson, R.M. (2002) Estimating the human health risk from possible B.S.E. infection of the British sheep flock. *Nature*, **415**, 420–4.

12. *The TSE Regulations* (2002) HMSO, London.

13. *The European Communities Act* (1972) HMSO, London.

14. Poultry meat, farmed game bird meat and rabbit meat (Hygiene and Inspection) Regulations as Amended (1995) HMSO, London.

15. *Welfare of Animals Regulations* (1995) HMSO, London.

16. Wild Game Meat (Hygiene and Inspection) Regulations (1995) HMSO, London.

17. Minced Meat and Meat Preparation (Hygiene) Regulations (1995) HMSO, London.

18. *Meat Product Regulations* (2003) HMSO, London.

19. Meat Products (Hygiene) Regulations as amended (1994) COP No. 17. and Guidance to be read in conjunction with COP, HMSO, London.

20. Bassett, W.H. (1998) *Environmental Health Procedures*, 5th edn, Spon Press, London.

21. Beef Labelling (Enforcement) Regulations (2000) HMSO, London.

22. Hudson, W.R. Mead, G.C. and Hinton, M.H. (1996) The relevance of abattoir hygiene assessments to microbrial contamination of British beef carcasses. *Veterinary Record*, **139**(24), 587.

23. Food Safety (General Food Hygiene) (Butchers Shops) Amendment Regulations (2000) HMSO, London.

24. Meat (Enhanced Enforcement Powers) Regulations 2003.

FURTHER READING

Gracey, J.F. (1998) *Meat Plant Operations*, Chalcombe Publications, Lincoln.

Gracey, J.F. and Collins, D.S. (1992) *Meat Hygiene*, 9th edn, Baillière Tindall, London.

Meat Hygiene Service (1995) *Meat Hygiene Service Operations Manual, Volume 1 – Procedures*, MHS, York.

Meat Hygiene Service (1995) *Meat Hygiene Service Operations Manual, Volume 2 – Legislation and Codes of Practice*, MHS, York.

MAFF (June 1998) *Bovine Spongiform Encephalopathy in Great Britain – A Progress Report*, MAFF, London.

Wilson, W.G. (1997) *Wilson's Practical Meat Inspection*, 6th edn, Blackwell Scientific Publications, Oxford.

Part Eight
Environmental Protection

34 Introduction to environmental protection

Chris Megainey

INTRODUCTION

Environmental protection is one of the three pillars of sustainable development, along with social and economic considerations. It contributes to both the conservation of natural resources and to improvements in public health. This chapter aims to explain how the work of the environmental health professional fits into to the wider environmental protection agenda.

For many years, local authorities and other agencies tended to focus on controlling stationary sources of pollution, such as power stations, chemical works and domestic fireplaces. They sought to mitigate the effects of these sources through the application of emission standards of one sort or another. Legislation – from the Alkali Act 1863, through the Clean Air Acts of the 1950s and 1960s, right up to the Environmental Protection Act of 1990 – required the regulator to reduce emissions, often in isolation from consideration of the effects of the source on the environment.

Nevertheless, this approach was often highly effective. The Clean Air Acts helped to save many lives by bringing to an end the perilous smogs of the post-war years. Industrial pollution regulation has played a major part in raising the profile of environmental issues in the boardroom, as well as bringing about dramatic reductions in emissions.

But a number of factors have combined over recent years to move environmental control philosophy in a rather different direction. The tremendous growth in the power of computers has assisted in the development and widening of understanding of the way in which emissions from a particular site impact on the environment, both locally and over greater distances.

Partly as a result of this phenomenon, the evidence base on which decisions can be made has also improved dramatically. Furthermore, regulation of pollution from stationary industrial sources, along with the decline of heavy industry, has caused these polluters to decline in significance. On the other hand, the massive growth in the use of road transport since the Second World War has led to the introduction of many millions of new mobile sources of pollution.

The results of these changes are various. The greater understanding of environmental impacts has led to a shift from source-based pollution control to a more strategic, effects-based approach. Industry has largely, although by no means wholly, been supplanted by the car in the perception of the public as the environmental issue that needs to be resolved most urgently. There is also greater awareness of the environmental impacts of activities previously considered relatively benign, such as agriculture.

Environmental regulation has therefore needed to develop accordingly and become more sophisticated. The National Air Quality Strategy [1–3] is a good example of this new approach. The aim of the strategy is not to put in place prescriptive rules, but to enable the most appropriate tools to be employed, often at a local level, to bring air quality throughout the United Kingdom up to health-based standards by 2005. The strategy does not deal with different types of pollution source in isolation, but requires that they should all be addressed in a proportionate way.

European Directives on water and ambient noise have adopted a similarly holistic approach.

However, source-based industrial pollution control is far from being irrelevant. The Integrated Pollution Prevention and Control Directive (96/61/EC L257 10/10/96), which requires reductions in emissions to all media from the major polluting industries, is one of the most important and demanding pieces of European environmental legislation to be adopted. The constantly tightening standards for vehicle emissions and fuel quality continue to play a key role in improving air pollution.

Working towards effects-based objectives as well as emission limits for individual sources of pollution presents new challenges for policy-makers, regulators and business. Decisions need to be taken to ensure that different sectors are treated equitably and to ensure that the most efficient measures are taken rather than the easiest ones. Nevertheless, pollution control issues are now so complex that without such a co-ordinated approach success will be hard to achieve.

THE DRIVERS FOR ENVIRONMENTAL PROTECTION

It has long been recognized that public concern plays a major role in the development of environmental legislation. It was, for example, public complaints in the 1860s about emissions from alkali works that resulted in a parliamentary inquiry, which in turn led to the first major piece of industrial pollution control legislation, the Alkali etc. Works Regulation Act 1863. The great London smog of 1952, which was responsible for 4000 deaths, aroused intense public outrage. The outcome was a detailed study of air pollution by the Beaver Committee, as a consequence of which the Clean Air Act of 1956 was enacted.

As scientific knowledge about the effects of our actions on public health, ecosystems and global climate has increased, media interest and public concern has inevitably intensified. Incidents such as the escape of toxic gases at Bhopal in 1984 and the nuclear reactor failure at Chernobyl in 1986 would have been always regarded as human tragedies. However, because of the increased engagement with environmental issues around that time, they were also seen in the context of their longer term effects, not only on the health of people living nearby, but also on the viability of ecosystems. As evidence emerges that issues such as global warming present real and tangible threats to people's way of life, public anxiety escalates.

As a result of this increase in concern, campaign groups such as Friends of the Earth and Greenpeace have gained a higher profile and their activities have served to generate yet more interest and political pressure. The disposal of the Brent Spar oil storage buoy in 1995 would have been unlikely to have been such a major media issue had it occurred 20 years earlier. This heightened public concern about environmental issues has, of course, given rise to a change in the attitude of governments.

Increasingly, politicians see a need for a strategic, proactive approach to environmental protection that aims to avert disasters rather than deal with them once they have occurred. The willingness of world leaders to participate in the United Nations Conference on Environment and Development (known as the Earth Summit) in Rio de Janeiro in 1992 and, subsequently, the World Summit on Sustainable Development in 2002 has demonstrated the higher profile that environmental matters now command.

The forces that underpin environmental protection are, however, more complex than this. In particular, business now realizes that its impact on the environment can affect it in a number of ways. This perception is clearly in part a consequence of the increased interest in the environment shown by governments and international organizations and the consequent tightening of environmental legislation. The introduction of fiscal incentives to encourage more environmentally responsible behaviour, such as the landfill tax and producer responsibility regimes, is also starting to take effect.

However, at least as important is the concern that many businesses have about the effect that their environmental performance has on their image and, ultimately, their share price and profitability. Companies that have been associated with environmental damage or poor practice, have increasingly been the subject of campaigns organized by pressure groups and public disapproval. From DIY shops to car manufacturers and oil companies, many businesses are now promoting their environmental integrity alongside the quality of their products.

Increased access to environmental information by the public has also had an effect. For example, in the United States, the Toxic Release Inventory (TRI) has been a major stimulus for change. All manufacturing processes are required to report their usage and releases of hazardous substances to the US Environmental Protection Agency. The TRI lists the companies in each state that release the greatest mass of pollution. Clearly, no company wishes to appear on these lists.

Another driver of environmental performance is quality management. A large number of companies are now accredited to the internationally recognized quality standard ISO 9001. Many companies will not deal with suppliers who do not have this accreditation, which should only be held by organizations that comply with relevant legislation, including environmental law. ISO 14001 operates in a similar way to ISO 9001, but is concerned specifically with environmental management. Although relatively few companies are currently accredited to ISO 14001, it has the potential to be a significant force in the move towards better environmental practice.

There is also the question of continued availability of resources. Some businesses, particularly those that are intensive users of natural resources, now realize that their longer term futures depend on the continued existence of those resources and that their commercial future is bleak if they fail to adapt their behaviour. Unfortunately, if understandably, this awareness is rather more apparent in rich countries than it is in the developing world.

INTERNATIONAL ACTION

It is increasingly accepted that many types of pollution can only be tackled at a regional or international level. Emissions of acid gases from power stations and other industrial sources in the United Kingdom may cause ecological damage hundreds or even thousands of miles away. Ozone is one of the key air pollutants of concern because of its adverse health effects and role in causing summer smogs. Yet, ozone in southern England is likely to be formed by a complex chemical reaction from nitrogen oxides and volatile organic compounds emitted largely in continental Europe.

Other important environmental issues require a global approach. An obvious example is climate change, which is caused by carbon dioxide and other greenhouse gases emitted from every country in the world. Stratospheric ozone depletion is another case in point.

Rio and beyond

The most high profile global events to focus on environmental protection issues to date have been the Earth Summit in Rio de Janeiro in 1992 and the World Summit on Sustainable Development that took place ten years later in Johannesburg. The Earth Summit was a direct result of the groundbreaking report *Our Common Future* [4], produced by the World Commission on Environment and Development in 1987, which identified, for the first time, the concept of sustainable development. The Summit was attended by heads of state, senior politicians and officials from more than 160 countries. The outcomes of the summit included the following initiatives:

- Production of a Framework Convention on Climate Change that aimed to stabilize emissions of greenhouse gases, such as carbon dioxide, at 1990 levels by 2000.

- Drawing up of a Convention on Biological Diversity that aimed to protect and preserve endangered species both on land and in the oceans.
- Production of Agenda 21, a detailed document setting out guidance for governments on the establishment of environmental policies that embrace the concept of sustainable development.
- A statement of principles for the management, conservation and sustainable development of all of the world's forests.

However, many commentators felt that progress towards sustainable development in the years following the Earth Summit was disappointing. The Johannesburg Summit therefore aimed to achieve a stronger focus on implementation rather than philosophical debate. The Summit reaffirmed sustainable development as a central element of the international agenda and looked to give new impetus to global action to fight poverty and protect the environment. Key commitments and targets included: [5]

- halving, by the year 2015, the proportion of people without access to safe drinking water;
- diversifying energy supply and substantially increasing the global share of renewable energy sources in order to increase its contribution to total energy supply;
- aiming, by 2020, to use and produce chemicals in ways that do not lead to significant adverse effects on human health and the environment;
- improving access by developing countries to alternatives to ozone-depleting substances by 2010, and assisting them to comply with the phase out schedule agreed under the Montreal Protocol.

In order to increase the likelihood of effective delivery of these commitments and targets, the Summit also resulted in a number of major initiatives to support them:

- The United States announced $970 million in investments over the next three years on water and sanitation projects.

- Nine major G7 electricity companies (the E7) signed a range of agreements with the United Nations to facilitate technical co-operation for sustainable energy projects in developing countries.
- Canada and Russia announced that they intended to ratify the Kyoto Protocol.
- The United Kingdom announced that it was doubling its assistance to Africa to £1 billion a year and raising its overall assistance to all countries by 50%.

However, as UN Secretary-General Kofi Annan said at the conclusion of the Summit, the true test of what the Johannesburg Summit achieves will be the actions taken in its wake. 'This is not the end', said Mr Annan 'It's the beginning.'

The role of the United Nations

The United Nations plays a leading role in promoting and co-ordinating action to protect and improve the environment at a global level. Its Department for Economic and Social Affairs was responsible for the Johannesburg Summit, while the United Nations Environment Programme, based in Nairobi, provides the Secretariat for a range of environmental protection conventions such as:

- The Basel Convention, dealing with the transport of and trade in hazardous wastes
- The Convention on the International Trade in Endangered Species
- The Montreal Protocol on ozone depleting substances
- The Kyoto Protocol on Climate Change

The United Nations Economic Commission for Europe (UNECE) has a wider membership than the EU and includes the United States, Canada, Russia and a number of Eastern European countries. It has overseen a number of protocols with the purpose of reducing transboundary pollution. These include the following:

- the Helsinki Protocol on the reduction of sulphur emissions. This required emissions of sulphur dioxide to be cut by 30%;

- the Sofia Protocol on the control of emissions of nitrogen oxides;
- the Aarhus Convention on access to environmental information;
- the ESPOO Convention on environmental impact assessment.

The European approach

The European Economic Community was set up by the Treaty of Rome in 1957 with the fundamental objective of setting up a 'common market' within which goods could be traded freely. The United Kingdom joined in 1973. Since 31 December 1992, there has been a 'single market' for goods, services, human resources and capital. In order for the single market to operate, there was a clear need for the harmonization of legislation to ensure that fair competition exists between member states and that trade barriers are not caused by differing standards, including environmental controls, for products and services. However, free trade is not the only motive for European action. There are many areas of environmental protection where action can be taken much more efficiently at the European level than by member states acting independently.

Community activity in the environmental field began in earnest with the European Commission's first environmental action programme in 1973. This described measures to be taken to reduce pollution and nuisances and to improve the environment. Further action programmes followed in 1977, 1983 and 1987. During this period, the Community adopted around 200 pieces of environmental legislation, mostly concerned with limiting pollution through the introduction of minimum standards, generally for water and air pollution and waste management.

In 1993, the fifth environmental action programme was approved, entitled *Towards Sustainability*. The issues covered ranged wider than previously and included climate change, acidification as well as air and water pollution and waste management. The overarching theme was sustainable development. The programme, which ran to 2000, focused on the need for all stakeholders in the environment, including governments, industry and the public, to take responsibility for environmental protection. It also stressed the need for environmental issues to be integrated into all areas of EU work.

A review of the fifth action programme was completed in 1998. Five areas were highlighted as being in need of increased effort:

- improving the basis for environmental policy;
- developing sustainable production and consumption patterns;
- encouraging practical ways of improving shared responsibility and partnership;
- promotion of local and regional initiatives;
- environmental themes.

The Treaty of Amsterdam, which was agreed in 1997, made clear that sustainable development was one of the Community's aims and that a high degree of environmental protection was one of its priorities. An overarching theme of the Treaty was an increased emphasis on the rights of individuals. The sixth environmental action programme [6], adopted in 2002, highlights four key areas – climate change, nature and biodiversity, environment and health and the management of natural resources and waste. It seeks to improve the effectiveness of Community action through improving the application of legislation on the ground, working better with the market and citizens and ensuring that other Community policies take greater account of environmental considerations. Through integrated product policy, the Commission and Member States aim to develop a more sustainable market by way of a focus on the whole lifecycle of goods.

SOME ENVIRONMENTAL PROTECTION CONCEPTS

Sustainable development

Sustainable development is much more than an environmental protection ideal. Rather, the need to protect the environment is just one strand of what will be needed to achieve sustainability.

There are various definitions of the term 'sustainable development' and rather more interpretations. It was first introduced in 1987 by the World Commission on Environment and Development in its report, '*Our Common Future*', often referred to as the Brundtland Report after its chairwoman, Groharlem Brundtland. The definition of sustainable development in this report was: 'meeting the needs of the current generation without compromising the ability of future generations to meet their own needs'.

The achievement of sustainable development will be dependant on meeting four broad objectives:

- Social progress which recognizes the needs of everyone
- Effective protection of the environment
- Prudent use of natural resources
- Maintenance of high and stable levels of employment.

Precautionary principle

Under the precautionary principle, a lack of firm scientific evidence should not prevent action being taken to deal with a risk of harm to the consumer, the public or the environment. However, two conditions should be satisfied before the principle can be applied: first, that a potentially negative effect has been identified; and second, that it is impossible to fully assess the risk because of lack of conclusive or precise data.

In determining whether to apply the precautionary principle, the policy maker should consider the level of risk that it is acceptable to the exposed population and the possible consequences of inaction. The European Commission has set out its views on the use of the Principle in a Communication [7].

Best available techniques and best available techniques not entailing excessive cost

The concept of best available techniques (BAT) attempts to define a balance between costs and benefits for industrial processes which fall under the Integrated Pollution Prevention and Control (IPPC) Directive. The Directive provides the following definition of BAT.

- 'Best' shall mean most effective in achieving a high general level of protection for the environment as a whole.
- 'Available' techniques shall mean those developed on a scale that allows implementation in the relevant industrial sector, under economically viable conditions, taking into consideration the costs and advantages, whether or not the techniques are used or produced inside the member state in question, as long as they are reasonably accessible to the operator.
- 'Techniques' shall include both the technology used and the way in which the installation is designed, built, maintained, operated and decommissioned.

The BAT concept is elaborated further for particular industrial processes in the European Commission's BAT Reference (BREF) documents, often supplemented by domestic guidance.

As the IPPC Directive comes into force for different industrial sectors, BAT is gradually replacing the earlier best available techniques not entailing excessive cost (BATNEEC) principle that was introduced under the 1984 Air Framework Directive. Guidance issued by the Secretary of State for the Environment, Transport and the Regions provides the following interpretation of BATNEEC [8].

- 'Best' should mean the most effective in preventing, minimizing and rendering harmless emissions. There may be more than one set of techniques that can be termed 'best'.
- 'Available' should mean procurable by any operator of the class of process in question. It need not imply that the technique is in general use, but it does require general accessibility.
- 'Techniques' includes both the process and how the process is operated, including the design of the process, staff numbers, supervision, training and working methods.
- 'Not entailing excessive cost' suggests that, although the presumption is that the best available techniques should be used, economic considerations may modify this judgement if the cost of applying them would be excessive in relation to the degree of environmental protection achieved.

In practice, what is required under BAT and BATNEEC may not differ greatly, although the former is clearly a more rigorous standard.

Best practicable environmental option

Best practicable environmental option (BPEO) is a complex concept that can be interpreted in a number of ways. The initial impetus behind BPEO came from the Royal Commission on Environmental Pollution, a group of distinguished independent advisers appointed by the UK Government. The Royal Commission's 12th report [9], issued in 1988, proposed the following definition of the term: 'A BPEO is the outcome of a systematic consultative and decision making procedure which emphasises the protection and conservation of the environment across land, air and water'. The BPEO procedure establishes, for a given set of objectives, the option that provides the most benefit or least damage to the environment as a whole, at acceptable cost, in the long term as well as the short term.

In its report, the Royal Commission stated that pollution should not be considered in terms of each individual receiving medium. Rather, there is a complex balance between emissions to air, water and land, and pollution can potentially be transferred between the three media. BPEO is an assessment based on the precautionary principle, promoting the adoption of clean techniques and technologies. The greatest area of difficulty in the evaluation process is the need to compare a number of very different forms of pollution and to reach a judgement about which are the least harmful to the environment as a whole.

The BPEO principle was central to the Integrated Pollution Control regime. However, it did not find favour with other Member States during the negotiation of the Integrated Pollution Prevention and Control Directive and will play a less important role under the PPC system.

The 'polluter pays' principle

The 'polluter pays' principle is central to the use of economic measures in environmental protection. It emerged from the United Nations Conference on the Human Environment in 1972 and was incorporated in the European Community's First Action Programme on the Environment in 1973. The principle recognizes that environmental resources are limited and that if their consumption is to be reflected in the market, the cost of preventing pollution or minimizing environmental damage should be borne by those causing the pollution.

There are three main options for putting the 'polluter pays' principle into practice:

- the setting of environmental standards, with the cost of achieving those standards being wholly borne by the polluter;
- the instigation of a system of pollution charges or taxes on the product or process that results in environmental damage;
- the issuing of tradable pollution permits that allow the release of a certain amount of pollution.

The application of the 'polluter pays' principle generally results in an increase in production costs, which are likely to be passed on to the consumer. Although this may seem unfair, it is the purchaser of the goods who demands their production and is ultimately responsible for the pollution. It therefore seems reasonable that the consumer should be faced with the choice of paying the increased cost or using less environmentally damaging products.

Nevertheless, the allocation of a value to pollution poses a series of problems. The increased level of environmental awareness and concern would seem to represent a preference for cleaner air and water and a safer, more attractive environment. A demand for these concepts can therefore be readily identified, and can be compared with the demand that exists for other goods and services. However, many environmental resources cannot be readily costed. It is very difficult, for example, to put a price on the presence of a certain species of fish in a particular river. Furthermore, as environmental improvements are not generally purchased in the same way that other goods are, there is not a clearly defined market for them.

It is therefore important that the allocation of economic values, monetary and otherwise, be further developed. This will help to reinforce the notion that the utilization of environmental resources carries costs.

KEY ENVIRONMENTAL PROTECTION ISSUES

Introduction – what is pollution?

Releases into the environment can result in a wide range of effects. In order to consider these effects further, it is helpful to think first about what the term 'pollution' actually means. Several definitions have been proposed. Part I of the Environmental Protection Act 1990 defines pollution in terms of the capability of a substance to cause harm to man or other living organisms. The term 'harm' is further defined as: 'harm to the health of living organisms, interference with ecological systems of which they form part or offence caused to any sense of man or harm to his property.' Thus, the term pollution takes into account physical risk, ecosystem damage, damage to property and the rather more subjective concept of offence to the senses.

However, it is important to realize that environmental protection covers a great deal more territory than simply controlling and preventing pollution in a narrow sense. For example, it encompasses resource conservation and, notably in the case of climate change, the way we will be able to live our lives in the future.

Climate change

Trace gases in the earth's atmosphere, including carbon dioxide, water vapour and methane, absorb the sun's heat and cause the atmosphere to be around 3°C warmer than it would otherwise be. Such gases are therefore commonly referred to as greenhouse gases. A significant proportion of the emissions of these gases are natural. Carbon dioxide is formed during respiration and is produced by volcanoes and forest fires. Methane is formed during the decomposition of organic matter.

However, particularly since the advent of industrialization, man's activities have increasingly added to the total releases of greenhouse gases. Vast quantities of carbon dioxide are produced by the combustion of fossil fuels, particularly by the generation of electricity and the use of various methods of transportation. Methane is emitted from intensive farming, the production, transmission and use of natural gas, and landfill sites.

The raised levels of these gases in the atmosphere increase the amount of heat that is trapped and there is now a broad consensus of scientific opinion that the temperatures are rising as a direct result of this effect. The Intergovernmental Panel on Climate Change has predicted a rise in mean global temperature of about 1°C by 2025 and 3°C by the end of the twenty-first century [10].

While an increase in temperatures may at first seem to be a not undesirable effect, there could be a significant impact on agriculture, and changes in weather patterns may well lead to more storms with a concomitant rise in damage to property and loss of life. The melting of regions of the polar ice caps may cause sea levels to rise, with a consequent loss of low-lying areas of land.

Climate change is a global problem that requires a global solution. At the Earth Summit in 1992, an international treaty was agreed, under which 184 signatory states pledged to stabilize their emissions of greenhouse gases at 1990 levels by 2000. The United Kingdom is likely to have been one of a small number of developed countries to meet this target – its greenhouse gas emissions are estimated to have been 13.5% below 1990 levels in 2000.

At the follow-up conference on climate change in Kyoto in December 1997, a more ambitious agreement was reached, requiring parties to control greenhouse gas emission by an amount that varied according to their particular circumstances and level of development. For example, EU countries agreed to reduce emissions by 8% by 2010, again compared with 1990 levels. Under a subsequent EU agreement in 1998, this 8% reduction was shared out between Member States. The United Kingdom agreed to reduce emissions by 12.5%, and this will now become its legally binding target under the Kyoto Protocol. Targets for other Member States ranged from a 21% reduction for Germany and Denmark to a 27% increase for Portugal.

Stratospheric ozone depletion

The stratosphere contains a layer of ozone about 20 km above the earth's surface. This ozone absorbs some of the ultraviolet radiation received from the sun that is thought to be the cause of conditions such as skin cancer and eye cataracts. Ultraviolet radiation can also damage crops and aquatic life.

Since the 1970s, it has been increasingly realized that the ozone layer is being depleted and that holes are developing in certain areas, notably in the polar regions. Research has led scientists to conclude that organic chemicals containing chlorine play an important part in this process. Among the most significant contributors to ozone depletion are chlorofluorocarbons (CFCs), which are also greenhouse gases.

Again, this is an international issue that can only be addressed effectively by international action. In 1987, the Montreal Protocol was agreed under the auspices of the United Nations Environmental Protection Programme (UNEP) and was subsequently ratified by more than 100 countries. Under the protocol, as revised and reinforced at various times (most recently in Beijing in 1999), CFC production by developed countries was phased out by 1 January 1996, with a similar commitment for less developed nations to do the same by 2010. Consumption was to be cut by 75% by January 1994. Production and consumption targets have also been set for other substances that play a part in ozone depletion, including halons, carbon tetrachloride, 1,1,1-trichloroethane and hydrochlorofluorocarbons (HCFCs).

Air quality

For hundreds of years, air pollution has been regarded as a cause of poor health. Emissions of acid gases, principally sulphur dioxide, can also cause damage to buildings. Other types of pollutants, such as odorous substances and coarse particulates, often cause concern because of their effect on amenity.

The need for further action to improve air quality was emphasized in a report by the Department of Health's Committee on the Medical Effects of Air Pollution [11]. This report contained estimates of the number of deaths and hospital admissions due in part to poor air quality. The suggestion that up to 24 000 deaths a year are brought forward by air pollution should act as a spur to encourage creative approaches to urban air quality problems.

Local air quality

Many deaths were attributed to the London smogs of the 1950s and as a result steps were taken to reduce the emissions of particulates and sulphur dioxide caused by the burning of coal and oil. These measures included the promotion of cleaner fuels and fireplaces and the requirement for industrial chimneys to be of an adequate height to ensure that effective dispersion of emissions took place. Because of these measures, along with the decline of manufacturing industry and the huge growth in road transport, motor vehicles have become the major source of air pollution in many busy urban areas.

Fine particulates remain a cause of concern, as although much smaller amounts are emitted from stationary sources than was the case at the time of the great smogs, substantially greater amounts are now produced by road traffic, particularly diesel engines. However, nitrogen dioxide has largely replaced sulphur dioxide as the other great air pollution challenge in urban areas.

The European Air Quality Framework Directive (96/62/EC L296 21/11/96) and its 'daughter' Directives set out target levels for various ambient air pollutants. In the United Kingdom, the Air Quality Regulations 2000 set statutory air quality objectives for seven pollutants, to be achieved between 2003 and 2008. The pollutants are fine particles, nitrogen dioxide, sulphur dioxide, benzene, 1,3-butadiene, lead and carbon monoxide. Non-statutory objectives have been set for ozone and polycyclic aromatic hydrocarbons, as well as nitrogen oxides and sulphur dioxide (to protect vegetation and ecosystems).

Other than the Air Quality Framework Directive, the main European measures for addressing local air quality are the Directives, which have, since the early 1970s, set increasingly tight emission standards for vehicles, along with requirements for substantial reduction in the levels of impurities, such as sulphur, in road fuels. Industrial air pollution is now largely covered by the IPPC Directive, although other measures set standards from particular types of plant, such as waste incinerators, large combustion plant and processes involving the use of organic solvents.

Transboundary air pollution

In the years immediately after the Second World War, as evidence began to accumulate on the effects of air pollutants on health, various policy decisions were made in an attempt to combat urban smogs. Domestic smoke control was one of the most successful actions. Other measures included the increasing of industrial chimney heights, and the removal of the most polluting processes, notably coal-fired power stations, from built-up areas and their replacement with much larger stations in rural locations.

In terms of local air quality, these policies largely achieved their aims. However, by the 1970s, scientists began to realize that these large new power stations were not a panacea. Huge plumes of acid gases were now being emitted at a sufficient height to travel hundreds of miles. Prevailing south-westerly winds were taking them to Scandinavia where acidification by sulphur and nitrogen oxides was causing substantial damage to lakes and vegetation. Steps have been taken to reduce emissions of acid gases through measures such as the EC Large Combustion Plant Directive (88/609/EEC L336 7/12/88). Nevertheless, further action will be required if the problem of acidification is to be eliminated.

Other types of pollution may cause harm many miles from the source of the pollutant. Ozone is formed by a complex reaction involving the effects of sunlight on nitrogen dioxide and hydrocarbons. Thus, high levels of ozone may be experienced many miles from the point of release of these pollutants and effective action can only be taken at an international level. Emissions of heavy metals and persistent organic pollutants, such as dioxins, can be transported by a variety of mechanisms and may cause damage to ecosystems thousands of miles away. For these pollutants, global action is required for effective control.

The UNECE Gothenburg Protocol and the European national emissions ceiling Directive set national total emission limits for sulphur dioxide, nitrogen oxides, volatile organic compounds (VOCs) and ammonia. In 2002, the Department of the Environment, Food and Rural Affairs (DEFRA) issued a consultation paper [12] setting out the UK Government and Devolved administrations' plans to abate acidification, eutrophication and ground-level ozone.[1]

Water quality

The importance of clean water supplies to public health and the availability of food has been recognized for many centuries. However, it was the industrial revolution and the consequent rise in discharges of wastes to water that brought about the first concerted action to tackle water pollution in the United Kingdom. The Clauses Acts of 1847 provided a framework for the prevention of industrial and domestic releases to water, while the Salmon Fisheries Act of 1861 represented the first attempt to preserve aquatic life.

More recently, it has been recognized that water pollution can seriously damage the viability of ecosystems. Thus, standards have been introduced that not only protect resources that contribute directly to public health and nutrition but also allow a greater number of species to prosper in rivers and other waters. An example is the European Directive on Freshwater Fish (78/659/EEC L222 14/8/78). Other measures, such as the Bathing Water Directive (76/160/EEC L631 5/2/76), which among other things, sets standards for Blue Flag beaches, focus on the public health impact of recreational activities.

Furthermore, as industrial pollution sources have both declined in number and been cleaned up, other sources, notably agriculture, have

---

[1]See the National Emissions Ceiling Regulations 2002 which were implemented on 10 January 2003.

attracted more attention. Nitrate pollution of waters, caused largely by run-off from fields that have been fertilized by manure of other N-containing fertilizers, can cause eutrophication of waters and harm to aquatic life. The European Nitrates Directive (91/676/EEC L375 31/12/91) was an attempt to introduce controls to alleviate this problem.

More recently, European policy makers have taken the view that a more co-ordinated, strategic approach needs to be taken to water pollution control rather than dealing with each issue in isolation. A result of this has been the Water Framework Directive (2000/60/EC L327 22/12/00), which requires all designated waters to be of a good standard by 2015. Within this overall framework, daughter directives will deal with particular issues, for example, the perceived need for more stringent groundwater standards.

Marine protection

Marine protection is a huge topic to which justice cannot be done here. The protection of bathing and coastal waters is, of course, in part a function of the way in which inland water quality is managed. However, there are many other issues of interest from an environmental protection perspective, ranging from the disposal of waste at sea through pipeline discharges, drilling and mining and overfishing to the acute problems caused by accidental spills. DEFRA has recently published an overview of the way in which these issues are being taken forward by the UK Government [13].

Waste management

As economies and consumption grow worldwide, so too does the amount of material that arises as waste. Unless the waste has a readily realizable value, the default option has tended to be disposal, generally by landfill. However, landfill requires a lot of space, often old mineral workings, and enough suitable sites are not always available near to large settlements.

It has also become increasingly apparent that buried waste cannot simply be forgotten about. As organic material decomposes, methane is formed, which is highly flammable. Properties built on old landfill sites have effectively been blighted by the risk of explosion. Liquid wastes and decomposing material can also leach out of the site into underground aquifers, potentially polluting supplies of drinking water and harming aquatic life. Under the European Landfill Directive (99/31/EC L182 16/7/99), landfill sites must now be engineered and restored in a manner that minimizes these problems, pushing up costs. In the United Kingdom, the undesirability of landfill as a means of waste disposal has been emphasized by the imposition of a landfill tax, which further increases costs and causes users to face up to more of the external costs of sites.

Thus, over the past 20 years, other disposal techniques have been explored more widely. Since the 1970s, there has been a move towards the incineration of waste, particularly in urban areas. Using this method, the volume of waste can be reduced substantially and many materials that could present hazards in landfill sites are rendered inert by combustion. There is also more possibility of separating out materials that are valuable or can be recycled. Some incinerators have been equipped with heat recovery plant to enable some of the energy present in waste to be used by nearby factories or district heating schemes.

However, incineration has not been without its own difficulties. Some of the early plants were problematic and subject to widespread local complaints of smoke and particulate fall-out. Increased understanding of the formation and effects of persistent organic pollutants such as dioxins focused public concern on incinerator emissions, particularly as most municipal plants were in densely populated areas. Stringent standards set in European directives and integrated pollution control authorizations raised costs and proved to be unachievable in many cases, with incinerators being closed or rebuilt.

Thus, the increased difficulty and costs associated with disposal methods, coupled with increased public interest in sustainability, have led to an increased level of interest in recycling and reusing waste materials. Approximately half of the domestic waste produced in the United Kingdom can be recovered.

More emphasis is also being placed on the responsibility of the producers of goods to play a leading part in waste reduction. Producer responsibility is an extension of the 'polluter pays' principle and aims to encourage business to recover value from products at the end of their life. Much European Waste legislation over recent years has focussed on the extension of producer responsibility to different sectors, starting with packaging and moving on to end-of-life vehicles and waste electrical and electronic goods.

Contaminated land

The contamination of land by previous use has become a complex and important issue. Contamination may have arisen from previous waste disposal activities, as in the case of methane and leachate releases from former landfill sites. In other circumstances, it may be related to past industrial activities, for example old gas works where the contaminants may include heavy metals and cyanides. It has been estimated that there are up to 100 000 contaminated sites in the United Kingdom covering 120 000 acres (48 600 ha), or 0.4% of the total land area. There are also at least 1400 former waste tips generating methane at potentially dangerous levels [14].

   Contamination of soil with toxic substances can give rise to health effects, particularly where houses are built on the land. Soil ingested by adults while gardening or children while playing can cause harm. Vegetables grown in contaminated soil may contain the chemicals that remain in the ground. Pollutants present in soil may be broken down naturally over time by a variety of chemical or microbiological processes. However, they may also be leached out, particularly where the soil is light and sandy, thus potentially causing contamination of groundwater. Where neither breakdown nor leaching occurs, high concentrations of pollutants may accumulate in the upper layers of the soil. Restoration of contaminated land to 'as new' condition is generally not possible. However, it is often possible to carry out remedial works to allow the land to be reused for some purpose, although this may sometimes be non-residential, particularly where contamination has been severe.

Chemicals

The production and use of chemicals is an emotive issue and a long history of control measures has failed to entirely quell public and NGO concern. Under European law, 'Existing' chemicals (in legal terms those on the market between 1971 and 1981) are subject to a process of data collection and, in the case of priority chemicals, risk assessment and reduction. In the case of 'new' substances (those first marketed after 1981), manufacturers must carry out prescribed test and provide information on production techniques and a risk assessment as well as proposing guidelines for classification and labelling. Under the European Marketing and Use Directive (76/769/EEC L262 27/9/76), the Commission can propose restrictions and bans on any substances with dangerous properties.

Genetically modified organisms

The release of genetically modified organisms (GMOs) into the environment has become a focus for public concern and activism. The subject is a challenging one which requires policymakers to find a balance between opportunities to produce food more efficiently and the need to ensure that unacceptable damage to biodiversity is avoided. The GMO issue, like climate change, therefore gets to the heart of the sustainable development agenda.

   The UN Cartagena Protocol on biodiversity introduced procedures to control transboundary movements of GMOs and is of particular benefit to developing countries that do not have their own legislation in this field. The Protocol has been signed by the EC and its Member States and European legislation that will allow ratification is currently in preparation.

Noise

Noise is sometimes defined as unwanted sound. As such, perception of noise is bound to be at least partly subjective. What is uplifting music to one person may well be an unwanted racket to another. Thus, simple measurement of noise levels may not always be the most appropriate indicator of the likely effects.

Nevertheless, it is clear that noise can cause feelings of annoyance and frustration, perhaps leading to stress. Among the sources of noise that cause particular irritation, if not the highest sound levels, are amplified music and barking dogs. However, in many parts of the United Kingdom, the noise climate is dominated by transportation, notably busy roads and those carrying fast-moving traffic such as motorways, railways and aircraft. There is concern that high levels of noise may cause cognitive dysfunction, leading to learning difficulties for children and others who are taught in urban areas and near to airports.

In recent years, there has been increasing interest in the subject of ambient noise (sometimes referred to as environmental noise). In 1996, the European Commission issued a Green Paper [15] which estimated that 20% of European citizens were exposed to noise levels that health experts considered unacceptable. There were already a number of directives to limit noise from particular sources, such as vehicles and machinery. However, the Green Paper proposed a more strategic approach that focussed on those receiving the noise.

Following the Green Paper, a Directive on the assessment and management of environmental noise was proposed. The Directive (2002/49/EC L189 18/7/02), which was eventually finalized in 2002, requires strategic noise maps to be produced by 2007 for all conurbations with a population greater than 250 000 and for all major roads, railway lines and airports. Action plans will have to be drawn up by 2008 for the management of noise in these areas.

Light pollution

It is only relatively recently that light pollution has been regarded as a significant problem. Concern exists at two levels: some people are worried that the increased use of lighting generally constitutes an unacceptable intrusion into nature, particularly as it has led to the night sky being less visible. People are also concerned about specific sources of light, notably security lights, industrial and commercial lighting, sports facilities and highway lighting. Domestic security lighting is the cause of the greatest number of complaints to local authorities [16]. In these circumstances, problems are often due to excessively powerful bulbs or their inconsiderate location.

There are no statutory powers designed to deal with light pollution in the United Kingdom. Consideration was given to adding floodlighting to the list of matters that can be treated as a statutory nuisance during the passage of the Bill that became the Environmental Protection Act 1990. However, this proposal was rejected by the House of Lords on the basis that the incidence of the problem was considered to be too low and that definition of the term would be difficult.

Indoor air pollution

Public debate has tended to focus on air quality in streets and other public spaces. In fact a significant proportion of most people's exposure to many pollutants occurs in the home and in the workplace. The latter is a health and safety matter and outside the scope of this chapter. However, the generation of pollution in the home is a further issue that needs to be taken into account when considering the effects of harmful substances on the population [17].

Smoking is one obvious source of indoor air pollution, and can give rise to emissions of fine particulates, polyaromatic hydrocarbons (PAHs) and dioxins. Furniture made from fibreboard has been associated with exposure to formaldehyde, although steps have been taken by the manufacturers to minimize this problem. Natural sources, such as dust mites, can also cause health effects. The problem has been exacerbated by modern luxuries such as carpets and central heating, and reduced ventilation brought about by the desire to insulate properties.

A related issue is air pollution in vehicles. People tend to assume that while they are in their cars they are protected from poor air quality outside. In fact, research suggests that cars travel in a tunnel of pollution emitted by the vehicle in front, which is only diluted to a small extent before being sucked in through the car's ventilation system [18]. Thus, drivers and passengers in cars are typically exposed to two to three times more pollution than cyclists and pedestrians, and generally more than passengers on buses.

Electromagnetic radiation

Mobile phones and base stations

Mobile phone use and ownership has increased enormously over the last few years. A large proportion of the United Kingdom's population now regards the mobile phone as essential to work and play. However, there has also been an increase in concern regarding the possible health effects of both the phones themselves and the base stations and masts that are necessary to enable the mobile networks to function.

In 2000, the Independent Expert Group on Mobile Phones (IEGMP), also known as the Stewart Group after its chair, Sir William Stewart, produced a comprehensive report [19] reviewing these issues. In his foreword to the Report, Sir William stated that:

> The balance of evidence does not suggest that mobile phone technologies put the health of the general population of the UK at risk. There is some preliminary evidence that outputs from mobile phone technologies may cause, in some cases, subtle biological effects although, importantly, these do not necessarily mean that health is affected.

The Group proposed that a precautionary approach be adopted until more robust scientific evidence becomes available, as it was not possible to say that exposure to RF radiation, even at very low levels, was totally without adverse health effects. It made a number of recommendations about siting of base stations and also recommended that all base stations, including those with small masts under 15 metres high, should be subject to the normal planning process. The Government responded positively to the report, although, at the time of writing, not all of the recommendations had yet been put into practice.

Overhead power lines

High voltage power lines can lead to increased exposure to electromagnetic radiation. A great deal of research continues to be carried out to determine whether there is any evidence linking ill-health, particularly childhood leukaemia, to power lines and other sources of electromagnetic fields. The National Radiological Protection Board (NRPB) currently takes the view that there is no conclusive evidence of health effects at levels to which people are normally exposed [20]. However, the possibility remains that intense and prolonged exposure to electomagnetic fields can increase risks. The NRPB is currently consulting on the possibility of taking a precautionary approach to this issue [21].

Emergency planning

The events in America on 11 September 2001 and subsequent terrorist attacks elsewhere in the world have put increased emphasis on the need for a comprehensive and effective emergency planning system for the United Kingdom. Legislation expected to be passed in 2004 is likely to place increased emphasis on this and create a firmer situation involving the statutory duties of local authorities and others to produce plans.[2]

The role of the environmental health officer in these situations is both important and diverse and needs to be fully reflected in such plans. A useful reference to this is the publication *The Role of EHOs in Emergency Planning* produced by the CIEH as a Professional Guidance Note in September 2000.

Liveability

Despite the increasing wealth of evidence to demonstrate the effects of, for example, climate change or air pollution on sustainability and public health, surveys of public opinion regularly show that many people are more concerned about more tangible damage to their local environment. Litter, dog fouling and graffiti, like neighbour noise, are clearly capable of not only making people feel less good about their neighbourhood, but also interfering materially with their overall quality of life.

---

[2]See the Civil Contingencies Bill 2004.

Governments are increasingly aware of the need to balance their attempts to address global and regional action with measure to protect and enhance the local environment. In the United Kingdom, for example, local authorities are being encouraged to clear dumped refuse and litter more quickly and to employ environmental 'wardens' to enforce offences such as dog fouling. It remains to be seen whether this process will discourage anti-social behaviour or merely mitigate its effects.

DELIVERY OF ENVIRONMENTAL PROTECTION IN THE UNITED KINGDOM

Sustainable development (also see Chapters 1 and 9)

The Government strategy for achieving sustainable development was set out in 1999 in *A better quality of life* [22]. Since then, annual reports have been produced setting out progress against the fifteen headline indicators identified by the original strategy and describing action taken thus far to promote sustainability.

In order to maintain momentum following the publication of its sustainable development strategy, the Government established the Sustainable Development Commission. The Commission's objectives are to:

- review the extent to which sustainable development is being achieved and identify risks to progress;
- identify unsustainable trends that will not be corrected by current policies and recommend appropriate action;
- enhance understanding and awareness of the concept of sustainable development;
- encourage and stimulate good practice.

Sustainable development at regional and local levels

Regional Sustainable Development Frameworks have been prepared for each English region by Government Offices, Regional Development Agencies, Regional Chambers and other bodies working jointly. These Regional Frameworks inform the development of the Community Strategies that all local authorities are required to prepare through Local Strategic Partnerships. The Community Strategies have subsumed the Local Agenda 21 process that took forward, at a local authority level, delivery of the Agenda 21 initiative agreed at Rio.

Access to environmental information

Public access to information about the environment is a cornerstone of sustainable development as it helps citizens to engage in decision-making processes. The UNECE Aarhus Convention sets out a framework for openness and public participation which has been fleshed out European level through a Directive. New Environmental Information Regulations came into force in the United Kingdom in 2003.

Climate change

The United Kingdom's climate change programme [23] was published in November 2000. It sets out the United Kingdom's policies for meeting its targets under the Kyoto Protocol and the Government's own target of a 20% cut in carbon dioxide emissions by 2010 (compared to a 1990 baseline).

The programme includes measures and policies designed to:

- improve the energy efficiency of businesses;
- encourage more efficient means of power generation;
- reduce emissions from transport;
- improve energy efficiency in the home;
- increase the energy efficiency of new buildings;
- ensure that the public sector sets a good example.

Two of the Government' key policies for improving energy efficiency in the business sector are the climate change levy and the UK emissions trading scheme. The climate change levy is a tax on the use of energy for non-domestic purposes. Certain types of businesses (mostly heavy industry) can receive an 80% discount from the levy through climate change agreements, under which they undertake to achieve energy-related targets.

Under the UK emissions trading scheme, 34 organizations have entered a legally binding agreement with the Government to reduce their emissions of greenhouse gases. These organizations, along with the 6000 companies with climate change agreements, can buy allowances to meet their targets, or sell any excess reductions.

Adaptation

Alongside efforts to control emissions of greenhouse gases and thus curb climate change, it is increasingly recognized that action is necessary to deal with its consequences. The Government is now incorporating a recognition of the possible implications of climate change into many of its activities – the precautionary principle being put into practice. An example is the strengthening of planning guidance to take account of the increased risk of flooding. Policies on subjects as diverse as building regulations, water resources and health now incorporate a climate change perspective.

Each Government Department is considering the implications of climate change for its work – for example, in 2003, DEFRA published an assessment of impacts across its remit [24]. Similar issues arise for local government, particularly in the areas of planning and investment.

Integrated pollution prevention and control and local air pollution prevention and control

Part I of the Environmental Protection Act 1990 set up two new systems of industrial pollution control. The most potentially polluting processes were prescribed for integrated pollution control (IPC) by the Environment Agency and the Scottish Environmental Protection Agency (SEPA) in Scotland. Processes with less potential to pollute, and for which the most significant emissions are to air, were prescribed for Local Air Pollution Control (LAPC) by local authorities and SEPA in Scotland.

However, as a result of the IPPC Directive and the subsequent UK implementing legislation, these systems underwent a major overhaul in 2000. The

new regime consisted of three tiers of controls (two in Scotland):

- IPPC – operated by the Environment Agency/SEPA, and essentially covering industrial (Part A) installations which fell under the IPPC Directive and had previously been subject to IPC (and LAPC in Scotland);
- Local Authority Integrated Pollution Prevention and Control (England and Wales only) – operated by local authorities and covering (Part A2) installations that fell under the IPPC Directive and had previously been subject to LAPC;
- Local Air Pollution Prevention and Control – operated by local authorities/SEPA and covering (Part B) installations that did not fall under the IPPC Directive but had previously been subject to LAPC.

A large number of industrial and commercial processes are not prescribed for control under Part I of the Act, and are therefore subject only to the statutory nuisance provisions of Part III of the Act and the Clean Air Act 1993.

There are two fundamental principles in Part I of the Act. These are:

- processes prescribed by regulation must not be operated without an authorization from the relevant enforcement agency;
- the authorization must contain specific conditions that are designed to achieve the objectives of the Act; the principal objective is that the process must operate using the best available techniques to prevent, where practicable, or minimize emissions of prescribed substances, and to render harmless any substance that may be emitted.

The transition from the old control systems to the new PPC regime is taking place on a gradual, sectoral, basis up to 2007, in line with the timetable in the Directive. The permitting system is similar in many respects to the integrated pollution control regime in the United Kingdom, on which it was partially modelled. However, there are a number of notable differences. The range of processes covered is broader and includes, for example, intensive livestock farming and large-scale food production. The Directive also requires a wider range of environmental issues to be taken into account, including noise, energy efficiency, heat, raw material usage, accident prevention and site remediation following final closure of the plant.

Ambient air pollution

The UK National Air Quality Strategy

The 1997 UK National Air Quality Strategy heralded a radical new approach to the improvement of air quality in the United Kingdom. At the heart of the strategy were health-based air quality standards and objectives for eight pollutants of particular concern: benzene, 1,3-butadiene, carbon monoxide, lead, nitrogen dioxide, ozone, sulphur dioxide and fine particulates (PM_{10}). The air quality standards were set at a level where the pollutants will cause no significant health effects in the population. A revised strategy was published in 2000, taking into account the enhanced evidence base amassed over the previous three years. In 2003, the Government and the devolved administrations published an addendum to the strategy, which introduced tighter objectives for particles, benzene and carbon monoxide, as well as a new objective for polycyclic aromatic hydrocarbons. The strategy envisages that the current objectives should be achieved between 2003 and 2008.

National and international measures currently in place should ensure that the objectives are met in many parts of the United Kingdom. For industry, these include the pollution prevention and control systems. However, it has increasingly been recognized that road transport is a significant contributor to air pollution, particularly in towns and cities. The Auto-Oil Directives are ensuring that new vehicles conform to more stringent emission standards, while the drive towards cleaner fuels continues.

However, in some areas, notably urban centres, further local action will be required if the objectives are to be achieved. The Government therefore gave local authorities new powers and duties to manage air quality under the Environment Act 1995. Each authority was required to review and assess air quality within its area and to predict, on the basis of this exercise, whether the objectives were likely to be met. Where the local authority considered that the objectives will not be achieved, it was under a duty to declare an air quality management area covering the predicted area of exceedance. For each air quality management area, an action plan must be devised in conjunction with other stakeholders, such as the Environment Agency, the Highways Agency and, where appropriate, the county council.

Control of air pollution from road transport

The most important means of controlling emissions from vehicles is through the type of approval system administered by the Vehicle Certification Agency. All new vehicles must meet emission standards specified in EU directives. These standards have become increasingly stringent since the early 1970s and the recent Auto-Oil Directives of the mid-1990s set standards for 2000 along with even more challenging limits for 2005. The Directives also set quality standards for fuels.

In-service vehicle testing is achieved largely through the annual MOT test for light vehicles and as part of the plating test for trucks and buses. The Vehicle Inspectorate carries out roadside testing on an *ad hoc* basis in conjunction with the police, usually as part of a general roadworthiness examination. Local authorities have also recently been given powers to carry out roadside emission tests, with on the spot penalties for offenders. Drivers who leave engines running in stationary vehicles can also be penalized.

Traffic authorities (county councils and metropolitan districts) can use Traffic Regulation Orders to prevent or restrict vehicular access to defined areas either permanently, periodically or temporarily. Either all or certain types of vehicles could be excluded, so it would be possible, for example, to allow only low emission vehicles into a defined area. It would also be feasible, provided that sufficient warning was given, to restrict or prevent access when air pollution exceeded specified levels.

Water pollution

With the formation of the Environment Agency and SEPA in 1996, the responsibility for protecting the cleanliness and purity of watercourses transferred to the new agencies from the National Rivers Authority. The agencies also took on responsibility for controlling emissions to water from processes prescribed for IPC.

Control of point discharges to water are currently undertaken in three ways:

- the Environment Agency/SEPA controls the discharge of substances from processes prescribed for IPC through the process authorization;
- the Environment Agency/SEPA also controls the discharge of pollutants to controlled waters, including rivers, watercourses, lakes and ground water;
- the water companies control the discharge of trade effluent to sewers through trade effluent discharge consents and agreements.

However, it is increasingly recognized that many of the more serious water pollution problems are due in large part to diffuse pollution sources, notably agriculture. Over half of England and Wales have been declared nitrate vulnerable zones, and farmers in these areas must obey rules about the amount of manure they can spread and the times of year that spreading can take place. However, emissions from diffuse sources will need to be reduced further if forthcoming European standards, such as those under the Water Framework Directive, are to be met. In 2003, DEFRA published a discussion document on diffuse water pollution from agriculture [25] that examines these issues in some detail. It is likely that this will be followed up with a consultation document exploring options for reducing diffuse emissions, including regulatory, voluntary and economic measures.

Audits and inspections of water companies in England and Wales, and enforcement action to ensure compliance with the Water Quality Regulations – including investigation of customer complaints and incidents which affect drinking water supplies – are the responsibility of the Drinking Water Inspectorate.

Waste management

Since the formation of the Environment Agency and SEPA in April 1996, most waste regulation functions have transferred from local authorities to these agencies. The central waste management delivery mechanism, in terms of the standard of facilities at least, is waste management licensing system, set up under Part II of the Environmental Protection Act 1990 through the Waste Management Licensing Regulations 1994. The licensing regime is supplemented by the duty of care placed by the act on businesses producing or handling waste to ensure that it is managed properly.

The waste management licensing system is currently being reviewed to ensure that it continues to be effective in the light of recent changes. The foremost of these is the coming into force of the Landfill Directive, which sets rigorous standards for landfill sites and which will bring to an end the longstanding UK practice of co-disposing hazardous and non-hazardous waste. This Directive, along with the landfill tax, is likely to bring about a reduction in landfill capacity in the United Kingdom. Another significant change will be the proposed extension of waste management controls to agricultural and mining and quarrying wastes.

Movements of hazardous wastes are also regulated by the Environment Agency/SEPA under the Special Waste Regulations 1996. These Regulations are also being reviewed in an attempt to streamline the controls to facilitate the inclusion of many waste streams newly designated as hazardous under the latest version of the European Hazardous Waste List.

The Secretary of State is required, under the Environment Act 1995, to produce a National Waste Strategy. Following a series of consultations, this was finally published in 2000 [26]. The Strategy confirms the Government's support for a waste hierarchy that places waste minimization at the top of its priorities, above recycling and energy recovery, with disposal, for example by landfill, the least desirable option. The strategy also sets challenging recycling targets for local authorities.

The concept of producer responsibility is an important tool for delivering improved rates of recycling. The Packaging Regulations set recovery rates for packaging used by businesses and the Government has adopted a marked-based approach to compliance with recovery credits able to be bought and sold, depending on whether companies are exceeding or falling short of the targets. The Government is currently considering how best to implement the two recent European producer responsibility directives, which require recycling or recovery of end-of-life vehicles and waste electrical and electronic goods.

Contaminated land

Until recently, the main means of ensuring that contaminated land is not used for unsuitable purposes was the development control system. However, as records of former contaminating uses, some of which may have taken place hundreds of years ago, were by no means complete, this did not provide sufficiently rigorous protection.

The first steps to improve legislation were taken in Part II of the Environmental Protection Act 1990. This placed a duty of care on any body handling waste. Operators of landfill sites were now responsible for the site until it has been declared safe and a certificate of completion has been issued. Provisions were also made for the establishment of registers of contaminated land but these were never commenced, largely through fear of widespread blight, and were repealed under the Environment Act 1995.

The framework for a revised system of control was put in place by the Environment Act. New controls have now come into force under which local authorities are the main regulators and are required to produce a strategy for inspecting their area. The Environment Agency also has responsibilities for providing site-specific advice to local authorities, dealing with defined 'special sites' and monitoring and reporting on progress made.

The IPPC Directive (and the PPC regime in the United Kingdom) should help to avoid further serious contamination of land. As part of the environmental assessment, consideration should be given to whether pollution control options present a risk of polluting the site. The operator of the installation should plan ahead for decommissioning and restoration of the site after plant closure.

Noise control

The control of point sources of noise is exercised by local authorities primarily under the provisions of Part III of the Environmental Protection Act 1990. There is a growing problem of neighbour noise, particularly in towns and cities, exacerbated by declining tolerance and higher expectations. The Noise and Statutory Nuisance Act 1994 introduced new controls over noise in public places. The Noise Act 1996 contains additional adoptive powers under which local authorities can deal with night-time noise.

The IPPC Directive will, for the first time, require noise to be taken into account when setting standards in industrial pollution control permits. The Government is currently considering how this might be most appropriately achieved.

In order to start preparations for implementation of the Directive on environmental noise, the Government consulted, in 2001, on a proposed national ambient noise strategy. The strategy is likely to involve mapping, a process with which local authorities in London and Birmingham have already made significant progress.

Radioactive substances

Control of the release of radioactive substances is now the responsibility of the Environment Agency, which enforces the provisions of the Radioactive Substances Act 1993. This legislation covers the registration of sites, the keeping of radioactive substances and the licensing of the discharge of radioactive waste to air, water and solid waste disposal routes.

Radiological standard setting

The main body responsible for setting standards relating to radiological protection is the International

Commission on Radiological Protection (ICRP), which is a non-governmental scientific organization. In the United Kingdom, the NRPB advises the government on the suitability of proposed standards. The current ICRP recommendations have been adopted by the UK government and consist of three central requirements:

- justification – no practice shall be adopted unless its introduction produces a net benefit;
- optimization – all exposures shall be kept as low as reasonably achievable (ALARA), taking into account economic and social factors;
- limits – the dose equivalent to individuals shall not exceed the limit recommended for the relevant circumstances by the ICRP.

Radiological monitoring

The Chernobyl disaster showed that an overseas radiological incident could affect the United Kingdom and that the national response plan for accidents occurring domestically was not suitable in such circumstances. As a result, the then Department of the Environment established the Radioactive Incident Monitoring Network (RIMNET), which consists of a series of automatic gamma ray dose rate monitoring stations throughout the United Kingdom. This provides: an independent warning of any overseas incidents; a monitoring database against which elevated levels may be judged; and an information handling network to determine the effects of such accidents.

A number of local authorities have also instigated environmental radiation programmes. Many of these rely on periodic gamma ray dose rate monitoring by portable instruments, although some authorities have embarked upon more far-reaching work involving sampling and analysis of water, soil and food. In order to ensure that the data obtained by local authorities are quality assured and are fed into RIMNET in the case of an accident, the local authority associations have formed the Local Authorities Radiation and Radioactivity Monitoring Advice and Collation Centre (LARRMAC). This organization is responsible for providing guidance on monitoring protocols, and also for collating local authority monitoring data and co-ordinating its input into the government database.

Development control

One important theme that applies to all environmental protection activities is the importance of planning. Enforcement of planning controls under the Town and Country Planning Acts is a local authority responsibility, and environmental matters are central to the decision-making process. One of the more recent developments in the planning field has been the emergence of environmental impact assessment (EIA) for certain categories of development proposal, particularly industrial processes with significant pollution potential. Planning is an important component of Community Strategies and local air quality management.

Planning Policy Guidance (PPG) notes provide a useful source of advice on the interface between planning and other concerns. PPG23, issued in 1994, considers the interface between planning and pollution control, and is currently being reviewed [27]. In general terms, it suggests that planning should not be used as a substitute for the proper use of pollution control legislation unless pollution control issues are also land-use planning issues.

Environmental assessment and strategic environmental assessment

Environmental assessment, also referred to as environmental impact assessment, is a systematic procedure for collecting, analysing and presenting data to ensure that the likely effects of a new development on the environment are fully understood and taken into account before construction work begins. The concept was given the force of law by EU Directive 85/337/EC, which was adopted in 1985 and amended in 1997.

The Directive contains two annexes: Annex I lists the types of projects for which environmental assessment is mandatory; and Annex II lists projects that 'shall be made subject to an assessment where member states consider that their characteristics so require'. The Directive was implemented in the United Kingdom primarily through the Town and

Country Planning (Assessment of Environmental Effects) Regulations 1988. Environmental assessment is now an integral part of the planning process.

In December 1995, the European Council of Ministers reached a common position on a proposal for a directive that amends the 1985 directive. The new directive, 96/61/EC L257, provides more guidance about the information to be provided by developers, and requires public consultation to take place prior to approval of the project rather than before commencement of work. The list of project types that are subject to mandatory environmental assessment has also been expanded to include 20 categories of development, while Annex II to the Directive has also been widened to take in a total of 50 project types.

The project types listed in Annex II closely reflect the processes that are subject to the IPPC Directive, but also encompass certain types of tourism and leisure development. Additional criteria for determining whether an Annex II project requires an environmental assessment include: whether it is likely to have significant environmental effects when judged against defined thresholds; and whether it is likely to have a significant effect on an area designated as a special environmental protection zone under other EU legislation.

The aim of an environmental assessment is that the developer should gather, make available and assess relevant information to enable full public scrutiny of the project to take place. It should also enable the predicted environmental effects and proposed mitigation measures to be properly evaluated by the planning authority before a decision on the development is taken. In the case of industrial processes, the environmental assessment will, in many cases, be complementary to the application for authorization under either the integrated pollution control or local air pollution control systems.

The environmental assessment is the full information collection and analysis process. It must be supported by an environmental statement, which is the developer's own assessment of the likely environmental impact of the project. The exact content of an environmental statement is largely dependent on the type of project.

However, all environmental statements should include information on:

- the nature of the proposed project, including the site location and the design, size and scale of the development;
- emissions and wastes;
- the likely effects on various receptors, including humans, plants, animals, water, soil, landscape and so on.

Where significant environmental effects are identified, the statement should include a description of the measures proposed to avoid, reduce or remedy those effects. This should include an outline of the alternatives considered and the reasons why the proposed mitigation methods were chosen.

An environmental assessment and statement is a public document and should therefore include a non-technical summary. Another important element of environmental assessment is continuation beyond the construction period to investigate the actual environmental effects and a comparison with those that were predicted. This will enable the accuracy of the statement to be gauged and also allow the assessment process and techniques to be refined for future use.

Strategic environmental assessment

The strategic environmental assessment Directive (2001/42/EC L197 21/7/01), which was adopted in 2001, aims to ensure that the environmental consequences of plans and programmes are identified and assessed during their preparation and before their adoption. Its aim is to increase transparency and thus help to achieve sustainable development.

Environmental management

Increasing public and regulatory pressure are encouraging many companies to demonstrate a high level of environmental protection – merely avoiding prosecution is often not enough. This trend has been complemented by the increasing insistence of buyers that suppliers should demonstrate quality management, for example, by following the procedures set down in the ISO 9001 standard.

In order to fulfil the resultant demand for an environmental quality standard, ISO developed ISO 14001, which is based on similar principles to ISO 9001 but with a clear focus on the environment. The EU has developed this approach further through the Eco-Management and Audit Scheme (EMAS), which is a management tool for businesses that allows them to evaluate, report and improve their environmental performance.

If an organization wishes to to receive an EMAS registration, it must:

- conduct an environmental review of its activities;
- establish an effective environmental management system based on the review;
- carry out an environmental audit, including an assessment of compliance with regulatory requirements;
- provide a statement of its environmental performance and future improvements.

The take up of EMAS has been slightly disappointing, perhaps because it seems rather daunting to many companies. In order to address some of these issues, the British Standards Institute has recently finalized BS8555, which provides a route to environmental accreditation for small and medium-sized businesses. In addition, some industry sectors, have developed their own schemes. For example, farm assurance schemes, although principally designed to improve food hygiene, are starting to include environmental standards.

The use of economic measures

A number of different economic measures can be used to reflect and encourage environmental protection. These include charges and levies, tradable quotas, and liability compensation schemes. The principle behind these economic measures is that if a market-based approach to the environment is taken, consumers and producers are given a clear indication of the cost of using environmental resources. This is clearly in line with the 'polluter pays' principle.

It is, however, important to recognize that even if economic instruments are widely used, there is still likely to be a role for regulation, enforcement and compliance with standards. Indeed, the two approaches can be combined. The recovery of the costs of authorizing and inspecting industrial processes under the integrated pollution control and local air pollution control systems is an example of the 'polluter pays' principle being incorporated into a command and control system of environmental protection.

The various economic measures available are considered in more detail below.

Levies and deposit refund schemes

Where a material or product is costly or dangerous to handle at the end of its operational life, for example, tyres, one possibility is that the consumer should meet the cost of this disposal in accordance with the 'polluter pays' principle. This can be achieved by charging a levy on the original product, which is then returned through a central scheme to organizations undertaking the final disposal of the waste. Another method is to use a deposit or refund, which encourages consumers to return materials or products that can be recycled. This type of scheme has been in use in continental Europe for many years for items such as glass bottles.

Charges based on damage to the environment

The application of a charge that reflects the environmental costs of using resources or releasing emissions or wastes has much to recommend it. It would encourage industry to review its operations and allow it to choose to pay the charge or invest in cleaner production methods. The major difficulty with such a scheme is the assignment of an appropriate value to environmental damage.

Charges or taxes on products and materials

This option involves adding an environmental charge or tax directly to the price of materials or services offered to reflect the environmental damage caused. The most widely considered of these has been the use of a carbon tax to encourage a reduction in the consumption of fossil fuels.

On a smaller scale, the application in the United Kingdom of a fuel duty escalator (the programmed increase of fuel duty above the rate of inflation) is an example of such a measure in practice. A variation on this theme is the use of differential rates of taxation to encourage the use of cleaner techniques or materials. Such measures have been used to promote the uptake of unleaded petrol and low sulphur diesel and road fuel gases.

Tradable quotas

This approach is based on the determination of an acceptable ceiling for total emissions of a pollutant and then the issuing of tradable quotas for discrete amounts of discharge that can be bought and sold by operators of industrial processes. This has the attraction of being economically and highly efficient as the market determines the value of the emissions. On the other hand, unless such a system is applied in tandem with regulation, control is lost over where the permitted emissions are released. Thus, this approach is more appropriate for pollutants that have regional or global effects than for those that have an impact on the local environment.

Liability and compensation

This philosophy is based on the theory that if a polluter is liable for environmental damage caused, more careful behaviour and reduced emissions will result. One of the difficulties of this approach is that it can be difficult to obtain insurance for strict liability to environmental risk. However, the European Commission has proposed an environmental liability Directive, which will render many businesses strictly liable for environmental damage caused by their activities. It seems likely that this Directive may be adopted in 2004.

CONCLUSION

As public concern about the environment continues to increase and knowledge of the effects of human activities improves, further steps will need to be taken to combat pollution, locally, nationally and globally. Although there may be a continuing move towards economic instruments as a means of protecting the environment, effective regulation and enforcement are likely to be at the heart of pollution control for the foreseeable future. The wealth of environmental strategies that are being put in place will only be effective through partnership between government, local authorities, industry, commerce and the public at large.

REFERENCES

1. Department of the Environment, Transport and the Regions (1997) *The United Kingdom National Air Quality Strategy*, Stationery Office, London.
2. Department of the Environment, Transport and the Regions (2000) *The Air Quality Strategy*, Stationery Office, London.
3. Department of the Environment, Food and Rural Affairs (2003) *An Addendum to the Air Quality Strategy*, Stationery Office, London.
4. World Commission on Environment and Development (1987) *Our Common Future*, Oxford University Press, Oxford.
5. United Nations (2002) *Report of the World Summit on Sustainable Development*, UN, New York.
6. The European Parliament and the European Council (2002) *Environment 2010: Our Future, Our Choice – The Sixth Environmental Action Programme of the European Communities*, EC, Brussels.
7. The European Commission (2000) *Communication from the Commission on the Precautionary Principle* COM(2000)1, The European Commission, Brussels.
8. Department of the Environment (1991) *General Guidance Note GG1(91): An Introduction to Local Authority Air Pollution Control*, HMSO, London.
9. Royal Commission on Environmental Pollution (1988) *Best Practicable Environmental Option, 12th Report*, HMSO, London.
10. UK Climate Change Impacts Review Group (1991) *The Potential Effects of Climate*

Change in the United Kingdom, First Report, Stationery Office, London.

11. Committee on the Medical Effects of Air Pollution (1997) *Quantification of the Effects of Air Pollution on Health in the United Kingdom*, Department of Health, London.

12. DEFRA (2002) *National Strategy to Combat Acidification, Eutrophication and Ground-Level Ozone – Consultation Paper*, DEFRA, London.

13. DEFRA (2002) *Safeguarding our Seas: A Strategy for the Conservation and Sustainable Development of our Marine Environment*, DEFRA, London.

14. Department of the Environment (1994) *Paying for our Past*, HMSO, London.

15. European Commission (1996) *Action Against Noise*, European Commission, Brussels.

16. The Chartered Institute of Environmental Health (1993) *Light Pollution Survey*, CIEH, London.

17. See also WHO (1998) *Assessment of Exposure to Indoor Air Pollutants*, TSO, London.

18. Environmental Transport Association (1997) *Road User Exposure to Air Pollution – Literature Review*, ETA, Weybridge.

19. Sir William Stewart (Chairman) (2000) *Mobile Phones and Health: A Report from the Independent Expert Group on Mobile Phones*, IEGMP Secretariat, Chilton.

20. Advisory Group on Non-Ionising Radiation (2001) *ELF Electromagnetic Fields and the Risk of Cancer*, National Radiological Protection Board, Chilton.

21. National Radiological Protection Board (2003) *Consultation Document – Proposals for Limiting Exposure to Electromagnetic Fields*, NRPB, Chilton.

22. Her Majesty's Government (1999) *A Better Quality of Life: A Strategy for Sustainable Development in the UK*, TSO, London.

23. DETR and the Devolved Administrations (2000) *Climate Change – The UK Programme*, DETR, London.

24. DEFRA (2003) *The Implications of Climate Change for Defra*, DEFRA, London.

25. DEFRA (2003) *Diffuse Water Pollution from Agriculture: The Government's Strategic Review – Stakeholder Discussion Paper*, DEFRA, London.

26. Department of the Environment, Transport and the Regions and the Welsh Assembly (2000) *Waste Strategy 2000*, The Stationery Office, London.

27. Department of the Environment (1994) *Planning Policy Guidance Note PPG23: Planning and Pollution Control*, HMSO, London.

FURTHER READING

National Society for Clean Air and Environmental Protection (2003) *NSCA Pollution Handbook*, NSCA, Brighton.

The websites of DEFRA (www.defra.gov.uk), the Environment Agency (www.environment-agency.gov.uk) and the European Commission (www.europa.eu.int) contain a wealth of information. For regular updates on the world of environmental protection, both within the United Kingdom and further afield, the ENDS Report and ENDS Daily are unparalleled.

35 Noise

Michael Squires and
John F. Leech

INTRODUCTION

We live in an environment full of sound, some sounds carry information, others may give us pleasure, while some are intrusive and are known as noise, which is unwanted sound. Environmentally the most annoying noise probably comes from barking dogs, domestic noise, transportation, construction and demolition works, and plant such as air conditioning or freezer motors.

Noise policy in the United Kingdom is being driven by European Union (EU) directives and the Ambient Noise Strategy [1] has been adopted by the UK Government. This strategy is designed to target the five perceived elements of noise nuisance:

- The behaviour and attitudes of the noisemaker
- The perception and resulting actions of the noise sufferer
- The passage or path of the sound between the two
- Any equipment or devices used in the production of the noise and
- The possible routes to conflict resolution – court action, mediation and so on.

It is recognized that control of the above will sit between central and local government.

Phase 1 of the Ambient Noise Strategy (2002–5) aims to establish:

- the ambient noise climate in this country. In simple terms the number of people affected by different levels of noise (i.e. road, rail, airports and industry) and the location of the people affected;
- the adverse effects of ambient noise, particularly regarding people's quality of life. Special consideration will be needed in regard to tranquillity;
- the techniques available to take action to improve the situation where it is bad or to preserve it where it is good; and
- the methodology to be used to undertake economic analysis.

Phase 2 (2004–6) aims to 'evaluate and prioritise' the need for action identified in Phase 1 bearing in mind benefits in terms of cost and environmental, economic and social issues.

Phase 3 (2007) The government will consider the way forward to complete the ambient noise strategy. It is intended that revisions to the strategy should be on a five-year cycle.

Local Authorities will have a key role in the development and implementation of the strategy alongside transport authorities.

Throughout Europe, it has been estimated that, around 20% of the population of the EU, approximately 80 million people, are exposed to noise levels that are considered unacceptable and likely to affect health [2]. A further 170 million people live in areas where noise levels are likely to cause serious annoyance during the daytime [2].

The Wilson Committee Final Report published in 1963 [3] can still be considered to be the most comprehensive study of noise in the United Kingdom. The committee was appointed in April 1960 'to examine the nature, sources and effects of the problem of noise and to advise what further measures can be taken to mitigate it'. Much of what the committee recommended later became incorporated in BS 4142:1967, which was revised in 1990 and again in 1997 [4]. The report highlighted road traffic, railways, aircraft, construction and entertainment/advertising as the sources responsible for the majority of unwanted sound affecting the population. Road traffic was considered the prime source of noise affecting people. The Wilson report so provides the widely accepted definition of noise: 'sound which is undesired by the recipient'.

A number of surveys have been undertaken to determine noise levels in the United Kingdom and the effects of environmental noise on people at home. The Building Research Establishment (BRE) carried out a survey in 1990 [5] at involved measuring noise levels over a 24-hour period outside 1000 dwellings in England and Wales. In addition to the measurement of sound levels, the survey also recorded the sources of noise identified at each site.

Data published in 1993 indicated that 7% of the dwellings were exposed to noise levels above 68 dB $L_{A10,\ 18\ hour}$, the level above which sound insulation compensation must be paid in the case of new road developments. Furthermore, at over 50% of the sites the daytime levels ($L_{Aeq,\ 16\ hour}$) exceeded 55 dB(A). The World Health Organization (WHO) [6] has suggested that general daytime outdoor noise levels of less than 55 dB L_{Aeq} are desirable to prevent any significant community annoyance. The main source of noise identified at the sample sites was road traffic noise, which was noted at 92% of the sites (Table 35.1).

Table 35.1 Noise sources outside dwellings: England and Wales, 1990

Noise source	Percentage of sample sites at which noise source was recorded
Roads	92
Aircraft	62
Animals and birds	57
Trees rustling	18
Children	18
Domestic	16
Railways	15
Farm equipment	10
Construction work	5
Industry	4
Motorways	2

Source: Sargent, J.W. and Fothergill, L.C. (1993) A survey of environmental noise levels in the UK. *Proceedings of Noise*, 93(7) 161–6.

A noise attitude survey [7] was carried out in 1991 by the BRE as part of the Department of the Environment's noise research programme. One adult from each of 2373 randomly selected households was surveyed. The results of the survey indicated that road traffic noise (28%) was the most widespread form of noise disturbance, followed by neighbour noise (22%), aircraft noise (16%) and train noise (4%). This survey is considered to have produced responses that, for the first time, fully reflected the extent and nature of the adverse consequences of noise.

The degree to which noise affects people depends on a number of factors.

1. *Frequency.* Generally, the higher the frequency, the greater the annoyance.
2. *Loudness.* Generally, the louder the noise, the greater the nuisance. It has been found that dislike of noise is more related to loudness than to any other easily measured characteristic.
3. *Time of day.* Sounds acceptable at 6 p.m. may not be acceptable at 6 a.m.
4. *Unexpectedness.* The effect of this is more apparent on some than on others, but all the evidence points to the effect being of short duration.

5. *Uncertainty of direction and unfamiliarity.* It is a natural reaction to experience a sense of unrest until the direction and source of a noise is established.

6. *Irregularity and duration.* Continuous impulsive or rhythmic noise is more irritating than a 'smooth' noise: with intermittent noise much will depend on the frequency of the repetition, for example, a noise 4 times a day is not as annoying as one heard 40 times a day.

7. *Necessity.* If the noise-generating activity is trivial or thought by the complainant to be unnecessary, his or her annoyance is increased.

8. *General state of health, sensitivity and emotional attitude of a person towards noise.* Poor health or emotional instability lowers the tolerance level of acceptability.

9. *Level of background noise.* The difference in loudness between a noise and the background noise is very important. The greater the difference, the greater the annoyance is likely to be – BS 4142: 1997 [4].

10. *Economic link.* A factor that might also affect a person's reaction to a noise is his involvement with the source of the noise. It is noticeable when investigating complaints of noise arising from a factory, that those who work there or occupy houses belonging to the company are much less voluble in their complaints. This may also explain why only a few isolated complaints are sometimes received in respect of noise that could rightly be regarded as a serious nuisance.

These factors indicate that the assessment of noise is subjective; sound, on the other hand, can be measured objectively (see section on sound and vibration). The effects of noise on people are various, often interrelated, and vary from person to person.

Noise can induce acute physiological responses such as increases in blood pressure and pulse rate. The quality of the noise seems to be of importance: uncontrollable noise from a neighbour's voice is reported as being more irritating than impersonal sounds such as traffic and machine noise [8]. There is evidence that the adverse effects of noise depend, to some extent, on individual characteristics that have been termed 'sensitivity' and predisposition to 'annoyance'.

Nuisance from noise is much more likely when sound insulation is inadequate, and is therefore commonly found in low-cost housing [9]. Moreover, there is evidence to suggest that self-reported nervousness, irritability and poor concentration are related to exposure to traffic noise [10]. Noise at night interferes with sleep patterns, shortens the time spent in deep sleep and reduces rapid eye movement (REM) sleep [11]. This pattern, if prolonged, can give rise to anxiety, headaches, irritability and chronic fatigue [12]. Moreover, daytime fatigue can impair motor co-ordination [12].

LEGAL FRAMEWORK

Prior to 1960, local authorities had to rely on local Acts and by-laws made under the Local Government Act 1933 and on common law to control noise. The findings of the Wilson Committee [3] resulted in the passing of the Noise Abatement Act 1960. This Act stated that noise or vibration that was a nuisance should be a statutory nuisance for the purposes of Part III of the Public Health Act 1936. Furthermore, restrictions were introduced on the operation on highways etc. of loudspeakers.

Concern about noise problems continued to develop rapidly during the 1960s, and in 1970 the government formed the Noise Advisory Council (NAC) with a brief: 'to keep under review the progress made generally in preventing and abating the generation of noise, to make such recommendations to ministers with responsibility in this field and to advise on such matters as they refer to council'.

The NAC was formed to keep the noise abatement activities of central and local government under review. To help the NAC achieve its aim, working groups concentrating on different aspects of noise abatement policy were set up. The NAC was disbanded in 1981 but it produced a number of reports and publications, some of which will be referred to later. The work of the NAC assisted the government in drawing up Part III of the Control of Pollution Act 1974, which repealed the Noise Abatement Act 1960. The 1974 Act formed the main framework of noise control legislation and provided powers for dealing with noise from construction sites, noise in streets and for drawing up codes of practice on particular noise problems such as burglar alarms (see section on Codes of Practice). It also empowered local authorities to set up noise abatement zones (see section on Noise abatement zones).

Although still applicable in Scotland, the noise nuisance provisions of the Act in England and Wales are now covered by Part III of the Environmental Protection Act (EPA) 1990 (see Chapter 6). The Noise and Statutory Nuisance Act 1993 amended the EPA and extended the duties of local authorities to deal with noise nuisance in the street, introduced discretionary powers to allow the operation of loudspeakers in the street, and provided adoptive powers relating to the control of the installation and operation of burglar alarms (see section on Codes of Practice). The Health and Safety at Work, Etc. Act 1974, generally covers occupational noise by requiring employers to provide a safe place and safe system of work. The Noise Act 1996 received Royal assent in July 1996 and has now been fully implemented. The Act covers noise nuisance from domestic premises (see section on Codes of Practice).

As with other areas of environmental health work, the EU has adopted directives relating to noise abatement which the United Kingdom must implement. Directives (see below) stating maximum permitted noise levels for motor vehicles, construction plant and subsonic aircraft and detailing measurement procedures for all these sources have been adopted. The Noise at Work Regulations 1989 implemented Directive 86/188/EEC on the protection of workers from the risk related to exposure to noise at work.

EU noise programme

Although environmental noise continues to be one of the main local environmental problems in Europe, action to tackle it has generally been given a low priority. In order to remedy this, the 1993 **Fifth Environmental Action Programme** detailed a number of targets for noise exposure to

be achieved by 2000. The review of the Fifth Action Programme (COM(95)647) announced the development of a noise abatement programme for action to meet these targets. (Now see European Parliament and Council Decision 2179/98/EC on the review of the EC programme of policy and action in relation to the environment and sustainable development, (OJ) L275 10.10.98.) The **European Commission Green Paper –** *Future Noise Policy* (COM(96)540(final)) adopted on 5 November 1996, aims to stimulate public discussion on the future approach to noise policy and is seen as the first step in the development of a noise abatement programme.

In the past, EU environmental noise policy has concentrated on legislation to fix the maximum sound levels for vehicles, aircraft and machines in order to ensure that new vehicles and equipment, at the time of manufacture, comply with the noise limits laid down in directives.

Although this has resulted in significant reductions of noise from individual sources, little has been achieved in relation to improvements in exposure to environmental noise. Therefore a new framework for noise policy is proposed in the green paper.

1. A proposal for a directive providing for the harmonization of methods of assessment of noise exposure and the mutual exchange of information. The proposal could include recommendations on noise mapping and the provision of information on noise exposure to the public. In a second stage consideration could be given to the establishment of target values and the obligation to take action to reach the targets.

2. The next phase of action to reduce road traffic noise will address tyre noise and look at the possibilities of integrating noise costs into fiscal instruments, amending EU legislation on roadworthiness tests to include noise, and the promotion of low noise surfaces through EU funding.

3. More attention needs to be paid to rail noise where some member states are planning national legislation and where there is considerable opposition to the expansion of rail capacity due to excessive noise. In addition to supporting research in this field, the Commission will investigate the feasibility of introducing legislation

setting emission limit values, negotiated agreements with the rail industry on targets for emission values, and economic instruments such as variable track charges.

4. In air transport the Commission is also looking at a combination of instruments. These include greater stringency in emission values and the use of economic instruments to encourage the development and use of lower noise aircraft, as well as the contribution that the local measures such as land use planning can make.[1]

5. The Commission plans to simplify the existing legislation setting emission limits for a limited range of outdoor equipment and will propose a framework directive covering a wide range of equipment including construction machinery, garden equipment and others and incorporate the existing seven directives. The principal feature of the new legislation will be the requirement to label all equipment with the guaranteed noise level. Limit values will only be proposed for equipment for which there is already noise legislation and a limited range of highly noisy equipment.

It is hoped that this green paper will help give noise abatement a higher priority in EU policy-making.

THE NOISE REVIEW WORKING PARTY

Although controls exist to deal with noise, it is still a major problem. In December 1989, the Secretary of State for the Environment announced a review of the controls on environmental noise. In February 1990, the Noise Review Working Party was formed with the following terms of reference:

The noise review will consider aspects of noise control, with particular reference to Part III of the Control of Pollution Act 1974. Topics of interest will be burglar and car alarms, traffic noise, noise and planning (with reference to Department of the Environment Circular 10/73) and neighbourhood noise. The review will consider the possible need to increase the powers of local authorities to deal with noise nuisance. There will also be an examination of the current state of research regarding measurement of noise and its effects.

The review was extended to include aircraft noise.

The Working Party Report [13] was published in October 1990 and made 53 recommendations. Some of those in relation to noise nuisance were included in the Environmental Protection Act 1990 (Section 79) and others were incorporated into the government's white paper on the environment [14] published in September 1990. The recommendations and the government's response to them are considered in relevant sections later in the chapter.

SOUND AND VIBRATION

Characteristics of sound and units of measurement

Sound is propagated by a source vibrating, which causes any elastic medium such as air to fluctuate. If a stone is thrown into a pond, waves can be seen radiating outwards across the surface of the pond. Sound propagates in three dimensions, not just two as with the water.

The sound waves that travel outwards from the source are really minute positive and negative fluctuations of pressure caused by the air being alternatively compacted and rarefied by the vibrating source. The lowest pressure difference detectable by the average person is 0.00002 Newtons per metre squared (2×10^{-5} N/m^2 [Pa]), while the largest pressure difference detectable without pain is 20 N/m^2 (Pa).

A sound level meter is a sound pressure meter measuring these fluctuations in pressure. Because the range of pressures detectable by the ear is so great, that is 20 Pa is one million times greater than 2×10^{-5} Pa, a logarithmic scale is used for convenience. The decibel (dB) is the unit used to measure the sound pressure variation; it is a ratio

---

[1]Now see the Aerodrames (Noise Restrictions) (Rules and Procedures) Regulations 2003 which Implement Directive 2002/30/EC (OJ L8528.3.02, p. 40).

of measured pressure squared to a reference pressure squared:

$$\log_{10} \frac{(\text{pressure})^2}{(2 \times 10^{-5})^2} \text{ bels}$$

or

sound pressure level (SPL)

$$= 10 \log_{10} \frac{(\text{sound pressure})^2}{(\text{reference pressure})^2}$$

where reference pressure $= 2 \times 10^{-5}$ N/m². Decibels, being logarithmic, cannot be added or subtracted arithmetically, that is two noise sources each producing 80 dB do not combine to produce 160 dB but 83 dB. The method of addition of SPLs is indicated in Table 35.2.

If the difference between two sources is 10 dB or more, then the SPL remains that of the higher. Some examples of sound pressure levels are shown in Table 35.3.

Table 35.2 Addition of sound pressure levels

Difference in dB between two sources	0	1	2	3	4	5	6	7	8	9	10
Add to highest level	3	2.5	2	2	1.5	1	1	1	0.5	0.5	0

Table 35.3 Typical sound pressure levels

Sound pressure (N/m²)	Sound source	SPL (dB)
0.00002	Silence (threshold of hearing)	0
0.0002	TV studio	20
0.002	Soft whisper	30
	Public library	40
0.02	Normal conversation at 1 m	60
	Radio set (loud)	70
0.2	Small car at 7.5 m	80
	Heavy lorry at 7.5 m	90
2.0	Pneumatic chipper	100
20.0	Boilermaker's shop	120

Frequency

Having considered pressure waves moving out from the source, frequency or the rate of repetition needs to be understood. With a continuous sound, a succession of air compactions will pass out from the source and the distance between each compaction will be the same – this is called the wavelength. The velocity of sound in air at sea level and at 20°C is 340 m/sec. If we take this as a working figure, we can see that:

$$\text{frequency} = \frac{\text{speed of sound (C)}}{\text{wavelength}}$$

or

$$\text{speed of sound} = \text{wavelength} \times \text{frequency}.$$

It is of interest to note that the velocity of sound is dependent on the medium through which it travels (Table 35.4), and it can be shown that it is dependent on the medium's modulus of elasticity (Young's modulus [E]) and its density.

The human perception of sound is dependent upon the frequency of the sound. The ear's best response is between 1000 and 4000 Hz. Generally, in noise control we are looking at a combination of frequencies rather than a single frequency or pure tone. For ease of measurement, the frequency spectrum can be divided into octave or one-third octave bands. Octave bands are ranges of frequency; each octave is designated by a number which is the square root of 2 multiplied by the lower frequency, that is 88–176 Hz has a centre frequency of 125 Hz.

Table 35.4 Velocity of sound in various materials

Material	Approximate velocity of sound (m/sec)
Air	340
Lead	1220
Water	1410
Brick	3000
Wood	3400
Steel	5200

One-third octave bands are exactly that, that is each band is one-third of an octave. Acoustic engineers involved in the design of silencing for a factory may well need to measure frequency by narrow band analysis, in which band widths can be much smaller, to identify particular problems.

'A' weighting

An A-weighting filter is incorporated into sound level meters allowing direct measurement of this parameter. The A-weighted frequency response approximately follows the 40 phon equal loudness curve and correlates closely with people's subjective response to noise. A-weighting is now used internationally in standards to rate noise (Fig. 35.1). There are other weighting curves (B, C, D and E), but all have specialized uses, the A weighting being by far the most frequently used.

Sound power level

Sound power level (SWL) is a measure of the total energy of the source; it is independent of the environment. The sound pressure level can be calculated at any distance from the source if the SWL of the source is known:

Sound power level (SWL)

$$= \frac{\text{(Sound power)}}{\text{(reference power)}} \text{ dB}$$

Convention is to use sound intensity rather than pressure, and intensity is expressed in watts per square metre. The reference sound power is 10 W.

Sound pressure is related to unit area. Therefore to determine the SWL, the sound pressure over a total area enclosing the source has to be found.

Therefore: SWL = SPL + 10 $\log_{10}$ A dB where A is the area in metres squared.

We know that A = $4\pi r^2$ where r is the distance from the source in metres.

So: SWL = SPL + 10 $\log_{10} 4\pi r^2$
    = SPL + 20 $\log_{10}$ r + 11.

Loudness

Loudness is a subjective characteristic of sound, that is a listener's impression of the amplitude of sound. Loudness is primarily related to sound pressure, but sound of a constant pressure can be made to appear louder or quieter by changing its frequency. In practice, an increase in noise level of 10 dB(A) at any point in the decibel scale approximates to a doubling of loudness. The unit used to indicate loudness is the phon which is defined in BS 3045:1981 [15] as a dimensionless unit used to express the given loudness of a sound or noise.

The loudness level in 'x' phons is judged by the average person to be equal in loudness to 'x' decibels at 1000 Hz, for example any sound judged to be equal in loudness to a pure tone of 50 dB at 1000 Hz has a loudness level of 50 phons. The phon scale is logarithmic. As phons are dimensionless units that allow for variation of the sensitivity of the ear with frequency, they do not give a direct indication of a change in loudness level, for example a sound of 60 phons is not generally twice as loud as a sound of 30 phons. The sone scale of loudness gives a direct relationship between different loudness levels.

The sone is defined in BS 3045:1981 [15] as 'a unit of loudness on a scale designed to give scale

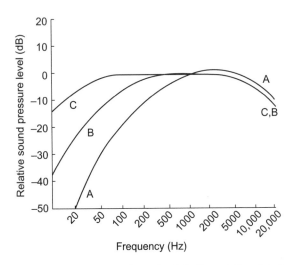

Fig. 35.1 International standard A, B, and C weighting curves for sound level meters.

numbers approximately proportional to the loudness'.

The relationship between phons and sones is shown in Table 35.5.

A noise of 64 sones (100 phons) is twice as loud as a noise of 32 sones (90 phons). Units of loudness do not usually feature in noise control measures.

NOISE MEASUREMENT

Noise indices

In practice in the real environment, noise levels are almost always fluctuating widely, are often multi-sourced, and the sources are often transient or mobile; a simple instantaneous measurement is therefore meaningless. As a result of these technical and social difficulties, surveys have been made to quantify the relationship between noise and its effect on the quality of life.

Different indices are used for different noise resources; they all relate the noise to time either directly or statistically. The indices most likely to be used by environmental health staff are $L_{Aeq,T}$, $L_{EP,d}$, L_{A10}, L_{A90} and $L_{Ar,T}$. Others sometimes used include noise and number index (NNI) and traffic noise index (TNI).

$L_{Aeq,T}$ (A-weighted equivalent continuous sound level with respect to time)

This is a measure of the energy of a noise. It is defined as 'the value of the A-weighted sound

Table 35.5 Relationship between phones and sones

Phones	Sones
40	1
50	2
60	4
70	8
80	16
90	32
100	64
110	128
120	256

pressure level in decibels of continuous steady sound that within a specified time interval, T, has the same mean-square sound pressure as a sound that varies with time'.

$$L_{Aeq,T} = 10 \log_{10}\left\{ 1/T \int_{t1}^{t2} (PA^2(t)PO^2)dt \right\}$$

This index is now widely used for environmental measurement, being quoted in BS 4142:1997 [4] as well as in the Health and Safety at Work, Etc. Act 1974. It is also increasingly being used for aircraft and railway noise.

$L_{EP,d}$

The $L_{EP,d}$ is the index used in the Noise at Work Regulations 1989 (SI 1989 No. 1790) and is the 'daily personal exposure to noise'. It is arrived at by measuring the A-weighted sound pressure level and the duration of exposure. One method of measuring it is to use an integrating sound level meter complying with BS EN 60804:1994 [14] and measuring the L_{Aeq}. It can also be measured on a simple sound level meter and calculated using the nomogram shown in Noise Guide No. 3 [16].

L_{A10} (18 hour)

This index of traffic noise is used as the basis for measurement in the Noise Insulation Regulations 1973 and 1975 (SI 1977/1763) made under the Land Compensation Act 1973. It is the sound level in dB(A) exceeded for 10% of the time worked on an hourly basis from 6 a.m. to midnight (18 hours) on a normal weekday.

$L_{A90,T}$ (background noise level)

The $L_{A90,T}$ is the A-weighted sound pressure level of the residual noise in decibels exceeded for 90% of a time 'T'. It is the level quoted in BS 4142:1997 [2] for background noise.

$L_{Ar,T}$ (rating level)

The $L_{Ar,T}$ is the equivalent continuous A-weighted sound pressure level in decibels at the measurement position produced by the specific noise source under investigation over a given reference time interval plus any adjustments for the character of the noise. It is quoted in BS 4142:1997 [4].

Noise and number index

This index of aircraft noise measurement is based on the combined effect of the loudness of individual aircraft and the number of flights in a specified period. The NNI is not relevant to airports with fewer than 80 aircraft per day, each of which needs to produce an average peak noise level of more than 80 PNdB. NNI was introduced following a survey at Heathrow Airport; it has now been superseded by $L_{Aeq,T}$.

Traffic noise index

This index resulted from a survey of noise levels and annoyance in Greater London in 1966 [15]. It is based on A-weighted sound pressure levels:

$$TNI = L_{A90} + 4(L_{A10} - L_{A90}) - 30.$$

Noise monitoring equipment

Sound level meters

A sound level meter is an instrument designed to measure sound pressure in an objective reproducible manner, and consists of a microphone, a processing unit and a visual display or read-out facility (Fig. 35.2).

Sound measuring equipment must currently meet the following standards:

1. BS EN 60651:1994 (specification for sound level meters) [17].
2. BS EN 60804:1994 (integrating/averaging sound level meters) [18].

The standards specify four degrees of precision for sound level meters.

1. Type 0 (intended as a laboratory reference standard).
2. Type 1 (for laboratory and field use where the acoustic environment can be closely specified and/or controlled).

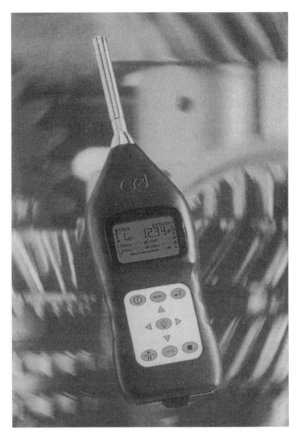

Fig. 35.2 Sound level meter (400 series). (Courtesy of CEL Instruments Ltd, 35–37 Bury Mead Road, Hitchin, Hertfordshire, SG5 1RT.)

3. Type 2 (for general field applications).
4. Type 3 (for field noise surveys).

Many of the regulations made under legislation enforceable by environmental health departments specify equipment complying with type 2 or better.

Environmental noise analysers

An alternative method of making long-term environmental noise surveys is to use either the environmental noise analyser (Fig. 35.3) or a graphic recorder. These instruments are self-contained, reasonably robust and by the use of microprocessor control can be programmed for prolonged hands-off monitoring in the field. Noise analysers produce a complete written history of

Fig. 35.3 Sound level meter (type 2260). (Courtesy of Brüel & Kjaer (UK) Ltd, Harrow Weald Lodge, 92 Uxbridge Road, Harrow Weald, Middlesex, HA3 6BZ.)

statistical parameters over the measurement period, and graphic recorders give the operator an analogue trace of the time/noise signal in addition to their ability to produce sequential set period L_{Aeq} bar charts.

Calibration of noise monitoring equipment

All sound-measurement equipment should be calibrated before and after each series of measurements using a calibrator or pistonphone complying with class 2, BS 7189:1989 [19]. If carrying out prolonged measurements, intermediate checks on calibration should be made.

Every two years, all equipment should be sent to an accredited laboratory for a compliance test; calibrators and pistonphones must comply with BS 7189:1989[19], and other equipment with BS EN 60804:1994 [18] or BS EN 60651:1994 [17] type 1 or 2.

VIBRATION

The Environmental Protection Act 1990 Part 3, Section 79 (statutory nuisances and clean air) defines noise as including vibration. Vibration is the movement of a solid body about an axis.

An oscillating metal sheet can be seen to display three physical properties: first, it moves over a distance, which is called its displacement; second, it has a velocity, which is at its peak as it crosses its axis and is zero at its point of maximum deflection; and, thirdly, it has positive and negative acceleration. These three physical properties are linked mathematically, and can each be measured individually or together.

There are two main types of vibration: deterministic and random. Deterministic vibration can be determined by mathematical formulae, while random vibration has to be determined by statistical means.

Generally, the root mean square (RMS) values are used for measurement as these reflect the time history, and peak measurements reflect an instantaneous moment only.

$$X_{RMS} = \sqrt{1/T} \int_0^{At} X^2(t)dt.$$

Investigating officers may wish to measure vibration to determine whether damage could arise, or if a nuisance is being caused. Measurements for nuisance confirmation are likely to be taken within a building, and it is worth noting that the worst effects will normally be felt at mid-floor spans on upper floors. BS 6472:1984 [20] specifies the measurements that should be made to determine human reaction to vibration.

NOISE CONTROL TECHNIQUES

Attenuation

Sound propagates outwards from its source; if the source is comparatively small in relationship with the distance to the receiver it can be considered a point source. Large sources, such as factories, listened to from close by, or sources such as roads, are considered as line or plane sources.

With a point source the intensity of sound is inversely proportional to the square of the distance from the source to the receiver (Fig. 35.4). In the case of a line source, the intensity is inversely proportional to this distance.

A point source gives an attenuation of 6 dB(A) per doubling of distance. A line source gives an attenuation of 3 dB(A) per doubling of distance. In practice quite large sources can be treated as points in origin.

In addition to the attenuation due to distance, many other factors can alter the sound level between the source and receiver. Meteorological changes, such as temperature gradient and wind, can play a part, and so it is recommended that no readings are taken with a wind speed above 24 kph (15 mph). The ground surface also plays an important part in altering the sound level – soft, absorbent vegetation attenuates sound quicker than hard concrete surfaces.

Barriers are often introduced primarily to reduce noise. This technique is often used where motorways pass through noise-sensitive areas (Fig. 35.5).

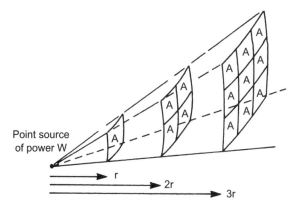

Fig. 35.4 The dispersion of sound from a sound point. It can be seen that the intensity is inversely proportional to the square of the distance between source and receiver, that is it attenuates 6 dB per doubling of distance. (Courtesy of Brüel & Kjaer (UK) Ltd, Harrow Weald Lodge, 92 Uxbridge Road, Harrow Weald, Middlesex, HA3 6BZ, *Acoustic Noise Measurement*.)

Mass law

Within a building the airborne sound insulation offered by a wall is quoted in terms of the sound reduction index (SRI), which is dependent on the transmitted sound power and the incident sound power:

$$SRI = \frac{Wi}{Wt}$$

where Wi = incident sound power and Wt = transmitted sound power.

The mass law states that within the mass controlled region, the SRI increases by 6 dB(A) for each doubling in frequency or given mass per unit area, or for each doubling of mass (thickness) at a given frequency.

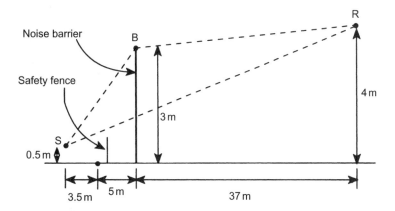

Fig. 35.5 Path of noise transmission. S = source; B = barrier; R = reception point. (Courtesy of the Department of Transport and the Welsh Office, *Calculation of Road Traffic Noise*, HMSO, London.)

$$\text{SRI} = 20 \log_{10} (\text{fM}) - 47 \text{ dB}$$

where f = frequency of the incident sound in Hz and M = mass per unit area of the wall in g/m^2.

Noise in buildings

Sound generated within a room reaches the listener both directly from the source, and also from reflections from the walls and furniture of the room. A room with hard surfaces readily reflects sound, and the reflections or echo take a comparatively long time to die away. This is a reverberant or live room. A room with many absorbent surfaces is quiet, the sound dying away quickly. It is an acoustically dead room.

Reverberation time (RT) is the time it takes for the sound level within a room to decrease by 60 dB. Sabine's formula shows the relationship between the absorption of a given room and the RT.

$$\text{RT} = \frac{0.161\text{V}}{\text{Sa}}$$

where V = the room's volume, S = the room's total surface area, and a = the absorption coefficient.

The Building Regulations 2000

The vast majority of noise complaints dealt with by local authority environmental health departments relate to domestic noise and to address this Part E of the Building Regulations has been amended to allow for up to 10% testing of new properties and conversions. This includes hostel and hotel type accommodation. The local authority may specify which units are to be tested, the developer must arrange and pay for the testing to be carried out.

For conversions of houses to hotel and hostel accommodation, the new regulations will come into force on 1 July 2003, for newly built houses and flats, it is expected that the amended regulations will come into force on 1 July 2004. This delay will give the House Builders Federation (HBF) the opportunity to formulate 'Robust Standard Details' for sound insulation of walls and floors. If the HBF can show that their Robust Standard Details provide consistantly good performance routine testing may be deemed unnecessary, the Building Regulations Advisory Committeee will determine if this has been achieved.

Enclosures

Noisy machines or activities can often be installed, or carried out, in enclosures or partial enclosures. BS 5228:1984 [21] Part I gives examples of the use of enclosures on construction sites. Enclosures used in factories nearly always have to have openings for access or ventilation but, even then, using sheet material with a weight of 10 kg/m^2 lined with 25 mm absorbent lining, it is possible to achieve a reduction in noise of about 20 dB.

AV mounts

The mounting of machinery on antivibration (AV) mounts can have a positive effect in reducing structure-borne noise. Care must be exercised in the choice of mounts as vibration produced by a machine at a lower frequency than its mounted resonant frequency will not be isolated. Vibrations at the resonant frequency may be amplified. Mountings should have internal damping. The displacement of all mounts must be equal to avoid the machine starting to rock.

Examples of practical noise control

Noise reduction at source is always the best answer when possible. The ideal solution is to stop making the noise. Is there a quieter way of doing the job? Is maintenance up to scratch? Blunt saws, worn bearings or transmissions can increase noise from a machine. Where two gear wheels of the same diameter, one with twice as many teeth as the other, are operating at the same rotational velocity, the one with twice the number of teeth will produce its noise at twice the frequency of the other because the source of the noise is a succession of impacts as the teeth mesh.

Machines can be enclosed or screened; in general, the closer the screen is to the machine, the greater the reduction in noise. Machines can be re-sited indoors or behind a building. An enclosure should be airtight and made of a heavy material that has stiffness designed into it. The more expensive commercially available machine enclosures often make use of a composite material made up of a dense material encased in steel. The inside of an enclosure should be lined with an acoustically absorbent material to reduce reflections and thus lower the internal noise levels.

Fan noise is a frequent source of complaint. If the noise is tonal in character, it can be beneficial to operate a large diameter fan at a slower speed. However, because the noise now generated will be at a lower frequency, it will travel further away and may widen the area of complaint. A better solution may be to install a silencer behind the original fan.

PLANNING AND NOISE

The right to enjoy the environment without being subjected to excessive noise intrusion has been established at common law. Obviously, prevention is better than cure and consideration should therefore be given to the likely effects of noise at the development control stage.

In order to address the issue of planning and noise and provide guidance for local authorities, the government introduced Department of the Environment Circular 10/73 [22]. This circular was directed primarily at the control of development and acknowledged that a great deal could be achieved by positive planning to reduce the risk of noise disturbance. It emphasized the need for close co-operation between planning, environmental health and highways authorities in the control of developments to contain and, where possible, reduce the impact of noise.

The circular gave guidance to local authorities on the use of their planning powers, and stated that new noise-sensitive development should not be allowed in areas exposed to unacceptable levels of noise. For example, the circular presumed that sites were not suitable for residential development if they were subjected to a traffic noise in excess of 70 dB(A) on the L_{10}(18 hour) scale unless the

properties were screened from the noise, or sound insulated to the standard prescribed in the Noise Insulation Regulations 1975. It also made references to noise from roads, aircraft and industry, and suggested criteria upon which local authorities should base their policies.

The Noise Review Working Party Report 1990 [13] considered the issue of planning and noise and the following recommendations were made.

1. Circular 10/73 should be revised as quickly as possible to include areas of concern such as aircraft noise including helicopters, noise from railways, sporting activities, recreation and entertainment. It should also distinguish between industrial noise from fixed installations, and noise from mineral workings and waste disposal facilities. Furthermore, it should make a requirement that all applications except for housing should include information relating to the possible noise implications of the project.
2. The establishment of action levels to assist in the assessment for noise-sensitive developments.
3. BS 4142:1997 should be more extensively revised to take account of peaks of noise and of noisy events of short duration, particularly at night.
4. More research in respect of community response to various types of industrial noise is required.
5. The provisions of Section 60 of the Control of Pollution Act 1974 should be extended to mineral extraction sites, oil and gas sites and waste disposal sites in conjunction with a specific code of practice for these sites.

The government issued Planning Policy Guidance (PPG) 24 *Planning and Noise* [23] in 1994 to replace Circulars 10/73 [22] and 1/85 [24]. The PPG takes on board the comments and recommendations of the working party and provides more comprehensive guidance about a wider range of noise sources, namely noise from traffic, aircraft, railways, industrial and commercial developments, construction and waste disposal sites, and sporting, entertainment and recreational activities. Furthermore, in order to assist local authorities in the appraisal of new noise-sensitive developments adjacent to a noise source, the concept of noise exposure categories was introduced. Four categories were created and recommended noise limits for each category for the development of dwellings and schools exposed to the various noise sources are detailed in an appendix to the document. The categories are:

A. For proposals in this category, noise need not be considered as a determining factor in granting permission, although the noise level at the high end of the category should not be regarded as a desirable level.
B. In this category authorities should increasingly take noise into account when determining planning applications, and require noise control measures.
C. There should be a strong presumption against granting planning permission for proposals in this category. Where permission is given, that is if no other suitable quieter sites are available, conditions should be imposed to ensure an adequate level of insulation against external noise.
D. For proposals in this category planning permission should normally be refused.

Where a planning authority is disposed to give planning permission for a noisy development, conditions should be attached to any permission granted. Local authorities should refer to the model conditions detailed in PPG 24.

Local authorities should impose conditions only where they are satisfied that they are:

1. necessary
2. relevant to planning
3. reasonable
4. enforceable
5. relevant to the development
6. precise.

A review of the implementation of PPG 24 by Goswell may be found in [25].

In April 1993, the Department of the Environment published *Mineral Planning Guidance* Note 11 (MPG 11) [26]. This provides advice for both mineral planning authorities and the industry on how the planning system can be used to keep noise emissions from surface mineral workings within environmentally acceptable limits without imposing unreasonable burdens on the mineral operators. The main provisions of MPG 11 are summarized below.

1. The use of a modified form of BS 5228 [21] on the prediction of noise emissions is recommended. The model takes account of barrier and soft ground attenuation, weather conditions, noise reflection and mobile plant.
2. Two methods for setting noise limits for mineral site are recommended. One method is based on BS 4142 [4], the other on absolute noise limits, which can be incorporated into planning conditions.
3. Advice is given on how to monitor noise levels from surface mineral sites effectively.
4. Advice is given on noise abatement methods, which can be the subject of planning conditions and/or incorporated into good practice by the operator.

Rating noise

BS 4142:1997 [4] provides a method for rating industrial noise affecting mixed residential and industrial areas. This should be used for assessing the calculated or measured noise levels from new or modified premises, or the measured noise levels from existing premises. This is a useful method when considering the likely environmental effects of new or modified industrial developments.

The revised standard recommends the use of $L_{Aeq,T}$ as the specific noise level of the noise source under investigation, and $L_{A90,T}$ for background noise levels. The method is not applicable for assessing noise where background noise levels are below a sound pressure level of 30 dB(A), or for assessing noise levels measured inside buildings. An assessment of the noise for complaint purposes is made by subtracting the measured background noise level from the rating level ($L_{Ar,T}$), which is the specific noise level plus any adjustment for the character of the noise. A difference of 10 dB or more suggests that complaints are likely. As the difference decreases the likelihood of complaint diminishes; at a difference of −10 dB there is a positive indication that complains are unlikely.

ENVIRONMENTAL ASSESSMENT

An environmental assessment (EA) is a technique for gathering expert quantitative analysis and qualitative assessments of a proposed development's likely environmental effects. A description of the project and measures to be taken to minimize any adverse effects on the environment, including, for example, a noise assessment, might be required by the planning authority when considering applications to which these regulations apply. Environmental health departments have an important role to play in advising planning departments and their committees. Equipment should be available within the department to predict, monitor and assess the likely environmental effects of proposed developments. Where possible, one officer from the department should liaise with the planning department, all applications should be screened, and comments and any conditions considered necessary should be forwarded to the planning department as quickly as possible. A representative from the department should also attend and advise at committee meetings where planning decisions are taken.

NEIGHBOURHOOD NOISE[2]

This term can be used to describe noise from a number of sources, for example industrial noise, transport noise, entertainment noise, noise in the street and noise from neighbours. The 1991 BRE survey [5] indicated that 22% of the adult population is bothered by neighbourhood noise or, more precisely, by neighbour noise. Neighbour noise complaints generally relate to hi-fi equipment, barking dogs, do-it-yourself (DIY) activities, voices, car repairs and domestic activities.

Neighbourhood noise is not a problem that is solely confined to large towns and cities. In the country, people can be affected by noise sources such as industrial operations such as quarrying, agricultural activities, and leisure activities such as clay pigeon shooting. In towns and the country alike, noise from sources such as well-established

industry or agricultural activities is often tolerated until new residents, with higher environmental expectations, move into the area.

Noise nuisances

Noise nuisance arising from fixed premises can be dealt with by the nuisance procedures of the Environmental Protection Act 1990, except in Scotland where Sections 57–59 of the Control of Pollution Act 1974 still apply. These Acts were amended by the provisions of the Noise and Statutory Nuisance Act 1993. This Act extended the list of statutory nuisances to include 'noise that is prejudicial to health or a nuisance and is emitted from or caused by a vehicle, machinery or equipment in the street'. The aim of the Act is to control nuisances caused by vehicle repairs, persistent DIY, vehicle alarms and noise from refrigeration units and generators on stationary vehicles. It is a defence against a statutory nuisance action to prove that the 'best practicable means' have been employed to control the noise, as defined in Section 79 of the Environmental Protection Act 1990 and Section 72 of the Control of Pollution Act 1974 (see Chapter 6). This defence is only available in the case of the new provisions where the vehicle, equipment or machinery is used for industrial, trade or business purposes. In the case of entertainment noise where, for example, a public entertainment licence is required, local authorities may, when issuing the licence, attach noise conditions. In these cases, nuisance action can still be taken if necessary.

Loudspeakers in streets

The Noise and Statutory Nuisance Act also amended Section 62 of the Control of Pollution Act 1974. This section of the Act bans the use of loudspeakers in streets between 9 p.m. and 8 a.m., except for emergency use by the fire, police and ambulance services. The EPA was amended to give discretionary power to local authorities to consent to the operation of loudspeakers for non-advertising

---

[2]The Fireworks Act 2003, to be implemented in 2004, enables the government to set a noise limit of 120 dB on fireworks, ban the use of fireworks during anti social hours and impose a human requirement a people who see fireworks.

purposes during the prohibited hours. It is suggested that this will allow charitable and other entertainment events to operate loudspeakers after 9 p.m. with the approval of the local authority. The use of loudspeakers in the street for advertising, entertainment, business or trade is not allowed at any time, except in the case of vehicles selling perishable foodstuffs, which may use loudspeakers between noon and 7 p.m. The EPA was also amended to enable the Secretary of State to alter these times should he so choose.

Codes of practice

Section 71 of the 1974 Act empowers the Secretary of State to issue codes of practice or approve codes produced by other bodies for minimizing noise. For example, BS 5228:1984 [21], which deals with the control of noise from construction sites (see section on Construction sites), was prepared by the British Standards Institution (BSI) and approved by the Secretary of State. Codes of practice on noise from ice-cream van chimes, the control of noisy parties, burglar alarms and model aircraft have been produced by the Department of the Environment. Work has been done on producing codes for clay pigeon shooting, pop concerts, motorbike scrambling, power boats, water skiing and audible bird scarers. The Noise Review Working Party Report [13] considers that codes of practice are of great importance as a means of providing advice and guidance for those involved in the control of environmental noise. It is important to remember that any code produced, while it may be taken into account by the courts, does not have the force of regulations.

As mentioned above, the Department of the Environment has issued a code of practice dealing with noise from burglar alarms. The Noise and Statutory Nuisance Act provides an adoptive power for local authorities, which introduces controls over the installation and operation of these types of alarms. There is no specified date when these provisions will come into force; the Secretary of State is given the power to make a Commencement Order and issue regulations to bring them into force.

Noise Act 1996 – neighbourhood noise

The Neighbour Noise Working Party was set up in 1994 following concerns about the ever increasing number of domestic noise complaints and the effectiveness of domestic noise controls. The recommendations of the working party have been implemented by the introduction of the Noise Act 1996 and the publication of two guides [27,28]. The noise management guide [27] aims to encourage local authorities to review noise policies and practice, and encourages service consistency while taking account of local needs. Furthermore, basic minimum standards of service to fulfil local authority statutory responsibilities are identified. The guide on police/local authority co-operation [28] aims to encourage effective local arrangements for dealing with noise complaints.

The Noise Act 1996 covers noise from domestic premises. Sections 2–9 are adoptive and have been implemented by the Night Time Noise Regulations 1997. Local authorities adopting these provisions must take reasonable steps to investigate neighbour noise problems between 11 p.m. and 7 a.m.

If an offence is committed the offender may be prosecuted. The Act also gives authorized officers powers to enter a dwelling and confiscate equipment and clarifies the seizure powers detailed in Section 81(3) of the Environmental Protection Act 1990.

The Government has stated its intention to amend the Noise Act 1996 to make it less prescriptive and non-adoptive. This should make the night-noise offence more widely available to deal with night-time noise complaints by local authorities.[3]

The Noise Emission in the Environment by Equipment for Outdoor use Regulations 2001 (SI 2001/1701) means that from 3 January 2002 all equipment covered by these regulations sold in the United Kingdom must be tested and meet these regulations or risk being prohibited from being distributed within the Community market.

The Crime and Disorder Act 1998

This Act contains provisions that allow local authorities to issue an anti-social behaviour order on anyone who is causing 'harassment, alarm or

---

[3]Section 48 of the Antisocial Behaviour Act 2003 makes these amendments.

distress', which includes noise pollution. Information on the use of these powers is given in a Guidance Note issued by the Home Office in March 1999.

The Home Office has proposed new powers in the Antisocial Behaviour Bill 2003. It is the intention that a Local Authority (usually an EHO) will be able to serve a notice requiring the noise to be abated within a short fixed period (suggested 10 minutes) after which if the noise is not abated the EHO can serve a fixed penalty (£100) on the spot fine or serve an abatement notice as they would at present.

A fixed penalty notice will be dealt with in a similar way to a police speeding ticket or a parking ticket and the Local Authority will deal with the paper work, it will not go before the court unless the recipient decides not to pay within the prescribed period.

This proposed legislation will also allow the Local Authority the power to close premises with immediate effect if a public nuisance is being committed. This will give the EHO the same power as the police. This proposed legislation effectively moves nuisance from civil to criminal law and will necessitate much closer co-operation between the EHO and the police.

ENTERTAINMENT – PUBS AND CLUBS

The Institute of Acoustics has published a *Good Practice Guide on the Control of Noise From Pubs and Clubs*. The document is aimed at the regulator, the operator and the planning authority and sets out the responsibilities both of Local Authorities and of businesses.

It provides guidance on the assessment and control of noise at sensitive premises both for local authority staff and the management of venues. It covers music, singing, public address systems, patrons, children's play areas, beer gardens, skittle alleys, car parks, deliveries/collections material handling, and plant and machinery.

It had been the intention to include objective noise criteria, however, it was felt that the numbers developed had not been subject to a sufficient and thorough enough validation process to publish at this time, it is hoped that the objective criteria will be published at a later date to be used in conjunction with this document.

NOISE FROM INDUSTRIAL PROCESSES COVERED BY INTEGRATED POLLUTION AND PREVENTION CONTROL

Noise and vibration is now included in the Integrated Pollution and Prevention Control (IPPC) (see Chapter 36) regime and operators of processes that fall within A1 and A2 categories, will have to consider this when applying for a permit under this legislation. Noise and vibration are defined as emissions as set out in the Pollution Prevention and Control Regulations. The definition of pollution includes 'emissions which may be harmful to human health or the quality of the environment, cause offence to human senses or impair or interfere with amenities and other legitimate uses of the environment'.

IPPC uses 'BAT'(best available technique) when setting emisson limits and in determining relevant conditions to control emissions. The aim of BAT should be to ensure that people beyond the boundary of any permitted installation are not given any reasonable cause for annoyance.

It will be for the courts to decide if BAT equates with normal nuisance test or is taken to be a sterner guideline.

CONSTRUCTION SITES

Part III of the Control of Pollution Act 1974 (Sections 60 and 61) gives local authorities the power to exercise certain controls over noise emanating from construction and demolition sites. The Act recognizes the need to protect residents and people working in the vicinity of these sites.

Local authorities may, regardless of whether a statutory nuisance has been caused or is likely to be caused, serve a notice on a developer or contractor requiring the implementation of measures to control noise on the site. These powers remain the same irrespective of the character of the area. However, local authorities can temper any conditions they impose to take account of the locality.

The conditions imposed could include limitation of hours worked per day, the positioning of noisy machinery on the site, the designation of vehicular access routes about the site, and the provision of enclosures for noisy machinery or operations. Local authorities should also ensure that best practicable means are employed by the contractor.

Developers or contractors can also apply to the local authority for a prior consent for their proposed works, and can appeal against any notice served by the local authority or against the failure to agree the details of a prior consent.

Once a prior consent is agreed or a notice served, it is a defence for a developer or contractor to prove that he was complying with the conditions therein in any subsequent nuisance action.

BS 5228:1984 [21] was produced to give guidance to local authorities in the methods of predicting, measuring and assessing noise on construction and open sites and its impact on those exposed to it. The standard is in four parts:

1. **Part 1.** Code of practice for basic information and procedures for noise control.
2. **Part 2.** Guide to legislation for noise control applicable to construction and demolition, including road construction and maintenance.
3. **Part 3.** Code of practice for noise control applicable to surface coal extraction by opencast methods.
4. **Part 4.** Code of practice for noise control applicable to piling operations.

In setting noise conditions, the local authority should take the following into account.

1. The site location: proximity of noise sensitive premises.
2. Existing ambient noise levels: the larger the increase in noise over the ambient noise level, the more likelihood of complaints.
3. Duration of site operations: the longer the perceived duration of site operations the more likely noise is to provoke complaint. Residents

may be willing to accept higher noise levels if they can anticipate an early cessation of activity on the site.

4. Hours of work: residential property and offices may have different priorities regarding the times when noisy work should be carried out on a site. Night-time work would keep domestic residents awake. However, the same activity in the day may interfere with speech in a nearby office. Noise conditions imposed on evening work may need to be much stricter than daytime limits – BS 5228:1984 [21] suggests as much as 10 dB(A) lower. Night-time working necessitates very careful consideration: the times when people are going to sleep and just before they rise appear to be particularly sensitive.
5. Attitude of the site operator: local people will be more amenable to noise if they consider that the site operator is, in his planning and management, doing all he can to keep noise to a minimum.
6. Noise characteristics: the presence of impulsive noise or noise with tonal characteristics may be less acceptable than actual sound level readings would lead one to believe.

NOISE ABATEMENT ZONES

The Control of Pollution Act 1974 (Sections 63–67), gave local authorities the power to establish noise abatement zones (NAZs) in order to control noise increases and to achieve noise reductions in existing noisy situations, as well as preventing the increase in background noise due to new developments. Under the Act, local authorities can designate all or part of their area as an NAZ. The order will specify classes of premises to which it applies, for example, industrial premises, places of entertainment, commercial premises, etc. In practice there is no reason why named premises could not be included in the schedule. Once the order has been implemented, the local authority is obliged to measure noise levels from the specific premises along the perimeter of the premises. The Control of Noise (Measurement and Register) Regulations 1976 make provision with respect to the methods to be used by local authorities when measuring noise levels, and the maintenance of a public register. Detailed guidance on the introduction of such areas may be found in Department of the Environment Circular 2/76 [29].

While these sections of the Act gave local authorities the power to control noise increases within the community, it is a power that is not widely used. It is very time consuming and labour intensive to collect the initial noise data and also to maintain the public register in which the measured noise levels must be recorded.

The government recognized these problems, and in its white paper on the environment [13] stated its intention to devise simpler and more practical procedures for dealing with NAZs. The Noise Review Working Party Report [12] recommended that the system be simplified by using procedures issued in a code of practice rather than under the existing legislation. Such a method would allow local authorities to adjust the requirements to local circumstances, and it is felt that this may result in a revival of NAZs. In response to the recommendation of the Noise Review Working Party, the Department of the Environment requested the BRE to undertake a review of the NAZ system. This was completed during 1992. The BRE surveyed local authorities and found that 37 had designated a total of 58 NAZs [30]. However, of these, only 28 of the zones had complete noise level registers and 40 of the NAZs were now either inactive or abandoned. While the survey concluded that NAZs were unpopular, it also indicated that they could be of use for preventing excessive noise from a small number of premises. As yet there have been no new policy developments in response to the findings of this report.

RAILWAY NOISE

Generally speaking, complaints relating to noise from this source are less common than those relating to other noise sources. The 1991 BRE survey [6] indicated that 4% of the adult population is bothered by railway noise. This is somewhat surprising when one considers that the noise generated can often be higher than levels from sources that frequently give rise to complaint. This could be due to the fact that the public tends to tolerate noise from this source and believes that nothing can be done to reduce the levels of noise. Furthermore, by comparison, far fewer people live adjacent to railways than, for example, next to major roads.

However, many local authorities are faced with an increasing demand for housing land such that locations adjacent to existing railway lines are now being considered for housing use. The increased use of existing lines and the construction of new routes suggest that railway noise is more likely to cause more widespread annoyance in the future. The opening of the Channel Tunnel and the completion of the high-speed rail links has had a major effect in this respect in the south-east.

A national survey of the community response to railway noise was undertaken by Fields and Walker of the Institute of Sound and Vibration Research (Southampton University) [31]. This study combined a social survey (1453 respondents) with a noise measurement survey in 75 study areas in the United Kingdom. The study concluded that railway noise is a problem for those living adjacent to railway lines, and this was quantified further to suggest that 2% of the nation's population is bothered by noise from this source. Furthermore, the equivalent continuous noise level over a 24-hour period ($L_{Aeq,\ 24\ hour}$) was closely related to people's reaction to railway noise, more so than any other accepted noise indices examined. It further concluded that approximately 170 000 people are exposed to railway noise above 65 dB(A)$L_{eq,24\ hour}$, and that maintenance noise is rated as a bigger problem than passing train noise. Above 45 dB(A)$L_{eq,\ 24\ hour}$ there was found to be a steady increase in annoyance with increasing $L_{Aeq,\ 24\ hour}$. There is no simple threshold below which people were not annoyed.

In the absence of any statutory guidelines, various standards have been adopted by local authorities when considering the development of land adjacent to railway lines. In some cases a 24-hour L_{eq} is set, while in others separate day-time and night-time levels are set. For example, some authorities prohibit noise-sensitive development where the 24-hour L_{eq} level exceeds

60 dB(A), while others may require noise insulation double glazing.

In March 1990, the government set up an independent committee to recommend a noise insulation standard for dwellings near new railway lines. The Mitchell Committee published its report in February 1991. In November 1991, the Minister for Public Transport confirmed that noise insulation regulations for new railway lines would be issued.

The Noise Insulation (Railways and Other Guided Transport Systems) Regulations 1996 and a Technical Memorandum [32] describe the procedures for predicting and measuring noise levels from moving railway vehicles. The purpose of the regulations is to introduce provisions for new railway lines similar to those contained in the Noise Insulation Regulations 1975 (as amended).

The regulations apply when it is not possible to reduce noise below the trigger levels (see below). It is expected that bodies constructing new railway lines will take all the necessary steps to reduce noise by careful design and the use of screens and landscaping.

Eligible properties, dwellings and other residential properties that will be entitled to noise insulation treatment must:

1. be within 300 m of the new or altered system; and
2. be subject to a noise increase of at least 1 dB(A) as a result of vehicles using that new system; and
3. be subject to noise levels from vehicles using the new, altered and any relevant existing system of not less than 68 dB $L_{Aeq\ 18\ hour}$ between 6 a.m. and midnight or 63 dB $L_{Aeq\ 6\ hour}$ between midnight and 6 a.m.; and
4. the railway noise level must have increased the noise level by at least 1 dB(A).

The regulations confer duties or powers on railway authorities to carry out insulation work or pay grant, set out the machinery for offering and accepting insulation work or grant payment, specify the extent of insulation work that should be carried out and detail appeal procedures.

The Technical Memorandum is divided into three main sections: a general method of predicting noise levels at a distance from a railway; procedures to deal with the prediction of railway noise in situations such as stations and sidings; and procedures and requirements for the measurement of railway noise where prediction is not possible.

ROAD TRAFFIC NOISE

The noise limits for road-going motor vehicles are prescribed in regulations made under the Road Traffic Act 1972, the Motor Vehicle (Construction and Use) Regulations 1986 (as amended) and the Motor Vehicle Type Approval (Great Britain) Regulations (Table 35.6). The Motor Vehicle Type.

Approval Regulations reflect EU directives and have been amended to encompass EU directive 92/97/EEC, and subsequently directive 96/20/EEC. This reduced the limits for new vehicles, cars, dual purpose and light goods vehicles with

Table 35.6 Maximum noise limits for motor vehicles

Vehicles	Current (db(A))	Proposed (dB(A))
Goods vehicles over 3.5 tonnes		
Over 150 kW engine power	84	80
75–150 kW engine power	83	78
Below 75 kW engine power	81	77
Goods vehicles below 3.5 tonnes		
Goods vehicles 2–3.5 tonnes	79	77
Goods vehicles below 2 tonnes	78	76
Buses		
With more than nine seats, over 3.5 tonnes GVW and over 150 kW engine power	83	80
With more than nine seats, over 3.5 tonnes GVW but below 150 kW engine power	80	78
With more than nine seats, but below 3.5 tonnes GVW	78–79	
Cars	77	74

Table 35.7 Motorcycle noise limits

Engine capacity (cc)	1st stage (dB(A))	2nd stage (dB(A))	Date of entry for type	
			1st stage	2nd stage
<80	77	75	1.10.88	1.10.93
80–175	79	77	1.10.89	31.12.94
>175	82	80	1.10.88	1.10.93

petrol engines from 1 October 1995, and for other engined new vehicles from 1 October 1996.

Motorcycles

Motorcycle noise is controlled in the United Kingdom by the Road Vehicle (Construction and Use) (Amendment No. 3) Regulations 1989 and the Motorcycles (Sound Level Measurement Certificates) (Amendment) Regulations 1989. These regulations implement EU Directive 78/1015/EEC as amended by Directive 87/56/EEC (Table 35.7).

Local authority involvement in road traffic noise

The local authority responsible for highways in an area may, using the Road Traffic Regulation Act 1967, re-route through traffic out of a residential area, which could overcome a particular noise problem. The Heavy Commercial Vehicles (Controls and Regulations) Act 1973 requires the highway authority to prepare written proposals for the management of lorries. This allows the introduction of improved traffic schemes.

The Land Compensation Act 1973 Part II enables householders to claim a grant from the highway authority to provide noise insulation if they experience significantly increased noise levels from a new or substantially improved road. The avenue that opens the possibility of noise insulation is set out in the Noise Insulation Regulations 1975 (as amended) for England and Wales (similar legislation/regulations apply in Scotland).

The Noise Insulation Regulations 1975 make grants available for secondary glazing of habitable rooms of a dwelling if it is estimated that the L10 (18 hour) will be raised to above 68 dB(A) due to a new road or substantially improved road during the 15 years following its opening (this is taken to mean at least the addition of a new carriageway), with a contribution of at least 1 dB(A) from traffic using the new road or 5 dB(A) from improvements to an existing road. Grants are not available to provide insulation where the new or increased noise is due to a traffic management scheme. The Noise Insulation Regulations 1975 refer to the *Calculation of Road Traffic Noise* [33].

The Noise Review Working Party [13] considered transportation noise and made a number of comments and recommendations. These included whether the current standard of 68 dB(A) L10 (18 hour) for noise insulation work remains appropriate and the possible extension of the existing rights to compensation against increased traffic noise to include those affected by permanent traffic management schemes.

In response to this report, the government has stated that it is considering covering noise in the annual MOT test, and will seek to tighten international noise regulations for vehicles. Furthermore, consideration will be given to the extension of compensation to those affected by permanent traffic management schemes and to an examination into improved design and specification in road construction to reduce traffic noise.

AIRCRAFT NOISE

According to a BRE survey [6], 16% of the adult population is bothered by aircraft noise, compared with 28% who are bothered by road traffic noises. Results from a survey (based on 1000 responses) by the Local Authority Aircraft Noise Monitoring Group (LAANMG) undertaken in 1992 indicated that as many as 72% of residents living near Heathrow airport claim that their sleep is disturbed by night-time aircraft movements. Other findings of the survey centred on the London boroughs of Hounslow and Richmond upon Thames and the Royal Borough of Windsor and Maidenhead included:

- 77% of residents are affected by aircraft noise, compared with a national average of 7%

- 69% ranked aircraft noise as the most annoying source of noise, compared with 32% for road traffic noise
- if the results of aircraft noise disturbance are representative of other local authorities around Heathrow, it can be estimated that several million people have their sleep disturbed to some degree
- 60% of respondents considered that their local authority should be able to take enforcement action against both the airline operator and the airport for noise infringement.

Aircraft engines are very noisy, especially jet engines, and even though much work continues to be done to reduce sound levels by, for example, producing quiet-engined aircraft, an acute problem still exists for many people as the above survey indicates.

Powers available to control environmental noise in the Control of Pollution Act 1974 and Environmental Protection Act 1990 deal with noise from fixed sources, and specifically exclude noise from aircraft other than noise from model aircraft. Section 76 of the Civil Aviation Act 1982 exempts aircraft from statutory action being taken in respect of noise nuisance. Sections 78 and 79 of this Act provide power for the Secretary of State for Transport to apply operational controls and enforce noise standards on aircraft and give directions to airport owners in relation to noise insulation grant schemes. For example, operational restrictions relating to night flights are applied at Heathrow, Stansted and Gatwick airports, and similar restrictions apply between April and October at Manchester and Luton airports.

Standards set by the International Civil Aviation Organization are enacted in the United Kingdom, and the Air Navigation (Noise Certification) Order 1986 implements the most recent international recommendations in respect of the control of aircraft noise at source.

The NNI (see p. 653) was formulated and adopted as an index of disturbance from aircraft noise. There was considerable support for the replacement of the NNI, and the government changed the index for measuring aircraft to L_{eq}

(16 hour) dB(A) based on measurements over 16 hours between 7 a.m. and 11 p.m.

In recent years, there has been a considerable increase in the number of non-commercial flights. There are in the region of 7000 British-registered general aviation aircraft flying from about 280 small airfields in the United Kingdom; these aircraft are generally used for private flying. This, together with the increased use of helicopters, which can land and take off from very small areas, has created a new aircraft problem. This problem was identified by the Noise Review Working Party [12], and in response to this report the government has confirmed that it intends to examine the need for further action on controlling noise from helicopters landing outside airfields, and from light or recreational aircraft.

In August 1991, the Department of Transport issued a consultation paper [34] that set out the government's proposal on the above issues. In March 1993, the Department of Transport published the government's conclusions [35] following the responses to the consultation paper. The new system was intended to build on the existing, mainly voluntary, scheme of control at most civil airports and airfields. The main points in the government's proposals were:

- to commission and consult on guidance to create a national framework to assist preparation of noise amelioration schemes
- to encourage aerodromes to review existing noise amelioration measures and their enforcement, and arrangements for local accountability
- to open discussions with the British Airports Authority and local consultative committees about making Heathrow, Gatwick and Stansted airports more responsible and locally accountable for their noise control measures
- to introduce a new enabling power for aerodromes to establish and enforce noise control arrangements, including for ground noise
- to introduce a new power of designation to replace existing Sections 5 and 78–80 of the Civil Aviation Act 1982

- designated aerodromes to prepare a noise amelioration scheme, consult locally and agree it with the 'lead' local authority; disputed points to be settled by the Secretary of State
- to introduce new powers of enforcement to enable local authorities to take action against designated aerodromes that do not enforce schemes
- introduce a 'call-in' power for the Secretary of State to approve schemes
- these new powers to be capable of being applied to all aerodromes from the largest airports to the smallest private sites, including those used by helicopters.

Primary legislation will be required to introduce the new designation power. The timetable for this legislation has not yet been set out.[4]

OCCUPATIONAL NOISE (SEE ALSO CHAPTER 21)

The Noise at Work Regulations 1989 came into force on 1 January 1990 and are designed to protect people at work from suffering damage to their hearing. These regulations implement the requirements of EU Directive 86/188/EEC. The regulations stipulate three 'action levels':

1. a first action level of 85 dB(A)
2. a second action level of 90 dB(A)
3. a peak action level of 200 pascals (equivalent to 140 dB ref. 20 Pa).

If the noise level in a workplace is above the first action level, then the employer shall ensure that a 'competent' person makes a noise assessment that is sufficient to:

1. identify which of the employees are so exposed; and
2. provide them with such information with regard to the noise to which those employees may be exposed that will facilitate compliance with the duties under the Health and Safety at Work, Etc. Act 1974 and the Noise at Work Regulations 1989.

When the level lies between the first and second action level, the employer must provide suitable and sufficient ear protectors to employees who ask for them. These must be maintained in good condition by the employer. However, there is no duty on either the employer or the employee to ensure that are worn.

Where the exposure is above the second action level, the employer must provide suitable and sufficient ear protectors capable of keeping risk down to no more than that expected from the action levels. Employers and employees have a duty to ensure that they are worn.

Areas in the workplace that are identified as being above the second action level must be identified and designated an 'ear protection zone'. These zones should be clearly marked with signs complying with BS 5378: 1980 [36], and it should be ensured that everyone entering the zones is wearing ear protection. The peak action level is most likely to be encountered where cartridge operated tools, shooting guns or similar loud, explosive noisy devices are used. Workers exposed above this level will also be exposed to levels above 90 dB(A) $L_{EP,d}$.

A 'competent person' does not have to be a qualified acoustic engineer. However, he should be capable of working unsupervised and have a good basic understanding of what information needs to be obtained and how to make the necessary measurements.

The Institute of Acoustics (IoA) provides training courses leading to the Certificate of Competence in Work-Place Assessment. Training on noise is also available in modular form as part of more general courses, that is, National Diploma in Occupational Safety and Health of the National Examination Board in Occupational Safety and Health (NEBOSH), the Certificate of Operational Competence in Comprehensive Occupational Hygiene of the British Examining and Registration Board in Occupational Hygiene (BERBOH), and the IoA Diploma in Acoustics and Noise Control. One method of measuring personal exposure to workplace noise is by the use of a noise dose meter (Fig. 35.6).

[4]But see the Aerodrames (Noise Restrictions) (Rules and Procedures) Regulations 2003.

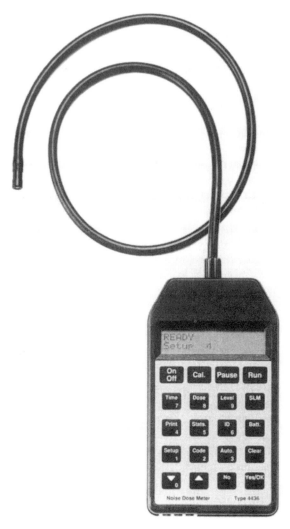

Fig. 35.6 Noise dose meter. (Courtesy of Brüel & Kjaer (UK) Ltd, Harrow Weald Lodge, 92 Uxbridge Road, Harrow Weald, Middlesex, HA3 6BZ.)

REFERENCES

1. EU Environmental Noise Directive 2002/49/EC DEFRA News release 20 December 2001.
2. Commission of European Communities (1996) EC Green Paper COM(96)540.
3. Wilson, A. (1963) *Noise, Final Report*, HMSO, London.
4. British Standards Institution (1997) *Method for Rating Industrial Noise Affecting Mixed Residential and Industrial Areas*, BS 4142:1997, BSI, Milton Keynes.
5. Sargent, J.W. and Fothergill, L.C. (1993) A survey of environmental noise levels in the UK. *Proceedings of Noise 93*, 7, 161–6.
6. World Health Organization (1980) *Noise: Environmental Health Criteria 12*, WHO, Geneva.
7. Grimwood, C.J. (1993) A national survey of the effects of environmental noise on people at home. *Proceedings of the Institute of Acoustics*, **15**, (Part 8), 69–76.
8. Mant, D.C. and Muir Gray, J.A. (1986) *Building Regulation and Health. Building Research Establishment Report*, Building Research Establishment, Watford.
9. Gray, P.G. and Cartwright, A. (1958) Noise in Three Groups of Flats with Different Floor Insulations, Res. Paper No. 27, National Building Studies, HMSO, London.
10. Ising, H. *et al.* (1980) Health effects of traffic noise. *International Archives of Occupational and Environmental Health*, **47**, 179.
11. Gloag, D. (1980) Noise and health: public and private responsibility. *British Medical Journal*, **281**, 1404.
12. Oswald, I. (1975) *Sleep*, Penguin Books, Harmondsworth.
13. Department of the Environment (1990) *Report of the Noise Review Working Party*, HMSO, London.
14. Government White Paper (1990) *This Common Inheritance – Britain's Environmental Strategy*, HMSO, London.
15. British Standards Institution (1981) *Method of Expression of Physical and Subjective Magnitude of Sound or Noise in Air*, BS 3045:1981, BSI, Milton Keynes.
16. Health and Safety Executive (1990) *Noise Assessment Information and Control*, Noise Guide No. 3, HSE Books, Sudbury, Suffolk.
17. British Standards Institution (1994) *Specification for Sound Level Meters* BS EN 60651:1994 BSI Milton Keynes.

18. British Standards Institution (1994) *Integrating/Averaging Sound Level Meters*, BS EN 60804:1994, BSI, Milton Keynes.

19. British Standards Institution (1989) *Sound Calibrators*, BS 7189:1989, BSI, Milton Keynes.

20. British Standards Institution (1984) *Guide to Human Exposure to Vibration in Buildings*, BS 6472:1984, BSI, Milton Keynes.

21. British Standards Institution (1984) *Noise Control on Construction and Open Sites*, BS 5228:1984, BSI, Milton Keynes.

22. Department of the Environment (1973) *Planning and Noise*, Circular 10/73, HMSO, London.

23. Department of the Environment (1994) *Planning and Noise*, Policy Guidance Note (PPG) 24, HMSO, London.

24. Department of the Environment (1985) *The Use of Conditions in Planning Permission*, Circular 1/85, HMSO, London.

25. Goswell, S. (1999) Making PPG24 work. *Environmental Health Journal*, 107(04), 119–23.

26. Department of the Environment (1990) *Mineral Planning Guidance: The Control of Noise at Surface Mineral Workings*, MPG 11, HMSO, London.

27. Chartered Institute of Environmental Health (1997) *Noise Management Guide – Guidance on the Creation and Maintenance of Effective Noise Management Policies and Practices for Local Authorities and their Officers*, CIEH, London.

28. Chartered Institute of Environmental Health/Association of Chief Police Officers (1997) *Noise – Good Practice Guidance for Police and Local Authority Co-operation*, CIEH/ACPO, London.

29. Department of the Environment (1976) *The Control of Pollution Act 1974. Implementation of Part 3: Noise*, Circular 2/76, HMSO, London.

30. Grimwood, C. (1992) Noise abatement zones – a method of preventing deterioration in environmental noise levels? *Proceedings of the Institute of Acoustics*, **14**, part 4.

31. Fields, J.M. and Walker, J.G. The response to railway noise in residential areas in Great Britain, *Journal of Sound and Vibration*, 85(2), 177–25.

32. Department of Transport (1995) *Calculation of Railway Noise*, HMSO, London.

33. Department of Transport and the Welsh Office (1988) *Calculation of Road Traffic Noise*, HMSO, London.

34. Department of Transport (1991) *Consultation Paper: Control of Aircraft Noise*, HMSO, London.

35. Department of Transport (1993) *Review of Aircraft Noise Legislation, Announcement of Conclusions*, HMSO, London.

36. British Standards Institute (1980) *Safety Signs and Colour Specification for Colour and Design*, BS 5378:1980, BSI, Milton Keynes.

FURTHER READING

Health and Safety Executive (1990) *Noise at Work Assessment, Information and Control*, Noise Guides 3–8, HMSO, London.

National Society for Clean Air and Environmental Protection (1994) *1994 Pollution Handbook*, NSCA, Brighton.

Open University *The Control of the Acoustic Environment*, OU, Milton Keynes.

Penn, C.N. (1979) *Noise Control*, Shaw and Sons, London.

Sharland, I. (1972) *Woods Practical Guide To Noise Control*, Cambridge University Press, Cambridge.

Williams, M. (1988) *Noise and Vibration Measurements for Environmental Health Officers*, Brüel and Kjaer (UK) Ltd, Harrow.

36 Air pollution

Michael J. Gittins

INTRODUCTION

Air pollution can have significant effects on human and animal health; damage vegetation; cause the destruction of building materials, fabrics and decoration; and bring about climatic change. The need to ensure that adequate measures are taken to effect its control is a vital part of sustainable development.

Although different countries face different problems there are few parts of the world where some action is not necessary. This chapter concentrates on the management of air quality in the United Kingdom, but also provides some information on international issues such as global warming, although these are discussed also in Chapter 34.

DEFINITION OF AIR POLLUTION

Air pollution is taken to be 'the presence in the atmosphere of substances or energy in such quantities and of such duration as to be liable to cause harm to human, plant, or animal life, or damage to human-made materials and structures, or changes in the weather and climate, or interference with the comfortable enjoyment of life or property or other human activity' [1].

SUSTAINABILITY

During the past few years, there has been a move towards a more proactive approach to the way we live, based on the concept of **sustainable development** [2] (see also p. 635). This concept is not entirely new to the United Kingdom. In 1989, the government published *Sustaining Our Common Future* [3], which set out policy aims and measures for the United Kingdom that were specifically directed towards sustainable development. The theme was also incorporated into the government's white paper *This Common Inheritance – Britain's Environmental Strategy* [4].

A Commission on Sustainable Development, under the aegis of the United Nations (UN), was established in 1993 as one outcome of the UN Conference on Environment and Development (the Earth Summit) held in Rio de Janeiro in 1992. The commission's role was to monitor progress in the implementation of the agreements reached there. A year later Agenda 21 was agreed, a comprehensive and forward looking action plan for the twenty-first century, which required national sustainable development strategies. (see Chapter 9.)

Agenda 21 is a global plan that recognizes the different needs of developed and developing countries. Because economic development in one nation will affect the economy of other countries, it is necessary to consider the impact of home policy, and practice, on other parts of the world. This will include the impact of the United Kingdom on the developing world and areas such as central and eastern Europe.

Decisions about levels of protection should be based on the best possible economic and scientific analysis. Where appropriate, actions should be based on the so-called '**precautionary principle**' It is also important that the private sector takes full account of environmental costs – the '**polluter pays principle**' (see p. 627).

INSTRUMENTS TO CONTROL POLLUTION

The Organization for Economic Co-operation and Development (OECD) has carried out considerable research into the various approaches that can be adopted to control pollution [5]. While there is a need for a regulatory approach, many countries use economic instruments as a secondary means of achieving control. Many economists have advocated the use of fiscal measures as these provide environmental policy-making with flexible, effective and efficient options in realizing its objectives. The range of fiscal measures available is discussed on p. 642.

EFFECTS OF AIR POLLUTION

Although some of the effects of air pollution are well understood, there are issues, especially those related to human health, where the picture is far from clear. Continuing research identifies an increasingly complex series of relationships. The principal issues were summarized by the Department of the Environment [6] as follows:

- harmful gases or particles may be inhaled by people or animals, or may attack skin, causing ill health or death;
- gases may damage leaves and shoots of plants, reducing amenity and the yields of crops and trees;

- particles of substances that settle out onto soil or vegetation may cause damage or contaminate human or animal food;
- air pollution can screen out sunlight, corrode structures and be a nuisance in many ways, especially through smells and the settlement of airborne dust.

Increasingly, interest is growing in the use of small population health statistics to enable perceived worries to be critically analysed and quantified. Investigations must be based on properly designed studies.

The level at which the ecosystems are judged to be at risk is called the '**critical load**'. The Department of the Environment formed a Critical Loads Advisory Group, which has produced maps of critical loads for soils and fresh waters. Further study is necessary to relate the response of ecosystems to measured acid depositions.

Effects on health

Although it is generally agreed that air pollution has a detrimental effect on health, its actions are generally insidious. In only a limited number of cases does it produce a specific disease or symptoms: photochemical smog can cause eye irritation in some major cities, for example, Los Angeles; chest diseases can be associated with occupational exposure to dust. Epidemiological studies show connections between air pollution and morbidity and mortality in chronic bronchitis sufferers, there is a demonstrable association between atmospheric sulphate levels and ill health, and there is growing concern at the increase in child asthma (although a a variety of causes have been associated with this phenonomem). Many factors contribute to this relationship; it is impossible to consider any one in isolation. Effects will depend on:

- general health of the subject
- duration of exposure
- concentration of pollutant
- nature of pollutant.

The following pollutants have been specifically identified by the government in its UK air quality management strategy [7]:

- benzene
- 1,3-butadiene
- carbon monoxide
- lead
- nitrogen dioxide (NO_2)
- ozone
- particulates
- sulphur dioxide.

The sources and health significance of these are given below.

Benzene

Benzene is an aromatic volatile organic compound (VOC) that is a minor constituent of petrol (about 2% by volume). The main sources of benzene in the atmosphere in Europe are the distribution and combustion of petrol. Of these, combustion by petrol vehicles is the single largest source (70% of total emissions), while the refining, distribution and evaporation of petrol from vehicles accounts for approximately a further 10% of total emissions. Benzene is emitted in vehicle exhaust not only as unburned fuel but also as a product of the decomposition of other aromatic compounds. Benzene is a recognized genotoxic human carcinogen [8].

1,3-butadiene

1,3-butadiene, like benzene, is a VOC emitted into the atmosphere principally from fuel combustion by petrol and diesel vehicles. Unlike benzene, however, it is not a constituent of the fuel but is produced by the combustion of olefins. This 1,3-butadiene is also an important chemical in certain industrial processes, particularly the manufacture of synthetic rubber. It is handled in bulk at a small number of industrial locations. Other than in the vicinity of such locations, the dominant source of 1,3-butadiene in the atmosphere is road traffic [9]. The health effect which is of most concern in relation to 1,3-butadiene exposure is the induction of cancers of the lymphoid system and blood-forming tissues, lymphomas and leukaemias. Like benzene, 1,3-butadiene is a genotoxic

carcinogen, and so no absolutely safe level can be defined.

Carbon monoxide

Carbon monoxide (CO) is a toxic gas that is emitted into the atmosphere as a result of combustion processes. It is also formed by the oxidation of hydrocarbons and other organic compounds. In European urban areas, CO is produced almost entirely (75%) from road traffic emissions. Even at levels found in ambient air CO may reduce the oxygen-carrying capacity of the blood. People who have an existing disease which affects the delivery of oxygen to the heart or brain (e.g. coronary artery disease (angina)) are likely to be at particular risk if these delivery systems are further impaired by carbon monoxide. It survives in the atmosphere for a period of approximately one month but is eventually oxidized to carbon dioxide (CO_2) [10].

Lead

Particulate metals in air result from activities such as fossil fuel combustion (including vehicles), metal processing industries and waste incineration. (There are currently no EU standards for metals other than lead, although several are under development.) Lead is a cumulative poison to the central nervous system, and is particularly detrimental to the mental development of children. Lead is the most widely used non-ferrous metal and has a large number of industrial applications. Its single largest industrial use worldwide is in the manufacture of batteries (60–70% of the total consumption of some 4 million tonnes) and it is also used in paints, glazes, alloys, radiation shielding, tank lining and piping. Tetraethyl lead has been used, for many years, as an additive in petrol. In the past, most airborne emissions of lead in Europe have originated from petrol-engine motor vehicles but the virtual phasing out of this fuel additive has resulted in significant reductions in the general levels of atmospheric lead.

Nitrogen dioxide

Nitrogen oxides are formed during high temperature combustion processes from the oxidation of nitrogen in the air or fuel. The principal source of nitrogen oxides – nitric oxide (NO) and nitrogen dioxide (NO_2), collectively known as NOx – is road traffic, which is responsible for approximately half the emissions in Europe. NO and NO_2 concentrations are therefore greatest in urban areas where traffic is heaviest. Other important sources are power stations, heating plants and industrial processes. Nitrogen oxides are released into the atmosphere, mainly in the form of NO, which is then readily oxidized to NO_2 by reaction with ozone. Elevated levels of NOx occur in urban environments under stable meteorological conditions, when the airmass is unable to disperse.

NO_2 has a variety of environmental and health impacts. It is a respiratory irritant, may exacerbate asthma and possibly increase susceptibility to infections. In the presence of sunlight it reacts with hydrocarbons to produce photochemical pollutants such as ozone. In addition, nitrogen oxides have a lifetime of approximately 1 day with respect to conversion to nitric acid. This nitric acid is in turn removed from the atmosphere by direct deposition to the ground, or transfer to aqueous droplets (e.g. cloud or rainwater), contributing to acid deposition [11].

Ozone

Ground-level ozone (O_3), unlike other primary pollutants mentioned above, is not emitted directly into the atmosphere, but is a secondary pollutant produced by a reaction between NO_2, hydrocarbons and sunlight. Ozone can irritate the eyes and air passages, causing breathing difficulties, and may increase susceptibility to infection [12]. It is a highly reactive chemical, and is capable of attacking surfaces, fabrics and rubber materials. Ozone is also toxic to some crops, vegetation and trees. Whereas NO_2 participates in the formation of ozone, nitrogen oxide destroys ozone to form oxygen (O_2) and NO_2. For this reason, ozone levels are not as high in urban areas (where high levels of NO are emitted from vehicles) as in rural areas. This is because as the nitrogen oxides and hydrocarbons are transported out of urban areas, the ozone-destroying NO is oxidized to NO_2, which participates in ozone formation.

Sunlight provides the energy to initiate ozone formation; near-ultraviolet radiation dissociates stable molecules to form reactive species known as free radicals. In the presence of nitrogen oxides, these free radicals catalyse the oxidation of hydrocarbons to CO_2 and water vapour. Partially oxidized organic species such as aldehydes, ketones and CO are intermediate products, with ozone being generated as a by-product. Since ozone itself is split up by sunlight to form free radicals, it promotes the oxidation chemistry, and so catalyses its own formation. Consequently, high levels of ozone are generally observed during hot, still, sunny, summertime weather in locations where the airmass has previously collected emissions of hydrocarbons and nitrogen oxides (e.g. urban areas with traffic). Because of the time required for chemical processing, ozone formation tends to be downwind of pollution centres. The resulting ozone pollution or 'summertime smog' may persist for several days and be transported over long distances.

Particulates

Airborne particulate matter varies widely in its physical and chemical composition, source and particle size. Concern about the potential health impacts has increased very rapidly over recent years [13]. Government concern was initially focused on PM10 particles (the fraction of particulates in air that are less than 10 µm in diameter) but interest is developing in PM2.5 particles (the fraction that are less than 2.5 µm in diameter). A major source of fine primary particles is combustion processes, particularly diesel combustion, where transport of hot exhaust vapour into a cooler exhaustpipe or flue can lead to spontaneous nucleation of 'carbon' particles before emission. Secondary particles are typically formed when low volatility products are generated in the atmosphere, for example the oxidation of sulphur dioxide to sulphuric acid. The atmospheric lifetime of particulate matter is strongly related to particle size, but may be as long as 10 days for particles of about 1mm in diameter.

The principal source of airborne PM10 matter in European cities is road traffic emissions, particularly from diesel vehicles. As well as creating dirt, odour and visibility problems, PM10 particles are associated with health effects including increased risk of heart and lung disease. In addition, they may carry surface-absorbed carcinogenic compounds into the lungs.

There is clear evidence of associations between concentrations of particles, similar to those encountered currently in the United Kingdom, and changes in a number of indicators of damage to health [14]. These range from changes in lung function, through increased symptoms and days of restricted activity, to hospital admissions and mortality. The consistency of the associations demonstrated by the studies undertaken in the United States and Europe is notable, especially with regard to mortality, although the reported effects on health of day-to-day variations in concentrations of particles are small in comparison with other uncontrolled factors, for example seasonal variations or variations in temperature. There is no clear evidence that these associations are restricted to specific types of particles.

The government has indicated a commitment to securing long-term abatement of vehicle emissions, particularly from diesel engines. In view of the possibility of local exceedences from specific sources, it is appropriate to reduce emissions from all other sources progressively, including industry, quarrying, construction and domestic coal burning [15].

Sulphur dioxide

Sulphur dioxide (SO_2) is a corrosive acid gas that combines with water vapour in the atmosphere to produce acid rain. Both wet and dry depositions have been implicated in the damage and destruction of vegetation and in the degradation of soils, building materials and watercourses. SO_2 in ambient air is also associated with asthma and chronic bronchitis [16]. The principal source of this gas is from power stations burning fossil fuels, which contain sulphur. Major SO_2 problems now tend to occur only in cities where coal is still widely used for domestic heating, in industry and in power stations. Since the decline in domestic coal burning in cities and in power stations overall, SO_2 emissions have diminished steadily and, in most European

countries, they are no longer considered to pose a significant threat to health. Of particular concern in the past was the combination of SO_2 and particulates.

Air quality information

The government has produced proposals on the way in which information is to be given to the public on the significance of air quality. Details are summarized in Table 36.1.

Mixed air pollutants

The Department of Health has published a study of the medical aspects of air pollution episodes in which the concentration of more than one pollutant was above the background level [17]. It concluded that predicting the combined effects of exposure to combinations of pollutants was difficult. It pointed to the fact that the level of understanding of the mechanisms of possible interactive effects is less well developed than the understanding of the health effects of individual pollutants. The advisory group that produced the report identified three common types of air pollution episodes:

- type 1: summer smog, the pollution mixture with the main or indicator pollutant being ozone
- type 2: vehicle smog, the indicator pollutant being oxides of nitrogen
- type 3: winter smog, the indicator pollutant being SO_2, with contribution from oxides of nitrogen.

Elevated concentrations of particulates may occur during all three types of air pollution episodes. A main concern has been whether the biological effects of mixtures of air pollution differ from those of the individual pollutants in isolation. Potentiation of damage (**synergism**) has been seen following exposure of animals to some combinations of pollution. However, there is a paucity of information on the toxicological effects of complex mixtures of pollutants, particularly those that relate to the three main types of pollution episodes seen in the United Kingdom.

Table 36.1 Air quality banding

Air pollutant	Bands			
	Low	*Moderate*	*High*	*Very high*
Sulphur dioxide (ppb, 15 minutes average)	Less than 100	100–199	200–399	400 or more
Ozone (ppb)	Less than 50 (8 hours running average)	50–89 (hourly average)	90–179 (hourly average)	180 or more (hourly average)
Carbon monoxide (ppm 8 hours running average)	Less than 10	10–14	15–19	20 or more
Nitrogen dioxide (ppb, hourly average)	Less than 150	150–299	300–399	400 or more
Fine particles ($\mu g/m^3$ 24 hours running average)	Less than 50	50–74	75–99	100 or more

Notes:
Low: Effects unlikely to be noticed even by individuals who know they are sensitive to air pollants.
Moderate: Standard threshold, mild effects, unlikely to require action, may be noticed by sensitive individuals.
High: information threshold, significant effects may be noticed by sensitive individuals and action to avoid or reduce these effects may be needed.
Very high: Alert threshold, the effects on sensitive individuals described for 'high' levels of pollution may worsen.

Pollen and fungal spores, or their fragments, are common aeroallergens during the period between the late spring and early autumn. They are responsible for attacks of hay fever or deterioration in asthma in sensitized individuals. Elevated aeroallergen levels may occur in conjunction with type 1 air pollution episodes. A limited number of laboratory studies in which human subjects have been exposed to relatively high concentrations of air pollution have demonstrated an enhanced reaction to aeroallergens.

It is probable that all three main types of pollution mixtures encountered during air pollution episodes in the United Kingdom could cause small mean reductions in lung function in normal individuals. The evidence for this is probably strongest for ozone-related pollution, where reduction in lung function and development of symptoms is more likely in those who take strenuous exercise. There is no evidence that any of the three episode types commonly seen in the United Kingdom cause symptoms or adverse health effects in people who are otherwise well. The elderly, especially those with chronic cardiovascular disorders, have

been identified as being at increased risk during episodes of air pollution. It has, however, proved difficult to demonstrate convincingly that patients with asthma are vulnerable to air pollution at levels encountered in the United Kingdom, except possibly those with severe asthma.

The Department of Health has also attempted to quantify the effects of air pollution on health in the United Kingdom [18]. It suggested that the deaths of between 12 000 and 24 000 vulnerable people may be brought forward because of, and between 14 000 and 24 000 hospital admissions and readmissions may be associated with, short-term air pollution episodes each year. Evidence relating to the long-term health effects of exposure to such episodes of air pollution as currently occur in the United Kingdom is not available.

The government recognizes that it is necessary to be able to provide information on air pollution episodes as they develop. Use is currently made of national and local weather forecasts, text services and the websites. The advisory group [17] identified criteria that need to be fulfilled if health advice provided to the public by forecast is to be beneficial. They include: the type of episode expected; the indicator pollutant and any associated pollutants; geographical location; severity; and the likely duration. Current evidence does not suggest that the use of smog masks is to be recommended!

Effects on animals

Air pollution has a similar effect on animals as it does on humans, at least where there is a common physiology. Of particular interest is the effect of airborne fluorine on herbivores as there is no parallel mechanism for human exposure (although humans do suffer from mottling of dental enamel if exposed to a significant intake). The main source of fluoride emission is from processes that involve the heating of minerals, such as the firing of heavy clay, the manufacture of glass, burning coal or aluminium smelting. Fluorine is a protoplasmic poison with an affinity for calcium – it interferes with the normal calcification process. Symptoms are hypoplasia of dental enamel and, at higher levels of exposure, abnormal growth rate of bones (which can cause lameness). Advanced fluorosis in cattle is indicated by anorexia, diarrhoea, weight loss, low fertility and reduction in milk production.

Effects on vegetation

Deposition has the effect of reducing the rate of photosynthesis by reducing the amount of light reaching leaves. Stomata and folial pores can be obstructed, reducing transpiration. Oxidant gases can cause leaf collapse, chlorosis, growth alterations and a reduction in nitrogen fixing. It is possible to diagnose potential sources of damage to plants by a careful study of their foliage. Due account must also be taken of frost or insect damage, drought, mould and disease.

Effects on materials

The principal mechanisms that cause damage are abrasion, deposition and removal (the damage caused by regular cleaning), direct chemical attack and electrochemical corrosion. The rate at which damage is caused will be influenced by moisture, temperature, sunlight, air movement and the chemical nature of the pollutant. Moisture encourages oxidation, permits electrolytic action and accelerates chemical action. The rate of chemical action is temperature-dependent, but the significance of water accumulation, which occurs when surfaces fall below dew-point, must not be ignored. Sunlight is itself damaging and aids the formation of photochemical oxidants. Particular chemicals in the atmosphere will cause particular effects on materials. CO_2 in solution will dissolve limestone but has little effect on sandstone because of the less reactive nature of silica. Most metals are vulnerable to acid attack, particularly in the presence of moisture. Textiles suffer soiling effects but other materials, such as leather and paper, can be embrittled by the absorption of SO_2 from the atmosphere. Rubber is particularly affected by ozone at levels as low as 0.02 ppm.

Research on effects

It is important that existing research and published information be used at the earliest stage in the design of an investigation. One criticism of local authorities is their reluctance to acknowledge or utilize work previously carried out. While there is a limit to the resources available to local government for research activities, there are opportunities for collaborative studies with other agencies such as universities.

A definitive source of information is the Environmental Health Criteria series published by the World Health Organization (WHO) in collaboration with the United Nations Environment Programme and the International Labour Organization. The main objective of this series is to establish the scientific basis for assessment of risk to human health and the environment from exposure to chemicals. The format of these studies is consistant; a review of physical and chemical properties, analytical method, sources of human and environmental exposure, environmental transport and distribution, environmental levels, human exposure, effects on organisms in the environment, metabolism, effects on experimental animals, effects on humans and evaluations of health and environmental risks. Each study is comprehensively referenced and runs into about 230 titles at the time of writing.

Further information on environmental epidemiology can be found in Chapter 15.

GLOBAL AIR POLLUTION ISSUES

The politics associated with the control of air pollutants at an international level is complex (see also Chapter 34). Economic theory would suggest that the minimum action required would be to reduce air pollution to the point at which the costs would be covered by the benefits. In practice this is not easily achieved. Cost benefit analysis is a developing science, and ascribing a monetary value to issues such as the health-related effects of exposure to a particular pollutant is a complicated calculation. The inclusion of probability and the implications of long- and short-term strategies in the calculations makes assessment even more difficult. A further complication is the transboundary nature of some problems – expenditure in one nation may have a limited advantage there and yet achieve effective amelioration of a situation in another part of the world. For this reason the extent to which pollution control can be implemented will depend on social pressure and political decisions.

The three principal issues of global concern are:

- acid rain
- global warming
- the ozone layer.

Acid rain

The term acid rain was first coined about a century ago by the United Kingdom's first pollution inspector, Robert Angus, to describe the polluted rain in Manchester, his home town. He had realized that the air there was not only dirty but also acidic, and was attacking vegetation and building materials. His observations were largely forgotten until the 1960s when Scandinavian scientists linked pollution blown from the United Kingdom and mainland Europe to the acidification of streams and lakes and the reduction in fish stocks

in Scandinavia. The mechanism that brought this about was not easily understood – acids exist naturally in soils yet did not seem to cause similar effects. Rain is slightly acidic in its natural state: precipitation reacts with CO_2 in the atmosphere to form weak carbonic acid, with a pH of about 5.6. In central Europe the rain is more acidic, with an average pH of about 4.1. Rain in individual storms can have a pH of less than 3; water droplets in fog can be even more acidic [19].

SO_2 makes a significant contribution to acid rain. Natural sources include rotting vegetation, plankton and volcanic activity. Globally about half of the SO_2 in the atmosphere comes from such sources, but in Europe the figure is only 15%. The balance comes from fuel combustion. In the atmosphere, SO_2 reacts with water vapour; it is first oxidized to a sulphate ion (SO_4), which subsequently combines with hydrogen to form sulphuric acid. When in a large mass, for example a plume from a power station chimney, the process may be much slower, resulting in SO_2 reaching ground level in an unconverted state. The speed of the reaction in the atmosphere may be increased by the presence of particles of metal (such as iron and manganese), which act as catalysts.

Nitrogen oxides (NO_2 and NO) also contribute to acid rain. The main sources of emission are power stations and vehicle exhausts. In the atmosphere NO converts to NO_2, which in turn is oxidized to nitric acid (HNO_3). The effects of acid rain are most clearly observed in areas where thin soil covers granite or sandy rock. Such soils are unable to neutralize acids and so gradually become increasingly acidic. There is evidence that lakes in Scandinavia, Scotland and Canada have become more acidic since the Industrial Revolution in the nineteenth century; this is revealed by close examination of sediments for the presence of diatoms. Nitrate ions would accumulate in the same way if they were not taken up by plants.

Fish, particularly salmonoidae, have disappeared from lakes and rivers in Scandinavia and some Scottish lochs are also fishless. Fish death is rarely caused by acidity; more often the cause is aluminium poisoning. Aluminium salts are abundant in the soil but are normally locked there in an insoluble form. Sulphate ions can dissolve aluminium, which is carried into water courses. Aluminium interferes with the way in which fish gills operate, causing them to be clogged by mucus, which reduces their ability to absorb oxygen.

Acid rain has also been associated with the decline of forests in Europe. In the mid-1970s, fir trees began to loose their needles. In this case the cause was essential nutrients, such as magnesium and calcium, being dissolved and washed out of the soil. SO_2 also damages plants directly.

In order to reduce acid rain it is necessary to reduce the use of fossil fuels that contain sulphur, reduce the amount of sulphur permitted in fuel, or require the application of gas cleaning techniques. The EC Large Combustion Plant Directive 88/609 EEC lays down stringent standards for new and existing plant.

Global warming

The earth's atmosphere traps heat from the sun that would otherwise escape into space. This is demonstrated by the fact that although the earth and moon are at similar distances from the sun, the earth's average surface temperature is 15°C while that of the moon is −18°C. This warming of the surface is known as the **greenhouse effect** and is essential for maintaining the earth in a condition suitable for life. The sun primarily radiates its energy in the visible spectrum, its surface temperature determining the nature of the radiation and where in the spectrum its energy output peaks.

Having received energy the earth achieves a state of equilibrium, radiating the same amount of energy back into space. The black-body spectrum corresponding to the earth's surface temperature is in the infra-red range and has a wavelength of 4–100 μm. Some of the outgoing radiation is trapped in the atmosphere, close to the ground. Water vapour strongly absorbs radiation with wavelengths 4–7 μm, and CO_2 absorbs radiation in the range 13–19 μm. (Radiation in the range 7–13 μm escapes into space.) The trapped radiation warms the lower part of the troposphere. This warmed air radiates energy – largely in the infra-red range – in all directions. Some energy finds its way back to the earth's surface causing additional heating.

Research using samples of air trapped in bubbles drilled from the Arctic ice sheet shows that the natural concentration of CO_2 has been about 270 ppm since the end of the last ice age, about 10 000 years ago. In 1957, the concentration was 315 ppm and in 1996 it was 360 ppm [20]. Most of the extra carbon assumed to cause the increased CO_2 levels has come from burning fossil fuels (although burning tropical forests will also have made a contribution). Burning of 1 tonne of carbon produces about 4 tonnes of CO_2. In 1980, some 5 gigatonnes of carbon, in fossil fuel, was burnt.

About half of the CO_2 released into the atmosphere is absorbed by natural sinks, such as vegetation and the biomass in the soil and the sea. There is concern that a point could be reached when the capacity for this absorption becomes saturated. If this were to happen, the rate of accumulation of carbon dioxide in the atmosphere would double.

It is widely accepted that since the nineteenth century the amount of CO_2 and other anthropogenic gases in the atmosphere has increased dramatically, and that the average temperature of the world has increased, if somewhat erratically. There are various estimates of the consequences of this situation. The Intergovernmental Panel on Climate Change has projected that there could be a warming of 1.6°C by 2050, which would result in a 0.25 m rise in sea levels [21].

Any increase in the temperature of the earth increases the amount of water vapour in the atmosphere, so further increasing heating. In addition, chlorofluorocarbons (CFCs), methane and nitrous oxide contribute to the effect. These gases absorb radiation in the range 7–13 μm. (Although emissions of CFCs may be reducing in response to concern about ozone depletion, they remain significant. A single molecule of a common CFC has the same warming effect as 1000 molecules of CO_2.)

International agreement is being sought on the action to be taken to reduce emissions of greenhouse gases. The United Kingdom has committed itself to a reduction of 20% below the 1990 level by 2010 [22].

At the conference of the Parties to the UN Framework Convention on Climate Change, held in Kyoto, Japan, in December 1997, some 160 industrial nations agreed a protocol to cut emissions of greenhouse gases by 6% between 2008 and 2012. The protocol required countries to keep on reviewing their reduction targets into the twenty-first century, culminating in a 60% cut in greenhouse gases.

It allowed for conventional trading mechanisms, including joint implementation. This would permit an industrial nation to sign an agreement with another industrialized country, or, more importantly, with so-called economies in transition, such as those in the former Soviet block, to carry out '**carbon mitigation**' schemes. Joint initiatives include building a clean power station to replace a dirty one, or planting trees to soak up CO_2. Under joint agreements the rich country providing the clean technology could offset its carbon pollution against the savings made overseas. Other schemes include '**carbon-trading**', whereby factories in industrial countries with rising emissions would be able to buy carbon credits from factories in other industrial countries that have reduced emissions. Countries will also be able to offset emissions using sinks, such as forests, which will absorb CO_2. Implementation of the protcol has proved a major political challenge for many of the signatories. The US congress refused to approve its adoption. In 2001, targets were revised downward by two-thirds of the original goals.

The ozone layer

The earth is the only planet in the solar system that has an atmosphere rich in oxygen. This is because oxygen is constantly being renewed by the process of photosynthesis. If this were not the case, oxidization would lock up all of the O_2 as oxides, such as CO_2, metal oxides and water.

Currently, the earth's atmosphere comprises (by weight) about 75% nitrogen, 23% oxygen O_2, 1.3% argon and 0.4% CO_2. The balance of 0.3% is comprised of trace gases. As the earth radiates heat into the atmosphere (in the infra-red range) it is absorbed by water vapour, CO_2, and other molecules in the atmosphere, causing warming. By a height of about 11 km, the temperature has dropped to $-60°C$. From this level to a height of about 50 km, the atmosphere becomes warmer with increasing altitude due to heating associated with the presence of ozone. The increasing warmth is caused by the absorption of ultraviolet radiation from the sun. (The boundary between these two zones varies in height from about 8 km over the poles to 16 km over the equator [23].)

The process by which ultraviolet radiation is absorbed is of considerable significance to life on earth. Essentially, ordinary molecules of O_2 are broken apart by ultraviolet radiation, which has a wavelength below 242 nm to form free oxygen atoms. Some of these react with ordinary O_2 to form ozone, liberating heat and warming the stratosphere. Ozone reacts with ultraviolet radiation with wavelengths between 230 and 290 nm, to form free oxygen atoms and O_2. These and other processes result in the dynamic balance of ozone levels in the stratosphere.

It is recognized that levels of ozone in the stratosphere have been falling since the 1950s [23]. In 1982, a hole was detected in the ozone layer over Halley Bay, Antarctica, by the British Antarctic Survey. More recently, a similar phenomenon has been detected over the Arctic. The mechanism that has been identified as the cause of **ozone depletion** is chemical reactions involving chlorine compounds released as a result of human activity. The sources of chlorine that caused most concern were CFCs. These organic liquids have a low boiling point and used to be widely used as a heat-exchange fluid in refrigerators, to make bubbles in foamed plastic and as a propellant in aerosols.

Widespread concern about the depletion of stratospheric ozone is based on the anticipated increases in the levels of ultraviolet radiation that will reach the surface of the earth. Increased exposure to ultraviolet radiation will increase the likelihood of deterious effects on biological systems. In humans these effects include sunburn and skin cancer, together with possible eye damage and changes in immune function. Similar effects can be expected in land and aquatic animals. There are concerns that ultraviolet radiation could modify the development of fauna and flora, compromising aquatic and terrestrial food chains. Natural and synthetic polymers, used by the construction, transport and agricultural industries, are degenerated by exposure to sunlight, ultraviolet radiation being particularly damaging. Ground level air quality is likely to deteriorate as a result of increased ultraviolet radiation due to the formation of photochemical smog and increased surface levels of ozone [24].

In 1987, the Montreal Protocol came into force. This was an agreement, initially between 31 countries, under the auspices of the United Nations Environmental Programme. Its effect is to control the production of the five main CFCs (11, 12, 113, 114 and 115) and three halons (1211, 1301 and 2402). Controls on carbon tetrachloride and 1,1,1-trichloroethylene have also been agreed. The need to control hydrochlorofluorocarbons (HCFCs) has now been recognized. Although HCFCs break down more easily in the lower atmosphere, some chlorine will still reach the stratosphere. It has been agreed that such substances must be used responsibly, and that they should be phased out by 2020 or 2040 at the latest [25].

UK AIR POLLUTION

Background

From the earliest point in the development of civilization, humans have burnt biological and fossil fuels to generate heat. The process releases a range of products of combustion that can give rise to air pollution. It is only when the quantities of these substances become sufficiently concentrated as to cause harm to human, plant or animal life that concerns have been identified. Human activity has caused such effects as a result of an increasingly urbanized style of living. As early as the thirteenth century, the use of sea coal in London was prohibited as being 'prejudicial to health'. In the mid-fifteenth century it was recognized that there was an association between atmospheric pollution and human health. John Evelyn, the diarist, described how the citizens of London were affected by smoke.

Writing in 1661, he said:

> Her inhabitants breathe nothing but an impure and thick Mist, accompanied with a fuliginous and filthy vapour, which renders them obnoxious to a thousand inconveniences, corrupting their Lungs, and disordering the entire habit of their Bodies; so that Catharrs, Phthisicks, Coughs, and Consumptions rage more in this City, than in the whole Earth besides. [26]

His observations, although profound, were largely ignored for some three centuries.

It was not until the Industrial Revolution that there was a widespread increase in the rate at which fuels were abstracted and consumed. From that time forward the effects of human activity have increasingly been seen on air quality. At the time, the widely prevailing view appears to have been that such pollution was the inevitable result of industrialization; that if the health of some sectors of the population was being affected, then that was the price that had to be paid for progress.

Not until the middle of the twentieth century was there widespread acceptance that air pollution caused unacceptable damage and that preventative action was necessary. Our understanding of the effects of air pollution on health has developed from studies of major air pollution incidents. Three episodes in particular focused attention on the problem: the Meuse Valley, Belgium, in 1930; Donora, USA, in 1948; and London, UK, in 1952. By studying these incidents it was possible to estimate the number of excess deaths associated with increased levels of pollution. Table 36.2 shows excess deaths associated with the major air pollution events this century [27].

It was the London smog episode in 1952 that acted as a catalyst in the United Kingdom and led to parliamentary action. A Committee on Air Pollution was established under the chairmanship of Sir Hugh Beaver to 'examine the nature, causes and effects of air pollution and the efficacy of the present preventative measures; to consider what further preventative measures are practicable; and to make recommendations'. After due deliberation, the production of two reports, [28,29], and

Table 36.2 Major air pollution episodes in the twentieth century

Date	Place	Excess deaths
December 1930	Meuse Valley, Belgium	63
October 1948	Donora, USA	20
December 1952	London, UK	4700
November 1953	New York, USA	250
January 1956	London, UK	480
December 1957	London, UK	300–800
November to December 1962	New York, USA	46
December 1962	London, UK	340–700
December 1962	Osaka, Japan	60
January to February 1963	New York, USA	200–405
November 1966	New York, USA	168
December 1991	London, UK	100–180

Source: Based on Elsom, D.M. [27].

considerable parliamentary debate, the Clean Air Act 1956 was enacted.

The main effect of that legislation was to reduce the burning of coal in domestic fireplaces, and to require industry and commerce to use coal more efficiently. The availability of natural gas has had a significant effect on fuel utilization, causing further reductions in the use of coal in all sectors of the energy market in the United Kingdom. Gas is currently being used for the generation of electricity. This trend, together with the year on year increase in the level of traffic on the roads, has resulted in vehicle exhaust becoming the chief cause for concern.

UK sources

Table 36.3 shows the relative contribution made by different sectors to the total national emission of the pollutants identified in the UK National Air Strategy [30]. (Ozone is not emitted directly: the pollutants that lead to its formation are nitrogen oxides and volatile organic compounds.) These figures do not necessarily reflect the relative contribution of these sources in any particular area of the country, or any particular 'hot spot'. There are great variations between urban and rural areas, and between residential, commercial and industrial areas [30]. The main sources of the pollutants are listed here.

Road transport

Road transport produces all the pollutants covered by the air pollution strategy [30] except sulphur dioxide. The proportion of emission varies according to traffic levels, weather conditions and other local sources.

Energy generation

Power generation plants are the dominant sources of sulphur dioxide, some NO_2 particulates and other pollutants.

Industrial processes

There are many sources. For example, VOCs evaporate from many liquid fuels, paints, printing inks and cleaners. Lead is widely used in the manufacture of vehicle batteries. Particulates are also given off from mineral extraction, construction and other processes.

Domestic sources

Emissions from domestic fires are much reduced as a result of domestic smoke control and natural gas, but domestic emissions of sulphur dioxide and smoke particles may still be a problem in some areas of the country. Domestic use of paint,

Table 36.3 The main sources of air pollution – national emission figures

Pollutant (Year)	PM10 (1994)	NOx (1993)	CO (1993)	SO$_2$ (1993)	Lead (1994)	Benzene (1993)	VOC (1994)	Butadiene (1993)
Transport	29%	59%	90%	5%	64%	68%	35%	77%
Combustion (industry)	35%	24%	3%	92%	4%	5%	2%	—
Chemicals and fuel	—	—	1%	—	—	11%	25%	18%
Other industrial	24%	14%	—	—	27%	16%	25%	—
Waste	—	—	1%	—	4%	—	—	5%
Domestic	14%	3%	5%	3%	1%	—	13%	—

Source: Department of the Environment, Transport and the Regions [30].

varnishes and other products is a significant source of volatile compounds.

Current UK policies

The principal mechanisms for controlling air quality in the United Kingdom are as follows.

Smoke control areas

Since 1956, there has been a progressive introduction of smoke control areas throughout the country. The effect of this has been to reduce the extent to which coal is burnt on open fireplaces significantly.

Authorization of specified industrial processes

Processes with pollution potential are regulated under a system of **integrated pollution control** (IPC) (see p. 699). Operators are required to adopt the **best practicable environmental option** (BPEO) for controlling emissions to air, water and land. Further processes are regulated under **local air pollution control** (LAPC) by local authorities. On 1 August 2000, a new regieme, Pollution Prevention and Control (PPC) was introduced. It will impose additional levels of control on all IPC processes and some subject to LAPC. Control is based on the principle of the **best available techniques not entailing excessive costs** (BATNEEC) (see p. 626). Authorizations are agreed on a plant by plant basis, allowing account to be taken of local circumstances.

Emission standards for fuels and new road vehicles

European vehicle emission and fuel quality standards were introduced in the early 1970s, but until the late 1980s, road transport pollution was not reduced because of the rise in the number of vehicles. Standards agreed in the EU since the early 1990s have substantially reduced emissions from new vehicles. For example, the 1993 standards required new petrol engine cars to be fitted with a catalytic converter. Such converters, if fully operational, can reduce emissions by up to 80%. By 2003, the fleet has been largely replaced by vehicles conforming to the new standards, which has resulted in a dramatic fall in emissions. Because the number of vehicles on the road continues to increase, the effect of this benefit has been diminished. It is estimated that there will be a 10% increase in vehicle numbers by the year 2013.

One unqualified success in emission improvement has been achieved by the reduction in the use of lead additives in petrol. (A result of this policy has been a significant reduction in the blood levels of people living near roadways.)

European developments

The EU has been legislating to control emissions of air pollutants and to establish air quality objectives for the last two decades. Directive 96/62/EC on ambient air quality assessment and management, the so called Air Quality Framework

Directive, sets a strategic framework for tackling air quality in a consistent way by setting European wide-limit values for twelve air pollutants in a series of daughter directives. The first daughter directive, covering sulphur dioxide, nitrogen oxides, particles and lead was agreed in April 1999 – Table 36.4.

EU policy is being further developed by proposals to integrate air quality policies with policies aimed at curbing the emissions of pollutants which impact on air quality. The Commission proposes that a more integrated approach should be taken to policy making, bringing together the process of setting health-based objectives with the process of developing strategies for controlling emissions from a range of sources. The Commission suggests that this single framework should set out the action proposed over a five-year period to co-ordinate policy on future air quality daughter directives, strategies dealing with specific aspects of air pollution and measures to control emissions from activities which lead to air pollution.

The UK Sustainable Development Strategy

The UK Sustainable Development Strategy, *A better quality of life* [31] was published in May 1999. It sets out the Government's vision of how to achieve sustainable development through:

- social progress which meets the needs of everyone;
- effective protection of the environment;
- prudent use of natural resources; and
- maintenance of high and stable levels of economic growth and employment.

It refers throughout to a series of indicators. This new set of around 150 indicators will be at the core of future reports on progress towards meeting the Government's sustainable development objectives. One of these headline indicators relates to ambient air quality. This indicator monitors the average number of days per monitoring site on which pollution levels are above the national air quality standards for five pollutants, as a way of

Table 36.4 Air quality daughter directive limit values for the protection of health

Pollutant	Limit value		Date to be achieved by
	Concentration	*Measured as*	
Lead	0.5 mg/m^3	annual mean	1 January 2005
Nitrogen dioxide	200 mg/m^3 not to be exceeded more than 18 times a year	1-hour mean	1 January 2010
	40 mg/m^3	Annual mean	1 January 2010
Particles (PM$_{10}$)[a]	50 mg/m^3 not to be exceeded more than 35 times a year	24-hour mean	1 January 2005
	40 mg/m^3	Annual mean	1 January 2005
Sulphur dioxide	350 mg/m^3 not to be exceeded more than 24 times a year	1-hour mean	1 January 2005
	125 mg/m^3 not to be exceeded more than 3 times a year	24-hour mean	1 January 2005

Note:
a The Directive also includes indicative limit values for PM$_{10}$ for 2010, which are subject to review.

monitoring progress towards meeting the sustainable development objectives.

National air quality strategy

The implementation of a national air quality strategy (NAQS) is a statutory duty contained in Part IV of the Environment Act 1995. It is based on the concept of meeting air quality objectives by 2005. Section 82 of the Act states:

1. Every local authority shall from time to time cause a review to be conducted of the quality of air at that time, and the likely future quality within the relevant period, of air within the authority's area.
2. Where a local authority causes a review under subsection (1) above to be conducted, it shall also cause an assessment to be made of whether air quality standards and objectives are being achieved, or are likely to be achieved, within the authority's area.
3. If, on an assessment under subsection (2) above, it appears that any air quality standards or objectives are not being achieved, within the local authority's area, the local authority shall identify any parts of its area in which it appears that those standards or objectives are not likely to be achieved within the relevant period.

An introduction to the function of local authorities in relation to the implementation of the government's national air quality strategy is given in a Circular 15/97 [32]. The Circular states:

National policies on air pollution are expected to deliver a significant improvement in air quality throughout the country. The strategy recognizes, however, that there is an important local dimension to air quality. In some locations, air quality problems occur because of local factors such as density of traffic, geography, topography, etc. Such hotspots require a more focused approach to ensure their elimination.

A series of guidance notes has been published that provides information on: the framework within which air quality should be reviewed and assessed [33]; the development of local air quality action plans [34]; the relationship between air quality and traffic management [35]; and the relationship between air quality and land use planning [36]. Technical guidance has been published on monitoring for air quality [37]; preparation of atmospheric inventories [38]; selection and use of dispersion models [39]; and pollution-specific guidance [40].

Review is defined [33] as the consideration of the levels of pollutants in the air for which prescribed objectives have been set, and the estimation of future levels. The **assessment of air quality** is the consideration of whether the levels estimated for the end of 2005 are likely to exceed the levels set in the prescribed objectives.

The first step in the local air quality management regime has been to provide the benchmark for action and the mechanism by which it can be measured. With the exception of ozone pollution, objectives have been set (Tables 36.5 and 36.6).

Air quality standards have been set[1] based on scientific and medical evidence on the effects of the particular pollutants on human health: they represent minimum or no significant risk levels. (They do not take into account an assessment of costs and benefits or issues of technical feasibility.) The air quality objectives represent the government's judgement of achievable air quality on the evidence of costs, benefits and technical feasibility. The recommendations of the Expert Panel on Air Quality Standards (EPAQS) [8,10,13,16] have been used as the standards to base objectives on. Where EPAQS did not make recommendations, the relevant information from WHO was used.

Air quality objectives provide a framework for determining the extent to which policies should aim to improve air quality. They also provide a measure for each pollutant of interest by which progress towards achieving improved air quality can be measured. Deciding whether objectives will be achieved by the relevant date, without local action, is likely to be a major challenge for some local authorities. To assist them in this task, the government has provided some information

---

[1]See the Air Quality (England) Regulations 2000 [41] and similar regulation of Scotland and Wales.

Table 36.5 Objectives for protecting human health to be included in Regulations for the purposes of local air quality management (The Air Quality (England) Regulations 2000)

Pollutant	Objective		Date to be achieved by
	Concentration	Measured as	
Benzene	16.25 μg/m^3 (5 ppb)	Running annual mean	31 December 2003
1,3-butadiene	2.25 μg/m^3 (1 ppb)	Running annual mean	31 December 2003
Carbon monoxide	11.6 mg/m^3 (10 ppm)	Running 8-hour mean	31 December 2003
Lead	0.5 μg/m^3	Annual mean	31 December 2004
	0.25 μg/m^3	Annual mean	31 December 2008
Nitrogen dioxide[a]	200 μg/m^3 (105 ppb) not to be exceeded more than 18 times a year	1-hour mean	31 December 2005
	40 μg/m^3 (21 ppb)	Annual mean	31 December 2005
Particles (PM$_{10}$)	50 mg/m^3 not to be exceeded more than 35 times a year	24-hour mean	31 December 2004
	40 μg/m^3	Annual mean	31 December 2004
Sulphur dioxide	350 μg/m^3 (132 ppb) not to be exceeded more than 24 times a year	1-hour mean	31 December 2004
	125 μg/m^3 (47 ppb) not to be exceeded more than 3 times a year	24-hour mean	31 December 2004
	266 μg/m^3 (100 ppb) not to be exceeded more than 35 times a year	15-minute mean	31 December 2005

Notes:
a The objectives for nitrogen dioxide are provisional.

[8,10,13,16]. In the case of CO_2, benzene, 1,3-butadiene, sulphur dioxide and lead, national targets are likely to be achieved, with the possible exception of locations in the vicinity of local sources of emission. In the case of nitrogen dioxide, a reduction of 5–10% above that achieved by national measures will be required to ensure that the objectives are achieved. PM$_{10s}$ are a particular concern. It is clear that the objectives will not be achieved by national measures alone as transboundary pollution from Europe accounts for a significant proportion of the annual mean concentrations.

A national air quality objective has not been set for ozone because of its transboundary nature. Local action will have an impact on the reduction of ozone precursors at a local level, but the effects of that action are unlikely to be experienced in the same locality. The responsibility for achieving the ozone objective is to be met at national and international levels.

Local air quality management

First stage review

Local authorities should have undertaken a first stage review and assessment for all pollutants of

Table 36.6 National objectives *not* to be included in regulations for the purposes of Local Air Quality Management

Pollutant	Objective		Date to be achieved by
	Concentration	Measured as	
Objectives for the protection of human health			
Ozone[a]	100 μg/m³ (50 ppb) not to be exceeded more than 10 times a year	Daily maximum of running 8-hour mean	31 December 2005
Objectives for the protection of vegetation and ecosystems			
Nitrogen oxides[b]	30 μg/m³ (16 ppb)	Annual mean	31 December 2000
Sulphur dioxide	20 μg/m³ (8 ppb)	Annual mean	31 December 2000
	20 μg/m³ (8 ppb)	Winter average (1 October to 31 March)	31 December 2000

Notes:
a The objective for ozone is provisional.
b Assuming NO_X is taken as NO_2.

concern. This process will have provided information on existing or anticipated significant sources of pollution of concern in their area, and consider whether people are likely to be exposed to levels that exceed the objectives. Consideration must have been given to the likelihood of sources of pollution outside the authority's area. This information will have formed the foundation of the report that the authority will have to produce on its review and assessment. Precise guidance has been produced [37] on the exact nature of the information to be collated. The first stage should have been be sufficiently comprehensive to demonstrate that the authority:

- had considered all sources of pollutants of concern that could have a significant effect on the locality
- had conducted a thorough search of existing information, and, where information is unavailable, has made attempts to collect missing information
- had used the information collected as the basis for a decision about whether to proceed to a second stage review and assessment.

There have been suggestions that some authorities, using consultants to collect this information, do not have details of the original data, which has caused difficulties in moving to the second and third stage of the review.

Second stage review and assessment

The aim of the second stage is to provide further screening of pollution levels in a local authority's area. It does not require that every area of exceedence has to be estimated, but locations where the highest concentrations can be expected should be identified and studied. If these studies indicate that objectives will not be exceeded there is no need for further action. If there is a likelihood of objectives being exceeded, the authority must proceed to the third stage of the process.

Third stage review and assessment

The third stage requires the authority to undertake an accurate and detailed review and assessment of current and future air quality. This is likely to involve more sophisticated modelling and monitoring techniques. The government expects local authorities to complete their reviews and assessments by the end of 1999.

Designation of quality management areas

Air quality management areas (AQMAs) must be designated where it appears that prescribed air quality objectives will not be met within the relevant period. No firm guidance has been provided on the geographical extent of an AQMA. Section 84(1) of the Environment Act 1995 requires that the assessment of air quality in AQMAs is carried out on a continuing basis. If it appears that the objectives are to be met without the need for formal action, the AQMA should be revoked. The local authority is required to produce a report on air quality within 12 months of the designation of an AQMA. That report must be available to the public and be subject to consultation.

Air quality action plans

Local authorities are required to produce an action plan for any AQMA that they have designated. That plan should be completed within one year of the air quality report. Action plans must reflect the fact that air quality may be influenced by factors far beyond the boundary of a local authority, and that solutions to air quality problems in one area may lie some distance away. In some circumstances action will need to complement and support the action of other local authorities. Detailed guidance on the way in which action plans should be provided has been published [34]. Action plans should set out a local air quality strategy that complements other local authority strategies. It is important that the principles of sustainable development form the core of all policy-making, and are clearly acknowledged within the action plan. Account must also be taken of transport planning, transport management, land use planning, energy and waste management, enforcement, economic development and other local functions. This will require a multidisciplinary approach. In addition to partnership with other local authorities, there is a need to develop partnership with other bodies, business interests and environmental organizations at the earliest stage in the development of the plan.

Environmental health departments, with their experience in health promotion, are well placed to take a leading role in the promotion of air quality issues. Many local authorities have established good practice in the use of car sharing schemes, the use of bicycles for business purposes, and flexible working hours. They should promote such initiatives and try to act as a catalyst to encourage similar action by other organizations.

The following measures are available to local authorities in the improvement of air quality, and their relevance should be considered in the plan:

- traffic management
- land use planning
- control of emissions from authorized processes under Part I of the Environmental Protection Act 1990
- control of industrial emissions under the provisions of the Clean Air Act 1993
- smoke control areas
- control of statutory nuisances under Part III(I) of the Environmental Protection Act 1990
- vehicle emission testing and control of stationary vehicles.

Once the action plan has been produced it must be made available to the public and a consultation process carried out.

Consultation and liaison

The preferred approach to consultation is that interested parties are given an opportunity to comment on the production of the action plan. Perhaps this is best achieved by a two-stage process: preliminary discussions resulting in a draft action plan, and secondary discussions that would lead to general agreement on the final action plan.

Section 11 of the Environment Act 1995 requires that consultation must be carried out with the following bodies:

- the Secretary of State
- the relevant Environment Agency

- in England and Wales, the relevant highway authority
- all neighbouring local authorities
- the relevant county council (if appropriate)
- any relevant National Parks Authorities
- other relevant public authorities as the local authority considers appropriate
- bodies representative of local business interests
- other bodies that the authority considers appropriate.

The consultation should relate to:

- an air quality review
- an air quality assessment
- a further air quality assessment in the AQMA
- the preparation or revision of an action plan.

In addition to their statutory consultees, local authorities should include relevant community and environmental groups in their consultation process. Implementation of the plans is more likely to be effective if the community understands the issues, and accepts and supports the proposals. Public meetings and widespread publicity of information, in an easily understandable form, will help this process.

Consultees should have access to all the data collected by the local authority in the review and assessment process. The Environment Act 1995 makes provision for public access to such information. Although that legislation does not require all information to be published, it must be borne in mind that the Environmental Information Regulations 1992 impose a duty on local authorities in relation to the provision of environmental information. Both the draft and the final action plan should include:

- details of the pollutants to be addressed and an indication of their source
- the involvement of other local authorities in securing the action plan's objectives
- proposals from the county council, if applicable, for inclusion in the plan
- the timescale over which each of the proposed measures in the plan are to be implemented

- an indication of where action by other people or agencies is required to secure the plan's objectives, and the arrangements being put in hand to secure the co-operation of those agencies.

Air quality and traffic management

A detailed review of national and international concerns associated with traffic is contained in the 18th report by the Royal Commission on Environmental Pollution [42]. Details of the implications of traffic in the context of air quality management are contained in guidance produced by Department of the Environment, Transport and the Regions (DETR) [35].

Road traffic is a major contributor to the pollutants identified in the national quality air strategy [30], with the exception of sulphur dioxide. Emissions of lead from this source should be negligible by 2005 as a result of a ban on the sale of leaded petrol from 1 January 2000.

Although national action to tighten vehicle emission standards will result in improvements, these are unlikely to be sufficient to achieve the air quality objectives for particulates and oxides of nitrogen in all areas. Traffic management can contribute to a reduction in CO_2, benzene and 1,3-butadiene levels.

Between 1984 and 1994, the number of cars increased from just over 16 million to about 20.5 million. The distances they drove also increased significantly [35]. Figure 36.1 shows the overall national traffic growth estimates from 1965 to 2025.

Measures to reduce vehicle emissions

EU Auto-Oil Directives will progressively introduce more stringent vehicle emission and fuel quality standards for vehicles and fuels from 2000. In addition to setting emission standards for new vehicles, checks are carried out as part of the MOT test. The Vehicle Inspectorate has also increased its programme of kerb-side emission tests.

In 1996, the UK National Cycling Strategy was launched. This set the target of doubling the number of cycle trips by 2002 and doubling them

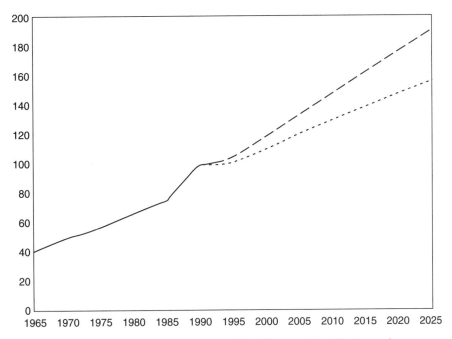

Fig. 36.1 Motor vehicle traffic from 1965 to 2025. ——— All traffic; ----- all traffic, lower forecast; ————— all traffic, upper forecast. Index, 1994 = 100. Reproduced from Department of the Environment, Transport and the Regions [35]; Crown copyright; reproduced with the permission of the Controller of Her Majesty's Stationery Office.

again by 2012. The DETR has established a group to develop a UK strategy for encouraging walking as a means of transport in its own right and as a link between other transport modes.

The government is introducing fiscal measures to encourage the use of less polluting fuels and low particulate emission commercial vehicles. Its stated intention (now see DETR (1998) White Paper, *A New Deal for Transport – Better for Europe*, DETR, London) is to bring about:

- a reduction in mileage
- changes to smoother, more fuel-efficient driving styles
- a reduction in the number of large, high fuel-consuming cars.

Traffic management

Strategies to reduce the environmental impact of transport have the following components:

- reducing the growth in the number and length of motorized journeys

- encouraging alternative means of travel that have less environmental impact
- reducing reliance on the private car.

Nevertheless, the importance of personal choice and the demands of economic competitiveness must be kept in mind [35].

The likely impact a traffic management scheme will have on local air quality management will depend on a variety of factors such as:

- the size of the scheme and the number of vehicles it affects
- the extent to which it reduces traffic levels, overall or at peak periods
- whether it redirects traffic to areas where fewer people are exposed to emissions
- whether it results in smoother traffic flow at a relatively constant speed
- whether it offers advantages to particular types of vehicles [35].

Journey distance/time has a significant effect on emissions. Before the engine of a vehicle has fully warmed up it cannot work at optimum performance. This is particularly the case in relation to pre-1997 petrol cars fitted with three-way catalysts. It takes a typical catalyst 1–4 km before it becomes effective. (From 1 January 1997 catalysts had to operate after 1–2 minutes.) As 40% of journeys are shorter than 5 km, there are a large number of occasions when cars fitted with catalysts are no better than those without. Policies to encourage alternative means of transport are important.

The way in which a vehicle is driven has a significant effect on emission levels. A speed of 25–55 mph (40–88 kph) leads to minimum pollution, but safety has to be considered. Traffic management measures that improve traffic flow by reducing start/stop driving can reduce total emissions.

Other measures available to local authorities include:

- traffic regulation orders prohibiting, restricting or regulating vehicle traffic
- permits restricting access to specific classes of vehicles, for example, residents, essential vehicles and orange badge holders; such schemes can be self-enforcing by the use of smart cards
- physical measures, such as the creation of narrow gateways to urban centres, which discourage car access
- reallocation of highway space by the introduction of bus or cycle lanes
- high occupancy vehicle lanes, which create a lane that can only be used by vehicles with two or more occupants
- pedestrian/vehicle restriction areas
- parking controls
- traffic control systems
- speed limits
- traffic calming
- promotion of public transport.

In addition, there are other potential methods of traffic management. Section 87 of the Environment Act 1995 empowers the Secretary of State to make regulations to prohibit or restrict the access of prescribed vehicles to prescribed areas. This could enable private cars to be banned from urban centres when there was a particular concern about air quality. Such provisions already exist in France. On 30 September 1997, it was decided to reduce the number of cars allowed into Paris because of very high ozone levels. Exclusion was based on the vehicle registration number, odd numbers having access on one day, even numbers the next; in addition, speed limits were reduced. Public transport was provided free. Cars carrying more than three people, taxis, delivery vans, hearses and emergency vehicles were exempt from the ban, as were vehicles running on liquefied petroleum gas (LPG), liquefied natural gas (LNG) and electricity.

Air quality and land use planning

It is important that due account is taken of the environmental impact of new development. Environmental health departments should be consulted by planners about whether certain land uses are appropriate, but planning policy guidance PPG23 [43] makes it clear that planning law cannot be used as a mechanism for controlling pollution. Guidance from the DETR [36] indicated that local authorities should have regard to the NAQS when undertaking their full range of functions, including regulating the development and use of land in the public interest. As part of a package of measures, land use planning forms an important component of air quality management. The DETR guidance states: 'By guiding the location of new development, reducing the need to travel and promoting transport choices, land use planning is an important part of an integrated strategy' [36]. However, this statement is in conflict with PPG23, which specifically excludes from planning control the indirect consequences of development such as pollution arising from vehicle use. It is important that planning, environmental health, economic development and transport planning departments work in partnership to achieve a multi-disciplinary approach to the consideration of land use proposals [36]. Two specific issues require consideration:

- that the land use planning system makes an appropriate contribution to the national air quality strategy

- that air quality considerations are properly considered along with other material considerations in the planning process.

Other PPG documents should be studied to ensure a general understanding of the thrust of government thinking on the subject. Of specific relevance is guidance on transport [44], town centres and retail development [45], and industrial and commercial development [46].

Unitary authorities are required to prepare unitary development plans. Non-unitary authorities must prepare four types of development plans depending on whether they are county or district authorities: counties prepare structure, minerals and waste plans; districts prepare local plans. Such plans are an important means of promoting environmental protection through integrated land use policies, and should take account of any constraints on development that result from the need to comply with any statutory environmental quality standards or objectives. This includes air quality objectives. Such plans also need to take into account trends in air quality over time and identify, where necessary, constraints on development in particular areas arising from the cumulative impact of existing and future development.

The need for planning authorities to liaise with other enforcement authorities, including the Environment Agency is reinforced in current guidance [36]. Where an environmental statement is required, the effect a project is likely to have on air quality should be considered. Environmental impact assessment (see also p. 640) is required under Directive 85/337/EEC and is implemented by the Town and Country Planning (Assessment of Environmental Effects) Regulations 1988 [47]. The local planning authority must consider all information in the environmental statement, including details of the effects, which must be regulated by pollution control authorities. If planning approval is granted, the planning authority must be satisfied that any remaining pollution concerns for which there is insufficient data can be dealt with under appropriate pollution control legislation.

Although planning legislation cannot be used as a substitute for pollution control legislation, there are occasions when conditions need to be imposed to address the impact of emissions. Such conditions must satisfy the following tests, which are included in Circular 11/95 [48], which gives details of the use of planning conditions. Conditions should be:

- necessary
- relevant to planning
- relevant to the permitted development
- enforceable
- precise
- reasonable in all other respects.

The circular gives clear guidance on the way in which conditions can be tested to satisfy compliance with these criteria.

Where planning concerns cannot be overcome by conditions it is possible, by virtue of section 106 of the Town and Country Planning Act 1990, for the developer and the planning authority to enter into an agreement that is similar in effect to a planning condition. Where the installation of new plant or the alteration of existing plant at a site authorized under the Environmental Protection Act 1990 does not require planning permission, it is still necessary for the Environment Agency and the local authority to take account of the impact of the proposal.

Modelling

The modelling of air pollution is an attempt to use physical principles to calculate how gases move in the atmosphere and are influenced by physical features, for example the way in which pollutants are transported and influenced by wind speed and turbulence. Turbulent mixing is a random process that varies with the stability of the atmosphere.

Modelling seeks to establish a mathematical relationship between known levels of discharge (based on emission inventories) and observed meteorology to determine the concentration of specific substances in the atmosphere. Monitoring can be used to test the validity of models by comparing real data with predicted results. The value of a model is that it should allow the prediction of future events; for example, the possible build up of pollution under particular atmospheric conditions, or the implications of possible air quality management strategies. A good model should be capable of anticipating a short-term episode or predict future long-term trends.

The DETR has produced technical guidance on the use and selection of dispersion models [39]. It advises that the use of models should not necessarily be at the first stage of the review and assessment process, that simple screening models can be used at the second stage, and that more sophisticated models may be necessary at the third stage.

Screening models generally only allow consideration of a single source. If predictions are at or near air quality objectives further investigation is necessary. For example, the Department of Transport manual [49] requires only vehicle flow, speed, fraction of heavy goods vehicles and distance from the road to the receptor to enable a calculation to be made. Intermediate models are generally computer-based and require more information, including meteorological data. Advanced models require high quality data but can integrate emissions from a large number of point, line and area sources and take topography into account.

When considering the use of modelling it is important to balance the resources required to allow the calculations to take place with the scale of the problem. Attention must also be paid to the accuracy of the technique.

The accuracy of modelling is of particular importance when associated with environmental impact assessments on which significant planning decisions may be taken. The US National Commission on Air Quality found that models over-predicted or underpredicted actual concentrations by a factor of two [50]. Large errors in dispersion models occur for a number of reasons:

- imperfect understanding of dispersion
- errors in emission estimates
- inappropriate dispersion model structure
- errors in meteorology.

But as the quality of information improves and understanding of dispersion develops, these errors should be reduced. Similarly, as mathematical manipulation of data becomes more sophisticated, so calculations can become more refined.

Three main sources of air pollution may be encountered when assessing air quality:

- road traffic
- industrial, commercial and domestic sources, for example, emissions from a process
- fugitive emissions, for example, dust from mineral extraction processes.

Dispersion models consider releases in four ways:

- point sources, for example, significant industrial chimneys
- line sources, for example, roads
- area sources, based on the accumulation of lesser point or line sources, for example, housing or industrial estates
- volume sources, for example, aircraft take-off and landings.

The Meteorological Office has carried out a review model of predictions for several towns, plotting prediction error against town size [51]. The study revealed that there was under-reporting in smaller towns. Two most likely causes currently being investigated are:

- limitations in the models used in forecasting
- emission data being inadequate in small towns.

This, the study concluded, emphasizes the point that air quality review, models and measurements should be used together for a better understanding of local problems.

All dispersion models require data that describes the amount of pollutants being released. In most circumstances the predicted concentration is directly proportional to the emission: if the emission doubles so will the predicted concentration. For this reason the more accurate the data used to feed the model the more reliable the output.

In the case of point sources, the emission rate describes how much pollution is released per unit time (generally grammes per second). This information can be derived from:

- measurements of the concentration of pollution in the stack
- plant manufacturer's specifications
- emission limits
- estimates based on fuel consumption.

Consideration must be given to the effective chimney height, that is the height to which the plume rises. In the case of line sources, the emission rate is given in grammes per metre per second. The emission of a section of road will depend on:

- the length of road
- the volume of traffic
- the vehicle speed
- the vehicle mix
- the gradient.

Dispersion from roads will be influenced by local topography (e.g. street canyon effects, elevated highways or cuttings). The output from dispersion models is reported as concentration values, which are expressed as averages over specified time periods.

It is important that the appropriate period is used for the pollutants of interest. In the case of benzene, 1,3-butadiene, lead and NO_2, the objectives are described by annual means of running annual means. (Calculation requires the averaging of hourly rates over a 12-month period.) In the case of carbon monoxide, NO_2, PM_{10} and sulphur dioxide, shorter periods are used. Dispersion models must be run taking account of various meteorological conditions in order to determine the worst case. The selection of appropriate models for assessment must take account of the likelihood of air quality failing to reach prescribed objectives. The resources required to purchase the necessary computer hardware and software to operate the more advanced models can run into tens of thousands of pounds. The staff requirements to obtain the information for the model to operate and to ensure proper and regular updating is similarly demanding.

A national mapping methodology has been developed, and used to generate UK-wide maps of annual mean benzene, 1,3-butadiene, NO_2 and

PM_{10} concentrations at background locations for both current and future years. These methods have been supplemented by more detailed modelling studies using the dispersion modelling techniques.

Atmospheric emission inventories

An atmospheric emission inventory is an organized collection of data that relates to the characteristics of processes or activities that release pollutants into the atmosphere across the study area. The DETR has published technical guidance on how inventories should be established [38].

Emission inventories provide:

- an overview of the intensity of emissions, allowing the identification of hot spots and the production of maps that will help to identify where air quality objectives are not being met
- guidance on appropriate locations for monitoring
- information on those sources that should be targeted for reductions in emissions
- a basis for dispersion modelling
- a mechanism to assess trends in air quality based on the effects of possible changes in emissions.

There are a number of sources of information on emissions within the United Kingdom: the UK Emissions Factor Databases gives representative emissions from a range of industrial and transport activities; the National Atmospheric Emission Inventory provides estimates for each pollutant on a 1-km square grid across the whole of the United Kingdom; and the DETR has produced local inventories for a number of major conurbations including Greater Manchester, the West Midlands and West Yorkshire.

Before beginning to establish an inventory a number of decisions must be taken:

- its use
- the pollutants and sources covered (local air quality management requires that account be taken of eight pollutants)
- the requirements of the database and software capable of carrying out possible subsequent modelling

- the geographic scope of the inventory and the spatial resolution
- the maintenance of the inventory
- the mechanism for achieving public access to the inventory.

Three types of source have to be considered:

- line sources, for example, roads and railways
- point sources, for example, factories
- area sources, arising from small-scale diffuse activities, for example, areas of housing.

There are few situations where it is possible to obtain continuous measurements of emissions, generally emission inventories are based on estimates. The first step in the process is to quantify the sources of pollution in the area of interest, after which recognized techniques can be used to calculate the likely emissions.

Various formats are available for storage and manipulation of inventories using standard computer software, such as spreadsheets or databases. In addition, a more sophisticated approach could be adapted by the use of geographic manipulation. It is important to consider at this stage whether modelling is likely to be carried out at some later date; if so, the format selected for the inventory must be capable of being transferred into the model.

The following sources of pollution must be considered as part of the process of air quality review and assessment:

- road traffic
- other mobile sources, for example, railways and aircraft
- point sources, for example, industrial processes
- area sources.

To give an indication of the approach that is adopted, emissions from traffic can be based on four subsources:

- exhaust from vehicles on major roads (for which there will be considerable information on traffic flow)
- exhaust from vehicles on minor roads

- additional emissions associated with vehicle movements, such as the number of cold starts
- other emissions, such as refuelling.

Detailed information on all of the considerations in relation to each of the major sources, together with a number of example showing how calculations are carried out, is given in the technical guidance from the DETR [38].

It would seem prudent to base any inventory on the determination of the pollutants identified in the UK NAQS. Other substances may be of interest, but the resources involved in collecting information must be carefully considered if there is no mechanism available for control. In other countries, where other substances are of interest, the inventory must reflect those requirements. Once the basis of an inventory has been determined, its expansion is only influenced by the identification of the emissions and their quantification.

The major problem with all inventories is the resources that have to be continuously invested to update the database so as to ensure that it provides a contemporary picture of air pollution in the area of interest.

Monitoring

The primary objectives of the UK air monitoring systems as described by the Department of Health [17] are to:

- assess population/ecosystem exposure and resulting effects
- meet present and future statutory requirements under EU directives
- provide a sound scientific basis for policy monitoring
- identify long-term trends and major pollution sources
- provide information to the public.

The DETR has provided technical information on monitoring [37]. Quality assurance and control must be an essential part of any air monitoring system. All measurements must meet defined levels of quality, with a stated level of confidence. The results must be suitable for the purpose for which they were collected [52]. The quality assurance and control system must ensure that:

- measurements are accurate, precise and credible
- data is representative of ambient conditions
- results are comparable and traceable
- measurements are consistent over time
- data is captured effectively
- the best use is made of resources.

In order to ensure that all these points are taken into account, clear procedures should be produced that identify: the process of site and equipment selection; site operation and maintenance; calibration procedures; and data handling. The use of procedures must not become a process for generating unnecessary paperwork; it is a means of ensuring that there is a clear understanding of operation methods at all levels within the organization. Having agreed operational procedures, management must ensure that staff are fully aware of their content, are trained in implementation, and realize that modifications to procedures must be formally acknowledged. Procedures must be written in such a way that they can be audited. There should be a structured audit process to evidence the effectiveness of the quality control system.

When establishing or reviewing an air monitoring network it is necessary to determine the resources that will be required in relation to capital, maintenance and operational costs and staffing requirements.

The design of a monitoring programme will depend on the size of the area throughout which information is required and the use that will be made of the data. Consideration of the number and general location of monitoring sites should include the following factors:

- significant pollution sources in the area
- target populations/ecosystems
- meteorology
- topography
- dispersion patterns based on local models
- existing information on air quality.

A specification should be drawn up that identifies the factors that have to be considered when determining the location of individual monitoring units. It should include:

- availability of mains electricity
- availability of telephone link for telemetry (if required)
- availability for the duration of the monitoring programme
- ease of access
- free from influence by local sources of pollution
- free from influence by local topographical features
- acceptable security/risks of vandalism
- availability of planning permission
- public safety.

In order to allow the accurate analysis of the air sample under consideration, the following matters must be taken into account:

- stability of the pollutant being examined
- minimal reaction with the sampling line
- sample delivered to equipment at an adequate rate
- acceptable pressure drop through sample line
- susceptibility of equipment to thermal shock
- ease of calibration and maintenance.

Instrument selection should ensure: consideration of monitoring and data quality objectives; suitable time resolution of measurement; satisfactory after-sales support and reliability.

A variety of techniques of varying levels of complexity are available for the measurement of concentrations of pollution in the air. The main methods fall into the following categories:

- passive sampling, for example, diffusion tubes
- active sampling, for example, bubblers and filters
- automatic point monitoring
- long-path/remote monitoring.

Passive sampling methods provide simple and cost-effective methods of screening air quality, giving a good indication of average pollution concentrations. The low cost of tubes permits sampling at a number of points, enabling the identification of hot spots. Diffusion tubes are available for NO_2, sulphur dioxide and benzene.

Active semi-automatic samplers collect pollution samples by chemical or physical means requiring subsequent laboratory analysis. A known volume of air is bubbled through a collector, such as a chemical solution or filter, for a known period of time. Subsequent analysis allows an indication of the pollution level to be assessed.

Automatic real-time point analysis produces high-resolution measurements for discrete periods of time. The guidance from the DETR [37] includes pollution-specific information on monitoring methods and issues; advice is presented in a clear and structured manner. The standard and objective for each of the pollutants in the UK air quality management strategy is set out, and is followed, where available, by information on reference methods of monitoring. A monitoring strategy is discussed in the guidance, providing suggestions on how monitoring should be carried out. It also contains details on methodology, quality assurance and control and data quality assurance in relation to both non-automatic and automatic techniques. Caution should be shown before buying equipment using innovative technology as it is not unknown for the promises made by a manufacture to fail to be met in the field.

When expensive equipment is to be purchased it is likely that invitations to tender will be required. The process of drafting a specification requires identification of all of the requirements in relation to the equipment to be obtained. A properly written specification enables the purchaser to demonstrate the defined operational parameters in the event of a subsequent dispute with the supplier.

Once a network has been established, data must be reviewed in a structured manner. Equipment can malfunction, power surges can affect data, and results can be corrupted by deliberate action or as the result of a very local release. (For example, PM_{10} measurements at one urban site increased by an order of magnitude as a result of local demolition work. An understanding of the cause of this change was important as an environmental lobby group tried to use the results to brand the authority as having the worst air in Europe.)

Data validation involves a rapid front-end screening process to identify or remove spurious data prior to initial use. Subsequently there should be a long-term review process prior to analysis and reporting. The validation should be carried out as soon as practicable after measurements have been made to allow investigation of anomalies. Many commercially available data telemetry monitoring systems allow for the identification of out-of-range or suspect data. Adherence to rigid data acceptance criteria and automatic rejection of flagged data do not necessarily guarantee high data quality. Such an approach could invalidate extreme (but valid) pollution measurements just because they lie outside pre-set limits [52].

Once data has been collected it must be analysed to allow it to be used for policy development and review, and as a source of information for the general public. WHO has provided useful guidance on recognized analytical methodology in relation to air quality [53].

A minimum level of data management could be regarded as the production of daily, monthly and annual summaries, involving simple statistical and geographical analysis. The use of geographical information systems should be considered, particularly when it is intended to combine pollution data with information from epidemiological and other geoco-ordinated social, economic or demographic sources [52].

Information must be presented in an appropriate manner to maximize understanding: data that is understood by those involved in the measurement may be incomprehensible to policy-makers and the general public. It may be necessary to produce data in more than one format to satisfy the various users.

UK LEGISLATIVE CONTROL (ALSO SEE [54])

Background

From the time that public health legislation was first enacted, the thrust of control has been directed towards the abatement of nuisance. It is for this reason that, generally, action has been taken to control emissions only when they have become the subject of complaint. Despite any duty to inspect the district 'from time to time', many local authorities have either been restrained or have restrained themselves from taking action until there has been a reasonable body of evidence to support the existence of nuisance. This has lead to accusations that controls have not been applied uniformly across the country.

An exception to this generalization relates to the control of smoke. It has long been realized that smoke can be offensive. Initially, the concerns were associated with coal smoke. The London smog of 1952 triggered the first of two Clean Air Acts (1956 and 1968). At that time much domestic and industrial heating was provided by the combustion of coal. The 1956 Act introduced the concept of domestic smoke control by following the pioneering work of a few far-sighted councils that had introduced their own smokeless zones. Industrial emissions, largely associated with hand-fired boilers, were limited by restrictions on the periods during which smoke could be emitted and the density of such smoke. Powers were also provided to limit grit and dust emissions from furnaces, although regulations related to coal-fired boilers were never extended to other plant, such as incinerators or cupolas.

It was not until the Environmental Protection Act 1990 came into force that specific powers to control processes were given to local authorities.

Almost all areas of the country that have previously been recognized as having poor air quality have been made the subject of smoke control orders. The procedure necessary to establish an order is complex, and is beyond the scope of this chapter. (For detailed procedures, see [54].) It must, however, be recognized that there may be a need for further smoke control areas to be declared as a result of the national air quality strategy. Some information is provided on enforcement action and on the implications of smoke control orders to commercial and industrial premises.

It is important that anyone involved in pollution control work is able to make observations of dark and black smoke, but such activities are more likely to be associated with the burning of trade or industrial waste than emissions from chimneys serving conventional furnaces. The convenience of gas has seen a massive change in fuel utilization over the past 20 years. This practice has made its own contribution to the reduction in the concentration of particulates in the atmosphere. The wisdom of burning natural gas in large combustion plants, such as power stations, is open to question.

While there is a need to be familiar with a range of combustion processes and installations, it is not practicable to provide such information here.

The enactment of the Environmental Protection Act 1990 has focused interest on the uniform control of a wide range of other furnaces. It is now as important to understand the principles of, for example, metal smelting, glass manufacture and mineral processing. These subjects are considered here, but reference to contemporary specialist works is necessary to achieve an adequate comprehension of the relevant technology.

Environmental Protection Act 1990

The 1990 Act laid down procedures for the application of IPC to a range of processes that have considerable pollution potential, and provided a new approach to air pollution control to be implemented by local authorities. Plant was required to apply BATNEEC (p. 626) to control emissions. From April 1991, all new and substantially changing major installations were subject to IPC. Existing plant was brought under IPC in a phased programme, starting with large combustion plants in 1991–2. Similarly, local authority control was introduced on a phased basis: new and existing processes were required to comply with the BATNEEC conditions contained in the relevance guidance rate. Further controls will be introduced under integrated pollution prevention and control (see, p. 699).

Definitions

Section 1 of the Act provides a series of definitions relevant to the interpretation of Part I. These are set out later.

Environment consists of all, or any, of the air, water, and land; and the medium air includes the air within buildings and the air within other natural or man-made structures above or below ground (Section 1(2)).

Pollution of the environment means pollution of the environment due to the release (into any environmental medium) from any process of substances that are capable of causing harm to humans or to any other living organism supported by the environment (Section 1(3)).

Harm means harm to health of any living organisms or other interference with ecological systems of which they form part and, in the case of humans, includes offence caused to any of his or senses or harm to his property; and harmless has a corresponding meaning (section 1(4)).

Process means any activities carried on in Great Britain, whether on premises or by means of mobile plant, that are capable of causing pollution of the environment; 'prescribed processes' means processes prescribed under Section 2(1).

In the context of **substances**:

1. **activities** means industrial or commercial activities or activities of any other nature whatsoever (including, with or without other activities, the keeping of a substance)
2. **Great Britain** includes so much of the adjacent territorial sea as is, or is treated as, relevant territorial waters for the purposes of Chapter I of Part III of the Water Act 1989 (now the Water Resources Act 1991) or, with respect to Scotland, Part II of the Control of Pollution Act 1974
3. **mobile plant** means plant that is designed to move or to be moved whether on roads or otherwise.

The **enforcing authority** in relation to England and Wales is the chief inspector or the local authority by which, under Section 4, the functions conferred or imposed by this part 'are for the time being exercised in relation respectively to releases of substances into the environment or the air'; **local enforcing authority** means the local authority (Section 1(7)). Subsection 8 of Section 1 defines the enforcing authority in relation to Scotland.

Authorization means an authorization for a process (whether on premises or by means of mobile plant) granted under Section 6; a reference

to the conditions of an authorization is a reference to the conditions to which at any time the authorization has effect (Section 1(9)).

A substance is **released** into any environmental medium whenever it is released directly into that medium whether it is released into it within or outside Great Britain. Release includes:

1. in relation to air, any emission of the substance into air
2. in relation to water, any entry (including any discharge) of the substance into water
3. in relation to land, any deposit, keeping or disposal of the substance in or on land; and for this purpose **water** and **land** shall be construed in accordance with subsections 11 and 12.

For the purpose of determining into what medium a substance is released:

1. any release into

   (a) the sea or the surface of the sea-bed
   (b) any river, watercourse, lake, loch or pond (whether natural or artificial or above or below ground) or reservoir or the surface of the river-bed or of other land supporting such waters
   (c) ground waters

2. any release into

   (a) land covered by water falling outside paragraph 1 above or the water covering such land
   (b) the land beneath the surface of the sea-bed or other land supporting waters falling within paragraph 1a above is a release on to land

3. any release into a sewer (within the meaning of the Water Industry Act 1991) shall be treated as a release to water. But a sewer and its contents shall be disregarded in determining whether there is pollution of the environment at any time.

In subsection 11, **ground water** means any waters contained in underground strata or in:

1. a well, borehole or similar work sunk into under-ground strata, including any adit or

passage constructed in connection with the well, borehole or work facilitating the collection of water in the well, borehole or work

2. any excavation into underground strata where the level of water in the excavation depends wholly or mainly on water entering it from the strata.

Substance shall be treated as including electricity or heat (Section 1(13)).

Prescribed substance means any substance of a description in regulations made under Section 2(5), or in the case of a substance of a description prescribed only for release in circumstances specified under Section 2(6)(b) means any substance of that description that is released in those circumstances.

Integrated pollution control

The concept of IPC was developed by the Royal Commission on Environmental Pollution in its fifth report [55]. In essence, BPEO (see p. 627) has to be applied to the control of solid, liquid or gaseous wastes. Where choice exists about the sector of the environment into which waste should be discharged, a decision has to be made on the means by which environmental damage can be minimized.

In the case of larger works, control is executed by the Environment Agency, which considers the most appropriate abatement technology on the basis of IPC. Process-specific guidance notes have been produced (see pp. 701–703).

Local air pollution control

The main features of the LAPC system are outlined later. Prescribed processes designated for local control must not be operated without an authorization from the local enforcing authority in whose area they are located. (In the event of an application being refused and an appeal being made, existing processes may continue to operate pending the appeal decision.) Mobile plant must be authorized by the local enforcing authority in whose area the operator has his principal place of business. Operators of prescribed processes must

submit a detailed application for authorization to the local enforcing authority.

Local authorities are legally obliged to include conditions in any authorization they issue that are designed to ensure that the process is operated using BATNEEC in order to prevent and minimize emissions of prescribed substances, and to render harmless any substance that may be emitted. Conditions must also secure compliance with the other objectives specified in Section 7(2) of the Act.

In addition to any specific conditions included in an authorization, all authorizations implicitly impose a duty on the operator to use BATNEEC in relation to any aspect of the process that is not covered by the specific conditions. This is the so-called 'residual' BATNEEC duty. The Secretary of State has arranged for process guidance notes (PGs) to be issued to every local enforcing authority on all the main categories of process coming under local enforcing authority control. These notes, which are listed later on in this chapter, contain the Secretary of State's views on the techniques appropriate in order to achieve the BATNEEC objective, as well as information in relation to any of the other objectives in Section 7(2) of the Act. They are likely to be of interest to operators of prescribed processes as well as to local authorities. Eighty PG notes have been published by the Stationery Office. These are intended to cover all the main categories of process prescribed for local air pollution control. A first four-year review of these notes has been substantially completed, and revised notes published. Some notes have been cancelled.

Operators can appeal against refusal of an application, against the conditions included in an authorization, and against the various forms of notice that may be served by a local enforcing authority. Appeals will not put notices into abeyance, except in the case of revocation notices. Local authorities can issue enforcement, variation, prohibition and revocation notices to ensure that appropriate standards of control are met and raised in line with new techniques and new awareness of environmental risk. Prohibition notices are a mechanism for stopping a process if there is an imminent risk of serious pollution of the environment.

All applications for authorization (except in relation to small waste oil burners and mobile plant) must be advertised locally and full details (except information that is commercially confidential or would prejudice national security) must be made available so that the public can comment before the process is authorized to start operation or to undergo a substantial change.

Public registers must be set up by each local enforcing authority giving details of all IPC and local enforcing authority air pollution control processes in its area. These must include specified particulars of applications, authorizations, notices, directions issued by the Secretary of State, appeal decisions, monitoring data, etc. The information relating to IPC processes will be supplied to local authorities by the national agencies (Environment Agency or Scottish Environment Protection Agency) as appropriate. Information is to be kept from the register only on grounds of national security or commercial confidentiality.

Local authorities are obliged to levy fees and charges in accordance with a scheme prescribed by the Secretary of State. The scheme is reviewed annually.

Local authorities have powers of entry, inspection, sampling, investigation and seizure of articles or substances that are a cause of imminent danger or serious harm.

Power to make regulations

Section 2 of the Act enables the Secretary of State for the Environment, Transport and the Regions to prescribe, by regulation, the processes that will require authorization.

The Secretary of State for the Environment, Transport and the Regions is also empowered to prescribe substances that may be released to the environment. Regulations may:

1. prescribe the substance that may be released, subject to control, to each environmental medium
2. prescribe the concentrations, periods of release and other limitations.

The Secretary of State may also make regulations establishing standards, objectives or requirements in relation to particular prescribed processes or substances:

1. in relation to releases from prescribed processes into any environmental medium, prescribe standards for:

   (a) the concentration, amount or the amount in any period of that substance that may be so released
   (b) any other characteristic of that substance in any circumstances in which it may be so released

2. prescribe standard requirements for the measurement or analysis of, or release of, substances for which limits have been set under paragraph 1 above
3. in relation to any prescribed process, prescribe standards or requirements in relation to any aspect of the process.

Power is given to the Secretary of State to make plans for:

1. establishing limits for the total amount, or the total amount in any period, of any substance that may be released into the environment
2. allocating quotas with respect to the releases of substances to people carrying on processes in respect of which such limit is established
3. establishing limits of the descriptions above so as to progressively reduce pollution of the environment
4. the progressive improvement in the quality objectives and quality standards established by regulation.

General guidance notes

Five general guidance notes, which explain the main controls and procedures, have been issued by the DETR. One 'upgrading guidance' (UG) note has also been issued. They are as follows.

GG1 – Introduction to Part I of the Act. This includes a general explanation of BATNEEC, guidance on interpreting 'substantial' change, an explanation of what is meant by existing process, a copy of the charging scheme, and the variation notice procedures.

GG2 – Authorizations. This contains advice for local authorities on drawing up authorizations and includes an outline authorization and a range of specimen conditions.

GG3 – Applications and registers. This explains the procedures for making an application and the public register requirements. It includes a suggested application form, flow chart of the main stages in the application procedures, and a list of the names and addresses of the statutory consultees.

GG4 – Interpretation of terms used in PG notes. As its title suggests, this provides additional guidance on some of the terminology commonly found in the PG notes.

GG5 – Appeals. This summarizes the procedures for making appeals and for the handling of appeals at a hearing or by exchange of written representations.

UG-1 Revisions/additions to existing process and general guidance notes: No 1. This note contains amendments to PGs 1/1, 1/3, 1/5, 1/10, 3/2, 5/1, 5/2, 5/4, 5/5, and minor corrections to 6/3, 6/21, 6/22, 6/23 and 6/26, as well as additional general guidance. All the PG note amendments and corrections have been superseded by the four-year review of the PG notes, apart from those in relation to PG1/10.

Air quality guidance notes

Air quality guidance notes (AQs) give information to local authorities about the LAPC system. They are published by DETR.

Process guidance notes

All such notes are headed 'Environmental Protection Act 1990, Part 1; Processes Prescribed for Air Pollution Control by Local Authorities; Secretary of State's Guidance'. The heading then specifies the name of the individual process, as listed later. Most notes have been subject to a four-year review conducted by the DETR, with appropriate consultation, between mid-1994

and mid-1996. The broad aim was to issue any amendments that arose out of the review at around the fourth anniversary of the publication date of the original note. The full list of extant notes is given later. Notes that have been cancelled as a result of the review are listed separately.

COMBUSTION

PG1/1(95) – waste oil burners, less than 0.4 MW net rated thermal input.

PG1/2(95) – waste oil or recovered oil burners, less than 3 MW net rated thermal input.

PG1/3(95) – boilers and furnaces, 20–50 MW net rated thermal input.

PG1/4(95) – gas turbines, 20–50 MW net rated thermal input.

PG1/5(95) – compression ignition engines, 20–50 MW net rated thermal input.

PG1/10(92) – waste derived fuel combustion processes less than 3 MW.

PG1/11(96) – reheat and heat treatment furnaces, 20–50 MW net rated thermal input.

PG1/12(95) – combustion of fuel manufactured from or comprised of solid waste in appliances between 0.4 and 3 MW net rated thermal input.

PG1/13(96) – processes for the storage, loading and unloading of petrol at terminals.

PG1/14(96) – unloading of petrol into storage at service stations.

PG1/15(96) – odorizing natural gas and liquefied petroleum gas.

NON-FERROUS METALS

PG2/1(96) – furnaces for the extraction of non-ferrous metal from scrap.

PG2/2(96) – hot dip galvanizing processes.

PG2/3(96) – electrical and rotary furnaces.

PG2/4(96) – iron, steel and non-ferrous metal foundry processes.

PG2/5(96) – hot and cold blast cupolas.

PG2/6(96) – aluminium and aluminium alloy processes.

PG2/7(96) – zinc and zinc alloy processes.

PG2/8(96) – copper and copper alloy processes.

PG2/9(96) – metal decontamination processes.

MINERAL PROCESSING

PG3/1(95) – blending, packing, loading and use of bulk cement.

PG3/2(95) – manufacture of heavy clay goods and refractory goods.

PG3/3(95) – glass (excluding lead glass) manufacturing processes.

PG3/4(95) – lead glass manufacturing processes.

PG3/5(95) – coal, coke, coal product and petroleum coke processes.

PG3/6(95) – processes for the polishing or etching of glass or glass products using hydrofluoric acid.

PG3/7(95) – exfoliation of vermiculite and expansion of perlite.

PG3/8(96) – quarry processes.

PG3/12(95) – plaster processes.

PG3/13(95) – asbestos processes.

PG3/14(95) – lime processes.

PG3/15(96) – mineral drying and roadstone coating processes.

PG3/16(96) – mobile crushing and screening processes.

PG3/17(95) – china and ball clay processes including the spray drying of ceramics.

CHEMICAL INDUSTRY

PG4/1(95) – processes for the surface treatment of metals.

PG4/2(96) – processes for the manufacture of fibre reinforced plastics.

INCINERATORS

PG5/1(95) – clinical waste incineration processes under 1 tonne an hour.

PG5/2(95) – crematoria.

PG5/3(95) – animal carcass incineration processes under 1 tonne an hour.

PG5/4(95) – general waste incineration processes under 1 tonne an hour.

PG5/5(91) – sewage sludge incineration processes under 1 tonne an hour (no revision of 1991 note planned; no plant under 1 tonne an hour currently known to exist).

OTHER INDUSTRIES

PG6/1(00) – animal byproduct.

PG6/2(95) – manufacture of timber and wood-based products.

PG6/3(97) – chemical treatment of timber and wood-based products.

PG6/4(95) – processes for the manufacture of particleboard and fibreboard.

PG6/5(95) – maggot breeding processes.

PG6/7(97) – printing and coating of metal packaging.

PG6/8(97) – textile and fabric coating and finishing processes.

PG6/9(96) – manufacture of coating.

PG6/10(97) – coating manufacturing processes.

PG6/ 11(97) – manufacture of printing ink.

PG6/12(91) – production of natural sausage casings, tripe, chitterlings and other boiled green offal products (due to be reviewed and republished).

PG6/13(97) – coil coating processes.

PG6/14(97) – film coating processes.

PG6/15(97) – coating in drum manufacturing and reconditioning processes.

PG6/16(97) – printworks.

PG6/17(97) – printing of flexible packaging.

PG6/18(97) – paper coating processes.

PG6/19(97) – fish meal and fish oil.

PG6/20(97) – paint application in vehicle manufacturing.

PG6/21(96) – hide and skin processes.

PG6/22(97) – leather finishing processes.

PG6/23(97) – coating of metal and plastic.

PG6/24(96) – pet food manufacturing.

PG6/25(97) – vegetable oil extraction and fat and oil refining.

PG6/26(96) – animal feed compounding.

PG6/27(96) – vegetable matter drying.

PG6/28(97) – rubber processes.

PG6/29(97) – di-isocyanate processes.

PG6/30(97) – production of compost for mushrooms.

PG6/31(96) – powder coating (including sheradizing).

PG6/32(97) – adhesive coating processes.

PG6/33(97) – wood coating processes.

PG6/34(97) – respraying of road vehicles.

PG6/35(96) – metal and other thermal spraying processes.

PG6/36(97) – tobacco processing.

PG6/40(94) – coating and recoating of aircraft and aircraft components.

PG6/41(94) – coating and recoating of rail vehicles.

PG6/42(94) – bitumen and tar processes.

Technical guidance notes

The following notes were issued by HM Inspectorate of Pollution prior to the establishment of the Environment Agency and are published by the Stationery Office.

1. *Sampling Facility Requirements for the Monitoring of Particulates in Gaseous Releases to Atmosphere*, Technical Guidance Note M1.
2. *Monitoring Emissions of Pollutants at Source*, Technical Guidance Note M2.
3. *Standards for IPC Monitoring, Part 1: Standards Organizations and the Measurement Infrastructure*, Technical Guidance Note M3.
4. *Standards for IPC Monitoring, Part 2: Standards in Support of IPC Monitoring*, Technical Guidance Note M4.
5. *Guidelines on Discharge Stack Heights for Polluting Emissions*, Technical Guidance Note D1.
6. *Pollution Abatement Technology for the Reduction of Solvent Vapour Emissions*, Technical Guidance Note A2, revised edition with minor amendments, October 1994.
7. *Pollution Abatement Technology for Particulate and Trace Gas Removal*, Technical Guidance Note A3.

Note: Technical Guidance Note A1 is not relevant to APC.

Integrated pollution prevention and control

Integrated pollution prevention and control (IPPC) is replacing, as from the 1 August 2000, the integrated pollution control system. EU Directive 96/61/EC, concerning integrated pollution prevention and control, was agreed in September 1996. Its objective is to prevent emissions into air, water and soil wherever practicable, taking into account waste management, and where not, minimizing them. The chief differences between IPPC and the existing IPC and LAPC regimes are:

- both IPC and LAPC are concerned with regulating pollution from industrial processes, but the former is geared to preventing or minimizing emissions to air, water or land, while the latter was aimed at preventing or minimizing air pollution from processes not thought to give rise to significant air, water or land pollution.
- IPPC considers emissions to air, water and land, and other environmental effects together in an integrated system of control, the aim being to ensure a high level of protection for the environment as a whole. It covers most of the installations presently regulated under IPC and around 1500 of the 13 000 currently regulated under LAPC.

The general classes of processes within the directive can be summarized as follows: energy industries, production and processing metals, mineral industries, chemical industry, waste management and other activities such as intensive rearing houses for animals. Installations used exclusively for research, development and testing are not covered.

The timetable contained in the regulations specifies that processes move from LAPC to LAPPC in three batches, as follows:

12 months beginning 1 April 2003: combustion, cement and lime, asbestos, other minerals, ceramics, incineration, timber
12 months beginning 1 April 2004: ferrous and non-ferrous metals, coating, manufacture of coating materials, rubber
12 months beginning 1 April 2005: gasification etc., glass, organic chemicals, storage of bulk chemicals, tar and bitumen, animal and vegetable.

Six general principles govern the basic obligations of the operators of processes:

- appropriate pollution prevention by use of best available technology (BAT) (see also p. 626)
- no significant pollution caused
- waste production avoided
- energy efficiency
- accident prevention
- site remediation.

The main air polluting substances that will be taken into account when fixing emission limits are:

- sulphur dioxide and other sulphur compounds
- oxides of nitrogen and other nitrogen compounds
- carbon monoxide
- volatile organic compounds
- metals and their compounds
- dust
- asbestos (suspended particles, fibres)
- chlorine and its compounds
- fluorine and its compounds
- arsenic and its compounds
- cyanides
- substances and preparations that have been proved to possess carcinogenic or mutagenic properties, or properties that may affect reproduction via the air.
- polychlorinated didenzodioxins and polychlorinated dibenzofurans.

There is also a list of substances that can cause pollution to water. BAT is defined. In essence the term has the same meaning as BATNEEC. The following considerations have to be taken into account by member states when considering BAT:

- use of low waste technology
- use of less hazardous substances
- recovery and recycling
- comparable processes found to be successful
- technological advance and change in scientific knowledge
- nature, effects and volumes of emission
- commissioning dates for new or existing installations
- time for introduction of BAT
- consumption of raw materials (including water)
- need to prevent or reduce to a minimum the overall impact of emissions
- need to prevent accidents and minimize consequences for the environment
- information published by the Commission.

Existing processes will be given up to eight years to carry out the necessary upgrades, unless they undergo a substantial change.

The process of IPC and LAPC has set a precedent in the United Kingdom for the provision of information to the regulatory authority. As indicated earlier, existing processes will have to meet the full conditions of an operating permit if the process undergoes a substantial change.

The Commission will exchange information on 'BAT, associated monitoring and developments in them'. It has already undertaken to produce a series of BAT reference documents (BREFs). The BREFs are to be taken into account in determining BAT at a national level, but they will not be prescriptive documents setting standards that have to be followed. They will, however, lay the foundations for a consistent approach across the EU. Member states whose controls are generally below the BREF levels are likely to have to produce persuasive reasons for their decisions, to satisfy public concern if for no other reason.

The public is entitled to the following information about businesses that operate under a permit granted by the directive:

- permits – on an individual basis
- monitoring data – by process
- inventory of emissions – by source.

Records will be held by the regulatory agency and will be available for public inspection. The intention is that the roles of granting permits, etc. and taking enforcement action will be divided between the Environment Agency and local authorities in England and Wales. In Scotland, they will be the sole responsibility of the Scottish Environment Protection Agency (SEPA), with the exception of matters affecting offshore installations. A fee will be payable to the authorities for the exercise of some of their functions as under the present regimes.

New installations will be regulated under IPPC from the day on which it comes into force, while existing installations will be phased into the new regime on a sector by sector basis.

Clean Air Act 1993

This Act was introduced to consolidate the Clean Air Acts 1956 and 1968 and certain related legislation. The following parts are of particular significance.

Section 1. Prohibition of dark smoke from chimneys

Section 1 prohibits, subject to conditions, emissions of dark and black smoke from chimneys serving boilers and industrial plant. Details of the restrictions are determined by the Dark Smoke (Permitted Period) Regulations 1958 (SI 1958 No. 498). Two separate circumstances are considered: continuous and intermittent emissions. The limit for a continuous emission of dark smoke, caused otherwise than by soot-blowing, is 4 minutes.

The determination of intermittent emissions is limited to aggregate emissions for a period of up to 8 hours. In the case of one chimney, emissions of dark smoke shall not exceed 10 minutes in the aggregate period of 8 hours or, if soot-blowing is carried out, for not longer than 14 minutes in aggregate in that period. The periods of 10 and 14 minutes shall increase in the case of a chimney serving two furnaces to 18 and 25 minutes, respectively; in the case of a chimney serving three furnaces 24 and 34 minutes, respectively; and in the case of a chimney serving four or more furnaces 29 and 41 minutes, respectively. The maximum aggregate emission of black smoke is 2 minutes in any period of 30 minutes. Aggregate emissions are considered by carrying out an observation over a period of time and accumulating incremental emissions.

In any proceedings taken under this legislation, account must be taken of the statutory defences contained in Section 1. They are as follows:

1. that the contravention complained of was solely due to the lighting up of a furnace that was cold, and that all practicable steps had been taken to prevent or minimize the emission of dark smoke;
2. that the contravention complained of was solely due to some failure of a furnace or the apparatus used in connection with a furnace, that that failure could not have been foreseen, or, if foreseen, could not have been provided against, and that the contravention could not reasonably have been prevented by action taken after the failure occurred;
3. that the contravention complained of was solely due to the use of unsuitable fuel, that suitable fuel was unobtainable, that the least unsuitable fuel that was available was used, and that all practicable steps had been taken to prevent or minimize the emission of dark smoke as the result of the use thereof;
4. that the contravention complained of was due to a combination of two or more of the causes specified in paragraphs 1–3 above and that the other conditions specified in those paragraphs are satisfied in relation to those causes respectively.

When considering the defences it is important to bear in mind the date of the primary legislation. In the mid-1950s coal was the most widely used industrial fuel, hence the allowance made for furnaces being lit from cold, the emphasis on the availability of suitable fuel and the generous periods allowed for emission. If the 1958 regulations had been updated, it would have been reasonable to have imposed more stringent standards that reflected contemporary combustion technology.

When an emission of dark smoke is observed, it is necessary to advise the owner of the boiler in plant 'as soon as may be', and confirm the fact, in writing, before the end of four days of becoming aware of the offence in accordance with Section 51 of the 1993 Act.

While procedures for notification are not prescribed, it is apparent that certain information must be obtained at the earliest opportunity after the alleged offence has been observed. Immediate examination of the plant is vital in order that comprehensive evidence can be gathered about the likely cause of the emission, and the possible relevance of the statutory exemption. It is for individual local authorities to determine the way in which investigations should be carried out, and to balance the competing pressures between requiring the investigating officer to seek immediate contact with senior management, a time-consuming exercise in a large organization, and the immediate visit to the boiler-house or furnace.

Familiarity with plant and the fundamental principles of combustion technology are important. It seems reasonable to begin an investigation by identifying the particular boiler or furnace that was causing the emission. Details should be recorded of the description of the plant and the fuel used to fire it. An attempt has to be made to gather concise information of the operating cycle, the nature of the fuel and the condition of the fuel and combustion air supplies. It is important to take account of information not only on the possibility of there having been a breakdown, but also on the immediate and longer term steps taken to mitigate the problem. What is the general state of maintenance? Does there appear to be a planned programme of maintenance, or does it seem that mechanical and electrical engineering works are only undertaken when absolutely necessary? These factors are important in determining the relevance of defences. While information can be considered during the preliminary stages of an investigation in an informal manner, it is necessary to issue a caution when it has been decided that there has been an offence, and to limit the use of statements to information obtained after the caution was given. Evidence must be based on contemporaneous notes.

Provisions similar to those related to industrial boilers and furnaces restrict emissions from vessels, provided that they are in waters as defined by Section 44 of the Clean Air Act 1993. Details are to be found in the Dark Smoke (Permitted Periods) (Vessels) Regulations 1958 (SI 1958 No. 878).

Section 2. Prohibition of dark smoke from industrial or trade premises

Section 2 prohibits emissions of dark smoke from trade or industrial premises. Subject to specific exemptions for prescribed matter, the prohibition is absolute: no permitted periods are provided. 'Industrial or trade premises' are defined, in subsection (6) as 'premises used for any industrial or trade purpose or premises not so used on which

matter is burnt in connection with any trade or industrial process'. This legislation is used to control smoke emissions from 'bonfires', including those on demolition sites.

Materials conditionally exempt from the provisions of this legislation are detailed in the Clean Air (Emission of Dark Smoke) (Exemption) Regulations 1969 (SI 1969 No. 1263) as follows.

1. Timber and any other waste matter (other than natural or synthetic rubber or flock or feathers) which results from the demolition of a building or clearance of a site in connection with any building operation or work of engineering construction (within the meaning of Section 176 of the Factories Act 1961). (Conditions A, B and C.)
2. Explosive (within the meaning of the Explosives Act 1875) which has become waste; and matter which has become contaminated by such explosive. (Conditions A and C.)
3. Matter which is burnt in connection with:

   (a) research into the cause or control of fire; or
   (b) training in fire-fighting.

4. Tar, pitch, asphalt and other matter which is burnt in connection with the preparation and laying of any surface, or which is burnt off any surface in connection with resurfacing, together with any fuel used for any such purpose. (Condition C.)
5. Carcasses of animals or poultry which:

   (a) have died, or are reasonably believed to have died, because of disease;
   (b) have been slaughtered because of disease; or
   (c) have been required to be slaughtered pursuant to the Diseases of Animals Act 1981. (Conditions A and C, unless the burning is carried out by or on behalf of an inspector (within the meaning of the Diseases of Animals Act 1981).)

6. Containers which are contaminated by any pesticide or by any toxic substance used for veterinary or agricultural purposes; and in this paragraph 'container' includes any sack, box, package or receptacle of any kind. (Conditions A, B and C.)

The three conditions are as follows.

1. Condition A: that there is no other reasonably safe and practicable method of disposing of the matter.
2. Condition B: that the burning is carried out in such a manner as to minimize the emission of dark smoke.
3. Condition C: that the burning is carried out under the direct and continuous supervision of the occupier of the premises concerned or a person authorized to act on his behalf.

As in the case of emission of smoke from chimneys, it is necessary to investigate the circumstances of the emission immediately after completing the investigation so as to be able to test the relevance of the exemptions.

Several issues arise from the exemptions that are worthy of brief consideration. There is growing concern about the way in which material is squandered; while it may be permissible to burn waste from building and demolition operations, more thought needs to go into finding methods of reusing such materials. While the legislation permits the burning of contaminated containers, it does not authorize smoke emissions associated with the burning of pesticides or similar hazardous substances. The disposal of such material should be carried out in an environmentally responsible manner and in accordance with the Control of Substances Hazardous to Health Regulations 1998 (COSHH) requirements (see Chapter 23).

Exemptions are not available for emissions of dark smoke produced for effect purposes associated with film-making or other 'entertainment' activities.

Section 3. Meaning of 'dark smoke'

Dark smoke is defined in Section 3 of the 1993 Act as smoke which, 'if compared in the appropriate manner with a chart of the type known at the date of the passing of the Act as the Ringelmann Chart, would appear to be as dark as or darker than shade 2 on that chart'. Black smoke is defined in the Dark Smoke (Permitted Period) Regulations 1958 (SI 1958 No. 498) as smoke which, 'if compared . . . with . . . the Ringelmann Chart, would appear to be as dark as or darker than shade 4 on the chart'.

Section 3(2) continues: 'for the avoidance of doubt it is hereby declared that . . . the court may be satisfied that smoke is or is not dark . . . notwithstanding that there has been no actual comparison with a Ringelmann Chart'. This does not seem to suggest that any officer can stand up in court testifying the intensity of a smoke emission without having carried out the necessary comparison; some familiarity with the procedure is necessary.

Smoke observations

The method used for the assessment of the intensity of the shade of a smoke emission is simple: it is based on comparison with a shaded chart. The Ringelmann Chart is described by BS 2742:1969 [56]. The chart comprises a white card onto which are printed, in black, grids that proximate to shades of obscuration. The 20% shade is represented by a grid in which the ratio of printing ink to plain card is 1 : 4. Similarly the 40%, 60% and 80% shades are represented by print to paper ratios of 2 : 3, 3 : 2 and 4 : 1. It is important to recognize the possibility that the characteristics of the chart could be influenced by the fading of the ink or the discoloration of the paper, whether by soiling or tinting caused by protective film.

Clear guidance is given in BS 2742 on how to use the chart. The position from which the observation is made should be chosen carefully. The chart should be used in daylight conditions and held in a vertical plane facing the observer. Where possible, it should be in line with the point of emission and so placed that the chart and smoke have similar sky backgrounds. The chart must be at a sufficient distance from the observer that the lines appear to merge until each square is a uniform shade of grey; for most observers this distance is in excess of 18 m.

Observations should be carried out, as far as practicable, under conditions of uniform illumination from the sky. If the sun is shining the chart should be at right angles to the line of vision, not in front or behind the observer. The white square provides a useful indication of illumination of the chart; it also enables soiling of the paper to be recognized. Observations at steep angles should be avoided. The darkness of the smoke, at the point where it leaves the chimney or bonfire, should be compared with the chart; the number of the shade that appears to match most closely the darkness of the smoke determined, and the duration of the emission measured. The darkness of smoke that is intermediate between two shades may be estimated to the nearest quarter Ringelmann number in favourable conditions. In order to use the chart, it seems necessary to use at least two people: one to support it and the other to make the observation. This has resource implications.

A miniature chart has been produced [56] to the same precision as the Ringelmann chart. Hatching has been replaced by shades of grey colour, similar to those used on paint colour charts. It is designed to be used about 1.5 m from the observer's eye; conditions for observation are as previously described. As the chart is printed on slightly translucent card, it should be backed with a sheet of white opaque material or be inserted in a holder.

Section 4. Requirement that new furnaces shall be so far as practicable smokeless

Section 4 requires all new furnaces (other than domestic boilers rated at less than 16 kW) to be, so far as practicable, capable of operating continuously without emitting smoke when burning the fuel for which they were designed.

'Furnace' is not defined in the Act, but can be interpreted as any enclosed or partially enclosed space in which solid, liquid or gaseous fuel is burned, or in which heat is produced.

Section 4 also requires that the local authority be advised of proposals to install new furnaces, other than domestic boilers. If requested, the local authority can be required to consider whether the unit is capable of substantially smokeless combustion; there are few circumstances when an application cannot be approved. Unless the officer has delegated authority, the decision has to be made by the authority.

Section 5. Emission of grit and dust from furnaces

Section 5 provides for regulations to be made limiting emission of grit and dust from furnaces. The only regulations so far enacted are the Clean Air (Emission of Grit and Dust from Furnaces) Regulations 1971, which apply to:

1. boilers
2. indirect heating appliances (in which combustion products are not in contact with the material being heated)
3. furnaces in which the combustion gases are in contact with the material being heated but the material does not contribute to grit and dust emissions.

The regulations lay down the permitted limits based on heat output and heat input depending on the type of furnace under consideration.

Grit is defined by the Clean Air (Emission of Grit and Dust from Furnaces) Regulations 1971 as particles exceeding 76 μm in diameter; dust by BS 3405:1983 [57] as particulates of between 1 and 75 μm in diameter; and fume as airborne solid matter smaller than dust.

Section 6. Arrestment plant for new non-domestic furnaces

Section 6 requires that new furnaces be provided with plant for the arresting of grit and dust that has been approved by the local authority. The plant to which this provision applies is:

1. furnaces that burn pulverized fuel
2. furnaces that burn solid matter at a rate of more than 45.4 kg an hour
3. furnaces that burn liquid or gaseous fuel at a rate equivalent to 366.4 kW or more.

Section 10. Measurement of grit, dust and fumes by occupiers

This section provides for the measurement of grit and dust. If the local authority is satisfied that there are grounds for concern, it may serve a notice requiring measurements to be made. If the plant burns less than 1.02 tonnes per hour (other than pulverized fuel), or less than 8.21 MW per hour of liquid or gaseous fuel, the operator can, and probably will, serve a counternotice requiring the authority to carry out the tests at its own expense, and to repeat the measurements 'from time to time' until the counternotice is withdrawn. This is a curious concept in the light of policy broadly accepting that the 'polluter must pay'. Section 12 enables a notice to be served requiring information on fuel or waste burned on furnaces in relation to this provision. Procedural details are contained in the Clean Air (Measurement of Grit and Dust from Furnaces) Regulations 1971 (SI 1971 No. 162).

The local authority must give at least six weeks' notice of requiring adaptations to be made to the chimney and the provision of necessary equipment to enable measurements to be made, in accordance with the method described in BS 3405:1983 [57]. Once the stack has been adapted, the local authority may require, giving at least 28 days' notice, tests to be carried out in accordance with the method set out in *Measurement of Solids in Flue Gases* [58].

Before making the measurements, the operator of the plant must give the local authority at least 48 hours' notice of the date and time of the commencement of the tests. The local authority may be present while testing is taking place. Results must be sent to the local authority within 14 days in a report that includes the following information:

1. the date of test(s)
2. the number of furnaces discharging into the stack at that time
3. the results of the measurements.

Further tests may be required to be made if the local authority is of the opinion that true emission levels cannot otherwise be determined, but not at intervals of less than 3 months.

Even if the plant is of sufficiently limited capacity for the operator to serve a counternotice on the local authority, it is still necessary for scaffolding and services to be provided at the expense of the operator. Measurements must be made by the authority, or by consultants acting on its behalf.

Section 14. Applications for approval of height of chimneys of furnaces

Control of chimney height for all significant combustion plant under local authority control is contained in the provisions of Section 14. It applies to boilers and plant that:

1. burn pulverized fuel
2. burn solid matter at a rate of 45.5 kg/hour
3. burn liquid or gaseous matter at a rate of 366.4 kW

and these may not be used unless the height of the chimney has been approved by the local authority. Details of the proposals are to be sent to the local authority and must be adequate to allow the necessary calculations to be carried out. The local authority has 28 days in which to determine the application. If it is anticipated that it will take a longer period, there must be agreement between the two parties. Failure to determine the application, and to convey the decision in writing to the applicant within 28 days, will allow approval to be assumed on the basis of the application, without conditions; it is important that the procedure associated with the processing of applications be well managed and that regard be taken of the resolutions of the authority in relation to delegated power of officers.

A chimney height shall not be approved unless the local authority is satisfied that it is sufficient to prevent, so far as is practicable, the smoke, grit, dust, gases or fumes becoming prejudicial to health or a nuisance, having regard to:

1. the purpose of the chimney
2. the position and description of buildings near it
3. the levels of the neighbouring ground
4. any other matters requiring consideration in the circumstances.

A simple method for the calculation of chimney heights is provided in *Chimney Heights, Third Edition of the 1956 Clean Air Act Memorandum* [59]. Its relevance is limited to chimneys serving plant with a gross heat input of between 0.15 MW and 150 MW, including stationary diesel generators. It does not deal with direct fired heating systems that discharge into the space being heated, gas turbines or incinerators (which require separate treatment depending on the pollutants to be emitted).

The method first provides an uncorrected chimney height based on gross thermal input, in the case of low sulphur fuels, or sulphur dioxide emission, in other cases. Adjustment is made based on the district in which the development is taking place, and the need to increase the height to correct for effects of surrounding buildings. The method must be treated with some caution. The memorandum made it clear that there will be cases in which local knowledge and experience suggest that the results obtained for the calculations should be varied.

Although the intermediate stages of the calculation should be performed with a reasonable degree of accuracy, the final result should be rounded to the nearest metre. In circumstances where the memorandum will not provide adequate guidance, for example, where a chimney discharges through a roof with complicated structures, or in difficult topography, specialist advice should be sought.

The memorandum also gives guidance on the location of outlets for **fan-diluted emissions**. Discharges at the uncorrected chimney height (U), are acceptable, even if they are below the roof level of the building, subject to the following conditions.

1. The emission velocity must be at least 75/F m/s, where 'F' is the fan dilution factor. $F = V/V_o$, where V is the actual volume of flue gas and V_o is the stoichiometric combustion volume. For natural gas V_o is 0.26 Q m^3/s.
2. The outlet must not be within 50 U/F of a fan assisted inlet (except for intakes of combustion air).
3. The outlet must not be within 20 U/F of an openable window on the emitting building.
4. The distance to the nearest building must be at least 60 U/F.
5. The lower edges of all outlets must be at least 3 m above ground level, with the exception of inputs of less than 1 MW, where 2 m is permissible.

6. The outlets must be directed at an angle above the horizontal – preferably 30° – and must not be under a canopy.
7. Flue gas should not be emitted into an enclosed, or almost totally enclosed, well or courtyard.

Under certain circumstances, defined in the Clean Air (Heights of Chimneys) (Exemption) Regulations 1969 (SI 1969 No. 411), it is not necessary to apply for approval for a chimney. The provisions largely relate to the use of temporary plant.

Sections 18–24. Smoke control areas

These sections lay down the procedures under which local authorities may establish and enforce the provisions of smoke control orders. The concept of smoke control is to prohibit the emission of smoke from domestic and industrial chimneys. The measure was introduced to remedy serious problems largely associated with the domestic combustion of coal. It has had a profoundly beneficial effect on air quality throughout the country, although it is arguable that similar improvements may have taken place in response to other social pressures, such as the widespread switch to gas-fired domestic heaters and boilers. It was never envisaged that the entire nation would be covered by smoke control orders – almost all of the areas where poor air quality has been reported have been made the subject of orders; where they have not, the procedure to introduce them is largely under way. For this reason, a detailed description of the steps to be taken to introduce an order is not provided here.

Owners or occupiers of private dwellings within a smoke control area may recover seven-tenths of the 'reasonable costs' of conversion of all fireplaces normally used to burn coal. Guidance of 'reasonable costs' was published in circulars by the then Department of the Environment. Discretionary powers exist to enable the payment of additional monies, up to the full 'reasonable costs' and to assist charities and other religious buildings. Originally, the local authority recovered a portion of its expenditure directly from the treasury; now the government contribution is included within the Housing Investment Programme allocation.

Financial contributions have never been available to industrial or commercial premises to enable them to meet the terms of an order.

Subsection (2) of Section 20 of the 1956 Act states, subject to exemption and limitation, 'if, on any day, smoke is emitted from a chimney of any building within a smoke control area, the occupier of the building shall be guilty of an offence'. At any time that an order is made, certain buildings or classes of fireplaces may be exempt from the terms of the order, but the local authority has no power to exempt other buildings at a later date unless a new order is made. The effect of this is of particular significance to industrial and commercial plant. Coal, oil and wood cannot be burnt unless on an exempt fireplace – exempt fireplaces consist of either generic plant, such as 'any fireplace specifically designed or adapted for the combustion of liquid fuel', or an appliance named in one of the Exempted Fireplaces Orders, made by regulation, under Section 21 of the Act.

The use of statutory instruments to define appliances is a clumsy and inappropriate approach; although the only option allowed by the Act – each statutory instrument defines not just the appliance but the manner in which it is operated – it follows that compliance with the manufacturer's instructions is necessary, but that the relevant instructions are those produced at the time of the passing of the statute. This does not allow documentation to keep up to date with technical progress.

Few incinerators have been the subject of statutory exemption; it is unlikely that incineration plant currently in use in premises is the subject of specific exemption unless the order was recently made, or is vague in its wording. It follows that almost all incinerators within a smoke control area must meet the 'no smoke' standard.

Although smoke control deliberately sought to prohibit the combustion of coal on open domestic fires, it can still be used on a small number of 'exempt fireplaces'. Use of other solid fuel was encouraged, with a recognition that smokeless combustion could not always be achieved. For administrative simplicity 'smokeless fuel' can be used, regardless of any possible emission, provided that it is specifically identified, by regulation, as an

authorized fuel. The lists of exempted fire-places and authorized fuels continue to grow, and for this reason details must be obtained from an examination of current legislation.

If a chimney is observed emitting smoke, notification must be made within four days as previously described. Investigation should be undertaken to make sure that the appliance does not enjoy the benefit of an exemption or that the fuel use has not been authorized.

Both acquisition and delivery of fuel, other than authorized fuel, in a smoke control area is illegal under Section 27 of the Clean Air Act 1993, although this does nothing to stop the retail sale of coal and other unauthorized fuels, a significant loophole in the legislation.

Section 45. Exemption for purposes of investigations and research

Section 45 enables the local authority to give whole or partial exemption from provision of the Act. If an application is made but not determined to the satisfaction of the applicant, appeal may be made to the Secretary of State for the Environment.

Other provisions

The Act also contains provisions that deal with other matters, including the following:

- Section 30, regulations about motor fuel
- Section 31, regulations about sulphur content of oil fuel for furnaces or engines
- Section 33, cable burning
- Section 34, research and publicity
- Section 35, obtaining information
- Section 36, notices requiring information about air pollution
- Section 42, colliery spoilbanks
- Section 43, railway engines
- Section 46, crown premises, etc.

Smoke nuisances

Details of procedures that enable action to be taken in respect of statutory nuisances from smoke etc. are contained in Part III of the Environmental Protection Act 1990 (see Chapter 7).

Investigation of complaints of air pollution

Investigation methodology

There are four components to the investigation of complaints. They may seem obvious, but this procedure is not always recognized in practice:

1. determination of the alleged problem
2. design of the investigation
3. consideration of remedial action
4. execution.

It may be considered to be stating the obvious to suggest that there is a particular need to identify the nature of a complaint, but this is vital if subsequent action is to be effective. It is not unknown for resources to be squandered on irrelevant issues. A carefully constructed interview with the complainant will ensure that the appropriate issues are identified from the first stage of the enquiry.

The design and implementation of investigations should always be considered, although many routine issues can be executed without the need for a significant level of preparation. In even the simplest case there should be a procedure – a standard approach that has been laid down and is available to all staff in an organization. This does not imply that each action is the subject of a multi-page document scripting the content of interviews, but there are clear advantages in management determining the manner in which the service should be delivered. This concept forms the basis of quality assurance and quality management. If implemented, it minimizes ambiguity within the organization: staff have an absolute understanding of what is required of them; 'customers' receive a similar or identical service; all legislation is applied in an equitable manner. Once the concept of quality management is widely used by individual local authorities, it should be possible to extend the principles across authority boundaries, thus unifying actions regionally and nationally. The use of a computerized database allows steps to be integrated into a procedure and for appropriate correspondence to be generated as and when necessary. It also allows work to be better monitored, including the time targets at various stages of the operation.

While the activities associated with a simple problem can be broken down into component steps, there are occasions when this approach is inappropriate. No management system is likely to anticipate every eventuality. In such circumstances, the need to define the objectives of the investigation is even more clear. Before expending considerable time on the collection of information, it is worthwhile reviewing the possible remedies and the legal framework. Only by being aware of the final destination it is possible to sketch out the route.

If a problem is felt to affect a significant number of people, it may be necessary to carry out a survey in the community. An early decision to make is the advisability of seeking to interview all potential householders in the prescribed area of interest. To do so can lead to the accusation that concerns have been generated by local authority action; failure to do so results in biased evidence. In order for action to be justified, it is necessary to be able to demonstrate that there is a genuine effect. There is a tendency for the defence to balance the evidence of complainants with the opinions of other residents who have a comparable experience of a situation. Clearly it is possible to take into account the existence of economic ties – employees of an organization are generally more tolerant than others who have no direct involvement. The crucial issue is the proportion of the exposed population that can be regarded as independent witnesses who register the existence of a nuisance.

The use of diaries to enable complainants to record the timing of events is helpful, especially if both quantitative and qualitative information is required. Results can be collated to show spatial and temporal distributions. Meteorological information can be obtained to relate observation to possible sources. The use of local meteorological detail, including impressions from respondents, should not be overlooked, especially if the nearest meteorological observation point is distant from the area under study. Local topography can influence wind conditions. Collation should not only highlight contradictory statements, but also consider the possibility of collusion. There is a limit to the possibility of independent observers reporting identical timing for events or descriptions of experiences.

If questionnaires are to be used, much effort has to be invested in their design and content. There is considerable advice published on the way in which questions should be posed and responses invited. (If questionnaires submitted to local authorities by various researchers are typical, then the understanding of this science seems less than adequate.) Questions must be precise and unambiguous. The question setter must carefully consider the precise information needed before construction of a step-by-step model for its retrieval. Questionnaires must be piloted, if only within a critical group of colleagues, to eliminate the worst errors or omissions. Questions to be considered are:

1. Who should be studied? Are particular sub-groups of the population at risk? How should control groups be selected?
2. Who should be measured? Can specific agents be measured? Is there a single pathway, for example, via inhalation, or have several routes to be considered simultaneously? How are health effects to be assessed?
3. Where has the study to take place? Should geographic position, meteorology, etc. be taken into account when selecting the locality? Are there existing monitoring stations or sets of data relating to the environmental factors in question? Is there a need to involve other agencies?
4. When should the study be carried out? Are seasonal effects likely to be important? Is the available time-span sufficient to provide a satisfactory estimate of long-term exposures? Should exposure be averaged over months or years, or are short-term peaks relevant in some cases?

Before a survey is begun, a clear protocol must be produced and agreed with all participating professionals to ensure the subsequent acceptability of findings.

There is no point in embarking on an investigation unless there is a clear outcome in terms of informal intervention, statutory remedy or the production of useful data. This principle applies in relation to all aspects of work, from the most simple case of complaint to complex epidemiological research. It is equally important that the public, whether a single complainant or a community, is made aware of actions and findings. Criticism frequently levelled against both local and central government is the way in which officials cling to the need for secrecy. Barriers are being broken down by progressive local authorities, pressure groups and even reported redirection of government policy. Information is best shared when the authority publicizes its activities. It is not sufficient to base a freedom of information policy on the principle that those seeking data will be satisfied; interested parties should be aware of the information held.

Choice of instruments for local surveys

SIMPLE COLLECTORS OF DEPOSITED MATERIAL Particulate matter can be collected by a variety of receptors allowing crude comparisons to be made between the mass collected at different sites under similar conditions. The following collectors are in general use:

- adhesive plastic sheets mounted horizontally or vertically
- greased glass plates
- Petri dishes
- other receptacles, such as plastic washing-up bowls.

Physical examination of the material may provide information on the source. It is possible to develop considerable expertise if particles are examined under an optical microscope, particularly if it is possible to use polarized light. Multi-volume catalogues of photographs of magnified particles are available for comparison. Alternatively, particles could be collected from various local emitters to build up a library of reference material. This is a time-consuming exercise that is likely to be appropriate in only a limited number of cases.

DEPOSIT GAUGES Various deposit gauges have been developed to carry out long-term assessment of deposited matter. They are of limited value as collectors, and in all cases collect precipitation, which may influence the nature of the deposited matter. Their main contribution is for the examination of long-term trends. The following are in general use.

- The British Standard Deposit Gauge (BS 1747:1969, Part 1) [60].
- The ISO deposit gauge.
- The CERL directional deposit gauge, which consists of four vertical pipes mounted on a frame. Each pipe has a rectangular opening cut into its wall. The upper end of each pipe is closed while the lower discharges into a receptacle. The stand is arranged so that the openings face to the cardinal points or towards, away from and at right angles to a suspect source.
- Frisbees.

GAS DETECTION TUBES A number of companies manufacture gas detection tubes that can be used for the detection and measurement of specific gases in the atmosphere. A range of detector tubes is produced, each being sensitive to a single chemical or range of chemicals. The tubes are sealed at either end and not exposed to the atmosphere until ready for use. A known volume of gas is aspirated through the tube by the operation of hand-held bellows or a syringe. The air first passes through a pre-filter designed to remove other chemicals that would interfere with the reaction, then through sensitized crystals which undergo colour change proportional to the concentration of the subject gas. Concentration is assessed by the depth of penetration of the colour change and the volume of gas tested. Such equipment is robust and accurate, within defined ranges, provided that the tube used is within its 'shelf date'. Often this is the only technique readily available for local authority use for specific surveys.

DIFFUSION TUBES Diffusion tubes that allow the assessment of specific gases by a process of adsorption can be used. They are left out for prolonged periods of time enabling investigation of particular substances without excessive expenditure. They have been used for the investigation of NO_2 in a number of cities, enabling, for example, the production of NO_2 contours across London.

AIR POLLUTION FROM CHEMICAL CONTAMINATION

There are occasions when members of the public are exposed to substances in the atmosphere as a result of chemical contamination incidents. Local authorities have a considerable role in the production of emergency plans [61]. (There is a specific responsibility for site plans to be drawn up under the Control of Major Accident Hazard Regulations 1999 (COMAH), but there is also a need for plans to be produced in relation to the range of events that can reasonably be anticipated.) Such plans must consider the interests of other bodies including the emergency services, site operators, the Health and Safety Executive and the Health Protection Agency.

It is likely that the environmental health department would play a major role in the immediate response to such a release, tracking the path, considering health implications and working with the emergency services on the possibility of evacuation. Officers involved in on-site action must ensure that they put neither themselves nor any other individuals at an unacceptable risk of exposure.

Some general guidance to responses has been published [62], and this forms a useful basis both for emergency plans and as a working document during an incident. It draws particular attention to the assistance available from the National Poisons Information Service.

One important measure sometimes overlooked at the height of an incident is the need for environmental samples of ambient air and deposition, if they can be safely collected. This is important when investigating the identity of possible emissions, the subsequent implications of the event and possible exposure levels. The emergency plan should include a provision for resources to be available at all times to facilitate such sampling.

As important as the need for environmental sampling is the need for biological samples when human exposure is suspected. It is vital that such samples are taken at the earliest opportunity because of the way in which the body naturally excretes ingested material. Such action should be taken by the health authority and used, if necessary, to assess health effects in the acute phase and for long- and short-term surveillance.

Few chemical incidents occur each year. There would seem to be an opportunity for lessons to be learned from one event to be recorded and collated. Unfortunately there does not seem to be a nationwide process in the United Kingdom to facilitate this.

REFERENCES

1. Elsom, D.M. (1992) *Atmospheric Pollution – A Global Problem*, 2nd edn, Blackwell, Oxford.
2. World Commission on Environment and Development (1987) *Our Common Future*, Oxford University Press, Oxford, p. 8.
3. Department of the Environment (1989) *Sustaining Our Common Future*, HMSO, London.
4. Cmnd 1200 (1990) *This Common Inheritance. Britain's Environmental Strategy*, HMSO, London.
5. Organization for Economic Co-operation and Development (1989) *Economic Instruments for Environmental Protection*, OECD, Paris.
6. Department of the Environment (1974) *The Monitoring of the Environment in the United Kingdom*, HMSO, London.
7. Department of the Environment, Transport and the Regions (1997) *Review and Assessment of Air Quality for the Purposes of Part IV of the Environment Act 1995*, DETR, London.
8. Department of the Environment (1994) *Expert Panel on Air Quality Standards. Benzene*, HMSO, London.
9. Department of the Environment (1994) *Expert Panel on Air Quality Standard. 1,3-Butadiene*, HMSO, London.
10. Department of the Environment (1994) *Expert Panel on Air Quality Standards. Carbon monoxide*, HMSO, London.
11. Department of the Environment (1996) *Expert Panel on Air Quality Standards. Nitrogen dioxide*, HMSO, London.
12. Department of the Environment (1994) *Expert Panel on Air Quality Standards. Ozone*, HMSO, London.
13. Department of the Environment (1995) *Expert Panel on Air Quality Standards. Particulates*, HMSO, London.
14. Department of Health (1995) *Committee on the Medical Effects of Air Pollution. Non-biological Particles and Health*, HMSO, London.
15. Department of Health (1997) *Committee on the Medical Effects of Air Pollution. Handbook on Air Pollution and Health*, HMSO, London.
16. Department of the Environment (1995) *Expert Panel on Air Quality Standards. Sulphur dioxide*, HMSO, London.
17. Department of Health (1995) *Advisory Group on the Medical Aspects of Pollution Episode. Health Effects of Exposure to Mixed Pollutants*, HMSO, London.
18. Department of Health (1998) *The Quantification of Health Effects in the United Kingdom*, HMSO, London.
19. Pearce, F. (1987) Inside science. Acid rain. *New Scientist*, 5 November.
20. Gribbin, J. and Gribbin, M. (1996) Inside science. The greenhouse effect. *New Scientist*, 6 July.
21. Department of the Environment (1996) *Review of the Potential Effects of Climate Change in the United Kingdom. United Kingdom Climate Change Impacts Review Group. Second Report*, HMSO, London.
22. DETR (1998) *UK Climate Change Programme – A Consultation Paper*, DETR, Wetherby.
23. Gribbin, J. (1988) Inside science. The ozone layer. *New Scientist*, 5 May.
24. Department of the Environment (1996) *The Potential Effects of Ozone Depletion in the United Kingdom. United Kingdom UVB Measurements and Impacts Review Group*, HMSO, London.
25. Department of the Environment (1991) *The Ozone Layer*, HMSO, London.
26. Evelyn, J. (1661) *Fumifugium: or the Inconvenience of the Aer and Smoke of*

London Dissipated, reprinted by the National Society for Clean Air in 1961, The Dorset Press, Dorset.

27. Elsom, D.M. (1987) *Atmospheric Pollution*, HMSO, London.

28. Cmnd 9011 (1953) *Committee on Air Pollution. Interim Report*, HMSO, London.

29. Cmnd 9322 (1954) *Committee on Air Pollution. Report*, HMSO, London.

30. Department of the Environment, Transport and the Regions (1997) *The United Kingdom National Air Quality Strategy*, HMSO, London.

31. DETR (1999) The UK Sustainable Development Strategy, *A better quality of life*, HMSO, London.

32. Department of the Environment, Transport and the Regions (1997) *Part IV The Environment Act 1995, Local Air Management*, Circular 15/97, HMSO, London.

33. Department of the Environment, Transport and the Regions (1997) *Framework for Review and Assessment of Air Quality*, LAQM.G1(97), HMSO, London.

34. Department of the Environment, Transport and the Regions (1997) *Developing Local Air Quality Strategies and Action Plans: the Principal Considerations*, LAQM.G2(97), HMSO, London.

35. Department of the Environment, Transport and the Regions (1997) *Air Quality and Traffic Management*, LAQM.G3(97), HMSO, London.

36. Department of the Environment, Transport and the Regions (1997) *Air Quality Management and Land Use Planning*, LAQM.G4(97), HMSO, London.

37. Department of the Environment, Transport and the Regions (1998) *Monitoring for Air Quality Reviews and Assessments*, LAQM.TG1(98), HMSO, London.

38. Department of the Environment, Transport and the Regions (1998) *Preparation and Use of Atmospheric Emission Inventories*, LAQM.TG2(98), HMSO, London.

39. Department of the Environment, Transport and the Regions (1998) *Selection and Use of Dispersion Models*, LAQM.TG3(98), HMSO, London.

40. Department of the Environment, Transport and the Regions (1998) *Review and Assessment: Pollution Specific Guidance*, LAQM.TG4(98), HMSO, London.

41. The Air Quality (England) Regulations 2000, HMSO, London.

42. Royal Commission on Environmental Pollution (1994) *Eighteenth Report. Transport and the Environment*, HMSO, London.

43. Department of the Environment (1994) *PPG23 Planning and Pollution Control*, HMSO, London.

44. Department of the Environment (1994) *PPG13 Transport*, HMSO, London.

45. Department of the Environment (1996) *Revised PPG6 Town Centres and Retail Developments*, HMSO, London.

46. Department of the Environment (1992) *PPG4 Industrial and Commercial Development and Small Firms*, HMSO, London.

47. Statutory Instrument 1988 No. 1199 Town and Country Planning (Assessment of Environmental Effects) Regulations 1988, HMSO, London.

48. Department of the Environment (1995) *The Use of Conditions in Planning Permissions*, Circular 11/95, HMSO, London.

49. Department of Transport (1994) *Design of Roads and Bridges. Environmental Assessment under EC Directive 85/337*, HMSO, London.

50. US National Commission on Air Quality (1981) *To Breath Clean Air*, USNCAQ, Washington, DC.

51. Middleton, D.P. (1996) Clean air and environmental protection. *Physical Models of Air Pollution for Air Quality Reviews*, **26**(2), 28–31.

52. Bower, J. (1997) Air quality management, in *Issues in Environmental Science and Technology* (eds R.E. Hester and R.M. Harrison), Volume 8, Royal Society of Chemistry, Letchworth, pp. 41–65.

53. World Health Organization (1980) *GEMS: Analysing and Interpreting Air Monitoring Data*, WHO, Geneva.

54. Bassett, W.H. (2002) *Environmental Health Procedures*, 6th edn, Spon Press, London.

55. Royal Commission on Environmental Pollution (1976) *Fifth Report. Air Pollution Control: An Integrated Approach*, HMSO, London.

56. British Standards Institution (1969) *Notes on the Ringelmann and Miniature Smoke Charts*, BS 2742:1969, BSI, London.

57. British Standards Institution (1983) *Method for Measurement of Particulate Emissions including Grit and Dust*, BS 3405:1983, BSI, London.

58. Hawksley, P.G.W., Badzioch, S. and Blackett, J.H. (1961) *Measurement of Solids in Flue Gases*, British Coal Utilization Retailers Association, Leatherhead.

59. Department of the Environment (1981) *Chimney Heights, Third Edition of the 1956 Clean Air Act Memorandum*, HMSO, London.

60. British Standards Institution (1969) *Methods for the Measurement of Air Pollution*, BS 1747:1969, BSI, London.

61. Wisner, B. and Adams, J. (2002) *Environmental Health in Emergency Disasters: A Practical Guide*, WHO, Geneva.

62. Schofield, S., Cummins, A. and Murray, V. (1995) *Toolkit for Handling Chemical Incidents*, Medical Toxicology Unit, London.

37 Waste management

Jeff Cooper

INTRODUCTION

The major influences on the way the UK government legislates for the control of solid waste is, and increasingly will be, the European Union (EU), together with its obligations under international treaties. These factors will become even more important as the number of member states (MSs) of the EU increases and the EU institutions establish uniform standards of solid waste pollution control in order to achieve market equality. At present, the enforcement of such standards is weak and is likely to remain so for the immediate future. Therefore, while the United Kingdom may not emulate the standards of solid waste management adopted by Germany, Denmark or The Netherlands due to their different difficulties with landfill, it will undoubtedly remain in advance of the new accession countries for some time to come.

Above all the United Kingdom's need to conform to the requirements of the Landfill Directive (p. 722) will be instrumental in determining waste management policies and practices during the period to 2020. With the Landfill Directive's requirement to restrict the amount of bio-degradable waste to just 35% of the total sent to landfill in 1995 by 2020 (by tonnage not proportion) at the latest, the Directive has had a profound influence on the Government's plans and the waste strategies developed for each of the devolved

administrations: England, Scotland, Wales and Northern Ireland. Its influence will therefore start to extend into every aspect of waste management of local authorities and all businesses.

BACKGROUND TO WASTE COLLECTION, DISPOSAL AND ADMINISTRATION

The collection of waste

Prior to the Control of Pollution Act 1974 (COPA), domestic solid waste was officially known as 'house refuse' for which there was no legal definition. However, it was accepted that it was the sort of refuse that arose from the ordinary domestic occupation of a dwelling. The main Act dealing with domestic refuse before the COPA was the Public Health Act 1936. Under Section 72, a local authority could, and if required by the minister had to, undertake the removal of household refuse with respect to the whole or any part of its area.

Surprisingly, perhaps, it was only with the introduction of the Collection and Disposal of Waste Regulations 1988, which enforced the provisions of Sections 12–14 of the COPA, that for the first time every collection authority was under a duty to collect household waste in its area.

The variation in the standards of refuse collection services provided by local authorities was examined in great detail by the Working Party on

Refuse Storage and Collection. Its report [1], published in 1967, revealed great diversity in the frequency of collection, types of material that were collected or not accepted for collection, and in the types of waste receptacle and equipment used for collection. Concern was also expressed about the lack of provision for the disposal of bulky waste, garden waste and cars, which were often disposed of in quiet country lanes, ditches or any other spot convenient for the increasingly affluent and mobile population.

The need for facilities for people to dispose of these types of waste was recognized by a Private Member's Bill introduced by Duncan Sandys MP, which became the Civic Amenities Act 1967 and was subsequently revised as the Refuse Disposal (Amenity) Act 1978. These Acts have permitted local authorities to establish civic amenity (CA) sites, more sensibly referred to as household waste disposal sites, tidy tips and by a number of other terms, including (re-use and) recycling centres, as the vast majority of sites now incorporate these facilities. Indeed, the duty to manage the site in return for salvage rights can be of financial advantage to the local authority. While most CA sites were managed by Waste Disposal Authorities (WDAs), Section 51 of the Environmental Protection Act 1990 (EPA) requires their management to be contracted out by the WDAs as well as the disposal of waste from these facilities.

The disposal of waste

Reports of the irresponsible disposal of toxic waste became steadily more prominent in the press during the 1960s, and the fear of contamination of water sources prompted government action. In 1964, the Minister of Housing and Local Government, together with the Secretary of State for Scotland, appointed the Technical Committee on the Disposal of Toxic Wastes. Its terms of reference were:

To consider present methods of disposal of solid and semi-solid toxic wastes from the chemical and allied industries, to examine suggestions for improvement, and to advise what, if any, changes are desirable in current practice, in the facilities available for disposal and in control arrangements, in order to ensure that such wastes are disposed of safely and without risk of polluting water supplies and rivers.

The recommendations of the Key Report [2] in 1970, the companion report of that of the Working Party on Refuse Disposal produced in 1971, and the earlier Report on the Working Party on Refuse Storage and Collection 1967 were being incorporated into a comprehensive piece of environmental legislation when several incidents of fly-tipping of drums of toxic chemicals in the Midlands prompted a rapid legislative response.

The Deposit of Poisonous Waste Act 1972 was enacted in 20 days, and was the first piece of legislation anywhere in the world to protect the environment from the dumping of hazardous waste. While it should have been repealed and replaced by the COPA in 1974, the 1972 Act lasted until 1980 when the special waste consignment note system was brought in under the Control of Pollution (Special Wastes) Regulations 1980. These regulations were designed primarily to comply with internationally agreed obligations under the EU Directive on dangerous and toxic wastes (78/319/EEC). The consignment note system was similar to the notification requirements of the 1972 Act to provide cradle to grave notification to ensure that the Environment Agency (EA) is able to monitor movements of special waste and the techniques of disposal. The present and future arrangements are outlined on p. 760.

The administration of waste disposal

Until 1965, responsibility for the disposal of house waste rested with the same authority that collected it, namely the metropolitan borough, county borough, or the urban or rural district council. However, this was recognized as increasingly unsatisfactory in metropolitan areas. Generally, waste disposal had to be undertaken through incineration, where standards of emission control were almost non-existent, or through the use of transfer stations where spillage *en route* accounted for a proportion of the disposal. In 1963, the

London Government Act transferred responsibility for waste disposal to the Greater London Council (GLC), which effectively became the first WDA. The new and enlarged London boroughs retained responsibility for collection of house waste, but were directed by the GLC on its ultimate destination by delivery to a landfill site, incinerator or transfer station. Even at its abolition in 1986 the establishment of an environmentally sound system of waste disposal by the GLC had not totally been achieved.

This system for waste management was extended to the rest of England under the Local Government Act 1972, but in Wales, Scotland and Northern Ireland responsibility for waste disposal remained with the much enlarged district councils rather than being transferred to the upper tier of local government administration (and that position has been retained through subsequent reorganizations of local government). Under the COPA these new local government bodies and English county councils became WDAs after they came into existence on 1 April 1974.

In 1976, under the COPA, for the first time a system of day-to-day regulation of waste disposal and waste transfer facilities was added to the planning conditions and Public Health Act 1936 powers that had been the only means of controlling waste disposal since 1947. Under the Town and Country Planning Act 1947, any change of land use, including use of land for disposal of waste, required planning permission, although use of land for waste disposal prior to 1 January 1948 meant that land could continue to be used for disposal of waste. The local government bodies that administered these waste regulation duties were the new WDAs created under the COPA.

The abolition of the metropolitan counties and the GLC in 1986 prompted a number of changes in the arrangements for waste disposal and waste regulation, with both subsequently being superseded by the provisions of the Environmental Protection Act 1990 and the Environment Act 1995, respectively. The post-1986 arrangements led to a series of WDAs, most formed voluntarily by metropolitan district councils covering a former metropolitan county area, or groups of London boroughs, which dealt with both the disposal of waste and the administration of waste regulation. However, under Section 32 of the EPA, the direct administration of waste disposal operations of all local authorities in England and Wales was progressively privatized, either through the formation of arm's length local authority waste disposal companies (LAWDCs) or through sale of assets, formation of joint ventures or handing over of management waste disposal facilities to private sector waste management companies. Of the original LAWDCs, most have now been subject either to management buy-out or sold to waste management companies.

The formation of the EA for England and Wales and the Scottish Environment Protection Agency (SEPA) under the Environment Act 1995 on 1 April 1996 brought together the National Rivers Authority (NRA), Her Majesty's Inspectorate of Pollution (HMIP) and the waste regulation departments of the local authorities.

EUROPEAN PERSPECTIVES

The EU has had and will continue to have a predominant influence on environmental legislation in the United Kingdom. However, it is commonly acknowledged that the first European statement on waste management, the Framework Directive 75/442/EEC, was closely modelled on the COPA. The directive on dangerous and toxic wastes (78/319/EEC) was one of the first daughter directives to be formulated. Both these directives have been amended by 91/156/EEC and 91/689/EEC, respectively. In addition, there are several other directives affecting waste management in the United Kingdom, covering, for example, disposal of polychlorinated biphenyls (PCBs) 96/59/EEC, air pollution from waste incinerators 89/369/EEC, 89/429/EEC and 2000/76/EC, and a European Regulation on the transfrontier shipment of waste 259/93/EC.

In the early 1990s, a number of waste streams were identified by the EU for special attention, some because of their environmental effects and others, such as packaging, mainly because of their political significance. The Packaging and Packaging Waste Directive 94/62/EC approved on 23 December 1994 set targets for the recovery and recycling of packaging waste and was amended in 2003 with higher targets for recycling.

The packaging directive

The directive stemmed from the introduction of legislation covering packaging waste introduced in Germany in 1991, which has set the agenda for the whole of Europe. In the United Kingdom, the packaging directive has been implemented through the producer responsibility provisions of the Environment Act 1995, Sections 93–95. The United Kingdom has opted for shared producer responsibility to raise the levels of recovery and recycling of packaging wastes from an estimated 30% of the 8.8 million tonnes of packaging used in 1996.

While most countries take the view that shared producer responsibility for packaging waste will involve at least a partnership between the consumer, local authorities and industry, the United Kingdom has a much more specific and narrower definition. Shared producer responsibility for packaging waste in the United Kingdom refers only to the industries that produce or use packaging. When local authorities and consumers are drawn in it will be to help the packaging producers fulfil their obligations, but there is no regulatory compulsion.

Shared responsibility has been instituted through imposing specific responsibilities on all businesses that passed on more than 50 tonnes of packaging or packaging materials and had a turnover of more than £5 million per annum (reducing to those with a turnover of more than £2 million from 2000). The share in the achievement of the United Kingdom's recycling and recovery targets accounted for by business is dependent on their place in the packaging chain (Table 37.1).

In addition, in order to determine the recovery and recycling targets of businesses, progressively increasing levels of recovery and recycling have been set by the Producer Responsibility

Table 37.1 Breakdown for responsibility by packaging activity

Activity	Share of responsibility (%)
Raw material manufacturer	6
Converter	9
Packer/filler	37
Seller	48

Table 37.2 UK businesses' recycling and recovery targets 2004–2008

Material	2004	2005	2006	2007	2008
Glass	49	55	61	66	71
Aluminium	26	28	30.5	33	35.5
Steel	52.5	55	58	60	61.5
Paper/fibreboard	65	66	68	69	70
Plastic	21.5	22	22.5	23	23.5
Wood	18	19	20	20.5	21
Recovery	63	65	67	69	70

Obligations (Packaging Waste) Regulations 1997 as amended. (Table 37.2).

Businesses can either arrange for the recovery and recycling of packaging waste themselves, in some cases through agents acting on their behalf, or through joining a compliance (collective or exempt) scheme, thereby placing responsibility on the scheme to arrange for the recovery and recycling to be undertaken on their behalf.

The End-of-Life Vehicles Directive

The End-of-Life Vehicles (ELV) Directive was one of the producer responsibility directives which were developed as one of the priority waste streams initiatives of the European Commission in the early 1990s which eventually passed into European legislation in 2000 through the ELV Directive (2000/53/EC). In the United Kingdom producer responsibility for the treatment and processing of the ELV will not come into effect until 2007 (in common with many other MSs) and as a result there may be an increase in the extent of abandonment of vehicles, which will therefore become the responsibility of local authorities. Measures to mitigate the potential adverse effects of the directive when last owners will have to pay £50–80 to a dismantler to handle their ELV and provide a CoD (certificate of destruction) were implemented in 2004. These included the continuous registration of vehicles.

The Waste Electrical and Electronic Equipment Directive (to be implemented by Summer 2004)

The Waste Electrical and Electronic Equipment (WEEE) Directive (2002/96/EC) is intended to pass the costs of treatment and processing for recovery and recycling of WEEE onto manufacturers and importers (producers) of EEE. Segregation of WEEE will not be compulsory but MSs are required to encourage consumers to undertake separation for re-use, recovery and recycling of WEEE. The extent of local authority involvement will depend on the extent to which local authorities are required to offer facilities for consumers to separate the WEEE items and dispose of them separately. Retailers will have a responsibility for taking back WEEE items on a like for like replacement basis.

There is also the Restriction of Hazardous Substances in Certain Electrical and Electronic Equipment (RoHS) Directive (2002/95/EC) which will preclude the use of heavy metals and certain other materials in new EEE.

The landfill directive

While there were lengthy discussions on the production of a landfill directive in the early 1990s, no directive resulted, mainly due to disagreement between the European Commission and the European Parliament. The introduction of a further proposed landfill directive in 1997 led to the Landfill Directive 2000/76/EC.

Overall, the various articles in the directive will seek to move waste 'up the hierarchy' away from landfill by:

- reducing the landfilling of biodegradable waste;
- the pretreatment of all waste prior to landfill;
- banning the co-disposal of hazardous industrial wastes and household wastes.

The main difficulty is that for a considerable period the United Kingdom has sought to support the 'flushing bioreactor' model of sustainable landfill, whereby through the promotion of enhanced degradation of biologically active waste materials a landfill could be stabilized within a generation, approximately 30 years. However, there has been no instance of the use of this concept in practice and thus it is difficult to assess whether a landfill could be chemically and physically stabilized within 30 years.

One of the main provisions in the landfill directive is the progressive reduction in the amount of biodegradable municipal waste (BMW) allowed into landfills. This is designed to reduce the potential for maximizing the generation of landfill (mainly methane) gases. The reductions are to:

- 75% of 1995 amounts by 2010
- 50% of 1995 amounts by 2013
- 35% of 1995 amounts by 2020.

Britain and some other MSs were allowed to take four years longer than other MSs to reach these targets because they landfilled more than 80% of their municipal solid waste (MSW) in 1995. The Landfill (England and Wales) Regulations 2002 (SI 2002 No. 1559) and the Landfill (Scotland) Regulations (SSI 2003 No.208) introduced most of the requirements of the directive, including:

- Classifying landfills for hazardous, non-hazardous and inert wastes
- Banning explosive, corrosive, oxidizing or highly flammable wastes from July 2002
- Banning infectious clinical waste from July 2002
- Banning whole tyres from landfill from July 2003
- Banning shredded tyres from July 2006.

However, administrative arrangements for the reduction of biodegradable waste to landfill were left to the Waste and Emissions Trading Act 2003. Under this act local authorities (WDAs) can arrange their waste disposal to meet the landfill directive targets in the most economic way by trading their allowances for biodegradable waste reductions to landfill. Those that undertake reductions greater than their allowance can trade their excess reductions with authorities which cannot or decide not to limit their BMW to landfill and who will have to pay for the privilege. In this way Government aims to keep the cost of conforming with the BMW reduction targets to a minimum.

The calculation of BMW and for reductions to landfill were the subject of detailed research concluding in late 2003 which helped to provide the targets for each WDA.

Economic and other waste policy instruments

While throughout Europe the landfill or waste disposal tax is the most popular economic instrument, there is a wide range of other waste policy instruments. These include: deposit refunds, product taxes, recycling targets, changing responsibilities, mandatory collection, subsidies, price support, recycling credits and bans.

These policy instruments are often used in combination and thus the UK approach to the EC Packaging and Packaging Waste Directive involves a combination of changing responsibilities and recycling targets. Industry is being forced to take on an enhanced responsibility for packaging waste, partly away from local authorities in the case of household packaging wastes, to reach the recovery and recycling target levels laid down in the Producer Responsibility Obligations (Packaging Waste) Regulations 1997.

The landfill tax

The United Kingdom's landfill tax was introduced on 1 October 1996 and is applied to all wastes going to landfill. It was applied at two rates, a lower rate of £2 per tonne for inactive (inert) waste, such as brick or concrete waste, and a higher rate of £7 per tonne applying to all other wastes. The landfill tax was raised to £10 per tonne (with inactive rate remaining at £2 per tonne) in March 1998 rising by £1 per tonne until 2004/05 after which it is planned to raise it at £3 per tonne to a limit of £35 in 2011.

The effect of the landfill tax was a reduction in the deposit of waste, especially that from construction and demolition. More of this waste is being sorted and processed on building sites or at transfer stations and recycling facilities.

The Aggregates Levy

The Aggregates Levy was introduced on 1 April 2002. It is payable on sand, gravel and crushed rock at a rate of £1.60 per tonne. It is expected that this will raise £385 m pa. The current rate of exploitation of these aggregates is around 240 m tonnes per annum (tpa), a figure which is much below the high levels of 360 m tpa reached in the 1980s.

As with the Landfill Tax the revenue raised through the Aggregates Levy is used for a reduction of 0.1% in employer's National Insurance Contributions and funds a £35 million Sustainability Fund. The Waste and Resources Action Programme (WRAP) set up in 2001 has been provided with the bulk of the funds raised in order to stimulate the recovery and recycling of construction and demolition (C&D) waste.

Climate Change Levy (CCL)

The CCL affects industry by increasing the cost of fossil fuel energy but again the costs are offset against Employers' National Insurance contributions. The aim of the CCL is to help ensure industry improves its energy efficiency to meet the United Kingdom's commitments under the Kyoto Protocol.

From the perspective of waste management the significant factor is that substantial rebates of CCL can be negotiated through the change in sources of fuel and/or raw materials in order to improve energy efficiency. Thus, for example, there is an added incentive for glass manufacturers to use cullet from waste sources and for cement manufacturers to use a wide variety of alternative fuels from waste sources, including: tyres, solvents and selected packaging products and production residues.

THE ENVIRONMENTAL PROTECTION ACT 1990 (EPA)

Under the COPA, responsibility for waste regulation and disposal of waste collected by local authorities rested with the WDAs. The EPA split the existing English and Welsh WDAs (county councils in the English non-metropolitan areas, district councils in Wales and a variety of arrangements for English metropolitan authorities (see above)) into three bodies:

1. that part of the WDA that operated waste disposal facilities became a LAWDC, where the council did not decide to sell off whatever assets it held;
2. the regulatory aspects were taken on by the waste regulation authority (WRA);
3. that part of the WDA arranging contracts for waste disposal retained the name of waste disposal authority.

The purpose of the new WDAs was to arrange for the disposal of any waste the collection authority had in its possession. District (and borough) councils continued as waste collection authorities (WCAs), and kept their function of collecting household waste, with discretionary power to collect other types of controlled waste.

These different authorities and the WCAs are defined in Section 30 of the EPA, and the transition process to the new LAWDCs was outlined in Section 32. Each authority had an individually agreed timetable for setting up an LAWDC, establishing a private company in partnership with an existing waste disposal contractor, or selling any waste disposal facilities it may have owned.

Further changes in the organization of both waste disposal and waste regulation were inevitable

as the reorganization of local government took place and because the Environment Agency was formed in April 1996. The responsibility for contracting for waste collection and disposal services remained with the new local authorities. Unitary authorities have both waste collection and disposal responsibilities but often work in close collaboration with a neighbouring WDA for their waste disposal services.

The EA brought together those functions performed by the NRA and HMIP, and the waste regulation responsibilities of local government, within a non-departmental public body to provide a 'one-stop shop' for pollution prevention.

The duty of care

One of the fundamental changes introduced by the EPA was the concept of a duty of care under which 'it shall be the duty of any person who imports, produces, carries, keeps, treats or disposes of controlled waste, or, as a broker, has control of such waste, to take all such measures applicable to him in that capacity as are reasonable in the circumstances'. Waste producers and others responsible for waste have 'to prevent the escape of the waste' on to unlicensed land or in contravention of the conditions of a waste management licence. It also requires the holder of the waste to transfer it only to an authorized person such as:

1. the WCA;
2. the holder of a waste management licence;
3. a registered carrier under Section 2 of the Control of Pollution (Amendment) Act 1989.

Section 34 'does not apply to an occupier of domestic property as respects the household waste produced on the property'.

Guidance on what actions are 'reasonable in the circumstances' is contained in a code of practice produced by the Department of the Environment [3]. See also [4].

Under the Environmental Protection (Duty of Care) (England) (Amendment) Regulations 2003 (SI 2003/63) WCAs have the power to serve a notice on businesses requiring them to furnish the WCA with their duty of care records.

This is designed to help officers of the WCA to check whether businesses are transferring their waste in accordance with the law. Also it will help WCAs investigate fly-tipping incidents by, for example, being able to track back evidence through information deposited, such as a letter headed paper.

The Department of the Environment, Food and Rural Affairs (DEFRA) will revise the statutory guidance on the Duty of Care during 2004.

The carriage of waste

The Control of Pollution (Amendment) Act 1989 (COP(A)A) enabled the Secretary of State to make regulations requiring any persons carrying controlled waste in the course of their business, or otherwise with a view to profit, to be registered.

The Controlled Waste (Registration of Carriers and Seizure of Vehicles) Regulations 1991 (SI 1991/1624) came into force on 14 October 1991 and the initial application period for registration for existing carriers expired on 31 May 1992.

Those exempt from registration are:

1. WCAs/WDAs;
2. people who carry only their own waste (except where building or demolition waste is concerned; all people involved in the carriage of waste from construction, building repair/improvement or demolition must be registered);
3. train operating companies, ferry operators, etc.;
4. a charity or recognized voluntary body.

One of the aims of the registration of carriers is to curb fly-tipping activities. Therefore, to ensure compliance with the duty of care it is an offence to transfer controlled waste to an unauthorized person, an authorized person being a registered carrier for these purposes.

Registration for people and companies is with the Environment Agency Area Office in the place where they have their principal place of business. A fee is payable with the application for registration, which lasts three years.

The EA must decide whether an applicant should be registered to carry controlled waste. It may refuse to register a carrier if the applicant (or an associate) fails to comply with the application requirements or has unspent environmental convictions for prescribed offences. The prescribed offences relate primarily to environmental matters but also include the offence of non-possession of a vehicle operator's licence. The registration may be revoked if the carrier commits a prescribed offence. Carriers whose applications are refused or whose registration is revoked may appeal to the Secretary of State for the Environment. Carriers who are registered receive a Certificate of Registration that is standardized throughout Great Britain and bears a unique registration number. They may also purchase certificated copies. Carriers stopped by the police or by EA officers may be required to produce documentary proof of registration within seven days.

The COP(A)A also empowers the EA to seize vehicles suspected of being used for the illegal deposit of waste where it proves impossible to trace those in charge of the vehicle(s) at the time of the offence. If not claimed, the vehicle and its load can be sold.

The Transfrontier Shipment of Waste Regulations 1994 provides a system of prior notification and authorization for the movement of waste consignments between states in the EU and in and out of the EU.

The definition of waste

One of the few aspects of the COPA not changed by the EPA was the definition of controlled waste. The definition used under Section 30 of the COPA is re-stated in Section 75 of the EPA, whereby:

(2) **Waste** includes –

(a) any substance which constitutes a scrap material or an effluent or other unwanted surplus substance arising from the application of any process; and

(b) any substance or article which requires to be disposed of as being broken, worn out, contaminated or otherwise spoiled;

but does not include a substance which is an explosive within the meaning of the Explosives Act 1875.

In addition, under Section 75(3), 'Any thing which is discarded or otherwise dealt with as if it were waste shall be presumed to be waste unless the contrary is proved'.

Controlled waste comprises household, industrial and commercial waste 'or any such waste'.

Household waste means waste from:

1. domestic property, that part of a building used wholly for the purposes of living accommodation;
2. a caravan;
3. a residential home;
4. a university, school or other educational establishment;
5. a hospital or nursing home.

Industrial waste comes from

1. any factory (within the meaning of the Factories Act 1961);
2. premises connected with transport services;
3. premises used for gas, water, electricity or sewage services;
4. premises used for postal or telecommunications services.

Commercial waste means waste from premises used for trade and business or for sport, recreation or entertainment, but excludes household and industrial waste and mine or quarry waste and agricultural waste.

Municipal waste was introduced as a legal phrase into the United Kingdom for the first time through the Landfill (England and Wales) Regulations 2002 (SI 2002 No. 1559) and is defined as 'waste from households as well as other waste which because of its nature or composition is similar to waste from households'.

The DEFRA view of municipal waste is that it confined just to waste collected by the WCA or through WCA contractors. The view in some other MSs is that all similar wastes fit within that definition. This would include at least all commercial waste.

The Secretary of State has reserve powers to specify types of waste as falling into specific categories of waste, so that under the Collection and Disposal of Waste Regulations 1988, for example, clinical waste is classified as industrial waste (see below). The Controlled Waste Regulations 1992 (amended 1993 and 1994) provide descriptions of wastes that are to be treated as being (or not being) household, industrial or commercial waste, and prescribe the cases in which a charge may be made for the collection of household waste. Certain types of agricultural and mining and quarrying wastes are expected to become controlled waste in 2004 to meet the requirements of the Waste Framework Directive 91/556/EEC.

Hazardous Waste. Section 17 of the COPA, which provided enabling powers to the Secretary of State to make regulations on special waste, was retained by Section 62 of the EPA. The control of special wastes is regulated by the Special Waste Regulations 1996. This is due to be replaced by the Hazardous Waste Regulations 2004 which will introduce a new system of control for producers of hazardous waste. Producers of hazardous waste will have to be registered with the EA. Because the European Waste Catalogue defines a wide range of wastes as hazardous, including: end-of-life vehicles, cathode ray tubes (televisions and computer monitors) and fluorescent tubes a large number of businesses and all local authorities will be classed as producers. The EA has produced guidance on the system for hazardous waste classification and control [5].

Article 1 of Directive 75/442/EEC provided that '"waste" means any substance or object which the holder disposes of or is required to dispose of pursuant to the provisions of national law in force'. This provision was given effect in Great Britain by the definition of waste in Section 30(1) of the Control of Pollution Act 1974 and the re-enactment in Section 75 of the Environmental Protection Act 1990. These provisions do not define authoritatively what is and is not waste. What they do is to include within the ordinary meaning of the word 'waste' certain substances about which there might otherwise have been doubt. Other MSs adopted their own national definition of waste.

The government decided that the application of a single definition of waste would best serve the interests of environmental protection and efficient waste management, and the Control of Pollution Act and the Environmental Protection Act definitions (see above) were therefore supplemented by the Section 75 definition.

Waste is defined in Schedule 22 of the Environment Act 1995. This amends Section 75 of the EPA to define waste as any substance or object, listed in a new Schedule 2B to the EPA, which the holder discards, or intends to discard, or is required to discard. This was a new definition and is based on the definition of waste in the Waste Framework Directive as amended by Directive 91/156/EEC. Schedule 2B to the EPA lists the following categories of waste:

1. Production or consumption residues not otherwise specified in this part of this schedule (Q1).
2. Off-specification products (Q2).
3. Products whose date for appropriate use has expired (Q3).
4. Materials spilled, lost or having undergone other mishap, including any materials, equipment, etc. contaminated as a result of the mishap (Q4).
5. Materials contaminated or soiled as a result of planned actions (e.g. residues from cleaning operations, packing materials, containers, etc.) (Q5).
6. Unusable parts (e.g. reject batteries, exhausted catalysts, etc.) (Q6).
7. Substances that no longer perform satisfactorily (e.g. contaminated acids, contaminated solvents, exhausted tempering salts, etc.) (Q7).
8. Residues of industrial processes (e.g. slags, still bottoms, etc.) (Q8).
9. Residues from pollution abatement processes (e.g. scrubber sludges, baghouse dusts, spent filters, etc.) (Q9).
10. Machining or finishing residues (e.g. lathe turnings, mill scales, etc.) (Q10).
11. Residues from raw material extraction and processing (e.g. mining residues, oil field slops, etc.) (Q11).

12. Adulterated materials (e.g. oils contaminated with PCBs, etc.) (Q12).
13. Any materials, substances or products whose use has been banned by law (Q13).
14. Products for which the holder has no further use (e.g. agricultural, household, office, commercial and shop discards, etc.) (Q14).
15. Contaminated materials, substances or products resulting from remedial action with respect to land (Q15).
16. Any materials, substances or products which are not contained in the above categories (Q16).

(*Note*: the references in brackets refer to the number of the corresponding paragraph in Annex I to the directive.)

In the spring of 2004 the DEFRA introduced a consultation paper to bring within the definition of waste certain types of agricultural wastes, such as pesticide containers, fertilizer bags and other non-natural wastes generated on agricultural premises that are currently outside the definition of waste. When the regulations are implemented, probably in 2005, this will preclude farms burning or burying a wide range of wastes previously dealt with on site and close the farm dumps estimated to be used on 70 000 of the 170 000 agricultural holdings in England and Wales.

Illegal deposit of waste

The key section of the EPA is Section 33. It makes fly-tipping illegal and requires that any other waste disposal activity is undertaken only under and in compliance with the conditions of a waste management licence. Section 33(1) states:

a person shall not:

(a) deposit controlled waste, or knowingly cause or knowingly permit controlled waste to be deposited in or on any land unless a waste management licence authorizing the deposit is in force and the deposit is in accordance with the licence;
(b) treat, keep or dispose of controlled waste or knowingly cause or knowingly permit

controlled waste to be treated, kept or disposed of:

(i) in or on any land, or
(ii) by means of any mobile plant except under and in accordance with a waste management licence;

(c) treat, keep or dispose of controlled waste in a manner likely to cause pollution of the environment or harm to human health.

A contravention of the main offence under Section 33(1), or a contravention of a condition of a waste management licence, is an offence.

Section 33 introduces the concepts of 'pollution to the environment' and 'harm to human health'. The definitions are given in Section 29 and reiterate the definitions used in Part I of the Act with specific reference to waste on land. Penalties for offences under Section 33 are £20 000 and/or six months' imprisonment on summary conviction and an indictment and unlimited fine and/or imprisonment for two years. For offences involving special waste the latter prison term is raised to a maximum of five years.

As with many other aspects of environmental legislation anyone can initiate a prosecution for an offence under Section 33, in that the previous COPA limitation restricting prosecution to either the Director of Public Prosecutions, through the Crown Prosecution Service, or the Waste Disposal (Regulation) Authority has been lifted. Nevertheless, the vast majority of prosecutions are pursued by the EA. In Scotland, the Procurator Fiscal takes the decision to prosecute or not based on information received from the SEPA.

Waste management licences (WMLs)

The COPA provisions with respect to licensing waste disposal sites, Sections 4–10, have been strengthened in Sections 35–44 of the EPA. Waste management licences are granted by the EA for facilities of treating, keeping and disposing (including depositing) 'any specified description of controlled waste', and also include the treatment and disposal of waste by mobile plant. Under the

EPA, the matters that the EA must take into consideration before issuing a licence are different from the COPA. Under Section 36(3), the EA has a duty not to reject a licence unless this action is necessary to prevent

1. pollution of the environment
2. harm to human health
3. serious detriment to the amenities of the locality.

The last proviso applies only to land covered by established use certificates or certificates of lawful use. These three reasons are very wide ranging and have replaced the two criteria used by Part I of the COPA, whereby a licence could only be refused 'for the purpose of preventing pollution of water or danger to public health'.

In addition to these criteria with respect to the site, the licence applicant has also to be assessed by the EA as 'a fit and proper person' to hold such a licence. Guidance on this matter is provided by Section 74. A person is judged on his ability both to carry out the activities authorized by the licence, and to fulfil the conditions of the licence. A person would fail this test on three counts:

1. that he or another relevant person has been convicted of a relevant offence;
2. that the management of the site covered by the licence is not in the hands of a technically competent person;
3. that a licence applicant has no intention of making financial provision adequate to discharge the obligations arising from the licence.

The definition of a fit and proper person, however, is not absolute. The EA has considerable discretion over whether the conviction by a person of a relevant offence makes a person 'fit and proper' (FAPP). The EA has the duty to have regard to guidance from the Secretary of State over this matter. This can include guidance on the qualifications and experience that a licence holder requires. This would, for example, include the Certificate of Competence developed under the auspices of the Waste Management Industry Training Advisory Board (WAMITAB). Regulations 4 and 5 of the 1994 regulations (see below) deal with this. In

July 2003, the Agency simplified the procedure for assessing FAPP, and in particular limiting the financial provision required by licence holders to guarantee clearance of waste and restoration of their site should they leave the site without proper surrender of the licence.

For the planning aspects of WMLs, see [6].

Licence modification, revocation and suspension

The power of the EA to modify licence conditions under Section 7 of the COPA is continued under Section 37 of the EPA. The holder of the licence can also apply for a modification. Under the EPA, the Secretary of State can also direct the EA to modify licence conditions. This could, for example, be used to implement provisions of any new EU directive, such as a landfill directive. Where an application to modify conditions is rejected or if the holder is unhappy about a modification imposed by the EA, the holder can appeal to the Secretary of State.

Section 38 replaces Section 7 of the COPA permitting the EA to revoke or suspend the licence or revoke certain parts of it. The latter may be needed at a landfill site, for example to retain in force conditions relating to gas and leachate management but prevent further tipping. There are two grounds for revocation under Section 38(1):

1. that the licence holder has ceased to be a fit and proper person due to conviction for a relevant offence;
2. that continuation of the activities authorized by the licence would cause pollution of the environment, harm to human health, or would be seriously detrimental to the amenities of the locality affected.

However, there is the prerequisite that a revocation can only be made where the pollution, harm or detriment cannot be avoided by modifying the conditions of the licence.

Revocations under Section 7(4) of the COPA have been rare because of the onerous task of having to prove both that the damage was so serious that revocation was imperative, and that modification of condition would not have resolved the problem. Similar potential also applies to a new provision, suspension of licences under Section 38(6) of the EPA, but oddly reference is made to **serious** pollution of the environment and **serious** harm to human health in that subsection.

Under the COPA, the WDA could also revoke a licence under Section 9. This power is contained in Section 42 of the EPA. If it appears to the EA that a condition is not being complied with, a compliance notice can be served on the holder. This specifies a time period for compliance which, unless fulfilled, enables the EA to revoke the licence entirely, partially or to suspend it.

Appeals and compensation

Under the provisions outlined above relating to licence modifications and revocations, the licence holder can appeal to the Secretary of State against the EA's action (Section 43). While the appeal is being determined, the EA's action is held in abeyance.

Surrender of licences

Section 39 provides for the surrender of a licence, but only if the surrender is accepted by the EA. This is a change from the COPA where an operator could surrender the licence whenever he felt like it, possibly leaving enormous problems with gas and leachate control, for example.

Under the EPA, when a licence holder wishes to surrender a licence the EA has the duty to satisfy itself that the licensed site will not cause pollution of the environment or harm to human health. On the surrender of a licence, the EA issues a certificate of completion [7].

Management of completed disposal facilities

With the granting of a certificate of completion on the surrender of a licence, the EA absolves the former licence holder of any further responsibility for the land. The liability for any further work needed will therefore fall on the EA once the licence has been surrendered. Under Section 61(9), should the EA identify a site that has created a significant environmental problem but where it has already granted a certificate of completion, it can undertake remedial work but cannot attempt to recover the cost from the landowner, unlike in all other circumstances. Given the fact that landfill gas emissions can continue for several decades, and leachate can likewise cause problems for long periods, the EA will be reluctant to issue certificates of completion for most landfill sites.

Faced with a refusal to issue a certificate of completion, the licence holder can appeal to the Secretary of State. A successful appeal against the wishes of the EA may therefore hold considerable financial implications under Section 61(9), noted above.

Fees and charges

Section 41 permits the Secretary of State to introduce fees and charges for applications for WMLs and their retention. (See the Waste Management Licensing (Fees and Charges) Scheme.) Such fees would be charged by them and paid to the EA. Indeed, under Section 41(7), failure to pay the fees may render a licence holder liable to revocation of the licence. The EA cannot set its own fees. While it is the duty of the Secretary of State to make regulations prescribing the fees and the charging scheme, it is done 'with the approval of the Treasury' (Section 41(2)).

Fee levels are revised annually and from April 2003 there is a small variable element determined by the score an operator achieves under the OPRA (Operator Performance Risk Assessment) system introduced by the Agency in 2000.

The Waste Management Licensing Regulations 1994

These regulations brought into operation the new concepts of waste licensing introduced by the EPA, outlined above, and became fully effective on 1 May 1994. Waste Management Paper No. 4 [8] and DoE Circular 11/94 [9] both give guidance on and support the new system. The Waste Management Paper has been given statutory status and provides formal guidance to the EA. Its revision was undertaken in late 2003 and at the same time a more fundamental review of Waste Management Licensing was

pursued by DEFRA covering not only licences for waste management facilities but also the exemptions under Schedule 3 to the Waste Management Licensing Regulations 1994.

Removal of unlawful waste deposits

The power to require the removal of waste unlawfully deposited on land under Section 16 of the COPA is continued under Section 59 of the EPA. It permits the EA or WCA to serve notice on the occupier of any land to remove a particular waste and/or to eliminate or reduce its consequences. However, the occupier can apply to the magistrates' court for the notice to be quashed within 21 days of the service of the notice. If the court is satisfied that the appellant neither deposited nor caused or permitted the deposit of waste, or that there is a material defect in the notice, the notice can be quashed. There is also provision under Section 59(7) and (8) for the EA or WCA to remove the waste or reduce its consequences and then recover the cost by charging the person who deposited, knowingly caused or knowingly permitted the deposit of waste. Recouping costs where material is fly-tipped is clearly only possible when someone has been successfully prosecuted under Section 33.

Waste management planning

One of the main concepts that emerged in the 1990s and that continues to influence the planning of waste into the new millennium is sustainability. Following the Earth Summit in Rio de Janeiro in 1992, the UK government produced *Sustainable Development: the UK Strategy* [10] in January 1994. Chapter 23 provided a sustainable framework for waste:

- to minimize the amount of waste produced;
- to make best use of the waste that is produced;
- to minimize pollution from waste.

It also defined a hierarchy of waste management options:

1. reduce
2. reuse

3. recover

  (a) materials recycling
  (b) composting
  (c) energy recovery

4. disposal.

This followed the priorities of the EU's Fifth Environmental Action Programme, although in 1996 the EU's hierarchy was adjusted slightly to place materials recycling above energy recovery, where this represents the best practicable environmental option (BPEO).

Further details of the government's priorities were announced in *Making Waste Work: a Strategy for Sustainable Waste Management in England and Wales*, published in December 1995 [11]. These were reinforced in the first statutory national waste plans established under the EPA, Section 92. *Waste Strategy 2000* was published in June 2000 as a waste strategy for England and Wales but Wales produced its own strategy in May 2002.

The key targets in *Waste Strategy 2000* are:

- to reduce the amount of industrial waste going to landfill to 85% of 1998 levels by 2005;
- to recover 40% of municipal waste by 2005, 45% by 2010 and 67% by 2015;
- to recycle or compost by 17% of household waste by 2003/04 and 25% by 2004/05;
- to achieve the reduction of BMW to landfill to 75% of 1995 levels by 2010 and 50% in 2013 and 35% by 2020.

The national target for recycling and composting 25% of household waste by 2005 (originally set as a target for 2000 by government in 1992) and the 2003/04 target of 17% are statutorily enforced through the best value performance indicators (BVPIs) originally designed to provide valuable information on the key services delivered locally. Local authorities were set individual targets in 2000 based on their recycling performance in 1998/99 for 2003/04. Those with the best performance had to achieve 33% recycling rates for 2003/04.

The data for the two years since the BVPIs were first set for 2000/01 has provided valuable information helping local authorities to assess where they are and to set meaningful targets for further improvements.

That part of the BVPI system which involves waste management and resource recovery in 2003 included:

- percentage of the total tonnage of household waste arisings recycled;
- percentage of the total tonnage of household waste arisings sent for composting;
- percentage of the total tonnage of household waste arisings used to recover heat, power and other energy sources;
- percentage of the total tonnage of household waste arisings landfilled;
- kilograms of household waste collected per head;
- cost of waste collection per household;
- cost of waste disposal per tonne for MSW;
- percentage of people satisfied with the cleanliness standard in their area;
- percentage of people satisfied with:

  (a) household waste collection
  (b) waste recycling
  (c) waste disposal

- percentage of population served by a kerbside collection of recyclable waste.

There is also the possibility for local authorities to promote waste minimization, although the need to achieve recycling targets reduces the incentive. The Waste Minimization Act 1998 gave local authorities new powers to reduce waste at source through:

- the need to change the perception of the waste hierarchy;
- more creative use of economic incentives to fund recycling;
- increased public involvement in the decision-making process.

In Scotland, the SEPA is responsible for drawing up a strategic overview of waste management planning. SEPA published its draft strategy in February 1997. After extensive consultation it published a draft strategy during 1999 and the final strategy in early 2000.

Provision for detailed local waste planning started under Section 2 of the COPA, whereby WDAs were required to establish what arrangements were needed to treat or dispose of controlled waste within their area. Waste management planning responsibilities were subsequently handed over to the WRAs under Section 50 of the EPA. With the WRAs subsumed into the EA in April 1996, Section 50 was repealed, and this left waste recycling plans as the only statutory local waste plans. While some local authorities have produced waste management plans, these have no statutory basis, although there are also plans that the local planning authorities are required to produce under planning legislation.

Waste recycling plans[1]

The duties of WCAs with regard to the reclamation of household and commercial waste arising in their area are outlined in Section 49(1) of the EPA. The WCA is required:

(a) to carry out an investigation with a view to deciding what arrangements are appropriate for dealing with the waste by separating, baling or otherwise packaging it for the purposes of recycling;

(b) to decide what arrangements are, in the opinion of the authority, needed for that purpose;

(c) to prepare a statement ('the plan') of the arrangements made and proposed to be made by the authority and other persons for dealing with waste in those ways;

(d) to carry out from time to time further investigations with a view to deciding what changes are needed;

(e) to make any modification of the plan that the authority thinks appropriate in consequence of any such further investigation.

Having drawn up its draft plan, the WCA is required to submit a copy to the Secretary of State for him to ensure that the authority has fulfilled its requirement with regard to Section 49(3), which specifies the type of information required to be included. Having complied with any directions

that the Secretary of State may impose, the WCA has the duty under Section 49(5):

(a) to take such steps as in the opinion of the authority will give adequate publicity in its area for the plan or modification;

(b) to send to the waste disposal authority and waste regulation authority for the area, which includes its area, a copy of the plan or particulars for the modification.

Unlike the procedure with the waste disposal plan, there is no requirement for consultation with the public or any other authority prior to its production. However, because of the requirement to undertake public consultation under the Best Value regime most authorities have extensive consultation procedures.

Despite the fact that the WCA must undertake the planning for waste reclamation it is the WDA that retains the final say over recycling. Among its other duties, the WDA 'shall give directions to the waste collection authorities within its area as to the persons to whom and places' at which the controlled waste they collect is to be delivered (EPA Section 51(4)). In addition, Section 48(4) requires that 'where a waste disposal authority has made with a waste disposal contractor arrangements, as respects household waste or commercial waste in its area or any part of its area, for the contractor to recycle the waste, or any of it, the waste disposal authority may, by notice served on the waste collection authority, object to the waste collection authority having the waste recycled; and the objection may be made as respects all the waste, part only of the waste or specified descriptions of the waste'. There is no provision in the EPA for WCAs to appeal against an objection by the WDA. Therefore, were a WCA to produce a recycling plan incorporating source segregation, this could be overturned by the WDA.

A further barrier to waste reclamation is introduced under the EPA for those authorities that are both WCAs and WDAs. Specifically, under Section 48(6) and (7) of the EPA:

6. A waste collection authority may, subject to subsection 7 below, provide plant and

---

[1]The requirement for local authorities to produce recycling plans was repealed by the Waste and Emissions Trading Act 2003 and will be replaced by joint municipal waste strategies.

equipment for costing and baling of waste retained by the authority . . . .

7. Subsection (6) does not apply to an authority which is also a waste disposal authority . . . .

Informal guidance provided by the minister when this restriction was debated during the third reading suggested that equipment beyond that normally associated with the fulfilment of a WCA's duties would be regarded as falling within the restriction of Section 48(6) and (7).

While a timetable for the production of waste recycling plans is not given in the EPA, the Secretary of State retains the power to require them within a specified period of time. In March 1997, the government requested WCAs to submit their recycling plans to the DETR by December 1998 and there have been subsequent revisions required by the DEFRA.

Guidance to WCAs in the preparation of waste recycling plans is available in the Waste Management Paper No. 28 produced in July 1991 with revised guidance published in March 1998 [12], which lays down clear guidelines on the authority's requirements with respect to the information to be included in this plan under Section 49(3):

(a) the kinds and quantities of controlled waste which the authority expects to collect during the period specified in the plan;

(b) the kinds and quantities of controlled waste which the authority expects to purchase during that period;

(c) the kinds and quantities of controlled waste which the authority expects to deal with in the ways specified in subsection (1)(a) during that period;

(d) the arrangements which the authority expects to make during that period with waste disposal contractors;

(e) the plant and equipment which the authority expects to provide;

(f) the estimated costs or savings attributable to the methods of dealing with the waste in the ways provided for in the plan.

However, there is no duty for the WCA to enhance its current waste reclamation activities, where these exist.[2]

One positive measure to encourage the reclamation of waste is contained in Section 52(1), which provides that where a WCA undertakes such reclamation, the WDA makes payments to the WCA 'of such amount representing its net saving of expenditure on the disposal of the waste as the authority determines'. Guidance and regulations issued in February 1992 and March 1994 provide for the means of calculating the amount to be paid by the WDA to the WCA from April 1992, and from April 1994 when payments doubled. The basis of payment from April 1994 is the long-run marginal cost of each WCA's highest cost method of waste disposal. While payments between WDAs and WCAs are mandatory, those to others undertaking waste reclamation measures that make savings in waste disposal and/or collection costs are discretionary, although the Secretary of State can introduce regulations to make such payments obligatory [13].

This measure, which is an advance on the position under the COPA and dealt with in the DoE Circular 13/88 [14], was felt to be necessary because while most source segregation reclamation is undertaken by WCAs, the main benefit accrues to the WDAs through savings in waste disposal costs.

Recycling plans are currently the only statutory waste management plans. In addition there are municipal waste plans which are not statutory but are important in that they can be prepared by WCAs jointly with other WCAs and together with WDAs in order to prepare waste management operational plans for a wider area. Regional waste management plans for each English planning region are also being developed to provide a wider spatial planning area for the provision of major waste management facilities. These should be concluded in 2005 for the first time.

In 2003, the Household Waste Recycling Act imposed a duty on Waste Authorities to prepare sustainable waste strategies and to report annually on progress in improving recycling rates. The WCAs are required to collect at least 2 types of recyclable waste together or separated from the rest of the household waste by 31 December 2010, unless the cost of doing so would be unreasonably high or comparable alternative arrangements are available.

---

[2]But see the Waste Emmissions and Trading Act 2003.

Waste collection and waste receptacles

Sections 45–47 of the EPA largely replicate the duties and powers of WCAs under sections 12–14 of the COPA, the regulations for which were finally introduced as part of the Collection and Disposal of Waste Regulations 1988. The ability of WCAs to establish kerbside source segregation collection schemes, as opposed to the pre-1992 door-to-door schemes, has been permitted under Section 46(1), whereby 'the authority may . . . require the occupier to place the waste for collection in receptacles of a kind and number specified. However, agreement has to be given by the highway or roads authority and provided that arrangements have been made as to the liability for any damage arising out of their being so placed' (Section 46(5)(b)).

Under Section 45, the WCA has a duty to arrange for collection of household waste except for premises that are isolated or inaccessible so that collection costs would be 'unreasonably high', and that adequate arrangements for its disposal can reasonably be expected to be made. There is also a duty for the WCA to collect or arrange collection of commercial waste when requested. As regards industrial waste, the WCA may arrange for its collection if requested by the occupier, but a WCA can only exercise this power with the consent of the WDA (Section 45(2)). For commercial and industrial waste, the WCA has a duty to recover the costs for both its collection and disposal, 'unless in the case of charge in respect of commercial waste the authority considers it inappropriate to do so'. Section 45 also deals with the duties of WCAs with regard to the emptying of cesspools and privies.

While Section 46 deals with the provision of receptacles for household waste, Section 47 covers those for commercial or industrial waste. While the authority has considerable discretion about the number and type of containers to be used and their positioning for the collection of waste and recyclables, under Section 46(7) an occupier can appeal to a magistrates' court that the authority's request is unreasonable or that 'the receptacles in which household waste is placed for collection from the premises are adequate'.

However, this appeal procedure must be initiated within 21 days of a notice being served on the occupier or the end of the notice period. A similar provision exists for receptacles from industrial and commercial premises, but the grounds for objection are different in that while the requirement of unreasonable provision is repeated the other ground is that 'the waste is not likely to cause a nuisance or be detrimental to the amenities of the locality' (Section 47(7)(b)).

REFUSE ANALYSIS

The analysis of household waste is concerned with the physical characteristics and composition of that waste, for example, its density, output per household and the amount of paper. Data obtained from such analyses help to predict the changing nature of waste, and is a valuable aid in determining methods of collection, transportation and disposal.

Since 1974, the Department of the Environment has co-ordinated UK waste analysis data. Table 37.3 shows the national average composition of household refuse, although it may vary substantially between different local authority areas.

Reliable waste analysis can be used for organizing collection rounds, calculating storage requirements of blocks of flats, and long-term forecasting to enable strategies to be planned and research organized. In seeking improved efficiency and higher standards of waste recycling, treatment and disposal, it is essential that reliable data are

Table 37.3 UK average composition of household refuse

Type	Percentage
Screenings – 2 cm	10
Vegetable and putrescible	20
Paper and board	33
Metals	8
Textiles and man-made fibres	4
Glass	10
Plastics	7
Unclassified	8

Source: Department of Environment [15].

Table 37.4 Distribution of waste disposal in the United Kingdom in 1990

Type of waste	Amount (millions of tonnes)
Agricultural waste	80.00
Mining waste	51.00
Quarrying waste	57.00
Industrial waste	45.00
Municipal solid waste	35.00
Sewage sludge	36.00
Hazardous waste	4.00
Radioactive waste	0.02

available on the raw material emanating from the waste industry. There are numerous examples where waste analysis, correctly carried out, has enabled treatment and disposal plant design to be adequately assessed. Similarly, there are many cases where lack of analysis data has not only led to under- or gross over-design, but also to the wrong treatment system being selected. Table 37.4 shows the distribution of waste disposed in the United Kingdom in 1990. The EA undertook a national survey of household waste in 1996 and for commercial and Industrial wastes in 1999 [16]. Both of these surveys will be updated in early 2004 using data from late 2003.[3]

REFUSE STORAGE

It is important that during periods of storage, refuse on any type of premises should be kept dry. An accumulation of moist or wet refuse is not only offensive and difficult to remove, but forms an excellent breeding medium for flies.

Domestic storage

The vast majority of domestic residences now have one of two alternative types of refuse storage facility. The traditional galvanized and plastic dustbins are gradually being phased out and replaced by plastic refuse sacks or 240 litre wheeled bins manufactured from plastic materials.

With increasing concern about the dangers of needle stick injury and sharp, protruding refuse in a plastic sack, allied to the litter problems generated by such containers, the wheeled bin has become increasingly popular. Although provision of such a receptacle usually requires the householder to place the bin at the curtilage of the property or at the kerbside, the merit of having almost three times the capacity of a traditional dustbin and nearly six times the capacity of a refuse sack is seen as a considerable advantage, both in containing waste generation in one place and dramatically reducing the litter problem. It also had a significant role in assisting the national household growth rate to increase by 3% a year on average from the late 1990s onwards because householders had fewer barriers to their waste production.

High-rise flats

British Standard 5906 details the requirements for the gravity transport system of waste disposal in multistorey buildings. It is specific on dimensions, materials, finishes and ventilation requirements. Waste is deposited on each floor in hoppers and transported via gravity chutes to containers in the basement. The British Standard gives information on these chutes and hoppers, and the location, dimension, construction, lighting, ventilation and cleansing of the storage chambers. Where there is likely to be excessive accumulations and bulk cylindrical containers are inadequate, a refuse compactor may be allocated to receive refuse.

Removal distances

Access for removal is very important, and consideration must be given to the distance of carrying to the external removal point. According to BS 5906, these distances should not be more than **25 m** or, when wheeling a container, not more than **10 m**. Care must be taken in the design of refuse storage areas, that the gradient from the storage point to the point of collection access should not **exceed 1 : 14**.

Commercial and industrial waste storage

Where there is a large volume of waste accumulation, refuse skips or bulk containers of 1100 l

---

[3]See DEFRA (August 2003) *Municipal Waste Management Survey 2001/2002*. DEFRA, London.

capacity and above are frequently used. Where collection frequency is limited or there is a problem of space, refuse compactors may be installed. These can reduce the volume of waste down to 20% of its original bulk, which also helps to control smell and litter, and reduces fire risk. Waste that is compacted is not likely to attract vermin and insects and gives lower transport costs. British Standard 5832 deals with the most popular type of compactor waste containers, which have a capacity of 10 m³. Provision for the installation of the compactor should be made at the design stage so that space, electrical fittings and access to, plus removal of, compacted refuse is planned in. Compacted refuse is heavy and may need wheeled transport, and so the siting of the compactor must take account of steps and ramp access.

External storage

Sufficient storage capacity should be provided for the agreed span of time between collections, and consideration should be given to providing an enclosed compound for refuse containers, which encourages hygiene around the building, controls litter, discourages vandalism and helps to avoid the problem of fly-tipping. Enclosed refuse storage compounds must have natural and artificial lighting, appropriate ventilation, smooth, cleanable floor surfaces, hose points and drainage facilities, be easily accessible for deposit and removal of refuse, and have a lock-up entry. The containers must usually be accessible from the ground floor, with vehicular access not more than 25 m from the material without collectors having to pass through a building. Rear access is an advantage, especially where traffic is heavy and parking is a problem.

All access roads for refuse vehicles must have suitable foundations and surfaces to withstand the maximum weights of the vehicles likely to be used. Manhole covers, gully gratings and the like should be of a heavy duty construction; overhead service cables and pipes should be not less than 7 m from ground level; and there should be sufficient room to allow manoeuvrability of heavy goods vehicles in the 17–24 tonne range. If skips or large compactors are used, consideration must be given to

high overhead clearance for the emptying of such equipment.

The screening of waste storage areas by planting creates a good impression. Landscaping should be practical with low maintenance requirements, and should be impact resistant.

RECYCLING

Recycling of waste (see also is undertaken in order to:

1. provide a cheaper alternative to virgin raw materials in the manufacturing process;
2. ensure that diminishing natural resources are not used up;
3. reduce waste disposal costs;
4. increase the life of diminishing waste disposal facilities.

The EPA imposes a duty on waste collection authorities to prepare 'recycling plans', accessible to the public, outlining what action the authority is taking in respect of recycling. The government considers that 50% of all waste is potentially recyclable and hopes that by 2005 all councils will be recycling 50% of this fraction, that is, 25% of overall waste.

Two methods of recycling are currently used in the United Kingdom: recycling at source, and recycling at the disposal point. Many local authorities base batteries of containers at strategic locations around their area where members of the public can bring recyclable commodities such as glass, cans, plastics and paper, to central collection points. Increasingly, authorities are introducing a second wheeled bin or a box or bags so that householders can separate potentially recyclable materials at source.

This waste is then taken to a recycling centre where it can either be manually or mechanically sorted. In some authorities where recycling is not undertaken at source, facilities are available to extract certain materials at the waste treatment and/or disposal point, for example electromagnetic separation of steel cans. If the recycling of household waste material is to become the preferred means for disposal of our waste, it is imperative that suitable markets for reclaimed waste are found.

Waste paper

There are many grades of waste paper, the highest being produced by paper mills themselves, almost all of which is recycled. Publishers and printers also create waste, which is usually worth recycling. However, it is lower grades of waste paper that are normally created in the home and made available for local authority collection, mainly newpapers and magazines used in the United Kingdom's three newsprint mills or mixed papers used in board mills. Over half the paper and board produced in the United Kingdom is made from waste paper, and over three-quarters of it is used to make packaging such as cardboard boxes and cartons for cereals, soap powders and other products. Waste paper is not used for cartons that come into direct contact with food because of possible hygiene problems.

Plastics

Recycling of plastics from household waste is still in its early stages of development in the United Kingdom. The most commonly available material is polyethylene, which is increasingly recycled, especially from supermarkets. It is used as a shrink-wrapping for pallets of goods and for trays of packaged goods. Plastic bottles made from PET used for carbonated soft drinks and an increasing range of goods and HDPE and used for milk containers are also recycled, and are often reprocessed to make things such as drainage pipes and garden posts. Polypropylene, used in the making of crates and battery cases, can be recycled into drainage pipes. PET can be reused in the clothing industry and also for bottles for non-food products. Mixed plastic waste can be reprocessed to make products such as road signs, motorway acoustic insulation walls, fencing, pallets and outdoor furniture. Residual plastics, which are in such tiny quantities that it is not possible to collect them, can be allowed to pass through the waste stream, where they may help to fuel incinerators and thereafter

the product can be reclaimed for use as heat or energy.

Ferrous metals

All ferrous metals are attracted by magnets and it is therefore not critical for this material to be separated at source if the waste is ultimately to go through an incinerator plant with over-band electromagnetic separation. However, larger materials, for example, white goods (cookers and washing machines), can normally be segregated at the disposal point for recycling. At present, about 90% of food and drinks cans are made of ferrous metal and are best extracted from mixed waste.

Since the beginning of 2002 all fridges and freezers which contain ozone depleting substances (ODS) have been subject to extra control over their disposal in order to remove the ODS from the insulating foam which contains sustantially more ODS compared to the cooling circuit. It was only on 6 June 2001 that it became clear that Article 16 (b) of the European Community Regulation No. 2037/2000 on ODS would apply to the foam insulation used in domestic refrigerators and freezers. Therefore, from 1 January 2002 no fridge or freezer could be processed through a shredder (the normal route for disposal/recovery of these items of WEEE) unless the 'controlled substances', mainly CFCs or HCFCs are removed from the foam insulation.

Facilities for the processing, removal and destruction of foam insulation in household fridges and freezers had been introduced in Germany and a number of other countries. Initially, the United Kingdom exported the fridges to Germany or the Netherlands, both of which had surplus processing capacity, or stored the fridges prior to the establishment of processing facilities within the United Kingdom. There is now sufficient UK processing capacity for all waste fridges and freezers generated and some of this capacity is being used for WEEE.

Aluminium

Much of the aluminium waste is obtained from old cars and white goods, with only very small quantities from household waste. Aluminium currently makes up less than 1% of household waste; it is used for milk bottle tops, foil containers and about half of all drinks cans. In 1992, a £28 million aluminium can smelting facility was opened in Warrington, which is capable of reclaiming 50 000 tonnes of all-aluminium cans in a year. It relies on imports of scrap aluminium cans from other countries for full production.

Glass

Glass bottles have traditionally been refilled rather than recycled, for example, door-to-door milk delivery, but inevitably there are casualties in this system. It is difficult to separate glass from other household waste mechanically, so it is best to keep it separate and deliver it to bottle banks. Glass is heavy and dense, and so needs to be collected in large quantities to be worth transporting to a treatment plant for cleaning and recycling. Most waste glass (cullet) can be recycled, and bottle banks are normally placed near shopping centres or household waste centres. Every effort should be made to keep different coloured glass separate because over two-thirds of the glass produced in the United Kingdom is clear, and mixed cullet can only be used in considerable volumes in green glass furnaces.

Since 2001 increasing amounts of green and mixed coloured glass have been used as an aggregate substitute. However, this recycling option relies on additional money being added by packaging compliance schemes through the PRN (packaging waste recovery note) system to supplement the lower value of glass cullet when used in this application.

Compost

The food and garden waste that ends up in household waste containers can be used to make compost. This waste makes up on average between one-fifth and one third of all household waste and is called the putrescible fraction. It must be separated by the householder. If separated at a treatment centre that 'compost' can only be used for landfill cover or other low level applications. Separating the kitchen waste compostable fraction,

which is currently not widely practised in the United Kingdom, has not been permitted except in special circumstances because of the concern to protect the health of farm animals, particularly from classical swine fever and foot and mouth disease. The English Animal By-Products Regulations and the EU Animal By-Products Regulation restrict the ways in which animal wastes and catering wastes can be treated and spread on land.

The advantages of composting in a comprehensive waste management system is that the rest of the waste can be burned more efficiently because much of the moisture is removed with the compostables, and even if the rest of the waste goes direct to landfill, such segregation helps to reduce the amount of BMW deposited and the landfill gas that is generated.

REFUSE COLLECTION

Organization

Labour

The collection of refuse from domestic, commercial and industrial premises is undertaken by local authorities themselves, using their direct labour organizations, or by commercial waste collection and disposal companies. There has been a significant increase in the involvement of the latter organizations since the introduction of compulsory competitive tendering to the UK waste management industry in the late 1980s.

Irrespective of the organization undertaking the work, the system universally favoured is the task system, where a defined number of premises are cleared each day and a bonus is paid on top of the basic wage for completion of that day's work. The system is favoured over a continuous system because, in a climate of increasing concern about security of premises and linked properties, it is important for occupiers to know precisely when they should make the receptacles available for collection. Rounds are routed to ensure the minimum of unproductive travel, usually geared to start and finish near to the depot and/or ultimate disposal point. Collection rounds are subject to periodic review, as the vehicles and equipment are constantly upgraded and to reflect any changes in the demographic nature of the authority, the working practices and population movement. There has been a move away from incentive bonus schemes, which have tended to encourage uncontrolled working practice to ensure early finishes on each day.

The implementation of compulsory competitive tendering has tended to load the working day in order to minimize the number of vehicles utilized on the service, thus reducing high capital cost elements for incorporation in tender bids. The system of task and finish has rapidly diminished. The introduction of the Best Value arrangements which replaced compulsory competitive tendering in 1998 have added to the complexity of tender preparation but have not changed the emphasis on efficiency.

It is normal for household refuse to be collected once per week but, where authorities are currently experimenting with recycling schemes, there is some slight movement towards fortnightly collection, although this is resented by the public and therefore politically unpopular. Usually, the new service comprises fortnightly collections of dry recyclable and compostable material and fortnightly collection of other waste. Such schemes have to be carefully monitored from an environmental point of view to ensure that there is no public health risk or nuisance. The advent of the wheeled bin system, together with a kerbside collection scheme, has dramatically reduced the number of crew members on each vehicle. Future developments tend to suggest a move towards double-shift working and potentially seven-day collections in order to further reduce the number of high-cost vehicles needed. Another advantage of the advent of the wheeled bin system is that the lifting device on the collection vehicles enables domestic waste and commercial waste to be collected at the same time, thus reducing the duplication of travelling time and overlapping of rounds when these services were segregated.

Transport

The traditional transport system for refuse collection is now compaction vehicles, ranging from 16-tonne single rear axle to 24-tonne twin rear axle vehicles. Where the geographical nature of the round is such that large vehicles cannot negotiate the area, then smaller derivatives are used, but in order to ensure maximum economic payloads and work content, these should operate within a short distance of the disposal point, or should have a transfer facility to ensure that there are no long, unproductive hauls to the tipping site. The optimum arrangement is to have two or, at the most, three trips to the disposal point, allowing rest periods and lunch breaks to coincide wherever possible, thus ensuring faster working when the work is resumed. The vehicles used tend to be compaction vehicles with power-operated pressure plates producing compaction ratios in the region of 4 : 1. Typical payloads of a 17-tonne gross vehicle weight (GVW) vehicle would be in the order of 8 tonnes per load. Although there are few variations between the major manufacturers of refuse collection vehicles, there are specialist variants for unusual situations, that is, a smaller, non-HGV type vehicle may be used in rural locations where there is considerable interpremises travel, narrow-bodied vehicles are used in inner cities where access and manoeuvrability is limited, and some authorities are experimenting with one-man crews using automatic lifting devices and/or side loading lifting devices on the vehicles. Figure 37.1 shows the operation of a rotary drum compaction system capable of dealing with both sacks and wheeled bins. This type of compactor is ideal for the collection of compostable materials of all types.

Where it is impossible to operate close to a disposal point or a transfer station, a system known as the relay system of working is introduced (where there is a ratio of vehicles to gangs of men in the order of 2 : 1 or 3 : 2). Drivers within the relay system are not considered part of the loading team, but as one vehicle moves out to make the journey to the disposal site, a new vehicle comes in to enable the crew to continue working. Starting and finishing times may have to be staggered to allow this system to work. Where the wheeled bin system is in operation, a lifting rig is fitted to the rear of the vehicle. Other lifting devices can be fitted to the rear of large-capacity compression vehicles to enable them to lift refuse compactors, refuse skips and a variety of bulk containers. For certain industrial applications, front-end loading

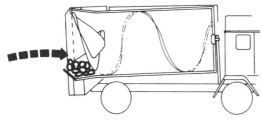

Refuse is loaded into the rear hopper.

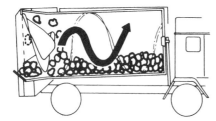

Refuse is carried forward, tumbled and broken down by drum rotation. Weight is always distributed along full length of drum.

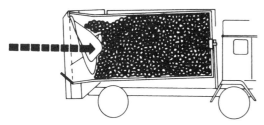

With all free space in drum filled as loading continues, the load is compacted by the action of the stationary helical compression cone against the rotating refuse.

For discharge the rear door is opened and the drum rotated in reverse direction to 'screw-out' refuse.

Fig. 37.1 Rotary compaction refuse collection vehicle. (Courtesy of Laird (Anglesey) Ltd.)

compression vehicles are used, where a pair of forks operate hydraulically in front of the cab, lifting containers over the top of the vehicle and tipping the refuse into the roof of the body. These vehicles are often used where access is particularly difficult.

Health and safety in refuse collection

There are serious health and safety implications involved in the operation of refuse compaction vehicles. Between 1977 and 1984, 28 fatal and 64 major injuries and 320 accidents were notified to the Health and Safety Executive (HSE). Of these total accidents, 77% occurred during the collection round. Guidance Note PM52 [17] gives recommendations on the design and operation of vehicles, including operational controls and safe systems of work.

LITTER

In the United Kingdom, the demand for action against litter is hardening as exemplified by its prominence in the waste management objectives detailed in the government's white paper *This Common Inheritance – Britain's Environmental Strategy*:

> The government aims to encourage the best use of valuable raw materials, and safe and efficient disposal of waste. This means:

- minimizing waste
- recycling materials and recovering energy
- tight controls over waste disposal
- action against litter. [18]

The national anti-litter organization, the Tidy Britain Group, formerly called the Keep Britain Tidy Group but renamed to reflect reality and since 2003 part of ENCAMS which also has other campaigns within its remit, such as Going for Green. There remains the paradox that despite the increased funding for the Tidy Britain Group and the ever widening legislative provision against littering in the early 1990s onwards, most people perceive that the problem is becoming worse. Therefore, in 2004 the Government introduced the Local Environment Quality Bill[4] complementing

---

[4]Not introduced by January 2004.

provisions in the Ant-Social Behaviour Act 2003, partly in order to tighten up on the problem of local social disruption, one of the outcomes being environmental deterioration due to irresponsible discarding of waste.

Although Part IV of the EPA did not repeal the whole of the Litter Act 1983, it did take over its main provision (Section 1), creating an offence of leaving litter (Section 87 of the EPA). The 1983 Act itself consolidated the previous Litter Acts of 1958 and 1971, and repealed the Dangerous Litter Act 1971. The essence of Part IV is provided in Section 87:

1. If any person throws down, drops or otherwise deposits in, into or from any place to which this section applies, and leaves anything whatsoever in such circumstances as to cause, or contribute to, or tend to leave to, the defacement by litter of any place to which this section applies, he shall, subject to subsection 2 below, be guilty of an offence.
2. No offence is committed under this section where the depositing and leaving of the thing was:

   (a) authorized by law, or
   (b) done with the consent of the owner, occupier or other person having control of the place in or into which that thing was deposited.

The maximum fine for littering was raised to £2500 from £400 under the 1983 Act, somewhat academic given that fines in 1993 averaged £48 although the level of costs was often greater. Section 88, however, extends to all litter authorities the fixed penalty notice system[5] originally introduced by Westminster City Council through the City of Westminster Act 1988. These powers cannot be exercised by county councils (Section 89(a)), except for areas designated by the Secretary of State, but do include National Park Committees and Boards and the Broads Authority, which are not principal litter authorities under Section 86(2). These provisions were introduced on 13 February 1991.

Members of the public can take court action to compel authorities to remove litter from public land in their control (Section 91, see later), although the five days' notice requirement (Section 91(5)) prior to the institution of proceedings does provide ample time for remedial action to be taken. There is also encouragement to institute proceedings in that the authority would have to reimburse the expenses of the complainant in bringing the complaint and proceedings before the court where magistrates are satisfied that the complaint is justified.

Local authorities have the power to establish litter control areas under Section 90, which would apply to private landowners subject to the Secretary of State setting the necessary regulations (Sections 90(1)). Also street litter control notices can be issued by the local authority (Section 93) on occupiers of premises to curb litter or refuse on the street or adjacent land.

The most significant change in litter control under the EPA is that local authorities have to reach a quality standard for cleanliness, or absence of litter, from an area (see the *Code of Practice* later). This is in contrast to the traditional specification of service input for street cleansing and litter removal services (Section 89(7)).

The final section of Part IV, Section 99, gives district and borough councils powers with regard to abandoned shopping and luggage trolleys. The extent of their powers to curb the measure of abandoned trolleys is given in Schedule 4, which includes the power to seize and remove trolleys and to exact payment for their return or to arrange for their disposal after a period of six weeks.

The EPA places a new duty on the Crown, local authorities designated statutory undertakers and the owners of some other land to keep land to which the public has access clear of litter and refuse, as far as is practicable. (In the case of designated statutory undertakers, the duty may apply additionally to land to which the public does not have access.) The duty also applies to the land of designated educational institutions.

The code of practice

The Act requires the Secretary of State, under Section 89(7), to issue a Code of Practice to which

---

[5]Section 119 of the Local Government Act 2003 allows local authorities to retain the amounts of fine received and use them to pay their litter control functions.

those under the duty are required to have regard [19]. The original code was issued in 1991 and revised in 1999. The objective of this Code of Practice is 'to provide guidance on the discharge of the duties under Section 89 by establishing reasonable and generally acceptable standards of cleanliness which those under the duty should be capable of meeting'.

It will immediately be apparent that this code, in its approach to litter clearance, is innovative in at least two ways. First, it attempts, by defining standards of cleanliness which are achievable in different types of location and under differing circumstances, to ensure uniformity of standards across Great Britain.

Second, the code is concerned with **output standards** rather than **input standards** – that is to say, it is concerned with **how clean** land is, rather than **how often it is swept**. Indeed, this code does not suggest cleaning frequencies at all – it simply defines certain standards which are achievable in different situations. This may mean that an area which all but escapes littering will seldom need to be swept whereas a litter blackspot may need frequent attention. It will be seen then, that the code offers considerable scope for local authorities and others to target their resources to areas most in need of them, rather than simply sweeping a street because of the dictates of an arbitrary rota. Expressed in its simplest terms: 'if it isn't dirty, don't clean it'.

Statutory duties

Section 89(1) of the Act places on the Crown and (in England and Wales) county, district and London borough councils, the Common Council of the City of London, and the Council of the Isles of Scilly and (in Scotland) district or island councils, and joint boards (collectively known as 'principal litter authorities') a duty to ensure that all land in their direct control which is open to the air and to which the public has access is kept clear of litter and refuse, so far as is practicable. In addition, where the duty extends to roads, they must also be kept clean – again, so far as is practicable.

Section 86(9) transfers the responsibility for cleaning all roads except motorways (which remain with the Secretary of State) from the highways authorities to the district and borough councils. This duty may be transferred to the highway authority by the Secretary of State. See, for example, the Highway Litter Clearance and Cleaning (Transfer of Responsibility) Order 1998.

A similar duty is placed on designated statutory undertakers. One difference between the duty as it applies to statutory undertakers and the duty applying to principal litter authorities is that the duty on the statutory undertakers might cover some land in the direct control of a statutory undertaker **to which the public has no right of access** (such as railway embankments).

The duty also applies to land in the open air and which is in the direct control of the governing body or local education authority of designated educational institutions.

Similar duties may be imposed by principal litter authorities (other than county councils or joint boards) on owners of other land by designating their land as a 'litter control area'.

The Act allows the Secretary of State to specify descriptions of animal faeces to be included within the definition of refuse (see the Litter (Animal Droppings) Order 1991). He may also, by regulation, prescribe particular kinds of things which, if on a road, are to be treated as litter or refuse. (See the Controlled Waste Regulations 1992.)

Practicability

The caveat in the summary of the duty concerning practicability is very important. It is inevitable that on some occasions circumstances may render it impracticable (if not totally impossible) for the body under the duty to discharge it. It will be for the courts to decide, in all cases brought before them, whether or not it was impracticable for a person under the duty to discharge it, but certain circumstances are foreseeable in which the discharge of the duty may be considered by the courts to be impracticable.

Enforcement

In the great majority of cases those under the duty will wish to achieve the highest possible standards of cleanliness. However, the Act makes provision for the occasion when a body under the duty may not discharge it adequately. Under Section 91, a citizen aggrieved by the presence of litter or refuse on land to which the duty applies may, after giving five days' written notice, apply to the magistrates' court (or, in Scotland, the sheriff) for a 'litter abatement order' requiring the person under the duty to clear away the litter or refuse from the area which is the subject of the complaint. Failure to comply with a litter abatement order may result in a fine (with additional fines accruing for each day the area remains littered). Any person contemplating enforcement action should not just consider the presence of litter but is advised to consider whether the body in question is complying with the standards in the code before notifying them, since, under Section 91(11), the code is admissible in evidence in any court proceedings brought under that section.

Similarly, local authorities can act against any other body under the duty which appears to them to be failing to clear land of litter and refuse.

STREET CLEANSING

There are many different types of street cleansing systems depending on the type of location, be it urban, rural, heavily industrialized, inland, seaside, seasonal population changes, etc. No particular system is wrong or right, it is very much a case of 'horses for courses', and tailoring the best option for each particular situation.

It is true to say that there has been a shift away from the manual emphasis of street cleansing towards mechanical systems in the last decade.

Street orderly barrow?

The simplest conventional form of street cleansing is the road sweeper with a street orderly barrow. This is unquestionably the most thorough method of cleansing, but by virtue of the considerable element of walking, and the limited capacity of the barrows involved, the distance travelled will not be great in a working day. The merits of this system are local knowledge, community identity and thoroughness.

Pedestrian-controlled electric vehicles

In inner town areas where access is restricted but where large accumulations of sweepings and waste are anticipated, pedestrian-controlled vehicles are often used. Their capacity may be as much as that of eight bins each of 4 ft^3 (1.13 m^3). As an alternative, the vehicle may have a side-and-end-loading tipping body, and the facility exists for a crew to work in a gang of three or four people, thus covering a larger area.

Cabacs

The next progression up the range is the cabac-type vehicle, which comprises a body similar to the pedestrian-controlled vehicle, but with a cab at the front. The cab will hold up to five drivers/sweepers. The range of these vehicles is considerably greater and speeds of up to 40 mph (65 kph) can be achieved. There are no exhaust emissions causing atmospheric pollution, and the vehicles are quiet in operation, which is particularly advantageous when working in residential areas early in the morning. Cabacs are also a useful mobile amenity block, providing shelter from the rain and warmth, hand-washing facilities, and clothes lockers. By virtue of their greater range than that of pedestrian-controlled vehicles, and the reduced likelihood of fatigue among the crew, they can be used to move between areas, particularly from shopping parade to shopping parade, and have a much wider range of options.

Mechanical/suction sweeping machines

The conventional mechanical sweeping vehicle is used extensively on trunk, principal and main arterial roads. The machines are available in single sweep or dual sweep options to cater for single carriageway and dual carriageway. The single version is left-hand drive to enable the driver to position himself with good sight of the channel that he is sweeping. These machines sweep at between 2 and 5 mph (3 and 8 kph). Smaller, scaled-down versions can be used in pedestrian precincts and areas of restricted manoeuvrability.

Miscellaneous vehicles

There are a number of miscellaneous vehicles that can be used for street cleansing purposes, for example, refuse collection vehicles or light vans, once they have fulfilled their useful life on a regular round, can often be used to service manual street cleansing crews, picking up bags or emptying street cleansing bins from strategic locations. In some of the larger cities in the United Kingdom, street orderly boxes are set into the pavement, and at first glance sweepers appear to be sweeping litter and detritus down a street gully. These boxes are often emptied by a nightshift street cleansing crew.

Street-washing is common in some large cities in the United Kingdom. It can be undertaken by attachments fitted to mechanical sweepers or gully emptiers, or in connection with a bowser and reel or stand-pipes. This practice is not however, as universal as it was 20 years ago, although open-air markets often still use it.

Seasonal variations

Winter brings snow, frost and ice as additional hazards, spring and summer the growth of weeds on footways and channels, and autumn leaf deposits.

Pavement gritters, knapsack and road sweepers, mounted weed control kits and pedestrian-operated or vehicle-mounted leaf blowing units are all recently developed equipment designed to improve operator performance in the field of street cleansing.

Other non-elemental factors that must be taken into account are tourism and student populations, often requiring seasonal staff to cope with fluctuating service demands and workload.

Street cleansing – prevention or cure?

The whole question of street cleansing and its 'bedfellow', litter abatement, hinges on the question, prevention or cure? Unquestionably, the cure is easier to manage and simpler to perform, but is more expensive to achieve. However, in the mid-1990s a new approach to the problem of an untidy country emerged. The Tidy Britain Group's 'People and Places Programme' is aimed at increasing the awareness of all sections of the community of littering and its effect on society. The programme is designed to change attitudes to the problem and promote a long-term systematic approach to litter abatement and environmental improvement. Nationally, around 160 councils have adopted the programme.

WASTE DISPOSAL AND TREATMENT

The principal methods are:

- incineration
- controlled landfilling
- composting.

Development of other techniques is being promoted because of the opposition to the more conventional methods of recovery and disposal but ultimately most of these newer techniques rely on the pre-treatment of wastes in order to utilize these three conventional methods. Therefore, mechanical-biological treatment (MBT), for example, provides processing options for separating of certain recyclable wastes, processing wastes for fuel use and treating organic wastes by aerobic or anaerobic methods, mainly as a pre-treatment technique prior to landfill.

Considerations that influence the adoption of a particular method of disposal are:

- the physical characteristics of the district
- the situation of disposal sites
- daily yield and average composition of refuse
- ultimate disposal of residual products
- capital outlay and running costs.

Incineration

Incineration is not popular, losing favour in the 1960s mainly on the grounds of cost but latterly due to public opposition. However, because of a lack of landfill space and increasing distances for haulage, landfill tax and the restrictions on BMW deposit under the landfill directive an increasing number of areas are considering adopting incineration as their means of disposal. Plants using the modern techniques for emission control should cause no nuisance, as confirmed by an assessment of the health impacts of waste management published by the government in late 2003. The limits on emission of pollutants are becoming more stringent under EU directives and were initially implemented through Part I of the Environmental Protection Act 1990. The former EU controls on the incineration of waste in 89/369/EEC and 89/429/EEC are being replaced and extended by the Waste Incineration Directive 2000/76/EC (WID). The Waste Incineration (England and Wales) Regulations 2002 came into effect on 28 December 2002. The government decided to bring all incinerators and all the other operations covered by the WID, such as co-incineration of waste in cement kilns and waste oil combustion under the Pollution Prevention and Control (PPC) regime.

The WID sets stringent emission limits for pollutants such as dioxins and acid gases. It also covers matters such as discharges to water and the recycling of ash. Compliance will be achieved through conditions in the PPC permit. These limits increase the cost of disposal by this method. Incineration will remain viable only when it is used to generate energy. In the United Kingdom, only incinerators that reclaim energy from waste continued to operate after 1996.

In theory it is possible to utilize the incinerator bottom ash (IBA) from the incineration process, especially when the ferrous metal content has been removed by magnetic extraction and non-ferrous metals by eddy-current separation; in normal circumstances the quality of the IBA is such that only a very small proportion would be used as a low quality fill. However, with both the landfill tax and aggregates tax and improved control of incinertors which allowed maximum burn-out of combustible wastes there has been increasing prospects of processing IBA for aggregate use.

Until the commissioning trials started for the South East London Combined Heat and Power (SELCHP) plant in December 1993 there had not been a municipal waste incinerator built in the United Kingdom since 1978. Incinerators dealing with other types of waste, such as special and clinical wastes and sewage sludge, however, continued to be constructed during that period. Also in 1993 a power plant fired by 60 000 tonnes a year of whole tyres started operating in Wolverhampton but this was closed due to technical problems in June 2000 and the year before a plant burning chicken litter was built in East Anglia. The SELCHP plant is illustrated in Fig. 37.2.

As with the SELCHP plant, the provision of enhanced payments for the electricity generated by these plants through the non-fossil fuel obligation (NFFO) payments system under the Electricity Act 1989 was a substantial additional inducement to the building of incinerators. The NFFO system has now been superceded by the ROC (renewables obligation certificates) system which forces electricity companies to meet increasing proportions of their electricity sales from renewable sources.

A variety of combustion systems are used for waste incineration, some being more suited to liquid chemical wastes or sewage sludge and others being most suited to municipal waste. The main types in use, using the terminology from the 1993 Royal Commission on Environmental Pollution's (RCEP's) Seventeenth Report [20] are:

- mass burn with excess air combustion
- ashing rotary kiln
- slagging rotary kiln
- fluidized bed
- pulsed hearth
- multiple hearth
- liquid injection.

The RCEP report suggested that the minimum economic threshold for development of an MSW incinerator is around 200 000 tpa. The SELCHP plant in Deptford, south London, has a capacity of 420 000 tonnes, while the Coventry plant, which also has energy recovery and was completed in 1976, has a capacity of 160 000 tonnes a year. More recent developments have mainly been located at sites which previously housed old incineration plant and generally under 200 000 tpa capacity.

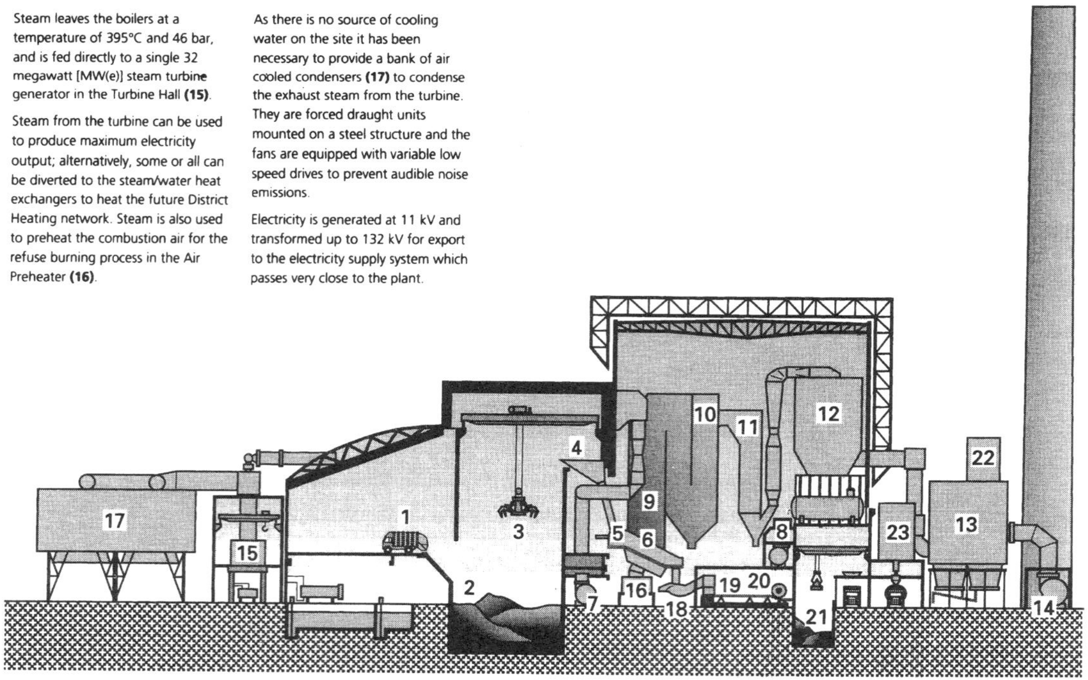

HOW THE PLANT WORKS

Refuse collection vehicles (1) tip the solid waste - without pre-sorting into the storage pit (2) from where it is transferred by over-head cranes (3) to the feed hopper (4) of the stoker feed chute Hydraulic ram feeders (5) provide controlled charging of refuse onto the surface of the Martin reverse-acting stoker grate (6).

The forced draught fan (7) supplies primary air via the under-grate air zones to the burning refuse layer on the stoker grate (6). This fan and the overfire air fan (8) draw the air from the refuse storage pit area thus preventing the egress of unpleasant odours from the plant.

The heart of the System for waste combustion is the stoker grate itself (6). The grate surface is sloped downwards from the feeder end towards the residue discharge end and is comprised of alter-nate steps of fixed and moving grate bars. The moving grate steps perform slow stirring strokes against the grate slope. This ensures that the burning refuse layer is continually rotated and mingled to form an even depth of bed and red hot mass is pushed back to the front end of the grate. In this way an intense fire builds up immediately at the front end of the grate, with all combustion phases (such as drying, ignition and combustion itself) taking place simultaneously.

Burned out residues are transferred at the bottom of the stoker grate, by an ash discharger (18) and by the residue handling sys-tem (19) and deposited in the residue pit (21). During the trans-fer, ferrous metals are removed by the magnetic separator (19).

The energy released by the process is recovered in the boiler. In this unit, the furnace walls and the division walls between the boiler sections (9) are of solidly welded membrane design. The superheater (10) is carefully sited in the multi-pass boiler whilst the economiser section (11) is in the fourth pass. The econo-miser is followed by the lime scrubber reactor (12) where a fine spray of a lime (from the lime storage silo (22)) and water mix-ture is introduced into the flue gases. This has the effect of neu-tralising acid gases contained in the flue gases and as the lime/salts cool, heavy metals condense onto the particulates. This particulate matter is removed from the gas stream by the bag-house filter (13) and the now clean and dedusted flue gases are ejected to atmosphere by the induced draft fan (14) via the 100 metre tall chimney.

Flue gas treatment residues are stored in the ash silo (23). Both these residues and the burned out ash residues and separated metals from the stoker grate are loaded onto transport within the building and removed from the site. Ash and flue gas treat-ment residues are landfilled and the metals are recycled.

POWER GENERATION

Steam leaves the boilers at a temperature of 395°C and 46 bar, and is fed directly to a single 32 megawatt (MW(e)) steam turbine generator in the Turbine Hall (15).

Steam from the turbine can be used to produce maximum electricity output; alternatively, some or all can be diverted to the steam/water heat exchangers mounted on a steel structure and the fans are equipped with variable low speed drives to prevent audible noise emissions.

As there is no source of cooling water on the site it has been necessary to provide a bank of air cooled condensers (17) to condense the exhaust steam from the turbine. They are forced draught units mounted on a steel structure and the exchangers to heat the future District Heating network. Steam is also used to preheat the combustion air for the refuse burning process in the Air Preheater (16).

Electricity is generated at 11 kV and transformed up to 132 kV for export to the electricity supply system which passes very close to the plant.

Fig. 37.2 Waste to energy plant, South East London Combined Heat and Power Ltd. (Diagram and description by courtesy of SELCHP, 7B Evelyn Court, Deptford Business Centre London, SE8 5AD.)

Mass burn incinerators

Mass burn incinerators are used to burn municipal solid waste but can also be used to burn suitable commercial and industrial wastes. The capacity of existing plants in the United Kingdom varies according to the number of furnace units incorporated into the design (up to five). The size of each unit can vary from around 10 tonnes per hour up to 30 tonnes per hour.

Waste is fed by crane from storage bunkers to the feed hopper, from where it flows down, usually aided by a mechanical stoker, into the combustion chamber. Air is introduced through a moving grate in the chamber, which agitates the waste promoting its thorough exposure to the air. Waste is fed on to the grate and during the initial drying stage, at 50–100°C, volatile compounds are released. These burn above the grate, where secondary air is introduced to facilitate complete gas phase combustion. The remaining waste moves down the grate and continues to burn slowly. After about an hour the residues are discharged from the end of the grate, where they are usually quenched. In most cases the ferrous fraction is magnetically extracted prior to deposit.

Although there is a considerable variety of designs for mass burn incinerators, the most efficient appear to be those with reciprocating or roller grates and the combustion chamber in the form of a vertical shaft.

Ashing rotary kiln

Ashing rotary kiln incinerators are primarily used to burn chemical waste. They are also suitable for the combustion of clinical waste and MSW, although they have not been used for the latter in the United Kingdom.

Rotary kilns used for clinical waste range from around 1 to 3 tonnes per hour, for chemical waste up to 10 tonnes per hour and for MSW up to 20 tonnes per hour.

Waste is fed into a refractory-lined drum which acts as the grate surface in the primary combustion stage. The waste is ignited on entry to the drum and the rotating action of the drum mixes it with air supplied via nozzles in the furnace wall.

A secondary combustion chamber is fitted with air inlets and after-burners, usually operating at 1100–1300°C, to complete the combustion of the gases.

Slagging rotary kiln

Slagging rotary kiln incinerators are most suitable for chemical waste. This type of incinerator is very similar in method of operation to the ashing rotary kiln but uses a higher combustion temperature (1400°C). This melts most inorganic waste and ash and therefore produces a liquid slag which traps particulate matter in the kiln before being solidified, usually in a quench tank but occasionally by air cooling. The vitrified slag has a very low leaching rate for heavy metals and organic compounds.

Fluidized bed

Fluidized bed incinerators are used in the United Kingdom mainly to burn dewatered sewage sludge cake, usually at about 3 tonnes an hour. They are also suitable for burning clinical waste, some chemical wastes and municipal waste that has been processed to effect size reduction. Of the units operational in the United Kingdom only the Dundee plant incinerates MSW.

Other systems

Multiple hearth, pulsed hearth and liquid injection incinerators are not used for the incineration of MSW.

The essential elements of an incineration plant (Fig. 37.3) are listed below.

1. **Facilities for handling, storing and mixing feedstock** without creating a hazard or a nuisance. In the diagram, vehicles enter the tipping hall (A) and unload waste into a bunker (B), from which a grab on a crane (C) transfers it to a hopper (D) leading to the grate.
2. **A combustion chamber** (F) is designed to achieve very efficient combustion in both the solid/liquid (**primary**) and gas/aerosol (**secondary**) phases. In the plant illustrated, the

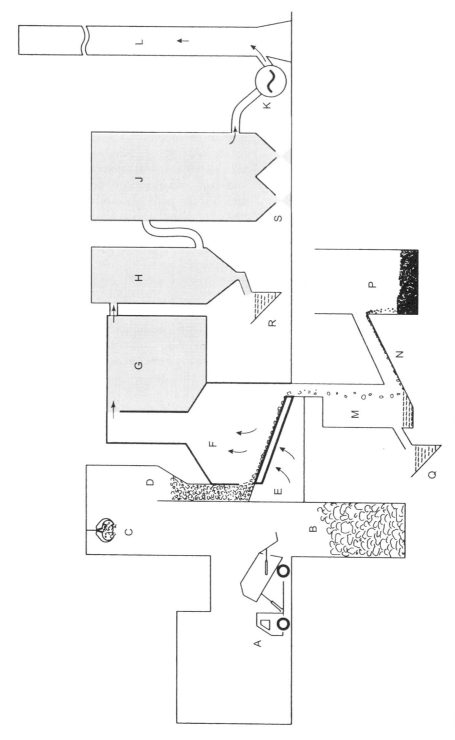

Fig. 37.3 An incineration plant for municipal waste (arrows indicate air or gas flow) (Source: [20]).

waste is agitated mechanically as it moves over the **grate** (E) in order to expose it to air and maintain a free flow through the grate bars.

3. **A heat recovery system** (G) cools the exhaust gases prior to cleaning them. This minimizes corrosion, thermal stress and the formation of undesirable compounds such as dioxins. It also gives the opportunity of using the energy recovered.

4. **A gas-cleaning system** typically consists of a scrubber (H) and an electrostatic precipitator (J).

5. **A fan** (K) draws the cleaned gases out of the plant so that they can be dispersed from the **stack** (L).

6. Facilities for the collection, handling and, where necessary, **treatment of solid residues**. Ash and clinker from the grate fall into a water-filled quench tank (M) and are moved by conveyor (N) to a bunker (P); fly ash is collected from the electrostatic precipitator (S).

7. Facilities for the collection and **treatment of liquid effluents** (Q, R) from the scrubber and quench tank.

Controlled landfilling

The term 'controlled landfilling' or 'sanitary land-fill' describes a system of disposing of refuse by depositing in a methodical, as distinct from a crude or indiscriminate, manner. The greater part of all controlled waste is disposed of in this manner and, if properly supervised, the system provides a satisfactory method of disposal. Mineral excavations, low-lying or open-cast sites are selected for the purpose, and, by the utilization of refuse in the manner described below, are reclaimed for useful purposes. In many districts, land reclamation alone justifies this method of refuse disposal. There has been a recent trend for disposal of waste over land so as to create a new and improved landform. There are three methods of landfill operation.

1. **Trench method**: this involves the excavation of a trench (which may be very large) into which waste is deposited. The excavated material is then used as cover. This technique is a variation on the cell method described below. It should be distinguished from the construction of trenches into already deposited solid waste for the disposal of liquid wastes and progressively banned up until July 2005. The trench method has found very limited application in the United Kingdom.

2. **Area method**: waste may be deposited in layers to form terraces over the available area. However, with this type of operation, excessive leachate generation may occur unless high-waste inputs are maintained, thereby providing adequate absorptive capacity to account for rainfall. This method has been used widely in the United Kingdom, but is no longer favoured since operational control may be difficult.

3. **Cell method**: this method involves the deposition of waste within preconstructed bunded areas (Fig. 37.4). It is now the preferred method since it encourages the concept of progressive filling and restoration. It now has widespread application.

The main points about the deposition of the crude refuse are as follows.

1. Where the base of the site cannot support the weight of vehicles, a preformed base is required.

2. Waste should be deposited at the top or base of a shallow sloping working face, and not over a vertical face. The face of the tip should slope at an incline that is no greater than 1 : 3.

3. The deposit of waste should be in thin layers. The use of a compactor enables a high density to be activated. Each layer should not exceed 0.3 m in thickness.

4. At the end of the working day, all exposed surfaces should be covered with an inert material to a depth of at least 0.15 m.

5. The management and workforce should be fully aware of the site safety regulations and of the need to observe them.

6. An effective system of litter control is essential.

7. Measures need to be taken to control birds, particularly gulls and crows, where sites receive quantities of putrescible matter. Control measures include:

(a) birdscarers
(b) distress calls

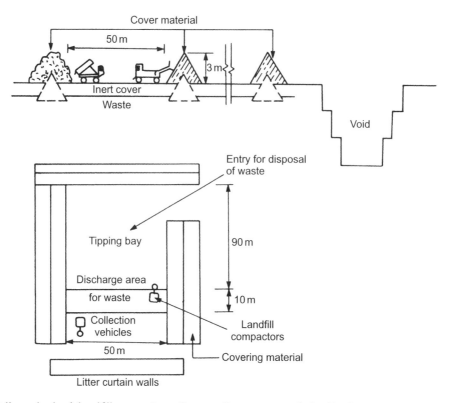

Fig. 37.4 Cell method of landfill operation. (Source: Department of the Environment (1986, revised 1990) *Landfilling Wastes*, Waste Management Paper No. 26, HMSO, London. Reproduced by kind permission of the Controller of Her Majesty's Stationery Office.)

(c) falcons

(d) nets.

8. Effective measures to ensure good pest control are necessary. Good compaction of the material will reduce the likelihood of infestation by both insects and rodents, as will the daily covering of waste. However, regular systems of both rodent and insect control need to be established on a programmed basis.

9. Waste should not be burnt on the landfill site.

Figure 37.5 shows a typical operational plan for a landfill site.

The main elements of the site restoration plan following the cessation of landfilling are as follows.

1. A plan for restoration should form an integral part of the landfill operation.

2. The intended final levels and contours should be indicated within the plan, as should the systems for leachate control.

3. There must be a clear and efficient system for the management of landfill gas.

4. Landfill site capping should be constructed of material with a permeability of 1×10^{-7} cm/s or less, and should extend over the whole site to increase water run-off. A basic aim of the cap is to minimize leachate production by minimizing water entry to the tipped area.

5. The cap should be covered with an appropriate thickness of soil to protect the cap from damage by, for example, agricultural machinery.

Figure 37.6 shows a section through a landfill site situated on clay strata and indicates both completed and operational parts of the site.

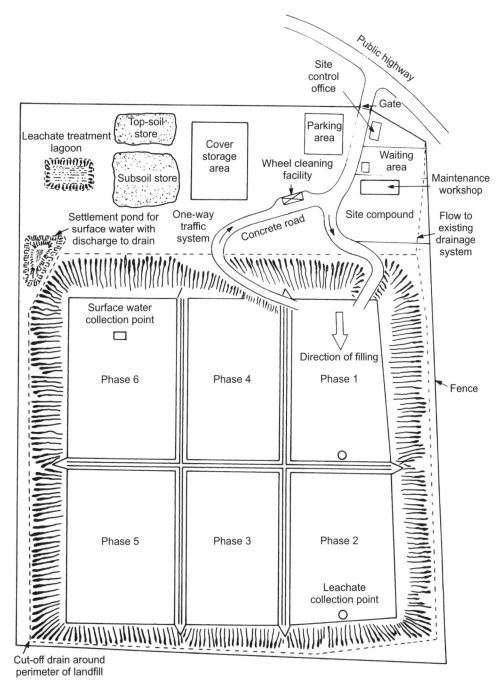

Fig. 37.5 Typical operational plan for landfill site. (Source: Department of the Environment (1986, revised 1990) *Landfilling Wastes*, Waste Management Paper No. 26, HMSO, London. Reproduced by kind permission of the Controller of Her Majesty's Stationery Office.)

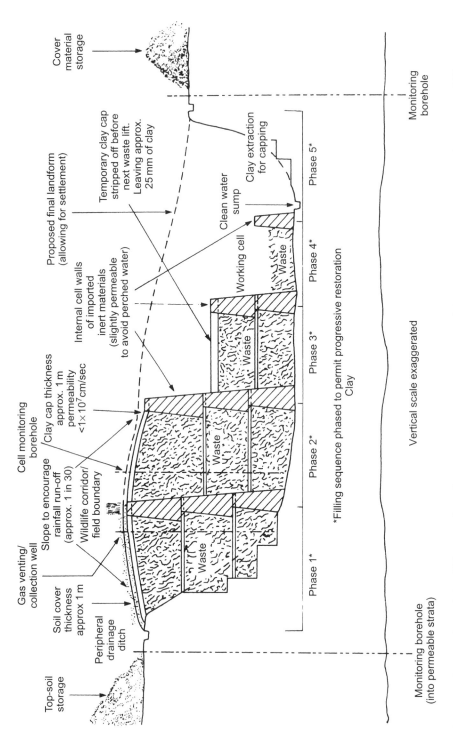

Fig. 37.6 Section through landfill situated on clay strata. (Source: Department of the Environment (1986, revised 1990) *Landfilling Wastes*, Waste Management Paper No. 26, HMSO, London. Reproduced by kind permission of the Controller of Her Majesty's Stationery Office.)

Department of the Environment Waste Management Paper No. 26 [21] provides a useful technical source of information on the problem and solutions in current practice. Waste Management Paper No. 26A [7] provides guidance on assessing the completion of licensed landfill sites. (See also [22].)

Types of landfill sites

There were two basic forms of landfill sites: dilute and disperse sites and containment sites. Under the Landfill Directive and the Landfill Regulations 2002 landfills must conform with the detailed technical requirements of the directive. These cover matters, such as lining systems, leachate and gas controls, siting of landfills and waste acceptance criteria. The few remaining dilute and disperse landfills will therefore be phased out.

By 16 July 2002, all landfill operators should have provided site conditioning plans to the EA in order to demonstrate that the sites could be brought up to the standards required. The Agency will be progressively permitting all landfill sites on the basis of assessment of risk to the environment. This work should be concluded by 2007 with hazardous waste landfill permitting completed by July 2004.

Dilute and disperse sites

The dilute and disperse site is operated on the basis that leachate that is generated slowly migrates through unsaturated strata. During its passage, chemical, physical, biological and microbiological activity reduces the pollutants present. When this leachate comes into contact with water it is quite weak. Further dilution takes place in the aquifer, thus reducing the polluting characteristics still further. This method has been used extensively in the United Kingdom.

In theory, the method appears to be safe, but there are a few cases where water has been polluted by the practice. Such sites are banned under the requirements of the landfill directive.

Containment sites

The containment site is a site where the base, sides and top (once landfilling finishes) are sealed with a suitable mineral or synthetic impermeable liner.

In the United Kingdom, clay is extensively used to seal sites. In the United States and some Continental European countries, synthetic plastic liners are used. Depending on the country and geology of the site, a single liner, a composite liner or a double liner is incorporated. Further research is being undertaken to evaluate the advantages and disadvantages of various liners in different conditions.

Decomposition of household waste after landfilling

Within a landfill site, aerobic conditions prevail initially, but anaerobic conditions are rapidly established. The biodegradation of various components of refuse in landfills is extremely complex, but three main stages can be distinguished (Fig. 37.7).

In the first stage, degradable waste is attacked by aerobic organisms, which are present in the oxygen in the air trapped in the waste, to form more simple organic compounds, carbon dioxide and water. Heat is generated and the aerobic organisms multiply.

The second stage commences when all the oxygen is consumed or displaced by carbon dioxide and the aerobic organisms die back. The degradation process is then taken over by organisms that can thrive in either the presence or absence of oxygen. These organisms can break down the large organic molecules present in food, paper and similar waste into more simple compounds such as hydrogen, ammonia, water, carbon dioxide and organic acids. During this stage, carbon dioxide concentrations can reach a maximum of 90%, but usually reach about 50%.

In the third and final stage, which is anaerobic, species of methane-forming organisms multiply and break down organic acids to form methane gas and other products (Fig. 37.8).

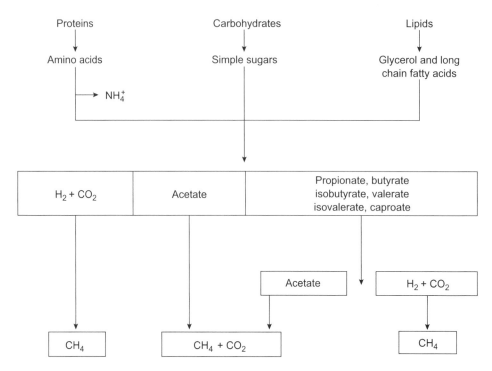

Fig. 37.7 Decomposition of materials occurring in household waste. (Source: Department of the Environment (1986, revised 1990) *Landfilling Wastes*, Waste Management Paper No. 26, HMSO, London. Reproduced by kind permission of the Controller of Her Majesty's Stationery Office.)

Composting

As a waste treatment technique in the United Kingdom, composting was almost eliminated in the 1970s. Up to that time several schemes were producing compost from a feedstock of ordinary household waste, usually using Dano drums or similar slowly rotating trommels where the waste was retained for between two days and one week before being placed in windrows for further processing.

The production of a growing medium made from mixed refuse became increasingly difficult to market and the cost of this method of waste processing was considerably higher than that of landfill. In contrast, in most Continental European countries the composting has continued to be used for treatment of up to 10% of household waste, although in most cases the resulting material has been used in land reclamation or as landfill cover.

While the average dustbin or black bag contains around 20–35% of putrescible and largely compostable waste, there is an even greater amount of readily compostable material available from residents' gardens and the councils' parks departments. Although traditionally composted, until very recently increasing amounts of this were going to landfill for disposal.

Research carried out on pilot collections of source-separated compostable waste in the United Kingdom suggests that at least half of the weight of compostable waste could be reclaimed. Monitoring of the 4000 households that participated in Leeds City Council's split wheeled bin trial, which started in December 1990, showed that of an estimated 4.4 kg of compostable waste produced each week by households, half was correctly separated for collection. Early results from East Hertfordshire District Council's 2200 household trial of brown wheeled 'bio-bins' showed a

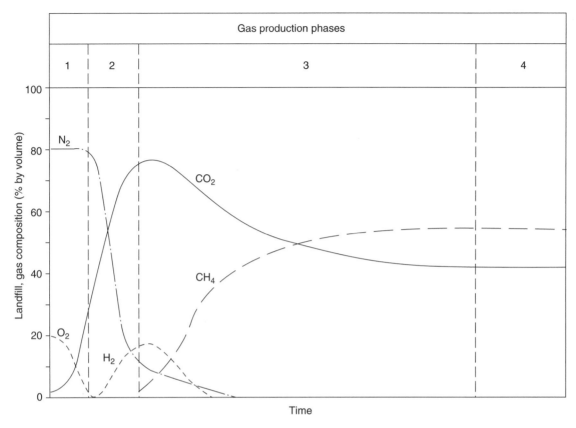

Fig. 37.8 Typical production pattern of landfill gas. (Source: Department of the Environment (1986, revised 1990) *Landfilling Wastes*, Waste Management Paper No. 26, HMSO, London. Reproduced by kind permission of the Controller of Her Majesty's Stationery Office.)

diversion rate of 40%, but the scheme started during the growing season, and a greater proportion of the population in its area has a garden compared with the initial trial area in Leeds.

A national collection of compostable waste would therefore yield around 1.5–2 million tpa of kitchen waste, plus an equal amount of garden waste, which is currently taken to civic amenity sites.

The development of composting systems in the United Kingdom is taking two main forms: the centralized composting of green wastes; and the promotion of home composting systems. The development of markets to accommodate increasing amounts of compost is an essential aspect of the development of centralized composting systems. The main priority of local authorities promoting home composting is the need for a comprehensive coverage of composting units in those areas chosen for this type of waste reduction strategy to ensure that there will be an effective diversion of the putrescible waste fraction.

In the early 1990s, a number of green waste composting systems were established, mainly using green wastes from parks departments and civic amenity waste disposal sites. In a limited number of cases selected commercial and industrial wastes were used to improve the mix of materials composted. In only a small minority of cases have centralized composting systems incorporated kitchen waste materials because of the potential problems of attracting vermin and the increased chance of odour and latterly the issue of the Animal By-Products Order/Regulation. In several mainland European countries these difficulties have been overcome by the building of enclosed composting facilities, which obviously increases the cost of this method of waste treatment.

The greatest barrier to the expansion of green waste composting schemes is the requirement to develop markets to take the growing media mulches and chippings produced throughout the year, as the market is highly concentrated into the spring months. See DETR (1999) *Report of the Composting Development Group on the Development and Expansion of Markets for Compost*, DETR, Wetherby.

In contrast, the establishment of home composting facilities attempts to effect the reduction of waste at source. By early 1994, around 25 UK local authorities had established home composting initiatives, each with more than 250 units. These included the free or heavily subsidized provision of conventional garden composters, rotating units and Green Cone digesters. In a minority of instances wormeries have also been offered, but these have greater operational limitations than other types of unit.

One of the main difficulties in assessing the success of home composting schemes is that there is often little data on what proportion of the population is already practising composting at home and what proportion of the materials generated in the home are composted. How many more people will be composting as a result of the provision of the new composting units is therefore not known.

To overcome some of these difficulties, the London Borough of Sutton offered composters in only one collection round where the record of waste collected had been measured over several years and therefore the effect of the provision of composters could be measured. Despite the generous discounts on the three types of unit offered, the participation rate was too low to assess accurately the effect of distributing the composters.

To overcome this problem, during 1994 the council planned to provide Green Cone digesters to all households in another part of the borough, with people having positively refused to accept the unit. This approach was not effective, however, because of a myriad of operational problems experienced with the Green Cones. Conventional composters

would have been more effective, but they are unsuitable for cooked items and meat products.

In some cases, such as in East Hertfordshire and Leeds, where there is separation of compostable material, alternate fortnightly collections of different fractions of household waste have been instituted. Potentially, where very high coverage of home composting units has been achieved, fortnightly collections might be considered in those circumstances. However, many residents, councillors and local authority officers are rightly wary of instituting such a change of service provision. Where fortnightly collections have been instituted, almost always where wheeled bin collection systems have been introduced, very few operational problems have been experienced or complaints received.

Solid fuel substitutes

A fairly recent development has been the manufacture of a fuel derived from waste by separating materials and compressing the paper, cardboard and plastics hard enough to form them into pellets. They have half to two-thirds of the calorific value of coal.

Some doubt has been expressed about its economical viability as well as its suitability due to the higher quality requirement on emissions of flue gases. The extensive use of the system is doubtful.

CLINICAL WASTE

General background

The World Health Organization (WHO), the European Commission and the United States Environmental Protection Agency (EPA) have all, in the course of the last few years, emphasized the need for clinical waste to be safely managed from the point of production via various methods of movement to its final disposal. The concern expressed by these bodies reflects the growing awareness of the threat to public health that can arise from the improper handling and disposal of such wastes. In *Management of Waste from Hospitals* [23] (published in 1985 and reflecting the conclusions of a working party in the summer of 1983), WHO classified health care waste into eight main categories, while the EPA, in *Guide for Infectious Waste Management* [24], recommended that health care facilities should prepare an infectious wastes management plan outlining policies and procedures for the management of infectious wastes, which it divided into six main categories plus four 'optional' categories. Details of these categories are included in *Clinical Waste – an Appraisal* [25].

Nationally

In the United Kingdom, the same concern about the need for safe management of clinical waste has been expressed by the Department of the Environment, the Department of Health, the Health and Safety Commission (HSC) and the Royal Commission on Environmental Pollution.

National awareness of the need for tighter control and regulation of clinical wastes increased at the beginning of the 1980s. Prior to this, items had been washed and cleaned or sterilized after use, but there was now a major conversion to the use of one-trip plastics and disposable materials. Disposable syringes, catheters, probes, urine bags and special clinical packs, together with one-use and discarded protective clothing, linen and dining equipment became the accepted substitute for much of the long-life equipment previously used in the medical and clinical field.

This trend, coupled with the switch to waste disposal in plastic bags in a whole range of colours, gave rise to a marked increase in the output of clinical wastes. The casual manner in which these wastes were mixed with general waste disposal streams also gave cause for concern, and in 1982 the HSC issued a guidance document entitled *The Safe Disposal of Clinical Waste* [26]. This defined and categorized clinical waste and gave advice on its handling, transport and disposal.

The HSC's definition of clinical waste, which has been generally accepted in the United Kingdom, is:

Waste arising from medical, nursing, dental, veterinary, pharmaceutical or similar practice, investigation, treatment, care, teaching or research which by nature of its toxic, infectious or dangerous content may prove a hazard or give offence unless previously rendered safe and inoffensive. Such waste includes human or animal tissue or excretions, drugs and medicinal products, swabs and dressings, instruments, and similar substances and materials.

The HSC document divided clinical waste into the following five categories.

1. **Group A**
   (a) Soiled surgical dressings, swabs and all other contaminated waste from treatment areas;
   (b) material other than linen from cases of infectious disease;
   (c) all human tissues (whether infected or not), animal carcasses and tissues from laboratories, and all related swabs and dressings.

2. **Group B**
   Discarded syringes, needles, cartridges, broken glass and any other sharp instruments.

3. **Group C**
   Laboratory and post-mortem room waste other than waste included in Group A.

4. **Group D**
   Certain pharmaceutical and chemical waste.

5. **Group E**
   Used disposable bed-pan liners, urine containers, incontinence pads and stoma bags.

In 1983 (revised 1995) the Department of the Environment published *Clinical Wastes: a Technical Memorandum on Arisings, Treatment and Disposal* (Waste Management Paper No. 25) [27] including a code of practice which estimated that the amount of clinical waste arisings from National Health and private hospitals in Great Britain in 1980 was some 33 000 tonnes. A comparable estimate was not available for clinical waste arisings from community health service practices, but the paper suggested it could be as much as 15% of clinical waste arisings at hospitals. It stated that waste from home treatment would in total represent the largest arisings of clinical wastes, but as individual arisings were usually small and irregular in time a precise figure could not be given, though three to five times the amount of arisings in hospitals was thought probable. It was noted that about 1.7 million cat and dog carcasses were also disposed of each year.

The paper stressed that the segregation of clinical waste from all other wastes was the key to proper disposal. It suggested that the segregation system set out in the HSC document should be adopted as a national standard and urged hospitals to ensure that their management structures for waste disposal were adequate, and that a waste disposal policy was formulated and implemented. It was emphasized that packaging for clinical wastes should be of a high integrity.

In 1987, the British Standards Institution (BSI) adopted a classification for the analysis of hospital wastes in BS 3316:1987 Part 4 [28] and grouped hospital waste into six categories.

The Royal Commission on Environmental Pollution in its eleventh report *Managing Waste: The Duty of Care* [3] recommended that all regional health authorities should prepare and implement waste disposal plans that matched the arisings and the disposal facilities in health care establishments. It also recommended that publicity should be given by government health departments through health authorities to the guidelines and code of practice prepared by the HSC and the Department of the Environment on the disposal of clinical waste so that community health care establishments were made aware of good practices, which could then be enforced. It was urged that publicity material prepared by the health departments should be made available for doctors and nurses in the community to give to patients who might have to dispose of clinical waste. It also recommended that Crown immunity should cease to apply to the NHS and that incineration and other waste disposal facilities operated by or on behalf of the NHS should, in consequence, be subject to exactly the same controls and standards as similar facilities operated by other organizations.

In its response to the Royal Commission in July 1986 [29], the government expressed its own concern about reported standards of waste handling within the NHS and stated that, while it had no current plans for removing Crown immunity, the situation would be kept under review. It considered it unnecessary for regional health authorities (RHAs) to duplicate, in effect, the work of the WDAs in preparing waste disposal plans, but urged the RHAs to co-operate with the WDAs in making their information available. The government also made it clear, in separate pronouncements, that it expected the NHS to meet and maintain the standards required by the Control of Pollution Act 1974, the Clean Air Acts 1956 and 1968 (now consolidated in the Clean Air Act 1993) and supporting regulations and that responsibility for emissions from NHS hospital chimneys would be considered as part of a forthcoming review of air pollution control (see Part I of the Environmental Protection Act 1990).

It will be clear from the foregoing that the 1980s heralded a heightened national awareness of the potential dangers from the mismanagement of clinical waste from all sources and saw the emergence of a new determination to tackle the problem.

Legislation

The general framework for the control of waste including clinical waste is included earlier in this chapter. However, there are specific requirements that need to be considered.

The Controlled Waste Regulations 1992

These regulations, applicable to both England and Wales, contain a legal definition of clinical waste. Clinical waste means:

1. any waste which consists wholly or partly of human or animal tissue, blood or other body fluids, excretions, drugs or other pharmaceutical products, swabs or dressings, or syringes, needles or other sharp instruments, being waste which unless rendered safe may prove hazardous to any person coming into contact with it; and
2. any other waste arising from medical, nursing, dental, veterinary, pharmaceutical or similar practice, investigation, treatment, care, teaching,

or research, or the collection of blood for transfusion, being waste which may cause infection to any person coming into contact with it.

Certain classes of waste are prescribed so that charges can be made for their collection and disposal. The regulations define clinical waste and then distinguish between the two basically different sources from which it emanates: clinical waste from a private dwelling or residential home (Schedule 2), which is categorized as household waste for which a charge for collection may be made, and clinical waste from any other source (Schedule 3), which is categorized as industrial waste.

Thus, a local collection authority has a duty to collect clinical waste in the first category if requested to do so, but has the power to make a charge for this service. The former Department of the Environment has stressed that it expects the collection authorities to take account of the social benefits, and the obvious undesirability of discouraging occupants from making the desired special arrangements, before deciding to make a charge. It also stresses that the regulations are not intended to disrupt existing arrangements, that is, collection by health authorities, where those are satisfactory and that the primary objective of the prescription is simply to promote public safety [14].

All other clinical waste, that is, that which does not emanate from a private dwelling or residential home, is categorized as industrial waste. The local collection authority does not have any duty to collect it, but may if requested to do so, and then only after first obtaining permission from the disposal authority. A disposal authority in England where it is separate from the collection authority may also collect clinical waste where it is prescribed as industrial waste. The collecting authority has a duty to make a reasonable charge for its collection and disposal.

All other waste from a hospital, unless otherwise prescribed, falls into the category of household waste. This means that the local collection authority has a duty to collect it free of charge.

Health and Safety at Work, Etc. Act 1974

This Act applies to all health care activities in the United Kingdom and thus its provisions apply in requiring the safe handling, storage, transportation and disposal of clinical waste arising from those activities. Briefly, the relevant sections of the Act require that a safe system of work (which should be in writing) exists, which is monitored to ensure that proper and adequate resources, training, equipment and information are provided to ensure the minimization of risk to health or safety of employees and public alike.

Segregation of waste

The HSC's *Safe Disposal of Clinical Waste* [26] recommends that the identification of various kinds of waste for disposal in a particular manner is best achieved by the use of easily identifiable colour-coded containers. Care should be taken to avoid confusion with other sorting systems that may be colour-coded, for example, for soiled, foul and infected linen in hospitals and other health care premises. All organizations involved with handling clinical waste are therefore strongly advised to adopt the colour-coding given in Table 37.5.

Table 37.5 Recommended colour-coding for containers for clinical waste

Colour of bag	Type of waste
Black	Normal household waste: not to be used to store or transport clinical waste
Yellow	All waste destined for incineration
Yellow with a black band	Waste, e.g. home nursing waste, which preferably should be disposed of by incineration, but which may be disposed of by landfill when separate collection and disposal arrangements are made
Light blue or transparent with light blue inscriptions	Waste for autoclaving (or equivalent treatment) before ultimate disposal

Source: Health and Safety Commission [26].

Disposal

Waste Management Paper No. 25 [27] recommends that incineration should be the preferred disposal route for the following clinical wastes whether they arise in the community or medical or veterinary establishments:

1. Group A: human tissues, limbs, placentae and infected carcasses and dialysis waste.
2. Group B: all sharps.
3. Group C: all pathology wastes unless they have been effectively autoclaved and are suitable for treatment and disposal by this method (Howie Code).
4. Group D: small amounts of solid medicines and injectables but not large arisings of pharmaceuticals whose disposal should be on the advice of the WDA, the NRA or River Purification Board and the Pharmaceutical Society of Great Britain. (Waste Management Paper No. 19 also provides advice [30].)

Waste Management Paper No. 25 (revised) [27] also expresses a preference that the following materials be incinerated, but notes that other disposal methods could be acceptable: Group A: soiled surgical dressings, swabs and other contaminated wastes.

It is now accepted that clinical waste should, wherever possible, be incinerated in incinerators designed to destroy all grades of clinical and hazardous waste.

In practical terms, landfill should only be used as an emergency back-up to incinerators, and even then only for certain types of non-infectious clinical waste. The dangers of infection from hepatitis B and human immunodeficiency virus (HIV) and the anxiety surrounding these matters has meant that the use of landfill for the disposal of clinical waste, even under carefully controlled conditions, has virtually ceased because of stringent restrictions on site licensing.

The third report of the Hazardous Waste Inspectorate in June 1988 [31] drew attention to the unsatisfactory performance of some NHS incinerators, which operate at temperatures that were too low to achieve totally sterile residues. For the future, in addition to maceration and high temperature incineration, there are other methods for the disposal of clinical wastes so as to render them non-hazardous and harmless. In Germany, a process is in operation whereby clinical waste is treated at the site of production; it is shredded, making the waste unrecognizable, then subjected to microwave treatment, which results in thermic disinfection. It is able to handle all types of clinical waste, including hypodermic needles, bandages, dialysis filters, etc., and the resulting disinfected granules can then be compacted into a container for shipment to a licensed landfill site for final disposal. The control of atmospheric emissions from hospital incinerators, including those handling clinical waste, is now exercised through the authorization procedures of Part I of the Environmental Protection Act 1990. In 1998, the EA produced a new version of Waste Management Paper No. 23 [32] bringing together strategies for best practice for the segregation, collection and disposal of clinical wastes.

STORAGE, COLLECTION AND DISPOSAL OF OTHER HAZARDOUS WASTE[6]

Special wastes

In the United Kingdom, the Hazardous Waste Directive (91/689/EEC) has been implemented through the Special Waste Regulations 1996 as amended by SI No. 2019 and SI No. 1997/251. Superficially, the Special Waste Regulations 1996 build on the Special Waste Regulations 1980. The EA, the SEPA and the NIHS published in 1998 *A Technical Guide on the Definition and Classification of Special Wastes*, in 2 volumes, HMSO, London. In preparation for the change to the Hazardous Waste Regulations 2004 a further Technical Guide was published in 2003 (5).

Under the 1980 regulations, wastes were classified by whether or not they contained listed substances so as to be 'dangerous to life' or had a low flash point ($<21°C$). Under the 1996 classification, and now under the 2002 CHIP3 regulations, substances can render a waste 'special' if

---

[6]Also see the Hazardous Waste Forum (December 2003) *Hazardous Waste – An Action Plan for its Reduction and Environmentally Sound Management*. DEFRA, London.

they possess a relevant hazardous property or properties (see Table 23.1 in Chapter 23).

There is a close link between the Hazardous Waste Directive, the Directive on the Classification, Packaging and Labelling of Dangerous Substances (67/584/EEC) and the Directive on the Classification, Packaging and Labelling of Dangerous Preparations (83/467/EEC), which are implemented in the United Kingdom through the Chemicals (Hazard Information and Packaging for Supply) Regulations 2002 (CHIP3) SI 2002 No. 1689. The link between special waste and CHIP3 is buried in Schedule 2 Part IV of the Special Waste Regulations 1996. This ties the classifications of hazardous substances in the waste to those listed in the Approved Supply List (ASL) or, if they are not listed, to classifications derived in line with methods given in the 'Approved Guide', both of which documents were prepared under the CHIP2 regulations.

Storage

There are only a small number of licensed facilities in the country for disposal of hazardous waste, and the cost of transport and disposal of hazardous waste has increased several-fold in recent years. This has resulted in an increasing number of storage facilities within the premises of the waste producers, as well as the installation of transfer stations and bulking stations, being licensed for this purpose by the EA and SEPA so that economical loads are transported for disposal.

Under the Special Waste Regulations 1996 the deposit on the premises on which it is produced of special waste pending its disposal elsewhere requires a licence, other than in the following quantities:

1. liquid waste of a total volume of not more than 23 000 l deposited in a secure container or containers; and either
2. non-liquid waste of a total volume of not more than 80 m³ deposited in a secure container or containers; or
3. non-liquid waste of a total volume of not more than 50 m³ deposited in a secure place or places.

These regulations are under review and will be changed in 2004.

A secure container or place is one designed or adapted so that, as far as it is practicable, waste cannot escape from it, and members of the public cannot have access to the waste contained within it. This means that safe storage of all hazardous waste is essential, even when the quantities are below the licensable limits. This is achieved by the proper segregation and labelling of all containers, and storing them in bunded areas or impervious areas so that no risk to water resources exists. The site is made secure by providing proper fencing and lockable gates.

Collection

The collection of hazardous waste is, and must be, carried out by operators who have knowledge of the properties of the waste and have experience of handling such waste so that they know what equipment and vehicles should be used. It is essential that two or more consignments of hazardous waste are not mixed during transportation. There have been serious incidents when two wastes, upon mixing, have resulted in a dangerous chemical reaction.

Some waste authorities, such as the Corporation of the City of London on behalf of most London boroughs, provide a specialist collection service for small quantities of hazardous chemicals. The advantage of such a service is that hazardous chemicals are then not so liable to be dumped in dustbins, streams or lay-bys where the risk to people and the environment is high.

Disposal

The following methods are used for the disposal of hazardous waste.

1. Landfilling. A significant number of hazardous wastes have previously been disposed of by co-disposal with household waste. This practice ceased in July 2004 after which hazardous wastes could be deposited in only a very few remaining sites permitted to accept hazardous wastes.

2. Incineration. Incineration of organic hazardous waste is the best method of disposal provided that the incinerator is capable of destroying such chemicals by proper design in terms of temperature, retention time and oxygen intake, and the operation is well controlled. Directive 94/67/EC dealing with the incineration of hazardous waste was implemented in Great Britain by the Environmental Protection (Prescribed Processes and Substances) (Amendment) (Hazardous Waste Incineration) Regulations 1998 but is now superceded by the Waste Incineration (England and Wales) Regulations 2002. Transition will be completed by 28 December 2005.

3. Chemical fixation. Hazardous waste, particularly inorganic waste, is capable of being mixed with suitable chemicals, such as cement, so that the polluting waste is bound chemically or physically within them and no polluting leachate is generated.

4. Co-disposal with sewage. Certain chemical wastes can be treated by controlled feeding into the flow of sewage works. The rate of deposit is controlled in such a manner that it does not kill micro-organisms in the treatment works. The micro-organisms then degrade these wastes into harmless components.

REFERENCES

1. *Report of the Working Party on Refuse Storage and Collection* (1967) HMSO, London.
2. Key Report (1970) *Report of the Technical Committee on the Disposal of Toxic Waste*, HMSO, London.
3. Department of the Environment (1996) *Managing Waste: The Duty of Care. A Code of Practice*, HMSO, London.
4. Department of the Environment (1991) *Environmental Protection Act 1990 Section 34, 'The Duty of Care'*, Circular 19/91, HMSO, London.
5. Environment Agency (2003) *Hazardous Waste: Interpretation of the Definition and Classification of Hazardous Waste*. Technical guidance WM2 EA Bristol (also available as a searchable pdf on request).
6. Department of the Environment, Transport and the Regions (1998) *Planning and Pollution Control*, PPG 23, Stationery Office, London.
7. Department of the Environment (1994) *Landfill Completion – a' Technical Memorandum Providing Guidance on Assessing the Completion of Licensed Landfill Sites*, Waste Management Paper No. 26A, HMSO, London.
8. Department of the Environment (1994) *Licensing of Waste Management Facilities*, Waste Management Paper No. 4, HMSO, London.
9. Department of the Environment (1994) *Environmental Protection Act 1990 Part 2. Waste Management Licensing*, Circular 11/94, HMSO, London.
10. Department of the Environment (1994) *Sustainable Development: the UK Strategy*, HMSO, London.
11. Department of the Environment (1995) *Making Waste Work: a Strategy for Sustainable Waste Management in England and Wales*, HMSO, London.
12. Department of the Environment (1991, revised 1998) *Recycling of Waste*, Waste Management Paper No. 28, HMSO, London.
13. Department of the Environment (1992) *The Environmental Protection (Waste Recycling Payments) Regulations*, Circular 4/92, HMSO, London.
14. Department of the Environment (1988) *The Control of Pollution Act 1974 – The Collection and Disposal of Waste Regulations*, Circular 13/88, HMSO, London.
15. Environment Agency (1998) *Proposals for a National Waste Survey*, EA, Bristol.
16. Environment Agency (1998) *An Action Plan for Waste Management and Regulation*, EA, Bristol.
17. Health and Safety Executive (1985) *Safety in the Use of Refuse Compaction Vehicles*, Guidance Note PM52, HMSO, London.

18. Government White Paper (1990) *This Common Inheritance – Britain's Environmental Strategy*, HMSO, London.

19. Department of the Environment (1991) *Code of Practice on Litter and Refuse*, HMSO, London.

20. Royal Commission on Environmental Pollution (1993) *Seventeenth Report, Incineration of Waste*, HMSO, London.

21. Department of the Environment (1986, revised 1990) *Landfilling Wastes*, Waste Management Paper No. 26, HMSO, London.

22. Department of the Environment (1991) *Landfill Gas – a Technical Memorandum Providing Guidance on the Monitoring and Control of Landfill Gas*, Waste Management Paper No. 27, HMSO, London.

23 World Health Organization (1985) *Management of Waste from Hospitals*, WHO, Geneva.

24. United States Environmental Protection Agency (1986) *Guide for Infectious Waste Management*, EPA, Washington, DC.

25. London Waste Regulation Authority (1989) *Clinical Waste – an Appraisal*, LWRA, London.

26. Health and Safety Commission (1982, revised 1994) *The Safe Disposal of Clinical Waste*, HMSO, London.

27. Department of the Environment (1983, revised 1995) *Clinical Wastes: a Technical Memorandum on Arisings, Treatment and Disposal*, Waste Management Paper No. 25, HMSO, London.

28. British Standards Institution (1987) BS 3316:1987 Part 4. Code of Practice for the design, specification, installation and commissioning of incinerators for the destruction of hospital waste, BSI, London.

29. Department of the Environment (1986) *The Government's Response to the 11th Report of the Royal Commission on Environmental Pollution*, HMSO, London.

30. Department of the Environment (1978) *Waste from Manufacturers of Pharmaceuticals, Toiletries and Cosmetics*, Waste Management Paper No. 19, HMSO, London.

31. Department of the Environment (1988) *Third Report of the Hazardous Waste Inspectorate*, HMSO, London.

32. Department of the Environment (1987) *Special Wastes – a Technical Memorandum Providing Guidance on their Definition*, Waste Management Paper No. 23, HMSO, London.

38 Radon in buildings

Alan Blythe and Jon Miles*

INTRODUCTION

Although radon was first discovered by the German chemist Friedrich Ernst Dorn in 1901, relatively modest importance was given to it in the United Kingdom until 1984, when the tenth report of the Royal Commission on Environmental Pollution [1] recognized the serious threat posed to health from radon, and the International Commission on Radiological Protection (ICRP) published new principles [2] for limiting public exposure.

The results of the National Radiological Protection Board's (NRPB's) national and regional surveys of radon in houses in 1982–5 [3,4] prompted the government to issue a ministerial statement in January 1987 [5], which detailed an action plan for tackling domestic radon.

Radon had at last 'come of age' and many more surveys – including three organized by the then Institution of Environmental Health Officers (now CIEH) [6–8] – ensued. New, more exacting standards for public protection have been developed and radon now has a justifiably high profile in the indoor air pollution arena.

BASIC PROPERTIES

Radon is a tasteless, odourless, invisible, noble gas that can only be detected using specialized equipment. It occurs naturally as three different isotopes: ^{222}Rn, ^{220}Rn and ^{219}Rn. It is a chemically inert, but alpha-radioactive, gas produced within both the uranium and thorium radioactive decay chains (Fig. 38.1). ^{222}Rn, commonly known as radon, is produced from the decay of ^{226}Ra in the ^{238}U decay chain; ^{220}Rn, commonly known as thoron, is produced from the decay of ^{224}Ra in the ^{232}Th decay chain. The dispersion of these gases is quite different owing to their different half-lives: 3.8 days for radon compared with 55 seconds for thoron.

Natural uranium also contains 0.7% of the isotope ^{235}U, which has a decay chain in which ^{219}Rn is produced. However, ^{219}Rn has a half-life of only 4 seconds and, together with its very low natural abundance, this means that it can be effectively ignored as a source of radiation exposure in the home.

In contrast to thoron, radon is able to move considerable distances from its parent during its lifetime. It can diffuse readily out of surface soil into the atmosphere and into the basements and living areas of houses. For this reason, it is the ^{238}U decay

*This chapter was originally written by Alan Blythe for the 18th edition of *Clay's Handbook*. Scientific and technical issues were updated for the 19th edition by Jon Miles.

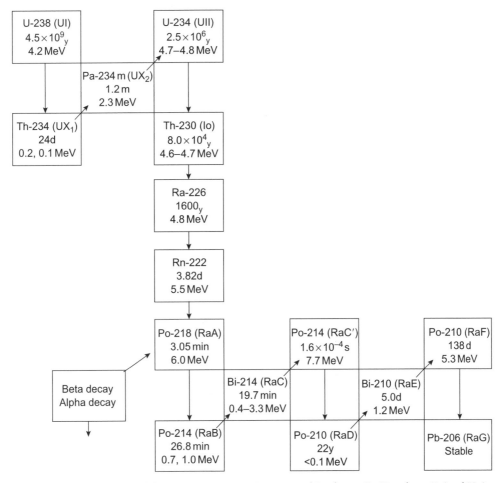

Fig. 38.1 Principal decay scheme of the uranium series. (Courtesy of Professor D. Henshaw, Bristol University.)

chain and radon that are considered more important in terms of environmental alpha-radioactivity than the ^{232}Th decay chain and thoron.

Radon contributes half of the average annual dose of ionizing radiation that the UK population receives (Fig. 38.2); in high radon areas this proportion may rise to a staggering 90%, or more, for exposures above the current Action Level (see later).

SOURCES OF RADON

All rocks and soils emit ionizing radiation to some extent because they contain traces of the alpha,

beta and gamma emitting nuclides within the two uranium decay series, the thorium decay series and the beta and gamma emissions of ^{40}K (radioactive potassium) [9].

Alpha particles are the least penetrating of all these types of radiation, the most energetic travelling only 10 cm in air. Beta radiation is absorbed by around 0.5 cm of rock or soil. Therefore, from the body of the earth itself and from within building materials it is only the gamma radiation that provides a direct background radiation dose to the occupants in houses, whereas gaseous radon diffuses into houses where it may be inhaled, providing an alpha-radiation dose to the body, chiefly the lungs.

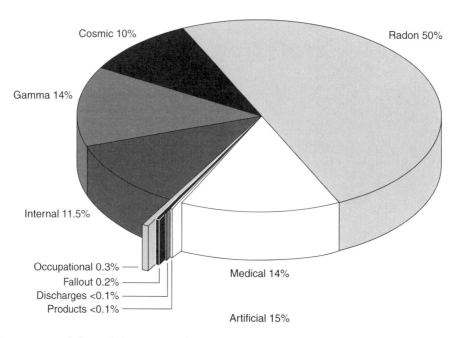

Fig. 38.2 Average annual dose of the UK population (2.6 mSv overall). (Courtesy of the National Radiological Protection Board.)

As uranium is a constituent of around a hundred minerals, it is not surprising to find traces of it in diverse geological formations [9]. The more soluble forms, such as pitchblende, coffinite and uraninite, may be found in alluvial deposits and sediments, while in igneous rocks uranium occurs in accessory minerals along grain boundaries and in crystal lattice defects. Uranium is normally found in concentrations of less than 1 part per million (ppm), but where it has been leached and precipitated in a reducing environment, for example in some sandstones and iron sulphites, it can attain concentrations of 0.05–0.2% uranium. Finer grain sediments, such as shales, concentrate uranium in organic matter at generally 30–60 ppm (much higher in some Upper Cambrian formations). Phosphatic sediments contain deposits of 10–60 ppm uranium, or more, while lignites formed under reducing conditions show greater variations in uranium content up to 1%. It should be remembered that uranium is mobile as organic complexed or colloidal particles, especially under oxidizing conditions in acid or carbonate-rich waters. See Table 38.1 for the content of uranium in different rock types.

The element thorium is present both as the precursor in the ^{232}Th decay chain, and as intermediate nuclides in the ^{238}U decay chain. The long half-lives of these nuclides means that the geological behaviour of thorium must be taken into account.

Granites are by no means the most radioactive rocks, and some shales contain one hundred times more uranium than the most highly radioactive granites. Acid volcanics cover large areas and have equivalent uranium and thorium contents to granite. The highly evolved and fissured potassium-rich granites in Cairngorm and Etive in Scotland emit relatively high levels of gamma radiation, as do the granite areas of the Lake District and the southwest of England. However, it should be appreciated that even higher levels of radioactivity may be associated with sedimentary rocks, such as phosphatic limestone of lower carboniferous age, and the black shales and marine shale bands in the coal measures.

Table 38.1 Uranium contents in ppm and typical ranges in different rock types

Rock type	Mean ppm	Range
Igneous		
Mafic	0.8	0.1–3.5
Diorite and quartz diorite	2.5	0.5–12
Silicic	4.0	1.0–22
Alkaline intrusive		0.04–20
Sedimentary		
Shale	3.0	1–15
Black shale		3–1250
Sandstone	1.5	0.5–4
Orthoquartzite	0.5	0.2–0.06
Carbonate	1.6	0.01–10
Phosphorite		50–2500
Lignite		10–2500

Source: Bowie and Plant [9].

Uranium that goes into solution is capable of migrating over long distances to be concentrated in both surface and ground waters, depending on the particular conditions that affect the rate of weathering. Radium is not as mobile in surface oxidizing conditions as uranium, but it can go into solution and migrate significant distances before being absorbed into other oxides or clay minerals, such as iron and magnesium. Wherever ^{226}Ra is found, its important immediate decay product, ^{222}Rn, will be produced. Since the half-life of ^{222}Rn is 3.8 days, it can diffuse easily through porous rocks, ground or surface waters, through faults and fissures and into thermal springs. As would be expected, radon levels generally relate to locally occurring radium levels [9] (Fig. 38.3).

HEALTH EFFECTS AND RISKS

When radon gas is breathed in, most of it is breathed out again without undergoing radioactive decay. But radon is accompanied by its short-lived decay products, some of which are also alpha particle emitters. These decay products are atoms of solid elements, and when these are breathed in, most of them will be deposited on the lungs and airways. Almost all of them decay within an hour or two after deposition, emitting radiation. It is the decay products of radon, rather than radon gas itself, that deliver almost all of the radiation dose that is attributed to radon.

Unsurprisingly, irradiating the lungs with alpha particles causes a risk of lung cancer. There remain uncertainties over the exact magnitude of the risk of exposure to radon, but it is clear that alpha particle irradiation of the bronchial epithelium by short-lived radon decay products is the most significant cause of radiation-induced lung cancer. Findings published in 1997 [10] indicate a progressively increasing risk of lung cancer with increasing indoor radon exposures, based on a meta-analysis of eight case-control epidemiological studies, including 4000 cases and 6000 controls. Later epidemiological studies, including one carried out in the high radon area of southwest England [11] have supported the earlier work. This research supports the international consensus of opinion on the health risks of radon and the utility of radon mitigation programmes around the world. In addition, it appears that there is a synergistic link between radon and smoking in increasing the risk of lung cancer, as shown in Table 38.2. It is unclear whether the joint effect is multiplicative or sub-multiplicative, but it is certainly greater than additive. In other words, the joint effect of radon and smoking is greater than the effect of either factor alone.

Using the results of such studies, the risk of people exposed to radon in their homes can be calculated. At the present radon Action Level of 200 Bq/m³ (see below), the estimated lifetime risk is 3%. This figure is calculated for the general population, including both smokers and non-smokers, but the risk is estimated to be higher for smokers (10%) than for non-smokers (1%), based on a multiplicative model. It may therefore be deduced that while the current Action Level for radon in homes may be quite acceptable in public protection terms for non-smokers, moderate smokers exposed to the UK average radon concentration of 20 Bq/m³ could still run a lifetime risk of premature death equivalent to their chances of dying in a motor vehicle accident (or, more significantly, 10 times that risk at the Action Level).

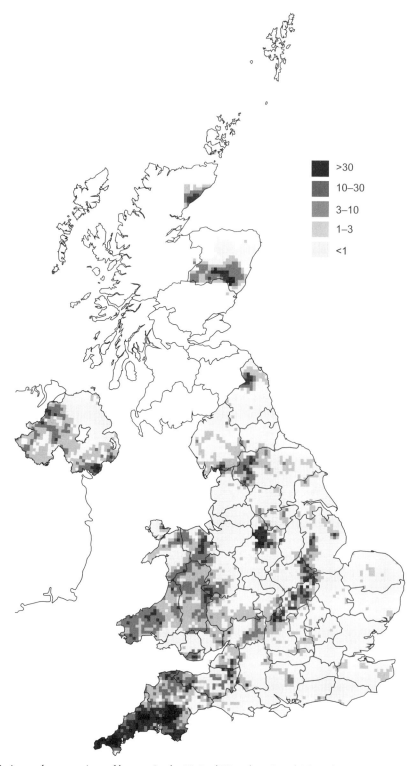

Fig. 38.3 Estimated proportion of homes in the United Kingdom in which radon concentrations exceed the Action Level in each 5-km Ordnance Survey grid square. (Courtesy of the NRPB.)

Table 38.2 Estimated lifetime risk of lung cancer from lifelong exposure to radon at home (%)

Average concentration (Bq/m³)	*Whole*	*Smokers*	*Non-Smokers*
20	0.3	1	0.1
100	1.5	5	0.5
200	3.0	10	1
400	6.0	20	2

Source: Documents of the NRPB (1990).

Perhaps this rather startling synergistic effect between smoking and radon should justify the offering of further special assistance and advice to inveterate smokers in high radon areas.

It is worth bearing in mind that these risks are calculated for lifetime exposure at the same radon level, though most people do not live in the same house for their whole life. The effect of people moving house between high and low radon houses is that the lung cancer risk caused by high radon levels is spread over a larger number of people than would be the case if they did not move house.

In the green paper *Our Healthier Nation* [12], the government recognized the concerns about the health effects of radon and stated:

100000 houses in the United Kingdom have high levels of radon gas. People want to be confident that if they are acting responsibly and protecting themselves from cancer by eating well and not smoking, then the Government and local authorities are actively engaged in reducing the health risk in those areas where radon gas in homes can increase the chances of developing lung cancer.

MONITORING AND EXPOSURE MECHANISMS

Monitoring techniques

Choice of measurement method

Because radon levels can vary greatly from hour to hour, and are usually higher at night than during the day, measurement methods that sample over a

few hours are a very poor guide to long-term average radon concentration, which is, after all, the significant factor we are seeking to assess.

Indoor radon levels also vary from day to day and month to month, depending on the weather and on the way occupiers use their houses. The monthly average radon levels vary with the seasons because people's homes are used in different ways at different times of the year: in winter, heating systems generate a relatively high negative pressure inside living areas, which tends to draw more radon-laden air into the living areas from the ground. At the same time, tightly closed doors, windows and draught stripping systems are all working to optimum effect in reducing air change and thereby innocently enhancing radon concentrations. Thus, radon concentrations tend to be at their highest in winter. Measurements made over this period would obviously give a falsely high result unless adjusted for this effect. Similarly, radon measurements made in summer are liable to give a falsely low reading because householders tend to open windows and doors and heating systems are not working at all. Hence, there is little or no negative internal air pressure in the house and higher air change rates dissipate the radon present to quite low levels. Because of these variations in radon levels, the best methods for assessing long-term average concentrations require the use of detectors, which average over periods of months, and adjust for typical seasonal variations.

In certain circumstances, for instance when a quick assessment is required prior to purchasing a property, screening measurements carried out over a few days have some value. They cannot, however, be expected to predict long-term average radon levels accurately. It is worth reiterating that the NRPB specifies monitoring to an approved standard **for at least three months** before the result can be regarded as being valid to justify application of remedial measures. The grant system, as embodied in the Housing and Local Government Act 1989, cites a similar three-month period for testing before a grant may be approved for homes above the current Action Level. NRPB has calculated seasonal adjustment factors for the three-month radon results taken at any time throughout the year, which may be utilized to estimate the average concentrations over the year. Measured concentrations are multiplied by the appropriate factors, as detailed in Table 38.3.

It has been reported that there are firms claiming to give householders an accurate assessment of radon concentration on the basis of a weekend's electronic meter measurements, or even on the basis of instantaneous samples. It is highly likely that such readings could generate false positives and negatives and may cause householders unnecessary anxiety and expense in the first case, and give false reassurance in the other.

For all these reasons, it is **essential** that environmental health departments support the use of accurate methods of radon measurement. NRPB runs a validation scheme for laboratories offering long-term radon measurements (see section on local authority radon programmes), and validated laboratories should be used before any decision is made to take remedial action to reduce radon levels (Table 38.4).

Direct 'active' instruments

These are sensitive electronic instruments that give a direct, almost instantaneous, reading related to the ambient radon or radon decay product concentration during the relatively short period of measurement.

Such instruments are most often used in mines and caves, where concentrations are usually stable, to determine relatively short-term occupational exposure. They are not normally used in houses, because variations in radon levels from hour to hour and day to day make instantaneous

Table 38.3 Seasonal adjustment of three-month radon results to obtain average concentrations over the whole year

Period of deployment	Adjustment factor
January to March	0.74
February to April	0.83
March to May	0.96
April to June	1.15
May to July	1.45
June to August	1.64
July to September	1.59
August to October	1.28
September to November	1.04
October to December	0.88
November to January	0.76
December to February	0.73

Note:
Measured concentrations are multiplied by the appropriate factor.

Table 38.4 Laboratories currently validated for radon measurements in homes

Laboratory	Telephone	Fax
DSTL Radiation Protection Services Institute of Naval Medicine Gosport, Hampshire PO12 2DL	02392–768152	02392–511065
Gammadata PO Box 36, Droitwich WR9 7ZB	01905–724840	01604–646456
Radon Centres Ltd Grove Farm, The Grove, Moulton Northampton NN3 7UF	01604–494118	01604–646456
NRPB Chilton, Didcot, Oxfordshire OX11 0RQ	01235–831600	01235–833891

Note:
Further details of the validation scheme are given in NRPB-M1140: *Validation scheme for laboratories making measurements of radon in dwellings: 2000 Revision*, available from the NRPB.

measurement results a poor guide to long-term average levels.

Radon gas monitors use a filter to exclude ambient radon decay products and sample the air continuously, averaging over a time period that is typically 1 hour. The measurement is repeated over successive time periods. Results are either printed onto paper or stored in a memory for later readout. This allows the variation in radon levels from hour to hour to be determined. They are sometimes used in workplaces to determine whether the radon concentration is lower during working hours than at other times.

Radon decay product monitors incorporate a small air pump, which draws air containing radionuclides through a filter where the decay products are deposited for counting. Pre-determined counting intervals are used, during which either the total number of alpha particles detected is counted, or the numbers of alpha particles of different energies are counted. Standard calculations allow the overall radon decay product concentration or the concentration of the individual decay products to be determined.

Passive track-etch detectors

These long-term monitors are now marketed by various companies and are currently recognized as the NRPB preferred method for accurately estimating radon exposure, and hence dose, over time.

The complete detector comprises a small plastic pot containing a postage stamp-sized rectangle of radon-sensitive plastic – a polymer of diethylene glycol bis (allyl carbonate) known as CR-39 or PADC. The tightly-lidded pot protects the sensitive plastic from physical damage and allows radon to diffuse inwards; alpha particles produced from radon decay invisibly damage the detecting plastic, in which chemical etching later reveals the damage as pits in its surface (Fig. 38.4). The number of these alpha 'tracks' correlates directly with radon concentration over the period of detector exposure. Ideally, measurements should be carried out for a minimum of three months and in different rooms in the house, so that natural variations can be integrated into the estimate of annual average radon concentration (Fig. 38.5).

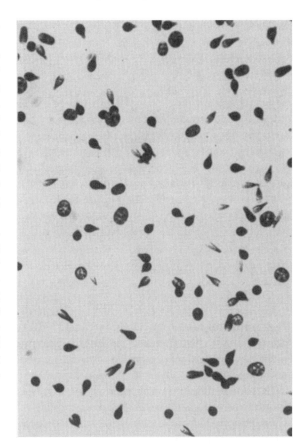

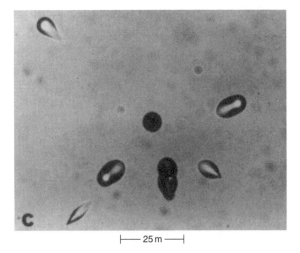

├── 25 m ──┤

Fig. 38.4 Radon track – etch pits. (Courtesy of Professor D. Henshaw, Bristol University.)

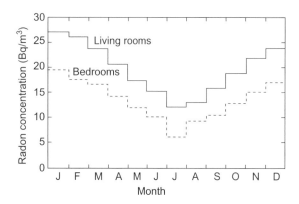

Fig. 38.5 Annual variation in radon including ground floor living rooms and first floor bedrooms. (Source: Wrixon *et al.* [13].)

Diffusion-barrier activated charcoal detectors

Perforated canisters of activated charcoal (25–100 g) are exposed to radon-laden air for the measurement period. Radon is adsorbed onto the charcoal and retained while it decays. The canisters must be kept sealed and dry until opened for exposure. Indoor measurements usually cover a relatively short period (2–7 days), which is strictly recorded for subsequent laboratory use. The detector is returned to the originating laboratory for measurement before too much of the radon decays.

The laboratory measures the gamma ray emissions from the radon decay products and calculates the original radon gas concentration. The results of these measurements can give an accurate estimate of the radon level over the period of exposure, particularly if the radon level is constant. While excellent 'rapid' screening results may be obtained with the method, charcoal canisters are not suitable for measurement of long-term mean concentrations.

Electrets

These are passive devices, properly described as electret ion chambers. The electret itself holds an electrostatic charge which is gradually neutralized by the ionization of the air by alpha particles emitted by radon and its decay products. Measuring the charge on the electret at the beginning and end of an exposure allows the average radon concentration to be calculated. In making this calculation, an allowance must be made for ionization caused by the natural background of gamma rays from rocks and building materials. Different types of electret are available, suitable for measurements over periods of a few days to a few months. Care must be taken with this device, as dropping it can cause a partial discharge of the electret, and an overestimation of the radon concentration.

Exposure mechanisms

The main factors that affect radon levels in buildings include varying combinations of the following.

1. The content of radon-producing radium in underlying ground. This is likely to be highest in fissured igneous rocks such as granite or porous rocks such as karstified limestone, but many other rock types can also cause elevated radon levels.
2. The presence of suitable routes for radon-laden air to seep into the building, for example, cracks, service entry points, poor construction and joints.
3. The presence of sufficient negative pressure inside the building to suck soil gas carrying radon inwards – usually caused by heating, wind effects and sometimes by inappropriate mechanical ventilation.
4. Restricted air change inside the building – energy conservation measures, such as double-glazing and draught stripping without the provision of trickle vents may fail to provide adequate dilution to dissipate incoming radon.
5. The habits of the occupants in terms of comfort standards and occupancy rates of different rooms in relation to the ground floor level.
6. The time of day or night and season – relative radon levels are usually higher at night (very early morning) and during the winter months.

The prevalence of these factors, acting singly or in various combinations, can cause wide differences in measured radon concentrations within houses in the same area or even in the same street.

It can, therefore, never be assumed that single-point sampling in defined locations (town, village, hamlet or street) is representative of houses or workplaces in that vicinity. The more tests undertaken, the more accurate the radon picture will become.

RADON SURVEYS AND REMEDIAL PROGRAMMES

NRPB and central government surveys

The first comprehensive survey of indoor radon levels throughout the United Kingdom was carried out in the 1980s by NRPB with support from the Commission of the European Communities [13]. Further surveys over the following years were mostly funded by the UK central government and carried out by NRPB. These surveys had two purposes: to identify houses with high radon levels, and to allow the mapping of high radon levels across the country. By 2003, radon levels had been measured in more than 400 000 homes. A complete radon map of England and Wales was published in 2002 [14] and of Northern Ireland in 1999 [15]. A radon map of limited parts of Scotland was published in 1993 [16], and work to complete the map of Scotland started in 2003. A joint radon map of the different parts of the United Kingdom is shown in Fig. 38.3.

The government-sponsored surveys identified a large number of homes – more than 40 000 – with radon concentrations at or above the Action Level. These householders were advised to reduce their radon concentrations, and provided with information on how this could be achieved. Unfortunately, research indicated that only 10–20% of the householders were taking this advice. In response to this disappointing statistic, the former Department for Environment, Transport and the Regions (now Department for Environment, Food and Rural Affairs – DEFRA) decided to alter the direction of the radon programme in England by refocusing at local authority rather than central government level. This programme is discussed here.

Local authority radon programmes

In 1987–90, the then Institution of Environmental Health Officers (IEHO now CIEH) organized three surveys of radon levels in almost 5500 houses in 27 English counties and 6 Welsh counties, finding a significant number of homes above the radon Action Level [6–8]. These surveys also served to highlight the fact that radon problems are not confined to granite and limestone areas, but occur in many parts of the country.

Since then, many local authorities have carried out their own surveys of radon in homes in their areas, which provided extra information over that available from the surveys funded by central government. Local authority surveys have normally included a provision that the results should be available to the authority. The centrally funded surveys provided summary results for each authority, but the results in individual houses were not available to local authorities as they were confidential to the householder. This caused difficulties for local authorities trying to tackle the radon problems in their area. This difficulty has been removed in the more recent radon programme organized by DEFRA, discussed in the next paragraph.

Environmental Health departments contemplating expansion of radon surveys in their districts should first consult DEFRA, the National Assembly for Wales, the Scottish Executive Development Department Housing Division or the Environment and Heritage Service of the Department of Environment for Northern Ireland, to find out whether central government funding will be available. Advice and help on planning and running surveys is available from NRPB.

Whether or not central funding is available, Environmental Health departments should be aware of the Defra Radon Good Practice Guide [17], which was drawn up in the light of pilot radon measurement and remediation campaigns carried out by Derbyshire Dales, Mendip and Cherwell District Councils in partnership with the government department. These pilot campaigns showed that enthusiastic local authority campaigns were much more successful than centrally organized campaigns in persuading householders to have radon measurements and to install effective radon remedial measures where necessary.

An evaluation report on these campaigns concluded:

> Response rates were generally high and remediation was successfully speeded up and increased in all three areas, with the total of remediated properties being raised by up to 100% on previously achieved numbers. Key factors in this success included local and pro-active delivery, effective targeting, appropriate timing of publicity, optimum use of technical expertise and sustained support and follow up. [18]

The campaigns targeted offers of a free measurement at householders in the areas of greatest risk, defined as 5% or greater probability of houses exceeding the Action Level. In addition, householders of homes already or subsequently identified as having a high radon level were offered free advice on remediation and a free re-measurement to test the efficiency of any works completed. To enable the local authorities to fulfil their role, intensive training sessions were organised by NRPB and the Building Research Establishment (BRE). DEFRA provided support on public relations and publicity.

The Radon Good Practice Guide provides detailed advice on developing an effective programme to identify high radon houses and persuade the householders to install effective remedial measures. It has been the basis for subsequent campaigns funded by DEFRA in 32 local authority areas in England. The campaigns have included the following phases:

1. Initial training by NRPB and BRE
2. Awareness raising
3. Targeting of first-time testers
4. Targeting of known high homes
5. Remediation phase
6. Exit strategy.

To avoid the earlier difficulties about the release of individual results to local authorities, care is taken to ensure that householders are fully aware that the main focus of the radon programme is the Environmental Health Department of the local council. All replies to initial letters, requests for tests and other queries are routed through the council and it is clearly stated in correspondence that the council will receive the results. Similar programmes, at various stages, are in operation in other parts of the United Kingdom.

It is important that the laboratories chosen to supply radon detectors for local authority surveys are capable of performing accurate measurements, and report the results in a suitable manner. NRPB runs a validation scheme for laboratories which ensures that those validated meet the required standards [19]. The laboratories currently validated are listed in Table 38.4. House renovation grants (see next section) may be available from local authorities to assist low-income households with radon remediation costs, subject to circumstances.

STANDARDS – ACTION LEVELS AND LIMITS, AFFECTED AREAS FOR HOMES AND OCCUPATIONAL EXPOSURE

Various national governments and international agencies – most notably the ICRP and the World Health Organization (WHO) – have sought over the years to establish and revise radiological protection standards as the state of knowledge concerning risk has grown.

Existing homes

There are various national and international recommendations for radon limits in existing homes (see Table 38.5), ranging from a 'low' of $150 \, Bq/m^3$ in the USA and Luxembourg, to a 'high' of $1000 \, Bq/m^3$ in France and Switzerland [20,21]. Note that the unique reference level of $20 \, Bq/m^3$ in the Netherlands is an exception, and is a target rather than an Action Level.

The UK's current Action Level of $200 \, Bq/m^3$ for existing homes was adopted by the government in January 1990 after the NRPB had revised its assessment of risk in line with new experimental and epidemiological evidence from the ICRP, the US National Research Council

Table 38.5 European (and some world) radon policies: radon reference ('Action' Levels) for dwellings

Country	Existing dwellings (Bq/m³)	Future dwellings (Bq/m³)	Proofing grants available
Australia	200	200	
Austria	400	200	
Belarus	200	200	
Belgium	400	400	Yes[c]
Canada	800	800	
Czech Republic	500	250	Yes
Denmark	400	200	
Estonia	400	200	
Finland	400	200	Yes
France	1000[a]		
Germany	250	250	
Greece	400	200	
Ireland	200	200	Yes
Latvia	300	300	
Lithuania	400	200	
Luxemburg	150	150	Yes[c]
Netherlands	20	20	
Norway	200	200	
Poland	400	200	
Russia	400	200	
Slovak Republic	500	250	
Slovenia	400	200	
Sweden	400[a] 200[b]	200[a]	Yes
Switzerland	1000[a] 400[b]	400[a]	
United Kingdom	200	200	Yes[c]
Yugoslavia		200	200
USA		150	Same as outdoors

Notes:
a Regulatory limit.
b Recommended upper limit for remediation ('Action Level').
c In theory, **but** means tested.

Source: Amended from Colgan, P.A. [20] and Akerblom, G. [21].

(NRC) and the United Nations Scientific Committee on the Effects of Atomic Radiation (UNSCEAR) [22].

Housing Health and Safety Rating System

Radon was not included in the 1990 Housing Fitness Standard, despite the environmental health profession's opinion that radon in sufficient concentration could indeed make a house uninhabitable. The Housing Health and Safety Rating System [23] (HHSRS) has been developed by the Office of the Deputy Prime Minister (ODPM) to replace the Housing Fitness Standard. The aim of the system is to provide a logical and practical means of assessing and grading dwellings from a health and safety perspective. The HHSRS covers 27 categories of risk, including radiation exposure from radon.

At the time of writing the HHSRS system is on trial. An Act of Parliament (expected 2003–4) will be required to implement it. The system is based on the assumption that an Equivalent Annual Risk of Death of 1 in 10 000 is an acceptable risk, while 1 in 1000 is unacceptable. Such risks are translated into HHSRS hazard scores: these two risks relate to hazard scores of 100 and 1000, respectively. Under this system, a hazard score of less than 100 should be considered acceptable, while a score of 1000 or more should be considered unacceptable.

Table 38.6 gives the HHSRS scores allocated to different radon concentrations, to be added to the HHSRS scores attributed to other risks. The rating system implies that radon concentrations over 220 Bq/m³ are unacceptable.

New homes

A narrower band of international recommendations applies to radon limits for new homes, ranging from 150 Bq/m³ in Luxembourg to 800 Bq/m³ in Canada. The NRPB's recommendation for new homes is that they 'should be so constructed that radon concentrations are as low as reasonably practicable and at least below 200 Bq/m³ [22]'.

Affected areas

This concept was introduced in 1990 by the NRPB [22] 'so as to concentrate effort on radon where it is most needed'. While recognizing that any such criteria are likely to be arbitrary, it proposed that parts of the country with 1% or more probability of present or future homes being

Table 38.6 Average annual likelihood and health outcomes for all persons aged 75 years or over, following exposure to radon

Measured radon level, Bq/m³	Average likelihood, 1 in	Spread of health outcomes				Average HHSRS scores
		Class I extreme (%)	Class II severe (%)	Class III serious (%)	Class IV moderate (%)	
800	277	90	10	0	0	3285
400	518	90	10	0	0	1757
200	1000	90	10	0	0	910
150	1322	90	10	0	0	688
100	1961	90	10	0	0	464
50	3902	90	10	0	0	233
25	7853	90	10	0	0	116

Note:
This age group was chosen because lung cancer risk rises with age and duration of exposure, and this age-group gives a high ('most vulnerable') estimate of the annual risk of lung cancer.

Source: [23].

above the Action Level of 200 Bq/m³ should be regarded as Affected Areas.

In 1990, Cornwall and Devon were formally declared Affected Areas based on the results of over 8000 radon tests in the two counties [24]. Different parts of the United Kingdom were declared as Affected Areas as measurement and mapping proceeded. The most current maps of Affected Areas at the time of writing are given in various NRPB publications [14–16]. Affected Area status encourages homeowners to have radon measurements made. Affected areas can only be defined by the NRPB, and are based on the results of radon measurements in homes.

Occupational exposure

Exposure to radon in workplaces is covered by the Ionizing Radiations Regulations 1999 (IRR99) [25,26], which replaced the 1985 regulations. An important element of the new regulations is the formal requirement for an assessment of risks. While this has always been an obligation on employers, it was previously required by the Management of Health and Safety at Work (MHSW) Regulations which address the whole range of hazards, among which radon is often not perceived as an obvious and important issue by

employers. Inclusion of risk assessment in IRR99 brings a sharper focus to the issue of radon, which will affect all employers with premises in which there might be significant exposures. The maps of radon levels in houses are the best guide to where high levels in workplaces are likely to be found.

If employers have premises in which there may be high radon levels, the first step in assessing the risk is measurement of radon concentrations in a representative sample of locations. If a valid result of more than 400 Bq/m³ is found in any area within a workplace, the IRR99 apply. The employer must inform their local health and safety inspector of the radon results. The employer then has a choice: either to reduce the radon concentrations to below 400 Bq/m³ throughout the premises, or to submit to the requirements of IRR99 to limit the radiation exposure of workers. Generally it will be less expensive and less burdensome to reduce the radon concentrations. If the employer does not reduce the concentrations below this threshold, then the annual radiation doses to employees in units of millisieverts (mSv) must be calculated based on radon concentrations and work patterns. The actions required depend on the doses calculated:

Below 1 mSv Officially, this is *work with radiation*, the lowest level of control recognized by the IRR99. The doses are self-limiting if the areas with high radon levels are rooms with low occupancies such as stores or basements. In consultation with a *radiation protection adviser* (RPA), radiation warning signs or key control might be advised. If usage of the areas changes, the risk assessment should be repeated to make sure that no additional controls are required.

From 1 mSv to 6 mSv The IRR99 describe this as a *supervised area*, and impose certain additional requirements to those above. The employer will need to appoint an RPA and a *radiation protection supervisor* (RPS), make sure that employees are informed and trained appropriately and conduct regular radon measurements. Alternatively, if radon levels are reduced, the need for formal controls will be reduced or eliminated.

Over 6 mSv The highest designation is a *controlled area*, which requires formal administrative and physical controls. Employees receiving doses over 6 mSv a year are among the most exposed radiation workers in the United Kingdom, and may need to be designated *classified workers* and have continuous dose monitoring and annual health checks. Reducing radon levels may be cheaper and easier than applying these stringent measures.

High Peak Borough Council reported their experience of implementing the regulations in commercial premises in 2002 [27]. Their programme consisted of the following steps:

1. Developing a Radon Practice Note, establishing procedures for dealing with radon at work during inspections of business premises.
2. Compiling a database of commercial premises in the high-risk areas.
3. Sending the HSE 'Radon at Work' leaflet and advice on testing to these businesses.
4. Following up 6 months later with a more formal letter to those businesses which had not responded.
5. Inspecting premises, and serving improvement notices where it was found that the premises had cellars or poorly ventilated workrooms and the employers had not carried out radon testing.

The council found that this programme was highly effective, and that all of the premises found to have excessive radon levels were remedied. Regular re-testing every few years is needed to ensure that remedial measures continue to be effective.

REMEDIAL MEASURES AGAINST RADON INGRESS

The principal motive force behind radon ingress to buildings is the negative pressure inside buildings caused by heating and/or wind effects. The under-pressure is usually no more than a few pascals in magnitude, but it will typically draw 1 or 2 m³ of air an hour out of the soil under a house, that air being replaced in the soil by air drawn into the open ground around the house. Restricted ventilation in modern homes (double-glazing, draught stripping and sealed flues), while conserving valuable heat, limits the ability of incoming fresh air to effect significant dilution and hence radon concentrations can increase. Radon is drawn inside buildings through a variety of entry routes, as shown in Fig. 38.6.

There are five main ways of reducing the amount of radon entering a house (Figs 38.7–38.10).

The first of these methods is the most effective; the choice between the other methods depends on the characteristics of individual houses.

1. *Install a radon sump system* A radon sump is a space about the volume of a bucket under a solid floor, connected by a pipe to a fan. The fan slightly reduces the air pressure underneath the house, counteracting the slight underpressure indoors which would otherwise draw soil air into the house (Figs 38.7–38.8). Radon in soil air which is drawn out of the sump by the fan is released into the atmosphere and dispersed harmlessly. This method can typically reduce indoor radon levels by a factor of ten.
2. *Improve ventilation under suspended timber floors* Improving ventilation under suspended timber floors is beneficial in reducing the likelihood of moist air causing timber to decay as well as reducing indoor radon levels. Installing extra airbricks under floors generally produces

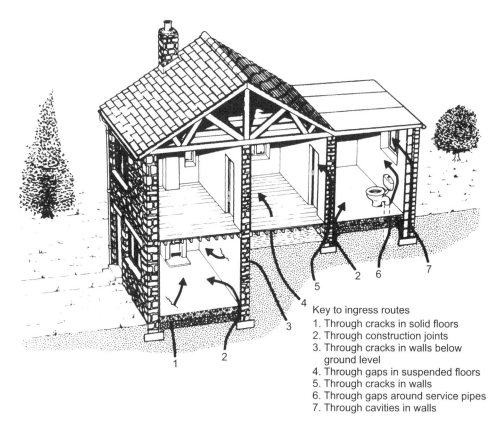

Key to ingress routes
1. Through cracks in solid floors
2. Through construction joints
3. Through cracks in walls below ground level
4. Through gaps in suspended floors
5. Through cracks in walls
6. Through gaps around service pipes
7. Through cavities in walls

Fig. 38.6 Entry of radon into buildings. (Source: DEFRA [29].)

modest reductions in indoor radon levels, and in most cases a fan will be needed to assist the air movement (Fig. 38.9).

3. *Use a positive supply ventilation system* Positive ventilation of houses involves the use of a fan installed in the loft to blow air into the house, reducing the underpressure that draws in radon. This measure is most successful in bungalows and double-glazed houses (Fig. 38.10).

4. *Seal cracks and gaps in floors* Sealing cracks and gaps in floors, though it seems an obvious way to reduce the entry of soil gas, is a method that is very difficult to implement successfully. Unless **all** cracks are sealed, it may have very little effect on radon levels. Nevertheless, when other radon remedial work is being carried out, it is worth sealing any major gaps in the floor to increase the effectiveness of systems such as sumps.

Timber floors should never be completely sealed, as this may cause them to rot.

5. *Change the way the house is ventilated* Changing the way a house is ventilated is generally the least successful of radon remedial measures. It can be successful only if permanent extra ventilators (such as trickle vents above windows) are installed at ground floor level, and left open when radon levels are highest, at night and in winter.

Table 38.7 compares the costs of some radon remedial measures with other home improvements [28].

The DEFRA publication *Radon: A Guide to Reducing Levels in Your Home* [29] provides valuable information; it explains the possibility of do it yourself (DIY) remedies for the less technical and non-arduous tasks. (See also BRE guides to radon remedial measures in existing dwellings,

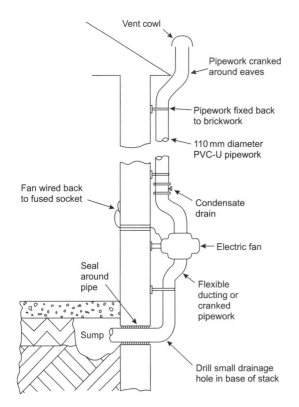

Fig. 38.7 Externally excavated sump. (Courtesy of BRE.)

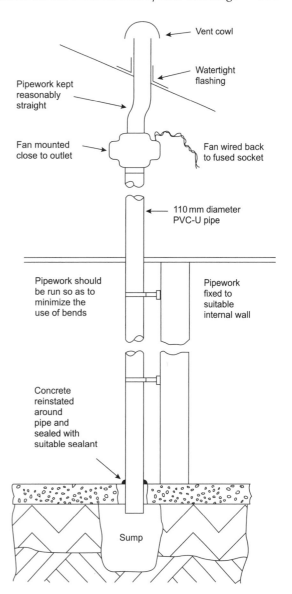

Fig. 38.8 Internal mini-sump. (Courtesy of BRE.)

which are aimed at builders who carry out remedial work and those who are competent at DIY [30,31].)

Whether carried out by the skilled amateur or reputable contractor, the essential element of all remedial work is that it must be performed to a high quality standard, as any shoddy workmanship would soon be revealed in a subsequent radon test, which is the only real way to check the effectiveness of radon-proofing measures. Also worth noting is that to qualify for a grant (see section on house renovation grants towards randon-proofing), a subsequent test should prove that radon reduction to at least below the Action Level has been attained.

GUIDANCE ON THE CONSTRUCTION OF NEW DWELLINGS

Guidance on how to avoid indoor radon problems when building new dwellings in high radon areas was originally introduced in 1988 as an extension of the Building Regulations. The guidance applied in certain high-risk areas of Cornwall and small parts of Devon. Since then, much more information on radon levels in houses around the United Kingdom has become available, and many more areas have been designated for radon protective measures in new houses.

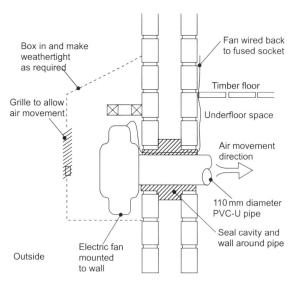

Fig. 38.9 Fan to provide increased ventilation under a timber floor. (Courtesy of BRE.)

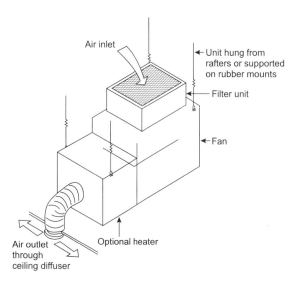

Fig. 38.10 Positive ventilation system. (Courtesy of BRE.)

The areas of the United Kingdom in which protective measures are required are set out in three parallel documents published by the BRE: BR211, Radon: Guidance on protective measures for new dwellings (which applies to England and Wales), BR376, Radon: Guidance on protective measures for new dwellings in Scotland, and BR413,

Table 38.7 Comparison of costs of radon remedial measures with other home improvements

Radon remedies	Typical costs
Fan-assisted radon sump	£500 – £1000
Positive ventilation system	£500 – £750
Mechanical ventilation under suspended floor	£350
Other home improvements	
Replace carpets	from £1000
Replace bathroom suite	from £700
Treat rising damp in one wall	£450

Source: [28].

Radon: Guidance on protective measures for new dwellings in Northern Ireland [32–34]. All three documents define two stages of protection, a first stage that is applied where the risk of high indoor radon concentrations is moderate and a second stage that is applied in the higher radon areas. There are significant differences between the guidance provided in the three documents: in Scotland and Northern Ireland, radon protective measures are required in areas where the probability of houses being above the UK radon Action Level is estimated to be 1% or more; in England and Wales the threshold is 3%. BR211, for England and Wales, is also more complex in that there are two sets of maps that define the areas in which the guidance applies. One set of maps is derived from grouping the results of house radon measurements by 5-km grid squares, and the other is derived from grouping the same results according to the geological units the houses are on. Both sets of maps must be taken account of when deciding whether new houses need radon protection.

The two stages of protection are based on the same principles in all parts of the United Kingdom, though the implementation differs to some extent because the Building Regulations are not identical for all parts of the United Kingdom. The first stage of protection requires a protective barrier to be installed right across the building, from the outside of each wall inwards. Joints in the barrier and service penetrations must be sealed to make the construction as gas-tight as possible.

The second stage of protection is applied in areas where 10% or more of houses are estimated to be above the Action Level, and requires that builders provide means of ventilation underneath the floors of houses. This can be achieved, for example, by using a suspended concrete (beam and block) floor, with vents allowing air exchange between the void under the floor and outside air. The vents allow for the addition of a fan later if necessary, but this type of construction has been found to be very successful in providing low indoor radon concentrations without the need for a fan. An alternative method of construction for use in these high radon areas is the use of a ground-supported concrete floor with a radon sump underneath, which can be connected to an extract fan if necessary. This method has the drawback that, unlike the beam and block method, it provides no reduction in radon levels unless a fan is in operation.

RADON IN WATER SUPPLIES

High indoor levels of radon are normally the result of soil air carrying radon into the building. In most cases, the radon originates in rock or soil nearby, but because it is soluble in water, it can also be carried over large distances by underground water movements. The highest concentrations of radon in water usually occur in ground water sources such as deep wells, boreholes and springs in areas of susceptible geology. Surface water generally has low concentrations.

Public water supplies derived from groundwater are often sufficiently delayed by storage and processing to allow radon to decline to acceptable levels before it reaches the consumer. However, in some cases it is necessary to take action to remove radon from UK public supplies. There may also be significant radon exposure to employees in the water industry, when radon is released during storage and processing.

It is important to recognize that radon in domestic water supplies almost certainly presents a smaller public health hazard than radon in air, both in terms of the numbers of people exposed to high levels, and in terms of the risks to the most exposed individuals. In general, those most at risk from radon in water are those who rely on a

Table 38.8 Amount of radon released into the indoor air by various domestic processes

Process	Amount of radon liberated in indoor air (%)
Washing clothes or dishes	95
Showering	66
Bathing	42
Drinking	35
Toilet flushing	35

Source: [35].

private supply from a borehole in an area where there are high radon concentrations indoors. These people receive doses from radon in two ways: by ingesting the radon when they drink the water, and by inhaling radon decay products produced from radon released to the air from the water from activities such as showering or bathing. Although the behaviour of ingested radon is not fully understood, it is clear that the great majority of the radiation dose is delivered to the stomach.

It has been estimated [35] that the domestic processes listed in Table 38.8 release amounts of radon into the indoor atmosphere that could equate to 1–2% (and normally not more than 10%) of the total radon entering the house. Generally, it has been estimated that for each 100 becquerels per litre (Bq/l) of radon in water, there is an added 10 Bq/m^3 of radon in air.

Standards for radon in water

The European Union (EU) has issued recommendations [36] on the protection of the public against exposure to radon in water supplies. The EU recommends that where water is supplied as part of a commercial or public activity, or distributed in public premises such as residential homes, schools and hospitals, then:

- Remedial action is deemed to be justified for concentrations of radon in water in excess of 1000 Bq/l.
- Member States should set a reference level for radon between 100 and 1000 Bq/l to be used for consideration whether remedial action is needed to protect human health.

Table 38.9 Normal radon concentrations in water

Surface water (average)	0.5–1 Bq/l
Ground water (average)	7.5–25 Bq/l
Ground water (Devon/Cornwall)	50–1000 Bq/l

Source: [37].

For an individual water supply, from which no water is supplied as part of any commercial or public activity, the EU recommends that a level of 1000 Bq/l should be used for consideration of remedial action.

The normal radon concentrations in ground water are given in Table 38.9. A survey by the British Geological Survey in West Devon found elevated levels of radon in some private water supplies: approximately 9% of the supplies tested were found to exceed the draft EU Recommended Action Level of 1000 Bq/l for such supplies, though the homes tested were not a representative sample [37].

Measurement techniques for radon in water

Samples of water for radon measurement must be taken in such a way as to avoid loss of radon to the air. When sampling well water, immerse a kilner jar or other container, with its lid, in a bucket of the water, and fit the lid and the seal under water. When sampling from a tap, any aerator should first be removed from the tap. The water should be run for 10 minutes, then allowed to flow into, and overflow, the sampling container. Seal the container and despatch to the sampling laboratory. Because radon decays with a 3.8-day half-life, the sample should be passed to the laboratory as quickly as possible.

The most commonly used measurement methods are listed below.

1. **Gamma spectroscopy** This method usually employs a sodium iodide crystal or other solid scintillator to detect the gamma radiation emitted by the radon decay products.
2. **Liquid scintillation** Water is injected into a glass vial, and toluene mixed with a scintillator is added. Radon dissolves in the toluene and the mixture separates to a transparent phase. The alpha particles emitted by radon and its decay products cause scintillations which are counted using a photomultiplier tube and suitable electronics.
3. **Electret** A small jar of water is placed in a large container that also holds an electret radon detector. The container is closed and the jar of water upset, to allow the radon to be released. After a fixed time the electret is read out to allow the radon concentration in the water to be calculated.

An indication of whether a house has a radon problem can also be obtained by using passive radon detectors to measure the radon concentration in air in rooms where water is released, such as shower rooms. Passive detectors may also be placed in the air space above the water in cisterns.

Methods of controlling radiation doses from radon in water

Two approaches are usually employed:

1. **Remove radon from the water before it reaches the tap**

   (a) **Storage** Store freshly drawn water for 13 days, which gives a 90% reduction in radon. This is not a practical method for normal use and can only be employed where large capacity storage systems are available.
   (b) **Aeration** Removal of radon is achieved by increasing the surface area of water exposed to air, either by spraying the water into a vessel or by bubbling air through a column of water. In either case it is important that the aeration apparatus is vented safely to the atmosphere.
   (c) **Filtration** This method utilizes an activated carbon absorption medium (granular activated carbon). It has the disadvantage that over time, long-lived radon decay products build up in the carbon.

2. **Dilute the radon after it has been released in the indoor air** This method requires good ventilation of bathrooms, laundries, showers

and kitchens. This is often sufficient to prevent appreciable radon build up. However, care should be taken not to depressurize the house using powerful extract fans, as this could increase the influx of soil gas carrying radon.

ADMINISTRATIVE AND FINANCIAL ISSUES

Confidentiality

As discussed above, earlier surveys of radon in houses funded by central government did not allow individual house radon results to be divulged to local authority environmental health officers, unless by prior approval. Later radon programmes funded by DEFRA have involved active participation by local authorities, and have ensured that the radon results were available to the relevant local authority.

Possession of these results by local authorities requires that the utmost discretion should be exercised. For most people, their home is their biggest asset: they often rely on its growing collateral for future prosperity. Any threat to its ultimate sales potential through radon blight would be reprehensible, and environmental health officers involved in radon survey work should act accordingly. Results – particularly high ones – should be made known only to the householder concerned and officers on a 'need to know' basis: when not in use they should be locked away and not freely discussed in public. However, knowledge of such results could and should form the basis for further surveys, even if the original householder chooses not to act to remedy the problem.

The acid test for confidentiality *vis-à-vis* public protection comes at the time of sale or resale, particularly where vendors have failed to agree to environmental health officers revealing radon test results on solicitors' property searches. Clearly legal advice should be sought from individual council solicitors, but the overriding principle must lean towards public protection. The potential health risk posed by a (possibly young) family unknowingly exposing themselves to high radon levels is such that the test **occurrence** should first be revealed to the purchaser's solicitor with a view to direct enquiry of the vendor's solicitor of the actual result. If, in spite of exhortations, it is still not forthcoming, the environmental health officer's professional duty is clearly to reveal the test result to the purchaser's solicitor, after due legal consideration.

Until such time as a mandatory radon test certificate is built into conveyancing procedures (possibly incorporated within the much-vaunted 'house log-book') this dilemma will remain.

House renovation grants towards radon-proofing

Sweden was the pioneer in providing state financial aid for radon-proofing works, introducing radon-proofing subsidies of up to 15 000 Kronor (£1200 at 2003 exchange rates) in 1985. Ireland began an innovative grant scheme for radon-proofing in 1998, under which affected householders may be paid up to £800 (50% of the cost of approved works), showing a real government commitment to eradicating this serious health threat from Ireland's housing stock. No fewer than seven countries already pay grants towards the cost of remediation in existing dwellings (Sweden, Ireland, Finland, the Czech Republic, Belgium, Luxembourg and the United Kingdom), although in some cases the grants are means tested.

Until July 2003, the provisions of the Housing Grants, Construction and Regeneration Act (1996) allowed local authorities to approve an application for a renovation (main) grant or a home repair assistance grant for the purposes of mitigating radon to below the Action Level. However, such grants were means-tested and totally discretionary, so that in the vast majority of cases house owners had to finance the radon-proofing themselves, forcing many to the DIY approach.

The CIEH consistently campaigned for mandatory grants for radon-proofing to ensure that householders get the message that radon has been officially recognized as a serious health risk and that radon mitigation really is a worthwhile and cost-effective cure. In 1993, the Department of the Environment commissioned Professor Terence Lee to look at why so few householders with high radon levels took remedial action. One of the conclusions of his summary report, published in 1994 [38], was that the most common reason for householders failing to take remedial action was 'an inability to afford [radon remediation] work'.

The housing grant system has recently changed. A general power in the Regulatory Reform (Housing Assistance) (England and Wales) Order 2002 removed the previous powers to give discretionary grants, with effect from July 2003. Disabled Facilities Grants remain as a mandatory provision. A key change is that it will now be up to Local Housing Authorities whether they means test any discretionary assistance, and if they do so what form the means test, and indeed the assistance, will take.

Under the Regulatory Reform Order, authorities will still be able to give 'assistance' for radon remedial works directly or indirectly (grants, loans, equity release, materials), but crucially that assistance will remain discretionary. Authorities can use the new power once they have adopted and published a policy setting out how they will use it. For local authorities in radon-affected areas it would seem sensible for them to cover radon remedial works, but they do not have to do so.

It is up to individual Housing Authorities to determine their own qualifying criteria for both eligibility for assistance and the standard to be achieved. Best practice guidance in this regard remains as follows: Radon-proofing works necessary to reduce radon below the prevailing Action Level should qualify for grant only if supported by the results of a survey carried out, for at least three months, by a laboratory validated by NRPB [19].

REFERENCES

1. Royal Commission on Environmental Pollution (1984) *Tackling Pollution – Experience and Prospects*, 10th Report, Cmnd 9149, HMSO, London.

2. International Commission on Radiological Protection (1984) Publication 39: Principles for limiting exposure of the public from natural resources of radiation, *Annals of the ICRP*, **14**(1).

3. Green, B.R.M.G., Brown, L., Cliff, K.D. *et al.* (1985) Surveys of natural radiation exposure in UK dwellings with passive and active measurement techniques. *Science of the Total Environment*, **45**, 459–66.

4. National Radiological Protection Board (1987) *Exposure to Radon Daughters in Dwellings*, NRPB GS.6, HMSO, London.

5. House of Commons (1987) *Hansard*, 27 January, wls 189–97.

6. Institution of Environmental Health Officers (1998) *Report of the IEHO Survey on Radon in Homes 1987/88*, IEHO, London.

7. Institution of Environmental Health Officers (1989) *Report of the IEHO Survey on Radon in Homes 1989*, IEHO, London.

8. Institution of Environmental Health Officers (1991) *Report of the IEHO Survey on Radon in Homes 1990*, IEHO, London.

9. Bowie, S.H.U. and Plant, J.A. (1983) *Natural Radioactivity in the Environment, Applied Environmental Geo-chemistry*, Academic Press, London.

10. Lubin, J.H. and Boice, J.D. (1997) Meta-analysis of residential radon case control studies. *Journal of the National Cancer Institute*, **89**, 49–57.

11. Sarah Darby, Elise Whitley, Paul Silcocks, Bharat Thakrar, Martyn Green, Paul Lomas, Jon Miles, Gillian Reeves, Tom Fearn, Richard Doll (1998) Risk of lung cancer associated with residential radon exposure in southwest England: a case-control study. *British Journal of Cancer*, **78**, 394–408.

12. Department of Health (1998) *Our Healthier Nation: A Contract for Health*, Cm 3852, Stationery Office, London.

13. Wrixon, A.D., Green, B.M.R., Lomas, P.R., Miles, J.C.H., Cliff, K.D., Francis, E.A., Driscoll, C.M.H., James, A.C. and O'Riordan, M.C. (1988) Natural radiation exposure in UK dwellings. NRPB R190.

14. Green, B.M.R., Miles, J.C.H., Bradley, E.J. and Rees, D.M. (2002) Radon Atlas of England and Wales, NRPB-W26.

15. Miles, J.C.H., Green, B.M.R. and Lomas, P.R. (1999) Radon affected areas: Northern Ireland – 1999 review. *Documents of the NRPB*, **10**(4), 1–8.

16. Miles, J.C.H., Green, B.M.R. and Lomas, P.R. (1993) Radon affected areas: Scotland. *Documents of the NRPB*, **4**(6), 1–8.

17. DEFRA Radon Good Practice Guide 2002. http:// www.defra.gov.uk/environment/radioactivity/ radon/pdf/radon_laguide.pdf

18. Thomas, A. and Hobson, J. (2000) Review and evaluation of the radon remedial pilot programme. DETR/RAS/00.004.

19. Miles, J.C.H. and Howarth, C.B. (2000) Validation scheme for laboratories making measurements of radon in dwellings: 2000 revision NRPB-M1140.

20. Colgan, P.A. (1996) Radon policy around the world. *Radiological Protection Bulletin*, **181**, 12–13.

21. Akerblom, G. (1999) Radon legislation and national guidelines. SSI Report 99:18.

22. Documents of the NRPB (1990) *Board Statement on Radon in Homes*, **1**(1), HMSO, London.

23. Housing Health and Safety Rating System papers at http://www.housing.odpm.gov.uk/ research/hhsrs/index.htm

24. Documents of the NRPB (1990) *Radon Affected Areas: Cornwall and Devon*, **1**(4), HMSO, London.

25. The Ionising Radiations Regulations 1999 SI3232 London, HMSO.

26. Work with ionising radiation (2000) HMSO, London. (This is the ACoP and guidance.)

27. Towers, M. and Nicholls, I. Radon in Commercial Premises: The High Peak Experience. Environmental Radon Newsletter 33, http://www.nrpb.org/publications/newsletters/ environmental_radon/index.htm

28. How much do remedial measures cost? Environmental Radon Newsletter 34, http:// www.nrpb.org/publications/newsletters/ environmental_radon/index.htm

29. *Radon, A Guide to Reducing Levels in Your Home* (2000) DEFRA, Wetherby.

30. BRE Report BR250: Surveying dwellings with high indoor radon levels: a BRE guide to radon remedial measures in existing dwellings (1993) CRC Ltd.

31. BRE Report BR343: Dwellings with cellars and basements: a BRE guide to radon remedial measures in existing dwellings (1998) CRC Ltd.

32. BRE publication BR211, Radon: Guidance on protective measures for new dwellings (1999) CRC Ltd.

33. BRE publication BR376, Radon: Guidance on protective measures for new dwellings in Scotland (1999) CRC Ltd.

34. BRE publication BR413, Radon: Guidance on protective measures for new dwellings in Northern Ireland (2001) CRC Ltd.

35. Nazaroff, W.W., Doyle, S.M., Nero, A.V. and Sextro, R.G. (1988) Release of entrained radon-222 from water to air by domestic use, in *Radon Entry Via Potable Water: Radon and its Decay Products in Indoor Air* (eds W.W. Nazaroff and A.V. Nero), Wiley & Sons, New York, pp. 131–57.

36. European Commission 2001. Commission Recommendation of 20 December 2001 on the protection of the public against exposure to radon in drinking water supplies (notified under document number C(2001)4580) (2001/928/Euratom). Official Journal of the European Communities L 344/85–88.

37. D.K. Talbot, J.R. Davis and M.P. Rainey (2000) Natural radioactivity in private water supplies in Devon. Report No: DETR/RAS/00.010. London: Department of the Environment, Transport and the Regions.

38. Lee, T. (1994) *Householder's Response to the Radon Risk: Summary Report*, HMSO, London.

39 The water cycle

David Clapham and
Nigel Horan

INTEGRATED WATER MANAGEMENT

The basic principle involved is management of the water cycle through the supply, distribution and waste control chain. As well as catchment control and the use of the river system, usage and final disposal of waste water are important parts of this cycle (Fig. 39.1) The Water Act 1989 (consolidated into the Water Industry Act 1991 and the Water Resources Act 1991), embodied the principle of managing the water cycle but did so by privatizing the water supply and sewage utilities. There were 10 water undertakers or companies (PLCs) set up under the Act. The original 29 statutory water companies were retained, but gained the power to become public companies. The Act made the regulatory functions and river basin management in England and Wales the responsibility of the National Rivers Authority (NRA). They were also given responsibility for the control of water pollution, management of water resources, fisheries, flood defence, land drainage and navigational functions. The NRA became part of the new Environment Agency (EA) on 1 April 1996, established by the Environment Act 1995, with eight regional areas in England and Wales:

- Anglian
- Midlands
- North-East
- North-West
- Southern
- South-West
- Thames
- Welsh.

Each region has the duty of appointing a regional fisheries, ecology and recreation advisory committee and in addition in Wales, the Secretary of State has a committee to advise him generally with respect to the EA. Two further national

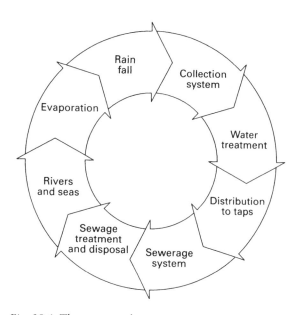

Fig. 39.1 The water cycle.

governmental bodies assist with the regulation of the water industry: the Office of Water Services (OFWAT) and the Drinking Water Inspectorate (DWI). The Director General of OFWAT has duties with respect to the economic and performance standards of the water industry. The Director General allocates each water undertaker a customer service committee to protect the interests of consumers and potential customers. Both the Secretary of State and the Director General are given power, by Section 18 of the Water Industry Act 1991,[1] to enforce the provisions on the water undertakers, both of the Act and its regulations. Concurrent powers are given to serve enforcement orders, although the bulk of the requirements, including the Water Supply (Water Quality) Regulations 2000 (SI 3184) are enforced by the Secretary of State. Failure to comply with an order (either provisional or final) may be followed by a 'special administration order' issued by the High Court under Section 24 of the 1991 Act, or by civil proceedings.

The DWI issues an annual report on drinking water quality in England and Wales. It has also reported at other times on specific issues, including nitrates, pesticides, *Cryptosporidium* and disinfection by-products. The role of the DWI is outlined in a free publication – *The Drinking Water Inspectorate – What Do We Do?* and on their website http://www.dwi.gov.uk/.

Provision of water by a water undertaker that is unfit for human consumption is an offence, although prosecutions may only be instituted by the Secretary of State or the Director of Public Prosecutions. The water companies have a general duty to maintain an efficient and economical system of water supply within their areas. They also have to supply water of adequate pressure for domestic purposes, subject to certain conditions. All water companies offer a free detection and leakage repair service for supply pipes and are taking other action to promote water conservation.

Public water supplies must comply with the standards laid down in the Water Quality (Water Supply) Regulations 2000 (SI 3184). The standards are based on the 1998 European Union Directive on the Quality of Water Intended for Human Consumption (98/83/EC-OJ L 330.05.12 98). This in turn was based on WHO guidelines from 1993 (see Further reading section). Power is available under the Water Quality Regulations for the Secretary of State to grant relaxations (authorizations) of drinking water standards. Details of the relaxations are to be sent to the relevant local authority and health authority, who are free to make objections [1]. The relaxations cannot be given for parameters that may cause ill health and will require improvement of the supply by a specified date. Local authorities have a general duty to keep themselves informed about the wholesomeness and sufficiency of all water supplies in their districts. If a local authority is aware of a public supply which is, or is likely to become, unwholesome or insufficient, it has a duty to notify the water undertaker. If the undertaker fails to take satisfactory action, then the local authority must notify the Secretary of State.

In respect of private supplies, local authorities have specific responsibility for monitoring. Where private supplies are unwholesome or insufficient, they have the power to require improvements by service of notice on the owners and occupiers of the premises The meaning of the term **wholesomeness** of water is defined for England and Wales by the 2000 Regulations (Part III). A separate definition is in the Private Water Supplies Regulations 1991(SI 2970). When these regulations are updated the definitions will probably become the same. Section 79 of the Water Industry Act 1991 gives local authorities wide powers to tailor their actions to the circumstances of particular supplies. Under Regulation 4 of the Private Water Supplies Regulations 1991, local authorities have a complementary power to grant relaxations (authorizations) for certain parameters, provided that no public health risk is involved. In order to carry out their duties under the Regulations and the Water Industry Act 1991, central advice has been made available to local authorities. Advice has been issued in circulars for England [2], Wales [3] and

---

[1]Also see Water Act 2003 to be implemented in 2004.

Scotland [4], as well as in the *Manual on Treatment of Small Water Supply Systems* [5].

In England and Wales, house plans that are deposited with the local authority under the Building Regulations must be rejected unless satisfactory arrangements are made for a water supply (Section 25 of the Building Act 1984). It should be noted, however, that the building regulations themselves do not apply to cold water supplies, and reference should be made to the Water Supply (Water Fittings) Regulations 1999 for installation standards. Local authorities also have powers under Section 69 of the Building Act 1984 to require a satisfactory water supply in an occupied house [1].

Disposal of waste water and effluent resulting from the use of water, including sewage, is also subject to a network of legal controls. Principal among these is Part IV of the Water Industry Act 1991, but some of the powers under the Public Health Acts 1936 and 1961 and the Building Act 1984, as well as other associated legislation, remain in force. Much of Part IV of the Water Industry Act 1991 is implemented by regulations, which, in turn, are subject to the influence of the requirements of European Union (EU) legislation.

Whilst the disposal of sewage is the responsibility of the water companies, a number of important powers and duties are also held by the EA and local authorities. The water companies have the duty, as sewerage undertakers, to provide, improve and extend a system of public sewers, ensuring the effective drainage of their area. They are also required to provide sewage disposal facilities, including trade effluent. Responsibility for sewage disposal is thus largely divided between the three main public and private organizations: the EA, the water companies, acting as 'sewerage authorities' and the local authorities.

Control of effluent discharges to the natural water environment is regulated by the EA. The Act provides for regional environmental protection advisory committees in England and Wales. The EA has River Quality Objectives in response to the EU Dangerous Substances Directive 76/464/EEC [6]. This directive requires EU member states to eliminate pollution of surface waters with 'List I' substances and to reduce pollution by 'List II' substances. List I substances are those substances which are the most toxic, persistent and bioaccumulative. List II substances display these characteristics to a lesser degree. EU standards for the list I substances are enforced through the Surface Waters (Dangerous Substances) (Classification) Regulations 1992 (SI 337). A further 21 list II substances are contained in the Groundwater Regulations 1998 (SI 2746).

The basis of the new system introduced by the Water Act 1989 and now contained in the Water Resources Act 1991, is the establishment of statutory water quality objectives (although none have been set as yet). Section 82 of the 1991 Act gives power to the Secretary of State to establish classification standards for all 'controlled waters', which include boreholes into ground water, inland waters, rivers, watercourses, docks, coastal waters and territorial waters. Classification of waters by the Surface Water (Classification) Regulations 1989 (SI 1148) under Section 104 do not, in themselves, create operational standards. Service of a notice under Section 105 of the Act by the Secretary of State on the EA triggers the standards by requiring their implementation by a specified date on particular waters.

In turn, the EA has a general duty to achieve and maintain the objectives, including monitoring the extent of pollution in controlled waters. Clearly, in exercising this duty there is a need for a close relationship with both the water utility companies (sewerage undertakers) and the local authorities. In the latter case, there is a particular need to ensure that any controlled waters abstracted for domestic use as private supplies comply with the River Quality Objectives in the 1992 classification regulations, as well as those in the 1991 Private Water Supplies Regulations (SI 2790).

Control powers over pollution are given to the EA, in the form of a prohibition notice under Section 86 of the Water Resources Act 1991, against the discharge of polluting matter into controlled waters, which has either been prohibited by notice or which 'contains a prescribed substance or a prescribed concentration of such a substance or derives from a prescribed process . . . which exceed the prescribed amounts'. In addition, by Section 87 of the 1991 Act, it is given power to consent to discharges and issue disposal licences, although not for any of the substances prohibited under Section 86.

The EA has powers to limit discharges from sewage works, industrial sources and individual polluters. There is, therefore a framework for the incremental improvement of the natural aquatic environment. It is open to the Secretary of State to introduce progressively more stringent water

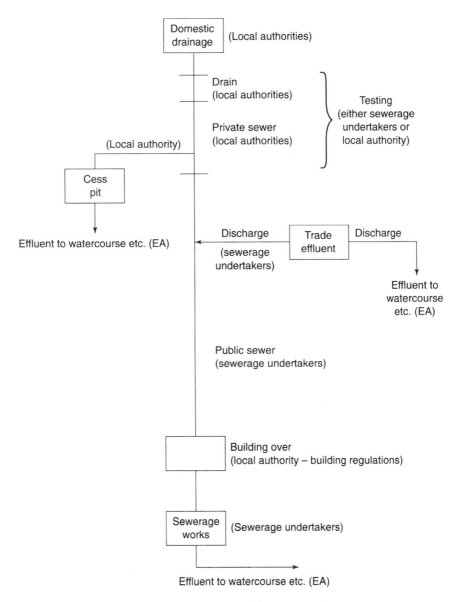

Fig. 39.2 Responsibility for control of the drainage sector of the water cycle.

quality objectives, which in turn will result in greater use of control powers by the EA.

It is an essential principle of the water cycle that local authorities are responsible for private drainage, including cesspools; sewerage undertakers are responsible for public sewers and sewage disposal works; and the EA controls final discharges to watercourses etc. (Fig. 39.2). Thus, if trade effluent is discharged to a public sewer, the discharge requires the consent of the sewerage undertakers, while if it is discharged direct to a natural water course, it is subject to control by the EA.

This completes the 'water cycle'. It is subject to a complex and interrelated legal code, some of which originated in the nineteenth century, and involves the complementary activities of both private water utility companies and public authorities. It is against the background of these legislative arrangements in England and Wales that the water cycle should be considered.

WATER COLLECTION

Physical properties of water

Water is a colourless, tasteless, odourless liquid. It is also one of the greatest solvents known to science. If there is sufficient contact time it will dissolve and take into solution all gases and most solids. It is therefore easily contaminated and pure water is not met in nature or in tap water. Uncontaminated water has a neutral pH of 7, that is, it is neither acidic nor basic. Water has many physical features that make it unique. It is naturally found as a liquid, a solid and a gas and is practically incompressible; the density of water is taken as the standard of comparison for all liquids and solids. The term 'pure' water nowadays is accepted as meaning water that is dietetically pure. A more useful term is wholesome or potable and for practical purposes this may be regarded as water that complies with the standards laid down in the Water Supply (Water Quality) Regulations 2000 (2001 for Wales SI 3911). Its high solvent properties must be remembered by all who have duties relating to its quality, storage and distribution.

Sources of water supply

The hydrogeological cycle is the continuous movement of water including evaporation from the land, rivers, seas and oceans via the diffusion of water vapour throughout the atmosphere, with its subsequent condensation and descent in the form of precipitation – rain, hail or snow. When precipitation reaches the earth, part percolates into the ground until it reaches an impermeable stratum, above which it accumulates in water-bearing rock known as aquifers. This water then moves slowly back to the sea. Some flows overland and forms rivers and lakes and part is absorbed by the soil itself and supports plant life. The remainder is re-evaporated either from the earth's surface or from plants (**transpiration**).

The sources from which it is practicable to derive water for public and domestic purposes are:

- rainwater (for isolated dwellings)
- surface water: lakes, rivers, etc.
- groundwater: springs, wells and boreholes
- sea water: by desalination or reverse osmosis treatment (not usual in the United Kingdom).

Supplies are broadly classified as:

- public supplies, distributed by statutory water undertakers;
- private supplies to a single user or a small group of users.

Rainwater

Rainwater may be used for domestic purposes in remote districts where groundwater is contaminated, particularly by sea water ('saline intrusion') or where there are no alternative sources. Where it is the sole source of supply allowance must be made in storage capacity to overcome any long periods of drought. As a practical source of domestic supply its use is limited since particular care is required to reduce any pollution from extraneous sources such as bird droppings from the roof or other surface the water is collected from. Care also has to be taken to ensure it is not contaminated by improper methods of storage. In any case, arrangements should be made for suitable treatment to ensure compliance with EU Directive 98/83/EC relating to the quality of water intended for human consumption.

Surface water

Surface water supplies are taken from rivers, lakes or man-made reservoirs. These often provide a consistent and manageable source for public supply, but it is subject to greater risk of contamination than groundwater and requires careful and suitable treatment. There is also the possibility of accidental or deliberate pollution by irresponsible industrialists or the agricultural community. To meet these two potential problems it is imperative that sufficient treatment plant is installed to enable compliance with both the compositional and microbiological standards set by the EU (Directive 98/83/EC). For supply purposes, they can be considered as two different types of source. Reservoirs and lakes store water for many months, during which time considerable improvements may occur in its quality. Bacterial decay takes place (due to a variety of factors) and the time available for settlement permits a reduction in suspended material (**turbidity**) and other constituents. On the other hand, some surface water conditions have caused the growth of toxic **blue–green algae** (which are perhaps more accurately known as **cyanobacteria**) [7]. The toxins they produce, particularly after the bacteria have died and start to decompose, are a risk to animals and humans ingesting the untreated water. The algae grow below the surface, but most cells contain a gas vesicle that enables them to move up and down to make the best use of ecological niches. When the conditions are right for growth (**eutrophic** water, usually at the end of a long summer) the cyanobacteria rise to the surface to form a layer of distinctive blue–green scum, although they can produce scums of other colours, such as the 'red tides' sometimes found in coastal waters. The use of algaecides, like copper sulphate, to reduce blooms of cyanobacteria is not always completely successful as when the cyanobacteria die they may release cyanotoxins into the water. Cyanotoxins are of three main types: neurotoxins, which damage the nervous system, hepatotoxins, which damage the liver and dermatotoxic alkaloids, which irritate the skin. Half to two-thirds of cyanobacterial blooms may be toxic, although toxicity varies from day to day. While small animals have died from ingesting water containing the blooms and cases of human illness are known, no confirmed report of human deaths has been recorded in the United Kingdom. Because of the dilution factors and efficient water treatment, carry-over of the toxins into the public water supply is highly unlikely but not thought to be absolutely impossible. The risk has not been quantified and no specific treatment has been advised. Granulated activated carbon and ozonation are thought to be effective.

Under certain conditions, impounded water may become stratified. This leads to higher levels of oxygen in the upper layers of the reservoir and **oxygen deficiency** near the bottom. This allows chemical reduction and the dissolution of various metals and other contaminants that may render the water difficult to treat. Water utilities are normally aware of the quality of the water they receive, however, and take steps to make sure their treament is appropriate. The huge volume of water held in reservoirs will also act as an effective buffer against small-scale accidental pollution. By good management, the quality of water abstracted from reservoirs can be reasonably constant in quality.

Water abstracted from rivers is generally of poorer quality than reservoirs and lakes and varies in quality throughout the year depending on the weather, particularly heavy rainfall. As a result, public water supply treatment has to be robust and surface water should only be considered for a private supply where no adequate ground water supply exists.

Groundwater

Water that has percolated through the ground can be used for drinking purposes, either via springs, boreholes or wells. The quality of the water varies depending on a variety of factors such as the depth

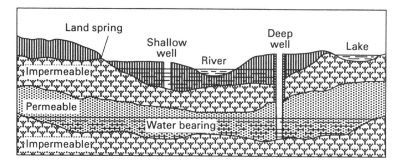

Fig. 39.3 Surface and underground water supplies.

it is taken from and the types of rock it has passed through. While subsequent treatment of the water supply can make it suitable for drinking, its initial condition influences the types of treatment necessary. Ground water, particularly from deep sources, often provides water of good microbiological quality although its quantity may not always be sufficient for public water supply demands.

Ground water can come from one of three sources:

1. **springs**;
2. **wells**: shallow, deep or artesian (Figs 39.3 and 39.4);
3. **boreholes**.

Springs

A spring is a simple outcropping of water that has percolated into a permeable subsoil and run along the top of an impermeable stratum to a point at which it reaches the surface (Fig. 39.3). Such a spring will vary in volume and contamination levels according to the rainfall in the immediate neighbourhood. It is obviously liable to pollution by direct percolation through the top soil. As a result, it is often little better than land drainage in quality and while it may be possible, after careful investigation, to authorize its use as an acceptable private supply, all too often such sources are unsatisfactory due to agricultural pollution and farm animal excreta. A spring supply issuing from

deep, water-bearing strata can produce both a consistent volume and a better quality supply. Whichever type of spring is used, animals should be excluded from the area by a stock-proof fence and any water overflowing the land after rain should be diverted by a suitable ditch. The immediate area where the spring arises should be protected by a suitable housing or **collection chamber** with a concrete apron to divert contamination away.

Wells

The practice of obtaining water from wells is universal and well water is an important source of supply in many developing countries. The necessity for care in the location and construction of wells can hardly be exaggerated. A well should be situated uphill from any possible sources of pollution. When water is extracted from a well sunk into any but the most open subsoil, for example gravel, the surface of the water about it will fall more or less into the shape of an inverted cone. This **cone of depression** may allow the ingress of polluting matter during pumping. To prevent this, protection of the source is needed, for example, by excluding animals from the vicinity (Figs 39.4 and 39.5). A well that taps water above the first impermeable stratum is known as a **shallow well** (Fig. 39.3). The term 'shallow' in this context is not a definite depth, but an indication of the stratum from

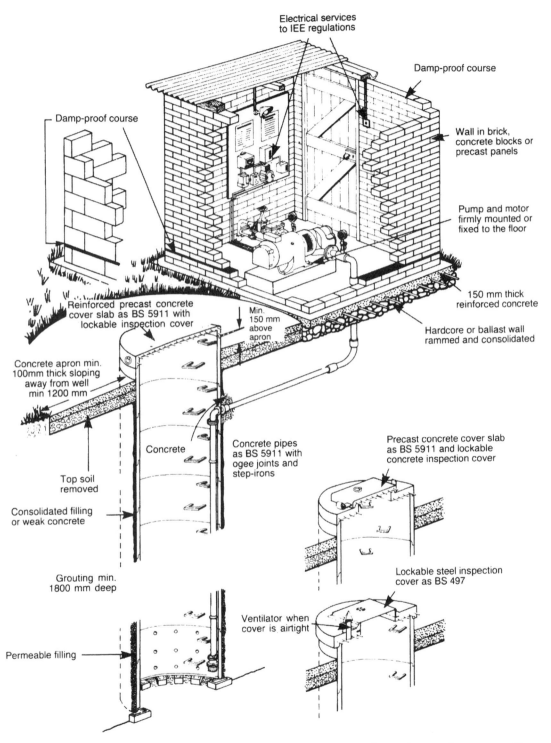

Electrical services
to IEE regulations

Damp-proof course

Damp-proof course

Wall in brick,
concrete blocks or
precast panels

Pump and motor
firmly mounted or
fixed to the floor

150 mm thick
reinforced concrete

Reinforced precast concrete
cover slab as BS 5911 with
lockable inspection cover

Min.
150 mm
above
apron

Hardcore or ballast wall
rammed and consolidated

Concrete apron min.
100mm thick sloping
away from well
min 1200 mm

Concrete

Concrete pipes
as BS 5911 with
ogee joints and
step-irons

Precast concrete cover slab
as BS 5911 and lockable
concrete inspection cover

Top soil
removed

Consolidated filling
or weak concrete

Lockable steel inspection
cover as BS 497

Grouting min.
1800 mm deep

Ventilator when
cover is airtight

Permeable filling

Fig. 39.4 Well and pump installation: typical arrangements to exclude pollution from surface sources. (Source: [5].)

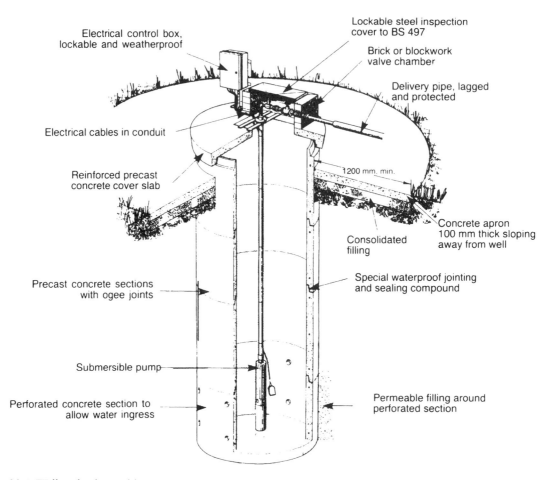

Electrical control box, lockable and weatherproof

Lockable steel inspection cover to BS 497

Brick or blockwork valve chamber

Delivery pipe, lagged and protected

Electrical cables in conduit

Reinforced precast concrete cover slab

1200 mm. min.

Consolidated filling

Concrete apron 100 mm thick sloping away from well

Precast concrete sections with ogee joints

Special waterproof jointing and sealing compound

Submersible pump

Perforated concrete section to allow water ingress

Permeable filling around perforated section

Fig. 39.5 Well and submersible pump installation: typical arrangements to exclude pollution from surface sources. (Source: [5].)

which it is abstracted. **Deep wells** (Fig. 39.3) are those derived from water-bearing strata below at least one impervious stratum. A deep well must be constructed so as to exclude subsoil water and contamination from aquifers above the level where the well takes its water. It should be watertight down to a point slightly below the level of the deep supply.

Well water (and sometimes spring water) from below impervious strata is sometimes under pressure because the natural surface of the watertable may be above the level of the stratum the water is contained in. The water will therefore rise in any well or borehole up to the level of the natural watertable, this is the **potentiometric surface** and may actually be above the surface of ground (Fig. 39.6). In this case it is known as an **artesian supply**. Similarly, a natural fault in the impounding stratum may give rise to a deep or artesian spring (Fig. 39.6).

Most of the old wells met with in practice are roughly lined with open jointed brickwork and this permits polluted water to enter from the top and sides. The compulsory closing of an unfit private well is a serious matter for those dependent upon it and in many cases, this may be avoided at

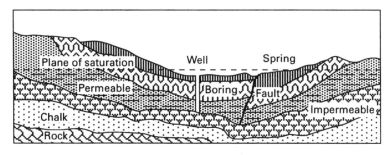

Fig. 39.6 Artesian water supplies.

reasonable expense by lining the well with concrete pipes grouted in cement or clay, although the drilling of a new borehole may be more cost-effective in the long-run. Provided the relining is combined with modern headworks and an adequate sloping concrete apron to prevent contaminating percolation, it may be possible to rehabilitate an old well in this way (Figs 39.7 and 39.8). However, it must be clearly established that the contamination is the result of inadequate construction; no reconstruction will improve an inadequate shallow supply affected by pollution sources such as a leaking cesspool or silage pit.

Boreholes

Boreholes are a modern version of the well and much of the information about wells applies to boreholes: siting is important, they should be protected from contamination, care should be taken to prevent ingress of water from higher aquifers into the source being tapped and they may be artesian. They are purposely drilled by specialist engineers and are lined to prevent contamination. Underground water from deep boreholes is normally considered to be the most safe, since the passage through the ground and the time taken to reach the source provides a natural filtration process. Because it may have travelled through a great amount of rock strata, the water from boreholes may be higher in dissolved metals and other natural minerals.

Shallow wells and springs are liable to local contamination and all private water supplies generally require treatment, particularly disinfection, to ensure their suitability for drinking purposes. Where no public supply is available, construction of new boreholes should have regard to the guidance given to local authorities in the *Manual on Treatment of Small Water Supply Systems* [5]. In many cases, where a suitable water-bearing substrata is available, a borehole with an electric pump may provide an adequate supply for several households. In all cases of new construction, particular care must be taken to protect the source by lining it to prevent percolation from surrounding ground at a shallow level and by provision of an impermeable apron to prevent seepage of contamination from above.

The European Union Directive 75/440/EEC, relating to the quality of surface water to be used for the abstraction of drinking water, has been in force since 1975 (see also next section). Such water may also be suitable for recognition as a natural mineral water (see the Natural Mineral Water, Spring Water and Bottled Drinking Water regulations 1999 (SI 1540) (as amended in 2003)).

The extent to which it is necessary to fence the area around a spring, borehole or shallow well as protection against surface contamination, particularly from animals, depends on the surrounding land and its topography, hydrographical conditions, the porosity and filtering efficiency of the soil, and the extent to which the water is liable to fall when ordinary pumping is in

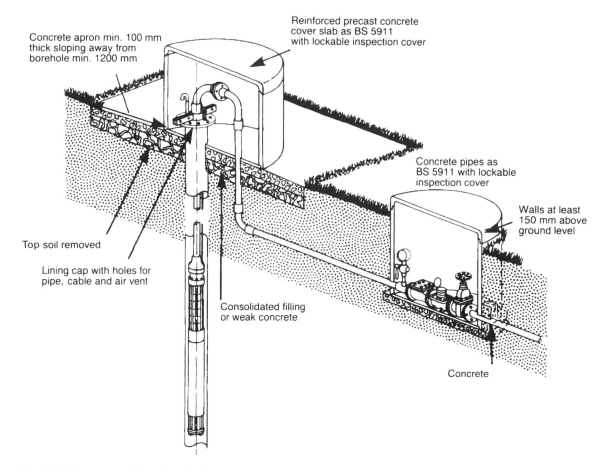

Reinforced precast concrete
cover slab as BS 5911
with lockable inspection cover

Concrete apron min. 100 mm
thick sloping away from
borehole min. 1200 mm

Concrete pipes as
BS 5911 with lockable
inspection cover

Walls at least
150 mm above
ground level

Top soil removed

Lining cap with holes for
pipe, cable and air vent

Consolidated filling
or weak concrete

Concrete

Fig. 39.7 Borehole well headworks. (Source: [5].)

progress. A grass crop on the protected area around a small water supply is desirable, as the roots form a filtering mat, and the grass tends to prevent surface drying and consequent cracking of the soil, which would permit seepage of contaminated surface water. Careful assessment of the supply should include taking microbiological and chemical samples in accordance with the Private Water Supplies Regulations 1991 (SI 2790). A sanitary survey, where all pollution sources in the area and defects in the supply are noted, investigated and where possible put right, is a vital component of the assessment of safety in private water supplies and should always be carried out in conjunction with the water sampling programme.

Acid rain and other atmospheric pollutants in water

A constituent of most natural waters is carbon dioxide (CO_2) (this is also the gas added to bottled water to make it fizzy). Sulphur dioxide (SO_2) is another gaseous contaminant of water. Both these gases, partly resulting from atmospheric pollution, are dissolved by rain falling through the atmosphere. They make the water slightly acid and this is the so called 'acid rain' that has been claimed to damage forests, herbage, natural lakes and wildlife when it passes over and through earth covered with vegetation. Some of these assertions are now being questioned as the damage may be much less severe than previously thought. This

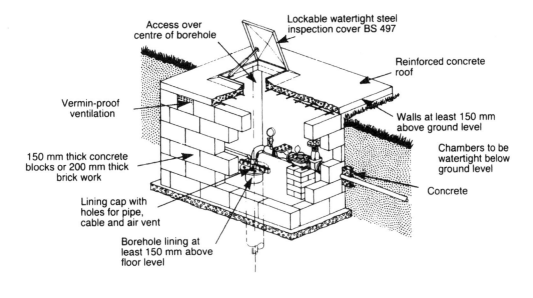

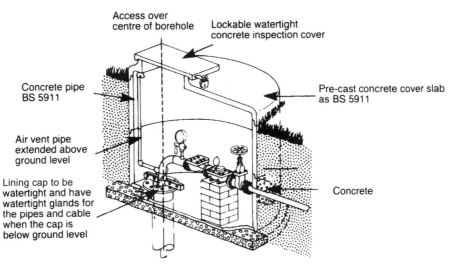

Fig. 39.8 Borehole well headworks. (Source: [5].)

acidity will also dissolve metals and other minerals from the rocks and soil as the water passes through them.

Hardness and softness

The nature of the underlying geological strata, as well as the particular chemical reactions that result, are crucial to the final composition of the water. In its simplest form, a dilute solution of carbonic acid increases the solvent properties of the water and enables it to dissolve calcium and magnesium from limestone or chalk strata and to hold them in solution as bicarbonates. Water in which these salts are dissolved is described as being 'hard'. Hardness, noted as the difficulty of raising a lather from soap, is generally described as being of two kinds: temporary hardness (which can be removed by boiling), due to the presence of bicarbonates of calcium and magnesium; and permanent hardness, due to the sulphates of these chemicals. It is conventional practice to express both types of hardness as the equivalent milligrams per litre (mg/l) of calcium carbonate ($CaCO_3$). Table 39.1 lists the commonly accepted levels of hardness in public water supplies against their content of calcium and magnesium salts expressed as $CaCO_3$. Water that does not pass through rock containing calcium or magnesium bicarbonates and sulphates will be soft water. It will be more acidic, contain more dissolved metals, particularly iron, manganese and aluminium and will tend to be aggressive to plumbing installations.

WATER TREATMENT

Contamination of public supplies

Whether good quality potable water reaches the consumer depends on its original quality, the treatment processes it is subject to and how it is protected at source, in storage and in distribution. Heightened public concerns over environmental pollution has increasingly focused on drinking water quality, sometimes unfairly. Run-off from surfaces, percolation through subsoil and natural and anthropogenic pollution of open bodies of water in streams, rivers, reservoirs, ponds and lakes, all expose water to contamination risks. The main sources of potential contamination are agricultural, industrial and sewage disposal. The EU Drinking-water Directive lists some 48

Table 39.1 Water hardness

Water type	Hardness (mg/l $CaCO_3$)
Soft	0–50
Moderately soft	50–100
Slightly hard	100–150
Moderately hard	150–200
Hard	200–300
Very hard	>300

parameters for water, of which a number relate to each source of contamination. In addition to obviously toxic substances such as heavy metals, there is an increasingly recognized risk associated with traces of such chemicals as organic solvents and hydrocarbons.

Nitrates

Increased use of artificial nitrogen fertilizers has raised the level of nitrate run-off from farmland. As a result, the level of nitrates in water supplies has increased, particularly in those parts of the United Kingdom subject to intensive aerable farming. Significant amounts of nitrates are also released when pasture land is ploughed up. Elevated levels of nitrates have been known to cause **methaemoglobinaemia** or **'blue baby syndrome'**, when associated with microbiological contamination, although it is exceedingly rare and the last cases in the UK were over 30 years ago. It is a rare condition that only affects bottle-fed babies and is caused by damage to the haemoglobin, which prevents oxygen transportation in the blood. It has also been suggested that there may be an elevated risk of stomach cancer from a high level of nitrates, but epidemiological studies have failed to support this contention. The EU Nitrates Directive (91/676) agreed in 1991, requires member states to introduce voluntary measures for agriculture designed to protect water against nitrate pollution. This has been subject to further EU regulation (1257/99). In October 1995, the Department of the Environment (DoE) designated an Initial 68 'nitrate vulnerable zones' in England and Wales. Further regulations on the grants available for farmers have been produced. (The Farm Waste Grant (Nitrate Vulnerable Zones) (England) Scheme 2003) (SI 908), the measures that farmers will be required to take (The Action Programme for Nitrate Vulnerable Zones (England and Wales) Regulations 1998) (SI 1202) and the ability to designate other areas (The Nitrate Vulnerable Zones (Additional Designations) (England) (No. 2) Regulations 2002) (SI 2164). Similar legislation has been produced for Scotland and Wales.

Metals

There was some concern that elevated levels of aluminium may be a precursor of **Alzheimer's disease**, a condition with symptoms of progressive dementia. There no longer appears to be this concern and aluminium in drinking water is considered an aesthetic parameter. It occurs naturally in many waters with low pH and also results from some treatment methods as aluminium salts can be used as floculants. An accident in 1988 in Camelford, Cornwall, occurred when a relief driver poured 20 tonnes of aluminium sulphate into the contact chlorine reservoir by mistake. This then entered the distribution system. There were many reported ill effects following the incident, but several epidemiological problems have made it difficult to decide how much the aluminium was to blame.

Lead has been recognized as a toxic water pollutant since Roman times, although acute lead poisoning from water, with immediate neurological effects, anaemia and muscle cramps is a thing of the past. Recent concerns have focused on the problem of subclinical lead poisoning, with a shift in educational ability of the very young and hyperactive behaviour. It is important to remember that lead has long-term cumulative effects and that it is therefore a multi-source problem, of which water is but one component.

Arsenic is another natural contaminant of waters with a low pH and although not a particular problem in most of the United Kingdom has caused serious skin cancer problems in Bangladesh, where it is a major contaminant of water from tube wells. Other metals, such as **iron** and **manganese** may cause discoloration in the water but at the levels found in the United Kingdom do not pose a risk to human health.

Industry

Industrial sources of pollution may include consented discharges into water courses, leaching of substances from contaminated land or licensed industrial tips and illegal dumping of industrial waste on land or into water courses. Agricultural sources of contamination include silage effluent, rural sanitary disposal and nitrates. In January 1995, the government established the UK Round Table on Sustainable Development. In 1997, the group published a report [8] on fresh water, making 12 recommendations, including calling for the government to publish a framework for fresh water policy and a review of the regulatory structure for the industry. These calls were repeated in the Round Table's Second Annual Report. The Group was disbanded in 2000.

A report on nitrates, pesticides and lead during the period 1991–94 was published by the DWI in January 1997 [9]. It indicated that for nitrates, the percentage of samples that exceeded the standard of 50 mg/l fell from 2.8% to 0.8%, those that exceeded the standard for individual pesticides of 0.1 µg/l fell from 2.8% to 1.2%, and those that exceeded the lead standard of 50 mg/l fell from 3.7% to 3.1% during the four-year period. In 2001, the percentage failure of the nitrate standard was 0.38%, for pesticides 0.01% and for lead 0.63%. Much of this improvement of samples taken at consumer taps was attributable to improved treatment processes.

Sewage discharges and Cryptosporidium

The microscopic protozoan parasite *Cryptosporidium* is endemic in many animals, particularly young farm animals. It is capable of causing enteric illness and diarrhoea in humans and can be fatal for the immuno-compromised. Several outbreaks and individual cases have been associated with water supplies [10,11]. About 2% of the cases of acute diarrhoea in the United Kingdom are caused by *Cryptosporidium*, although the majority of these are due to direct infection from pets and farm animals or person-to-person contact. As well as run-off from grazing areas, final discharges from sewage works into fresh water, even if meeting defined standards, may contaminate water supplies. This is of particular importance if river abstraction takes place close to the point of discharge, or the catchment area is at risk of contamination by sprayed or injected sewage sludge on to agricultural land. In the case of river or surface water abstraction, it is important that specific allowance be made for this risk in any water treatment process. Improperly controlled drainage from animal husbandry or silage may contaminate water courses, notably private supplies. While chlorination is important for eliminating bacteria and viruses that may survive the treatment process, normal levels of chlorination are ineffective against *Cryptosporidium*. The principal means of control at the treatment stage is filtration. Protection of the water catchment area is also crucial in reducing the incidence of the parasite [12]. Slow sand filters seem to be the most effective traditional means for the removal of *Cryptosporidium* oocysts; the effectiveness of other filters depends on the filter medium and pore size. Operational mistakes include failure to backwash filters effectively and the inefficient control of filters. Tertiary treatment by ultraviolet irradiation, membrane filtration (reverse osmosis) and ozonation have also proved useful. Infection of small water supplies is usually from agricultural sources, often as a result of periods of heavy rain.

The Group of Experts on Cryptosoridium in Water Supplies has reported three times [12–14] and concluded that although *Cryptosporidium* is occasionally present in the United Kingdom's drinking water supplies, water from public supplies is generally safe to drink. The second report included a recommendation, accepted by government, that laboratories isolating *Cryptosporidium* should notify local authorities and the Communicable Diseases Surveillance Centre of the Public Health Laboratory Service (PHLS) (which since April 2003 is called the Health Protection Agency) [13]. In the second report, the Group of Experts also drew attention to the policy of the DWI to investigate any incident of *Cryptosporidium* in the community that might be related to drinking water. The Group of Experts published their third report in 1999, following an outbreak of cryptosporidiosis from a groundwater source [14]. This report looked at groundwater contamination, particularly from a catchment control and risk assessment viewpoint, and updated advice on treatment and outbreak control. There are two important themes running through the recommendations. The first is about access to information and the second concerns the need for local authorities, water companies and

health authorities to work together to prevent the occurrence of outbreaks, rather than act after the event. There should be practice exercises to test these safety systems.

The Water Supply (Water Quality) (Amendment) Regulations 1999 (SI 1524) required water undertakers to risk assess all their treatment works sources for *Cryptosporidium*. Where there is a significant risk they need to install treatment to ensure less than 1 oocyst per 10 l and continuously moniter the final water to ensure that it meets this standard.

Contamination of private supplies

The operators and users of private water supplies need to be aware of the possible hazards associated with their use: microbiological, chemical and occasionally radiological [15].

Microbiological

Microbiological contamination may arise from improper siting, construction, protection and use, from the manure of animals deposited on land or from the unobserved seepage of farm drainage or septic tank effluent. Regular sanitary surveys and testing for **faecal coliform** indicator organisms is necessary to ensure that supplies remain satisfactory and that no harmful organisms gain access. Most private water supplies are contaminated with faecal material at least some time during the year so it is always beneficial to fit disinfection plant.

Chemical

Chemical contamination may involve aluminium, iron and manganese, particularly in soft water areas, nitrates from excessive fertilizer use, hardness from limestone and chalk, a variety of other organic and inorganic chemicals and, very rarely, pesticides from agriculture. Occasional analyses should be undertaken and treatment fitted where they are present in excessive amounts. A particular concern with many small domestic supplies is the pick-up of lead from old lead piping since, in many cases, shallow wells and springs have soft and acidic water, resulting in plumbosolvency.

Radiological

Occasionally, in granitic or karstic limestone areas, **radon** may be a problem. Treatment is simple however, by aeration units or activated carbon filters.

Monitoring

Responsibility for monitoring public water supplies is placed on the statutory water undertakers in England and Wales by the Water Supply (Water Quality) Regulations 2000 (SI 3184). Water companies must also determine, in respect of each of their supply zones, sufficient suitably located sampling points to provide a representative sampling pattern for each of the parameters laid down. Standard arrangements for patterns and the number of samples are detailed in the regulations. Water companies must notify local authorities and district health authorities of the results of monitoring by 30 June in the following year, and must inform them of any event that may be a risk to the health of people living in that local authority area. Local authorities have the power to require information about any such incident and may also have their own sampling arrangements.

As they are required to be aware of the water quality in their area, Local authorities are also given discretion to make their own arrangements to check public supplies. They are free to take such samples of water as they reasonably require, although according to DoE Circular 20/89 [16] implementing the Water Act 1989, the Secretary of State did not expect any increase in sampling in view of the availability of information from water companies.

Private water supplies are the responsibility of local authorities as specified in the Water Industry Act 1991 and the Private Water Supplies Regulations 1991, including the maintenance of sampling programmes. This responsibility is usually undertaken by the environmental health department, which must identify private supplies in its district, carry out the necessary sampling and where required, provide advice on improving the supplies to achieve the standards required by the regulations. The Act specifies the information that local authorities are able to obtain about private supplies and gives local authorities the power to serve a notice to have them improved (Section 80). The regulations define wholesomeness, detail the classification of supplies, the ability to relax various standards, what parameters must be tested for, the standards to be achieved and prescribe monitoring frequencies and sampling charges.

Private supplies include rural supplies, bottled water that is not 'natural mineral water', and water supplies in food production undertakings that do not come directly from the public main. This may include transport catering in long distance buses or trains. Powers exist in England and Wales, under Section 140 of the Public Health Act 1936, to cut off polluted private water supplies from wells, tanks, or water butts that are liable to contamination [1].

Sampling techniques

Under regulation 16 of the Water Supply (Water Quality) Regulations 2000, water undertakers must apply 'the appropriate requirements' when taking samples. Similarly, under regulation 38, local authorities are required to follow the same procedures. Steps should be taken during the taking, handling, transportation, storage and analysis of water (or causing it to be analysed) to ensure that the sample is:

1. representative of the quality of the water;
2. not contaminated when being sampled;
3. kept at such a temperature and in conditions that will ensure there is no material change in the sample;
4. analysed as soon as possible;
5. analysed by a laboratory with a verified quality control system.

In practice, sampling for chemical analysis is a relatively simple procedure and only requires a fairly large but properly representative quantity of water in a 1 l or 2 l laboratory bottle. Microbiological samples, on the other hand, should be taken with care, to avoid contamination of the sample. The tap should be wiped with a disinfectant swab both inside and out, the water run for several minutes to ensure the sample is taken from the mains supply outside the property, and the tap then 'flamed' for a minute or so with a portable gas blow lamp (provided that the tap is metal). Subsequently, the water should be run for a further 2–3 minutes before the sample is taken to ensure the removal of all possible contamination from the tap and internal plumbing. Care should be taken to prevent the water from becoming contaminated by the neck of the bottle and from splashes from the sink. The cap should be held only by the rim, avoiding touching the actual surface with the fingers, before replacing immediately the bottle is full. All bottles used for sampling chlorinated supplies should contain a small quantity of sodium thiosulphate, to dechlorinate the water, so that the sample is representative of the water at the time it was taken.

It is important to remember that results from any single sample only indicate the condition of the water at that particular time (a 'snapshot'). Sanitary assessments of private supplies will give a better picture of the potential quality of the water than single samples. It is only by regular, systematic sampling and maintenance of a database of results over a period of time that a clear picture of water quality may be seen.

Water quality standards

Public health objectives for the analysis of water samples are mainly:

- to determine its safety and suitability, particularly for domestic purposes;
- to ensure concentrations of contaminants are below the maximums laid down in European and UK legislation.

The standards also apply in the food industry if the wholesomeness of the finished product may be affected by the quality of the water used. Examples of situations where food or drink may be affected are where the water is used:

- as an ingredient;
- in the course of preparing food;

- for the washing of equipment etc. or for the personal cleanliness of food handlers.

Water must be of the necessary standard when it arrives at a food premises; if it is mains water, this is the responsibility of the water undertakers. If it is a private supply, the person in charge of the supply and the proprietor of the food business are responsible. In either case, it is the responsibility of the local authority to ensure they carry out their duties.

EU directives now cover:

- the quality requirements of surface water for the abstraction of drinking water (Directive 75/440/EEC);
- the quality of water intended for human consumption (the 'Drinking-water Directive' 98/83/EC);
- the exploitation and marketing of natural mineral waters (Directives 80/777/EEC and 96/70/EC).

These directives utilize 'guide' values, which are preferred levels, as well as maximum admissible concentrations. Similarly, in respect of softened water intended for human consumption, values have been fixed for certain parameters that must be equal to minimum required concentrations. These are termed 'MRCs'.

Quality of surface water

The Surface Water Directive (75/440/EEC) sets values for each of the different categories of water from the purest upland source (A1) to lowland polluted river supplies (A3), and indicates the treatment requirements for each. Table 39.2 gives the 'I' values, which are mandatory for the three types of water A1, A2 and A3. Directive standards are applied by the statutory water undertakers and are monitored by the EA.

Quality of drinking water

After abstraction and treatment, water intended for human consumption in England and Wales is subject to the Drinking-water Directive. The directive introduced a duty on member states to

Table 39.2 EC requirements for abstracted surface water (intended for human consumption)

Parameter	A1	A2	A3
Coloration (mg/l Pt Scale)	20	100	200
Temperature (°C)	25	25	25
Nitrate (mg/l NO_3)	50	50	50
Fluoride (mg/l F)	1.5	—	—
Dissolved iron (mg/l Fe)	0.3	2	—
Copper (mg/l Cu)	0.05	—	—
Zinc (mg/l Zn)	0.05	0.05	0.1
Arsenic (mg/l As)	0.05	0.05	0.1
Cadmium (mg/l Cd)	0.005	0.005	0.005
Total chromium (mg/l Cr)	0.05	0.05	0.05
Lead (mg/l Pb)	0.05	0.05	0.05
Selenium (mg/l Se)	0.01	0.01	0.01
Mercury (mg/l Hg)	0.001	0.001	0.001
Barium (mg/l Ba)	0.1	1	1
Cyanide (mg/l Cn)	0.05	0.05	0.05
Sulphates (mg/l SO_4)	250	250	250
Phenols (mg/l C_6H_5OH)	0.001	0.005	0.1
Dissolved or emulsified hydrocarbons (mg/l)	0.05	0.2	1
Polycyclic aromatic hydrocarbons (mg/l)	0.0002	0.0002	0.001
Total pesticides (mg/l)	0.001	0.0025	0.005
Ammonia	—	1.5	4

take measures necessary to ensure that water for human consumption is wholesome and clean. This means that it must be free from micro-organisms, parasites and substances that, by their number and concentration, constitute a danger to human health. Water for human consumption has to meet a number of microbiological parameters detailed in the annexes, with member states setting more stringent values or values for additional parameters where necessary.

The Water Industry Act 1991 requires local authorities to keep themselves informed about the quality of water supplies in their area. In practice, this is generally satisfied by a sampling programme, including access to information on water quality held by the water service company concerned. Mains water supplied for human consumption must be 'wholesome', as defined by reference to the standards set in the Water Supply (Water Quality) Regulations 2000, which in turn incorporate the relevant elements of the EU Directive.

The directive requirements cover the following water supplies:

- water for human consumption supplied by a statutory company as a 'public supply';
- water for consumption supplied by a non-statutory water company, including private supplies and bottled water not covered by the exclusions;
- water supplied for use in food production undertakings.

Exclusions from the Directive are, broadly, private supplies to domestic supplies to less than 50 people, bottled natural mineral waters, medicinal waters and non-potable use of water in food production undertakings. To comply with the 2000 regulations, steps must be taken to ensure that drinking water supplies meet the requirements of all parameters in the Directive. The requirements (Table 39.3) are in the form of MAC values. The regulations make no requirements in respect of guide level values (GLs), which are included in the Directive's standards.

Natural mineral waters

Before exploitation and bottling of natural mineral waters is permitted, quality criteria and rules established under the Natural Mineral Water, Spring Water and Bottled Drinking Water Regulations 1999 (SI 1540) amended by the Natural Mineral Water, Spring Water and Bottled Drinking Water (Amendment) (England) Regulations 2003 (SI 1666) must be satisfied. The waters may then be 'officially recognized', and are subject to specific requirements in respect of treatment and bottling (EU Directive 80/777/EEC amended by 96/70/EC). Rules pertaining to labelling of spring water and the quality of bottled water are included in these regulations. Similar regulations apply to Scotland and Wales. The Drinking Water in Containers Regulations 1994 were revoked by the 1999 Regulations. The major enforcement provisions and defences of the Food Safety Act 1990 also apply.

Purification of water

The methods used for purifying water depend on the volume and quality of the water to be treated. Where underground supplies of good quality are involved, simple treatment can be used consisting of chlorination, pH correction and occasionally, filtration. Surface supplies usually require more extensive treatment, which can include storage, screening, aeration, coagulation, settlement, filtration, pH correction and chlorination. For some waters, additional processes such as softening, activated carbon filtration, microstraining or ozonization are employed. Where the local health authority has requested it, fluoride may also be added.

Treatment processes

STORAGE   Storage of polluted water in reservoirs, particularly from rivers (**bankside storage**) has been a recognized treatment process for many years. Because of low nutrient levels and natural predation, during long periods of storage pathogenic bacteria decay to relatively low numbers. Suspended matter also settles out, often carrying with it harmful micro-organisms, metals and other substances. A significant reduction in nitrate levels may occur by the action of bottom muds (**the benthic layer**), and hardness values can change. A disadvantage may be the growth of cyanobacteria, if the water is warm and **eutrophic** (due to organic impurities in the incoming water when concentrations of nitrates and phosphates are higher). This gives problems at the settlement and filtration stages of treatment, or impart tastes and odours to the water. The water can also become stratified and deoxygenated at lower levels. Mixing devices are installed in some reservoirs to aerate the water and overcome this problem. Aeration devices can be fitted on the bottom of reservoirs, or pumps can be fitted with jetting nozzles to induce turbulence.

SCREENING   Virtually all intakes to water supply systems are screened by filters of varying sizes. The devices used vary in size from simple bar types to complex microstrainers. The nature of the screening is at least partly determined by the quality of the raw water entering the treatment works.

SLOW SAND FILTRATION   This process, which was one of the earliest and best has largely been superseded by chemical coagulation and rapid sand filters. A proportion of the water supplied by the water industry however, is still subject to slow sand filtration. Slow sand filters consist of large open filter beds constructed with impermeable walls and floors and containing about 1 m of fine sand laid on 0.3 m of shingle with suitable drainage tiles at the bottom. Filters work under a head of about 1 m of water, percolating through at a rate of 100–300 l/m^2 per hour. The bed acts partly as a fine strainer, but most of the purification occurs by biological action, involving the formation of a gelatinous film on the surface of the sand (the *Schmutzdecke*), which

Table 39.3 Water Supply (Water Quality) Regulations 1989, the Private Water Supplies Regulations 1991, and the Drinking Water in Containers Regulations 1994

Parameter	Units of measurement	Concentration of value (maximum unless otherwise stated)
Organoleptic		
1. Colour	mg/l Pt/Co scale	20
2. Turbidity (including suspended solids)	Formazin turbidity units	4
3. Odour (including hydrogen sulphide)	Dilution number	3 at 25°C
4. Taste	Dilution number	3 at 25°C
Physiochemical		
5. Temperature	°C	25
6. Hydrogen ion	pH value	9.5 5.5 (minimum)
7. Sulphate	mg SO_4/l	250
8. Magnesium	mg Mg/l	50
9. Sodium	mg Na/l	150
10. Potassium	mg K/l	12
11. Dry residues	mg/l	1500 (after drying at 180°C)
Undesirable substances		
12. Nitrate	mg NO_3/l	50
13. Nitrite	mg NO_2/l	0.1
14. Ammonium (ammonia and ammonium ions)	mg NH_4/l	0.5
15. Kjeldahl nitrogen	mg N/l	1
16. Oxidizability (permanganate value)	mg O_2/l	5
17. Total organic carbon	mg C/l	No significant increase over that normally observed
18. Dissolved or emulsified hydrocarbons (after extraction with petroleum ether); mineral oils	µg/l	10
19. Phenols	µg C_6H_5OH/l	0.5
20. Surfactants	µg/l (as lauryl sulphate)	200
21. Aluminium	µg Al/l	200
22. Iron	µg Fe/l	200
23. Manganese	µg Mn/l	50
24. Copper	µg Cu/l	3000
25. Zinc	µg Zn/l	5000
26. Phosphorus	µg P/l	2200
27. Fluoride	µg F/l	1500
28. Silver	µg Ag/l	10[a]
Toxic substances		
29. Arsenic	µg As/l	50
30. Cadmium	µg Cd/l	5
31. Cyanide	µg CN/l	50
32. Chromium	µg Cr/l	50
33. Mercury	µg Hg/l	1
34. Nickel	µg Ni/l	50
35. Lead	µg Pb/l	50
36. Antimony	µg Sb/l	10
37. Selenium	µg Se/l	10

Table 39.3 Continued

Parameter	Units of measurement	Concentration of value (maximum unless otherwise stated)
38. Pesticides and related products:		
(a) individual substances	µg/l	0.1
(b) total substances	µg/l	0.5
39. Polycyclic aromatic hydrocarbons	µg/l	0.2
Microbiological parameters		
40. Total coliforms	Number/100 ml	0
41. Faecal coliforms	Number/100 ml	0
42. Faecal streptococci	Number/100 ml	0
43. Sulphite-reducing *Clostridia*	Number/20 ml	1 Multiple tube method
44. Colony counts	Number/1 ml at 22°C or 37°C	No significant increase over that normally observed
Other parameters (the maximum concentrations are average values or concentrations over the preceding 12 months)		
45. Conductivity	µS/cm	1500 at 20°C
46. Chloride	mg Cl/l	400
47. Calcium	mg Ca/l	250
48. Substances extractable in chloroform	mg/l dry residue	1
49. Boron	µg B/l	2000
50. Barium	µg Ba/l	1000
51. Benzo 3,4 pyrene	ng/l	10
52. Tetrachloromethane	µg/l	3
53. Trichloroethene	µg/l	30
54. Tetrachloroethene	µg/l	10
Additional parameters for desalinated or softened waters		
55. Total hardness	mg Ca/l	60
56. Alkalinity	mg HCO$_3$/l	30.0

Note:
a If silver is used in a water treatment process, 80 may be substituted for 10.

contains a variety of micro-organisms feeding off the contamination in the water. The film slowly builds up in thickness until the rate of filtration is slowed to below a practical limit. At this point the filter is cleaned. Filters will not reach full purification efficiency until the *Schmutzdecke* has formed, so the quality of filtered water must be monitored carefully during the initial ripening phase to ensure its safety. In many works, raw water is first passed through primary filters to reduce suspended matter and hence the load on the slow sand filters, this permits an increased time between cleaning.

COAGULATION, FLOCCULATION AND SEDIMENTATION
These are the processes used for the removal of impurities in most water treatment works.

Coagulation is the process of aggregation of the colloidal-sized negatively charged particles in the raw water. It is brought about by the addition of coagulants, usually either aluminium or ferric sulphate or polyelectrolytes, which have positively charged ions. The effectiveness of the process is also dependent on the pH over a range of values of alkalinity (Table 39.4). Sometimes potassium permanganate is added to the water at this stage if manganese levels are high and lime (calcium hydroxide), if overacidity is a problem.

Flocculation follows coagulation and is the growth in size of the coagulated particles. The end result of this process is a relatively dense precipitate known as floc. These larger particles will settle out more readily or be more effectively removed by filtration than the original particulate and organic contents of the water.

Sedimentation is achieved in tanks, of which there are many designs. Most designs ensure that the floc remains in contact with the water stream for a long period of time. This is achieved by preventing immediate settlement and by continuously removing only a proportion of the floc as a sludge, since the floc is an efficient absorber of many impurities including trace metals and bacteria. Most tanks work on an upward flow principle. The water enters the tank at its base, rising through a carefully controlled 'floc-blanket', with the clarified water being removed via decanting troughs from the top of the tank. Up-flow rates vary from 1 to 3 m/h, but higher rates can be used with some designs.

RAPID SAND FILTRATION   Sedimentation is usually followed by rapid sand filtration to ensure removal of any floc carried through with the decanted water. Rapid gravity filters are up to 40 times as fast as slow sand filters and normally employ filtra-

tion rates up to 20 m/h with peak flows as high as 50 m/h. They do not however employ any biological activity and merely act as physical filters. The filter normally consists of a large open tank containing a bed of sand 1–1.2 m deep supported on layers of coarse sand and gravel. Water is fed on to the top of the sand and removed through drains in the floor. The difference in level between the water in the tank and the filtered water channel provides a 'head' to force the water through the filter. As solids are taken out, the filter becomes clogged and the head loss across it increases. As this reaches a maximum, usually when the bed has increased to 2 m, or when breakthrough of suspended material occurs, the filter is backwashed using various combinations of clean water and the injection of compressed air to break up the aggregates in the sand (Fig. 39.9). Care has to be taken following the backwashing operation to ensure the sand settles down properly and is ripened to full efficiency. The water from the backwashing and ripening process may go to waste or in some cases a proportion is reintroduced at the head of the works.

Table 39.4 Optimum coagulation pH over alkalinity values

Alkalinity (CaCO₃)	Optimum coagulation pH
Soft waters <50 mg/l	5.0–5.5
Medium hard waters 50–200 mg/l	6.0–7.2
Hard alkaline waters >200 mg/l	6.0–7.5

PRESSURE FILTERS   There are a number of different forms and designs of pressure filters, with various filter mediums. They are used where it is undesirable to break the pressure after pumping, for example from a bore-hole. These perform in the same way as a rapid gravity filter, but the filter shell is a pressure vessel, usually cylindrical in shape. Somewhat higher head losses can be allowed to develop before backwashing is required.

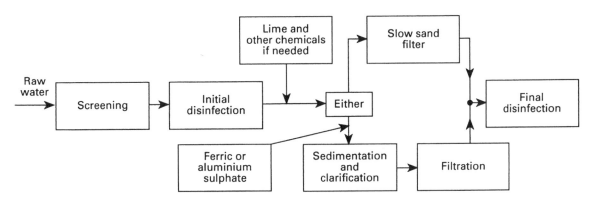

Fig. 39.9 The water treatment process.

Disinfection

CHLORINATION   Because it has proved to be so reliable and effective over the years, chlorine is used as a disinfectant almost universally in the water industry to ensure that water is free from harmful microorganisms. It is added either as a gas or as a solution of sodium hypochlorite, the latter being the norm for smaller supplies. Chlorine can also be generated on site for small supplies by the electrolysis of common salt solution. When the source is very pure, for example borehole water, the dose rarely exceeds 0.5 mg/l and the process is known as **marginal chlorination**. For most other supplies, **breakpoint chlorination**, or **super-chlorination** and **dechlorination**, are used. Most natural waters contain various microorganisms, organic contaminants, oxidizable material and ammonium salts and these exert a **chlorine demand** that must be satisfied if **free chlorine**, which is the most bactericidal form of chlorine, is to remain in the water. It is therefore usual to add just enough chlorine to overcome the chlorine demand and to maintain a small free-chlorine residual. This amount of free chlorine is 0.1–0.2 mg/l. This process is known as breakpoint chlorination. Super-chlorination is the name given to the process of adding excessive chlorine so that the breakpoint is well exceeded, and the residual is then controlled by the addition of a dechlorinating agent such as sulphur dioxide. Although it needs to be sufficient to ensure effective bactericidal activity, the amount of free residual should be low enough to leave no taste in the water.

ULTRAVIOLET IRRADAITION   Shortwave ultraviolet (UV) radiation at around 254 nm is effective against most microorganisms and may be used to disinfect water supplies. This UV spectrum is harmful to the eyes and skin and manufacturers' safety precautions must be adhered to. UV treatment has no smell associated with the treated water but there is no residual effect, so it is usually used for smaller supplies with short pipe runs. It is therefore popular for private water supplies. UV is also important in the control of *Legionella* and other organisms in recirculating water systems. UV filters (Fig. 39.10) are usually in the form of a

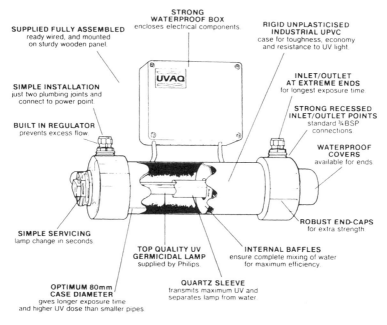

Fig. 39.10 Ultraviolet light sanitizing unit. (Courtesy of UVAQ, Unit 2, Mills Road, Chilton Industrial Estate, Sudbury, Suffolk.)

tubular unit with a central low-pressure mercury vapour discharge lamp and UV-emitting element, fitted into a quartz sleeve. Water flows through the unit with contact time maintained by a flow regulator device. The mercury tube gradually deteriorates and must be replaced every 6–12 months. The unit must also be maintained in a clean condition to ensure effectiveness. Several units have failed where a build-up of iron deposits have prevented the UV light reaching microorganisms. Multiple-tube, medium pressure units are also available where greater amounts of water are to be treated. UV is considered useful for the treatment of water containing *Cryptosporidium*.

OZONIZATION   Ozone, the triatomic form of oxygen (O_3), was first produced in large quantities by Von Siemens in 1857, using an electrical discharge tube. Water treatment plants throughout Europe have used this process for protozoa and virus elimination, as well as the removal of pesticides and other organic chemicals. Ozonization is relatively expensive when compared with chlorination and has no residual effect, but it has some advantages:

- ozone is better at removing viruses and protozoan parasites such as *Cryptosporidium* and *Giardia* than chlorine;
- it can be very effective in removing colour from moorland water;
- organic chemical, tastes and odours are removed;
- there is no smell associated with the treated water.

PH BALANCING   The addition of coagulants and chlorine to water tends to make it acidic. In addition, many upland waters are already acidic due to the presence of natural substances such as humic acids. Such water is often soft and aggressive and treatment is required to prevent corrosion. Copper pipes may corrode to give blue-green staining on ceramic ware, iron or galvanized pipes may corrode to cause water discoloration ('red water'), and with lead pipes it may cause an increased lead intake by the consumer. Acidity is normally corrected by dosing the supply with lime (calcium hydroxide), either as a slurry or more rarely as a solution prepared in lime saturators. Occasionally, sodium hydroxide (caustic soda) or sodium carbonate are used as an alternative to lime. A pH of 7 or slightly above is normally recommended.

REMOVAL OF IRON, MANGANESE AND ALUMINIUM Depending on the source of the water and the resulting level of contaminants, other treatment may be necessary. High manganese, aluminium and iron levels for instance, occur in upland surface waters that are low in pH. Balancing the pH with the addition of potassium permanganate or aeration will change the metal ions so that they precipitate out. This is followed by sedimentation or filtration, if necessary in two stages.

REMOVAL OF TASTES AND ODOURS   These may be reduced by the use of granular activated carbon, where the aromatic materials adhere to the charcoal or by ozonation.

FLUORIDATION   Sodium fluoride is added to the public supply at the request of the health authority in order to reduce the incidence of dental caries. This models the effect that has been demonstrated in those areas where calcium fluoride exists naturally. The addition of fluoride to public water supplies is contentious and has been debated by Parliament on a number of occasions. It is principally controlled by the Water (Fluoridation) Act 1985. Excessive levels of natural flouride (far above those found in public water supplies) have been known to cause mottling of teeth and brittleness of bones.

WATER DISTRIBUTION

The distribution, fitting and use of public water supplies is governed by The Water Supply (Water Fittings) Regulations 1999 (SI 1148). A British Standards Institution code of practice [17] gives general guidance on water installation to buildings and their curtilages. In general, pipes laid underground must be protected from damage and permeation by gas. They should not pass through sources of pollution, and not be constructed from materials that can cause contamination. As a principle, this means that planners should avoid having pipes passing through or close to drainage, manholes, cesspools, or similar and any alignment with gas mains should be carefully assessed. **Back syphonage** of water from installations, hoses and the like must to be avoided by the use of check valves, pipe interrupters, vacuum breaks, or air gaps between the mains supply and the point of use. Several waterborne outbreaks of disease have been caused by back syphonage causing contaminated water to be pulled into a water supply.

Storage cisterns

Domestic supplies of cold water for drinking purposes should be taken directly from the water main, through a service pipe, to the domestic taps used for drinking or cooking. There are however, circumstances in which it is necessary to have a domestic storage cistern; for instance, where mains supplies are intermittent or a pumped private supply is used. In addition, many industries that need large quantities of potable water for processing that require a storage cistern to ensure sufficient water is always available. In each of these cases the 1999 Water Fitting Regulations require a covered cistern, protected against ingress of insects and an adequate overflow warning pipe. Where an underground or partially below ground storage tank for potable water is required, this should be constructed of waterproof concrete. It must be designed, constructed and tested in accordance with BS 8007:1987 [18].

Water supply pipes

Pipes used for water distribution may be made from a variety of materials, including ductile iron, steel tube, wrought steel, asbestos cement or concrete. Internal linings containing coal tar are now prohibited because of the polynuclear aromatic hydrocarbons (PAHs) released into the water. Service pipes are usually of copper, plastic or occasionally, stainless steel. Lead, which was formerly used, is now prohibited for installation due to the ill health it causes. Copper pipes may not be connected to, or incorporated into, any lead pipe unless it is protected from galvanic action.

To achieve this, copper should not be connected or inserted 'upstream' of lead pipework.

DOMESTIC FILTRATION AND PURIFICATION

Although water supplied by the statutory water undertakers is consistently of more than adequate quality, circumstances sometimes arise, particularly in the case of private supplies, that make some form of additional domestic treatment either necessary or desirable. Treatment and filtration are somewhat interchangeable terms. Private water supplies often contain faecal material and physico-chemical contamination is usually higher than in public water supplies. Private supplies are also strongly influenced by periods of rain, particularly springs, surface supplies and wells, with high levels of contamination following heavy downpours. For this reason fitration and disinfection should be applied to all private supplies except those few that never show any signs of contamination. Because sampling programmes tend to be inefficient at finding contamination in private supplies, sanitary surveys or risk assessment surveys should be used to identify obvious sources of pollution and contamination pathways. These problems should then be rectified before consideration of treatment options. It is important that filtration should be related to the contamination that needs to be removed, as filter mediums vary in their selectivity and effectiveness at removing particular substances. Treatment units can be either point-of-use or point-of-entry. Point-of-use single tap devices may be appropriate where nitrate or heavy metal contamination is taking place but where microbiological contamination is a problem, whole house treatment with a point-of-entry unit is required to ensure safety for washing, brushing teeth, etc. as well as drinking.

Distillation and membrane filtration

Distillation and some membrane filtration units may be used to desalinate sea water or brackish waters. Distillation units may be fitted to the cold water supply under the sink and consist of a boiling unit with either water or air cooling for the condensation stage. They are very effective at removing impurities, but have a slow flow rate. Where contamination is caused by organic chemicals, distillation may carry over light fractions. These may be removed by an activated carbon filter. The resultant unit may be bulky and expensive to run.

Membrane filtration units use a semipermeable membrane and come in a variety of pore sizes. The name of the unit corresponds to the pore size. Reverse osmosis units are the finest and will filter out most molecules, including essential trace elements. The other sizes are nanofilters (pore size 10^{-9} m), ultrafiltration (10^{-7} m) and microfiltration (10^{-6} m). Filters with pore sizes larger than this are called microstrainers with a pore size of 25 or 35×10^{-6} m. Filtration is achieved by the differential pressure on each side of the membrane. The units are sometimes combined with other types of filters to produce a complete system that is very effective at removing contaminants. The water flow rate is low and a lot of water is wasted with typically 75% losses.

Filtration

A wide variety of point-of-use water filters are available for domestic use (Fig. 39.11) and their effectiveness is variable. They are basically two types: on-tap fittings and under-sink arrangements. On-tap filters fit directly on to the tap, in which case the limited size affects flow rate and effectiveness or they stand on an adjacent work surface. Under-sink units are permanently plumbed into the cold water feed and provide filtered water through a separate tap. Filtration units are generally fitted with replaceable or cleanable cartridges. Filter mediums may be of various materials, including plastic, metal gauzes and discs, ion exchange resin, activated carbon and silver.

Unglazed ceramic cartridges (also called ceramic candles) are effective at filtration, but require frequent cleaning and a high water pressure to produce even a low rate of flow. The fine pores of ceramic material are effective at mechanical filtration. Cleaning is achieved by scrubbing the cartridge with a stiff brush and immersion in boiling water. They can also contain activated carbon to improve filtration or a silver coating (silver is thought in some quarters to be bacteriostatic). Silver is not effective in this context for protozoan parasites.

Activated carbon filters are the most popular type of domestic filter, particularly for the removal of the taste of chlorine from public supplies. They are also sometimes combined with silver to reduce bacterial growth. The charcoal media has a large surface area and the process works by a combination of adsorbtion and physical filtration. However, they do not remove all impurities and do not remove hardness. Because they are very effective at removing organic contamination, the cartridges should be regularly replaced otherwise adventitious bacteria will grow on the organic material and start to enter the water supply. Many carbon filters are designed to work with treated public supplies and not contaminated private supplies.

Where hardness or nitrates are a problem, specific ion-exchange resins can be used. These can be very effective and replace the problem ions in the water with others, usually sodium or chloride. People who are on low sodium diets should be aware of this.

All filters should conform to BS 6920 [19] and other British Standards (CWA 14247:2001, BS EN12901 and 12902:1991) and should be installed in accordance with the Water Supply (Water Fittings) Regulations 1999, including a non-return valve, to prevent contamination of the mains supply. Cartridges need changing in line with manufacturers recommendations and units should be frequently cleaned to prevent a build-up of bacteria.

Emergency methods of domestic purification

Boiling is the best method of dealing promptly with drinking-water suspected of spreading disease. All pathogenic bacteria, viruses and protozoa will be killed by water brought to the boil. Statutory water companies will issue 'boil orders' or 'boil advisory notices' where there is a short-term risk from the mains supply. It is the primary emergency means of protection in the case of a suspect supply and if a boil order is in force its

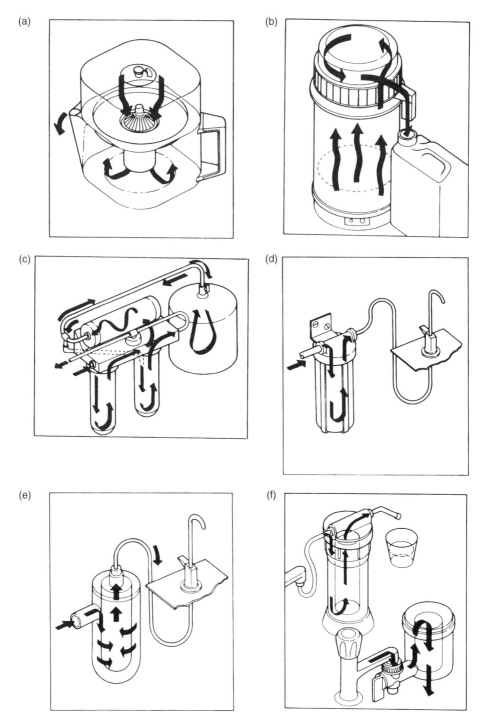

Fig. 39.11 (a) Jug filter. (b) Distillation unit. (c) Reverse osmosis under-sink filter. (d) Activated carbon under-sink filter. (e) Ceramic filter. (f) On-tap filters. (Reproduced with kind permission of the Consumers' Association Ltd from 'Water filters', *Which?* , August 1990.)

importance should be impressed on all concerned. It is not a practical proposition for any length of time and is difficult to implement for a large population. As a result, other means of treatment must be used for sanitizing water on a long term basis.

Chemical treatment by chlorination or ultraviolet light treatment is effective for reducing microorganisms in water. Chlorine is readily available as a solution of sodium hypochlorite, containing about 10% available chlorine or, occasionally, as calcium hypochlorite (bleaching powder), containing about 33% available chlorine. Both tend to decay over a period of time and should be kept in airtight containers, although any deficiency in strength may be counteracted by an increased dose. An effective sanitizing strength will achieve at least 1 μg/l of free chlorine in the water supply, although this may result in taste problems. A solution prepared from 31 g of bleaching powder, or 100 ml of sodium hypochlorite, to 1 l of water will be sufficient to treat 10 000 litres of water.

If the water is very impure, discoloured, or highly turbid it will reduce the effectiveness of chlorine and should first be filtered. This can be done by the addition of equal parts of alum (aluminium sulphate) and powdered chalk. It should then be allowed to stand before decanting to a separate tank for chlorination. A contact time of at least four hours is necessary before consumption. If primary filtration is carried out, this should precede chlorination, although on-tap or under-sink filtration may follow the process. In the case of heavily polluted water, it is important to test for available free chlorine after addition of the solution to the water.

Testing for chlorine

Many methods have been used over the years to test for residual chlorine. The main on-site test uses Palin's DPD (NN-diethyl-paraphenylene diamine) test method. By using a standard Lovibond, Tintometer or Hach comparator, with appropriate comparison discs or digital readout and DPD tablets, estimates can be made of free and combined residual chlorine, as well as total residual levels. It is sometimes useful to estimate chlorine levels at the same time as taking microbiological samples.

THE SEWERAGE SYSTEM

Introduction

The provision of an adequate systems of drains and sewers is fundamental to the effective drainage of an area. The duty of providing sewerage services was transferred to the sewerage undertakers under the Water Act 1989, now operating under the framework of the Water Industry Act 1991,[2] the first time in the United Kingdom that such provision has been the responsibility of the private sector.

Drains and sewers[3]

For an adequate understanding of the principles of drainage, it is important that the difference between 'drains' and 'sewers' should first be understood. The various definitions are contained in Section 343 of the Public Health Act 1936 and Section 219 of the Water Industry Act 1991.

Effectively, a **drain** is a single pipe from either one building or a number of buildings 'within the same curtilage'. The meaning of the word curtilage is sometimes subject to fine legal interpretation and regard for the precise circumstances of a particular case (see, for example, *Cook* v. *Mignion* (1979) JPL 305 and *Weaver* v. *Family Housing* (1975) 75 LGR 255), but broadly the word means the boundary or fence of a particular property.

A **sewer** is generally a pipe taking drainage from a number of drains. Such a sewer may be a **public sewer**, in which case it is 'vested' in a sewerage undertaker by the various powers under the two Acts. All sewers built before the introduction of the 1936 Public Health Act (October 1937) are public sewers. In other cases, where drains have been connected together to form a sewer (in order to simplify drainage, but the overall ownership remains conjointly with the property owners), the

---

[2]Also see amendments made by the Water Act 2003 to be implemented in 2004.
[3]Also see DEFRA (July 2003) *Review of Existing Private Sewers and Drains*. DEFRA, London.

sewer is a **private sewer**, being constructed solely for the use of those properties. In exactly the same way as a drain, it is to be maintained by, and at the expense of, the person or persons to whom it belongs. Private sewers and drains are generally laid in the gardens, courts, or yards of the various premises to which they belong. However, they remain private up to the point of connection with the public sewer, even under public pavements and highways.

Provision of sewerage

Under Section 94 of the Water Industry Act 1991, sewerage undertakers must 'provide, improve and extend . . . a system of public sewers'. They must also maintain and cleanse the sewers, as well as provide disposal facilities, including sewage disposal works. Their duty extends to making provision for trade effluent discharged under the powers of Part 3 of the 1991 Act. This duty to provide sewerage for their districts is similar to that placed on the predecessor authorities, but with the significant introduction of enforcement orders under Section 18 of the Water Industry Act 1991. In addition, regulations made under Section 95 of the same Act may prescribe performance standards (see the Water Supply and Sewerage Services (Customer Service Standards) Regulations 1989 (SI 1159) as amended in 1989).

Connections to sewers

Owners and occupiers of premises are given the right, by Section 34 of the Public Health Act 1936, subject to certain restrictions, to drain into public sewers. The actual connection to the sewer may be made by the sewerage undertakers if they elect to do so, under Section 36 of the same Act, or otherwise it may be made by the owner or occupier himself. In the latter case, owners or occupiers must give 'reasonable notice' to the statutory undertaker, which may supervise the works. In either case, and subject to appeals, the sewerage undertaker may refuse to allow the drain to communicate with the public sewer if 'the mode of construction or condition of the drain or (private) sewer is such that [it] would be prejudicial to their sewerage system'. Sewerage undertakers may also, at their expense, alter the existing drainage system

of premises where the drainage is sufficient for the premises, but is 'not adapted to the general sewerage system'. This is subject to giving adequate notice, the provision of 'equally effectual' drainage and rights of appeal to a magistrates' courts [1].

Sewer requisition

Section 98 of the Water Industry Act 1991 provides requisitioning arrangements for public sewers similar to those first introduced by the Water Act 1973. Essentially this allows owners, occupiers or local authorities (including new town and development corporations) to serve a notice on the appropriate sewerage undertaking company requiring it to provide a public sewer for domestic purposes in a particular area. In these circumstances, Section 99 of the 1991 Act requires the person or authority to pay the capital costs over a period of 12 years.

A new Section 101A was added to the provisions by the Environment Act 1995, making it the duty of a sewerage undertaker to provide a public sewer for domestic drainage under particular circumstances. These are broadly if:

- it is for a dwelling house constructed before 20 June 1995;
- the drains currently do not connect, directly or indirectly, with a public sewer;
- the drainage is having an adverse effect on the environment or amenity, and that it is appropriate to provide a public sewer.

A number of other factors must also be taken into account, including the geology of the area, the number of premises involved, the cost of providing a sewer, the nature and extent of any adverse effects and the practicability and cost of making alternative arrangements to overcome the effects.

Agency arrangements

The relationship between sewerage undertakers and local authorities was reinforced by Section 97 of the 1991 Act, which allowed the latter to carry out sewerage functions on behalf of the sewerage undertakers. Clearly, where agency agreements continue to exist, the operational arrangements and the various powers are consolidated within one authority, which facilitates local decision-making.

Miscellaneous provisions

Section 21 of the amended Public Health Act 1936 also gives sewerage undertakers the power to use highway drains and sewers for sanitary purposes by agreement with the highways authority. This power is also actionable by the local authorities. The powers in the Public Health Act 1936 over the adoption of sewers (Sections 17 and 28), construction standards (Section 9), purification of sewage (Sections 30 and 31), rights of owners and occupiers to drain into public sewers (Section 34), the ability to alter the drainage system of premises (Section 42) and the ability to make connections to public sewers (Section 36) were unaltered by the Water Act 1989 and the Water Industry Act 1991, except that they are also exercised by the sewerage undertakers.

Local authority powers

For a detailed description of local authority powers in relation to sewers and drains, together with flow diagrams, see *Environmental Health Procedures* [1].

New drains and sewers

Provision of drainage in new buildings is controlled by Section 21 of the Building Act 1984. When plans of a building or an extension are deposited with a local authority, under the Building Regulations 2000, they must be rejected if the arrangements for drainage, where necessary, are not satisfactory. Proposals for provision of drainage must include proper connection to the public sewer, cesspool or some other place, as required by the local authority or, on appeal, by the magistrates' court. Connection to a public sewer cannot be required unless the sewer is within 30.5 m and the owner is entitled to construct a drain through the intervening land. A local authority may, however, require a building to be drained into a sewer that is more than 30.5 m distant if it bears the cost of construction and maintenance of the drain from the sewer up to the point within that distance of the building.

Section 22 of the Building Act 1984 gives the local authority power to require construction of a private sewer to new buildings where they 'may be drained more economically or advantageously in combination'. Clearly, if there is to be any expectation that a private sewer is to be adopted as a 'public sewer' by the sewerage authority, it should accord with any constructional standards laid down by the sewerage authority.[4] The maintenance of private sewers is likely to cause interowner disputes and legal difficulties, particularly if a large number of properties are connected. For this reason, developers should be encouraged to enter into a formal adoption agreement under Section 18 of the Public Health Act 1936, ensuring that the sewer will be adopted provided that it is constructed to the agreed standard.

The Building Regulations 2000 schedule 1 part H1 states that 'any system which carries foul water from appliances within the building to a sewer, a cesspool or a settlement tank shall be adequate'. The regulations make similar provision at H3 for roof water. (See for the role of Approved Documents under the Building Regulations.)

Section 60 of the Building Act 1984 makes a specific requirement, enforceable by local authorities, for the provision of ventilation to a soil pipe serving a WC. The same section also places a prohibition on the use of rainwater pipes from a roof as soil pipes, or the use of surface water drains as a ventilation shaft for foul drainage.

If it is proposed to erect a building or extension over a drain, public sewer or disposal main, Section 18 of the Building Act 1984 requires the local authority to reject any plans submitted under the 2000 regulations if maintenance or access to the drain, sewer, etc. is to be interfered with. ('Disposal main' means an outfall pipe from a sewage works that is not a public sewer.) Consent may be given to erect a building under these circumstances, and the plans passed, provided that

---

[4]See Water Act 2003 for new requirements relating to the adoptable standards for lateral drains.

the authority is satisfied that it may 'properly consent'. Approval of such a construction may be subject to conditions and specific constructional requirements.

Plans that are deposited that show a proposal to build over a drain, sewer or disposal main must be notified to the sewerage undertaker by the council concerned. In turn, the undertaker may specify to the council 'the manner in which [it is] to exercise [its] functions' under Section 18. In practice this means that it will either require the local authority to reject the plans, or specify the conditions and/or constructional standards required. Clearly, where the local council is itself acting as agent for the sewerage undertaken under Section 21, it may determine these matters on behalf of the undertaker. The power relating to the rejection of plans also applies to an 'initial notice' given by an approved inspector under the Building Regulations 2000.

Defective drainage

Three main powers are given to local authorities to deal with defective existing private sewers and drains.

1. Under Section 17(3) of the Public Health Act 1961 they may serve a 48-hour notice to deal with a **stopped up drain, private sewer**, etc. If the notice is not complied with, they may carry out the necessary work themselves and reclaim the expenses from the persons concerned. This is subject to rights of appeal, particularly over the cost of the works and the apportionment of the costs if more than one property is concerned.

2. Under Section 17(1) of the Public Health Act 1961, they may serve a seven-day notice of their intention to deal with a **drain, private sewer, etc. that has not been maintained** and **kept in good repair** and that can be repaired for less than £250. They may recover the expenses from the 'person concerned' but, on appeal to the courts, will be required to justify their conclusion that the pipe or appliance was not properly maintained. The court may also enquire into the apportionment of the costs

charged to the person or persons on whom the notice was served and make an order concerning the expenses or apportionment (Section 17(7)).

3. Section 59 of the Building Act 1984, which replaced the almost identical powers formerly contained in Section 39 of the Public Health Act 1936, gives local authorities the duty of serving a notice in a wide variety of situations concerning existing **defective private drainage**. This applies to virtually all pipes and appliances, including cesspools, other than public sewers. The period allowed for appeal against notices served under this section is 21 days. As a consequence, a longer period, often 28 days, must be allowed for compliance. This delay may cause difficulty where a serious risk to public health exists.

Section 22 of the Public Health Act 1961 gives local authorities the power to **cleanse or repair** defective drainage etc. at the **request** of an owner or occupier and at his expense. In practice, this is a useful alternative to service of a notice, carrying out work in default and recovering the expenses, where the owner or occupier is willing to carry out the work. Indeed, many local authorities enclose a 'request authority' with the service of notice, to offer the opportunity of early remedy for what is often a serious public health problem.

Any **court, yard, etc. that is inadequately surfaced** so that surface water drainage is unsatisfactory may be dealt with by the local authority serving a notice on the owner under Section 84 of the Building Act 1984. This provision applies only in relation to a yard or court that is common to two houses, or a house and a commercial or industrial building. It does not apply to highways maintained at public expense.

In addition, powers contained in the 1936 Act to deal with overflowing and leaking cesspools (Section 50) and care of sanitary conveniences (Sections 51 and 52) are similarly enforced by unitary and district councils serving notice on owners, or occupiers, as appropriate.

The **testing** of drains or private sewers that are thought to be defective is an important prerequisite for serving notice to repair or cleanse them.

The power to carry out tests is contained in Section 48 of the Public Health Act 1936, and is given to the 'relevant authority', which in the case of drains and sewers communicating with a public sewer means the sewerage undertakers. It is not clear whether this restricts local authorities to testing drains and sewers that communicate with private disposal facilities, or whether it simply limits their ability to test the final run of drainage that connects to the public sewer. The different wording of subsection 1 of Section 48 should be noted, which refers to 'a drain or private sewer communicating directly or indirectly with a public sewer'. Once again, where an agency agreement exists between a local authority and the relevant sewerage undertaker, the matter is simplified.

It is not always necessary to expose and examine the drains to check for defects and, initially, the application of a non-pressure smoke test or colour test is sufficient. At the least, this will give an indication of any break in the drains and its approximate location. If necessary, remote television probes may be used to examine the interior of drainage systems, and it is now rarely necessary to expose drains before the nature and location of the defect has been established. If alterations or repairs are carried out to underground drains, other than in an emergency, the person carrying out the work must give the local authority at least 24 hours' notice, by Section 61 of the Building Act 1984. If any work involves the permanent disconnection or disuse of any drains, they must, by Section 62 of the same Act, be sealed off 'at such points as the local authority may reasonably require'. Section 63 of the 1984 Act also requires that a water closet, drain or soil pipe should be properly constructed or repaired. If conditions prejudicial to health or that are a nuisance are caused by such works, then the person carrying them out is guilty of an offence.

A desription of the various tests which may be applied to sewers and to drains is given in the following sections.

Local authorities are authorized, by Section 67 of the Building Act 1984, to lend temporary sanitary conveniences where works on drainage etc. necessitates disconnecting WCs etc. They may make charges for the loan, provided it exceeds seven days and is not a result of work made necessary by a defect in a public sewer.

TRADE EFFLUENT

Section 27 of the Public Health Act 1936 places a general prohibition on passing any matter into a drain or public sewer that is harmful to it or that might interfere with the disposal of sewage. There are also specific prohibitions on the discharge of chemical waste, steam, high-temperature liquids, petroleum and calcium carbide. This is particularly important when considering the means of arresting materials before they enter the drainage system (see next section).

The Public Health (Drainage of Trade Premises) Act 1937, originally enforced by local authorities, is now the responsibility of the EA, and its provisions have been consolidated in the Water Resources Act 1991 Part 4. Before discharging trade effluent into a public sewer, the occupier or owner of the trade premises concerned must seek a consent from the EA. The application must be in writing and must state:

- the nature or composition of the effluent;
- the maximum quantity to be discharged in any one day;
- the highest rate at which it is to be discharged.

Copies of any consents must be kept available for public inspection.

Important changes have been made to the provisions of the 1937 Act by Section 138 of the Water Industry Act 1991. This section makes provision for 'prescribed substances' present in the effluent in 'prescribed concentrations' and effluent derived from a 'prescribed process', or involving use of 'prescribed substances' in specified quantities or amounts. Regulations made under this section, the Trade Effluents (Prescribed Processes and Substances) Regulations 1992, have 'prescribed' the substances contained in list 1 of EU Directive 76/464/EEC on pollution caused by certain substances discharged into the aquatic environment. In addition, the regulations prescribe effluents derived from five types of processes, including asbestos processes, giving effect to Article 3 of EU Directive 87/217/EEC on the prevention and reduction of environmental pollution by asbestos.

Section 130 of the Water Industry Act 1991 requires the EA to refer any applications for consent, or any proposal to make an agreement to discharge trade effluent, to the Secretary of State. It is for him to determine whether the operation should be prohibited, or any conditions applied to a consent or agreement. Appeals by aggrieved persons are to be made to the Director General of Water Services. These are important modifications the 1937 Act, foreshadowing an evolving policy to protect the aquatic environment.

Sewelage systems

Two main systems are employed for the disposal of sewage, trade effluent and surface water run-off from roofs and hard surfaces. In the one case, rain or surface water is kept separate from soil and waste, and the two are conveyed to separate outfalls. Surface water is conveyed direct to a natural watercourse and foul sewage to a disposal works. This is known as the **separate**, or occasionally the dual, system.

Where surface water and foul water are conveyed together, this is referred to as a **combined system**. Combined systems have the virtue of simplicity but requires pipework of much larger capacity to deal with storm water. The need to allow for smaller, and therefore cheaper, capacity sewerage systems often leads to the construction of **storm water overflows**. These are usually constructed as part of an inspection chamber and vary in design; essentially they provide an overflow from the system at a high level, usually connecting with a natural watercourse.

At one time it was thought that the effluent was graduated at times of storms, with freshwater flowing over the top of the foul sewage, and the discharge was thought to consist almost entirely of rainwater. It is now accepted that this is not the case, and the overflows at times of storm are of highly diluted sewage. This, in turn, introduces the risk of contamination of watercourses by overflows. This problem can be particularly serious in the case of sewer blockages, which can lead to the discharge of large volumes of undiluted sewage through storm water overflows.

The advantage of separate systems is the avoidance of surcharging of sewers by vast volumes of surface water during storms. This, in turn, allows the construction of smaller capacity drains and sewers to deal only with foul matter. In practice, however, truly separate systems are difficult to maintain, and is often confusing, leading to wrong connections being made by developers. The contamination by tar used in roads and domestic driveways and, more particularly, by increasing amounts of oil and other deposits from motor vehicles, are rendering this system less and less effective as a means of preventing contamination of the natural water environment. In towns, often the only water sufficiently clean for direct discharge into a river is water from roofs. The essential principles of separate drainage require that this should be conveyed through pipes and gullies used exclusively for the purpose. In some instances there is a combination of the two systems called **partially separate**, when some rainwater is allowed into the foul water system. On balance, despite higher construction costs, the separate system should be favoured.

The discharge of foul water into a sewer provided for surface water or, except with the approval of the sewerage undertaker, the discharge of surface water into a sewerage for foul water, is prohibited by Section 34 of the Public Health Act 1936 (as amended). The connection of a waste pipe from a lavatory basin or a WC, for example, to a rainwater pipe discharging into a sewer provided for surface drainage only would be an offence under this section and might also be an offence under the Building Regulations 2000.

Separate systems should always be adopted in connection with any individual house or building that drains into a cesspool or private sewage disposal plant. Rainwater pipes from roofs and gullies (provided only for surface drainage) may be made to discharge into prepared soakaways in the gardens etc. subject to consent from the EA. In premises where sewage treatment is carried out on site, drains conveying surface water may discharge into the outlet of the purification plant provided that this is also agreed with the EA.

Drainage: design and construction – approved document H

The Schedule 1 part H of the Building Regulations 2000 gives broad requirements relating to the design and construction of new drainage and waste disposal. For example, in regard to rainwater drainage it is a requirement, and therefore a breach of the regulations if not complied with, that 'Any system which carries rainwater from the roof of a building to a sewer, soakaway, watercourse or some other suitable rainwater outlet shall be adequate'.

The method of achieving a compliance with these broad requirements is then at the discetion of the contractor in that this can either be done in accordance with Approved Documents issued in support of the regulations, or by other ways which will achieve compliance.

Approved Document H which came into effect on 1 April 2002, deals with drainage and waste disposal and is constructed in 6 parts:

- Foul water drainage
- Wastewater treatment systems and cesspools
- Rainwater drainage
- Building over sewers
- Separate systems of drainage
- Solid waste storage.

Reference should therefore be made to that document in relation to the more detailed aspects of the standards for design and construction. This includes reference to the approprite BS and BS EN standards.

DRAIN TESTING

The smoke test

This test is usually applied to existing drains with a view to revealing and locating leaks and detecting defects, which may be connected with rat infestations. It may be applied by the use of a smoke 'rocket' or by a specially constructed machine. A smoke rocket is essentially a large smoke-generating firework and its value is very limited. No pressure can be applied, and its main use is for tracing connections and demonstrating the presence or absence of obstruction, rather than as a test for soundness.

One of the best-known smoke testing machines and the method of applying a smoke test is illustrated in Fig. 39.12. The apparatus consists of a double-action bellows and a smoke box, which are enclosed by a light copper dome working in a water seal. More compact and lightweight equipment is also available, which includes a vertical pump connected to the smoke chamber and a manometer to indicate pressure.

Smoke-producing material, such as a specially prepared paper or oily cotton-waste, is placed in the smoke box and lighted. By slowly working the bellows, large volumes of pungent smoke are produced, and this is conveyed by means of a flexible tube into the drain through a gully, WC or access chamber, whichever is convenient, but preferably at the low end of the system.

In order to ensure that the drain in filled with smoke and not merely air locked, traps should be emptied and re-sealed one by one as the smoke appears through them. The last openings to be sealed are those of the ventilating pipes. When these are sealed, a slight pressure may be set up in the drain. The copper dome rises with the action of the bellows, and if the drain is quite sound, is maintained in a slightly elevated position.

Considerable care is needed in the application of this test. If any pressure, however slight, is exerted in excess of that which can be resisted by the head of the water seals of traps when depressed, smoke is forced through into the building and the test may be rendered misleading and, therefore, useless.

The colour test

This test is useful for tracing drain connections, leaks and gradients. A drain that is sound under a pressure test is not necessarily a good drain. The joints may contain surplus cement on the inside – a common defect in stoneware drains laid by unskilled people. The fall may be insufficient or may even be in the wrong direction. These are serious defects and may be detected and, to some extent, measured by the colour test.

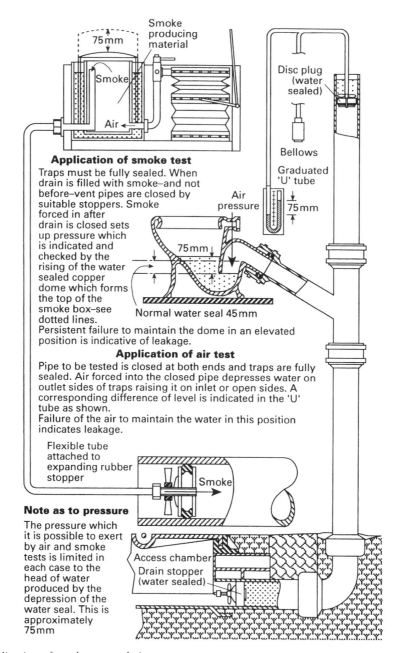

Application of smoke test

Traps must be fully sealed. When drain is filled with smoke–and not before–vent pipes are closed by suitable stoppers. Smoke forced in after drain is closed sets up pressure which is indicated and checked by the rising of the water sealed copper dome which forms the top of the smoke box–see dotted lines.

Persistent failure to maintain the dome in an elevated position is indicative of leakage.

Application of air test

Pipe to be tested is closed at both ends and traps are fully sealed. Air forced into the closed pipe depresses water on outlet sides of traps raising it on inlet or open sides. A corresponding difference of level is indicated in the 'U' tube as shown.

Failure of the air to maintain the water in this position indicates leakage.

Flexible tube attached to expanding rubber stopper

Note as to pressure

The pressure which it is possible to exert by air and smoke tests is limited in each case to the head of water produced by the depression of the water seal. This is approximately 75mm

Fig. 39.12 Application of smoke test and air test.

The test should be applied to the drain in sections, each section being first washed out with clean water. Coloured water, produced by the use of a dye sachet, is poured into the drain in small measured quantities with intervals of time between each. Any water retained by obstructions or improper levels is displaced by the coloured water, the amount of which it is found necessary to add before the first trace of it is discharged at the outlet represents approximately the amount of water normally retained by the drain.

The colour test is also useful for determining the source of any water that may be causing a nuisance, for example, water percolating through a basement wall that is normally dry. A drain passing near the wall may be suspected. For this purpose, a solution of fluorescene should be poured into the drain. This chemical when added to water possesses a bright green fluorescence, which is unmistakable and easily discerned even though it may be very highly diluted.

The air test

This is a very convenient and useful test for soil and waste pipes. The application of the test to a soil pipe is illustrated in Fig. 39.12. Approved Document H contains information about the requirements of this test.

All plugs used in the test must be water-sealed in order that any leakage of air through them may be detected. The test is delicate, and the pressure obtainable limited by the depth of water seals in the traps. A valuable and important feature of the test is that it indicates the weakest water seal in connection with a pipe under test. A pressure equal to 38 mm water gauge is usually applied and this should remain constant for 3 minutes.

The test may also be used for drains before they are submerged in water or the trench is filled in. Junctions are sealed off and expanding plugs screwed into each end of the length to be tested. The U tube is connected to the plug at one end of the drain, and air is pumped through the other until 100 mm water pressure is indicated. The cock at the pump is turned off, and if the water pressure drops by more than 25 mm in 5 minutes for a 100 mm gauge the drain is unsatisfactory.

A maximum drop of 12 mm is allowed for a 50 mm gauge (after allowing a suitable time for stabilization of the air temperature).

Unlike the water test, the air test is of equal severity regardless of the length of the drain tested. An objection sometimes raised against this test is that while it may indicate leakage, it does not locate it. This is only partially true. If leakage is indicated, suspected joints should be coated with a soap solution. By this means, escaping air is easily discerned.

The water test

Approved Document H requires that drains should be watertight and tested by either a water test or an air test as prescribed for new drainage systems and gives details of the conduct of this test.

The lower end of the drain is plugged by means of an expanding rubber stopper or, if this is difficult, by a flexible pneumatic stopper. A 1.5 m head of water above the invert of the pipe at the top end of the pipeline is used to create the pressure, and this can be obtained by temporarily jointing a 90° bend and a straight pipe of the same diameter as the pipeline to the head of the line and filling with water. Care must be taken to prevent airlocks in unventilated branches. This is done either by emptying the traps or by passing a ventilated pipe through the water seals.

The section of the drain should be filled and allowed to stand for two hours before being topped up. The leakage over 30 minutes should be measured by adding water up to the original level, and should not be more than 0.05 l for each metre run of drain, for a 100 mm diameter drain, equal to a 6.4 mm/m drop in water level. In the case of a 150 mm diameter drain, the loss should not exceed 0.08 l/m run, or 4.5 mm/m drop in water level.

To prevent damage, the head of water to the lower end of the drain should not exceed 4 m and it may be necessary to test the system in sections.

Television surveys

Increasingly, the testing of drains and sewers is being replaced by internal inspection, using specialized closed circuit television (CCTV) systems. As a result, it is now rarely necessary to expose drains or sewers before the precise nature and location of any defect has been established. Special cylindrical video cameras of stainless steel construction, watertight to over 100 psi, are used to examine the interior of pipes from as small as 50 mm through to 600 mm in diameter. They may be mounted on appropriately sized skids, which are pushed or pulled through the pipe with rods. Alternatively, the camera may be mounted on an electrically driven tractor unit, which can crawl along the pipe's interior. The camera transmits an image to a closed-circuit monitor, usually in a vehicle, where defects can be detected and noted. Video tapes and photographs may also be produced.

In its simplest form, CCTV pipe inspection units provide monochrome pictures from a single angle. A cable meter indicates the position of the defect by measuring the length that has been inserted into the drain or sewer. Other units may be highly sophisticated, providing high definition colour video reproduction, with a wide variety of control and measuring facilities, including the ability to negotiate acute bends. Pan, tilt and zoom facilities may be available on the lens, as well as an online computer analysis of the video signal to provide text data on the pipe dimensions and levels.

The cleansing and repair of drains and sewers

The methods available for cleansing and repairing underground pipework and removing blockages include the following.

1. **Rodding**. Rodding is still widely used and is particularly applicable to smaller pipes. Clearance of the blockage is attempted by bringing the end of a flexible rod into contact with the material responsible for the obstruction. Various devices are available for use at the tip of the rod, including spirals and rubber plungers, the latter using air compression in addition to mechanical pressure. The material responsible for the obstruction is not, however, removed, but is merely transported along the drainage system, and for this reason large quantities of water are required for flushing.
2. **Flushing**. The routine flushing of sewers and drains that are not laid to a self-cleansing velocity is essential, and flushing of other sewers may be necessary from time to time in order to remove sludge and grit. Flushing devices, such as the automatic siphon that is fed by mains water and discharges by siphonic action, are constructed as part of some systems. Where such systems are not installed, flushing can be accomplished by the use of hydrant water supplies via 100 mm diameter hoses, or with mobile flushing tanks.

3. **Hydraulically propelled devices**. Water is discharged into the sewer behind a sewer ball or jetting ball, which is slightly smaller in diameter than the pipe. The flow behind the device is driven through the gap between the device and the pipe wall at a high velocity, and a scouring action results. This method is used for dislodging sludge and grit.
4. **Winched devices**. This system can be used to remove debris and roots, but access via two manholes is required. A bucket or similar device is winched through the sewer by hand or power, and the debris is collected and removed.
5. **Jetting**. This is most effective in pipes of 150–375 mm in diameter. Water under pressure is fed through a hose to a nozzle containing a rosette of jets, and is sufficient to dislodge material on the pipe walls and debris lying along the invert. The system uses a purpose-built vehicle incorporating a water tank and including a suction unit for gully cleansing.
6. **Repairs**. As an alternative to conventional repairs to underground pipes by excavation and replacement of damaged sections, new *in situ* repair systems are now available. Essentially, the commercially available systems consist of installing a liner to the damaged pipe(s) and subsequent internal, remotely controlled cutting of lateral connections. The steps involved include use of CCTV cameras to carry out a preliminary survey of the pipe, to prepare the interior for the insertion and to control the installation. Pipe liners currently available include those made of preformed stainless steel or polyethylene. Alternatively, *in situ* linings may be formed by use of polyester and polyurethane felt tubing impregnated with thermosetting resin, cured on site with hot water. In this latter case, internal patches up to 3-m long may also be applied to pipe defects using the same technology and also controlled by CCTV monitoring.

THE NATURE OF SEWAGE

Sewage is a cloudy, aqueous solution containing mineral and organic matter as floating particles

Table 39.5 Typical chemical composition of dry weather domestic sewage

Mineral content dissolved and suspended	420 mg/l
Organic matter dissolved and suspended	620 mg/l
Total solids	1040 mg/l
5-day biochemical oxygen demand	310 mg/l

and, in suspension, colloidal material and substances in true solution. It also contains living organisms, particularly very large numbers of bacteria, viruses and protozoa. Its nature may be altered by design (combined system), or by infiltration through the pipe joints. Trade wastes also alters its composition (Table 39.5).

TREATMENT OF SEWAGE

Inland sewage disposal works usually provide full primary and secondary treatment, with discharge of final effluent mainly into rivers. Sewage from some coastal towns is still discharged directly into the sea, with only primary treatment. Sometimes the sewage is screened and macerated before discharge in order to hasten its breakdown by the salt water, wave action and sunlight. This unsatisfactory state of affairs has been changed by the Urban Waste Water Treatment Directive (91/271/EEC) [20] through the Urban Waste Water Treatment (England and Wales) Regulations 1994 (SI 2841). These regulations, enforced by the EA, provide for varying degrees of sewage treatment over a timescale dependent upon population size (Table 39.6). In addition, the dumping of sludge at sea by ships was prohibited after December 1998.

Discharges of waste waters entering collecting systems are now subject to requirements that will accelerate the installation of in-house waste water treatment systems by industrialists. A similar effect has resulted from the requirements over direct discharges of biodegradable industrial waste water.

Disposal schemes

The basic principles underlying the conventional treatment of sewage are physical and biological.

In the case of public disposal works, operated by statutory sewerage undertakers, the treatment may be extensive. Fig. 39.13 shows the sewage treatment process.

Physical

This involves the removal of solids usually by screening followed by detritus and sedimentation tanks. Screens may be fixed or moving. Fixed screens may be vertical or inclined at an angle and, in the larger works, are mechanically raked. Moving screens may be an endless belt of wire mesh or perforated plate, or a revolving drum where the sewage passes into the drum and out through the sides, which are perforated.

Grit, sand, etc. is removed by passing the sewage through a settlement tank (grit chamber), or sometimes by passing the sewage through a channel at a constant velocity.

This is followed by sedimentation, allowing much of the suspended solids to settle by passing the sewage slowly through tanks. These may be circular, rectangular with a horizontal flow, or square with an upward flow. The solids settle as a sludge which, in a properly designed system, consists of 40–60% of the solids originally contained in the sewage. The sludge from the rectangular tanks may be cleared manually by suction or mechanical scraper, and the vertical flow tank is mechanically scraped and discharged under hydrostatic head.

Biological

This consists of the oxidation of the organic matter and ammonia left in the sewage after settlement and is effected by the action of micro-organisms and in particular the bacteria. This part of the treatment is achieved either by attached growth processes (such as trickling filters or biological aerated filters), or suspended growth processes (such as activated sludge).

FILTRATION The use of the word filtration is rather misleading as the function of the filter is not to filter out suspended solids but to bring the settled sewage into contact with a durable medium, such as hard clinker or broken stone or plastic, on

Table 39.6 Urban Waste Water Treatment (England and Wales) Regulations 1994

Population size	Provision of sewerage date	Primary treatment date	Secondary treatment date
Over 15 000 generally	31/12/2000	31/12/2000	31/12/2000 (exceptionally by 31/12/2005)
Over 15 000 discharging into waters with high natural dispersion	31/12/2000	31/12/2000 (20% reduction in BOD; 50% reduction in suspended solids)	
Over 10 000 in a 'sensitive area'	31/12/98	31/12/98	31/12/98 (with strict treatment)
10 000–15 000	31/12/2005	31/12/2005	31/12/2005
2000–10 000 discharging to fresh waters and estuaries	31/12/2005	31/12/2005	31/12/2005
2000–10 000 discharging to coastal waters	31/12/2005	Appropriate treatment to achieve quality objectives	Appropriate treatment to achieve quality objectives
Populations of fewer than 2000 discharging to all waters		Appropriate treatment to achieve quality objectives	Appropriate treatment to achieve quality objectives

the surfaces of which develops a film containing micro-organisms that feed on or oxidize the impurities in the sewage. The process is assisted by the action of worms and insects. The micro-organisms require oxygen in order to remove organic matter so the filter bed must be adequately ventilated to ensure their activity. The media making up the filter beds should conform to BS 1438:1971 [21]. It is usually of 50 mm nominal size, except for the bottom 150 mm, which is 100–150 mm in size. The tanks may be either rectangular or circular, and vary from 1.4 to 2.0 m in depth. However, filters which employ plastic media can be much deeper and up to 6 m. Some mechanical means is provided for the dispersion of the liquor over the surface of the filter: in the case of rectangular beds, this is usually a tipping trough with channels or a moving arm siphon system. In circular beds, a rotary distributor fed via a dosing chamber and siphon are used. The dosing chamber balances the variation in flow and ensures a head sufficient to operate the rotary arms. Otherwise during periods of low flow (for instance at night-time) the arms

will not turn and all the flow is distributed at the same point on the filter bed; consequently ponding will result. The filter material is usually supported on perforated tiles or pipes to ensure aeration of the lower media.

The amount of settled sewage that can be treated on the filter, depends on the flow and the strength and is expressed as a hydraulic and organic loading rate. The hydraulic loading rate is usually about $1–4 \ m^3/m^2$ per day and the organic loading rate is about $0.08–0.32 \ kg \ BOD/m^3$ per day.

Another system aimed at the same effect is the alternate double filtration (ADF) system where settled sewage is applied at a high organic loading rate to one filter and the treated effluent afterwards passes over a second filter at a much lower organic loading rate. Periodically (weekly or even daily) the order of the filters is reversed. This system is very efficient and permits a smaller total volume of filter to be used, but requires good operation to ensure that ponding does not occur.

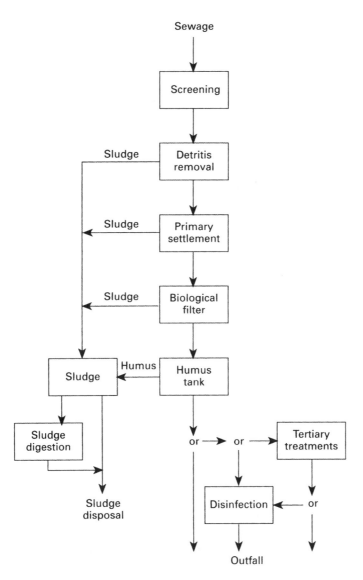

Fig. 39.13 The sewage treatment process.

Another form of biological treatment is the rotating biological contactor (RBC), which uses a medium in the form of discs or random elements packed in a perforated drum. This is very popular for small treatment works with population equivalents <2000. An RBC is usually followed by a tertiary reed bed system in order to ensure the final effluent can meet its consent for suspended solids.

ACTIVATED SLUDGE PROCESS   When large volumes of sewage have to be treated, the biological breakdown is achieved by the activated sludge process. In this process the settled sewage, is aerated in tanks by compressed air being blown into the sewage through porous plates (which produce fine bubbles) or pipes (which produce coarse bubbles) at the bottom of the tank. Alternatively, the sewage takes up atmospheric oxygen by being continually agitated by mechanical aerators rotating on or near the surface of the sewage. For small works, an oxidation ditch may be used as an alternative to tanks.

Two recent developments of the activated sludge process have proved very popular. A sequencing batch reactor (SBR) employs a number of tanks usually between 2 and 4 which ideally should treat screened, degritted sewage (i.e. without primary sedimentation). Each tank is filled with sewage in turn. After filling the sewage is then aerated for a fixed period and the aeration then turned off. This permits the solids to settle to the bottom of the tank leaving a clear treated effluent at the surface. This is then decanted to a watercourse. Usually up to 1.8–2.0 m of the tank depth can be decanted with each cycle. The tank then has an idle period until it is due to receive sewage. At times of high flow the idle time will be negligible and *vice versa*. A typical cycle is of 6 hours duration and comprises: Fill 60 minutes, Aerate, 180 minutes, Settle and Decant 100 minutes, Idle, 20 minutes. Thus, each tank undergoes four cycles each day. The advantages of the SBR process are that it requires only one type of tank constructing. In addition it does not need primary sedimentation, secondary sedimentation or sludge recyle facilities to be constructed.

A membrane bioreactor (MBR) is a modification of the activated sludge process in which solids separation is achieved without the requirement of a secondary clarifier. Instead this function is carried out by a membrane with a small pore size (typically as low as 0.4 μm), which retains the micro-organisms within the reactor and allows the treated, clarified effluent to pass to the next treatment train. An MBR is able to produce an extremely high-quality effluent which is low in BOD and contains negligible suspended solids. In addition the reactors can be configured to achieve the removal of ammonia (by nitrification), total nitrogen (by denitrification) and phosphorus by biological phosphorus removal. A further benefit of these reactors is that due to the low pore size of the membranes the effluent they produce is disinfected with an almost complete removal of faecal coliforms and faecal streptococci. Up to a 6-log removal of certain viruses is also achieved despite the fact that viruses are so much smaller than the membrane pore size. This disinfection capability makes MBRs an attractive option at coastal sites and sites discharging to inland recreational waters. Although there are a number of full-scale MBRs in the British Isles these are currently restricted to smaller sites. This is because of the high cost of the membranes themselves. The area of membrane needed for separation is proportional to the flow to the treatment plant and thus there are no economies of scale to be achieved as plant size increases. However, the cost of membranes is falling rapidly and it seems very likely that their application will continue to grow.

Sedimentation processes

The film that forms on the media in the biological filter periodically becomes detached and this, together with other fine particles still in suspension in the sewage, is allowed to settle in humus tanks. After passing through these tanks, the effluent should be fairly clear and innocuous and suitable for discharge into a watercourse. Typically it will have a biochemical oxygen demand (BOD) less than 20 mg/l and a suspended solids concentration less than 30 mg/l.

It is also necessary to remove the suspended biomass from treated effluent in suspended growth processes. This is achieved in secondary sedimentation tanks which are very similar both in design and mode of operation to humus tanks. The role of secondary sedimenation is to clarify the treated effluent and thus remove all the solids particles. In addition however, the tanks allow the sludge to accumulate at the base where the concentration of solids will increase. This means a smaller volume of sludge must be removed from the tanks for treatment and disposal. Final effluent from a modern activated sludge process would typically have BOD and suspended solids values less than 10 mg/l and ammonia less than 1 mg/l.

Tertiary treatment

When very high standards of purification are necessary, a third stage is adopted in which there is further sedimentation or filtration. However, modern treatment systems have rendered tertiary treatment unnecessary in many cases and it is now used primarily at smaller works for effluent polishing and at larger works for disinfection. The methods used include reed bed treatment, slow or rapid sand filtration or membrane systems. A final disinfection process may also be used at this stage and UV filters are used almost universally to achieve this, although chlorination has been practised at some sites. This is now considered unacceptable due to the discharge of chlorinated organic compounds that some people consider may pose a threat to aquatic life.

Microstraining by use of continuous membrane filtration (CMF) results in a large reduction in both bacteria and viruses. The technology has largely been developed in Australia, although pilot plants exist in both the USA and the United Kingdom. CMF is capable of producing a very high-quality effluent, but its application has so far been confined to small-scale sewage disposal works.

SLUDGE

The solid material deposited in the settlement tanks is known as sludge. It is a thickish, offensive-smelling liquid with a water content ranging from 60% to 95%. One of the most difficult and costly aspects of sewage treatment is final disposal of this sludge and there are many options available for its treatment. In the United Kingdom around 1.1 million dry tonnes of sludge are generated each year.

Thickening

The first stage in treatment is to remove as much water as possible in order to reduce the volume of material. As the water is removed the solids concentration increases. Any process which increases the solids concentration up to 5% is referred to as sludge thickening and this must be considered as the first step in any sludge processing. The two most common types of thickener are the picket fence thickener and the gravity belt thickener (GBT). The former comprises one or more circular tanks of concrete or steel, which are slowly stirrred

with a stirrer that resembles a picket fence. Thickened sludge is removed from the base and the water rises to the surface where it is removed and recirculated back to the treatment plant. A GBT involves the addition of polymer to the sludge which causes the solid particles to agglomerate and the liquid to separate out. The separated liquid passes through the belt, which is made of a coarse wire mesh to permit water to pass through.

Dewatering

This is the process that further increases the solids concentration up to around 40% are known as dewatering and improvments in dewatering technology has led the production of drier cakes. Dewatering is carried out using either centrifuges, belt presses or plate presses.

Any increase in solids concentration above 40% is achieved by sludge drying by use of thermal energy. The big advantage of drying is that it reduces the bulk of the dewatered cake by up to four fold and makes management of the final product easier.

Enhancement

A number of processes are employed to treat sewage sludge, which collectively are termed enhancement. Such processes do more than simply remove water and they fundamentally change the nature of the sludge itself. Anaerobic digestion and composting are two such processes which help to reduce the amount of organic material (and thus the amount of sludge solids to dispose of) and in doing so they render the product more stable. Digested and composted sludge reduces odour nuisance and depending on the temperature of the process, may bring about a significant reduction in the number of pathogens in the sludge. A number of other enhancements occupy a small niche and these include sludge vitrification, oil from sludge and gassification.

Final disposal

Final disposal of sludge is becoming increasingly difficult. Disposal to the North Sea accounted for

30% of our sludge but this route ceased in December 1998, despite over 100 years of close monitoring that had demonstrated no evidence of environmental degradation away from the disposal zone. This leaves only landfilling, recycling to agricultual land and incineration as major routes for ultimate disposal. Future EC regulations are likely to make the landfilling route increasingly threatened and so Water Companies are looking to agriculture and incineration. Recycling to agricultural land is regarded as the best practical environmental option (BPEO) but recent guidelines, known as the Safe Sludge Matrix, require additional treatment to ensure a 6-log removal of microbial pathogens are removed from the sludge before it can be used on agricultural land. This effectively requires that the sludge is sterile and will require a lot of investment in order to acquire the appropriate technology. These developments make incineration a more attractive long-term proposition for most Water Companies.

ODOUR CONTROL

Sewage disposal works, sewers and pumping stations are a consistent source of smell problems to their neighbours and cause considerable offence to residential areas in particular. As a result, any proposal to build new plant, particularly large works, is likely to be very sensitive, and will be subject to a full environmental impact assessment. Odour control must be an important part of that assessment and is best given careful consideration at the design stage. Modern plants are designed to minimize or eliminate smell and those parts of the works, which have the potential to generate most odour (namely the inlet works and sludge handling systems) are now generally fully enclosed that permits ventilation control of odours. A wide range of odour control technologies are now available including bioscrubbers, carbon adsorption and biofilters. Modern, well-operated treatment plants now have negligible odour beyond their boundaries.

A successful appeal in the High Court in 2003 may change an anomoly in nuisance law regarding sewage works causing a smell nuisance. The ruling means sewage works can now be classed as 'premises' under the Environmental Protection Act (EPA) 1990, overturning previous judgments that sewage works fall outside nuisance law. An abatement notice under Section 80 of the EPA was served in 2001 but in May 2002, magistrates quashed the notice on the grounds that the works were not a 'premises' under Section 79(1)(d) of the act. The water undertaker's barristers argued that the Public Health Act 1875 implicitly excluded sewage treatment work, referring to *R* v. *Parlby* 1889. The High Court judges have now ruled (June 2003) that this exclusion does not apply to later legislation, saying: 'The definition of statutory nuisance was completely recast by the 1990 act and there was no intention to exclude any particular premises from operation of Section 79.' The water undertaker intends to appeal against this decision so the ruling may not be final [22].

SMALL SEWAGE DISPOSAL SCHEMES

Small sewage treatment plant are defined as those serving population equivalents of less than 2000 and although there numbers are decreasing rapidly as sewage treatment becomes more centralized, they are still the most numerous type of plant in the British Isles. The attainment of a satisfactory standard of purification is possible if the plant is properly constructed and designed to meet the requirements of each particular case. Size, shape and construction are all important. Size must be relative to the strength and amount of crude sewage to be dealt with, which in turn depend upon the number of people occupying a building and the amount of water likely to be used by them. Where an adequate supply of water is available, a sewage flow of 180 l per person per day may be taken as the average to be dealt with.

A high degree of purification must be the prime object of any disposal scheme, and to be efficient it must be automatic, require a minimum of attention, and be free from objection, either of sight or smell. The design should be in accordance with BS 6297:1983 [23]. The principal points of construction and use of the different units are described here.

Septic tank

The purpose of this unit is to retain crude sewage until the soluble solids liquify, and the insoluble solids precipitate as sludge. During decomposition, solids float to the surface, forming a scum that should be removed only when necessary. A useful leaflet for householders with septic tanks is produced by CIRIA [24].

1. **Capacity.** In BS 6297:1983 [23] total capacity, where desludging is carried out at not more than yearly intervals, is given for general purposes by the formula:

   $$C = (180P + 2000) \text{ litres.}$$

   where C is the capacity of the tank (in litres) with a minimum value of 2720, and P is the design population, with a minimum of four.
2. **Design.** The tanks should be in series, either one tank divided into two by a partition or two separate tanks. The length of the first tank or section should be twice its width and provide about two-thirds of the total capacity. They are now only in use for populations of less than 50 people.
3. **Inlets and outlets.** These should be designed so that the discharge of sewage into the tank and the decanting of the clarified sewage is done with the minimum disturbance of the scum or sludge in the tank. For tanks up to 1200 mm width, 'dip' pipes, with the bottom arms about 450 mm below liquid level, are satisfactory. For tanks wider than 1200 mm, two dip pipes fed from a trough inlet are recommended, with a full width weir acting as an outlet. It is customary to have a scum board fixed about 150 mm from the weir; this board should stretch the full width of the tank and be about 450 mm wide, 300 mm of this being below the water level.
4. **Desludging.** The design should include facilities for desludging, and the floor of the tank should slope towards the inlet, providing a

'well' for the collection of the sludge as it is being pumped out or, if the site is sloping and makes it impossible, discharged by gravity through a valve.
5. **Covers.** These are unnecessary from the point of view of efficient working of the plant but should be provided to small tanks purely as a safety precaution. Larger tanks can be surrounded by a fence.

Settlement tanks

Above a population of 50 people a primary settlement tank is generally employed, either as a separate structure or more usually incorporated into the secondary treatment stage. Settlement tanks may be of the horizontal-flow or upward-flow type. Tanks are also available in either precast concrete or fibreglass reinforced plastics in a range of capacities from 2750 to 18 000 l, and as standard prefabricated tanks or in larger capacities when required. The principal function of the tank is effectively to produce quiescent conditions by reducing upward flow velocities between chambers, thus permitting maximum retention of sludge in the base.

Biological filter

The principles for biological treatment are the same as in large public sewage disposal schemes. In very small schemes, the filtration process is very much simplified. Care should be taken to ensure that distribution of sewage over the surface of the filter is even, or the effluent will form defined channels and purification will be proportionately incomplete. The design of the distributor is an important factor in the success or failure of a plant [23].

Humus or secondary settlement tank

The design should be similar to that of the septic tank, with a weir inlet and outlet with baffle and scum boards. The capacity should be not less than that suggested in BS 6297:1983 [23].

It may be desirable to make secondary tanks equal in size to primary tanks, otherwise the formulae for calculating them are as follows.

- for horizontal flow tanks:

$$C = 135P^{0.85}$$

where C is the gross capacity of the tank (in litres), and P is the design population with a minimum value of four.
- for upward flow tanks the gross capacity should not be less than that determined by the formula above, and the surface area should not be less than:

$$A = 0.075P^{0.85}$$

where A is the minimum area (in square metres) of the tank at the top of the hopper, and P is the design population with a minimum value of four.

Outfall

Purified sewage effluent from small sewage treatment plants may be disposed of by surface or subsoil irrigation or, subject to consent by the EA, by discharge into a receiving watercourse. Such a consent will prescribe conditions about the quality and quantity of the discharge and may include provisions relating to the siting and construction of the outlet, sampling point, etc. Rain and surface water should be kept separate from sewage where purification has to be effected on site, but a drain conveying this water may with advantage join the outlet humus tank.

A large number of prefabricated 'packaged' sewage disposal units are available for small communities. An example of this is the Klargester BioDisc, which is a complete self-contained sewage treatment plant designed for small communities in a range of sizes to serve from 5 to 150 people. Larger communities can be accommodated by arranging units in parallel. The complete treatment process is carried out in one totally enclosed compact unit.

In the example given, units are fabricated in glass reinforced plastics (GRP) and may be installed either above or below ground. Crude sewage is piped direct to the BioDisc, entering the baffled primary settlement zone via a deflector box that stills the flow. The heavier solids sink to the bottom of the compartment to disperse into the main sludge zone. Any floating solids are retained by means of a baffle. The effluent, with lighter solids still in suspension, passes into the 'biozone', which comprises a chamber with transverse baffles arranged so that the liquid must follow a serpentine path from zone inlet to outlet. The baffles also separate a series of slowly rotating circular discs into banks, so that the sewage passes through each bank in turn. Micro-organisms that are naturally present in the sewage adhere to the partially immersed discs to form a biologically active film, feeding upon the impurities and rendering them inoffensive. The organisms feed and multiply very rapidly in the presence of an ample supply of oxygen, and as each portion of the film on the rotating discs is alternately in contact with settled sewage and atmospheric oxygen, conditions are ideal for efficient purification.

The sludge from the primary and biozones collects and consolidates in the base of the unit. The unit can accommodate a large quantity of consolidated sludge before desludging is necessary (at intervals of, say, every 4–6 months), and this can be done by a gully emptier or similar suction unit.

Effluents – analyses and interpretation of reports

Sewage effluents discharging into streams must normally conform to standards set by the EA, usually based on the recommendations of the Royal Commission on Sewage Disposal. These most frequently refer to biochemical oxygen demand and suspended solids. The values achievable by a well-run works are fractions of the plant design capacity. These and other constituents have been interpreted as detailed below.

The Royal Commission on Sewage Disposal

A maximum BOD concentration of 20 mg/l in effluents was recommended, where the dilution in the river was at least 8 : 1. Although this value is now varied according to circumstances by the EA to meet environmental quality objectives, the majority of works have been designed to produce an effluent reaching at least this standard.

The Biochemical Oxygen demand (BOD) test is essentially a measure of biodegradable material present in an effluent and is dependent upon the conditions under which the test is applied. Furthermore, under the original conditions of the test, the result included the oxygen demand of nitrogenous materials. To overcome this, allylthiourea (ATU) is added to suppress nitrogenous oxidation.

Total oxygen demand and chemical oxygen demand are two further tests. They differentiate oxidizable organic and inorganic matter present.

In the case of suspended matter, a limit of 30 mg/l was recommended, where the dilution was 8 : 1, but in a similar manner to BOD, this value is now varied to suit the circumstances. Suspended solids are normally determined as total and volatile.

Sewage always contains chloride to an extent that depends upon the strength of the sewage, the presence of trade wastes and the content of the water supply. It remains unaltered throughout the treatment process. Medium sewage contains up to 100 mg/l of chloride whereas a strong sewage may contain up to 500 mg/l.

Nitrogen is present in a number of forms, particularly ammonia, nitrite and nitrate, representing different oxidation states. Ammonia arises from the aerobic or anaerobic decomposition of nitrogenous organic matter and since it is extremely toxic to fish and interferes with the chlorination of drinking water, it is undesirable in effluents discharged to these waters. Nitrates represent the final oxidation stage of ammonia, and their presence indicates a well-run plant. When a plant is operating ineffectively, there is likely to be oxygen deficiency at, for example, an overloaded filter, and the oxidation of ammonia will be inhibited.

For activated sludge plants, it has been considered uneconomic to carry the purification process beyond the clarification stage, and little or no nitrate may be produced, yet the effluent may be well clarified with low BOD and suspended matter.

EC Urban Waste Water Standards (see also the Urban Waste Water Treatment (England and Wales) Regulations 1994)

Royal Commission standards have now been replaced by the requirements of Directive 91/271/EEC [20]. In each case, the Directive lays down sampling frequencies, technical compliance standards for discharges and the percentage of failed samples allowable (see Tables 39.7–39.9). The minimum annual number of samples is determined according to the size of the treatment plant and must be collected at regular intervals during the year, as shown in Table 39.9. In many cases the EA requires standards in excess of those prescribed in the Directive where it is thought this is necessary to further enhance the quality of the receiving watercourse.

ALTERNATIVE SANITATION

Sometimes it is not possible to connect premises to the sewerage system and alternative methods of faecal waste disposal have to be used. Very little of this takes place but some form of closet system may, however, be the only practicable system in sparsely populated districts. The term for these systems is 'conservancy' and it often goes together with private water supplies. Absence of an adequate public water supply necessitates recourse to a private means of supply; it also necessitates conservancy. This produces the dual public health problem of insufficient safe water for domestic purposes and inadequate disposal of sewage; frequently this occurs on sites that are not suitable for a septic tank disposal system.

Provided that they are properly sited, are installed under skilled supervision, and are properly maintained, on-site sanitation systems do not necessarily imply a lowering of normal standards of hygiene. Nevertheless, such systems must always be regarded as potential dangers to health, for example, to water, by contamination of the earth, and to food, through contamination by flies.

Table 39.7 Requirements for discharges from urban waste water treatment plants subject to Articles 4 and 5 of EC Directive 91/271/EEC. The values for concentration or the percentage of reduction shall apply

Parameters	Concentration	Minimum percentage of reduction	Reference method of measurement
Biochemical oxygen demand (BOD) at 20°C without nitrification	25 mg/l O$_2$	70–90 40 under Article 4(2)	Homogenized, unfiltered, undecanted sample; determination of dissolved oxygen before and after 5-day incubation at 20°C ± 1°C, in complete darkness; addition of a nitrification inhibitor
Chemical oxygen demand (COD)	125 mg/l O$_2$	75	Homogenized, unfiltered, undecanted sample; potassium dichromate
Total suspended solids	35 mg/l (optional) 35 mg/l under Article 4(2) (more than 10 000 pe) 60 mg/l under Article 4(2) (2000–10 000 pe)	90 (optional) 90 under Article 4(2) (more than 10 000 pe) 70 under Article 4(2) (2000–10 000 pe)	Filtering of a representative sample through a 0.45 μm filter membrane; drying at 105°C and weighing; centrifuging of a representative sample (for at least 5 minutes with mean acceleration of 2800–3200 *g*), drying at 105°C and weighing

Note:
pe = population equivalent.

Table 39.8 Requirements for discharges from urban waste water treatment plants to sensitive areas that are subject to eutrophication as identified in Annexe IIA(a). One or both parameters may be applied depending on the local situation. The values for concentration or for the percentage of reduction shall apply

Parameters	Concentration	Minimum percentage of reduction (in relation to the influent)	Reference method of measurement
Total phosphorus	2 mg/l P (10 000–100 000 pe) 1 mg/l P (more than 100 000 pe)	80	Molecular absorption spectrophotometry
Total nitrogen	15 mg/l N (10 000–100 000 pe) 10 mg/l N (more than 100 000 pe)	70–80	Molecular absorption spectrophotometry

Note:
pe = population equivalent.

Table 39.9 Minimum number and frequency of treatment plant samples under EC Directive 91/271/EEC (British Standards Institution (1987))

Population equivalent	Number and frequency of samples
2000–9999 pe	12 samples during the first year; four samples in subsequent years if it can be shown that the water during the first year complies with the provisions of the directive; if one of the four samples fails, 12 samples must be taken in the year that follows
10 000 to 49 999 pe	12 samples
50 000 or more pe	24 samples

Notes:
pe = population equivalent.
For full details of sampling methodology and standards, consult Annexe I of Council Directive 91/271/EEC (British Standards Institution (1987).)

Cesspools

Cesspools are the cheapest means of storage of sewage outside a sewered area. A cesspool is not a means of disposal, it is merely the storage of sewage until it can conveniently be disposed of. A drain discharging into a cesspool should be reserved for foul water only and should be 'disconnected' by means of an intercepting trap and chamber. Surface drainage should be treated separately, for example, by discharge into soakaway pits. Cesspools should be constructed in accordance with BS 6297:1983 [23]. The cesspool should be on sloping ground, lower than nearby buildings, and regard should also be paid to prevailing winds due to possible odour problems. There should be no risk of pollution, particularly of water supplies, and the tank should be at least 15 m from any inhabited building.

Cesspools should have a minimum capacity of 18 000 l (18 m³) (Building Regulations 2000 approved document H2), which preferably should represent not less than 45 days usage for the building it serves. Regard should be had for the means of emptying before a decision is made about capacity, or, indeed, about whether to use a cesspool at all. Cesspools must be watertight and may be made of a variety of materials. Factory-made cesspools are available in GRP, polyethylene or steel. Prefabricated tanks should have a British Board of Agreement (BBA) certificate.

Under the Building Regulations 1991, cesspools have to be constructed so as to enable access for emptying and also to avoid contamination of water supplies. An existing cesspool that leaks or overflows is subject to the requirement of Section 50 of the Public Health Act 1936, and the service of a notice by the local authority, which can specify the works to be done to deal with any soakage or overflow. This includes regular emptying [1].

The emptying of cesspools by contractors or local authorities, rather than by individuals, enables the use of gully emptiers, or cesspool emptier's, as part of the public cleansing service. A cesspool emptying vehicle has a closed steel tank typically of about 4500 l in capacity, and is fitted with a 15-cm inlet valve. It should be provided with armoured rubber pipe of sufficient length to connect the tank with the cesspools in the area. The engine produces a vacuum equal to 50 cm in the cylinder. The valve is then opened and the contents of the cesspool are sucked through the pipe into the cylinder. The pipe should extend to the bottom of the cesspool so that the sludge is removed first, followed by the liquid sewage above it. There is no agitation of the sewage; the whole proceeding should be inoffensive and may be carried out at any time of day. The tank should be discharged either at a sewage disposal works or into a large capacity public sewer, through a proper disposal chamber. Indiscriminate disposal on to land is not acceptable.

Earth closets

Now little used, these depend upon the power of dry earth to 'neutralize' faecal matter. The receptacle is essentially a stout, galvanized iron bucket which, in order to ensure frequent emptying should not be of more than 55 l in capacity. It should be movable, but held in position by a suitable guide, and fitted closely into a suitable enclosure. The earth should be clean, dry, fertile top-soil sifted to exclude particles smaller than

6 mm. Sand or ashes are not suitable for the purpose. The fitness standard laid down in the Local Government and Housing Act 1989 requires the provision of a water closet. Any property provided with dry conservancy, including earth closets, is unfit and should be subject to action under housing legislation (see Chapter 17).

Chemical closet

Chemical closets have improved greatly in appearance over the years, although they are now more generally used for camping, caravanning and boating. They all operate with a deodorizing, liquefying or sterilizing liquid that is basically a solution of formalin. Some manufacturers claim waste matter that has been treated with their chemicals can be safely disposed of down a drain, but unless there are specific disposal points, such as on camp sites, at marinas, etc., the views of the controlling authority should be sought beforehand.

The closets are usually constructed of strong plastic or fibreglass and range from an enclosed bucket to semi-permanent recirculating units with filtration units and electric 'flushing' pumps. Similar built-in units are used in aircraft, long-distance coaches and in some trains (other trains however, merely dump the sewage effluent on the track).

Treatment and disposal of farm wastes

The development of intensive farming practices, together with increasing urbanization and rising expectations about the environment, have led to a rise in complaints about smell. Farming is an industry in its own right, and produces a volume of trade waste that must be disposed of without a risk of nuisance or danger to public health.

Character of the waste

Animal excreta is the main problem: it has been estimated that the total annual volume of excreta from all livestock (cattle, pigs, sheep, poultry) is 121 million tonnes [25]. Also, the demand by supermarkets for prepacked washed vegetables has led to a big increase in this work being undertaken

at farms: it has been estimated that washed carrots require 22 000 l of water per hectare of crop, which means that large quantities of polluted water have to be dealt with.

Methods of disposal

The seven usual methods adopted are detailed here.

1. **Dry handling.** The aim is to keep manure as dry as possible and the excreta, mixed with bedding, is taken to a midden or store with walls of earth, railway sleepers or maybe concrete. The manure is restricted to a depth of 1–1.5 m to help evaporation, and is retained for 6–9 months before being distributed on the land. The bedding absorbs much of the liquid, and some drains from the store into channels or pipes to be dealt with separately.
2. **Semi-dry handling.** The excreta is kept apart from any bedding; faeces and urine are mixed to give a slurry, which is spread on the land.
3. **Semi-liquid handling.** Initially the procedure is as for 2 above, but the faeces and urine are mixed with water (one part water to one part manure) to enable it to be spread by vacuum tanker.
4. **Irrigation.** The solid and liquid manure are washed down by water to a storage tank with enough water to give a mixture of one part manure to two parts water. Some system of agitation is required in the tank, and the liquid manure is spread on the land by pump, pipes or rain gun. The tank should be large enough to hold 10–14 days' diluted effluent.
5. **Discharge to sewers.** Being a trade waste, farm effluent could, subject to the consent and conditions laid down by the EA, be discharged into a public sewer. Regard would need to be paid to the 'strong' nature of the effluent and its effect upon any sewage treatment works.
6. **Oxidation in lagoons.** Unfortunately, due to the low temperature usually experienced in the United Kingdom, this has not proved very successful. The best results are obtained by having as large an area as possible with the slurry at a depth no greater than 1.5 m. If the slurry is too

deep, anaerobic action begins, creating an offensive smell.

7. **Soakaways.** Provided the soil is suitable and there is no danger of polluting underground water supplies, particularly by nitrates, soakaways or blind ditches can be used for draining off smaller amounts of the liquor, the deposited solid matter being removed periodically. The consent of the EA is required for this.

Camp sanitation

Among the public health issues associated with the use of camps of all kinds, none is more important than the provision of sanitary accommodation. Large permanent camps require piped water supplies; they also need a water-carriage system of drainage. Where, in addition, a sewage disposal plant is necessary, the types of plant already described are suitable and can be adapted to meet the needs of most sites.

In small camps and those of a temporary nature, some form of conservancy system alone is practicable. In open-air conditions, the provisions described in the following subsections can be quite satisfactory but in every case they must be sited, constructed and maintained under the direction and supervision of the environmental health officer of the district concerned. All latrines must be fitted with seats and covers and otherwise made fly-proof.

Earth latrines

The simplest method of disposing of human excrement in non-sewered districts is a dug latrine, but it is necessary to minimize all danger of spreading disease. Surface pollution must be avoided. Earth latrines, soakaways, etc. may be used with safety only where pollution of subsoil water is not a risk. Faecal organisms reaching the groundwater from a point source, for example, a latrine, do not travel evenly in all directions, but are carried with the groundwater flow. Pollution has been found to travel a distance of 25–30 m 'with the stream', but did not reach 3 m in any other direction. Usually, the ground water is flowing in a definite direction and, provided the direction is known, an earth

latrine or a soakaway can be located safely. In ordinary soils it may be assumed that the area outside a radius of 6 m, and extending round one-half of a circle 'up-stream' or above the latrine, is safe from danger of pollution. In chalk and similar formations, however, the water is in fissures (**karstic**), which form subterranean streams running long distances in all directions. No earth latrine may be used with safety in such formations.

The deep trench

In temporary camps where only field sanitation is practicable, a properly constructed and supervised earth latrine is a sanitary and satisfactory provision. The latrine in most common use is of the deep trench type which, owing to its size, is generally arranged for communal use and, because of that, is open to many objections.

The trench latrine should be 1 m wide and not less than 2.5 m deep. It should be provided with a well-constructed riser or seat, the openings of which should be arranged over the lateral centre of the trench, the front and back being constructed to prevent the ingress of surface water. Self-closing covers should be provided to the openings, and all necessary measures should be taken to render the trench fly-proof when it is closed.

The contents of a deep trench latrine should be covered with about 75 mm of fertile top soil (not the earth previously dug out) at frequent intervals. The nitrifying organisms in top soil rapidly neutralize and break down faecal solids, rendering them innocuous. The action is biological and quite efficient. Only where a latrine is sunk into dense watertight earth and contains foul liquid matter should a disinfectant be used.

The bored-hole latrine

This simple form of earth latrine has been adopted with marked success in the tropics, but is equally suitable for use elsewhere. By means of a hand-operated land auger of simple design, a hole 400 mm in diameter is bored in the earth to a depth of 4.5–6 m. Being circular and undisturbed, the walls of the hole are, in ordinary soil conditions, self-supporting.

Container closets

The best of these are the earth closet and the chemical closet already described; they are both suitable for camp purposes.

Sullage water

On no account should crude sullage (bath, shower, kitchen, etc. liquid waste) be discharged into a river, stream, ditch or lake, or over unprepared ground. The simplest means of disposal is by soil absorption through a soakage pit. Even in the most favourable formations, the ability of the soil to absorb sullage water is lessened as time goes on, and it may be reduced to a point at which complete disposal is not possible. In order to preserve soil absorptivity, grease and oils should be removed by passing sullage water through a grease-trap before it is discharged into a soakage pit. For field use, a grease-trap is designed to ensure that sullage water passing through it has a long journey at low velocity between inlet and outlet, so that grease the may separate from the water and float to the surface, where it is retained. Only two 'baffles' are necessary, one to form the inlet and the other the outlet chamber; these should be deep rather than shallow, and long and narrow rather than square. A length to breadth proportion of 3 : 1 is recommended. A capacity of 225 l is suitable for most general purposes.

Soakage pits

The purpose of a soakage pit is to receive waste liquids as and when they are produced, and to act as a reservoir from which they may soak continuously into the surrounding ground. The pit should be filled with coarse rubble to support the sides and cover, while leaving the maximum void. It should be covered with at least 300 mm of earth. Apart from geological conditions, efficiency is dependent upon two factors: water content and the size of the soakage surface. Generally, the water content should be equal to 1 day's production of sullage. The soakage surface is the area of the perimeter walls plus the base; the shape should provide the maximum perimeter by comparison with the volume, i.e., rectangular rather than square. Where the surface stratum is dense, it should, if possible, be pierced and a more permeable stratum be brought into use. A form of vertical drainage, useful for both sullage and surface water, is thereby obtained.

Chemical precipitation

In large camps or where the soil has become non-absorbent, sullage must be purified or, more correctly, clarified; it may then be disposed of in the same way as ordinary surface water. A simple method of purification that is used in military camps consists of treating sullage with two chemicals: ferrous sulphate and hydrated lime. The quantities required vary with the sullage, but effective precipitation depends upon obtaining a correct alkalinity (pH of 9); the lime is used for this purpose. It is added in equal parts to or slightly more than, the ferrous sulphate.

The sullage is collected in a tank. The ferrous sulphate, after being dissolved in water, is then added; the lime is similarly dealt with. Then the whole contents of the tank are thoroughly agitated. A heavy floc results, which precipitates rapidly, forming a closely packed sludge and leaving the supernatant water clear. The process does not completely remove grease; water from a cookhouse should therefore be passed though a grease-trap before treatment. The clarified water is comparable to an ordinary purified sewage effluent, except that the oxygen demand is high: about 10 parts per 100 000. This can be corrected by aeration, for example by causing the effluent to pass through an open channel and, if possible, over weirs before discharging into a stream.

A sullage purification plant is easily constructed. Two precipitating tanks are usually necessary, to be used in rotation; they can be formed from precast concrete rings or pipes placed on end in a suitable excavation, the bottoms being concreted in the form of an inverted frustum of a cone, into which the sludge can gravitate. The outlet for clarified effluent is fixed not less than 300 mm above the bottom of the tank and at a level from which the effluent can gravitate to a suitable place of disposal. Two outlets fixed at different levels afford useful flexibility of working.

Sludge is dealt with in shallow lagoons. It is usually at too low a level to run by gravity to these; the hydraulic head provided by the supernatant water when the tank is full is utilized to eject and lift it to the required level. It dehydrates rapidly, forming an innocuous residue that may be disposed of on land. The ferrous sulphate and lime contained in the sludge produced from the first 2 or 3 charges is used to assist in the precipitation of subsequent charges. For this reason, desludging is carried out only when the sludge level rises to within a few inches of the clear water outlet.

DISPOSAL OF EFFLUENT TO RIVERS AND THE SEA

The final stage in the water cycle is the drainage of water from the land environment to the sea. Sources of drainage include:

- natural drainage from hard surfaces and land
- fresh water drains
- storm water overflows
- effluent from sewerage systems.

Drainage may be via the rivers and their tributaries, or direct to the sea, but eventually all flow to the ocean. In the United Kingdom, as an yet incomplete network of statutory standards, voluntary standards and information influences both the quality of drainage and the resultant river and sea water.

Rivers

Powers were included in the Water Resources Act 1991 to enable OFWAT to declare statutory water quality objectives (SWQOs). It was intended that they would replace the non-statutory river quality objectives (RQOs), which were mostly adopted in the period between 1979 and 1981 by the former water authorities. In February 1995, the then DoE (now DEFRA) announced the intention to consult 'on a small set of SWQOs', resulting in a set of 10 proposals that were not then implemented.

Despite the consultation and the stated intention of the government in 1987 to have SWQOs in place by 1989, there is still no legally enforceable water quality targets for rivers, estuaries, lakes, canals or coastal waters in place at the time of writing.

If river water quality needs to be improved to comply with the existing RQOs, action will be required to regulate or prohibit individual discharges into the receiving waters. In effect, the Urban Waste Water Directive [20] allowed for two options to regulate discharges from sewage works:

1. uniform discharge limits for suspended solids, BOD and chemical oxygen demand (COD);
2. minimum percentage reductions in each of the values of the parameters during the treatment process.

Generally, the second method is more difficult to enforce, requiring monitoring of both the influent and the effluent. However, in December 1993, the DoE decided that the percentage reduction method was to be adopted as the basis for implementing the Urban Waste Water Directive in the United Kingdom. The resultant safeguards over river water quality, by using this method, do not necessarily provide the power to achieve the non-statutory RQOs. The Urban Waste Water Directive also requires that all significant discharges receive two stages of treatment (primary and secondary). In the United Kingdom, treatment to at least secondary level is to be provided for all the discharges to coastal waters designated as significant, serving as few as 2000 people. This goes beyond the requirements of the Directive.

The Directive requires special treatment for discharges to waters that are designated as sensitive. One type is called the Eutrophic Sensitive Area (Eutrophication leads to a lack of oxygen in the water and can be detrimental to wildlife). The Directive requires that member states review the designations of these every four years. Once a water has been identified, the larger sewage treatment works discharging into it must meet the Directive's standards for the removal of nutrients, unless it is demonstrated that the removal will have no effect on the degree of eutrophication.

There are 80 Eutrophic Sensitive Areas in England and Wales.

The Directive also requires member states to 'decide on measures to limit pollution from storm water overflows'. In the five-year period before 2000, a total of 1200 were improved, but more than 5500 overflows still required attention. At the rate of investment during the 1990s, the job would not have been completed until beyond 2020. The Government has accelerated this trend and asked that 85% of the overflows are brought up to standard by 2005 [26].

The sea

The sea is the final stage of the water cycle under human control before natural evaporation and the climate takes over. The disposal of sewage outfall effluent to sea, either directly or as a result of drainage from rivers, has caused concern for three main reasons:

1. its effects on bathing beaches and other recreational areas;
2. its impact on the sea environment and its ecosystems;
3. the concentration of polluting substances into the food chain of animals and humans.

Increased use of sea water for recreational purposes has led to the first of these being of the greatest concern to environmental organizations and the media.

Natural bathing water quality

Directive (76/160/EEC) (as amended) [27] concerning the quality of bathing water, commonly referred to as the Bathing Water Directive, was adopted by the Council of the European Communities in December 1975. The principal objectives of the Directive are to protect both human health and the environment by maintaining the quality of bathing waters. This Directive does not apply to swimming pools or waters intended for therapeutic purposes, but to those fresh or saline open waters where bathing is explicitly authorized by the competent authorities,

or is not prohibited, and is practised by a large number of bathers. In the United Kingdom, the term 'bathing water' was taken by the Department of the Environment, Transport and the Regions (DETR) to mean an area where bathing is traditionally practised by a large number of people as defined by the directive. Examples are sites with lifeguards or where changing huts, car parks or chalets are provided on a substantial scale for bathers. There is no definition in the legislation of what might constitute 'large numbers' of bathers, resulting in wide divergence within the EU over implementation. Luxemburg, for example, a very small land-locked country, declared more 'bathing waters' than the United Kingdom when the legislation was introduced. 'Bather' also appears to be interpreted rather narrowly in the United Kingdom to mean a swimmer, and the bathing season is taken to be from May to September.

Because of its responsibilities for controlling water pollution, the EA has been designated as the competent authority by the DETR. In turn, the EA is required to identify bathing waters for sampling and monitoring and to ensure that the waters comply with the standards set by the Directive. The results of the sampling programme undertaken by the EA at all EU designated beaches during the bathing season are supplied to the coastal local authorities. The EA publishes an annual report on bathing water quality in England and Wales.

With respect to coastal water quality, the local authority must inform the public of any health concern that could arise as a result of bathing. In 1990, the Minister of State for the Environment and Countryside announced, in response to a House of Commons Environment Select Committee report on pollution of beaches [28], that local authorities were to be encouraged to display bathing water quality information on beaches and other prominent sites around resorts.

Standards set by Directive 76/160/EEC [27] have been subject to considerable controversy and there have been calls for new imperative standards (especially in relation to viruses) to be put into effect. In its fourth report [28], the House of Commons Environment Select Committee added to the call for a review of the EU directive standards in the light of the then current epidemiological and viral scientific research.

Quality requirements for bathing water are detailed in the annexe to Directive 76/160/EEC [27] (see Tables 39.10–39.12). Some 19 physical, chemical and microbiological parameters are listed, of which 13 are 'I' (imperative) values and/or 'G' (guide) values. Coliform counts are important indicative values, although they are regarded by some researchers as too fragile to reflect survival rates of pathogens in sea water. Only in exceptional circumstances have derogations been permitted to the standards. To conform, 95% of samples for parameters where the 'I' value is given must meet the values set, and 90% of samples in other cases. Waivers may be granted because of exceptional weather or geographical conditions, etc. Minimum sampling frequencies are laid down and the Directive specifies how and where samples are to be taken. It does not however, specify handling procedures for samples before analysis. The **streptococcal** and *Salmonella* parameters only have to be checked when there is reason to suppose the presence of these organisms, although an amendment proposed by the European Commission early in 1994 would have introduced a new standard for faecal streptococci (now called enterococci). Based on the established 95% pass results, it was an 'I' standard of 400 organisms per 100 ml of water, and a 'G' standard of 100 organisms per 100 ml. Directive 76/160/EEC has been partially implemented in the United Kingdom by the Bathing Waters (Classification) Regulations 1991 (SI 1597), which are enforceable by the EA.

By 2002, DEFRA had designated 407 beaches in the United Kingdom as 'bathing waters', 401 complied with European mandatory coliform bacteria standards. There remain, however, large areas of water that are used extensively for **recreational activities**, including swimming, canoeing and windsurfing, that have not been designated. Most of these activities involve immersion in water, and must represent a risk to health if the water is seriously polluted. There is as yet no standard available to define acceptable limits for recreational water quality.

During the 1990s, the statutory sewerage undertakers made an increasing capital investment in new sewage treatment works. However, there are concerns that, where full primary, secondary and tertiary treatments are not provided, outfalls designed to protect bathing waters may simply divert sewage to areas of water recreational activity [29].

On 24 October 2002, the EC Commission produced a proposal for the revision of the Bathing Water Directive. It proposes:

1. An obligation to meet a much tighter bathing water quality standard than the existing Bathing Water Directive.
2. Very limited provisions for recreational waters with no standards for these waters.
3. Some management measures for bathing waters.
4. Improved provision of information at bathing waters.

Table 39.10 Quality requirements for bathing water: microbiological parameters

Microbiological parameters	G	I	Minimum sampling frequency
Total coliforms per 100 ml	500	10 000	Fortnightly[a]
Faecal coliforms per 100 ml	100	2000	Fortnightly[a]
Faecal streptococci per 100 ml	100	—	[b]
Salmonella per litre	—	[c]	[b]
Enteroviruses PFU 10 1	—	[c]	[b]

Notes:
a When sampling in previous years has produced result that are appreciably better than those in this annexe, and when no new factor likely to lower the quality of the water has appeared, the competent authorities may reduce the sampling frequency by a factor of two.
b Concentration to be checked by the competent authority when an inspection in the bathing area shows that the substance may be present, or that the water quality has deteriorated.
c Provision exists for exceeding the limits in the event of exceptional geographical or meteorological conditions.
G = guide; I = mandatory.

Source: Hatchett (2003).

Table 39.11 Quality requirements for bathing water: physiochemical parameters

Physiochemical parameters	G	I	Minimum sampling frequency
pH	—	6–9[a]	b
Colour	—	No abnormal change in colour	Fortnightly
Mineral oils (mg/l)	—	No film visible on the surface of the water and no odour	Fortnightly b
Surface-active substances reacting with methylene blue (mg/l)	≤ 0.3	No lasting foam	b Fortnightly c
Phenols (mg/l) (phenol indices)	≤ 0.3	—	b
	—	No specific colour	Fortnightly c
	≤ 0.005	≤ 0.05	b
Transparency (m)	2	1 (0)	Fortnightly c
Dissolved oxygen (% saturated O$_2$)	80–120	—	
Tarry residues and floating materials such as wood, plastic articles, bottles, containers of glass, plastic, rubber or any other substance; waste or splinters	Absence		Fortnightly c
Ammonia (mg/l NH4)			d
Nitrogen Kjeldahl (mg/l N)			d

Notes:

a Provision exists for exceeding the limits in the event of exceptional geographical or meteorological conditions.

b Concentration to be checked by the competent authority when an inspection in the bathing area shows that the substance may be present, or that the water quality has deteriorated.

c When sampling in previous years has produced results that are appreciably better than those in this annexe, and when no new factor likely to lower the quality of the water has appeared, the competent authorities may reduce the sampling frequency by a factor of two.

d These parameters must be checked by the competent authorities when there is a tendency towards the eutrophication of the water.

G = guide; I = mandatory.

Source: [22].

Health risks from bathing water

There was for many years a view in the United Kingdom that sewage-contaminated bathing water did not represent a serious risk to public health. This was fostered to a large extent by a PHLS study [30] and a Medical Research Council report [31], both of which concluded in 1959 that infection from sea water bathing could be discounted for all practical purposes, unless the pollution was so gross as to be aesthetically revolting. The 10th report of the Royal Commission on Environmental Pollution [32], published in 1984, showed that many bathing waters and beaches in the United Kingdom suffered from an undesirable degree of contamination by sewage. It also said that the risk of infection by serious disease was small, but that the visible presence of faecal and other offensive materials led to a serious loss of amenity and was unacceptable. Views expressed by the World Health Organization (WHO) [33] and the US Environmental Protection Agency [34] stated that there is potential risk to health from bathing

Table 39.12 Quality requirements for bathing water: other parameters

Other substances regarded as indicators of pollution	G	I	Minimum sampling frequency
Pesticides (mg/l) (parathion, HCH, dieldrin)			a
Heavy metals (mg/l) Arsenic (As) Cadmium (Cd) Chrome VI (VI) Lead (pb) Mercury (Hg)			a a
Cyanides (mg/l CN)			
Nitrates and phosphates (mg/l NO, PO^4)			a

Notes:
a Concentration to be checked by the competent authority when an inspection in the bathing area shows that the substance may be present, or that the water quality has deteriorated.
G = guide; I = mandatory.

Source: [22].

in sewage-contaminated sea water. In *The State of the Marine Environment* [35], an Environmental Programme report published in 1990, the United Nations also supported this view. In a section dealing with human health effects, the report concludes:

> Gastrointestinal infection due to swimming in sewage-polluted sea water is the most widespread health effect in estuarine and coastal areas with high population densities. Recent epidemiological studies in the USA and in the Mediterranean have cast new light on the causal relationship between bathing in sea water contaminated with pathogens of faecal origin and disease among the bathers. The relationship is particularly strong in the case of children under five. Earlier views that there is no demonstrable link between human disease and bathing in sea water can no longer be supported.

Bathing water studies

A number of studies have been undertaken in the search for a definitive 'health' standard for bathing water, Cabelli in the USA [36] concluded that bathing in modestly polluted sea waters produced a significant elevation of minor infections of the ear, nose and throat as well as gastrointestinal infections in bathers compared with non-bathers. A UK-wide was carried out by the Water Research Centre (WRC) [37] and a total of 14 separate, interrelated studies involved over 17 000 people. The final report to the Department of the Environment was published in 1994 [37]. Conclusions included a noted increase in minor symptoms after contact with sea water, which were not related to the microbiological quality of the water. However, relative increases in diarrhoea seemed to be related to the mean counts of coliforms and enteroviruses. During a controlled clinical trial, symptoms of gastroenteritis in bathers were related to the counts of **enterococci** in the water at about chest level. Overall, the report concludes: '. . . findings suggest that activity in sea water meeting these Imperative standards does not pose any significant risks to health'. It is worth noting that the randomized clinical trials conclude:

1. A significant dose–response relationship was identified between faecal streptococci (per 100 ml) measured at chest depth and gastroenteritis ($p < 0.001$). The relationship was independent of the site studied.
2. Non-water-related risk factors did not confound the relationship and no significant interaction between the confounders and the water quality index was found.
3. The threshold of risk was objectively defined as 32 faecal streptococci per 100 ml at chest depth. The resulting model ($p = 0.012$) is shown in Fig. 39.14.

Clean beaches campaign

The Blue Flag Awards for clean beaches is a European campaign initiated by the Foundation for Environmental Education in Europe (FEEE) to

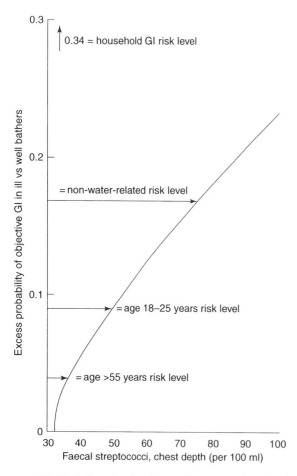

Fig. 39.14 Policy implications. GI = gastrointestinal illness. 1. The model allows the prediction of the probability of gastroenteritis (i.e. the risk of illness) at a give faecal streptococcal level. This probability can be compared with the risk of illness attributable to the other risk factors, such as household illness. 2. The results clearly indicate that the current mandatory standards specified in Directive 76/160/EEC (Hatchett (2003)) may not be appropriate. Consideration should be given to changing both the recommended sampling depth and the microbial indicators used to assess compliance of EU marine waters. 3. The results provide the necessary scientific information for the construction of standards or objectives for marine recreational waters used by 'normal' adult bathers. (Source: Public Health Laboratory Service (1959).)

reward efforts that meet the standards of the Blue Flag charter. The four basic tests are water quality, beach cleanliness and management, environmental education and information.

Water quality criteria are met by full compliance with EU bathing water imperative and guideline standards, with no industrial or sewage discharges. Recommended emergency plans and monitoring regimes are included. There must be no gross pollution of beaches by faeces or other wastes and no litter or oil pollution. The banning of dogs, together with monitoring, are also criteria for determining beach quality. Environmental education is achieved by the provision of current information on water quality, which must be published and updated regularly. It is also required that immediate public warning is given if, for any reason, the beach becomes unsafe, or is grossly polluted. Beach management and safety criteria are to include the provision of adequate litter bins, which are regularly maintained and emptied, beach clean-ups after peak days, no driving or camping on beaches, no dumping and safe access. There are also recommended standards for sanitary facilities, beach guards and first-aid facilities and telephones. A definitive guide to bathing beaches is produced by the Marine Conservation Society [38].

REFERENCES

1. Bassett, W.H. (2002) *Environmental Health Procedures*, 6th edn, Spon Press, London.
2. Department of the Environment (1991) *Private Water Supplies*, Circular 24/91, HMSO, London.
3. Department of the Environment (1991) *Private Water Supplies*, Circular 68/91, HMSO, London.
4. Scottish Office (1992) *Private Water Supplies*, Circular 20/92, HMSO, Edinburgh.
5. Drinking Water Inspectorate/WRc/-NSF Ltd (2002) *Manual on Treatment of Small Water Supply Systems*, WRc-NFS Ltd (London and Medmenham).
6. European Union (1976) *The Dangerous Substances Directive*, 76/464/EEC.
7. Chorus, I. and Bartram, J. (editors) (1999). *Toxic Cyanobacteria in Water: A Guide to Their Public Health Consequences, Monitoring*

and Management. E & FN Spon on behalf of WHO (Geneva).

8. Department of the Environment (1996) *Report on Freshwater by the Round Table on Sustainable Development*, HMSO, London (available from Portland House, Floor 23, Stag Place, London, SW1E 5DF).

9. Drinking Water Inspectorate (1995/96) *Nitrates, Pesticides and Lead*, DWI, Floor 2/A1, Ashdown House, 123 Vicitoria Street, London, SW1E 6DE.

10. Willocks, L., Crampin, A., Milne, L., Seng, C., Susman, M., Gair, R., Moulsdale, M., Shafi, S., Wall, R. Wiggins R. and Lightfoot, N. (1998) A Large Outbreak of Cryptosporidiosis Associated with a Public Water Supply from a Deep Chalk Borehole, *Communicable Disease and Public Health*, 1(4), 239–43.

11. Duke, L.A., Breathnach, A.S., Jenkins, D.R., Harkis, B.A. and Codd, A.W. (1996) A mixed outbreak of *Cryptosporidium* and *Campylobacter* infection associated with a private water supply. *Epidemiological Infections*, 116, 303–8.

12. Group of Experts on Cryptosporidium In Water Supplies (The Badenoch Report) (1990) *Cryptosporidium in Water Supplies*, Report of the Group of Experts, HMSO, London.

13. Group of Experts on Cryptosporidium in Water Supplies (The Second Badenoch Report) (1995) *Cryptosporidium in Water Supplies*, Second Report of the Group of Experts, HMSO, London.

14. Group of Experts on Cryptosporidium in Water Supplies (The Bouchier Report) (1999) *Clostrosporidium in Water Supplies*. Third Report of the Group of Experts, HMSO, London.

15. Drinking Water Inspectorate (1998) Private Water Supplies: *A Guide on the Protection of Private Water Supplies*, DWI, London.

16. Department of the Environment (1989) *The Water Act 1989*, Circular 20/89, HMSO, London.

17. British Standards Institution (1997) *Specification for Design, Installation, Testing and Maintenance of Services Supplying Water for Domestic Use within Buildings and their Curtilages*, BS 6700:1997, BSI, London.

18. British Standards Institution (1987) *Code of Practice for the Design of Concrete Structures for the Retaining of Aqueous Liquids*, BS 8007:1987, BSI, London.

19. British Standards Institution (1996) *Suitability of Non-metallic Products for use in Contact with Water Intended for Human Consumption with Regard to their Effect on the Quality of Water*, BS 6920, BSI, London.

20. European Union (1991) Council Directive of 21 May 1991 concerning urban waste water treatment (91/271/EEC). *Official Journal of the European Communities*, L135, 30 May 1991.

21. British Standards Institute (1971) Specifications for Biological Percolating Filters BS 1438:1971, BSI, London.

22. Hatchett, W. (2003) Stinking sewers could face fines. *Environmental Health News*, 18(22), 1.

23. British Standards Institution (1983) *Design and Installation of Small Sewage Treatment Works and Cesspools*, BS 6297:1983, BSI, London.

24. CIRIA (1998) *Septic Tank Systems – A User's Guide*, CIRIA, London.

25. Riley, C.T. and Jones, K.B.C. (1970) *Origins and Nature of Farm Wastes*, Symposium on Farm Wastes, University of Newcastle upon Tyne.

26. Environment Agency Online Article (2003) http://www.environment-agency.gov.uk/subjects/waterquality Accessed 11 June 2003

27. European Union (1975) Council Directive of 8 December 1975 Concerning the Quality of Bathing Water (76/160/EEC). *Official Journal of the European Communities*, L31, 5 February, 1976.

28. House of Commons Environment Committee (1990) *Pollution of Beaches*, 4th Report, HMSO, London.

29. Department for Environment, Food and Rural Affairs Online Article (2003) http://www.defra.gov.uk/ Accessed 11 June 2003.

30. Public Health Laboratory Service (1959) Sewage contamination of coastal bathing waters in England and Wales. *Journal of Hygiene*, **57**.
31. Medical Research Council (1959) *Sewage Contamination of Bathing Beaches in England and Wales*, Memorandum No. 37, HMSO, London.
32. Royal Commission on Environmental Pollution (1984) *Tackling Pollution – Experience and Prospects*, Cmnd 9149, HMSO, London.
33. World Health Organization (1989) *Microbiological/Epidemiological Studies on the Correlation between Coastal Recreational Water Quality and Health Effects*, revised protocol submitted by WHO Secretariat, ICP/CEH083/10, WHO, Geneva.
34. US Environmental Protection Agency (1986) *Ambient Water Quality Criteria for Bacteria*, EPA 440/5-84-002, USEPA, Washington, DC.
35. United Nations Environment Programme (1990) *The State of the Marine Environment*, UNEP, UN, Geneva.
36. Cabelli, V.J. (1983) *Health Effects Criteria for Marine Recreational Waters*, EPA-600/1-80-031, US Environmental Protection Agency, Health Effects Research Laboratory, Research Triangle Park, North Carolina.
37. Water Research Centre (1994) *Health Effects of Sea Bathing*, Final Report to the DoE, WM1 9021, DoE 3412/2 January, HMSO, London.
38. Marine Conservation Society (1994) *The Reader's Digest Good Beach Guide*, David & Charles, Exeter. (New edition published every year and available on the World Wide Web: www.readersdigest.co.uk.)

FURTHER READING

Dawson, A. and West, P. (eds) (1993) *Drinking Water Supplies: A Microbiological Perspective*, HMSO, London.

World Health Organization (1993–7) *Guidelines for Drinking Water Quality*, 2nd edn, Vols 1–3, HMSO, London.

40 Contaminated land

Alan Higgins

INTRODUCTION

The study of contaminated land is a complex issue and in a book of this type can only be covered in outline. Environmental Health Practitioners should see this chapter as a starting point for further reading and study.

Various estimates have been made of the extent of potential contamination in the United Kingdom. Expert estimates referred to in the report *Contaminated Land* published by the Parliamentary Office of Science and Technology in 1993 [1] ranged from 50 000 to 100 000 sites across the United Kingdom with the extent of land involved put at between 100 000 and 200 000 hectares. Only a small proportion of these would represent an immediate risk to human health. More recently the Environment Agency has estimated that some 300 000 hectares have been affected by industrial or natural contamination [2]. The estimated cost of cleaning up contaminated sites is calculated to be billions of pounds.

This legacy of the industrial development of the United Kingdom has been an issue that has taxed our politicians and regulatory bodies significantly over the last 15 years.

In 1990, the House of Commons Environment Committee published their report on Contaminated Land [3] in which, amongst other things, they recommended legislation to establish registers of contaminative uses. This was subsequently enacted as S.143 of the Environmental Protection Act 1990 but never implemented, principally as a consequence of concerns regarding blight arising from the registers and the absence of any mechanism for cleaning up contaminated sites. The provisions were subsequently repealed.

Following a review of the legal powers [4], and the publication of a consultation document [5] the Government published a policy document – the Framework for Contaminated Land (1994) – [6] that led to the enactment of the Environment Act 1995, containing powers for dealing with existing and future land contamination.

The policy document endorsed a number of principles for the control and treatment of existing contamination and requires remedial action only where:

- the contamination poses unacceptable actual or potential risk to human health or the environment, and
- that the polluter should be responsible and pay for past contamination, otherwise known as the 'polluter pays' principle, and
- there are appropriate and cost effective means of cleaning up the contamination, taking into account the actual or intended use of the site, otherwise known as the 'suitable for use' principle.

The 'suitable for use' approach takes into account the risks of contamination and the variability of those risks dependent on the level and type of contamination. In addition, the use to which the site is being or is intended to be put and a range of other factors such as the underlying geology of the site are also considered. The approach underpins any regulatory action to eliminate conditions on any site causing an unacceptable risk to human health or the environment. It also means that any permission for a new use takes account of the need to assess contamination risks and that requirements for remediation are limited by the proposed land use.

When determining the appropriate remediation in association with a planning application a broader range of issues must be considered including amenity, odour and long-term management [7].

However, although there is a framework for regulating land contamination, dealing with the sudden discovery of contamination on an occupied site, particularly if it is residential or for some other sensitive use, is tremendously difficult both technically and also in terms of the political and public relations issues. Some instances can be catastrophic as in the methane explosion that destroyed a bungalow in Loscoe, Derbyshire in the 1970s.

Local Authorities are the key body in dealing with contaminated land in a regulatory and planning context and Environmental Health Practitioners are key to ensuring the Local Authorities receive consistent and authoritative technical advice.

TYPES OF LAND CONTAMINATION

Many human activities cause contamination and this has been particularly true during the period of rapid industrialization of the last two centuries. The Government has published profiles of industrial uses encompassing a range of industries and their possible effects in terms of contamination [8]. Examples of the industrial uses that can give rise to contamination are given in Table 40.1.

Of the industrial uses listed in Table 40.1 a number of studies have indicated that waste disposal, mining, metal/engineering works and chemical/gas works form the major areas of

potentially contaminative uses [2]. However, the most significant areas of concern dealt with by Local Authorities are likely to be small intensive sites now in housing areas such as asphalt works, hatters, petrol stations.

Although industrial uses are a primary source of land contamination and will be predominantly found in urban areas, rural areas have their own problems particularly with the use of fertilizers, pesticides, herbicides and fungicides if used excessively. A further problem may arise from the practice of spreading sewage sludge on agricultural land where excess use may result in accumulation of heavy metals contained within the sludge. Other uses of potential concern in rural areas will be abattoirs, scrapyards and uncontrolled filling of quarries/ponds etc.

In some areas potentially harmful substances may occur naturally and so any man-made contribution has to be viewed in the context of the naturally occurring background. Soils derived from igneous rocks generally give higher levels of metals. Some rock types contain uranium and thorium whose radioactive decay gives rise to radon gas, a carcinogen, which can accumulate in buildings. Methane gas, which is combustible and potentially explosive, can also arise from the natural breakdown of organic material especially in marshy or peat areas or where there are strata containing naturally occurring organic material.

Potential hazards

It is possible to classify contaminants in a number of ways:

- **Inorganic substances** such as toxic metals and cyanides, oxidants, corrosive substances such as acids and alkalis and materials such as asbestos which are harmful because of their physical nature.
- **Organic contaminants** such as coal tar, phenols, petroleum products, solvents and chlorinated compounds such as polychlorinated biphenyls(PCBs).
- **Gases and volatiles**, particularly landfill gas which is a potentially explosive mixture of methane and

Table 40.1 Industrial uses giving rise to potential land contamination

- Coal and mineral mining and preparation
- Smelters, foundries, steel works and metal processing and finishing installations
- Heavy engineering and engineering works, for example, car manufacture, ship building
- Electrical and electronic equipment manufacture and repair
- Gasworks, coal carbonization plants, power stations
- Oil refineries, petroleum storage and distribution sites
- Manufacture of asbestos, cement, lime and gypsum
- Manufacture of organic chemicals including pesticides, pharmaceuticals, detergents and cosmetics
- Rubber industry including tyre manufacture
- Munitions production and testing sites
- Glass making and ceramics manufacture
- Textile industry including tanning
- Paper and pulp manufacture and printing works
- Timber treatment
- Food processing industry
- Railway depots, dockyards, garages, road haulage depots, airports and petrol filling stations
- Sewage works and farms
- Landfill, incineration of waste, waste handling and recovery and recycling
- Burial of diseased livestock
- Scrap yards
- Dry cleaning premises

Note:
This list is not exhaustive and many other historic and current land uses can given rise to ground contamination [1].

carbon dioxide but also may contain a range of other potentially harmful gases and volatiles such as hydrogen sulphide. Landfill gas may also give rise to odour problems.

- **Micro-organisms** such as anthrax which may present a disease risk to humans.

The presence of different chemicals may also result in synergistic effects as a result of chemical interactions, much of which is difficult to predict.

It is also helpful sometimes to classify contaminants by the nature of their potentially adverse affects:

- **Toxicity** to humans, other animals and to plants (**phytotoxicity**). Toxic effects may be subdivided into acute effects which arise from short-term exposure, or chronic effects which arise after long-term exposure. The nature of the effect also depends on the type of exposure that is, whether by inhalation, ingestion or via skin absorption.

- **Carcinogenicity or mutagenicity** – a wide variety of contaminants have the potential to induce cancers in humans. Some substances may cause birth defects (teratogenicity).
- **Corrosivity** – ranging from mild skin irritation to permanent physical damage of human tissue through inhalation, ingestion or skin contact. Corrosion of building materials can weaken building structures.
- **Combustibility** – as well as the dangers of fire, toxic or asphyxiant gases may be released.
- **Flammability and explosiveness.**
- **Asphyxiation.**
- **Physical damage** to humans or buildings through ground instability.
- **Surface and groundwater pollution** including pollution of drinking water sources [1].

Table 40.2 sets out hazards from some commonly encountered contaminants but these only represent a small number of the potential contaminants that are dealt with more comprehensively in a range of

Table 40.2 Hazards from some commonly encountered contaminants [1]

Contaminant	Sources	Hazards
Inorganics **Toxic metals** (e.g. arsenic, cadmium, copper lead, mercury, selenium)	Metal mines, iron and steel works, foundries and smelters, metal finishing works, engineering works, scrap yards, agricultural use (e.g. arsenic in herbicides and pesticides), paints, pharmaceutical manufacturers, sewage works and farms, incinerators and sewage sludge spreading	Toxic to humans by inhalation, ingestion or skin contact. May also be corrosive or irritate skin or other tissues. Arsenic and some salts of nickel and zinc are carcinogenic. Groundwater pollution may occur if metals are leached from site. Chromium salts are highly soluble in water
	Plating works, heat-treating works, photography, pigment manufacture, gas works and waste sites	Chemical attack on building materials (concrete, metals, plastic, rubber and brickwork). Corrosive to human tissue
Phytotoxic metals (e.g. boron, chromium, nickel, zinc)	As above	Damaging to plant life at relatively low concentrations
Aggressive substances (e.g. acids and alkalis)	Waste from fertilizer and chemical industries, metal surface preparation and finishing, plastics manufacture, paper and glass industries, gasworks	Toxic by ingestion, inhalation or absorption through skin. Inhalation of sulphur compounds leads to inflammation of respiratory tract. Sulphur compounds are also phytotoxic and corrosive to building materials
Finely divided materials (e.g. asbestos)	Industrial buildings, waste disposal sites, breakers yards (particularly of ships or defence equipment)	Carcinogenic and irritant if inhaled
Organics **Coal tar** (a complex mixture of over 10 000 compounds, including phenols)	Coal carbonization processes, e.g. old gas works	Toxic by inhalation. Some components (e.g. benzene and toluene) have narcotic properties, some are carcinogenic; strong irritant; phytotoxic; attacks plastics; combustion risk

Contaminant	Sources	Hazards
Phenols (a mixture of compounds with characteristic odour)	Coal carbonization processes – present in coal tar and ammoniacal liquors from gasworks. Wastes from pharmaceutical, dye and indicator industries	Toxic by inhalation, ingestion and skin contact Corrosive to skin. Attacks building materials, plastic and rubber piping and can lead to contamination of water supplies. Phytotoxic
Petroleum products (including benzene, toluene and xylenes)	Petrol retail sites, distribution terminals and refineries	Fire, asphyxiation and explosion. Benzene is carcinogenic. Pollution of groundwater
Solvents (including hydrocarbons, halogenated hydrocarbons, alcohols and esters)	Wastes from printing, oil extraction, degreasing, metal finishing and dry cleaning industries	Fire and explosion. Toxic by inhalation. Pollution of groundwater
Chlorinated compounds (excluding solvents, e.g. dioxins, polychlorinated biphenyls)	PCBs used in transformers and capacitors and as coolants and hydraulic fluids. Dioxins result from incomplete combustion of chlorine and organic compounds material, e.g. in waste incineration plant	Toxic, highly persistent (accumulate in food chain). Possibly carcinogenic
Others **Landfill gas** (methane and carbon dioxide)	Landfill sites	Explosive. Asphyxiant (carbon dioxide is also toxic). Phytotoxic due to reduction of oxygen around plant roots
Combustible substances (e.g. coal and coke dust, refuse)	Gasworks, power stations, railway land, landfill sites	Underground fires
Micro-organisms	Sewerage farms/works, burial of diseased livestock, slaughter houses, tanneries, cemeteries, hospital and research laboratory waste	Risk of disease to humans and animals

guidance set out later in this chapter but most important of these is the Contaminated Land Exposure Assessment (CLEA) Model [9].

The following are some commonly identified risks from land contamination to health, buildings and the environment [1]:

Human health

(a) **Uptake of contaminants by food plants grown in contaminated soil.** The accumulation of heavy metals such as cadmium and lead in the edible portions of some food plants may make their consumption over a long period unsafe. Uptake by the plant depends on the concentrations of the metals in the soil, the chemical forms in which metals are present, and the soil pH. The plant species and the proportion of home-grown food in the diet also influence the importance of this hazard.

(b) **Ingestion and inhalation.** Metals may be ingested directly by young children playing on contaminated soil (e.g. hand to mouth contact), or by eating plants which have absorbed metals or are contaminated with soil or dust. Ingestion may also occur via water supplies, contaminated either by leaching of soil contaminants into groundwater or by migration of organic compounds through plastic pipework laid in contaminated soil. Metals and some organic materials may also be inhaled from dusts and soils.

(c) **Skin contact.** Soil containing tars, oils and corrosive substances, for example, phenols may cause irritation to the skin through direct contact. Some contaminants, for example, phenols, may be absorbed into the body through the skin or through cuts and abrasions.

(d) **Fire and Explosion.** Materials such as coal and coke particles, oil, tar, pitch, rubber, plastic and domestic waste are all combustible. If they are heated, for example, by contact with buried power cables, by careless disposal of hot ashes or by lighting of fires on the site surface, they may ignite and continue to burn underground. Such fires may pose a threat to human health through production of toxic gases or flammable gases such as methane, which if allowed to accumulate in confined spaces could cause explosions.

Buildings

(a) **Fire and explosion.** Underground fires (as above) may result in ground subsidence posing a threat to the structural integrity of buildings. Accumulation of flammable gases in confined spaces within or beneath buildings also leads to a significant risk of explosion.

(b) **Chemical attack on building materials and building services.** Some substances such as sulphate attack concrete structures. The presence of acids, oily and tarry substances and other organic compounds may accelerate the corrosion of metals and attack plastics, rubber and other polymeric materials used in pipework (e.g. underground water pipes or storage containers) and service conduits or as jointing seals and protective coatings to concrete and metals.

(c) **Physical.** Old blast furnace and steel-making slags may expand if disturbed, for several decades after deposition.

Natural environment

(a) **Phytotoxicity (the prevention or inhibition of plant growth).** Certain metals such as boron, copper, nickel and zinc, which are essential for plant growth at low levels, are phytotoxic at higher concentrations. Methane and other gases may also give rise to phytotoxic effects by depleting the oxygen content of the soil in the root zone.

(b) **Contamination of water resources.** Soil has a limited capacity to absorb, degrade or attenuate the effects of pollutants, and when this is exceeded, polluting substances may be leached out of the soil into surface and groundwater. Some contaminants such as hydrocarbons can be present in a liquid-free product which will readily migrate through the ground.

(c) **Ecotoxicological effects.** The presence of contaminants in the soil may also affect microbial, animal and plant populations. Ecosystems or individual species, both on the original sites of contamination and in rivers or areas affected by migration from the site, may be affected.

LANDFILL GAS

Landfill gas is a particular concern to Local Authorities as it is mainly generated from landfill sites and often in urban areas these are in close proximity to housing and other development. Landfill gas comprises methane, carbon dioxide and other trace gases and if allowed to accumulate in confined spaces can achieve explosive concentrations. Although there have only been two incidents of explosions caused by accumulations of methane in the United Kingdom in the last 30 years, the potential for loss of life and severe injury means that this issue has to be taken extremely seriously. Methane, which is the main explosive content of landfill gas, can also arise from naturally occurring organic materials in some silts or alluvial deposits. Investigation in respect of landfill gas needs to be conducted over a long period of time, particularly targeting falling and low pressure to be meaningful and may require expert advice and support. There is a range of guidance available to assist evaluation and control measures:

CIRIA 149 'Protecting Development from Methane' [10].
CIRIA 150 'Methane Investigation Strategies' [11].
CIRIA 151 'Interpreting Measurements of Gas in Ground' [12].
CIRIA 152 'Risk Assessment for Methane and other Gas on the Ground' [13].
Wilson & Card February 1999 'Ground Engineering, Reliability and Risk in Gas Protection Design' [14].

LEGISLATIVE FRAMEWORK FOR DEALING WITH CONTAMINATION

The United Kingdom has a comprehensive framework for dealing with both past and present land contamination. That framework comprises of a number of elements:

(a) Legislation to deal with past contamination

Water Resources Act 1991 – Pollution of Controlled Waters. Allows the Environment Agency to serve a 'works notice' to prevent the pollution of controlled waters or carry out works themselves in urgent circumstances. There is potential for overlap between this provision and the Part IIA provisions where the pollution is being caused by contaminated land. Guidance requires the Part IIA provisions to be used where contaminated land is involved but also requires full consultation between the Agency and the Local Authority before either take action which affects controlled waters involving contaminated land.

Environmental Protection Act, Part IIA – Power to control and remediate contaminated land. Part IIA of the 1990 Act was added by the 1995 Environment Act which also established the Environment Agency. It established a regime for the identification and remediation of contaminated land. In this context contaminated land is defined as

'any land that appears to the Local Authority in whose area it is situated to be in such a condition, by reason of substances in, on or under the land, that:

(a) significant harm is being caused or there is a significant possibility of such harm being caused, or:
(b) pollution of controlled water is being, or is likely to be caused.'

(Harm means harm to the health of living organisms or other interference with the ecological systems of which they form a part and in the case of man, includes harm to his property.) Examples of significant harm are given in the statutory guidance [15].

In order to determine whether or not harm is likely to arise as a result of hazardous materials on land, Part IIA of the Environmental Protection Act 1990 uses the concept of a 'pollutant linkage' – that is, a linkage between a contaminant and a receptor by means of a pathway. An example of a pollutant linkage is given below:

HAZARD	+	PATHWAY	+	RECEPTOR	= RISK
(e.g. Methane from a landfill)		(e.g. Granular Strata and sub floor void)		(e.g. Occupied Residential Premises)	

In this instance, methane being generated either by a landfill site or a naturally occurring substance, that is, the contaminant, is finding its way via a pathway into a building where it can accumulate to an explosive degree and hence cause harm to a receptor, in this instance human beings or the building.

In order for land to be defined as contaminated within the Part IIA regime, all three elements of contaminant, pathway and receptor must be present. It therefore follows that by removing one of these elements one can introduce a control mechanism into contaminated land and this is the basis upon which remediation is carried out and is dealt with later on in this chapter.

The effect of this definition is to limit the number of sites that are likely to require action under the Act to a relatively small number. This leaves a range of sites having some degree of contamination but not falling within the definition; derelict sites and other brown field sites to be dealt with by the planning and regeneration processes, through use of planning conditions requiring appropriate desk study, site investigation and remediation.

Unitary or second tier local authorities are the primary regulators for the regime. They have a duty to inspect their districts to identify contaminated land, to determine whether any particular site is contaminated land and to carry out enforcement duties in respect of such sites except those designated as special sites. Special sites are defined by regulation and are the responsibility of the Environment Agency although the Local Authority is initially responsible for determining a special site as contaminated. Generally, they are sites that are already the responsibility of the Agency under other legislation. Local Authorities were required to publish their strategies for carrying out their responsibilities under Part IIA by June 2001.

Enforcing authorities are required to:

- determine who is responsible for remediation of contaminated land (the 'appropriate' person);
- to consult with all stakeholders including the Environment Agency;
- to decide what remediation is required and when it should take place either by voluntary agreement or through an enforcement procedure;

- to determine liability and apportion costs when enforcement action is taken;
- to keep records of formal actions taken.

The Government's intention is that enforcing authorities, wherever possible, should seek agreement on remediation with the landowner and any other responsible persons and should take enforcement action only where this is not possible. Detailed formal guidance on the procedures to be followed under Part IIA has been published by the Government and should be studied by Environmental Health Practitioners before commencing any informal or formal action [16].

An excellent source of reference on Part IIA is 'The Local Authority Guide to the Application of Part IIA of the Environmental Protection Act 1990' which covers the procedural issues in great detail in relation to the legislation. This document has been published as a joint collaboration with the Local Government Association, Department of the Environment, Food and Rural Affairs (DEFRA), the Environment Agency and the Chartered Institute of Environmental Health [17].

The Environment Agency has a number of roles under Part IIA:

- to assist local authorities in identifying contaminated land especially where pollution of controlled waters is involved;
- to provide site-specific guidance to local authorities in respect of remediation;
- to act as an enforcing authority for special sites;
- to publish periodic reports and carry out technical and scientific research on contaminated land.

In addition, under the terms of a Memorandum of Understanding with the Local Government Association [18], they will provide a range of information to assist local authorities in identifying all contaminated land, for example, information on waste licences.

Part IIA excludes sites which are subject to other legislation such as waste consents, authorization under the integrated pollution prevention and control regime, fly tipping, radioactivity, or discharge consents under the Water Resources Act 1990 administered by the Environment Agency.

(b) Prevention of future contamination

The Environmental Protection Act 1990, Part II – Waste Management. Waste disposal is primarily controlled by the licensing of waste carriage and waste disposal sites. Where there is site licence in force Part IIA does not normally apply except in circumstances where any harm or pollution of controlled water arises from something other than a breach of the site licence conditions.

It is important to note that remediation of contaminated sites will require a licence under the waste provisions where waste disposal or recovery is part of the remediation process.

Pollution and Prevention Control Act 1999. Deals with the control of prescribed processes to prevent, minimize or counteract pollution of the air, land or water. It is predominantly enforced by the Environment Agency although local authorities have some responsibilities for a limited range of processes. The legislation established a regime that takes account of pollution as part of the process rather than applying 'end of pipe' solutions. Processes that produce land contamination are required to clean up sites to an agreed standard, often the standard that applied before the process commenced, before a licence can be surrendered.

Such sites are outside the Part IIA controls as they have their own enforcement regime mainly controlled by the Environment Agency.

Planning and development control. Land contamination is a material consideration for the purposes of town and country planning. Planning authorities must consider land contamination in respect of applications for development control and within development plans. Policy Planning Guidance [6] published by the Government requires all parties to the planning process to be aware of the potential for land contamination, and to take any necessary steps as part of the redevelopment process to remove any unacceptable risks, applying the suitable for use approach. Planning authorities are able to apply conditions to development control applications covering such issues as site investigation, remediation and maintenance of monitoring, reporting and record keeping in respect of land contamination. The emphasis within the guidance is that developers

are primarily responsible for dealing with land contamination issues but encourages informal pre-application discussions. Planning Officers will normally seek guidance from Environmental Health Practitioner colleagues when considering applications involving land contamination issues. Developers need to provide appropriate and sufficient validation information to satisfy the local Planning Authority that all works have been completed before a recommendation can be made to discharge a planning condition.

(c) Other related legislation

Environmental Protection Act 1990, Part III – Control of Statutory Nuisances. Until the implementation of Part IIA the main regulatory mechanism for dealing with contaminated land. Contaminated land is now specifically excluded from the statutory nuisance regime.

Food and Environment Protection Act 1985 – Food Safety. Provides powers to control agricultural activities to protect consumers from exposure to contaminated food.

Health and Safety at Work Act 1974. Provides controls on risks to the public from work-related activities such as might arise when excavating or investigation of contaminated land. Liaison with the Health and Safety Executive is important to avoid duplication of controls and to ensure protection of the public and the workforce. There are a number of regulatory controls that are of particular importance:

- Control of Substances Hazardous to Health (COSHH) Regulations, 2002.
- Management of Health & Safety at Work Regulations, 1999.
- The Workplace (Health, Safety & Welfare) Regulations, 1992.
- The Provision & Use of Work Equipment Regulations, 1998.
- The Personal Protective Equipment Regulations, 1992.
- The Construction, Design & Management Regulations, 1994.
- Transport & Packaging Regulations.

Finance Act 1996 – Landfill Tax. Allows for exemption from the tax when material is being removed from land to prevent harm or facilitate development. Requires a specific application for exemption before works commence.

Building Regulation. To deal with health and safety issues in relation to the actual buildings themselves, which may include requirements in respect of land contamination.

LOCAL AUTHORITY ROLES

Local Authorities in England and Wales have a number of diverse and sometimes conflicting roles in relation to contaminated land:

Regulation

Contaminated Land Controls under Part IIA of the Environmental Protection Act 1990. Local authorities are the primary regulator for the purposes of dealing with land where significant harm is being caused to humans or the environment. These controls are dealt with in more detail elsewhere in this chapter. Local authorities are required to have a published strategy for dealing with contaminated land.

Planning Controls. For the purposes of the Town & Country Planning Act 1990, and the Planning & Compensation Act 1991, the potential for contamination is a material planning consideration to be taken account of during the normal course of development and redevelopment. It is the primary mechanism for regenerating contaminated and derelict land. The establishment of good working relationships between Environmental Health Practitioners and Planning Officers is fundamental to good control.

Building Control. Linked to the planning role where redevelopment or new development takes place. The Building Regulations require that precautions be taken to avoid danger to health and safety caused by substances on or in the ground to be covered by the building. Any requirements as to the land surrounding the building are to be dealt with under planning conditions, although it is possible that changes proposed to the Building

Regulations may bring areas beyond the building blueprint within these controls.

Land owner

Many Local Authorities have substantial land holdings and in many instances some of these land holdings will be contaminated, particularly where they have been used previously for waste disposal or other industrial uses, for example, industrial estates. Local authorities often hold land such as allotments or playing fields that have been placed on land previously used for waste disposal. The possibility of any risk arising to users of such sites will need to be considered.

Local authorities will have to give consideration not only to land that they currently own, but also to land that they have previously sold or are considering selling to ensure that they have discharged their liability under the contaminated land provisions.

Under the 'polluter pays' principle, enacted within Part IIA of the Environmental Protection Act, previous ownership of a site which has been subject to contamination during the Local Authorities ownership may result in a continuing liability unless measures have been taken to discharge that liability. Ensuring that land sales take place under conditions of full disclosure as to the condition of the land is one means of helping to discharge liability.

Waste management

All purpose local authorities have responsibility for domestic waste collection and disposal. Second tier authorities have responsibility for collection only.

The provisions of the Environmental Protection Act 1990 Part II – Waste on Land, require the Environment Agency to enforce provisions which impose a duty of care on all waste holders for the proper disposal of controlled waste including domestic waste. It also makes provision for the granting of waste management licences for the treating, keeping or disposal of controlled waste. This area of regulation transferred from Local Authorities to the Environment Agency in April 1996. Local authorities will need to ensure their compliance with legislative provisions, although in practice many have passed these responsibilities to private contractors.

Economic development

Local Authorities have a statutory responsibility to promote economic development in their areas and they do this in a number of different ways:

- through the planning process, and particularly by way of the development plan provisions for industrial and commercial development;
- by building partnerships with businesses and financial interests to promote particular developments;
- by assisting land assembly, particularly in respect of flexible use of their own land resources, but also by compulsory purchase to promote particular areas for business development;
- by providing services to businesses to encourage them to move to a particular area;
- by facilitating applications for grants through various government and European programmes available to promote economic development particularly in deprived areas.

The process invariably requires a high degree of facilitation, enabling and negotiating skills and involves consultation with the local community in order to arrive at solutions that not only provide employment opportunity, but also balance the other interests of the community, particularly in respect of environmental matters.

The redevelopment of brown field sites has become an increasing priority in urban areas with the government reluctant to support development of green field sites and the shortage of land in urban areas. Conflict between the economic development and the regulatory roles of local authorities can be avoided by having a clear and properly consulted strategy on contaminated land use preferably endorsed within the local planning mechanisms.

Information

The Local Authority provides information to the general public on a range of issues including information related to the environment and contaminated land through a number of mechanisms.

Land Charge Searches – providing information held on statutory actions taken by the Local Authority in response to formal inquiries where land purchases are progressing. Local authorities are only required to provide information on contaminated land issues where they take formal action under the Environmental Protection Act 1990. However, some purchasers will follow up with specific questions on other information held by the Local Authority in respect of land contamination which may be considerably more than that held in respect of statutory action.

Environmental Information Regulations 1992 provide that except in specific circumstances local authorities must make available any environmental information that they hold. This would include any information in relation to land contamination.

Local authorities will also have to consider how much information they make available or require when involved in land transactions themselves either buying or selling to ensure they discharge liabilities in respect of land contamination and do not unknowingly acquire liabilities.

IDENTIFYING AND DETERMINING CONTAMINATED LAND

The strategy

The technical guidance [15] gives detailed requirements for local authorities in respect of establishing and reviewing their strategies for identifying contaminated land. In order to establish a strategy in respect of contaminated land, the Local Authority needs to have a clearly identified framework for action which will have a number of principle elements to it:

- Clearly identified aims, objectives and priorities.
- Appropriate timescales for action.
- Identification and prioritization of sites, including:
  - obtaining and evaluating information on actual harm or pollution of controlled waters;

– identifying receptors and assessing the possibility or likelihood that they are being or could be affected by a contaminant;

– obtaining and evaluating existing information on the possible presence of contaminants and their effects.

• The process of identification and prioritization has been the subject of guidance from the Government which local authorities will have to take into account when dealing with contaminated land issues.

• Managing information requirements.

• Process for dealing with sites that fall outside of action under Part IIA, that is, principally those sites that are suitable for use at the present time but are likely to be re-developed in the future.

• Liaison with other agencies and other owners and occupiers of land.

• Responding to members of the public, businesses and voluntary organizations.

Identification and prioritization

In order to consider its priorities for work to establish a detailed programme of inspection in a particular area, local authorities will need to have some mechanism for identifying those areas which will have the highest priority for them. They will need to collect a range of information and the information sources set out in Table 40.4 will be useful, both at this stage and at the desk study stage. Some of the factors that they will need to take into account in identifying priority areas are as follows:

• Evidence of actual significant harm or pollution of controlled waters.

• Nature and location of Part IIA receptors.

• Extent to which relevant receptors may be exposed to contaminants.

• Extent of existing information on land contamination.

• History/scale/nature of industrial or other activities.

• Extent to which remedial action has already been taken or is planned through redevelopment to deal with contamination.

• Extent of other regulatory interest in the land.

Linking contamination to potential receptors will be the major methodology to try and identify potentially contaminated sites and Table 40.3 gives some indication of Part IIA receptors and their locations.

The geographical co-incidence of potential contamination and sensitive receptors will confirm that two parts of a potential pollutant linkage are in place and will allow the authority to start to define inspection areas. There is a third element that will need to be verified and that is the presence of a relevant pathway.

Some authorities have developed priority matrices to try and ensure that resources are allocated to meet the most severe problems. These may take the form of a relatively simplistic ordering of sites in decreasing order of sensitivity with the most sensitive at the top and the least sensitive at the bottom, for example,

• Residential with gardens
• Allotments
• Schools
• Amenity areas
• Commercial/industrial
• Paved car parking areas.

This may be made more sophisticated by introducing other factors and giving them a weighting, for example, the distance from a gassing landfill site. By combining the scores allocated to each element this can result in a scoring method where the high scores will represent the most risk and therefore the sites that should be investigated first. There are ranges of variations on this approach which can be found in the various contaminated land guidance and also in health and safety guidance documents in relation to risk management [15,19,20].

Risk management process

In order to achieve an objective process for identifying, describing and evaluating the risks associated with contaminated land and deciding the best ways of controlling or reducing those risks to an acceptable level one must adopt a risk management approach. The process involves hazard identification, hazard assessment, risk estimation, risk

Table 40.3 Potentially sensitive receptors [1]

Receptors	Land use types
Human beings	• Allotments • Residential with gardens • Residential without gardens • Schools or nurseries • Recreational/parks, playing fields, open space • Commercial/industrial
Ecological systems or living organisms forming part of a system within statutorily protected locations	• SSSI's • National nature reserves • Marine nature reserves • Areas of special protection for bids • European sites (SCAs, SPAs, candidate SACs and SPAs) • Ramsar sites • Nature reserves
Property in the form of buildings and structures	• Ancient monuments • Buildings
Property in other forms (crops, livestock, home grown produce, owned of domesticated animals, wild animals subject to shooting or fishing rights)	• Agricultural land • Allotments and gardens • Forestry areas • Other open spaces, rivers, lakes etc.
Controlled waters	• Surface waters • Drinking water abstractions • Source protection zones • Groundwaters – private abstractions • Groundwaters – major aquifers

evaluation and risk control. Its principal advantages are that it is structured, objective, comprehensive and it explicitly considers uncertainties. The approach provides a rational, transparent and defensible basis for discussion of a proposed course of action, something which regulators will find essential in their discussions with landowners and developers. The first stage of this process involves the site investigation which will

• determine the nature and extent of any contamination;
• identify any hazards posed by the contamination;
• identify and describe pathways and targets;
• allow assessment of current and potential risks;
• determine whether remedial action is needed;
• inform decisions on the nature, extent and performance requirements of any remedial works.

Detailed guidance is available from the Construction Industry Research & Information Association (CIRIA) to assist in constructing a comprehensive site investigation approach which will deliver the required outcomes [21,22]. The principle phases of investigation are:

• Preliminary investigation – site reconnaissance and desk study.
• Exploratory investigation.
• Detailed investigation and assessment.

There may be a requirement for further supplementary investigations to clarify any particular technical matters or to collect information relevant to the application of the selected remedial option.

It may be at any point during this process that the Local Authority will determine that they have sufficient evidence to define a site as contaminated and then they will be looking to the 'appropriate person' (the person legally responsible for carrying out any remedial action) to carry out the remaining elements of the site investigation in order to define a remedial process. Although the local authorities' responsibility for site investigation ends once they are able to determine whether a site is contaminated in the context of Part IIA, they will still have to have a detailed involvement in the assessment of any proposed remedial action in order to satisfy themselves that the site will no longer be contaminated at the completion of any remedial works.

It is therefore important that the objectives for any site investigation are clearly stated. For a local authority it may be to provide sufficient information on potential hazards, pathways, targets and other site characteristics to make a determination as to whether or not a site can be determined as contaminated land under Part IIA of the Environmental Protection Act 1990.

However, for a person or body responsible for designing appropriate remedial measures a more appropriate objective might be to provide sufficient information on potential hazards, pathways, targets and other site characteristics (e.g. engineering constraints) to permit an assessment of risks and allow decisions to be made on the need for, and nature of, any remedial work including any immediate action to protect public health or the environment.

The defined objective will assist in determining the extent and nature of any site investigation that has to be carried out.

Preliminary site investigation

The preliminary site investigation comprises a site reconnaissance and a formal desk study. This should provide initial information on the actual and probable nature and location of contamination and other hazards. It will also provide the information needed to determine the site-specific investigation objectives and procedures for any detailed site investigation as well as assisting in identifying the health and safety and environmental requirements for on-site work.

The initial element of the preliminary site investigation is the desk study which is a review of the information about the particular site, including the history, its geological and hydrogeological setting and any present or future receptors. A range of information sources is useful in preparing the desk study and some sources of information are set out in Table 40.4.

The desk study should provide a range of information about the site layout, the design and construction of any buildings, the nature and quantities of materials handled on the site, the nature of surrounding land use, physical features, previous history particularly focusing on industrial uses or spillages and leakages and information about the geology and hydrogeology of the site. The information about the nature of the surrounding land uses is important to determine issues in relation to receptors.

Once the desk study has been completed the report can be used to focus the work of the site reconnaissance which involves a visit to the site, armed with the information from the desk study, so that any potential hazards can be identified and a strategy for the reconnaissance agreed. It is at this stage where sites are obviously seriously contaminated that a local authority may well be in a position to make a determination in respect of contaminated land. However, in the absence of this the site reconnaissance will provide visual confirmation of the desk study and could allow a limited amount of sampling to help inform the detailed strategy for the detailed investigation design.

Amongst other things that the site reconnaissance may identify will be any obvious immediate hazards to public health or safety, including:

- Any areas of discoloured soil, polluted water, distressed vegetation or significant odours.
- The location and condition of buildings, roads, fences etc. and any deviations from those shown on available plans.
- Location of services, sewers, outfalls of any surface water and any standing water or water in any rivers, streams or canals.
- The location and condition of any surface deposits, made ground, any signs of any settlement, subsidence or disturbed ground.

Table 40.4 Information sources likely to be used in the desk study [20]

Site records	Drawings, production logs, maintenance records, supplier records, environmental audits, compliance records, COSHH assessment records, R&D activities
Company records	Environmental audit reports, archival information, title deeds
Plant personnel	Plant Manager, Safety Manager, Product Manager, Employees
Maps	O.S., town maps, geological maps, hydrogeological maps, thematic maps, groundwater vulnerability maps, tithe maps
Photographic material	Ground level, aerial (e.g. military photo-reconnaissance)
Local literature	Library, local history departments, local newspapers, local specialist societies, clubs etc.
Directories	Trade, Street etc.
The regulatory authorities	Local councils (e.g. environmental health, trading standards, planning, Waste Regulation Authorities, Hazardous Substances Authorities), Environment Agency, River Purification Boards
Local knowledge	Residents, former employees etc.
Technical literature	Research and review papers, specialist guidance
The fire and emergency services	Fire certification documents, accident reports, petroleum tank records
Other organizations	British Coal, Opencast Executive, British Geological Survey, Water, Gas and Power Companies etc.

- Any evidence of seepage through river or canal banks.
- Anything that may be useful during any subsequent investigations, including depots, offices, labs etc. and any boreholes left from any previous investigations.
- Site constraints which could limit the site investigation e.g. wide of access gates.
- Results from limited sampling of both gas, surface deposits and surface water.

It is always useful as part of the site reconnaissance to make a photographic record of site conditions, layout and any individually important features.

The preliminary site investigation will hopefully assist in drawing conclusions about the probable distribution of contamination and whether it is appropriate to zone the site for the purposes of subsequent investigation. The detailed investigation will be based upon the conclusions of the preliminary site investigation.

Exploratory investigation

Exploratory investigation may be carried out to test the conclusions formed from the preliminary site investigation as to the presence, nature and extent of contamination. It will not of itself usually be sufficient to prove the absence of contamination; a more detailed investigation will be necessary to determine this question. It may be necessary for the purposes of determining the site as contaminated land to carry out an exploratory investigation which will involve the need to take the minimum amount of samples to determine whether the conclusions reached about contamination in the preliminary investigation are proven.

If this is the case then moving onto a more detailed investigation may not be necessary for the purposes of the Local Authority in determining the site as contaminated land for the purposes of Part IIA.

Detailed investigation

A more detailed site investigation may be carried out by the Local Authority where they have not been able to demonstrate that the land is either uncontaminated or contaminated land by the preliminary and exploratory site investigations. Where contaminated land has been determined then the Local Authority may well require the landowner or some other person responsible for the contamination to carry out a detailed site investigation. The scope of any detailed investigation will be highly site-specific. There is significant guidance available on the design of site investigations from CIRIA, British Standards and the Environment Agency [21,23–25].

A number of issues need to be taken into account when implementing a site investigation.

Project management   The normal practice is for specialist advisers to be appointed to conduct site investigations as generally the knowledge and expertise to carry out this type of work is unavailable within local authorities or from landowners or persons responsible for contamination who may have to carry out such investigations. Such specialist advisors should operate a quality management system and demonstrate knowledge and expertise in a range of issues including risk assessment, health and safety, soil and water quality, modelling techniques, remediation techniques, analytical and testing aspects of investigation, environmental legislation, contract procedures and validation techniques. They should also be able to demonstrate previous experience in this field, be familiar with publicly available guidance and also be able to access additional specialist expertise or advice which may be required. A detailed site investigation will comprise a number of elements.

Quality management   The standing of any report produced as a result of a site investigation will be tested by its quality management standards. It should be able to demonstrate that the key aspects of the site investigation process, including such matters as handling of samples, storage and preparations of samples, methods of analysis and testing, have been subject to a comprehensive system of progress and performance checks. The participation by specialist service providers in accreditation schemes for quality management systems, will assist in lending confidence to the process. The development of the Environment Agency's Monitoring Certification Scheme (MCERTS) will be important in this context [26].

Environmental protection requirements   It is essential to ensure that during the course of any site investigation of contaminated sites that environmental protection measures are taken to protect uncontaminated lower strata and groundwater during drilling operations and surface waters during excavation activities. Controlled disposal of surplus materials and proper cleaning of equipment and site vehicles, protection of ground surfaces from spillages, will also be necessary.

Sampling strategies   The type, number, method, position and analysis of samples will be determined by a range of variable factors in relation to the site, as well as the information derived from the preliminary and exploratory site investigations. Each sampling strategy will be site-specific and will be dependent upon a conceptual model which has been developed in relation to the site from the information already available. The conceptual model will draw on the available information in relation to contamination sources, the potential pathways and the potential receptors to identify 'pollutant linkages' which may exist. There may be a number of these and therefore the model will need to take account of this. Assumptions will be made about the likely areas, types and mobility of pollutants, the impact that they may have on potential receptors and how these assumptions can be tested through the sampling strategy and subsequent risk assessment.

The conceptual model   This may suggest that parts of the site are uncontaminated, that there are known points and sources of contamination, that contamination is uniformly distributed or that there are an unknown number of point sources within the site, or that there is actually little rational pattern to the distribution of contamination, either horizontally or vertically. The strategy will also need to take into account the potential for contaminant migration into or out of the site.

Guidance on sampling strategies is provided both in the CIRIA documents [22] and the British Standards [23,24].

Sample collection   Table 40.5 sets out the main exploratory methods of the collection of soil and groundwater samples. It is important to note that methods to collect samples should not contaminate the sample, or allow the absorption or escape of contaminants. Containers should be kept clean to avoid cross contamination between the samples.

Analytical and testing strategies   It is not economically feasible to analyse a large number of samples for all possible contaminants and therefore some means of rationalizing the analytical strategy is usually required. This will be based on the objectives of the investigation, the findings of the desk study, previous investigation findings and observations made on site and visual inspection of the samples. Depending on the site-specific circumstances a comprehensive approach to analysis is likely to involve testing for ubiquitous contaminants such as lead, zinc and arsenic oils, substances that are specific to the past and present use of the site and the use of screening techniques to detect adventitious substances whose presence cannot be easily predicted. The analysis technique used for determining more complex contaminants such as hydrocarbons must be appropriate for the type of hydrocarbons likely to be present. The analysis will also need to take account of any statutory requirements. The use of industry profiles to determine likely contaminants is appropriate.

There is a range of other methods that are non-intrusive which can image the ground through a variety of techniques without changing the ground itself or any contaminant profiles which from a health and safety perspective can be useful.

These techniques can help to focus the intrusive work in appropriate areas, particularly if coupled with the information from the desk study. The six main methods used are:

- Seismic
- Gravity
- Electrical
- Electromagnetic
- Magnetic
- Ground penetrating radar.

There are other non-intrusive techniques such as satellite imagery, aerial photography and thermal imaging which may be considered.

Site investigation report

A typical site investigation report should contain the following elements [22]:

- A summary and description of the work carried out stating the main aims and findings, together with their implications and brief account of the conclusions and recommendation.
- A table of contents – this should include a list of tables, figures and appendices.
- An introduction which should set out the conclusions from the preliminary report and include elements such as the details, location and history of the site including a brief description of its current state. Any constraints on the site investigation should be noted.
- General matters – this should relate to the general site conditions while the work was in progress and any matters affecting the outcome of the investigation.
- Photographs and drawings are usefully included.
- Sampling – an outline of the strategy adopted along with appropriate plans of the sampling points etc.
- A description of materials encountered (inclusion of trial pit, borehole and strata sample records).
- Any on-site testing or measurements and the results obtained should be provided.
- Analysis and off-site testing – issues related to the sample preparation, analytical and test methods should be described. All analytical and test results should be included in the report.
- A narrative description and discussion of results.
- Data evaluation. Assessment of data in terms of adequacy, type, quantity and quality.
- Hazard assessment.
- Risk estimation.
- Risk evaluation.
- Conclusions and recommendations.

Table 40.5 Methods of exploration [20]

Methods	Advantages	Disadvantages
Surface sampling	• Ease of sampling • Allows assessment of immediate hazards	Only very shallow sampling possible
Augering	• Allows examination of soil profile and collection of samples at pre-set depths	• Limited depths achievable • Ease of use very dependent on soil type • Can lead to cross contamination if not done with care
Trial pits and trenches	• Allows detailed examination of ground conditions • Ease of access for discrete sampling purposes • Rapid and inexpensive	• Limitation of depth of exploration • Greater exposure of media to air and greater risk of changes to contamination • Greater potential health and safety impacts • More potential disruption/damage to site • May generate wastes for disposal • More potential for escape of contaminants to air/water (need to protect ground surface) • May need to import clean material to site for backfilling (to ensure clean surface)
Boreholes	• Permits greater sampling depth • Provides access for permanent sampling/monitoring points • Less potential for adverse effects on health and safety, or above ground environment (but note potential risks to groundwater) • Smaller volumes of waste to dispose of • May permit integrated sampling for contamination, geotechnical and gas/water sampling	• More costly and time consuming • More amenable to visual inspection • Limited access for discrete sampling purposes • Depending on the technique may be disturbance to samples and therefore loss of contaminants • Potential for contamination to an underlying aquifer • Potential for groundwater flow between strata within aquifer
Driven probes	• Minimal disturbance of site – no need to remove material from the hole • Some soil properties can be determined during penetration • Sample can be sleeved to prevent contamination • Undisturbed samples can be recovered • Variety of measuring devices can be installed once hole is formed • Fewer health and safety, and above-ground environmental implications	• Limited opportunity to inspect strata • High mobilization costs for most powerful equipment

Table 40.5 Continued

Methods	Advantages	Disadvantages
	• Can be used where there is limited access/space available	
Soil gas sampling	• Can detect volatile organics • Allows contamination plumes to be mapped • Avoids difficulties of handling and analysing samples for volatiles	• Can detect only volatiles and semi-volatiles • Semi-quantitative data only • Limited use in clay rich soils
Groundwater Sampling Standpipe/well	• Permits water levels, water quality and permeability to be measured and pump tests to be carried out	• Cannot provide accurate water quality, levels or pressure data in sequences of strata of different permeability • May permit cross-flow within and between acquifers
'nested' wells and similar installations	• Permits sampling from different depths	• More expensive
Piezometer	• Permits monitoring of water pressures and permeability to be tested, particularly over specific zones	• Not usually suitable for groundwater quality testing

Risk assessment

This will be an important tool in relation to many sites where there is no gross pollution but pollution is likely to contribute to chronic risks. In this context the source–pathway–target approach will form the basis of any judgement in terms of risk estimation. Soil guideline values, developed in the CLEA model [9] and other environmental standards, will assist this process.

In developing a conceptual model to assess risk, site-specific risks must be identified based on the 'source–pathway–receptor' approach. This effectively requires that a source or sources of a substance or substances with the potential to cause harm or water pollution must be present on or in the land; a receptor or target, that is, the presence of humans or other receptors that can be harmed by the pollutant must be present or have access to the land; and there must be a pathway by means of which the receptor can actually be exposed to the potential pollutant if the source, pathway and the receptor have a 'pollutant linkage'.

It is only when this linkage is present that there is a potential for land to be considered as contaminated. There may be more than one 'pollutant linkage' and the conceptual model for the site must identify all the pollutant linkages of potential concern.

The Local Authority has to take into account the definitions of significant harm which are contained in the Act and in the Guidance. 'Harm' is defined in Part IIA as meaning: 'harm to the health of living organisms or other interference with the ecological systems of which they form part, and in the case of man, includes harm to his properties'. Further detailed guidance [16] is given on the question of whether the harm is significant and whether the possibility of significant harm being caused is significant.

The guidance states that the following types of harm are to be regarded as significant harm:

Chronic or acute toxic effect, serious injury or death to humans.

Irreversible or other adverse change in the functioning of an ecological system.

Substantial damage to or failure of buildings, plant and equipment.

Disease, other physical damage, to, or death of livestock or crops.

The possibility of the harm caused being significant will be dependent on a number of factors including the nature and degree of harm, the timescale within which the harm might occur and the vulnerability of the receptors to which the harm might be caused.

Other issues that might be taken into account, particularly where there is a low probability of harm, are the number of people that are likely to be affected, whether the harm is likely to be irreversible and whether it could be caused by a single incident (such as a fire or an explosion).

The assessment of risk is a complex process as one must assess all the potential exposure pathways by which a receptor, particularly humans, may come into contact with a contaminant, and if there are a range of contaminants for each pathway it is necessary to predict the quantities involved and the frequency and duration of contact. For example, if there is a release of toxic gas to the air, factors to be taken into account should include the assumed amount of gas released, dilution factors including weather conditions, how many breaths a person would take during their exposure to the gas and what the overall effect as far as the population as a whole is concerned, taking into account the effects upon vulnerable elements of the population, that is, children and the elderly.

In order to assist the process of risk management and also to try and limit the resource implications of such a process a tiered approach has been developed to risk assessment of increasing complexity to enable the risk assessment to be undertaken in the most cost effective manner.

The approach is set out below in Fig. 40.1.

The model follows the three-tiered approach to risk assessment.

Tier 1 – Risk screening

This involves the development of an understanding of the contaminated site in its environmental setting, including the identification of all the possible sources of risk, the pathways and the potential receptors. Consideration is given as to the sensitivity of receptors and initial selection of appropriate environmental benchmarks for each group of receptors, as well as some consideration of the potential impacts on each receptor via a simple qualitative or quantitative approach. If no pathways exist between potential receptors and the contamination or if none of the contamination is likely to cause harm to any specific receptor then the output from the risk screening is that no further action would need to be taken. The use of environmental benchmarks to provide a standard for initial assessment is essential.

When considering environmental benchmarks it can be the case that benchmarks for specific contaminants are unavailable within UK guidance. This is particularly true of some of the volatile component of landfill gas. In these instances it is appropriate to consider other guidance values and the normal hierarchical approach to this is to first of all consider UK standards, then EU standards, then WHO standards and then other national standards. It is important when considering other standards to understand the circumstances in which they have been produced, particularly the specific purpose and the political context, that is, standards in the United States are likely to be more stringent than elsewhere, partly because the United States has been looking at this issue for longer and partly because of their more stringent approach to environmental legislation. However, they are also likely to be extensively peer reviewed and tested in the courts.

Tier 2 – Simple risk assessment

Should the initial screening identify potential pathways then there is a need for the next stage of the risk model to be implemented, the simple risk

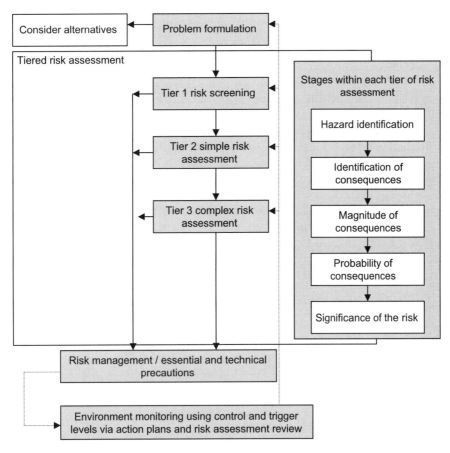

Fig. 40.1 Tiered approach to risk assessment (from DETR *et al.*, 2000) [27,28].

assessment. This should consist of quantitative calculations using conservative input parameters, assumptions and methods where it is possible to define the potential sources, pathways and receptor issues with sufficient certainty that they can be represented within the calculations.

Tier 3 – Complex risk assessment

It is unlikely that a simple risk assessment will generally be applicable in sensitive locations. In more sensitive locations where there is a degree of uncertainty regarding any of the source, pathway or receptor information the more complex risk assessment is carried out tailored to the characteristics of the specific contamination issues

at the site supported by specific information derived from site monitoring, site investigations and detailed modelling of contaminant behaviour.

Such assessments will take account of both short and long-term exposure risks as well as acute and chronic effects.

Environmental standards or benchmarks are used to risk screen each particular hazard. These standards can take a number of forms dependent upon the type of hazard, whether it is liquid, solid or gaseous.

The CLEA model has been developed to provide soil guideline values as one form of environmental benchmark which can be used as a quantitative guideline for concentrations of contaminants in soil in order to allow for a consistent approach to risk assessment of contaminated sites. These guideline values are based not only on the toxicological information on health effects but also on an element of judgement in relation to what is an acceptable risk. The primary purpose of soil guideline values is as an intervention level value in the regulatory framework for assessment of risks in relation to land use. They have been developed on the basis of the 'suitable for use' approach. The CLEA model is one element in determining whether to take action on a particular site and assisting in defining remediation objectives. Comprehensive guidance has been produced to assist in the understanding and use of soil guideline values [29]. It must be adhered to in using the soil guideline values as they are limited in their application to certain land use types and specific circumstances.

Risk assessment is a specialized area and Environmental Health Practitioners will need an understanding of how to carry out risk screening and simple risk assessments. Complex risk assessments should be carried out by specialists, although Environmental Health Practitioners will need sufficient knowledge to understand and review the process.

Detailed guidance is available to assist in this understanding as set out in Table 40.6 [9].

A further refinement as to the hierarchy of guidance on land contamination is set out in Table 40.7.

REMEDIATION

Remedial action is primarily taken to reduce the risks posed by the presence of contamination to an acceptable level. The remedial action on any particular site will be determined by a risk assessment process which takes account of the acceptable contaminant levels for each medium, exposure routes or pathways and the receptors in relation to that particular site. This may be expressed in terms of quantitative contamination levels for near surface soils, replacement material, growth media and landscape areas and any soils that represent an ongoing source of contamination for

surface or groundwater associated with the site. The remediation will need to take account of the proposed or existing use of the site and will need to be economically feasible and require the minimum of post treatment monitoring and maintenance. In addition issues such as the time taken for carrying out any remediation option and its compatibility with the existing use of the site may need to be considered.

Any remediation process will need to ensure avoidance of any long-term legal liability and its acceptability to funding bodies. Finally, account will need to be taken of any regulatory requirements and community acceptance of any proposals. Remediation principles/objectives for Part IIA will be different to remediation principles and objectives for planning. Planning will deal with a broader range of issues related to contamination, not just risk, for example, amenity issues (odour), practicality of maintenance, longevity of measures. Part IIA measures will focus on the removal of risk, the simplest example of which may be a change of land use to a less sensitive use, for example, fencing of a site to limit access for potential receptors.

There is no universally agreed classification for remediation methods for contaminated land but they can be broadly divided into:

- **Civil engineering methods** which use conventional civil engineering techniques, either to remove the contaminant source or to modify contaminant pathways without necessarily removing, destroying or modifying the source.
- **Process-based methods** that involve the application of physical, chemical or biological processes to remove, destroy or modify contaminant sources or the pathways along which they may be released.

There follows a brief description of some of the remediation techniques available. More detailed information can be found in the 12 volumes produced by CIRIA SP102-112, 122 and 124 [21]. Collectively, they cover all aspects of remediation of contamination. Each volume deals with a specific subject and promotes a systematic approach to the management of contaminated land.

Table 40.6 Assessment of risk to human health from land contamination. Key reports from DEFRA and the Environment Agency

CLR7 *Assessment of Risks to Human Health from Land Contamination: An Overview of the Development of Soil Guideline Values and Related Research* (DEFRA and Environment Agency 2002a). CLR7 serves as an introduction to the other reports in this series. It sets out the legal framework, in particular the statutory definition of contaminated land under Part IIA of the Environmental Protection Act (EPA) 1990; the development and use of Soil Guideline Values; and references to related research.

CLR8 *Priority Contaminants for the Assessment of Land* (DEFRA and Environment Agency, 2002b). Identifies priority contaminants (or families of contaminants), selected on the basis that they are likely to present on many current or former sites affected by industrial or waste management activity in the United Kingdom in sufficient concentrations to cause harm; and that they pose a risk, either to human health, buildings, water resources or ecosystems. It also indicates which contaminants are likely to be associated with particular industries.

CLR9 *Contaminants in Soil: Collation of Toxicological Data and Intake Values for Humans* (DEFRA and Environment Agency, 2002c). This report sets out the approach to the selection of tolerable daily intakes and Index Doses for contaminants to support the derivation of Soil Guideline Values.

CLR TOX1-10 (DEFRA and Environment Agency, 2002d). These reports detail the derivation of tolerable daily intakes and Index Doses for the following contaminants, which are arsenic, benzo (a) pyrene, cadmium, chromium, inorganic cyanide, lead, phenol, nickel, mercury and selenium.

CLR10 *The Contaminated Land Exposure Assessment Model (CLEA): Technical Basis and Algorithms* describes the conceptual exposure models for each standard land-use that are used to derive the Soil Guideline Values. It sets out the technical basis for modelling exposure and provides a comprehensive reference to all default parameters and algorithms used.

CLR GV 1–10 (DEFRA and Environment Agency, 2002e). These reports set out the derivation of the Soil Guideline Values for the following contaminants, which are arsenic, benzo (a) pyrene, cadmium, chromium, cyanide (free, simple and complex inorganic compounds), lead, phenol, nickel, mercury (inorganic compounds) and selenium.

CLR 11 *Model Procedures for the Management of Contaminated Land* (DEFRA and the Environment Agency, in preparation). This report incorporates existing good technical practice, including the use of risk assessment and taking appropriate action to deal with contamination, in a way that is consistent with UK policy and legislation.

Civil engineering processes

Excavation

Excavation may be the methodology that precedes disposal off-site, disposal on the site and on or off-site treatment. In all cases such disposal will need to be to a suitably licensed facility or area. It is a methodology that is applicable to a wide range of contaminants which can be carried out over a relatively short time period. The main constraint for this method is the cost of handling, transporting and disposing of large volumes of contaminated material. There may also be difficulties with suitable replacement fill materials and the health and safety implications of handling contaminated material, both for site workers and the general public.

On-site disposal

This approach may overcome some of the potential disadvantages of off-site disposal and therefore may be applicable to very large schemes allowing for controlled disposal directly supervised by clients. It does however require controls under both the planning and waste management legislation and may require engineering works to prepare the site. In addition, there are possible

Table 40.7 Hierarchy of guidance on land contamination [8]

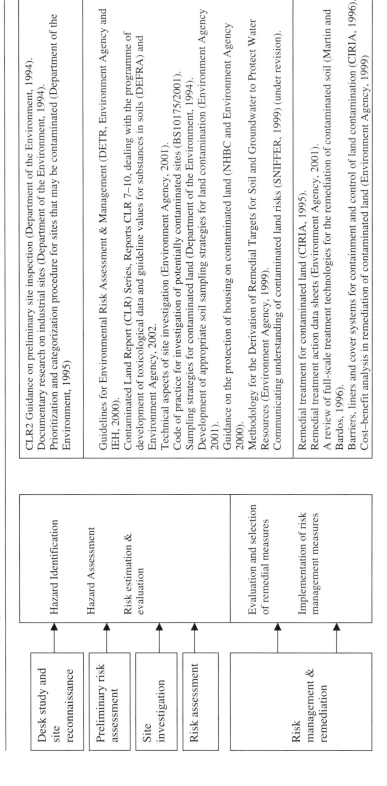

Stage	Sub-stage	Guidance
Desk study and site reconnaissance	Hazard Identification	CLR2 Guidance on preliminary site inspection (Department of the Environment, 1994). Documentary research on industrial sites (Department of the Environment, 1994). Prioritization and categorization procedure for sites that may be contaminated (Department of the Environment, 1995)
Preliminary risk assessment	Hazard Assessment	Guidelines for Environmental Risk Assessment & Management (DETR, Environment Agency and IEH, 2000).
Site investigation	Risk estimation & evaluation	Contaminated Land Report (CLR) Series, Reports CLR 7–10, dealing with the programme of development of toxicological data and guideline values for substances in soils (DEFRA) and Environment Agency, 2002. Technical aspects of site investigation (Environment Agency, 2001). Code of practice for investigation of potentially contaminated sites (BS10175/2001). Sampling strategies for contaminated land (Department of the Environment, 1994).
Risk assessment		Development of appropriate soil sampling strategies for land contamination (Environment Agency 2001). Guidance on the protection of housing on contaminated land (NHBC and Environment Agency 2000). Methodology for the Derivation of Remedial Targets for Soil and Groundwater to Protect Water Resources (Environment Agency, 1999). Communicating understanding of contaminated land risks (SNIFFER, 1999) (under revision).
Risk management & remediation	Evaluation and selection of remedial measures / Implementation of risk management measures	Remedial treatment for contaminated land (CIRIA, 1995). Remedial treatment action data sheets (Environment Agency, 2001). A review of full-scale treatment technologies for the remediation of contaminated soil (Martin and Bardos, 1996). Barriers, liners and cover systems for containment and control of land contamination (CIRIA, 1996). Cost–benefit analysis in remediation of contaminated land (Environment Agency, 1999)

long-term restrictions on land use and potential for continuing liability. Ongoing licence requirements in the long-term may make it unsuitable for certain issues, for example, housing.

Covering systems

This involves placing a specified depth of clean material possibly incorporating break layers and/or membranes on the surface of the contaminated ground. Such covering systems may be used in conjunction with vertical and horizontal barriers and may require trenches of clean material to contain building services. The process is applicable to a wide range of contaminants and physical conditions and can be economic, even on large sites where other options become too expensive to implement. It is a well-established approach and is widely used, although there may be issues of long-term degradation and failure due to accidental or unauthorized disturbance of which account has to be taken. The depth of capping must be robust in the long-term for proposed/existing land use and maintenance of future control.

In-ground barriers

In-ground barriers physically isolate contaminated ground from surrounding media and may take the form of physical or chemical barriers. They come in a variety of different types of system ranging from sheet steel piling to slurry trench walls, chemical grouting and ground freezing. They are potentially economic where it is not feasible to remove or treat large volumes of contaminated materials and they can be used to control solid, liquid or gaseous contamination hazards. They can be used in combination with other remedial methods. The biggest constraint is that there is a lack of information on long-term performance and durability and therefore monitoring is required to check long-term performance.

Hydraulic measures

Involves management of the local hydrological regime to prevent or reduce contact between the contaminated ground mass and surface and groundwater bodies and the reduction or containment of contaminated groundwater. This may support other remediation methods such as excavation or physical barriers to provide a comprehensive approach. It may be a more economic solution than physical barriers for some applications and it is applicable to most contaminants present in a water-soluble form. Its basic constraint is that it requires long-term monitoring, maintenance and adjustment. It may be difficult to ensure it is maintained over a period of time. It will require detailed characterization of local and regional hydrogeological regimes which will require specialist input.

Process-based methods

These methods are generally derived from existing technologies used in the oil and chemicals industry, waste treatment, or minerals processing. They can be applied both *ex situ* and *in situ*, with the advantage that *in situ* applications can be applied where the cost and potential health, safety and environmental impacts associated with the excavation and transportation of contaminated materials are too great. In practice, *ex situ* methods are used far more widely because of the greater operator control and the much greater degree of uncertainty in applying *in situ* techniques.

The main forms of process-based treatment available are – thermal, physical, chemical, biological and stabilization/solidification.

Thermal treatment methods

Thermal treatment uses heat to remove or destroy most of the organic contaminants by:

- Thermal desorption where contaminants are removed by evaporation at high temperature (600°C) from the feedstock and subsequently the volatile components are treated at a higher temperature.
- Incineration – this is the application of high temperatures (800–1200°C) to destroy or detoxify contaminants. Exhaust gases must be treated.

- Vitrification – involves the application of high temperatures (1000–1700°C) to destroy some contaminants and trap others in a glass product.

In practice all of these treatment methods are well established internationally, although their use in the United Kingdom is limited. They offer the prospect of complete destruction of some contaminants. Generally the processes can be carried out using a mobile plant.

Some methodologies are highly energy intensive and therefore extremely costly and there may be requirements for pre-treatment. Regulatory control over atmospheric emissions and mobile plant will be required.

Physical treatment methods

Physical treatment methods separate contaminated and uncontaminated material by exploiting differences in physical properties, for example, density, particle size, by applying external physical force, for example, abrasion or by altering physical characteristics to enable segregation to occur, for example, flotation. The main physical methods are classified in four main groups – particle separation techniques, soil washing systems, solvent extraction systems and other physical separation techniques. The systems are applicable to a wide range of contaminants and use existing well-established technologies. Mobile plant is available. However, they do produce secondary waste streams for treatment or disposal and they are generally difficult to apply to peaty soils or soils with a high clay content. Some of the methodologies will have health and safety implications. There will be a need for regulatory approval for operation of the process.

Chemical treatment methods

These methods use chemical reactions to destroy or modify contaminants present in solids and liquids. The range of chemical processes that can be used include oxidation/reduction/ dechlorination/PH adjustment and hydrolysis. Chemical treatment processes can be applied

in situ, although effective contact between the contaminant and reagents can be difficult to achieve and control. Experience in their application to contaminated soils is limited. It is used where it is possible to be very specific about particular contaminants to be removed. It does require a good mixing contact for many of the processes to work effectively. Some of the chemical reagents' by-products may have public health and environmental implications which may require regulatory approval for the operation of any process.

Biological treatment methods

These methods use the natural processes of micro-organisms such as bacteria and fungi to transform, destroy, fix or mobilize contaminants. Both *ex situ* and *in situ* systems are available but because they use living organisms to be effective they must have optimal process conditions in terms of temperature, oxygen concentrations (or absence of oxygen), nutrient balance, PH etc. They do offer the potential for the complete destruction/ detoxification of certain contaminants and can be highly specific in their action. However, some variations do require long treatment periods and many organic compounds and most inorganic contaminants are not amenable to biological treatment. Some contaminants may actually inhibit bacterial degradation.

Stabilization solidification methods

These methods operate by solidifying contaminated materials, converting contaminants into a less mobile chemical form by binding them with an insoluble matrix presenting a minimal surface area to leaching agents. *Ex situ* and *in situ* treatment variations are available. Typically, they use reagents such as Portland cement/lime/flyash/ clay/silicates/polymers and proprietary additives. These methods are proven for certain inorganic contaminants and require a relatively short treatment period. Mobile plant is available. However, they only contain contaminants and all methods require good mixing/penetration. The process requires long-term monitoring to check performance.

Remediation process selection

In selecting a remediation process consideration will need to be given to the applicability of the process, its effectiveness, any limitations which the process may have, the development status of the process – is it well established or an emerging technology – its availability, operational requirements and the timescale involved. Cost will be a significant factor and is probably the single most important non-technical factor which will influence the selection process.

INFORMATION REQUIREMENTS

The statutory requirements within Part IIA require Local Authorities to maintain registers which include information in relation to remediation notices that have been served, remediation statements and declarations, any appeals, charging notices or convictions, and notifications regarding what has been claimed to have been done by way of remediation.

This statutory requirement is fairly limited as sites that will fall within the definition of contaminated land under Part IIA will be relatively few and even fewer where formal action will be taken.

There will be a far wider range of sites that have some degree of contamination that are fit for the purpose for which they are being used for the time being. Local Authorities will need to consider how they hold information in respect of these sites in order to deal with their obligations in terms of redevelopment and the planning process.

This has implications in terms of how information is held, how long it is held and what sort of public access there should be to this information. The Freedom of Environmental Information Regulations 1992 require that most of this information is readily accessible for those who wish to seek it.

Much of the information being held is of a highly technical nature and when reviewed by specialist advisors for developers, for example, it should not present too many problems, although consideration will have to be given to what charges are made for such services. However, a more difficult problem arises with access for the general public who may have a limited understanding of the information with which they are presented and this has the potential to blight land values where house sales are involved. The preparation of a simple fact sheet summarizing available information on sites where there are regular enquiries may help to overcome this problem.

Additionally, local authorities are required to respond to local land charge enquiries where properties are being sold and purchased. The specific requirements of the normal land charge questionnaire would require that the Local Authority only respond where there has been actual formal action taken under Part IIA. Some solicitors may follow up with more specific questions regarding other information held regarding site contamination and it is at this point that simple summary sheets of the issues affecting any particular site may be useful to provide information and also point to the more detailed information held on the site.

Some local authorities are adopting the use of computer-based geographic information systems which enables information on sites to be held in a readily accessible form with the benefit of historical information such as historic maps in respect of each site. The cost of these systems is rapidly reducing but the resource implications in establishing a comprehensive record base are significant.

OTHER CONSIDERATIONS

The Part IIA regime for contaminated land raises a number of practical issues for all those concerned in the regulatory roles.

Consistency of approach

This will be vital in establishing a level playing field for action thus avoiding some of the problems of blight that will otherwise arise and are likely to bring the process into disrepute.

How Local Authorities handle the development of their strategy for identifying contaminated land and what sort of enforcement approach they take in respect of remediation notices can clearly have significant impacts on economic development in particular areas.

Since the introduction of the Part IIA regime in April 2000 the number of sites actually determined as contaminated have been relatively few, only 33 by 31 March 2002 of which 11 were designated as special sites [2].

This may have been because Local Authorities have been concentrating on their inspection strategies and the number of sites will increase as this phase is left behind. It may also be because the legislation is complex and Local Authorities are pursuing a more informal approach to remediation. The sites have been relatively small, mainly contaminated with organic compounds and metals, predominantly from the fuel/oil storage, construction and waste management industries [2].

Local authority liaison groups have been set up across the country to enable local authority Environmental Health Practitioners to exchange information, identify training needs and develop a common approach to the complex issue of contaminated land.

Training

Training for dealing with contaminated land issues is essential for Environmental Health Practitioners as this is a complex area of work and keeping up to date with new government guidance and technical advances is an essential pre-requisite to ensuring that evaluation of sites is done effectively.

There have been joint initiatives between the Government, Local Government Association, Environment Agency and professional associations such as the Chartered Institute of Environmental Health to promote procedural training on the Part IIA regime and also specific training on the technical areas such as site characterization, remediation and risk assessment. Locally, local authority liaison groups have promoted training and this has often taken the form of workshops looking at actual case studies. There are also a number of academic institutions that provide training courses which are particularly useful for practitioners new to the

subject. This is supplemented by a number of private providers who run ad hoc courses.

A group of professional and technical associations have developed a registration scheme for Specialists in Land Condition (SiLC) which aims to provide a recognizable standard for professionals working in the contaminated land field. This may be helpful in selecting consultants to provide specialist advice and support.

Relationships with the Environment Agency

The Local Government Association and the Environment Agency have agreed to a protocol of land contamination as part of the Memorandum of Understanding between these organizations [18]. This clearly identifies the roles and responsibilities of local authorities and the Agency as well as processes for information exchange, consultation, co-operation and transparency. The Agency have provided all local authorities with a package of information to assist in preparing their strategies, including information on catchment areas, catchment management plans, licensed landfill sites and bathing waters, licensed water abstraction points, consents to discharge and sites for waste management licences, IPC and IPPC authorizations, nuclear sites, and information on water and river quality objectives.

The Agency has also been closely involved in organizing and delivering elements of the procedural training on Part IIA as well as some joint training on the technical aspects related to land contamination. At a local level agency representatives sit on local authority liaison groups and more specifically have provided advice on site specific situations. Liaison between local authorities and the Agency at an early stage on potential special site has also been an important factor. The Agency have also established a National Groundwater & Contaminated Land Centre as a source of expertise and advice.

The Agency have also published a summary of information on local authority strategies [2] and will also have to produce an Annual Report on contaminated land in England and Wales.

Resources

Resourcing of the contaminated land regime comes in a number of forms.

- **Revenue funding.** To support Part IIA work within local authorities. The government included some revenue funding within the rate support grant in 1999/2000 to support the implementation of the new regime. For many authorities the specific funding has only been sufficient to fund part posts. The amount of resource made available for contaminated land will be dependent upon the degree of priority that the Local Authority give to the issue. In some authorities where there is a significant problem resources attached to it may have to be significant to support both the planning and economic development processes and will come from general revenue funds.
- **Capital funding to support site investigation and remediation.** The government has a Supplementary Credit Approval scheme to support the remediation of local authority sites and sites determined under Part IIA where it is not possible, or, for reasons of hardship, not practical to take formal action against an appropriate person. There is a limited amount of funding available and it is accessed via submission to DEFRA.
- **Capital funding for the redevelopment of sites.** Predominantly, this will come through developers but may be supported by a range of other funding, either through government or the EU, specifically aimed at areas of economic deprivation which are often those areas that have the most complex issues surrounding land contamination. Regional Economic Development Agencies may also be a source of funding.

Guidance

There is a significant range of existing guidance which comes in a number of different forms:

- Statutory guidance – produced in the context of the Part IIA regime which must be adhered to when taking formal action under Part IIA.

- Advisory guidance which has been produced to support statutory guidance.
- Technical guidance. A wide range of technical advice and guidance has been produced, some of which has been referenced in the course of this chapter, some of which is referenced at the end of this chapter. It is produced by Government, by the Environment Agency, by professional associations and bodies such as CIRIA, other statutory and voluntary agencies such as British Standards, and by professional institutes such as the Chartered Institute of Environmental Health (CIEH).

Environmental Health Practitioners who wish to work in the area of contaminated land will need to become familiar with and be able to reference to the relevant guidance.

New guidance is being produced on a range of issues by the Environment Agency and DEFRA, so it is important to keep up-to-date.

REFERENCES

1. Parliamentary Office of Science & Technology (1993) *Contaminated Land.*
2. Environment Agency (2002) *Dealing with Contaminated Land in England & Wales.*
3. House of Commons Environment Select Committee (1990) *Contaminated Land.*
4. HMSO (1992) *The Government's Response to the House of Commons Select Committee on the Environment on Contaminated Land.*
5. HMSO (1994) *Paying for Our Past.*
6. HMSO (1994) *Framework for Contaminated Land.*
7. HMSO (1994) PPG23 *Planning & Pollution Control.*
8. Department of the Environment (1995) *Industry Profiles.*
9. DEFRA/Environment Agency (2002) *The Contaminated Land Exposure Assessment Model*, CLR10.
10. CIRIA 149, Protecting Development from Methane.
11. CIRIA 150, Methane Investigation Strategies.

12. CIRIA 151, Interpreting Measurements of gas in ground.
13. CIRIA 152, Risk Assessment for Methane & Other Gas in Ground.
14. Wilson & Card, Ground Engineering (1999) *Reliability and Risk in Gas Protection Design.*
15. DETR (2001) *Contaminated Land Inspection Strategies*, Technical Advice Note for Local Authorities.
16. DETR Part IIA (2000) *Contaminated Land, Environmental Protection Act 1990.*
17. Local Government Association, DEFRA, Environment Agency (2001) *CIEH, The Local Authority Guide to the Application of Part IIA of the Environmental Protection Act 1990.*
18. Local Government Association and Environment Agency (2001) *Memorandum of Understanding and Protocol on Land Contamination.*
19. DOE (1995) *Prioritisation and Categorisation Procedure for Sites which may be Contaminated.*
20. HSE (1991) *Protection of Workers and the General Public during the Development of Contaminated Land.*
21. Construction Industry Research & Information Association (CIRIA) (1995) Special Publications 102–12, 122 & 124, Remedial Treatment for Contaminated Land, Various Topics.
22. CIRIA Sp103 (1995) Remediation Treatment for Contaminated land Volume III: Site Investigation and Assessment.
23. BSI (2000) Investigation of Potentially Contaminated Sites, DD175.
24. BSI (1999) Code of Practice for Site Investigations BS5930.
25. DOE Report by Applied Environmental Research Centre Ltd (1994) *Guidance on Preliminary Site Inspection of Contaminated Land*, vols I & II.
26. Source Testing Association Guide (2002) *Environment Agency Monitoring and Certification Scheme (MCERTS)*, pp. 4–10.
27. DETR/CIRIA (2001) *Contaminated Land Risk Assessment.*
28. DEFRA/Environment Agency (2002) *Model Procedures for the Management of Contaminated Land CLR11.*
29. DEFRA/Environment Agency (2002) *Guidelines Values for Contamination in Soils, CLR10 GV1-10* (covers a range of contaminants).

FURTHER READING

The Local Authority Guide to the Application of Part IIA [16] has a detailed listing of key publications on contaminated land which is a useful source of reference and further reading. As well as the detailed listing of publications from a wide range of statutory, professional and private organizations, it also includes a listing of web addresses for reference.

41 Pest control

Veronica Habgood

INTRODUCTION

The control of pests is a long-established public health function of local authorities. A pest may be defined as a creature that, in a particular situation, is seen as undesirable, whether for health and hygiene purposes, or for aesthetic or economic reasons. Pest control is the term applied to activities designed to identify, reduce or eliminate pest populations in any given situation.

Control of pests through local authority intervention is primarily concerned with preventing a risk to public health in whatever situation a pest may be encountered. Diseases caused by bacteria, viruses, protozoa and fungi may be transmitted actively through a bite or a sting, or passively via contaminated food or from contaminated food preparation surfaces and equipment. Insects and rodents are the most common pests associated with risks to public health. Rats and mice have been shown to transmit salmonellosis and leptospirosis (Weil's disease) to humans; cockroaches and pharaoh's ants can transmit pathogenic bacteria; fleas and bedbugs may cause infection at the site of a bite, in addition to transmitting pathogenic organisms to a host's blood. Feral pigeons and seagulls may be carriers of *Salmonellae* and their presence, particularly close to food premises, is undesirable.

Within local authorities, the function of pest control is most often the responsibility of the environmental health service because of the intrinsic link to health and hygiene. The execution of the pest control function, however, varies considerably between local authorities. For many local authorities, the pest control function amounts solely to the provision of advice together with the enforcement of the provisions of the Prevention of Damage by Pests Act 1949, the Public Health Act 1936 and the Food Safety Act 1990, and an assortment of minor legislation. Many local authorities have now contracted out their pest control function; others still provide a full advisory and treatment service. The policy on charging for treatments varies between local authorities and with the type of pest infestation.

LEGAL PROVISIONS

A miscellany of legislation of direct and indirect relevance to pest control has been produced. Generally, from a public health stance, the legal provisions provide for:

- the duties of local authorities;
- the prevention of risk to health, and nuisance;
- health and safety in the use and storage of pesticides;
- applications of relevance to particular pests.

The following is a summary of the most relevant provisions in respect of pest control.

Prevention of Damage by Pests Act 1949

The Prevention of Damage by Pests Act 1949 is primarily concerned with the control of rats and mice and the prevention of loss of food through infestation. Infestation is defined as the presence of rats, mice, insects or mites in such numbers or under such conditions that there is a potential risk of substantial loss or damage to food.

A duty is placed on local authorities under Section 2 to ensure that their district is kept free from rats and mice. To this end, local authorities are required to:

- carry out inspections from time to time;
- destroy rats and mice on any land that the local authority occupies, and keep that land free, so far as is practicable, from rats and mice;
- enforce the provisions of the Act in respect of the duties of the owners and occupiers of land.

A duty is also placed on the owner or occupier of land to notify the local authority in writing when substantial numbers of rats and mice are living on their land, excluding agricultural land. The expression 'substantial numbers' is not defined. The law essentially places the responsibility for maintaining land free from rats and mice on owners and occupiers.

Where local authorities are aware of circumstances where action should be taken to destroy rats or mice on land, or where there is a need to keep the land free from rats or mice, they may serve a notice on the owner or occupier of the land requiring steps to be taken within a reasonable period of time to effect specified treatment, structural repair or other work. The local authority can carry out the work in default of the person(s) on whom the notice was served, and recover the costs.

Other provisions of the Act are concerned with infestations in connection with the business of manufacture, storage, transportation or sale of food. These provisions are administered by the Department of Environment, Food and Rural Affairs (DEFRA), which may delegate powers to the local authority. Further provision in respect of the infestation of food can be found in the Prevention of Damage by Pests (Infestation of Food) Regulations 1950.

Public Health Act 1936

The provisions of Sections 83–85 of the Public Health Act 1936 deal with action available to local authorities in the case of filthy and verminous premises, articles or individuals. The expression 'verminous' includes reference to the eggs, larvae and pupae of insects and parasites. In the case of premises that are considered to be verminous or in such a filthy or unwholesome condition as to be prejudicial to health, the local authority has the power to serve a notice on the owner or occupier of the premises specifying works that are to be effected to remedy the condition of the premises, or to remove and destroy the vermin. Works may be carried out in default of the owners or occupiers with the costs being recovered from them. Filthy or verminous premises are frequently characterized by an accumulation of material that can make access to a premises difficult and that may present a physical or fire risk to the occupants or those of adjoining premises.

The application of these powers is not uncommon in many local authorities. Referral may often come from social care organizations, which encounter individuals or households in such conditions in the course of their intervention activities. Resolution of many cases requires a tenacious and firm but sympathetic approach, and the environmental health department working in partnership with others, such as social care organizations. Some local authorities will employ contractors to effect any work carried out in default; others will have their own 'dirty squad'.

In the case of filthy or verminous articles, similar provisions are available to the local authority to require that those articles be cleaned, purified, disinfected or destroyed, or, if appropriate, removed from the premises to prevent injury or danger of injury to the health of any person.

In the case of a verminous person and his clothing, either the local authority or the county council may remove that person, with his consent, to a cleansing station. A court order may be obtained from a magistrates' court requiring the person's removal to a cleansing station if he refuses to consent to removal. County councils or local authorities may provide cleansing stations for the purpose of exercising these functions, although many local authorities no longer have access to such facilities within their area.

Public Health Act 1961

Section 37 of the Public Health Act 1961 prohibits any person who trades or deals in household articles (a 'dealer') from preparing for sale, selling, offering or exposing for sale, or depositing with any person for sale or preparation for sale any household article known to be verminous. Any household article being prepared for sale, offered or exposed for sale or deposited for sale in any premises can be disinfected or destroyed on or off those premises under the authorization of the proper officer of the local authority. Clearly, this has implications for the second-hand furniture market.

Food Safety Act 1990

The provisions of the Food Safety Act 1990 are not directly concerned with pests and pest control. However, under Section 8 it is an offence to sell food that is unfit for human consumption or so contaminated that it would not be reasonable to expect it to be used for human consumption. Food contaminated as a result of a pest infestation may therefore be construed as being unfit for human consumption, although each case should be considered individually.

The Food Safety (General Food Hygiene) Regulations 1995 make some specific references to the control of pests. In Chapter I of Schedule 1, the layout, design, construction and size of food premises must be such as to protect against external sources of contamination, such as pests; in Chapter II, windows opening to the outside environment must be fitted with cleanable insect-proof screens, where necessary; in Chapter VI, refuse stores must be designed and managed to prevent against access by pests, and in Chapter IX of the same schedule, adequate procedures must be in place, generally, to ensure that pests are controlled.Further guidance can be found in the Food Safety (General Food Hygiene) Regulations 1995 Industry Guide to Good Hygiene Practice: Catering Guide (HMSO,

1995) available from: http://www.doh.gov.uk/pub/docs/doh/catering.pdf.

Remedial activities can be effected through the service of an improvement notice issued under Section 10 of the Food Safety Act 1990.

Other legal provisions

Powers are given to local authorities under Section 74 of the Public Health Act 1936 to deal with nuisance or damage caused in built-up areas through the congregation of house doves, pigeons, sparrows or starlings. No bird that has an owner can be seized or destroyed. All reasonable precautions must be taken to ensure that birds are destroyed humanely, and nothing may be done that is contrary to Part I of the Wildlife and Countryside Act 1981.

Provisions exist to deal with the presence of rodents, insects and vermin on ships and aircraft, where their presence may be a threat to public health. The Public Health (Aircraft) Regulations 1979 and the Public Health (Ships) Regulations 1979 are applicable in these circumstances. Powers are exercisable by the director of public health, not the local authority environmental health function (see p. 111).

Where rats are threatened by or infected with plague, or are dying in unusual numbers, the Public Health (Infectious Diseases) Regulations 1988 require the local authority to report the situation to the chief medical officer for England (or Wales) and to take all necessary measures for destroying rats in its area and to prevent rats from gaining entry to buildings.

The Public Health (International Trains) Regulations 1994 make provision for the 'deratting' of international trains leaving the United Kingdom in the unlikely event that rats from a plague control area are, or are suspected of being, on board (also see p. 117).

There is a wide range of other legislation, dealing largely with the control of particular species. These requirements are concerned with ensuring the humane destruction of wild animals, health and safety for pest control operatives and the public, protection of the environment and limitations on the control of certain species.

PEST CONTROL AND PREVENTION

Rather than waiting for an infestation to occur, emphasis should be placed on preventative measures. These should be introduced at the design and construction stage of a building and continued through the application of good housekeeping and regular maintenance. Where infestations do occur, treatment should take the form of pest eradication and remedial measures to eliminate the risk factors contributing to the infestation. The eradication of infestation in domestic premises is an established function of local authorities, although many services are now contracted out; commercial premises normally employ a contractor directly. For high-risk commercial premises, such as healthcare and food premises, a regular pest control service, offered by a contractor is necessary. Such a service provides for continual monitoring and treatment where necessary and is designed to prevent infestations of significance. Contractors may be members of either the British Pest Control Association (see http://www.bpca.org.uk) or the National Pest Technicians Association (see http://www.npta. org.uk).

PLANNED PEST CONTROL

All pest control activities can be expensive, whether they take the form of the eradication of pests, the provision and maintenance of measures to prevent an infestation or the implementation of legal controls.

Where pests are to be eradicated through the application of chemical, physical or biological measures, a planned, co-ordinated approach is essential to maintain health and safety, reduce poor results and prevent the ill-considered choice of pesticide and its associated undesirable effects. Planned pest control can be considered in five stages using the mnemonic RAPID:

- Recognition that an infestation exists, its extent, nature and identification of the species present.
- Appreciation of all the factors that may influence the effectiveness of the treatment.
- Prescription of the exact measures that need to be taken, whether proofing, hygiene control or

the application of pesticides, together with the relevant health and safety measures.
- Implementation of the prescribed measures by trained personnel.
- Determination of the effectiveness of control, and follow-up where appropriate.

Health, safety and environmental considerations

All pesticides are inherently dangerous, not just to their target organism but also to humans, domestic animals, non-target groups and the environment. Their safe storage, handling and application are therefore necessary.

The responsibility for the general regulation of pesticides is split between the Pesticide Safety Directorate, an executive agency of DEFRA, and the Health and Safety Executive's Biocides and Pesticides Assessment Unit. Environmental protection matters are the concern of the environment agencies.

The sale, supply, storage, advertisement and use of pesticides is governed by Part III of the Food and Environment Protection Act 1985, supported by the Control of Pesticides Regulations 1986 and the Biocidal Products Regulations 2001. The Biocidal Products Regulations 2001 are concerned with pesticides in non-agricultural use and provide for the approval of the active substances in pesticides, biocidal products and their labelling and packaging. The Regulations derive from the Biocidal Products Directive (98/08/EC) and will eventually apply to all biocidal products, superseding the Control of Pesticides Regulations 1986. The competent authority responsible for biocidal products is the Biocides and Pesticides Assessment Unit. A pesticide may only be used where an approval is current. Details of approvals are published monthly in 'The Pesticides Register', available from The Stationery Office, and on the website of the Biocides and Pesticides Assessment Unit (see http://www.hse.gov.uk/hthdir/noframes/ bluebook/bluebook.htm) and that of the Pesticides Safety Directorate (see http://www.pesticides. gov.uk/raid_info/prod_inf.htm). Provision is made in the Food and Environment Protection Act 1985 for the publication of Approved Codes of Practice for the purpose of providing sound practical advice, primarily for commercial users of pesticides. It is generally considered to be good practice to provide a householder with simple information about a pesticide product used in their home to treat an infestation. More generally, however, the Pesticides Act 1998 provides for wider public access to information concerning pesticides and legislation to this effect is anticipated.

Persons handling pesticides must also have due regard to the provisions of the Health and Safety at Work etc. Act 1974, and in particular to the Control of Substances Hazardous to Health Regulations 2002. Guidance and Approved Codes of Practice have been produced to support those working with pesticides to meet the requirements of the legislation (see 'Further Reading' at the end of this Chapter). Guidance on preventing pollution during the handling, use and disposal of pesticides is provided by the environment agencies in PPG9: Prevention of Pollution by Pesticides (2000) and through the Environment Agency's website (see http://www.environment-agency.gov.uk).[1]

PEST CONTROL THROUGH DESIGN AND CONSTRUCTION

Of primary importance in preventing or reducing the impact of an infestation is the provision of advice with regard to the design and construction of buildings. This can be an effective proactive tool. Advice may be given by the local authority or the pest control contractor. Consideration must be given to the risk of infestation according to the use to which a building may be put, the effect of building location and the acceptability of infestation. Any form of infestation in health care premises would pose an unacceptable risk; food premises would be considered high risk; and domestic, detached or semi-detached buildings, low risk.

In general, buildings should be designed to prevent access and harbourage, and building materials should be such that they are unsuitable for nesting purposes. Intervention at the planning approval stage, or during the early design stage of renovation work, will enable good practice in design and construction to be adopted.

External walls should be designed and constructed so as to ensure that there are no holes greater than 5 mm and that access to any wall cavity that may offer harbourage is prevented. Airbricks should be protected by wire mesh or possess openings smaller than 5 mm. Smooth-faced finishes deter climbing. Internal walls, partitions and ceiling cavities should prevent access to other parts of the building and be designed to prevent harbourage. Hollow spaces behind skirting boards, architraves, decorative moulding and panels are to be avoided. The use of silica aerogel within stud partitions will eliminate insect pests through desiccation. Insulation materials may be used for nesting by rodents; rigid foams have shown less susceptibility to damage than semi-flexible foams.

When closed, doors should not permit access or a gnawing edge. Doors closing on to a level threshold will ensure this. Self-closing doors are recommended so as to prevent doors being accidentally left open and thus to reduce the opportunity for rodent access.

Birds, particularly feral pigeons, can be discouraged from alighting and roosting by ensuring that the number of ledges is reduced to a minimum, and any remaining surfaces are inclined at least 45° from the horizontal. Access to roof spaces can be denied through careful attention to design and construction detail.

Ductwork, trunking and service pipes can offer easy access to all parts of a building unless they are closely built in or the openings sealed. Widespread infestation by cockroaches or pharaoh's ants of many system-built buildings may arise through the migration of insects via communal building components and fittings. Of particular concern are district heating systems, where an ideal warm environment that is conducive to the survival of insects is presented. Cables, pipes, sanitaryware and ducting passing into or out of a building should also be tightly fitted to prevent access or egress of insects. A diverse range of pipe and wire rat guards is available to prevent ingress via soil and rainwater pipework.

---

[1]See also HSE Information Sheet MISC 515 – *Urban Rodant Control and the Safe Use of Rodenticidis by Professional Uses.* (November 2003) HSE Books, Sudbury, Suffolk.

In multi-occupied buildings, communal facilities such as refuse chutes and lifts can provide access to all levels of a building. Rodents may damage the service components through gnawing. In refuse collection bin rooms, food scraps may encourage infestation. A self-closing, tight-fitting metal door, or a timber door with metal kick plates will act as a deterrent.

Food premises

Design and construction considerations beyond those described may be applicable in food premises and other high-risk buildings. The elimination of voids, the creation of space around fittings, and the coving or splaying of junctions at walls, floors and ceilings will reduce the potential for harbourage and facilitate effective cleaning. Fly screens at windows are effective in preventing the entry of flying insects, but their design must be conducive to effective cleaning and in any case should not permit the entry of other material into a food room.

RODENTS

Rodents are mammals with a characteristic gap between their front and back teeth known as the diastema. The rodents most commonly encountered by environmental health departments are *Rattus norvegicus* (the brown, common or sewer rat) and *Mus domesticus* (the house mouse). *Rattus rattus* (the black or ship rat) is less common in the United Kingdom, and its presence may be confined to port areas. Grey squirrels (*Sciurus carolinensis*) may be of significance in certain areas of the United Kingdom.

Public health significance

Rats and mice will readily infest both domestic and commercial premises, particularly where there is stored food. Entry may be gained through poor design, construction and maintenance of the building fabric, or via containers of food when few precautions are taken during the transportation and movement of goods. Stored food may be eaten or contaminated; packaging, the building fabric, fixtures and fittings may be damaged or soiled through gnawing and defecation, causing economic loss and nuisance; and localized subsidence may occur as a consequence of burrowing activity. An infestation may be indicative of poor standards of hygiene and housekeeping, particularly in food premises, coupled with a lack of awareness of suitable preventive measures.

Rodents may also be involved in the transmission of disease. Work at the University of Oxford has identified a variety of zoonoses of wild brown rats (*Rattus norvegicus*) that can give rise to such diseases as Q fever, leptospirosis, listeriosis, typhus, Lyme disease, cryptosporidiosis and toxoplasmosis [1]. Once established with shelter and a supply of food, both rats and mice will readily breed, giving rise to significant populations within a relatively short space of time. The National Rodent Survey 2003 by the National Pest Technician's Association [2] shows a steady increase in the number of brown rat (*Rattus norvegicus*) infestations, particularly during the summer months. A number of reasons are proposed for this [2]: mild winters, permitting greater survival and causing breeding cycles to continue; reduction in funding for sewer baiting and sewer repairs; more litter, fly-tipping and tolerance of poor rubbish standards; greater availability of food in urban areas; drains being left open by developers and builders; a reduction in funding for the pest control function within local authorities and the charging policies of local authorities.

Characteristics

The **brown rat** (*Rattus norvegicus*) (Fig. 41.1(a)) is generally brownish-grey in colour, with a paler greyish belly. The tail is thick and shorter than the head and body, and is nearly always pale below and dark above. An adult may weigh on average 340 g. The snout is blunt and the ears small and furry. Droppings may be grouped or scattered and are ellipsoidal or spindle-shaped.

The brown rat is a burrowing animal, and will live indoors, outdoors or in sewers. It enjoys both rural and urban environments. It may frequently

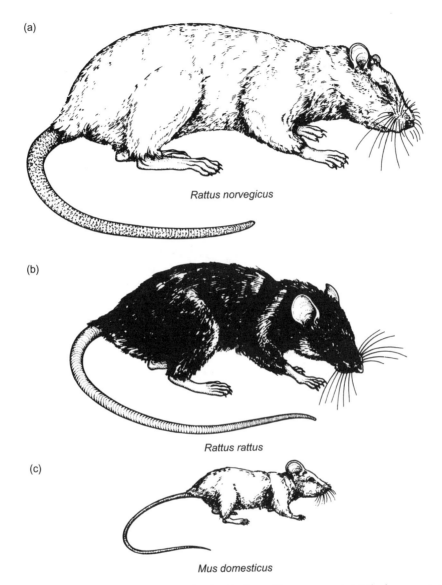

(a)

Rattus norvegicus

(b)

Rattus rattus

(c)

Mus domesticus

Fig. 41.1 (a) The brown rat (*Rattus norvegicus*); (b) the black rat (*Rattus rattus*); (c) the house mouse (*Mus domesticus*). (Source: Burgess, N.R.H. (1990) *Public Health Pests*, Chapman & Hall, London, p. 139, Fig. 16.1.)

be found at landfill sites and railway embankments, and possesses climbing and swimming skills. The diet of a brown rat is that of an omnivore but with a preference for cereals and a need for water. It will rarely venture far from a nest site in search of food – up to 660 m in the case of adult males. Foraging for food takes place mainly at night. All brown rats exhibit a cautious reaction to new objects, and this is an important consideration

in developing an approach to the control of an infestation.

By contrast, the **black rat** (*Rattus rattus*) (Fig. 41.1(b)) has a black or dark brown body with a pale, sometimes white, belly. The tail is thin and longer than the head and body. An adult weighs up to 300 g. The snout is pointed, and the ears large, translucent and furless. Droppings are scattered and banana- or sausage-shaped.

The black rat is non-burrowing and is rarely found in sewers. It has superior climbing skills. The diet is omnivorous, with a preference for fruit and vegetables. In common with brown rats, a cautious reaction is extended to new objects.

The **house mouse** (*Mus domesticus*) (Fig. 41.1(c)) is brownish-grey with a thin tail that is much longer than the head and body. Adults weigh up to 25 g. The snout is pointed and the ears small, with fine hairs. Droppings are scattered, thin and spindle-shaped.

The house mouse rarely burrows but has good climbing skills. It may be found both indoors and outdoors and is ubiquitous. The diet is omnivorous, with a preference for cereals. Behaviour is erratic, and there is a transient reaction to new objects.

Identifying an infestation

Evidence of the presence of rats or mice can be established without necessarily sighting an animal. Typical signs, both inside and outside premises, include damage to building materials, packaging and food from gnawing. Tooth marks may be evident, and will help to indicate if the rodent is a rat or mouse: mice tend to nibble from the centre of a grain, rats often leave half grains or small pieces of debris. Holes, which may be the entrance to a nest, will typically be about 80 mm in diameter in the case of rats, and 20 mm in the case of mice. These holes may appear in the ground or in floors, walls and the base of doors. Footprints may be evident in dusty environments. Rats are creatures of habit and will regularly use the same run from one place to another. The run will exhibit characteristic 'smear' marks, as the grease and dirt from the rat's fur makes contact with surfaces. Outside, soil and vegetation will become flattened. Droppings will aid identification of the types of infestation, and whether or not the infestation is current. A soft, wet appearance is indicative of fresh droppings, becoming dry and hard after a few days. Old droppings have a dull appearance.

Control principles

Rodents require food, water and shelter to survive. Preventive measures to repel an infestation in the first instance can be achieved through attention to design, construction and maintenance of buildings, in addition to good housekeeping. The latter is especially important in food and other high-risk premises. Effective cleaning of all parts of premises and equipment is essential, together with storage of food in rodent-proof containers, maintenance of refuse storage and collection points in a clean condition, and regular inspection of premises for anything that may encourage rodents or offer harbourage.

Where an infestation has been established, the use of a rodenticide will be required until control of the infestation has been achieved. Remedial work in the form of repair and proofing, coupled with a revision of hygienic practices will help to prevent reinfestation. Rodenticides are poisons used to kill rodents following a single or multi-dose of poisoned bait. The chemicals most commonly used are anticoagulants, which interfere with the production of prothrombin, which clots blood quickly when blood vessels are damaged. The animals therefore die from internal or external haemorrhaging. Over time, sublethal ingestion of anticoagulants has given rise to resistance to some of the more common formulations, particularly, warfarin. Multi-dose anticoagulants include difenacoum, coumatetralyl, bromadiolone and warfarin; single-dose anticoagulants include brodifacoum and flocoumafen. Other rodenticides interfere with the rodent's metabolism: alpha-chloralose, effective against mice where temperatures do not exceed 16°C, reduces the body temperature, resulting in death from hypothermia; calciferol causes a fatal disruption of calcium metabolism.

Rodenticides will only be effective if ingested and therefore tend to be combined with a food that is appealing to the rodent. Foods commonly used are cereals and grains. Pellets, pastes and sachets containing poisoned bait may also be employed. Rodenticidal dusts spread along runs will be picked up on the feet and fur of passing animals and ingested during preening.

The behavioural characteristics of rodents must be taken into account when laying poisoned bait. Locations should be selected carefully, having regard to the evidence of infestation. Laying unpoisoned bait until the rodents are feeding readily will help to overcome a cautious reaction and 'bait shyness'. Once feeding is established, poisoned bait can be laid. Care should be taken to ensure that humans and other animals cannot gain access to the poisoned bait, and that foodstuffs will not be contaminated. Use may be made of bait trays or boxes.

The bait should be checked every two to three days, and topped up according to the manufacturer's instructions. Untouched bait should be removed. The infestation may take from a few days to three or four weeks to be eradicated, particularly where multi-dose rodenticides are in use.

Where no further 'takes' have been recorded for a week, it is likely that the infestation has been eradicated, and all bait should be removed. It may be expedient to maintain permanent baiting sites on farmland, or where effective proofing measures are impracticable.

Other treatments that may be applied in particular circumstances include trapping and gassing.

Trapping

Trapping can be used to eliminate small infestations or as a temporary means of preventing reinfestation. Approved 'break-back' traps are recommended. They should be placed on runs or at the entrance to harbourage. Traps have limited application.

Gassing

Gassing is used outdoors to kill rats in burrows. Extreme care must be exercised when using this technique. Tablets are placed in the burrows and the exit holes are sealed. On contact with moisture, a gas is liberated, which kills the rats in the sealed burrow.

Control of rats in sewers

The practice of sewer baiting to control rat populations in the sewerage system has decreased in recent years. The responsibility for sewer-baiting rests with the sewerage undertakers (water companies) created by the Water Act 1989. Sewerage undertakers make their own arrangements for sewer-baiting; work may be contracted out to local authorities or undertaken by the sewerage undertaker or its subsidiary company.

Test baiting may be used to assess the extent of populations and to target treatment. Access to the sewers is gained via manholes. Test bait or poisoned bait is deposited on the benching alongside the invert. Alternatively, where no benching is present or the angle is too steep, a bait tray can be fixed to the side of the manhole. A rope leading from the bait tray to the benching enables rats to reach the bait easily. Usually, one manhole in each direction from an affected area is baited. In many systems, however, infestations are such that all manhole points are baited.

See also The Robens Centre for Public and Environmental Health (1998) *Rat Control, Underground Drainage and Public Health*, The Robens Centre, University of Guildford, Surrey.

Grey squirrels

Grey squirrels (*Sciurus carolinensis*) are common residents in urban areas and will give rise to complaints from time to time. The body is about 250 mm long with a 220 mm long bushy tail. The winter coat is grey with a white underside; the summer coat shorter and brownish-grey above. Grey squirrels have no public health significance, but may be a nuisance due to the damage caused to trees, fruits and the fabric of a building. The animals are determined and ingenious, and any proofing measures must take this into account. Destruction of squirrels may be only a temporary measure, since the area is likely to be recolonized.

Approved cage and spring traps can be used throughout the year. Poison baiting can be employed both indoors and outdoors, subject to the statutory provisions of the Control of Pesticides Regulations 1986, the Biocidal Products Regulations 2001 and the Grey Squirrels Warfarin Order 1973. The traditional method of squirrel control is shooting and drey-poking, which is usually carried out during the winter months.

COCKROACHES

There is a large number of species of cockroach throughout the world but only two are commonly found in the United Kingdom: *Blatta orientalis* (the **oriental cockroach**) is found throughout the United Kingdom, usually in warm indoor environments such as restaurants, hospitals, prisons and other institutional premises; *Blattella germanica* (the **German cockroach**) favours similar environments, particularly kitchens, bakeries, district heating systems and other warm, moist areas. It is sometimes known as the 'steamfly'. The life cycle of the cockroach is one of incomplete metamorphosis. Females produce egg cases (oothecae) that contain eggs. The eggs hatch into nymphs, which resemble small versions of the adult. The life cycle progresses through a number of nymphal stages, depending on the species, before the cockroaches become fully grown and sexually mature.

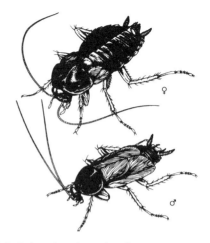

Fig. 41.2 Oriental cockroach (*Blatta orientalis*). (Source: Burgess, N.R.H. (1990) *Public Health Pests*, Chapman & Hall, London, p. 41, Fig. 6.3.)

Public health significance

Both adults and nymphal stages will feed on a variety of organic foods, including food intended for human consumption, refuse and material in drains. Regurgitation of gastric fluids on to food and indiscriminate defecation contaminate fresh foods and surfaces. Scavenging occurs over wide areas, and tends to take place at night. Cockroaches are unable to survive more than a few weeks without supply of water or high water-content foods. Both the oriental and German cockroach are known to carry pathogenic organisms such as *Salmonellae* and *Staphylococci*, although evidence to suggest transmission of disease is scant. Their presence in premises will frequently give rise to feelings of revulsion, particularly where large numbers are exposed.

Characteristics

Blatta orientalis is dark brown to black in colour and 20–25 mm in length. Males and females differ in appearance, the females possessing small, vestigial wings (Fig. 41.2). Males pass through seven nymphal stages, the females, 10. Between five and 10 dark brown oothecae are produced during

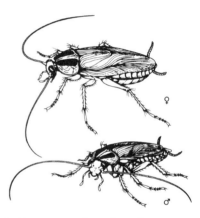

Fig. 41.3 German cockroach (*Blattella germanica*). (Source: Burgess, N.R.H. (1990) *Public Health Pests*, Chapman & Hall, London, p. 36, Fig. 6.2.)

adult life, each containing about 15 eggs. Adults can live for up to 300 days. The preferred temperature range is 20–29°C. The climbing ability of oriental cockroaches is not particularly effective, and the species is most likely to be found on horizontal surfaces or rough vertical surfaces.

By contrast, **Blattella germanica** is mid-brown in colour with two distinctive dark longitudinal bands in front of the wings (Fig. 41.3). It is up to 15 mm in length. Males pass through five nymphal stages, the females, seven. Between four and eight oothecae are produced during adult life, each containing about 35 eggs. The female carries the ootheca until just before the eggs hatch. Adult males can live for 120 days; the females' life is much shorter. German cockroaches prefer temperatures within the range 15–35°C, and a relative humidity of around 80%.

German cockroaches move quickly and are adept climbers. During the daytime they will seek harbourage on both horizontal and vertical surfaces.

Identifying an infestation

Since both species of cockroach will not readily emerge during daytime, and night inspection may be impractical, careful observation is necessary to establish the extent of any infestation. Visual evidence will include the presence of egg cases and faecal spotting. A strong sour smell is frequently noticeable. Pyrethrum sprays can be used to 'flush out' cockroaches from less accessible areas such as behind equipment, pipework and voids and crevices within the structure. Cockroach traps with a sticky surface can be left out overnight to determine the extent of infestation, and can also be used to monitor the effectiveness of treatment.

Control principles

Preventive measures should be employed to reduce the likelihood of an infestation in the first instance. Consideration should be given to the design, construction and maintenance of the building fabric, paying particular attention to the avoidance of voids and crevices. The adoption of good hygienic practices in food premises will discourage infestation. Thorough cleaning and removal of food debris should be undertaken.

Where an infestation has arisen, successful control will only be achieved through a planned programme of treatment, coupled with the adoption of preventive measures. The favoured method of treatment is through use of a residual insecticide, applied as a spray. The active ingredient may be fenitrothion, bendiocarb or pyrethrins. Boric acid and hydramethylnon may be incorporated in scatter baits or pastes for use in less accessible areas.

Residual sprayed insecticides should be applied to wall and floor surfaces and around places likely to offer harbourage, using a suitable spray nozzle configuration. Cockroaches will be readily 'flushed' into other areas of a building or adjoining premises. Treatment should therefore commence at the most distant point of the infestation, working towards the centre of it. Dust injection of inaccessible voids may be used additionally. Follow-up treatment is necessary in the case of the oriental cockroach because the oothecae are resistant to the effects of insecticides. A residual effect should be maintained for up to three months after the initial treatment, to control emerging nymphs. Treatment for German cockroaches should consider harbourage areas at height because of their superior climbing ability. Residual insecticides have the disadvantage of dead and dying insects being in evidence. These should be removed and suitably disposed of on a daily basis.

In some situations, the use of insecticides is undesirable. Insect growth inhibitors have recently been introduced with some success, particularly in block treatments in system-built premises. Juvenile hormones applied to the nymphal stages in the cockroach's life cycle inhibit metamorphosis into the sexually mature adult, thus preventing reproduction. The treatment may take a number of weeks to be effective.

FLEAS

Adult fleas are ectoparasites of warm-blooded animals. They tend to be host-specific, but will readily feed on other species in the absence of their primary host. The number of local authority treatments for fleas has increased greatly in the past 10 years. Most treatments are carried out in respect of the **cat flea** (*Ctenocephalides felis*). The **human flea** (*Pulex irritans*) is present throughout the United Kingdom, but with improved standards of personal and domestic hygiene it is becoming an increasingly rare occurrence. The **rat flea** (*Xenopsylla cheopis*) will infest the black rat (*Rattus rattus*), acting as the vector for plague. Bird fleas (*Ceratophyllus* spp.) are generally host-specific and rarely feed on human blood.

Public health significance

Fleas are known to act as vectors of human disease such as plague and typhus and may transmit the dog tapeworm (*Dipylidium caninum*). Human and cat fleas, however, are more likely to give rise to an irritant reaction, where bites are scratched and become swollen and infected. The irritation is thought to be due to a reaction to the anti-blood-clotting agent contained in the insect's saliva.

Characteristics

All fleas are laterally compressed, which allows them to move easily through their host's hair. The hind legs are long and well developed for jumping and the eye is black and prominent. A row of stiff, backward-facing spines run along the back. Both human and cat fleas are about 2 mm long and dark brown or mahogany in colour. The cat flea can be distinguished from the human flea by the presence of a prothoracic comb (spines) (Fig. 41.4). The life cycle is a complete metamorphosis.

The cat flea lays oval, white, translucent eggs that are about 1 mm long on the hair of the host. These readily fall off on to bedding, carpets and upholstery. After two to three days, the eggs hatch and white larvae, up to 5 mm long, emerge. The larvae feed off animal debris, and the excreta of adults. After three larval stages, lasting up to 4 weeks, the larvae spin a cocoon and pupate. The adult flea is ready to emerge within a week, but may not do so until the vibration of a passing blood meal is sensed. In this way, fleas can remain dormant in the cocoon for up to 12 months. This explains how heavy infestations of fleas arise in premises that have been vacant for a period of time. The fleas feed on blood from the host. Cat fleas will also feed from dogs, small rodents and humans.

In common with the cat flea, the human flea will lay pearly white, oval eggs of about 0.5 mm in length in the clothing or bedding of its host. The larvae, which emerge two or three days later, are white and bristly, and feed on dust, fluff, shed skin, hair, dandruff and faecal pellets of adults. Pupation follows in the same manner as that of the cat flea. The adult flea will remain dormant in the cocoon until a suitable host passes nearby. Human fleas will live on blood from the host, but spend more time on the host's bedding or clothing, than on the host itself.

Identifying an infestation

The fleas may be seen on the host animal or on bedding or clothing. More commonly, humans will be alerted to the presence of fleas as a result of being bitten. The bites of cat fleas tend to be confined to the lower legs and ankles, whereas the bites of human fleas tend to be concentrated around the waist and abdomen.

Control principles

Both adults and larvae can be readily controlled provided that both the host and the environment are treated. This generally involves the concurrent use of a contact insecticide (e.g. pyrethrin) and an insect growth regulator (e.g. methoprene),

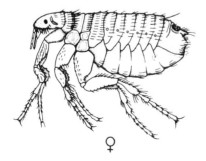
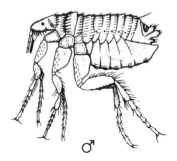

Fig. 41.4 Cat flea (*Ctenocephalides felis*). (Source: Burgess, N.R.H. (1990) *Public Health Pests*, Chapman & Hall, London, p. 80, Fig. 9.3.)

together with efficient vacuuming of carpets and upholstery, and replacement or thorough washing of the host's bedding or clothing. Aerosols containing both a contact insecticide and an insect growth inhibitor are readily available for public use, although many people prefer the infestation to be dealt with by a trained pest control operator.

BEDBUGS

Bedbugs (*Cimex lectularius*) are ectoparasites of humans, feeding largely during night-time on human blood. During the hours of daylight, they inhabit cracks and crevices in furniture, pictures and wallpaper seams. Recent research [3] suggests an increase in the number of infestations. The reasons for this increase have not been established, although increased travel and tourism has been cited as a possible cause [3].

Characteristics

Adults are about 6 mm in length, with flattened bodies and a red-brown coloration when unfed (Fig. 41.5). Their bodies show a rich mahogany colour as they become engorged with blood, growing up to six times their original size following a blood meal. Adults can survive for 6 months between feeds. A characteristic almond smell and the presence of faecal spotting are evident during an infestation. Eggs are yellowish-white and laid in crevices. A minimum room temperature of 14°C is required before the eggs will hatch; the optimum being about 25°C. The nymphs pass through five developmental stages, each requiring one or more blood meals. The life cycle is 6–7 weeks, but may be considerably longer under adverse conditions.

Identifying an infestation

Infestations are generally associated with areas of social deprivation and poor standards of personal hygiene although hotels are increasingly the site of infestations [3]. However, bedbugs can easily be introduced into any home following the introduction of infested second-hand furniture and other

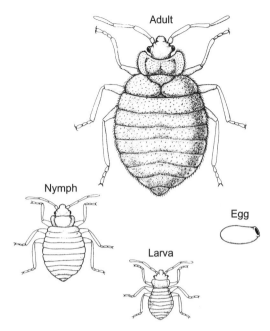

Fig. 41.5 Bedbug (*Cimex lectularius*) – life cycle. (Source: Burgess, N.R.H. (1990) *Public Health Pests*, Chapman & Hall, London.)

effects. Much stigma is still attached to the presence of bedbugs in a home, although they have not been implicated in the transmission of disease. Bites will tend to be evident on parts of the body exposed at night, and produce swelling, unpleasant irritation and possible secondary infection.

Control principles

Treatment with a residual insecticide containing pyrethrins will control an infestation. The insecticide should be applied to all surfaces and furniture over which the bedbugs will crawl. Bedding and clothing can be washed and tumble-dried, which will kill bedbugs and eggs; mattresses may be steam-treated at a cleansing station or disposed of.

ANTS

Pharaoh's ants (*Monomorium pharaonis*) are a yellow-brown colour and approximately 2 mm in length (Fig. 41.6). They are associated with

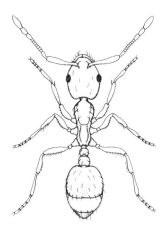

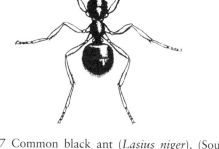

Fig. 41.6 Pharoah's ant (*Monomorium pharaonis*). (Source: Burgess, N.R.H. (1990) *Public Health Pests*, Chapman & Hall, London, p. 99, Fig. 12.3.)

Fig. 41.7 Common black ant (*Lasius niger*). (Source: Burgess, N.R.H. (1990) *Public Health Pests*, Chapman & Hall, London, p. 98, Fig. 12.2.)

indoor environments, particularly hospitals, prisons, housing estates with district heating systems and other warm institutional premises. Under optimum temperatures (30°C), the ants will breed prolifically and readily form new colonies when the original one is under threat. Workers will scavenge widely in search of food, and may be found in refuse rooms, the vicinity of drains, food rooms, sterile rooms and in hospital wards. Their mandibles are strong enough to chew through packaging and plastics.

Control principles

Control is difficult, and colonies may persist for years in some premises. Conventional treatment involves the use of residual insecticidal sprays containing bendiocarb and poisoned baits containing hydramethylnon. Use of such treatments may not affect the queen, who will continue to produce eggs. Juvenile hormone baits containing methoprene are very effective since the growth regulating hormone will be introduced to the nest, sterilizing the queen and suppressing the development of larvae into adults. Treatment must be thorough, ensuring that all colonies are subject to control. Effective control may not be achieved for up to 20 weeks.

The **common black ant** (*Lasius niger*) may be dark brown or black in colour and 3–5 mm in length (Fig. 41.7). It generally lives outside, but may invade buildings in search of food. Although omnivorous, ants have a predilection for sweet foods. Nests may be found by observing the trail of ants. Common nesting sites are around foundations, under paving slabs, the edges of flowerbeds and lawns, often where there is sandy soil. To control ants, an insecticidal spray or dust containing bendiocarb is effective. Proprietary insecticidal sprays available to the public will only provide temporary relief in heavier infestations.

FLIES

The habits and lifestyle of many flies lead to contamination of food and implicate them as carriers of enteric disease. Flies will feed indiscriminately on faeces, rotting food, refuse and fresh foods. Material from these sources will adhere to leg and body surfaces, and may be deposited during a subsequent feeding stop. During feeding, flies regurgitate the contents of their gut over the food, releasing organisms from an earlier meal. Defecation occurs randomly both during a meal and at rest. A number of flies are of public health significance in the United Kingdom.

Houseflies

The **common housefly** (*Musca domestica*) has a well-differentiated head, thorax and abdomen and two broad wings. Adult flies are 6–9 mm in length with a 13–15 mm wingspan (Fig. 41.8). The abdomen is grey and black in the male, but more yellowish and black in the female. The female lays 120–150 eggs at a time on organic matter. These are white and about 1 mm long. Within eight to 48 hours the eggs hatch into tiny larvae. These maggots feed voraciously and pass through the three larval stages in a minimum of four to 5 days. The larvae then pupate, the pupa hardening and changing in colour from yellow through red to brown and finally to black. This stage takes three to 5 days under optimum conditions, but may take several weeks in cold or adverse conditions.

The adult fly is attracted to breeding sites that will provide food and warmth for larvae. Decaying animal and vegetable matter, human and animal faeces and even fresh foods are favoured sites.

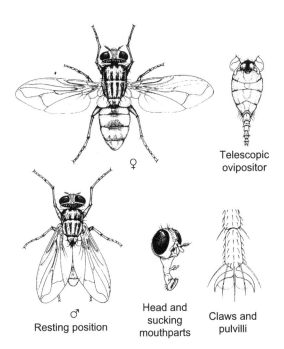

♀

Telescopic ovipositor

♂
Resting position

Head and sucking mouthparts

Claws and pulvilli

Fig. 41.8 Common house fly (*Musca domestica*). (Source: Burgess, N.R.H. (1990) *Public Health Pests*, Chapman & Hall, London.)

The **lesser housefly** (*Fannia canicularis*) is similar in appearance and lifestyle to the common housefly, breeding mainly in refuse and the soil of chicken runs, but not usually travelling between waste matter and human food.

Bluebottles or **blow flies** (*Calliphora* spp.) are widely distributed. Their life cycle is similar to that of the common house fly, preferring decaying animal matter, kitchen refuse containing meat or fish, or fresh meat on which to deposit or 'blow' eggs. The adult fly is 10–15 mm long, with a 25 mm wingspan, and has a distinctive dark blue shiny body, large compound eyes and a loud buzzing flight.

Greenbottles (*Lucilia sericata*) have a similar life cycle and habits to those of bluebottles, but are smaller in size, being about 10 mm long, with a coppery metallic green colouration.

Flesh flies (*Sarcophaga* spp.) are larger than bluebottles and have a grey chequered colouring. They feed on carrion and decaying animal matter, and may be found in the vicinity of dustbins during the summer months, but rarely venture indoors.

Fruit flies (*Drosophila* spp.) are small flies about 2 mm long, with a wingspan of 3–4 mm and a greyish-yellow body. They are associated with fermenting matter such as decaying fruit and vegetables, yeasts and vinegar. Large numbers in food premises are a nuisance and may give rise to contamination of food.

Cluster flies

Cluster flies can give rise to a nuisance where large numbers congregate around buildings prior to hibernation in the autumn, or on leaving their hibernation in spring. They will use roof spaces or cavity walls for shelter. Common species include the cluster fly (*Pollenia rudis*), the autumn fly (*Musca autumnalis*), the green cluster fly (*Dasyphora cyanella*), the yellow swarming fly (*Thaumatomya notata*) and the window fly (*Anisopus fenestralis*).

Principles of control

Control is most easily effected through the removal of organic matter and the maintenance of refuse areas in a clean and tidy condition. Drains and gullies should be free from organic debris. Where appropriate, fly screens can be installed at opening windows. Self-closing doors or the use of heavy-duty plastic door strips can prevent flies from entering food rooms. Electronic flying insect killers attract flying insects to an electrified grid with ultraviolet light. The dead insects are caught in a catch-tray, which must be emptied regularly. These devices should not be sited over open food or food equipment. Sticky flypapers have some application in storage and refuse areas, although they may be considered aesthetically unpleasant, and should be changed frequently.

Both 'knock-down' and residual chemical treatments containing pyrethroids can be applied, although some resistance has been noted. Fly baits based on sugar and incorporating a housefly pheromone and an active ingredient can be successful. To break the life cycle, a larvicide, diflubenzuron, has been approved for use in agricultural fly control. The application of any insecticide in food premises should be used only as a back up to physical controls. Where treatment is carried out, all food and equipment coming into contact with food should be removed or protected from the insecticide and dead insects.

INSECTS OF STORED PRODUCTS

This is a large group of insects comprising beetles, weevils and moths, which readily attack food during manufacture, processing, storage or transportation. All have a four-stage life cycle, and most damage is done by the larvae that live in the food, contaminating it with waste products and secretions. Adults further contaminate food through excrement, empty pupae and dead bodies. These pests are not vectors of disease.

Beetles

The **larder** or **bacon beetle** (*Dermestes lardarius*) is a member of the hide beetle species. It can be a serious pest in food premises. The adult is about 7–9 mm in length with a dark body and a distinctive light band across the body (Fig. 41.9). Larvae are dark brown and covered with tufts of hair. The larvae will eat any material of animal origin, including meat, bone, hide, fur and wool. Their presence in food premises is indicative of poor hygiene.

The **flour beetle** (*Tribolium confusum*) and **rust-red flour beetle** (*Tribolium castaneum*) are commonly found in flour mills and animal feed mills, but will also feed on other stored foods such as nuts, dried fruit and spices. The larvae of these species are almost identical; adults differ in the shape of their antennae. Both beetles may reach 4 mm in length and under favourable conditions can live for 18 months. Adults produce bitter secretions which taint foods.

The **saw-toothed grain beetle** (*Oryzaephilus surinamensis*) is commonly found in bulk grain stores, but will also attack rice, dried fruits and nuts. The adult is about 3 mm in length and is of a dull brown colour, with distinctive serrated ridges on the thorax (Fig. 41.10). The larvae are about 5 mm long and pale yellow. Established infestations can be widespread and difficult to control.

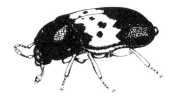

Fig. 41.9 Larder beetle (*Dermestes lardarius*). (Source: Burgess, N.R.H. (1990) *Public Health Pests*, Chapman & Hall, London, p. 107, Fig. 13.6.)

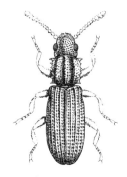

Fig. 41.10 Saw-toothed grain beetle (*Oryzaephilus surinamensis*). (Source: Burgess, N.R.H. (1990) *Public Health Pests*, Chapman & Hall, London.)

Flat grain beetles (*Cryptolestes* spp.) feed largely on cereals and cereal products. Adults are 1.5–3.5 mm long, shiny mahogany brown and have long antennae. The size and flattened shape of these beetles enables them to survive within machinery. Warm environments are preferred, with optimum temperatures of 30–35°C. Species are often found in conjunction with weevils and saw-toothed grain beetles.

The **biscuit beetle** (*Stegobium paniceum*) is a pest found in both commercial and domestic food stores. Adults are 2–3 mm in length and mid- to dark brown in colour (Fig. 41.11). Infestation can be widespread because of the ability of adults to fly. The preferred food of larvae is cereal products and dried vegetable material such as that found in packeted soup. Larvae have the ability to chew through most packaging.

A related species is the **cigarette** or **tobacco beetle** (*Lasioderma serricorne*). This beetle will survive in a similar habitat to the biscuit beetle, but is less common in the United Kingdom. In tropical regions, extensive damage may be caused to tobacco and cigars.

The **spider beetle** (*Ptinus tectus*) is common in the food industry, infesting grain, flour, spices, dog biscuits, nuts and dried fruits. In the domestic situation it is associated with old birds' nests and may cause damage to clothing and fabrics. The adult is 2–4 mm in length, mid-brown in colour, and the rounded body is covered in fine hairs (Fig. 41.12).

Yellow mealworm beetles (*Tenebrio molitor*) are usually associated with birds' nests; when found in food premises they are indicative of neglected hygiene practices. Adults may reach 15 mm, and are a shiny dark brown (Fig. 41.13). The life cycle may take a year to complete. Preference is for cereals or cereal products, but the beetles will scavenge on dead insects, birds or rodents. The **dark mealworm beetle** (*Tenebrio obscurus*) is a similar beetle, but less common in the United Kingdom. The **lesser mealworm beetle** (*Alphitobius diaperinus*) may be found on imported products such as oilseed, rice bran and cereals. It requires warmth and is more frequently found in piggery and poultry units.

Fig. 41.11 Biscuit beetle (*Stegobium paniceum*). (Source: Burgess, N.R.H. (1990) *Public Health Pests*, Chapman & Hall, London, p. 102, Fig. 13.1.)

Fig. 41.12 Spider beetle (*Ptinus tectus*). (Source: Burgess, N.R.H. (1990) *Public Health Pests*, Chapman & Hall, London.)

Fig. 41.13 Yellow mealworm beetle (*Tenebrio molitor*). (Source: Burgess, N.R.H. (1990) *Public Health Pests*, Chapman & Hall, London.)

The **grain weevil** (*Sitophilus granarius*) can be recognized by the presence of a prominent snout at the front of the head (Fig. 41.14). Adults are 3–4 mm in length and dark brown to black.

Eggs are laid inside grains of cereal and the larva remains in the grain, feeding and pupating, before the adult emerges through a small exit hole.

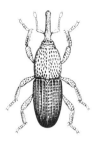

Fig. 41.14 Grain weevil (*Sitophilus granarius*). (Source: Burgess, N.R.H. (1990) *Public Health Pests*, Chapman & Hall, London.)

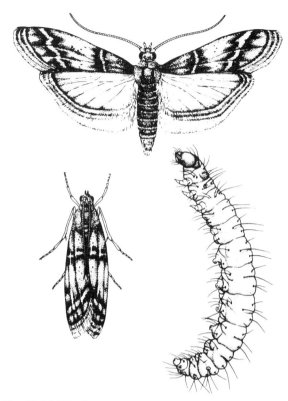

This activity produces large quantities of dust and faecal material known as 'frass' and may cause significant economic loss. Infestation can go undetected for a period of time, and can be introduced into new areas when grain is moved. The **rice weevil** (*Sitophilus oryzae*) and the **maize weevil** (*Sitophilus zeamais*) are similar to each other, but can be distinguished from the grain weevil by being less shiny and having four distinct orange patches on the wing cases. They are imported in grain and cereals.

Moths

The larvae of a number of species of moth will readily feed on cereals, dried fruits, spices, chocolates and nuts. The **warehouse** or **cocoa moth** (*Ephestia elutella*) is the foremost moth pest of stored food in the United Kingdom. A wide range of foods is attacked, including cocoa beans and chocolate products. The larvae will contaminate food through faecal pellets and from trailing strands of silk produced as they move through the food. The silken threads are difficult to remove and in heavy infestations may hang in festoons from packaging. The larvae are creamy white with dark spots on each segment, and are up to 12 mm long. The moths are 14–16 mm long with pale buff or grey wings (Fig. 41.15).

The **tropical warehouse moth** (*Ephestia cautella*), also known as the almond or dried currant moth, is frequently imported into the United Kingdom. The **mill moth** or **Mediterranean flour**

Fig. 41.15 Warehouse or cocoa moth (*Ephestia elutella*). (Source: Burgess, N.R.H. (1990) *Public Health Pests*, Chapman & Hall, London, p. 118, Fig. 14.4.)

moth (*Ephestia kuehniella*) prefers cereals and may be found in flourmills and bakeries. The larval silk may block chutes and choke sieves and milling machinery. The **Indian meal moth** (*Plodia interpunctella*) is imported in foods such as peanuts, cocoa beans and dried fruit. The larvae may be yellowish and do not have dark spots on the segments. Adult moths have reddish-brown wing tips.

Mites

Mites that infest stored products are pale fawn to brown in colour and are not usually visible as individuals without magnification. They enjoy humid conditions and feed on the moulds that form on food products. The **flour mite** (*Tyroglypus farinae*) and the **cheese mite** (*Tyrophagus casei*) are the most common species.

Heavy infestations may give rise to an allergic dermatitis in people handling infested products.

Controlling insects of stored products

To reduce the spread of an infestation where insects may be inadvertently introduced into premises, good housekeeping practices are essential:

- inspection of incoming goods and separation of new stock from old;
- effective stock rotation;
- regular inspection of goods stored for extended periods;
- frequent cleaning of storage areas and removal of any spillages;
- maintenance of the building fabric and suitable ventilation, where appropriate.

Where an infestation has arisen, insecticides are usually employed. Badly damaged or heavily infested products may have to be destroyed. A sprayed residual insecticide or an insecticidal dust containing an appropriate active ingredient such as pirimiphos-methyl or fenitrothion can deal with an infestation. However, fumigation by trained operators may be the only satisfactory treatment for some infestations.

WOOD-BORING INSECTS

The term 'woodworm' is used generally to describe any beetle whose larval stage attacks timber. In all cases, eggs are laid in cracks and crevices of dead wood, fence posts, seasoned wood, door and window frames, structural timbers and furniture. The larvae hatch and burrow into the wood, tunnelling randomly, excreting a 'bore dust' or 'frass' characteristic of the species. In the final larval stage, a pupation chamber is constructed close to the surface, where the larva pupates. Adult beetles emerge some weeks later and leave the timber through an exit hole. The size and shape of the exit holes and the presence of 'bore dust' close to the infested timber will help to determine the species of beetle. The life cycle can take between two and ten years, depending on the species.

The **common furniture beetle** (*Anobium punctatum*) is 3–5 mm in length with a dull medium-brown coloration and a 'humped' thorax (Fig. 41.16). It leaves exit holes of about 2 mm in diameter. This beetle is widespread in the United Kingdom and may cause extensive damage, seriously threatening the structural integrity of floors and roof timbers and ruining the appearance of furniture.

The **death watch beetle** (*Xestobium rufovillosum*) is 5–9 mm long and has a dark brown mottled appearance (Fig. 41.17). The life cycle may take up to 10 years to complete, and the emerging adult leaves exit holes about 4 mm diameter. This beetle is found in the southern two-thirds of the United Kingdom, and favours hardwoods such as oak and willow. Serious structural weakening can occur over a period of time.

The **powder post beetle** (*Lyctus brunneus*) (Fig. 41.18) is about 6 mm long and leaves exit holes about 1 mm in diameter. It commonly feeds on the sapwood of felled hardwoods and produces large quantities of powder from the timber. The **house longhorn beetle** (*Hylotrupes bajalus*)

Fig. 41.16 Common furniture beetle (*Anobium punctatum*). (Source: Burgess, N.R.H. (1990) *Public Health Pests*, Chapman & Hall, London.)

Fig. 41.17 Death watch beetle (*Xestobium rufovillosum*). (Source: Burgess, N.R.H. (1990) *Public Health Pests*, Chapman & Hall, London.)

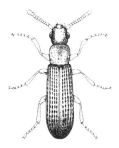

Fig. 41.18 Powder post beetle (*Lyctus brunneus*). (Source: Burgess, N.R.H. (1990) *Public Health Pests*, Chapman & Hall, London, p. 112, Fig. 13.11.)

Fig. 41.19 House longhorn beetle (*Hylotrupes bajalus*). (Source: Burgess, N.R.H. (1990) *Public Health Pests*, Chapman & Hall, London.)

(Fig. 41.19) is about 16 mm long and leaves ovalshaped exit holes of about 3 mm by 6 mm. It will infest seasoned softwoods and can cause extensive structural damage during its four to five year life cycle.

Two **wood-boring weevils** (*Pentarthrium huttoni* and *Euphyrum confine*) will also attack timber, particularly that which has been damaged through fungal action. Adult weevils are 2.5–4.5 mm long and dark brown in appearance, with a typical weevil 'snout'. Exit holes are similar in size and shape to those of *Anobium punctatum*. The life cycle can take eight months to complete. The adult weevils are commonly found in basements, cellars and other sub ground floor areas of buildings.

Control principles

Infestation can be prevented through the use of timbers that have been impregnated under pressure with a residual insecticide. Surface treatments will discourage females from laying eggs but will have no effect on larvae in the timber. Frequent examination of older, stored furniture will enable early recognition of an infestation and permit remedial and preventive action to be taken.

In all cases of infestation by wood-boring insects, treatment should be carried out by specialist personnel. Unsound structural timbers may need to be removed and destroyed, and replaced with sound, seasoned timbers treated with a residual insecticide. Treatment of timber or furniture *in situ* generally involves the application of a residual insecticide by brushing, spraying or injection. Lindane has routinely been used in the past, but there has been a move towards photostable pyrethroids such as permethrin and cypermethrin. Small infested items can be fumigated.

OTHER PESTS OF PUBLIC HEALTH SIGNIFICANCE

Feral birds

The presence of large numbers of **pigeons** (*Columba livia var*) or **starlings** (*Sturnus vulgaris*) in urban areas frequently gives rise to complaints of nuisance. Fouling of pavements and buildings where the birds roost and nest, noise and the blockage of gutters and rainwater pipes with feathers, nests and dead birds are common complaints. There is little evidence to substantiate the claim that these birds transmit disease to man, although pigeons have been shown to be infected with ornithosis and salmonellosis. Pigeons, in particular, may gain access to food premises, contaminating food and machinery.

Control principles

Pigeons are attracted to urban areas by the presence of food dropped deliberately or accidentally by the public. Limiting the availability of food through a prohibition on feeding of pigeons and strict regulation over the storage and collection of refuse containing food waste will go some way towards discouraging these birds. Under the

Wildlife and Countryside Act 1981, both pigeons and starlings may be taken or killed by authorized persons, provided that approved methods are used.

Trapping is effective for pigeons, but less so in the case of starlings. A number of suitable designs are available. All permit birds to enter the trap freely, but deny exit. The traps may be placed on the ground but in urban areas they are more commonly put on the flat roof of low-rise buildings. Bait, usually a mixture of maize and wheat or a proprietary mix, is placed inside and outside the trap. Water should also be provided inside the trap. The trap should be left open for a period of about a week to allow the birds to become accustomed to its presence. When the birds are feeding freely, the trap is closed. The trap should be checked daily and all unringed birds humanely destroyed, usually by cervical dislocation. Ringed birds should be returned to their owners.

Stupefying or narcotic baits approved for use under the Food and Environment Protection Act 1985 can be used in urban areas, but may present a public reaction if unconscious or dying birds are evident. Pre-baiting is carried out for up to 28 days, after which alpha-chloralose is added. Narcotized bait is usually laid before dawn and left for a few hours before being cleared away.

A search must then be made for all affected birds. Ringed birds and non-target species should be allowed to recover; pigeons should be destroyed humanely.

Repellents and scaring devices can be employed to prevent roosting and perching on buildings. Netting can be applied to buildings, but will only be viable if the mesh size is suited to the size of the bird. Gels can be applied to ledges and windowsills. Over time, however, their performance is impaired by birds attempting to land and flattening the gel or covering it with droppings. Spring-tensioned wire positioned around ledges is really only suitable for pigeons. Acoustic methods of scaring include sirens, ultrasound and recordings of the distress calls of the target species. Some success has been claimed using birds of prey. Shooting can be used effectively for small numbers of birds, but is likely to give rise to an adverse public reaction.

Brown-tail moth

The **brown-tail moth** (*Euproctis chrysorrhoea*) may give rise to localized concern among the public and tends to be confined to the southern part of the United Kingdom. The adult is of no public health significance; the larvae, however, are covered in fine hairs, which can cause severe irritation and skin rash. The caterpillars emerge from the silky tent in which they have hibernated in April/May, and feed on fruit trees, blackthorn, hawthorn and oak. The larvae are covered in thick tufts of brown hair with two white lines of hair tufts on their backs and two orange warts. The hairs are easily detached and can make contact with human skin. Many of the hairs are barbed and resist washing and brushing off. Hairs in the eyes can cause serious discomfort. The larval stage is usually completed by June.

Control principles

Control may be exercised through the cutting out and burning of the overwintering tents of the larvae between November and March. Alternatively, insecticidal sprays can be applied in mid-September or mid-May. Sprays based on permethrin may be approved for control of these moths.

Carpet beetle

The **carpet beetle** (*Anthrenus verbasci*) is found in domestic premises in carpets, clothing, stuffed specimens, animal furs and skins, and may be associated with birds' nests. The adult beetles are 1.5–4 mm long, with dark bodies, mottled with patches of lighter coloration, giving rise to distinctive wavy patterns (Fig. 41.20). The larvae are segmented and dark in colour, with tufts of bristles that give them the common name 'woolly bear'. The larvae cause extensive damage to non-synthetic fabrics and animal fur and skins. Adults feed on pollen and nectar.

Control principles

Control involves the removal and destruction of heavily infested materials, together with thorough

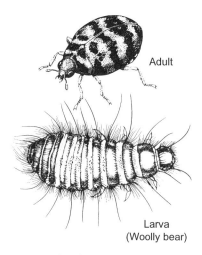

Adult

Larva
(Woolly bear)

Fig. 41.20 Carpet beetle (*Anthrenus verbasci*). (Source: Burgess, N.R.H. (1990) *Public Health Pests*, Chapman & Hall, London, p. 103, Fig. 13.2.)

Fig. 41.21 Common clothes moth (*Tineola bisselliella*). (Source: Burgess, N.R.H. (1990) *Public Health Pests*, Chapman & Hall, London.)

Fig. 41.22 Fur beetle (*Attagenus pellio*). (Source: LGMB (1992) *Pest Control: a reference manual for pest control staff; invertebrates*, LGMB, Luton, p. 68.)

vacuuming. A residual insecticide containing fenitrothion, bendiocarb or permethrin, may be applied.

Clothes moths

The term 'clothes moth' is applied generally to species of moth that commonly damage natural products of animal origin such as wool, fur and feathers. The three common species are the **common clothes moth** (*Tineola bisselliella*) (Fig. 41.21), the **brown house moth** (*Hofmannophila pseudospretella*) and the **white-shouldered house moth** (*Endrosis sarcitrella*). All adult moths are 8–10 mm long, shun light and are not strong flyers. The larvae are white, and through feeding cause holes and damage to blankets, clothing and carpets.

Control principles

Control is achieved through careful storage of articles with naphthalene crystals or mothballs. Dichlorvos strips can be used in enclosed situations. A residual insecticide can be applied to the structure of premises, or to packaging, where approved for use. Where an infestation has occurred, heavily infested articles should be removed and destroyed.

Fur beetle

The **fur beetle** (*Attagenus pellio*) lives in birds' nests and is frequently found in the domestic situation. The life cycle and habits of this beetle are similar to those of the carpet beetle. In appearance, however, there are differences. The adult beetle is about 5 mm long and black, with distinctive white spots: one on each of the wing cases, and three at the base of the prothorax (Fig. 41.22). The larvae may be up to 12 mm long with a long 'tail' of silky hairs. Control is the same as that for carpet beetles.

Lice

All lice have well-developed mandibles for biting and piercing skin, enabling them to suck the blood of their host. Three lice are of public health significance: the **head louse** (*Pediculus humanus capitis*), the **body louse** (*Pediculus humanus corporis*) and the **crab** or **pubic louse** (*Phthirus pubis*).

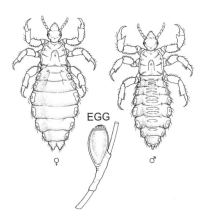

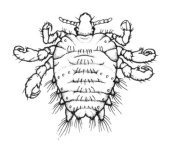

Fig. 41.24 Crab louse (*Phthirus pubis*). (Source: Burgess, N.R.H. (1990) *Public Health Pests*, Chapman & Hall, London, p. 92, Fig. 11.2.)

Fig. 41.23 Human louse (*Pediculus humanus capitis*). (Source: Burgess, N.R.H. (1990) *Public Health Pests*, Chapman & Hall, London.)

Characteristics

Head and body lice are 2–4 mm in length and pale grey to light brown in colour. The crab louse is about 2 mm in length and greyish. All lice are darker in colour following a feed. Head lice (Fig. 41.23) are widely found in the United Kingdom and primarily affect children. Despite their prevalence, there is still much embarrassment associated with an infestation. Eggs are laid at the base of the hair and hatch, leaving the pale-coloured egg casing, known as a 'nit' on the hair. The nymphs feed on blood until sexual maturity. Transmission is through physical contact.

Body lice live in the clothing of the host, moving to the host to feed. Survival is reliant on the same clothing being worn at no more than three-day intervals. Transmission is via infested clothing or physical contact. Body lice may transmit typhus, trench fever and relapsing fever.

Crab lice (Fig. 41.24) favour the coarser body hair found in the pubic areas and armpits, and spread through intimate contact. All lice will cause irritation, and scratching may give rise to secondary infection.

Control principles

Control of head and crab lice can be achieved through application of insecticidal shampoos or lotions although low-grade resistance to permethrins has been identified [4]. Mechanical removal using a fine-toothed comb can be effective, particularly if coupled with the use of an insecticidal product. In 1998, a working group convened by the Public Health Medicine Environment Group produced a good-practice document on the control of head lice. It is available on http://www.phmeg.org.uk/Documents/Headlice/phmeghl.htm. In the case of body lice, infested clothing should be destroyed or disinfested at a cleansing station.

Mosquitoes

About 30 species of mosquito are present in the United Kingdom, occupying different aquatic habitats such as coastal salt waters, brackish inland waters, stagnant ponds and water-filled hollows in trees and logs. The two main mosquito groups in the United Kingdom are the **anophelines** and the **culicines**. In general, all have slender bodies, long legs and a well-developed proboscis. The length of the body is dependent on the species, but will range from 7 to 15 mm (Figs 41.25 and 41.26).

Characteristics

Eggs are laid on water, and hatch within a few hours. The larvae will breathe oxygen by moving to the surface of the water. The larvae feed on organic matter and micro-organisms in the water or on the surface. After four larval stages, a larva pupates, forming a comma-shaped pupa which can propel itself using paddles at the bottom of the abdomen. The adult mosquito emerges

Fig. 41.25 Anopheline mosquito. (Source: Burgess, N.R.H. (1990) *Public Health Pests*, Chapman & Hall, London, p. 71, Fig. 8.4.)

Fig. 41.26 Culicine mosquito. (Source: Burgess, N.R.H. (1990) *Public Health Pests*, Chapman & Hall, London, p. 71, Fig. 8.4.)

from the pupa on to the surface of the water. Only females bite and suck blood; the males feed on the nectar of flowering plants. Females are attracted to a host by heat and exhaled carbon dioxide.

A blood meal is required before viable eggs can be laid. During feeding, a small amount of anticoagulant saliva will be injected into the host to prevent the blood from clotting. The irritation, swelling and erythema associated with a mosquito bite are antibody reactions to this anticoagulant.

Public health significance

The mosquito has significant public health importance in tropical and subtropical regions, the female being the vector of malaria, yellow fever, filariasis, dengue fever and forms of viral encephalitis. In temperate regions the effect of a mosquito bite causes discomfort and possible secondary infection as a consequence of scratching. No link has been established between the transmission of the human immunodeficiency virus (HIV) and a mosquito bite.

Control principles

Effective mosquito control relies on knowledge of the species involved. However, in general, mosquito control should be aimed at both the larval and adult stages of the life cycle. Breeding sites can be removed through emptying natural or man-made containers of water, draining puddles, ditches and small pools, and channelling water to increase the flow. The larval stages can be eliminated in a number of ways. The application of an agent such as light oil or lecithin will reduce the surface tension of the water, preventing the larvae from obtaining oxygen. These agents spread readily over a large area, and can be applied aerially. Consideration must be given to the ecological effect. Larvicides containing permethrin, pirimiphos-methyl or chlorpyrifos, which are applied to breeding sites, do not have approval for use in the United Kingdom. The use of methoprene, the insect growth regulator, and a biological control agent, the bacterium *Bacillus thuringiensis israelensis*, have been used for larval control in developed countries. Adult mosquitoes can be eliminated using 'knock-down' agents or residual insecticides. Hand-held aerosols can be used in a domestic environment. Individual protection can be achieved through the application of suitable insect repellents to exposed parts of the body or the use of mosquito coils or candles containing citronella to deter the insects. In tropical areas, sleeping accommodation can be protected by a mosquito net impregnated with a suitable insecticide. Windows and doors can be fitted with mosquito proofing.

Plaster beetles

Plaster beetles (*Lathridiidae* spp.) are small dark brown beetles 1.5–2.5 mm long that have a predilection for damp and humid conditions (Fig. 41.27). They can be significant when infestation occurs in food premises. Control over humidity or removal of damp conditions will remove any problems, although a residual insecticide can be applied.

Psocids

Psocids or **booklice** (*Psocoptera* spp.) are generally found in high humidity situations where they

Fig. 41.27 Plaster beetle (*Lathridiidae* spp.). (Source: Burgess, N.R.H. (1990) *Public Health Pests*, Chapman & Hall, London.)

Fig. 41.28 Silverfish (*Lepisma saccharina*). (Source: Burgess, N.R.H. (1990) *Public Health Pests*, Chapman & Hall, London.)

feed on moulds. Damp wallpaper, bookbindings and food packets are favoured, and damage may occur where significant numbers are present. These insects are 1–1.5 mm long, wingless and greyish in colour, although the colour may reflect their food. Infestations spread readily through the movement of infested books or foodstuffs. Their presence in domestic situations frequently gives rise to complaints.

Control principles

Control can be achieved through improving hygiene and environmental conditions. Infested food packets should be destroyed and storage areas treated with a residual insecticide.

Silverfish

Silverfish (*Lepisma saccharina*) are commonly found in damp situations such as bathrooms, kitchens and pantries. They may be up to 20 mm long, with carrot-shaped wingless bodies and three bristly tails (Fig. 41.28). The colour is a silvery-grey. They feed on protein-rich gums and binding pastes in books and packeted foods, wallpaper paste and fine textiles. The life cycle may take up to a year to complete. Silverfish are of no public health significance, although severe infestations may be a nuisance.

Firebrats (*Thermobia domestica*) are similar to silverfish, but require a warm, dry environment and have a preference for starchy foods.

Control principles

Control of silverfish and firebrats can be achieved through attention to hygiene or the application of residual insecticides based on synthetic pyrethroids.

Wasps

The **common wasp** (*Vespula vulgaris*) is frequently found nesting in roof spaces, cavity walls, trees or in the ground. Nests are constructed each year from chewed wood pulp, which is converted into a paper-like substance and formed into many cells within an outer layer. The nest will be expanded during the season to accommodate the growing colony, which may reach many thousands. The wasps seen outside the nest are workers. They are 10–20 mm in length, with a distinctive yellow and black banding and a 'wasp-waist' (Fig. 41.29). As cold weather approaches, the new queens find a suitable place for overwintering and the rest of the colony dies. The queens emerge from hibernation in spring and form new colonies. Wasps are usually a nuisance in late summer when the workers forage for sweet substances and, if provoked, will sting.

Control principles

Control is effected through the application of insecticidal dust at the entrance to the nest. Workers will then carry the insecticide into the

Fig. 41.29 Common wasp (*Vespula vulgaris*). (Source: Burgess, N.R.H. (1990) *Public Health Pests*, Chapman & Hall, London.)

nest, spreading it to other wasps in the colony. Pyrethroids tend to excite wasps, and active ingredients such as bendiocarb are generally used. Operatives should be protected against aggressive behaviour by wasps if the nest should be disturbed. Care must be taken to avoid treatment having an effect on beneficial wasp and bee species. General application of residual insecticides to loft spaces may present a risk to bats.

INSECT IDENTIFICATION

To aid the identification of less common public health insect pests, the Department of Entomology at the Natural History Museum offers an identification and advisory service for which a charge is made. For further information see http://www.nhm.ac.uk/entomology/insident/index.html.

ACKNOWLEDGEMENTS

Many thanks to Dr Burgess and Helen Hadjidimitriadon, the artist, for their kind permission to reproduce some of the drawings from Burgess, N.R.H. (1990) *Public Health Pests*, Chapman & Hall, London.

REFERENCES

1. Webster, J.P. (1996) Wild brown rats (*Rattus norvegicus*) as a zoonotic risk on farms in England and Wales. *Communicable Disease Report*, **6**(3), R46–R49.
2. NPTA (2003) National Rodent Survey 2003. http://www.npta.org.uk/rodent-report%20htm.htm
3. Boase, C. (2001) Bedbugs – Back from the Brink. *Pesticide Outlook*, August, 159–62.
4. Burgess, I. (2001) Head Lice. *Pesticide Outlook*, June, 109–13.

FURTHER READING

Bassett, W.H. (2002) *Environmental Health Procedures*, 6th edn, Spon Press, London.

British Standards Institute (1992) *BS 5502: 1992, Buildings and Structures for Agricultural Use Part 30: Code of Practice for Control of Infestation.*

Building Research Establishment (1996) *Reducing the Risk of Pest Infestations in Buildings*, BRE, Aylesbury.

Burgess, N.R.H. (1990) *Public Health Pests*, Arnold, London.

Cornwell, P.B. (1979) *Pest Control in Buildings: a Guide to the Meaning of Terms*, 2nd edn, Rentokil, London.

HSE (1990) *Recommendations for Training Users of Non-Agricultural Pesticides*, HSE Books, Sudbury, Suffolk.

HSE (1995) *The Safe Use of Pesticides for Non-Agricultural Purposes. Control of Substances Hazardous to Health Approved Code of Practice*, L9, HSE Books, Sudbury, Suffolk.

HSE (1996) *Approved Code of Practice for the Control of Substances Hazardous to Health in Fumigation Operations*, L86, HSE Books, Sudbury, Suffolk.

HSE (1999) *Safe Use of Pesticides on Farms and Holdings*, HSE Books, Sudbury, Suffolk.

Judge, L. (1996) The brown-tailed moth (*Euproctis chrysorrhoea*). *The Safety and Health Practitioner*, May, 22–23.

Mullen, G. and Durden, L. (Eds) (2002) *Medical and Veterinary Entomology*, Academic Press.

Local Government Management Training Board (1992) *Pest Control: a Reference Manual for Pest Control Staff*, LGMB, Luton.

National Britannia Library (1998) *Insect Pests of Food Premises*, National Britannia Ltd, Caerphilly.

Munro, J.W. (1966) *Pests of Stored Products*, Hutchinson, London.

Sprenger, R.A. (2002) *Hygiene for Management*, Highfield Publications, Doncaster.

42 The control of animals

Andrew Griffiths

REASONS FOR CONTROLLING ANIMALS

There are a number of reasons why legislation exists to control animals. The key driver is to prevent the transmission of disease from animals to humans, but further reasons include the need to control and prevent nuisance and to protect animals from cruelty and ill treatment. Furthermore, action to control animals such as dogs can result in considerable community benefits; local authorities are frequently subject to considerable pressures to invest resources in the control of fouling by dogs. A further aspect of the protection of the welfare of animals is the beneficial health effects of companion animals.

Specific controls are available to control dogs, pet shops, boarding establishments, riding establishments and zoos. This chapter does not address the issue of pest control (dealt with in Chapter 41) or the control of animals in slaughterhouses (Chapter 33) or zoonoses which cause food poisoning (Chapter 13).

Zoonoses

Mankind shares a world with animals; it is unsurprising therefore that we also share some diseases. Zoonoses, or zoonotic diseases (of which approximately three thousand are known) are infectious diseases transmissible between humans and other animals (including parasitic diseases);

many thousands of zoonoses have been identified. The risk to humans depends on the kind of disease and the type of exposure. Some zoonoses are notifiable including anthrax, leptospirosis or rabies (Chapter 12).

Zoonoses are diseases that are transmissible between humans and animals. While the reported instances of transmission of disease between domesticated animals and humans are not that frequent, they nevertheless represent a significant disease prevalence. In addition, it is likely that the incidence of zoonotic infections is higher than that reported as many of the less well-understood infections may not be accurately diagnosed.

Animal diseases that can infect humans (Table 42.1)

A number of zoonoses, which are discussed in more detail in other chapters, are listed below:

- Food poisoning organisms such as *Salmonella* and *Campylobacter* are technically zoonoses (see Chapter 13).
- Other zoonoses are those transmitted in a working environment such as ringworm (a fungal skin infection), which can be contracted by those working with cattle, and orf (a contagious pustular dermatitis), commonly found in people working with sheep (Chapters 23 and 33).

Table 42.1 Animal diseases that can infect humans

Disease	Common source	Clinical effects	Mode of acquisition	Prevention
Bacterial diseases				
Campylobacter	Cattle, sheep, pigs, dogs, rodents, poultry	Acute gastro-enteritis, nausea, headache, diarrhoea	Direct contact with food contaminated with animals' faeces	Avoid contact with infected animals and faeces contaminated food
Cat-scratch diseases	Cats, dogs, fomites	Fever, primary skin papule, regional lymphadenopathy	Direct contact with infected animals	Avoidance of animal scratches and puncture wounds
Salmonellosis	Cattle, cats dogs, horses poultry, turtles	Chills, fever, headache, diarrhoea, vomiting	Direct contact with animal or its faeces, food contamination from infected animals	Improved food processing and preparation
Viral diseases				
Encephalitis	Horses, rodents	Lethargy, fever, headache, disorientation	Mosquito or tick bite	Protective clothing, insect repellents
Rabies	Cats, dogs, racoons, skunks, bats, foxes	Fever, headache, agitation, confusion, seizures, excessive salivation, death	Animal bite, contact with infected tissue, body fluids or faeces	Avoid contact with suspected animals, local wound care, pre- and post-exposure immunization/ vaccination
Chlamydial diseases				
Psittacosis (Ornithosis)	Pigeons, turkeys, parakeets, parrots	Fever, headache, pneumonia	Inhaled from infected birds, carcasses, secretions and contaminated facilities	Avoid contact with infected birds, control of disease with antibiotics
Fungal disease				
Ringworm	Cats, cattle	Skin lesions	Direct contact	Avoid close contact with infected animals. Children and individuals with immune suppressed system are more susceptible
Parasitic diseases				
Tosocariasis (Visceral larval migrans)	Dogs, racoons, cats	Eye disease, brain disease	Ingestion and contact with infected ovum of parasites	Treat pets, avoid faecal contaminated soil and sandboxes

Table 42.1 Continued

Disease	Common source	Clinical effects	Mode of acquisition	Prevention
Toxoplasmosis	Cats, sheep, undercooked meat	Fever, lymph-adenopathy, abortion, still-birth, mental retardation	Ingestion of infected meats, oocysts in faecal contaminated soil	Proper disposal of cat faeces, cook meat well, avoid contaminated soil (especially pregnant women and immune compromised individuals)
Scabies	Dogs, racoons	Itching skin, lesions	Direct contact with infected animals	Treat pets, avoid contact with infected animals

- The organism *Leptospira interrogans* causes Weil's disease, a form of leptospirosis; the organism can be picked up from rat and cattle urine (Chapter 14).

Nuisance

Nuisances that are associated with animals are principally noise and smell. Such nuisances are more likely to give rise to complaints when animals which are usually associated with farms and the countryside (such as cockerels and geese) are kept in urban areas. Probably the most common nuisance from animals is barking by dogs. Domestic animals that are kept in large numbers in towns and cities can also give rise to considerable nuisance. Noise nuisance from barking dogs is a key issue for planning authorities when considering applications for the building of dog boarding establishments (dealt with in more detail later in this chapter).

Nuisances from animals can be a statutory nuisance under the Environmental Protection Act 1990 Section 79 ('Any animal kept in such a state or manner as to be prejudicial to health or a nuisance'). An abatement notice served under the Environmental Protection Act can specify 'the execution of such works, and the taking of such other steps, as may be necessary (for the abatement of the nuisance)'.

The noise from barking dogs is a frequent complaint to environmental health departments, and one that is often difficult to resolve. Dogs that are

not properly trained or controlled, or that are kept in unsuitable environments may cause noise by barking. Dog breeding premises and animal boarding establishments have been the focus of well-publicized allegations of noise nuisance, some of which have been the subject of inquiry by the Local Government Ombudsman. Prolonged and persistent barking may constitute a statutory noise nuisance, which can be dealt with under Part III of the Environmental Protection Act 1990 (see Chapter 6). A dog warden can often provide advice concerning training to reduce the incidence of barking and the use of proprietary anti-barking devices.

Accumulations of dog faeces on premises may also give rise to a statutory nuisance or risk to public health and can be dealt with similarly under Part III of the Environmental Protection Act 1990.

Noise from cockerels and other birds

One specific type of nuisance that has found its way to the courts in recent years is that of noise from cockerels, geese, waterfowl, etc. Notices under S79 of the Environmental Protection Act have been successfully served on owners of such birds requiring the abatement of a statutory (noise) nuisance. Notices have required defendants to provide dark housing (and other measures) to prevent crowing etc. between 10.00 and 06.00 hours. Other practical measures that have been successfully executed include the lowering of the height of housing for cockerels to prevent them from stretching their necks – a necessary precursor to crowing. One celebrated case was that of 'Corky the Devon cockerel'. A nuisance notice issued by the local authority was initially quashed but an injunction obtained by a neighbour was upheld and required that steps be taken to prevent the bird from causing a noise nuisance between 12.00 and 07.00 hours.

A further common cause of nuisance involving animals is that of smell. Complaints relating to smell are common when animals normally associated with rural areas are kept in towns and cities. Domestic pets that are inadequately cared for or kept in large numbers can also cause a statutory nuisance by reason of smell.

Health and safety

There are two aspects to the need for health and safety in respect of the control of animals, namely in the work environment and in the community. The former is dealt with through occupational health and safety legislation (and to a certain extent through zoo licensing legislation) for the protection of animal handlers and keepers and the latter through the Dangerous Wild Animals (DWA) Act 1976 (dealt with later in the chapter).

A number of incidents have occurred when keepers employed in zoos have been fatally injured by animals in their care. Several such incidents have been caused by elephants and considerable pressure has been applied by several organizations in recent years (including the Royal Society for the Prevention of Cruelty to Animals (RSPCA)) to ban the keeping of elephants in zoos. Lions, tigers and other big cats have also been responsible for deaths and serious injuries. The basic principles of health safety in respect of safe systems of work and the responsibilities of employer and employee are relevant to the prevention of death and injury caused by animals.

Cruelty

Legislation to prevent cruelty is only of indirect relevance to this chapter. Local authorities will however have cause from time to time to work with, or with reference to, animal welfare organizations most of which do not possess enforcement powers, the key

exception being the RSPCA. Legislation requiring the licensing of animal establishments of various kinds (dealt with later in this chapter) is designed to protect the welfare of the animals concerned as well as the protection of the health safety and welfare of people coming into contact with them.

The basis of animal protection legislation is the Protection of Animals Act 1911 (1912 in Scotland). Cruelty is defined widely and in general terms but can be summarized as 'the infliction of unnecessary pain'; it can be caused by both action and omission. It has been held by the courts that some activities involving animals while legal, such as intensive rearing, can nevertheless be considered as cruel. In addition to the issue of protecting animals, research in the United States has shown that abuse of animals by individuals has frequently led to abuse of humans.

Community and personal benefits

Action to control and protect the welfare of animals has been shown to have a beneficial effect on communities as a whole.

Much research is available that identifies the health benefits to mankind from a relationship with animals. The need is to relate this to the requirement for socially acceptable and responsible actions on behalf of the animal's keepers.

Owning animals extends the life span of keepers, reduces their blood pressure, and provides for a reduction in stress. Animals are taken round factories in Japan to help keep workers relaxed and horse riding helps the rehabilitation into society of the injured, disabled and those with learning difficulties.

The role of Guide Dogs is well known; less well known are 'Hearing Dogs' and 'Pets as Therapy', where dogs visit hospitals to help children, particularly, to recover. Companion animals help people to relate to others via dog walking and social discourse. Exercise is gained from some pets, and most provide companionship.

LICENSING[1]

Some form of licensing, certification or registration covers most commercial and professional activities involving animals. Notable exceptions are circuses,

---

[1]Also see Bassett, W.H. (2002) *Environmental Health Procedures*, 6th edn, Spon Press, London.

Table 42.2 Activities for which licences are required

Activity	Legislation
1. Pet shops	Pet Animals Act 1951
2. Animal boarding establishments	Animal Boarding Establishments Act 1963
3. Riding establishments	Riding Establishments Act 1964
4. Dog breeding establishments	Breeding of Dogs Act 1973
5. Zoos	Zoo Licensing Act 1981
6. Exhibiting, or training performing animals	Performing Animals (Regulation) Act 1925
7. Keeping a dangerous wild animal	Dangerous Wild Animals Act 1976
8. Scientific procedures involving protected animals	Animals (Scientific Procedures) Act 1986
9. Slaughterhouses	Fresh Meat (Hygiene and Inspection) Regulations 1995
10. Killing, taking or selling badgers	Protection of Badgers Act 1992
11. Killing, taking or selling types of wild birds	Wildlife and Countryside Act 1981
12. Slaughtering animals	The Welfare of Animals (Slaughter or Killing Regulations 1995)
13. Establishments for the conduct of scientific procedures	Animals (Scientific Procedures) Act 1986
14. Transport of cattle, sheep, pigs and horses for more than 8 hours	Welfare of Animals (Transport) Order 1997

markets, farms and animal sanctuaries (which take in stray or unwanted animals).

Activities for which licences are required are summarized in Table 42.2 (items 1 to 7 are discussed in more detail in this chapter).

Pet shops

The Pet Animals Act 1951 (amended in 1983) requires that a person keeping a pet shop holds a licence from the local authority. Pet shops are defined as premises (including private dwellings) at which the business of selling animals as pets is conducted. Two key exemptions are the keeping or selling of pedigree animals (or the offspring of such animals) and the sale of excess or unwanted breeding stock.

Pedigree animal means an animal of any description which is by its breeding eligible for registration with a recognized club or society keeping a register of animals of that description.

The selling or keeping of animals as pets (in respect of cats and dogs) relates to selling or keeping wholly or mainly for domestic purposes and (in respect of other animals) for ornamental purposes.

Pet animals include any type of vertebrate. The 1983 Act removed the definition of premises but prohibits the sale in any part of a street or public place or at a stall or barrow in a market.

Licence conditions

Local authorities can attach any appropriate condition to a licence in respect of the following:

(a) Size, temperature, lighting, ventilation and cleanliness
(b) Food and drink plus visits at suitable intervals (as appropriate)
(c) Mammals not to be sold at too young an age
(d) Prevention of the spread of infection (between animals)
(e) Appropriate precautions in case of fire and other emergencies.

The Local Government Association has produced, in conjunction with the Chartered Institute of Environmental Health and others, model licence conditions for pet shops [1].

Local authority officers, vets or other practitioners authorized in writing have powers of entry to licensed pet shops, but no such powers exist to enable local authority officers to gain access to premises suspected of operating an unlicensed pet shop.

Animal boarding establishments

The Animal Boarding Establishments Act 1963 requires that a person keeping an establishment for cats or dogs holds a licence issued by the local authority. An animal boarding establishment is defined as premises (including private dwellings) at which a business of providing accommodation for other people's animals is conducted. The Act does not apply if the provision of accommodation for other people's animals is not the main activity of a business. Similarly, the Act does not apply if a requirement under the Animal Health Act 1981 has been imposed at the premises concerned.

Licence conditions

Local authorities can attach any appropriate condition to a licence in respect of the following:

(a) Suitability of accommodation as respects construction, size, numbers of occupants, exercise facilities, temperature, lighting, ventilation and cleanliness.
(b) Adequate and suitable food, drink and bedding and adequate exercise and visits for animals.
(c) Prevention of spread of infection, including the provision of isolation facilities.
(d) Appropriate precautions to prevent animals in the case of fire or other emergencies.
(e) The maintenance of a register, available for inspection, to record details of animals and their owners together with arrival and departure dates.

The Chartered Institute of Environmental Health has produced model licence conditions for both dog and cat boarding establishments [2,3].

Local authority officers, vets or other practitioners authorized in writing have powers of entry to licensed animal boarding establishments, but no such powers exist to enable local authority officers to gain access to premises suspected of operating an unlicensed boarding establishment.

Riding establishments

The Riding Establishment Acts 1964 and 1970 require that a person keeping a riding establishment be licensed by the local authority. A riding establishment is defined as the carrying on of a business of keeping horses for hire for riding or instruction in riding. The Act does not apply to Ministry of Defence establishments, the police, the Zoological Society of London and universities providing veterinary courses.

When deciding whether or not to grant a licence, a local authority must take a number of matters into consideration summarized below:

1. whether the licence applicant is suitable and qualified; and
2. the need for securing the following:

   (a) that paramount consideration will be given to the condition of horses so that they are maintained in good health and that horses for hire or for riding instruction are suitable for the purpose;
   (b) that the feet of all animals are properly trimmed and that, if shod, their shoes are properly fitted and in good condition;
   (c) that suitable accommodation for horses will be available – both new buildings and those converted for use as stables;
   (d) for horses maintained at grass, that adequate pasture, shelter and water will always be available (with supplementary feeds provided when necessary);
   (e) that horses will be adequately supplied with suitable food, drink and bedding material and will be adequately cared for;
   (f) that all reasonable precautions will be taken to prevent the spread of infection (including the maintenance of veterinary first-aid equipment and medicines);
   (g) that appropriate steps will be taken for protecting and removing horses in case of fire and that details of the licence holder (and instructions for removing horses in the event of fire) are prominently displayed outside the premises;
   (h) that adequate accommodation will be provided for forage, bedding, stable equipment and saddlery.

Licence conditions

Local authorities have to specify appropriate conditions to achieve the objectives set out above. The following matters are also required under the Act:

(a) a horse found to be in need of veterinary attention until the licence holder has obtained a veterinary certificate that the horse is fit for work;
(b) horses must not be hired out for riding without supervision, unless the licence holder is satisfied the hirer is suitably competent;
(c) the business shall not be left in the charge of anyone under 16;
(d) the licence holder shall hold an insurance policy insuring against liability for injuries sustained by hirers of horses as a result of the hire;
(e) the licence holder shall keep a register at the premises, available for inspection, of all horses aged three years and under.

Powers of entry

Local authority officers and veterinary surgeons authorized by the authority have powers of entry to inspect licensed riding establishments, premises subject to a licence application and, unlike powers of entry in respect of pet shops and animal boarding establishments, premises suspected of being used as a riding establishment.

Dog breeding

The Breeding of Dogs Acts 1973 and 1991 require that a person keeping a breeding establishment for dogs holds a licence from the local authority. A breeding establishment for dogs is defined as the carrying on at any premises a business of breeding dogs for sale. The Acts do not apply to the keeping of dogs at premises in pursuance of a requirement under the Animal Health Act 1981.

When deciding whether or not to grant a licence, a local authority must take a number of matters into consideration summarized below:

(a) that the dogs will be kept in suitable accommodation as respects construction, numbers (of animals), size, temperature, lighting, ventilation, cleanliness and exercise facilities;
(b) that the dogs will be supplied with adequate food, drink and bedding materials and be adequately exercised and visited;
(c) that all reasonable precautions will be taken to prevent and control the spread of infection;
(d) that appropriate steps will be taken for the protection of dogs in the event of fire or other emergencies;
(e) that appropriate steps will be taken to ensure dogs are provided with adequate food, water and bedding whilst in transit.

Conditions

When granting a licence a local authority must specify appropriate conditions to achieve the objectives set out above.

Zoos

The Zoo Licensing Act 1981 provides powers to regulate the conduct of zoos through licensing. It is unlawful to operate a zoo without a licence or in contravention of licence conditions. Licences can last from four to six years. Licence applications to local authorities can be refused on the following grounds:

• the health and safety of the public;
• inadequate standards of accommodation, staffing or management;
• the applicant, director or keeper gas has been convicted under a range of animal welfare legislation;
• absence of planning permission.

The Standards of Modern Zoo Practice [4] consists of guidance on standards of practice that zoos should meet. The standards include:

• the need to ensure that captive animals are kept in accordance with five principles:
  – provision of food and water
  – provision of suitable environment
  – provision of animal health care
  – provision for opportunity to express most normal behaviour
  – provision of protection from fear and distress

- the requirement that zoos engage in conservation, education and research;
- the establishment of an ethical review process to assess a zoo's animal husbandry practices in a critical and objective way;
- the introduction of a pre-inspection process giving the inspector a better understanding of the zoo before visiting the site;
- the categorization of dangerous animals that lists species according to whether they must be kept separate from visitors and those which may come into contact with visitors under supervision.

Under regulations made under the 1981 Act, which came into force in 2003, local authorities can close, or partially close, zoos which fail to participate in conservation activities. In the event of closure, local authorities have powers to ensure that animals are disposed of in a manner that takes account of their conservation and welfare needs.

In 1985, the Health and Safety Commission published an Approved Code of Practice and Guidance Note [5] which set out practical guidance with respect to Sections 2, 3, 4 and 7 of the Health and Safety at Work etc. Act 1974.

Dangerous wild animals

Before granting a licence under the DWA Act, local authorities have to be satisfied that

- the public will be properly protected and kept free from nuisance;
- the applicant is suitable;
- the animals will be kept secure in suitable accommodation (construction, temperature, lighting, ventilation) and will be properly fed and watered;
- the animals will
  - be properly protected in case of fire or other emergency;
  - be subject to precautions to control infectious diseases;
  - have adequate exercise facilities.

The Act requires local authorities to specify conditions in the licence including where the animal is kept, that the keeper should be the licence holder, that the licence holder is adequately insured against liability for damage caused by the animal and other matters necessary to secure the objectives of the conditions. Local authorities can authorize competent persons to enter licensed premises to inspect them and the relevant animal(s). Local authorities can seize, retain or destroy animals that are not properly licensed.

Circuses

A circus is defined as any place where animals are kept or introduced mainly or wholly for the purpose of performing tricks or manoeuvres (DWA Act 1976 and Zoo Licensing Act 1981).

The All Parliamentary Group for Animal welfare set up a Circus Working Group in 1996 to examine current legislation, practice and theory relating to the welfare and management of all circus animals including domestic animals. The Group reported in 1998 [6] and its conclusion about legislation was as follows:

> It is apparent that a wide range of legislation has application in the circus industry. However, the Working Group has found that no current legislation adequately addresses the welfare needs of animals in live entertainment. It is recommended that the consideration be given to the introduction of annual pre-season registration, and a 'log-book' to record relevant safety documentation, together with details of inspection findings requiring repair or improvement, be carried with each circus.

The government recently indicated that it had no plans to introduce additional statutory measures for travelling circuses, but would assist the Association of Circus Proprietors (ACP) in preparing a voluntary Code of Practice [7].

The document was circulated by the Department for Environment, Food and Rural Affairs (DEFRA) but it is not a DEFRA Code of Practice and is not intended to provide government approved guidelines for the care of circus animals.

DEFRA ministers believe that some degree of self-regulation and guidance, if supported by local authorities, is likely to improve the welfare of animals performing in circuses. The Department indicated that concerns about the welfare of animals in circuses, expressed in the consultation process on the draft Animal Welfare Bill (see end of chapter), would be taken into account in the preparation of the draft Bill.

According to the ACP, its animal welfare policy is designed to ensure that its member circuses provide a 'scientifically and ethically sound environment' for their animals. The Code's animal husbandry requirements provide for:

- Regular inspections and spot checks of animals in ACP circuses.
- Detailed guidelines of how animals must be cared for.
- Veterinary knowledge of wild animals to keep them in good condition.

The Performing Animals (Regulations) Act 1925 and 1968 and the Protection of Animals Act 1954 regulate animal trainers working in circuses. Under the 1925 Regulations, all people training or exhibiting performing animals must be registered with the local authority. If a complaint is made on grounds of cruelty and upheld in court, the person may have his or name removed from the register and be disqualified from the register.

One-day sales

The sale of animals at fairs, markets, clubs, shows and other one-day sales are increasingly commonplace. Animals sold at one-day sales are often birds or reptiles, but, since the 1980s, such events for the sale of koi carp have become a regular feature. Some events are organized by clubs, ostensibly for members only while others are clearly public markets. The use of legislation by local authorities to control such activities has been inconsistent and, in some cases, inappropriate. Some authorities have in the past issued licences under the Pet Animals Act for such events.

The Pet Animals Act 1951 states that a person keeping a pet shop must be licensed. The keeping of a pet shop means carrying on at premises of any nature (including private dwellings) a business of selling pets. 'Animal' includes any description of vertebrate. Pets must not be sold in streets or public places.

The 1983 Amendment Act prohibited the sale of pets in a public place such as a market. It states: 'Any person carrying out the business of selling animals as pets will be guilty of an offence if it takes place from a stall or barrow in a market or public place.'

Discussions have revolved around the hiring of public halls for 'Fairs', in which stallholders gather to sell animals. Questions have also been raised about the legality of selling animals in shopping malls. An example of this is licensing the sale of 'aquababies', which are small Perspex cubes containing live fish, from kiosks or barrows in the public thoroughfare within shopping malls.

Legal opinion holds that such sales are excluded from the licensing ability of local authorities. The issue of a licence in such circumstances might be regarded as *ultra vires*.

Some definitions of relevance to the issue of one-day sales follow.

Public place

1. Licensing Act 1902: 'Any place to which the public have access whether on payment or otherwise.'
2. Indecent Displays (Control) Act 1981: 'Any place to which the public have or are permitted to have access (whether on payment or otherwise).'
3. Environmental Protection Act 1990. Part VIII, s.149 (11): 'Any highway and any other place to which the public are entitled or permitted to have access.'
4. Dangerous Dogs Act 1991, s.10 (2): 'Any street, road or other place to which the public have, or are permitted to have, access.'

Premises

The main purpose of the 1951 Act is to allow for the licensing of 'Premises' used for the business of selling animals as pets. It is clear that premises can, and should, be licensed by local authorities. The 1983 amendment made a distinction between premises and the practice of selling animals as pets in any part of a street or public place, or at a stall or barrow in a market.

A clear distinction has been drawn between 'premises' and 'public place'. Pet shops that comply with the requirements of the Act should be licensed, but other attempts at sale should not be so licensed, that is, the sale of pet animals to the public in a public place cannot be licensed under the provisions of the Pet Animals Acts 1951 and 1983.

Market

The Oxford English Dictionary defines a market as 'The gathering of people for the purchase and sale of provisions, livestock etc., especially with a number of different vendors – an open space or covered building used for this.'

'Chartered markets' are covered for different circumstances and have particular rules but are not seen as directly relevant to this review.

Common law definition of a market: 'A concourse of buyers and sellers.'

Enforcement

It is likely to be those trading rather than the person hiring out the hall against whom enforcement action is taken. It is widely held that the purpose of the Pet Animals Act 1951, as amended, was to afford protection to animals using the objects specified in Section 1. Discretion is available in the application of conditions.

Pet shops should be fixed, defined premises within which standards of care may be monitored and maintained, with named individuals responsible for the care of those animals. Temporary fairs and the like involve numerous dealers and changeable environmental conditions within which the animals are held awaiting sale.

It is difficult to ensure that adequate welfare conditions are achieved at temporary fairs and auctions; local authorities are empowered to take enforcement action against vendors at such events.

Member only events

Many events are claimed to be for the benefit of members only; such events would fall outside the restrictions on the sale of animals in public places. There are loopholes available here for exploitation; the following factors will help determine whether or not an event is private within the requirements and spirit of the Acts.

(a) the organization responsible for the show must be a bona fide club;
(b) members of the public cannot gain admittance without membership;
(c) admittance by joining the club at the door on the day of the show should not be allowed;
(d) membership fees should be reasonably substantial and not just a nominal figure (a guide of £10–£12 has been suggested);
(e) clear benefits of membership of the club are identified in the membership documentation;
(f) no sales of pet animals in the course of a business, take place at the event.

The application of the Human Rights Act 1998

Some organizers of one-day animal sales, have cited the Human Rights Act in defence of such events, claiming that action to curtail such sales infringes the right to free assembly.

It is clearly unlawful for local authorities to act in a way which is incompatible with Convention rights, subject to certain exceptions. Exceptions include a situation where authorities are compelled to act in a certain way by legislation. There is currently no certainty in the application of the Convention rights under the 1998 Act to pre-existing legislation.

CONTROL OF DOGS

Stray dogs

It is estimated that on any given day in the United Kingdom there can be up to 500 000 uncontrolled dogs on the streets; at least 200 000 are registered each year as strays [8]. These dogs can cause road accidents, kill or maim livestock, bite or attack members of the public and deposit tonnes of excrement daily.

Statutory duties for local authorities in respect of stray dogs were introduced under the Environmental Protection Act 1990. These provisions became enforceable on 1 April 1992, since when the police have no longer exercised their discretionary powers in rounding up strays. However, members of the public are still able to take stray dogs to a police station, pursuant to the Dogs Act 1906.

Environmental Protection Act 1990, Sections 149–151

Every local authority is required to appoint an officer whose function is to administer the duties required under the Environmental Protection Act 1990. These duties can be discharged through the appointment of dog wardens, who will be responsible for day-to-day activities in connection with the seizure and detention of stray dogs. Dog wardens may be local authority employees or contractors.

When a dog that is believed to be a stray is found in a public place, the officer has a duty to seize and detain the animal. If the dog is on private land, the officer must receive the consent of the owner or occupier of the land before seizing the dog. The term 'stray dog' is not defined, but implies that there is no person in charge of the dog for the time being. A duty is placed on any person who finds a stray dog to return it to the owner or take it to the police or local authority, where particulars of the dog and the name and address of the finder will be taken. The person finding the dog may keep it for at least one month, or hand it over to the local authority or police to be dealt with.

Where the owner of the dog can be identified through information on a collar tag or microchip implant, the officer must notify that person by notice that the dog has been seized and where it is being held. The notice must give the person seven days to collect the dog and pay the necessary charges, otherwise the dog is liable to be disposed of or destroyed. A public register containing prescribed particulars of seized dogs is required to be kept by the officer.

Such particulars, prescribed in the Environmental Protection (Stray Dogs) Regulations 1992, include a brief description of the dog, including any distinguishing characteristics; information recorded on the collar and tag; the date, time and place of seizure; date and service of any notice; the name and address of any person claiming to be the owner to whom the dog was returned; and the date of return. The register must also record where the dog is disposed of; the date of disposal; the method of disposal (destruction, gift or sale); the place of sale and the price fetched; and the name and address of the person receiving, purchasing or effecting the destruction of the dog.

Seized dogs may be detained in kennels provided by the local authority or a voluntary organization, or private kennels with an arrangement with the local authority. Arrangements must be made for receiving and dealing with dogs found or reported outside usual working hours. The charge to the owner on collection of a stray dog is based on the cost per day of kennelling the dog together with a sum of £25, prescribed by the Environmental Protection (Stray Dogs) Regulations 1992. There is no automatic entitlement to the return of a dog unless the full amount is paid; however, local authorities do have the discretion to determine that a dog be returned to the owner without full payment of the costs.

Dogs must be kept for a minimum of seven days following seizure or service of a notice on the owner, whichever is the longer. Dogs that are injured or in poor health can be destroyed before the end of seven days to avoid suffering. In this instance, the advice of a veterinary surgeon should be sought. Following the seven-day detention period, if a stray has not been reclaimed the officer must determine the most appropriate means of disposal.

The local authority may make its own arrangements for selling or giving the dog to a person who, in the officer's opinion, will care for the dog properly, or, commonly, may have an arrangement with a voluntary organization, which will attempt to find a suitable owner. Any person who has purchased or received a dog in good faith has the ownership of that dog vested in him such that the original owner has no rights to reclaim the dog. Alternatively, the dog may be destroyed in a humane manner. Guidance to local authorities on dealing with stray dogs has been produced by the Association of District Councils [9].

Control of Dogs Order 1992

This Order, made under the Animal Health Act 1981, requires every dog on a public highway or in a place of public resort, to wear a collar identifying its owner. Exceptions include guide dogs, dogs used in emergency rescue work, dogs being used by the armed forces, police or Customs and Excise, and dogs used for sporting purposes. Any dog found on a highway or in a public place without a collar may be seized and treated as a stray under the Dogs Act 1906 or Sections 149–151 of the Environmental Protection Act 1990.

Identification of stray dogs

Traditionally, stray dogs have been identified through information contained on a collar tag, but there is increasing support for the use of microchip implants. Not only does this provide a permanent means of identification, it also helps to overcome breeding fraud. Microchipping has increased significantly in the United Kingdom, and is also the method preferred in some European countries and for certain 'dangerous' breeds worldwide. Both the National Canine Defence League and London's Battersea Dogs' Home report an increased number of strays being returned to their owners as microchipping becomes more widespread [10,11].

DANGEROUS DOGS

Until 1991, the law concerned with controlling attacks on the public existed within the Dogs Act 1871, which was concerned with dogs that were dangerous and not kept under proper control, and the Town Police Clauses Act 1847, which made it an offence for a ferocious dog to be at large. In 1989, these provisions were extended by the Dangerous Dogs Act 1989. This provided for the making of a court order requiring that a dog be handed over for destruction and the owner disqualified from keeping a dog in the future if a dog was deemed to be dangerous and was not being kept under proper control.

Dangerous Dogs Act 1991

This Act provides for even more stringent control. Section 1 applies to certain prescribed breeds of dog bred for fighting or possessing the characteristics of a type of dog bred for fighting. To date, the breeds prescribed are:

- pit bull terrier
- tosa (Japanese fighting dog)
- dogo Argentino
- fila Braziliero.

The provisions of the Act are enforced by the police or an authorized officer of the local authority. Additionally, the local authority's role in controlling stray dogs will inevitably bring them into contact with prescribed breeds. The Act prohibits the breeding, sale, exchange, offering as gifts or importation of prescribed dogs. Furthermore, no one is permitted to have one of these breeds in his possession unless it has been exempted. Exempted dogs are registered and the owners issued with a certificate of exemption.

To gain a certificate of exemption, an owner must arrange for the dog to be neutered, permanently implanted with a transponder/microchip implant, tattooed with the dog's exemption reference number, covered by third party insurance, kept in secure conditions at home and muzzled and held on a lead by someone aged at least 16 years when in a public place. Additionally, the owner must provide the police with the name, age and gender of the dog, together with the address at which the dog is kept. The transponder is implanted in the scruff of the dog's neck and an electronic reading device passed within 100–1300 mm of the transponder will produce a digital display that will uniquely identify the dog on the Index of Exempted Dogs.

Identification of prescribed breeds

One of the major difficulties encountered by those concerned with the legislation is the identification of the prescribed breeds, particularly, the pit bull terrier and dogs with pit bull terrier characteristics. The former Department of the Environment has provided guidance [12] and the RSPCA has a small number of qualified expert witnesses. Standards laid down by the American Dog Breeders' Association [13] include reference to behavioural characteristics in the identification of dangerous dogs. A High Court ruling in 1993 [14] determined that these behavioural characteristics could be taken into account during any proceedings to confirm whether a dog fell within the prescribed description.

The onus of proving that a dog is not a prescribed breed or exhibits characteristics of a prescribed breed rests with the plaintiff. The difficulty in positively identifying a prescribed breed has led to a series of appeals and dogs being kept kennelled at police or local authority expense for periods in excess of 15 months.

Dogs dangerously out of control

Any dog that is dangerously out of control in a public place is dealt with under Sections 4 and 4A of the Dangerous Dogs Act 1991. Proceedings may be taken against the owner or person temporarily in charge of the dog. 'Dangerously out of control' means that there are grounds for believing that the dog will injure a person, whether or not it actually does so. An aggravated offence is deemed to have been committed where the dog injures any person. In the case of a dog that is allowed to enter a place that is not a public place and where it is not permitted to be, for example, a private garden, the owner or person in charge of the dog is guilty of an offence.

Seizure of dangerous dogs

Powers are available under Sections 4 and 5 of the Dangerous Dogs Act 1991 enabling the police or local authority to seize a prescribed breed that is in a public place and that is not exempt, or, if exempt, is not muzzled and kept on a lead, or is a prescribed 'specially dangerous dog', or is a dog that is dangerously out of control.

In general, the Dangerous Dogs Act 1991 provides for the destruction of a seized dog. The original provisions were extremely contentious and vigorous campaigning for their amendment took place. This pressure forced some relaxation through the Dangerous Dogs (Amendment) Act 1997, and destruction orders are no longer automatically made following a conviction for an offence. The amendments allow some discretion for exempted breeds. The court may also disqualify the offender from having custody of a dog for whatever period it thinks fit.

In cases where an owner is convicted of an offence under the Dangerous Dogs Act 1991, and a court order is made for a dog's destruction, the dog will not be destroyed until the end of the appeal period, and in the event of an appeal not until the appeal has been withdrawn or determined.

Dog wardens

To fulfil the statutory duties under the Environmental Protection Act 1990 and the Dangerous Dogs Act 1991, local authorities have generally provided a dog warden service through the direct employment of dog wardens or through a contractual arrangement. In most cases, the duties of a dog warden will comprise both statutory and non-statutory functions. These functions include:

- the seizure and detention of stray dogs, dangerous dogs and associated duties under the Environmental Protection Act 1990 and the Dangerous Dogs Act 1991, as amended;
- the enforcement of orders and by-laws relating to the fouling of public areas, including 'poop-scoop' schemes; identification of dogs under the Control of Dogs Order 1992; the control of dogs on highways;
- the investigation of complaints relating to the keeping of dogs, particularly in relation to general welfare and nuisance from barking or the accumulation of faeces;

- the provision of advice in respect of the licensing of pet shops, animal boarding establishments and dog breeding premises;
- health education, through the provision of information to the public promoting the dog warden service; the production and distribution of publicity material concerned with the keeping of dogs; education of the public through talks and presentations.

The emotive nature of dog-related issues necessitates the appointment of someone who is sensitive to the sentiments of dog-lovers but is nonetheless capable of acting in an enforcement capacity. The job relies on fostering good public relations and building up a close working relationship with the police and animal welfare agencies. Although no formal qualifications are required, a recently introduced National Vocational Qualification in Animal Welfare and Management has become available and is suitable for those employed as a dog warden [15].

Dog control

A variety of legal controls are available to local authorities to provide for control to deal with matters such as defecation and noise, and these are discussed next.

Fouling in public places

Complaints concerning dog fouling in public places are commonplace in most local authorities. Although unpleasant aesthetically, the major issue associated with dog fouling is the potential risk to human health. All dogs may carry parasitic worms in their intestines, including the roundworm *Toxocara canis*. The worms are most commonly found in young dogs. Eggs from the roundworm are disseminated into the environment via the faeces of the dog. These eggs may remain viable for a period of years. Humans become infected through ingestion of viable eggs. Children are most at risk because of the nature of play, which may bring them into contact with contaminated ground. Much attention is therefore focused on the maintenance of dog-free children's play areas in public parks and on beaches. Each year, a number of

children are identified as exhibiting the clinical symptoms of toxocariasis. These symptoms are a form of blindness or eye disease and physical damage to the viscera caused by migrating larvae which hatch from the eggs. Surveys have shown that asymptomatic infection with *Toxocara canis* is common, indicating that up to one million people in the United Kingdom have been infected at some time or other [16]. *Toxocara canis* cannot develop into an adult roundworm in the human.

Toxocara canis and other worm infestations in dogs can be readily controlled through the administration of 'worming' preparations, together with the control of defecation. Individuals can reduce the risk of infection by maintaining good standards of hygiene in the home. Such measures include discouraging dogs from licking people; maintaining separate food bowls, utensils, toys and blankets for pets; not allowing dogs to sleep on beds, and washing one's hands after handling dogs.

Public Health Act 1875, Section 164; Open Spaces Act 1906, Sections 12 and 15

By-laws can be made under these provisions to control the entry of dogs into prescribed areas of parks, recreation grounds, beaches and promenades and to maintain dogs on leashes in prescribed areas at prescribed days and hours. There is also provision to require owners to clear up faeces within prescribed dog exercise areas. Model by-laws have been produced for local authorities to adopt subject to confirmation by the Home Office.

Litter (Animal Droppings) Order 1991

This Order applies the provisions of Part IV of the Environmental Protection Act 1990 in respect of litter and refuse to dog faeces on land that is not heath or woodland or used for grazing animals. The Environmental Protection Act 1990 requires local authorities, certain Crown authorities, designated statutory transport undertakers and occupiers of certain other land, to keep their land clear of litter and refuse. The effect of the order is not to create an offence in respect of dog fouling, but merely to ensure that any faeces are removed.

Dogs (Fouling of Land) Act 1996

The approach to dealing with fouling in public places has largely concentrated on education of pet owners, supported by provisions under Section 235 of the Local Government Act 1972, which have enabled local authorities to adopt by-laws requiring the removal of canine faeces ('poop-scoop' by-laws). These by-laws have been enthusiastically adopted by many local authorities. The Dogs (Fouling of Land) Act 1996 simplifies the law and will gradually replace the 'poop-scoop' by-laws. It applies to land open to the air to which the public have access, with some exceptions such as agricultural land and woodland, and includes land adjacent to a carriageway, where the speed limit does not exceed 40 miles per hour. Under the Act, local authorities may, by order, designate land to which the Act applies. The effect of this is to create an offence where a dog defecates and the owner fails to remove the faeces. Powers exist to authorized officers of the local authority to issue a fixed penalty notice.[2] The 'poop-scoop' by-laws made under the Local Government Act 1972 cease to have effect once land has been designated, and in any case after August 2008.

Designation of land under the Dogs (Fouling of Land) Act 1996 requires careful planning and effective publicity, together with capital expenditure for the provision of suitable receptacles, education of the public geared towards toilet training of dogs and high-profile enforcement action, although in many cases, the capital investment will have been made when the local authority adopted the 'poop-scoop' by-laws. Designation under the Dogs (Fouling of Land) Act 1996 may, however, offer some local authorities scope for a renewed publicity and awareness campaign. Many contractors offering dog warden services are able to supply a full range of suitable equipment and publicity material.

Control of rabies

Rabies, or hydrophobia, is a viral zoonosis transmitted via the saliva of a rabid animal. The saliva is introduced by a bite, or more rarely through a scratch or break in the skin. Rabies occurs worldwide and is endemic in many European countries.

Canine and feline animals are the main vectors in Europe, particularly the fox and dog. The spread of rabies across continental Europe, however, has been stalled as a consequence of an aerial vaccination programme. No cases of indigenous rabies have been reported in England and Wales since 1902; the only reported cases having been contracted abroad. There is much concern at the risk of rabies being introduced into the United Kingdom through the accidental or intentional importation of rabid animals aboard ships, aircraft and, more recently, via the Channel Tunnel link. The occupations most at risk are those of the animal handlers at quarantine kennels and zoos, port health inspectors, veterinary surgeons, animal health inspectors and dog wardens.

Rabies (Importation of Dogs, Cats and Other Mammals) Order 1974

Statutory powers provide both a proactive and reactive approach to dealing with rabies. The Rabies (Importation of Dogs, Cats and Other Mammals) Order 1974, as amended, prohibits the landing in Great Britain of any animal brought from outside Great Britain, except in accordance with the terms of a licence that has been issued in advance. Animals landed within the terms of the order are subject to a quarantine period of six months at the owner's expense. Certain qualified exceptions do apply to dogs originating in European Union (EU) states and that are offered for sale in the United Kingdom by virtue of the Rabies (Importation of Dogs, Cats and Other Mammals) (Amendment) Order 1994.

The quarantine provisions have recently been reviewed [17,18], with the key recommendation that the law be relaxed under certain circumstances. More widespread use of microchip implants lends itself to improved record-keeping concerning a dog's origin, health and vaccination record and may obviate the need for routine quarantine for every imported dog.

Rabies (Control) Order 1974

The Rabies (Control) Order 1974 provides for measures to be taken in the event of an outbreak of

---

[2]Section 119 of the Local Government Act 2003 allows local authorities to retain the sums received from these fines and use them to pay for dog control functions.

rabies. Such measures include the declaration of an infected area; the seizure and destruction of the suspect animal; restriction on the movement of animals in to and out of an infected area; action to control dogs, cats and other animals through leashing or muzzling; seizure, detention or destruction of animals not kept under proper control; compulsory vaccination of animals; prohibition on events and activities where animals are brought together; and the destruction of foxes within the infected area.

Rabies outbreak contingency plans have been produced by animal health authorities, that is, London boroughs, metropolitan authorities, county councils and unitary authorities. These plans set out the action to be taken in the event of an outbreak of rabies in their area, including publicity, liaison with other statutory bodies, manpower, equipment and similar considerations.

Public Health (International Trains) Regulations 1994

With regard to the Channel Tunnel link, the operators introduced various measures during the design and construction of the tunnel to minimize the likelihood of foxes, dogs and rodents gaining access to it. However, there is always the possibility of 'stowaway animals' on trains, and the Public Health (International Trains) Regulations 1994 address this situation. A 'stowaway animal' is defined as 'any animal on board an international train, except one which is being lawfully transported through the tunnel, or smuggled through the tunnel'. The effect of this statutory instrument is to require any member of the train crew to report the presence of a stowaway animal to the train manager. The train manager then has a duty to report that sighting to the local authority environmental health service at the next designated stopping place for that service. Where the animal is a rabies suspect, either the train manager or the local authority will advise the Ministry of Agriculture, Fisheries and Food, which will be responsible for any necessary rabies control measures. Where the stowaway animal is not a rabies suspect, the local authority should be able to require the deratting, decontamination or disinfestation of the train.

Pet travel scheme

The Pet Travel Scheme (PETS) is the system that allows pet animals from certain countries to enter the United Kingdom without quarantine as long as they meet the prescribed rules. It also means that people in the United Kingdom can, having taken their pets to these countries, bring them back without the need for quarantine.

The Scheme:

- only applies to pet cats and dogs ('pets') including guide dogs and hearing dogs;
- is limited to pets coming into the United Kingdom from certain countries and territories;
- only operates on certain sea, air and rail routes to England.

PETS was introduced on 28 February 2000 for dogs and cats travelling from certain European countries. The Scheme was extended to Cyprus, Malta and certain Long Haul countries and territories on 31 January 2001. Bahrain joined on 1 May 2002. The United States and Canada joined on 11 December 2002.

Animals which meet the PETS rules can enter (or re-enter) the United Kingdom without having to undergo six months quarantine. Animals which do not meet all the rules must be licensed into quarantine. They might then be able to obtain early release if they can be shown to comply with the necessary PETS requirements.

There are a series of key steps that have to be taken in order to comply with PETS rules:

Microchipping – pets have to be microchipped to enable proper identification.
Vaccination – pets have to be vaccinated against rabies.
Blood test – after vaccination, pets must have their blood tested to ensure that the vaccine has given them a satisfactory level of protection against rabies.
PETS certificate – following the completion of the above steps, an official PETS certificate can be obtained from a government authorized vet.
Treatment against ticks and tapeworm – pets must be treated against ticks and a tapeworm between

24 and 48 hours before they are checked for a journey to the United Kingdom. Any qualified vet can carry out the treatment. The vet must also issue an official certificate of treatment to show that this treatment has been carried out.

Declaration of residency – pet owners have to sign a declaration that the pet has not been outside any of the PETS qualifying countries in the six months before it enters the United Kingdom.

When the pet is brought to the United Kingdom, it must travel on an approved route with an approved travel company. Requirements of PETS qualifying countries vary according to whether they are in Europe, Canada/the United States or classified as 'long haul' countries.

Animal Welfare Bill[3]

In 2002, the DEFRA announced plans for a new Animal Welfare Bill. The proposals cover the welfare of all farmed, wild or exotic animals in captivity, domestic animals and animals in entertainment and sport and include:

- a two-tier structure of offences. The first to maintain the present ban on cruelty but with the addition of a statutory duty to promote the welfare of all animals kept by humans;
- powers to make regulations to ban mutilations of animals such as the docking of dogs' tails;
- powers under which Ministers can make regulations to prevent cruelty and promote welfare to enable the Government more easily to keep animal welfare legislation up to date and make necessary changes.

Further proposals in the plans include raising the age at which children can buy pets unaccompanied by an adult and the licensing of animal sanctuaries.

DEFRA also made a commitment to look again at the Council of Europe Convention for the Protection of Pet Animals, the main components of which could be included in a new bill. The Convention, among other things, would ensure that proper account is taken of welfare needs when setting breeding standards, in co-operation with breed societies.

The proposed Bill will not be about the welfare of wild animals in general, which is already covered by legislation; or hunting, which is being dealt with separately. Nor will it seek to place restrictions on shooting for sport.

The laws protecting domestic or captive animals which the Department was looking to consolidate and update include:

- Protection of Animals Act 1911
- Performing Animals (Regulation) Act 1925
- Pet Animals Act 1951
- Cockfighting Act 1952
- Abandonment of Animals Act 1960
- Animal Boarding Establishments Act 1963
- Riding Establishments Acts 1964 and 1970
- Breeding of Dogs Acts 1973 and 1991
- Protection Against Cruel Tethering Act 1988
- Breeding and Sale of Dogs (Welfare) Act 1999

At the time of writing the Bill had not progressed into statute law.

REFERENCES

1. Local Government Association *et al.* (1998) *Model Licence Conditions for Pet Shops*, LGA, London.
2. Chartered Institute of Environmental Health (1995) *Model Licence Conditions and Guidance for Dog Boarding Establishments*, CIEH, London.
3. Chartered Institute of Environmental Health (1995) *Model Licence Conditions and Guidance for Cat Boarding Establishments*, CIEH, London.
4. DETR (March 2000) *Guidance on Modern Zoo Practice*, DETR, London.
5. HSC (1985) Approved Code of practice and Guidance, Zoos – *Safety, Health and Welfare Standards for Employees and Persons at Work*, HSE Books, London.

---

[3]But now see DEFRA (December 2003) *Consultation on the Animal Health and Welfare Strategy Implementation Plan for England*. DEFRA, London. A final strategy will be published in spring 2004.

6. Chartered Institute of Environmental Health *et al.* (1998) *A Report into the Welfare of Circus Animals in England and Wales*, CIEH, London.

7. Association of Circus Proprietors (ACP) (2002) *Standards for the Care and Wellbeing of Circus Animals on Tour*, ACP, Blackburn, Lancashire.

8. Moore, S.R. and Dhaliwal, P. (1992) Campaign on Dogs. *Environmental Health*, **100**(6), 152–4.

9. Association of District Councils (1992) *Dealing with Stray Dogs*, ADC, London (through LGA, London).

10. Fred's Fact File http://www.okima.com/dogs/facts.html

11. Brown, C. (1997) Micro-chip – the high-tech, low-cost solution to strays. *Environmental Health*, **105**(8), 219–20.

12. Department of the Environment/Welsh Office/Home Office (1992) *Environmental Protection Act 1990, Part IV. Control of Stray Dogs. Circular 6/1992*, DoE, London.

13. UK looks to adopt US standards on Pit Bull Terriers (1992) *Environmental Health News*, **8**(28), 4.

14. *R v Knightsbridge Crown Court ex parte Dunne* (1993) Q.B.D and *Brock v Director of Public Prosecutions* (1993) The Times Law Reports, 23 July 1993.

15. Anon. (1997) Taking the lead on the Isle of Wight. *Environmetal Health*, **105**(1), 22.

16. Gillespie, S.H. (1993) Human Toxicaris. *CDR Review*, 3(10), R140–3.

17. Advisory Group on Quarantine (1998) *Quarantine and Rabies: A Reappraisal*, MAFF, London.

18. Savage, C. (1990) Quarantine Wars. *Environmental Health*, **107**(01), 8–10.

FURTHER READING

Malcolm, R. and Pointing, J. (2002) *Statutory Nuisance, Law and Practice*, Oxford University Press.

McCracken, R. *et al.* (2001) *Statutory Nuisance*, Butterworths, London.

Radford, M. (2001) *Animal Welfare Law in Britian*, Oxford University Press.

Bassett, W.H. (2002) *Environmental Health Procedures*, 6th edn, Spon Press, London.

British Medical Association (1995) *The BMA Guide to Rabies*, Radcliffe Medical Press, Oxford and New York.

Chartered Institute of Environmental Health (1991) *Dogs – Control by Local Authorities: Report of Dogs Survey 1991*, CIEH, London.

Department of the Environment/Welsh Office (1989) *Action on Dogs: The Government's Proposals for Legislation. A Consultation Paper*, HMSO, London.

Department of the Environment/Welsh Office (1996) *The Dogs (Fouling of Land) Act 1996. Circular 18/1996*, DoE, London.

Department of Health (2000) *Memorandum on Rabies*, HMSO, London.

Dog Owner's Guide (USA) http://www. canismajor.com/dog/guide.html

Health and Safety Executive (1993) *The Occupational Zoonoses*, HSE, London.

Home Office/Department of the Environment/Welsh Office/Scottish Office (1990) *The Control of Dogs: A Consultation Paper*, HMSO, London.

k9netuk – The Complete UK Dog World. http://www.k9netuk.com/contents.html

National Canine Defence League. http://www.k9netuk.com/ncdl/index.html

People and Dogs Society. http://www.gurney.co.uk/pads/index.htm

Price Waterhouse (1991) *Resource Implications of s. 149–151 Environmental Protection Act 1990 (Dog Control) for the Department of the Environment*, Price Waterhouse, London.

Tidy Britain Group (1993) *Local Authority Survey – Control of Dog Fouling*, Tidy Britain Group at the Pier, Wigan.

Royal Society for the Prevention of Cruelty to Animals (2002) *Partners in Animal Welfare (Local Authority Good Practice Guide)* RSPCA, Horsham, Sussex.

Feline Advisory Bureau (2002) *Boarding Cattery Manual*, Feline Advisory Bureau, Tisbury, Wiltshire.

Index